Lecture Notes in Computer Science 4708

Lecture Notes in Computer Science 4708

Commenced Publication in 1973
Founding and Former Series Editors:
Gerhard Goos, Juris Hartmanis, and Jan van Leeuwen

Luděk Kučera   Antonín Kučera (Eds.)

Mathematical Foundations of Computer Science 2007

32nd International Symposium, MFCS 2007
Český Krumlov, Czech Republic, August 26-31, 2007
Proceedings

Springer

Volume Editors

Luděk Kučera
Charles University
Faculty of Mathematics and Physics, Department of Applied Mathematics
Malostranské nám. 25, 118 00 Praha 1, Czech Republic
E-mail: ludek@kam.mff.cuni.cz

Antonín Kučera
Masaryk University
Faculty of Informatics, Department of Computer Science
Botanická 68a, 602 00 Brno, Czech Republic
E-mail: tony@fi.muni.cz

Library of Congress Control Number: 2007932973

CR Subject Classification (1998): F.1, F.2, F.3, F.4, G.2, E.1

LNCS Sublibrary: SL 1 – Theoretical Computer Science and General Issues

ISSN 0302-9743
ISBN-10 3-540-74455-X Springer Berlin Heidelberg New York
ISBN-13 978-3-540-74455-9 Springer Berlin Heidelberg New York

Springer is a part of Springer Science+Business Media

springer.com

Printed in Germany

Typesetting: Camera-ready by author, data conversion by Scientific Publishing Services, Chennai, India
Printed on acid-free paper SPIN: 12112838 06/3180 5 4 3 2 1 0

Preface

This volume contains the proceedings of the 32nd International Symposium on Mathematical Foundations of Computer Science (MFCS 2007). The purpose of the MFCS symposia is to encourage high-quality research in all fields of theoretical computer science. This year's conference was held in Český Krumlov, Czech Republic, during August 26–31.

The conference program of MFCS 2007 consisted of 61 contributed papers selected by the Program Committee from a total of 167 submissions. All submissions were read and evaluated by at least four referees, and the resulting decision was based on electronic discussion which often included help from outside experts. A selection of contributed papers will appear in the journal *Theoretical Computer Science.*

Complementing the contributed papers, the program of MFCS 2007 included invited lectures by Vašek Chvátal (Montreal, Canada), Anuj Dawar (Cambridge, UK), Kurt Mehlhorn (Saarbrücken, Germany), Luke Ong (Oxford, UK), and Leslie Valiant (Cambridge, USA). We are grateful to the invited speakers for accepting our invitation and sharing their knowledge and skills with all MFCS 2007 participants.

As the editors of these proceedings, we would like to thank everyone who contributed to the success of the symposium. First of all, we thank the authors of all submitted papers for considering MFCS 2007 as an appropriate platform for presenting their work. Since the number of submissions was very high, many good papers could not be accepted. We also thank the Program Committee members for their demanding and responsible work, the referees for careful reading of all the submissions, and the staff at Springer for the professional support in producing this volume.

June 2007

Luděk Kučera
Antonín Kučera

Organization

The 32nd International Symposium on Mathematical Foundations of Computer Science (MFCS 2007) was held in *Hotel Růže*, Český Krumlov, Czech Republic, during August 26–31, 2007. The conference was organized by the Faculty of Mathematics and Physics, Charles University, Prague in cooperation with other institutions in the Czech Republic.

Organizing Committee

Milena Zeithamlová *(Action M Agency)*
Blanka Puková *(Action M Agency)*
Martin Mareš *(Charles University)*
Luděk Kučera *(Charles University)*

Program Committee

Parosh Abdulla *(Uppsala University, Sweden)*
Jos Baeten *(Eindhoven University of Technology, The Netherlands)*
Narsingh Deo *(University of Central Florida, USA)*
Josep Díaz *(Universitat Politecnica de Catalunya, Spain)*
Yefim Dinitz *(Ben-Gurion University of the Negev, Israel)*
Javier Esparza *(Technische Universität München, Germany)*
Fedor Fomin *(University of Bergen, Norway)*
Pierre Fraigniaud *(CNRS and University Paris 7, France)*
Juraj Hromkovič *(ETH Zürich, Switzerland)*
Giuseppe F. Italiano *(Università di Roma "Tor Vergata", Italy)*
Kazuo Iwama *(Kyoto University, Japan)*
Michael Kaufmann *(Universität Tübingen, Germany)*
Barbara König *(Universität Duisburg-Essen, Germany)*
Petr Kolman *(Charles University, Czech Republic)*
Rastislav Královič *(Comenius University, Slovak Republic)*
Antonín Kučera *(Co-chair, Masaryk University, Czech Republic)*
Luděk Kučera *(Co-chair, Charles University, Czech Republic)*
Alberto Marchetti-Spaccamela *(Università di Roma "La Sapienza", Italy)*
Martin Mareš *(Charles University, Czech Republic)*
Friedhelm Meyer auf der Heide *(Universität Paderborn, Germany)*
Madhavan Mukund *(Chennai Mathematical Institute, India)*
Mogens Nielsen *(University of Aarhus, Denmark)*
Reinhard Pichler *(Technische Universität Wien, Austria)*
Rajeev Raman *(University of Leicester, UK)*
José Rolim *(University of Geneva, Switzerland)*

Davide Sangiorgi *(University of Bologna, Italy)*
Philippe Schnoebelen *(École Normale Supérieure de Cachan, France)*
Jerzy Tyszkiewicz *(Warsaw University, Poland)*
Uzi Vishkin *(University of Maryland, USA)*
Peter Widmayer *(ETH Zürich, Switzerland)*
James Worrell *(Oxford University, UK)*
Christos Zaroliagis *(CTI & University of Patras, Greece)*

Referees

Farid Ablayev
Isolde Adler
Manindra Agrawal
Luca Allulli
Kazuyuki Amano
Vikraman Arvind
Albert Atserias
Jiří Barnat
Amos Beimel
Wolf Bein
Amir Ben-Amram
Daniel Berend
Stefan Berghofer
Nayantara Bhatnagar
Vittorio Bilò
Davide Bilò
Johannes Blömer
Peter van Emde Boas
Hans Bodlaender
Filippo Bonchi
Guillaume Bonfante
Vincenzo Bonifaci
Paul Bonsma
Ahmed Bouajjani
Torben Braüner
Václav Brožek
Sander Bruggink
Henning Bruhn
Jakub Bystroň
H.-J. Böckenhauer
Peter Bürgisser
Toon Calders
Alberto Caprara
Arturo Carpi
Olivier Carton
Didier Caucal
Jakub Černý
Ho-Lin Chen
Jianer Chen
Paul Christophe
Josef Cibulka
Maxime Crochemore
Felipe Cucker
Flavio D'Alessandro
Peter Damaschke
Valentina Damerow
Samir Datta
Anuj Dawar
Francien Dechesne
Bastian Degener
Stéphane Demri
Jörg Derungs
Volker Diekert
Martin Dietzfelbinger
Yannis Dimopoulos
Arnaud Durand
Bruno Durand
Ron Dutton
Martin Dyer
Miroslav Dynia
Tomáš Ebenlendr
Thomas Erlebach
Zoltán Ésik
Serge Fehr
Maribel Fernández
Jiří Fiala
Andrzej Filinski
Jean-Christophe Filliâtre
Jiří Fink
Lev Finkelstein
Michele Flammini
Abraham Flaxman
Vojtěch Forejt
Enrico Formenti
Lance Fortnow
Gudmund Frandsen
Kimmo Fredriksson
Keith Frikken
Stefan Funke
Joaquim Gabarró
Anna Gambin
Alfredo Garcia
Michael Gatto
Cyril Gavoille
Markus Geyer
Beat Gfeller
Dan Ghica
Kristian Gjøsteen
Bart Goethals
Paul Goldberg
Avi Goldstein
Rajeev Goré
Fabrizio Grandoni
Petr Gregor
Gregory Gutin
Alex Hall
Magnus Halldorsson
Xin Han
Sariel Har-Peled
Ichiro Hasuo
Tobias Heindel
Harald Hempel
Miki Hermann
Ulrich Hertrampf
Petr Hliněný

Michael Hoffmann
Jan Holeček
Markus Holzer
Tomas Hruz
Yinghua Hu
Cornelis Huizing
Pierre Hyvernat
Costas Iliopoulos
Riko Jacob
Florent Jacquemard
Mark Jerrum
Ravi Kant
Christos Kapoutsis
Jan Kára
Jarkko Kari
Neeraj Kayal
Iordanis Kerenidis
Uzma Khadim
Stefan Kiefer
Pekka Kilpelainen
Martin Klazar
Bartek Klin
Johannes Köbler
Roman Kolpakov
Elisavet Konstantinou
Spyros Kontogiannis
Peter Korteweg
Takeshi Koshiba
Michal Koucký
Elias Koutsoupias
Jakub Kozik
Richard Královič
Maksims Kravcevs
Steve Kremer
Danny Krizanc
Peter Krusche
Manfred Kufleitner
Werner Kuich
Viraj Kumar
Michal Kunc
Petr Kůrka
Piyush Kurur
Anna Labella
Ugo Dal Lago
Giovanni Lagorio
Jim Laird
Gad Landau
Ivan Lanese
Sophie Laplante
Sławomir Lasota
Søren Lassen
Luigi Laura
Ranko Lazić
Emmanuelle Lebhar
Katharina Lehmann
Pietro Di Lena
Nutan Limaye
Guohui Lin
Giuseppe Liotta
Maciej Liskiewicz
Bruce Litow
Kamal Lodaya
Martin Loebl
Christoph Löding
Markus Lohrey
Daniel Lokshtanov
Sylvain Lombardy
Alex Lopez-Ortiz
Martin Lotz
Antoni Lozano
Michael Luttenberger
Ian Mackie
Meena Mahajan
Peter Mahlmann
Johann A. Makowsky
Christos Makris
Federico Mancini
Nicolas Markey
Hendrik Maryns
Luděk Matyska
Jens Maue
Elvira Mayordomo
Ernst Mayr
Jacques Mazoyer
Tyrrell B. McAllister
Andrew McGregor
Pierre McKenzie
Klaus Meer
Daniel Meister
Stéphane Messika
Jan Midtgaard
Matus Mihalak
Zoltan Miklos
Peter Bro Miltersen
Tobias Moemke
Luminita Moraru
Andrzej Murawski
Christophe Morvan
Nysret Musliu
Veli Mäkinen
Takayuki Nagoya
Masaki Nakanishi
Shin-ichi Nakano
K. Narayan Kumar
Alan Nash
Pavel Nejedlý
Jaroslav Nešetřil
Cyril Nicaud
Rolf Niedermeier
Harumichi Nishimura
Dirk Nowotka
Marc Noy
Marc Nunkesser
Jan Obdržálek
Alexander Okhotin
Luke Ong
Jaroslav Opatrný
Simona Orzan
Michiel van Osch
S.P. Suresh
Ondřej Pangrác
Dana Pardubská
Mike Paterson
Dirk Pattinson
Andrzej Pelc
Paolo Penna
Martin Pergel
Libor Polák
John Power
Sanjiva Prasad
Rudy Raymond
J. Radhakrishnan
M. Sohel Rahman
C.R. Ramakrishnan
R. Ramanujam

Srinivasa Rao
Jean-François Raskin
Stefan Ratschan
Bala Ravikumar
Ran Raz
Vojtěch Řehák
Eric Remila
Chloé Rispal
Mike Robson
Piet Rodenburg
Laurent Rosaz
Salvador Roura
Hana Rudová
Daniel Russel
Jan Rutten
Harald Räcke
David Šafránek
Jacques Sakarovitch
Louis Salvail
Piotr Sankowski
Saket Saurabh
Zdeněk Sawa
Francesco Scarcello
Guido Schaefer
Gunnar Schomaker
Florian Schoppmann
Lutz Schroeder
Stefan Schwoon
Sebastian Seibert
Helmut Seidl
Maria Serna
Olivier Serre
Anil Seth
Nikolay Shilov
Amir Shpilka
Jakob Grue Simonsen
Alex Simpson
Sitabhra Sinha
Christian Sohler
Ana Sokolova
Robert Špalek
Jiří Srba
Srikanth Srinivasan
Ludwig Staiger
Yannis Stamatiou
Martin Staněk
Ian Stark
Bjoern Steffen
Benjamin Steinberg
Krzysztof Stencel
Jan Strejček
K.V. Subrahmanyam
K.G. Subramanian
Stefan Szeider
Andrzej Szepietowski
Siamak Taati
Tadao Takaoka
Jean-Marc Talbot
Seiichiro Tani
Greg Tener
Véronique Terrier
Dimitrios Thilikos
Seinosuke Toda
Marc Tommasi
Jacobo Torán
Tayssir Touili
Kostas Tsichlas
Dekel Tsur
Emilio Tuosto
Pavel Tvrdik
Paweł Urzyczyn
Tarmo Uustalu
Tomas Valla
Pavel Valtr
Victor Vianu
Elias Vicari
Yngve Villanger
Tjark Vredeveld
Tomasz Waleń
Fang Wei
Pascal Weil
Michael Weiss
Carola Wenk
Mark Weyer
Ronald de Wolf
Alexander Wolff
David Wood
Thomas Worsch
Ondřej Zajíček
Ayal Zaks
Hans Zantema
Guochuan Zhang
Jie Zheng
Martin Ziegler

Sponsoring Institutions

European Association for Theoretical Computer Science

Previous MFCS

MFCS symposia have been organized in the Czech Republic, Poland, and Slovak Republic since 1972. The previous meetings are listed below.

1972 Jablonna (Poland)
1973 Štrbské Pleso (Czechoslovakia)
1974 Jadwisin (Poland)
1975 Mariánské Lázně (Czechoslovakia)
1976 Gdańsk (Poland)
1977 Tatranská Lomnica (Czechoslovakia)
1978 Zakopane (Poland)
1979 Olomouc (Czechoslovakia)
1980 Rydzyna (Poland)
1981 Štrbské Pleso (Czechoslovakia)
1984 Praha (Czechoslovakia)
1986 Bratislava (Czechoslovakia)
1988 Karlovy Vary (Czechoslovakia)
1989 Porabka-Kozubnik (Poland)
1990 Banská Bystrica (Czechoslovakia)
1991 Kazimierz Dolny (Poland)
1992 Praha (Czechoslovakia)
1993 Gdańsk (Poland)
1994 Košice (Slovak Republic)
1995 Praha (Czech Republic)
1996 Kraków (Poland)
1997 Bratislava (Slovak Republic)
1998 Brno (Czech Republic)
1999 Szklarska Poreba (Poland)
2000 Bratislava (Slovak Republic)
2001 Mariánské Lázně (Czech Republic)
2002 Warsaw-Otwock (Poland)
2003 Bratislava (Slovak Republic)
2004 Praha (Czech Republic)
2005 Gdańsk (Poland)
2006 Bratislava (Slovak Republic)

Table of Contents

Invited Papers

Random Graphs

Rewriting

Approximation Algorithms

Iteration and Recursion

Algorithms I

Automata

Complexity II

Protocols

Graphs II

Networks

Algorithms II

Languages

Graphs III

Quantum Computing

Isomorphism

Equilibria

Games

Algebra and Strings

Algorithms III

Words and Graphs

How To Be Fickle

Vašek Chvátal

Canada Research Chair in Combinatorial Optimization
Concordia University, Montreal, Canada

Abstract. The backbone of many popular algorithms for solving NP-hard problems is *implicit enumeration.* There, the input problem is split into two or more subproblems by assigning some variable a new fixed value in each new subproblem – this is called *branching* – and the procedure is repeated recursively until the subproblems become easy enough to be dealt with directly.

Unfortunate branching choices may have disastrous consequences: once you have branched, there is no turning back and you may be doomed to painfully replicate all your subsequent moves. In this sense, branching is like marrying in the Roman Catholic church: Branch in haste, repent at leisure.

To some, the ominous prospect of irrevocable matrimony may supply the motivation for utmost care in choosing a spouse; others may prefer to choose spouses carelessly and to make divorce easy.

An implementation of the former plan is the prototype of *strong branching,* first introduced in Concorde, a computer code for the symmetric traveling salesman problem: use first quick and coarse criteria to eliminate some of a large number of initial candidates in a first round, then slower and finer criteria to eliminate some of the remaining candidates in the next round, and so on in an iterative fashion.

Two of the most popular implementations of the latter plan are *dynamic backtracking* of Matthew Ginsberg and *GRASP* of João Marques Silva and Karem Sakallah; one of the most neglected ones is *resolution search.* In reviewing resolution search, I will go into its origins, details, and extensions beyond the original presentation in 1995. In particular, I will point out that it can bridge the gap between certain local search heuristics and so-called exact algorithms.

L. Kučera and A. Kučera (Eds.): MFCS 2007, LNCS 4708, p. 1, 2007.

Finite Model Theory on Tame Classes of Structures

Anuj Dawar

University of Cambridge Computer Laboratory, William Gates Building,
J.J. Thomson Avenue, Cambridge, CB3 0FD, UK
Anuj.Dawar@cl.cam.ac.uk

Abstract. The early days of finite model theory saw a variety of results establishing that the model theory of the class of finite structures is not well-behaved. Recent work has shown that considering subclasses of the class of finite structures allows us to recover some good model-theoretic behaviour. This appears to be especially true of some classes that are known to be algorithmically well-behaved. We review some results in this area and explore the connection between logic and algorithms.

1 Introduction

Finite model theory is the study of the expressive power of various logics—such as first-order logic, second-order logic, various intermediate logics and extensions and restrictions of these—on the class of finite structures. Just as model theory is the branch of classical mathematical logic that deals with questions of the expressive power of languages, so one can see finite model theory as the same study but carried out on finite interpretations. However, finite model theory is not simply that as it has evolved its own specific methods and techniques, its own significant questions and a core of results specific to the subject that all make it quite distinct from model theory. These methods, questions and results began to coalesce into a coherent research community in the 1980s, when the term *finite model theory* came into common use. The core of the subject is now well established and can be found in books such as [17,33,23]. Much of the motivation for the development of finite model theory came from questions in computer science and in particular questions from complexity theory and database theory. It turns out that many important questions arising in these fields can be naturally phrased as questions about the expressive power of suitable logics (see [1,29]). Moreover, the requirement that the structures considered are available to algorithmic processing leads to the study of such logics on specifically finite structures. Such considerations have provided a steady stream of problems for study in finite model theory.

In his tutorial on finite model theory delivered at LICS in 1993, Phokion Kolaitis [31] classified the research directions in finite model theory into three categories that he called *negative*, *conservative* and *positive*. In the first category are those results showing that theorems and methods of classical model theory

L. Kučera and A. Kučera (Eds.): MFCS 2007, LNCS 4708, pp. 2–12, 2007.

fail when only finite structures are considered. These include the compactness theorem, the completeness theorem and various interpolation and preservation theorems. In the second category are results showing that certain classical theorems and methods do survive when we restrict ourselves to finite structures. One worth mentioning is the result of Gaifman [22] showing that any first-order sentence is equivalent to a Boolean combination of local sentences. This has proved to be an extremely useful tool in the study of finite model theory. A more recent example in the vein of conservative finite model theory is Rossman's result [38] that the homomorphism preservation theorem holds in the finite (a topic we will return to later). Finally, the third category identified by Kolaitis is of results exploring concepts that are meaningful only in the context of finite structures. Among these are work in descriptive complexity theory as well as 0-1 laws.

Much early work in finite model theory focussed on the negative results, as researchers attempted to show how the model theory of finite structures differed from that of infinite structures. The failure of compactness and its various consequences led to the conclusion that the class of finite structures is not model-theoretically well behaved. Indeed, Jon Barwise once stated that the class of finite structures is not a natural class, in the sense that it is difficult to define (in a formal logic) and does not contain limit points of sequences of its structures. However, recent work in finite model theory has begun to investigate whether there are subclasses of the class of finite structures that may be better behaved. We call such classes *tame*. It is impossible to recover compactness in any reasonable sense in that any class that contains arbitrarily large finite structures but excludes all infinite ones will not have reasonable compactness properties. Thus, interesting subclasses of the class of finite structures will not be natural in the sense of Barwise, but as we shall see, they may still show interesting model-theoretic behaviour. The subclasses we are interested in are motivated by the applications in computer science. It is often the case in a computational application where we are interested in the expressive power of a logic that the structures on which we interpret the logic are not only finite but satisfy other structural restrictions. Our aim is to understand how such restrictions may affect the model-theoretic tools available.

Preservation Theorems. Consider classical preservation theorems, which relate syntactic restrictions on first-order formulas with semantic counterparts. A key example is the extension preservation theorem of Łoś and Tarski which asserts that a first-order formula is preserved under extensions on all structures if, and only if, it is logically equivalent to an existential formula (see [27]). One direction of this result is easy, namely that any formula that is purely existential is preserved under extensions, and this holds on any class of structures. The other direction, going from the semantic restriction to the syntactic restriction makes key use of the compactness of first-order logic and hence of infinite structures. Indeed, this direction is known to fail in the case of finite structures as it was shown by Tait [41] that there is a first-order sentence whose finite models are closed under extensions but that is not equivalent *on finite structures* to an existential sentence. Thus, we can consider the extension preservation question relativised to a class of structures $\mathcal{C}$ as: if a first-order sentence φ is preserved under extensions on $\mathcal{C}$, is it equivalent on $\mathcal{C}$ to an existential sentence? If we replace $\mathcal{C}$ by a class $\mathcal{C}'$ that is contained in $\mathcal{C}$, we are weakening both the hypothesis and the consequent of the question. Thus, one cannot deduce the truth or otherwise of the preservation theorem on $\mathcal{C}'$ from that on $\mathcal{C}$. The question arises anew for every class $\mathcal{C}$. The extension preservation theorem for various classes of finite structures $\mathcal{C}$ is explored in [3].

A related preservation result of classical model theory is the *homomorphism preservation theorem* which states that a first-order formula is preserved under homomorphisms on all structures if, and only if, it is logically equivalent to an existential positive formula. For many years it was an open question whether this preservation theorem was true in restriction to the class of finite structures. The question was finally settled by Rossman [38] who showed that it is indeed true in this case. This provides a rare example of a preservation theorem that sits in the *conservative* rather than the *negative* category in Kolaitis' classification of results. Once again, for every class $\mathcal{C}$ of finite structures, the question of whether the homomorphism preservation theorem holds on $\mathcal{C}$ is a new question. The preservation property is established for a large variety of classes in [4].

Descriptive Complexity. In the *positive* research direction, the most prominent results are those of descriptive complexity theory. The paradigmatic result in this vein is the theorem of Fagin [19] which states that a class of finite structures is definable in existential second-order logic if, and only if, it is decidable in NP. Similar, descriptive, characterisations were subsequently obtained for a large number of complexity classes (see [29]). In particular, Immerman [28] and Vardi [42] showed that LFP—the extension of first-order logic with a least fixed point operator—expresses exactly those classes of finite *ordered* structures that are decidable in P (a similar result is shown by Livchak [35]). Whether or not there is a logic that expresses exactly the polynomial time properties of finite structures, without the assumption of order, remains the most important open question in descriptive complexity. It was shown by Cai, Fürer and Immerman [8] that LFP + C, the extension of LFP with a counting mechanism, does not suffice. However, it turns out that on certain restricted classes of structures, LFP + C is sufficient to express all properties in P. We will see examples of this below.

2 Tame Classes of Structures

We consider classes of finite structures defined in terms of restrictions on their underlying adjacency (or Gaifman) graphs. The adjacency graph of a structure $\mathbb{A}$ is the graph $G\mathbb{A}$ whose vertices are the elements of $\mathbb{A}$ and where there is an edge between vertices a and b if, and only if, a and b appear together in some tuple of some relation in $\mathbb{A}$. The restrictions we consider on these graphs are obtained from graph structure theory and algorithmic graph theory. They are restrictions which have, in general, yielded interesting classes from the point of view of algorithms. Our aim is to explore to what extent the classes are also well-behaved in terms of their model-theoretic properties. From now on, when we say that a class of structures $\mathcal{C}$ satisfies some restriction, we mean that the collection of graphs $G\mathbb{A}$ for $\mathbb{A} \in \mathcal{C}$ satisfy the restriction.

The restrictions we consider and their interrelationships are depicted in Figure 1.

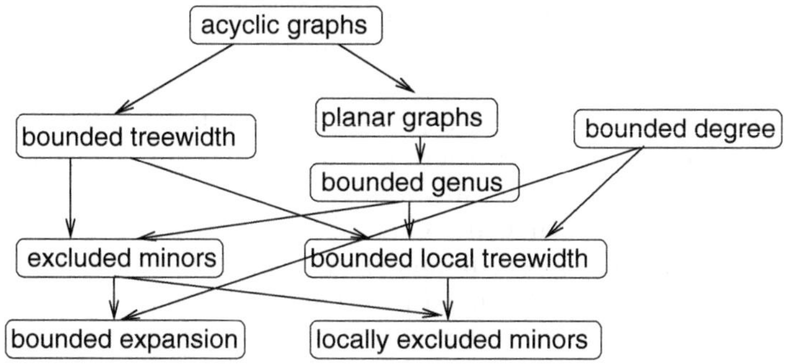

Fig. 1. Relationships between tame classes

Among the restrictions given in Figure 1, that of acyclicity and planarity are of a different character to the others in that they apply to single graphs. We can say of graph G that it is acyclic or planar. When we apply this restriction to a class $\mathcal{C}$, we mean that all structures in the class satisfy it. The other conditions in the figure only make sense in relation to classes of graphs. Thus, it makes little sense to say of a single finite graph that it is of bounded degree (it is necessarily so). When we say of a class $\mathcal{C}$ that it is of bounded degree, we mean that there is a uniform bound on the degree of all structures in $\mathcal{C}$.

The arrows in Figure 1 should be read as implications. Thus, any graph that is acyclic is necessarily planar. Similarly, any class of acyclic graphs has bounded treewidth. The arrows given in the figure are *complete* in the sense that when two restrictions are not connected by an arrow (or sequence of arrows) then the first does not imply the second and separating examples are known in all such cases.

The restrictions of acyclicity, planarity and bounded degree are self-explanatory. We say that a class of graphs $\mathcal{C}$ has bounded genus if there is a fixed orientable surface S such that all graphs in $\mathcal{C}$ can be embedded in S (see [37]). In particular, as planar graphs are embeddable in a sphere, any class of planar graphs has bounded genus. The treewidth of a graph is a measure of how tree-like it is (see [16]). In particular, trees have treewidth 1, and so any class of acyclic graphs has treewidth bounded by 1. The measure plays a crucial role in the graph structure theory developed by Robertson and Seymour in their proof of the graph minor theorem. We say that a graph G is a minor of H (written $G \prec H$) if G can be obtained from a subgraph of H by a series of edge contractions (see [16] for details). We say that a class of graphs $\mathcal{C}$ excludes a minor if there is some G such that for all $H \in \mathcal{C}$ we have $G \not\prec H$. In particular, this includes all classes $\mathcal{C}$ which are closed under taking minors and which do not include all graphs. If G is embeddable in a surface S then so are all its minors. Since, for any fixed integer k, there are graphs that are not of genus k, it follows that any class of bounded genus excludes some minor.

The notion of bounded local treewidth was introduced as a common generalisation of classes of bounded treewidth and bounded genus. A variant, called the diameter width property was introduced in [18] while bounded local treewidth is from [21]. Recall that the r-neighbourhood of an element a in a structure $\mathbb{A}$, denoted $N_{\mathbb{A}}^r(a)$, is the substructure of $\mathbb{A}$ induced by the set of elements at distance at most r from a in the graph $G\mathbb{A}$. We say that a class of structures $\mathcal{C}$ has bounded local treewidth if there is a nondecreasing function $f : \mathbb{N} \to \mathbb{N}$ such that for any structure $\mathbb{A} \in \mathcal{C}$, any a in $\mathbb{A}$ and any r, the treewidth of $N_{\mathbb{A}}^r(a)$ is at most $f(r)$. It is clear that any class of graphs of bounded treewidth has bounded local treewidth (indeed, bounded by a constant function f). Similarly, any class of graphs of degree bounded by d has local treewidth bounded by the function d^r, since the number of elements in $N_{\mathbb{A}}^r(a)$ is at most d^r. The fact that classes of bounded genus also have bounded local treewidth follows from a result of Eppstein [18].

We say that a class of structures $\mathcal{C}$ locally excludes minors if there is a nondecreasing function $f : \mathbb{N} \to \mathbb{N}$ such that for any structure $\mathbb{A} \in \mathcal{C}$, any a in $\mathbb{A}$ and any r, the clique $K_{f(r)}$ is not a minor of the graph $GN_{\mathbb{A}}^r(a)$. This notion is introduced in [11] as a natural common generalisation of bounded local treewidth and classes with excluded minors. Classes of graphs with bounded expansion were introduced by Nešetřil and Ossona de Mendez [40] as a common generalisation of classes of bounded degree and proper minor-closed classes. A class of graphs $\mathcal{C}$ has bounded expansion if there is a function $f : \mathbb{N} \to \mathbb{N}$ such that for any graph $G \in \mathcal{C}$, any subgraph H of G and any minor H' of H obtained from H by contracting neighbourhoods of radius at most r, the average degree in H' is bounded by $f(r)$. In particular, classes that exclude a minor have bounded expansion witnessed by a constant function f.

3 Logic and Algorithms on Tame Classes

The interest in tame classes of structures from the point of view of algorithms is that it is often the case that problems that are intractable in general become tractable when a suitable restriction on the structures is imposed. For instance, for any class of graphs of bounded treewidth, there are linear time algorithms for deciding Hamiltonicity and 3-colourability and on planar graphs there is a polynomial time algorithm for the MAX-CUT problem. On the other hand, many problems remain hard as, for instance, 3-colourability is NP-complete even on planar graphs.

What is of interest to us here is that in many cases the good algorithmic behaviour of a class of structures can be explained or is linked to the expressive power of logics. This is especially the case with so-called *meta-theorems* that link definability in logic with tractability. Examples of such meta-theorems are Courcelle's theorem [10] which shows that any property definable in monadic second-order logic is decidable in linear time on classes of bounded tree-width and the result of Dawar *et al.* [13] that first-order definable optimization problems admit polynomial-time approximation schemes on classes of structures that exclude a minor. Also among results that tie together logical expressiveness and algorithmic complexity on restricted classes, one can mention the theorem of Grohe and Mariño [26] to the effect that LFP + C captures exactly the polynomial-time decidable properties of classes of structures of bounded treewidth. In this section, we take a brief tour of some highlights of such results.

Acyclic Structures. To say that the adjacency graph $G\mathbb{A}$ of a structure $\mathbb{A}$ is acyclic is to say that all relations in $\mathbb{A}$ are essentially unary or binary and the union of the symmetric closures of the binary relations is a forest. One interesting recent result on such classes of structures is that of Benedikt and Segoufin [6] that any first-order sentence that is order-invariant on trees is equivalent to one without order. This contrasts with a construction of Gurevich (see [1, Exercise 17.27]) that shows that there is a first-order sentence that is order-invariant on the class of finite structures but is not equivalent to any first-order sentence without order. The theorem of Benedikt and Segoufin can be seen as a special case of interpolation. The general version of Craig's interpolation theorem (see [27]) is known to fail on the class of finite structures and even on the class of finite acyclic structures.

Another important respect in which acyclic structures are well-behaved is that while the validities of first-order logic on finite structures are not recursively enumerable, the validities on acyclic structures are decidable. Indeed, it is well-known that even monadic second-order logic (MSO) is decidable on trees (see [7] for a treatment). Moreover, by Courcelle's theorem mentioned above, we know that the problem of deciding, given a formula φ of MSO and an acyclic structure $\mathbb{A}$, whether or not $\mathbb{A} \models \varphi$ is decidable by an algorithm running in time $O(f(|\varphi|)|\mathbb{A}|)$ for some computable function f. We express this by saying that the *satisfiability* problem for the logic (also often called the *model-checking* problem) is *fixed-parameter tractable.* It has also been known, since results of Immerman and Lander and Lindell that LFP + C captures polynomial time on trees [30,34].

Finally, it has been proved that the homomorphism and extension preservation theorems hold on the class of acyclic structures (see [4] and [3] respectively). Indeed these preservation properties hold of any class of finite acyclic structures which is closed under substructures and disjoint unions, but may fail for other subclasses.

Bounded Treewidth. Let $\mathcal{T}_k$ denote the class of all structures of treewidth at most k. It is known that many of the properties of acyclic structures that make it a well-behaved class also extend to $\mathcal{T}_k$ for values of k larger than 1. However, it is not known if the order-invariance result of Benedikt and Segoufin is one of these properties. This remains an open question. Monadic second-order logic is as tame on $\mathcal{T}_k$ as it is on $\mathcal{T}_1$ since it is known that the satisfiability problem is decidable [9] and the satisfaction problem is fixed-parameter tractable [10].

It has been shown that $\mathcal{T}_k$ has the homomorphism preservation property [4] as well as the extension preservation property [3]. The former holds, in fact, for all subclasses of $\mathcal{T}_k$ that are closed under substructures and disjoint unions, but this is not true of extension preservation. Indeed, it is shown in [3] that extension preservation fails for the class of all planar graphs of treewidth at most 4, which is a subclass of $\mathcal{T}_4$.

We have mentioned above that Grohe and Mariño [26] proved that LFP + C captures polynomial time computation on $\mathcal{T}_k$ for any k. Recently, this has been shown to be optimal, in the following sense. For any nondecreasing function $f : \mathbb{N} \to \mathbb{N}$, let $\mathcal{T}_f$ denote the class of structures where any structure $\mathbb{A}$ of at most n elements has treewidth at most $f(n)$. Then, we can show [15] that as long as f is not bounded by a constant, there are polynomial time properties in $\mathcal{T}_f$ that are not expressible in LFP + C. Note, this does not preclude the possibility that LFP + C capture P on subclasses of $\mathcal{T}_f$ of unbounded treewidth. Indeed, just such a possibility is realised by the result of Grohe that LFP + C captures P on planar graphs [24] and more generally on graphs of bounded genus [25].

Bounded Degree Structures. Bounding the maximum degree of a structure is a restriction quite orthogonal to bounding its treewidth and yields quite different behaviour. While graphs of maximum degree bounded by 2 are very simple, consisting of disjoint unions of paths and cycles, structures of maximum degree 3 already form a rather rich class. That is, if $\mathcal{D}_k$ is the class of structures with maximum degree k, then the MSO theory of $\mathcal{D}_3$ is undecidable as is its first-order theory. Indeed, the first-order theory of planar graphs of degree at most 3 is also undecidable [12]. Furthermore, the satisfaction problem for MSO is intractable as one can construct sentences of MSO which express NP-hard problems on planar grids. However, it is the case that the satisfaction problem for first-order logic is fixed-parameter tractable on $\mathcal{D}_k$ for all k. This was shown by Seese [39].

The question of devising a logic in which one can express all and only the polynomial-time properties of bounded degree structures is an interesting one. The graph isomorphism problem is known to be solvable in polynomial time on graphs of bounded degree [36], and indeed, there is a polynomial-time algorithm for canonical labelling of such graphs [5]. It follows from general considerations about canonical labelling functions (see [17, Chapter 11]) that there is some logic that captures exactly P on $\mathcal{D}_k$, for each k. However, we also know, by the construction of Cai, Fürer and Immerman [8] that LFP + C is too weak a logic for this purpose. It remains an open question to find a "natural" logic that captures P on bounded degree classes.

On the question of preservation properties, both the homomorphism and extension preservation theorems have been shown to hold, not only on $\mathcal{D}_k$, but also on subclasses closed under substructures and disjoint unions [4,3].

Excluded Minor Classes. Classes with excluded minors are too general a case for good algorithmic behaviour of MSO. This logic is already undecidable, and its satisfaction problem intractable, on planar graphs. Indeed, first-order logic is also undecidable on planar graphs. However, it has been shown that the satisfaction problem for first-order logic is fixed-parameter tractable on any class of structures that excludes a minor [20]. While the extension preservation theorem fails in general on such classes, and was even shown to fail on planar graphs [3], the homomorphism preservation property holds of all classes which exclude a minor and are closed under taking substructures and disjoint unions [4]. It remains an open question whether one can construct a logic that captures P on excluded minor classes. Grohe conjectured [25] that LFP+C is actually sufficient for this purpose. Indeed, he proved that LFP + C captures P on all classes of bounded genus.

Further Extensions. Frick and Grohe showed that the satisfaction problem for first-order logic is fixed-parameter tractable, even on classes of structures of bounded treewidth [21]. This result was recently extended to classes of graphs that locally exclude a minor [11] by an algorithmic analysis of the graph structure theorem of Robertson and Seymour. It is an open question whether or not it can also be extended to classes of graphs of bounded expansion. The model-theoretic and algorithmic properties of classes of graphs of bounded expansion and that locally exclude minors are yet to be studied in detail and a number of open questions remain.

4 Preservation Theorems

Among the results in the last section, we looked at classes of structures where the homomorphism and the extension preservation theorems are known to hold. Indeed, the homomorphism preservation theorem survives all the restrictions we considered, while the extension preservation is available in some. We now take a brief look at the methods used to establish the homomorphism and extension preservation theorems in the tame classes where they have been shown.

The key idea in these proofs is to establish an upper bound on the size of minimal models of a first-order sentence that has the relevant preservation property. For instance, suppose φ is a sentence that is preserved under extensions on a class of structures $\mathcal{C}$. Then, we say that a structure $\mathbb{A}$ is a minimal model of φ in $\mathcal{C}$ if $\mathbb{A} \models \varphi$ and no proper induced substructure of $\mathbb{A}$ is a model of φ. It is then immediate that the models of φ in $\mathcal{C}$ are exactly the extensions of minimal models. It is not difficult to show that φ is equivalent to an existential sentence on $\mathcal{C}$ if, and only if, it has finitely many minimal models. The same holds true for sentences preserved under homomorphisms if we take minimal models, not with respect to induced substructures, but allowing substructures that are not induced (see [4] for details). The preservation properties for tame classes mentioned above are then proved by showing that from every sentence φ we can extract a bound N such that all minimal models of φ have at most N elements. This bound is obtained by considering structural properties that a minimal model must satisfy.

It can be shown that if φ is preserved under homomorphisms on a class $\mathcal{C}$ (closed under disjoint unions and substructures) then there are positive integers d and m such that no minimal model of φ in $\mathcal{C}$ contains a set of m elements that are pairwise distance d or greater from each other. This result is essentially obtained from a construction of Ajtai and Gurevich [2] and is a consequence of Gaifman's locality theorem for first-order logic. A more involved construction, again based on Gaifman's theorem establishes this density property also for formulas preserved under extensions. An immediate consequence is the preservation theorem for certain classes we call wide. A class of structures $\mathcal{C}$ is *wide* if for all d and m, there is an N such that every structure in $\mathcal{C}$ with at least N elements contains a set of m elements that are pairwise distance at least d from each other. For instance, any class of bounded degree is easily seen to be wide.

The construction of Ajtai and Gurevich shows further that for any sentence φ preserved under homomorphisms on $\mathcal{C}$, and for every positive integer s, there are d and m such that no minimal model of φ in $\mathcal{C}$ contains a set of m elements that are pairwise distance d or greater from each other, even after s elements are removed from it. This leads to a definition of classes that are almost wide: $\mathcal{C}$ is *almost wide* if there is an s such that for all d and m there is an N such that in every structure $\mathbb{A}$ in $\mathcal{C}$ with at least N elements, one needs to remove at most s elements to obtain a set of m elements that are pairwise distance at least d from each other. A combinatorial construction is needed to prove that classes of graphs that exclude a minor are almost wide (see [4] and also [32]). Almost wideness is not sufficient in itself to establish the extension preservation property (as is witnessed by the class of planar graphs). However, we can strengthen the requirement of closure under disjoint unions to closure under unions over "bottlenecks" (see [3]) and obtain a sufficient condition. This leads, in particular, to the proof that the extension preservation theorem holds for the classes $\mathcal{T}_k$.

It is not clear if classes of structures of bounded expansion or with locally excluded minors are almost wide. However, they can be shown to satisfy a weaker condition. Say a class of structures $\mathcal{C}$ is *quasi-wide* if for all d there is an s such that for all m, there is an N such that if $\mathbb{A} \in \mathcal{C}$ has N or more elements, then there is a set B of at most s elements in $\mathbb{A}$ such that $\mathbb{A} \setminus B$ contains a set of m elements that are pairwise at least distance d from each other. It can be shown that classes of structures of bounded expansion and that locally exclude minors are quasi-wide. Furthermore, it seems that a strengthening of the Ajtai-Gurevich lemma can establish the homomorphism preservation theorem for quasi-wide classes that are closed under disjoint unions and minors [14].

5 Conclusion

The class of all finite structures is not a model-theoretically well-behaved class. Recent work has investigated to what extent considering further restricted classes may enable us to discover interesting model-theoretic properties. The restrictions that have been found that yield tame classes are also those that yield good algorithmic behaviour. The interaction between logical and algorithmic properties of these classes remains an active area of investigation. Besides preservation theorems, many model-theoretic properties of these classes remain to be explored. In the absence of the Compactness Theorem, which is the bedrock of the model theory of infinite structures, the methods used on tame classes of finite structures are varied and often combinatorial in nature. However, methods based on locality appear to play a central role.

References

1. Abiteboul, S., Hull, R., Vianu, V.: Foundations of Databases. Addison-Wesley, Reading (1995)
2. Ajtai, M., Gurevich, Y.: Datalog vs first-order logic. J. of Computer and System Sciences 49, 562–588 (1994)
3. Atserias, A., Dawar, A., Grohe, M.: Preservation under extensions on well-behaved finite structures. In: Caires, L., Italiano, G.F., Monteiro, L., Palamidessi, C., Yung, M. (eds.) ICALP 2005. LNCS, vol. 3580, pp. 1437–1449. Springer, Heidelberg (2005)
4. Atserias, A., Dawar, A., Kolaitis, P.G.: On preservation under homomorphisms and unions of conjunctive queries. Journal of the ACM 53, 208–237 (2006)
5. Babai, L., Luks, E.M.: Canonical labeling of graphs. In: Proc. of the 15th ACM Symp. on the Theory of Computing, pp. 171–183. ACM Press, New York (1983)
6. Benedikt, M., Segoufin, L.: Towards a characterization of order-invariant queries over tame structures. In: Ong, L. (ed.) CSL 2005. LNCS, vol. 3634, pp. 276–291. Springer, Heidelberg (2005)
7. Börger, E., Grädel, E., Gurevich, Y.: The Classical Decision Problem. Springer, Heidelberg (1997)
8. Cai, J-Y., Fürer, M., Immerman, N.: An optimal lower bound on the number of variables for graph identification. Combinatorica 12(4), 389–410 (1992)
9. Courcelle, B.: The monadic second-order logic of graphs ii: Infinite graphs of bounded width. Theory of Computing Systems 21, 187–221 (1989)
10. Courcelle, B.: Graph rewriting: An algebraic and logic approach. In: van Leeuwan, J. (ed.) Handbook of Theoretical Computer Science. Formal Models and Sematics (B), vol. B, pp. 193–242. Elsevier, Amsterdam (1990)
11. Dawar, A., Grohe, M., Kreutzer, S.: Locally excluding a minor. In: Proc. 22nd IEEE Symp. on Logic in Computer Science (2007)
12. Dawar, A., Grohe, M., Kreutzer, S., Schweikardt, N.: Model theory makes formulas large. In: ICALP'07: Proc. 34th International Colloquium on Automata, Languages and Programming. LNCS, Springer, Heidelberg (2007)
13. Dawar, A., Kreutzer, S., Grohe, M., Schweikardt, N.: Approximation schemes for first-order definable optimisation problems. In: Proc. 21st IEEE Symp. on Logic in Computer Science, pp. 411–420 (2006)
14. Dawar, A., Malod, G.: forthcoming
15. Dawar, A., Richerby, D.: The power of counting logics on restricted classes of finite structures. In: CSL 2007: Computer Science Logic. LNCS, Springer, Heidelberg (2007)
16. Diestel, R.: Graph Theory, 3rd edn. Springer, Heidelberg (2005)
17. Ebbinghaus, H-D., Flum, J.: Finite Model Theory. 2nd edn., Springer, Heidelberg (1999)
18. Eppstein, D.: Diameter and treewidth in minor-closed graph families. Algorithmica 27, 275–291 (2000)
19. Fagin, R.: Generalized first-order spectra and polynomial-time recognizable sets. In: Karp, R.M. (ed.) Complexity of Computation, SIAM-AMS Proceedings, vol. 7, pp. 43–73 (1974)
20. Flum, J., Grohe, M.: Fixed-parameter tractability, definability, and model checking. SIAM Journal on Computing 31, 113–145 (2001)
21. Frick, M., Grohe, M.: Deciding first-order properties of locally tree-decomposable structures. Journal of the ACM 48, 1184–1206 (2001)

22. Gaifman, H.: On local and non-local properties. In: Stern, J. (ed.) Proceedings of the Herbrand Symposium Logic Colloquium '81, pp. 105–135. North-Holland, Amsterdam (1982)
23. Grädel, E., Kolaitis, P.G., Libkin, L., Marx, M., Spencer, J., Vardi, M.Y., Venema, Y., Weinstein, S.: Finite Model Theory and Its Applications. Springer, Heidelberg (2007)
24. Grohe, M.: Fixed-point logics on planar graphs. In: Proc. 13th IEEE Annual Symp. Logic in Computer Science, pp. 6–15 (1998)
25. Grohe, M.: Isomorphism testing for embeddable graphs through definability. In: Proc. 32nd ACM Symp.Theory of Computing, pp. 63–72 (2000)
26. Grohe, M., Mariño, J.: Definability and descriptive complexity on databases of bounded tree-width. In: Beeri, C., Bruneman, P. (eds.) ICDT 1999. LNCS, vol. 1540, Springer, Heidelberg (1998)
27. Hodges, W.: Model Theory. Cambridge University Press, Cambridge (1993)
28. Immerman, N.: Relational queries computable in polynomial time. Information and Control 68, 86–104 (1986)
29. Immerman, N.: Descriptive Complexity. Springer, Heidelberg (1999)
30. Immerman, N., Lander, E.S.: Describing graphs: A first-order approach to graph canonization. In: Selman, A. (ed.) Complexity Theory Retrospective, Springer, Heidelberg (1990)
31. Kolatis, P.G.: A tutorial on finite model theory. In: Proceedings of the Eighth Annual IEEE Symp. on Logic in Computer Science, LICS 1993, pp. 122–122 (1993)
32. Kreidler, M., Seese, D.: Monadic NP and graph minors. In: Gottlob, G., Grandjean, E., Seyr, K. (eds.) Computer Science Logic. LNCS, vol. 1584, pp. 126–141. Springer, Heidelberg (1999)
33. Libkin, L.: Elements of Finite Model Theory. Springer, Heidelberg (2004)
34. Lindell, S.: An analysis of fixed-point queries on binary trees. Theoretical Computer Science 85(1), 75–95 (1991)
35. Livchak, A.: The relational model for process control. Automated Documentation and Mathematical Linguistics 4, 27–29 (1983)
36. Luks, E.M.: Isomorphism of graphs of bounded valence can be tested in polynomial time. Journal of Computer and System Sciences 25, 42–65 (1982)
37. Mohar, B., Thomassen, C.: Graphs on Surfaces. Johns Hopkins University Press (2001)
38. Rossman, B.: Existential positive types and preservation under homomorphisisms. In: 20th IEEE Symposium on Logic in Computer Science, pp. 467–476 (2005)
39. Seese, D.: Linear time computable problems and first-order descriptions. Math. Struct. in Comp. Science 6, 505–526 (1996)
40. Nešetřil, J., de Mendez, P.O.: The grad of a graph and classes with bounded expansion. International Colloquium on Graph Theory, 101–106 (2005)
41. Tait, W.W.: A counterexample to a conjecture of Scott and Suppes. Journal of Symbolic Logic 24, 15–16 (1959)
42. Vardi, M.Y.: The complexity of relational query languages. In: Proc. of the 14th ACM Symp. on the Theory of Computing, pp. 137–146 (1982)

Minimum Cycle Bases in Graphs Algorithms and Applications

Kurt Mehlhorn

Max-Planck-Institut für Informatik, Saarbrücken, Germany

A cycle basis of a graph is a family of cycles which spans all cycles of the graph. In an undirected graph, a cycle is simply a set of edges with respect to which every vertex has even degree. We view cycles as vectors indexed by edges. The entry for an edge is one if the edge belongs to the cycle and is zero otherwise. Addition of cycles corresponds to vector addition modulo 2 (symmetric difference of the underlying edge sets). In this way, the cycles of a graph form a vector space and a cycle basis is simply a basis of this vector space. The notion for directed graphs is slightly more involved.

The weight of a cycle is either the number of edges in the cycle (in unweighted graphs) or the sum of the weights of the edges in the cycle (in weighted graphs). A minimum cycle basis is basis of total minimum weight.

The analysis of the cycle space has applications in various fields, e.g., electrical engineering [Kir47], structural analysis [CHR76], biology and chemistry [Gle01], surface reconstruction [GKM$^+$], and periodic timetabling [Lie06]. Some of these applications require bases with special properties [LR07]. In the first part of the talk, I will discuss applications of cycle basis.

In the second part, I turn to construction algorithms. The first polynomial time algorithms for constructing minimum cycle bases in undirected graphs are due to Horton [Hor87] and de Pina [dP95]. Faster realizations of the latter approach are discussed in the papers [BGdV04, KMMP04, MM]. Both approaches can be generalized to directed graphs [LR05, KM05, HKM06, Kav05]. Approximation algorithms are discussed in [KMM07].

Integral cycle basis are required for the application to periodic timetabling. Finding minimal integral or fundamental bases is NP-complete. Construction and approximation algorithms are described in [Lie03, Lie06, Kav, ELR07].

References

[BGdV04] Berger, F., Gritzmann, P., de Vries, S.: Minimum cycle basis for network graphs. Algorithmica 40(1), 51–62 (2004)

[CHR76] Cassell, A.C., Henderson, J.C., Ramachandran, K.: Cycle bases of minimal measure for the structural analysis of skeletal structures by the flexibility method. In: Proc. Royal Society of London Series A, vol. 350, pp. 61–70 (1976)

[dP95] de Pina, J.C.: Applications of Shortest Path Methods. PhD thesis, University of Amsterdam, Netherlands (1995)

[ELR07] Elkin, M., Liebchen, Ch., Rizzi, R.: New length bounds for cycle bases. Technical report, TU Berlin (June 2007)

[GKM$^+$] Gotsman, C., Kaligosi, K., Mehlhorn, K., Michail, D., Pyrga, E.: Cycle Basis of Graphs and Sampled Manifolds (submitted for publication)

L. Kučera and A. Kučera (Eds.): MFCS 2007, LNCS 4708, pp. 13–14, 2007.

[Gle01] Gleiss, P.M.: Short Cycles, Minimum Cycle Bases of Graphs from Chemistry and Biochemistry. PhD thesis, Fakultät Für Naturwissenschaften und Mathematik der Universität Wien (2001)

[HKM06] Hariharan, R., Kavitha, T., Mehlhorn, K.: A Faster Deterministic Algorithm for Minimum Cycle Basis in Directed Graphs. In: Bugliesi, M., Preneel, B., Sassone, V., Wegener, I. (eds.) ICALP 2006. LNCS, vol. 4051, pp. 250–261. Springer, Heidelberg (2006)

[Hor87] Horton, J.D.: A polynomial-time algorithm to find the shortest cycle basis of a graph. SICOMP 16, 358–366 (1987)

[Kav] Kavitha, T.: A simple approximation algorithm for integral bases. personal communication

[Kav05] Kavitha, T.: An $O(m^2n)$ randomized algorithm to compute a minimum cycle basis of a directed graph. In: Caires, L., Italiano, G.F., Monteiro, L., Palamidessi, C., Yung, M. (eds.) ICALP 2005. LNCS, vol. 3580, Springer, Heidelberg (2005)

[Kir47] Kirchhoff, G.: Über die Auflösung der Gleichungen, auf welche man bei der Untersuchungen der linearen Verteilung galvanischer Ströme geführt wird. Poggendorf Ann. Phy. Chem. 72, 497–508 (1847)

[KM05] Kavitha, T., Mehlhorn, K.: A Polynomial Time Algorithm for Minimum Cycle Basis in Directed Graphs. In: Diekert, V., Durand, B. (eds.) STACS 2005. LNCS, vol. 3404, Springer, Heidelberg (2005)

[KMM07] Kavitha, T., Mehlhorn, K., Michail, D.: New Approximation Algorithms for Minimum Cycle Bases of Graphs. In: Thomas, W., Weil, P. (eds.) STACS 2007. LNCS, vol. 4393, Springer, Heidelberg (2007)

[KMMP04] Kavitha, T., Mehlhorn, K., Michail, D., Paluch, K.: A Faster Algorithm for Minimum Cycle Bases of Graphs. In: Díaz, J., Karhumäki, J., Lepistö, A., Sannella, D. (eds.) ICALP 2004. LNCS, vol. 3142, Springer, Heidelberg (2004)

[Lie03] Liebchen, Ch.: Finding short integral cycle bases for cyclic timetabling. In: Di Battista, G., Zwick, U. (eds.) ESA 2003. LNCS, vol. 2832, pp. 715–726. Springer, Heidelberg (2003)

[Lie06] Liebchen, Ch.: Periodic Timetable Optimization in Public Transport. PhD thesis, TU Berlin (2006)

[LR05] Liebchen, C., Rizzi, R.: A greedy approach to compute a minimum cycle basis of a directed graph. Information Processing Letters 94(3), 107–112 (2005)

[LR07] Liebchen, C., Rizzi, R.: Classes of cycle bases. Discrete Applied Mathematics 155(3), 337–355 (2007)

[MM] Mehlhorn, K., Michail, D.: Minimum Cycle Bases: Faster and Simpler (submitted for publication)

Hierarchies of Infinite Structures Generated by Pushdown Automata and Recursion Schemes

C.-H.L. Ong

Oxford University Computing Laboratory
Wolfson Building, Parks Road, Oxford OX1 3QD, England
users.comlab.ox.ac.uk/luke.ong/

Abstract. Higher-order recursion schemes and higher-order pushdown automata are closely related methods for generating infinite hierarchies of infinite structures. Subsuming well-known classes of models of computation, these rich hierarchies (of word languages, trees, and graphs respectively) have excellent model-checking properties. In this extended abstract, we survey recent expressivity and decidability results about these infinite structures.

A class of infinite-state systems of fundamental importance to software verification are *pushdown automata*. It is an old idea that first-order imperative programs with recursive procedures can be accurately modelled by pushdown automata (see e.g. [1] for a precise account). Viewed abstractly, a pushdown automaton (PDA) is a finitely-presentable infinite-state transition graph in which a state (vertex) is a reachable configuration that carries a finite but unbounded amount of information, namely, the contents of the stack. Müller and Schupp [2] have proved that the monadic second-order (MSO) theory of the transition graph of a pushdown automaton is decidable, and the automata-theoretic technique using two-way alternating automata [3] provides a direct (elementary) model-checking decision procedure for temporal properties.

1 Pushdown Automata and Safe Recursion Schemes

There is in fact an infinite hierarchy of *higher-order pushdown automata*. First introduced by Maslov [4,5] as accepting devices for word languages, order-0 and order-1 pushdown automata are, by definition, the finite-state and standard pushdown automata respectively. An order-2 PDA has an order-2 stack, which is a stack of order-1 (i.e. ordinary) stacks; in addition to the usual (first-order actions) *push* and *pop*, it has an order-2 $push_2$ that duplicates the top 1-stack, and an order-2 pop_2 that pops the entire top 1-stack. As n varies over the natural numbers, the languages accepted by order-n PDA form an infinite hierarchy. In *op. cit.* Maslov showed that the hierarchy can be defined equivalently as languages generated by *higher-order indexed grammars*, generalising indexed grammars in the sense of Aho [6]. Yet another characterisation of Maslov's hierarchy was given by Damm and Goerdt [7,8]: they studied *higher-order recursion*

L. Kučera and A. Kučera (Eds.): MFCS 2007, LNCS 4708, pp. 15–21, 2007.

schemes that satisfy the constraint of *derived types*, and showed that the word languages generated by order-n such schemes coincide with those accepted by order-n PDA. The low orders of the Maslov hierarchy are well-known – orders 0, 1 and 2 are respectively the regular, context-free and indexed languages, though little is known about languages at higher orders.

Higher-order PDA as a generating device for (possibly infinite) ranked trees was first studied by Knapik, Niwiński and Urzyczyn in a TLCA'01 paper [9]. As in the case of word languages, an infinite hierarchy of trees are thus defined. Lower orders of the pushdown hierarchy are well-known classes of trees: orders 0, 1 and 2 are respectively the regular [10], algebraic [11] and hyperalgebraic trees [9]. In a follow-up paper in FoSSaCS'02 [12] Knapik *et al.* considered another method for generating such trees, namely, by higher-order (deterministic) recursion schemes that satisfy the constraint of *safety*[1]. A major result in that work is the equi-expressivity of the two methods as generators of trees; another is that all trees in the pushdown hierarchy have decidable monadic second-order (MSO) theories.

In a MFCS'02 paper, Caucal [14] introduced a tree hierarchy and a graph hierarchy, which are defined by mutual induction using a pair of transformations, one from trees to graphs, and the other from graphs to trees. The order-0 graphs are by definition the finite graphs. Inductively the order-n trees are defined as the unravelling of the order-n graphs; and the order-$(n+1)$ graphs are defined as the *inverse rational mappings* of the order-n trees. Since these transformations preserve the decidability of MSO theories, all infinite structures belonging to the two hierarchies have decidable MSO theories. Caucal's hierarchies turn out to be closely related to higher-order PDA: Carayol and Wöhrle [15] have shown that the order-n graphs in Caucal's graph hierarchy are exactly the ε-closure of the configuration graphs of order-n pushdown systems.

To summarise, the hierarchies of infinite structures generated by higher-order pushdown automata are of considerable interest to infinite-state verification:

1. *Expressivity*: They are rich and very general; as tabulated below, the lower orders of the hierarchies are well-known classes of models of computation

	Pushdown (= Safe Recursion Scheme) Hierarchies of		
Order	*Word Languages*	*Trees*	*Graphs*
0	regular	regular	finite
1	context-free	algebraic [11]	prefix-recognisable [16]
2	indexed [6]	hyper-algebraic [9]	
⋮	⋮	⋮	⋮

   though little is known about the structures at higher orders.

[1] As a syntactic constraint, *safety* is equivalent to Damm's *derived types* [7]; see de Miranda's thesis [13] for a proof.

2. *Robustness*: Remarkably, order-n pushdown automata are equi-expressive with order-n recursion schemes that satisfy the syntactic constraint of safety, as generators of word languages [7,8] and of trees [12] respectively.
3. *Excellent model-checking properties*: The infinite structures that are generated have decidable MSO properties. The criterion of MSO decidability is appropriate because the MSO logic is commonly regarded as the gold standard of specification languages for model checking: standard temporal logics such as LTL, ETL, CTL, CTL*, and even modal mu-calculus can all be embedded in MSO; moreover any obvious extension of the logic would break decidability.

2 Recursion Schemes and Collapsible Pushdown Automata

Several questions arise naturally from the preceding equi-expressivity and decidability results.

1. Syntactically awkward, the *safety* constraint [12] seems unnatural. Is it really necessary for MSO decidability? Precisely, do trees that are generated by (arbitrary) recursion schemes have decidable MSO theories?
2. Can the expressivity of (arbitrary) recursion schemes be characterised by an appropriate class of automata (that contains the higher-order pushdown automata)?
3. Does safety constrain expressivity? I.e. is there a tree or a graph that is generated by an unsafe, but *not* by any safe, recursion scheme?

Recent work has provided answers to the first two of these questions, and a partial answer to the third. Using new ideas and techniques from *innocent game semantics* [17], we have proved [18]:

Theorem 1 (MSO decidability). *The modal mu-calculus model checking problem for ranked trees generated by order-n recursion schemes is n-EXPTIME complete, for each $n \geq 0$. Hence these trees have decidable MSO theories.*

To our knowledge, the hierarchy of trees generated by (arbitrary) recursion schemes is the largest, generically-defined class of ranked trees that have decidable MSO theories, subsuming earlier results such as [10,11] and also [12,19,20,21] etc. A novel ingredient in the proof of Theorem 1 is a certain *transference principle* from the tree $[\![G]\!]$ generated by the recursion scheme G – the *value tree* – to an auxiliary *computation tree* $\lambda(G)$, which is in essence an infinite λ-term obtained by unfolding the recursion scheme *ad infinitum*. The transference relies on a strong correspondence theorem between *paths* in the value tree and what we call *traversals* in the computation tree, established using innocent game semantics [17]. This allows us to prove that a given alternating parity tree automaton (APT) has an accepting run-tree over the value tree iff it has an accepting traversal-tree over the computation tree. The second ingredient is the simulation of an accepting traversal-tree by a certain set of annotated paths over the computation tree.

Higher-order recursion schemes are essentially closed terms of the simply-typed lambda calculus with general recursion, generated from uninterpreted first-order function symbols. A fundamental question in higher-type recursion is to characterise the expressivity of higher-order recursion schemes in terms of a class of automata. Thus the results of Damm and Goerdt [8], and of Knapik *et al.* [12], may be viewed as partial answers of the question. An exact correspondence with recursion schemes has never been established before.

Another partial answer was recently obtained by Knapik, Niwiński, Urzyczyn and Walukiewicz. In an ICALP'05 paper [19], they proved that order-2 homogeneously typed (but not necessarily safe) recursion schemes are equi-expressive with a variant class of order-2 pushdown automata called *panic automata.* In recent joint work [22] with Hague, Murawski and Serre, we have given a complete answer to the question. We introduce a new kind of higher-order pushdown automata (which generalises *pushdown automata with links* [23], or equivalently panic automata, to all finite orders), called *collapsible pushdown automata* (CPDA), in which every symbol in the stack has a link to a (necessarily lower-ordered) stack situated somewhere below it. In addition to the higher-order stack operations $push_i$ and pop_i, CPDA have an important operation called *collapse*, whose effect is to "collapse" a stack s to the prefix as indicated by the link from the top_1-symbol of s. A major result of [22] is the following:

Theorem 2 (Equi-expressivity). *For each $n \geq 0$, order-n recursion schemes and order-n collapsible pushdown automata define the same trees.*

Thus order-n CPDA may be viewed as a machine characterisation of order-n recursively-defined lambda-terms, and hence also of order-n innocent strategies (since innocent strategies are a universal model of higher-order recursion schemes). In one direction of the proof, we show that for every (tree-generating) order-n pushdown automaton, there is an order-n recursion scheme that generates the same tree. In the other direction, we introduce an algorithm (as implemented in the tool HOG [24]) translating an order-n recursion scheme G to an order-n CPDA A_G that computes exactly the traversals over the computation tree $\lambda(G)$ and hence paths in the value tree $[\![G]\!]$.

The Equi-Expressivity Theorem has a number of useful consequences. It allows us to translate decision problems on trees generated by recursion schemes to equivalent problems on CPDA and *vice versa.* Chief among them is the Modal Mu-Calculus Model-Checking Problem (equivalently the Alternating Parity Tree Automaton Acceptance Problem); another is the Monadic Second-Order (MSO) Model-Checking Problem. We observe that these problems – concerning infinite structures generated by recursion schemes – reduce to the problem of solving a parity game played on a *collapsible pushdown graph* i.e. the configuration graph of a corresponding collapsible pushdown system (CPDS).

The transfer of techniques goes both ways. Another result in our work [22] is a self-contained (without recourse to game semantics) proof of the solvability of parity games on collapsible pushdown graphs by generalising *standard* techniques in the field:

Theorem 3 (Solvability). *For each $n \geq 0$, solvability of parity games over the configuration graphs of order-n collapsible pushdown systems is n-EXPTIME complete.*

The Theorem subsumes a number of well-known results in the literature [25,26,19]. By appealing to the Equi-Expressivity Theorem, we obtain new proofs for the decidability (and optimal complexity) of model-checking problems of trees generated by recursion schemes as studied in [18].

Finally, in contrast to higher-order pushdown graphs (which do have decidable MSO theories [14]), we show in [22] that the MSO theories of collapsible pushdown graphs are undecidable. Hence collapsible pushdown graphs are, to our knowledge, the first example of a natural class of finitely-presentable graphs that have undecidable MSO theories while enjoying decidable modal mu-calculus theories.

3 Practical Relevance to Semantics and Verification

Recursion schemes are an old and influential formalism for the semantical analysis of both imperative and functional programs [27,7]. Indeed one of the first models of "Algol-like languages" (i.e. higher order procedural languages) was derived from the pushdown hierarchy of word languages (see Damm's monograph [7]). As indicated by the recent flurry of results [12,14,19,18,22], the hierarchies of (collapsible) pushdown automata and recursion schemes are highly relevant to infinite-state verification. In the light of the mediating algorithmic game semantics [17,28], it follows from the strong correspondence (Theorem 2) between recursion schemes and collapsible pushdown automata, that the collapsible pushdown hierarchies are accurate models of computation that underpin the computer-aided verification of higher-order procedural languages (such as Ocaml, Haskell, F#, etc.) — a challenging direction for software verification.

Acknowledgements. Some of the results cited above are based on (as yet unpublished) joint work [22] with Matthew Hague, Andrzej Murwaski and Olivier Serre. A full account will be published elsewhere.

References

1. Jones, N.D., Muchnick, S.S.: Complexity of finite memory programs with recursion. Journal of the Association for Computing Machinery 25, 312–321 (1978)
2. Muller, D.E., Schupp, P.E.: The theory of ends, pushdown automata, and second-order logic. Theoretical Computer Science 37, 51–75 (1985)
3. Kupferman, O., Vardi, M.Y.: An automata-theoretic approach to reasoning about infinite-state systems. In: Emerson, E.A., Sistla, A.P. (eds.) CAV 2000. LNCS, vol. 1855, Springer, Heidelberg (2000)
4. Maslov, A.N.: The hierarchy of indexed languages of an arbitrary level. Soviet mathematics Doklady 15, 1170–1174 (1974)
5. Maslov, A.N.: Multilevel stack automata. Problems of Information Transmission 12, 38–43 (1976)

6. Aho, A.: Indexed grammars - an extension of context-free grammars. J. ACM 15, 647–671 (1968)
7. Damm, W.: The IO- and OI-hierarchy. Theoretical Computer Science 20, 95–207 (1982)
8. Damm, W., Goerdt, A.: An automata-theoretical characterization of the OI-hierarchy. Information and Control 71, 1–32 (1986)
9. Knapik, T., Niwiński, D., Urzyczyn, P.: Deciding monadic theories of hyperalgebraic trees. In: Abramsky, S. (ed.) TLCA 2001. LNCS, vol. 2044, pp. 253–267. Springer, Heidelberg (2001)
10. Rabin, M.O.: Decidability of second-order theories and automata on infinite trees. Transactions of the American Mathematical Society 141, 1–35 (1969)
11. Courcelle, B.: The monadic second-order logic of graphs IX: machines and their behaviours. Theoretical Computer Science 151, 125–162 (1995)
12. Knapik, T., Niwiński, D., Urzyczyn, P.: Higher-order pushdown trees are easy. In: Nielsen, M., Engberg, U. (eds.) ETAPS 2002 and FOSSACS 2002. LNCS, vol. 2303, pp. 205–222. Springer, Heidelberg (2002)
13. de Miranda, J.: Structures generated by higher-order grammars and the safety constraint. PhD thesis, University of Oxford (2006)
14. Caucal, D.: On infinite terms having a decidable monadic theory. In: Diks, K., Rytter, W. (eds.) MFCS 2002. LNCS, vol. 2420, pp. 165–176. Springer, Heidelberg (2002)
15. Carayol, A., Wöhrle, S.: The Caucal hierarchy of infinite graphs in terms of logic and higher-order pushdown automata. In: Pandya, P.K., Radhakrishnan, J. (eds.) FST TCS 2003: Foundations of Software Technology and Theoretical Computer Science. LNCS, vol. 2914, pp. 112–123. Springer, Heidelberg (2003)
16. Caucal, D.: On infinite transition graphs having a decidable monadic theory. In: Meyer auf der Heide, F., Monien, B. (eds.) ICALP 1996. LNCS, vol. 1099, pp. 194–205. Springer, Heidelberg (1996)
17. Hyland, J.M.E., Ong, C.H.L.: On Full Abstraction for PCF: I. Models, observables and the full abstraction problem, II. Dialogue games and innocent strategies, III. A fully abstract and universal game model. Information and Computation 163, 285–408 (2000)
18. Ong, C.H.L.: On model-checking trees generated by higher-order recursion schemes. In: Proceedings of the 21st Annual IEEE Symposium on Logic in Computer Science (LICS'06), pp. 81–90. IEEE Computer Society Press, Los Alamitos (2006)
19. Knapik, T., Niwiński, D., Urzyczyn, P., Walukiewicz, I.: Unsafe grammars and panic automata. In: Caires, L., Italiano, G.F., Monteiro, L., Palamidessi, C., Yung, M. (eds.) ICALP 2005. LNCS, vol. 3580, pp. 1450–1461. Springer, Heidelberg (2005)
20. Aehlig, K., Miranda, J.G.d., Ong, C.H.L.: The monadic second order theory of trees given by arbitrary level two recursion schemes is decidable. In: Urzyczyn, P. (ed.) TLCA 2005. LNCS, vol. 3461, pp. 39–54. Springer, Heidelberg (2005)
21. Aehlig, K.: A finite semantics for simply-typed lambda terms for infinite runs of automata. In: Ésik, Z. (ed.) CSL 2006. LNCS, vol. 4207, pp. 104–118. Springer, Heidelberg (2006)
22. Hague, M., Murawski, A.S., Ong, C.H.L., Serre, O.: Collapsible pushdown automata and recursion schemes. Technical report, Oxford University Computing Laboratory, p. 56 (Preprint, 2007), downloadable from `users.comlab.ox.ac.uk/luke.ong/publications/cpda-long.pdf`

23. Aehlig, K., de Miranda, J.G., Ong, C.H.L.: Safety is not a restriction at level 2 for string languages. In: Sassone, V. (ed.) FOSSACS 2005. LNCS, vol. 3441, pp. 490–501. Springer, Heidelberg (2005)
24. Blum, W.: A tool for constructing structures generated by higher-order recursion schemes and collapsible pushdown automata (2007), `web.comlab.ox.ac.uk/oucl/work/william.blum/`
25. Walukiewicz, I.: Pushdown processes: games and model-checking. Information and Computation 157, 234–263 (2001)
26. Cachat, T.: Higher order pushdown automata, the Caucal hierarchy of graphs and parity games. In: Baeten, J.C.M., Lenstra, J.K., Parrow, J., Woeginger, G.J. (eds.) ICALP 2003. LNCS, vol. 2719, pp. 556–569. Springer, Heidelberg (2003)
27. Nivat, M.: On the interpretation of recursive polyadic program schemes. Symp. Math. XV, 255–281 (1975)
28. Murawski, A.S., Ong, C.H.L., Walukiewicz, I.: Idealized Algol with ground recursion and DPDA equivalence. In: Caires, L., Italiano, G.F., Monteiro, L., Palamidessi, C., Yung, M. (eds.) ICALP 2005. LNCS, vol. 3580, pp. 917–929. Springer, Heidelberg (2005)

Evolvability

Leslie G. Valiant*

School of Engineering and Applied Sciences
Harvard University
valiant@deas.harvard.edu

Abstract. Living organisms function according to complex mechanisms that operate in different ways depending on conditions. Evolutionary theory suggests that such mechanisms evolved as result of a random search guided by selection. However, there has existed no theory that would explain quantitatively which mechanisms can so evolve in realistic population sizes within realistic time periods, and which are too complex. In this paper we suggest such a theory. Evolution is treated as a form of computational learning from examples in which the course of learning is influenced only by the fitness of the hypotheses on the examples, and not otherwise by the specific examples. We formulate a notion of evolvability that quantifies the evolvability of different classes of functions. It is shown that in any one phase of evolution where selection is for one beneficial behavior, monotone Boolean conjunctions and disjunctions are demonstrably evolvable over the uniform distribution, while Boolean parity functions are demonstrably not. The framework also allows a wider range of issues in evolution to be quantified. We suggest that the overall mechanism that underlies biological evolution is *evolvable target pursuit*, which consists of a series of evolutionary stages, each one pursuing an evolvable target in our technical sense, each target being rendered evolvable by the serendipitous combination of the environment and the outcome of previous evolutionary stages.

1 Introduction

We address the problem of quantifying how complex mechanisms, such as those found in living cells, can evolve into existence without any need for unlikely events to occur. If evolution merely performed a random search it would require exponential time, much too long to explain the complexity of existing biological structures. Darwin made this observation eloquently in the context of the evolution of the eye and suggested selection as the critical controlling principle. He called the supposition that the eye could evolve "··· absurd in the highest possible degree" were it not for the fact that eyes "vary ever so slightly" and might therefore evolve over time by selection [5].

This paper describes a quantitative theory of the possibilities and limitations of what selection can achieve in speeding up the process of acquiring complex

* This work was supported by grants NSF-CCR-03-10882, NSF-CCF-04-32037 and NSF-CCF-04-27129.

L. Kučera and A. Kučera (Eds.): MFCS 2007, LNCS 4708, pp. 22–43, 2007.

mechanisms beyond mere exhaustive search. In particular, we show that, in a defined sense, selection for a given beneficial behavior can provably support the evolution of certain specific classes of mechanisms, and provably not support that of certain other classes.

We approach this problem by viewing mechanisms from the viewpoint of the mathematical functions they realize. In particular the subject matter of the field of computational learning theory [3, 16, 20, 23] can be viewed as that of delineating limits on the complexity of functions that can be acquired with feasible resources without an explicit designer or programmer. A primary instance studied there is the acquisition of a recognition algorithm for a function given just positive and negative examples of it. The quantitative study of computational learning over the last two decades has shown that certain classes of recognition mechanisms can indeed be learned in a feasible amount of time, while others encounter apparently intractable computational impediments.

Our goal here is to give a quantitative theory of the evolution of mechanisms. What we formalize is concerned with four basic notions. First, since the biology of cells consists of thousands of proteins and operates with circuits with complex mechanisms, we seek mechanisms that can evaluate *many-argument functions*. This permits the behavior of circuits to vary in complex ways depending on the particular combination of values taken by a large number of input parameters. For example, such a function may determine the expression level of a protein in terms of the expression levels of all the other proteins. Second, any particular many-argument function has a measure of *performance* that is determined by the values of the function on inputs from a probability distribution over the conditions that arise. By applying the function to a variety of conditions the organism will enjoy a cumulative expected benefit that is determined by this performance. Third, for any function only a limited number of variants can be explored per generation, whether through mutations or recombination, since the organisms that can exist at any time have a *limited population*. Fourth, there is the requirement that mechanisms with significant improvements in performance evolve in a *limited number of generations*.

We show that our notion of evolvability is a restricted case of PAC learnability. This offers a unifying framework for the fields of evolution and cognition. The behavior of a biological organism is clearly affected both by the results of evolution and those of learning by the individual. Distinguishing between the effects of nature and nurture on behavior has proved problematic, and it will perhaps help to have a unifying viewpoint on them.

While evolvability as we define it is a form of learnability, it is a constrained form. In PAC learning, an update to a hypothesis may depend arbitrarily, as permitted by polynomial time computation, on the particular examples presented, on the values computed on them by the hypothesis, and on the values of the functions to be learned on those examples. In the more restricted Statistical Query (SQ) model of Kearns [14] the updates can depend only on the result of statistical queries on the distribution of examples. However, these queries can ask about the percentage of examples that have certain properties and hence the learning algorithm may act differentially for the different inputs. In evolution, we assume that the updates depend only on the aggregate performance of the hypothesis on a distribution of examples or experiences, and cannot differentiate at all among different kinds of examples. This last restriction expresses the idea that the relationship between the genotype and phenotype may be extremely complicated, and the evolution algorithm does not understand it. We shall observe that the function classes that are evolvable, in our sense, form a subset of those that are SQ learnable, and it is an open question whether the containment is proper.

As far as the evolvability of specific classes of functions we have both positive and negative results. On the positive side we show that the classes of monotone Boolean conjunctions and disjunctions are evolvable over the uniform distribution of inputs for the most natural representation of these functions. On the negative side we observe that the class of Boolean parity functions is not evolvable over the uniform distribution, as it is not SQ learnable. Since the latter class is known to be learnable we can conclude that evolvability is more constrained than learnability.

Our intention is to leave little doubt that functions in classes that are provably evolvable in the defined sense, do correspond to mechanisms that can logically evolve into existence over realistic time periods and within realistic populations, without any need for combinatorially unlikely events to occur. Previous quantitative theories of evolution had aims other than that of quantifying the complexity of the mechanisms that evolved. The major classical thrust has been the analysis of the dynamics of populations [10, 21, 28]. A more recent development is the study of evolutionary algorithms [1, 27], a field in which the goal is to develop good computer algorithms inspired by evolution, usually for optimization, but not necessarily those that model biological evolution. For example, these algorithms may choose mutations depending on the current inputs or experiences, may act on exponentially small increases in performance, and may be forced to start from a fixed configuration. We note also that the term evolvability has been used in a further different sense, that of measuring the intrinsic capacity of genomes to produce variants [26].

We observe that while most discussions of evolution emphasize competition and survival rather than evolution towards targets, such discussions have not yielded quantified explanations of which mechanisms can evolve.

2 Many-Argument Functions

The structures or circuits that are the constituents of living cells have to respond appropriately to wide variations in the internal and external conditions. We shall represent the conditions as the values of a number of variables $x_1, \cdots, x_n$, each of which may correspond, for example, to the output of some previously existing circuit. The responses that are desirable for an organism under the various combinations of values of the variables $x_1, \cdots, x_n$ we view as the values of an *ideal* function $f(x_1, \cdots, x_n)$. This f will have a low value in circumstances when such a low value is the most beneficial, and a high value in those other circumstances when a high value is the most beneficial. It will be beneficial to evolve a circuit that behaves as closely as possible to this f. For simplicity we shall consider the variables x_i and the functions f to take just two possible values, a low value of -1, and a high value of +1. One can, for example, think of x_i having value +1 or -1 according to whether the ith protein is being expressed or not, and f having value +1 or -1 according to whether a further protein is being expressed. Then f is an ideal function if for each combination of the conditions represented by the x_i the value of f is the best one for the organism.

We shall consider in this paper two particular function classes. The first, the *parity* functions, will be shown in this paper not to be evolvable. The second, *conjunctions*, will be shown to be evolvable in the monotone case over the uniform distribution.

A parity function is odd or even. An odd (even) parity function f over $x_1, \cdots, x_n$ has value +1 iff an odd (even) number of the variables in a subset S of the variables have value +1. For example, an odd parity function over $\{x_1, x_2, x_3, x_4\}$ with $S = \{x_1, x_3, x_4\}$ has value +1 if $x_1 = x_2 = x_3 = x_4 = +1$, and value -1 if $x_1 = x_2 = x_3 = +1$ and $x_4 = -1$.

We shall show that for evolving arbitrary parity functions over n variables either the number of generations or the population size of the generations would need to be exponentially large in terms of n.

In contrast we shall also show that there are classes with similarly substantial structure that are evolvable in a strong sense. An example of such a class is that of monotone conjunctions, defined as conjunctions of a subset S of the literals $\{x_1, x_2, \cdots, x_n\}$. An example of a conjunction is the function that is true if and only if $x_1 = +1$, and $x_4 = +1$. We abbreviate this function as $x_1 x_4$.

We denote by X_n the set of all 2^n combinations of values that the variables $x_1, \cdots, x_n$ can take. For both of the above defined function classes the set S is unknown to the evolution algorithm and the challenge for it is to approximate it from among the more than polynomially many possibilities. We define D_n to be a probability distribution over X_n that describes the relative frequency with which the various combinations of variable values for $x_1, \cdots, x_n$ occur in the context of the organism. Evolution algorithms that work for all distributions would be particularly compelling.

Definition 1. *The* performance *of function $r : X_n \to \{-1, 1\}$ with respect to ideal function $f : X_n \to \{-1, 1\}$ for probability distribution D_n over X_n is*

$$Perf_f(r, D_n) = \sum_{x \in X_n} f(x) r(x) D_n(k).$$

The performance is simply a measure of the correlation between the ideal function f and a hypothesis r we have at hand. The value will always be a real number in the range [-1, 1]. It will have value 1 if f is identical to r on points with nonzero probability in D_n. It will have value -1 if there is perfect anti-correlation on these points.

The interpretation is that every time the organism encounters a condition, in the form of a set of values x of the variables $x_1, \cdots, x_n$, it will undergo a benefit amounting to +1 if its circuit r agrees with the ideal f on that x or a penalty -1 if r disagrees with f. Over a sequence of life experiences (i.e. different points in X_n) the total of all the benefits and penalties will be accumulated. Organisms or groups for which this total is high will be selected preferentially to survive over organisms or groups with lower such totals.

The performance $Perf_f(r, D_n)$ may be viewed as a fitness landscape over the genomes r. Our analysis discusses the viability of this landscape for evolution in terms of the nature of the ideal function f and the distribution D_n of inputs for f. Instead of speculating on the nature of fitness landscapes that may be encountered in the world, we instead discuss which landscapes correspond to evolvable ideal functions.

An organism or group will be able to test the performance of a function r by sampling a limited set $Y \subseteq X_n$ of size *s(n)* of inputs or experiences from D_n, where, for simplicity we assume that the Y are chosen independently for the various r. These experiences may be thought of as corresponding to one or more per organism, so that $s(n)$ certainly upper bounds the population size.

Definition 2. *For a positive integer s, ideal function $f : X_n \to \{-1, 1\}$ and probability distribution D_n over X_n the* empirical performance $Perf_f(r, D_n, s)$ *of function $r : X_n \to \{-1, 1\}$ is a random variable that makes s selections independently with replacement according to D_n and for the multiset Y so obtained takes value*

$$s^{-1} \sum_{x \in Y} f(x)r(x).$$

In our basic definition of evolvability we insist that evolution be able to proceed from any starting point. Otherwise, if some reinitialization process were permitted, then proceeding to the reinitialized state from another state might incur an arbitrarily large decrease in performance.

The evolution algorithm for monotone conjunctions that we describe in detail in Section 5 behaves as follows. The learning of a target function x_1x_4 will be achieved by an algorithm that maintains a conjunction of a number of literals. Clearly the aim is to evolve the function x_1x_4 which has performance +1 since it is identical with the target, and is the only conjunction with that performance. The mutations will consist of *adding or deleting a single literal* for the current conjunction, or *swapping one literal for another*. Since there are then about n^2 possible mutations at each step it is feasible to explore the space of all mutations with a population of (polynomial) size, namely n^2.

3 Definition of Evolvability

Given the existence of an ideal function f the question we ask is whether it is possible to evolve a circuit for a function r that closely approximates f. Roughly, we want to say that a class C of ideal functions f is evolvable if any f in the class C satisfies two conditions. (i) From any starting function r_0 the sequence of functions $r_0 \Rightarrow r_1 \Rightarrow r_2 \Rightarrow r_3 \Rightarrow \cdots$ will be such that r_i will follow from r_{i-1} as a result of a single step of mutation and selection in a moderate size population, and (ii) after a moderate number i of steps r_i will have a performance value significantly higher than the performance of r_0, so that it is detectable after a moderate number of experiences. The conditions will be sufficient for evolution to start from any function and progress towards f, predictably and inexorably.

While the ideal function may be viewed as an abstract mathematical function, the hypothesis r needs to be represented concretely in the organism and should be viewed as a *representation* of a function. We generally consider C to be a class of ideal functions and R a class of representations of functions from which the organism will choose an r to approximate the f from C. We shall assume throughout that the representation R is *polynomial evaluatable* in the sense that there is a polynomial $u(n)$ such that given the description of an r in R and an input x from X_n, the value of $r(x)$ can be computed in $u(n)$ steps. This reflects the one assumption we make about biology, that its processes can be simulated in polynomial time on a computer. For brevity, and where it introduces no confusion, we shall denote by r both the representation as well as the function that that representation computes. We denote by C_n, R_n, and D_n, the restrictions of C, R, and D to n variables, but sometimes omit these distinctions where the meaning is clear. Also, we shall denote by ε the error parameter of the evolution, which describes how close to optimality the performance of the evolved representation has to be. We shall be prepared to expend resources that are polynomial in n, the number of arguments of the ideal function, and also in $1/\varepsilon$. (To be more precise n is the maximum of the input size and the size of the smallest representation of an ideal target function, but we will, for simplicity assume here that the representations needed are of size polynomial in the number of variables, which removes this distinction.) Hence our resource bounds will be polynomial functions, such as $p(n, 1/\varepsilon)$ in the following definition.

Definition 3. *For a polynomial $p(\cdot\ ,\ \cdot)$ and a representation class R a p-neighborhood N on R is a pair M_1, M_2 of randomized polynomial time Turing machines such that on input the numbers n and $\lceil 1/\varepsilon \rceil$ a number n in unary and a representation $r \in R_n$ act as follows: M_1 outputs all the members of a set $Neigh_N(r) \subseteq R_n$, that contains r and has size at most $p(n, 1/\varepsilon)$. If M_2 is then run on this output of M_1, it in turn outputs one member of $Neigh_N(r)$, with member r_1 being output with a probability $Pr_N(r, r_1) \geq 1/p(n, 1/\varepsilon)$.*

The interpretation here is that for each genome the number of variants, determined by M_1, that can be searched effectively is not unlimited, because the population at any time is not unlimited, but is polynomial bounded. But a significant number of experiences with each variant, generated by M_2, must be available so that differences in performance can be detected reliably.One possible implementation, clearly, is to regard R as the set of possible genomes, restrict mutations to a fixed constant number of base pairs, and regard the genome length as a polynomial in the relevant n. We consider exponentially many such variants to be impractical, while modest polynomial bounds such as n or n^2 are feasible. As in other areas of algorithmic analysis natural polynomially bounded processes usually have reasonably modest polynomial bounds, and hence such results are meaningful [11, 19]. The theory, as presented here, aims to distinguish between polynomial and exponential resources, insisting as it does that population sizes, numbers of generations, and numbers of computational steps all have to be upper bounded by a polynomial in the number of variables on which a circuit depends, and in the inverse of the error. Clearly, using more careful analysis finer distinctions can also be made. We note that estimates of actual mutation rates in various organisms are available [6, 17, 18].

Definition 4. *For error parameter ε, positive integers n and s, an ideal function $f \in C_n$, a representation class R with $p(n, 1/\varepsilon)$-neighborhood N on R, a distribution D, a representation $r \in R_n$ and a real number t, the* mutator *$Mu(f, p(n, 1/\varepsilon), R, N, D, s, r, t)$ is a random variable that on input $r \in R_n$ takes a value $r_1 \in R_n$ determined as follows. For each $r_1 \in Neigh_N(r)$ it first computes an empirical value of $v(r_1) = Perf_f(r, D_n, s)$. Let Bene be the set $\{r_1 \mid v(r_1) \geq v(r) + t\}$ and Neut be the set difference $\{r_1 \mid v(r_1) \geq v(r) - t\}$ – Bene. Then*

(i) if *Bene* $\neq \phi$ then output $r_1 \in$ *Bene* with probability

$$Pr_N(r, r_1) / \sum_{r_1 \in Bene} Pr_N(r, r_1)$$

(ii) if *Bene* $= \phi$ then output an $r_1 \in$ *Neut*, the probability of a specific r_1 being

$$Pr_N(r, r_1) / \sum_{r_1 \in Neut} Pr_N(r, r_1).$$

In this definition a distinction is made between beneficial and neutral mutations as revealed by a set of s experiments. In the former the empirical performance after the mutation exceeds that of the current representation r by an additive tolerance of at least t, a quantity which will, in general, be large enough, in particular some inverse polynomial, that it can be reliably distinguished from a zero increase in performance. In neutral mutations no significant increase in performance is expected, but it is expected that the performance is not worse than that of the current r by more than t. If some beneficial mutations are available one is chosen according to the relative probabilities of their generation by N as allowed by machine M_2 in Definition 3. If none is available then one of the neutral mutations is taken according to the relative probabilities of *their* generation by N. Since in Definition 3 we insist that $r \in Neigh_N(R), r$ will always be empirically neutral, by definition, and hence *Neut* will be nonempty.

Definition 5. *For a mutator $Mu(f, p(n, 1/\varepsilon), R, N, D, s, r, t)$ a t-*evolution step *on input $r_1 \in R_n$ is the random variable $r_2 = Mu(f, p(n, 1/\varepsilon), R, N, D, s, r_1, t)$. We then say $r_1 \to r_2$ or $r_2 \leftarrow Evolve(f, p(n, 1/\varepsilon), R, N, D_n, s, r_1, t)$.*

We say that polynomials $tl(x,y)$ and $tu(x,y)$ are *polynomially related* if for some $\eta > 1$ for all x, y $(0 < x, y < 1)$ $(tu(x,y))^\eta \leq tl(x,y) \leq tu(x,y)$. We now define an evolution sequence as a sequence of t-evolution steps where the t at each step is bounded between two polynomially related quantities $t\ell(1/n, \varepsilon)$, $tu(1/n, \varepsilon)$ and computable in polynomial time by a Turing machine T that takes $r \in R, n$ and ε as inputs.

Definition 6. *For a mutator* $Mu(f, p(n, 1/\varepsilon), R, N, D, s, r, t)$ *a* $(t\ell, tu)$-evolution sequence *for* $r_1 \in R_n$ *is a random variable that takes as values sequences* $r_1, r_2, r_3, \ldots$ *such that for all* i $r_i \leftarrow Evolve(f, p(n, 1/\varepsilon), R, N, D, s, r_{i-1}, t_i)$, *where* $t\ell(1/n, \varepsilon) \leq t_i \leq tu(1/n, \varepsilon)$, *tl and tu are polynomially related polynomials, and* t_i *is the output of a TM* T *on input* r_{i-1}, n *and* ε.

We shall find that if we want to evolve to performance very close to one, say $1-\varepsilon$, we shall need numbers of experiments s or numbers of generations g that grow inversely with ε, and the tolerances t that diminish with ε. We therefore regard these as functions of n and ε, and denote them by $s(n, 1/\varepsilon), g(n, 1/\varepsilon), tl(1/n, \varepsilon)$ and $tu(1/n, \varepsilon)$.

Definition 7. *For polynomials* $p(n, 1/\varepsilon)$, $s(n, 1/\varepsilon)$, $t\ell(1/n, \varepsilon)$ *and* $tu(1/n, \varepsilon)$, *a representation class* R *and* $p(n, 1/\varepsilon)$*-neighborhood* N *on* R, *the class* C *is* $(t\ell, tu)$-evolvable by $(p(n, 1/\varepsilon), R, N, s(n, 1/\varepsilon))$ over distribution D *if there is a polynomial* $g(n, 1/\varepsilon)$ *and a Turing machine* T, *which computes a tolerance bounded between tl and tu, such that for every positive integer* n, *every* $f \in C_n$, *every* $\varepsilon > 0$, *and every* $r_0 \in R_n$ *it is the case that with probability greater than 1 -* ε, *a* (tl, tu)*-evolution sequence* $r_0, r_1, r_2, \cdots$, *where* $r_i \leftarrow Evolve(f, p(n, 1/\varepsilon), R, N, D_n, s(n, 1/\varepsilon), r_{i-1}, T(r_{i-1}, n, \varepsilon)$, *will have* $Perf_f(r_{g(n,1/\varepsilon)}, D_n) > 1 - \varepsilon$.

The polynomial $g(n, 1/\varepsilon)$, the generation polynomial, upper bounds the number of generations needed for the evolution process.

Definition 8. *A class* C *is* evolvable by $(p(n, 1/\varepsilon), R, N, s(n, 1/\varepsilon))$ over D *iff for some pair of polynomially related polynomials* $t\ell, tu$, C *is* (tl, tu)*-evolvable by* $(p(n, 1/\varepsilon), R, N, s(n1/\varepsilon))$ *over* D.

Definition 9. *A class* C *is* evolvable by R over D *iff for some polynomials* $p(n, 1/\varepsilon)$ *and* $s(n, 1/\varepsilon)$, *and some* $p(n, 1/\varepsilon)$*-neighborhood* N *on* R, C *is evolvable by* $(p(n, 1/\varepsilon), R, N, s(n, 1/\varepsilon))$ *over* D.

Definition 10. *A class* C *is evolvable over* D *if for some* R *it is evolvable by* R *over* D.

Definition 11. *A class* C *is* evolvable *if it is evolvable over all* D.

Our definition of evolvability is closely related to that of learnability, but includes the extra ingredients that each step of learning (i) chooses from a polynomial size set of hypotheses, (ii) tolerates at most a small decrease in performance, and further (iii) the choice of the next hypothesis from among the candidate hypotheses is made on the basis of their aggregate performance on inputs, and not differentially according to the values of the various inputs.

Proposition 1. *If C is evolvable by R over D then C is learnable by R over D. In particular, if C is evolvable by R then C is learnable by R.*

Proof. If C is evolvable over D then, by definition, for some polynomials $p(n, 1/\varepsilon)$, $s(n, 1/\varepsilon)$, $g(n, 1/\varepsilon), t\ell(1/n, \varepsilon)$ and $tu(1/n, \varepsilon)$, some polynomial evaluatable representation R and some $p(n, 1/\varepsilon)$-neighborhood N on R, C is $(t\ell, tu)$-evolvable by $(p(n, 1/\varepsilon), R, N, s(n, 1/\varepsilon))$ over distribution D with generation polynomial $g(n, 1/\varepsilon)$. The main observation here is that we can replicate this evolution algorithm exactly in terms of the PAC learning framework. At each stage the evolution algorithm takes fixed size samples of $s(n, 1/\varepsilon)$ labelled examples from the distribution, computes for its current hypothesis the empirical performance, and from that generates the next hypothesis in a polynomially bounded fashion. But computing this performance is equivalent to computing the fraction of examples on which the hypothesis predicts correctly. Hence the access required to examples is that of random labelled examples from D, and every step is a polynomial time computational process. All this is permitted within the PAC model. Also the final hypothesis of the evolution model satisfies the requirements of the learning model since it ensures that the performance is at least $1 - \varepsilon$, and hence accurate on at least $1 - \varepsilon/2$ of D. □

We can strengthen the above statement by observing that evolvability implies learnability in the more restricted sense of statistical queries defined by Kearns [12]. In that model oracles provide not individual examples but estimates, to within inverse polynomial additive error, of the fraction of examples that satisfy polynomially evaluatable properties. This is clearly sufficient for the proof of Proposition 12, which therefore supports also the following.

Proposition 2. *If C is evolvable by R over D then it is efficiently learnable from statistical queries using R over D.*

Evolvability for all D is a very strong and desirable notion. As mentioned previously, it guarantees evolution independent of any assumptions about the distribution. It also means that evolution can continue even if the D changes. Of course, a change in D can cause a reduction in the value of Perf for any one r, and hence may set back the progress of evolution. However, the process of finding improvements with respect to whatever the current D is will continue. It remains an open problem as to whether such distribution-free evolution is possible for a significant class of functions.

Note also that the representation class R may represent a class of functions that differs from C. For example an R richer than C may be helpful. Alternatively, a weaker class may still produce good enough approximations and may have better properties for evolution. In general if we wish to identify or emphasize the class R that supports an evolution algorithm we say that *C is evolvable by R for D*, or *C is evolvable by R*.

The purpose of the main definition above of evolvability is to capture the notion of evolution towards a target under stable conditions. In biological evolution other phenomena are involved also, and, we believe, many of these can be discussed in the language of our formalism by appropriate variants. One restriction is *evolvability with initialization.* In that case in Definition 7, instead of requiring convergence from any starting point $r_0 \in R_n$, we require only that there is convergence from one fixed starting point r_0 for all targets $f \in C_n$. The more general definition given is more robust, allowing for successive phases of evolution, each one with a different target ideal function, for example. The evolved representation for one phase can then serve as the starting representation for the next, without a decrease in performance at any step. In evolution with initialization, the steps of going from the end of one phase to a reinitialized new state may suffer an arbitrary performance decrease. In our definition of evolvability we seek to avoid allowing any mechanism that would provide for such initialization by a back door. We therefore insist that the tolerance be bounded between two polynomially related functions. Allowing the tolerance to be arbitrarily large would allow initialization in one step via an arbitrarily large drop in performance.

Another variation is *variable population evolution.* In this the sample size s may vary. In particular, if it is made small than random variations in the empirical performance may make a low performance mutation appear as neutral or even beneficial, and be adopted. This permits reinitializations, for example, for a subsequent phase of evolution with initialization. In biology evolution in small populations is believed to play a special role.

A further variant is *evolvability with optimization.* Here we insist that in Definition 4 the representation r_1 selected is any one with empirical performance within tolerance t of the best empirical performance in $\mathrm{Neigh}_N(r)$. However, it is easy to see that this variant is no more powerful than the main definition. One can simulate the search for the best representation, as required in one step of the optimized evolution, in no more than $6/t$ basic steps of looking for a representation with an empirical additive improvement of at least $t/2$, each step using a new sample. (Note that the actual performance can increase cumulatively by at most 2. Using the Hoeffding Bound (Fact 2) one can show that the cumulative empirical performance increase on the different samples can be limited to 3 with overwhelming probability.) For this simulation we change the representation to $i \cdot r^M \cdot r^P$ where $i \leq 6/t$ is an integer denoting which basic step we are in, $r^M \in R$ is the representation that generates the mutations (using the M_1 of Definition 3) for each of the up to $6/t$ basic steps, and $r^P \in R$ is the one with best performance found so far. (In other words r^P is the function this representation is computing, but the representation also has a memory of r^M from which it can generate new mutations in R, that may not be generatable from r^P alone.) After $i = 6/t$ basic steps the final r^P is adopted as the starting r^M and r^P of the next step of the optimized evolution. Note that the constructed representation in this reduction is a *redundant* representation in the sense that there are many representations that correspond to the same function r^P. It illustrates the power of storing history, namely R^M, in addition to the active part, R^P.

Proposition 3. *If C is evolvable with optimization by R over D, then C is evolvable by R over D. If C is evolvable with initialization and optimization by R over D, then C is evolvable with initialization by R over D.*

A simpler variant is that of *fixed-tolerance* evolvability, obtained if the bounds tl, tu on the tolerance are the same.

We note that the aspect of evolution that is outside the control of the evolution algorithm itself is the population size. Thus evolvability guarantees inexorable convergence only if the population is appropriate. Our algorithms require only that the population, as represented by s, be large enough. The variable population variant defined earlier permits schedules of varying population sizes.

4 Limits to Evolvability

The obvious question arises as to whether the converse of Proposition 12 holds: does learnability imply evolvability? Our next observation answers this in the negative, saying as it does that for a certain function class there is a distribution that defeats all combinations of representations and neighborhoods.

We define Lin_n to be the set of odd parity functions $f(x_1, \cdots, x_n)$ over $\{-1, 1\}^n$. Each such f corresponds to some subset of the variables $x_{i[1]}, \cdots, x_{i[k]} \in \{x_1, \cdots, x_n\}$. The function f has value 1 if and only if an odd number of the variables $\{x_{i[1]}, \cdots, x_{i[k]}\}$ have value 1. Clearly there are 2^n functions in Lin_n. We define U to be the uniform distribution over $\{-1, 1\}^n$. We note that the functions in Lin are easy to compute, and further the class is known to be learnable not only for U but for all distributions [9, 12].

Proposition 4. *Lin is not evolvable for U by any representation R.*

Proof. Kearns [14] shows that Lin is not efficiently learnable from statistical queries over U using any representation. The result then follows from Proposition 12 above. □

The class Lin may appear to be biologically unnatural. That is exactly the prediction of our theory, which asserts that evolution cannot be based on the evolvability of such a class.

An important class of functions that is known to be learnable is that of linear halfspaces $\{\mathbf{a.x} \leq b \mid \mathbf{a} \in \mathbf{R}^n, b \in \mathbf{R}\}$ in n-dimensional space $\mathbf{R}^n$. This class is learnable for all distributions by the natural representation of linear halfspaces if the coefficients $\mathbf{a}$, b are represented as rational numbers with n digits of accuracy, by virtue of the existence of polynomial time algorithms for linear programming [3]. However, if both the class and its representation is restricted to {0,1} coefficients then we have the following.

Proposition 5. *If C is the class of Boolean Threshold Functions* $\{\mathbf{a.x} \leq b \mid \mathbf{a} \in \{0, 1\}^n, b \in \mathbf{R}\}$ *in* $\mathbf{R}^n$ *and R is the given representation of it, then C is not evolvable by R, unless NP = RP.*

Proof. In [20] it is shown that this class is not learnable by its natural representation unless NP = RP. (The proof there shows that an NP-complete problem, integer programming, can be mapped to instances of learning Boolean threshold functions for a certain distribution to accuracy better than $1/n$.) The result follows from Proposition 12 above. □

There appear to be at least four impediments that can be identified to evolvability in our sense, the first three of which derive from general impediments to learnability, while the last is particular to evolvability: (i) A purely information theoretic impediment [8]: the complexity of the mechanism that is to evolve exceeds the number of experiences. (ii) A representational limit such as Proposition 16 above, where learnability by a fixed representation would imply solving a computational problem that is believed to be hard. (iii) An intrinsic complexity limitation [15]: the function class is so extensive that learning it by *any* representation would imply an efficient algorithm for a problem believed to be hard to compute. (iv) Limits such as Proposition 15 above, that show that for information theoretic reasons evolvability cannot proceed because no empirical test of a polynomial number of hypotheses in a neighborhood can guarantee sufficient convergence in performance. Note that impediments (i) and (iv) are absolute, requiring no unproven computational assumptions.

5 Some Provably Evolvable Structures

We now describe some basic classes of Boolean functions and distributions that are provably evolvable. Here disjunction or Boolean "or" is denoted by +, conjunction or Boolean "and" by the multiplication sign, and Boolean negation of a variable x_i by x'_i. In general we shall have n variables $x_1, \cdots, x_n$. A *q-disjunction* is a disjunction of $k \leq q$ of the n variables or their negations, while a *q-conjunction* is a conjunction of $k \leq q$ of the n variables or their negations. Thus a q-disjunction is $y_{i[1]} + \cdots + y_{i[k]}$ where $1 \leq i[1], \cdots, i[k] \leq n$ and $y_{i[j]} \in \{x_1, \cdots, x_n, x'_1, \cdots, x'_n\}$, and a q-conjunction is $y_{i[1]} \cdots y_{i[k]}$. The uniform distribution over $\{-1, +1\}$ will be denoted again by U. A conjunction or disjunction is *monotone* if it contains no negated literals. We note that Ros (Section B2.8 in [1]; [22]) has analyzed evolutionary algorithms for learning conjunctions and disjunctions. However, a step of his algorithm is allowed to depend not just on the value of his current hypothesis on an input, but on more detailed information such as the number of bits on which the hypothesis and input differ. Such dependence on the input condition we consider unrealistic for evolution, and is outside our model. With regard to the literature on evolutionary algorithms [27] we also note that there the functions being evolved are often real rather than Boolean valued, and that provides more feedback to the process.

Fact 1. *Over the uniform distribution U for any conjunction $Pr_U(y_{i[1]} \cdots y_{i[k]} = 1) = 2^{-k}$ and for any disjunction $Pr_U(y_{i[1]} + \cdots + y_{i[k]} = 1) = 1 - 2^{-k}$.*

For our probabilistic arguments below it will be sufficient to appeal to the following:

Fact 2 *(Hoeffding [13]). The probability that the mean of s independent random variables each taking values in the range $[a, b]$ is greater than or less than the mean of their expectations by more than δ is at most $exp(-2s\delta^2/(b-a)^2)$, where $exp(x)$ denotes e^x.*

Fact 3 *(Coupon Collector's Problem). Suppose that there is a bucket of n balls and M is a subset of m of them. Then after $j = CC(n, m, \eta) = n(\log_e m + \log_e(1/\eta))$ samples with replacement the probability that some member of the chosen set M has been missed is less than η.*

Proof. This probability is upper bounded by $m(1-1/n)^j < m(1-1/n)^{jn/n} = me^{-j/n} < \eta$. □

We note that an evolution algorithm for a representation R needs to have defined at each step (i) the neighborhood N, (ii) the tolerance t and (iii) the sample sizes, so that the mutator random variable can be evaluated at that step.

Theorem 1. *Monotone conjunctions and disjunctions are evolvable over the uniform distribution for their natural representations.*

Proof. We first note that for our definition of evolvability it is sometimes advantageous for a local search procedure to introduce literals that do not appear in the ideal function f. For example, suppose $f = x_1x_2x_3$ and we start with a hypothesis 1, the conjunction of zero literals. Then the hypothesis will disagree with f on 7/8 of the distribution, and the introduction of the literal x_4 will be an improvement, reducing this probability from 7/8 to 1/2.

If we are evolving to accuracy ε and have n variables we let $q = \lceil \log_2(dn/\varepsilon) \rceil$ for some constant d. We choose the effective representation class R to be monotone q-conjunctions.

We first assume that both the ideal function f and the initial representation r_0 have at most q literals. We denote by r^+ and r^- the sets of conjunctions consisting of the literals of r with one literal added, and with one taken away, respectively. In the case that r has the maximum number q of literals then r^+ is empty. In the case that $|r| = 0, r^-$ is empty. Also we define r^{+-} to be the conjunctions consisting of the literals in r with one further literal added and then one literal taken away. Clearly $r \in r^{+-}$. We then choose the neighborhood structure N to be such that

$$Neigh_N(r) = r^+ \cup r^- \cup r^{+-}.$$

Finally r and the members of r^+ and r^- will each have equal probabilities, so that their total is 1/2, while the remaining members of r^{+-} will also have equal probabilities, again totaling 1/2. Clearly N is a $p(n, 1/\varepsilon)$-neighborhood structure where $p(n, 1/\varepsilon) = O(n^2)$.

The construction will ensure that every mutation in N either causes an improvement in performance of at least 2^{-2q} or causes no improvement (and a possible degradation.) We choose the tolerance $t(1/n, \varepsilon) = 2^{-2q-1}$ and the number of samples $s(n, 1/\varepsilon) = t^{-3}$. It will then follow that the empirical test will, except with exponentially small probability, correctly identify an available mutation that has true improvement $2t$, distinguishing it from one that gives no improvement in performance. This can be seen by substituting $a = -1, b = 1, \delta = t$ and $s = \delta^{-3}$ in the Hoeffding Bound above to obtain that the probability of s trials each with expected improvement 2δ will produce a mean improvement of less than $t = \delta$ is at most $\exp(-2s\delta^2/(b-a)^2) = \exp(-(dn/\varepsilon)^2)$. A similar argument also shows that the same test will not mistakenly classify a mutation with no performance increase with one with an increase of $2t$. In the same way the same tolerance will distinguish a mutation with a nonnegative performance increase from one whose performance decreases by at least $2t$.

In a run of the evolution algorithm there will be $g(n, 1/\varepsilon)$ stages, and in each stage up to $p(n, 1/\varepsilon)$ mutations will be tested, where g and p are polynomials. We will want that in all $p(n, 1/\varepsilon)g(n, 1/\varepsilon)$ empirical tests the probability of even one failure to make such a distinction be less than $\varepsilon/2$. But $p(n, 1/\varepsilon)g(n, 1/\varepsilon)\exp(-(dn/\varepsilon)^2) < \varepsilon/2$ for all n, ε for a suitable constant d.

Suppose that W is the true set of m literals in the ideal conjunction, and that the current hypothesis r is the conjunction of a set V of k literals. We claim that:

Claim 1. *For (a) – (g) suppose that $k \leq q$. Then*

(a) If $k < q$ then adding to r any literal z in W - V will increase the performance of r by at least 2^{1-q}.
(b) Removing from r any literal z in $V \cap W$ will decrease the performance of r by at least 2^{1-q}.
(c) Adding to r a literal in W - V and removing from r a literal in V - W will increase the performance by at least 2^{-q-m}.
(d) Adding to r some literal not in W, and removing from r a literal in $V \cap W$ will decrease the performance of r by at least 2^{-q-m}.
(e) Adding to r a literal in W - V and removing from r a literal in $V \cap W$ will leave the performance unchanged, as will also adding to r a literal not in W, and removing one in V - W.
(f) If r contains all the literals in W, then removing a z in V - W will increase the performance of r by at least 2^{1-q}.
(g) If r contains all the literals in W then adding a z in V - W will decrease the performance of r by at least 2^{-q}.
(h) If $m > q$ then adding a z to an r of length at most q - 2 will increase performance by at least 2^{1-q}, and removing a z from an r of length at most $q-1$ will decrease performance by at least 2^{1-q}.

To verify the above eight claims suppose that r is $y_{i[1]} \cdots y_{i[k]}$.

(a) Consider a z in $W - V$. Then conjoining z to $y_{i[1]} \cdots y_{i[k]}$ will change the hypothesis from $+1$ to the value of -1 on the points satisfying $z' y_{i[1]} \cdots y_{i[k]}$. Clearly, the ideal function f takes value -1 at all these points since $z = -1$. These points will, by Fact 1, have probability $2^{-(k+1)} \geq 2^{-q}$. Hence, the performance will improve by at least twice this quantity, namely 2^{1-q}.

(b) Suppose that $z = y_{i[1]}$. Removing it from r will change the hypothesis from -1 to the value of +1 on the points satisfying $z'y_{i[2]} \cdots y_{i[k]}$. Clearly, the ideal function takes value -1 at all these points. These points will, by Fact 1, have probability $2^{-k} \geq 2^{-q}$ and hence the performance will degrade by at least twice this quantity.

(c) Suppose the added literal is z and the removed literal is $y_{i[1]}$. Then the hypothesis changes (i) from 1 to -1 on the points where $z'y_{i[1]} \cdots y_{i[k]} = 1$, and (ii) from -1 to +1 on the points such that $zy'_{i[1]}y_{i[2]} \cdots y_{i[k]} = 1$. Now (i) changes from incorrect to correct at all such points, and applies with probability $2^{-(k+1)}$. Also (ii) applies to a set of points with the same total probability $2^{-(k+1)}$, but the change on some of the points may be from correct to incorrect. To show that the net change caused by (i) and (ii) in combination is beneficial as claimed it is sufficient to observe that (ii) is nondetrimental on a sufficient subdomain. To see this we consider the literals Z in W that are missing from r but other than z, and suppose that there are u of these. Then on the domain of points $zy'_{i[1]}y_{i[2]} \cdots y_{i[k]} = 1$ that specify (ii) we note that on the fraction 2^{-u} of these the correct value of the ideal function is indeed 1. Hence the improvement due to (i) is not completely negated by the degradation due to (ii). The improvement in performance is therefore at least $2^{-u-k} \geq 2^{-m-q}$.

(d) Suppose the added literal is z and the removed literal is $y_{i[1]}$. Then the hypothesis changes (i) from 1 to -1 on the points where $z'y_{i[1]} \cdots y_{i[k]} = 1$, and (ii) from -1 to +1 on the points such that $zy'_{i[1]}y_{i[2]} \cdots y_{i[k]} = 1$. Now (ii) is an incorrect change at every point and applies with probability $2^{-(k+1)}$. Also (i) applies to a set of points with the same total probability $2^{-(k+1)}$. To show that the net change caused by (i) and (ii) in combination is detrimental to the claimed extent, it is sufficient to observe that (i) is detrimental on a sufficient subdomain. To see this we consider the literals Z in W that are missing from r, and suppose that there are u of these. Then on the domain of points $z'y_{i[1]}y_{i[2]} \cdots y_{i[k]} = 1$ that specify (i) we note that on the fraction 2^{-u} of these the correct value of the ideal function is indeed 1. Hence (i) suffers a degradation of performance on a fraction 2^{-u} of its domain, and hence the rest cannot fully compensate for the degradation caused in (ii). The combined decrease in performance is therefore at least $2^{-u-k} \geq 2^{-m-q}$.

(e) Suppose the added literal is z and the removed literal is $y_{i[1]}$. Then the hypothesis changes (i) from 1 to -1 on the points where $z'y_{i[1]} \cdots y_{i[k]} = 1$, and (ii) from -1 to +1 on the points such that $zy'_{i[1]}y_{i[2]} \cdots y_{i[k]} = 1$. Now (ii) is an incorrect change at every point and applies with probability $2^{-(k+1)}$, and (i) applies to a set of points with the same total probability $2^{-(k+1)}$ but is a correct change at every point. The second part of the claim follows similarly. Again each of the two conditions holds with probability 2^{-k-1}. But now if there are u literals in W missing from r, then over each of the two conditions stated in (i) and (ii) function f is true on a fraction z^{-u}. Hence the effect of the two changes is again to cancel and keep the performance unchanged.

(f) Suppose that $z = y_{i[1]}$. Removing z from $y_{i[1]} \cdots y_{i[k]}$ will change the value of the hypothesis from -1 to +1 on the points satisfying $z' y_{i[2]} \cdots y_{i[k]}$. But all such points have true value +1 if r contains all the literals in W. Hence this gives an increase in performance by an amount $2^{1-k} \geq 2^{1-q}$.

(g) Consider a z in $V - W$. Then conjoining z to $y_{i[1]} \cdots y_{i[k]}$ will change the hypothesis from +1 to -1 on the points satisfying $z' y_{i[1]} \cdots y_{i[k]}$. But all such points have true value +1 if r contains all the literals in W. Hence conjoining z will cause a decrease in performance by an amount $2^{-k} \geq 2^{-q}$.

(h) If $m > q$ then the hypothesis equals -1 on a large fraction of at least $1 - 2^{-2-q}$ of the points. A conjunction of length $k \leq q - 2$ will equal -1 on $1 - 2^{-k} \leq 1 - 2^{2-q}$ points, and a conjunction of length $k \leq q - 1$ on $1 - 2^{-k} \leq 1 - 2^{1-q}$ of the points. Hence the fraction of points on which the -1 prediction will be made increases by $(1 - 2^{-k-1}) - (1 - 2^{-k}) = 2^{-k-1} \geq 2^{1-q}$ if $k \leq q - 2$ and a literal is added, and decreases by $(1 - 2^{-k}) - (1 - 2^{-k+1}) = 2^{-k} \geq 2^{1-q}$ with the removal of one, if $k \leq q - 1$. If $m > q$ then the corresponding increase/decrease in the fraction of points on which predictions are correct is at least 2^{-q}, since the fraction of predicted -1 points changes by twice this quantity, and the true +1 points amount to at most a half this quantity.

To prove the proposition, we are first supposing that the number m of literals in the ideal function is no more than q. Then the intended evolution sequences will have two phases. First, from any starting point of length at most q the representation will increase the number of its literals that are in W by a sequence of steps as specified in Claims (a) and (c). Interspersed with these steps there may be other steps that cause similar inverse polynomial improvements, but add or remove non-ideal literals. Once the conjunction contains all the literals of the ideal conjunction, it enters into a second phase in which it contracts removing all the non-ideal literals via the steps of Claim (f).

The assertions of the above paragraph can be verified as follows. Claims (a) and (c) ensure that *as long as some ideal literal is missing from r*, beneficial mutations, here defined as those that increase performance by at least 2^{-2q} will be always available and will add a missing ideal literal. Further, Claims (b), (d) and (e) ensure that mutations that remove or exchange ideal literals will be deleterious, reducing the performance by at least 2^{-2q}, or neutral, and hence will not be executed. Some beneficial mutations that add or remove non-ideal literals may however occur. However, since each literal not already in the conjunction, will be generated by N with equal probability as a target for addition or swapping in, the Coupon Collector's model (Fact 3) can be applied. If the ideal conjunction contains m literals then after $\mathrm{CC}(n, m, \varepsilon/2) = O((n \log n + n \log(1/\varepsilon))$ generations all m will have been added or swapped in, except with probability $\varepsilon/2$.

Once r contains all the ideal literals then the only beneficial mutations are those that remove non-ideal literals (Claim (f)). Adding non-ideal literals (Claim (g)), replacing an ideal literal by a non-ideal literal (Claims (d)), or replacing a non-ideal literal by a non-ideal literal (Claim (e)) are all deleterious or neutral. Hence in this second phase after O(n) steps the ideal conjunction with perfect performance will be reached.

We conclude that in the case that the number of literals in both r_0 and the ideal conjunction is at most q then the evolution will reach the correct hypothesis in the claimed number of stages, except with probability ε, which accounts for both the empirical tests being unrepresentative, as well as the evolution steps failing for other reasons.

In case the initial conjunction is of length greater then q we allow for a prologue phase of evolution of the following form. We define the neighborhood of each such long conjunction to be all the conjunctions obtained by removing one or none of its literals, each one being generated with equal probability. Clearly the removal of a literal from a hypothesis having $k > q$ literals will change its value on at most $2^{-k} < 2^{-q} = 2\varepsilon/(dn)$ of the distribution and hence the performance, if it decreases, will decrease by no more than $2\varepsilon/(dn)$. Hence if we set the tolerance to $t = 4\varepsilon/(dn)$ then a mutation that decreases the number of literals will be available, and will be detected as a neutral mutation as long as its empirical performance is not less than its true performance by more than $\delta = 2\varepsilon/(dn)$. The probability of this happening is small, as can be seen by substituting $a = -1, b = 1, \delta = t/2$, and $s = \delta^3$ in the Hoeffding bound (Fact 2), yielding $\exp(-dn/4\varepsilon)$. After this process runs its course, which will take $O(n)$ stages except with exponentially small probability, a short conjunction of length at most q will be reached.

Finally we consider the alternative case that the number m of literals in the ideal conjunction is more than q. If r_0 has at most q literals then in the evolution beneficial steps (h) that add literals will be always available until the hypothesis becomes of length $q - 1$. Further, steps (h) that remove literals will be never taken since these are deleterious. Once length $q - 1$ is achieved the length can change only between length $q - 1$ and q, and the performance will be at least $1 - 2(2^{1-q} + 2^{-m}) = 1 - 5.2^{-q} = 1 - 5\varepsilon/(dn)$. In the alternative case that r_0 has more than q literals, the prologue phase will be involved as before until length at most q will be reached, and the previous condition joined.

The result for conjunctions therefore follows with $g(n, 1/\varepsilon) = O(n \log(n/\varepsilon))$. We note that $s = \Omega((n/\varepsilon)^6)$ is sufficient for both phases.

The claimed result for disjunctions follows by Boolean duality: Given an expression representing a Boolean function, by interchanging "and" and "or" operators and negating the inputs will yield the negation of the original function. □

The algorithm as described above may be applied to conjunctions with negated variables, but will then fail sometimes. For example, if the starting configuration contains many literals that are negations of literals in the ideal function, then it may have high performance because it predicts -1 everywhere. However, it would appear difficult in that case to find an improved hypothesis by local search.

If initialization is allowed then the above results can be obtained much more easily, and then also allow negations.

Proposition 6. *Conjunctions and disjunctions are evolvable with initialization over the uniform distribution.*

The reader can verify this by considering conjunctions with initial representation 1. If the hypothesis is of length k and consists only of literals that occur in the true conjunction of length $m > k$, then adding a literal from the true conjunction will increase performance by 2^{-k}, while adding one not in the true conjunction will increase performance by $2^{-k} - 2^{1-h}$. If $m = k$ then adding a literal will decrease performance by at least 2^{-k}. Then if we let $q = \log_2(d/\varepsilon)$ for an appropriate constant d, choose tolerance 2^{-q-1}, have a neighborhood that either adds a literal or does nothing, and stop adding new literals if the conjunction reaches length q, then evolution with optimization will proceed through conjunctions of literals exclusively from W until performance at least $1-\varepsilon$ is reached. It follows that this evolution algorithm will work with optimization and initialization, and hence, by Proposition 14 it is evolvable with initialization alone, but for a redundant representation.

6 Discussion

We have introduced a framework for analyzing the quantitative possibilities of and limitations on the evolution of mechanisms. Our definition of evolvability has considerable robustness. It can be weakened in several ways, separately and in combination, to yield notions that impose less onerous requirements. First, one can entertain the definition for just one specific distribution as we did for our positive results in Section 5. The question whether significant classes are provably evolvable for all distributions is one of the more important questions that our formulation raises. Second, the requirement of having the performance be able to approach arbitrarily close to the best possible can be relaxed. This permits processes where computations are feasible only for obtaining approximations to the best possible. Third, the starting point need not be allowed to be arbitrary. There may be a tradeoff between the robustness offered by allowing arbitrary starting points, and the complexity of the mechanisms that can evolve. Wider classes may be evolvable in any of these less onerous senses than in the most robust sense. We can equally study, in the opposite direction, the quantitative tradeoffs obtained by constraining the model more, by disallowing, for example, neutral mutations or redundant representations, or by insisting on a fixed tolerance. We note that our algorithm for conjunctions as describe does exploit neutral mutations. Also, it uses a fixed tolerance for the main phase, and a different tolerance in the prologue.

Our result that some structures, namely monotone conjunctions and disjunctions are evolvable over the uniform distribution, we interpret as evidence that the evolution of significant algorithmic structure is a predictable and analyzable phenomenon. This interpretation is further supported by the observation that the theory, analogously to learning theory, analyzes only the *granularity* of the structure that can evolve in a single phase with a single ideal function. If multiple phases are allowed with different ideal functions in succession then arbitrarily complex structures can evolve. For example, in response to various initial ideal functions some set of conjunctions and disjunctions may evolve first. At the next phase the outputs of these functions can be treated as additional basic variables, and a second layer of functionality can evolve on top of these in response to other ideal functions. This process can proceed for any number of phases, and build up circuits of arbitrary complexity, as long as each layer is on its own beneficial. We call this *evolvable target pursuit.*

Clearly much work remains to be done before we can characterize the classes that are evolvable. A further step would be to characterize which classes of evolution algorithms are themselves evolvable.

We note that our model makes no assumptions about the nature of a mutation, other than that it is polynomial time computable by a randomized Turing machine. Thus the biological phenomena found in DNA sequences of point mutations, copying of subsequences, and deletion of subsequences, are all easily accommodated in the model.

The proof of Proposition 14 hints at one possible purpose of redundancy that is also believed to occur widely in biology. In that construction two near identical copies of a subsequence are maintained, one of which acts as a reservoir that records history and offers expanded possibilities for future mutations.

The idea that diversity in the gene pool of a population serves the purpose of protecting a species against unpredictable changes in the environment can also be expressed in the model. We would represent the hypotheses of all N members of a population by a hypothesis that concatenates them all but has a distinguished first member. The total hypothesis would still be of polynomial size if N is. The distinguished first member determines the performance while the rest form a reservoir to facilitate future mutations. In a mutation the subhypotheses would be cyclicly shifted by an arbitrary amount so that any one of them can come into first place, and only this first one would undergo mutation. In this way the diverse gene pool of a population can be represented. In one phase of evolution the hypotheses that have their first subhypotheses best fitted to the then current environment would win, but they would retain diversity in their reservoir. Of course, once we regard the genome of a population as a single genome, then there may be useful operations on them beyond cyclic shifts, such as operations that splice together parts of the individual subgenomes. The latter operations correspond to recombination.

In our model large populations are useful when small improvements in performance need to be detected reliably. Small populations can also have a role in permitting deleterious mutations to be adapted, which would not be in large populations.

It is natural to ask what is the most useful view of the correspondence between our view of circuits and what occurs in biology. What do the nodes and the functions that evolve in our model correspond to? It seems plausible to suppose that at least some of the nodes correspond to the expression of particular proteins. Then the regulatory region associated with each protein coding region would correspond to the function evaluated at that node. Possibly such regions may have to be subdivided further into nodes and functions. The fact that there are highly conserved regions in the genome that code for proteins, and also some that do not [2, 6], is consistent with this viewpoint.

In the case that a node corresponds to the expression of a fixed protein the interpretation of the ideal functions in our model is particularly simple. Suppose the genome has an evolution algorithm for the class C of functions, such as disjunctions. Then the ideal function f simply expresses for each combination of other variables the best choice of whether to, or whether not to (or how much to) express that protein. Evolution will then be guaranteed towards f provided f lies within C.

Modularity, in biology or engineering, is the idea that systems are composed of separate components that have identifiable separate roles. If evolvability is severely constrained to limited classes of functions as our theory suggests, then systems that evolve would be constrained to be modular, and to consist of many identifiable small modules. Hence modularity in biology would be a consequence of the limitations of evolvability.

A unified theory for learning and evolution is of potential significance to the studies of cognition and of its emulation by machine. A major challenge in understanding cognition is that in biological systems the interplay between the knowledge that is learned through experience by an individual and the knowledge inherent in the genes, is complex, and it is difficult to distinguish between them. In attempts to construct computer systems for cognitive functions, for example for vision, this challenge is reflected in the difficulty of providing an effective split between the preprogrammed and the learning parts. The unification of learning and evolution suggests that cognitive systems can be viewed as pure learning systems. The knowledge and skills a biological organism possesses can be viewed as the accumulation of what has been learned by its ancestors over billions of years, and what it has learned from its individual experience since conception. Robust logic [24] is a mathematical framework based on learning that aims to encompass cognitive tasks beyond learning, particularly reasoning. The pragmatic difficulty of finding training data for systems to be built along such principles has been pointed out [25]. By acknowledging that the training data may also need to cover knowledge learned through evolution one is acknowledging what happens in existing cognitive systems, namely the biological ones. It is possible that learning is the only way of guaranteeing sufficient robustness in large-scale cognitive systems. In that case it would follow that the construction of cognitive systems with human level performance should be conceptualized as a learning task that encompasses knowledge acquired in biological systems through evolution as well as experience.

We have shown that with regard to the acquisition of complex mechanisms evolvability can be viewed as a restricted form of learnability. While evolvability may be technically the more constrained, it is not inherently more mysterious.

Acknowledgements

The author is grateful to Daniel Fisher and Martin Nowak for stimulating discussions and encouragement, to Brian Jacobson, Loizos Michael and Rocco Servedio for their technical comments on earlier drafts of this paper, and to Gill Bejerano and Richard Karp for some helpful observations. This work was supported by grants NSF-CCR-0310882, NSF-CCF-0432037 and NSF-CCF-0427129 from the National Science Foundation.

References

[1] Bäck, T., Fogel, D.B., Michalewicz, Z. (eds.): Handbook of Evolutionary Computation. Oxford Univ. Press, Oxford (1997)
[2] Bejerano, G., et al.: Ultraconserved elements in the human genome. Science 304, 1321–1325 (2004)
[3] Blumer, A., Ehrenfeucht, A., Haussler, D., Warmuth, M.K.: Learnability and the Vapnik Chervonenkis dimension. J. ACM 36(4), 929–965 (1989)
[4] Bürger, R.: The Mathematical Theory of Selection, Recombination, and Mutation. Wiley, Chichester (2000)
[5] Darwin, C.: On the origin of species by means of natural selection. John Murray, London (1859)
[6] Dermitzakis, E.T., et al.: Conserved non-genic sequences - an unexpected feature of mammalian genomes. Nature Reviews Genetics 6, 151–157 (2005)
[7] Drake, J.W., et al.: Rates of spontaneous mutation. Genetics 148, 1667–1686 (1998)
[8] Ehrenfeucht, A., Haussler, D., Kearns, M., Valiant, L.G.: A general lower bound on the number of examples needed for learning. Inf. and Computation 82(2), 247–261 (1989)
[9] Fischer, P., Simon, H.U.: On learning ring-sum expressions. SIAM J. Computing 21(1), 181–192 (1992)
[10] Fisher, R.A.: The Genetical Theory of Natural Selection. Oxford University Press, Oxford (1930)
[11] Garey, M.R., Johnson, D.S.: Computers and Intractability: a Guide to the Theory of NP-Completeness. Freeman, San Francisco (1979)
[12] Helmbold, D., Sloan, R., Warmuth, M.K.: Learning integer lattices. SIAM J. Computing 21(2), 240–266 (1992)
[13] Hoeffding, W.: Probability inequalities for sums of bounded random variables. J. Amer. Stat. Assoc. 58, 13 (1963)
[14] Kearns, M.: Efficient noise tolerant learning from statistical queries. J.ACM 45(6), 983–1006 (1998)
[15] Kearns, M., Valiant, L.G.: Cryptographic limitations on learning Boolean formulae. J. ACM 41(1), 67–95 (1994)
[16] Kearns, M., Vazirani, U.: An Introduction to Computational Learning Theory. MIT Press, Cambridge (1994)
[17] Kimura, M.: Evolutionary rate at the molecular level. Nature 217, 624–626 (1968)
[18] Kumar, S., Subramanian, S.: Mutation rates in mammalian genomes. Proc. Nat. Acad. Sci. 99, 803–808 (2002)
[19] Papadimitriou, C.H.: Computational Complexity. Addison-Wesley, Reading, Mass (1994)
[20] Pitt, L., Valiant, L.G.: Computational limitations on learning from examples. J. ACM 35(4), 965–984 (1988)
[21] Roff, D.A.: Evolutionary Quantitative Genetics. Chapman & Hall, New York (1997)

[22] Ros, J.P.: Learning Boolean functions with genetic algorithms: A PAC analysis. In: Whitley, L.D. (ed.) Foundations of Genetic Algorithms, pp. 257–275. Morgan Kaufmann, San Mateo, CA (1993)

[23] Valiant, L.G.: A theory of the learnable. C. ACM 27(11), 1134–1142 (1984)

[24] Valiant, L.G.: Robust logics. Artificial Intelligence Journal 117, 231–253 (2000)

[25] Valiant, L.G.: Knowledge infusion. In: Proc. 21st National Conference on Artificial Intelligence, AAAI06, pp. 1546–1551 (2006)

[26] Wagner, G.P., Altenberg, L.: Complex adaptations and the evolution of evolvability. Evolution 50(3), 967–976 (1996)

[27] Wegener, I.: Theoretical aspects of evolutionary algorithms. In: Orejas, F., Spirakis, P.G., van Leeuwen, J. (eds.) ICALP 2001. LNCS, vol. 2076, pp. 64–78. Springer, Heidelberg (2001)

[28] Wright, S.: Evolution and the Genetics of Populations, A Treatise. University of Chicago Press, Chicago (1968-78)

Expander Properties and the Cover Time of Random Intersection Graphs*

Sotiris E. Nikoletseas[1,2], Christoforos Raptopoulos[1,2], and Paul G. Spirakis[1,2]

[1] Computer Technology Institute, P.O. Box 1122, 26110 Patras, Greece
[2] University of Patras, 26500 Patras, Greece
nikole@cti.gr, raptopox@ceid.upatras.gr, spirakis@cti.gr

Abstract. We investigate important combinatorial and algorithmic properties of $G_{n,m,p}$ random intersection graphs. In particular, we prove that with high probability (a) random intersection graphs are expanders, (b) random walks on such graphs are "rapidly mixing" (in particular they mix in logarithmic time) and (c) the cover time of random walks on such graphs is optimal (i.e. it is $\Theta(n \log n)$). All results are proved for p very close to the connectivity threshold and for the interesting, non-trivial range where random intersection graphs differ from classical $G_{n,p}$ random graphs.

1 Introduction

Random graphs are interesting combinatorial objects that were introduced by P. Erdös and A. Rényi and still attract a huge amount of research in the communities of Theoretical Computer Science, Algorithms, Graph Theory and Discrete Mathematics. This continuing interest is due to the fact that, besides their mathematical beauty, such graphs are very important, since they can model interactions and faults in networks and also serve as typical inputs for an average case analysis of algorithms.

There exist various models of random graphs. The most famous is the $G_{n,p}$ random graph, a sample space whose points are graphs produced by randomly sampling the edges of a graph on n vertices independently, with the same probability p. Other models have also been quite a lot investigated: $G_{n,r}$ (the "random regular graphs", produced by randomly and equiprobably sampling a graph from all regular graphs of n vertices and vertex degree r) and $G_{n,M}$ (produced by randomly and equiprobably selecting an element of the class of graphs on n vertices having M edges). For an excellent survey of these models, see [2,4].

* This work was partially supported by the IST Programme of the European Union under contact number IST-2005-15964 (AEOLUS) and by the Programme PENED under contact number 03ED568, co-funded 75% by European Union – European Social Fund (ESF), 25% by Greek Government – Ministry of Development – General Secretariat of Research and Technology (GSRT), and by Private Sector, under Measure 8.3 of O.P. Competitiveness – 3rd Community Support Framework (CSF).

L. Kučera and A. Kučera (Eds.): MFCS 2007, LNCS 4708, pp. 44–55, 2007.

In this work we study important properties (expansion properties and the cover time) of a relatively recent model of random graphs, namely the random intersection graphs model introduced by Karoński, Sheinerman and Singer-Cohen [16,27]. Also, Godehardt and Jaworski [12] considered similar models. In the $G_{n,m,p}$, to each of the n vertices of the graph, a random subset of a universal set of m elements is assigned, by independently choosing elements with the same probability p. Two vertices u, v are then adjacent in the $G_{n,m,p}$ graph if and only if their assigned sets of elements have at least one element in common.

Importance and Motivation. First of all, we note that (as proved in [17]) any graph is a random intersection graph. Thus, the $G_{n,m,p}$ model is very general. Furthermore, for some ranges of the parameters m, p ($m = n^{\alpha}, \alpha > 6$) the spaces $G_{n,m,p}$ and $G_{n,\hat{p}}$ are equivalent (as proved by Fill, Sheinerman and Singer-Cohen [11], showing that in this range, the total variation distance between the graph random variables has limit 0).

Second, random intersection graphs may model real-life applications more accurately (compared to the $G_{n,\hat{p}}$ Bernoulli random graphs case). In fact, there are practical situations where each communication agent (e.g. a wireless node) gets access only to some ports (statistically) out of a possible set of communication ports. When another agent also selects a communication port, then a communication link is implicitly established and this gives rise to communication graphs that look like random intersection graphs. Even epidemiological phenomena (like spread of disease) tend to be more accurately captured by this "proximity-sensitive" random intersection graphs model. Other applications may include oblivious resource sharing in a distributed setting, interactions of mobile agents traversing the web etc.

Regarding the properties we study, we believe that their importance is evident. So we just mention the fact that expander graphs are basic building blocks in optimal network design. Also, at a combinatorial/algorithmic level, it is well known that random walks whose second largest eigenvalue is sufficiently less than 1 are "rapidly mixing", i.e. they get close (in terms of the variation distance) to the steady state distribution after only a polylogarithmic (in the number of vertices/states) number of steps (see e.g. [26]); this has important algorithmic applications e.g. in efficient random generation and counting of combinatorial objects. Finally, the cover time of a graph is one of its most important combinatorial measures which also captures practical quantities like the expected communication time in a network of mobile entities, infection times in security applications etc.

Related Work. Random intersection graphs have recently attracted a growing research interest. The question of how close $G_{n,m,p}$ and $G_{n,\hat{p}}$ are for various values of m, p has been studied by Fill, Sheinerman and Singer-Cohen in [11]. In [19], new models of random intersection graphs have been proposed, along with an investigation of both the existence and efficient finding of close to optimal independent sets. The authors of [10] find thresholds (that are optimal up to a constant factor) for the appearance of hamilton cycles in random intersection graphs. The efficient construction of hamilton cycles is studied in [24]. Also, by using a sieve method, Stark [28] gives exact formulae for the degree distribution of an arbitrary fixed vertex of $G_{n,m,p}$ for a quite wide range of the parameters of the model. In [21], the authors use a coupling technique to bound the second eigenvalue of random walks on instances of symmetric random intersection graphs $G_{n,n,p}$ (i.e. in random intersection graphs with $m = n$), when p is near the connectivity threshold. The upper bound proved holds for almost every instance of the symmetric random intersection graphs model. We should note that in this paper we deal with the case $m = n^{\alpha}$, for $\alpha < 1$, which is very different from the symmetric case, as in the first case each label is selected by a large number of vertices (which allows for much tighter concentration bounds that help in the analysis).

In [15] the author proves that with high probability (whp) the cover time (that is the *expected time* to visit all the vertices of the graph) of a simple random walk on a Bernoulli random graph $G_{n,\hat{p}}$ is quite close to optimal when $\hat{p} = \omega\left(\frac{\ln n}{n}\right)$. Also, he proves that by further increasing the value of $\hat{p}$, the same bound that holds for the cover time holds whp for the actual time needed for the random walk on $G_{n,\hat{p}}$ to visit all the vertices of the graph. His results are improved by Cooper and Frieze in [5], who prove that when $\hat{p} = \frac{c \log n}{n}$, $c > 1$, the cover time of $G_{n,\hat{p}}$ is asymptotic to $c \log\left(\frac{c}{c-1}\right) n \log n$.

Geometric proximity between randomly placed objects is also nicely captured by the model of random geometric graphs (see e.g. [7,8,23]) and important variations (like random scaled sector graphs, [9]). In [3], the cover time of random geometric graphs near the connectivity threshold is found almost optimal, by showing that the effective resistance of the graph is small. Other extensions of random graph models (such as random regular graphs) and several important combinatorial properties (connectivity, expansion, existence of a giant connected component) are performed in [18,22].

Our Contribution. As proved in [11], the spaces $G_{n,m,p}$ and $G_{n,\hat{p}}$ are equivalent when $m = n^{\alpha}$, with $\alpha > 6$, in the sense that their total variation distance tends to 0 as n goes to ∞. Also, the authors in [24] show that, when $\alpha > 1$, for any monotone increasing property there is a direct relation (including a multiplicative constant) of the corresponding thresholds of the property in the two spaces. So, it is very important to investigate what is happening when $\alpha \leq 1$ where the two spaces are statistically different. In this paper, we study the regime $\alpha < 1$. In particular

(a) We first prove that $G_{n,m,p}$ random intersection graphs are c-expanders (i.e. every set S of at most $n/2$ vertices is connected to at least $c|S|$ other vertices outside S) with high probability. This is shown for $p = \frac{\ln n + g(n)}{m}$, where $g(n) \to \infty$ arbitrarily slowly, i.e. p is *just* above the connectivity threshold. [1] Note that [21] has no equivalent results to this one.

[1] The connectivity threshold for $\alpha \leq 1$ is proved to be $\tau_c = \frac{\ln n}{m}$ in [27].

(b) We then show that random walks on the vertices of random intersection graphs are whp rapidly mixing (in particular, the mixing time is logarithmic on n). This is shown for p very close to the connectivity threshold τ_c of $G_{n,m,p}$, with $m = n^\alpha, \alpha < 1$. We interestingly note that the c-expansion property shown in (a) cannot ensure "small" rapid mixing. For example imagine the following graph pointed out to us by Noga Alon [1]: two cliques of size $n/2$ each, connected by a perfect matching of their vertices is a c-expander but has mixing time $\Omega(n)$. To get our result on the mixing time we had to prove an upper bound on the second eigenvalue of $G_{n,m,p}$, that holds with high probability, through coupling arguments of the original Markov Chain describing the random walk and another Markov Chain on an associated random bipartite graph whose conductance properties we show to be appropriate. The attentive reader can easily understand that although the general technique used to prove the results of section 4 is similar to the technique used in [21], the proofs are quite different. More specifically, in the case of $G_{n,m,p}$, with $m = n^\alpha, \alpha < 1$, the concentration results of section 2 (and especially the first part of Lemma 1) can be used to give an elegant proof of Lemma 5 (which cannot be applied in the symmetric case considered in [21]).

(c) Finally, we show that the cover time of such graphs (in the interesting, non-trivial range mentioned above and for p close to the connectivity threshold) is whp $\Theta(n \ln n)$, i.e. optimal up to multiplicative constants. To get this result we had to prove a technically involved intermediate result relating the probability that our random walk on $G_{n,m,p}$ has not visited a vertex v by time t with the degree of v. Note that [21] has no equivalent results to this one. Also, to prove the results of section 6, one needs to prove an extra preliminary result (namely Lemma 3) that does not appear in [21].

2 Notation, Definitions and Properties of $G_{n,m,p}$

Let $Bin(n, p)$ denote the Binomial distribution with parameters n and p. We first formally define the *random intersection graphs model.*

Definition 1 (Random Intersection Graph). *Consider a universe $\mathcal{M} = \{1, 2, \ldots, m\}$ of elements and a set of vertices $V(G) = \{v_1, v_2, \ldots, v_n\}$. If we assign independently to each vertex v_j, $j = 1, 2, \ldots, n$, a subset S_{v_j} of $\mathcal{M}$ choosing each element $i \in \mathcal{M}$ independently with probability p and put an edge between two vertices v_{j_1}, v_{j_2} if and only if $S_{v_{j_1}} \cap S_{v_{j_2}} \neq \emptyset$, then the resulting graph is an instance of the random intersection graph $G_{n,m,p}$. In this model we also denote by L_l the set of vertices that have chosen label $l \in M$. The degree of $v \in V(G)$ will be denoted by $d_G(v)$. Also, the set of edges of $G_{n,m,p}$ will be denoted by $e(G)$.*

Consider now the bipartite graph with vertex set $V(G) \cup \mathcal{M}$ and edge set $\{(v_j, i) : i \in S_{v_j}\} = \{(v_j, i) : v_j \in L_i\}$. We will refer to this graph as the bipartite random graph $B_{n,m,p}$ associated to $G_{n,m,p}$.

In this section we assume that $m = n^\alpha$, for some $\alpha < 1$. This is the interesting regime where $G_{n,m,p}$ differs from $G_{n,\hat{p}}$ (see also "Our Contribution" in the previous section). Let $\tau_c \stackrel{def}{=} \frac{\ln n}{m}$ be the connectivity threshold for $G_{n,m,p}$ in that case. Also we assume that $p = 4\tau_c$. We prove that

Lemma 1. *The following hold with high probability in $G_{n,m,p}$ when $\alpha < 1$ and $p = 4\frac{\ln n}{m}$*

(a) *For every label $l \in \mathcal{M}$ we have $(1-\epsilon)np \leq |L_l| \leq (1+\epsilon)np$ for any $\epsilon \in [n^{-(1-\alpha)/2}, 1)$.*
(b) *For every vertex $v \in V$ we have $|S_v| \in (1 \pm \sqrt{4/5})4\ln n$.*

Proof. See full paper [20]. □

Note that Lemma 1 implies that the minimum degree in $G_{n,m,p}$ when $\alpha < 1$ and p just above the connectivity is whp at least $\Omega(n^{1-\alpha}\ln n)$. In fact, we prove the following

Lemma 2. *The following hold with high probability in $G_{n,m,p}$ when $\alpha < 1$ and $p = 4\frac{\ln n}{m}$*

(a) *The degree of any vertex $v \in V$ satisfies $d_G(v) \in (1 \pm n^{-\epsilon'})4|S_v|n^{1-\alpha}\ln n$ for some positive constant ϵ' bounded away from 0 and $1-\alpha$.*
(b) *The number of edges of the $G_{n,m,p}$ graph satisfies $|e(G)| \in (1 \pm \epsilon'')\frac{1}{2}16n^{2-\alpha}\ln^2 n$, for some small constant $\epsilon'' > 0$.*
(c) *There are no vertices $x \neq y \in V(G)$ such that $|S_x \cap S_y| \geq \lceil \frac{3}{\alpha} \rceil$.*

Proof. See full paper [20]. □

Let now $D(k)$ be the number of vertices $v \in V(G)$ that have $|S_v| = k$, i.e. $D(k) = |\{v \in V : |S_v| = k\}|$. Consider also the following partition of the set $\{0, 1, \ldots, m\}$.

$$
\begin{aligned}
M_2 &= \{k \in (1 \pm \sqrt{4/5})\ln n : E[D(k)] > \ln n\} \\
M_1 &= \{k \in (1 \pm \sqrt{4/5})\ln n : E[D(k)] \leq \ln n\} \\
M_0 &= \{\{0, 1, \ldots, m\} \backslash \{M_2 \cup M_1\}\}
\end{aligned}
$$

We can then prove the following lemma that will be useful for upper bounding the cover time.

Lemma 3. *For the $G_{n,m,p}$ with $\alpha < 1$ and $p = 4\tau_c$ the following hold with high probability*

1. *For every $k \in M_0$, $D(k) = 0$*
2. *For every $k \in M_1$, $D(k) \leq \ln^3 n$ and*
3. *For every $k \in M_2$, $D(k) \leq 2E[D(k)]$.*

Proof. See full paper [20]. □

3 Expansion Properties of Random Intersection Graphs

We first give the following definition:

Definition 2 (c-expanders). *Let c be a positive constant. A graph $G = (V(G), E(G))$ is a c-expander if every set $S \subseteq V(G)$ of at most $n/2$ vertices is connected to at least $c|S|$ vertices outside S.*

In this section we assume that $p = \frac{\ln n + g(n)}{m}$, that is, p is *just* above the connectivity threshold τ_c. In the following, let $S_X = \bigcup_{v \in X} S_v$, for $X \subseteq V$ and let $L_Y = \bigcup_{l \in Y} L_l$, for $Y \subseteq \mathcal{M}$. We prove the following:

Lemma 4. *Assume that $m = n^\alpha$, $\alpha < 1$ and $p = \frac{\ln n + g(n)}{m}$, for some function $g(n) \to \infty$ (arbitrarily slowly). With high probability $G_{n,m,p}$ is a c-expander, for some constant $c > 0$.*

Proof. See full paper [20]. □

4 Bounds for the Second Eigenvalue and the Mixing Time

In this section we give an upper bound on the second eigenvalue (i.e. the eigenvalue with the largest absolute value less than 1) of $G_{n,m,p}$, with $\alpha < 1$ and $p = 4\tau_c$, that holds for almost every instance. This will imply a logarithmic mixing time.

Let $\tilde{W}$ be a Markov Chain on state space V (i.e. the vertices of $G_{n,m,p}$) and transition matrix given by

$$\tilde{P}(x,y) = \begin{cases} \sum_{l \in S_x \cap S_y} \frac{1}{|S_x| \cdot |L_l|} & \text{if } S_x \cap S_y \neq \emptyset \\ 0 & \text{otherwise.} \end{cases}$$

Note that this Markov Chain comes from observing the simple random walk on the $B_{n,m,p}$ graph associated with $G_{n,m,p}$ every two steps. This means that $\tilde{W}$ is reversible and we can easily verify that its stationary distribution is given by

$$\tilde{\pi}(x) = \frac{|S_x|}{\sum_{y \in V} |S_y|}, \qquad \text{for every } x \in V.$$

Now let W denote the random walk on $G_{n,m,p}$ and let P denote its transition probability matrix. It is known that W is reversible and its stationary distribution is given by $\pi(x) = \frac{d_G(x)}{2|e(G)|}$, for every $x \in V$.

Notice now that $P(x,y) > 0 \Leftrightarrow \tilde{P}(x,y) > 0$. By using Theorem 2.1 of [6], we can show that if λ_1 (respectively $\tilde{\lambda}_1$) is the second largest eigenvalue of P (respectively $\tilde{P}$), then

$$\lambda_1 \leq 1 - \frac{\beta}{A}(1 - \tilde{\lambda}_1) \tag{1}$$

where β is such that $\tilde{\pi}(x) \geq \beta\pi(x)$, for all $x \in V$, and

$$A = \max_{(x,y):P(x,y)>0} \left\{ \frac{\tilde{\pi}(x)\tilde{P}(x,y)}{\pi(x)P(x,y))} \right\}.^2$$

Because of (a) of Lemma 2 and the fact that $|e(G)| = \sum_{y \in V} d_G(v)$, there exist two positive constants $\beta < 1 < \beta'$, such that $\beta\pi(x) \leq \tilde{\pi}(x) \leq \beta'\pi(x)$, for all $x \in V$ (These two constants can be quite close to 1 as they depend on the value of ϵ'' of Lemma 2). Also, by (a) and (c) of Lemma 2, for any (x,y) such that $P(x,y) > 0$, we have

$$\frac{\tilde{P}(x,y)}{P(x,y)} = \frac{\sum_{l \in S_x \cap S_y} \frac{1}{|S_x| \cdot |L_l|}}{\frac{1}{d_G(x)}} \leq (1+\gamma')|S_x \cap S_y| \leq \gamma$$

for some positive constants γ', γ. This means that A is upper bounded bounded by some constant. Thus, we have established that $\lambda_1 \leq 1 - \zeta_1(1 - \tilde{\lambda}_1)$, for $\zeta_1 = \frac{\beta}{\beta'\gamma}$.

We now show that $\tilde{\lambda}_1$ is whp bounded away from 1 by some constant (which will mean that λ_1 is also bounded away from 1).

Let $\hat{W}$ denote the random walk on the $B_{n,m,p}$ bipartite graph that is associated to $G_{n,m,p}$. Let also $\hat{P}$ denote its transition probability matrix and let $\hat{\lambda}_i, i = 1, \ldots, m+n$, its eigenvalues and $\hat{x}_i, i = 1, \ldots, m+n$, their corresponding eigenvectors. Note that

$$\hat{P}^2 = \begin{bmatrix} \tilde{P} & \emptyset \\ \emptyset & Q \end{bmatrix}$$

where Q is some transition matrix. But for any $i = 1, \ldots, m+n$, we have $\hat{P}^2\hat{x}_i = \hat{\lambda}_i^2\hat{x}_i$. So, in order to give the desired upper bound to $\tilde{\lambda}_1$, we need to show that $\hat{\lambda}_1$ is whp bounded away from 1 by some constant (because we already have that $\tilde{\lambda}_1 \leq \hat{\lambda}_1^2$). In order to do so, we use the notion of *conductance* $\Phi_{\hat{W}}$ of the walk $\hat{W}$ that is defined as follows:

[2] The original theorem is as follows: For each pair $x \neq y$ with $\tilde{P}(x,y) > 0$, fix a sequence of steps $x_0 = x, x_1, x_2, \ldots, x_k = y$ with $P(x_i, x_{i+1}) > 0$. This sequence of steps is called a *path* γ_{xy} of length $|\gamma_{xy}| = k$. Let $\mathcal{E} = \{(x,y) : P(x,y) > 0)\}$, $\tilde{\mathcal{E}} = \{(x,y) : \tilde{P}(x,y) > 0)\}$ and $\tilde{\mathcal{E}}(z,w) = \{(x,y) \in \tilde{\mathcal{E}} : (z,w) \in \gamma_{xy}\}$, where $(z,w) \in \mathcal{E}$. Then

$$\lambda_1 \leq 1 - \frac{\beta}{A}(1 - \tilde{\lambda}_1)$$

where β is such that $\tilde{\pi}(x) \geq \beta\pi(x)$, for all $x \in V$, and

$$A = \max_{(z,w) \in \mathcal{E}} \left\{ \frac{1}{\pi(x)P(x,y))} \sum_{\tilde{\mathcal{E}}(z,w)} |\gamma_{xy}|\tilde{\pi}(x)\tilde{P}(x,y) \right\}.$$

In our case we have taken $\gamma_{x,y} = \{x_0 = x, x_1 = y\}$ for every $(x,y) \in \tilde{\mathcal{E}}$ which simplifies our formula.

Definition 3. *Consider the bipartite random graph $B_{n,m,p}$ that is associated to $G_{n,m,p}$. The vertex set of $B_{n,m,p}$ is $V(B) = V(G) \cup \mathcal{M}$. For every $x \in V(B)$, let $d_B(x)$ be the degree of x in B. For any $S \subseteq V(B)$, let $e_B(S : \overline{S})$ be the set of edges of S with one end in S and the other in $\overline{S} = V(B) \backslash S$, let $d_B(S) = \sum_{v \in S} d_G(v)$ and $\hat{\pi}(S) = \sum_{v \in S} \hat{\pi}(v)$. Then*

$$\Phi_{\hat{W}} = \min_{\hat{\pi}(S) \leq 1/2} \frac{|e_B(S : \overline{S})|}{d_B(S)}.$$

We now prove the following

Lemma 5. *With high probability, the conductance of the random walk on $B_{n,m,p}$ satisfies $\Phi_{\hat{W}} \geq \zeta_2$, where ζ_2 is some positive constant.*

Proof. See full paper [20]. □

By a result of [13,25], we know that $\hat{\lambda}_1 \leq 1 - \frac{\Phi^2_{\hat{W}}}{2}$ and so $\hat{\lambda}_1$ is (upper) bounded away from 1. By the above discussion, we have proved the following

Theorem 1. *With high probability, the second largest eigenvalue of the random walk on $G_{n,m,p}$, with $m = n^\alpha, \alpha < 1$ and $p = 4\tau_c$, satisfies $\lambda_1 \leq \zeta$, where $\zeta \in (0, 1)$ is a constant that is bounded away from 1.*

Such a bound on λ_1 implies (as shown in Proposition 1 of [26]) a logarithmic mixing time. Thus we get

Theorem 2. *With high probability, there exists a sufficiently large constant $K > 0$ such that if $\tau_0^{(G)} = K \log n$, then for all $v, u \in V(G)$ and any $t \geq \tau_0^{(G)}$,*

$$|P^{(t)}(u, v) - \pi(v)| = O(n^{-3})$$

where $P^{(t)}$ denotes the t-step transition matrix of the random walk W on $G_{n,m,p}$, with $m = n^\alpha, \alpha < 1$ and $p = 4\tau_c$. We will refer to $\tau_0^{(G)}$ as the mixing time *of $G_{n,m,p}$.*

5 A Useful Lemma

In order to give bounds to the cover time of $G_{n,m,p}$, for $m = n^\alpha$, $\alpha < 1$ and p four times the connectivity threshold we first prove a lemma that the probability that the random walk on $G_{n,m,p}$ has not visited a vertex v by time t with the degree of v. Before presenting the lemma we give some notation.

Let G be an instance of the random intersection graphs model and let $H(v) = G - \{v\}$. We will sometimes write H instead of $H(v)$ when v is clear from the context. Let $\tau_0^{(H)}$ denote the mixing time of H, namely the time needed for the random walk on H to get closer than $O(n^{-3})$ to its steady state distribution (see also definition of $\tau_0^{(G)}$ in Theorem 2). Note that because of the definition of

H and by Lemma 1, the removal of v from G does not affect its mixing time[3] and so $\tau_0^{(G)} \sim \tau_0^{(H)} \leq \tau_0 \stackrel{def}{=} \Theta(\log n)$ for any v whp. We will denote by $\mathcal{W}_{u,H}$ the random walk on H that starts at vertex $u \in V(H)$. Let also $\mathcal{W}_{u,H}(t)$ be the random walk generated by the first t steps. For $u \neq v \in V$, let $A_t(v)$ be the event that $\mathcal{W}_{u,G}(t)$ has not visited v.

Lemma 6. *Let G be an instance of $G_{n,m,p}$, with $m = n^\alpha, \alpha < 1$ and $p = 4\frac{\ln n}{m}$, that satisfies Lemma 1 and has $\tau_0^{(H)} \leq \tau_0 = \Theta(\log n)$ for every $H = H(v)$ (note that almost every instance of $G_{n,m,p}$ in this range satisfies these requirements). Let δ_v be the minimum degree of the neighbours of $v \in V(G)$. Then, for every $v \in V(G)$,*

$$Pr(A_t(v)) \leq \left(1 - \left(\frac{\delta_v - 1}{\delta_v} - o(1)\right)\frac{d_G(v)}{2|e(G)|}\right)^{t-\tau_0} Pr(A_{\tau_0}(v)).$$

Proof. See full paper [20]. □

6 An Upper Bound on the Cover Time

Let G be an instance of $G_{n,m,p}$, where $m = n^\alpha, \alpha < 1$ and $p = 4\frac{\ln n}{m}$. Fix an arbitrary vertex u. Let $T_G(u)$ be the time that the random walk W_u on G needs to visit every vertex in $V(G)$. The following theorem shows that the cover time on G is optimal assuming that G is a "typical" instance of the $G_{n,m,p}$ model in this range, i.e. an instance that satisfies Lemmata 1, 2, 3 and has $\tau_0^{(H)} \leq \tau_0 = \Theta(\log n)$ for every $H = H(v)$ (the last assumption assures us that Lemma 6 can be applied). Note that almost every instance of $G_{n,m,p}$ in this range is "typical", since these requirements are satisfied whp.

Theorem 3. *The cover time C_u of the random walk starting from u is almost surely at most $\Theta(n \ln n)$.*

Proof. We will denote by U_t the number of vertices that have not been visited by W_u at step t. Clearly, the cover time of W_u satisfies

$$C_u = E[T_G(u)] = \sum_{t=0}^{\infty} Pr(T_G > t) = \sum_{t=0}^{\infty} Pr(U_t > 0) \leq \sum_{t=0}^{\infty} \min\{1, E[U_t]\}$$

by Markov's inequality. So, for any $t_0 > 0$,

$$C_u \leq t_0 + \sum_{t \geq t_0+1} E[U_t] = t_0 + \sum_{t \geq t_0+1} \sum_{v \in V(G)} Pr(A_t(v)) \tag{2}$$

where in the last equality we used the linearity of expectation and the fact that the events $\{v \in U_t\}$ and $A_t(v)$ are the same.

[3] In fact the same analysis of section 4 can be applied unchanged to $H(v)$.

We set $t_0 = 5n \log n$. Since $\delta_v \geq n^{1-\alpha} \ln n$, for every $v \in V(G)$, by setting $Pr(A_{\tau_0}(v))$ equal to 1 in Lemma 6 and using the well known inequality $1+x \leq e^x$, for any real x, we have that for all $t \geq t_0$,

$$Pr(A_t(v)) \leq \exp\left\{-\frac{td_G(v)}{2|e(G)|}\left(1 - O\left(\frac{\tau_0}{\delta_v}\right)\right)\right\} \leq \exp\left\{-(1-B)\frac{t|S_v|}{4n \ln n}\right\}$$

for some small constant $B > 0$. Note that for the final inequality we used the fact that $\tau_0 = o(\delta_v)$.

Now equation (2) becomes

$$C_u \leq 5n \ln n + \sum_{v \in V(G)} \sum_{t \geq t_0+1} e^{(1-B)\frac{-t|S_v|}{4n \ln n}} \leq 5n \ln n + 5n \ln n \sum_{v \in V(G)} \frac{1}{|S_v|} e^{-\frac{5(1-B)}{4}|S_v|}$$

$$\leq 5n \ln n + 5n \ln n \left(\sum_{v:|S_v| \in M_1} \frac{1}{|S_v|} e^{-|S_v|} + \sum_{v:|S_v| \in M_2} \frac{1}{|S_v|} e^{-|S_v|}\right) \quad (3)$$

Because of Lemma 1 and Lemma 3, the first sum is clearly $o(1)$ (just notice that for any v such that $|S_v| \in M_1$ we have that $|S_v| = \Theta(\ln n)$ and $D(|S_v|) \leq \ln^3 n$). For the second sum, by Lemma 3 we have

$$\sum_{v:|S_v| \in M_2} \frac{1}{|S_v|} e^{-\frac{|S_v|}{5}} \leq \sum_{k=1}^{m} D(k) \frac{1}{k} e^{-k} \leq \sum_{k=1}^{m} 2n \binom{m}{k} p^k (1-p)^{m-k} \frac{1}{k} e^{-k}$$

$$\leq 7n \frac{1}{mp} \sum_{k=1}^{m} \binom{m+1}{k+1} p^{k+1} (1-p)^{m-k} e^{-(k+1)}$$

$$\leq 7n \frac{1}{mp} \left(1 - p + pe^{-1}\right)^{m+1} = o(1).$$

By (3) this means that $C_u \leq \Theta(n \ln n)$ for any fixed vertex u. □

Since the cover time $C = \max_{u \in V(G)} C_u$, and it is known that $C \geq (1 - o(1)) n \ln n$, we have proved

Theorem 4. *The cover time of an instance of $G_{n,m,p}$, with $m = n^\alpha, \alpha < 1$ and $p = 4\frac{\ln n}{m}$, is $C = \Theta(n \ln n)$ with high probability.*

7 Conclusions and Future Work

In this work, we investigated the expansion properties, the mixing time and the cover time of $G_{n,m,p}$ random intersection graphs for the non-trivial regime where $m = n^\alpha$, for $\alpha < 1$ and p very close to the connectivity threshold. We showed that the mixing time is logarithmic on the number of vertices and that the cover time is asymptotically optimal. Our analysis can be pushed further (although not without many technical difficulties) to provide even tighter results. However, the cover time and expansion properties in the case $\alpha = 1$ remains an open problem. It is worth investigating other important properties of $G_{n,m,p}$, such as dominating sets, existence of vertex disjoint paths between pairs of vertices etc.

Acknowledgement. We wish to warmly thank Noga Alon for pointing out to us the particular graph (two cliques of size $n/2$ with a perfect matching between them) that despite being a c-expander is not rapidly mixing. Also, for providing material on his relevant research to us.

References

1. Alon, N.: Personal communication (January 2007)
2. Alon, N., Spencer, J.: The Probabilistic Method, 2nd edn. Wiley & Sons, Inc., Chichester (2000)
3. Avin, C., Ercal, G.: On the Cover Time of Random Geometric Graphs. In: Caires, L., Italiano, G.F., Monteiro, L., Palamidessi, C., Yung, M. (eds.) ICALP 2005. LNCS, vol. 3580, pp. 677–689. Springer, Heidelberg (2005)
4. Bollobás, B.: Random Graphs, 2nd edn. Cambridge University Press, Cambridge (2001)
5. Cooper, C., Frieze, A.: The Cover Time of Sparse Random Graphs. Random Structures and Algorithms 30, 1–16 (2007)
6. Diaconis, P., Saloff-Coste, L.: Comparison Theorems for Reversible Markov Chains. The Annals of Applied Probability 3(3), 696–730 (1993)
7. Díaz, J., Penrose, M.D., Petit, J., Serna, M.: Approximating Layout Problems on Random Geometric Graphs. Journal of Algorithms 39, 78–116 (2001)
8. Díaz, J., Petit, J., Serna, M.: Random Geometric Problems on $[0, 1]^2$. In: Rolim, J.D.P., Serna, M.J., Luby, M. (eds.) RANDOM 1998. LNCS, vol. 1518, pp. 294–306. Springer, Heidelberg (1998)
9. Díaz, J., Petit, J., Serna, M.: A Random Graph Model for Optical Networks of Sensors. In: Jansen, K., Margraf, M., Mastrolli, M., Rolim, J.D.P. (eds.) WEA 2003. LNCS, vol. 2647, pp. 186–196. Springer, Heidelberg (2003)
10. Efthymiou, C., Spirakis, P.G.: On the Existence of Hamilton Cycles in Random Intersection Graphs. In: Caires, L., Italiano, G.F., Monteiro, L., Palamidessi, C., Yung, M. (eds.) ICALP 2005. LNCS, vol. 3580, pp. 690–701. Springer, Heidelberg (2005)
11. Fill, J.A., Sheinerman, E.R., Singer-Cohen, K.B.: Random Intersection Graphs when $m = \omega(n)$: An Equivalence Theorem Relating the Evolution of the $G(n, m, p)$ and $G(n, p)$ models, `http://citeseer.nj.nec.com/fill98random.html`
12. Godehardt, E., Jaworski, J.: Two models of Random Intersection Graphs for Classification. In: Opitz, O., Schwaiger, M. (eds.) Studies in Classification, Data Analysis and Knowledge Organisation, pp. 67–82. Springer, Heidelberg (2002)
13. Jerrum, M., Sinclair, A.: Approximate Counting, Uniform Generation and Rapidly Mixing Markov Chains. Information and Computation 82, 93–133 (1989)
14. Jerrum, M., Sinclair, A.: The Markov Chain Monte Carlo Method: an Approach to Approximate Counting and Integration. In: Hochbaum, D. (ed.) Approximation Algorithms for NP-hard Problems, PWS, pp. 482–520 (1996)
15. Jonasson, J.: On the Cover Time of Random Walks on Random Graphs. Combinatorics, Probability and Computing 7, 265–279 (1998)
16. Karoński, M., Scheinerman, E.R., Singer-Cohen, K.B.: On Random Intersection Graphs: The Subgraph Problem. Combinatorics, Probability and Computing journal 8, 131–159 (1999)

17. Marczewski, E.: Sur deux propriétés des classes d' ensembles. Fund. Math. 33, 303–307 (1945)
18. Nikoletseas, S., Palem, K., Spirakis, P.G., Yung, M.: Short Vertex Disjoint Paths and Multiconnectivity in Random Graphs: Reliable Network Computing, In: The Proceedings of the 21st International Colloquium on Automata, Languages and Programming (ICALP), Lecture Notes in Computer Science vol. In: Shamir, E., Abiteboul, S. (eds.) ICALP 1994. LNCS, vol. 820, pp. 247–262. Springer, Heidelberg (1994)
19. Nikoletseas, S., Raptopoulos, C., Spirakis, P.G.: The Existence and Efficient Construction of Large Independent Sets in General Random Intersection Graphs. In: Díaz, J., Karhumäki, J., Lepistö, A., Sannella, D. (eds.) ICALP 2004. LNCS, vol. 3142, pp. 1029–1040. Springer, Heidelberg (2004)
20. Nikoletseas, S., Raptopoulos, C., Spirakis, P.G.: Expander Properties and the Cover Time of Random Intersection Graphs, DELIS technical report, `http://delis.upb.de/paper/DELIS-TR-0491.pdf`.
21. Nikoletseas, S., Raptopoulos, C., Spirakis, P.G.: The Second Eigenvalue of Random Walks on Symmetric Random Intersection Graphs. In: Proceedings of the 2nd International Conference on Algebraic Informatics (CAI 2007) (2007)
22. Nikoletseas, S., Spirakis, P.G.: Expander Properties in Random Regular Graphs with Edge Faults. In: Mayr, E.W., Puech, C. (eds.) STACS 95. LNCS, vol. 900, pp. 421–432. Springer, Heidelberg (1995)
23. Penrose, M.: Random Geometric Graphs. Oxford Studies in Probability (2003)
24. Raptopoulos, C., Spirakis, P.G.: Simple and Efficient Greedy Algorithms for Hamilton Cycles in Random Intersection Graphs. In: Deng, X., Du, D.-Z. (eds.) ISAAC 2005. LNCS, vol. 3827, pp. 493–504. Springer, Heidelberg (2005)
25. Sinclair, A.: Algorithms for Random Generation and Counting: a Markov Chain Approach, PhD Thesis, University of Edimburg (1988)
26. Sinclair, A.: In: Birkhauser (ed.) Algorithms for Random Generation and Counting (1992)
27. Singer-Cohen, K.B.: Random Intersection Graphs, PhD thesis, John Hopkins University (1995)
28. Stark, D.: The Vertex Degree Distribution of Random Intersection Graphs. Random Structures & Algorithms 24(3), 249–258 (2004)

Uncover Low Degree Vertices and Minimise the Mess: Independent Sets in Random Regular Graphs*

William Duckworth[1] and Michele Zito[2]

[1] Mathematical Sciences Institute
Australian National University
Canberra, ACT 0200, Australia
Billy.Duckworth@maths.anu.edu.au

[2] Department of Computer Science
University of Liverpool
Liverpool L69 3BX, UK
M.Zito@csc.liv.ac.uk

Abstract. We present algorithmic lower bounds on the size s_d of the largest independent sets of vertices in random d-regular graphs, for each fixed $d \geq 3$. For instance, for $d = 3$ we prove that, for graphs on n vertices, $s_d \geq 0.43475n$ with probability approaching one as n tends to infinity.

1 Introduction

Given a graph $G = (V, E)$, an *independent set* is a subset $\mathcal{I}$ of V which spans no edge. In this paper we are interested in finding (by algorithmic means) independent sets of the largest possible cardinality. Let $\alpha(G)$ be the size of the largest independent sets in G divided by $|V(G)|$. The problem has a long history. It is one of the first optimization problems whose decision version was shown to be NP-complete [12]. Since then many results have appeared either proving that an optimal structure can be found in polynomial time on special graph classes [1,13] or showing that particular polynomial-time heuristics return solutions that are not too small for particular classes of graphs [5,6,8,16] or else proving that finding heuristics returning solutions significantly close to the optimal ones is at least as hard as solving the optimization problem exactly [15].

The algorithmic results in this paper are valid *asymptotically almost surely* (*a.a.s.*), i.e. with probability approaching one as $|V(G)|$ tends to infinity, under the assumption that the input structure is presented according to a pre-specified probability distribution. The *maximum independent set problem* (MIS) has been studied thoroughly in several of such random structures. For graphs generated according to the well-known $G(n, p)$ model (n vertices, edges appear independently with probability p) it has been proven [7,10] that as long as pn tends to

* Part of this work was carried out during a visit of the second author to the Australian National University (ANU). The authors wish to thank the Mathematical Sciences Institute at the ANU for its support.

L. Kučera and A. Kučera (Eds.): MFCS 2007, LNCS 4708, pp. 56–66, 2007.

Table 1. A.a.s. bounds on $\alpha(G)$ for random d-regular graphs

d	l.b.	u.b.	α_d
3	0.4328	0.4554	0.4348
4	0.3901	0.4163	0.3921
5	0.3566	0.3844	0.3593
6	0.3296	0.3580	0.3330
7	0.3071	0.3357	0.3106

infinity $\alpha(G(n,p))n \sim 2\log np/\log 1/(1-p)$ (although no polynomial time algorithm is known which returns, even just a.a.s., an independent set containing more than half that many vertices). For random d-regular graphs the situation is less satisfactory. For large d, $\alpha(G)$ is close to $\frac{2\log d}{d}$ [11]. However if d is a small fixed constant (say 3 or 5 or even 100), only lower and upper bounds are known. The best known bounds are reported in the second and third column of Table 1 for d up to 7. The upper bounds are derived using Markov's inequality [18]. The lower bounds are algorithmic [19]. It is quite interesting that for the past 12 years, nobody has been able to improve these bounds (in fact the upper bounds are even older than that). More to this, the existence of greedy algorithms that return independent sets asymptotically larger than those promised by Wormald's algorithm [19], is an open problem raised ten years ago by Frieze and McDiarmid [9]. Notice that combinatorial methods like the one studied in [2], although successful for other packing problems [3,4], failed to shed any light on the exact location of $\alpha(G)$ [2, Chap. IV].

In this paper we argue that careful algorithm design may be of significant help in improving the current situation. We propose a couple of heuristics that, when followed by an algorithm, lead to improvements on the performances of Wormald's algorithm for all values of d with only minimal additional running time costs. Wormald [19] showed that a simple process (termed "neighbourly" algorithm in that paper) that repeatedly picks vertices of minimum positive degree, adds them to the independent set that is being built and then removes all edges at distance at most one from the chosen vertex, builds fairly large independent sets a.a.s. if $G \in \mathcal{G}(n,d\text{-reg})$. It turns out that, in some cases, it is more convenient to add to the independent set one of the neighbours v of the initially chosen vertex u, rather than u itself. More precisely, v should be chosen if this is guaranteed to create a number of low degree vertices (*sparsification* principle) or if it leads to the removal of very few edges (*minimal mess* principle). A detailed description of our algorithm is in Section 3.

We contend that our solution is simple to analyse: the proof of our results relies on a standard application of the differential equation method, spelled out in [20] (the method is general enough to allow the analysis of even more sophisticated algorithms). Furthermore, it seems plausible that similar principles may lead to improvements for other optimization problems.

In Section 2 we present the model of random regular graphs that we use and a statement of our main result. In Section 3 we describe our new algorithm.

The final part of the paper is devoted to the proof of our result: we first briefly describe the differential equation method, then we fill in all the details needed in the specific case.

2 Model and Main Result

Let $\mathcal{G}(n,d\text{-reg})$ denote the uniform probability space of d-regular graphs on n vertices. Notation $G \in \mathcal{G}(n,d\text{-reg})$ will signify that the graph G is selected according to such model.

A construction that gives the elements of $\mathcal{G}(n,d\text{-reg})$ is the *configuration model* (see, for example, [17, Chapter 9]). Let n urns be given, each containing d balls. A set F of $dn/2$ unordered pairs of balls is chosen uniformly at random (or *u.a.r.*). Let Ω be the set of all such pairings. Each $F \in \Omega$ corresponds to an d-regular (multi)graph with vertex set $V = \{1, \ldots, n\}$ and edge set E formed by those sets $\{i, j\}$ for which there is at least one pair with one ball belonging to urn i and the other ball belonging to urn j. Let Ω^* be the set of all pairings not containing an edge joining balls from the same urn or two edges joining the same two urns. Since each simple graph corresponds to exactly $(d!)^n$ such pairings, a random pairing $F \in \Omega^*$ corresponds to an d-regular graph G without loops or multiple edges chosen u.a.r.

Notice that a random pairing can be picked by choosing pairs one after the other. Moreover, the first point in a pair may be selected using any rule whatsoever, as long as the second point is chosen u.a.r. from all the remaining unpaired points. This property implies the existence of two equivalent ways of describing the algorithm presented in this paper. The description in Section 3 works on a previously generated (random) graph. However it would be equally easy to define a process that, by working on configurations, at the same time generates the graph and simulates our algorithm (this approach might help understanding our analysis).

In this paper we prove the following result:

Theorem 1. *For every integer n and $d \geq 3$, if $G \in \mathcal{G}(n,d\text{-reg})$ then $\alpha(G) \geq \alpha_d$ a.a.s. where the values of α_d are obtained through the method described in Section 4 and are reported, for $d \leq 7$ in the fourth column of Table 1.*

The proof of Theorem 1 is based on the definition of an algorithm and on the fact that, for each constant value of d, the algorithm dynamics can be described with sufficient precision by a random process that, for large n, behaves in a predictable way.

3 Greedy Algorithms for Independent Sets

If $G = (V, E)$ is a graph then $\Gamma(v) = \{u \in G : \{u, v\} \in E(G)\}$, for each $v \in V(G)$. The degree of a vertex v of G is the size of $\Gamma(v)$. Let G be a graph whose maximum degree is bounded by some fixed constant $d > 0$. We call the *isolation of* u the process of deleting all edges at distance at most one from vertex u (where the distance of an edge $\{u, v\}$ from a vertex w is the minimal length of a shortest path connecting w to u or v). For each $q \in \{1, \dots, d\}$ define Op_q to be the process of picking a vertex u of degree q in G u.a.r., and isolating it unless the minimal degree x in $\Gamma(u)$ is at most q, there is a vertex v of degree x in $\Gamma(u)$ and one of the following conditions hold (in such case v gets isolated):

1. there are at least two vertices of degree x in $\Gamma(u)$, OR
2. there is a single vertex v of degree x in $\Gamma(u)$ AND
   (a) $q = 2$ and the minimum degree in $\Gamma(v) \setminus u$ is larger than that in $\Gamma(u) \setminus v$ OR
   (b) $2 < q < d - 1$, the minimum degree in $\Gamma(v) \setminus u$ is larger than q, and the sum of all degrees in $\Gamma(v) \setminus u$ is smaller than that in $\Gamma(u) \setminus v$.

For each $q \in \{1, \dots, d\}$ Op_q obeys the sparsification (cases 1. and 2.(a)) and minimal mess principles (case 2.(b)) described in Section 1. We may then consider the following process:

> **Careful minimum degree process.** While there are still edges in G, define q as the minimum positive degree in G and perform Op_q on G, adding to $\mathcal{I}$ the vertex that has been (deliberately) isolated in the process.
>
> If there is no edge left then return $\mathcal{I}$ and stop.

Note that other vertices (apart from u or v) may get isolated while performing Op_q. They are not taken into account by our analysis. Denote by G_t, for each integer t, the graph obtained from G by removing all edges at distance at most one from any of the first t vertices added to $\mathcal{I}$ (of course $G_0 \equiv G$).

The differential equation method [20] allows us to estimate the size of $\mathcal{I}$ at the end of the process from (approximate) knowledge of the dynamics of the vector $(|V_1|, \dots, |V_d|)$ (where $V_i = V_i(t) = \{v \in G_t : |\Gamma(v)| = i\}$, for each $i \in \{1, \dots, d\}$). It turns out that the spawning of vertices of smaller and smaller degree can be described by well-known probabilistic tools (e.g. [14]). This implies that a.a.s. the process proceeds by picking vertices of degree $d-1$ and $d-2$ during an initial phase, then $d-2$ and $d-3$ during a second phase, and so on eventually running out of vertices of positive degree during phase $d-2$ while picking mainly vertices of degree two or one, and by that time all graph has been explored. The main technical drawback of this approach is the fact that the minimum degree in G_t is a random variable and a number of conditions must be checked to ensure that the proposed analysis method works smoothly. However, there is a different, but equivalent (see [20]), process that avoids such problems. In the forthcoming description d is a fixed integer greater than two, ϵ a small positive real number, and function $p(q, x, \mathbf{y})$ a discrete probability measure defined precisely in Section 4.

> **Algorithm** CarefulGreedy$_{d,\epsilon}(G)$
> **Input:** a graph $G = (V, E)$ on n vertices and maximum degree d.
> $\mathcal{I} \leftarrow \emptyset$;
> **for** $t \leftarrow 0$ **to** $\lceil \epsilon n \rceil$ perform Op$_d$;
> **while** $E \neq \emptyset$
> compute a probability distribution $p(q, \frac{t}{n}, \frac{|V_1|}{n}, \ldots, \frac{|V_d|}{n})$,
> for $q \in \{1, \ldots, d\}$;
> choose $q \in \{1, \ldots, d\}$ with probability $p(q, \frac{t}{n}, \frac{|V_1|}{n}, \ldots, \frac{|V_d|}{n})$;
> perform Op$_q$ on G_t;
> $t \leftarrow t + 1$;
> **return** $\mathcal{I}$.

A *step* of this algorithm is a complete iteration of the algorithm main while loop. Assuming that each vertex adjacencies are retrievable in time $O(d)$ and that all vertex degrees are computed before the main loop is executed and then updated as edges get removed, it is easy to believe that the algorithm time complexity is linear in the size of the input graph.

Initially $V_d = V$ and $V_i = \emptyset$ for $i \leq d - 1$. The main difference between CarefulGreedy$_{d,\epsilon}(G)$ and the minimum degree process described before is that, here, for $t > \lceil \epsilon n \rceil$, the choice to perform Op$_q$, for $q \in \{1, \ldots, d - 1\}$ is based on a probability distribution $p(q, x, \mathbf{y})$, rather than being dictated by the value of the minimum degree in G_t. The general definition of $p(q, x, \mathbf{y})$, valid when G is a random d-regular graph, will be given in Section 4. Depending on the particular probability distribution $p(q, x, \mathbf{y})$ that is used at a given step, the algorithm will be in one of a number of different *phases*, defined formally in the next section. The outcome of our analysis implies that the algorithm processing goes through successive phases. In phase $j \in \{1, 2, \ldots\}$ the process performs only Op$_{d-j}$ or Op$_{d-j-1}$. In this sense algorithm CarefulGreedy$_{d,\epsilon}$ simulates the careful minimum degree process described above.

4 Analysis Method

In order to obtain estimates on the size of the independent set returned by algorithm CarefulGreedy$_{d,\epsilon}(G)$ we use the differential equation method proposed by Wormald [20]. Given the input graph, our algorithm peels off a number of edges (upper bounded by an expression depending only on d) from the graph G_t and updates the structure $\mathcal{I}$ ($\mathcal{I}_t$ will denote the content of $\mathcal{I}$ before G_t is further processed) that is being built. Let $Y_i(t) = |V_i(t)|$ for $i \in \{1, \ldots, d\}$ and $Y_{d+1}(t) = |\mathcal{I}_t|$. In what follows, for $i \in \{1, \ldots, d + 1\}$ and $q \in \{1, \ldots, d - 1\}$, functions $f_{i,q}$ in $\mathbb{R}^{d+2}$ will be such that the expected change to $Y_i(t)$, conditioned on the history of the process up to step t and following one occurrence of Op$_q$ during step $t + 1$ is asymptotically $f_{i,q}(\frac{t}{n}, \frac{Y_1(t)}{n}, \ldots, \frac{Y_{d+1}(t)}{n}) + o(1)$, whenever $Y_q(t) > 0$. Let $\nu_{d-s}(x, \mathbf{y}) = \frac{f_{d-s-1,d-s}(x,\mathbf{y})}{f_{d-s-1,d-s}(x,\mathbf{y}) - f_{d-s-1,d-s-1}(x,\mathbf{y})}$. Assuming that these functions are continuous and bounded in

$$\mathcal{D}_\epsilon = \{(x, y_1, \ldots, y_{d+1}) : 0 \leq x \leq d,\ 0 \leq y_i \leq d \text{ for } 1 \leq i \leq d + 1,\ y_d \geq \epsilon\}$$

we may consider the following $d-1$ distinct systems of differential equations

$$\begin{aligned} \tfrac{\mathrm{d}y_i}{\mathrm{d}x} &= \nu_{d-s}(x,\mathbf{y}) f_{i,d-s-1}(x,\mathbf{y}) + (1-\nu_{d-s}(x,\mathbf{y})) f_{i,d-s}(x,\mathbf{y}) \quad && s \in \{1,\ldots,d-2\} \\ \tfrac{\mathrm{d}y_i}{\mathrm{d}x} &= f_{i,1}(x,\mathbf{y}) && s = d-1 \end{aligned} \tag{1}$$

Note that, if we run the careful minimum degree process, during phase $d-s$, the expression $f_{d-s-1,d-s}(x,\mathbf{y})$ (resp. $-f_{d-s-1,d-s-1}(x,\mathbf{y})$) would be approximately the expected number of vertices of degree $d-s-1$ created by one Op_{d-s} (resp. removed by an instance of Op_{d-s-1}). Thus, in a sense, $\nu_{d-s}(x,\mathbf{y})$ (resp. $1-\nu_{d-s}(x,\mathbf{y})$) represent, on average, the proportion of times the minimum degree process performs Op_{d-s} (resp. Op_{d-s-1}) during phase $d-s$. Provided

(F1) the functions $f_{i,q}$ are rational with no pole in $\mathcal{D}_\epsilon$ and
(F2) there exist positive constants C_1, C_2, and C_3 such that for each $i \in \{1,\ldots,d\}$, everywhere in $\mathcal{D}_\epsilon$, $f_{i,q} \geq C_1 y_{i+1} - C_2 y_i$ (for $q \neq i$), and $f_{i,q} \leq C_3 y_{i+1}$ for all q (see [20])

each system in (1), coupled with a suitably defined initial condition, admits a unique solution over $[x_{s-1}, x_s]$ (for $s \in \{1,\ldots,d-1\}$), where

> $x_0 = 0$ and x_s is defined as the infimum of those $x > x_{s-1}$ for which at least one of the following holds:
> (C1) $f_{d-s-1,d-s-1}(x,\mathbf{y}) \geq 0$ or
> $f_{d-s-1,d-s}(x,\mathbf{y}) - f_{d-s-1,d-s-1}(x,\mathbf{y}) \leq \epsilon$ and $s < d-1$;
> (C2) the component $d-s$ of the solution falls below zero or
> (C3) the solution is outside $\mathcal{D}_\epsilon$ or ceases to exist. (2)

Let $\tilde{\mathbf{y}} = \tilde{\mathbf{y}}(x) = (\tilde{y}_1(x),\ldots,\tilde{y}_{d+1}(x))$ be the function defined inductively as follows:

> For each $i \in \{1,\ldots,d+1\}$, $\tilde{y}_i(0) = \frac{Y_i(0)}{n}$. For $s \geq 1$, $\tilde{\mathbf{y}}$ is the solution to (1) over $[x_{s-1}, x_s]$, with initial condition $\mathbf{y}(x_{s-1}) = \tilde{\mathbf{y}}(x_{s-1})$. (3)

We may now state the result which bounds from below $\alpha(G)$.

Theorem 2. *Let d be a positive integer with $d \geq 3$, and ϵ an arbitrarily small positive real number. For $q \in \{1,\ldots,d-1\}$, let $f_{i,q}$, for each $i \in \{1,\ldots,d+1\}$ be the functions referred to in the description above and defined in Sections 4.1 and 4.2. Then there exists a positive integer m such that the algorithm CarefulGreedy$_{d,\epsilon}(G)$ a.a.s. returns a structure of size $n\tilde{y}_{d+1}(x_m) + o(n)$ where functions $\tilde{y}_1,\ldots,\tilde{y}_{d+1}$ are defined in (3) and $x_0,\ldots,x_m$ in (2) when $G \in \mathcal{G}(n,d\text{-reg})$.*

The values of m and $\tilde{y}_{d+1}(x_m)(=\alpha_d)$ referred to in Theorem 2 were found solving the various systems numerically using `Maple`'s Runge-Kutta Fehlberg method (a very primitive solver written in C for the case $d = 3$ is enclosed in the Appendix). The distributions used in CarefulGreedy$_{d,\epsilon}(G)$ satisfy the following definition: $p(d-s-1, x, \mathbf{y}) = \nu_{d-s}(x, \mathbf{y})$, $p(d-s, x, \mathbf{y}) = 1 - p(d-s-1, x, \mathbf{y})$ and $p(q, x, \mathbf{y}) = 0$ otherwise, when $x \in [x_{s-1}, x_s]$, for each $s \in \{1, \ldots, m\}$. The intervals $[x_{s-1}, x_s]$ define the process *phases.* For $x \in [x_{s-1}, x_s]$ the definition of $p(q, x, \mathbf{y})$ implies that only Op$_{d-s}$ and Op$_{d-s-1}$ have a positive probability of being performed during one step of algorithm CarefulGreedy$_{d,\epsilon}(G)$.

The proof of Theorem 2 is carried out invoking Theorem 1 in [20]. The important point to stress is that the argument has two quite separate components. The definition of a number of functions and numerical quantities (satisfying certain conditions) related to the particular algorithm and the proof that everything works and the solutions of (1) actually give information on $|\mathcal{I}|$ after the execution of CarefulGreedy$_{d,\epsilon}(G)$. As long as we are able to define the various $f_{i,q}$ and verify conditions (F1) and (F2), we do not need to be concerned with the second part of the argument (which mirrors the proof of Wormald's general result).

Before digging into the details of the specific cases we introduce a few notations. In what follows for integers a and b, $\delta_{a,b}$ is equal to one (resp. zero) if $a = b$ (resp. otherwise). Given a vertex u, the probability of creating a vertex of degree $i-1$ in $\Gamma(u)$ when removing an edge incident to u is asymptotically $P_i = \frac{iY_i}{\sum_j jY_j}$. In what follows S_a^b will denote the sum of all P_i's for $a \leq i \leq b$ (with $S_a^b = 0$ if $a > b$). Furthermore let $\chi_c(a) = (S_a^b)^c - (S_{a+1}^b)^c$. For large n, $\chi_c(a)$ approximates the probability that the minimum degree in a given set of c vertices is a, given that all degrees are between a and b. The expected change in Y_i due to the degree changes in $\Gamma(u)$ following the removal of an edge incident to u can be approximated by $\rho_i = P_{i+1} - P_i$ with $P_{d+1} = 0$. Similarly, if $e = \{u, v\}$ the expected change in Y_i due to the removal of e and of any other edge incident to v is asymptotically $\mu_i = -P_i + \rho_i \sum_{z=2}^{d} P_z(z-1)$. Finally, if $\mathcal{P}$ is some boolean condition, define $\diamond_r^i(\mathcal{P}) = (r-1)\rho_i - \delta_{i,r}$ (resp. $\delta_{i,r-1} - \delta_{i,r}$) if $\mathcal{P}$ is true (resp. false).

4.1 The Simple Case $d = 3$

Before describing the general case, it may be useful to follow an informal description of the analysis on cubic graphs.

For $d = 3$, we may assume that, at the beginning of a step, vertex u has degree either one or two (the initial $\lceil \epsilon n \rceil$ steps will make sure that this assumption is valid). Algorithm CarefulGreedy$_{3,\epsilon}(G)$ behaves exactly like Wormald's algorithm except in the case when Op$_2$ is performed and the two neighbours of u both have degree two. In such case our algorithm chooses a random neighbour of u rather than u itself. Thus, if $f_{i,q}^W$ denotes the function $f_{i,q}$ associated with Wormald's algorithm (a precise definition is given in formula (2.12) of [20]) then $f_{i,q} = f_{i,q}^W + \delta_{q,2}(P_2)^2(\delta_{i,1} + \mu_i - 2\rho_i)$, for $i \in \{0, \ldots 3\}$ and $q \in \{1, 2\}$ (whereas $f_{4,q} = f_{4,q}^W$). Of course each $f_{i,q}$ satisfies conditions (F1) and (F2) that imply

Theorem 2 (this follows from the properties of $f^W_{i,q}$ in [20]). For $d = 3$ it turns out that $m = 1$. Condition (C2) eventually becomes true exactly when the vector (x, y_1, y_2, y_3, y_4) hits the boundary of $\mathcal{D}_\epsilon$. At that point the process stops and $\tilde{y}_4 \simeq 0.4347531298$, which agrees with the value for α_3 in Table 1.

4.2 Arbitrary $d \geq 4$

We next state the result characterising the dynamics of $(Y_1, \ldots, Y_{d+1})$ for arbitrary $d \geq 4$ following an instance of Op_q, for $q \in \{1, \ldots, d-1\}$.

Lemma 1. *Let $d \geq 4$ and $\epsilon > 0$. For each $q \in \{1, \ldots, d-1\}$, conditioned on the history of algorithm* $\mathrm{CarefulGreedy}_{d,\epsilon}(G)$ *up to step t, the expected change to $Y_i(t)$ following one occurrence of* Op_q *is asymptotically*

$$-\delta_{i,q} \ + (S^d_{q+1})^q \ \times \textstyle\sum_{k=q+1}^d q \frac{P_k}{S^d_{q+1}}((k-1)\rho_i - \delta_{i,k})$$

$$+ \textstyle\sum_{x=1}^q \Big[(\chi_q(x) - qP_x(S^d_{x+1})^{q-1}) \ ((x-1)\mu_i - \delta_{i,x}) \ - \ (qP_x(S^d_x)^{q-1} - \chi_q(x))$$

$$\times(\delta_{i,x} - \delta_{i,x-1}) - q(\chi_{q-1}(x) - (q-1)P_x(S^d_{x+1})^{q-2}) \textstyle\sum_{k=x+1}^d P_k(\delta_{i,k} - \delta_{i,k-1})\Big]$$

$$+ \ qP_1(S^d_2)^{q-1}(-\delta_{i,1} + (q-1)\textstyle\sum_{k=2}^d((k-1)\rho_i - \delta_{i,k})\frac{P_k}{S^d_2})$$

$$+ \ \textstyle\sum_{x=2}^q qP_x\Big\{\sum_{z=x+1}^{d-1}\Big[\ \sum_{m=1}^{q+(d-q)\delta_{q,2}}\Big[-\delta_{i,x}\chi_{q-1}(z)\chi_{x-1}(m)$$

$$+ \ (q-1)\chi_{x-1}(m) \times ((S^d_z)^{q-2} \diamond^i_z \ (m \leq z)P_z + \chi_{q-2}(z)\textstyle\sum_{r=z+1}^d \diamond^i_r(m \leq z)P_r)$$

$$+ \ (x-1)\chi_{q-1}(z) \cdot ((S^d_m)^{x-2} \diamond^i_m \ (m > z)P_m + \chi_{x-2}(m)\textstyle\sum_{s=m+1}^d \diamond^i_s(m > z)P_s)\Big]$$

$$+ \ \textstyle\sum_{m=q+(d-q)\delta_{q,2}+1}^d \sum_{\boldsymbol{j}:j_z>0}\binom{q-1}{j_z,\ldots,j_d}P_z^{j_z}\ldots P_d^{j_d}\sum_{\boldsymbol{k}:k_m>0}\binom{x-1}{k_m,\ldots,k_d}P_m^{k_m}\ldots P_d^{k_d}\gamma(\boldsymbol{j},\boldsymbol{k})\Big]$$

$$+ \ P_d^{q-1}((x-1)\rho_i - \delta_{i,x} + ((d-1)\rho_i - \delta_{i,d})(q-1))\Big\}$$

where $\gamma(\boldsymbol{j},\boldsymbol{k}) = \sum_{r=z}^d \diamond^i_r(\mathcal{P}) \cdot j_r - \delta_{i,x} - \sum_{r=m}^d \diamond^i_r(\neg\mathcal{P}) \cdot k_r$ and $\mathcal{P} \equiv \sum_{r=z}^d rj_r < \sum_{r=m}^d rk_r$, if $i \leq d$. Finally $f_{d+1,q} = 1$ for all q.

Remark. The first line in the asymptotic expression for $f_{i,q}$ $(i \leq d)$ refers to the case when the minimum degree around u is at least $q+1$. The subsequent sum deals with the case when there are at least two vertices of minimum degree $x \leq q$ in $\Gamma(u)$. The remainder of the expression covers the case when there is a single vertex of minimum degree $x \leq q$ in $\Gamma(u)$.

Proof. We sketch the definition of $f_{i,q}$ for each q in the given range and $i \leq d$. The stated expression (more convenient from the numerical point of view) can then be obtained through simple algebraic manipulations.

For arbitrary, fixed $d \geq 4$, each step of algorithm $\mathrm{CarefulGreedy}_{d,\epsilon}(G)$ may perform Op_q for $q \in \{1, \ldots, d-1\}$. More importantly the execution of such an operation generates a number of alternative events whose probabilities are approximately multinomial, under the assumption that $G \in \mathcal{G}(n, d\text{-reg})$. Hence, for each $i \in \{1, \ldots, d\}$ and $q \in \{1, \ldots, d-1\}$ function $f_{i,q}$ satisfies (a.a.s.) the following definition:

$$f_{i,q} = -\delta_{i,q} + \sum_j \binom{q}{j_{q+1}, \ldots, j_d} P_{q+1}^{j_{q+1}} \ldots P_d^{j_d} \left(\sum_{k=q+1}^d ((k-1)\rho_i - \delta_{i,k}) j_k\right) +$$

$$+ \sum_{x=1}^q \Big\{ \sum_{j: j_x > 1} \binom{q}{j_x, \ldots, j_d} P_x^{j_x} \ldots P_d^{j_d} \Big((x-1)\mu_i - \delta_{i,x} +$$

$$- \sum_{k=x}^d (\delta_{i,k} - \delta_{i,k-1})(j_k - \delta_{k,x}) \Big) + g_{i,q,x} \Big\}$$

where the first sum is over all sequences $(j_{q+1}, \ldots, j_d)$ of non-negative integers adding up to q, the second sum on the second line is over all $(j_x, \ldots, j_d)$ with the further restriction that j_x must be positive (x represents the minimum degree in $\Gamma(u)$), and $g_{i,q,x}$ is the expected change to Y_i conditioned on performing Op_q and on the existence of a single vertex of minimum degree x around u (this is case 2. in the description of the equivalent minimum degree process). Function $g_{i,d-1,x}$ has a very simple definition, since if we perform Op_{d-1} we are essentially just replicating Wormald's algorithm. To define $g_{i,q,x}$ for $q < d-1$ we need to condition on the degree structure in $\Gamma(v) \setminus u$ and $\Gamma(u) \setminus v$. This enables us to analyse algorithms based on the two proposed optimization principles. In particular if $x = 1$ then $\Gamma(v) \setminus u$ is empty and therefore (just following Wormald's algorithm) $g_{i,q,1} = \sum \binom{q}{1, j_s, \ldots, j_d} P_1 \ldots P_d^{j_d} (-\delta_{i,1} + \sum_{k=2}^d ((k-1)\rho_i - \delta_{i,k}) j_k)$. For $x \geq 2$,

$$g_{i,q,x} = qP_x \Big\{ \sum_{z=x+1}^{d-1} \Big[h_{i,q,x,z} + \sum_{m \neq z} \sum_{j: j_z > 0} \binom{q-1}{j_z, \ldots, j_d} P_z^{j_z} \ldots P_d^{j_d} \times$$

$$\times \sum_{k: k_m > 0} \binom{x-1}{k_m, \ldots, k_d} P_m^{k_m} \ldots P_d^{k_d} \gamma_{i,q,x,z,m} \Big] +$$

$$+ \quad P_d^{q-1} ((x-1)\rho_i - \delta_{i,x} + ((d-1)\rho_i - \delta_{i,d})(q-1)) \Big\}$$

where $h_{i,q,x,z}$ describes the case when the minimum degree in $\Gamma(u) \setminus v$ and $\Gamma(v) \setminus u$ are the same and $\gamma_{i,q,x,z,m}$ the expected updates necessary in any other case. If $z \neq m$ the algorithm's rule is quite simple: v is added to $\mathcal{I}$ if the minimum degree in $\Gamma(v) \setminus u$ is larger than that in $\Gamma(u) \setminus v$. Then $\gamma_{i,q,x,z,m} = \sum_{r=z}^d \diamond_r^i (m \leq z) j_r + \sum_{s=m}^d \diamond_s^i (m > z) k_s$. Finally, for $q > 2$, function $h_{i,q,x,z}$ is computed conditioning on each possible pair of sequences $(j_z, \ldots, j_d)$ and $(k_z, \ldots, k_d)$ adding up to $q-1$ and $x-1$ respectively, and having $j_z > 0$ and $k_z > 0$. Following the minimum mess principle, vertex v is added to $\mathcal{I}$ if $\sum_{r=z}^d r j_r > \sum_{r=m}^d r k_r$.

To get to the expression in the Lemma statement we use repeatedly well-known multinomial identities like

$$\sum \binom{q}{j_x, \ldots, j_d} P_x^{j_x} \ldots P_d^{j_d} = (S_x^d)^q \quad \text{or} \quad \sum \binom{q}{j_x, \ldots, j_d} j_k P_x^{j_x} \ldots P_d^{j_d} = qP_k (S_x^d)^{q-1},$$

remembering that the case $q = 2$ needs slightly different computation. □

References

1. Alekseev, V.E.: Polynomial algorithm for finding the largest independent sets in graphs without forks. Discrete Applied Mathematics 135(1–3), 3–16 (2004)
2. Assiyatun, H.: Large Subgraphs of Regular Graphs. PhD thesis, Department of Mathematics and Statistics - The University of Melbourne (2002)
3. Assiyatun, H., Wormald, N.: 3-star factors in random d-regular graphs. European Journal of Combinatorics 27(8), 1249–1262 (2006)
4. Assiyatun, H.: Maximum induced matchings of random regular graphs. In: Akiyama, J., Baskoro, E.T., Kano, M. (eds.) IJCCGGT 2003. LNCS, vol. 3330, pp. 44–57. Springer, Heidelberg (2005)
5. Baker, B.S.: Approximation algorithms for NP-complete problems on planar graphs. Journal of the Association for Computing Machinery 41(1), 153–180 (1994)
6. Berman, P., Fujito, T.: On approximation properties of the independent set problem for low degree graphs. Theory of Computing Systems 32(2), 115–132 (1999)
7. Bollobás, B., Erdős, P.: Cliques in random graphs. Mathematical Proceedings of the Cambridge Philosophical Society 80, 419–427 (1976)
8. Chen, Z.-Z.: Approximation algorithms for independent sets in map graphs. Journal of Algorithms 41(1), 20–40 (2001)
9. Frieze, A., McDiarmid, C.: Algorithmic theory of random graphs. Random Structures and Algorithms 10, 5–42 (1997)
10. Frieze, A.M.: On the independence number of random graphs. Discrete Mathematics 81(2), 171–175 (1990)
11. Frieze, A.M., Łuczak, T.: On the independence and chromatic number of random regular graphs. Journal of Combinatorial Theory B 54, 123–132 (1992)
12. Garey, M.R., Johnson, D.S.: Computer and Intractability, a Guide to the Theory of NP-Completeness. Freeman and Company (1979)
13. Gavril, F.: Algorithms for minimum coloring, maximum clique, minimum covering by cliques and maximum independent set of a chordal graph. SIAM Journal on Computing 1(2), 180–187 (1972)
14. Harris, T.E.: The Theory of Branching Processes. Springer, Heidelberg (1963)
15. Håstad, J.: Clique is hard to approximate within $n^{1-\varepsilon}$. Acta Mathematica 182, 105–142 (1999)
16. Hunt, H.B., Marathe, M.V., Radhakrishnan, V., Ravi, S.S., Rosenkrantz, D.J., Stearns, R.E.: NC-approximation scheme for NP- and PSPACE-hard problems for geometric graphs. Journal of Algorithms 26, 238–274 (1998)
17. Janson, S., Łuczak, T., Ruciński, A.: Random Graphs. John Wiley and Sons, Chichester (2000)
18. McKay, B.D.: Independent sets in regular graphs of high girth. Ars Combinatoria 23A, 179–185 (1987)
19. Wormald, N.C.: Differential equations for random processes and random graphs. Annals of Applied Probability 5, 1217–1235 (1995)
20. Wormald, N.C.: Analysis of greedy algorithms on graphs with bounded degrees. Discrete Mathematics 273, 235–260 (2003)

Appendix

We enclose a short C program that solves the single system relevant for $d = 3$. The final print statement outputs α_3. While the code is by no mean perfect (in particular it contains no checking on $f_{i,q}$ or $\tilde{y}$) the fact that it gives an answer very close to the one returned by Maple's solver may be taken as further evidence of the robustness of our numerical results.

```
#include <stdio.h>

double p (int i, double y[]) {return i*y[i]/(y[1]+2*y[2]+3*y[3]);}
double R (int i, double y[]) {return (i<3?p(i+1,y):0.0) - p(i,y);}
double M (int i, double y[]) {return -p(i,y)+R(i,y)*(p(2,y)+2*p(3,y));}

double f (int i, int q, double y[]) {
  if (i==4) return 1.0;
  else return -(i==q)+q*M(i,y)-(q==2)*p(2,y)*p(2,y)*(2*R(i,y)-M(i,y)-(i==1));
}

double F(int i,double y[]) {
  double p2=-f(1,1,y)/(f(1,2,y)-f(1,1,y));
  return p2*f(i,2,y)+(1-p2)*f(i,1,y);
}

main(int argc, char *argv[]) {
  int i,l;
  double h = 0.00000001,w[5],mid[5];

  for(i=0;i<5;i++) w[i]=0.0+(!(3-i));
  for (l=0;;l++) {
    for (i=0;i<5;i++) mid[i]=w[i]+(h*F(i,w)/2.0);
    for (i=0;i<5;i++) w[i]=w[i]+h*F(i,mid);

    if ( (f(1,1,w)>0||f(1,2,w)-f(1,1,w)<=h) || (w[2]<=0.0) ) {
      printf("|I| = %11.10f\n",w[4]); break;
    }
  }
}
```

Transition Graphs of Rewriting Systems over Unranked Trees

Christof Löding and Alex Spelten

RWTH Aachen, Germany
{loeding,spelten}@informatik.rwth-aachen.de

Abstract. We investigate algorithmic properties of infinite transition graphs that are generated by rewriting systems over unranked trees. Two kinds of such rewriting systems are studied. For the first, we construct a reduction to ranked trees via an encoding and to standard ground tree rewriting, thus showing that the generated classes of transition graphs coincide. In the second rewriting formalism, we use subtree rewriting combined with a new operation called flat prefix rewriting and show that strictly more transition graphs are obtained while the first-order theory with reachability relation remains decidable.

Keywords: Infinite graphs, reachability, rewriting, unranked trees.

1 Introduction

One of the main trends in verification is the field of infinite state model checking, in which procedures (and limits to their applicability) are developed to check systems with infinite state spaces against formal specifications (for a survey on infinite graphs cf. [19]).

In automatic verification, checking whether a system can reach an undesirable state or configuration translates to the reachability problem "Given a finite representation of an infinite graph G and two vertices u, u' of G, is there a path from u to u'?". From this point of view, an important task in the development of a theory of infinite graphs is to identify classes of infinite graphs where such elementary problems like reachability are decidable.

The strong formalism of monadic second-order logic (MSO) subsumes temporal logic (cf. [11]) and thus allows to express reachability properties. A well-known representative of graph classes with decidable MSO theory is the "pushdown hierarchy" introduced by Caucal [6]. Although this hierarchy is very rich and contains a lot of graphs, grid-like structures are not captured.

In order to compensate this weakness, a different approach of generating transition systems is to employ ranked trees (or terms) as the basic objects of the rewriting formalism, as already considered in [2]. Thereby, the internal structure of the trees is not of primary interest, but the different rewriting operations that can be applied on trees. Consequently, the vertices of the generated infinite graphs are represented by ranked trees, while the edge relation is induced by (simple) tree operations.

L. Kučera and A. Kučera (Eds.): MFCS 2007, LNCS 4708, pp. 67–77, 2007.

Among attractive subclasses of rewriting systems, an interesting and practical subclass is made up by the ground tree rewriting systems (which contain the infinite grid as transition graph; for an extensive analysis cf. [15]). "Ground rewriting" means that no variables occur in the rules, thus in ground tree (or term) rewriting systems (GTRSs), only explicitly specified subtrees can be replaced by other explicitly specified subtrees. Though in general MSO is undecidable for transition graphs of GTRSs, there is a decidable logic that allows to express reachability problems: first-order logic with the reachability relation [10]. In [13, 15] the structure of the transition graphs of GTRSs and their relation to other classes of infinite graphs, in particular to pushdown graphs, was studied. Furthermore, in [14, 15] several variants of the reachability problem for this class of graphs were investigated and a decidable logic was defined for the class of (regular) ground tree rewriting graphs.

For many applications however, the modelling of system states, messages, or data by ranked trees is not the most intuitive approach (if not impossible as e.g. the modelling of associative operations), since every symbol is of a fixed arity. Thus, our aim is to investigate a possible generalization of the idea of ground tree rewriting systems to the case of unranked trees. Briefly, unranked trees are finite labeled trees where nodes can have an arbitrary but finite number of children, and no fixed rank is associated to any label.

In this paper, we investigate to which extent results for ground tree rewriting systems are transferable to the unranked case. Note however, that the direct adaptation of this rewriting principle is not of interest: When starting from a fixed initial tree and applying a finite set of rewrite rules with constant trees, the resulting trees are of bounded branching and hence can be traced to the case of ground tree rewriting over ranked trees. Another natural approach to handle unbounded branching of unranked trees is to encode unranked trees as binary trees. Using this formalism, we show that there is a class of rewriting systems over unranked trees, which will be called *partial subtree rewriting systems*, that generates the same class of infinite graphs as ground tree rewriting systems over ranked trees.

However, encodings are problematic as they alter locality and path properties. This means that this approach of compensating unbounded branching via a dispersal into subtrees blurs a decisive point, namely the separation of two types of unboundedness: one is derived from the arbitrariness of "hierarchy levels" (represented by the height of the tree) while the other unboundedness refers to the number of data on the same hierarchy level. Pursuing the latter aspect, we define a new class of rewriting systems over unranked trees, the *subtree and flat prefix rewriting systems*, which combine ground tree rewriting with prefix word rewriting on the "flat front" of a tree. Here, a flat front indicates a successor sequence wherein all nodes are leaves. With this approach related to ground tree rewriting, we obtain a class of infinite graphs which has a decidable first-order theory with reachability predicate. Furthermore, analogous to regular ground tree rewriting systems over ranked trees, a regular variant of these rewriting systems is considered.

After introducing the basic terminology in Section 2, the class of transition graphs of partial subtree rewriting systems is treated in Section 3. We show that this class coincides with the class of transition graphs of ground tree rewriting systems over ranked trees. Section 4 introduces (regular) subtree and flat prefix rewriting systems, relates the classes of transition graphs to the previous ones, and investigates the decidability of the reachability problem over the transition graphs of (regular) subtree and flat prefix rewriting systems. Furthermore, it is shown that the structure consisting of the set of unranked trees and the relations reachability and one-step reachability is automatic for a suitable definition over unranked trees and thus has a decidable first-order theory (cf. [1]). Section 5 concludes with a short summary and points to further aspects of interest.

2 Preliminaries

It is assumed that the reader is familiar with the basic notions of automata theory and regular languages (for an introduction cf. [12], for automata on ranked trees cf. [7], and on unranked trees cf. [3]).

An *unranked tree* over an alphabet Σ is a mapping from a nonempty finite domain $dom_t \subseteq \mathbb{N}^*$ to Σ, where dom_t is prefix closed and it holds that if $xi \in dom_t$ then $xj \in dom_t$ for $x \in \mathbb{N}^*$, $i \in \mathbb{N}$, and $j \leq i$. In an unranked tree, each node may have an arbitrary but finite number of successors. If the root of a finite tree t is labeled by $a \in \Sigma$ and has k successors at which the subtrees $t_1, \ldots, t_k$ are rooted, then t can be written as the term $a(t_1, \ldots, t_k)$. The set of unranked trees over an alphabet Σ is denoted by T_Σ.

The *subtree* $t_{\downarrow x}$ of t is the tree rooted at node $x \in dom_t$ (i.e. $t_{\downarrow x}(u) = t(xu)$ for $xu \in dom_t$). The *height* of a tree is defined as $ht(t) := max\{|x| \mid x \in dom_t\}$; if a tree t is of height 1, the word derived from the front (i.e. the sequence of leaves read from left to right) of t is called the *flat front.*

A *hedge* as introduced by Courcelle [9] is a (possibly empty) finite ordered sequence of trees. The width of a hedge is defined as the number of trees that are contained in the sequence; consequently, a tree is a hedge of width 1.

A *nondeterministic bottom up tree automaton (N↑TA)* on unranked trees is of the form $\mathcal{A} = (Q, \Sigma, \Delta, F)$ over an unranked alphabet Σ, with a finite set Q of states, a set $F \subseteq Q$ of final states, and a finite set of transitions $\Delta \subseteq REG(Q) \times \Sigma \times Q$, where $REG(Q)$ denotes the class of regular word languages over Q, which are given for single transitions e.g. by a nondeterministic finite (word) automaton (NFA). A *run* of $\mathcal{A}$ on t is a mapping $\rho : dom_t \rightarrow Q$ such that for each node $x \in dom_t$ there is a transition $(L, t(x), \rho(x)) \in \Delta$ such that the sequence $q_1 \cdots q_n$ of states formed by the run at the successors of x is a word in L. Thus, an N↑TA employs NFAs that read the successor sequence of a node, and decide with this word and the label of the current node which state to assign to the current node.

As usual, a run is *accepting* if the root is labeled with a final state, and the accepted language $T(\mathcal{A})$ contains all trees for which there is an accepting run. If there is a run labeling the root with state q then we write $\mathcal{A} : t \rightarrow^* q$.

We also use the equivalent model N↓TA as well as an extended model (denoted by ε-N↑TA) with ε-transitions from the set $Q \times Q$, each with the standard semantics.

A tree is called *ranked* if every symbol $a \in \Sigma$ is assigned a unique arity $rk(a) \in \mathbb{N}$, and each node labeled with a has exactly $rk(a)$ successors.

A *ground tree rewriting system (GTRS)* over ranked trees is defined as a tuple $\mathcal{R} = (\Sigma, \Gamma, R, t_{in})$, with ranked alphabet Σ, transition alphabet Γ, finite set R of rules of the form $s \overset{\sigma}{\hookrightarrow} s'$ with $s, s' \in T_\Sigma$, $\sigma \in \Gamma$, and initial tree $t_{in} \in T_\Sigma$. A rule $s \overset{\sigma}{\hookrightarrow} s' \in R$ is applicable to a tree t if there is a node $x \in dom_t$ with $s = t_{\downarrow x}$, and the resulting tree is $t' = t[x|s']$, where the subtree rooted at node x is replaced by the tree s'. In this case, t' is *derived* from t by the rule $s \overset{\sigma}{\hookrightarrow} s'$ and we write $t \to_{\mathcal{R}}^{\sigma} t'$. The tree language that is generated by $\mathcal{R}$ is denoted $T(\mathcal{R}) = \{t \in T_\Sigma \mid t_{in} \to_{\mathcal{R}}^{*} t\}$; the focus of this paper will be the structure induced by the rewriting system with respect to the tree language. This is a directed edge labeled *transition graph* $G_{\mathcal{R}} = (V_{\mathcal{R}}, E_{\mathcal{R}}, \Gamma)$, of $\mathcal{R}$ with $V_{\mathcal{R}} = T(\mathcal{R})$, and $(t, \sigma, t') \in E_{\mathcal{R}}$ iff $t \to_{\mathcal{R}}^{\sigma} t'$. Note that the vertex set $V_{\mathcal{R}}$ is defined as the set of trees that are reachable from t_{in} by repeated application of the rewrite rules. The class of transition graphs of GTRSs is denoted by GTRG. For an extensive survey on GTRG cf. [15].

One way of dealing with unranked trees is to encode them by ranked trees. We use here a formalism proposed in [18], and employed as an encoding in [4], that uses only one binary symbol corresponding to an operation for constructing unranked trees. The *extension operator* $@ : T_\Sigma \times T_\Sigma \to T_\Sigma$ extends a given tree t by t' by adjoining t' as the next sibling of the last child of t: $a(t_1, \ldots, t_n) @ t' = a(t_1, \ldots, t_n, t')$, respectively for case $n = 0$: $a @ t' = a(t')$. Furthermore, we can also adjoin a hedge instead of a single tree t' in the intuitive way. Note that every unranked tree can be generated uniquely from trees of height 0 using the extension operator: $a(t_1, \ldots, t_n) = [(\cdots (a @ t_1) @ t_2) \cdots @ t_n]$, and thus, this formalism can be used as an encoding of unranked trees into binary ones (by assigning rank 0 to each symbol of the unranked alphabet and rank 2 to the extension operator @).

3 Partial Subtree Rewriting Systems

As mentioned in the Introduction, the direct transfer of the ground tree rewriting principle to unranked trees would result in bounded branching, therefore new rewriting principles have to be considered. The first rewriting principle considered aims at an easy transfer of nice properties of GTRSs. Therefore, unranked trees are encoded into ranked ones via the extension operator encoding as introduced in Section 2. Subtrees of the tree obtained after the encoding are hereby mapped to *partial subtrees* in the corresponding unranked tree; where if $a(t_1, \ldots, t_n)$ is a subtree, then $a(t_1, \ldots, t_i)$ is a partial subtree for each $0 \leq i \leq n$. The rewriting system therefore is defined such that exactly those partial subtrees are replaced.

The set $T_{\Sigma,\xi}$ is the set of all unranked trees over Σ with one occurrence of the variable ξ as leaf and rightmost child of the root, i.e. the set of trees of the form $\bar{t}\,@\,\xi$ with $\bar{t} \in T_\Sigma$.

A *partial subtree rewriting system (PSRS)* over unranked trees in T_Σ is of the form $\mathcal{R} = (\Sigma, \Gamma, R, t_{in})$, with an unranked alphabet Σ, a transition alphabet Γ, a finite set of rules R, and an initial tree t_{in}. The set R consists of subtree rewrite rules over trees of $T_{\Sigma,\xi}$ of the form: $r \overset{\sigma}{\hookrightarrow} r'$ with $r, r' \in T_{\Sigma,\xi}$ and $\sigma \in \Gamma$.

A tree t' is derived from t ($t \rightarrow^{\sigma}_{\mathcal{R}} t'$), if there is a node $x \in dom_t$, a hedge h over Σ, and a rule $r \overset{\sigma}{\hookrightarrow} r' \in R$, such that $r[\xi|h] = t_{\downarrow x}$, and $t[x|r'[\xi|h]] = t'$ (cf. Figure 1). The class of transition graphs of PSRSs is denoted by PSRG.

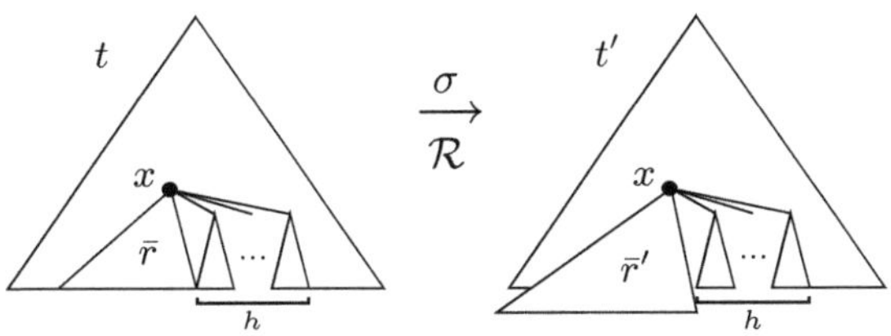

Fig. 1. Application of rewrite rule $r \overset{\sigma}{\hookrightarrow} r'$ according to the definition of PSRSs

With these definitions it can be shown that PSRSs over unranked trees and GTRSs over ranked trees generate the same transition graphs up to isomorphism.

Theorem 1. *Partial subtree rewriting systems generate the same class of transition graphs as ground tree rewriting systems (PSRG = GTRG).*

Proof (Sketch). When unranked trees are encoded into ranked ones by the extension operator encoding, subtrees of the ranked encoding correspond exactly to the partial trees of $T_{\Sigma,\xi}$ by construction. Thus applying a rule of a PSRS corresponds to rewriting an entire subtree in the ranked tree obtained after the encoding. The technical details of the construction of a GTRS for a given PSRS can be found in [17].

Since ranked trees can be viewed as unranked trees, and since each symbol has a unique rank, the construction of a PSRS $\mathcal{R}$ for a GTRS $\mathcal{S} = (\Sigma_r, \Gamma, S, t_{in})$ over ranked alphabet Σ_r is straightforward. The ranks of the symbols are simply omitted, the initial tree is kept, and the given rules of the GTRS are endorsed by extending the trees in the rules with the variable ξ to obtain trees in $T_{\Sigma,\xi}$. Consequently, with the same initial tree for both rewriting systems, the variable ξ can only be substituted by the empty hedge, thus resulting in isomorphic transition graphs. □

This class equivalence of GTRG and PSRG induces that by disregarding the inner structure of the vertices (unranked vs. ranked trees), the transition graphs are of identical structure.

Corollary 2. *The first-order theory with reachability is decidable for PSRG.*

Additionally, several other decidability and undecidability results for GTRG can be transferred to PSRG (cf. [2, 15]).

4 Subtree and Flat Prefix Rewriting Systems

Previously, unbounded branching was coped with via a dispersal into the unboundedness of depth of a tree. In order to respect the nature of these two different types of unboundedness, we now consider a new rewriting formalism. Towards a compromise between known principles and meeting this requirement, we combine standard subtree substitution with flat prefix substitution. That means, for subtrees of height 1, a prefix of the successor sequence of this subtree can be replaced by another sequence, enabling us to exploit properties of prefix rewriting over words.

Note that these prefix rewrite rules can be regarded as a kind of synchronization: they can only be applied at a node x if all subtrees rooted at its successors have a certain property, namely are of height 0. This kind of structural control is not available for the previously considered rewriting systems, and thus yields a new class of transition graphs.

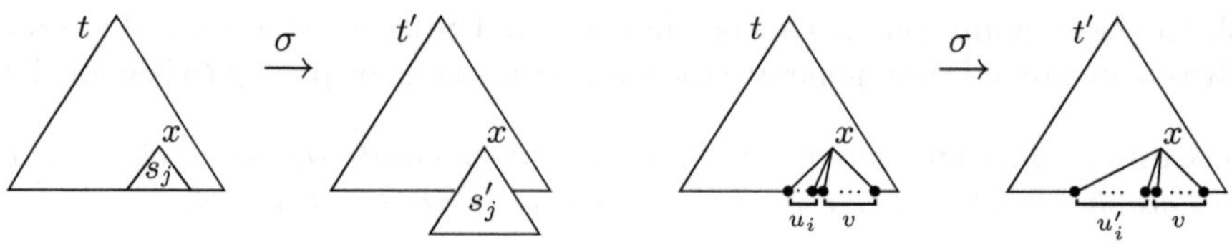

(a) Subtree substitution (b) Flat prefix substitution

Fig. 2. Application of rewrite rules of according to the definition of SFPRSs

A *subtree and flat prefix rewriting system (SFPRS)* over unranked trees in T_Σ is of the form $\mathcal{R} = (\Sigma, \Gamma, R, t_{in})$, with a finite unranked alphabet Σ, a finite transition alphabet Γ, an initial tree t_{in}, and a finite set R of rules of two types:

1. subtree substitution (cf. Figure 2(a))
   with rules of the form $r_j : s_j \overset{\sigma}{\hookrightarrow} s'_j$ for $j \in J$, $s_j, s'_j \in T_\Sigma$, $\sigma \in \Gamma$, and
2. flat prefix substitution at the flat front of the tree (cf. Figure 2(b))
   with rules of the form $r_i : u_i \overset{\sigma}{\hookrightarrow} u'_i$ for $i \in I$, $u_i, u'_i \in \Sigma^+$, $\sigma \in \Gamma$,

with $I \cup J = \{1, \ldots, |R|\}$ and $I \cap J = \emptyset$. The class of transition graphs of SFPRSs is denoted by SFPRG.

A tree t' is derived from t ($t \rightarrow^{\sigma}_{\mathcal{R}} t'$) by applying a subtree rewrite rule r_j, if there is a node $x \in dom_t$ with $t_{\downarrow x} = s_j$ such that $t[x|s'_j] = t'$ (cf. Figure 2(a)).

A tree t' is derived from t by applying a prefix rewrite rule r_i, if there is a node $x \in dom_t$ with $ht(t_{\downarrow x}) = 1$ and $\mathit{flatfront}(t_{\downarrow x}) = u_i v$, a tree $s \in T_\Sigma$ with $ht(s) = 1$ and $s(\varepsilon) = t(x)$, $\mathit{flatfront}(s) = u_i' v$, such that $t[x|s] = t'$ for some $v \in \Sigma^*$ (cf. Figure 2(b)).

Naturally, the definition of SFPRSs can be extended to *regular* SFPRSs by introducing regular sets of trees resp. words in the rules to obtain an even larger class of transition graphs. Conversely, SFPRSs can be regarded as the special case of singletons in the rules of regular SFPRSs. Note that all negative results in this paper are shown for SFPRSs while correspondingly, all positive results are shown for regular SFPRSs and thus hold for both classes of transition graphs.

4.1 Classification of Transition Graph Classes

Towards a classification of the transition graph classes PSRG and (regular) SFPRG, consider the SFPRS $\mathcal{R}_0 = (\Sigma, \Gamma, R, t_{in})$ with $\Sigma = \{a, c, e\}$, $\Gamma = \{0, 1\}$, $R = \{r_1 : c \overset{1}{\hookrightarrow} \begin{smallmatrix} e \\ | \\ e \end{smallmatrix}, r_2 : e \overset{1}{\hookrightarrow} \begin{smallmatrix} e \\ | \\ e \end{smallmatrix}, r_3 : c \overset{0}{\hookrightarrow} cc\}$ $(I = \{3\}, J = \{1, 2\})$, and $t_{in} = \begin{smallmatrix} & a & \\ / & & \backslash \\ c & & c \end{smallmatrix}$, whose transition graph is depicted in Figure 3.

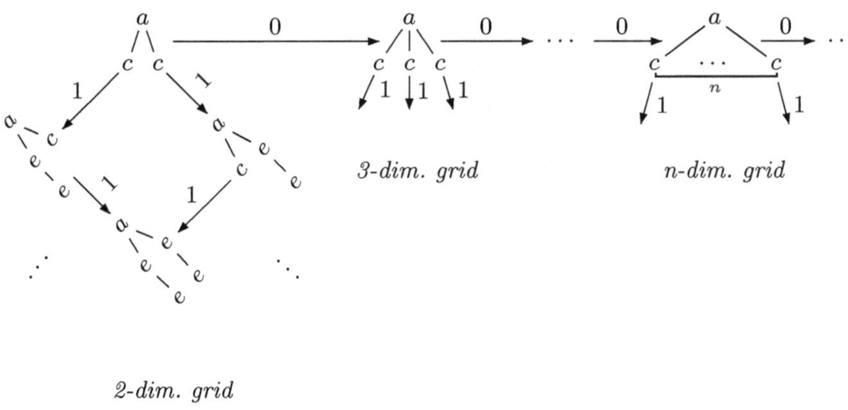

Fig. 3. Transition graph of SFPRS $\mathcal{R}_0$

Note that the 0-transition r_3 can only be applied at the trees of the vertices on the top line in Figure 3, since these are the only trees that have a subtree of height 1 with flat front cw for $w \in \Sigma^+$. For the transition graph this means that after traversing a 1-edge, no 0-edges are available any more.

Lemma 3. *The transition graph of SFPRS $\mathcal{R}_0$ cannot be generated by a GTRS.*

Proof (Sketch). It can be shown that using a GTRS, an enabled 0-transition cannot be disabled by an aritrary number of 1-transitions leading to different nodes.

Towards a contradiction: consider a vertex of the top row of Figure 3 with n out edges with label 1 and one out edge with label 0. For the tree at this vertex in a corresponding transition graph of a GTRS $\mathcal{S}$, there have to be n different 1-transitions which rewrite the subtree available for the applicable 0-transition in order to prevent a 0-transition afterwards. However, the number of nodes in the tree where these 1-transitions have to be applied in order to fulfill this requirement is bounded by the number of rewrite rules of $\mathcal{S}$ and the height of the trees of the left hand sides of the rewrite rules of $\mathcal{S}$. For n large enough this is a contradiction (for details, we refer the reader to [17]). □

Note that Lemma 3 is already true if we omit rule r_2 from SFPRS $\mathcal{R}_0$. Conversely, every ground tree rewriting system can always be conceived as a SFPRS with subtree rewrite rules only. With the same initial tree and the same subtree rewrite rules, omitting the ranks of the symbols does not provide more substitution possibilities. Since the classes of transition graphs GTRG and PSRG are equivalent, one obtains the following.

Proposition 4. *The class of transition graphs of PSRSs is strictly included in the class of transition graphs of SFPRSs: PSRG $\subsetneq$ SFPRG.*

Thus, undecidability results for PSRSs carry over to (regular) SFPRSs. These include the reachability problems: constrained reachability, universal reachability, and universal recurrence (cf. [15]).

Additionally, since this is the case for ground tree rewriting systems, the monadic second-order logic of SFPRSs is undecidable. This can also be observed directly from the transition graph of $\mathcal{R}_0$, since it includes the two-dimensional grid, whose monadic second-order logic is undecidable (as proven e.g. in [16]).

We would like to point out that the increase of expressive power of SFPRSs over PSRSs results from the fact that prefix rewrite rules can only be applied to flat fronts. Due to this restriction it is not possible to transfer these rules to standard rewriting rules over encodings.

4.2 Reachability Via Saturation

The main contribution of this paper is the decidability of the reachability problem for transition graphs of (regular) SFPRSs. This is done by an adaption of the well-known saturation algorithm which e.g. solves the reachability problem for semi-monadic linear rewriting systems over ranked trees (cf. [8]) by calculating the set $\mathrm{pre}^*_{\mathcal{R}}(T) = \{t \in T_\Sigma \mid \exists t' \in T : t \rightarrow^*_{\mathcal{R}} t'\}$ of trees from which the set T can be reached. Thereby, the rewrite rules of a (regular) SFPRS are simulated by adding transitions to an ε-N↑TA that recognizes the union of the target set T and all trees that correspond to a left hand side of the rewrite rules similar to the construction in [15]. In the very same manner, the set $\mathrm{post}^*_{\mathcal{R}}(T) = \{t \in T_\Sigma \mid \exists t' \in T : t' \rightarrow^*_{\mathcal{R}} t\}$ of trees which are reachable from the set T can be obtained by pursuing the same strategy for the reversed rewriting system, i.e. the left and right hand sides of the rules are simply swapped.

However, due to the different natures of the two types of rules of (regular) SFPRSs, and the employment of word automata in ε-N↑TAs over unranked trees, the saturation is based on an interleaving of two saturation algorithms on different levels of automata. In detail, for a prefix rewrite rule the saturation is basically realized by adding ε-transitions on the level of word automata that recognize the sequence of labels of the successors of a node, while for subtree rewrite rules, the saturation is realized by adding ε-transitions on the level of tree automata. The crucial interleaving aspects include that by the application of subtree rewrite rules new flat fronts may be introduced, which also need to be saturable.

For an elaborate example, the full construction, and the formal correctness proof, we refer the reader to [17]. The automaton resulting from the saturation accepts exactly those trees from which the target set is reachable and thus we obtain the following theorem.

Theorem 5. *Given a (regular) SFPRS $\mathcal{R}$, and a regular set T of unranked trees, the sets $pre^*_{\mathcal{R}}(T)$ and $post^*_{\mathcal{R}}(T)$ are again regular.*

As emptiness for unranked tree automata is decidable, we obtain the following corollary.

Corollary 6. *The reachability problem for (regular) SFPRSs: "Given a (regular) SFPRS $\mathcal{R}$, vertex t, and regular set T of vertices, is there a path from t to a vertex in T?" is decidable.*

4.3 First-Order Theory Via Automatic Structures

In addition to the decidability of the reachability problem for (regular) SFPRSs, we now address the first-order theory for these rewriting systems. We will show that the first-order theory enriched with the predicates reachability and one-step reachability remains decidable and thus obtain a proper superclass of (regular) GTRG with the same decidability properties.

We show that the structure consisting of the universe T_Σ, the one-step reachability relation, and the reachability relation is tree-automatic for a suitable definition over unranked trees, thus exploiting the feature that any (tree-) automatic structure has a decidable first-order theory (cf. [1]). Due to space restrictions, we stick to an informal description of the automaton that works on the convolution of two trees.

Briefly, the convolution $t = \langle t_1, t_2 \rangle$ encodes two trees $t_1, t_2 \in T_\Sigma$ such that the automaton reading the new tree t has access to both original ones. This is realized by labeling the node of t with pairs of symbols from t_1 and t_2 such that the successor sequences of the nodes line up on the right and are padded with a filling symbol $\Box$ on the left where necessary. For example, the trees $t_1 = a(bc)$ and $t_2 = d(efg)$ are convolved into $t = \langle t_1, t_2 \rangle = [a,d]([\Box, e], [b, f], [c, g])$.

In the construction of the automaton recognizing the reachability relation $\rightarrow^*$, we embark on a strategy similar to one for pushdown systems (cf. [5]): Given two trees $t_1, t_2 \in T_\Sigma$, we guess the set of "minimal" points of the rewriting steps in $t_1 \rightarrow^*_{\mathcal{R}} t_2$ and then check whether the first component of the convolution can be rewritten into the left side of the applied rule while the second component can be rewritten from the right side of the rule.

For (regular) SFPRSs this means that we start with an automaton $\mathcal{B}$ working on the convolution $t = \langle t_1, t_2 \rangle$ of two trees, which recognizes the identity function, i.e. the set $\{\langle t_1, t_2 \rangle \mid t_1 = t_2\}$. The automaton then nondeterministically guesses the set of minimal (w.r.t. the prefix ordering) nodes at which a rewrite rule was applied. Furthermore, $\mathcal{B}$ also guesses which rule was applied for each of these nodes.

For a subtree rewrite rule, $\mathcal{B}$ checks whether the projections of the subtree $t_{\downarrow x}$ belong to the regular sets *pre** resp. *post** of the applied rule. Theorem 5 yields automata $\mathcal{A}_{pre^*}$, $\mathcal{A}_{post^*}$ over Σ for these sets, and with a straightforward automaton construction, we can add transitions to $\mathcal{B}$ such that if the projections of $t_{\downarrow x}$ to the components belong to the corresponding sets, $\mathcal{B}$ accepts subtree $t_{\downarrow x}$ of the convolution. A similar but slightly more involved strategy works for flat prefix rewrite rules.

The construction of an automaton for the one-step reachability relation $\rightarrow$ is straightforward. The automaton works in a similar way but ensures that exactly one rule was applied.

Since both relations are automatic, we obtain the following theorem.

Theorem 7. *The first-order theory enriched by the relations reachability and one-step reachability is decidable for (regular) SFPRSs.*

5 Summary and Outlook

We showed that using rewriting systems over unranked trees one can generate a class of infinite graphs that coincides with the class of transition graphs of ground tree rewriting systems over ranked trees. The rewriting principle of these PSRSs consists of substituting unranked trees partially, which corresponds to ground tree rewriting over an encoding of unranked trees as ranked ones. Due to the class equivalence, several decidability results over the transition graphs of GTRSs over ranked trees can be transferred to those of PSRSs.

Furthermore, (regular) SFPRSs over unranked trees were introduced, which add flat prefix rewriting to the known paradigm of subtree substitution. The class of transition graphs of SFPRSs was shown to strictly include the class of transition graphs of PSRSs, which allows to transfer several undecidability results. Additionally, we described a saturation algorithm which yields the decidability of the reachability problem over (regular) SFPRG. We have also shown that the structure consisting of the set T_Σ of unranked trees and the relations reachability and one-step reachability is automatic for a suitable definition over unranked trees, and thus we can conclude that the first-order theory with these reachability relations is decidable for (regular) SFPRSs. Thus, the class of (regular) SFPRSs contains strictly more graphs than GTRG, but has the same decidable properties.

In general, other rewriting principles over unranked trees have yet to be investigated. One aspect could be to use other word rewriting techniques in combination with subtree substitution. Another interesting point of application is to define and investigate an adaption of (semi) monadic rewriting systems to unranked trees, which were introduced for ranked trees in [8].

Finally, we would like to thank Arnaud Carayol for his useful comments on the decidability proof for the first-order theory.

References

[1] Blumensath, A.: Automatic Structures. Diploma thesis, RWTH Aachen, Germany (1999), `http://www-mgi.informatik.rwth-aachen.de/Publications/pub/blume/AutStr.ps.gz`

[2] Brainerd, W.: Tree generating regular systems. Inf. and Contr. 14(2), 217–231 (1969)

[3] Brüggemann-Klein, A., Murata, M., Wood, D.: Regular tree and regular hedge languages over unranked alphabets. Unfinished technical report, Hongkong University (April 2001), `http://citeseer.ist.psu.edu/451005.html`

[4] Carme, J., Nieren, J., Tommasi, M.: Querying unranked trees with stepwise tree automata. In: van Oostrom, V. (ed.) RTA 2004. LNCS, vol. 3091, pp. 105–118. Springer, Heidelberg (2004)

[5] Caucal, D.: On the regular structure of prefix rewriting. TCS, 106(1), 61–86 (1992)

[6] Caucal, D.: On infinite terms having a decidable theory. In: Diks, K., Rytter, W. (eds.) MFCS 2002. LNCS, vol. 2420, pp. 165–176. Springer, Heidelberg (2002)

[7] Comon, H., Dauchet, M., Gilleron, R., Jacquemard, F., Lugiez, D., Tison, S., Tommasi, M.: Tree Automata Techniques and Applications (Unpublished electronic book, 1997), `http://www.grappa.univ-lille3.fr/tata`

[8] Coquidé, J.-L., Dauchet, M., Gilleron, R., Vágvölgyi, S.: Bottom-up tree pushdown automata: classification and connection with rewrite systems. TCS, 127(1), pp. 69–98 (1994)

[9] Courcelle, B.: A representation of trees by languages. TCS, 7, 25–55 (1978)

[10] Dauchet, M., Tison, S.: The theory of ground rewrite systems is decidable. In: Proc. LICS 1990, pp. 242–248. IEEE CSP, Los Alamitos (1990)

[11] Emerson, E.A.: Temporal and modal logic. In: van Leeuwen, J. (ed.) Handbook of TCS, pp. 995–1072. Elsevier, Amsterdam (1990)

[12] Hopcroft, J., Motwani, R., Ullman, J.: Introduction to Automata Theory, Languages, and Computation, 2nd edn. Addison-Wesley, Boston (2001)

[13] Löding, C.: Ground tree rewriting graphs of bounded tree width. In: Alt, H., Ferreira, A. (eds.) STACS 2002. LNCS, vol. 2285, pp. 559–570. Springer, Heidelberg (2002)

[14] Löding, C.: Model-checking infinite systems generated by ground tree rewriting. In: Nielsen, M., Engberg, U. (eds.) ETAPS 2002 and FOSSACS 2002. LNCS, vol. 2303, pp. 280–294. Springer, Heidelberg (2002)

[15] Löding, C.: Infinite Graphs Generated by Tree Rewriting. PhD thesis, RWTH Aachen, Germany (2003)

[16] Seese, D.: Entscheidbarkeits- und Definierbarkeitsfragen der Theorie, netzartiger Graphen-I. Wiss. Zeitschrift HU Berlin, XXI(5), 513–517 (1972)

[17] Spelten, A.: Rewriting Systems over Unranked Trees. Diploma thesis, RWTH Aachen, Germany (2006), `http://www-i7.informatik.rwth-aachen.de/download/papers/spelten/sp06.pdf`

[18] Takahashi, M.: Generalizations of regular sets and their application to a study of context-free languages. Inf. and Contr. 27(1), 1–36 (1975)

[19] Thomas, W.: A short introduction to infinite automata. In: Kuich, W., Rozenberg, G., Salomaa, A. (eds.) DLT 2001. LNCS, vol. 2295, pp. 130–144. Springer, Heidelberg (2002)

Rewriting Conjunctive Queries Determined by Views

Foto Afrati

National Technical University of Athens, Greece
afrati@softlab.ntua.gr

Abstract. Answering queries using views is the problem which examines how to derive the answers to a query when we only have the answers to a set of views. Constructing rewritings is a widely studied technique to derive those answers. In this paper we consider the problem of existence of rewritings in the case where the answers to the views uniquely determine the answers to the query. Specifically, we say that a view set $\mathcal{V}$ *determines* a query Q if for any two databases D_1, D_2 it holds: $\mathcal{V}(D_1) = \mathcal{V}(D_2)$ implies $Q(D_1) = Q(D_2)$. We consider the case where query and views are defined by conjunctive queries and investigate the question: If a view set $\mathcal{V}$ determines a query Q, is there an equivalent rewriting of Q using $\mathcal{V}$? We present here interesting cases where there are such rewritings in the language of conjunctive queries. Interestingly, we identify a class of conjunctive queries, CQ_{path}, for which a view set can produce equivalent rewritings for "almost all" queries which are determined by this view set. We introduce a problem which relates determinacy to query equivalence. We show that there are cases where restricted results can carry over to broader classes of queries.

1 Introduction

The problem of using materialized views to answer queries [19] has received considerable attention because of its relevance to many data-management applications, such as information integration [6,11,16,18,20,26], data warehousing [25],[5] web-site designs [14], and query optimization [10]. The problem can be stated as follows: given a query Q on a database schema and a set of views $\mathcal{V}$ over the same schema, can we answer the query using only the answers to the views, i.e., for any database D, can we find $Q(D)$ if we only know $\mathcal{V}(D)$? Constructing rewritings is a widely used and extensively studied technique to derive those answers [17].

A related fundamental question concerns the information provided by a set of views for a specific query. In that respect, we say that a view set $\mathcal{V}$ *determines* a query Q if for any two databases D_1, D_2 it holds: $\mathcal{V}(D_1) = \mathcal{V}(D_2)$ implies $Q(D_1) = Q(D_2)$ [24]. A query Q can be thought of as defining a partition of the set of all databases in the sense that databases on which the query produces the same set of tuples in the answer belong to the same equivalence class. In the same sense a set of views defines a partition of the set of all databases. Thus, if

L. Kučera and A. Kučera (Eds.): MFCS 2007, LNCS 4708, pp. 78–89, 2007.

a view set $\mathcal{V}$ determines a query Q, then the views' partition is a refinement of the partition defined by the query. Thus, the equivalence class of $\mathcal{V}(D)$ uniquely determines the equivalence class of $Q(D)$. Hence, a natural question to ask is: if a set of views determines a query is there an equivalent rewriting of the query using the views? In this paper we consider the case where query and views are defined by conjunctive queries (CQ for short) and investigate decidability of determinacy and the existence of equivalent rewriting whenever a view set determines a query.

The existence of rewritings depend on the language of the rewriting and the language of the query and views. Given query languages $\mathcal{L}$, $\mathcal{L}_{\mathcal{V}}$, $\mathcal{L}_Q$ we say that a language $\mathcal{L}$ is *complete* for $\mathcal{L}_{\mathcal{V}}$-to-$\mathcal{L}_Q$ rewriting if whenever a set of views $\mathcal{V}$ in $\mathcal{L}_{\mathcal{V}}$ determines a query Q in $\mathcal{L}_Q$ then there is an equivalent rewriting of Q in $\mathcal{L}$ which uses only $\mathcal{V}$. We know that CQ is not complete for CQ-to-CQ rewriting [22]. However there exist interesting special cases it is complete [24,22].

In this paper we consider subclasses of CQs and investigate a) decidability of determinacy, b) special cases where CQ or first order logic is complete for rewriting and c) the connection between determinacy and query equivalence. In more detail, our contributions are the following:

1. We show that CQ is complete for the cases a) where the views are full (all variables from the body are exported to the head) and b) where query has a single variable and view set consists of a single view with two variables.
2. We show that determinacy is decidable for chain queries and views.
3. We identify a class of conjunctive queries, CQ_{path}, which is almost complete for CQ_{path}-to-CQ_{path} rewriting. This is the first formal evidence that there are well behaved subsets of conjunctive queries.
4. Query rewritings using views is a problem closely related to query equivalence. Hence it is natural to ask what is the connection between determinacy and query equivalence. We investigate this question and introduce a new problem which concerns a property a query language may have and is a variant of the following: For a given query language, if Q_1 is contained in Q_2 and Q_2 determines Q_1, then are Q_1 and Q_2 equivalent? We solve special cases of it such as for CQ queries without self joins.
5. We make formal the observation that connectivity can be used to simplify the problem of determinacy and as a result of it we provide more subclasses with good behavior.

1.1 Related Work

In [24], the problem of determinacy is investigated for many languages including first order logic and fragments of second order logic and a considerable number of cases are resolved. The results closer to our setting show that if a language $\mathcal{L}$ is complete for UCQ-to-UCQ (i.e., unions of CQs) rewriting, then $\mathcal{L}$ must express non-monotonic queries. Moreover, this holds even if the database relations, views and query are restricted to be unary. This says that even Datalog is not complete for UCQ-to-UCQ rewritings. Datalog is not complete even for $CQ^{\neq}$-to-CQ rewritings. In [22,24], special classes of conjunctive queries and views are identified for which the language of conjunctive queries is complete: when views are unary or Boolean and when there is only one path view. It is shown that determinacy is undecidable for views and queries in the language of union of conjunctive queries [24].

Determinacy and notions related to it are also investigated in [15] where the notion of subsumption is introduced and used to the definition of complete rewritings and in [7,8] where the concept of lossless view with respect to a query is introduced and investigated both under the sound view assumption (a.k.a. open world assumption) and under the exact view assumption (a.k.a. closed world assumption) on regular path queries used for semi-structured data. Losslessness under the CWA is identical to determinacy. There is a large amount of work on equivalent rewritings of queries using views. It includes [19] where it is proven that it is NP-complete to decide whether a given CQ query has an equivalent rewriting using a given set of CQ views, [12] where polynomial subcases were identified. In [23], [4], [13] cases were investigated for CQ queries and views with binding patterns, arithmetic comparisons and recursion, respectively. In some of these works also the problem of maximally contained rewritings is considered. Intuitively, maximally contained rewritings is the best we can do for rewritings in a certain language when there is no equivalent rewriting in the language and want to obtain a query that uses only the views and computes as many certain answers [1] as possible. In [21] the notion of p-containment and equipotence is introduced to characterize view sets that can answer the same set of queries. Answering queries using views in semi-structured databases is considered in [7] and references therein.

2 Preliminaries

2.1 Basic Definitions

We consider queries and views defined by conjunctive queries (CQ for short) (i.e., select-project-join queries) in the form:

$$h(\bar{X}) : -g_1(\bar{X}_1), \ldots, g_k(\bar{X}_k).$$

Each subgoal $g_i(\bar{X}_i)$ in the *body* is a *relational atom*, where predicate g_i defines a *base relation* (we use the same symbol for the predicate and the relation), and every argument in the subgoal is either a variable or a constant. A variable is called *distinguished* if it appears in the head $h(\bar{X})$.

A *relational structure* is a set of atoms over a *domain* of variables and constants. A relational atom with constants in its arguments is called a *ground atom.* A *database instance* or *database* is a finite relational structure with only ground atoms. The body of a conjunctive query can be also viewed as a relational structure and we call it *canonical database of query* Q and denote D_Q; we say that in D_Q the variables of the query are *frozen* to distinct constants. A query Q_1 is *contained* in a query Q_2, denoted $Q_1 \sqsubseteq Q_2$, if for any database D on the base relations, the answer computed by Q_1 is a subset of the answer by Q_2, i.e., $Q_1(D) \subseteq Q_2(D)$. Two queries are *equivalent*, denoted $Q_1 \equiv Q_2$, if $Q_1 \sqsubseteq Q_2$ and

$Q_2 \sqsubseteq Q_1$. Chandra and Merlin [9] show that a conjunctive query Q_1 is contained in another conjunctive query Q_2 if and only if there is *containment mapping* from Q_2 to Q_1. A containment mapping is a *homomorphism* which maps the head and all the subgoals in Q_2 to Q_1. A CQ query Q is *minimized* if by deleting any subgoal we obtain a query which is not equivalent to Q. We denote by $\mathcal{V}(D)$ the result of computing the views on database D, i.e., $\mathcal{V}(D) = \bigcup_{V \in \mathcal{V}} V(D)$, where $V(D)$ contains atoms $v(t)$ for each answer t of view V.

Definition 1 (*equivalent rewritings*). *Given a query Q and a set of views $\mathcal{V}$, a query P is an* equivalent rewriting *of query Q using $\mathcal{V}$, if P uses only the views in $\mathcal{V}$, and for any database D on the schema of the base relations it holds: $P(\mathcal{V}(D)) = Q(D)$.*

The *expansion* of a CQ query P on a set of CQ views $\mathcal{V}$, denoted P^{exp}, is obtained from P by replacing all the views in P with their corresponding base relations. Existentially quantified variables (i.e., nondistinguished variables) in a view are replaced by fresh variables in P^{exp}. For conjunctive queries and views a conjunctive query P is an equivalent rewriting of query Q using $\mathcal{V}$ iff $P^{exp} \equiv Q$.

2.2 Determinacy

For two databases D_1, D_2, $\mathcal{V}(D_1) = \mathcal{V}(D_2)$ means that for each $V_i \in \mathcal{V}$ it holds $V_i(D_1) = V_i(D_2)$.

Definition 2 (*views determine query*). *Let query Q and views $\mathcal{V}$. We say that $\mathcal{V}$* determines *Q if the following is true: For any pair of databases D_1 and D_2, if $\mathcal{V}(D_1) = \mathcal{V}(D_2)$ then $Q(D_1) = Q(D_2)$.*

Let $\mathcal{L}$ be a query language or a set of queries (it will be clear from the context). We say that a subset $\mathcal{L}_1$ of $\mathcal{L}$ *contains almost all* queries in $\mathcal{L}$ if the following holds: Imagine $\mathcal{L}$ as a union of specific sets of queries, called *eq-sets* such that each eq-set contains exactly all queries in $\mathcal{L}$ that are equivalent to each other (i.e., every two queries in a particular eq-set are equivalent). Then $\mathcal{L}_1$ contains all queries in $\mathcal{L}$ except those queries contained in a finite number of eq-subsets.

Definition 3 (*(almost) complete language for rewriting*). *Let $\mathcal{L}_Q$ and $\mathcal{L}_\mathcal{V}$ be query languages or sets of queries. Let $\mathcal{L}$ be query language.*

We say that $\mathcal{L}$ is complete for $\mathcal{L}_\mathcal{V}$-to-$\mathcal{L}_Q$ rewriting *if the following is true for any query Q in $\mathcal{L}_Q$ and any set of views $\mathcal{V}$ in $\mathcal{L}_\mathcal{V}$: If $\mathcal{V}$ determines Q then there is an equivalent rewriting in $\mathcal{L}$ of Q using $\mathcal{V}$. We say that $\mathcal{L}$* is complete for rewriting *if it is complete for $\mathcal{L}$ -to-$\mathcal{L}$ rewriting.*

We say that $\mathcal{L}$ is almost complete for $\mathcal{L}_\mathcal{V}$-to-$\mathcal{L}_Q$ rewriting *if there exists a subset $\mathcal{L}_{Q1}$ of $\mathcal{L}_Q$ which contains almost all queries in $\mathcal{L}_Q$ such that the following holds: $\mathcal{L}$ is complete for $\mathcal{L}_\mathcal{V}$-to-$\mathcal{L}_{Q1}$ rewriting. We say that $\mathcal{L}$* is almost complete for rewriting *if it is almost complete for $\mathcal{L}$ -to-$\mathcal{L}$ rewriting.*

It is easy to show that if there is an equivalent rewriting of a query using a set of views then this set of views determine the query. The following proposition states some easy observations.

Proposition 1. *Let query Q and views $\mathcal{V}$ be given by minimized conjunctive queries. Suppose $\mathcal{V}$ determines Q.*

Let Q' be query resulting from Q after deleting one or more subgoals. Let D_Q and $D_{Q'}$ be the canonical databases of Q and Q' respectively. Then the following hold: **a**) $\mathcal{V}(D_Q) \neq \mathcal{V}(D_{Q'})$. **b**) *For any database D, the constants in the tuples in $Q(D)$ is a subset of the constants in the tuples in $\mathcal{V}(D)$.* **c**) *All base predicates appearing in the query definition appear also in the views (but not necessarily vice versa).* **d**) $\mathcal{V}(D_Q) \neq \emptyset$.

Canonical Rewriting. Let D_Q be the canonical database of Q. We compute the views on D_Q and get view instance $\mathcal{V}(D_Q)$ [3,2]. We construct *canonical rewriting* R_c as follows. The body of R_c contains as subgoals exactly all unfrozen view tuples in $\mathcal{V}(D_Q)$ and the tuple in the head of R_c is as the tuple in the head of query Q. Here is an example which illustrates this construction.

Example 1. We have the query $Q : q(X,Y) : -a(X,Z_1), a(Z_1,Z_2), b(Z_2,Y)$ and views $\mathcal{V}$: $V_1 : v_1(X,Z_2) : -a(X,Z_1), a(Z_1,Z_2)$ and $V_2 : v_2(X,Y) : -b(X,Y)$. Then D_Q contains the tuples $\{a(x,z_1), a(z_1,z_2), b(z_2,y)\}$ and $\mathcal{V}(D_Q)$ contains the tuples $\{v_1(x,z_2), v_2(z_2,y)\}$. Thus, R_c is: $q(X,Y) : -v_1(X,Z_2), v_2(Z_2,Y)$.

The following proposition can be used when we want to show that there is no equivalent CQ rewriting of a query using a set of views.

Proposition 2. *Let Q and $\mathcal{V}$ be conjunctive query and views and R_c be the canonical rewriting. If there is a conjunctive equivalent rewriting of Q using $\mathcal{V}$ then R_c is such a rewriting.*

2.3 Cases for Which CQ Is Complete for Rewriting

Theorem 1. *CQ is complete for $\mathcal{L}_\mathcal{V}$-to-$\mathcal{L}_Q$ rewriting in the case where $\mathcal{L}_\mathcal{V}$ and $\mathcal{L}_Q$ are subclasses of conjunctive queries in either of the following cases:*

1. *$\mathcal{L}_Q = CQ$ and $\mathcal{L}_\mathcal{V}$ contains only queries with no nondistinguished variables.*
2. *Binary base predicates, one view in the view set, $\mathcal{L}_Q$ contains only queries with one variable and $\mathcal{L}_\mathcal{V}$ contains only queries with one non-distinguished variable.*

3 Chain and Path Queries

In this section we consider chain and path queries and views.

Definition 4. *A CQ query is called* chain query *if it is defined over binary predicates and also the following holds: The body contains as subgoals a number of binary atoms which if viewed as labeled graph (since they are binary) they form a directed simple path and the start and end nodes of this path are the arguments in the head. For example, this is a chain query: $q(X,Y) : -a(X,Z_1)$, $b(Z_1,Z_2), c(Z_2,Y)$.*

Path queries *are chain queries over a single binary relation. Path queries can be fully defined simply by the length of the path in the body (i.e., number of subgoals in the body). Hence we denote by* P_k *the path query of length* k*. We denote the language of all chain queries by* CQ_{chain} *and the language of all path queries by* CQ_{path}.

3.1 Chain Queries – Decidability

In the case of chain queries and views, we show that the following property fully characterizes cases where a set of views determine a query (Theorem 2), hence for this class determinacy is decidable.

Definition 5. *Let* Q *be a binary query over binary predicates. We say that* Q *is* disjoint *if the body of* Q *viewed as an undirected graph does not contain a (undirected) path from one head variable of* Q *to the other.*

Theorem 2. *Let query* Q *and views* $\mathcal{V}$ *be chain queries. Then the following hold:*

1. $\mathcal{V}$ *determines* Q *iff the canonical rewriting of* Q *using* $\mathcal{V}$ *is not disjoint.*
2. *First order logic is complete for* CQ_{chain}*-to-*CQ_{chain} *rewriting.*
3. *It is decidable whether a set of views determines a query.*

3.2 Path Queries – CQ Is Almost Complete for Rewriting

In this section we will prove the following theorem and we will also get a complete characterization for path queries and two path views as concerns CQ being complete for this class of queries and views.

Theorem 3. CQ_{path} *(and hence CQ) is almost complete for* CQ_{path}*-to-*CQ_{path} *rewriting. Hence* CQ_{path} *is almost complete for rewriting.*

The above theorem is a consequence of Lemma 2. In order to acquire some intuition we present first some intermediate results.

Theorem 4

1. CQ_{path} *(and hence CQ) is complete for* $\{P_2, P_3\}$*-to-*CQ_{path} *rewriting.*
2. CQ_{path} *(and hence CQ) is complete for* $\{P_3, P_4\}$*-to-*CQ_{path1} *rewriting, where* CQ_{path1} *is* CQ_{path} *after deleting* P_5.

*Proof (*of Part 1*).* The proof of part 1 is easy: The view set does not determine query P_1 for the following reason: Take a database which is empty and another database which contains a single tuple, then in both databases, the views compute the empty set while the query computes the empty set only in the former database. All other path queries have easy equivalent CQ_{path} rewritings. ⊣

It is interesting to note (as another counterexample that CQ is not complete for rewriting) that viewset $\{P_3, P_4\}$ determines the query P_5 because the following formula is a rewriting of $P_5(X, Y)$ (it is not a CQ however):

$$\phi(X,Y) : \exists Z[P_4(X,Z) \wedge \forall W((P_3(W,Z) \rightarrow P_4(W,Y))]$$

However there is no CQ rewriting of P_5 using $\{P_3, P_4\}$.

We generalize the result in Theorem 4 for two views P_k and P_{k+1}. The following theorem is a complete characterization of all path queries with respect to viewset $\{P_k, P_{k+1}\}$.

Theorem 5. *Let $\mathcal{QP}_{k+2}$ be the set of all path queries except the set of queries*

$$\mathcal{QPP}_{k+2} = \bigcup_{n=1}^{n=k-2} \{P_{nk+n+1}, P_{nk+n+2}, \ldots, P_{(n+1)k-1}\}$$

Then the following hold:

1. *CQ_{path} (and hence CQ) is complete for $\{P_k, P_{k+1}\}$-to-$\mathcal{QP}_{k+2}$ rewriting.*
2. *CQ is not complete for $\{P_k, P_{k+1}\}$-to-$\mathcal{QPP}_{k+2}$ rewriting.*

Proof. (Sketch) First we use Theorem 2 to prove that all path queries except queries $P_1, \ldots, P_{k-1}$ are determined by $\{P_k, P_{k+1}\}$. We only need to show that there is in the expansion of the canonical rewriting an undirected path from head variable X to head variable Y which ends in a forward edge. Inductively, for query P_m ($m \geq k$) there is such a directed path which ends in a forward edge. For query P_{m+1}, we augment the undirected path of P_m by taking a backward edge for P_k and then a forward edge for P_{k+1}.

Then we use similar argument as in the case of the viewset $\{P_2, P_3\}$ to prove that none of the queries $P_1, \ldots, P_{k-1}$ are determined by $\{P_k, P_{k+1}\}$. Finally we prove that, for each path query in $\mathcal{QP}_{k+2}$, the canonical rewriting is an equivalent rewriting. ⊣

The following theorem is a corollary of Theorem 5 and Theorem 7 generalizes for any two views P_k, P_m. The proof of Theorem 7 is a consequence of Lemma 1.

Theorem 6. *CQ_{path} (and hence CQ) is almost complete for $\{P_k, P_{k+1}\}$-to-CQ_{path} rewriting.*

Theorem 7. *Let k, m be positive integers. Then, CQ_{path} (and hence CQ) is almost complete for $\{P_k, P_m\}$-to-CQ_{path} rewriting.*

Lemma 1. *Let P_n be a query and let viewset be $\{P_k, P_m\}$. Then the following hold.*

1. *If $n \geq km$ and the greatest common divisor of k and m divides n then there is a CQ_{path} equivalent rewriting of the query using $\{P_k, P_m\}$.*
2. *If the greatest common divisor of k and m does not divide n then $\{P_k, P_m\}$ does not determine the query.*

Finally the following lemma generalizes Lemma 1 for any number of views:

Lemma 2. *Let P_n be a query and let viewset be $\mathcal{V}=\{P_{k_1}, P_{k_2}, \ldots, P_{k_K}\}$. Then there is a positive integer n_0 which is a function only of $k_1, k_2, \ldots, k_K$ such that for any $n \geq n_0$ the following statements are equivalent.*

1. *There is no equivalent rewriting in CQ of P_n using $\mathcal{V}$.*
2. *The canonical rewriting of P_n using $\mathcal{V}$ is disjoint.*
3. *$\mathcal{V}$ does not determine P_n.*

4 Determinacy and Query Equivalence

The problem that we investigate in this paper relates determinacy to query rewriting. The simplest way to produce an equivalent rewriting of a query Q is when we have only one view and the view is equivalent to the query. Hence, a natural related problem is: If Q_1 is contained in Q_2 and Q_2 determines Q_1, are Q_1 and Q_2 equivalent? The following simple example shows that this statement does not hold: Let $Q_1 : q_1(X,X) : -a(X,X)$ and $Q_2 : q_2(X,Y) : -a(X,Y)$. Obviously Q_1 is contained in Q_2. Also Q_2 determines Q_1 because there is an equivalent rewriting of Q_1 using Q_2, it is $R : q(X,X) : -q_2(X,X)$. But Q_1 and Q_2 are *not* equivalent.

We add some stronger conditions: Suppose in addition that there is a containment mapping that uses as targets all subgoals of Q_1 and this containment mapping maps the variables in the head one-to-one. Still there is a counterexample:

Example 2. We have two queries:
$Q_1 :\ q_1(X,Y,Z,W,A,B) : -r(Y,X), s(Y,X), r(Z,W),$
$s(Z,Z_1), s(Z_1,Z_1), s(Z_1,W), s(A,A_1), s(A_1,A_1), s(A_1,B).$
and
$Q_2 :\ q_2(X,Y,Z,W,A,B) : -r(Y,X), s(Y,X), r(Z,W),$
$s(Z,Z_1), s(Z_1,Z_2), s(Z_2,W), s(A,A_1), s(A_1,A_1), s(A_1,B).$
Clearly Q_1 is contained in Q_2 and Q_2 determines Q_1 because there is an equivalent rewriting of Q_1 using Q_2:
$R : q_1'(X,Y,Z,W,A,B) : -q_2(X,Y,Z,W,A,B), q_2(X_1,Y_1,Z_1,W_1,Z,W).$
Moreover there is a homomorphism from Q_2 to Q_1 that uses all subgoals of Q_1 and is one-to-one on the head variables. But Q_1 and Q_2 are not equivalent.

Finally we add another condition which we denote by $Q_2(D_1) \subseteq_s Q_2(D_2)$, where D_1, D_2 are the canonical databases of Q_1, Q_2 respectively.

We need first explain the notation $Q(D_1) \subseteq_s Q(D_2)$ which in general expresses some structural property of databases D_1 and D_2 with respect to Q and this property is invariant under renaming. We say that $Q(D_1) \subseteq_s Q(D_2)$ holds if there is a renaming of the constants in D_1, D_2 such that $Q(D_1) \subseteq Q(D_2)$. For an example, say we have query $Q : q(X,Y) : -r(X,Y)$ and three database instances $D_1 = \{r(1,2), r(2,3)\}$, $D_2 = \{r(a,b), r(b,c)\}$ and $D_3 = \{r(a,b), r(a,c)\}$. Then it holds that $Q(D_1) \subseteq_s Q_1(D_2)$ and $Q(D_1) \subseteq_s Q(D_2)$ because there is a renaming of D_2 (actually here D_1, D_2 are isomorphic) such that $Q(D_1) \subseteq Q_1(D_2)$ and $Q(D_1) \subseteq Q(D_2)$. But the following does not hold: $Q(D_3) \subseteq_s Q(D_2)$.

We may also allow some constants in D_1, D_2 that are special as concerns renaming. Although we need incorporate these constants in the notation, we will keep (slightly abusively) the same notation here since we always mean the same constants. By $Q_2(D_1) \subseteq_s Q_2(D_2)$ we mean in addition that (i) the frozen variables in the head of the queries are identical component-wise, i.e., if in the head of Q_1 we have tuple $(X_1, \ldots, X_m)$ then in the head of Q_2 we also have same tuple $(X_1, \ldots, X_m)$ and in both D_1, D_2 these variables freeze to constants $x_1, \ldots, x_m$ and (ii) we are not allowed to rename the constants $x_1, \ldots, x_m$.

Now we introduce a new problem which relates determinacy to query equivalence:

Determinacy and query equivalence: Let Q_1, Q_2 conjunctive queries. Suppose Q_2 determines Q_1, and Q_1 is contained in Q_2. Suppose also that the following hold: a) there is a containment mapping from Q_2 to Q_1 which (i) uses as targets all subgoals of Q_1 and (ii) maps the variables in the head one-to-one, and b) $Q_2(D_1) \subseteq_s Q_2(D_2)$, where D_1, D_2 are the canonical databases of Q_1, Q_2 respectively. Then are Q_1 and Q_2 equivalent? If the answer is "yes" for any pair of queries Q_1, Q_2 where Q_1 belongs to CQ class CQ_1 and Q_2 belongs to CQ class CQ_2, then we say that *determinacy defines CQ_2-to-CQ_1 equivalence.*

This problem seems to be easier to resolve than the determinacy problem and Theorem 8 is formal evidence of that.

Theorem 8. *Let CQ_1, CQ_2 be subsets of the set of conjunctive queries. For the following two statements it holds: Statement (A) implies statement (B).*

A) CQ is complete for CQ_2-to-CQ_1 single view rewriting.
B) Determinacy defines CQ_2-to-CQ_1 equivalence.

In [22] it is proven part A of the above theorem for one path view. A consequence of it and Theorem 8 is the following:

Theorem 9. *Determinacy defines CQ_{path}-to-CQ equivalence.*

The *determinacy and query equivalence* question remains open. Theorem 10 settles a special case where we have replaced condition (b) with a stronger one. Theorem 11 is a consequence of Theorem 10.

Theorem 10. *Let Q_1, Q_2 be conjunctive queries. Suppose Q_2 determines Q_1, and Q_1 is contained in Q_2. Suppose also that the following hold: a) there is a containment mapping that uses as targets all subgoals of Q_1 and this containment mapping maps the variables in the head one-to-one, and b) $Q_2(D_1)$ contains exactly one tuple, where D_1 is the canonical database of Q_1. Then Q_1 and Q_2 are equivalent.*

Theorem 11. *Consider queries in either of the following cases: a) Q_1 has no self joins (i.e., each predicate name appears only once in the body) or b) Q_1 contains a single variable.*

Suppose CQ query Q_2 determines Q_1 and Q_1 is contained in Q_2. Then Q_1 and Q_2 are equivalent.

5 Connectivity

In this section, we present a case where good behavior for determinacy can carry over to a broader class of queries. Specifically we relate determinacy to connectivity in the body of the query. The following example shows the intuition.

Example 3. We have query: $Q : Q(X) : -r(Y, X), s(Y, X), s_1(Z, Z_1), s_2(Z_1, Z)$ and views $\mathcal{V}$: $v_1(X, Y) : -r(Y, X)$ and $v_2(X, Y) : -s(Y, X), s_1(Z, Z_1), s_2(Z_1, Z)$. First observe that all variables contained in the last two subgoals of Q are not contained in any other subgoal of Q and neither they appear in the head of Q. In this case we say that subgoals $s_1(Z, Z_1), s_2(Z_1, Z)$ form a connected component (see definitions below). Moreover, let us consider the canonical rewriting (which happens to be an equivalent rewriting) of Q using these two views $R_1 : Q(X) : -v_1(X, Y), v_2(X, Y)$. Observe that none of the variables in the two last subgoals of the query appear in the rewriting (we conveniently retain the same names for the variables). In this case, we say in addition that the subgoals $s_1(Z, Z_1), s_2(Z_1, Z)$ form a semi-covered component wrto the views (see definition below). We conclude the observations on this example by noticing that the following query and views a) are simpler and b) can be used "instead" of the original query and views. Query $Q'(X) : -r(Y, X), s(Y, X)$ and views $\mathcal{V}$: $v'_1(X, Y) : -r(Y, X)$ and $v'_2(X, Y) : -s(Y, X)$. They were produced from the original query and views by a) deleting the semi-covered subgoals from the query and b) deleting an isomorphic copy of the semi-covered subgoals from view v_2 (see Lemma 3 for the feasibility of this). Then the canonical rewriting of Q' using $\mathcal{V}'$ is isomorphic to R_1, specifically it is: $R'_1 : Q'(X) : -v'_1(X, Y), v'_2(X, Y)$ and is again an equivalent rewriting. In this section, we make this observation formal, i.e., that in certain cases, we can reduce the original problem to a simpler one.

Definition 6 (Connectivity graph of query). *Let Q be a conjunctive query. The nodes of the* connectivity graph *of Q are all the subgoals of Q and there is an (undirected) edge between two nodes if they share a variable or a constant.*

A *connected component* of a graph is a maximal subset of its nodes such that for every pair of nodes in the subset there is a path in the graph that connects them. A *connected component of a query* is a subset of subgoals which define a connected component in the connectivity graph. A query is *head-connected* if all subgoals containing head variables are contained in the same connected component.

Definition 7 (semi-covered component). *Let Q and $\mathcal{V}$ be CQ query and views. Let G be a connected component of query Q. Suppose that every variable or constant in G is such that there is no tuple in $\mathcal{V}(D_Q)$ (D_Q is the canonical database of Q) that contains it. Then we say that G is a* semi-covered component *of Q* wrto *$\mathcal{V}$.*

Lemma 3. *Let Q and $\mathcal{V}$ be conjunctive query and views. Suppose $\mathcal{V}$ determines Q. Let G_Q be a connected component of Q which is semi-covered wrto $\mathcal{V}$. Then there is a view in $\mathcal{V}$ which contains a connected component which is isomorphic to G_Q.*

As a consequence of Lemma 3, we can identify the semi-covered components of the query in the views definitions as well. Hence, we define the *semi-covered-free pair*, $(Q', \mathcal{V}')$, of a pair $(Q, \mathcal{V})$ of query and views: Q' results from Q by deleting all semi-covered components wrto $\mathcal{V}$ and each view in $\mathcal{V}'$ results from a view in $\mathcal{V}$ by deleting the components isomorphic to the semi-covered components of the query. Then the following holds:

Theorem 12. *Let CQ_1, CQ_2 be subsets of the set of conjunctive queries such that each query in either of them is head-connected. Let CQ_c be a conjunctive query language. Let CQ_{1f}, CQ_{2f} be subsets of the set of conjunctive queries such that for each query Q in CQ_1 (CQ_2 respectively) there is a query in CQ_{1f} (CQ_{2f}, respectively) which is produced from Q by deleting a connected component. Then the following holds:*

Language CQ_c is complete for CQ_1-to-CQ_2 rewriting iff it is complete for CQ_{1f}-to-CQ_{2f} rewriting.

The following is a corollary of Theorem 12 and results from Section 3:

Theorem 13. *Let P_k^a be a query with two variables in the head whose body contains i) a path on binary predicate r from one head variable to the other and ii) additional subgoals on predicates distinct from r and using variables distinct from the variables that are used to define the path. We call the language of such queries CQ_{apath}.*

Suppose we have query Q and views $\mathcal{V}$ that are in CQ_{apath}. Then it holds: CQ_{path} (and hence CQ) is almost complete for CQ_{apath}-to-CQ_{apath} rewriting.

6 Conclusion

The case about finding well behaved subclasses of conjunctive queries is of interest and is far from closed. We include some suggestions that are close to the research presented in this paper. For chain queries, we don't have a full characterization as concerns subclasses for which CQ is complete. We don't know whether determinacy defines CQ-to-CQ equivalence. Decidability of determinacy for conjunctive queries remains open.

Acknowledgments. Many thanks to Jeff Ullman for insightful discussions and for providing Example 2. Thanks also to the anonymous reviewers for their very useful comments.

References

1. Abiteboul, S., Duschka, O.M.: Complexity of answering queries using materialized views. In: PODS (1998)
2. Afrati, F., Li, C., Ullman, J.D.: Generating efficient plans using views. In: SIGMOD (2001)

3. Afrati, F., Li, C., Ullman, J.D.: Using views to generate efficient evaluation plans for queries. JCSS, to appear
4. Afrati, F.N., Li, C., Mitra, P.: Rewriting queries using views in the presence of arithmetic comparisons. Theor. Comput. Sci. 368(1-2) (2006)
5. Agrawal, S., Chaudhuri, S., Narasayya, V.R.: Automated selection of materialized views and indexes in sql databases. In: Proc. of VLDB (2000)
6. Bayardo Jr., R.J., et al.: Infosleuth: Semantic integration of information in open and dynamic environments (experience paper). In: SIGMOD (1997)
7. Calvanese, D., De Giacomo, G., Lenzerini, M., Vardi, M.Y.: Lossless regular views. In: PODS, ACM, New York (2002)
8. Calvanese, D., De Giacomo, G., Lenzerini, M., Vardi, M.Y.: View-based query query processing: On the relationship between rewriting, answering and losslessness. In: International Conference on Database Theory (ICDT) (2005)
9. Chandra, A.K., Merlin, P.M.: Optimal implementation of conjunctive queries in relational data bases. In: STOC (1977)
10. Chaudhuri, S., Krishnamurthy, R., Potamianos, S., Shim, K.: Optimizing queries with materialized views. In: ICDE (1995)
11. Chawathe, S.S., et al.: The TSIMMIS project: Integration of heterogeneous information sources. In: IPSJ (1994)
12. Chekuri, C., Rajaraman, A.: Conjunctive query containment revisited. In: Afrati, F.N., Kolaitis, P.G. (eds.) ICDT 1997. LNCS, vol. 1186, Springer, Heidelberg (1996)
13. Duschka, O.M., Genesereth, M.R.: Answering recursive queries using views. In: PODS (1997)
14. Florescu, D., Levy, A., Suciu, D., Yagoub, K.: Optimization of run-time management of data intensive web-sites. In: Proc. of VLDB (1999)
15. Grumbach, S., Tininini, L.: On the content of materialized aggregate views. In: PODS (2000)
16. Haas, L.M., Kossmann, D., Wimmers, E.L., Yang, J.: Optimizing queries across diverse data sources. In: Proc. of VLDB (1997)
17. Halevy, A.Y.: Answering queries using views: A survey. VLDB Journal 10(4)
18. Ives, Z., Florescu, D., Friedman, M., Levy, A., Weld, D.: An adaptive query execution engine for data integration. In: SIGMOD (1999)
19. Levy, A., Mendelzon, A., Sagiv, Y., Srivastava, D.: Answering queries using views. In: PODS (1995)
20. Levy, A., Rajaraman, A., Ordille, J.J.: Querying heterogeneous information sources using source descriptions. In: Proc. of VLDB (1996)
21. Li, C., Bawa, M., Ullman, J.: Minimizing view sets without losing query-answering power. In: Van den Bussche, J., Vianu, V. (eds.) ICDT 2001. LNCS, vol. 1973, Springer, Heidelberg (2000)
22. Nash, A., Segoufin, L., Vianu, V.: Determinacy and rewriting of conjunctive queries using views: A progress report. In: Schwentick, T., Suciu, D. (eds.) ICDT 2007. LNCS, vol. 4353, Springer, Heidelberg (2006)
23. Rajaraman, A., Sagiv, Y., Ullman, J.D.: Answering queries using templates with binding patterns. In: PODS (1995)
24. Segoufin, L., Vianu, V.: Views and queries: Determinacy and rewriting. In: PODS, ACM Press, New York (2005)
25. Theodoratos, D., Sellis, T.: Data warehouse configuration. In: Proc. of VLDB (1997)
26. Ullman, J.D.: Information integration using logical views. In: Afrati, F.N., Kolaitis, P.G. (eds.) ICDT 1997. LNCS, vol. 1186, Springer, Heidelberg (1996)

Approximation Algorithms for the Maximum Internal Spanning Tree Problem

Gábor Salamon*

Department of Computer Science and Information Theory
Budapest University of Technology and Economics
1117 Budapest, Magyar tudósok körútja 2., Hungary
gsala@cs.bme.hu

Abstract. We consider the MAXIMUMINTERNALSPANNINGTREE problem which is to find a spanning tree of a given graph with a maximum number of non-leaf nodes. From an optimization point of view, this problem is equivalent to the MINIMUMLEAFSPANNINGTREE problem, and is NP-hard as being a generalization of the HAMILTONIANPATH problem. Although there is no constant factor approximation for the MINIMUMLEAFSPANNINGTREE problem [1], MAXIMUMINTERNALSPANNINGTREE can be approximated within a factor of 2 [2].

In this paper we improve this factor by giving a $\frac{7}{4}$-approximation algorithm. We also investigate the node-weighted case, when the weighted sum of the internal nodes is to be maximized. For this problem, we give a $(2\Delta - 3)$-approximation for general graphs, and a 2-approximation for claw-free graphs. All our algorithms are based on local improvement steps.

Keywords: Approximation algorithm, Spanning tree leaves, Hamiltonian path.

1 Introduction

In this paper we consider a generalization of the Hamiltonian path problem: the MAXIMUMINTERNALSPANNINGTREE (MAXIST) problem, in which the goal is to find a spanning tree of a given graph with a maximum number of internal nodes (non-leaves). Note that a Hamiltonian path is a spanning tree with $|V|-2$ internal nodes. If the goal is to find an optimum solution then this problem is clearly equivalent to the MINIMUMLEAFSPANNINGTREE problem. However, from an approximation point of view, the two problems behave differently. On one hand, Lu and Ravi [1] proved that there is no constant factor approximation for the MINIMUMLEAFSPANNINGTREE problem. On the other hand, by slightly modifying the depth-first search algorithm, one can get a spanning tree that is either a Hamiltonian path or a tree with independent leaves. Such a tree always

* Research is supported by Grant No. 67651 of the Hungarian National Science Fund (OTKA).

L. Kučera and A. Kučera (Eds.): MFCS 2007, LNCS 4708, pp. 90–102, 2007.

has at least half as many internal nodes as the optimal one. This yields a linear-time 2-approximation for the MaxIST problem [2].

In this paper we improve the approximation ratio by giving a $\frac{7}{4}$-approximation algorithm for the MaxIST problem. We also consider the node-weighted case, when the aim is to find a spanning tree maximizing the weighted sum of internal nodes. For the MaximumWeightedInternalSpanningTree (MaxWIST) problem we present a $(2\Delta - 3)$-approximation for general graphs, and a 2-approximation for claw-free graphs—here Δ stands for the maximum degree. Our algorithms are based on local improvement steps. They work in arbitrary graphs; however, the above approximation ratios are guaranteed only when the input graph has no degree 1 node. The removal of this extra condition from the analysis is subject of further research.

Let us mention that for the MaximumLeafSpanningTree problem—when the aim is to maximize the number of leaves instead of the number of non-leaves—the best known approximation factor is 2 and is due to Solis-Oba [3].

2 Notation and Basic Definitions

By a graph $G = (V, E)$ we mean an undirected simple connected graph. Throughout this paper we suppose that G has no degree 1 nodes. Let $T = (V, E')$ be a spanning tree of G. The elements of E' are called *tree edges*, the edges in $E \setminus E'$ are called *non-tree edges*. We say that a node x is a *G-neighbor* of a node y if $(x, y) \in E$, and that x is a *T-neighbor* of y if (x, y) is a tree edge. A set $X \subseteq V$ is *G-independent* if it spans no edges of G, and a node x is *G-independent from* a set $Y \subseteq V$ if x has no G-neighbors in Y. We use $d_G(x)$ and $d_T(x)$ to denote the number of G-neighbors and T-neighbors of x, respectively. A node x is a *leaf* of T, a *forwarding node* of T, or a *branching* of T if $d_T(x) = 1$, $d_T(x) = 2$, or $d_T(x) \geq 3$, respectively. Forwarding nodes and branchings are called *internal nodes* of T. We denote the set of leaves by $L(T)$ and the set of internal nodes by $I(T)$. For nodes x and y, we denote by $P_T(x, y)$ the unique path in T between nodes x and y. The branching $b(l)$ of a leaf l is the one being closest to l in T. Note that if T is not a Hamiltonian path then each leaf l has a unique branching $b(l)$. The path $P_T(l, b(l))$ is called the *branch* of l and is denoted by $br(l)$. The node $b^-(l)$ is the T-neighbor of $b(l)$ being in the branch of l. If l is a leaf and (l, x) is a non-tree edge then $x^{\to l}$ is the predecessor of x and $b(l)^{\to x}$ is the successor of $b(l)$ along the path $P_T(l, x)$. The branch of l is *short* if $b^-(l) = l$, otherwise the branch of l is *long*. A leaf l is called short (long) if its branch is short (long). $L_s(T)$ stands for the set of short leaves and $L_g(T)$ for the set of long leaves. A leaf l is called *(x-)supported* if there is a non-tree edge (l, x) with $x \notin br(l)$. If l is a long leaf and there exists a non-tree edge (l, x) such that $x \in br(l)$ then $x^{\to l}$ is called an *l-leafish node*. In this case, the node x is called the *base* of $x^{\to l}$. The set of l-leafish nodes is denoted by $F(l)$. Note that not every long leaf has a leafish node in its branch. We denote by $L_p(T)$ the set of long leaves of T having no leafish nodes in their branch. For a set $X \subseteq V$, the graph $G[X]$ is the subgraph of G spanned by X. The *trunk* of tree T is $T[V \setminus \cup_{l \in L(T)}(br(l) \setminus \{b(l)\})]$. The maximum node-degree in G is denoted by Δ. For the sake of simplicity, we use $X + x$, and $X - x$ instead of $X \cup \{x\}$, and $X \setminus \{x\}$, respectively.

3 Maximizing the Number of Internal Nodes

In this section we present a local search algorithm for finding a spanning tree with many internal nodes. The algorithm starts by creating a DFS-tree. Then it applies local improvement rules as long as possible in order to reduce the number of leaves. Finally we obtain either a Hamiltonian path or a locally optimal spanning tree (a LOST). We show that a LOST provides a $\frac{7}{4}$-approximation for the MAXIST problem.

To prove this approximation ratio we show a primal problem formulation that has a solution of value $|I(T)|$ corresponding to T. We also show that any dual solution has a value of at least $|I(T^*)|$—the number of internal nodes of an optimum spanning tree. We give two different dual solutions to establish the approximation factor. Both dual solutions are based on a set S of forwarding nodes of T that are G-independent from the leaves.

The structure of the improvement rules is as follows. A precondition part determines when the specific rule can be executed, and an action part defines the replacement of some (1 or 2) edges of T by non-tree edges to obtain a spanning tree T' having less leaves than T.

We define a LOST to be a spanning tree T in which none of these improvement rules can be applied, that is, the precondition of the rules are not satisfied. If T has many short branches then the violated preconditions of Rules 1–6, if T has many long branches then the violated preconditions of Rules 1 and 7–14 ensure a set S that is G-independent from the leaves and whose size is big enough to yield the approximation ratio.

Rule 1. Precondition: T has two leaves l_1 and l_2 such that $(l_1, l_2) \in E(G)$. **Action:** Let $E(T') = E(T) + (l_1, l_2) - (b(l_1), b^-(l_1))$. (See Fig. 1(a)).

Rule 2. Precondition: T has an x-supported leaf l such that $d_T(x^{\to l}) > 2$. **Action:** Let $E(T') = E(T) + (l, x) - (x, x^{\to l})$. (See Fig. 1(b)).

Rule 3. Precondition: T has an x-supported leaf l_1 and a leaf l_2 such that $d_T(x^{\to l_1}) = 2$, and that $(l_2, x^{\to l_1})$ is a non-tree edge. **Action:** Let $E(T') = E(T) + (l_1, x) - (x, x^{\to l_1})$. Apply Rule 1 on leaves l_2, $x^{\to l_1}$. (See Fig. 1(c)).

Rule 4. Precondition: T has an x-supported leaf l such that $d_T(b(l)^{\to x}) > 2$. **Action:** Let $E(T') = E(T) + (l, x) - (b(l), b(l)^{\to x})$. (See Fig. 1(d)).

Rule 5. Precondition: T has an x-supported leaf l_1 and a leaf l_2 such that $d_T(b(l)^{\to x}) = 2$, and that $(l_2, b(l)^{\to x})$ is a non-tree edge. **Action:** Let $E(T') = E(T) + (l_1, x) - (b(l_1), b(l)^{\to x})$. Apply Rule 1 on leaves l_2, $b(l)^{\to x}$. (See Fig. 1(e)).

Rule 6. Precondition: T has a short leaf l, and an edge (x, y) such that (l, x) and (l, y) are both non-tree edges. **Action:** Let $E(T') = E(T) + (l, x) + (l, y) - (x, y) - (l, b(l))$. (See Fig. 1(f)).

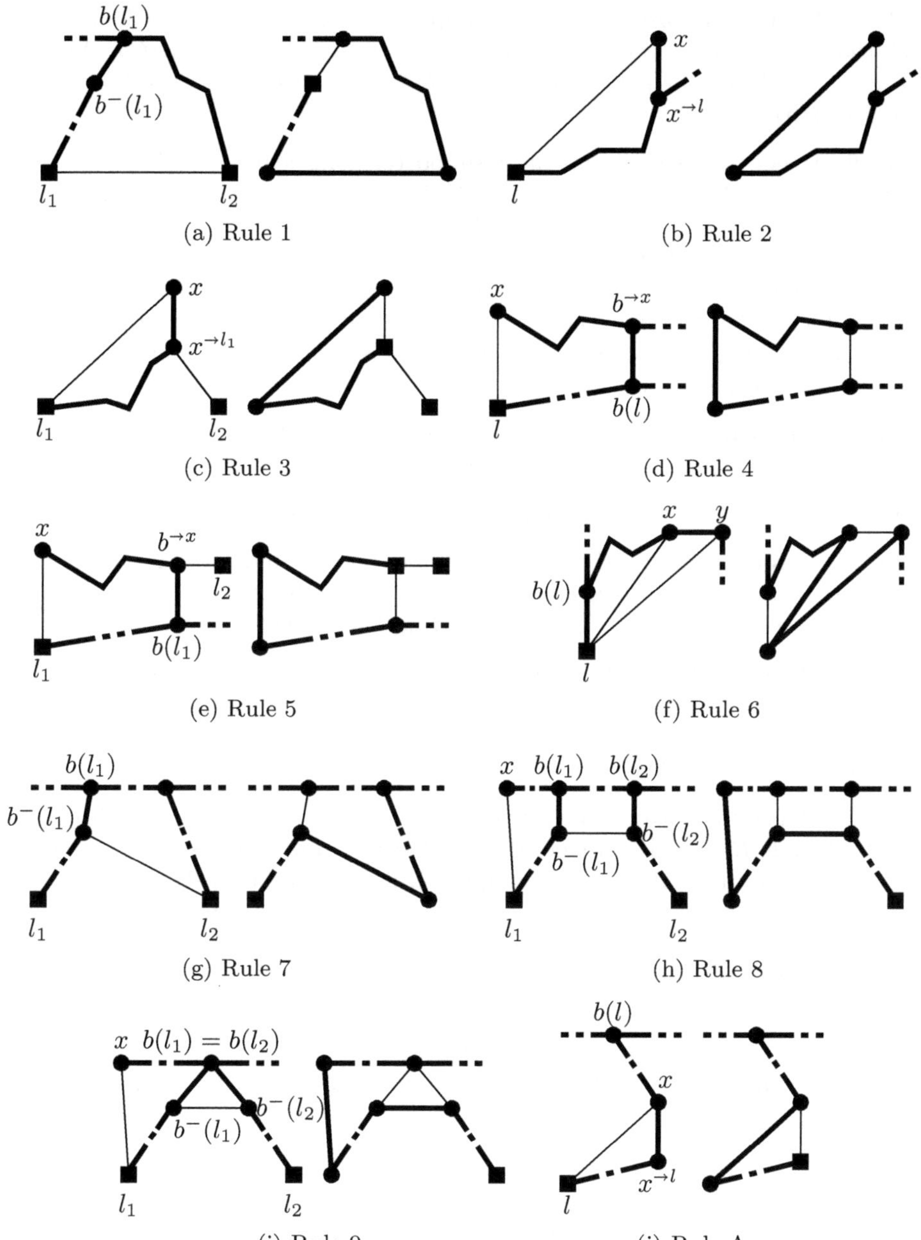

Fig. 1. Local improvement steps for creating a LOST (squares represent leaves, circles represent internal nodes)

Rule 7. Precondition: T has a long leaf l_1 and a leaf l_2 such that $(b^-(l_1), l_2) \in E(G)$. **Action:** Let $E(T') = E(T) + (b^-(l_1), l_2) - (b^-(l_1), b(l_1))$. (See Fig. 1(g)).

Rule 8. Precondition: T has an x-supported long leaf l_1 and a long leaf l_2 such that $x \notin br(l_2) - b(l_2)$, $b(l_1) \neq b(l_2)$, and $(b^-(l_1), b^-(l_2)) \in E(G)$. **Action:** Let $E(T') = E(T) + (l_1, x) + (b^-(l_1), b^-(l_2)) - (b(l_1), b^-(l_1)) - (b(l_2), b^-(l_2))$. (See Fig. 1(h)).

Rule 9. Precondition: T has an x-supported long leaf l_1 and a long leaf l_2 with $b(l_1) = b(l_2)$ such that $d_T(b(l_1)) \geq 4$, $x \notin br(l_2)$, and $(b^-(l_1), b^-(l_2)) \in E(G)$. **Action:** Let $E(T') = E(T) + (l_1, x) + (b^-(l_1), b^-(l_2)) - (b(l_1), b^-(l_1)) - (b(l_2), b^-(l_2))$. (See Fig. 1(i)).

The following rule differs from the above ones as—while applied on the long branch of a leaf l—it changes neither the trunk nor any other branch of T. Only the branch of l is modified such that one of its leafish nodes becomes leaf and l becomes an internal node.

Rule A. Precondition: T has a leaf l and an l-leafish node $x^{\to l}$ with base x. **Action:** Let $E(T') = E(T) + (l, x) - (x, x^{\to l})$. (See Fig. 1(j)).

We can use Rule A to decrease the number of leaves as follows.

Rule 10. Precondition: T has two leaves l_1 and l_2, and an l_1-leafish node u such that (u, l_2) is a non-tree edge. **Action:** Apply Rule A on u to make it a leaf and l_1 an internal node. Then apply Rule 1 on leaves l_2, u.

Rule 11. Precondition: T has two leaves l_1 and l_2, an l_1-leafish node u, and an l_2-leafish node v such that (u, v) is a non-tree edge. **Action:** Apply Rule A on u and on v to make both of them a leaf while l_1 and l_2 an internal node. Then apply Rule 1 on leaves u, v.

Rule 12. Precondition: T has two leaves l_1 and l_2, and an l_1-leafish node u such that $(u, b^-(l_2))$ is a non-tree edge. **Action:** Apply Rule A on u to make it a leaf and l_1 an internal node. Then apply Rule 7 on leaves u, l_2.

The following rules do not change the number of leaves. They can be applied only on pairs of branches that contain no leafish nodes. Rule 13 decreases the size of $L_p(T)$ and Rule 14 decreases the length-sum of branches without increasing $|L_p(T)|$.

Rule 13. Precondition: T has two leaves $l_1, l_2 \in L_p(T)$ and two non-tree edges (l_1, x) and (l_2, y) such that $x \in br(l_2)$ and $y \in br(l_1)$. **Action:** Let $E(T') = E(T) + (l_1, x) - (x, x^{\to}(l_1))$. (See Fig. 2(a)).

Rule 14. Precondition: T has two leaves $l_1, l_2 \in L_p(T)$ such that $b(l_1) = b(l_2)$, $d_T(b(l_1)) = 3$ and $(b^-(l_1), b^-(l_2))$ is a non-tree edge. **Action:** Let $E(T') = E(T) + (b^-(l_1), b^-(l_2)) - (b(l_2), b^-(l_2))$. (See Fig. 2(b)).

Definition 1. *A spanning tree T is a* locally optimal spanning tree (LOST) *if none of Rules 1–14 can be applied on it.*

Now we build an approximation algorithm for the MAXIST problem using the above improvement rules.

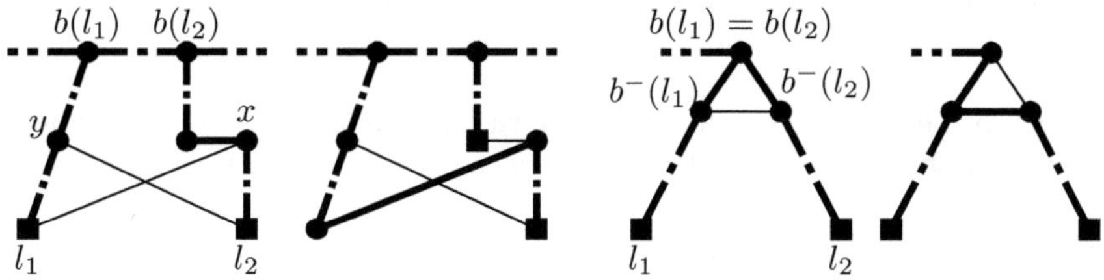

(a) Rule 13 (b) Rule 14

Fig. 2. Local improvement steps for creating a LOST, cont'd

Algorithm LOST. Create a DFS tree. Then apply Rules 1–14 as long as possible. If several rules can be applied, execute the one with the lowest number.

Theorem 1. *Algorithm LOST is an $\mathcal{O}(|V|^4)$ time $\frac{7}{4}$-approximation for the* MaxIST *problem in graphs that have no degree 1 nodes.*

The proof of the running time is omitted here because of space restrictions. We only mention that each application of Rules 1–12 decreases the number of leaves, thus these rules are applied at most $\mathcal{O}(|V|)$ times. Rules 13 and 14 do not change the number of leaves but decrease the sum $\sum_{l \in L_{\mathrm{p}}(T)} |br(l)|$. Hence these rules are applied at most $\mathcal{O}(|V|^2)$ times. The approximation ratio is established by the following lemma.

Lemma 1. *Let T be a LOST of a graph G that has no degree 1 nodes, and let T^* be a spanning tree of G with a maximum number of internal nodes. Then $\frac{|I(T^*)|}{|I(T)|} \leq \frac{7}{4}$.*

Proof. First observe some basic properties of T which are immediate consequences of the definition of a LOST.

Property 1. $L(T)$ forms a G-independent set. (As Rule 1 is no more applicable on T).

Property 2. Let l be an x-supported leaf. Then $d_T(x^{\rightarrow l}) = d_T(b(l)^{\rightarrow x}) = 2$. Furthermore, both $x^{\rightarrow l}$ and $b(l)^{\rightarrow x}$ are G-independent from every leaf $l_2 \neq l$. (As Rules 2–5 are no more applicable on T).

Property 3. Let l be a short leaf. Then no two G-neighbors of l are T-neighbors. (As Rule 6 is no more applicable on T).

Property 4. Let l_1 be a long leaf. Then no leaf $l_2 \neq l_1$ is a G-neighbor of $b^-(l_1)$. (As Rule 7 is no more applicable on T).

Property 5. Let l_1 and l_2 be leaves. Then the l_1-leafish nodes are G-independent from l_2. (As Rule 10 is no more applicable on T).

Property 6. Let l_1 and l_2 be leaves. Then the l_1-leafish nodes are G-independent from the l_2-leafish nodes. (As Rule 11 is no more applicable on T).

Property 7. Let l_1 and l_2 be leaves. Then the l_1-leafish nodes are G-independent from $b^-(l_2)$. (As Rule 12 is no more applicable on T).

Property 8. Let l_1 and l_2 be long leaves such that their branches do not contain leafish nodes, and $(b^-(l_1), b^-(l_2))$ is a non-tree edge. Then $b(l_1) = b(l_2)$ and $d_T(b(l_1)) = 3$. (As Rules 8, 9, and 13 are no more applicable on T.) Moreover, as Rule 14 is no more applicable, T must have exactly 3 leaves. From this point we suppose that T has at least 4 leaves. (Since it is easy to see that a LOST with 3 leaves satisfies the approximation factor of 7/4.) As a result, $l_1, l_2 \in L_{\mathrm{p}}(T)$ implies $(b^-(l_1), b^-(l_2)) \notin E(G)$.

To prove the approximation ratio we use a primal-dual linear programming approach. Let us recall a formulation of the spanning tree polyhedron [4]:

$$\mathcal{SP}(G)=\{x \mid \forall S\subseteq V : x(S) \leq |S|-1, -x(V)\leq -(|V|-1), \forall e \in E : 0 \leq x(e)\},$$

where $x(S) = \sum_{e\in E(G[S])} x(e)$ is the sum of x over all edges spanned by S.

Let us consider the following linear program ($\delta(v)$ is the set of edges incident to v):

$$\begin{array}{lll} \text{maximize} & \displaystyle\sum_{v\in V} z(v) & \\ \text{subject to} & x \in \mathcal{SP}(G) & \\ & \displaystyle -\sum_{e\in\delta(v)} x(e) + z(v) \leq -1 & \text{for all } v \in V \\ & 0 \leq z(v) \leq 1 & \text{for all } v \in V \end{array}$$

We obtain a solution $\mathbf{P}$ of this primal problem by setting $x(e) = 1$ when e is an edge of T, and $z(v) = 1$ when v is an internal node of T. All other variables are 0. Since T is a spanning tree, it is easy to see that this solution is feasible and has a value of $\mathrm{val}(\mathbf{P}) = |I(T)|$.

The dual of the above program is:

$$\begin{array}{lll} \text{minimize} & \displaystyle\sum_{S\subseteq V}(|S|-1)y(S) - (|V|-1)t - \sum_{v\in V} w(v) + \sum_{v\in V} r(v) & \\ \text{subject to} & \displaystyle\sum_{e\in E(G[S])} y(S) - t - \sum_{e\in\delta(v)} w(v) \geq 0 & \text{for all } e \in E \qquad (1) \\ & w(v) + r(v) \geq 1 & \text{for all } v \in V \\ & y(S), t, w(v), r(v) \geq 0 & \text{for all } S \subseteq V, v \in V \end{array}$$

We consider two different dual solutions corresponding to T, of which the first one is used in case of many short branches, and the second one is used in case of many long branches.

Let l be a short leaf. We define the set

$$Q(l) = \{x^{\to l} \mid \exists x : (l, x) \in E(G) \setminus E(T)\} \cup \{b(l)^{\to x} \mid \exists x : (l, x) \in E(G) \setminus E(T)\}.$$

Note that $Q(l) \neq \emptyset$, since as $d_G(l) \geq 2$. Let $Q = \cup_{l \in L_s(T)} Q(l)$.

The first dual solution $\mathbf{D}_1$ is constructed as follows. Let $y(V) = 1$, $y(Q) = 1$, $w(v) = 1$ for each $v \in L(T) \cup Q$, and $r(v) = 1$ for each $v \in V \setminus (L(T) \cup Q)$. All other variables are set to 0.

To see the feasibility of this solution, it is enough to check (1) for all edges of G. As $y(V) = 1$, only the edges of $G[L(T) \cup Q]$ could violate the inequality. However, by Property 1, there is no edge spanned by $L(T)$, and by Properties 2 and 3, there is no edge between $L(T)$ and Q. Thus $y(Q) = 1$ ensures the feasibility.

Let us define $c_1 = \frac{|Q|}{|I(T)|}$. The value of this dual solution is

$$\begin{aligned}\mathrm{val}(\mathbf{D}_1) = |V| - 1 + |Q| - 1 - |L(T)| - |Q| + |V| - |L(T)| - |Q| \\ = 2(|I(T)| - 1) - |Q| < (2 - c_1)|I(T)|.\end{aligned}$$

To construct the second dual solution $\mathbf{D}_2$ we denote the set of l-leafish nodes by $F(l)$, and define the set of leafish nodes to be $F = \cup_{l \in L_g(T)} F(l)$. Recall that $L_p(T)$ denotes the set of long leaves that have no leafish node in their branch. We use B_p^- to denote the set $\cup_{l \in L_p(T)} b^-(l)$. We immediately obtain $|B_p^-| = |L_p(T)|$.

Dual variables are set as: $y(V) = 1$, $y(F(l)+l) = 1$ for each $l \in L_g(T) \setminus L_p(T)$, $y(\{l, b^-(l)\}) = 1$ for each $l \in L_p(T)$, $w(v) = 1$ for each $v \in (L(T) \cup F \cup B_p^-)$, $r(v) = 1$ for each $v \in V \setminus (L(T) \cup F \cup B_p^-)$. All other variables are set to 0.

To see the feasibility of this solution, it is enough to check (1) for all edges of G. As $y(V) = 1$, only the edges of $G[(L(T) \cup F \cup B_p^-)]$ could violate the inequality. However, by Properties 1 and 4–8, the graph $G[(L(T) \cup F \cup B_p^-)]$ has no edge between different branches of T. On the other hand, the edges of $G[(L(T) \cup F \cup B_p^-)]$ within a single branch of T are also covered by some set S with $y(S) = 1$.

Let us define $c_2 = \frac{|V| - |L_s(T)|}{|I(T)|}$. Then as $|I(T)| = |V| - |L_s(T)| - |L_g(T)|$ we obtain $|L_g(T)| = (c_2 - 1)|I(T)|$.

The value of the solution is

$$\begin{aligned}\mathrm{val}(\mathbf{D}_2) = |V| - 1 + |F| + |L_p(T)| - |L(T)| - |F| - |L_p(T)| \\ + |V| - |L(T)| - |F| - |L_p(T)| < 2|I(T)| - |F| - |L_p(T)| \\ \leq 2|I(T)| - |L_g(T)| = (3 - c_2)|I(T)|.\end{aligned}$$

Here we have used that $|L_g(T)| \leq |F| + |L_p(T)|$.

Let $\mathbf{OPT}$ be the optimal LP-solution. Then $\mathrm{val}(\mathbf{P}) \leq |I(T^*)| \leq \mathrm{val}(\mathbf{OPT}) \leq \min(\mathrm{val}(\mathbf{D}_1), \mathrm{val}(\mathbf{D}_2))$. This gives

$$\frac{|I(T^*)|}{|I(T)|} \leq \min(2 - c_1, 3 - c_2). \tag{2}$$

Now let $N(L_s(T))$ be the set of G-neighbors of short leaves. Observe that by the definition of Q, each element of $N(L_s(T))$ has a T-neighbor in Q. Moreover, by Property 2, all nodes of Q are forwarding nodes of T. Thus

$$|N(L_s(T))| \leq 2|Q| = 2c_1|I(T)|. \tag{3}$$

Let us remark that the condition $d_G(l) \geq 2$ is used here to ensure that the set $Q(l)$ is not empty, for each leaf l. This latter fact is necessary to upper bound $|N(L_s(T))|$ by a function of $|Q|$.

Let us recall that the *scattering number* [5] of a non-complete graph is

$$\mathrm{sc}(G) = \max\left\{\mathrm{comp}(G[V \setminus X]) - |X| : \emptyset \subset X \subset V, \mathrm{comp}(G[V \setminus X]) \geq 2\right\}, \tag{4}$$

where $\mathrm{comp}(G[V \setminus X])$ is the number of components of $G[V \setminus X]$.

Salamon and Wiener [6] proved that $\mathrm{sc}(G) + 1 \leq \mathrm{ml}(G)$, where $\mathrm{ml}(G)$ is the minimum number of leaves in a spanning tree.

As the short leaves of T are G-independent, $G[V - N(L_s(T))]$ has at least $|L_s(T)|$ components implying that $\mathrm{sc}(G) \geq |L_s(T)| - |N(L_s(T))|$.

Thus, by (3) and (4), we have

$$\begin{aligned} |I(T^*)| = |V| - \mathrm{ml}(G) &\leq |V| - \mathrm{sc}(G) \\ &\leq |V| - |L_s(T)| + |N(L_s(T))| \leq (c_2 + 2c_1)|I(T)|. \end{aligned} \tag{5}$$

If $c_1 \geq \frac{1}{4}$ or $c_2 \geq \frac{5}{4}$ then by (2), otherwise by (5), we obtain $\frac{|I(T^*)|}{|I(T)|} \leq \frac{7}{4}$. □

4 Maximum Weighted Internal Spanning Tree

Let $G = (V, E)$ be a graph without degree 1 nodes and with a non-negative weight-function $c : V \longrightarrow \mathbb{Q}_+$ on its nodes. The MaxWIST problem aims to find a spanning tree T of G that maximizes the sum $c(I(T)) = \sum_{v \in I(T)} c(v)$. Obviously, this problem is NP-hard, as it contains the unweighted version, the MaxIST problem. In this section we present a $(2\Delta - 3)$-approximation algorithm for general graphs which is further refined to get a 2-approximation algorithm for claw-free graphs. The algorithms are based on local improvement steps.

4.1 General Graphs

Let us consider an arbitrary spanning tree T of G. In order to get a good approximation we apply local improvement rules as long as possible. Each such rule either creates an internal node from a leaf or it replaces a leaf with one of strictly smaller weight. The weighted sum of leaves decreases in both cases. Similarly to the unweighted case, a rule has a precondition and an action part.

Rule 15. Precondition: T has an x-supported leaf l such that $d_T(x^{\to l}) > 2$.
Action: Let $E(T') = E(T) + (l, x) - (x, x^{\to l})$ (Fig. 3(a)).

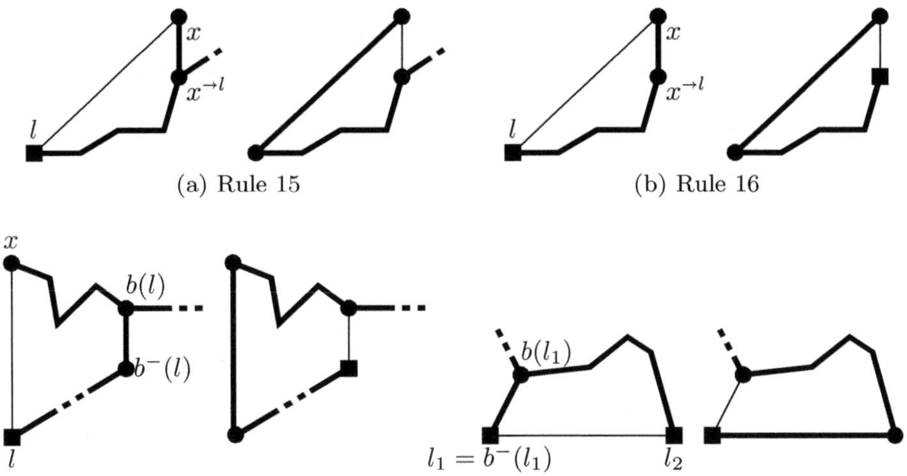

Fig. 3. Local improvement rules for creating a WLOST

Rule 16. Precondition: T has a leaf l, and there is a node x such that (l, x) is a non-tree edge, $d_T(x^{\to l}) = 2$, and $c(x^{\to l}) < c(l)$. **Action:** Let $E(T') = E(T) + (l, x) - (x, x^{\to l})$ (Fig. 3(b)).

Rule 17. Precondition: T has an x-supported leaf l such that $c(b^-(l)) < c(l)$ **Action:** Let $E(T') = E(T) + (l, x) - (b(l), b^-(l))$ (Fig. 3(c)).

Rule 18. Precondition: T has a short leaf l_1 and a leaf l_2 such that $(l_1, l_2) \in E(G)$. **Action:** Let $E(T') = E(T) + (l_1, l_2) - (b(l_1), b^-(l_1))$ (Fig. 3(d)).

Definition 2. *A spanning tree T is a* weighted locally optimal spanning tree (WLOST) *if none of Rules 15–18 can be applied on it.*

We can construct an approximation algorithm for the MaxWIST problem using the above rules.

Algorithm WLOST. Create an arbitrary spanning tree. Then apply Rules 15–18 as long as possible. If several rules can be applied, execute the one with the lowest number.

Theorem 2. *Algorithm WLOST is an $\mathcal{O}(|V|^4)$ time $(2\Delta - 3)$-approximation for the* MaxWIST *problem in graphs that have no degree 1 nodes.*

The proof of the running time is omitted here due to space restrictions. The following lemma proves the approximation ratio.

Lemma 2. *Let T be a WLOST of a node-weighted graph $G = (V, E, c)$ that has no degree 1 nodes. Then $(2\Delta - 3)c(I(T)) \geq c(V)$.*

Proof. First, observe some basic properties of T that are immediate consequences of the definition of a WLOST.

Property 9. If l is a leaf of T and (l, x) is a non-tree edge then $d_T(x^{\to l}) = 2$ and $c(x^{\to l}) \geq c(l)$. (As Rules 15 and 16 are no more applicable on T).

Property 10. If l is a supported leaf of T then $c(b^-(l)) \geq c(l)$. (As Rule 17 is no more applicable on T).

Property 11. If l is a short leaf of T then l is G-independent from all other leaves of T. (As Rule 18 is no more applicable on T).

We define a mapping $f : L(T) \longrightarrow I(T)$ as follows: for a leaf l, let x be the node such that (l, x) is a non-tree edge and the length of path $P_T(l, x)$ is the maximum possible. Such an x always exists as $d_G(l) \geq 2$. If $br(l)$ is a short branch or $x \in br(l)$ then let $f(l) = x^{\to l}$. Otherwise, let $f(l) = b^-(l)$. This means that each long leaf l is mapped to its own branch, thus the images of long leaves are disjoint. By the above properties, it is easy to see that for every leaf l, we have $c(f(l)) \geq c(l)$, and $d_T(f(l)) = 2$.

The mapping f can be used to establish the approximation ratio. Unfortunately, several short leaves can be mapped to the same node y. The following proposition is used to upper bound the number of such leaves.

Proposition 1. *For any node $y \in V$, we have $|\{l : y = f(l)\}| \leq 2(\Delta - 2)$.*

Proof. Let $l_1, l_2, \ldots, l_r$ be leaves of T with $f(l_i) = y$ $(1 \leq i \leq r)$. As shown above, $d_T(y) = 2$. Let y_1 and y_2 be the T-neighbors of y. Recall that a long leaf is mapped to a node of its own branch. This implies that neither y_1 nor y_2 is a leaf of T. Supposing the contrary, namely e.g. $d_T(y_1) = 1$, y must be in the long branch of y_1, and at least $r - 1$ of the branches $br(l_i)$ $(1 \leq i \leq r)$ must be short. Moreover, by the definition of the mapping f, graph G must have edges (l_i, y_1) that is a contradiction to Property 11. Hence, at least 2 edges incident to y_1 (and y_2, respectively) are tree edges, and so at most $\Delta - 2$ are non-tree edges. This shows that $r \leq 2(\Delta - 2)$, yielding the proposition. □

Using Proposition 1, and the fact that the images of long leaves are G-independent, we obtain

$$\sum_{v \in L(T)} c(v) = \sum_{l \in L_g(T)} c(l) + \sum_{l \in L_s(T)} c(l)$$
$$\leq \sum_{l \in L_g(T)} c(f(l)) + 2(\Delta - 2) \sum_{l \in L_s(T)} c(f(l)) \leq 2(\Delta - 2) \sum_{v \in I(T)} c(v). \quad (6)$$

Hence, $\sum_{v \in V} c(v) \leq (2\Delta - 3) \sum_{v \in I(T)} c(v)$, proving the approximation ratio. □

4.2 Claw-Free Graphs

In this subsection we extend the above algorithm by an additional improvement step in order to get a 2-approximation algorithm for the MAXWIST problem in claw-free graphs—that is, graphs without induced $K_{1,3}$. Observe that (6) would guarantee an approximation ratio of 2 if we could find a WLOST without short branches. The new improvement rule does exactly this job, namely it converts short branches to long ones. Throughout this subsection, G is supposed to be claw-free. First we point out a property of WLOSTs of claw-free graphs.

Let T be a WLOST of G, and l be a short leaf of T. Furthermore, let $l = x_0, x_1, x_2, \ldots, x_k$ are the T-neighbors of $b(l)$. Then by Property 9, none of $x_1, x_2, \ldots, x_k$ is a G-neighbor of l. Thus, nodes $x_1, x_2, \ldots, x_k$ must span a complete subgraph of G, since otherwise $b(l), l, x_i, x_j$ would induce a $K_{1,3}$ for some i, j. As a result—using Property 9—all the nodes x_i are internal nodes of T for $i \geq 1$. Thus we have

Property 12. If l is a short leaf of T and the T-neighbors of $b(l)$ are $x_0 = l, x_1, \ldots, x_k$, then $x_1, x_2, \ldots, x_k$ are all internal nodes of T and they induce a complete subgraph of G.

Using this property, we can now give the additional improvement rule to decrease the number of short branches, while not changing the set of leaves.

Rule 19. Precondition: T has a short leaf l, and the T-neighbors of $b(l)$ are $x_0 = l, x_1, \ldots, x_k$ such that for some $1 \leq i \leq k$ the node x_i is a branching, or x_i has a T-neighbor $v_i \neq b(l)$ which is an internal node. **Action:** Let $E(T') = E(T) \setminus \{(b(l), x_j)\}_{j=1..k, j\neq i} \cup \{(x_i, x_j)\}_{j=1..k, j\neq i}$.

Definition 3. *A WLOST is called a* refined WLOST (RWLOST) *if Rule 19 cannot be applied on it.*

Using the above Rule 19, we can improve Algorithm WLOST to obtain a better approximation ratio for claw-free graphs.

Algorithm RWLOST. Create an arbitrary spanning tree. Then apply Rules 15–18 and Rule 19 as long as possible. If several rules can be applied, execute the one with the lowest number.

Theorem 3. *Algorithm WLOST is an $\mathcal{O}(|V|^4)$ time 2-approximation for the* MAXWIST *problem in claw-free graphs that have no degree 1 nodes.*

The proof is omitted here because of space restrictions. We only mention that the core idea is to prove that an RWLOST has no short branches. Then (6) can be used to show the approximation ratio.

Acknowledgment. Author thanks to anonymous referees, Gábor Wiener and Dániel Marx for their valuable comments that helped a lot to improve the presentation of the paper.

References

1. Lu, H.I., Ravi, R.: The power of local optimization: Approximation algorithms for maximum-leaf spanning tree (DRAFT). Technical Report CS-96-05, Department of Computer Science, Brown University, Providence, Rhode Island (1996)

2. Salamon, G., Wiener, G.: On finding spanning trees with few leaves (submitted 2006)
3. Solis-Oba, R.: 2-approximation algorithm for finding a spanning tree with maximum number of leaves. In: Bilardi, G., Pietracaprina, A., Italiano, G.F., Pucci, G. (eds.) ESA 1998. LNCS, vol. 1461, pp. 441–452. Springer, Heidelberg (1998)
4. Schrijver, A.: 50: Shortest spanning trees. In: Combinatorial optimization. In: Polyhedra and efficiency, vol. B, pp. 855–876. Springer, Heidelberg (2003)
5. Zhang, S., Wang, Z.: Scattering number in graphs. Networks 37, 102–106 (2001)
6. Salamon, G., Wiener, G.: Leaves of spanning trees and vulnerability. In: The 5th Hungarian-Japanese Symposium on Discrete Mathematics and Its Applications, pp. 225–235 (2007)

New Approximability Results for 2-Dimensional Packing Problems*

Klaus Jansen[1] and Roberto Solis-Oba[2]

[1] Institut für Informatik, Universität zu Kiel, Kiel, Germany
[2] Department of Computer Science, University of Western Ontario, London, Canada

Abstract. The strip packing problem is to pack a set of rectangles into a strip of fixed width and minimum length. We present asymptotic polynomial time approximation schemes for this problem without and with 90° rotations. The additive constant in the approximation ratios of both algorithms is 1, improving on the additive term in the approximation ratios of the algorithm by Kenyon and Rémila (for the problem without rotations) and Jansen and van Stee (for the problem with rotations), both of which have a larger additive constant $O(1/\varepsilon^2)$, $\varepsilon > 0$.

The algorithms were derived from the study of the rectangle packing problem: Given a set R of rectangles with positive profits, the goal is to find and pack a maximum profit subset of R into a unit size square bin $[0,1] \times [0,1]$. We present algorithms that for any value $\epsilon > 0$ find a subset $R' \subseteq R$ of rectangles of total profit at least $(1-\epsilon)OPT$, where OPT is the profit of an optimum solution, and pack them (either without rotations or with 90° rotations) into the augmented bin $[0,1] \times [0,1+\epsilon]$.

1 Introduction

Recently, there has been a lot of interest in two-dimensional packing problems, like strip packing [2,12,15,16], two-dimensional bin packing [1], and rectangle packing [2,4,5,9,10]. These problems play an important role in diverse applications like cutting stock, VLSI layout, image processing, internet advertisement, and multiprocessor scheduling. Let $R = \{R_1, \ldots, R_n\}$ be a set of n rectangles; rectangle R_i has width $w_i \in (0,1]$, length $\ell_i \in (0,1]$ and profit $p_i \in \mathbb{R}^+$. The *rectangle-packing problem* is to pack a maximum profit subset $R' \subseteq R$ into a square bin B, $[0,1] \times [0,1]$. We only allow orthogonal packings, i.e., the rectangles must not overlap and their sides must be parallel to the sides of the bin. In the *strip packing* problem the goal is to pack all the rectangles R into a strip of unit width and minimum length. We consider two variants of these problems: without rotations and with 90° rotations.

The rectangle packing problem (with and without rotations) is known to be strongly NP-hard even for the restricted case of packing squares with unit profits

* Research supported in part by the EU project AEOLUS contract number 015964, the Natural Science and Engineering Research Council of Canada grant R3050A01, and by the DAAD German Academic Exchange Service.

L. Kučera and A. Kučera (Eds.): MFCS 2007, LNCS 4708, pp. 103–114, 2007.

[14]. Jansen and Zhang [10] designed a polynomial time approximation scheme (PTAS) for packing squares with unit profits in a rectangular bin and a $(2+\epsilon)$-approximation algorithm for packing rectangles with arbitrary profits [9], for any $\epsilon > 0$. Recently, Fishkin et al. [5] gave an algorithm for the *rectangle packing problem with resource augmentation*, that packs a subset of rectangles with profit at least $(1-\epsilon)OPT$ into an augmented square bin $[0, 1+\epsilon] \times [0, 1+\epsilon]$, where OPT is the maximum profit of any subset of rectangles that can be packed into a unit size square bin. For the rectangle packing problem restricted to squares, the best known algorithm [7] has performance ratio $\frac{5}{4}+\epsilon$, for any $\epsilon > 0$.

The strip packing problem (without and with rotations) is also strongly NP-hard [6]. The currently best approximation algorithms for the strip packing problem without rotations have absolute performance ratio 2 [15,16] and asymptotic performance ratio $1+\epsilon$, for any $\epsilon > 0$ [12]. For strip packing with rotations the best approximation algorithm has absolute performance ratio 2 [16] and asymptotic performance ratio $1+\epsilon$, for any $\epsilon > 0$ [8]. The asymptotic fully polynomial time approximation schemes (AFPTAS) by Kenyon and Rémila [12] and Jansen and van Stee [8] compute, respectively, strip packings without rotations and with 90° rotations of total length at most $(1+\epsilon)OPT + O(1/\epsilon^2)$ for any $\epsilon > 0$.

The first main result presented in this paper is the following:

Theorem 1. *Let $R = \{R_1, \ldots, R_n\}$ be a set of rectangles with positive profits. There are polynomial time algorithms that select and pack (without or with 90° rotations) a subset $R' \subseteq R$ of rectangles into a rectangular bin $[0,1] \times [0, 1+\epsilon]$, for any constant $\epsilon > 0$. The profit of R' is at least $(1-\epsilon)OPT$, where OPT is the maximum profit of any subset of rectangles that can be packed (either without rotations or with 90° rotations) into a unit size square bin $[0,1] \times [0,1]$.*

An interesting feature of our algorithms is that they only need to augment the length of the bin, while its width does not need to be changed. We consider this result to be an important step towards the solution of other 2-dimensional packing problems without resource augmentation. A related result was recently obtained by Bansal and Sviridenko [1] for two dimensional bin packing, where the goal is to pack a set of rectangles into the minimum number of unit size square bins. They designed an algorithm that packs the rectangles into $OPT + O(1)$ bins of size $[0,1] \times [0, 1+\epsilon]$, for any value $\epsilon > 0$, where OPT is the minimum number of unit size square bins needed to pack the rectangles. This algorithm rounds up, both, the widths and lengths of large rectangles in a similar fashion as in [12]. This rounding causes an increase in the dimensions of the bins and in the number of bins. For our rectangle packing problem we cannot round the width and length of the large rectangles, as this would require us to increase both dimensions of the bin.

Using our rectangle packing algorithm, we obtain the following results:

Theorem 2. *There are asymptotic polynomial time approximation schemes (APTAS) for the strip packing problems without rotations and with 90° rotations that produce solutions of length at most $(1+\epsilon)OPT_{SP} + 1$, where OPT_{SP} is the value of the optimum solution for each problem.*

Our additive constant 1 greatly improves on the $O(1/\epsilon^2)$ additive constant of [12] and [8]. Unfortunately, this is obtained at the expense of a higher running time. A similar situation happened with the bin packing problem (special case of strip packing with unit length rectangles): Karp and Karmarkar [11] designed an algorithm using $(1+\epsilon)OPT + O(1/\epsilon^2)$ bins and Fernandez de la Vega and Lueker [3] gave a slower algorithm that uses only $(1+\epsilon)OPT + 1$ bins.

We also show that the strip packing problem with 90° rotations for instances with optimum value $OPT_{SP} \geq 1$ has a PTAS.

Theorem 3. *There is a polynomial time approximation scheme (PTAS) for the strip packing problem with* 90° *rotations on instances with optimum value* $OPT_{SP} \geq 1$.

In this paper we describe the algorithms for the variant without rotations. For the case with 90° rotations we refer to the full version.

2 Near-Optimum Packings with a Simple Structure

We show that there are near-optimum solutions with a simple structure, described in Corollary 1. This greatly simplifies the search space for the problem, as our algorithm only needs to consider feasible solutions with that structure.

2.1 Partitioning the Set of Rectangles

Let $\epsilon' = \min\{1/2, \epsilon/4\}$, where ϵ is the desired precision for the solution. We assume that $1/\epsilon'$ is an even integer. Consider an optimum solution S^* for the rectangle packing problem. Let R^* be the subset of rectangles chosen by S^*, and let $OPT = \sum_{R_i \in R^*} p_i$.

Let $\sigma_1 = \epsilon'$ and $\sigma_k = (\sigma_{k-1})^{9/\sigma_{k-1}^2}$ for all integers $k \geq 2$. For each $k \geq 2$, we define the following sets: $\mathcal{R}_k^* = \{R_i \in R^* \mid w_i \in (\sigma_k, \sigma_{k-1}] \text{ or } \ell_i \in (\sigma_k, \sigma_{k-1}]\}$. Each R_i belongs to at most two of these sets, so $\sum_{k \geq 2} \sum_{R_i \in \mathcal{R}_k^*} p_i \leq 2OPT$. Hence, there must be an index $\tau \in \{2, \ldots, 4/\epsilon' + 1\}$ such that *profit*$(\mathcal{R}_\tau^*) \leq (\epsilon'/2)OPT$, where *profit*$(\mathcal{R}_\tau^*) = \sum_{R_i \in \mathcal{R}_\tau^*} p_i$. Let

$$\delta = \sigma_\tau \quad \text{and} \quad s = 9/\delta^2 + 1. \tag{1}$$

We partition the rectangles in three groups according to their lengths: $\mathcal{L} = \{R_i \in R \mid \ell_i > \delta\}$, $\mathcal{H} = \{R_i \in R \mid \ell_i \leq \delta^s\}$, and $\mathcal{M}_\ell = \{R_i \in R \mid \ell_i \in (\delta^s, \delta]\}$. Then, we consider the widths of the rectangles and partition them into three additional groups: $\mathcal{W} = \{R_i \in R \mid w_i > \delta\}$, $\mathcal{N} = \{R_i \in R \mid w_i \leq \delta^s\}$, and $\mathcal{M}_w = \{R_i \in R \mid w_i \in (\delta^s, \delta]\}$. The rectangles in $\mathcal{L}$, $\mathcal{H}$, $\mathcal{W}$, and $\mathcal{N}$ are called long, short, wide, and narrow, respectively (thus, there are long-wide, long-narrow, short-wide, and short-narrow rectangles). We also define $\mathcal{L}^* = \mathcal{L} \cap R^*$, $\mathcal{H}^* = \mathcal{H} \cap R^*$, $\mathcal{M}_\ell^* = \mathcal{M}_\ell \cap R^*$, $\mathcal{W}^* = \mathcal{W} \cap R^*$, $\mathcal{N}^* = \mathcal{N} \cap R^*$, and $\mathcal{M}_w^* = \mathcal{M}_w \cap R^*$.

Since $\delta = \sigma_\tau$ and $\delta^s = \sigma_{\tau+1}$, then $\mathcal{M}_\ell^* \cup \mathcal{M}_w^* = \mathcal{R}_\tau^*$. Discard the rectangles in $\mathcal{M}_\ell^* \cup \mathcal{M}_w^*$ from the optimum solution, creating a gap $(\sigma_{\tau+1}, \sigma_\tau]$ between the

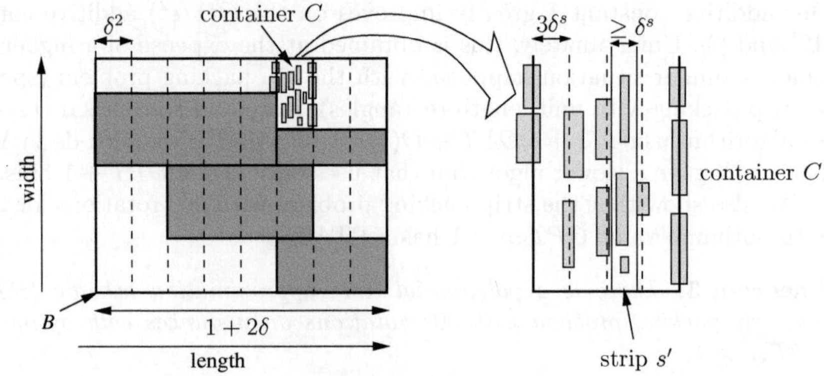

Fig. 1. Shifting the rectangles and discarding short rectangles crossing slot boundaries. Long rectangles appear in darker shade.

lengths of the rectangles in $\mathcal{L}^*$ and $\mathcal{H}^*$, and between the widths of the rectangles in $\mathcal{W}^*$ and $\mathcal{N}^*$. This separation between $\mathcal{L}^*$ and $\mathcal{H}^*$, and $\mathcal{W}^*$ and $\mathcal{N}^*$ is critical as it will allow us to deal independently with rectangles from different groups.

Even when an optimum subset R^* of rectangles is not known, we can still assume that the value of τ is known, because there is only a constant number, $4/\epsilon'$, of possible values for τ. Thus, our algorithm tries all these values (increasing its time complexity by only a constant factor); among all the solutions computed, the algorithm chooses one with maximum profit.

2.2 Rounding the Lengths of the Long Rectangles

We round up the length of each long rectangle $R_i \in R^*$ to the nearest multiple of δ^2. Then, we set the origin of a Cartesian system of coordinates at the left-bottom corner of the bin. In S^* we shift the rectangles horizontally to the right until all the long rectangles have their corners placed at points (x, y) where the x coordinates are multiples of δ^2 (See Figure 1).

Since long rectangles have length at least δ, these transformations increase the length of the packing by at most $\frac{1}{\delta}(2\delta^2) = 2\delta$. Accordingly, let us increase the length of the bin B to $1 + 2\delta$. This rounding and shifting limits the set of possible lengths and positions for the long rectangles in R^*.

2.3 Containers for Short Rectangles

Draw vertical lines spaced by δ^2 across the bin (the left and right sides of each long rectangle lie along two of these lines). These lines split the bin into at most $(1 + 2\delta)/\delta^2$ vertical strips, that we call *slots*. A *container* C is the rectangular region in a slot whose upper boundary is either the lower side of a long rectangle or the upper side of the bin, and whose lower boundary is either the upper side of a long rectangle or the lower side of the bin (see Figure 1).

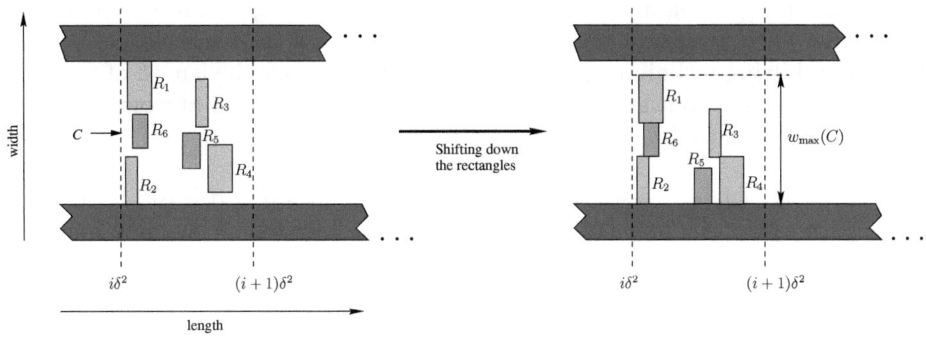

Fig. 2. (a) A container C. $R_1 - R_6$ are short-wide rectangles. (b) Container after shifting down the short-wide rectangles.

Consider a container C in S^* as shown in Figure 1. Note that the two vertical sides of a container might be crossed by short rectangles. We remove those rectangles through the following shifting technique. First, allocate all small rectangles crossing the border of two containers C and C' to the rightmost container. Then, divide C into $\delta^2/(3\delta^s) > 1/(4\epsilon')$ vertical strips of length $3\delta^s$. There must be a strip s' for which the total profit of the short rectangles completely contained in the strip is at most $\frac{\epsilon'}{4}OPT(C)$, where $OPT(C)$ is the total profit of the rectangles assigned to C in S^*. Remove all short rectangles completely contained in s', creating in C an empty vertical gap of length at least δ^s. Now, we move all small rectangles crossing the left boundary of C to this empty gap (See Figure 1). By performing the above process over all containers we loose profit at most $\frac{\epsilon'}{4}OPT$, but now no short rectangle crosses the boundary of a container.

For the sequel we only consider containers with at least one short-wide rectangle inside them. Notice that any packing has at most $O(1/\delta^3)$ such containers. Consider a container C. Remove all short-narrow rectangles from C. Compute the y-coordinate of the bottom side of each short-wide rectangle packed in C and then sort the rectangles in non-decreasing order of these y-coordinates. Take the rectangles in this order and shift them down until they touch either the bottom of C or the top of another rectangle. This yields a new feasible packing (see Figure 2). Let $w_{\max}(C)$ be the width of this new packing. Note that $w_{\max}(C)$ is the sum of the widths of at most $1/\delta$ short-wide rectangles.

Consider again the original packing for the short rectangles in C. Round the width of C down to the nearest value of the form $w_{\max}(C) + i\delta^s$, for i integer. This limits the number of possible widths for the containers; however, now some of the rectangles previously packed in C might not fit anymore. To ensure that all these rectangles still fit, we increase the length of the container.

Lemma 1. *Let S_c^* be the set of rectangles packed by S^* in C. S_c^* can be packed in a container of width w' and length $\delta^2 + 2\delta^4$, where w' is the width of C rounded down to the nearest value of the form $w_{\max}(C) + i\delta^s$, for an integer $i \leq n$.*

By Lemma 1 every container C storing short-wide rectangles can be replaced by a container C' of length at most $\delta^2 + 2\delta^4$ and width of the form $w_{\max}(C) + i\delta^s$. The same shifting technique described above can decrease the length of C' down to δ^2 while losing profit at most $4\delta^2 OPT(C)$. If we do this on all containers, we loose in total profit of value at most $4\delta^2 OPT \leq (\epsilon'/4)OPT$.

Corollary 1. *There is a set R^+ of rectangles of total profit at least $(1-\epsilon')OPT$ and a packing S^+ for them in a bin of width 1 and length $1 + 2\delta$ such that*

- *every long rectangle in R^+ has its length rounded up to the nearest multiple of δ^2 and its left side is at a position x that is a multiple of δ^2, and*
- *each container C storing at least one short-wide rectangle has length δ^2 and width $w_{\max}(C) + i\delta^s$, where $w_{\max}(C)$ is the sum of widths of at most $1/\delta$ short-wide rectangles and $i \leq n$ is a non-negative integer.*

3 Rectangle Selection

As described in Section 2.1, our algorithm tries all possible values $\{2, 3, \ldots, 4/\epsilon'+1\}$ for τ. Consider one of these and define the sets $\mathcal{L}$, $\mathcal{M}_\ell$, $\mathcal{H}$, $\mathcal{W}$, $\mathcal{M}_w$, and $\mathcal{N}$ as described above. Assume that for this choice of τ, *profit*$(\mathcal{M}_\ell \cup \mathcal{M}_w) \leq (\epsilon'/2)OPT$. Set $\mathcal{M}_\ell \cup \mathcal{M}_w$ is discarded.

3.1 Selecting Long Rectangles

Let R^* be the set of rectangles selected by S^*. We round up the length of each long rectangle to the nearest multiple of δ^2. For any constant $K > 0$, let L_K^* be the set of the K long rectangles in R^* of largest profit. Let $\bar{L}_K^* = (R^* \cap \mathcal{L}) \setminus L_K^*$, and let $\bar{L}_{Ki}^*$ be the subset of $\bar{L}_K^*$ formed by rectangles of length $i\delta^2$, for $i = 1/\delta, 1/\delta + 1, \ldots, 1/\delta^2$. Let $w(\bar{L}_{Ki}^*)$ be the total width of the rectangles in $\bar{L}_{Ki}^*$.

For any constant K, the set L_K^* is, of course, not known. However, since $|L_K^*|$ is constant, our algorithm can construct all $O(n^K)$ subsets of K long rectangles from R; clearly, one of these sets must be equal to L_K^*. For each possible L_K^*, we can find good approximations for $w(\bar{L}_{Ki}^*)$ and $\bar{L}_{Ki}^*$ as follows.

- If $|\bar{L}_{Ki}^*| \leq 1/\delta^4$, our algorithm will simply try all $O(n^{1/\delta^4})$ different subsets of at most $1/\delta^4$ long rectangles of length $i\delta^2$. One of these sets will be $\bar{L}_{Ki}^*$.
- If $|\bar{L}_{Ki}^*| > 1/\delta^4$, the algorithm considers all subsets of R with $1/\delta^4$ rectangles of length $i\delta^2$. One of them will coincide with the set S_{Ki}^* of $1/\delta^4$ widest rectangles in $\bar{L}_{Ki}^*$. Let $R_\ell^* \in S_{Ki}^*$ have lowest profit (at most δ^4*profit*$(\bar{L}_{Ki}^*)$) and use as approximations for $w(\bar{L}_{Ki}^*)$ values of the form $\varpi + kw(R_\ell^*)$, where $\varpi = \sum_{j=1}^{1/\delta^4} w(R_j^*)$, $k \in \{0, \ldots, n - 1/\delta^4\}$, and $w(R_j^*)$ is the width of R_j^*. Note that $w(\bar{L}_{Ki}^*) \in [\varpi + xw(R_\ell^*), \varpi + (x+1)w(R_\ell^*)]$ for some integer $x \leq n - \delta^{-4}$. If we remove R_ℓ^* from $\bar{L}_{Ki}^*$, the total width of the rectangles in $\bar{L}_{Ki}^*$ is at most $\varpi + xw(R_\ell^*)$ and their profit is at least $(1-\delta^4)$*profit*$(\bar{L}_{Ki}^*)$.

Corollary 2. *For each $i = 1/\delta, 1/\delta+1, \ldots, 1/\delta^2$, we can find in polynomial time a set Λ_i of $O(n^{1/\delta^4+1})$ values of the form $\varpi + xw_\ell$, where ϖ is the sum of widths of at most $1/\delta^4$ rectangles of length $i\delta^2$, w_ℓ is the width of a rectangle in S of minimum profit, and $x \in \{0, 1, \ldots, n - 1/\delta^4\}$. There is a value $\varpi^* + x^* w_\ell^* \in \Lambda_i$, such that $\varpi^* + x^* w_\ell^* \leq w(\bar{L}_{Ki}^*) \leq \varpi^* + (x^* + 1)w_\ell^*$, where w_ℓ^* is the width of a rectangle of profit at most δ^4*profit$(\bar{L}_{Ki}^*)$.

For each set $\bar{L}_{Ki}^*$ there are at most $O(n^{1/\delta^4+1})$ possible values for $w(\bar{L}_{Ki}^*)$ in Λ_i. We try all these values, and since there are fewer than $1/\delta^2$ sets $\bar{L}_{Ki}^*$, the total number of possible widths that we need to try for them is $O\left((n^{1+1/\delta^4})^{1/\delta^2}\right)$, which is (a huge) polynomial in n. For each possible $w(\bar{L}_{Ki}^*)$ we need to select a subset of rectangles of at most this width and high profit to pack in our solution. To do this selection we use the knapsack algorithm of [13] on the set of rectangles of length $i\delta^2$: the desired precision used in the knapsack algorithm is δ, the width of each rectangle is used as its size, and the guessed sum of widths is the knapsack's capacity. Among all $O(n^{1/\delta^4+1})$ sets selected, at least one of them must have profit at least $(1 - \delta)(1 - \delta^4)$*profit*$(\bar{L}_{Ki}^*)$ and total width no larger than $w(\bar{L}_{Ki}^*)$. Therefore, one of these selections will include rectangles of total profit at least $(1-\delta)(1-\delta^4)\sum_{i=1}^{1/\delta^2}$ *profit*$(\bar{L}_{Ki}^*) \geq (1-2\delta)\sum_{i=1}^{1/\delta^2}$ *profit*$(\bar{L}_{Ki}^*)$.

3.2 Selecting Short-Wide and Short-Narrow Rectangles

By Corollary 1 there is a near optimum solution S^+ with at most δ^{-3} containers where short-wide rectangles are packed. The length of each container C is δ^2 and its width is of the form $w_{\max}(C) + i\delta^s$. Our algorithm builds packings with $0, 1, \ldots, \delta^{-3}$ containers, and for each container we try all its $O(n^{1/\delta+1})$ possible widths. Clearly, one of these sets of containers is identical to that in S^+. Consider such a choice of containers and a selection S of long rectangles such that (a) the profit of S is at least $(1-\epsilon)$ times the profit of the long rectangles in S^+ and (b) the total width of the long rectangles of length $i\delta^2$ in S is no larger than the total width of the corresponding long rectangles in S^+, for all $i = 1/\delta, 1/\delta+1, \ldots, 1/\delta^2$.

Let A_{sn} be the area of the bin of width 1 and length $1 + 2\delta$ minus the area of the rectangles in S and the area of the containers. We choose a set of short-narrow rectangles to pack outside containers by using the knapsack algorithm of [13] with precision ϵ, and using the area of each rectangle as its size and A_{sn} as the capacity of the knapsack. The profit of this set of rectangles is at least $(1-\epsilon)$ times the profit of short-narrow rectangles packed by S^+ outside containers.

Next, we need to choose short-wide and short-narrow rectangles to be packed inside the containers. To do this we use the algorithm in [4]; this algorithm can select and pack a near-optimum profit set of rectangles into a container of width at least $1/(\epsilon')^4$ times its length. The algorithm in [4] was designed to pack rectangles in a single container, but a straightforward extension allows it to consider a constant number of containers. The total profit of the short rectangles selected by the knapsack algorithm and by the algorithm in [4] is at least $(1-\epsilon')$ times the total profit of the short rectangles in S^+.

4 Positioning Long Rectangles and Containers

To determine the positions for the rectangles, we first split the bin into vertical slots of length δ^2. Let L be the set of rectangles selected by our algorithm. Let C be one of the containers selected by our algorithm and let S_C be the set of short-wide and short-narrow rectangles that the algorithm packed in C. Replace in L all rectangles in S_C by a new rectangle R_C of the same width and length as C. The same is done for all containers, so L consists only of long rectangles, rectangles for containers, and short-narrow rectangles.

Let $L' \subseteq L$ be the rectangles corresponding to the containers and the K long rectangles of highest profit in L. A *slot assignment* for L' is a mapping $f : L' \rightarrow 2^M$ where $M = \{1, \ldots, (1+2\delta)/\delta^2\}$ is the set of slots. For each $R_j \in L'$, $f(R_j)$ is a consecutive set of γ_j slots, where $\gamma_j\delta^2$ is the length of R_j. Since the number of different mappings f is polynomial, $O(n^{(1+2\delta)/\delta^2})$, we consider all mappings f, and for each one we try to find a packing for L that is consistent with f. If no packing is found, then L is discarded and a different set of rectangles is selected as described above. At the end, the packing with the largest profit is produced.

Consider a packing S for L' consistent with f. A *snapshot* is any set of rectangles from L' that is intersected by a horizontal line (see Figure 3). Every $R_j \in L'$ appears in a sequence of consecutive snapshots $SHOT(\alpha_j), \ldots, SHOT(\beta_j)$. We index the snapshots from the bottom to the top of the bin as shown in Figure 3. For example, in Figure 3, R_2 appears in snapshots 2 and 3, so $\alpha_2 = 2$ and $\beta_2 = 3$.

Partition $L \setminus L'$ into two groups: the short-narrow rectangles L_{sn}, and the long rectangles L_{lo}. For packing S, let $m'(i)$ be the slots occupied by the rectangles from L' in $SHOT(i)$. $M \setminus m'(i)$ is the set of *free slots.* Let $\mathcal{F}$ be the family formed by all possible subsets of free slots.

For each set $F \in \mathcal{F}$ of free slots, a *configuration* (SN, Π) is a tuple where SN is a subset of F and Π is a partition of $F \setminus SN$ into sets of consecutive slots.

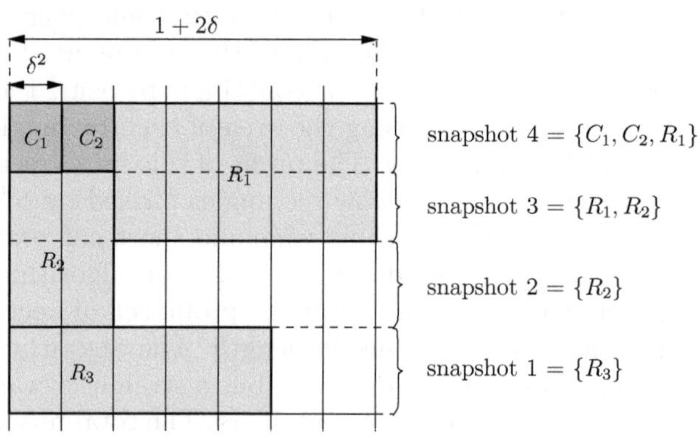

Fig. 3. Packing for a set of rectangles and containers, and the induced snapshots

Let $c_{F,i}$, $i = 1, \ldots, n_F$, denote all possible configurations for F, and let n_F be the number of these configurations. Given a configuration $c_{F,i} = (SN_{F,i}, \Pi_{F,i})$, $SN_{F,i}$ is reserved to pack rectangles from L_{sn}; every subset of slots $F' \in \Pi_{F,i}$ of cardinality ℓ is reserved for packing long rectangles from L_{lo} of length $\ell\delta^2$. Let $n_{F,i}(\ell)$ be the number of subsets of cardinality ℓ in $\Pi_{F,i}$.

To pack L_{sn} and L_{lo}, we first use a linear program to assign them to slots. In this linear program, a variable $x_{F,i}$ is used for each $c_{F,i}$ to denote the total width of the snapshots where free slots are allocated according to $c_{F,i}$. Hence, the area reserved by $c_{F,i}$ to pack short-narrow rectangles is $|SN_{F,i}|\delta^2 x_{F,i}$. Let $W_i(L_{lo})$ be the total width of the long rectangles of length $i\delta^2$ in L_{lo}, for all $i = 1/\delta, \ldots, 1/\delta^2$, and let A_{sn} be the total area of the short-narrow rectangles in L_{sn}. Since $|L'| \le K + \delta^{-3}$, the number of snapshots is at most $g = 2(K + \delta^{-3})$.

For each $R_i \in L'$ we try all possible values for α_i and β_i. Since L' has a constant number of rectangles, there is only a constant number (at most $g^{2(K+\delta^{-3})}$) of possible assignments of starting and ending snapshots for the rectangles. Let f be a slot assignment for L' and let α, β be assignments of starting and ending snapshots. The following linear program allocates rectangles to slots and snapshots. Variable t_i is the sum of widths of the first i snapshots; e_F is the total width of the snapshots where the set of slots not occupied by L' is F.

$$
\begin{array}{lll}
\mathbf{LP}(\mathbf{f}, \alpha, \beta): & t_0 = 0, t_g \le 1 & \\
& t_i \ge t_{i-1} & i = 1, \ldots, g \\
& t_{\beta_j} - t_{\alpha_j - 1} = w_j & \forall R_j \in L' \\
& \sum_{i:F=M\setminus m'(i)} (t_i - t_{i-1}) = e_F & \forall F \in \mathcal{F} \\
& \sum_{i=1}^{n_F} x_{F,i} \le e_F & \forall F \in \mathcal{F} \\
& \sum_{F\in\mathcal{F}} \sum_{i=1}^{n_F} n_{F,i}(\ell) x_{F,i} \ge W_\ell(L_{lo}) & \ell = 1, \ldots, |M| \\
& \sum_{F\in\mathcal{F}} \sum_{i=1}^{n_F} |SN_{F,i}|\delta^2 x_{F,i} \ge A_{sn} & \\
& x_{F,i} \ge 0 & \forall F \in \mathcal{F}, i = 1, \ldots, n_F.
\end{array}
$$

If $LP(f, \alpha, \beta)$ has no feasible solution, then we discard f, α, and β.

5 Generating a Packing

Let (t^*, e^*, x^*) be a feasible solution for $LP(f, \alpha, \beta)$. For simplicity, let us remove all snapshots $[t_i^*, t_{i+1}^*)$ of zero width. Let g^* be the number of remaining snapshots. Without loss of generality we may assume that all the configurations in which the set of free slots is $F = \{1, \ldots, M\}$ appear in the last snapshot $[t_{g^*-1}^*, t_{g^*}^*)$; otherwise, we can simply shift these configurations there.

5.1 Packing the Long Rectangles

Each $R_i \in L'$ is placed in the snapshots $f(R_i)$ so its bottom is at distance $t_{\alpha_{i-1}}^*$ from the bottom of the bin. Notice that no rectangle from L' is split. To pack the other long rectangles L_{lo}, consider each snapshot $[t_a^*, t_{a+1}^*)$, starting with $[t_0^* = 0, t_1^*)$. For each $[t_a^*, t_{a+1}^*)$, $m'(a+1)$ is the set of slots used by L'; we consider all the configurations $c_{F,i}$ with $x_{F,i} > 0$ corresponding to $F = M \setminus m'(a+1)$,

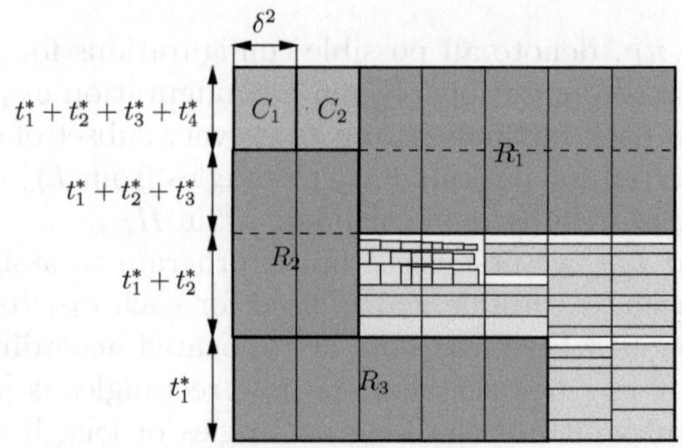

Fig. 4. Packing long rectangles and short narrow rectangles into the snapshots. $R_1, R_2, R_3 \in L'$; C_1 and C_2 are containers.

ordered so that all configurations $c_{F,i} = (SN, \Pi)$ with the same set SN appear in consecutive positions. This will ensure that a contiguous block of $|SN|$ slots will be available inside the snapshot to process short-narrow rectangles.

Let $\mathcal{R}_\ell = \{R_{\ell,1}, \ldots, R_{\ell,n_\ell}\} \subseteq L_{lo}$ be the long rectangles of length $\ell\delta^2$, for every $\ell = 1, \ldots, 1/\delta^2$. Let y^*_{a+1} be the width of snapshot $[t^*_a, t^*_{a+1})$. Take the first configuration $c_{F,i} = (SN_{F,i}, \Pi_{F,i})$ in the above ordering, for which $x^*_{F,i} > 0$. Select for each set $X \in \Pi_{F,i}$, the first not-yet (completely) packed rectangle $R_{\ell,j} \in \mathcal{R}_\ell$ with $\ell = |X|$. This rectangle is packed inside snapshot $[t^*_a, t^*_{a+1})$ in the set of slots X. $R_{\ell,j+1}, R_{\ell,j+2}, \ldots$ are packed in the slots X inside snapshot $[t^*_a, t^*_{a+1})$ until either their total width is at least $y^* = \min(x^*_{F,i}, y^*_{a+1})$, or all rectangles in $\mathcal{R}_\ell$ are packed. If the total width of the rectangles is larger than y^*, the last rectangle is split so that the width of the rectangles is exactly y^*.

This process is repeated for all sets $X \in \Pi_{F,i}$. If $x^*_{F,i} < y^*_{a+1}$, we set $y^*_{a+1} \leftarrow y^*_{a+1} - x^*_{F,i}$; then we consider the next configuration $c_{F,i'}$ with $x_{F,i'} > 0$ and pack long rectangles as described above. Otherwise, we set $x^*_{F,i} \leftarrow x^*_{F,i} - y^*_{a+1}$ and continue packing according to configuration $c_{F,i}$ in the next snapshot $[t^*_{a+1}, t^*_{a+2})$.

5.2 Packing the Short-Narrow Rectangles

All configurations $c_{F,i}$ with the same component $SN_{F,i}$ within an interval $[t^*_a, t^*_{a+1})$ leave a reserved area of total width $|SN_{F,i}|\delta^2$ for short-narrow rectangles. This reserved area gets split by L' and L_{lo} into at most $|M|/2 + 1$ rectangular blocks $B_1, B_2, \ldots, B_k$. Our algorithm leaves empty those blocks B_j of width $b < 4\delta^{s-1}$. Let B_j have length $d\delta^2$ and width $b \geq 4\delta^{s-1}$. Take short-narrow rectangles off L_{sn} an put them in a set S until their total area is at least $d\delta^2 b$. We use the First Fit Decreasing Width (FFDW) algorithm [2] to pack S into block B_j.

Lemma 2. *[2] Let S' be a set of rectangles, each of length and width at most Δ. FFDW can pack these rectangles in a rectangular bin of length 1 and width $FFDW(S') \leq AREA(S')(1 + \Delta) + \Delta$.*

Since all rectangles in S have width and length at most δ^s, FFDW can pack S into a bin of length $d\delta^2$ and width at most $AREA(S)(1 + \delta^s)/(d\delta^2) + \delta^s =$

$(d\delta^2 b + \delta^{2s})(1+\delta^s)/(d\delta^2) + \delta^s \le b + 3\delta^s$. Hence, we use FFDW to pack all the rectangles of S into a block of length $d\delta^2$ and width $b + 3\delta^s$. Then, divide the block into horizontal strips of width $4\delta^s$, partitioning S into at least $1/\delta$ disjoint groups (if a rectangle from S intersects two strips, we consider that it belongs to the strip that is above). We remove the lowest profit group, which has profit at most $\delta \times \mathit{profit}(S)$, so the remaining rectangles fit in block B_j.

5.3 Analysis of the Algorithm

Observe that our algorithm might not produce a valid packing for the rectangles in L_{lo} since a subset, SPLIT, of them might have been split into several pieces. Let us remove the subset SPLIT from the bin, thus obtaining a valid packing. The total profit lost by doing this is at most $3\epsilon' OPT$ (for details see our full paper). Short-narrow rectangles were packed in the blocks B_i as described in Section 5.2, but we do not pack anything in blocks B_i of width smaller than $4\delta^{s-1}$. The total space wasted in these blocks is at most δ. Hence, short-narrow rectangles of total area at most δ might not get packed. These rectangles are packed to the right of the bin using the FFDW algorithm. This increases the length of the bin by at most 2δ, so the total length of the bin is at most $1 + 4\delta$.

Now, we prove Theorem 1 for the case without rotations. Our algorithm produces a large number of packings, the best of which is finally selected. Consider this largest profit packing and the iteration of the algorithm in which it is computed. In Section 3.1 we select a set L_K of long rectangles of profit at least $(1-2\delta)p_L^+$, where p_L^+ is the profit of the long rectangles in S^+. In Section 3.2 we select and pack in the containers short rectangles of profit at least $(1-\epsilon')p_C^+$, where p_C^+ is the total profit of the short rectangles packed in containers in S^+. Then, a set of short-narrow rectangles of profit at least p_{sn}^+ is chosen, where p_{sn}^+ is the total profit of the short-narrow rectangles in S^+ packed outside containers.

In Section 5.1 the long rectangles selected are packed in the bin and a subset of profit at most $\epsilon' p_L^+$ is discarded. Thus, the total profit of the rectangles packed in the bin $[0,1] \times [0, 1+4\delta]$ is at least $(1-2\delta)p_L^+ + (1-\epsilon')p_C^+ + p_{sn}^+ - \epsilon' p_L^+ - \delta p_{sn}^+ \ge (1-3\epsilon')(p_L^+ + p_C^+ + p_{sn}^+) = (1-3\epsilon')\mathit{profit}(S^+) \ge (1-4\epsilon')OPT = (1-\epsilon)OPT$, where OPT is the profit of an optimum solution. The complexity of the algorithm is $O\left(n^{(4/\delta^2)^{5/\delta^2}}\right)$, where δ is a constant as defined in (1).

6 APTAS for Strip Packing Without Rotations

Use the algorithm of Steinberg [16] to pack R in a strip of length $v \le 2OPT_{SP}$. Set $\varepsilon \leftarrow \epsilon/5$. Consider all values $v' = v/2, (1+\varepsilon)v/2, (1+2\varepsilon)v/2, \ldots, v$. Note that for one of these values v^*, $OPT_{SP} \le v^* \le (1+\varepsilon)OPT_{SP}$. For each value v' we scale the lengths of the rectangles by $1/v'$ and define the profit of each rectangle as its area; then, we use our rectangle packing algorithm to pack a subset of scaled rectangles into a bin of unit width and length.

Observe that when $v' = v^*$ there is a way of packing all scaled rectangles into the augmented bin and, thus, by Theorem 1 for this value of v' our algorithm must be able to pack most of the rectangles; the rectangles that our algorithm cannot pack have total area at most ε. Find v^*, the smallest value v' such that in the packing produced by our algorithm the total area of the set S of un-packed rectangles is at most ϵ. Then, use the algorithm of Steinberg to pack S in a strip of width 1 and length $\epsilon + 1/v^*$, and append this to the solution produced above.

The total length of the packing produced by this algorithm is at most $v^*(1 + 2\varepsilon + 1/v^*) = (1+2\varepsilon)v^* + 1 \leq (1+\varepsilon)(1+2\varepsilon)OPT_{SP} + 1 \leq (1+5\varepsilon)OPT_{SP} + 1 = (1+\epsilon)OPT_{SP} + 1$, for $\varepsilon \leq 1$.

References

1. Bansal, N., Sviridenko, M.: Two-dimensional bin packing with one dimensional resource augmentation. Discrete Optimization 4, 143–153 (2007)
2. Coffman, E.G., Garey, M.R., Johnson, D.S., Tarjan, R.E.: Performance bounds for level-oriented two-dimensional packing algorithms. SIAM Journal on Computing 9, 808–826 (1980)
3. de la Vega, W.F., Lueker, G.S.: Bin packing can be solved within 1+epsilon in linear time. Combinatorica 1(4), 349–355 (1981)
4. Fishkin, A.V., Gerber, O., Jansen, K.: On weighted rectangle packing with large resources. In: Conference Theoretical Computer Science (TCS 2004), pp. 237–250 (2004)
5. Fishkin, A.V., Gerber, O., Jansen, K., Solis-Oba, R.: Packing weighted rectangles into a square. In: Jedrzejowicz, J., Szepietowski, A. (eds.) MFCS 2005. LNCS, vol. 3618, pp. 352–363. Springer, Heidelberg (2005)
6. Garey, M.R., D. S., Johnson.: Computers and Intractability: A guide to the theory of NP-completeness. W. H. Freeman and Company, San Francisco (1979)
7. Harren, R.: Approximating the orthogonal knapsack problem for hypercubes. In: Bugliesi, M., Preneel, B., Sassone, V., Wegener, I. (eds.) ICALP 2006. LNCS, vol. 4051, pp. 238–249. Springer, Heidelberg (2006)
8. Jansen, K., van Stee, R.: On strip packing with rotations. In: ACM Symposium on Theory of Computing. STOC 2005, pp. 755–761 (2005)
9. Jansen, K., Zhang, G.: On rectangle packing: maximizing benefits. In: ACM-SIAM Symposium on Discrete Algorithms. SODA 2004, pp. 197–206 (2004)
10. Jansen, K., Zhang, G.: Maximizing the number of packed rectangles. In: Hagerup, T., Katajainen, J. (eds.) SWAT 2004. LNCS, vol. 3111, pp. 362–371. Springer, Heidelberg (2004)
11. Karmarkar, M., Karp, R.M.: An efficient approximation scheme for the one-dimensional bin-packing problem. In: IEEE Symposium on Foundations of Computer Science. FOCS 1982, pp. 312–320 (1982)
12. Kenyon, C., Remila, E.: A near-optimal solution to a two-dimensional cutting stock problem. Mathematics of Operations Research 25, 645–656 (2000)
13. Lawler, E.: Fast approximation algorithms for knapsack problems. Mathematics of Operations Research 4, 339–356 (1979)
14. Leung, J.Y.-T., Tam, T.W., Wong, C.S., Young, G.H., Chin, F.Y.L.: Packing squares into a square. Journal Parallel and Dist. Computing 10, 271–275 (1990)
15. Schiermeyer, I.: Reverse-fit: a 2-optimal algorithm for packing rectangles. In: van Leeuwen, J. (ed.) ESA 1994. LNCS, vol. 855, pp. 290–299. Springer, Heidelberg (1994)
16. Steinberg, A.: A strip-packing algorithm with absolute performance bound two. SIAM Journal on Computing 26, 401–409 (1997)

On Approximation of Bookmark Assignments*

Yuichi Asahiro[1], Eiji Miyano[2], Toshihide Murata[2], and Hirotaka Ono[3]

[1] Department of Social Information Systems, Kyushu Sangyo University, Fukuoka 813-8503, Japan
asahiro@is.kyusan-u.ac.jp

[2] Department of Systems Innovation and Informatics, Kyushu Institute of Technology, Fukuoka 820-8502, Japan
miyano@ces.kyutech.ac.jp, hide_m@theory.ces.kyutech.ac.jp

[3] Department of Computer Science and Communication Engineering, Kyushu University, Fukuoka 819-0395, Japan
ono@csce.kyushu-u.ac.jp

Abstract. Consider a rooted directed acyclic graph $G = (V, E)$ with root r, representing a collection V of web pages connected via a set E of hyperlinks. Each node v is associated with the probability that a user wants to access the node v. The access cost is defined as the expected number of steps required to reach a node from the root r. A bookmark is an additional shortcut from r to a node of G, which may reduce the access cost. The bookmark assignment problem is to find a set of bookmarks that achieves the greatest improvement in the access cost. For the problem, the paper presents a polynomial time approximation algorithm with factor $(1-1/e)$, and shows that there exists no polynomial time approximation algorithm with a better constant factor than $(1-1/e)$ unless $\mathcal{NP} \subseteq \mathcal{DTIME}(N^{O(\log\log N)})$, where N is the size of the inputs.

1 Introduction

1.1 Motivation and Formulation

The world wide web is one of the most famous and the hugest repository of information, which consists of web pages and hyperlinks. The hyperlink is a reference from one page to another, and hence it gives us a non-linear searchability; a user can travel to access desired pages by starting from a home page, clicking correct hyperlinks, and displaying pages one after another. In order to display the target (desired or favorite) pages by clicking as few hyperlinks as possible, one of the most popular approaches is to add a "shortcut" linkage, usually called a *bookmark*.

The above natural solution for improving web access can be formulated as the following optimization problem, called the k-Bookmark Assignment Problem (k-BAP), which has been introduced in [19]: Let $G = (V, E)$ be a rooted,

* This work is partially supported by Grant-in-Aid for Scientific Research on Priority Areas No. 16092222 and 16092223, and by Grant-in-Aid for Young Scientists (B) No. 17700022, No. 18700014 and No. 18700015.

L. Kučera and A. Kučera (Eds.): MFCS 2007, LNCS 4708, pp. 115–124, 2007.

connected, directed acyclic graph (DAG for short) with root r, representing a web hierarchical structure. Let p_v be the *access probability* associated to a node $v \in V$, where $\sum_{v \in V} p_v = 1$ and each $p_v \geq 0$. For a node v, let $d(v)$ denote the *distance* from the root r to the node v, i.e., the length of the shortest directed path. The *expected number of steps* to reach a node from the root r is defined by $\sum_{v \in V} p_v \cdot d(v)$. A *bookmark to a node* v is an additional directed edge $(r, v) \notin E$. Only for simplicity, throughout the paper, we identify the bookmark (r, v) with the node v itself. Let $B = \{b_1, b_2, \ldots, b_k\}$, $b_i \in V$, be a set of bookmarks and also let $G \oplus B$ be the graph resulting from the assignment of B. The distance from the root r to a node v in $G \oplus B$ is denoted by $d_B(v)$. Thus, the expected number of steps to reach a node of $G \oplus B$ from the root is also defined by $\sum_{v \in V} p_v \cdot d_B(v)$. For simplicity, let $p_v \cdot d(v)$ and $p_v \cdot d_B(v)$ be represented by $c(v)$ and $c_B(v)$, respectively, in the following. A *gain* $g(B)$ of bookmarks B is defined as

$$g(B) = \sum_{v \in V} p_v \left(d(v) - d_B(v) \right) = \sum_{v \in V} \left(c(v) - c_B(v) \right),$$

and thus the problem is formulated as follows:

k-Bookmark Assignment Problem (k-BAP):
INSTANCE: A directed acyclic graph $G = (V, E)$, the access probability p_v for each node $v \in V$, and a positive integer $k \leq |V|$.
GOAL: Find a bookmark set $B \subseteq V$ of size k maximizing $g(B)$.

1.2 Previous and Our Results

The problem has been considered also in the context of locating web proxies in the Internet. For restricted network topologies, such as paths and trees, polynomial time exact algorithms are successfully developed: For paths, an $O(k|V|^2)$ time algorithm is proposed by Li, Deng, Golin, and Sohraby in [10] and for trees Li, Golin, Italiano, Deng and Sohraby proposed an $O(k^2|V|^3)$ time algorithm in [11]. In particular, Czyzowicz, Kranakis, Krizanc, Pelc, and Vergas Martin showed that if the input graph is a perfect binary tree and $k \leq \sqrt{|V|+1}$, the optimal set of bookmarks satisfies a good property [4] and a faster $O(|V|)$ time algorithm can be developed based on that property [19]. On the other hand, Vergas Martin shows that k-BAP is $\mathcal{NP}$-hard for directed acyclic graphs in [19]. As far as the authors know, there is no result with respect to approximation for k-BAP.

In this paper, we consider (in)approximability of k-BAP, and obtain the following results.

(Approximability). There is a polynomial time approximation algorithm with factor $(1 - 1/e)$.

(Inapproximability). There is no polynomial time approximation algorithm with a better constant factor than
- $(1 - 1/e)$ unless $\mathcal{NP} \subseteq \mathcal{DTIME}(N^{O(\log \log N)})$, where N is the size of the inputs, and

– $(1 - \delta)$ under a weaker assumption, i.e., $\mathcal{P} \neq \mathcal{NP}$, where δ is a fixed small positive constant.

An algorithm is called an *approximation algorithm with factor* α, or an α *factor approximation algorithm* for k-BAP, if $g(\mathsf{ALG})/g(\mathsf{OPT}) \geq \alpha$ holds for any instance G, where ALG and OPT are sets of bookmarks obtained by the algorithm and an optimal algorithm, respectively. α is called *approximation ratio.*

1.3 Related Work

There is a large amount of literature on the problem of improving the accessibility of the most popular pages by adding hyperlinks over the underlying structure (e.g., [3,6,13,17,18]). A variant of a bookmark is called a *hotlink*, which is defined as a hyperlink that links an *arbitrary* page to its descendant page. The problem variant is known as the Hotlink Assignment Problem (HAP), which has been introduced by Bose et al. in [3]. It is known that HAP is $\mathcal{NP}$-hard for arbitrary DAGs [3] but unknown for trees whether it is $\mathcal{NP}$-hard or not. Many researchers have studied on HAP in perfect binary tree structures [3,6]. Matichin and Peleg present a polynomial time 1/2 factor approximation algorithm for DAGs in [13], and also a 1/2 factor approximation algorithm for trees under some realistic access model in [14]. It is important to note that no result on inapproximability has been obtained so far for HAP.

1.4 Organization

In Section 2, we present the polynomial time approximation algorithm with factor $(1-1/e)$. The inapproximability of k-BAP is discussed in Section 3. Then we conclude in Section 4.

2 Approximation Guarantee of a Greedy Algorithm

Consider the following greedy algorithm for a DAG $G = (V, E)$, which selects k bookmarks in total by iteratively assigning a new bookmark that obtains the maximum gain:

Algorithm Greedy:
    $B = \emptyset$
    for $i = 1$ to k do
        select a bookmark $b \in V \setminus B$ that maximizes $g(B \cup \{b\})$
        $B = B \cup \{b\}$
    endfor
    output B

Theorem 1. *Algorithm* Greedy *achieves an approximation ratio of* $(1 - \frac{1}{e})$ *for* k*-BAP, and its running time is* $O(k|V||E|)$.

Proof. First we show the approximation ratio. The basic framework of the proof is the same as in [20]. That is, we utilize a result of Nemhauser et al. [15], which analyzes the approximation ratio of a greedy algorithm for a certain maximization problem whose cost is defined as a monotone submodular function. The submodularity and its monotonicity are defined as follows:

Definition 1. *Let S be a finite set, and $f : 2^S \to \mathbb{R}$ be a function with $f(\emptyset) = 0$. f is called* submodular *if for any sets $X, Y \subseteq S$,*

$$f(X \cup Y) + f(X \cap Y) \leq f(X) + f(Y).$$

f is called monotone *if for any set $X \subset S$ and $s \in S \setminus X$*

$$f(X \cup \{s\}) - f(X) \geq 0.$$

The result of Nemhauser et al. is that the problem of selecting k-element subsets maximizing a monotone submodular function can be approximated within a constant factor $(1 - \frac{1}{e})$ by a greedily picking algorithm [15]. By Proposition 2 (monotonicity) and Lemma 1 (submodularity) shown later, the gain function g is monotone submodular, and hence the approximation ratio of the above algorithm Greedy is also $(1 - \frac{1}{e})$.

As for the running time, Greedy selects k bookmarks by evaluating the values of $g(B \cup \{b\})$'s for each $b \in V$. Computing $g(B \cup \{b\})$ requires to solve the shortest path problem for the DAG $G \oplus (B \cup \{b\})$, which takes $O(|E|)$ time [5]. Since the number of possible bookmarks in each iteration is at most $|V|$ and the number of iterations is k, the total running time is $O(k|V||E|)$. □

Now we show that the gain function g is monotone and submodular. First we briefly note the monotonicity of g. The following property obviously holds.

Proposition 1. *For $S' \subseteq S \subseteq V$ and $u \in V$, $d_{S'}(u) \geq d_S(u)$ holds.* □

Then, from the above proposition, we can see that g is monotone in a straightforward way:

Proposition 2 (monotonicity). *For $S' \subseteq S \subseteq V$, $g(S') \leq g(S)$ holds.* □

We next prove the submodularity of the gain function g. Let $V(X)$ be a set of descendant nodes that are reachable from a set X of nodes in G.

Proposition 3. *For two subsets of nodes $X, Y \subseteq V$*

(i) $V(X) \cup V(Y) = V(X \cup Y)$
(ii) $V(X) \cap V(Y) \supseteq V(X \cap Y)$

Proof. (i) It is easy to see that $V(S') \subseteq V(S)$ holds for $S' \subseteq S \subseteq V$. Thus, $V(X) \cup V(Y) \subseteq V(X \cup Y)$ holds. We show that $V(X \cup Y) \subseteq V(X) \cup V(Y)$ also holds in the following: First, note that for $\forall u \in V(X \cup Y)$, there exists a node $v \in X \cup Y$ such that u is reachable from v, i.e., for $\forall u \in V(X \cup Y)$, u belongs to $V(X) \cup V(Y)$. It follows that $V(X \cup Y) \subseteq V(X) \cup V(Y)$ holds. As a result, $V(X \cup Y) = V(X) \cup V(Y)$ holds.

(ii) By a similar discussion, we can show that $V(X) \supseteq V(X \cap Y)$ and $V(Y) \supseteq V(X \cap Y)$ hold, which means that $V(X) \cap V(Y) \supseteq V(X \cap Y)$ is satisfied. □

Using Propositions 1, 2, and 3, we can show the submodularity of the gain function g:

Lemma 1 (submodularity). *The gain function $g : 2^V \to \mathbb{R}$ is submodular, i.e., for any subsets of nodes $X, Y \subseteq V$, $g(X \cup Y) + g(X \cap Y) \leq g(X) + g(Y)$ holds.*

Proof. Note that the following holds for a node set $S \subseteq V$ by definition:

$$g(S) = \sum_{v \in V(S)} (c(v) - c_S(v)).$$

Thus,

$$\begin{aligned}
& g(X) + g(Y) - (g(X \cup Y) + g(X \cap Y)) \\
= & \sum_{v \in V(X)} (c(v) - c_X(v)) + \sum_{v \in V(Y)} (c(v) - c_Y(v)) \\
& - \left(\sum_{v \in V(X \cup Y)} (c(v) - c_{X \cup Y}(v)) + \sum_{v \in V(X \cap Y)} (c(v) - c_{X \cap Y}(v)) \right) \\
= & \left(\sum_{v \in V(X) \cup V(Y)} c_{X \cup Y}(v) + \sum_{v \in V(X) \cap V(Y)} c(v) \right) \qquad (1) \\
& - \left(\sum_{v \in V(X)} c_X(v) + \sum_{v \in V(Y)} c_Y(v) \right) \qquad (2) \\
& - \sum_{v \in V(X \cap Y)} (c(v) - c_{X \cap Y}(v))
\end{aligned}$$

holds, where the last equality comes from Proposition 3 (i). By definition, the following two equations are satisfied:

$$\begin{aligned}
\sum_{v \in V(X)} c_X(v) &= \sum_{v \in V(X) \setminus V(Y)} c_X(v) + \sum_{v \in V(X) \cap V(Y)} c_X(v) \\
\sum_{v \in V(Y)} c_Y(v) &= \sum_{v \in V(Y) \setminus V(X)} c_Y(v) + \sum_{v \in V(X) \cap V(Y)} c_Y(v).
\end{aligned}$$

Also, as for the union $X \cup Y$ of bookmarks,

$$\begin{aligned}
& \sum_{v \in V(X) \cup V(Y)} c_{X \cup Y}(v) \\
= & \sum_{v \in V(X) \setminus V(Y)} c_{X \cup Y}(v) + \sum_{v \in V(Y) \setminus V(X)} c_{X \cup Y}(v) + \sum_{v \in V(X) \cap V(Y)} c_{X \cup Y}(v)
\end{aligned}$$

holds. Thus, the above two terms (1) and (2) can be replaced by the following terms:

$$\sum_{v \in V(X) \setminus V(Y)} (c_{X \cup Y}(v) - c_X(v)) + \sum_{v \in V(Y) \setminus V(X)} (c_{X \cup Y}(v) - c_Y(v))$$
$$+ \sum_{v \in V(X) \cap V(Y)} (c(v) + c_{X \cup Y}(v) - c_X(v) - c_Y(v)).$$

Note that even if we add bookmarks to nodes in Y, the length of the shortest path from r to a node v does not change if $v \in V(X) \setminus V(Y)$. Thus,

$$\sum_{v \in V(X) \setminus V(Y)} (c_{X \cup Y}(v) - c_X(v)) = 0.$$

From a similar reason,

$$\sum_{v \in V(Y) \setminus V(X)} (c_{X \cup Y}(v) - c_Y(v)) = 0.$$

As a result,

$$g(X) + g(Y) - (g(X \cup Y) + g(X \cap Y))$$
$$= \sum_{v \in V(X) \cap V(Y)} (c(v) + c_{X \cup Y}(v) - c_X(v) - c_Y(v)) \quad (3)$$
$$- \sum_{v \in V(X \cap Y)} (c(v) - c_{X \cap Y}(v)). \quad (4)$$

Since $c_{X \cup Y}(v) = \min\{c_X(v), c_Y(v)\}$ for $v \in V(X) \cap V(Y)$ from Proposition 3-(i), the term (3) can be replaced to the following:

$$\sum_{v \in V(X) \cap V(Y)} (c(v) - \max\{c_X(v), c_Y(v)\}). \quad (5)$$

Since $c_{X \cap Y}(u) \geq c_X(u)$ and $c_{X \cap Y}(u) \geq c_Y(u)$ hold for any vertex u from $X \cap Y \subseteq X$ and $X \cap Y \subseteq Y$, respectively, the inequality $c_{X \cap Y}(v) \geq \max\{c_X(v), c_Y(v)\}$ holds for $v \in V(X \cap Y)$. Hence the following inequality is true for the term (4):

$$\sum_{v \in V(X \cap Y)} (c(v) - c_{X \cap Y}(v)) \leq \sum_{v \in V(X \cap Y)} (c(v) - \max\{c_X(v), c_Y(v)\}). \quad (6)$$

From (5), (6), and Proposition 3-(ii), we can obtain the following:

$$g(X) + g(Y) - (g(X \cup Y) + g(X \cap Y))$$
$$\geq \sum_{v \in (V(X) \cap V(Y)) \setminus V(X \cap Y)} (c(v) - \max\{c_X(v), c_Y(v)\})$$
$$\geq 0.$$

Therefore, the gain function g is submodular. □

Before concluding this section, we mention a possibility of speeding the algorithm up. In Greedy, we need to compute the shortest path problem $O(|V|)$ times for each iteration. Although the shortest path of a directed acyclic graph can be computed in $O(|E|)$ time, it still might be time-consuming and actually many parts of the computation may be redundant; since most of the graph structure is preserved even if we add a bookmark, the shortest path computation can reuse the previous computation. This idea can be implemented by using several results from dynamic topological sort algorithms (cf., [12,1]), although we do not give the detail here because the main scope of the paper is the approximation factor.

Another issue of the running time is on the restriction of graph classes. One of the most common subclasses of directed acyclic graphs is a tree. In the case of trees, we actually can reduce the running time by the following: We prepare $q(u)$ and $w(u)$ for each node u to store the distance of u from the root r and $w(u) = \sum_{v \in V(u)} p_v$, respectively. Then, in each iteration of the for-loop, the algorithm processes the following operations:

1. Pick a node b which has the maximum gain $q(b) \cdot w(b)$ as a new bookmark,
2. Replace $q(u)$ with $q(u) - q(b) + 1$ for each node $u \in V(b)$, and
3. Replace $w(u)$ with $w(u) - w(b)$ for each node u on the simple path from r to b (exclusive).

It is easy to see that the above procedure maintains q and w correctly, and each step can be done in $O(|V|)$ time. Hence the total running time of Greedy for trees is reduced to $O(k|V|)$. It is much faster than the running time of the known exact algorithm in [11], $O(k^2|V|^3)$, although Greedy obtains only approximate solutions.

3 Lower Bounds

We show that the $(1 - \frac{1}{e})$ approximation ratio proved in the previous section is the best possible for k-BAP, in the sense that no approximation algorithm with factor $(1 - \frac{1}{e} + \varepsilon)$ exists for any $\varepsilon > 0$, unless $\mathcal{NP} \subseteq \mathcal{DTIME}(N^{O(\log\log N)})$. The hardness of approximation is shown via a *gap-preserving reduction* [2] from the Unit Cost Maximum Coverage Problem (UCMCP):

Unit Cost Maximum Coverage Problem (UCMCP):

INSTANCE: A collection of sets $S = \{S_1, S_2, \ldots, S_m\}$ with associated unit cost $c_i = 1$ for each $i = 1, 2, \ldots, m$ defined over a domain of elements $X = \{x_1, x_2, \ldots, x_n\}$ with associated unit weight $w_j = 1$ for each $j = 1, 2, \ldots, n$, and a positive integer $\ell \leq |S|$.

GOAL: Find a collection of sets $S' \subseteq S$ such that the total cost of sets in S' does not exceed ℓ, and the total weight of elements covered by S' is maximized.

Theorem 2 ([8]). *No approximation algorithm with approximation ratio better than $(1 - \frac{1}{e})$ exists for UCMCP unless $\mathcal{NP} \subseteq \mathcal{DTIME}(N^{O(\log\log N)})$.*

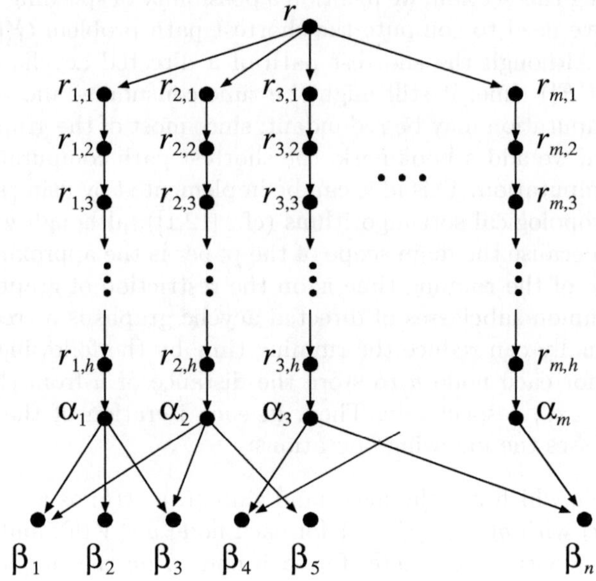

Fig. 1. Gap-preserving reduction

Let $OPT_{mcp}(I)$ denote the weight of elements covered by a collection of sets output by an optimal algorithm for the instance I of UCMCP. Also, let $OPT_{bap}(G)$ be the gain of bookmarks output by an optimal algorithm for the graph G of k-BAP.

Lemma 2. *There is a gap-preserving reduction from UCMCP to k-BAP that transforms an instance I of UCMCP to a graph $G = (V, E)$ of k-BAP such that*

(i) *if $OPT_{mcp}(I) = \mathtt{max}$, then $OPT_{bap}(G) \geq \frac{h}{n}\mathtt{max}$, and*
(ii) *if $OPT_{mcp}(I) \leq (1 - \frac{1}{e})\mathtt{max}$, then $OPT_{bap}(G) \leq (1 - \frac{1}{e} + \frac{\ell}{h \cdot \mathtt{max}})\frac{h}{n}\mathtt{max}$, where h is a sufficiently large integer such that $h = O(|I|^q)$ for some constant q.*

Proof. Consider an instance I of UCMCP; a collection of sets $S = \{S_1, S_2, \ldots, S_m\}$ defined over a domain of elements $X = \{x_1, x_2, \ldots, x_n\}$, and a positive integer ℓ. Then, we construct the following directed graph, illustrated in Figure 1. Let $V_R = \{r, r_{1,1}, r_{1,2}, \ldots, r_{1,h}, r_{2,1}, r_{2,2}, \ldots, r_{m,h-1}, r_{m,h}\}$ be a set of $1 + mh$ nodes associated with $|S| = m$. Also, let $V_S = \{\alpha_1, \alpha_2, \ldots, \alpha_m\}$ be a set of m nodes corresponding to the m sets, S_1 through S_m, and $V_X = \{\beta_1, \beta_2, \ldots, \beta_n\}$ be a set of n nodes corresponding to the n elements, x_1 through x_n. Here, V_R, V_S, and V_X are pairwise disjoint, and let $V = V_R \cup V_S \cup V_X$. The set E of directed edges is defined as follows: $E_1 = \{(r, r_{i,1}), (r_{i,1}, r_{i,2}), \ldots, (r_{i,h}, \alpha_i) \mid i = 1, 2, \ldots, m\}$, $E_2 = \{(\alpha_i, \beta_j) \mid x_j \in S_i$ for each i and each $j\}$, and $E = E_1 \cup E_2$. As for the access probabilities of nodes, we define $p_{\beta_j} = \frac{1}{n}$ for $j = 1, 2, \ldots, n$, $p_r = 0$, and $p_{r_{i,1}} = p_{r_{i,2}} = \cdots = p_{r_{i,h}} = p_{\alpha_i} = 0$ for $i = 1, 2, \ldots, m$. Finally, we set $k = \ell$. Clearly this reduction can be done in polynomial time since $h = O(|I|^q)$.

(i) Suppose that $OPT_{mcp}(I) = \mathtt{max}$ and the optimal collection of sets is $OPT = \{S_{i_1}, S_{i_2}, \ldots, S_{i_\ell}\}$ where $\{i_1, i_2, \ldots, i_\ell\} \subseteq \{1, 2, \ldots, m\}$. Then, if we select a set $B = \{\alpha_{i_1}, \alpha_{i_2}, \ldots, \alpha_{i_k}\}$ of k $(= \ell)$ nodes as bookmarks, then we can obtain the gain of $\frac{h}{n}\mathtt{max}$ since $d(\alpha_{i_j}) - d_B(\alpha_{i_j}) = h$ for each j and the nodes α_{i_j}'s are connected with exactly $\mathtt{max}$ leaves β's of probability $\frac{1}{n}$.

(ii) Next suppose that $OPT_{mcp}(I) \le (1 - \frac{1}{e})\mathtt{max}$ and again the optimal collection of sets is $OPT = \{S_{i_1}, S_{i_2}, \ldots, S_{i_\ell}\}$ where $\{i_1, i_2, \ldots, i_\ell\} \subseteq \{1, 2, \ldots, m\}$. If a set $B = \{\alpha_{i_1}, \alpha_{i_2}, \ldots, \alpha_{i_k}\}$ of k $(= \ell)$ nodes is selected as bookmarks, then the gain is at most $(1 - \frac{1}{e})\frac{h}{n}\mathtt{max}$. For example, the replacement of α_{i_j} with, say, $r_{i_j,h}$ decreases the gain by $\frac{1}{n}$ because $d(r_{i_j,h}) - d_B(r_{i_j,h}) = h - 1$. On the other hand, if we select, say, β_1 instead of α_{i_j} as a bookmark, then the gain possibly increases at most by $(h+1) \cdot \frac{1}{n} - h \cdot \frac{1}{n} = \frac{1}{n}$, i.e., at most $\frac{1}{n}$ gain per such replacement. As a result, the gain on G is at most $(1 - \frac{1}{e})\frac{h}{n}\mathtt{max} + \frac{k}{n} = (1 - \frac{1}{e} + \frac{\ell}{h \cdot \mathtt{max}})\frac{h}{n}\mathtt{max}$. This completes the proof. □

The following theorem is obtained by the above lemma:

Theorem 3. *No $(1 - \frac{1}{e} + \varepsilon)$ factor approximation algorithm exists for k-BAP unless $\mathcal{NP} \subseteq \mathcal{DTIME}(N^{O(\log\log N)})$, where ε is an arbitrarily small positive constant.*

A similar gap-preserving reduction from the Maximum k-Vertex Cover Problem [7,9,16] gives us the following hardness of approximation under the different weak assumption:

Theorem 4. *No $(1-\delta)$ factor approximation algorithm exists for k-BAP unless $\mathcal{P} = \mathcal{NP}$, where δ is a fixed small positive constant.*

Proof. Let $\delta_0 < 1$ be the approximation hardness factor of the Maximum k-Vertex Cover Problem. By using similar ideas in [2,9,7], we can provide the gap-preserving reduction with $\delta = \frac{1305}{1349} \cdot (1 - \delta_0)$. Details are omitted. □

4 Conclusion

In this paper we have considered the problem of assigning bookmarks from the root to the nodes of a DAG in order to maximize the gain in the expected cost. Then we have shown that there is a polynomial time approximation algorithm with factor $(1 - 1/e)$ and the factor is the best possible under the assumption that $\mathcal{NP}$ is not in $\mathcal{DTIME}(N^{O(\log\log N)})$.

As for further researches, there is a gap between the inapproximability result we have shown under the assumption that $\mathcal{P} \ne \mathcal{NP}$, and the approximability result of $(1 - 1/e)$ ratio. To reduce the time complexity for trees is another interesting topic.

References

1. Alpern, B., Hoover, R., Rosen, B.K., Sweeney, P.F., Zadeck, F.K.: Incremental evaluation of computational circuits. In: Proc. SODA'90, pp. 32–42 (1990)
2. Arora, S., Lund, C.: Hardness of Approximation. In: Hochbaum, D.S. (ed.) Approximation Algorithms for NP-hard problems, pp. 399–446. PWS publishing company (1995)
3. Bose, P., Kranakis, E., Krizanc, D., Vergas Martin, M., Czyzowicz, J., Pelc, A., Gasieniec, J.: Strategies for hotlink assignments. In: Lee, D.T., Teng, S.-H. (eds.) ISAAC 2000. LNCS, vol. 1969, pp. 23–34. Springer, Heidelberg (2000)
4. Czyzowicz, J., Kranakis, E., Krizanc, D., Pelc, A., Vergas Martin, M.: Assigning bookmarks in perfect binary trees. Ars Combinatoria, vol. LXXXII (2007)
5. Cormen, T.H., Leiserson, C.E., Rivest, R.L., Stein, C.: Introduction to Algorithms, 2nd edn. MIT Press, Cambridge (2001)
6. Fuhrmann, S., Krumke, S.O., Wirth, H.-C.: Multiple hotlink assignment. In: Brandstädt, A., Le, V.B. (eds.) WG 2001. LNCS, vol. 2204, pp. 189–200. Springer, Heidelberg (2001)
7. Feige, U., Langberg, M.: Approximation algorithms for maximization problems arising in graph partitioning. J. Algorithms 41(2), 174–211 (2001)
8. Khuller, S., Moss, A., Naor, J.: The budgeted maximum coverage problem. IPL 70(1), 39–45 (1999)
9. Langberg, M.: Approximation Algorithms for Maximization Problems arising in Graph Partitioning. M. Sc. thesis, Weizmann Institute of Science (1998)
10. Li, B., Deng, X., Golin, M.J., Sohraby, K.: On the optimal placement of web proxies in the Internet: Linear Topology. In: Proc. HPN'98, pp. 485–495 (1998)
11. Li, B., Golin, M.J., Italiano, G.F., Deng, X., Sohraby, K.: On the optimal placement of web proxies in the Internet. In: Proc. INFOCOMM'99, pp. 1282–1290 (1999)
12. Marchetti-Spaccamela, A., Nanni, U., Rohnert, H.: Maintaining a topological order under edge insertions. IPL 59(1), 53–58 (1996)
13. Matichin, R., Peleg, D.: Approximation algorithm for hotlink assignments in web directories. In: Dehne, F., Sack, J.-R., Smid, M. (eds.) WADS 2003. LNCS, vol. 2748, pp. 271–280. Springer, Heidelberg (2003)
14. Matichin, R., Peleg, D.: Approximation algorithm for hotlink assignment in the greedy model. In: Kralovic, R., Sýkora, O. (eds.) SIROCCO 2004. LNCS, vol. 3104, pp. 233–244. Springer, Heidelberg (2004)
15. Nemhauser, G., Wolsey, L., Fisher, M.: An analysis of the approximations for maximizing submodular set functions. Mathematical Programming 14, 265–294 (1978)
16. Petrank, E.: The hardness of approximation: gap location. Computational Complexity 4, 133–157 (1994)
17. Perkowitz, M., Etzioni, O.: Towards adaptive web sites: conceptual framework and case study. Computer Networks 31, 1245–1258 (1999)
18. Pessoa, A.A., Laber, E.S., de Souza, C.: Efficient algorithms for the hotlink assignment problem: the worst case search. In: Fleischer, R., Trippen, G. (eds.) ISAAC 2004. LNCS, vol. 3341, pp. 778–792. Springer, Heidelberg (2004)
19. Vergas Martin, M.: Enhancing Hyperlink Structure for Improving Web Performance. PhD thesis, School of Computer Science, Carleton University (2002)
20. Vohra, R.V., Hall, N.G.: A probabilistic analysis of the maximal covering location problem. Discrete Applied Mathmatics 43(2), 175–183 (1993)

Height-Deterministic Pushdown Automata

Dirk Nowotka[1] and Jiří Srba[2,*]

[1] Institut für Formale Methoden der Informatik
Universität Stuttgart, Germany
[2] BRICS**, Department of Computer Science
Aalborg University, Denmark

Abstract. We define the notion of height-deterministic pushdown automata, a model where for any given input string the stack heights during any (nondeterministic) computation on the input are a priori fixed. Different subclasses of height-deterministic pushdown automata, strictly containing the class of regular languages and still closed under boolean language operations, are considered. Several such language classes have been described in the literature. Here, we suggest a natural and intuitive model that subsumes all the formalisms proposed so far by employing height-deterministic pushdown automata. Decidability and complexity questions are also considered.

1 Introduction

Visibly pushdown automata [3], a natural and well motivated subclass of pushdown automata, have been recently introduced and intensively studied [8,2,4]. The theory found a number of interesting applications, e.g. in program analysis [1,9] and XML processing [10]. The corresponding class of visibly pushdown languages is more general than regular languages while it still possesses nice closure properties and the language equivalence problem as well as simulation/bisimulation equivalences are decidable [3,11]. Several extensions [7,5] have been proposed in order to preserve these nice properties while describing a larger class of systems. These studies have been particularly motivated by applications in the field of formal verification. However, unlike the natural model of visibly pushdown automata, these extensions are rather technical and less intuitive.

In this paper we suggest the model of height-deterministic pushdown automata which strictly subsumes all the models mentioned above and yet possesses desirable closure and decidability properties. This provides a uniform framework for the study of more general formalisms.

The paper is organized as follows. Section 2 contains basic definitions. Section 3 introduces height-deterministic pushdown automata, or *hpda*. It studies the languages recognized by real-time and deterministic *hpda*, and proves a number of interesting closure properties. Section 4 shows that these classes properly contain the language class of [7] and the classes defined in [3] and [5].

* Partially supported by the research center ITI, project No. 1M0021620808.
** Basic Research In Computer Science, Danish National Research Foundation.

L. Kučera and A. Kučera (Eds.): MFCS 2007, LNCS 4708, pp. 125–134, 2007.

2 Preliminaries

Let $\Sigma = \{a, b, c, \ldots\}$ be a finite set of *letters*. The set Σ^* denotes all finite words over Σ. The *empty word* is denoted by λ. A subset of Σ^* is called a *language*. Given a nonempty word $w \in \Sigma^*$ we write $w = w_{(1)}w_{(2)} \cdots w_{(n)}$ where $w_{(i)} \in \Sigma$ denotes the *i-th letter* of w for all $1 \leq i \leq n$. The *length* $|w|$ of w is n and $|\lambda| = 0$. By abuse of notation $|\cdot|$ also denotes the *cardinality* of a set, the *absolute value* of an integer, and the *size* of a pushdown automaton (see definition below). We denote by $\bullet w$ the word $w_{(2)}w_{(3)} \cdots w_{(n)}$, and define further $\bullet a = \lambda$ for every $a \in \Sigma$ and $\bullet\lambda = \lambda$. Finally, we let L^c abbreviate $\Sigma^* \setminus L$ for $L \subseteq \Sigma^*$.

Finite State Automata. A finite state automaton (*fsa*) R over Σ is a tuple $(S, \Sigma, s_0, \varrho, F)$ where $S = \{s, t, \ldots\}$ is a finite set of *states*, $s_0 \in S$ is the *initial state*, $\varrho \subseteq S \times \Sigma \times S$ is a set of *rules*, and $F \subseteq S$ is the set of final states. We call R a *deterministic finite state automaton* (*dfsa*) if for every $s \in S$ and every $a \in \Sigma$ there is exactly one $t \in \Sigma$ such that $(s, a, t) \in \varrho$, i.e., the relation ϱ can be understood as a function $\varrho \colon S \times \Sigma \to S$. Given a nonempty $w \in \Sigma^*$ we write $s \xrightarrow[R]{w} t$ (or just $s \xrightarrow{w} t$ if R is understood) if either $w \in \Sigma$ and $(s, w, t) \in \varrho$ or there exists an $s' \in S$ such that $(s, w_{(1)}, s') \in \varrho$ and $s' \xrightarrow{\bullet w} t$. We say that R *recognizes* the language $\mathcal{L}(R) = \{w \in \Sigma^* \mid s_0 \xrightarrow[R]{w} t, t \in F\}$. A language is *regular* if it is recognized by some *fsa*. The class of all regular languages is denoted by *REG*.

Finite State Transducers. A finite state transducer (*fst*) T from Σ^* to a monoid M (in this paper we have either $M = \Sigma^*$ or $M = \mathbb{Z}$), is a tuple $(S, \Sigma, M, s_0, \varrho, F)$ where $(S, \Sigma \times M, s_0, \varrho', F)$ is an *fsa* and $\varrho = \{(s, a, m, t) \mid (s, (a, m), t) \in \varrho'\}$. Given $w \in \Sigma^*$ and $m \in M$, we write $s \xrightarrow[T]{w,m} t$ (or $s \xrightarrow{w,m} t$ if T is understood) if either $w \in \Sigma$ and $(s, w, m, t) \in \varrho$ or if there exists an $s' \in S$ such that $(s, w_{(1)}, m_1, s') \in \varrho$, $s' \xrightarrow{\bullet w, m_2} t$ and $m = m_1 \oplus m_2$, where $\oplus$ is the operation associated with the monoid M. Given $L \subseteq \Sigma^*$, the *image of* L *under* T, denoted by $T(L)$, is the set of elements m such that $s_0 \xrightarrow[T]{w,m} t$ for some $t \in F$ and $w \in L$.

Pushdown Automata. A pushdown automaton (*pda*) A over an alphabet Σ is a tuple $(Q, \Sigma, \Gamma, \delta, q_0, F)$ where $Q = \{p, q, r, \ldots\}$ is a finite set of *states*, $\Gamma = \{X, Y, Z, \ldots\}$ is a finite set of *stack symbols* such that $Q \cap \Gamma = \emptyset$, $\delta \subseteq Q \times \Gamma \times (\Sigma \cup \{\varepsilon\}) \times Q \times \Gamma^* \cup Q \times \{\bot\} \times (\Sigma \cup \{\varepsilon\}) \times Q \times \Gamma^*\{\bot\}$ is a finite *set of rules*, where $\bot \notin \Gamma$ (empty stack) and $\varepsilon \notin \Sigma$ (empty input word) are special symbols, $q_0 \in Q$ is the *initial state*, and $F \subseteq Q$ is a set of *final states*. The *size* $|A|$ of a *pda* A is defined as $|Q| + |\Sigma| + |\Gamma| + \{|pXq\alpha| \mid (p, X, a, q, \alpha) \in \delta\}$. We usually write $pX \overset{a}{\underset{A}{\longmapsto}} q\alpha$ (or just $pX \overset{a}{\longmapsto} q\alpha$ if A is understood) for $(p, X, a, q, \alpha) \in \delta$. We say that a rule $pX \overset{a}{\underset{A}{\longmapsto}} q\alpha$ is a *push*, *internal*, or *pop* rule if $|\alpha| = 2, 1,$ or 0, respectively. A *pda* is called *real-time* (*rpda*) if $pX \overset{a}{\longmapsto} q\alpha$ implies $a \neq \varepsilon$. A *pda* is called *deterministic* (*dpda*) if for every $p \in Q$, $X \in \Gamma \cup \{\bot\}$ and $a \in \Sigma \cup \{\varepsilon\}$ we have (i) $|\{q\alpha \mid pX \overset{a}{\longmapsto} q\alpha\}| \leq 1$ and (ii) if $pX \overset{\varepsilon}{\longmapsto} q\alpha$ and $pX \overset{a}{\longmapsto} q'\alpha'$ then $a = \varepsilon$. A real-time deterministic pushdown automaton is denoted by *rdpda*.

The set $Q\Gamma^*\bot$ is the set of *configurations* of a *pda*. The configuration $q_0\bot$ is called *initial*. The transition relation between configurations is defined by: if

$pX \overset{a}{\longmapsto} q\alpha$, then $pX\beta \overset{a}{\longrightarrow} q\alpha\beta$ for every $\beta \in \Gamma^*$. A transition $p\alpha \overset{\varepsilon}{\longrightarrow} q\beta$ is called *ε-transition* or *ε-move*. The *labelled transition system* generated by A is the edge-labeled, directed graph $(Q\Gamma^*\bot, \bigcup_{a\in\Sigma\cup\{\epsilon\}} \overset{a}{\longrightarrow})$. Wherever convenient we use common graph theoretic terminology, like *(w-labeled) path* or *reachability*. Given $w \in \Sigma^*$, we write $p\alpha \overset{w}{\underset{A}{\Longrightarrow}} q\beta$ (or just $p\alpha \overset{w}{\Longrightarrow} q\beta$ if A is understood) if there exists a finite w'-labeled path from $p\alpha$ to $q\beta$ in A such that $w' \in (\Sigma\cup\{\varepsilon\})^*$ and w is the projection of w' onto Σ. We say that A is *complete* if $q_0\bot \overset{w}{\underset{A}{\Longrightarrow}} q\alpha$ for every $w \in \Sigma^*$. We say that A *recognizes* the language $\mathcal{L}(A) = \{w \in \Sigma^* \mid q_0\bot \overset{w}{\Longrightarrow} p\alpha, p \in F\}$. A language recognized by a *pda* (*dpda*, *rpda*, *rdpda*) is called (deterministic, real-time, real-time deterministic) *context-free* and the class of all such languages is denoted by *CFL*, *dCFL*, *rCFL*, and *rdCFL*, respectively.

Pushdown automata may reject a word because they get stuck before they read it completely, or because after reading it they get engaged in an infinite sequence of ε-moves that do not visit any final state. They may also scan a word and then make several ε-moves that visit both final and non-final states in arbitrary ways. Moreover, in a rule $pX \overset{a}{\longmapsto} q\alpha$ the word α can be arbitrary. For our purposes it is convenient to eliminate these "anomalies" by introducing a normal form.

Definition 1. *A pushdown automaton* $A = (Q, \Sigma, \Gamma, \delta, q_0, F)$ *is* normalized *if*

(i) A is complete;
(ii) for all $p \in Q$*, all rules in* δ *of the form* $pX \overset{a}{\longmapsto} q\alpha$ *either satisfy* $a \in \Sigma$ *or all of them satisfy* $a = \epsilon$*, but not both;*
(iii) every rule in δ *is of the form* $pX \overset{a}{\longmapsto} q\lambda$*,* $pX \overset{a}{\longmapsto} qX$*, or* $pX \overset{a}{\longmapsto} qYX$ *where* $a \in \Sigma \cup \{\varepsilon\}$*.*

States which admit only ε-transitions (see property (ii)), are called *ε-states*.

Lemma 1. *For every pda (dpda, rpda, rdpda) there is a normalized pda (dpda, rpda, rdpda, respectively), that recognizes the same language.*

3 Height Determinism

Loosely speaking, a *pda* is height-deterministic if the stack height is determined solely by the input word; more precisely, a *pda* A is height-deterministic if all runs of A on input $w \in (\Sigma \cup \{\varepsilon\})^*$ (here, crucially, ε is considered to be a part of the input) lead to configurations of the same stack height. Given two height-deterministic *pda* A and B, we call them synchronized if their stack heights coincide after reading the same input words (again, this includes reading the same number of ε's between two letters). The idea of height-determinism will be discussed more formally below.

Definition 2. *Let A be a pda over the alphabet* Σ *with the initial state* q_0*, and let* $w \in (\Sigma\cup\{\varepsilon\})^*$*. The set* $N(A, w)$ *of* stack heights reached by A after reading w *is* $\{|\alpha| \mid q_0\bot \overset{w}{\underset{A}{\longrightarrow}} q\alpha\bot\}$*. A* height-deterministic *pda (hpda) A is a pda that is*

(i) normalized, and
(ii) $|N(A,w)| \leq 1$ *for every* $w \in (\Sigma \cup \{\varepsilon\})^*$.

A language recognized by some hpda is height-deterministic context-free. *The class of height-deterministic context-free languages is denoted by hCFL.*

Note that every normalized *dpda* is trivially an *hpda*.

Definition 3. *Two hpda A and B over the same alphabet* Σ *are* synchronized, *denoted by* $A \sim B$, *if* $N(A,w) = N(B,w)$ *for every* $w \in (\Sigma \cup \{\varepsilon\})^*$.

Intuitively, two *hpda* are synchronized if their stacks increase and decrease in lockstep at every run on the same input. Note that $\sim$ is an equivalence relation over all *hpda*. Let $[A]_\sim$ denote the equivalence class containing the *hpda* A, and let A-*hCFL* denote the class of languages $\{\mathcal{L}(A) \mid A \in [A]_\sim\}$ recognized by any *hpda* synchronized with A.

In the following subsections we will study some properties of general, real-time, and deterministic *hpda*.

3.1 The General Case

Let us first argue that height-determinism does not restrict the power of *pda*.

Theorem 1. *hCFL = CFL.*

The basic proof idea is that for any context-free language L a *pda* A can be constructed such that $\mathcal{L}(A) = L$ and for every non-deterministic choice of A a different number of ε-moves is done.

Proof. Let $L \in$ *CFL*. There exists an *rpda* $A = (Q, \Sigma, \Gamma, \delta, q_0, F)$ with $\mathcal{L}(A) = L$. We can assume that A is normalized by Lemma 1. Certainly, $|N(A,w)| \leq 1$ for every $w \in \Sigma^*$ does not hold in general. However, we can construct a *pda* $A' = (Q', \Sigma, \Gamma, \delta', q_0, F)$ from A such that a different number of ε-moves is done for every non-deterministic choice of A after reading a letter. In this way every run of A' on some input w is uniquely determined by the number of ε-moves between reading letters from the input. Hence, $|N(A,w)| \leq 1$ for every $w \in (\Sigma \cup \{\varepsilon\})^*$ (condition (ii) of the Definition 2) is satisfied.

Formally, over all $p \in Q$ and $X \in \Gamma \cup \{\bot\}$ and $a \in \Sigma$, let m be the maximum number of rules of the form $pX \overset{a}{\longmapsto} q\alpha$ for some q and α. For every $q\alpha$ appearing on the right-hand side of some rule, we introduce m new states $p^1_{q\alpha}, p^2_{q\alpha}, \ldots, p^m_{q\alpha}$ and for every $X \in \Gamma \cup \{\bot\}$ and $1 \leq i < m$ we add the rules

$$p^i_{q\alpha}X \overset{\varepsilon}{\longmapsto} p^{i+1}_{q\alpha}X \qquad \text{and} \qquad p^m_{q\alpha}X \overset{\varepsilon}{\longmapsto} q\alpha \ .$$

Now, for all $p \in Q$, $X \in \Gamma \cup \{\bot\}$ and $a \in \Sigma$, let

$$pX \overset{a}{\longmapsto} q_1\alpha_1,\ pX \overset{a}{\longmapsto} q_2\alpha_2,\ \ldots, pX \overset{a}{\longmapsto} q_n\alpha_n$$

be all rules under the action a with the left-hand side pX; we replace all these rules with the following ones:

$$pX \overset{a}{\longmapsto} p^1_{q_1\alpha_1}X,\ pX \overset{a}{\longmapsto} p^2_{q_2\alpha_2}X,\ \ldots, pX \overset{a}{\longmapsto} p^n_{q_n\alpha_n}X \ .$$

Note that A' is normalized if A is normalized, and that $\mathcal{L}(A') = \mathcal{L}(A)$. □

Theorem 2. *Let A be any hpda. Then $REG \subseteq A$-hCFL.*

In particular, if R is a complete dfsa then there exists an hpda $B \in A$-hCFL such that $\mathcal{L}(B) = \mathcal{L}(R)$ and $|B| = \mathcal{O}(|A|\,|R|)$. Moreover, if A is deterministic, then B is deterministic.

Proof. Let $L \in REG$, and let R be a *dfsa* recognizing L. W.l.o.g. we can assume that R is complete, that is, for every $a \in \Sigma$ and state r in R there is a transition $r \xrightarrow[R]{a} r'$. We construct a *pda* B as the usual product of (the control part of) A with R: for all $a \in \Sigma$, B has a rule $(q,r)X \overset{a}{\underset{B}{\longmapsto}} (q',r')\alpha$ if and only if $qX \overset{a}{\underset{A}{\longmapsto}} q'\alpha$ and $r \xrightarrow[R]{a} r'$; for every state r of R, B has an ε-rule $(q,r)X \overset{\varepsilon}{\underset{B}{\longmapsto}} (q',r)\alpha$ if and only if $qX \overset{\varepsilon}{\underset{A}{\longmapsto}} q'\alpha$. The final states of B are the pairs (q,r) such that r is a final state of R. Clearly, we have $|B| = \mathcal{O}(|A|\,|R|)$. Moreover, every run of B on some $w \in \Sigma^*$ ends in a final state (q,r) if and only if R is in r after reading w, and hence, $\mathcal{L}(B) = L$.

Next we show that B is *hpda*. Firstly, condition (ii) of Definition 2 and completeness (Definition 1(i)) clearly hold. Secondly, every state of B either admits only ε-transitions or non-ε-transitions but not both (Definition 1(ii)) since $(p,r)X \overset{\varepsilon}{\underset{B}{\longmapsto}} (q,r)\alpha$ and $(p,r)Y \overset{a}{\underset{B}{\longmapsto}} (q',r')\beta$ implies $pX \overset{\varepsilon}{\underset{A}{\longmapsto}} q\alpha$ and $pY \overset{a}{\underset{A}{\longmapsto}} q'\beta$, contradicting the normalization of A. Finally, Definition 1(iii) follows trivially from the fact that A is normalized. It remains to prove $A \sim B$, however, this follows easily because the height of B's stack is completely determined by A. □

Note that the *pda* B in Theorem 2 is real-time (deterministic) if A is real-time (deterministic). The following closure properties are easily proved using classical constructions.

Theorem 3. *Let A be any hpda. Then A-hCFL is closed under union and intersection.*

In particular, let A and B be two hpda with $A \sim B$.

(i) The language $\mathcal{L}(A) \cup \mathcal{L}(B)$ is recognized by some hpda C of size $\mathcal{O}(|A|+|B|)$ such that $A \sim C \sim B$.
(ii) If A and B are deterministic, then the language $\mathcal{L}(A) \cup \mathcal{L}(B)$ is recognized by some deterministic hpda C of size $\mathcal{O}(|A|\,|B|)$ such that $A \sim C \sim B$.
(iii) The language $\mathcal{L}(A) \cap \mathcal{L}(B)$ is recognized by some hpda C of size $\mathcal{O}(|A|\,|B|)$ such that $A \sim C \sim B$. If A and B are deterministic, then C is deterministic.

Moreover, we have in all cases that if both A and B are rpda, then C is an rpda.

3.2 The Real-Time Case

Let *rhpda* denote a real-time *hpda*, and let *rhCFL* denote the class of languages generated by *rhpda*. We remark that *rhpda* contain visibly pushdown automata introduced in [3] but not vice versa as shown in Example 1 below. A visibly pushdown automaton A (*vpda*) over Σ is an *rpda* together with a fixed partition of $\Sigma = \Sigma_c \cup \Sigma_i \cup \Sigma_r$ such that if $pX \overset{a}{\underset{A}{\longmapsto}} qYX$ then $a \in \Sigma_c$ and if $pX \overset{a}{\underset{A}{\longmapsto}} qX$ then $a \in \Sigma_i$ and if $pX \overset{a}{\underset{A}{\longmapsto}} q\lambda$ then $a \in \Sigma_r$. By *vCFL* we denote the class of languages generated by *vpda*.

Example 1. Consider the language $L_1 = \{a^n b a^n \mid n \geq 0\}$ which is not recognized by any *vpda*; see also [3]. Indeed, a *vpda* recognizing L_1 would have to either only push or only pop or only change its state whenever the letter a is read, but then the two powers of a in an input word from a^*ba^* could not be compared for most inputs. However, the obvious *rdpda* that pushes the first block of a's into the stack, reads the b, reads the second block of a's while popping the first block from the stack, and compares whether they have the same length, is a *rhpda* that accepts L_1. □

On the other hand, it is easy to see that not every language accepted by an *rpda* can also be accepted by a *rhpda*. For example, the language of all palindromes over Σ is in *rCFL* but not in *rhCFL*. This follows from the fact that this language does not belong to *rdCFL*, and from the fact that *rdCFL* = *rhCFL*, which is proved below in Theorem 4. All together, we get the following hierarchy.

$$REG \subsetneq vCFL \subsetneq rhCFL = rdCFL \subsetneq rCFL = hCFL = CFL$$

The next theorem shows that real-time *hpda* can be determinised. The proof of this theorem uses the same basic technique as for determinising *vpda* [3].

Theorem 4. *rhCFL* = *rdCFL*.

In particular, we can construct for every rhpda A a deterministic rhpda B such that $\mathcal{L}(A) = \mathcal{L}(B)$ *and* $A \sim B$ *and B has* $\mathcal{O}\left(2^{n^2}\right)$ *many states and a stack alphabet of size* $\mathcal{O}\left(|\Sigma|2^{n^2}\right)$ *where n is the number of pairs of states and stack symbols of A.*

It follows from Theorem 4 and the closure of *rdCFL* under complement that a complement A^c exists for every *rhpda* A. However, the following corollary more precisely shows that A^c can be chosen to satisfy $A^c \sim A$.

Corollary 1. *rhCFL is closed under complement.*

In particular, for every rhpda A there exists an rhpda B such that $\mathcal{L}(B) = \mathcal{L}(A)^c$ *and* $A \sim B$ *and* $|B| = 2^{\mathcal{O}(|A|^2)}$.

The emptiness problem can be decided in time $\mathcal{O}(n^3)$ for any *pda* of size n; see for example [6]. In combination with the previous results we get the bound on the equivalence problem.

Theorem 5. *Language equivalence of synchronized rhpda is decidable.*

In particular, let A and B be two rhpda with $A \sim B$, *and let* $n = |A|$ *and* $m = |B|$. *We can decide* $\mathcal{L}(A) \stackrel{?}{=} \mathcal{L}(B)$, *in time* $2^{\mathcal{O}(n^2+m^2)}$.

3.3 The Deterministic Case

Contrary to the real-time case, arbitrary *hpda* cannot always be determinised, as shown by Theorem 1. For this reason we investigate the synchronization relation $\sim$ restricted to the class of deterministic pushdown automata. Certainly, *dhCFL* = *dCFL* since every *dpda* can be normalized by Lemma 1 and then it is trivially height-deterministic. However, we lay the focus in this section on the closure of each equivalence class of $\sim$ under complement. Therefore, we denote a deterministic *hpda* by *dhpda*. The class of languages recognized by some *dhpda* synchronized with the *dhpda* A is denoted by A-*dhCFL*.

First, we show that, as in the real-time case, every *dhpda* can be complemented without leaving its equivalence class. The proof is, however, more delicate due to the presence of ε-rules. In fact, the normalization of Definition 1 has been carefully chosen to make this theorem possible.

Theorem 6. *Let A be any dhpda. Then A-dhCFL is closed under complement.*

In particular, for every dhpda B there exists a complement dhpda B^c such that $B^c \sim B$ and $|B^c| = \mathcal{O}(|B|)$.

Proof. Let $B = (Q, \Sigma, \Gamma, \delta, q_0, F)$. Let $Q' \subseteq Q$ be the set of all ε-states of B and let $Q'' = Q \setminus Q'$. We construct B^c by first defining an *dhpda* B' equivalent to B such that a word is accepted if and only if it can be accepted with a state in Q'', that is, a state which allows only non-ε-moves. Then the set of accepting states is a subset of states in Q'' that do not accept $\mathcal{L}(B)$. This gives the complement of B.

We will define a *dhpda* B' such that $B \sim B'$ and $\mathcal{L}(B') = \mathcal{L}(B)$ and every accepting path in the transition system generated by B' ends in a state in $Q' \cup (Q'' \cap F)$, that is, when B' accepts a word w, then B' shall end in a final state after reading w with a maximal (and finite by property (i) in Definition 1) number of ε moves after reading the last letter of w. Note that the completeness property of B in Definition 1 implies that B is always in a state in Q'' after reading w followed by a maximal number of ε-transitions.

Let $B' = (Q \times \{0,1\}, \Sigma, \Gamma, \vartheta, q_0', F')$ with $F' = Q \times \{1\}$, and $q_0' = (q_0, 1)$ if $q_0 \in F$ and $q_0' = (q_0, 0)$ otherwise. The set of rules ϑ is defined as follows:

- $((p,i), X, e, (q,1), \alpha) \in \vartheta$ if $(p, X, e, q, \alpha) \in \delta$ and $q \in F$,
- $((p,i), X, a, (q,0), \alpha) \in \vartheta$ if $(p, X, a, q, \alpha) \in \delta$ and $q \notin F$, and
- $((p,i), X, \varepsilon, (q,i), \alpha) \in \vartheta$ if $(p, X, \varepsilon, q, \alpha) \in \delta$ and $q \notin F$.

where $e \in \Sigma \cup \{\varepsilon\}$ and $i \in \{0,1\}$ and $a \in \Sigma$. We have now $\mathcal{L}(B') = \mathcal{L}(B)$. Indeed, we have two copies, indexed with 0 and 1, of B in B' and whenever an accepting state is reached in B then it is reached in the 1-copy of B in B' (the first two items in the definition of ϑ above) and B' is in an accepting state and both B and B' accept the word read so far. The set of accepting states of B' is only left when the next letter is read from the input and B reaches a non-accepting state (the third item in the definition of ϑ above). Otherwise, B' remains in the respective copy of B (first and fourth item in the definition of ϑ above). Clearly, $B' \sim B$.

Now, $B^c = (Q \times \{0,1\}, \Sigma, \Gamma, \vartheta, q_0', Q'' \times \{0\})$. □

The equivalence checking problem for two synchronized *dhpda* is, like in the real-time case, decidable.

Theorem 7. *Language equivalence of synchronized dhpda is decidable.*

In particular, for any dhpda A and B such that $A \sim B$, we can decide whether $\mathcal{L}(A) \stackrel{?}{=} \mathcal{L}(B)$ in time $\mathcal{O}\left(|A|^3\,|B|^3\right)$.

4 Other Language Classes — A Comparison

In this section height-deterministic context-free languages are compared to two other recent approaches of defining classes of context-free languages closed under boolean operations. In [5], Caucal introduced an extension of Alur and Madhusudan's visibly pushdown languages [3], and proved that it forms a boolean algebra. The second class is the one introduced by Fisman and Pnueli in [7]. We show in this section that *rhCFL* (which is a proper subclass of *dhCFL*) properly contains these two classes.

4.1 Caucal's Class

Caucal's class is defined with the help of a notion of synchronization, just as our *hCFL* class.[1] Before we can define Caucal's synchronization, we need some preliminaries.

A *fst* is *input deterministic*, if $(s, a, m, t) \in \varrho$ and $(s, a, n, t') \in \varrho$ implies that $m = n$ and $t = t'$. Caucal considers input deterministic transducers from Σ^* to $\mathbb{Z}$ (the additive monoid of integers) where every state accepts, i.e., transducers whose transitions are labeled with a letter from Σ and an integer. When the transducer reads a word over Σ, it outputs the sum of the integers of the transitions visited. Notice that if a transducer T is input deterministic then the set $T(w)$ is a singleton, i.e., a set containing one single integer. By abuse of notation, we identify $T(w)$ with this integer. We let $|T(w)|$ denote the absolute value of $T(w)$.

Given an input deterministic *fst* T from Σ^* to $\mathbb{Z}$ and an *rpda* A over Σ with initial state q_0, we say that A is a *T-synchronized pda* (*T-spda*) if $q_0\bot \xrightarrow[A]{w} p\alpha\bot$ implies $|\alpha| = |T(w)|$ for every $w \in \Sigma^*$ and every configuration $p\alpha$ of A. Let *wSCFL* denote the class of all languages that are recognized by some *T-spda* for some T. (See also Caucal's introduction of *wSCFL* in [5]).

Theorem 8. *wSCFL* $\subsetneq$ *rhCFL.*
In particular, the language

$$L_3 = \{a^m b^n w \mid m > n > 0, |w|_a = |w|_b, w_{(1)} = a \textit{ if } w \neq \lambda\}$$

belongs to rhCFL but not to wSCFL.

4.2 Fisman and Pnueli's Class

We define the class of M-synchronized *pda*, which is the formalism used by Fisman and Pnueli in their approach to non-regular model-checking [7].

Let $M = (\Delta, \Gamma, \delta)$ be a *1-rdpda*, let $R = (Q, \Sigma \times \Gamma, q_0, \varrho, F)$ be a *dfsa*, and let $\phi\colon \Sigma \to \Delta$ be a substitution. The *cascade product* $M \circ_\phi R$ is the *rdpda* $(Q, \Sigma, \Gamma, \delta', q_0, F)$ with $qX \xmapsto{a} \varrho(q, (a, X))\delta(\phi(a), X)$ for all $q \in Q$, $a \in \Sigma$ and $X \in \Gamma \cup \{\bot\}$. An *rdpda* A is called *M-synchronized* (*M-spda*) if there exists

[1] In fact, Caucal's class was the starting point of our study.

a substitution ϕ and a *dfsa* R such that $A = M \circ_\phi R$. Let *1SCFL* denote the class of all languages that are recognized by some M-*spda* for some *1-rdpda* M. See also Fisman and Pnueli's introduction of *1SCFL* in [7].

Theorem 9. *1SCFL* $\subsetneq$ *rhCFL.*
In particular, the language

$$L_4 = \{a^n b a^n \mid n \geq 0\} \cup \{a^n c a^{2n} \mid n \geq 0\}$$

belongs to rhCFL but not to 1SCFL.

5 Conclusion

We have introduced several (sub)classes of the class of context-free languages that are closed under boolean operations. Our key technical tools are height-deterministic pushdown automata (*hpda*) and synchronization between *hpda*. These notions are inspired by and generalize Caucal's work on real-time synchronized pushdown graphs [5]. In fact, our results can be seen as an extension of Caucal's ideas to pushdown automata with ϵ-transitions. This extension has turned out to be rather delicate. Both Theorem 2 (*REG* $\subseteq$ *A-hCFL*) and Theorem 6 (*A-hCFL* is closed under complement) depend crucially on the normalization of Definition 1 which had to be carefully chosen. In a sense, one of the contributions of the paper is to have worked out the right notion of normalization. We have also showed that language equivalence of real-time height-deterministic pushdown automata is decidable in EXPTIME.

Both this paper and Caucal's have been also inspired by Alur and Madhusudan's work on visibly pushdown automata, initiated in [3]. From an automata-theoretic point of view, we have extended the theorem of [3], stating that visibly pushdown automata are closed under boolean operations, to deterministic *hpda*. This is rather satisfactory, because deterministic *hpda* recognize all deterministic context-free languages, while visibly *pda* are far from it. Remarkably, the extension is be achieved at a very low cost; in our opinion, height-deterministic *pda* are, at least from the semantical point of view, as natural and intuitive as visibly *pda*.

Acknowledgments. The authors are deeply indebted to Javier Esparza who contributed to this work in many ways. We also thank to the anonymous referees for their useful remarks.

References

1. Alur, R., Etessami, K., Madhusudan, P.: A temporal logic of nested calls and returns. In: Jensen, K., Podelski, A. (eds.) TACAS 2004. LNCS, vol. 2988, pp. 467–481. Springer, Heidelberg (2004)

2. Alur, R., Kumar, V., Madhusudan, P., Viswanathan, M.: Congruences for visibly pushdown languages. In: Caires, L., Italiano, G.F., Monteiro, L., Palamidessi, C., Yung, M. (eds.) ICALP 2005. LNCS, vol. 3580, pp. 1102–1114. Springer, Heidelberg (2005)
3. Alur, R., Madhusudan, P.: Visibly pushdown languages. In: ACM Symposium on Theory of Computing (STOC'04), pp. 202–211. ACM Press, New York (2004)
4. Bárány, V., Löding, C., Serre, O.: Regularity problems for visibly pushdown languages. In: Durand, B., Thomas, W. (eds.) STACS 2006. LNCS, vol. 3884, pp. 420–431. Springer, Heidelberg (2006)
5. Caucal, D.: Synchronization of pushdown automata. In: Ibarra, O.H., Dang, Z. (eds.) DLT 2006. LNCS, vol. 4036, pp. 120–132. Springer, Heidelberg (2006)
6. Esparza, J., Hansel, D., Rossmanith, P., Schwoon, S.: Efficient algorithms for model checking pushdown systems. In: Emerson, E.A., Sistla, A.P. (eds.) CAV 2000. LNCS, vol. 1855, pp. 232–247. Springer, Heidelberg (2000)
7. Fisman, D., Pnueli, A.: Beyond regular model checking. In: Hariharan, R., Mukund, M., Vinay, V. (eds.) FST TCS 2001: Foundations of Software Technology and Theoretical Computer Science. LNCS, vol. 2245, pp. 156–170. Springer, Heidelberg (2001)
8. Löding, Ch., Madhusudan, P., Serre, O.: Visibly pushdown games. In: Lodaya, K., Mahajan, M. (eds.) FSTTCS 2004. LNCS, vol. 3328, pp. 408–420. Springer, Heidelberg (2004)
9. Murawski, A., Walukiewicz, I.: Third-order idealized algol with iteration is decidable. In: Sassone, V. (ed.) FOSSACS 2005. LNCS, vol. 3441, pp. 202–218. Springer, Heidelberg (2005)
10. Pitcher, C.: Visibly pushdown expression effects for XML stream processing. In: Proceedings of Programming Language Technologies for XML (PLAN-X), pp. 5–19 (2005)
11. Srba, J.: Visibly pushdown automata: From language equivalence to simulation and bisimulation. In: Ésik, Z. (ed.) CSL 2006. LNCS, vol. 4207, pp. 89–103. Springer, Heidelberg (2006)

Minimizing Variants of Visibly Pushdown Automata

Patrick Chervet and Igor Walukiewicz*

LaBRI, Université de Bordeaux and CNRS
351, Cours de la Libération, F-33 405, Talence cedex, France

Abstract. The minimization problem for visibly pushdown automata (VPA) is studied. Two subclasses of VPA are introduced: call driven automata, and block automata. For the first class, minimization results are given unifying and generalizing those present in the literature. Unfortunately, this class shares the drawback of all other classes for which a minimization result is known: it is exponentially less succinct than VPA. The second class, block automata, is introduced to address this problem. These automata are as succinct as VPA. A minimization procedure for them is presented that requires one additional parameter to be fixed. An example of an exponential gain in succinctness is given.

Introduction

The class of visibly pushdown languages is the class of languages defined by pushdown automata where an input letter determines a stack action of the automaton. It seems that this class was first studied by Melhorn under the name of input driven automata. In [11] he shows that the parsing problem is in $\mathcal{O}(\log^2(n))$. Alur and Madhusudan [4] exhibit many good properties of this class. It is closed under boolean operations and it contains some interesting previously studied classes as: parenthesis languages [10] and balanced grammars [5]. Visibly pushdown languages have several different characterizations. One is via syntactic congruences in a style of Myhill-Nerode congruence for regular languages [2]. This characterization permits to obtain a canonical visibly pushdown automaton for a given language. Unfortunately, this canonical automaton is not always the minimal visibly pushdown automaton for the language.

In this paper we study the minimization problem for deterministic VPA. Our research is motivated by the presence of two different subclasses in the literature: SEVPA [2] and MEVPA [8]. These are two subclasses of VPA for which some minimization results are known. We introduce two new classes: call driven automata (CDA), and their expanded version (eCDA). The class CDA is a superset of both SEVPA and MEVPA; while eCDA is included in these two classes. We prove a minimization result for eCDA and show how it can be used to get known and new minimization results for the other three classes. This gives a unified picture of the previous studies of the minimization problem.

* This work was supported by ANR grant ANR-06-SETIN-001.

L. Kučera and A. Kučera (Eds.): MFCS 2007, LNCS 4708, pp. 135–146, 2007.

The drawback of all of the above results is that translation from deterministic VPA to automata in one of these classes may incur exponential blowup. This happens due to structural constraints on the form of automata. We propose a new subclass of VPA, called block VPA (BVPA), which is much better in this respect. The translation from VPA to a BVPA results in at most quadratic blowup. Thus a minimization result for BVPA would give an approximative minimization of VPA. Unfortunately, we are able to show minimization of BVPA only when some additional parameter is fixed. The advantage of this result is that minimizations of eCDA and SEVPA are its special cases. It makes also evident that minimizing VPA is related to optimizing the one parameter that we exhibit.

Since the paper of Alur and Madhusudan [4], VPA has appeared in several contexts: XML [15,7], verification [9,1], learning [8], semantics of programing languages [13]. It is in particular this last paper that motivated the current work. In that paper an algorithm is given for constructing a VPA describing the semantics of a given program expression. This way comparing program expressions is reduced to VPA equivalence. For the feasibility of the translation from programs to VPA it is essential to be able to reduce the size of intermediate automata during construction. This leads directly to the problem of minimization. It is worth noting that in other applications mentioned above minimization can also play an important role.

Let us comment on the importance and feasibility of minimization in general. Small canonical ways of representing objects are omnipresent. Consider two examples from verification: Binary Decision Diagrams [12] (BDD's) are nothing else but minimal automata for languages of bit strings; difference bounded matrices [6] (DBM's) are canonical representations of sets clock valuations. Good representations are rare. For example, nondeterministic automata are in principle more succinct than deterministic automata, still it seems very difficult to obtain, with a reasonable computational effort, a nondeterministic automaton of size close to a minimal one. Similar problems appear with two way deterministic automata, or even with two-pass deterministic automata that read the word first from left to right and then from right to left. In this context having minimal automata even for a restricted class of VPA is rather unexpected. The general minimization of VPA seems at least as difficult as minimization of two-pass automata.

The plan of the paper is as follows. We start with basic definitions on VPA. In the following section we introduce the new classes CDA and eCDA. We also show the minimization result for eCDA and point out how it implies a minimization result for CDA. Section 3 discusses relations with MEVPA and SEVPA. Finally, we present BVPA, discuss their properties and present a minimization procedure for them. The missing proofs can be found in the full version of the paper [14].

1 Visibly Pushdown Automata

A *visibly pushdown alphabet* $\hat{\Sigma} = (\Sigma_{call}, \Sigma_{ret}, \Sigma_{int})$ consists of three disjoint finite sets: Σ_{call} a set of *calls*, Σ_{ret} a set of *returns*, and Σ_{int} a set of *internal actions*.

For any such $\hat{\Sigma}$, let Σ denote $\Sigma_{call} \cup \Sigma_{ret} \cup \Sigma_{int}$. In the following we will use: $c, c_1, \ldots$ for elements of Σ_{call}; $r, r_1, \ldots$ for elements of Σ_{ret}; $i, i_1, \ldots$ for elements of Σ_{int}.

Definition 1. *A* visibly pushdown automaton *(VPA) is a pushdown automaton* $\mathcal{A} = \langle Q, \hat{\Sigma}, q_0, \Gamma, \delta, Q_F \rangle$, *where* Q *is a finite set of states,* $\hat{\Sigma} = (\Sigma_{call}, \Sigma_{ret}, \Sigma_{int})$ *is a visibly pushdown alphabet,* $q_0 \in Q$ *is an initial state,* Γ *is a (finite) stack alphabet,* Q_F *is a set of final states and* $\delta = \delta_{call} \cup \delta_{ret} \cup \delta_{int}$ *is a transition function, such that:* $\delta_{call} : Q \times \Sigma_{call} \to Q \times \Gamma$, $\delta_{ret} : Q \times \Sigma_{ret} \times \Gamma \to Q$ *and* $\delta_{int} : Q \times \Sigma_{int} \to Q$.

A stack over Γ will be represented by a finite word over Γ with the top on the left of the word. We will write $\gamma\sigma$ to denote a stack with the top letter γ and the rest of the stack σ. A *configuration* is a pair (q, σ) where q is a state and σ is a stack.

An *execution* of a VPA $\mathcal{A}$ as above on a word $w = a_1 \cdots a_k$ from Σ^* is a sequence of configurations $(q_0, \sigma_0), \ldots, (q_k, \sigma_k)$ where σ_0 is the empty stack ε, and for every $j \in [1, k]$:

- if a_j is a call then $\delta_{call}(q_j, a_j) = (q_{j+1}, \gamma)$ and $\sigma_{j+1} = \gamma\sigma_j$,
- if a_j is an internal action then $\delta_{int}(q_j, a_j) = q_{j+1}$ and $\sigma_j = \sigma_{j+1}$,
- if a_j is a return then $\delta_{ret}(q_j, a_j, \gamma) = q_{j+1}$ and $\sigma_j = \gamma\sigma_{j+1}$.

Intuitively, on reading a call the automaton is obliged to do a push operation and moreover it cannot look at the top of the stack. On internal actions the automaton cannot change the stack, neither it can look at the top of it. When reading a return, the automaton has to do a pop, but this time it can use the information on the top of the stack. We will write $q \xmapsto{c/\gamma} q'$, $q \xmapsto{i} q'$, and $q \xmapsto{r/\gamma} q'$ for push, internal, and pop transitions, respectively.

An execution $(q_0, \sigma_0), \ldots, (q_k, \sigma_k)$ is *accepting* if q_k is a final state ($q_k \in Q_F$). A word $w \in \Sigma^*$ is *recognized* by an automaton $\mathcal{A}$ if the unique execution of $\mathcal{A}$ on w is accepting. The *language recognized by* $\mathcal{A}$, denoted $L(\mathcal{A})$, is the set of all words of Σ^* recognized by $\mathcal{A}$. A language L over an alphabet $\hat{\Sigma}$ is a *VPL* if there is a visibly pushdown automaton over $\hat{\Sigma}$ recognizing L.

If $\mathcal{A}$ is a VPA and δ is its transition function, we will write $\delta(u)$ to denote the state reached by $\mathcal{A}$ after the reading of $u \in \Sigma^*$. We will sometimes also use $\to_{\mathcal{A}}$ to denote the transition function of $\mathcal{A}$.

Remark 1. A visibly pushdown automaton is a deterministic pushdown automaton with one important restriction that input letters determine stack actions. The restrictions disallowing to look at the top of the stack when doing push or internal actions are not essential if recognizability is concerned as one can always remember the top of the stack in a state. One can also consider nondeterministic visibly pushdown automata, but we will not do it here.

Remark 2. Without a loss of generality, one can assume that $\Gamma = Q \times \Sigma_{call}$ and that in a state q when reading a call c the automation pushes (q, c) on the stack. This works as (q, c) is the maximal information the automaton has when doing the push. In the following we will use this form of Γ when convenient.

Definition 2 (Matched calls and returns). *Let $\hat{\Sigma}$ be a pushdown alphabet, and u be a word in Σ^*.*

The word u is matched calls *if every call has a matching return, i.e. if for every suffix u' of u the number of call symbols in u' is at most the number of return symbols of u'.*

Similarly, the word u is matched returns *if every return has a matching call, i.e. if for every prefix u' of u the number of return symbols in u' is at most the number of call symbols in u'.*

The word u is well-matched *if it is matched calls and matched returns. Observe that being well matched means being well bracketed when call symbols are considered as opening brackets and return symbols are considered as closing brackets.*

Let then $MC(\hat{\Sigma})$, $MR(\hat{\Sigma})$ and $WM(\hat{\Sigma})$ be respectively the set of matched calls, matched returns and well-matched words. A language L is well-matched *if $L \subseteq WM(\hat{\Sigma})$.*

Remark 3. In this paper we consider only well-matched languages. In this case it is not restrictive to disallow return actions with empty stack: to fit with the classical definition of VPA it is enough to add a sink where every return transition with empty stack can go.

Given a VPL L over a visibly pushdown alphabet $\hat{\Sigma} = (\Sigma_{call}, \Sigma_{ret}, \Sigma_{int})$, let us define the equivalence relation $\approx_L$ on well-matched words:

$$w_1 \approx_L w_2 \quad \text{if for all } u, v \in \Sigma^* : uw_1v \in L \text{ iff } uw_2v \in L.$$

Observe that $\approx_L$ is a congruence with respect to the concatenation.

Theorem 1. *[2] A language $L \subseteq WM(\hat{\Sigma})$ is a VPL iff $\approx_L$ has finitely many equivalence classes.*

2 Call Driven Automata and Their Minimization

In this section we introduce the class of call driven automata. A call driven automaton is a special kind of visibly pushdown automaton where we require that a call letter determines uniquely the state to which the automaton goes. We will show later that this subclass of VPA is larger than previously introduced subclasses: SEVPA and MEVPA.

Definition 3. *A VPA $\mathcal{A} = \langle Q, \hat{\Sigma}, q_0, \Gamma, \delta, Q_F \rangle$ is a* Call Driven Automaton (CDA) *if there is a function $Target : \Sigma_{call} \mapsto Q$ and an equivalence relation $\overset{Q}{\leftrightarrow}$ on Q such that:*

- *for all $c \in \Sigma_{call}$, $q_0 \overset{Q}{\not\leftrightarrow} Target(c)$*
- *if $q \xrightarrow{i} q'$ then $q \overset{Q}{\leftrightarrow} q'$,*
- *if $q \xrightarrow{c/(q,c)} q'$ then $q' = Target(c)$,*
- *if $q \xrightarrow{r/(q',c)} q''$ then $q' \overset{Q}{\leftrightarrow} q''$.*

This definition essentially says that the set of states is divided into equivalence classes of $\overset{Q}{\leftrightarrow}$. An internal transition has to stay in the same equivalence class. A call transition goes to the class determined by the call letter. A return transition has to go back to the class from which the matching call was done. The first condition says that the equivalence class of q_0 is not the target of any call, so being in this class we know that the stack is empty.

An interesting subclass of CDA, which we call expanded CDA, is obtained by requiring that each call leads to a different equivalence class. We will show in the next subsection that eCDA are relatively straightforward to minimize. Moreover, minimization of CDA can be done via minimization of eCDA (cf. Theorem 2).

Definition 4. *A VPA $\mathcal{A}$ is an* expanded Call Driven Automaton *(eCDA) if it is a CDA for some function Target and equivalence relation $\overset{Q}{\leftrightarrow}$ such that if $Target(c) \overset{Q}{\leftrightarrow} Target(c')$ then $c = c'$.*

2.1 Minimization of eCDA

Due to the restriction on their structure, minimization of eCDA resembles very much minimization of finite automata. Later, we will show that minimizations of other subclasses of visibly pushdown automata can be obtained via reduction to minimization of eCDA. As usual, by the size of an automaton we mean the number of its states.

Theorem 2. *Let $\hat{\Sigma}$ be a visibly pushdown alphabet. For every VPL $L \subseteq WM(\hat{\Sigma})$ there is a unique (up to isomorphism) minimum-size eCDA recognizing L.*

Moreover, for a given eCDA in cubic time it is possible to find a minimal equivalent eCDA.

Proof. The proof uses the same method as minimization of SEVPA [2,3] but it is notationaly simpler, due to the simpler structure of eCDA. The idea is to construct first a syntactic eCDA $\mathcal{A}$ recognizing L, and then to prove its minimality by showing for any eCDA recognizing L a surjective homomorphism from it to $\mathcal{A}$. The construction uses congruence relations which are coarser than $\approx_L$, and hence of finite index (cf Theorem 1). So the construction resembles Myhill-Nerode minimization for finite automata. As the more complicated construction for SEVPA has the cubic complexity, the same is true here.

2.2 Minimization of CDA

An obvious question is whether one can minimize CDA in the same way as eCDA. The answer is negative as there is no unique minimal CDA for the language (see Example 1 below). Without much work though we can obtain an approximative minimization of CDA, i.e., minimization up to the factor $|\Sigma_{call}|$. The construction uses the minimization of eCDA.

Lemma 1. *Given a CDA of size n, an equivalent eCDA of size $\mathcal{O}(n \times |\Sigma_{call}|)$ can be constructed in linear time.*

The construction only needs to duplicate states so that the equivalence relation $\overset{Q}{\leftrightarrow}$ relates no two states in the range of *Target*. Given a CDA, this lemma together with the Theorem 2 allows us to find the smallest equivalent eCDA. Moreover, due to the Lemma 1, the smallest equivalent eCDA is of size $\mathcal{O}(n \times |\Sigma_{call}|)$, where n is the minimal size of an equivalent CDA. This gives us:

Corollary 1. *Given a CDA in a cubic time it is possible to find a CDA of size* $|\Sigma_{call}| \times n$ *where* n *is the size of a minimal equivalent CDA.*

The following example shows a language L such that there is no unique minimal CDA recognizing L. We take for L the VPL $c_1(1 \cdot 1)^*r + c_2 1^* r$ over the visibly pushdown alphabet $(\{c_1, c_2\}, \{r\}, \{1\})$. The two automata presented in Figure 1 have the same number of states and transitions. While L is a regular language, of course the same problem arises for some languages that are not regular.

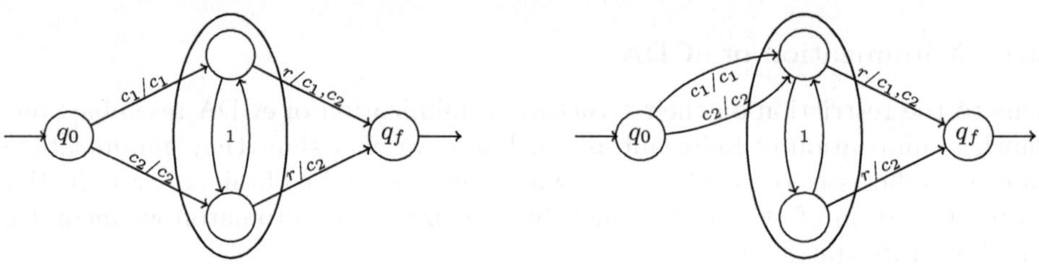

Fig. 1. Two non-isomorphic minimal CDA recognizing L (each automaton also has a sink that is not represented)

3 Comparison Between Different VPA Subclasses

In this section we discuss the relations between CDA and two other classes of VPA introduced in the literature: single entry visibly pushdown automata (SEVPA) [2], and multi-entry visibly pushdown automata (MEVPA) [8]. Not only these two classes are included in CDA, but we can recover minimization results for them from our basic minimization result for eCDA.

A SEVPA is a CDA when every equivalence class of $\overset{Q}{\leftrightarrow}$ has a single entry, i.e., only one state in a class is a target of a push transition.

Definition 5. *A CDA with Target and* $\overset{Q}{\leftrightarrow}$ *is a* single entry visibly pushdown automaton (SEVPA) *if, for all call actions* c *and* c'*:*

$$\textit{if } Target(c) \overset{Q}{\leftrightarrow} Target(c') \textit{ then } Target(c) = Target(c').$$

Multi-entry automata (MEVPA) represent another kind of restriction on CDA. In general we can assume that the stack alphabet of a visibly pushdown automaton is $Q \times \Sigma_{call}$ (cf. remark 2) as a push transition depends only on the current state and a letter being read. In MEVPA we assume that the symbol pushed on the stack depends only on the state and not on the input letter.

Definition 6. *A CDA* $\mathcal{A} = \langle Q, \hat{\Sigma}, q_0, \Gamma, \delta, Q_F \rangle$ *is a* Multi Entry Visibly Pushdown Automaton *(MEVPA) when for all transitions* $q_1 \xrightarrow{c_1/\gamma_1} q_1'$ *and* $q_2 \xrightarrow{c_2/\gamma_2} q_2'$*: if* $q_1 = q_2$ *then* $\gamma_1 = \gamma_2$.

By definition, CDA includes all other classes: eCDA, SEVPA, MEVPA. Also by definition, eCDA is included in SEVPA. The class eCDA is also included in MEVPA. Indeed, if $(\mathcal{A}, \mathit{Target}, \overset{Q}{\leftrightarrow})$ is an eCDA then each state of $\mathcal{A}$ is related in $\overset{Q}{\leftrightarrow}$ to at most one state $\mathit{Target}(c)$, so the automaton always knows what is the last call read, and does not need to put this information on the stack. This gives the following graph of inclusions between the subclasses.

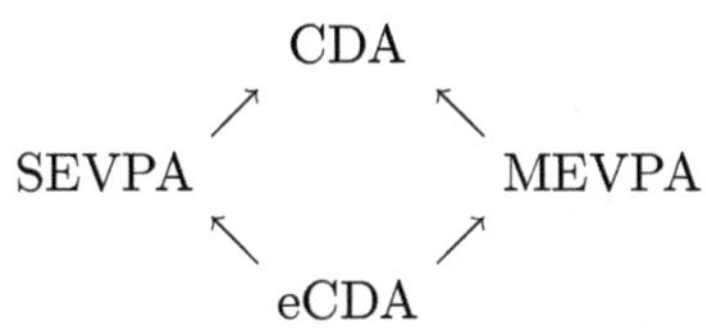

Due to Theorem 2 and Lemma 1, for every VPA there is an equivalent automaton in every one of the four classes above. The question is, given a VPA, how small an equivalent CDA can be. The following example shows that the blow-up can be exponential; which is also an upper bound.

Example. Consider an alphabet $\hat{\Sigma} = (\{c\}, \{r\}, \{a_1, \ldots, a_k\})$ with one call symbol c and one return symbol r. Let $L_k = a_1 c L_{a_1} r + \cdots + a_k c L_{a_k} r$, where $L_{a_i} \subseteq \{a_1, \ldots, a_n\}^*$ is the set of words where the number of a_i is even. By counting equivalence classes one can show that the minimal eCDA recognizing L_k is of size bigger than 2^k. Lemma 1 gives the bound $2^k/k$ for CDA . On the other hand, L_k can be recognized by a VPA of size in $\mathcal{O}(k^2)$ (see figure 2 below).

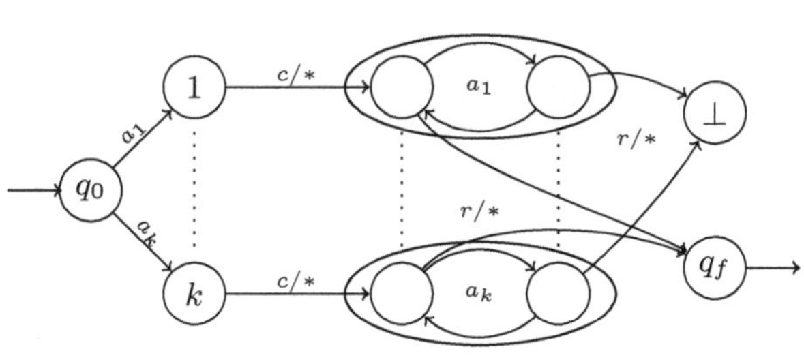

Fig. 2. VPA of size in $\mathcal{O}(k)$ recognizing L_k

4 Other Results on MEVPA and SEVPA

It turns out that we can reprove known minimization results on MEVPA and SEVPA using the result about eCDA. The main idea is to use an appropriately chosen injective morphism $\Phi : \Sigma^* \to \Lambda^*$ where Λ is some alphabet. The idea of the construction is presented in the schema in Figure 3. Given an automaton $\mathcal{A}$ from a class $\mathcal{C}$, one constructs an eCDA $\overline{\mathcal{A}}$ recognizing $\Phi(L)$. Then one finds the minimal eCDA $\overline{\mathcal{B}}$ equivalent to $\overline{\mathcal{A}}$, and finally translates $\overline{\mathcal{B}}$ to an automaton $\mathcal{B}$ recognizing L. If Φ is suitably chosen then the size of $\overline{\mathcal{A}}$ can be bounded by a small polynomial in the size of $\mathcal{A}$ and usually the translation from $\overline{\mathcal{B}}$ to $\mathcal{B}$ does not introduce extra states.

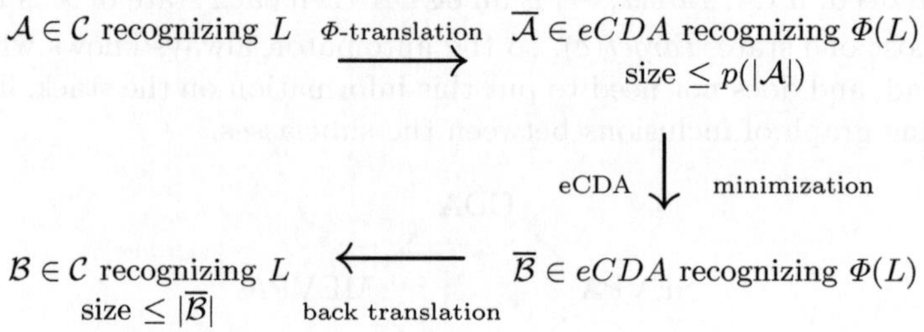

Fig. 3. Translation method: $p(n)$ is a fixed polynomial

4.1 MEVPA

We explain how to obtain a minimization of MEVPA using eCDA.

Theorem 3 (Kumar, Madhusudan & Viswanathan). *[8] Let $\hat{\Sigma}$ be a visibly pushdown alphabet. For every VPL $L \subseteq WM(\hat{\Sigma})$, there is a unique (up to isomorphism) minimum-size equivalent MEVPA. Moreover, it can be constructed in a cubic time from a given MEVPA recognizing L.*

Proof. We apply the translation method for the homomorphism Φ that is an identity on all letters but on the call letters where we put $\Phi(c) = 1c$. The idea of the construction is that each call transition on a letter c is split into a 1 transition that does the push followed by internal c transition. An eCDA $\overline{\mathcal{A}}$ for the transformed language can be obtained by adding one entry state E, which is the target of 1 transition, and modifying c transitions appropriately (see figure 4).

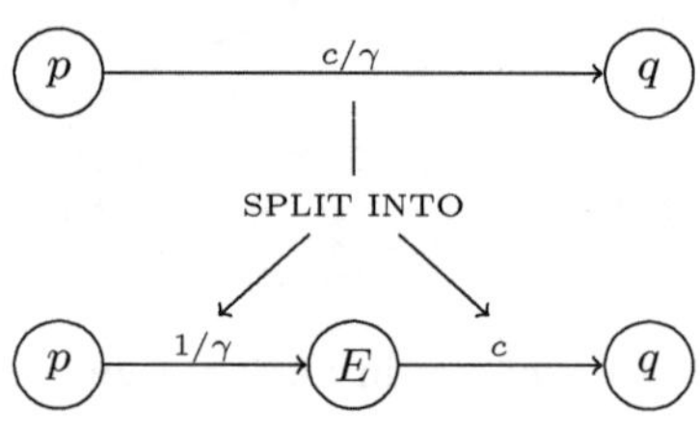

Fig. 4. $c \rightarrow 1c$ translation

We then apply eCDA minimization to $\overline{\mathcal{A}}$, and collapse back the previously split translations of the resulting automaton. Finally, we prove that the MEVPA $\mathcal{B}$ obtained this way is the unique minimal equivalent MEVPA.

4.2 SEVPA

In [2] Alur, Kumar, Madhusudan and Viswanathan give a minimization of SEVPA provided some additional structure is preserved. We show how to obtain this result using the method of translations. Before doing this we present two remarks. The first shows that preserving the structure can result in exponential blowup. The second points out that our result on CDA gives approximative minimization that avoids the blow-up.

For a SEVPA $\mathcal{A} = \langle Q, \hat{\Sigma}, q_0, \Gamma, \delta, Q_F \rangle$ with *Target* and $\overset{Q}{\leftrightarrow}$ we define an equivalence relation $\overset{\Sigma}{\leftrightarrow}$ on call actions Σ_{call} by:

$$c \overset{\Sigma}{\leftrightarrow} c' \text{ iff } Target(c) = Target(c').$$

We then call $\mathcal{A}$ a $\overset{\Sigma}{\leftrightarrow}$*-SEVPA*. Equivalence relation $\overset{\Sigma}{\leftrightarrow}$ fixes bit more of the structure of automaton. One can consider eCDA as SEVPA with $\overset{\Sigma}{\leftrightarrow}$ that is the identity relation.

Remark 4. It may happen that minimal SEVPA for one $\overset{\Sigma}{\leftrightarrow}$ relation can be much bigger than an equivalent SEVPA for some other relation. Consider visibly pushdown alphabet $\hat{\Sigma} = (\{c_1, \ldots, c_k\}, \{r\}, \{a_1, \ldots, a_k\})$, and a language $L_k = a_1 c_1 L_{a_1} r + \cdots + a_k c_k L_{a_k} r$, where $L_{a_i} \subseteq \{a_1, \ldots, a_k\}^*$ is the set of words with even number of occurrences of a_i. When $\overset{\Sigma}{\leftrightarrow}$ is the identity relation, the minimal size of a $\overset{\Sigma}{\leftrightarrow}$-SEVPA recognizing L is in $\mathcal{O}(k)$. If on the other hand $\overset{\Sigma}{\leftrightarrow}$ is a complete relation where every pair of call letters is related then the size of the minimal $\overset{\Sigma}{\leftrightarrow}$-SEVPA is bigger than $\mathcal{O}(2^k)$.

Remark 5. Lemma 1 implies that taking $\overset{\Sigma}{\leftrightarrow}$ to be the identity relation is not far from optimal. Indeed, every SEVPA is a CDA, so the minimal eCDA for a given language is only $|\Sigma_{call}|$ bigger than the minimal equivalent SEVPA. As noted above an eCDA is a $\overset{\Sigma}{\leftrightarrow}$-SEVPA when $\overset{\Sigma}{\leftrightarrow}$ is the identity relation. So modifying $\overset{\Sigma}{\leftrightarrow}$ relation we can gain at most a $|\Sigma_{call}|$ factor.

The following result is the main minimization result of [2]. It can be proved via translation with the homomorphism Φ that is an identity on all the letters but Σ_{call} for which we let $\Phi(c) = ct$ where $t = Target(c)$. Note that this theorem is also a particular case of the BVPA minimization we will study in Section 5.

Theorem 4 (Alur, Kumar, Madhusudan & Viswanathan). *[2] Let $\hat{\Sigma}$ be a visibly pushdown alphabet and let $\overset{\Sigma}{\leftrightarrow}$ be an equivalence relation over Σ_{call}. For every VPL $L \subseteq WM(\hat{\Sigma})$ there is a unique (up to isomorphism) minimum-size $\overset{\Sigma}{\leftrightarrow}$-SEVPA recognizing L. Moreover it can be constructed in a cubic time given a $\overset{\Sigma}{\leftrightarrow}$-SEVPA for the language.*

5 Block Visibly Pushdown Automata

In the previous section we have observed that there is an exponential lower bound for the translation from VPA to CDA. This motivates the quest for a new subclass of VPA which admits good properties with respect to minimization. Here we introduce the class of blocks visibly pushdown automata (BVPA).

One important property is that for every VPA there is an equivalent BVPA of quadratic size. The idea of BVPA is that its states are divided in equivalence classes of some $\overset{Q}{\leftrightarrow}$ and only call transitions can change the classes. But this time a call does not determine a class to which the automaton has to go.

Definition 7 (Block Visibly Pushdown Automaton). *A VPA* $\mathcal{A} = (Q, \hat{\Sigma}, q_0, \Gamma, \delta, Q_F)$ *is a Block Visibly Pushdown Automaton (BVPA) if there is a function* $Target : Q \times \Sigma_{call} \mapsto Q$ *and an equivalence relation* $\overset{Q}{\leftrightarrow}$ *on* Q *such that:*

- *for all* $(q, c) \in Q \times \Sigma_{call}$, $q_0 \overset{Q}{\not\leftrightarrow} Target(q, c)$
- *if* $q \xrightarrow{i} q'$ *then* $q \overset{Q}{\leftrightarrow} q'$,
- *if* $q \xrightarrow{c/(q,c)} q'$ *then* $q' = Target(q, c)$,
- *if* $q \xrightarrow{r/(q',c)} q''$ *then* $q' \overset{Q}{\leftrightarrow} q''$.
- *if* $Target(q, c) \overset{Q}{\leftrightarrow} Target(q', c')$ *then* $Target(q, c) = Target(q', c')$

To conver a VPA to a BPA one can make a copy of $\mathcal{A}$ for each state q of $\mathcal{A}$. This copy will simulate $\mathcal{A}$ afer reading a call pointing at q.

Proposition 1. *Given a VPA of size* n*, there is an equivalent BVPA of size* $\mathcal{O}(n^2)$*, that can be computed in quadratic time.*

Due to Proposition 1, every VPA has an equivalent BVPA of quadratic size. So approximative minimization of BVPA is as difficult as approximative minimization of VPA. We propose here a weaker minimization in the same sense as SEVPA minimization preserving $\overset{\Sigma}{\leftrightarrow}$ relation (cf. Section 4.2). Here again, our minimization will keep the structure of call transitions fixed. The automata studied in this section are not call driven any more. So we need a new way to characterize the structure of call transitions. Let $\mathcal{A}$ and $\mathcal{B}$ be two BVPA recognizing the same language L. $\mathcal{A}$ and $\mathcal{B}$ will have the same structure if when reading $u \in L$, the two automata pass through the "same" blocks simultaneously.

We use $Pref(L)$ for the set of prefixes of a language L. Take a BVPA recognizing L: $\mathcal{A} = \langle Q, \hat{\Sigma}, q_0, \Gamma, \delta, Q_F \rangle$ together with $Target$ and $\overset{Q}{\leftrightarrow}$. Let T be the range of $Target$. The *associated partition* is the partition of $K_L = \{uc | u \in MR(\hat{\Sigma}), c \in \Sigma_{call}, uc \in Pref(L)\}$ defined by:

$$\text{for every } t \in T,\ K_t = \{uc | u \in MR(\hat{\Sigma}), c \in \Sigma_{call}, \delta_0(uc) = t, uc \in Pref(L)\}$$

Theorem 5. *Given a consistent BVPA, in a cubic time it is possible to find the unique (up to isomorphism) minimal equivalent BVPA with the same associated partition.*

The proof uses again the method of translations, but this time the function Φ is not a homomorphism. Φ adds to a word m, after each occurrence of a call symbol $c \in \Sigma_{call}$, the state reached by $\mathcal{A}$ during the execution of m when reading this c.

Remark 6. A $\overset{\Sigma}{\leftrightarrow}$-SEVPA is a BVPA whose associated partition $(K_t)_{t \in T}$ verifies: uc and $u'c'$ are in the same K_t iff $c \overset{\Sigma}{\leftrightarrow} c'$. So the previous theorem gives a minimization of $\overset{\Sigma}{\leftrightarrow}$-SEVPA as a special case.

We are now able, given a BVPA, to find the minimal equivalent BVPA of same associated partition. This is a first step in minimization out of the class of CDA. But this is not sufficient, as the example in section 3 shows that the blow-up between the minimal equivalent BVPA and the minimal equivalent BVPA with respect to a given associated partition can be exponential; which, again, is also an upper bound. Indeed, recall L_k is the VPL $L_k = a_1cL_{a_1} + \cdots + a_kcL_{a_k}r$ over $\hat{\Sigma} = (\{c\}, \{r\}, \{a_1, \ldots, a_k\})$, where $L_{a_i} \subseteq \{a_1, \ldots, a_n\}^*$ is the set of words where the number of a_i is even. Here K_{L_k} is $\{a_1c, \ldots, a_kc\}$. The minimal BVPA of associated to the trivial partition of (K_{L_k}) is again of size bigger than 2^k. The example of Figure 2 gives a BVPA of associated partition $(\{a_1c\}, \ldots, \{a_kc\})$ recognizing L_k and of size in $\mathcal{O}(k^2)$.

So the choice of the associated partition has a big influence on the size of the automaton. Nevertheless, in the case of SEVPA, even with an optimal choice of the equivalence relation $\overset{\Sigma}{\leftrightarrow}$ over Σ_{call} the minimal automata $\overset{\Sigma}{\leftrightarrow}$-SEVPA can be exponentially bigger than an equivalent VPA. In the case of BVPA the choice of an optimal associated partition gives a BVPA of quadratic size with respect to that of a minimal VPA.

6 Conclusion

Our results indicate that the class eCDA is the best choice for minimization if the call alphabet is small. The class CDA is more interesting than SEVPA because it includes SEVPA and moreover it permits a general minimization result without requiring to preserve additional structure. Class MEVPA is still interesting, but: for getting canonical machines eCDA is simpler, and for getting the smallest machines possible CDA is better because it is strictly larger. The problem with all these classes is that VPA are exponentially more succinct than CDA.

The class of BVPA seems quite promising. Our results show that minimization of VPA can be understood as fixing a partition of calls. At present we do not know how to calculate a good partition. Still in some contexts it may be easy to guess a partition that gives good results. Example 2 shows such a partition that is simple but takes us outside CDA.

The above results can be compared with the situation for regular languages. Deterministic automata are not the most succinct way of representing regular languages. Nondeterministic automata or even two-pass deterministic automata (which reads the input first from left to right and then from right to left) are exponentially more succinct. Still, as no minimization, even approximative, is known for other classes, deterministic automata are widely used.

References

1. Alur, R., Etessami, K., Madhusudan, P.: A temporal logic of nested calls and returns. In: Jensen, K., Podelski, A. (eds.) TACAS 2004. LNCS, vol. 2988, pp. 467–481. Springer, Heidelberg (2004)

2. Alur, R., Kumar, V., Madhusudan, P., Viswanathan, M.: Congruences for visibly pushdown languages. In: Caires, L., Italiano, G.F., Monteiro, L., Palamidessi, C., Yung, M. (eds.) ICALP 2005. LNCS, vol. 3580, pp. 1102–1114. Springer, Heidelberg (2005)
3. Alur, R., Kumar, V., Madhusudan, P., Viswanathan, M.: Congruences for visibly pushdown languages. Technical Report UIUCDCS-R-2005 -2565, UIUC, Technical report (2005)
4. Alur, R., Madhusudan, P.: Visibly pushdown languages. In: Proceedings of STOC, pp. 202–211. ACM Press, New York (2004)
5. Berstel, J., Boasson, L.: Balanced grammars and their languages. In: Brauer, W., Ehrig, H., Karhumäki, J., Salomaa, A. (eds.) Formal and Natural Computing. LNCS, vol. 2300, pp. 419–426. Springer, Heidelberg (2002)
6. Dill, D.: Timing assumptions and verification of finite-state concurrent systems. In: Sifakis, J. (ed.) Automatic Verification Methods for Finite State Systems. LNCS, vol. 407, pp. 197–212. Springer, Heidelberg (1990)
7. Kumar, V., Madhusdan, P., Viswanathan, M.: Visibly pushdown automata for streaming XML. In: Proceedings of WWW, pp. 1053–1062. ACM Press, New York (2007)
8. Kumar, V., Madhusudan, P., Viswanathan, M.: Minimization, learning, and conformance testing of boolean programs. In: Baier, C., Hermanns, H. (eds.) CONCUR 2006. LNCS, vol. 4137, pp. 203–217. Springer, Heidelberg (2006)
9. Loding, C., Madhusudan, P., Serre, O.: Visibly pushdown games. In: Lodaya, K., Mahajan, M. (eds.) FSTTCS 2004. LNCS, vol. 3328, Springer, Heidelberg (2004)
10. McNaughton, R.: Parenthesis grammars. Journal of the ACM 14, 490–650 (1967)
11. Mehlhorn, K.: Pebbling mountain ranges and its application of dcfl-recognition. In: de Bakker, J.W., van Leeuwen, J. (eds.) Automata, Languages and Programming. LNCS, vol. 85, pp. 422–435. Springer, Heidelberg (1980)
12. Michon, J-F., Champarnaud, J.-F.: Automata and binary decision diagrams. In: Champarnaud, J.-M., Maurel, D., Ziadi, D. (eds.) WIA 1998. LNCS, vol. 1660, pp. 742–746. Springer, Heidelberg (1999)
13. Murawski, A., Walukiewicz, I.: Third-order idealized algol with iteration is decidable. In: Sassone, V. (ed.) FOSSACS 2005. LNCS, vol. 3441, pp. 202–218. Springer, Heidelberg (2005)
14. Chervet, P., Walukiewicz, I.: Minimizing variants of visibly pushdown automata (2007), `http://www.labri.fr/perso/igw/publications.html`
15. Pitcher, C.: Visibly pushdown expression effects for XML streem processing. In: Programming Language Techonologies for XML, pp. 1–14 (2005)

Linear Circuits, Two-Variable Logic and Weakly Blocked Monoids

Christoph Behle[1], Andreas Krebs[1], and Mark Mercer[2]

[1] WSI - University of Tuebingen, Sand 13, 72076 Tuebingen, Germany
[2] McGill University, Montreal, Canada

Abstract. Following recent works connecting two-variable logic to circuits and monoids, we establish, for numerical predicate sets $\mathfrak{P}$ satisfying a certain closure property, a one-to-one correspondence between $FO[<,\mathfrak{P}]$-uniform linear circuits, two-variable formulae with $\mathfrak{P}$ predicates, and weak block products of monoids. In particular, we consider the case of linear TC^0, majority quantifiers, and finitely typed monoids. This correspondence will hold for any numerical predicate set which is $FO[<]$-closed and whose predicates do not depend on the input length.

1 Introduction

The computational power of boolean circuits are of great interest as they are among the most powerful classes of computation devices for which we can prove nontrivial lower bounds [7]. To understand the power granted by nonuniformity in this setting, we often consider circuit families which can be generated under bounded resources.

In the case of small depth circuits, we are particularly interested in circuit families whose structure can be described in terms of some restricted class of logical formulae (Barrington, Immerman, and Straubing [2]). Such circuit families can often be characterized in terms of logic. For instance, the languages recognized by $FO[+,*]$-uniform AC^0 circuits are exactly those which are expressible by $FO[+,*]$ formulae. Likewise, $FO + MOD[+,*]$ formulae correspond to ACC^0 circuits, and $FO + MAJ[+,*]$ formulae correspond to TC^0. This establishes a strong connection between circuit classes and logical formulae.

The class of languages recognized by logical formula can be also characterized in terms of algebra. For instance, the class of languages recognized by $FO[<]$ formula corresponds exactly to the class of *star-free* languages, which are exactly those which are recognized via morphisms to finite aperiodic monoids, or equivalently, block products of U_1. This gives us a three-fold connection between circuits, logic and algebra.

In the case of AC^0 and ACC^0, restricting to linear size corresponds in logic to a restriction to using only two variables. This was shown in [9], and corresponds in algebra to weakly-blocked monoids. In [21] Therién and Wilke gave for first order formulae over two variables with the order predicate an algebraic characterization as the variety **DA**. By an result of Straubing and Therién [20] this

L. Kučera and A. Kučera (Eds.): MFCS 2007, LNCS 4708, pp. 147–158, 2007.

variety, and thus $FO_2[<]$, can be characterized as weakly blocked monoids of U_1. Analogously, $FO+MOD_2[<]$ was shown in [19] to correspond to the variety $\mathbf{DO} \Box \mathbf{G_{sol}}$, which is the closure of **DA** under weak block products of abelian groups.

The notion of *finitely typed monoids* was introduced in [8] to obtain an algebraic characterization for $\mathrm{TC}^0 = \mathcal{L}(MAJ[<, Sq])$ in terms of finitely typed monoids. It is clear that general numerical predicates, as well as linear TC^0, need the use of infinite monoids. In this paper, we show that types can be used to algebraically characterize logical formulae for many types predicate sets in a uniform way. Second, we apply these results to give matching logical and algebraic characterizations to a broad class of uniformity conditions for linear TC^0, which include the $FO[<]$-closure of the predicate sets $\{<\}$, $\{<, +\}$, and $\{<, arb\}$.

In particular we show, subject to a closure property of $\mathfrak{P}$, that the following properties of a language L are equivalent: (1) that it is recognized by a $FO[<, \mathfrak{P}]$-uniform family of TC^0 circuits of linear size and linear fan-in, (2) that it can be described by a $FO + \widehat{MAJ}_2[<, \mathfrak{P}]$ formula, and (3) that it is recognized by a restricted type of morphism into a particular type of finitely typed group, constructed from weak block products of simpler groups. Recent results suggest that these characterizations can be used to prove lower bounds on linear sized circuits [5].

The remainder of the paper is structured as follows. In Sections 2 and 3 we review notions from circuits, logic, and algebra which we will require in the exposition. In Section 4 we state the main result of this paper, and in the remaining sections we prove three inclusions that yield our result.

2 Definitions

2.1 Logic

Following the conventions of Straubing's book [17], we express words $w \in \Sigma^*$ of length n as structures over the universe $[n] = \{1, \ldots, n\}$ in the following way. For each $\sigma \in \Sigma$ we have a unary relation Q_σ such that $Q_\sigma(x)$ is true when the value of w at the position x is σ. A formula ϕ over a set of free variables $\mathcal{V}$ is interpreted over $\mathcal{V}$-*structures*, which are strings $w = (w_1, \mathcal{V}_1)(w_2, \mathcal{V}_2) \ldots (w_n, \mathcal{V}_n)$ over $\Sigma \times 2^{\mathcal{V}}$, where the $\mathcal{V}_i$s are disjoint and $\bigcup_i \mathcal{V}_i = \mathcal{V}$. We define $\Sigma^* \otimes \mathcal{V}$ to be the set of all $\mathcal{V}$-structures over Σ^*, while we use $(\Sigma \times 2^{\mathcal{V}})^*$ to denote the set of arbitrary strings over $\Sigma \times 2^{\mathcal{V}}$. Let $L_{\phi,\mathcal{V}}$ be the set of all $\mathcal{V}$-structures modeling ϕ. Then for any first-order sentence ψ we can associate a language $L_\psi = L_{\psi,\emptyset}$.

A predicate is called *numerical* if its truth value does not depend on the input. (See Section 2.2) Let $\mathfrak{P}$ be a set of numerical predicates. A first-order formula over $\mathcal{V}$ is a first order formula built from the atomic formulae $\{Q_\sigma(x)\} \cup \{P \mid P \in \mathfrak{P}\}$ and free variables $\mathcal{V}$.

There are several cases in the literature where a new quantifier has been defined to obtain a correspondence between logic and algebra. For example, $Mod\, x\, \phi(x)$ [18] has been used to connect $FO+MOD$ formulae to ACC^0 circuits.

Likewise, TC^0 was shown to correspond to logical formulae using the majority quantifier $Maj\, x\, \phi(x)$, which is true iff for more than half of the positions x the formula $\phi(x)$ evaluates to true. This construction requires that we can use the logic to simulate *counting quantifiers* $\exists^{=y} x \phi(x)$ [10], which are true if and only if there are y many positions for x where $\phi(x)$ is true. But since a counting quantifier is defined with two variables, one has difficulties to apply this result in the case of two-variable logic. This leads to the following definition, which is equivalent in power to counting quantifiers and thus majority quantifiers if the number of variables is not restricted, but gives the right expressibility in the case of two variables to capture linear TC^0.

Definition 1 (Extended Majority Quantifier). *Let $\phi_1(x), \ldots, \phi_c(x)$ be formulae with one free variable. Then $Maj\ x\ \langle \phi_1(x), \ldots, \phi_c(x) \rangle$ is a formula. We define the semantics so that the formula is true if $w_{x=i} \models \phi_j$ for the majority of $(i, j) \in [n] \times [c]$. In other words,*

$$w \models Maj\ x\ \langle \phi_1, \ldots, \phi_c \rangle \Leftrightarrow 0 < \sum_{i=1}^{n} \sum_{j=1}^{c} \begin{cases} 1 & \text{if } w_{x=i} \models \phi_j \\ -1 & \text{otherwise} \end{cases}$$

In the case of $c = 1$ we have the old definition of the majority quantifier.

Definition 2. *$FO + \widehat{MAJ}_2[<, \mathfrak{P}]$ is the class of two-variable logical sentences over words which are constructed from atomic formulae, the order predicate, numerical predicates from the set $\mathfrak{P}$, and the extended majority quantifier.*

2.2 Numerical Predicates

A c-ary predicate P is called *numerical* if the truth value of $P(x_1, \ldots, x_c)$ depends only on the the numeric value of $x_1, \ldots, x_c$ and the length of the input word. An assignment to a c-ary predicate can be expressed as a $\mathcal{V}$-structure over a unary alphabet with $\mathcal{V} = \{x_1, \ldots, x_c\}$. A predicate is said to be expressible in logic $\mathcal{Q}[\mathfrak{P}]$ if the corresponding $\mathcal{V}$-structures are expressible in first order with quantifiers $\mathcal{Q}$ and predicates $\mathfrak{P}$. We can naturally represent such predicates as subsets of $\mathbb{N}^{c+1}$. For a predicate P we have the subset $\mathcal{P} = \{(i_1, \ldots, i_c, n) \mid a^n_{x_1=i_1,\ldots,x_c=i_c} \models P\}$.

Definition 3 (Shifting Predicates). *A numerical c-ary predicate P is a shift of a numerical predicate P', if there exist integers $v_1, \ldots, v_{c+1}$ such that $\mathcal{P} = \{(i_1, \ldots, i_c, n) \mid (i_1 + v_1, \ldots, i_c + v_c, n + v_{c+1}) \in \mathcal{P}'\}$.*

Now we define the closure properties of predicates we need in this paper. For a set $\mathfrak{P}$ of numerical predicates, we say that a numerical predicate P is *$FO[<]$-constructible from* $\mathfrak{P}$ if P can be expressed by a $FO[<, \mathfrak{P}]$ formula.

Definition 4. *We denote by $\overline{\mathfrak{P}}$ the smallest set of predicates that contains $\mathfrak{P}$ and is closed under $FO[<]$-constructions and shifting.*

In the case of $\{<\}$, $\{<, +\}$, $\{<, +, *\}$ we have that $\overline{\{<\}}$, $\overline{\{<, +\}}$, $\overline{\{<, +, *\}}$ are the $FO[<]$ closure of these predicate sets, i.e. the shifting closure does not introduce new predicates. Shifting may, in general, add extra predicates for predicates that depend on the length of the word.

2.3 Circuits

In this paper we consider circuits which compute functions $f : \Sigma^n \rightarrow \{0, 1\}$. Our circuits will consist of *majority gates* and *input query gates*. A majority gate is true when more than half of the inputs are true and an $Inp_\sigma(i)$ query gate will output true when the ith letter of the input is σ.

A family $\{C_n\}_{n\in\mathbb{N}}$ of such circuits can be said to recognize a language in the usual way. The complexity class TC^0 consists of those languages recognized by families of threshold circuits of constant depth and polynomial size. We define LTC^0 to be the class of languages recognized by TC^0 circuit families of linear size and linear fan-in.

We consider the class of LTC^0 circuits with a uniformity condition that is expressed in terms of first order formulae over words. As in [6], we need the following definition in order to construct a uniformity language that can be expressed by $FO[<]$ formulae: For $v = (v_1, \ldots, v_c) \in [n]^c$, the unary shuffled encoding $\langle v_1, \ldots, v_c\rangle$ of v is the word w of length n over alphabet $\{\alpha, \beta\}^c$ defined by $\pi_j(w_i) = \alpha \Leftrightarrow v_j \leq i$, where $\pi_j((a_1, \ldots, a_c)) = a_j$.

Definition 5 (Uniformity language). *Let $C = \{C_n\}$ be an LTC^0 circuit family. Fix $c \in \mathbb{N}$, a labeling of the gates of each C_n with tuples $(x_1, x_2) \in [n]\times[c]$, and a unique identifier from $[|\Sigma| + 1]$ for each possible type of gate (i.e. Inp_α or majority). Additionally, we require $(1, 1)$ to be the output gate of the circuit. Then a uniformity language of C is the set of all shuffled encodings $\langle x_1, x_2, y_1, y_2, t\rangle$ such that if t denotes majority gate, then the gate (x_1, x_2) is a majority gate and has gate (y_1, y_2) as an input gate, or if t denotes an Inp_σ gate, then (x_1, x_2) is an $Inp_\sigma(y_1)$ query gate (y_2 is arbitrary).*

Using the definition of an uniformity language we can easily define uniform circuits for our setting.

Definition 6 (Uniform LTC^0). *$FO[<, \mathfrak{P}]$-uniform LTC^0 is the class of languages recognizable by a family of LTC^0 circuits with a uniformity language expressible in $FO[<, \mathfrak{P}]$.*

3 Finitely Typed Groups

In this section we recall the definition of finitely typed groups introduced in [8]. The motivation for finitely typed groups arises from the fact that the syntactic monoid of the majority function is infinite, yet the majority gates have a finite output. Typed groups allow us to model majority gates as morphisms in a meaningful way.

Let T be a group. A *type* of T is a collection of disjoint subsets $\mathfrak{T} = \{T_i \mid i \in I\}$ of T for finite I. A *finitely typed group* is a group T equipped with a type $\mathfrak{T}$.

We call the elements of the boolean closure of $\mathfrak{T}$ the *extended types* of T. If the type set $\mathfrak{T}$ of T is understood we often simply write T instead of $(T, \mathfrak{T})$. Note that a finite monoid T can be regarded as a finitely typed monoid equipped with the type $\mathfrak{T} = \{\{t\} \mid t \in T\}$. The *direct product* $(S, \mathfrak{S}) \times (T, \mathfrak{T})$ of two finitely typed monoids $(S, \mathfrak{S})$ and $(T, \mathfrak{T})$ is the usual Cartesian product equipped with the type $\mathfrak{S} \times \mathfrak{T} = \{\mathcal{S} \times \mathcal{T} \mid \mathcal{S} \in \mathfrak{S},\ \mathcal{T} \in \mathfrak{T}\}$.

In the following we extend the notion of *block products* to finitely typed groups. Let $(S, \mathfrak{S})$, $(T, \mathfrak{T})$ be finitely typed monoids. Recall that the ordinary block product of S with T is defined as the bilateral semidirect product $S^{T \times T} * * T$ of $S^{T \times T}$, the set of all functions from $T \times T$ to S, with T, where the right (resp. left) action of T on $S^{T \times T}$ is given by $(f \cdot t)(t_1, t_2) = f(t_1, t\ t_2)$ $(t \cdot f)(t_1, t_2) = f(t_1 t, t_2)$, t, t_1, $t_2 \in T$, $f \in S^{T \times T}$. Note that this set may be uncountable in the case that S and T are infinite. As in [8], a discrete version of the block product is defined. We begin by defining a set of qualified functions:

Definition 7 (Type respecting functions). *A function $f(t_1, t_2) : (T, \mathfrak{T}) \times (T, \mathfrak{T}) \to S$, where S is any set, is called type respecting if it has a finite image and, for each $s \in S$, the preimage $f^{-1}(s)$ can be described by a finite boolean combination of conditions of the form $t_1 \cdot c_1 \in \mathcal{T}_1, c_2 \cdot t_2 \in \mathcal{T}_2, t_1 \cdot c_3 \cdot t_2 \in \mathcal{T}_3$ where c_1, c_2, c_3 are constants in T and $\mathcal{T}_1, \mathcal{T}_2, \mathcal{T}_3$ are types in $\mathfrak{T}$.*

The definition of the block product is the same as in the finite case but restraining the functions used to type respecting functions.

Definition 8 (Block product). *Let $(S, \mathfrak{S})$, $(T, \mathfrak{T})$ be finitely typed monoids and let V be the set of all type respecting functions with respect to T. The* finitely typed block product *$(X, \mathfrak{X}) = (S, \mathfrak{S}) \boxdot (T, \mathfrak{T})$ of $(S, \mathfrak{S})$ with $(T, \mathfrak{T})$ is defined as the bilateral semidirect product $V * * T$ of V with T (with respect to the actions given above). The type set $\mathfrak{X}$ of X consists of all types $\hat{\mathcal{S}} = \{(f, n) \in X \mid f(e_S, e_S) \in \mathcal{S}\}$, where $\mathcal{S} \in \mathfrak{S}$ and e_S is the neutral element of S. We also write $\pi_1 \mathcal{X}$, with $\mathcal{X} \in \mathfrak{X}$, for the type $\mathcal{S} \in \mathfrak{S}$, such that $\hat{\mathcal{S}} = \mathcal{X}$.*

Note that for finite M and M' equipped with the type sets as above, every function $f : M \times M \to M'$ will be type respecting. Thus we have the ordinary definition of block product as a special case.

As usual we write the operation in V additively to provide a more readable notation. Note that this does not imply that V is commutative. By definition of the bilateral semidirect product we have:

$$(*) \quad (f_1, m_1) \ldots (f_n, m_n) = (\sum_{i=1}^{n} m_1 \ldots m_{i-1} \cdot f_i \cdot m_{i+1} \ldots m_n, m_1 \ldots m_n).$$

The neutral element of $(S, \mathfrak{S}) \boxdot (T, \mathfrak{T})$ is $(\mathbf{e}, e_T)$ where $\mathbf{e}$ is the function mapping all elements to the neutral element of S and e_T is the neutral element of T.

We also have the equivalence:

$$(f_1, m_1) \ldots (f_n, m_n) \in \mathcal{X} \Leftrightarrow \sum_{i=1}^{n} f_i(m_1 \ldots m_{i-1}, m_{i+1} \ldots m_n) \in \pi_1 \mathcal{X},$$

where $\pi_1 \mathcal{X}$ is the base type as in Definition 8 above.

Definition 9. *We say that a finitely typed monoid* $(T, \mathfrak{T})$ *recognizes the language* $L \subseteq \Sigma^*$ *if there is a morphism* $h : \Sigma^* \to T$ *and a subset* $\{T_1, \ldots, T_k\} \subseteq \mathfrak{T}$ *of types of* T *such that* $L = h^{-1}(\bigcup_{i=1}^{k} T_i)$.

Now we turn our attention to how we can characterize predicates via morphisms.

Theorem 1. *For each binary numerical predicate* $P(x, y)$ *there exists a finitely typed group* $(T, \mathfrak{T})$ *and a distinguished element* $m \in T$ *with the following properties:*

1. *there is a morphism* $h : (\{a\} \times 2^{\{x,y\}})^* \to T$ *with* $h((a, \emptyset)) = m$ *and an extended type* $\mathcal{T}$ *such that* $a^n_{x=i,y=j} \models P(x, y)$ *if and only if* $h(a^n_{x=i,y=j}) \in \mathcal{T}$.
2. *for all extended types* $\mathcal{T}$ *over* $\mathfrak{T}$ *and all morphisms* $h : (\{a\} \times 2^{\{x,y\}})^* \to T$ *with* $h((a, \emptyset)) = m$ *the predicate corresponding to the language* $h^{-1}(\mathcal{T}) \cap \{a\}^* \otimes \{x, y\}$ *is in* $\overline{\{P\}}$.

We call m *the* incremental element.

If $\mathfrak{P}$ is a set of predicates that are unary or binary the previous theorem is also true if we transform an unary predicate $P(x)$ into a binary predicate $P'(x, x)$. In following we always assume all predicates in the two-variable logic are binary predicates.

Definition 10 (Predicate group). *The tuple of a finitely typed group* $(T, \mathfrak{T})$ *and incremental element* m *is called a predicate group of* P *if it satisfies the conditions of Theorem 1.*

In the following we denote by $(T_P, \mathfrak{T}_P)$ and m_P the predicate group and incremental element for the predicate P. We define now the algebraic variety which corresponds to $FO + M\widehat{A}J_2[<, \mathfrak{P}]$.

Definition 11. *Let* $\mathfrak{P}$ *be a set of predicates. We let* $\mathbf{W}_{\mathbb{Z}}(\mathfrak{P})$ *be the smallest variety closed under weak block products with* $\bigtimes_{k=1}^{c}((\mathbb{Z}, \mathbb{Z}^+) \boxdot (\bigtimes_{l=1}^{c'_k}(T_{kl}, \mathfrak{T}_{kl})))$ *for* $c, c'_1, \ldots, c'_c \in \mathbb{N}$ *, where* $(T_{kl}, \mathfrak{T}_{kl})$ *are predicate groups for predicates* P_{kl}*, i.e.*

$$G \in \mathbf{W}_{\mathbb{Z}}(\mathfrak{P}) \implies G \boxdot (\bigtimes_{k=1}^{c}((\mathbb{Z}, \mathbb{Z}^+) \boxdot \bigtimes_{l=1}^{c'_k}(T_{kl}, \mathfrak{T}_{kl}))) \in \mathbf{W}_{\mathbb{Z}}(\mathfrak{P}).$$

We now introduce *restricted elements* to ensure that the predicate groups that appear in the structure of groups of our variety cannot be "abused". If we do not restrict the class of allowable morphisms, then the typed monoids above can simulate counting quantifiers by using the predicate group to simulate a quantifier which should not be possible with two-variable majority logic. To assure the predicate groups are used in the designated way we start with the following definition:

Definition 12 (Restricted Element). *We define inductively the set of restricted elements:*

1. *All elements of* $(\mathbb{Z}, \mathbb{Z}^+)$ *are restricted.*
2. *For each predicate group* $(T_P, \mathfrak{T}_P)$ *only the incremental element* m_P *is restricted.*
3. *An element* $x \in A \times B$ *is restricted iff* $\pi_1(x)$ *and* $\pi_2(x)$ *are restricted.*
4. *An element* $x \in A \boxdot B$ *is restricted iff all elements in the image of* $\pi_1(x)$ *are restricted and* $\pi_2(x)$ *is restricted.*

Definition 13. *A morphism* $h : \Sigma^* \to G$ *is restricted if all elements of* $h(\Sigma)$ *are restricted.*

The following definition yields the characterization of the languages that we deal with in this paper.

Definition 14. *For a variety* $\mathbf{V}$, $\mathcal{H}_R^{-1}(\mathbf{V})$ *is the set of all languages that are recognized by a some* $(T, \mathfrak{T}) \in \mathbf{V}$ *with a restricted morphism.*

4 Results

The main theorem translates the well known connections between two-variable logic, weak blocked monoids, and linear size circuits [21,19,9] to the case of majority. By establishing a similar uniformity result as in [6], we can show how the predicates used in logic have their counterparts in algebra and circuits.

Theorem 2. *Let* $\mathfrak{P}$ *be a set of predicates closed under* $FO[<]$*-constructions and shifting. The following are equivalent:*

1. $L \in FO[<, \mathfrak{P}]$*-uniform* LTC^0,
2. $L \in \mathcal{L}(FO + \widehat{MAJ}_2[<, \mathfrak{P}])$,
3. $L \in \mathcal{H}_R^{-1}(\mathbf{W}_\mathbb{Z}(\mathfrak{P}))$.

Proof. First we show that we can express a circuit family by a logic formula (Theorem 4). Then we show that a language in this logic can by recognized by a restricted morphism (Theorem 5). Finally, we show how to construct a circuit family for a restricted morphism (Theorem 6).

It is unknown whether the TC^0 depth hierarchy is strict. In the next theorem we show a relation between circuit depth and quantifier depth:

Theorem 3. *Let* $\mathfrak{P}$ *be a set of predicates closed under* $FO[<]$*-constructions and shifting.* $FO[<, \mathfrak{P}]$*-uniform* LTC^0 *circuits form a hierarchy in the circuits depth iff* $FO + \widehat{MAJ}_2[<, \mathfrak{P}]$ *form a hierarchy in the quantifier depth.*

Proof. The proof of Theorem 4 translates a circuit of depth d into a formula of depth $d + c$ for a constant c. Similarly the proof for Theorem 5 translates a formula of quantifier depth d in a homomorphism into a group of weak block depth $d + c$. The construction of a circuit in Theorem 6 from a group of weak block depth d yields a circuit of depth $c \cdot d$.

5 Circuits to Logic

In this section we show how we can transform a circuit into a logical formula. We proceed inductively, starting with the input gates.

The following lemma helps us to express the uniformity:

Lemma 1. *Let ϕ be a formula in $FO[<,\mathfrak{P}]$ such that L_ϕ is the uniformity language of a family of LTC^0 circuits. Then the following predicates are in $\overline{\mathfrak{P}}$:*

1. *for all $x_2, y_2 \in [c]$ the binary predicate $C_{x_2,y_2}(x_1, y_1)$ which is true iff the gate labeled (x_1, x_2) is connected to (y_1, y_2) in C;*
2. *for all $\sigma \in \Sigma, x_2 \in [c]$ the binary predicate $Inp_{\sigma,x_2}(x_1, y_1)$ which is true iff the gate labeled (x_1, x_2) is an input gate that checks if there is an σ at position y_1 in the input; and*
3. *for all $x_2 \in [c]$ the unary predicate $M_{x_2}(x_1)$ which is true iff the gate labeled (x_1, x_2) is a majority gate.*

Now we show that, given a subset of positions by a formula $\phi(x)$, we can express if a formula $\psi(x)$ is true for the majority of these positions.

Lemma 2 (Relativization). *Let $\phi(x)$ and $\psi(x)$ be formulae in $FO+\widehat{MAJ}_2[<,\mathfrak{P}]$ with one free variable. Then there exists a sentence in $FO+\widehat{MAJ}_2[<,\mathfrak{P}]$ that is modeled by w iff*

$$|\{i \mid w_{x=i} \models \phi(x) \wedge w_{x=i} \models \psi(x)\}| > |\{i \mid w_{x=i} \models \phi(x) \wedge \neg w_{x=i} \models \psi(x)\}|.$$

Proof. The formula $Maj\ x\ \langle \phi(x) \wedge \psi(x), \neg\phi(x) \vee \psi(x)\rangle$ will do. If $\phi(x)$ is false, both formula add to 0 in the evaluation of the extended majority quantifier. If $\phi(x)$ is true, the contribution of the two formulae to the sum will be $+2$ or -2 depending on the value of $\psi(x)$.

Theorem 4. *If L is recognized by a $FO[<,\mathfrak{P}]$-uniform family of LTC^0-circuits, then L can be expressed as a formula in $FO+\widehat{MAJ}_2[<,\overline{\mathfrak{P}}]$.*

Proof. The construction we use is standard (see e.g. [17,6]) but must be modified to work with two variables. Let $(C_n)_{n\in\mathbb{N}}$ be the LTC^0-circuit family recognizing L. By the assumption there is an $FO[<,\mathfrak{P}]$ formula ϕ that recognizes the uniformity language of $(C_n)_{n\in\mathbb{N}}$. As shown above we can assume that we have the predicates $C_{x_2,y_2}(x_1,y_1)$, $M_{x_2}(x_1)$, and $Inp_{\sigma,x_2}(x_1,y_1)$ in $\overline{\mathfrak{P}}$.

We now recursively construct a sentence ψ in $FO+\widehat{MAJ}_2[<,\overline{\mathfrak{P}}]$ which describes the same language as $(C_n)_{n\in\mathbb{N}}$. We construct formulae $\psi^{(d)}_{x_2}$ such that $\psi^{(d)}_{x_2}(x_1)$ is true iff gate (x_1,x_2) outputs true and has depth at most d. For $d=0$, (x_1,x_2) outputs true iff it is an input gate which outputs true, so:

$$\psi^{(0)}_{x_2}(x_1) = \bigvee_{\sigma\in\Sigma} \exists\, y_1\ (Inp_{\sigma,x_2}(x_1,y_1) \wedge Q_\sigma(y_1)).$$

Now let $G^{(d)}_{x_2}(x_1) =$

$$\begin{aligned} Maj\ y_1\ \langle\ & C_{x_2,1}(x_1,y_1) \wedge \psi^{(d-1)}_1(y_1), \neg C_{x_2,1}(x_1,y_1) \vee \psi^{(d-1)}_1(y_1), \ldots, \\ & C_{x_2,c}(x_1,y_1) \wedge \psi^{(d-1)}_c(y_1), \neg C_{x_2,c}(x_1,y_1) \vee \psi^{(d-1)}_c(y_1) \quad \rangle. \end{aligned}$$

This is the essential step. Observe that $G_{x_2}^{(d)}(x_1)$ models a majority gate at depth d. By the proof of Lemma 2 it evaluates to true iff the number of true predecessors is larger than the number of false predecessors. With the help of the formula $G_{x_2}^{(d)}(x_1)$, we define: $\psi_{x_2}^{(d)}(x_1) = M_{x_2}(x_1) \wedge G_{x_2}^{(d)}(x_1) \vee \psi_{x_2}^{(0)}(x_1)$. Finally, we define ψ to be the value of the gate labeled $(1,1)$, thus $\psi = \psi_1^{(d)}(1)$ where d is the depth of the circuit family.

6 Logic to Algebra

We will show that we can replace a logic formula over two variables by applying the *weak block product principle* a finite number of times. This extends the construction of [20].

Definition 15 (weak block product principle). *Let $\alpha : \Sigma^* \to (T, \mathfrak{T})$ be a morphism, Γ be a finite alphabet and $r : T \times \Sigma \times T \to \Gamma$ be a function such that $r_\sigma(t_1, t_2) = r(t_1, \sigma, t_2)$ is a type respecting function for all $\sigma \in \Sigma$. Then we define a length-preserving mapping $\tau_{r,\alpha} : \Sigma^* \to \Gamma^*$ by $\tau_{r,\alpha}(v_1 \cdots v_n) = w_1 \cdots w_n$, where $w_i = r(\alpha(v_1 \cdots v_{i-1}), v_i, \alpha(v_{i+1} \cdots v_n))$. If α is a restricted morphism, then we say $\tau_{r,\alpha}$ is restricted.*

As in the usual case [20], $\tau_{r,\alpha}$ is not a morphism. Without loss of generality we can assume an innermost formula of quantifier depth one to always be of the form $Maj\ x\ \langle Q_{\sigma_i}(x) \wedge P_i(x,y)\rangle_{i=1,\dots,c}$. First predicates using only y can be moved out of the scope of the quantifier, and the formulae inside the quantifier can be assumed to have the form $Q_{\sigma_i}(x) \wedge P_i(x,y)$ since the predicate set is closed under boolean combinations. We now proceed by induction on the depth.

Lemma 3. *Let ϕ be a formula in $FO + \widehat{MAJ}_2[<, \mathfrak{P}]$ with an innermost formula ψ of quantifier depth one over the alphabet Σ, and $\Gamma = \Sigma \times \{0,1\}$. We let ϕ' be the formula over Γ, which is ϕ if we replace $Q_\sigma(y)$ by $Q_{(\sigma,0)}(y) \vee Q_{(\sigma,1)}(y)$ and $\psi(y)$ by $\bigvee_{\sigma\in\Sigma} Q_{(\sigma,1)}(y)$. Then there exists a morphism $\alpha : \Sigma^* \to (T, \mathfrak{T}) = (\mathbb{Z}, \mathbb{Z}^+) \boxdot \bigtimes_{l=1}^{c} (T_{P_l}, \mathfrak{T}_{P_l})$ and type respecting function $r : T \times \Sigma \times T \to \Gamma$ such that $\tau_{r,\alpha}^{-1}(L_{\phi'}) = L_\phi$.*

Lemma 4. *Let ϕ be a formula in $FO + \widehat{MAJ}_2[<, \mathfrak{P}]$ of quantifier depth $d > 1$ over the alphabet Σ. Then there exists a finite alphabet Γ and a restricted mapping $\tau_{r,\alpha} : \Sigma^* \to \Gamma^*$ and a formula ϕ' in $FO + \widehat{MAJ}_2[<, \mathfrak{P}]$ of quantifier depth $d-1$ such that $L_\phi = \tau_{r,\alpha}^{-1}(L'_\phi)$.*

Lemma 5. *Let $\tau_{r,\alpha}$ be a restricted mapping with $\alpha : \Sigma^* \to (T, \mathfrak{T})$ and let $L \subseteq \Gamma^*$ be a language recognized by a morphism to $(S, \mathfrak{S})$. Then $\tau_{r,\alpha}^{-1}(L)$ is recognized by a morphism to $(S, \mathfrak{S}) \boxdot (T, \mathfrak{T})$.*

Theorem 5. *For each $L \in FO + \widehat{MAJ}_2[<, \mathfrak{P}]$ there is a $(T, \mathfrak{T})$ in $\mathbf{W}_{\mathbb{Z}}(\mathfrak{P})$ and a restricted morphism h such that $L = h^{-1}(T)$ for an* extended type *T over $\mathfrak{T}$.*

Proof. Let ϕ be a $FO + \widehat{MAJ}_2[<, \mathfrak{P}]$ formula of depth d with $L = L_\phi$. By applying Lemma 4 inductively we get a chain of mappings:

$$\Sigma^* \xrightarrow{\tau_{r_1,\alpha_1}} \Gamma_1^* \xrightarrow{\tau_{r_2,\alpha_2}} \Gamma_2^* - \cdots \rightarrow \Gamma_{d-2}^* \xrightarrow{\tau_{r_{d-1},\alpha_{d-1}}} \Gamma_{d-1}^*$$

and a $FO + \widehat{MAJ}_2[<, \mathfrak{P}]$ formula $\phi^{(d-1)}$ of depth one such that $L = \tau_{r_1,\alpha_1}^{-1} \circ \cdots \circ \tau_{r_{d-1},\alpha_{d-1}}^{-1}(L_{\phi^{(d-1)}})$.

The remaining formula ϕ' is of depth 1 and has no free variable $\phi' = Maj\ x\ \langle P_1(x) \wedge Q_{\sigma_1}(x), \ldots, P_c(x) \wedge Q_{\sigma_c}(x) \rangle$. hence it is easy to apply the construction of Lemma 3 for the morphism α. Since we do not have a free variable y we replace a_x by a_{xy} in the construction that simulates a variable y at the position x but is ignored by the formula ϕ'.

Now we have a morphism h' and a type $\mathcal{T}$ such that $L_{\phi^{(d-1)}} = h'^{-1}(\mathcal{T})$. By applying Lemma 5 inductively to $\tau_{r_{d-1},\alpha_{d-1}}$ up to τ_{r_1,α_1}, we will get a morphism $h : \Sigma^* \rightarrow (\cdots((T \square S_{d-1}) \square S_{d-2}) \cdots) \square S_1$, and a type $\mathcal{X}$ with $L = h^{-1}(\mathcal{X})$.

7 Algebra to Circuits

In order to model a morphism by a circuit, we will first split the morphism into mappings.

Lemma 6. *Let $h : \Sigma^* \rightarrow (S, \mathfrak{S}) \boxdot (T, \mathfrak{T})$ and $L = h^{-1}(\mathcal{X})$ for some type $\mathcal{X}$ of $(S, \mathfrak{S}) \boxdot (T, \mathfrak{T})$. Then there is a finite alphabet Γ and a map $\tau_{r,\alpha} : \Sigma^* \rightarrow \Gamma^*$ with $\alpha : \Sigma^* \rightarrow (T, \mathfrak{T})$ and a morphism $h' : \Gamma^* \rightarrow (S, \mathfrak{S})$ such that $\tau_{r,\alpha}^{-1}(h'^{-1}(\mathcal{S})) = L$ for some $\mathcal{S} \in \mathfrak{S}$. If h is restricted, then $\tau_{r,\alpha}$ and h' are also restricted.*

So if L is recognized by restricted morphism into a group $\mathbf{W}_\mathbb{Z}(\mathfrak{P})$, then there is a set of mappings $\tau_1, \ldots, \tau_d$ such that $L = \tau_1^{-1} \circ \cdots \circ \tau_d^{-1}(h^{-1}(\mathcal{T}))$, where all the morphisms map to a group of the form $\bigtimes_{k=1}^{c} \left((\mathbb{Z}, \mathbb{Z}^+) \boxdot \bigtimes_{l=1}^{c'_k} (T_{P_l}, \mathfrak{T}_{P_l}) \right)$.

Lemma 7. *A $FO[<, P]$-uniform LTC^0 circuit can compute the function $\tau_{r,\alpha}$ where $\alpha : \Sigma^* \rightarrow (T, \mathfrak{T}) = \bigtimes_{k=1}^{c} \left((\mathbb{Z}, \mathbb{Z}^+) \boxdot \bigtimes_{l=1}^{c'_k} (T_{P_l}, \mathfrak{T}_{P_l}) \right)$ is restricted. We require here for each letter $\gamma \in \Gamma$ the corresponding output gates to be labeled by (i, γ).*

Theorem 6. *Let $(T, \mathfrak{T}) \in \mathbf{W}_\mathbb{Z}(\mathfrak{P})$ recognize L then L is in $FO[<, \mathfrak{P}]$-uniform LTC^0.*

Proof. Let $h : \Sigma^* \rightarrow (T, \mathfrak{T})$ be a restricted morphism with $L = h^{-1}(\mathcal{T})$, where $\mathcal{T} \in \mathfrak{T}$. By applying Lemma 6 inductively we get a chain of mappings τ_{r_k,α_k} and a morphism h':

$$\Sigma^* \xrightarrow{\tau_{r_1,\alpha_1}} \Gamma_1^* \xrightarrow{\tau_{r_2,\alpha_2}} \Gamma_2^* - \cdots \rightarrow \Gamma_{d-2}^* \xrightarrow{\tau_{r_{d-1},\alpha_{d-1}}} \Gamma_{d-1}^* \xrightarrow{h'} T'$$

where $T' = \times_{k=1}^{c}\left((\mathbb{Z},\mathbb{Z}^+) \boxdot \times_{l=1}^{c'_k}(T_{P_l},\mathfrak{T}_{P_l})\right)$ and there is a $\mathcal{T}' \in \mathfrak{T}'$ such that $L = \tau^{-1}_{r_1,\alpha_1} \circ \cdots \circ \tau^{-1}_{r_{d-1},\alpha_{d-1}}(h'^{-1}(\mathcal{T}'))$.

To recognize $h'^{-1}(\mathcal{T}')$ we construct τ_{r_d,α_d} with $r(t_1,\sigma,t_2) = 1$ iff $t_1 \cdot t_2 \in \mathcal{T}$, $r(t_1,\sigma,t_2) = 0$ otherwise and $\alpha = h$. Then $\tau_{r_d,\alpha_d} = 1^n$ iff $w \in L$ and 0^n otherwise. Hence we can apply Lemma 7 to construct a circuit with only one output gate.

Now for each τ_{r_k,α_k} we can construct a circuit as in Lemma 7, by connecting these circuits together and also append the circuit for h' that we just created, we get a circuit that recognized L. To see that this circuit has a uniformity language in $FO[<,\mathfrak{P}]$, we label the gates (x_1,x_2) that belong to τ_{r_k,α_k} with $(x_1,(k,x_2))$ and the gates (x_1,x_2) that belong to h' by $(x_1,(d,x_2))$. Since we now have that the uniformity language for the individual circuit layers is in $FO[<,\mathfrak{P}]$, also the uniformity language for all layers is in $FO[<,\mathfrak{P}]$. The interconnection between these circuits is $FO[<]$-uniform since we always connect a series of output gates labeled by a tuple $(y_1,(d_k,y_2))$ where y_2 is a fixed constant to an input gate $(x_1,(d_{k+1},x_2))$ where $x_1 = y_1$ and x_2 is a fixed constant.

8 Discussion

In this paper we extend the known connections between linear circuits, two-variable logic, and weakly blocked algebra from the case of linear AC^0 and linear ACC^0 to the case of linear TC^0. This algebraic characterization can be used to prove that the word problem over A_5 (known to be complete for NC^1 [1]) is not in uniform LTC^0 [5].

$FO_2[<]$ (resp. $FO + MOD_2[<]$) was linked to weakly blocked U_1 (resp. $\mathbb{Z}_p$) but no connection to circuits is known. On the other hand, $FO_2[arb]$ (resp. $FO + MOD_2[arb]$) corresponds to linear AC^0 (resp. linear ACC^0). We obtain a three-way correspondence for predicate sets respecting certain closure properties. Our proofs also hold for the case of FO_2 and $FO + MOD_2$: The group $(\mathbb{Z},\mathbb{Z}^+)$, which simulates the quantifier, can be substituted by U_1, or by U_1 and $\mathbb{Z}_p$ to get results for those cases. In this way we obtain the possibility to handle predicate sets between the order predicate and arbitrary numerical predicates, e.g. $\overline{\{<,+\}}, \overline{\{<,+,*\}}$.

We want to thank Klaus-Jörn Lange and Stephanie Reifferscheid for helpful comments.

References

1. Barrington, D.A.: Bounded-width polynomial-size branching programs recognize exactly those languages in NC^1. J. Comp. System Sci. 38, 150–164 (1989)
2. Barrington, D.A., Immerman, N., Straubing, H.: On uniformity within NC^1. J. Comp. System Sci. 41, 274–306 (1990)
3. Barrington, D., Immerman, N., Lautemann, C., Schweickardt, N., Thérien, D.: The Crane Beach Conjecture. In: Proc. of the 16th IEEE Symposium On Logic in Computer Science, pp. 187–196. IEEE Computer Society Press, Los Alamitos (2001)

4. Barrington, D., Thérien, D.: Finite Monoids and the Fine Structure of NC^1. Journal of ACM 35(4), 941–952 (1988)
5. Behle, C., Krebs, A., Reifferscheid, S.: A_5 not in FO+MOD+MAJ$_2$[reg], `http://www-fs.informatik.uni-tuebingen.de/publi/a5notinltc0.pdf` (to appear)
6. Behle, C., Lange, K.-J.: $FO[<]$-Uniformity. In: IEEE Conference on Compuatational Complexity (2006)
7. Furst, M., Saxe, J.B., Sipser, M.: Parity circuits and the polynomial-time hierarchy. In: Proc. 22th IEEE Symposium on Foundations of Computer Science, pp. 260–270 (1981)
8. Krebs, A., Lange, K.-J., Reifferscheid, St.: In: Diekert, V., Durand, B. (eds.) STACS 2005. LNCS, vol. 3404, Springer, Heidelberg (2005)
9. Koucký, M., Lautemann, C., Poloczek, S., Thérien, D.: Circuit lower bounds via Ehrenfeucht-Fraissé games. In: Proc. 21st Conf. on Compuatational Complexity (CCC'06) (2006)
10. Lange, K.-J.: Some results on majority quantifiers over words. In: Proc. of the 19th IEEE Conference on Computational Complexity, pp. 123–129. IEEE Computer Society Press, Los Alamitos (2004)
11. Lautemann, C., McKenzie, P., Schwentick, T., Vollmer, H.: The descriptive complexity approach to LOGCFL. J. Comp. System Sci. 62, 629–652 (2001)
12. Lawson, M.: Finite Automata. Chapman & Hall/CRC (2004)
13. Rhodes, J., Tilson, B.: The Kernel of Monoid Morphisms. J. Pure Applied Alg. 62, 227–268 (1989)
14. Roy, A., Straubing, H.: Definability of Languages by Generalized First-Order Formulas over (N,+). In: Durand, B., Thomas, W. (eds.) STACS 2006. LNCS, vol. 3884, Springer, Heidelberg (to appear 2006)
15. Ruhl, M.: Counting and addition cannot express deterministic transitive closure. In: Proc. of 14th IEEE Symposium On Logic in Computer Science, pp. 326–334. IEEE Computer Society Press, Los Alamitos (1999)
16. Schweikardt, N.: On the Expressive Power of First-Order Logic with Built-In Predicates. In: Dissertation, Universität Mainz (2001)
17. Straubing, H.: Finite Automata, Formal Logic, and Circuit Complexity. Birkhäuser (1994)
18. Straubing, H., Thérien, D., Thomas, W.: Regular languages defined by generalize quantifiers. Information and Computation 118, 289–301 (1995)
19. Straubing, H., Thérien, D.: Regular Languages Defined by Generalized First-Order Formulas with a Bounded Number of Bound Variables. In: Ferreira, A., Reichel, H. (eds.) STACS 2001. LNCS, vol. 2010, pp. 551–562. Springer, Heidelberg (2001)
20. Straubing, H., Thérien, D.: Weakly Iterated Block Products of Finite Monoids. In: Rajsbaum, S. (ed.) LATIN 2002. LNCS, vol. 2286, pp. 91–104. Springer, Heidelberg (2002)
21. Thérien, D., Wilke, T.: Over Words, Two Variables are as Powerful as One Quantifier Alternation. In: Proc. 30th ACM Symposium on the Theory of Computing, pp. 256–263 (1998)

Combinatorial Proof that Subprojective Constraint Satisfaction Problems are NP-Complete

Jaroslav Nešetřil* and Mark Siggers

Department of Applied Mathematics and Institute for Theoretical Computer Science (ITI), Charles University Malostranské nám. 25, 11800 Praha 1 Czech Republic
nesetril@kam.mff.cuni.cz, mhsiggers@gmail.com

Abstract. We introduce a new general polynomial-time construction- the *fibre construction*- which reduces any constraint satisfaction problem CSP($\mathcal{H}$) to the constraint satisfaction problem CSP($\mathcal{P}$), where $\mathcal{P}$ is any subprojective relational structure. As a consequence we get a new proof (not using universal algebra) that CSP($\mathcal{P}$) is NP-complete for any subprojective (and thus also projective) relational structure. This provides a starting point for a new combinatorial approach to the NP-completeness part of the conjectured Dichotomy Classification of CSPs, which was previously obtained by algebraic methods. This approach is flexible enough to yield NP-completeness of coloring problems with large girth and bounded degree restrictions.

1 Introduction and Previous Work

Many combinatorial problems can be expressed as Constraint Satisfaction Problems (CSPs). This concept originated in the context of Artificial Intelligence (see e.g. [20]) and is very active in several areas of Computer Science. CSPs includes standard satisfiability problems and many combinatorial optimization problems, thus are also a very interesting class of problems from the theoretical point of view. The whole area was revitalized by Feder and Vardi [9], who reformulated CSPs as homomorphism problems (or H-coloring problems) for relational structures. Motivated by the results of [28] and [13], they formulated the following conjecture.

Conjecture 1 (Dichotomy). *Every Constraint Satisfaction Problem is either P or NP-complete.*

Schaefer [28] established the dichotomy for CSPs with binary domains, and Hell-Nešetřil [13] established the dichotomy for undirected graphs; it follows from [9] that the dichotomy for CSPs can be reduced to the dichotomy problem for H-coloring for oriented graphs. This setting, and related problems, have motivated

* Supported by grant 1M0021620808 of the Czech Ministry of Education and AEOLUS.

L. Kučera and A. Kučera (Eds.): MFCS 2007, LNCS 4708, pp. 159–170, 2007.

intensive research in descriptive complexity theory. This is surveyed, for example, in [7], [13] and [11].

Recently the whole area was put in yet another context by Peter Jeavons and his collaborators, in [15] and [5], when they recast the complexity of CSPs into properties of algebras and polymorphisms of relational structures. Particularly, they related the complexity of CSPs to a Galois correspondence between polymorphisms and definable relations (obtained by Bodnarčuk et al. [1] and by Gaiger [10]; see [25] and [26]). This greatly simplified elaborate and tedious reductions of particular problems and led to the solution of the dichotomy problem for ternary CSPs [2] and other results which are surveyed, for example, in [5] and [12]. This approach to studying CSPs via certain algebraic objects yields, in particular, that for every *projective* structure H the corresponding CSP(H) is an NP-complete problem [16], [15]. The success of these general algebraic methods gave motivation for some older results to be restated in this new context. For example, [4] treats H-coloring problems for undirected graphs in such a way that the dichotomy between the tractable and NP-complete cases of H-coloring problem agrees with the general CSP Dichotomy Classification Conjecture stated in [6].

In this paper we propose a new approach to the dichotomy problem. We define a general construction- the *fibre construction-* which allows us to prove in a simple way that for every projective structure H, CSP(H) is NP-complete. In fact we define a *subprojective* structure and prove that for every subprojective relational structure H, CSP(H) is NP-complete. Though by an example of Ralph McKenzie [27], we know there are structures H that are not sub-projective for which CSP(H) is NP-complete, this is a first step in a combinatorial approach to the CSP Dichotomy Conjecture. In a later paper we extend this approach to include all structures that are known to be NP-complete, and possibly others. This will provide a new CSP Dichotomy Classification Conjecture, which is one of the main results yielded by algebraic methods. A discussion of this new Conjecture can be found in the full version of this paper, [23].

The fibre construction lends easily to restricted versions of CSPs, so allows us to address open problems from [8] and [17]. In particular, for any subprojective structure H, we show that CSP(H) is NP-complete for instances of bounded degree. Thus a fibre construction approach to a CSP Dichotomy Classification will reduce the Feder-Hell-Huang conjecture that NP-complete CSPs are NP-complete for instances of bounded degree to the CSP Dichotomy Classification Conjecture.

Our approach is motivated by the *Sparse Incomparability Lemma* [22] and *Müller's Extension Theorem* [21] (both these results are covered in [14]). These results are recalled and extended in Sect. 5. Strictly speaking, we do not need these results for our main results, Thm. 3 and Cor. 4, but they provided an inspiration for early forms of the fibre construction in [29] and [30] and for the general case presented here. Moreover, we do need these results to address the Dichotomy Conjecture for instances of large girth, extending results of [17].

The fibre construction is simple, and is a refinement of gadgets, or indicator constructions [13,14], using familiar extremal combinatorial results [21,22,24]). However, the simplicity becomes obscured by the notation when dealing with general relational structures. Thus we find it useful to prove, in Sect. 3, only a simple case of the fibre construction. The case we prove is simple, but contains all the essential ingredients of the general fibre construction.

In Sect. 2 we introduce all the definitions and state the main results: Thm. 3 and Cor. 4. In Sect. 3 we prove a simple case of Thm. 3. In Sect. 4 we consider further applications of the fibre construction. Sect. 5 contains an extension of the some of the motivating results of our construction. Finally, in Sect. 6, we consider the relation of the fibre construction to the Dichotomy Classification Conjecture of [6].

2 Definitions and Statement of Results

We work with finite relational structures of a given type (or signature). A *type* is a vector $K = (k_i)_{i \in I}$ of positive integers, called *arities*. A *relational structure* $\mathcal{H}$ of type K, consists of a finite vertex set $V = V(\mathcal{H})$, and a k_i-ary relation $R_i = R_i(\mathcal{H}) \subset V^{k_1}$ on V, for each $i \in I$. An element of R_i is called an k_i-tuple. Thus a (di)graph is just a relational structure of type $K = (2)$. Its edges (arcs) are 2-tuples in the 2-ary relation R_1.

Given two relational structures $\mathcal{G}$ and $\mathcal{H}$ of the same type, an *$\mathcal{H}$-coloring of* $\mathcal{G}$ is a map $\phi : V(\mathcal{G}) \to V(\mathcal{H})$ such that for all $i \in I$ and every k_i-tuple $(v_1, \ldots, v_{k_i}) \in R_i(\mathcal{G})$, $(\phi(v_1), \ldots, \phi(v_{k_i}))$ is in $R_i(\mathcal{H})$. Fix a relational structure $\mathcal{H}$ (sometimes called *template*). CSP($\mathcal{H}$) is the following decision problem:

Problem CSP($\mathcal{H}$)
Instance: A relational structure $\mathcal{G}$;
Question: Does there exists an $\mathcal{H}$-coloring of $\mathcal{G}$?

We write $\mathcal{G} \to \mathcal{H}$ to mean that $\mathcal{G}$ has an $\mathcal{H}$-coloring.

A relational structure $\mathcal{H}$ is a *core* if its only $\mathcal{H}$-colorings are automorphisms.

It is well known, (see, for example, [14]) that $\mathcal{G} \to \mathcal{H}$ if and only if $\mathcal{G}' \to \mathcal{H}'$, where $\mathcal{G}'$ and $\mathcal{H}'$ are the cores of $\mathcal{G}$ and $\mathcal{H}$ respectively. Therefore, in the sequel, we only consider relational structures that are cores.

All relational structures of a given type form a category with nice properties. In particular, this category has products and powers which are defined explicitly as follows:

Given a relational structure $\mathcal{H}$, and a positive integer d, the d-ary *power* $\mathcal{H}^d$ of $\mathcal{H}$ is the relational structure of the same type as $\mathcal{H}$, defined as follows.

- $V(\mathcal{H}^d) = \{(v_1, \ldots, v_d) \mid v_1, \ldots, v_d \in V(\mathcal{H})\}$.
- For $i \in I$, $((v_{1,1}, v_{1,2}, \ldots, v_{1,d}), \ldots, (v_{k_i,1}, \ldots, v_{k_i,d}))$ is in $R_i(\mathcal{H}^d)$ if and only if all of $(v_{1,1}, v_{2,1}, \ldots, v_{k_i,1}), \ldots, (v_{1,d}, \ldots, v_{k_i,d})$ are in $R_i(\mathcal{H})$.

An $\mathcal{H}$-coloring of $\mathcal{H}^d$ (i.e. a homomorphism $\mathcal{H}^d \to \mathcal{H}$) is called a *$d$-ary polymorphism* of $\mathcal{H}$. A d-ary polymorphism ϕ is called a *projection* if there exists some $i \in 1, \ldots, d$ such that $\phi((v_1, \ldots, v_d)) = v_i$ for any $v_1, \ldots, v_d \in V(\mathcal{H})$. Let $\mathrm{Pol}(\mathcal{H})$, $\mathrm{Aut}(\mathcal{H})$ and $\mathrm{Proj}(\mathcal{H})$ be the sets of polymorphisms, automorphisms and projections (of all arities) of $\mathcal{H}$. A relational structure $\mathcal{H}$ is *projective* if for every $\phi \in \mathrm{Pol}(\mathcal{H})$, $\phi = \sigma \circ \pi$ for some $\sigma \in \mathrm{Aut}(\mathcal{H})$ and some $\pi \in \mathrm{Proj}(\mathcal{H})$. (It is shown in [19] that almost all relational structures are projective.)

The following definition of graphs that are, in a sense, locally projective, is our principal definition.

Definition 2. *A subset S of $V(\mathcal{H})$ is called* projective *if for every $\phi \in \mathrm{Pol}(\mathcal{H})$, ϕ restricts on S to the same function as does $\sigma \circ \pi$ for some $\sigma \in \mathrm{Aut}(\mathcal{H})$ and some $\pi \in \mathrm{Proj}(\mathcal{H})$. S is called* non-trivial *if $|S| > 1$. A relational structure $\mathcal{H}$ is called* subprojective *if it is a core and it contains a non-trivial projective subset.*

Note that any subset of a projective set is again projective. A structure is projective if and only if the set of all its vertices is projective. It is easy to find subprojective structures which fail to be projective.

The main tool of the paper is the following general indicator construction which we call the *fibre construction.* This construction extends a construction first used in a in a Ramsey theory setting in [29], and then proved in [30] in the present form, for $\mathcal{H} = K_3$ and $\mathcal{P}$ being projective. A special case of it is proved in Sect. 3, the full proof is relegated to the full version of the paper.

Theorem 3. *Let $\mathcal{H}$ be any relational structure, and let $\mathcal{P}$ be any subprojective relational structure. Then there exists a polynomial time construction, the fibre construction, $\mathcal{M}^{\mathcal{P}}_{\mathcal{H}}$ which provides for any instance $\mathcal{G}$ of $\mathrm{CSP}(\mathcal{H})$, an instance $\mathcal{M}^{\mathcal{P}}_{\mathcal{H}}(\mathcal{G})$ of $\mathcal{P}$ such that*

$$\mathcal{G} \to \mathcal{H} \iff \mathcal{M}^{\mathcal{P}}_{\mathcal{H}}(\mathcal{G}) \to \mathcal{P}.$$

Note that $\mathcal{H}$ and $\mathcal{P}$ need not be of the same type. Since $\mathrm{CSP}(K_3)$ is NP-complete, taking $\mathcal{H}$ in to be K_3 gives the following result.

Corollary 4. *For any subprojective relational structure $\mathcal{P}$, the problem $\mathrm{CSP}(\mathcal{P})$ is NP-complete.*

The fibre construction also has immediate applications to restricted versions of CSP complexity.

The *degree* of a vertex v in a relational structure $\mathcal{G}$ is the number of tuples it occurs in in $\bigcup R_i$, and the maximum degree, over all vertices in $\mathcal{H}$, is denoted by $\Delta(\mathcal{G})$. $\mathcal{G}$ is called *b-bounded* if $\Delta(\mathcal{G}) \leq b$.

It is conjectured in [8] that for any relational structure $\mathcal{H}$, if $\mathrm{CSP}(\mathcal{H})$ is NP-complete, then there is some finite b such that $\mathrm{CSP}(\mathcal{H})$ is NP-complete when restricted to b-bounded instances.

In [30], this was shown to be true in the case of graphs and projective relational structures $\mathcal{H}$. Further explicit bounds were given on $b(\mathcal{H})$, which is the minimum b such that $\mathrm{CSP}(\mathcal{H})$ is NP-complete when restriced to b-bounded instances. In Sect. 4, we observe the following corollary of the proof of Thm. 3.

Corollary 5. *For any subprojective relational structure $\mathcal{P}$,*

$$b(\mathcal{P}) < (4 \cdot \Delta(\mathcal{P})^6).$$

This greatly improves the bound on $b(H)$ from [30] in the case of sub-projective graphs H. We intend, in a later paper to show that all non-bipartite graphs are subprojective, thus applying this better bound to all graphs.

Degrees and short cycles are classical restrictions for coloring problems. Recall that *girth* $g(G)$ of a graph G is the length of the shortest cycle in G. We then observe that the following result about sparse graphs follows from our extension (from Sect. 5) of the *Sparse Incomparability Lemma* [22], [24].

Theorem 6. *Let H be a subprojective graph, and ℓ a positive integer. Then the problem* CSP(H) *is NP-complete when restricted to graphs with girth $\geq \ell$.*

This addresses a problem of [17] where the question of CSPs when restricted to instances with large girth was studied. This result can be generalized futher to relational structures but we decided to stop here.

3 The Fibre Construction

So called indicator constructions are often used relate the conplexity of different CSPs. The basic idea is that one can reduce CSP(H) to CSP(H') by constructing in polynomial time, for any instance G of CSP(H), an instance G' of CSP(H'), such that

$$G \to H \iff G' \to H'.$$

If CSP(H) is NP-complete, then CSP(H') must also be NP-complete. See the proof of the H-coloring dichotomy in [13] for an intricate use of such constructions.

One of the difficulties with indicator constructions is that one uses many ad hoc tricks to find a constrution for a particular graphs H' or H. The fibre construction, Thm. 3, is an indicator construction that will suffice for all reductions.

In this section, we prove the following simple case of the fibre construction.

Proposition 7. *There exists a polynomial time construction which provides for and graph G, a graph $M(G)$ such that*

$$G \to C_5 \iff M(G) \to K_3.$$

Ours is not the most elegant known reduction of C_5-coloring to K_3-coloring, but it has the advantage that it can be easily generalized. After the proof, we discuss a couple of issues that we must deal with in the general case, Thm. 3. The proof of the general case can be found in the full version of the paper.

3.1 Notation

Given an indexed set $W^* = [w_1^*, \ldots, w_d^*]$ of vertices, *a copy* W^a *of the set* W^* will mean the indexed set $W^a = [w_1^a, \ldots, w_d^a]$. Given two copies W^a and W^b of the same set W^* we say that we *identify* W^a *and* W^b *index-wise* to mean we identify the vertices w_i^a and w_i^b for $i = 1, \ldots, d$. When we define a function f on W^*, we will assume it to be defined on any copy W^a of W^* by $f(w_\alpha^a) = f(w_\alpha^*)$ for all $\alpha = 1, \ldots, d$. We will often refer to a function f on an indexed set W^* as a *pattern* of W^*. In the case that the image of f is contained in the vertex set of some graph H we speak about H*-pattern* of W^*.

3.2 The Fibre Gadget

The construction consists of two parts. In the first part we build a fibre gadget M which depends only on C_5 and K_3. To build the fibre gadget M we need the following simple lemma which is motivated by a result of Müller, [21].

Lemma 1. *Let $\mathcal{P}$ be a subprojective relational structure with non-trivial projective subset S. Let W be an indexed set, and let $\Gamma = \{\gamma_1, \ldots, \gamma_d\}$ be a set of S patterns of W satisfying the following condition (*).*

> *For any pair $w \neq w' \in W$, there exists some $\gamma \in \Gamma$ for which $\gamma(w) \neq \gamma(w')$.*

Then there exists a relational structure $\mathcal{M}$, isomorphic to $\mathcal{P}^d$, with $W \subset V(\mathcal{M})$, such that the set of $\mathcal{P}$-colorings of $\mathcal{M}$, when restricted to W, is exactly

$$\{\alpha \circ \gamma \mid \alpha \in \mathrm{Aut}(\mathcal{P}), \gamma \in \Gamma\}.$$

Proof (Proof of Lemma 1). Put $\mathcal{M} = \mathcal{P}^d$ and for each $w \in W$, identify w with the vertex $(\gamma_1(w), \ldots, \gamma_d(w))$ of $\mathcal{M}$. By condition (*), these are distinct elements of $V(\mathcal{M})$.

Since S is a projective subset of $\mathcal{P}$, the only $\mathcal{P}$-colorings of $\mathcal{M} = \mathcal{P}^d$ restrict on S^d, which contains W, to $\alpha \circ \pi$ where α is a automorphism of $\mathcal{P}$ and π is a projection. But the projections restrict on W to exactly the maps of Γ, so the lemma follows.

The following lemma provides the fibre gadget M.

Lemma 2. *There exists a graph M containing two copies W^a and W^b of an indexed set W^*, and a set $F = \{f_x \mid x \in V(C_5)\}$ of distinct K_3-patterns of W^*, such that the following conditions are true upto some permutation of $V(K_3)$.*

i. Any K_3-coloring of M, restricted to W^a, (or to W^b) is in F.
ii. For any K_3-coloring ϕ of M, ϕ restricts on W^a to f_x and on W^b to f_y for some edge xy of C_5.
iii. For any edge xy (or yx) of C_5, there is a K_3-coloring ϕ of M that restricts on W^a to f_x and on W^b to f_y.

Moveover, $M \cong (K_3)^{10}$.

Proof. Let $V(K_3) = \{0, 1, 2\}$. Let $W^* = [w_x^* \mid x \in V(C_5)]$, and let $F = \{f_x \mid x \in V(C_5)]$, where f_x is the $\{0,1\}$-pattern (K_3-pattern) defined by

$$f_x(w_y^*) = \begin{cases} 1 & \text{x = y} \\ 0 & \text{otherwise.} \end{cases}$$

Let $W = W^a \cup W^b$, where W^a and W^b are disjoint copies of W^* and let $\Gamma = \{\gamma_{xy} \mid xy \in E(C_5)\}$ where γ_{xy} is the $\{0,1\}$-pattern of W defined by

γ_{xy} restricted to W^a is f_x and restricted to W^b is f_y.

Observe that γ_{xy} and γ_{yx} are distinct elements of Γ for every edge xy of C_5, so $|\Gamma| = 10$.

Apply Lem. 1 to W and Γ. The instance $\mathcal{M}$ of CSP(K_3) that it returns is clearly the graph M that we are looking for.

The name 'fibre gadget' comes from the relation of the vertices of W^* to the set of K_3-patterns F. We view $w \in W^*$ as a fibre in $V(K_3)^{|F|}$, whose i^{th} postition corresponds to its image under the ith pattern f_{x_i} of F.

3.3 The Fibre Construction

The fibre gadgets are put together with the following construction, which call the fibre construction.

Construction 8. *Let W^*, F, and M be as in Lem. 2. Let G be an instance of* CSP(K_3), *and construct $M(G)$ as follows. (See Fig. 1.)*

i. For each vertex v of G let W^v be a copy of W^*.
ii. For each edge uv of G let M_{uv} be a copy of M. Index-wise, identify W^u and W^v with the copies of W^a and W^b, respectively, in M_{uv}.

Thus $M(G)$ consists of $|V(G)|$ copies of W^* and $|E(G)|$ copies of M. All vertices are distinct unless identified above.

We can now prove Prop. 7.

Proof. For any graph G let $M(G)$ be the graph defined by the fibre construction, Const. 8. As $M(G)$ is made of $|E(G)|$ copies of M, which is independent of G, this is a polynomial time construction.

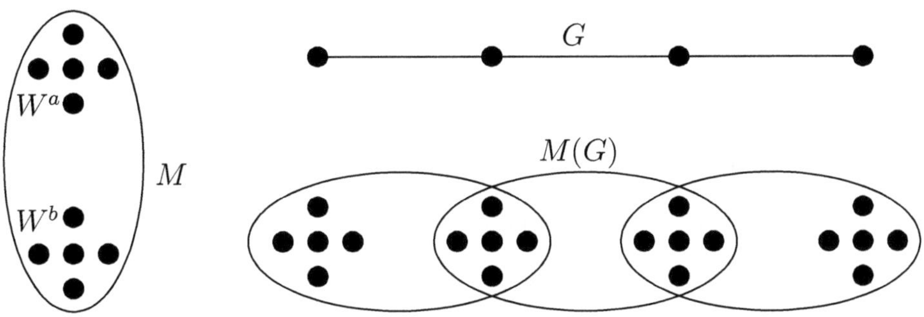

Fig. 1. Fibre Construction

Let ϕ be a K_3-coloring of $M(G)$. We show that this defines a C_5 colouring ϕ' of G. It is enough to show this for a compontent of $M(G)$. Now ϕ restricts on W^v, for each vertex v of G, to $\sigma \circ f$ for some permutation σ of $V(K_3)$ and some pattern f in F. Since the number of vertices of each color is constant over all patterns of F, and this property is not preserved under any permutaion of $V(K_3)$, the permutaion σ must be constant for all W^v. We assume that it is the identity permutation, so ϕ restricts on each W^v to some pattern f in F. Thus $\phi' : V(G) \to V(C_5)$ is well defined by letting $\phi'(v) = x$ where ϕ restricts on W^v to the pattern f_x. Moreover, by property (ii) of Lem. 2, ϕ' is a C_5-coloring of G.

On the other hand, given a C_5-coloring ϕ' of G we define a K_3-coloring ϕ of $M(G)$ as follows. For all vertices v of G, let ϕ be $f_{\phi'(v)}$ on the set W^v. For every edge uv of G, the sets W^u and W^v are already colored by ϕ, and we must extend this coloring to M^{uv}. Now ϕ restricts on W^u to $f_{\phi'(u)}$ and on W^v to $f_{\phi'(v)}$, and $\phi'(u)\phi'(v)$ is an edge of C_5, so by property (iii) of Lem. 2 ϕ can be extended to M^{uv}. Thus ϕ can be extended to a K_3-coloring of $M(G)$.

3.4 Remark

This outline gives only the idea of the general proof. There are several obstacles. For example, in the general case of relational structures, we will need a different fibre gadget for each kind of relation. And, of course, our relations need not be symmetric. In the general case, the set F from Lem. 2 will be S-patterns, instead of K_3-patterns. Also, for general $\mathcal{H}$, it may be more difficult to to ensure that Γ in the proof of Lem. 2 satisfies property (*) of Lem. 1. We thus use a more general version on Lem. 1 in which Γ need not satisfy (*), but which returns a $\mathcal{M}$ that is not necessarily isomorphic to $\mathcal{P}^d$.

These are just technicalities which can be handled with care.

4 Applications

4.1 Degree Bounded CSPs

We mentioned in the introduction, that because CSP(K_3) is NP-complete, taking $\mathcal{H} = K_3$, Cor. 4 follows from Thm. 3. In fact, CSP(K_3) is NP-complete for 4-bounded instances G.

For a 4-bounded graph G, the fibre construction would yield an instance $\mathcal{M}^{\mathcal{P}}_{K_3}(G)$ of CSP$(\mathcal{P})$ with maximum degree $(4 \cdot \Delta(\mathcal{P}))^6$. Thus Cor. 5, follows from the proof of Thm. 3.

4.2 Girth Restricted H-Coloring

The results in this subsection are for graphs.

The following lemma is proved in [21] in the case that P is a complete graph, and is proved in [24] without item (iii) in the case that P is projective. In both of these cases, $S = V(P)$.

Lemma 3. *Let P be a subprojective graph with projective subset S, and let $\ell \geq 3$ be an integer. Let W be an indexed set, and let $\Gamma = \{\gamma_1, \ldots, \gamma_d\}$ be a set of S patterns of W. Then there exists a relational structure M with $W \subset V(M)$, such that the following are true:*

i. The set of P-colorings of M, when restricted to W, is exactly

$$\{\alpha \circ \gamma \mid \alpha \in \mathrm{Aut}(P), \gamma \in \Gamma\}.$$

ii. M has girth at least ℓ.
iii. The distance, in M, between any two vertices of W is at least g.

Proof. This lemma follows from Thm. 11 which is a local form of the main result of [24]. The result will be stated in Sect. 5.

Using this in place of Lem. 1 in the fibre construction, we can ensure that the graph $\mathcal{M}_H^P(G)$ that is returned has girth ℓ. Thus Thm. 6 follows.

4.3 Conservative CSPs

A constraint satisfaction problem CSP($\mathcal{H}$) is *conservative* if $\mathcal{H}$ contains all possible unary relations. Such a CSP is also known as List $\mathcal{H}$-colouring.

In [3], Bulatov proves the following dichotomy for conservative CSPs.

Theorem 9. *[3] A conservative constraint satisfaction problem* CSP($\mathcal{H}$) *is NP-complete if there there is a set $B \subset V(\mathcal{H})$ of size at least 2 such that for any polymophism ϕ of $\mathcal{H}$, ϕ restricted to B is essentially unary. Otherwise,* CSP($\mathcal{H}$) *is polynomial time solvable.*

The difficult part of Bulatov's paper is the polynomial time solvable part of this result. The NP-complete part follows quickly from the algebraic approach of [15] and [5]. We observe that the NP-complete part is also immedieate from Cor. 4. Indeed, since we only consider cores $\mathcal{H}$, any essentially unary operation on B is $\alpha \circ \pi$ where π is a projection of B^d to B and α is an automorphism of $\mathcal{H}$. Thus B is a projective subset of $\mathcal{H}$.

5 Coloring Theorems - Combinatorial Background

The main motivation for our construction is a result of Müller [21], which is a special graph case of Lem. 1, except that it returns a graph M of arbitrary girth. The difficult part of the lemma is, of course, ensuring that M has arbitrary girth. He did this with a special case of Lem. 11. Müller's lemma was extended in [24], and the form here, is a localisation of their version.

Localising the notion of H-pointed graphs H, from [24], we get the following definition.

Definition 10. *Let H, H' be graphs. Subsets S of $V(H)$ and S' of $V(H')$ are said to be (H, H')*-pointed subsets *if for any two homomorphisms $g, g' : H \to H'$ which satisfy $g(x) = g'(x) \in S'$, whenever $x \neq x_0$ and $x \in S$ (for some fixed vertex $x_0 \in S$), then $g(x_0) = g'(x_0) \in S'$.*

Theorem 11. *For every graph H and every choice of positive integers k and l there exists a graph G together with a surjective homomorphism $c : G \to H$ with the following properties.*

i. $g(G) > l$;
ii. *For every graph H' with at most k vertices and there exists a homomorphism $g : G \to H'$ if and only if there exists a homomorphism $f : H \to H'$.*
iii. *For every (H, H')-pointed subsets $S \subset V(H), S' \subset V(H')$ with at most k vertices and for every homomorphism $g : G \to H'$ holds: if homomorphisms $f, f' : H \to H'$ satisfy $g = f \circ c$, then $f(x) = f'(x)$ for every $x \in S$.*

The proof of Thm. 11 is along the same lines as less general version proved in [24]. We essentially repeat their proof replacing the notion of H-pointed with its localization, (H, H')-pointed. This is routine but lengthy, so we omit the proof.

6 CSP Dichotomy Classification Conjecture

In [5], the universal algebra approach of [15] is extended to to show that CSP($\mathcal{H}$) is NP-complete for a large class of CSPs. A conjecture is made that CSP($\mathcal{H}$) is polynomial time solvable for all other CSPs $\mathcal{H}$. In [18], this conjecture is then transported to the language of posets.

An algebra $\mathcal{A} = (A, F)$ consists of a non-empty set A, and a set F of finitary operations on A. It is *finite* if A is finite. Given a relational structure $\mathcal{H}$, recall that Pol($\mathcal{H}$) is the set of polymorphisms of $\mathcal{H}$. This defines an algebra $\mathcal{A}_{\mathcal{H}} = (V(\mathcal{H}), \mathrm{Pol}(\mathcal{H}))$. We say that $\mathcal{A}_{\mathcal{H}}$ is NP-complete if CSP($\mathcal{H}$) is.

The following two definitions are borrowed directly from [5].

Definition 12. *Let $\mathcal{A} = (A, F)$ be an algebra and B a subset of A such that, for any $f \in F$ and for any $b_1, \ldots, b_d \in B$, where d is the arity of f, we have $f(b_1, \ldots, b_d) \in B$. Then the algebra $\mathcal{B} = (B, F|_B)$ is called a subalgebra of $\mathcal{A}$, where $F|_B$ consists of the restrictions of all operations in F to B.*

Definition 13. *Let $\mathcal{B} = (B, F_1)$ and $\mathcal{C} = (C, F_2)$ be such that $F_1 = \{f_i^1 \mid i \in I\}$ and $F_2 = \{f_i^2 \mid i \in I\}$, where both f_i^1 and f_i^2 are d_i-ary, for all $i \in I$. Then $\mathcal{C}$ is a homomorphic image of $\mathcal{B}$ if there exists a surjection $\psi : B \to C$ such that the following identity holds for all $i \in I$, and all $b_1, \ldots, b_{d_i} \in B$.*

$$\psi \circ f_i^1(b_1, \ldots, b_{d_i}) = f_i^2(\psi(b_1), \ldots, \psi(b_{d_i})).$$

Given an algebra $\mathcal{C} = (C, F)$, the *term operators* of $\mathcal{C}$ refer to the set of finitary operators of C that preserve the same relations on C as F does. Thus all operators in F are term operators. A d-ary operator f of F is *essentially unary* if $f = f' \circ \pi$ for some projection $\pi : C^d \to C$ and some non-constant function $f' : C \to C$. Because this f' is non-constant, if F has any essentially unary operators, then $|C| \geq 2$.

The following result is Cor. 7.3 in [5].

Theorem 14. *A finite algebra $\mathcal{A}$ is NP-complete if it has a subalgebra $\mathcal{B}$ with a homomorphic image $\mathcal{C}$, all of whose term operators are essentially unary.*

Further, they conjecture that CSP($\mathcal{H}$) is NP-complete for a relational structure $\mathcal{H}$, only if it is NP-complete by the above theorem.

We at first thought that any relational structure $\mathcal{H}$ such that $\mathcal{A}_{\mathcal{H}}$ has a subalgebra $\mathcal{B} = \mathcal{A}(B, F_1 = \mathrm{Pol}(\mathcal{H})|_B)$ with a homomorphic image $\mathcal{C} = \mathcal{A}(C, F_2)$, all of whose term operators are essenially unary, was subprojective. However, Ralph McKenzie [27] provided us with an explicit counterexample to this fact. Further, he showed that for any subprojective structure $\mathcal{H}$ that $\mathcal{A}_{\mathcal{H}}$ has a subalgebra which has a homomorphic image, all of whose term operators are essentially unary.

This shows that subprojective structures are certainly not the only structures yielding NP-complete CSPs. However, the fibre construction can be adapted to show NP-completeness for much more than subprojective structures. In a future paper we extend the fibre construction to show, at least, that all $\mathcal{H}$ such that $\mathcal{A}_{\mathcal{H}}$ has a subalgebra $\mathcal{B} = \mathcal{A}(B, F_1 = \mathrm{Pol}(\mathcal{H})|_B)$ with a homomorphic image $\mathcal{C} = \mathcal{A}(C, F_2)$, are NP-complete. A discussion of this extension appears in [23], the full version of this paper, and a full presentation will appear in a future paper.

References

1. Bodnarčuk, V.G., Kaluzhnin, L.A., Kotov, V.N., Romov, B.A.: Galois theory for Post algebras I - II (in Russian), Kibernetika, 3 (1969), 1-10 and 5 (1969), 1-9. English version: Cybernetics, 243-252 and 531-539 (1969)
2. Bulatov, A.: A dichotomy theorem for constraints on a three element set. FOCS'02 , 649–658 (2002)
3. Bulatov, A.: Tractable conservative Constraint Satisfaction Problems, ACM Trans. on Comp. Logic. LICS'03, 321–330 (2003)
4. Bulatov, A.: H-Coloring Dichotomy Revisited. Theoret. Comp. Sci. 1, 31–39 (2005)
5. Bulatov, A., Jeavons, P., Krokhin, A.: Classifying the complexity of constraints using finite algebras. SIAM J. Comput. 34(3), 720–742 (2005)
6. Bulatov, A., Jeavons, P., Krokhin, A.: Constraint satisfaction problems and finite algebras. In: Welzl, E., Montanari, U., Rolim, J.D.P. (eds.) ICALP 2000. LNCS, vol. 1853, pp. 272–282. Springer, Heidelberg (2000)
7. Creignou, N., Khanna, S., Sudan, M.: Complexity Classifications of Boolean Constraint Satisfaction Problems. SIAM Monographs on Discrete Mathematics and Applications, SIAM (2001)
8. Feder, T., Hell, P., Huang, J.: List Homomorphisms of Graphs with Bounded Degree (Submitted)
9. Feder, T., Vardi, M.: The computational structure of monotone monadic SNP and constraint satisfaction: a study through Datalog and group theory. SIAM J. Comput. 28(1), 57–104 (1999)
10. Geiger, D.: Closed systems of functions and predicates. Pacific. Journal of Math. 27, 95–100 (1968)

11. Hell, P.: Algorithmic aspects of graph homomorphisms. In: Survey in Combinatorics 2003, pp. 239–276. Cambridge University Press, Cambridge (2003)
12. Hell, P.: From Graph Colouring to Constraint Satisfaction: There and Back Again. In: Klazar, M., Kratochvil, J., Loebl, M., Matousek, J., Thomas, R., Valtr, P. (eds.) Topics in Discrete Mathematics. Dedicated to Jarik Nesetril on the Occasion of his 60th Birthday, pp. 407–432. Springer, Heidelberg (2006)
13. Hell, P., Nešetřil, J.: On the complexity of H-colouring. J. Combin. Theory B 48, 92–100 (1990)
14. Hell, P., Nešetřil, J.: Graphs and Homomorphisms. Oxford University Press, Oxford (2004)
15. Jeavons, P.G.: On the algebraic structure of combinatorial problems. Theoret. Comput. Sci. 200(1-2), 185–204 (1998)
16. Jeavons, P.G., Cohen, D.A., Gyssens, M.: Closure properties of Constraints. Journal of the ACM 44, 527–548 (1997)
17. Kostochka, A., Nešetřil, J., Smolíková, P.: Colorings and homomorphisms of degenerate and bounded degree graphs. Graph theory (Prague, 1998). Discrete Math 233(1-3), 257–276 (2001)
18. Larose, B., Zádori, L.: The Complexity of the Extendibility Problem for Finite Posets. SIAM J. Discrete Math. 17(1), 114–121 (2003)
19. Łuczak, T., Nešetřil, J.: A probabilistic approach to the dichotomy problem. SIAM J. Comput. 36(3), 835–843 (2006)
20. Montanari, U.: Networks of constraints: Fundamental properties and applications to picture processing. Information Sciences 7, 95–132 (1974)
21. Müller, V.: On colorings of graphs without short cycles. Discrete Math. 26, 165–176 (1979)
22. Nešetřil, J., Rödl, V.: Chromatically optimal rigid graphs. J. Comb. Th. B 46, 122–141 (1989)
23. Nešetřil, J., Siggers, M.: A New Combinatorial Approach to the CSP Dichotomy Classification (submitted, 2007)
24. Nešetřil, J., Zhu, X.: On sparse graphs with given colorings and homomorphisms. J. Combin. Theory Ser. B 90(1), 161–172 (2004)
25. Pippenger, N.: Theories of Computability. Cambridge University Press, Cambridge (1997)
26. Pöschel, R., Kalužnin, L.A.: Funktionen- und Relatrionenalgebren. DVW, Berlin (1979)
27. McKenzie, R.: Personal Communication
28. Schaefer, T.J.: The complexity of satisfiability problems. In: Proceedings of the 10th ACM Symposium on Theory of Computing (STOC'78), pp. 216–226 (1978)
29. Siggers, M.: On Highly Ramsey Infinte Graphs. (submitted, 2006)
30. Siggers, M.: Dichotomy for Bounded Degree H-colouring (submitted 2007)

NP by Means of Lifts and Shadows*

Gábor Kun and Jaroslav Nešetřil

Department of Mathematics, University of Memphis
373 Dunn Hall, Memphis, TN 38152
Department of Applied Mathematics (KAM) and
Institute of Theoretical Computer Science (ITI),
Charles University, Malostranské nám 22, Praha
kungabor@cs.elte.hu
nesetril@kam.mff.cuni.cz

Abstract. We show that every NP problem is polynomially equivalent to a simple combinatorial problem: the membership problem for a special class of digraphs. These classes are defined by means of shadows (projections) and by finitely many forbidden colored (lifted) subgraphs. Our characterization is motivated by the analysis of syntactical subclasses with the full computational power of NP, which were first studied by Feder and Vardi.

Our approach applies to many combinatorial problems and it induces the characterization of coloring problems (CSP) defined by means of shadows. This turns out to be related to homomorphism dualities. We prove that a class of digraphs (relational structures) defined by finitely many forbidden colored subgraphs (i.e. lifted substructures) is a CSP class if and only if all the the forbidden structures are homomorphically equivalent to trees. We show a surprising richness of coloring problems when restricted to most frequent graph classes. Using results of Nešetřil and Ossona de Mendez for bounded expansion classes (which include bounded degree and proper minor closed classes) we prove that the restriction of every class defined as the shadow of finitely many colored subgraphs equals to the restriction of a coloring (CSP) class.

Keywords: Digraph, homomorphism, duality, NP, Constraint Satisfaction Problem.

1 Introduction, Background and Previous Work

Think of 3-colorability of a graph G. This is a well known hard (and a canonical NP-complete) problem. From the combinatorial point of view there is a standard way how to approach this problem (and monotone properties in general): investigate minimal graphs without this property, denote by $\mathcal{F}$ the language of

* Part of this work was supported by ITI and DIMATIA of Charles University Prague under grant 1M0021620808, by OTKA Grant no. T043671, NK 67867, by NKTH (National Office for Research and Technology, Hungary), AEOLUS and also by Isaac Newton Institute (INI) Cambridge.

L. Kučera and A. Kučera (Eds.): MFCS 2007, LNCS 4708, pp. 171–181, 2007.

all such critical graphs and define the set $Forb(\mathcal{F})$ of all structures which do not "contain" any $F \in \mathcal{F}$. Then the language $Forb(\mathcal{F})$ coincides with the language of 3-colorable graphs. Unfortunately, in the most cases the set $\mathcal{F}$ is infinite. However the properties characterized by a finite set $\mathcal{F}$ are very interesting if we allow *lifts* and *shadows*.

Let us briefly illustrate this by our example of 3-colorability. Instead of a graph $G = (V, E)$ we consider the graph G together with three unary relations C_1, C_2, C_3 (i.e. colors of vertices) which *cover* the vertex set V; this structure will be denoted by G' and called a *lift* of G (thus G' has one binary and three unary relations). There are 3 *forbidden substructures*: For each $i = 1, 2, 3$ the single edge graph K_2 together with cover $C_i = \{1, 2\}$ and $C_j = \emptyset$ for $j \neq i$ form structure $\mathbf{F}'_i$ (where the signature of $\mathbf{F}'_i$ contains one binary and three unary relations). The language of all 3-colorable graphs is just the language $\Phi(Forb(\mathbf{F}'_1, \mathbf{F}'_2, \mathbf{F}'_3))$, where Φ is the forgetful functor which transforms G' to G. We call G the *shadow* of G'.

Clearly this situation can be generalized and one of the main results of this paper is Theorem 3 which states that every NP problem is polynomially equivalent to the membership problem for a class $\Phi(Forb(\mathcal{F}'))$. Here $\mathcal{F}'$ is a finite set of (vertex pair)-colored digraphs, $Forb(\mathcal{F}')$ is the class of all lifted graphs G' for which there is no homomorphism $F' \longrightarrow G'$ for an $F' \in \mathcal{F}'$. Thus $Forb(\mathcal{F}')$ is the class of all graphs G' with *forbidden* homomorphisms from $\mathcal{F}'$. (See Section 2 for definitions.) Theorems 4 and 5 provide similar results for forbidden colored subgraphs and for forbidden induced subgraphs (in both cases vertex colorings suffice).

We should add one more remark. We of course do not only claim that every problem in NP can be polynomially *reduced* to a problem in any of these classes. This would only mean that each of these classes contains an NP-complete problem. What we claim is that these classes have the *computational power* of the whole NP class. More precisely, to each language L in NP there exists a language M in any of these three classes such that M is *polynomially equivalent* to L, i.e. there exist *polynomial reductions* of L to M and M to L. E.g. assuming P $\neq$ NP there is a language in any of these classes that is neither in P nor NP-complete, since there is such a language in NP by Ladner's celebrated result [14].

The expressive power of classes $\Phi(Forb(\mathcal{F}'))$ corresponds to many combinatorially studied problems and presents a combinatorial counterpart to the celebrated result of Fagin [4] who expressed every NP problem in logical terms by means of an Existential Second Order formula.

The fact that the membership problem for classes $\Phi(Forb(\mathcal{F}'))$ and their injective and full variants $\Phi(Forb_{inj}(\mathcal{F}'))$ and $\Phi(Forb_{full}(\mathcal{F}'))$ have full computational power is pleasing from the combinatorial point of view as these classes cover well known examples of hard combinatorial problems: Ramsey type problems (where as in Theorem 3 we consider edge colored graphs), colorings of bounded degree graphs (defined by an injectivity condition as in Theorem 4) and structural partitions (studied e.g. in [8] as in Theorem 5). It follows that, in the full generality, one cannot expect dichotomies here. On the other side of the spectrum, Feder and Vardi have formulated the celebrated *Dichotomy conjecture* for all coloring problems (CSP).

Our main result is Theorem 9: we give an easy characterization of those languages $\Phi(Forb(\mathcal{F}'))$ which are coloring problems (CSP). This can be viewed as an extension of the duality characterization theorem for structures [6]. We demonstrate the power of this theorem while reproving some theorems about the local chromatic number. In contrast with this we show that the shadow $\Phi(Forb(\mathcal{F}'))$ of a vertex colored class of digraphs $Forb(\mathcal{F}')$ is always a CSP language when restricted to a bounded expansion class (this notion generalizes bounded degree and proper minor closed classes) [20]. Our main tools are *finite dualities* [23,6], *restricted dualities* [21], and the *Sparse Incomparability Lemma* [22,9]. The detailed proofs can be found in the full version of this paper [13].

2 Preliminaries

We consider finite relational structures although in most of the paper we only deal with digraphs, i.e. relational structures with just one binary relation. This itself is one of the main features of this note: oriented graphs suffice. Digraphs will be denoted by $\mathbf{A}, \mathbf{B}, \ldots$ (as we want to stress that they may be replaced by more general structures).

Let Γ denote a finite set we refer to as colors. A *Γ-colored graph (structure)* is a graph (or structure) together with either a coloring of its vertices or a coloring of all pairs of vertices by colors from Γ. Only in Theorem 3 we shall consider coloring of all pairs (but in Theorem 3 this will play an important role). Thus in the whole paper we shall undestand by a colored graph a graph with colored vertices. We denote colored digraphs (relational structures) by $\mathbf{A}', \mathbf{B}'$ etc. Following the more general notions in category theory we call $\mathbf{A}'$ a *lift* of $\mathbf{A}$ and $\mathbf{A}$ is called the *shadow* of $\mathbf{A}'$. Thus (vertex-) colored digraphs (structures) can be also described as *monadic* lifts. A homomorphism of digraphs (relational structures) preserves all the edges (arcs). A homomorphism of colored digraphs (relational structures) preserves the color of vertices (pairs of vertices), too. The *Constraint Satisfaction Problem* corresponding to the graph (relational structure) $\mathbf{A}$ is the membership problem for the class of all graphs (structures) defined by $\{\mathbf{B} : \mathbf{B}$ is homomorphic to $\mathbf{A}\}$. We call a mapping between two (colored) digraphs a *full homomorphism* if in addition the preimage of an edge is an edge. Full homomorphisms have very easy structure, as every full homomorphism which is onto is a retraction. The other special homomorphisms we will be interested in are *injective* homomorphisms.

Let $\mathcal{F}'$ be a finite set of colored relational structures (digraphs). By $Forb(\mathcal{F}')$ we denote the set of all colored relational structures (digraphs) $\mathbf{A}'$ satisfying $\mathbf{F}' \not\longrightarrow \mathbf{A}'$ for every $\mathbf{F}' \in \mathcal{F}'$. (If we use injective or full homomorphisms this will be denoted by $Forb_{inj}(\mathcal{F}')$ or $Forb_{full}(\mathcal{F}')$, respectively).

Similarly (well, dually), for the finite set of colored relational structures (digraphs) $\mathcal{D}'$ we denote by $CSP(\mathcal{D}')$ the class of all colored digraphs $\mathbf{A}'$ satisfying $\mathbf{A}' \longrightarrow \mathbf{D}'$ for some $\mathbf{D}' \in \mathcal{D}'$. (This is sometimes denoted by $\rightarrow \mathcal{D}$.) Now suppose that the classes $Forb(\mathcal{F}')$ and $CSP(\mathcal{D}')$ are equal. Then we say that the pair $(\mathcal{F}', \mathcal{D}')$ is a *finite duality*. Explicitly, a finite duality means that the following equivalence holds for every (colored) relational structure (digraph):

$$\forall \mathbf{F}' \in \mathcal{F}' \quad \mathbf{F}' \not\longrightarrow \mathbf{A}' \iff \exists \mathbf{D}' \in \mathcal{D}' \quad \mathbf{A}' \longrightarrow \mathbf{D}'.$$

We say that the structure $\mathbf{A}$ is *core* if every homomorphism $\mathbf{A} \longrightarrow \mathbf{A}$ is an automorphism. Every finite structure $\mathbf{A}$ contains (up to an isomorphism) a uniquely determined core substructure homomorphically equivalent to $\mathbf{A}$, see [23] [9]. The following result was recently proved in [6] and [23]. It characterizes finite dualities of digraphs (or more generally relational structures with a given signature).

Theorem 1. *For every finite set $\mathcal{F}$ of (relational) forests there exists (up to homomorphism equivalence) a finite uniquely determined set $\mathcal{D}$ of structures such that $(\mathcal{F}, \mathcal{D})$ forms a finite duality, i.e. $Forb(\mathcal{F}) = CSP(\mathcal{D})$. Up to homomorphism equivalence there are no other finite dualities.*

Let Φ denote the forgetful functor which corresponds to a Γ-colored relational structure (digraph) the uncolored one, i.e. it forgets about the coloring. We will investigate classes of the form $\Phi(Forb(\mathcal{F}'))$. We call the pair $(\mathcal{F}', \mathcal{D})$ *shadow duality* if $\Phi(Forb(\mathcal{F}')) = CSP(\mathcal{D})$. An example of shadow duality is the language of 3-colorable graphs discussed in the introduction (or, as can be seen easily, any CSP problem in general). Finite dualities became much more abundant when we demand the validity of the above formula just for all graphs from a given class $\mathcal{K}$. In such a case we speak about $\mathcal{K}$*-restricted duality*. It has been proved in [21] that so called *Bounded Expansion* classes (which include both proper minor closed classes and classes of graphs with bounded degree) have a restricted duality for every choice of $\mathcal{F}'$.

The study of homomorphism properties of structures not containing short cycles (i.e. with a large girth) is a combinatorial problem studied intensively. The following result has proved particularly useful in various applications. It is often called the *Sparse Incomparability Lemma*:

Theorem 2. *Let k, ℓ be positive integers and let $\mathbf{A}$ be a structure. Then there exists a structure $\mathbf{B}$ with the following properties:*

1. *There exists a homomorphism $f : \mathbf{B} \longrightarrow \mathbf{A}$;*
2. *For every structure $\mathbf{C}$ with at most k points the following holds: there exists a homomorphism $\mathbf{A} \longrightarrow \mathbf{C}$ if and only if there exists a homomorphism $\mathbf{B} \longrightarrow \mathbf{C}$;*
3. *$\mathbf{B}$ has girth $\geq \ell$.*

This result was proved by probabilistic method in [22] [24], see also [9]. The polynomial time construction of $\mathbf{B}$ is possible, too: in the case of binary relations (digraphs) this was done in [18] and for relational structures in [12].

3 Statement of Results

3.1 NP by Means of Finitely Many Forbidden Lifts

The class SNP consists of all problems expressible by an existential second-order formula with a universal first-order part [4]. The class SNP is computationally equivalent to NP. Feder and Vardi [5] have proved that three syntactically defined subclasses of the class SNP still have the full computational power of the class NP. We reformulate this result to our combinatorial setting of lifts and shadows.

Theorem 3. *For every language $L \in NP$ there exist a finite set of colors Γ and a finite set of Γ-colored digraphs $\mathcal{F}'$, where we color all pairs of vertices such that L is computationally equivalent to the membership problem for $\Phi(Forb(\mathcal{F}'))$.*

Theorem 4. *For every language $L \in NP$ there exist a finite set of colors Γ and a finite set of Γ-colored digraphs $\mathcal{F}'$, (where we color the vertices) such that L is computationally equivalent to the membership problem for $\Phi(Forb_{inj}(\mathcal{F}'))$.*

Theorem 5. *For every language $L \in NP$ there exist a finite set of colors Γ and a finite set of Γ-colored digraphs $\mathcal{F}'$, (where we color the vertices) such that L is computationally equivalent to the membership problem for $\Phi(Forb_{full}(\mathcal{F}'))$.*

3.2 Lifts and Shadows of Dualities

It follows from Section 3.1 that shadows of $Forb$ of a finite set of colored digraphs, this is classes $\Phi(Forb(\mathcal{F}'))$, where $\mathcal{F}'$ is a finite set, have the computational power of the whole NP. What about finite dualities? Are the shadow dualities also more frequent? The negative answer is expressed by Theorem 7 and shows a remarkable stability of dualities. Towards this end we first observe that every duality (of lifted structures) implies a shadow duality:

Theorem 6. *Let Γ be a finite set of colors and $\mathcal{F}'$ a finite set of Γ-colored digraphs (relational structures), where we color all of the vertices. Suppose that there exists a finite set of Γ-colored digraphs (relational structures) $\mathcal{D}'$ such that $Forb(\mathcal{F}') = CSP(\mathcal{D}')$. Then $\Phi(Forb(\mathcal{F}')) = CSP(\Phi(\mathcal{D}'))$.*

Theorem 6 may be sometimes reversed: Shadow dualities may be "lifted" in case that lifted graphs have colored vertices (this is sometimes described as *monadic lift*). This is non-trivial and in fact Theorem 7 may be seen as the core of this paper.

Theorem 7. *Let Γ be a finite set of colors and $\mathcal{F}'$ be a finite set of Γ-colored digraphs (relational structures), where we color all of the vertices. Suppose that $\Phi(Forb(\mathcal{F}')) = CSP(\mathcal{D})$ for a finite set $\mathcal{D}$ of digraphs (relational structures).*

Then there exists a finite set $\mathcal{D}'$ of Γ-colored digraphs (relational structures) such that $Forb(\mathcal{F}') = CSP(\mathcal{D}')$.

4 Proofs

The proofs of Theorems 3, 4 and 5 are in the full version of this paper [13]. We do not include them as they need some new definitions (and space) but nevertheless basically follow the strategy of [5].

Before proving Theorems 6 and 7 we formulate first a simple lemma which we shall use repeatedly:

Lemma 1. *(lifting) Let $\mathbf{A}, \mathbf{B}$ relational structures, homomorphism $f : \mathbf{A} \longrightarrow \mathbf{B}$, a finite set of colors Γ and $\Phi(\mathbf{B}') = \mathbf{B}$ be given. Then there exists a lift $\mathbf{A}'$, such that $\Phi(\mathbf{A}') = \mathbf{A}$ and the mapping f is a homomorphism $\mathbf{A}' \longrightarrow \mathbf{B}'$ (of colored structures).*

Proof (of Theorem 6). Suppose that $\mathbf{A} \in CSP(\Phi(\mathcal{D}'))$, say $\mathbf{A} \in CSP(\Phi(\mathbf{D}'))$. Now for a homomorphism $f : \mathbf{A} \longrightarrow \Phi(\mathbf{D}')$ there is at least one lift $\mathbf{A}'$ of $\mathbf{A}$ such that the mapping f is a homomorphism $\mathbf{A}' \rightarrow \mathbf{D}'$ (here we use Lifting Lemma 1). Since the pair$(\mathcal{F}', \mathcal{D}')$ is a duality $\mathbf{F}' \nrightarrow \mathbf{A}'$ holds for any $\mathbf{F}' \in \mathcal{F}'$ and thus in turn $\mathbf{A} \in \Phi(\mathrm{Forb}(\mathcal{F}'))$.

Conversely, let us assume that $\mathbf{A}' \in \mathrm{Forb}(\mathcal{F}')$ satisfies $\Phi(\mathbf{A}') = \mathbf{A}$. But then $\mathbf{A}' \in CSP(\mathcal{D}')$ and thus by the functorial property of Φ we have $\mathbf{A} = \Phi(\mathbf{A}') \in CSP(\Phi(\mathcal{D}'))$.

Proof (of Theorem 7). Assume $\Phi(\mathrm{Forb}(\mathcal{F}')) = CSP(\mathcal{D})$. Our goal is to find $\mathcal{D}'$ such that $\mathrm{Forb}(\mathcal{F}') = CSP(\mathcal{D}')$. This will follow as a (non-trivial) combination of Theorems 1 and 2. By Theorem 1 we know that if $\mathcal{F}'$ is a set of (relational) forests then the set $\mathcal{F}'$ has a dual set $\mathcal{D}'$ (in the class of covering colored structures; we just list all covering colored substructures of the dual set guaranteed by Theorem 1). It is $\Phi(\mathcal{D}') = \mathcal{D}$ by Theorem 6. So assume to the contrary that one of the structures, say $\mathbf{F}'_0$, fails to be a forest (i.e. we assume that one of the components of $\mathbf{F}'_0$ has a cycle). We proceed by a refined induction (which will allow us to use more properties of $\mathbf{F}'_0$) to show that $\mathcal{D}'$ does not exist. Let us introduce carefully the setting of the induction.

We assume shadow duality $\Phi(\mathrm{Forb}(\mathcal{F}')) = CSP(\mathcal{D})$. Let $\mathcal{D}$ be fixed throughout the proof. Clearly many sets $\mathcal{F}'$ will do the job and we select the set $\mathcal{F}'$ such that $\mathcal{F}'$ consists of cores of all homomorphic images (explicitly: we close $\mathcal{F}'$ under homomorphic images and then take the set of cores of all these structures). Among all such sets $\mathcal{F}'$ we take a set of minimal cardinality. It will be again denoted by $\mathcal{F}'$. We proceed by induction on the size $|\mathcal{F}'|$ of $\mathcal{F}'$.

The set $\mathrm{Forb}(\mathcal{F}')$ is clearly determined by the minimal elements of $\mathcal{F}'$ (minimal in the homomorphism order). Thus let us assume that one of these minimal elements, say $\mathbf{F}'_0$, is not a forest. By the minimality of $\mathcal{F}'$ we see that we have a proper inclusion $\Phi(\mathrm{Forb}(\mathcal{F}' \setminus \{\mathbf{F}'_0\})) \supset CSP(\mathcal{D})$. Thus there exists a structure $\mathbf{S}$ in the difference. But this in turn means that there has to be a lift $\mathbf{S}'$ of $\mathbf{S}$ such that $\mathbf{F}'_0 \longrightarrow \mathbf{S}'$ and $\mathbf{S} \not\rightarrow \mathbf{D}$ for every $\mathbf{D} \in \mathcal{D}$. In fact not only that: as $\mathbf{F}'_0$ is a core, as $\mathrm{Forb}(\mathcal{F}')$ is homomorphism closed and as $\mathcal{F}'$ has minimal size we conclude that there exist $\mathbf{S}$ and $\mathbf{S}'$ such that any homomorphism $\mathbf{F}'_0 \longrightarrow \mathbf{S}'$ is a monomorphism (i.e. one-to-one, otherwise we could replace $\mathbf{F}'_0$ by a set of all its homomorphic images - $\mathbf{F}'_0$ would not be needed).

Now we apply (the second non-trivial ingredient) Theorem 2 to structure $\mathbf{S}$ and an $\ell > |X(\mathbf{F}'_0)|$: we find a structure $\mathbf{S}_0$ with the following properties: $\mathbf{S}_0 \longrightarrow \mathbf{S}$, $\mathbf{S}_0 \longrightarrow \mathbf{D}$ if and only if $\mathbf{S} \longrightarrow \mathbf{D}$ for every $\mathbf{D} \in \mathcal{D}$ and $\mathbf{S}_0$ contains no cycles of length $\leq \ell$. It follows that $\mathbf{S}_0 \notin CSP(\mathcal{D})$. Next we apply Lemma 1 to obtain a structure $\mathbf{S}'_0$ with $\mathbf{S}'_0 \longrightarrow \mathbf{S}'$. Now we use that $\mathbf{S}'_0$ is a monadic lift and so does not contain cycles of length $\leq \ell$. Now for any $\mathbf{F}' \in \mathcal{F}'$, $\mathbf{F}' \neq \mathbf{F}'_0$ we have $\mathbf{F}' \not\rightarrow \mathbf{S}'_0$ as $\mathbf{S}'_0 \rightarrow \mathbf{S}'$ and $\mathbf{F}' \not\rightarrow \mathbf{S}'$. As the only homomorphism $\mathbf{F}'_0 \longrightarrow \mathbf{S}'$ is a monomorphism the only (hypothetical) homomorphism $\mathbf{F}'_0 \longrightarrow \mathbf{S}'$ is also monomorphism. But this is a contradiction as $\mathbf{F}'_0$ contains a cycle while $\mathbf{S}'_0$ has no cycles of length $\leq \ell$. This completes the proof.

5 Applications

5.1 Classes with Bounded Expansion

We study the restriction of classes $\Phi(Forb(\mathcal{F}'))$ to a class of digraphs with bounded expansion recently introduced in [20]. These classes are a generalization of proper minor closed and bounded degree classes of graphs. Using the decomposition technique of [20] [21] we can prove that any class $\Phi(Forb(\mathcal{F}'))$ (for a finite set $\mathcal{F}'$ of monadic lifts) when restricted to a bounded expansion class equals to a CSP class (when restricted to the same class).

Theorem 8. *Consider the finite set of colors Γ and the class $\Phi(Forb(\mathcal{F}'))$ for a finite set $\mathcal{F}'$ of Γ-colored digraphs. Let $\mathcal{C}$ be a class of digraphs of bounded expansion. Then there is a finite set of digraphs $\mathcal{D}$ such that $\Phi(Forb(\mathcal{F}')) \cap \mathcal{C} = CSP(\mathcal{D}) \cap \mathcal{C}$.*

Consider a monotone, first-order definable class of colored digraphs $\mathcal{C}$ which is closed under homomorphism and disjoint union. By a combination with recent results of [2] we also obtain (perhaps a bit surprisingly) that the shadow $\mathcal{C}$ is a CSP language of digraphs. It remains to be seen to which bounded expansion classes (of graphs and structures) this result generalizes.

5.2 The Classes MMSNP and FP - A Characterization

We conclude with an application to descriptive theory of complexity classes. Recall that the class of languages defined by monotone, monadic formulas without inequality is denoted by MMSNP (*Monotone Monadic Strict Nondeterministic*

Polynomial). (Feder and Vardi proved that the class MMSNP is computationally equivalent to the class CSP in a random sense [5], this was later derandomized by the first author [12].) Madeleine [16] introduced the class FP of languages defined similarly to our forbidden monadic lifts of structures.

It has been proved in [16] that the classes FP and MMSNP are equal. In fact the class MMSNP contains exactly the languages defined by forbidden monadic lifts.

Proposition 1. *A language of relational structures L is in the class MMSNP if and only if there is a finite set of colors Γ and a finite set of Γ-colored relational structures $\mathcal{F}'$ such that $L = \Phi(Forb(\mathcal{F}'))$.*

Madelaine and Stewart [17] gave a long process to decide whether an FP language is a finite union of CSP languages. We use Theorems 6 and 7 and the description of dualities for relational structures [6] to give a short characterization of a more general class of languages.

Theorem 9. *Consider the finite set of colors Γ and the language $\Phi(Forb(\mathcal{F}'))$ for a finite set $\mathcal{F}'$ of Γ-colored digraphs (relational structures).*

If no $\mathbf{F}' \in \mathcal{F}'$ contains a cycle then there is a finite set of digraphs (relational structures) $\mathcal{D}$ such that $\Phi(Forb(\mathcal{F}')) = CSP(\mathcal{D})$. If one of the lifts $\mathbf{F}'$ in a minimal subfamily of $\mathcal{F}'$ contains a cycle in its core then the language $\Phi(Forb(\mathcal{F}'))$ is not a finite union of CSP languages.

Proof. If no $\mathbf{F}' \in Forb(\mathcal{F}')$ contains a cycle then the set $\mathcal{F}'$ has a dual $\mathcal{D}'$ by Theorem 1, and the shadow of this set $\mathcal{D}'$ gives the dual set $\mathcal{D}$ of the set $\Phi(Forb(\mathcal{F}'))$ (by Theorem 6). On the other side if one $\mathbf{F}' \in Forb(\mathcal{F}')$ contains a cycle in its core and if $\mathcal{F}'$ is minimal (i.e. $\mathbf{F}'$ is needed) then $Forb(\mathcal{F}')$ does not have a dual. The shadow of the language $Forb(\mathcal{F}')$ is the language L and consequently this fails to be a finite union of CSP languages by Theorem 7.

Theorem 9 may be interpreted as stability of dualities for finite structures. While shadows of the classes $Forb(\mathcal{F}')$ are computationally equivalent to the whole NP, the shadow dualities are not bringing anything new: these are just shadows of dualities. In other words: the coloring problems in the class MMSNP are just shadow dualities. This holds for graphs as well for relational structures.

5.3 On the Local Chromatic Number

Now we apply Theorem 9 in the analysis of local chromatic number introduced in [3] (see also [26]): we say that a graph G is locally (a, b)-colorable if there exists a proper coloring of G by b colors so that every (closed) neighborhood of a vertex of G gets at most a colors. It follows from [3] that the class of all locally (a, b)-colorable graphs is of the form CSP($U(a, b)$) for an explicitely constructed graph $U(a, b)$. We conclude this paper with an indirect proof of this result with an application to complexity:

Proposition 2. *Let a, b be integers and consider the membership problem for the class of locally (a, b)-colorable graphs. This is actually a Constraint Satisfaction Problem which is NP-complete if $a, b \geq 3$ and it is polynomial time solvable else.*

Proof. Consider the color set $\Gamma = \{1, \ldots, b\}$ and the following set $\mathcal{F}'$ of Γ-colored undirected graphs. Let $\mathcal{F}'$ consist of all monochromatic edges (colored by any of the b colors) and all the stars with $a+1$ vertices colored by at least $a+1$ colors. The corresponding language is exactly the required one: a graph G is in the language iff it admits a proper Γ-coloring, this is no monochromatic edge is homomorphic to the colored graph, such that the neighbourhood of every vertex (including the vertex itself) has at most a different colors, i.e. no star with $a+1$ vertices of different color is homomorphic to it. Since $\mathcal{F}'$ consists of colored trees this will be a CSP language by Theorem 9.

Hell and the second author proved that CSP problems defined by undirected graphs are in P if the graph is bipartite and NP-complete else [9]. We do not determine which graph defines this particular CSP problem (of locally (a, b)-colorable graphs). But if $a, b \geq 3$ then we know that it contains the triangle if, so the problem is NP-complete. It is easy to see that this membership problem is in P if $a < 3$ or $b < 3$.

6 Summary and Future Work

We found a computationally equivalent formulation of the class NP by means of finitely many forbidden lifts of very special type. An ambitious project would be to find an equivalent digraph coloring problem for a given NP language really effectively (in human sense, our results provide a polynomial time algorithm). For example it would be nice to exhibit a vertex coloring problem that is polynomially equivalent to the graph isomorphism problem. In general this mainly depends on how to express the problem in terms of logic. The next class we seem to be able to deal with are coloring problems of structures with an equivalence relation. Another good candidate are lifts using linear order. This promises several interesting applications which were studied earlier in a different setting.

We also proved that shadow dualities and lifted monadic dualities are in $1-1$ correspondence. This abstract result has several consequences and streamlines some earlier results in descriptive complexity theory (related to MMSNP and CSP classes). The simplicity of this approach suggests some other problems. It is tempting to try to relate Ladner's diagonalization method [14] in this setting (as it was pioneered by Lovász and Gács [7] for NP$\cap$coNP in a similar context). The characterization of Lifted Dualities is beyond reach but particular cases are interesting as they generalize results of [23] [6] and as the corresponding duals present polynomial instances of CSP.

But perhaps more importantly, our approach to the complexity subclasses of NP is based on lifts and shadows as a combination of algebra, combinatorics and logic. We believe that it has further applications and that it forms a useful paradigm.

References

1. Atserias, A.: On Digraph Coloring Problems and Treewidth Duality. In: 20th IEEE Symposium on Logic in Computer Science (LICS), pp. 106–115 (2005)
2. Atserias, A., Dawar, A., Kolaitis, P.G.: On Preservation under Homomorphisms and Conjunctive Queries. Journal of the ACM 53(2), 208–237 (2006)
3. Erdős, P., Füredi, Z., Hajnal, A., Komjáth, P., Rödl, V., Seress, Á.: Coloring graphs with locally few colors. Discrete Math. 59, 21–34 (1986)
4. Fagin, R.: Generalized first-order spectra and polynomial-time recognizable sets. In: Karp, R. (ed.) Complexity of Computation, SIAM-AMS Proceedings, vol. 7, pp. 43–73 (1974)
5. Feder, T., Vardi, M.: The computational structure of monotone monadic SNP and constraint satisfaction: A study through Datalog and group theory. SIAM J. Comput. 28(1), 57–104 (1999)
6. Foniok, J., Nešetřil, J., Tardif, C.: Generalized dualities and maximal finite antichains in the homomorphism order of relational structures. In: KAM-DIMATIA Series 2006-766 (to appear in European J. Comb.)
7. Gács, P., Lovász, L.: Some remarks on generalized spectra. Z. Math. Log. Grdl. 23(6), 547–554 (1977)
8. Feder, T., Hell, P., Klein, S., Motwani, R.: Complexity of graph partition problems. In: 31st Annual ACM STOC, pp. 464–472 (1999)
9. Hell, P., Nešetřil, J.: Graphs and Homomorphism. Oxford University Press (2004)
10. Immerman, N.: Languages that capture complexity classes. SIAM J. Comput. 16, 760–778 (1987)
11. Kun, G.: On the complexity of Constraint Satisfaction Problem, PhD thesis (in Hungarian) (2006)
12. Kun, G.: Constraints, MMSNP and expander structures, Combinatorica (submitted, 2007)
13. Kun, G., Nešetřil, J.: Forbidden lifts (NP and CSP for combinatorists). In: KAM-DIMATIA Series 2006-775 (to appear in European J. Comb.)
14. Ladner, R.E.: On the structure of Polynomial Time Reducibility. Journal of the ACM 22(1), 155–171 (1975)
15. Luczak, T., Nešetřil, J.: A probabilistic approach to the dichotomy problem. SIAM J. Comp. 36(3), 835–843 (2006)
16. Madelaine, F.: Constraint satisfaction problems and related logic, PhD thesis (2003)
17. Madelaine, F., Stewart, I.A.: Constraint satisfaction problems and related logic, (manuscript, 2005)
18. Matoušek, J., Nešetřil, J.: Constructions of sparse graphs with given homomorphisms (to appear)
19. Nešetřil, J., Pultr, A.: On classes of relations and graphs determined by subobjects and factorobjects. Discrete Math. 22, 287–300 (1978)
20. Nešetřil, J., de Mendez, P.O.: Low tree-width decompositions and algorithmic consequences. In: STOC'06, Proceedings of the 38th Annual ACM Symposium on Theory of Computing, pp. 391–400. ACM Press, New York (2006)
21. Nešetřil, J., de Mendez, P.O.: Grad and Classes with bounded expansion III. - Restricted Dualities, KAM-DIMATIA Series 2005-741 (to appear in European J. Comb.)
22. Nešetřil, J., Rödl, V.: Chromatically optimal rigid graphs. J. Comb. Th. B 46, 133–141 (1989)

23. Nešetřil, J., Tardif, C.: Duality theorems for finite structures (characterising gaps and good characterization. J. Combin. Theory B 80, 80–97 (2000)
24. Nešetřil, J., Zhu, X.: On sparse graphs with given colorings and homomorphisms. J. Comb. Th. B 90(1), 161–172 (2004)
25. Rossman, B.: Existential positive types and preservation under homomorphisms. In: 20th IEEE Symposium on Logic in Computer Science (LICS'2005), pp. 467–476 (2005)
26. Simonyi, G., Tardos, G.: Local chromatic number, Ky Fan's theorem and circular colorings. Combinatorica 26, 589–626 (2006)
27. Vardi, M.Y.: The complexity of relational query languages. In: Proceedings of 14th ACM Symposium on Theory of Computing pp. 137–146 (1982)

The Complexity of Solitaire

Luc Longpré[1] and Pierre McKenzie[2,*]

[1] University of Texas at El Paso
[2] Université de Montréal

Abstract. Klondike is the well-known 52-card Solitaire game available on almost every computer. The problem of determining whether an n-card Klondike initial configuration can lead to a win is shown NP-complete. The problem remains NP-complete when only three suits are allowed instead of the usual four. When only two suits of opposite color are available, the problem is shown NL-hard. When the only two suits have the same color, two restrictions are shown in AC^0 and in NL respectively. When a single suit is allowed, the problem drops in complexity down to $AC^0[3]$, that is, the problem is solvable by a family of constant depth unbounded fan-in {AND, OR, MOD3}-circuits. Other cases are studied: for example, "no King" variant with an arbitrary number of suits of the same color and with an empty "pile" is NL-complete.

1 Introduction

Solitaire card games, called patience games outside of the United States, apparently originate from the fortune-telling circles of the eighteenth century [9]. Of the many hundred different solitaire card games in existence [8], to the best of our knowledge, only *FreeCell* [4] and *BlackHole* [5] have been studied from a complexity viewpoint. In both cases, determining whether an initial configuration can lead to a win was shown NP-complete.

Over the last decade, a particular variation of solitaire, the Klondike version, was popularized by Microsoft Windows (Figure 1). Earlier, Parlett [8] had described Klondike as the "most popular of all perennial favorites in the realm of patience, which is surprising since it offers the lowest success rate of any patience".

In a paper entitled *Solitaire: Man Versus Machine*, Yan, Diaconis, Rusmevichientong and Van Roy [10] report that a human expert (and distinguished combinatorialist, former president of the *American Mathematical Society*!) patiently recorded data on his playing 2000 games of *thoughtful solitaire*, that is, the intellectually more challenging Klondike in which the complete initial game configuration is revealed to the player at the start of the game. The expert averaged 20 minutes per game and was able to win the game 36.6% of the time. Pointing to the prohibitive difficulty of obtaining non-trivial mathematical estimates on the odds of winning this common game, Yan *et al.*, then describe heuristics developed using the so-called rollout method, and they report a 70% win rate.

* Supported by the NSERC of Canada and the (Québec) FQRNT.

L. Kučera and A. Kučera (Eds.): MFCS 2007, LNCS 4708, pp. 182–193, 2007.

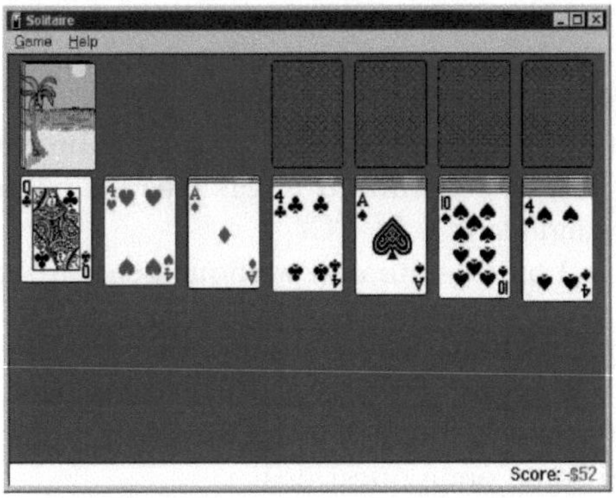

Fig. 1. Initial Klondike configuration on Microsoft Windows. In the terminology of [10], there are four empty *suit stacks*, seven *build stacks* containing 1, 2, 3, 4, 5, 6 and 7 cards respectively with only the top card facing up, and one *pile* containing the remaining cards facing down; another stack, the *talon*, will appear in the course of the game when cards from the pile are moved to the talon three by three.

Why is it so hard to compute the odds of winning at Klondike? In part, this question prompted our investigation of the complexity of the game. As well, after using the Minesweeper game [6] for several years as a motivating example for students, we looked for different NP-complete examples based on other widely popular and deceptively simple games.

The precise rules of Klondike are described in Section 2. To make the game amenable to a computational complexity analysis, the game is generalized to allow instances of arbitrary finite size. Hence an instance of the game involves a "deck" containing the "cards"

$$\begin{array}{c} 1\clubsuit, 2\clubsuit, 3\clubsuit, \ldots, n-1\clubsuit, n\clubsuit \\ 1\diamondsuit, 2\diamondsuit, 3\diamondsuit, \ldots, n-1\diamondsuit, n\diamondsuit \\ 1\heartsuit, 2\heartsuit, 3\heartsuit, \ldots, n-1\heartsuit, n\heartsuit \\ 1\spadesuit, 2\spadesuit, 3\spadesuit, \ldots, n-1\spadesuit, n\spadesuit \end{array}$$

and the game configurations are generalized appropriately. The problem of interest is to determine, given an initial game configuration, whether the game can be won. We show here that this is NP-complete.

Our NP-completeness proof bears resemblance to the proof that FreeCell is NP-complete [4]. In particular, the method of nondeterministically assigning truth values to card configurations is the same. However, our strategy gets by with only three card suits as opposed to the usual four used in the current proof that FreeCell is NP-complete. Many differences between FreeCell and Klondike further arise, for instance, when we argue the NP upper bound in the presence of "backward" moves (ie from a suit stack to a build stack) and when we consider restricted variants of the game. We highlight the following as our main results:

1. Klondike is NP-complete and remains so with only three suits available,
2. Klondike with a black suit and a red suit is NL-hard,
3. Klondike with any fixed number b of black suits is in NL,
4. flat Klondike (that is, without a pile) with an input-dependent number of black suits and without generalized Kings (see below) is NL-complete,
5. Klondike with a single suit is in $AC^0[3]$,
6. flat Klondike with 2 black suits and without generalized Kings is in AC^0.

Section 2 contains preliminaries and a precise description of the Klondike variation studied in this paper. Section 3 proves that Klondike is NP-complete and Section 4 considers restricting the usual four-suit game. Section 5 concludes with a discussion and some open questions.

2 Preliminaries

2.1 Complexity Theory

We assume familiarity with basic complexity theory, such as can be found in standard books on the subject, for example [7]. We recall the inclusion chain

$$AC^0 \subset AC^0[3] \subset AC^0[6] \subseteq L \subseteq NL = \text{co-NL} \subseteq P \subseteq NP.$$

Here, the complexity class AC^0 is the set of langages accepted by DLOGTIME-uniform unbounded-fan-in constant-depth $\{\wedge, \vee, \neg\}$-circuits of polynomial size. The larger class $AC^0[m]$ is the set of languages AC^0-Turing reducible to the MOD_m Boolean function, defined to output 1 iff m does not divide the sum of its Boolean inputs. The classes L and NL stand for deterministic and nondeterministic logarithmic space respectively. The classes P and NP are deterministic and nondeterministic polynomial time respectively. We adopt the definitions of [3] for constant-depth circuit uniformity (see also [2]).

If not otherwise stated, the hardness results in this paper are in terms of many-one logspace reducibility.

2.2 Klondike

We allow ourselves to borrow the following excellent description of Klondike found in [10, Section 2]:

> The goal of the game is to move all cards into the suit stacks, aces first, then twos, and so on, with each suit stack evolving as an ordered increasing arrangement of cards of the same suit. On each turn, the player can move cards from one stack to another in the following manner:
>
> 1. Face-up cards of a build stack, called a card block, can be moved to the top of another build stack provided that the build stack to which the block is being moved accepts the block (see Points 6 and 7 below for the meaning of acceptance). Note that *all* face-up cards on the source stack must be moved together. After the move, these

cards would then become the top cards of the stack to which they are moved, and their ordering is preserved. The card originally immediately beneath the card block, now the top card in its stack, is turned faceup. In the event that all cards in the source stack are moved, the player has an empty stack.
2. The top face-up card of a build stack can be moved to the top of a suit stack, provided that the suit stack accepts the card.
3. The top card of a suit stack can be moved to the top of a build stack, provided that the build stack accepts the card.
4. If the pile is not empty, a move can deal its top three cards to the talon, which maintains its cards in a first-in-last-out order. If the pile becomes empty, the player can redeal all the cards on the talon back to the pile in one card move. A redeal preserves the ordering of cards. The game allows an unlimited number of redeals.
5. A card on the top of the talon can be moved to the top of a build stack or a suit stack, provided that the stack to which the card is being moved accepts the card.
6. A build stack can only accept an incoming card block if the top card on the build stack is adjacent to and braided with the bottom card of the block. A card is adjacent to another card of rank r if it is of rank r + 1. A card is braided with a card of suit s if its suit is of a color different from s. Additionally, if a build stack is empty, it can only accept a card block whose bottom card is a King.
7. A suit stack can only accept an incoming card of its corresponding suit. If a suit stack is empty, it can only accept an Ace. If it is not empty, the incoming card must be adjacent to the current top card of the suit stack.

Yan *et al.*, coin the name *thoughtful solitaire* for the Klondike variation in which the player sees the complete game configuration, including the ranks of all the cards facing down, throughout the course of the game. The Klondike rules are otherwise unchanged.

Since we generalize Klondike to involve a variable number of cards, we adjust the notion of a game configuration to allow an arbitrary number of build stacks of arbitrary size. We will occasionally consider Klondike with a number of suits other than 4. In that case, the number of suit stacks is adjusted accordingly.

Card numbers and suit numbers are represented in binary notation. We assume any reasonable encoding of cards and game configurations that allows extracting individual card information in AC^0. In particular, the pile, talon and stacks are represented in table form so that the predicate "c is the ith card in the table" is AC^0-computable.

Definition. Problem $\text{SOLIT}(b, r)$:

Given: an initial b-black-suit and r-red-suit Klondike configuration involving the same number n of cards in every suit.
Determine: Whether the $(b+r)n$ cards can be placed on the $b+r$ suit stacks by applying the Klondike game rules starting from the given initial configuration.

In Section 3 we will be studying SOLITAIRE, by which we mean SOLIT(2, 2). In Section 4, we will consider Klondike restrictions, such as FLAT-SOLIT(b, r), by which we mean SOLIT(b, r) with an initial configuration having an empty pile and empty talon. We define the further restriction FLAT-SOLIT$_{\text{NoKing}}$(b, r) to mean FLAT-SOLIT(b, r) played with modified rules that forbid an empty stack from accepting a generalized King (ie we disallow refilling an empty build stack; this is equivalent to viewing the highest ranked cards as generalized Queens rather than generalized Kings). Using a "$*$", such as in FLAT-SOLIT($*, 0$), means that the number of suits corresponding to the $*$ is not fixed and depends on the input.

3 Klondike Is NP-Complete

Theorem 1. SOLITAIRE *is* NP*-complete.*

Proof. NP *upper bound.* Consider a winning N-card Klondike instance involving k build stacks. We need to argue that the length of a shortest winning sequence of movesis polynomial in N. We define three types of moves:

Type 1: moving a card out of the talon, or turning a build stack card face up
Type 2: moving a block of cards from one build stack to another
Type 3: moving a card from a build stack to a suit stack, or vice versa.

Let ℓ be a shortest winning sequence of moves, neglecting the pile-talon moves that are not of type 1. Such a sequence contains $N - k$ moves of type 1 since exactly k cards were visible at the outset. We claim that two successive moves of type 1 in this sequence are separated by $O(kN)$ moves of type 2 or type 3. The NP upper bound follows from the claim since moves of type 1 are irreversible and no obstacle remains after the $N - k$ moves of type 1; thus ℓ is $O(kN^2)$.

To see the claim, note first that a repetition-free sequence of type 3 moves is no longer than $2N$. Now any move of type 2 uncovers a card from some build stack B. Such a move of type 2 excludes any further type 2 move or type 3 move involving B until the next type 1 move occurs (unless B was emptied by the type 2 move, in which case a King could reactivate B at most once). Hence, between any two successive type 1 moves in a shortest sequence, there can be at most $(k - 1 + 4)(2N + 1)$ moves of types 2 or 3, proving the claim.

NP*-hardness.* We reduce from 3SAT. The main idea is to construct a pair of build stacks for each 3SAT formula variable. The top cards on these stacks will correspond to whether we want the variable to be true or false. Only one of these two top cards can be moved to another build stack. Moving a top card will uncover cards corresponding to the clauses that become true when the variable takes the Boolean value associated with the top card.

Let F be a 3CNF Boolean formula with n variables $v_1, ..., v_n$ and m clauses $c_1, ..., c_m$. Construct an initial configuration corresponding to this formula so that the configuration is winning if and only if the formula is satisfiable. For each variable v_i, associate cards $2i$ et $2i + 1$. For each clause c_j, associate cards $2n + 7j, ..., 2n + 7j + 6$.

For each variable v_i, construct 3 build stacks (called the *variable* stacks):

1. one with card $2i$♠ facing up on top and cards to be determined below,
2. one with card $2i$♣ facing up on top and cards to be determined below,
3. one with the sole card $(2i+1)$♡.

For each clause $c_j = (l_p, l_q, l_r)$, construct 3 build stacks (called *clause* stacks) with one card facing up on top and one card facing down below:

1. one stack with $(2n+7j+6)$♡ on top and $(2n+7j+5)$♣ below,
2. one stack with $(2n+7j+4)$♡ on top and $(2n+7j+3)$♣ below,
3. one stack with $(2n+7j+2)$♡ on top and $(2n+7j+1)$♣ below.

Also, if $l_p = v_i$, put $(2n+7j+1)$♠ facing down in stack $2i$♠, and if $l_p = \overline{v_i}$, put $(2n+7j+1)$♠ facing down in stack $2i$♣. If $l_q = v_i$, put $(2n+7j+3)$♠ facing down in stack $2i$♠, and if $l_q = \overline{v_i}$, put $(2n+7j+3)$♠ facing down in stack $2i$♣. If $l_r = v_i$, put $(2n+7j+5)$♠ facing down in stack $2i$♠, and if $l_r = \overline{v_i}$, put $(2n+7j+5)$♠ facing down in stack $2i$♣. Arrange for all clause (spade, face down) cards within any given build stack to occur in order of increasing card rank.

Finally, create the *critical* build stack, facing down, with cards $(2n+7j)$♡ in any order for $1 \leq j \leq m$, followed by all remaining cards in increasing order, starting with aces, and followed further by 3 generalized Kings $2n+7m+7$♡, $2n+7m+7$♣ and $2n+7m+7$♠. Regrouping all the generalized Kings at the bottom of the critical stack serves to prevent moves to an empty build stack during the core of the simulation. The pile, talon and suit stacks are empty. For each clause j, we will refer to the card $2n+7j$♡ as to the *critical clause-j card.*

Suppose that some assignment satisfies the formula. Here is how to win the game. In the assignment, if the variable v_i is false, put card $2i$♣ on card $(2i+1)$♡. If v_i is true, put card $2i$♠ instead. Then, move all the cards that were below the $2i$ cards. This is possible because all these cards are spade cards numbered $2n+7j+1$ or $2n+7j+3$ or $2n+7j+5$, and the red cards $2n+7j+2$ and $2n+7j+4$ and $2n+7j+6$ all sit facing up on top of their build stacks. If the formula is satisfiable, all the clauses have at least one literal set to true, so at least one clause j card will be released in this manner for each j.

Claim: For each j, a sequence of j-clause stack moves now exists such that

1. one clause-j stack can be made to accept the critical clause-j card, and
2. after this sequence, if a situation is reached such that all the cards ranked less than $2n+7j$ are placed on the suit stacks and a black card ranked $2n+7j$ sits on top of the critical stack, then all the cards ranked $2n+7j, 2n+7j+1, \ldots, 2n+7j+6$ can be placed on the suit stacks.

This claim implies a win as follows. Part (1) of the claim ensures that all the critical cards can be moved from the critical stack to the clause stacks. This releases the aces and allows moving all the cards ranked less that $2n+7$, including those that remained on the variable stacks, to the suit stacks. Part (2) of the claim together with an induction on j then yield the winning sequence of moves.

To prove the claim, fix j and let $2n + 7j + k\spadesuit$ for some $k \in \{1, 3, 5\}$ be the smallest clause-j card that got released from a variable build stack. If $k = 1$, then $2n + 7j + k\spadesuit$ was accepted by $2n + 7j + 2\heartsuit$ and the resulting stack accepts the critical clause-j card. If $k = 3$, then $2n + 7j + k\spadesuit$ was accepted by $2n + 7j + 4\heartsuit$, so the $2n + 7j + 2\heartsuit$ card can be displaced, thus uncovering $2n + 7j + 1\clubsuit$ which in turn accepts the critical clause-j card. Finally, if $k = 5$, then $2n + 7j + k\spadesuit$ was accepted by $2n + 7j + 6\heartsuit$; now $2n + 7j + 4\heartsuit$ can be displaced, followed by $2n + 7j + 3\clubsuit$, followed by $2n + 7j + 2\heartsuit$, again uncovering $2n + 7j + 1\clubsuit$ which accepts the critical clause-j card. This proves part (1) of the claim. To prove part (2) of the claim, it suffices to observe that although some $2n + 7j + k\clubsuit$ card(s) remain(s) under some $2n + 7j + k + 1\heartsuit$ card(s) after the above sequence of moves, the resulting stack configurations do not form an obstacle when the complementary cards in all suits are available in increasing order.

Conversely, suppose that the configuration produced from the formula is winning. Then the initial sequence of a winning sequence of moves must uncover the aces. This initial sequence cannot involve backward moves (ie from a suit stack to a build stack). This initial sequence must then first release every critical card $2n + 7j\heartsuit$. Each of these cards must be moved to a black $(2n + 7j + 1)$ card. For any given j, this cannot happen unless for some i, some clause-j card is released from one (and only one, since a single card $2i + 1$ is visible) of the two v_i-variable stacks. An assignment of variables v_i based on which of the two v_i-variable stacks was first released is, by construction, a satisfying assignment to our formula. □

4 Complexity of Klondike Restrictions

The proof of Theorem 1 used only $\{\clubsuit, \heartsuit, \spadesuit\}$. Furthermore, the initial configuration constructed had an empty pile and empty talon. Thus we have:

Theorem 2. SOLIT$(2, 1)$ *and* FLAT-SOLIT$(2, 1)$ *are* NP*-complete.*

Because the NP argument from Theorem 1 extends to the case in which an arbitrary number of (red and black) suits is allowed, we also have:

Theorem 3. SOLIT$(*, *)$ *and* FLAT-SOLIT$(*, *)$ *are* NP*-complete.*

Recall the "no King" game restriction, in which empty build stacks can never be filled. Because the Klondike instances constructed in the NP-hardness proof from Theorem 1 neither allow nor tolerate refilling an empty stack (except at the very end when all the cards have been released), we also have:

Theorem 4. FLAT-SOLIT$_{\text{NoKing}}(b, r)$ *is* NP*-complete for any* $b > r \geq 1$.

One might expect the remaining cases, namely the case of one black suit and one red suit, and the case in which all suits are black, to be trivial. This is not quite so. We begin with the latter.

A FLAT-SOLIT$(*, 0)$ instance w involves a set of nb cards $c_{1,1}, c_{1,2}, \ldots, c_{1,b}, c_{2,1}, \ldots, \ldots, c_{n,b}$ scattered within an arbitrary number of build stacks, where the card $c_{i,s}$ is the suit-s card of rank i. Since only black suits occur in w, the only actions possible are those that move a generalized King and its block to an empty build stack and those that move a card from a build stack to a suit stack. Even when all suits are black, the generalized King moves are powerful because the choice of which black King to move to an empty stack can be critical to the successful completion of the game. We do not yet fully understand the power of such moves. So we turn to FLAT-SOLIT$_{\text{NoKing}}(*, 0)$.

We say that a FLAT-SOLIT$_{\text{NoKing}}(*, 0)$ instance w is *nontrivial* if for every s, the suit-s cards occur in increasing order in every build stack. Clearly, no win is possible from a trivial w. When w is nontrivial, we define the directed graph $H(w)$ on the set of cards of w as follows: for $1 \leq i, j \leq n$ and $1 \leq s, t \leq b$, the arc $(c_{i,s}, c_{j,t})$ exists in $H(w)$ iff

1. $s = t$ and $j = i - 1$ (call this a *horizontal* edge), or
2. $s \neq t$ and the card $c_{i,s}$ is immediately beneath $c_{j,t}$ in some build stack (call this a *vertical* edge).

The following proposition is proved in the full paper.

Proposition 5. *Consider a nontrivial* FLAT-SOLIT$_{\text{NoKing}}(*, 0)$ *instance w. A win is possible from w iff $H(w)$ is cycle-free.*

Theorem 6. FLAT-SOLIT$_{\text{NoKing}}(*, 0)$ *is* NL*-complete.*

Proof. NL *upper bound.* Consider a FLAT-SOLIT$_{\text{NoKing}}(*, 0)$ instance w. We first check in AC^0 that w is nontrivial. If w is trivial then we reject immediately. Otherwise, Proposition 5 implies a co-NL = NL upper bound, because $H(w)$ is easily constructed in log space (in AC^0 in fact), and the total number bn of nodes in $H(w)$ together with the card numbers and suit numbers involved in w are $O(\log n)$-bit numbers.

NL*-hardness.* We reduce to FLAT-SOLIT$_{\text{NoKing}}(*, 0)$ the co-NL-complete problem of determining whether no path exists from node s to node $t \neq s$ in a directed graph G without self-loops and with edge set $E \subseteq \{1, \ldots, n\} \times \{1, \ldots, n\}$. The reduction is rendered delicate by the fact that several cards need to be assigned to each node in G: this is because a card can only occur once in a Klondike instance and furthermore, the "horizontal requirements" arising from the ranks of the cards are of course not compatible with the implicit ordering of the nodes in G. We now describe the flat Klondike instance w produced from G. It uses the cards $c_{i,u}$, $1 \leq i \leq 2n^2$, $1 \leq u \leq n$, arranged in $|E|(n-1) + 2$ build stacks as follows:

1. for each edge (i, j) in E and for each k, $0 \leq k < n - 1$, one stack with $c_{k(2n)+i,j}$ on top and with $c_{(k+1)(2n)+n+j,i}$ facing down below
2. one stack with $c_{2n^2,s}$ on top and with $c_{1,t}$ facing down below
3. one stack with the remaining cards facing down, in increasing order.

The reader can check that no card is mentioned twice in this (log space) construction. Furthermore, w is nontrivial, since only one build stack contains two cards of the same suit, and this stack is properly ordered. Hence the graph $H(w)$ is defined. It is a $n \times 2n$ grid with the horizontal edges forming n parallel lines running from left to right (card ranks decrease from left to right).

To see how the vertical edges operate, imagine the grid partitioned into columns $n, n-1, \ldots, 1$ of width $2n$. Each such column is further partitioned into a *target* region, of width n, and a *source* region, of width n. The vertical edges arising from $(i,j) \in E$ run from the source region in every column $k+1$ on line i to the target region in column k on line j. Observe then that the vertical $H(w)$ edges arising from E together with the horizontal $H(w)$ edges are incapable of forming a cycle in $H(w)$. The vertical edge $(c_{1,t}, c_{2n^2,s})$ is the only edge in $H(w)$ which connects a column (in fact, the rightmost entry in the source region of the nth column on line t) to a column situated to its left (in fact, to the leftmost entry in the target region of the first column on line s).

It follows that if a cycle exists in $H(w)$, then $(c_{1,t}, c_{2n^2,s})$ is part of it. Hence a path exists in $H(w)$ from the line s to the line t. This implies that a path existed from s to t in G.

Conversely, if a path $s = v_1, v_2, \ldots, v_m = t$ with $m \leq n-1$ exists in G, then a path can be traced from the column n on line s to the column $n-m$ on line t in $H(w)$ by appropriately combining neighbouring column traversals with horizontal displacements on the successive lines $v_1, v_2, \ldots, v_m$. A final horizontal displacement leads to $c_{1,t}$ and thus to $c_{2n^2,s}$, creating a cycle in $H(w)$.

Hence a cycle exists in $H(w)$ iff a path exists in G. By Proposition 5, a path exists in G iff no win is possible from w. This concludes the NL-hardness proof. □

We now relate the case of an arbitrary number of black suits to the case of a red suit and a black suit. We can show that when generalized King moves are disallowed, the all-blacks case reduces to the case of a red suit and a black suit.

Proposition 7
FLAT-SOLIT$_{\text{NoKing}}(*,0)$ AC^0*-reduces to* FLAT-SOLIT$_{\text{NoKing}}(1,1)$.

Proof. Let a FLAT-SOLIT$_{\text{NoKing}}(*,0)$ instance w involve the nb cards $c_{1,0}$, $c_{1,1}$, $\ldots$, $c_{1,b-1}$, $c_{2,0}, \ldots, \ldots, c_{n,b-1}$ where $c_{i,s}$ is the suit-$(s+1)$ card of rank i. The idea is to rename each $c_{i,s}$ as a ♡ card, and to use ♣ cards to restrict the release of the renamed cards in such a way as to enforce the rules that had to be followed in w when the original black cards were constrained by their respective suit stacks. Once the renamed images are released, all the auxiliary cards will be released to produce a win in the target {♣, ♡} instance.

This is done as follows. The FLAT-SOLIT$_{\text{NoKing}}(1,1)$ instance constructed will involve $3nb+1$ club cards and $3nb+1$ heart cards. First, for $0 \leq s < b$, we rename the suit-$(s+1)$ cards in the instance w as follows:

$$\begin{aligned} c_{1,s} &\rightarrow 3ns + 3n\heartsuit \\ c_{2,s} &\rightarrow 3ns + 3n - 3\heartsuit \\ &\cdots \\ c_{n-1,s} &\rightarrow 3ns + 6\heartsuit \\ c_{n,s} &\rightarrow 3ns + 3\heartsuit. \end{aligned}$$

Then, for $0 \leq s < b$, we add the n following build stacks, with the top card facing up and the bottom card (when present) facing down:

Below :		$3ns + 3n - 2\clubsuit$	$\cdots$	$3ns + 7\clubsuit$	$3ns + 4\clubsuit$
Top :	$3ns + 3n + 1\clubsuit$	$3ns + 3n - 1\clubsuit$	$\cdots$	$3ns + 8\clubsuit$	$3ns + 5\clubsuit$

Finally, the *critical* stack is set to the cards $3ns + 2\clubsuit$, $0 \leq s < b$, in any order, followed by the remaining cards $A\clubsuit, A\heartsuit, 2\heartsuit, 3\clubsuit, 4\heartsuit, 5\heartsuit, 6\clubsuit, \ldots, 3nb\clubsuit, 3nb{+}1\heartsuit$ in increasing order.

For any s, $0 \leq s < b$, until the $A\heartsuit$ becomes visible, no backward move is possible, and the cards $3ns + 3n\heartsuit$, $3ns + 3n - 3\heartsuit$, ..., $3ns + 6\heartsuit$ and $3ns + 3\heartsuit$ can only be placed in that order on the n build stacks designed to accept them. This holds for each s independently. Only after the $3ns + 3\heartsuit$ cards for $0 \leq s < b$ have found their ways to their mates $3ns + 4\clubsuit$ can the critical stack be freed of the b cards $3ns + 2\clubsuit$ sitting on top of it. In such an event, all the original build stacks arising from the renamed w cards are empty and only sorted build stacks remain, leading to a win. This happens iff the original w instance was winning. □

Corollary 8. FLAT-SOLIT$_{\text{NoKing}}(1, 1)$ *and* SOLIT$(1, 1)$ *are* NL*-hard.*

The simplest Klondike restrictions can be solved by constant depth circuits. We prove parts b) and c) of the following theorem in the full paper.

Theorem 9
a) *For any constant* b*,* FLAT-SOLIT$_{\text{NoKing}}(b, 0)$ *is in* AC^0
b) FLAT-SOLIT$(1, 0)$ *is in* AC^0
c) SOLIT$(1, 0)$ *is in* $\text{AC}^0[3]$.

Proof. Part a): By Proposition 5, we need to determine whether a cycle exists in the graph $H(w)$ of a nontrivial FLAT-SOLIT$_{\text{NoKing}}(b, 0)$ instance w. We first prove it for $b = 2$ and explain later how to generalize the proof.

When $b = 2$, we claim that $H(w)$ has a cycle iff there exist an edge $(i\clubsuit, k\spadesuit)$ and an edge $(j\spadesuit, \ell\clubsuit)$ in $H(w)$ such that $i < \ell$ and $j < k$. This condition is AC^0-testable.

We now prove the claim. Call a pair of edges $(i\clubsuit, k\spadesuit)$ and $(j\spadesuit, \ell\clubsuit)$ in $H(w)$ a *crossing* when $i < \ell$ and $j < k$. Clearly, a crossing together with the horizontal edges in $H(w)$ form a cycle. Conversely, let G be a cycle in $H(w)$. The cycle must have cards from both suits otherwise the instance is trivial. Consider the two parallel paths, one for $\clubsuit$ and one for $\spadesuit$, running from left to right and formed by the horizontal edges in $H(w)$. Let $i\clubsuit$ and $j\spadesuit$ be the rightmost (ie lowest ranked)

♣ and ♠ cards that belong to G. The edge in G leaving from i♣ must lead to a ♠, say k♠, otherwise the instance is trivial or i was not the rightmost. Similarly, the edge in G leaving from j♠ must lead to a ♣, say ℓ♣. It is not possible for both $i = \ell$ and $j = k$ to hold, since no configuration of the build stacks can simultaneously give rise to the edges (i♣, j♠) and (j♠, i♣). So assume with no loss of generality that $i < \ell$. Then $j < k$ otherwise the instance is trivial. So we have $i < \ell$ and $j < k$.

For the case $b > 2$, we claim that $H(w)$ has a cycle iff there exists a sequence of $d \leq b$ edges $(c_{a_1,s_1}, c_{b_2,s_2}), (c_{a_2,s_2}, c_{b_3,s_3}), (c_{a_3,s_3}, c_{b_4,s_4}), \ldots (c_{a_d,s_d}, c_{b_1,s_1})$ such that $a_i \leq b_i$ for $0 \leq i \leq d$. Clearly, these edges together with the horizontal edges in $H(w)$ creating paths from c_{b_i,s_i} to c_{a_i,s_i} form a cycle. Conversely, assuming a cycle G, consider the edges leaving from the rightmost card from G for each suit involved in G in the order they appear in G. These edges have the claimed property. This proves the claim and concludes part a). □

Finally we note an upper bound, proved in the full paper, that applies to the all-black instances:

Proposition 10. *For any constant* b, SOLIT$(b, 0)$ *is in* NL.

5 Conclusion

Figure 2 summarizes what we have learned in this work about the complexity of Klondike. Some gaps are obvious. In particular, the cases involving two suits beg for a more satisfactory characterization. The flat cases of a red suit and a black suit are especially puzzling. We suspect these cases to be in P, but could they possibly be hard for P? Are they in NL? Some simple Klondike cases involve the graphs $H(w)$ built around a grid with the horizontal lines representing the suit stack constraints. Could some of these be related to the grid graph reachability problems studied in [1]?

Our Klondike definition does not allow creating new build stacks in the course of the game, but the initial number of build stacks is not bounded. Does Klondike remain NP-hard if we insist on only seven build stacks initially, as in the usual 52-card Klondike?

	Flat Klondike, no King	Flat Klondike	Klondike
1 black	in AC^0	in AC^0	in AC^0[3]
b blacks	in AC^0	in NL	in NL
* blacks	NL-complete	NL-hard, in NP	NL-hard, in NP
1 black, 1 red	NL-hard, in NP	NL-hard, in NP	NL-hard, in NP
2 blacks, 1 red	NP-complete	NP-complete	NP-complete
* blacks, * red	NP-complete	NP-complete	NP-complete

Fig. 2. Our current knowledge of the complexity of Klondike. A "b" represents any fixed number and an "*" represents an input-dependent number.

Returning to one of our original motivations, we note that Klondike being NP-complete does not provide a mathematical justification that investigating the odds of winning in the case of a standard 52-card deck will be difficult. But the fact that Klondike is just another name for SAT can at least be seen as confirmation that the game does involve a good level of intricacy. We note that Figure 2 might suggest the following: start investigating the odds of winning in the apparently simpler game restrictions and then proceed onwards to the NP-complete cases.

Acknowledgement. We thank François Laviolette from Laval University who showed us in January 2007 how to handle backward moves to prove the Klondike NP upper bound in Theorem 1. The second author thanks Andreas Krebs and Christoph Behle in Tübingen for helpful discussions.

References

1. Allender, E., Barrington, D., Chakraborty, T., Datta, S., Roy, S.: Grid Graph Reachability Problems. In: Proc. 21st Annual IEEE Conference on Computational Complexity, pp. 299–313 (2006)
2. Barrington, D., Immerman, N., Straubing, H.: On uniformity within NC^1. J. Computer and System Sciences 41(3), 274–306 (1990)
3. Buss, S., Cook, S., Gupta, A., Ramachandran, V.: An optimal parallel algorithm for formula evaluation. SIAM J. Computing 21, 755–780 (1992)
4. Helmert, M.: Complexity results for standard benchmark domains in planning. Artificial Intelligence 143(2), 219–262 (2003)
5. Gent, I., Jefferson, C., Lynce, I., Miguel, I., Nightingale, P., Smith, B., Tarim, A.: Search in the Patience Game "Black Hole". In: AI Communications. pp. 1–15. IOS Press, Amsterdam, ISSN 0921-7126
6. R, Kaye, Minesweeper is NP-complete, The Mathematical Intelligencer, Springer Verlag, vol. 22, no. 2, pp. 9-15 (2000)
7. Papadimitriou, C.: Computational Complexity. Addison-Wesley, Reading (1994)
8. Parlett, D.: Solitaire: "Aces Up" and 399 Other Card Games, Pantheon (1979)
9. Parlett, D.: A History of Card Games, Oxford University Press, Oxford (1991)
10. Yan, X., Diaconis, P., Rusmevichientong, P., Van Roy, B.: Solitaire: Man versus Machine. In: Proc. Advances in Neural Information Processing Systems,NIPS vol. 17 (2004)

Adapting Parallel Algorithms to the W-Stream Model, with Applications to Graph Problems*

Camil Demetrescu[1], Bruno Escoffier[2], Gabriel Moruz[3], and Andrea Ribichini[1]

[1] Dipartimento di Informatica e Sistemistica, Università di Roma "La Sapienza", Rome, Italy
{demetres,ribichini}@dis.uniroma1.it
[2] Lamsade, Université Paris Dauphine, France
escoffier@lamsade.dauphine.fr
[3] MADALGO, BRICS, Department of Computer Science, University of Aarhus, Denmark
gabi@daimi.au.dk

Abstract. In this paper we show how parallel algorithms can be turned into efficient streaming algorithms for several classical combinatorial problems in the W-Stream model. In this model, at each pass one input stream is read and one output stream is written; streams are pipelined in such a way that the output stream produced at pass i is given as input stream at pass $i + 1$. Our techniques give new insights on developing streaming algorithms and yield optimal algorithms (up to polylog factors) for several classical problems in this model including sorting, connectivity, minimum spanning tree, biconnected components, and maximal independent set.

1 Introduction

Data stream processing has gained increasing popularity in the last few years as an effective paradigm for processing massive data sets. Huge data streams arise in several modern applications, including database systems, IP traffic analysis, sensor networks, and transaction logs [13, 23]. Streaming is an effective paradigm also in scenarios where the input data is not necessarily represented as a data stream. Due to high sequential access rates of modern disks, streaming algorithms can be effectively deployed for processing massive files on secondary storage [14], providing new insights into the solution of computational problems in external memory. In the classical read-only streaming model, algorithms are constrained to access the input data sequentially in one (or few) passes, using only a small amount of working memory, typically much smaller than the input size [14, 18, 19]. Usual parameters of the model are the working memory size s and the number of passes p that are performed over the data, which are usually

* Supported in part by the Sixth Framework Programme of the EU under contract number 001907 ("DELIS: Dynamically Evolving, Large Scale Information Systems"), and by the Italian MIUR Project "MAINSTREAM: Algorithms for massive information structures and data streams".

L. Kučera and A. Kučera (Eds.): MFCS 2007, LNCS 4708, pp. 194–205, 2007.

functions of the input size. Among the problems that have been studied in this model under the restriction that $p = O(1)$, we recall statistics and data sketching problems (see, e.g., [2, 11, 12]), which can be typically approximated using polylogarithmic working space, and graph problems (see, e.g., [5, 9, 10]), most of which require a working space linear in the vertex set size.

Motivated by practical factors, such as availability of large amounts of secondary storage at low cost, a number of authors have recently proposed less restrictive streaming models, where algorithms can both read and write data streams. Among them, we mention the W-Stream model and the StrSort model [1, 21]. In the W-Stream model, at each pass we operate with an input stream and an output stream. The streams are pipelined in such a way that the output stream produced at pass i is given as input stream at pass $i+1$. Despite the use of intermediate streams, which allows achieving effective space-passes tradeoffs for fundamental graph problems, most classical lower bounds in read-only streaming hold also in this model [8]. The StrSort model is just W-Stream augmented with a sorting primitive that can be used at each pass to reorder the output stream for free. Sorting provides a lot of computational power, making it possible to solve several graph problems using polylog passes and working space [1]. For a comprehensive survey of algorithmic techniques for processing data streams, we refer the interested reader to the extensive bibliographies in [4, 19].

It is well known that algorithmic ideas developed in the context of parallel computational models have inspired the design of efficient algorithms in other models. For instance, Chiang *et al.* [7] showed that efficient external memory algorithms can be derived from PRAM algorithms using a general simulation. Aggarwal *et al.* [1] discussed how circuits with uniform linear width and polylog depth (NC) can be simulated efficiently in StrSort, providing a systematic way of constructing algorithms in this model for problems in NC that use a linear number of processors. Examples of problems in this class include undirected connectivity and maximal independent set.

Parallel techniques seem to play a crucial role in the design of efficient algorithms in the W-Stream model as well. For instance, the single-source shortest paths algorithm described in [8] is inspired by a framework introduced by Ullman and Yannakakis [25] for the parallel transitive closure problem. However, to the best of our knowledge, no general techniques for simulating parallel algorithms in the W-Stream model have been addressed so far in the literature.

Our Contributions. In this paper, we show how classical parallel algorithms designed in the PRAM model can be turned into near-optimal algorithms in W-Stream for several classical combinatorial problems. We first show that any PRAM algorithm that runs in time T using N processors and memory M can be simulated in W-Stream using $p = O((T \cdot N \cdot \log M)/s)$ passes. This yields near-optimal trade-off upper bounds of the form $p = O((n \cdot \text{polylog } n)/s)$ in W-Stream for several problems, where n is the input size. Relevant examples include sorting, list ranking, and Euler tour. For other problems, however, this simulation does not provide good upper bounds. One prominent example concerns graph problems, for which efficient PRAM algorithms typically require $O(m+n)$ processors on graphs with n vertices and m edges. For those problems, this simulation method yields $p = O((m \cdot \text{polylog}\, n)/s)$ bounds, while $p = \Omega(n/s)$ almost-tight lower bounds in W-Stream are known for many of them.

To overcome this problem, we study an intermediate parallel model, which we call RPRAM, derived from the PRAM model by relaxing the assumption that a processor can only access a constant number of cells at each round. This way, we get the PRAM algorithms closer to streaming algorithms, since a memory cell in the working memory can be processed against an arbitrary number of cells in the stream. For some problems, this enhancement allows us to substantially reduce the number of processors while maintaining the same number of rounds. We show that simulating RPRAM algorithms in W-Stream leads to near-optimal algorithms (up to polylogarithmic factors) for several fundamental problems, including sorting, minimum spanning tree, biconnected components, and maximal independent set. Since algorithms obtained in this way are not always optimal – although very close to being so –, for some of the problems above we give better *ad hoc* algorithms designed directly in W-Stream, without using simulations.

Finally, we show that there exist problems for which the increased computational power of the RPRAM model does not help in reducing the number of processors required by a PRAM algorithm while maintaining the same time bounds, and thus cannot lead to better W-Stream algorithms. An example is deciding whether a directed graph contains a cycle of length two.

2 Simulating Parallel Algorithms in W-Stream

In this section we show general techniques for simulating parallel algorithms in W-Stream. We show in the next sections that our techniques yield near-optimal algorithms for many classical combinatorial problems in the W-Stream model. In Theorem 1 we discuss how to simulate general CRCW PRAM algorithms. Throughout this paper, we assume that each memory address, cell value, and processor state can be stored using $O(\log M)$ bits, where M is the memory size of the parallel machine.

Theorem 1. *Let A be a* PRAM *algorithm that uses N processors and runs in time T using space $M = \text{poly}(N)$. Then A can be simulated in* W-Stream *in $p = O((T \cdot N \cdot \log M)/s)$ passes using s bits of working memory and intermediate streams of size $O(M + N)$.*

Proof (Sketch). In the PRAM model, at each parallel round, every processor may read $O(1)$ memory cells, perform $O(1)$ instructions to update its internal state, and write $O(1)$ memory cells. A round of A can be simulated in W-Stream by performing $O((N \log M)/s)$ passes, where at each pass we simulate the execution of $\Theta(s/\log M)$ processors using s bits of working memory. The content of the memory cells accessed by the algorithm and the state of each processor are maintained on the intermediate streams. We simulate the task of each processor in a constant number of passes as follows. We first read from the input stream its state and the content of the $O(1)$ memory cells used by A and then we execute the $O(1)$ instructions performed. Finally, we write to the output stream the new state and possibly the values of the $O(1)$ output cells. Memory cells that remain unchanged are simply propagated through the intermediate streams by just copying them from the input stream to the output stream at each pass.

There are many examples of problems that can be solved near-optimally in W-Stream using Theorem 1. For instance, solving list ranking in PRAM takes $O(\log n)$ rounds and $O(n/\log n)$ processors [3], where n is the length of the list. By Theorem 1, we obtain a W-Stream algorithm that runs in $O((n \log n)/s)$ passes. An Euler tour of a tree with n vertices is computed in parallel in $O(1)$ rounds using $O(n)$ processors [15], which by Theorem 1 yields again a $p = O((n \log n)/s)$ bound in W-Stream. However, for other problems, the bounds obtained this way are far from being optimal. For instance, efficient PRAM algorithms for graph problems typically require $O(m + n)$ processors, where n is the number of vertices, and m is the number of edges. For these problems, Theorem 1 yields bounds of the form $p = O((m \cdot \text{polylog } n)/s)$, while $p = \Omega(n/s)$ almost-tight lower bounds are known for many of them.

In Definition 1 we introduce RPRAM as an extension of the PRAM model. It allows every processor to handle in a parallel round not only $O(1)$ memory cells, but an arbitrary number of cells. Since in W-Stream a value in the working memory might be processed against all the data in the stream, we view RPRAM as a natural link between PRAM and W-Stream, even though it may be unrealistic in a practical setting. We first introduce a generic simulation that turns RPRAM algorithms into W-Stream algorithms. We then give RPRAM implementations that lead to efficient algorithms in W-Stream for a number of problems where the PRAM simulation in Theorem 1 does not yield good results.

Definition 1. *An* RPRAM *(Relaxed* PRAM*) is an extended CRCW* PRAM *machine with N processors and memory of size M where at each round each processor can execute $O(M)$ instructions that:*

- *can read an arbitrary number of memory cells. Each cell can only be read a constant number of times during the round, and no assumptions can be made as to the order in which values are given to the processor;*
- *can write an arbitrary subset of the memory cells. The result of concurrent writes to the same cell by different processors in the same round is undefined. Writing can only be performed after all read operations have been done.*

Similarly to a PRAM*, each processor has a constant number of registers of size $O(\log M)$ bits.*

The jump in computational power provided by RPRAM allows substantial improvements for many classical PRAM algorithms such as decreasing the number of parallel rounds while preserving the number of processors or reducing the number of processors used while maintaining the same number of parallel rounds. We show in Theorem 2 that parallel algorithms implemented in this more powerful model can be simulated in W-Stream within the same bounds of Theorem 1.

Theorem 2. *Let A be an* RPRAM *algorithm that uses N processors and runs in time T using space $M = \text{poly}(N)$. Then A can be simulated in* W-Stream *in $p = O((T \cdot N \cdot \log M)/s)$ passes using s bits of working memory and intermediate streams of size $O(M + N)$.*

Proof (Sketch). We follow the proof of Theorem 1. The main difference is that a processor in the RPRAM model can read and write an arbitrary number of memory cells at each round, executing many instructions while still using $O(\log M)$ bits to maintain its internal state. Since the instructions of algorithm A performed by a processor during a round do not assume any particular order for reading the memory cells, reading memory values from the input stream can still be simulated in one pass. Replacing cell values read from the input stream with the new values written on the output stream can be performed in one additional pass.

3 Sorting

As a first simple application of the simulation techniques introduced in Section 2, we show how to derive efficient sorting algorithms in W-Stream. We first recall that n items can be sorted on a PRAM with $O(n)$ processors in $O(\log n)$ parallel rounds and $O(n \log n)$ comparisons [15]. By Theorem 1, this yields a W-Stream sorting algorithm that runs in $p = O((n \log^2 n)/s)$ passes. In RPRAM, however, sorting can be solved by $O(n)$ processors in constant time as follows. Each processor is assigned to an input item; in one parallel round it scans the entire memory and counts the numbers i and j of items smaller than and equal to the item the processor is assigned to respectively. Then each processor writes its own item into all the cells with indices between $i+1$ and $i+1+j$, and thus we obtain a sorted sequence.

Theorem 3. *Sorting n items in* RPRAM *can be done in $O(1)$ parallel rounds using $O(n)$ processors.*

Using the simulation in Theorem 2, we obtain the result stated below.

Corollary 1. *Sorting n items in* W-Stream *can be performed in $O(n \log n/s)$ passes.*

We obtain a W-Stream sorting algorithm that takes $p = O((n \log n)/s)$ passes, thus matching the performance of the best known algorithm for sorting in a streaming setting [18]. Since sorting requires $p = \Omega(n/s)$ passes in W-Stream, this bound is essentially optimal. However, both our algorithm and the algorithm in [18] perform $O(n^2)$ comparisons. We reduce the number of comparisons to the optimal $O(n \log n)$ at the expense of increasing the number of passes to $O((n \log^2 n)/s)$ by simulating an optimal PRAM algorithm via Theorem 1, as stated before.

4 Graph Problems

In this section we discuss how to derive efficient W-Stream algorithms for several graph problems using the RPRAM simulation in Theorem 2. Since efficient PRAM graph algorithms typically require $O(m+n)$ processors on graphs with n vertices and m edges [6], simulating such algorithms in W-Stream using Theorem 1 yields bounds of the form $p = O((m \cdot \text{polylog } n)/s)$, while $p = \Omega(n/s)$

almost-tight lower bounds in W-Stream are known for many of them. Graph connectivity is one prominent example [8]. Notice that, assigning each vertex to a processor, RPRAM gives enough power for each vertex to scan its entire neighborhood in a single parallel round. Since many parallel graph algorithms can be implemented using repeated neighborhood scanning, in many cases this allows us to reduce the number of processors from $O(m+n)$ to $O(n)$ while maintaining the same running time. By Theorem 2, this yields improved bounds of the form $p = O((n \cdot \text{polylog } n)/s)$.

4.1 Connected Components (CC)

A classical PRAM random-mating algorithm for computing the connected components of a graph with n vertices and m edges uses $O(m+n)$ processors and runs in $O(\log n)$ time with high probability [6, 20]. We first describe the algorithm and then we give an RPRAM implementation that uses only $O(n)$ processors which, by Theorem 2, leads to a nearly optimal algorithm in W-Stream.

PRAM Algorithm. The algorithm is based on building a set of star subgraphs and contracting the stars. It each parallel round it performs the following sequence of steps.

1. Each vertex is assigned the status of parent or child independently with probability 1/2;
2. For each child vertex u, determine whether it is adjacent to a parent vertex. If so, choose one such a vertex to be the parent $f(u)$ of u, and replace each edge (u, v) by $(f(u), v)$ and each edge (v, u) by $(f(v), u)$;
3. For each vertex having parent u, set the parent to $f(u)$.

The algorithm performs $O(\log n)$ parallel rounds with high probability [6].

RPRAM Implementation. We show how to implement each parallel round in RPRAM in $O(1)$ rounds using only $O(n)$ processors. We attach a processor to each vertex. We first assign each vertex the status of parent or child, and then for each vertex we scan its neighborhood to find a parent, if there exists one (in case of several parents, we break ties arbitrarily). Updating the parents according to the third step also takes one round in RPRAM. We obtain the result in Theorem 4.

Theorem 4. *Solving CC in* RPRAM *takes $O(n)$ processors and $O(\log n)$ rounds with high probability.*

By Theorem 2, this yields the following bound in W-Stream.

Corollary 2. *CC can be solved in* W-Stream *in $O((n \log^2 n)/s)$ passes with high probability.*

By the $p = \Omega(n/s)$ lower bound for CC in W-Stream [8], this upper bound is optimal up to a polylogarithmic factor. We notice that the same bound can be achieved deteministically by starting from the PRAM algorithm for CC in [22]. This bound can be further improved to $O((n \log n)/s)$ passes as shown in [8].

4.2 Minimum Spanning Tree (MST)

In this section, we first describe the PRAM algorithm in [6] for computing the MST of an undirected graph. We then give an RPRAM implementation that leads to an optimal algorithm (up to a polylog factor) in W-Stream by using the simulation in Theorem 2. Finally, we give an algorithm designed in W-Stream that outperforms the algorithm obtained via simulation.

PRAM Algorithm. The randomized CC algorithm previously introduced can be extended to find a minimum spanning tree in a (connected) graph [6]. It also takes $O(\log n)$ rounds with high probability and uses $O(m+n)$ processors. The algorithm is based on the property that given a subset V' of vertices, a minimum weight edge having one and only one endpoint in V' is in some MST. We modify the second step of the CC algorithm as follows. Each child vertex u determines the minimum weight incident edge (u, v). If v is a parent vertex, then we set $f(u) = v$ and flag the edge (u, v) as belonging to the spanning tree. This algorithm computes a MST and performs $O(\log n)$ rounds with high probability.

RPRAM Implementation. The updated second step runs in $O(1)$ rounds in RPRAM and uses $O(n)$ processors. Since the implementations of the other steps of the CC algorithm are unchanged and take $O(1)$ rounds and $O(n)$ processors, we obtain the result stated in Theorem 5.

Theorem 5. *MST is solvable in* RPRAM *using $O(n)$ processors and $O(\log n)$ rounds with high probability.*

Assuming edge weights can be encoded using $O(\log n)$ bits, we obtain the following bound in W-Stream by Theorem 2.

Corollary 3. *MST can be solved in* W-Stream *in $O((n \log^2 n)/s)$ passes.*

We now give a deterministic algorithm designed directly in W-Stream that improves the bounds achieved by using the simulation.

A Faster ad hoc *W-Stream Algorithm.* We again assume edge weights can be encoded using $O(\log n)$ bits. We build the MST by progressively adding edges as follows. We compute for each vertex the minimum weight edge incident to it. This set of edges E' is added to the MST. We then compute the connected components induced by E' and contract the graph by considering each connected component a single vertex. We repeat these steps until the graph contains a single vertex or there are no more edges to add. More precisely, we consider at each iteration a contracted graph where the vertices are the connected components of the partial MST so far computed. Denoting $G_i = (V_i, E_i)$ the graph before the ith iteration, the $(i+1)^{th}$ iteration consists of the following steps.

1. for each vertex $u \in V_i$, we compute a minimum weight edge (u, v) incident to u, and flag (u, v) as belonging to the MST (cycles that might occur due to weight ties are avoided by using a tie-breaking rule). Denote $E'_i = \{(u, v), u \in V_i\}$ the set of flagged edges.

2. we run a CC algorithm on the graph (V_i, E'_i). The resulted connected components are the vertices of V_{i+1}.
3. we replace each edge (u, v) by $(c(u), c(v))$, where $c(u)$ and $c(v)$ denote the labels of the connected components previously computed.

We now analyze the number of passes required in W-Stream. Let $|V_i| = n_i$. The first and the third steps require $O((n_i \log n)/s)$ passes each, since we can process in one pass $O(s/\log n)$ vertices. Computing the connected components also takes $O((n_i \log n)/s)$ passes, and therefore the ith iteration requires $O((n_i \log n)/s)$ passes. We note that at each iteration we add an edge for every vertex in V_i and thus $|V_{i+1}| \leq |V_i|/2$, i.e., the number of connected components is divided by at least two. We obtain that the total number of passes performed in the worst case is given by $T(n) = T(n/2) + O((n \log n)/s)$, which sums up to $O((n \log n)/s)$.

Theorem 6. *MST can be computed in $O((n \log n)/s)$ passes in* W-Stream.

By the $p = \Omega(n/s)$ lower bound for CC in W-Stream [8], this upper bound is optimal up to a polylog factor. To the best of our knowledge, no previous algorithm was known for MST in W-Stream.

4.3 Biconnected Components (BCC)

Tarjan and Vishkin [24] gave a PRAM algorithm that computes the biconnected components (BCC) of an undirected graph in $O(\log n)$ time using $O(m + n)$ processors. We give an RPRAM implementation of their algorithm that uses only $O(n)$ processors while preserving the time bounds and thus can be turned using Theorem 2 in a W-Stream algorithm that runs in $O((n \log^2 n)/s)$ passes. We also give a direct implementation that uses only $O((n \log n)/s)$ passes.

PRAM Algorithm. Given a graph G, the algorithm considers a graph G' such that vertices in G' correspond to edges in G and connected components in G' correspond to biconnected components in G. The algorithm first computes a rooted spanning tree T of G and then builds a subgraph G'' of G' having as vertices all the edges of T. The edges of G'' are chosen such that two vertices are in the same connected component of G'' if and only if the corresponding edges in G are in the same biconnected component. After computing the connected components of G'' the algorithm appends the remaining edges of G to their corresponding biconnected components. We now briefly sketch the five steps of the algorithm.

1. build a rooted spanning tree T of G and compute for each vertex its preorder and postorder numbers together with the number of descendants. Also, label the vertices by their preorder numbers.
2. for each vertex u, compute two values, $low(u)$ and $high(u)$, as follows.

$$low(u) = \min(\{u\} \cup \{low(w) | p(w) = u\} \cup \{w | (u, w) \in G \setminus T\})$$
$$high(u) = \max(\{u\} \cup \{high(w) | p(w) = u\} \cup \{w | (u, w) \in G \setminus T\}),$$

where $p(u)$ denotes the parent of vertex u.

3. add edges to G'' according to the following two rules. For all edges $(w, v) \in G \setminus T$ with $v + desc(v) \leq w$, add $((p(v), v), (p(w), w))$ to G'', and for all $(v, w) \in T$ with $p(w) = v$, $v \neq 1$, add $((p(v), v), (v, w))$ to G'' if $low(w) < v$ or $high(w) \geq v + desc(v)$, where $desc(v)$ denotes the number of descendants of vertex v.
4. compute the connected components of G''.
5. add the remaining edges of G to their biconnected components. Each edge $(v, w) \in G \setminus T$, with $v < w$, is assigned to the biconnected component of $(p(w), w)$.

RPRAM Implementation. We give RPRAM descriptions for all the five steps of the algorithm, each of them using $O(\log n)$ time and $O(n)$ processors. First, we compute a spanning tree of the graph using the RPRAM algorithm previously introduced. Rooting the tree and computing for each vertex the preorder and postorder numbers as well as the number of descendants are performed using list ranking and Euler tour [24], which take $O(\log n)$ time and $O(n)$ processors in PRAM, and thus in RPRAM. Since the second step takes $O(\log n)$ time using $O(n)$ processors in PRAM [24], the same bounds hold for RPRAM. We implement the third step in RPRAM in constant time and $O(n)$ processors, since it suffices a scan of the neighborhood for each vertex. For computing the connected components of G'' in the fourth step, we use the RPRAM algorithm previously introduced that takes $O(\log n)$ time and $O(n)$ processors. Finally, we implement the last step of the algorithm in RPRAM in $O(1)$ time and $O(n)$ processors by scanning the neighborhood for all vertices v and assigning the edges to the proper biconnected components. Since we implement all the steps of the algorithm in RPRAM in $O(\log n)$ rounds and $O(n)$ processors, we obtain the following result.

Theorem 7. *BCC can be solved in* RPRAM *using $O(n)$ processors in $O(\log n)$ rounds with high probability.*

By Theorem 2, this yields the following bound in W-Stream.

Corollary 4. *BCC can be solved in* W-Stream *in $O((n \log^2 n)/s)$ passes with high probability.*

We now show that we can achieve better bounds with an implementation designed directly in W-Stream.

A Faster ad hoc *W-Stream Algorithm.* We describe how to implement directly in W-Stream the steps of the parallel algorithm of Tarjan and Vishkin [24]. Notice that we have given constant time RPRAM descriptions for the third and the fifth step, thus by applying the simulation in Theorem 2 we obtain W-Stream algorithms that run in $O((n \log n)/s)$ passes. For computing the connected components in the fourth step, we use the algorithm in [8] that requires $O((n \log n)/s)$ passes. Therefore, to achieve a global bound of $O((n \log n)/s)$ passes, it suffices to give implementations that run in $O((n \log n)/s)$ passes for the first two steps. For the first step, we can compute a spanning tree within the bound of Theorem 6. Rooting the tree and computing the preorder and postorder numbers together with the number of descendants can be implemented in $O((n \log n)/s)$ passes using list ranking, Euler tour and sorting. Concerning the second step, we compute the *low* and *high* values by processing $\Theta(s/\log n)$ vertices at each pass, according to the postorder numbers.

Theorem 8. *BCC can be solved in* W-Stream *in* $O((n \log n)/s)$ *passes in the worst case.*

By the $p = \Omega(n/s)$ lower bound for CC in W-Stream [8], this upper bound is optimal up to a polylog factor. To the best of our knowledge, no previous algorithm was known for BCC in W-Stream.

4.4 Maximal Independent Set (MIS)

We give an efficient RPRAM algorithm for the maximal independent set problem (MIS), based on the PRAM algorithm proposed by Luby [17]. Using the simulation in Theorem 2, this leads to an efficient W-Stream implementation.

PRAM Algorithm. A maximal independent set S of a graph G is incrementally built through a series of iterations, where each iteration consists of a sequence of three steps, as follows. In the first step, we compute a random subset I of the vertices in G, by including each vertex v with probability $1/(2 \cdot deg(v))$. Then, for each edge (u, v) in G, with $u, v \in I$, we remove from I the vertex with the smallest degree. Finally, in the third step, we add to S the vertices in I, and then we remove from G the vertices in I together with their neighbors. The above steps are iterated until G gets empty. The algorithm uses $O(m + n)$ processors and $O(\log n)$ parallel rounds.

RPRAM Implementation. We implement the first step of each iteration in constant time and $O(n)$ processors in RPRAM, since it requires each vertex to compute its own degree. The second step can also be implemented in constant time, by having each vertex in I scan its neighborhood, and remove itself upon encountering a neighbor also in I with a larger degree. Finally, we implement the third step in constant time as well by scanning the neighborhood of each vertex that is not in I, and removing it from G if at least one of its neighbors is in I. Since the algorithm performs $O(\log n)$ iterations with high probability [17], we obtain the bound in Theorem 9.

Theorem 9. *MIS can be solved in* RPRAM *using* $O(n)$ *processors in* $O(\log n)$ *rounds with high probability.*

By Theorem 2, this yields the following bound in W-Stream.

Corollary 5. *MIS can be solved in* W-Stream *in* $O((n \log^2 n)/s)$ *passes with high probability.*

We now show that the bound in Corollary 5 is optimal up to a polylog factor.

Theorem 10. *MIS requires* $\Omega(n/s)$ *passes in* W-Stream.

Proof (Sketch). The proof is based on a reduction from the bit vector disjointness communication complexity problem. Alice has an n-bit vector A and Bob has an

n-bit vector B; they wish to know whether A and B are disjoint, i.e., $A \cdot B = 0$. They build a graph on $4n$ vertices v_i^j, where $i = 1, \cdots, n$ and $j = 1, \cdots, 4$. If $A_i = 0$, then Alice adds edges (v_i^1, v_i^2) and (v_i^3, v_i^4), whereas if $B_i = 0$, then Bob adds edges (v_i^1, v_i^3) and (v_i^2, v_i^4). The size of any MIS is $2n$ if $A \cdot B = 0$ and strictly greater otherwise.

5 Limits of the RPRAM Approach

In this section we prove that the increased power that RPRAM provides does not always help in reducing the number of processors to $O(n)$ and thus in obtaining W-Stream algorithms that run in $O((n \cdot \text{polylog } n)/s)$ passes. As an example, in Theorem 11 we prove that detecting cycles of length two in a graph takes $\Omega(m/s)$ passes.

Theorem 11. *Testing whether a directed graph with m edges contains a cycle of length two requires $p = \Omega(m/s)$ passes in* W-Stream.

Proof (Sketch). We prove the lower bound by showing a reduction from the bit vector disjointness two-party communication complexity problem. Alice has an m-bit vector A and Bob has an m-bit vector B; they wish to know whether A and B are disjoint, i.e., $A \cdot B = 0$. Alice creates a stream containing an edge $e(i) = (x_i, y_i)$ for each i such that $A[i] = 1$ and Bob creates a stream containing an edge $e^r(i) = (y_i, x_i)$ for each i such that $B[i] = 1$, where $x_i = i \text{ div } \lceil \sqrt{m} \rceil$ and $y_i = i \bmod \lceil \sqrt{m} \rceil$. Let G be the directed graph induced by the union of the edges in the streams created by Alice and Bob. Clearly, there is a cycle of length two in G if and only if $A \cdot B > 0$. Since solving bit vector disjointness requires transmitting $\Omega(m)$ bits [16], and the distributed execution of any streaming algorithm requires the working memory image to be sent back and forth from Alice to Bob at each pass, we obtain $s = \Omega(m)$, which leads to $p = \Omega(m/s)$.

Testing whether a digraph has a cycle of length two can be easily done in one round in RPRAM using $O(m)$ processors, by just checking in parallel whether there is any edge (x, y) that also appears as (y, x) in the graph. This leads to an algorithm in W-Stream that runs in $O((m \log n)/s)$ passes by Theorem 2.

References

[1] Aggarwal, G., Datar, M., Rajagopalan, S., Ruhl, M.: On the streaming model augmented with a sorting primitive. In: Proceedings of the 45th Annual IEEE Symposium on Foundations of Computer Science (FOCS'04), IEEE Computer Society Press, Los Alamitos (2004)

[2] Alon, N., Matias, Y., Szegedy, M.: The space complexity of approximating the frequency moments. J. Computer and System Sciences 58(1), 137–147 (1999)

[3] Anderson, R., Miller, G.: A simple randomized parallel algorithm for list-ranking. Information Processing Letters 33(5), 269–273 (1990)

[4] Babcock, B., Babu, S., Datar, M., Motwani, R., Widom, J.: Models and issues in data stream systems. In: Proceedings of the 21st ACM Symposium on Principles of Database Systems (PODS'02), pp. 1–16. ACM Press, New York (2002)

[5] Bar-Yossef, Z., Kumar, R., Sivakumar, D.: Reductions in streaming algorithms, with an application to counting triangles in graphs. In: Proc. 13th annual ACM-SIAM symposium on Discrete algorithms (SODA'02), pp. 623–632. ACM Press, New York (2002)
[6] Blelloch, G., Maggs, B.: Parallel algorithms. In: The Computer Science and Engineering Handbook, pp. 277–315 (1997)
[7] Chiang, Y., Goodrich, M., Grove, E., Tamassia, R., Vemgroff, D., Vitter, J.: External-memory graph algorithms. In: Proc. 6th Annual ACM-SIAM Symposium on Discrete Algorithms (SODA'95), pp. 139–149. ACM Press, New York (1995)
[8] Demetrescu, C., Finocchi, I., Ribichini, A.: Trading off space for passes in graph streaming problems. In: Proc. 17th Annual ACM-SIAM Symposium of Discrete Algorithms (SODA'06), pp. 714–723. ACM Press, New York (2006)
[9] Feigenbaum, J., Kannan, S., McGregor, A., Suri, S., Zhang, J.: On graph problems in a semi-streaming model. In: Díaz, J., Karhumäki, J., Lepistö, A., Sannella, D. (eds.) ICALP 2004. LNCS, vol. 3142, pp. 207–216. Springer, Heidelberg (2004)
[10] Feigenbaum, J., Kannan, S., McGregor, A., Suri, S., Zhang, J.: Graph distances in the streaming model: the value of space. In: Proceedings of the 16th ACM/SIAM Symposium on Discrete Algorithms (SODA'05), pp. 745–754 (2005)
[11] Feigenbaum, J., Kannan, S., Strauss, M., Viswanathan, M.: An approximate L^1 difference algorithm for massive data streams. SIAM Journal on Computing 32(1), 131–151 (2002)
[12] Gilbert, A., Guha, S., Indyk, P., Kotidis, Y., Muthukrishnan, S., Strauss, M.: Fast, small-space algorithms for approximate histogram maintenance. In: Proc. 34th ACM Symposium on Theory of Computing (STOC'02), pp. 389–398. ACM Press, New York (2002)
[13] Golab, L., Ozsu, M.: Data stream management issues: a survey. Technical report, School of Computer Science, University of Waterloo, TR CS-2003-08 (2003)
[14] Henzinger, M., Raghavan, P., Rajagopalan, S.: Computing on data streams. In: "External Memory algorithms". DIMACS series in Discrete Mathematics and Theoretical Computer Science 50, 107–118 (1999)
[15] Jájá, J.: An introduction to parallel algorithms. Addison-Wesley, Reading (1992)
[16] Kushilevitz, E., Nisan, N.: Communication Complexity. Cambr. U. Press (1997)
[17] Luby, M.: A simple parallel algorithm for the maximal independent set problem. SIAM Journal of Computing 15(4), 1036–1053 (1986)
[18] Munro, I., Paterson, M.: Selection and sorting with limited storage. Theoretical Computer Science 12, 315–323 (1980)
[19] Muthukrishnan, S.: Data streams: algorithms and applications. Technical report (2003), Available at `http://athos.rutgers.edu/~muthu/stream-1-1.ps`
[20] Reif, J.: Optimal parallel algorithms for integer sorting and graph connectivity. Technical Report TR 08-85, Aiken Comp. Lab, Harvard U., Cambridge (1985)
[21] Ruhl, M.: Efficient Algorithms for New Computational Models. PhD thesis, Massauchussets Institute of Technology (September 2003)
[22] Shiloach, Y., Vishkin, U.: An $o(\log n)$ Parallel Connectivity Algorithm. J. Algorithms 3(1), 57–67 (1982)
[23] Sullivan, M., Heybey, A.: Tribeca: A system for managing large databases of network traffic. In: Proceedings USENIX Annual Technical Conference (1998)
[24] Tarjan, R., Vishkin, U.: Finding biconnected components and computing tree functions in logarithmic parallel time. In: Proc. 25th Annual IEEE Symposium on Foundations of Computer Science (FOCS'84), pp. 12–20. IEEE Computer Society Press, Los Alamitos (1984)
[25] Ullman, J., Yannakakis, M.: High-probability parallel transitive-closure algorithms. SIAM Journal on Computing 20(1), 100–125 (1991)

Space-Conscious Compression

Travis Gagie and Giovanni Manzini*

Dipartimento di Informatica
Università del Piemonte Orientale
{travis,manzini}@mfn.unipmn.it

Abstract. Compression is most important when space is in short supply, so compression algorithms are often implemented in limited memory. Most analyses ignore memory constraints as an implementation detail, however, creating a gap between theory and practice. In this paper we consider the effect of memory limitations on compression algorithms. In the first part we assume the memory available is fixed and prove nearly tight upper and lower bounds on how much memory is needed to compress a string close to its k-th order entropy. In the second part we assume the memory available grows (slowly) as more and more characters are read. In this setting we show that the rate of growth of the available memory determines the speed at which the compression ratio approaches the entropy. In particular, we establish a relationship between the rate of growth of the sliding window in the LZ77 algorithm and its convergence rate.

1 Introduction

Data compression has come of age in recent years and compression algorithms are now vital in situations unforeseen by their designers. This has led to a discrepancy between the theory of data compression algorithms and their use in practice: compression algorithms are often designed and analysed assuming the compression and decompression operations can use a "sufficiently large" amount of working memory; however, in some situations, particularly in mobile or embedded computing environments, the memory available is very small compared to the amount of data we need to compress or decompress.

Even when compression algorithms are implemented to run on powerful desktop computers, some care is taken to be sure that the compression/decompression of large files do not take over all the RAM of the host machine. This is usually accomplished by splitting the input in blocks (bzip2), using heuristics to determine when to discard the old data (compress, ppmd), or by maintaining a "sliding window" over the more recently seen data and forgetting the oldest data (gzip). With the exception of the use of a sliding window (see Sect. 4), the validity of these techniques has not been established in a satisfying theoretical way.

In this paper we initiate the theoretical study of space-conscious compression algorithms. Although data compression algorithms have their own peculiarities,

* Both authors partly supported by Italian MUIR Italy-Israel FIRB Project "Pattern Discovery Algorithms in Discrete Structures, with Applications to Bioinformatics".

L. Kučera and A. Kučera (Eds.): MFCS 2007, LNCS 4708, pp. 206–217, 2007.

this study belongs to the general field of algorithmics in the streaming model (see, e.g., [1,10]), in which we are allowed only one pass over the input and memory sublinear (possibly polylogarithmic or even constant) in its size.

Our results. The first contribution of this paper is nearly tight upper and lower bounds on the compression ratio achievable by one-pass algorithms that use an amount of memory independent of the size of the input. The bounds are worst case and given in terms of the empirical k-th order entropy of the input string. More precisely we prove the following results:

(a) Let $\lambda \geq 1$, $k \geq 0$ and $\epsilon > 0$ be constants and let g be a function independent of n. In the worst case it is impossible to store a string s of length n over an alphabet of size σ in $\lambda H_k(s)n + o(n \log \sigma) + g$ bits using one pass and $O(\sigma^{k+1/\lambda-\epsilon})$ bits of memory.
(b) Given a $(\lambda H_k(s) + o(n \log \sigma) + g)$-bit encoding of s, it is impossible to recover s using one pass and $O(\sigma^{k+1/\lambda-\epsilon})$ bits of memory.
(c) Given $\lambda \geq 1$, $k \geq 0$ and $\mu > 0$, we can store s in $\lambda H_k(s)n + \mu n + O(\sigma^{k+1/\lambda} \log \sigma)$ bits using one pass and $O(\sigma^{k+1/\lambda} \log^2 \sigma)$ bits of memory, and later recover s using one pass and the same amount of memory.

While σ is often treated as constant in the literature, we treat it as a variable to distinguish between, say, $O(\sigma^{k+1/\lambda-\epsilon})$ and $O(\sigma^{k+1/\lambda} \log^2 \sigma)$ bits. Informally, (a) provides a lower bound to the amount of memory needed to compress a string up to its k-th order entropy; (b) tells us the same amount of memory is required also for decompression and implies that the use of a powerful machine for doing the compression does not help if only limited memory is available when decompression takes place; (c) establishes that (a) and (b) are nearly tight. Notice λ plays a dual role: for large k, it makes (a) and (b) inapproximability results — e.g., we cannot use $O(\sigma^k)$ bits of memory without worsening the compression in terms of $H_k(s)$ by more than a constant factor; for small k, it makes (c) an interesting approximability result — e.g., we can compress reasonably well in terms of $H_0(s)$ using, say, $O(\sqrt{\sigma})$ bits of memory. The main difference between the bounds in (a)–(b) and (c) is a $\sigma^\epsilon \log^2 \sigma$ factor in the memory usage. Since μ is a constant, $\mu n \in o(n \log \sigma)$ and the bounds on the encoding's length match. Note that μ can be arbitrarily small, but the term μn cannot be avoided (Lemma 4).

A second contribution of this paper is the proof of lower bounds similar to (a) and (b) for the case in which the input string is generated by a stationary ergodic k-th order Markov source $\mathcal{X}$. We show that, with high probability, a length-n string drawn from $\mathcal{X}$ cannot be stored in $\lambda H(\mathcal{X})n + o(n \log \sigma) + g$ bits using one pass and $o(\sigma^k \log \sigma)$ bits of memory (here λ, σ and g have the same meaning as in (a)). A symmetrical result holds for the space required for recovering a string emitted by $\mathcal{X}$ and compressed up to $\lambda H(\mathcal{X})n + o(n \log \sigma) + g$ bits. Note that an upper bound analogous to (c) automatically holds for strings generated by an ergodic k-th order Markov source $\mathcal{X}$ (the term H_k is simply replaced by $H(\mathcal{X})$).

The final contribution of the paper is a first step in the analysis of the power of compressors when the amount of available working memory grows with the size of the input. We model this behavior assuming that we are given an increasing function $f(t)$ and that after reading t characters the compression algorithm is allowed to use $\Theta(f(t))$ bits of memory. In this setting, the result (c) above implies (Lemma 7) that for any diverging function f (i.e. $\lim_{t\to\infty} f(t) = +\infty$) it is possible to compress every string up to its k-th order entropy *for any* $k \geq 0$. Given this state of affairs, it is clear that to understand the role played by the rate-of-growth function f, we must go deeper than simply considering whether the compression ratio approaches the k-th order entropy. We initiate this study with the analysis of LZ77 with a sliding window, which is the algorithm at the heart of the gzip tool. We show quantitatively that the rate-of-growth function f influences the convergence rate of the algorithm; that is, the speed at which the algorithm approaches the k-th order entropy. In particular, now treating σ as a constant, we prove that

(d) if LZ77 uses a sliding window that grows as $f(t) = t/\log^2 t$, then for any string s, the output size is bounded by $H_k(s)n + O((n \log\log n)/\log n)$ simultaneously for any $k \geq 0$;
(e) if LZ77 uses a sliding window that grows as $f(t) = \log^{1-\epsilon} t$, with $0 < \epsilon < 1$, then for any $n > 0$ we can build a string $\hat{s}$ of length n such that LZ77's output size is at least $H_0(s)n + \Omega\big((n \log\log n)/\log^{1-\epsilon} n\big)$ bits.

In other words, a faster growing sliding window yields a provably faster rate of convergence. To our knowledge, these are the first results relating the size of LZ77's sliding window and its rate of convergence in the worst case setting. In the probabilistic setting (see below) what it is known [14] is that using a window of *fixed* size W the rate of convergence of LZ77 is $\Theta((n \log\log W)/\log W)$.

2 Notation

In the following we use s to denote the string that we want to compress. We assume that s has length n and is drawn from an alphabet of size σ. Note that in Section 3 we measure memory in terms of alphabet size so σ is considered a variable; conversely, in Section 4 the memory depends on the input size n, so σ is considered a constant that remains hidden in the asymptotic notation.

For $i = 1, 2, \ldots, \sigma$, let n_i be the number of occurrences of the i-th alphabet symbol in s. The 0*-th order empirical entropy* of s is defined as $H_0(s) = -\sum_{i=1}^{\sigma}(n_i/|s|)\log(n_i/|s|)$ (throughout this paper we assume that all logarithms are taken to the base 2 and $0 \log 0 = 0$). It is well known that H_0 is the maximum compression we can achieve using a fixed codeword for each alphabet symbol. We can achieve a greater compression if the codeword we use for each symbol depends on the k symbols preceding it. In this case the maximum compression is bounded by the k-th order entropy $H_k(s)$ (see [6] for the formal definition). We use two properties of k-th order entropy in particular: $H_k(s_1)|s_1| + H_k(s_2)|s_2| \leq H_k(s_1 s_2)|s_1 s_2|$ and, since $H_0(s) \leq \log|\{a : a \text{ occurs in } s\}|$, we have $H_k(s) \leq \log \max_{|w|=k}\{j : w \text{ is followed by } j \text{ distinct characters in } s\}$.

We point out that the empirical entropy is defined *pointwise* for any string and can be used to measure the performance of compression algorithms as a function of the *string structure*, thus without any assumption on the input source. For this reason we say that the bounds given in terms of H_k are *worst case* bounds. Another common approach in data compression is to assume that the input string is generated by a Markov source $\mathcal{X}$. To measure the effectiveness of a compression algorithm in this setting its *average* compression ratio is compared with the entropy of the source $H(\mathcal{X})$. We call this the *probabilistic* setting, and we consider it in Sect. 3.3.

Some of our arguments are based on Kolmogorov complexity [8]; the Kolmogorov complexity of s, denoted $K(s)$, is the length in bits of the shortest program that outputs s; it is generally incomputable but can be bounded from below by counting arguments (e.g., in a set of m elements, most have Kolmogorov complexity at least $\log m - O(1)$). We use two properties of Kolmogorov complexity in particular, as well: if an object can be easily computed from other objects, then its Kolmogorov complexity is at most the sum of theirs plus a constant; and a fixed, finite object has constant Kolmogorov complexity.

In this paper we consider space-conscious compressors, that is, algorithms that are allowed to use a limited amount of memory during their execution. We assume that the algorithms are one-pass in the sense that they are allowed to read each input symbol only once. Hence, if an algorithm needs to access (portions of) the input more than once it must store it—consuming part of its precious working memory. In Section 5 we briefly comment on the possibility of extending our results to multi-pass algorithms. Being space-conscious ourselves, most of the proofs omitted; they can be found in a technical report [4].

3 Compressing with Memory Independent of Length

Move-to-front compression [2] is probably the best example of a compression algorithm whose space complexity is independent of the input length: keep a list of the characters that have occurred in decreasing order by recency; store each character in the input by outputting its position in the list (or, if it has not occurred before, its index in the alphabet) encoded in Elias' δ code, then move it to the front of the list. Move-to-front stores a string s of length n over an alphabet of size σ in $\big(H_0(s) + O(\log H_0(s))\big)\, n + O(\sigma \log \sigma)$ bits using one pass and $O(\sigma \log \sigma)$ bits of memory. Note that we can store s in $\big(H_k(s) + O(\log H_k(s))\big)\, n + O(\sigma^{k+1} \log \sigma)$ bits by keeping a separate list for each possible context of length k; this increases the memory usage by a factor of at most σ^k.

In this section we first use a more complicated algorithm to get a better upper bound: given constants $\lambda \geq 1$, $k \geq 0$ and $\mu > 0$, we can store s in $(\lambda H_k(s) + \mu)n + O(\sigma^{k+1/\lambda} \log \sigma)$ bits using one pass and $O(\sigma^{k+1/\lambda} \log^2 \sigma)$ bits of memory. We then prove that $\mu > 0$ is necessary and that we need to know k. We use the idea from these proofs to prove a nearly matching lower bound for compression: in the worst case it is impossible to store a string s of length n over an alphabet of size σ in $\lambda H_k(s)n + o(n \log \sigma) + g$ bits, for any function g independent of n, using one encoding pass and $O(\sigma^{k+1/\lambda-\epsilon})$ bits of memory. We prove a symmetric lower bound for decompression, and close with slightly weaker lower bounds for when the input comes from a stationary Markov source.

3.1 A Nearly Tight Upper Bound

The main drawback of move-to-front is the $O(\log H_0(s))$ in its analysis (or $O(\log H_k(s))$ using contexts of length k); we now show how we can replace this by any given constant $\mu > 0$. We start with the following lemma about storing an approximation Q of a probability distribution P in few bits, so that the relative entropy between P and Q is small. The relative entropy $D(P\|Q) = \sum_{i=1}^{\sigma} p_i \log(p_i/q_i)$ between $P = p_1, \ldots, p_\sigma$ and $Q = q_1, \ldots, q_\sigma$ is the expected redundancy per character of an ideal code for Q when characters are drawn according to P.

Lemma 1 ([3]). *Let s be a string of length n over an alphabet of size σ and let P be the normalized distribution of characters in s. Given s and constants $\lambda \geq 1$ and $\mu > 0$, we can store a probability distribution Q with $D(P\|Q) < (\lambda - 1)H(P) + \mu$ in $O(\sigma^{1/\lambda} \log(n + \sigma))$ bits using $O(\sigma^{1/\lambda} \log(n + \sigma))$ bits of memory.* □

Armed with this lemma, we adapt arithmetic coding [12] to use $O(\sigma^{1/\lambda} \log(n + \sigma))$ bits of memory with a specified redundancy per character:

Lemma 2. *Given a string s of length n over an alphabet of size σ and constants $\lambda \geq 1$ and $\mu > 0$, we can store s in $(\lambda H_0(s) + \mu)n + O(\sigma^{1/\lambda} \log(n + \sigma))$ bits using $O(\sigma^{1/\lambda} \log(n + \sigma))$ bits of memory.* □

We boost our space-conscious arithmetic coding algorithm to achieve a bound in terms of $H_k(s)$ instead of $H_0(s)$ by running a separate copy for each possible k-tuple, just as we boosted move-to-front compression:

Lemma 3. *Given a string s of length n over an alphabet of size σ and constants $\lambda \geq 1$, $k \geq 0$ and $\mu > 0$, we can store s in $(\lambda H_k(s) + \mu)n + O(\sigma^{k+1/\lambda} \log(n + \sigma))$ bits using $O(\sigma^{k+1/\lambda} \log(n + \sigma))$ bits of memory.* □

To make our algorithm use one pass and to change the $\log(n + \sigma)$ factor to $\log \sigma$, we process the input in blocks $s_1, \ldots, s_b$ of length $O(\sigma^{k+1/\lambda} \log \sigma)$. Notice each individual block s_i fits in memory — so we can apply Lemma 3 to it — and $\log(|s_i| + \sigma) = O(\log \sigma)$.

Theorem 1. *Given a string s of length n over an alphabet of size σ and constants $\lambda \geq 1$, $k \geq 0$ and $\mu > 0$, we can store s in $(\lambda H_k(s) + \mu)n + O(\sigma^{k+1/\lambda} \log \sigma)$ bits using one pass and $O(\sigma^{k+1/\lambda} \log^2 \sigma)$ bits of memory, and later recover s using one pass and the same amount of memory.* □

3.2 Lower Bounds

Theorem 1 is still weaker than the strongest compression bounds that ignore memory constraints, in two important ways: first, even when $\lambda = 1$ the bound on the compression ratio does not approach $H_k(s)$ as n goes to infinity; second, we need to know k. It is not hard to prove these weaknesses are unavoidable when using fixed memory, as follows.

Lemma 4. *Let $\lambda \geq 1$ be a constant and let g be a function independent of n. In the worst case it is impossible to store a string s of length n in $\lambda H_0(s)n+o(n)+g$ bits using one encoding pass and memory independent of n.*

Proof. Let A be an algorithm that, given λ, stores s using one pass and memory independent of n. Since A's future output depends only on its state and its future input, we can model A with a finite-state machine M. While reading $|M|$ characters of s, M must visit some state at least twice; therefore either M outputs at least one bit for every $|M|$ characters in s — or $n/|M|$ bits in total — or for infinitely many strings M outputs nothing. If s is unary, however, then $H_0(s) = 0$. □

Lemma 5. *Let λ be a constant, let g be a function independent of n and let b be a function independent of n and k. In the worst case it is impossible to store a string s of length n over an alphabet of size σ in $\lambda H_k(s)n + o(n \log \sigma) + g$ bits for all $k \geq 0$ using one pass and b bits of memory.*

Proof. Let A be an algorithm that, given λ, g, b and σ, stores s using b bits of memory. Again, we can model it with a finite-state machine M, with $|M| = 2^b$ and M's Kolmogorov complexity $K(M) = K(\langle A, \lambda, g, b, \sigma \rangle) + O(1) = O(\log \sigma)$. (Since A, λ, g, and b are all fixed, their Kolmogorov complexities are $O(1)$.)

Suppose s is a periodic string with period $2b$ whose repeated substring r has $K(r) = |r| \log \sigma - O(1)$. We can specify r by specifying M, the states M is in when it reaches and leaves any copy of r in s, and M's output on that copy of r. (If there were another string r' that took M between those states with that output, then we could substitute r' for r in s without changing M's output.) Therefore M outputs at least

$$K(r) - K(M) - O(\log |M|) = |r| \log \sigma - O(\log \sigma + b) = \Omega(|r| \log \sigma)$$

bits for each copy of r in s, or $\Omega(n \log \sigma)$ bits in total. For $k \geq 2b$, however, $H_k(s)$ approaches 0 as n goes to infinity. □

The idea behind these proofs is simple — model a one-pass algorithm with a finite-state machine and evaluate its behaviour on a periodic string — but, combining it with the following simple results, we can easily show a lower bound that nearly matches Theorem 1. (In fact, our proofs are valid even for algorithms that make preliminary passes that produce no output — perhaps to gather statistics, like Huffman coding [5] — followed by a single encoding pass that produces all of the output; once the algorithm begins the encoding pass, we can model it with a finite-state machine).

Lemma 6 ([3]). *Let $\lambda \geq 1$, $k \geq 0$ and $\epsilon > 0$ be constants and let r be a randomly chosen string of length $\lfloor \sigma^{k+1/\lambda-\epsilon} \rfloor$ over an alphabet of size σ. With high probability every possible k-tuple is followed by $O(\sigma^{1/\lambda-\epsilon})$ distinct characters in r.* □

Corollary 1. *Let $\lambda \geq 1$, $k \geq 0$ and $\epsilon > 0$ be constants. There exists a string r of length $\lfloor \sigma^{k+1/\lambda-\epsilon} \rfloor$ over an alphabet of size σ with $K(r) = |r| \log \sigma - O(1)$ but $H_k(r^i) \leq (1/\lambda - \epsilon) \log \sigma + O(1)$ for $i \geq 1$.* □

Consider what we get if, for some $\epsilon > 0$, we allow the algorithm A from Lemma 5 to use $O(\sigma^{k+1/\lambda-\epsilon})$ bits of memory, and evaluate it on the periodic string r^i from Corollary 1. Since r^i has period $\lfloor \sigma^{k+1/\lambda-\epsilon} \rfloor$ and its repeated substring r has $K(r) = |r| \log \sigma - O(1)$, the finite-state machine M outputs at least

$$K(r) - K(M) - O(\log |M|) = |r| \log \sigma - O(\sigma^{k+1/\lambda-\epsilon}) = |r| \log \sigma - O(|r|)$$

bits for each copy of r in r^i, or $n \log \sigma - O(n)$ bits in total. Because $\lambda H_k(r^i) \leq (1-\epsilon)\log \sigma + O(1)$, this yields the following nearly tight lower bound; notice it matches Theorem 1 except for a $\sigma^\epsilon \log^2 \sigma$ factor in the memory usage.

Theorem 2. *Let $\lambda \geq 1$, $k \geq 0$ and $\epsilon > 0$ be constants and let g be a function independent of n. In the worst case it is impossible to store a string s of length n over an alphabet of size σ in $\lambda H_k(s)n + o(n \log \sigma) + g$ bits using one encoding pass and $O(\sigma^{k+1/\lambda-\epsilon})$ bits of memory.* □

With a good bound on how much memory is needed for compression, we turn our attention to decompression. Good bounds here are equally important, because often data is compressed once by a powerful machine (e.g., a server or base-station) and then transmitted to many weaker machines (clients or agents) who decompress it individually. Fortunately for us, compression and decompression are essentially symmetric. Recall Theorem 1 says we can recover s from a $\big(\lambda H_k(s) + \mu\big)n + O(\sigma^{k+1/\lambda} \log \sigma)\big)$-bit encoding using one pass and $O(\sigma^{k+1/\lambda} \log^2 \sigma)$ bits of memory. Using the same idea about finite-state machines and periodic strings gives us the following nearly matching lower bound:

Theorem 3. *Let $\lambda \geq 1$, $k \geq 0$ and $\epsilon > 0$ be constants and let g be a function independent of n. There exists a string s of length n over an alphabet of size σ such that, given a $(\lambda H_k(s)n + o(n \log \sigma) + g)$-bit encoding of s, it is impossible to recover s using one pass and $O(\sigma^{k+1/\lambda-\epsilon})$ bits of memory.* □

3.3 Markov Sources

As many classic analyses assume the data comes from a Markov source, we close this section with versions of Theorems 2 and 3 that have slightly weaker bounds on memory usage — we show $o(\sigma^k \log \sigma)$ bits of memory are insufficient, instead of $O(\sigma^{k+1/\lambda-\epsilon})$ — but apply when the data are drawn from a such a source. (All other things being equal, upper bounds are stronger when proven in terms of empirical entropy, without any assumptions about the source; conversely, lower bounds are stronger when they hold even with such assumptions.) The proofs of these theorems are slightly different and involve de Bruijn sequences; a σ-ary de Bruijn sequence of order k contains every possible k-tuple exactly once and, so, has length $\sigma^k + k - 1$. This property means every such sequence has kth-order empirical entropy 0 and, equivalently, can be generated by a deterministic kth-order Markov source. Rosenfeld [13] proved there are $(\sigma!)^{\sigma^{k-1}}$ such sequences so, by Stirling's formula, a randomly chosen one d has expected Kolmogorov complexity $\mathrm{E}[K(d)] = \log(\sigma!)^{\sigma^{k-1}} - O(1) = |d| \log \sigma - O(|d|)$.

Theorem 4. *Let $\lambda \geq 1$ and $k \geq 0$ be constants and let g be a function independent of n. There exists a stationary ergodic kth-order Markov source $\mathcal{X}$ over an alphabet of size σ such that, if we draw a string s of length n from $\mathcal{X}$, then with high probability it is impossible to store s in $\lambda H(\mathcal{X})n + o(n \log \sigma) + g$ bits using one pass and $o(\sigma^k \log \sigma)$ bits of memory.* □

Theorem 5. *Let $\lambda \geq 1$ and $k \geq 0$ be constants and let g be a function independent of n. There exists a stationary ergodic kth-order Markov source $\mathcal{X}$ over an alphabet of size σ such that, if we draw a string s of length n from $\mathcal{X}$, then with high probability it is impossible to recover s from a $(\lambda H(\mathcal{X})n + o(n \log \sigma) + g)$-bit encoding of s using one pass and $o(\sigma^k \log \sigma)$ bits of memory.* □

4 Compressing with (Slowly) Growing Memory

In the previous section we have given upper and lower bounds to the amount of memory required to compress up to the k-th order entropy for a fixed k. It is well known that the best compressors, e.g. LZ77, LZ78, and BWT-based tools, are able to compress up to the k-th order entropy *for all $k \geq 0$ simultaneously.* That is, for any $k \geq 0$ and for any string s, their output is bounded by $\lambda H_k(s)n + g_k(n)$, with $g_k(n) \in o(n)$. Intuitively this means that these algorithms can take advantage of an "order-k regularity" for an arbitrarily large k. Unfortunately, Lemma 5 tells us that using memory indepedendent of n, it is impossible to compress up to $\lambda H_k(s)n$ for any $k \geq 0$.

For the above reasons, in this section we study compression algorithms in which the available memory grows with the size of the input. Given an increasing function f, we define the class C_f of one-pass compressors in which the working space grows according to f in the sense that when the algorithm has read t characters it is allowed to use a working space of size $\Theta(f(t))$ bits. Our first result shows that if $\lim_{t\to\infty} f(t) = \infty$ the algorithms in C_f can compress up to $\lambda H_k(s)n$ for any $k \geq 0$.

Lemma 7. *For any increasing and diverging function f there exists an algorithm in C_f achieving the compression ratio given in Theorem 1 for any $k \geq 0$.*

Proof. For a given k let n' be such that $f(n')$ is greater than the working space of the algorithm in Theorem 1. Consider now the procedure that outputs the first n' characters without compression and then executes the algorithm of Theorem 1. Since the space for the initial n' characters is just a constant overhead, for sufficiently long strings this procedure asymptotically achieves the space bound of Theorem 1 as claimed. □

The proof of Lemma 7 suggests that although any diverging working space suffices to get a compression ratio close to H_k for any $k \geq 0$, the rate of growth of the working space is likely to influence the rate of convergence, that is, the speed with which the compression ratio approaches the entropy. The quantitative study of this problem in the general setting appears to be a rather challenging task. In the following we initiate this study by exploring the relationship between working space and rate of convergence for the important special case of the algorithm LZ77 with a growing sliding window.

4.1 Window Size vs. Convergence Rate for LZ77

In the following we assume that the alphabet size is a constant (see comment at the beginning of Section 2). The LZ77 algorithm works by parsing the input string s into a sequence of words $w_1, w_2, \ldots, w_d$ and by encoding a compact representation of these words. For any non-empty string w let w^- denote the string w with the last character removed, and, if $|w| > 1$, let $w^{--} = (w^-)^-$. Assuming the words $w_1, w_2, \ldots, w_{i-1}$ have been already parsed, LZ77 selects the i-th word as the longest word w_i that can be obtained by adding a single character to a substring of $(w_1 w_2 \cdots w_i)^{--}$. Note that although this is a recursive definition there is no ambiguity. In fact, if $|w_i| > 1$ at least the first character of w_i belongs to $w_1 w_2 \cdots w_{i-1}$.

In the algorithm LZ77 with sliding window (LZ77$_{sw}$ from now on) the word w_i is selected using a sliding window of size L_i, that is, w_i^- must be a substring of $(z_i w_i)^{--}$ where z_i is the length-L_i suffix of $w_1 w_2 \cdots w_{i-1}$. In practical implementations the sliding window length is usually fixed (for example it is equal to 2^{15} in gzip) but for our analysis we will consider a sliding window which grows with the size of the parsed string. Once w_i has been found, it is encoded with the triplet (p_i, ℓ_i, α_i), where p_i is the starting position of w_i^- in the sliding window, $\ell_i = |w_i|$, and α_i is the last character of w_i. In the following we assume that encoding p_i takes $\log L_i + O(1)$ bits, encoding ℓ_i takes[1] $\log \ell_i + O(\log\log \ell_i)$ bits, and encoding α_i takes $\log \sigma + O(1)$ bits[2]. If we store the already parsed portion of the input in a suffix tree, the algorithm LZ77 runs in linear time and uses a working space of $\Theta(n \log n)$ bits. The same result holds for LZ77$_{sw}$ as well: the only difference is that we use a truncated suffix tree [7,11] to maintain the sliding window so the working space is $\Theta(L \log L)$ bits, where L is the maximum size of the sliding window.

For the algorithm LZ77 we know (see [6, Th. 4.1]) that for any $k \geq 0$ and for any string s:

$$|\mathsf{LZ77}(s)| \leq H_k(s)n + O\left(n \frac{\log\log n}{\log n}\right) \tag{1}$$

which implies that the convergence rate is $O((n \log\log n)/\log n)$.

In the following we say that LZ77$_{sw}$ uses an $f(t)$-size sliding window to denote that when t characters have been read, LZ77$_{sw}$ maintains a sliding window of size $\lceil f(t) \rceil$ (hence, for $f(t) = t$ we have the original LZ77 algorithm). We prove that for $f(t) = t/\log^2 t$ the convergence rate is still $O((n \log\log n)/\log n)$, whereas for $f(t) = \log^{1-\epsilon} t$, with $0 < \epsilon < 1$, the convergence rate is $\Omega((n \log\log n)/\log^{1-\epsilon} n)$.

---

[1] Since we cannot bound in advance the size of ℓ_i, we are assuming we code it using Elias' δ code.

[2] Other encodings are possible but we believe our analysis can be adapted to all "reasonable" encodings.

To bound the convergence rate of $\mathsf{LZ77}_{sw}$ with a sliding window growing as $(t/\log^2 t)$, we first bound the number of times the same word can appear in the parsing of the input string.

Lemma 8. *Let $g(t)$ denote an increasing and diverging function. The $\mathsf{LZ77}_{sw}$ algorithm with window size $f(t) = t/g(t)$ produces a parsing of the input string in which the same word appears at most $O(g(n)\log(n))$ times.* □

The next lemma relates the number of words in the parsing with the k-th order entropy, and Lemma 10 gives an upper bound to the total number of words.

Lemma 9 ([6, Lemma 2.3]). *Let $y_1, \ldots, y_d$ denote a parsing of a string s in which each word y_i appears at most M times. For any $k \geq 0$ we have*

$$d \log d \leq |s| H_k(s) + d \log\left(\frac{|s|}{d}\right) + d \log M + \Theta(d)\,. \qquad \square$$

Lemma 10. *Let $y_1, \ldots, y_d$ denote a parsing of a string s in which each word y_i appears at most M times. We have $d = O(n/\log(n/M))$.* □

Theorem 6. *The algorithm $\mathsf{LZ77}_{sw}$ with a sliding window growing as $f(t) = (t/\log^2 t)$ produces an output bounded by $nH_k(s) + O((n \log\log n)/\log n)$.*

Proof. Let $w_1 \cdots w_d$ denote the $\mathsf{LZ77}_{sw}$ parsing of s. Recall that for each word w_i $\mathsf{LZ77}_{sw}$ outputs a triple (p_i, ℓ_i, α_i) whose encoding is described above. Using elementary calculus it is easy to show that if $\mathsf{LZ77}_{sw}$ parses s into d words, the output size is bounded by

$$|\mathsf{LZ77}_{sw}(s)| \leq d \log n + d \log(n/d) + O(d \log\log n)\,.$$

Recall that by Lemma 8 each word appears at most $O\big(\log^3 n\big)$ times in the parsing. Since $d \log n = d \log d + d \log(n/d)$, using Lemma 9 we get

$$\begin{aligned} |\mathsf{LZ77}_{sw}(s)| &\leq d \log d + 2n \log(n/d) + O(d \log\log n) \\ &\leq H_k(s)n + 3d \log(n/d) + O(d \log\log n)\,. \end{aligned}$$

Finally, by Lemma 10 we have $d = O(n/\log n)$, hence

$$|\mathsf{LZ77}_{sw}(s)| \leq nH_k(s) + O\left(\frac{n \log\log n}{\log n}\right)$$

as claimed. □

Now we show that the algorithm $\mathsf{LZ77}_{sw}$ with a sliding window of size $o(\log t)$ has a convergence rate $\omega(n \log\log n/\log n)$. To this end, for any ϵ, with $0 < \epsilon < 1$, we consider the $\mathsf{LZ77}_{sw}$ algorithm with a sliding window of size $f(t) = \log^{1-\epsilon}(t)$. Fix $n > 0$ and let $b = 1 + \lceil \log^{1-\epsilon}(n) \rceil$. Note that $b-1$ is the maximum window size reached when compressing a string of length n. We define $\hat{s} = 0^j(10^{b-1})^h$ where $h = \lfloor n/b \rfloor$ and $j < b$ is such that $|\hat{s}| = n$.

Lemma 11. *We have* $H_0(\hat{s})n \leq (n/b)\log b + \Theta(n/b)$. □

Lemma 12. *Let* $\hat{s} = w_1 w_2 \dots w_d$ *denote the* $\mathsf{LZ77}_{sw}$ *parsing of* $\hat{s}$. *Then, if the word* w_i *contains the character 1, the word* w_{i+1} *contains only 0's.*

Proof. It is easy to see that if 1 appears in w_i it must be the last character of w_i. As a consequence, the next word will be $w_{i+1} = 0^\ell$ where ℓ is the current window size. □

Lemma 13. *For the* $\mathsf{LZ77}_{sw}$ *algorithm with a sliding window of size* $f(t) = \log^{1-\epsilon}(t)$ *we have*

$$|\mathsf{LZ77}_{sw}(\hat{s})| \geq 3(n/b)\log b - O(n/b).$$

Proof. Observe that when we have read $t \geq \sqrt{n}$ characters, the sliding window has size $\lceil \log^{1-\epsilon} t \rceil \geq ((\log n)/2)^{1-\epsilon}$. From that point on, encoding a position in the sliding window—which must be done for each word in the parsing—takes at least

$$(1-\epsilon)\log((\log n)/2) = (1-\epsilon)\log\log n - (1-\epsilon) \geq \log b - 2$$

bits. In addition, by the proof of Lemma 12 we see that each other word will have length equal to the window size; encoding each one of these lengths will again cost at least $\log b - 2$ bits. By Lemma 12, after we have read $\sqrt{n}$ characters there are still $2(n - \sqrt{n})/b$ words to be parsed. The above observations imply that their encoding takes at least $3(n/b)\log b - O(n/b)$ bits. □

Theorem 7. *For the* $\mathsf{LZ77}_{sw}$ *algorithm with a sliding window growing as* $f(t) = \log^{1-\epsilon}(t)$, *we can build an arbitrarily long string* $\hat{s}$ *such that*

$$|\mathsf{LZ77}_{sw}(\hat{s})| \geq H_0(\hat{s})n + \Omega\big((n \log\log n)/\log^{1-\epsilon} n\big).$$

Proof. By Lemmas 11 and 13 we have

$$|\mathsf{LZ77}_{sw}(\hat{s})| - H_0(\hat{s})n \geq \frac{2n\log b}{b} - O\Big(\frac{n}{b}\Big) = \frac{2n\log\log n}{\log^{1-\epsilon} n} - O\left(\frac{n}{\log^{1-\epsilon} n}\right).$$ □

Comparing Theorems 6 and 7 we see that the penalty we pay for using a smaller window is a slower convergence rate. Further work is needed to narrow the huge gap between the rate of growth of the sliding window in the two theorems. In particular, it would be interesting to determine the smallest rate of growth that guarantees an output size bounded by $H_k(s)n + O((n \log\log n)/\log n)$ as in (1).

5 Future Work

We plan to generalize the results in Section 3 to multipass algorithms. Munro and Paterson [9] introduced a model for multipass algorithms in which the data is "stored on a one-way read-only tape. [...] Initially the storage is empty and the tape is placed with the reading head at the beginning. After each pass the tape is rewound to this position with no reading permitted." Among other things, they proved sorting a set of n distinct elements in p passes takes $\Theta(n/p)$ memory locations (each of which can hold a single element).

It seems we can modify the algorithm in Theorem 1 so that, allowed p passes, during each pass it processes only those character following a $(1/p)$-fraction of the possible contexts; any character following a different context is ignored. This way, the algorithm might need only a $(1/p)$-fraction as much memory during each pass. On the other hand, consider a p-pass compression algorithm compressing the string r^i from Corollary 1: we can specify r by specifying the algorithm, the algorithm's memory configurations when it enters and leaves a particular copy of r in r^i during each pass (i.e., $2p$ configurations in all), and its output while reading that copy during each pass. Thus, allowing the algorithm in the proof of Theorem 2 to use p passes but only $O((1/p) \cdot \sigma^{k+1/\lambda-\epsilon})$ bits of memory seems not to affect the proof.

References

1. Babcock, B., Babu, S., Datar, M., Motwani, R., Widom, J.: Models and issues in data stream systems. In: Proceedings of the 21st Symposium on Principles of Database Systems, pp. 1–16 (2002)
2. Bentley, J.L., Sleator, D.D., Tarjan, R.E., Wei, V.K.: A locally adaptive data compression scheme. Communications of the ACM 29, 320–330 (1986)
3. Gagie, T.: Large alphabets and incompressibility. Information Processing Letters 99, 246–251 (2006)
4. Gagie, T., Manzini, G.: Space-conscious compression. Technical Report TR-INF-2007-06-02, Università del Piemonte Orientale (2007)
5. Huffman, D.A.: A method for the construction of minimum-redundancy codes. Proceedings of the IRE 40, 1098–1101 (1952)
6. Kosaraju, R., Manzini, G.: Compression of low entropy strings with Lempel–Ziv algorithms. SIAM Journal on Computing 29(3), 893–911 (1999)
7. Larsson, N.J.: Extended application of suffix trees to data compression. In: DCC '96: Proceedings of the Conference on Data Compression, Washington, DC, USA, p. 190. IEEE Computer Society Press, Los Alamitos (1996)
8. Li, M., Vitányi, P.: An Introduction to Kolmogorov Complexity and Its Applications, 2nd edn. Springer, Heidelberg (1997)
9. Munro, J.I., Paterson, M.S.: Selection and sorting with limited storage. Theoretical Computer Science 12, 315–323 (1980)
10. Muthukrishnan, S.: Data Streams: Algorithms and Applications. Now Publishers, See also: `http://www.nowpublishers.com/tcs/` (2005)
11. Na, J.C., Apostolico, A., Iliopoulos, C., Park, K.: Truncated suffix trees and their application to data compression. Theor. Comput. Sci. 304(1-3), 87–101 (2003)
12. Rissanen, J.: Generalized Kraft inequality and arithmetic coding. IBM Journal of Research and Development 20, 198–203 (1976)
13. Rosenfeld, V.R.: Enumerating De Bruijn sequences. MATCH Communications in Mathematical and in Computer Chemistry 45, 71–83 (2002)
14. Wyner, A.J.: The redundancy and distribution of the phrase lengths of the fixed-database Lempel-Ziv algorithm. IEEE Transactions on Information Theory 43, 1452–1464 (1997)

Small Alliances in Graphs*

Rodolfo Carvajal[1], Martín Matamala[1,2],
Ivan Rapaport[1,2], and Nicolas Schabanel[2,3]

[1] Departamento de Ingeniería Matemática, Universidad de Chile
{rocarvaj,mmatamal,rapaport}@dim.uchile.cl
[2] Centro de Modelamiento Matemático, Universidad de Chile
[3] LIP, École Normale Supérieure de Lyon, France
nicolas.schabanel@gmail.com

Abstract. Let $G = (V, E)$ be a graph. A nonempty subset $S \subseteq V$ is a (strong defensive) *alliance* of G if every node in S has at least as many neighbors in S than in $V \setminus S$. This work is motivated by the following observation: when G is a locally structured graph its nodes typically belong to small alliances. Despite the fact that finding the smallest alliance in a graph is NP-hard, we can at least compute in polynomial time $depth_G(v)$, *the minimum distance one has to move away from an arbitrary node v in order to find an alliance containing v.*

We define $depth(G)$ as the sum of $depth_G(v)$ taken over $v \in V$. We prove that $depth(G)$ can be at most $\frac{1}{4}(3n^2 - 2n + 3)$ and it can be computed in time $O(n^3)$. Intuitively, the value $depth(G)$ should be small for clustered graphs. This is the case for the plane grid, which has a depth of $2n$. We generalize the previous for bridgeless planar regular graphs of degree 3 and 4.

The idea that clustered graphs are those having a lot of small alliances leads us to analyze the value of $r_p(G) = \mathbb{P}\{S \text{ contains an alliance}\}$, with $S \subseteq V$ randomly chosen. This probability goes to 1 for planar regular graphs of degree 3 and 4. Finally, we generalize an already known result by proving that if the minimum degree of the graph is logarithmically lower bounded and if S is a large random set (roughly $|S| > \frac{n}{2}$), then also $r_p(G) \to 1$ as $n \to \infty$.

1 Introduction

The clustering coefficient of a vertex v, denoted by $c(v)$, indicates the extent to which neighbors of v are neighbors themselves [1]. More precisely, if the number of edges within the neighborhood of v is Γ and the degree of v is d, then $c(v) = \frac{2\Gamma}{d(d-1)}$. The average of $c(v)$ taken over all the nodes of a graph G gives the clustering coefficient of G. With this coefficient, Watts and Strogatz [1] were able to justify empirically the idea that small-world networks are locally connected while classical random graphs are not (with both families having a small diameter).

* Partially supported by Programs Conicyt "Anillo en Redes", Fondap on Applied Mathematics and Ecos-Conicyt.

L. Kučera and A. Kučera (Eds.): MFCS 2007, LNCS 4708, pp. 218–227, 2007.

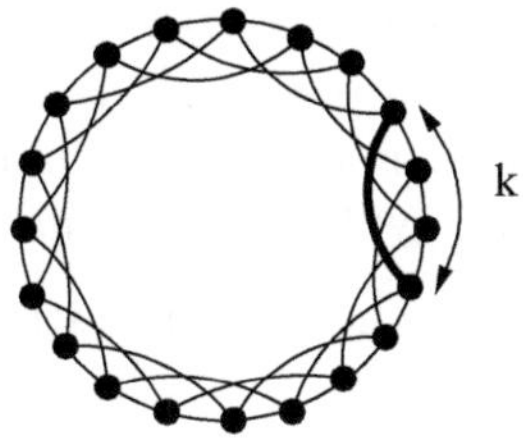

Fig. 1. A locally structured graph

A fundamental limitation of the clustering coefficient is shown in the graph of Fig. 1, consisting of a ring where every vertex has two additional neighbors at ring-distance k. When $k = 2$, the clustering coefficient is $\frac{1}{2} > 0$. However, when $k \geq 3$ it becomes 0, despite the fact that the structure of the graph remains the same.

This and other limitations have lead researchers to propose all kinds of generalizations. In [2], for instance, authors introduced a new definition intended to filter out the effect of degree correlations. In [3], instead of asking "how many of my neighbors are connected?", researchers started to ask "how closely related are my neighbors?". Roughly, this is the approach behind most of the new notions such as grid coefficient [4], meshedness coefficient [5], weighted clustering coefficient [6], high order clustering coefficient [7] and efficiency [8].

In the present work we take another approach, which is motivated by the following observation: the nodes of locally structured graphs typically belong to small *alliances*. In Fig. 1, these alliances are cycles of length k. More precisely, a subset of nodes $S \subseteq V$ is an alliance if each of its nodes has at least as many neighbors inside S than outside S. The formal definition was given in [9], where they used the term *strong defensive alliance* (which for simplicity we call alliance).

Our results. Let S_v be, among all the alliances containing the node v, a minimum one (with respect to the cardinality). Since finding $|S_v|$ is NP-hard, we exhibit in Sect. 2 a polynomial time algorithm that computes $depth_G(v)$, the minimum distance one has to move away from v in order to find an alliance containing v. We define $depth(G)$ as the sum of $depth_G(v)$ taken over all the nodes of G. We prove that the depth of G can be at most $\frac{1}{4}(3n^2 - 2n + 3)$ and it can be computed in time $O(n^3)$. For *bridgeless planar* cubic graph we obtain a better upper bound, namely $\frac{15}{2}n$. We consider this section the starting point of our research efforts by which we expect to find bounds for the depth of different graph classes.

We also take a probabilistic approach introducing another coefficient. We consider the probability of finding an alliance in a randomly chosen subset of nodes S (each one independently and with probability p). We prove that, as expected and in accordance with the previous result, this probability goes to 1 for planar regular graphs of degree 3 and 4.

It is known that in every graph $G = (V, E)$ of n nodes there exists an alliance of size at most $\lfloor \frac{n}{2} \rfloor + 1$ [9,10]. We prove a stronger result which says that, for graphs where the degree of every node is $\omega(\log(n))$, if the chosen set S is large enough (i.e, with $p > \frac{1}{2}$), then S will be an alliance with high probability.

Related work. The notion of alliance was first studied in [9], where the authors introduced various types of alliances which have been studied later, calculating and bounding their size on certain classes of graphs. Namely, these types of alliances are: defensive alliances [9,11], offensive alliances [12], global defensive/offensive alliances [13,14], dual or powerful alliances [15] and k-alliances [16,17,18,19].

The notion of alliance is very natural and, for that reason, it has appeared in other works in different contexts. In [20] the notion of *web community* was introduced: "a community is a set of sites that have more links to members of the community than to non-members". In [21] the authors refer to a "white block" as a subset W of an $(m \times n)$-torus composed of vertices "each of which has at least two neighbors in W". This set W is, of course, an alliance. It appeared when researchers were trying to bound the size of monopolies and coalitions in graphs [22,23]. A closely related line of research consists in trying to partition the graph into communities (alliances in this work). Here the key object is the partition itself and the measure of its quality. Newman in [24], together with a state-of-the-art survey and a complete list of references, provides an algorithm for partitioning based on the eigenspectrum of a matrix he calls modularity matrix.

Some terminology. Let $G = (V, E)$ be a (simple) undirected graph. We will usually assume $|V| = n$. Let $X \subseteq V$ and $v \in V$. Let $d_X(v)$ be the number of neighbors the node v has in X. In other words, $d_X(v) = |N_G(v) \cap X|$, where $N_G(v)$ is the (open) neighborhood of v. A nonempty subset $S \subseteq V$ is a strong defensive alliance [9] if for every vertex $v \in S$ it holds that $|N_G(v) \cap S| \geq |N_G(v) \cap \overline{S}|$. Note that this is equivalent to $d_S(v) \geq d_{\overline{S}}(v)$. In this work such a set S will simply be called an alliance. The eccentricity of a node v, denoted by $ecc_G(v)$, is the greatest distance between v and any other node in G.

2 The Depth of a Graph

Let v be a node of a graph $G = (V, E)$. Let $S_v \subseteq V$ denote a minimum size alliance containing v. Our work is motivated by the following observation: in locally structured graphs the value $|S_v|$ is typically small. Therefore, if we want to measure how locally structured a graph is, we should compute the average of $|S_v|$ taken over all the nodes. Unfortunately, calculating the size of each S_v turns out to be NP-hard. In fact, let us define the problem ALLIANCE as follows:

ALLIANCE
Instance: Graph G and $k \in \mathbb{N}$.
Question: Is there any alliance S in G such that $|S| \leq k$?

This problem is NP-complete [25]. For sake of completeness we present our own reduction in the Appendix. Despite the fact that the previous result implies that in practice *there is no efficient way to find* $|S_v|$, we can still do something. In fact, since we are looking for a measure of "clustering", it would be enough to compute the *depth* of v, the minimum distance one has to move away from v in order to find an alliance containing v:

$$depth_G(v) = \min\{ecc_S(v) : S \text{ alliance with } v \in S\},$$

where $ecc_S(v)$ is the S-eccentricity of v, the distance from v to the farthest node in S. We are going to present first an algorithm that, given $A \subseteq V$, outputs $m(A)$, the largest alliance contained in A.

ALLIANCE Input: $G = (V, E), A \subseteq V$. Output: $m(A)$.
$S \leftarrow A$
$S' \leftarrow \{v \in S : 2d_S(v) \geq d_G(v)\}$
while $S' \neq S$ **do**
  $S \leftarrow S'$
  $S' \leftarrow \{v \in S : 2d_S(v) \geq d_G(v)\}$
end while
return S

Proposition 1. *If the set $m(A)$ computed by algorithm **ALLIANCE** is not empty, then it is the largest alliance contained in A. The time complexity of **ALLIANCE** is $O(n^2)$.*

Proof. Let S be any set of vertices and let

$$S' = \{v \in S : 2d_S(v) \geq d_G(v)\}$$

Clearly, $S' = S$ if and only if S is an alliance. Moreover, an alliance is contained in S if and only it is contained in S'. Hence, the largest alliance contained in S (if any) is also contained in S'. Therefore, if **ALLIANCE** sets S' to $\emptyset$ during some iteration, then it will finish with $m(A) = \emptyset$. Otherwise, it stops with $m(A) = S' = S \neq \emptyset$, for some set S. For the time complexity notice that the construction of S' is $O(n)$ and there are at most n iterations. □

By using **ALLIANCE** we propose the following algorithm to compute the depth of a vertex.

DEPTH Input: $G = (V, E), v \in V$. Output: $depth_G(v)$.
$A \leftarrow N_G(v) \cup \{v\}$, $r \leftarrow 1$
while $r \leq n$ **do**
  if $v \in$ **ALLIANCE**(A) **then**
    return r
  end if
  $A \leftarrow A \cup N_G(A)$
  $r \leftarrow r + 1$
end while

Proposition 2. ***DEPTH*** *returns* $depth_G(v)$ *and its time complexity is* $O(n^3)$.

Proof. In order to prove the statement we prove that the depth of a vertex v corresponds to the smallest radius $r > 0$ such that the ball of radius r centered in v contains an alliance containing v, which is exactly the quantity returned by **DEPTH**.

Clearly, if S is an alliance contained in a ball of radius r centered in v, then the distance between v and any vertex in S is at most r. Hence the eccentricity of v in S is at most r. Therefore, the depth of v is at most r. Conversely, for sake of contradiction, let us assume that there is an alliance S containing v such that the eccentricity of v in S is less than r. Then the distance from v to any vertex in S is less than r. Hence, S is an alliance contained in a ball of radius smaller than r.

Since running **DEPTH** involves running **ALLIANCE** at most n times, the time complexity follows. □

The depth of a graph G is the sum of the depth of its vertices. It is denoted by *depth*(G). From Proposition 2, *depth*(G) can be computed in polynomial time. As we have already mentioned, it is known that every graph G with n vertices has an alliance of size at most $\lfloor \frac{n}{2} \rfloor + 1$ [9, 10]. In order to find an upper bound for the depth of G we prove now a slightly different result.

Proposition 3. *Every graph* $G = (V, E)$ *has an alliance* $S \subseteq V$ *such that* $|S| \in \{\lceil \frac{n}{2} \rceil, \lceil \frac{n}{2} \rceil + 1\}$.

Proof. Let us consider the set of all "almost balanced" cuts of $G = (V, E)$. More precisely, $\mathcal{C} = \{E(U, \overline{U}) \subseteq E \ : \ U \subseteq V, \ |U| \in \{\lceil n/2 \rceil, \ \lceil n/2 \rceil + 1\}\}$. Let $U_0 \subseteq V$ be such that $E(U_0, \overline{U_0})$ is a min-cut of $\mathcal{C}$. i.e, $|E(U_0, \overline{U_0})| = \min_{\tilde{E} \in \mathcal{C}} |\tilde{E}|$.

All we need to prove now is that U_0 is an alliance of G. Suppose that there is a node $u \in U_0$ such that $d_{U_0}(u) < d_{\overline{U_0}}(u)$. In that case, if we define $U_1 = U_0 \setminus \{u\}$, we would have

$$|E(U_1, \overline{U_1})| = |E(U_0, \overline{U_0})| - (d_{\overline{U_0}}(u) - d_{U_0}(u)) < |E(U_0, \overline{U_0})|.$$

In order to conclude we need to show that $E(U_1, \overline{U_1}) \in \mathcal{C}$. In fact, if $|U_0| = \lceil n/2 \rceil + 1$ then $|U_1| = \lceil n/2 \rceil$. On the other hand, if $|U_0| = \lceil n/2 \rceil$ then $|\overline{U_1}| \in \{\lceil n/2 \rceil, \ \lceil n/2 \rceil + 1\}$. □

Corollary 1. *The depth of any graph* G *is at most* $\frac{1}{4}(3n^2 - 2n + 3)$.

Proof. The depth of $\lceil \frac{n}{2} \rceil$ vertices is at most $\lceil \frac{n}{2} \rceil$. □

We do not know whether this upper bound is tight. Nevertheless, by forcing the graph to be bridgeless planar of degree at most 4, the upper bound decreases drastically.

First notice the following: the depth of every vertex in the $(m \times n)$-grid is 2, since every vertex belongs to a small alliance (a cycle of length 4). We are now going to generalize this and prove that the depth of any planar bridgeless graph of degree at most 4 is linear in n. This result goes in the right direction: the depth of a graph seems to be a good generalization of its clustering coefficient.

Let G be a bridgeless plane cubic graph. Then, each vertex v belongs to a cycle C which is the boundary of a face. We call these faces *facial cycles* in the sequel. Clearly, C is an alliance containing v, and therefore $depth_G(v) \leq |V(C)|/2$.

Proposition 4. *The depth a bridgeless planar cubic graph is at most* $\frac{15}{2}n$.

Proposition 4 is a consequence of the following lemma.

Lemma 1. *Let* $G = (V, E)$ *be a bridgeless plane cubic graph and let* $F(G)$ *denote the set of its faces. Then there exists a function* f *associating to each vertex a facial cycle containing it and such that no face is associated with more than five vertices.*

Proof. Consider the dual graph $G^* = (F(G), E^*)$. Since G is plane, cubic and bridgeless, then G^* is a plane graph with no loops and with no multiple edges. We are looking for a function f that assigns to each vertex of G a particular alliance to which it belongs. By duality, this is equivalent to look for a function f^* that assigns to each face h^* of G^* a particular vertex v^* of G^* with v^* lying in the boundary of h^*. Hence, in order to prove the lemma we have to make sure that in our construction (of f^*) at most 5 faces of G^* are labeled with the same vertex.

We proceed by induction on the number of vertices of G^*. If G^* has just one vertex then we label the unique face of the graph with this vertex. Suppose now that G^* has $n+1$ vertices. Since G^* is plane (with no loops and with no multiple edges) there must be a vertex v^* of degree at most five. Consider the graph $G' = G^* \setminus v^*$ (i.e, we delete the vertex and the incident edges). By the induction hypothesis one can solve the problem in G' without using v^*. The point occupied by v^* belongs to one face of G'. That face contains at most 5 faces of G^*. We label all of them with v^* and we get the result. □

Proof. (of Proposition 4). Let us consider the function f of the previous lemma. It follows:

$$\sum_{v \in V} |f(v)| = \sum_{h \in F(G)} |h||f^{-1}(h)| \leq 5 \sum_{h \in F(G)} |h| = 5 \times 2|E| = 5 \times 3|V|.$$

Therefore, $depth(G) = \sum_{v \in V} depth_G(v) \leq \frac{1}{2} \sum_{v \in V} |f(v)| \leq \frac{15}{2}|V|$. □

3 A Probabilistic Approach

Clustered graphs are those having a lot of small alliances. So a natural way of testing this is to compute the probability of finding an alliance in a small fraction of nodes (chosen randomly).

We can formalize this question. Let $p \in [0, 1]$. Let us denote $V_p(G)$ the outcome of selecting each node of V with probability p. Let us denote $r_p(G) = \mathbb{P}\{V_p(G) \text{ contains an alliance}\}$. The problem of computing $r_p(G)$ seems to be very difficult in general. But it can be tackled in some cases.

Proposition 5. *Let $G = (V, E)$ be a cubic planar graph. Let $0 < p < 1$. Then* $r_p(G) \geq 1 - (1 - p^6)^{\frac{n+4}{56}}$.

Proof. Let us assume that G is already embedded in the plane and let F be the set of faces of G. As any face is an alliance of G, we have that

$$r_p(G) \geq \mathbb{P}\{S : \exists f \in F, V(f) \subseteq S\}.$$

Let F' be any maximal set of vertex pairwise disjoint faces of size at most 6. The probability that a random set S does not contain a given face f in F' is $1 - p^{|V(f)|} \leq 1 - p^6$. Since the faces in F' are vertex disjoint the probability that S does not contain any face in F' is at most $(1 - p^6)^{|F'|}$. In order to conclude we will prove that $56|F'| \geq |V| + 4$.

Let a_i be the number of faces of size i and let b_i be the number of faces of size at most i. By maximality, any face f with size at most 6 intersects at least one element of F' and a face $f \in F'$ intersect at most 6 faces of size at most 6 not in F'. Therefore, $6|F'| \geq b_6 - |F'|$ and then $7|F'| \geq b_6$. From the definition of a_i, we get that $|F| = \sum_{i \geq 3} a_i$ and $2|E| = \sum_{i \geq 3} i a_i$. From Euler's Formula for cubic graphs, $2|E| = 6|F| - 12$, we get the following.

$$\sum_{i \geq 7} (i - 6) a_i + 12 = a_5 + 2a_4 + 3a_3 \tag{1}$$

Let c be a positive number. From equation 1 we deduce that if $a_5 + 2a_4 + 3a_3 < c|F|$, then $|F| - b_6 < c|F|$ and hence $b_6 > (1 - c)|F|$. Otherwise, if $a_5 + 2a_4 + 3a_3 \geq c|F|$ then $b_6 \geq \frac{c}{3}|F|$. By choosing $c = \frac{3}{4}$ we conclude that $b_6 \geq \frac{1}{4}|F|$. By using again Euler's formula and the upper bound $|F'| \geq \frac{b_6}{7}$ we conclude that

$$|F'| \geq \frac{b_6}{7} \geq \frac{1}{28}|F| = \frac{1}{28}(2 + |V|/2)$$

Therefore,

$$r_p(G) \geq 1 - (1 - p^6)^{|F'|} \geq 1 - (1 - p^6)^{\frac{|V|+4}{56}}. \qquad \square$$

We say that a sequence of graphs $(G_k)_{k \in \mathbb{N}}$ is an increasing sequence if the order (number of nodes) of the graphs grows with k.

Corollary 2. *Let $0 < p < 1$. Every increasing sequence $(G_k)_{k \in \mathbb{N}}$ of cubic planar graphs satisfies* $\lim_{k \to \infty} r_p(G_k) = 1$.

Remark 1. The previous result also holds for planar regular graphs of degree 4.

As we have already seen, every graph G with n nodes has an alliance of size at most $\lfloor \frac{n}{2} \rfloor + 1$. This alliance comes from a *very particular construction*, dealing with an "almost balanced" minimum cut of G. What if we choose randomly a large set of nodes? Is it going to contain an alliance with high probability?

Proposition 6. *Let $G = (V, E)$ be a graph with minimum degree d. Let $\frac{1}{2} < p < 1$. Then $r_p(G) \geq 1 - ne^{-\frac{p\delta^2}{2}d}$, where $\frac{1}{2} = p(1 - \delta)$.*

Proof. We apply the Chernoff bound in a standard way as explained in [26]. Let $X_v = 1$ if $v \in V_p(G)$ and $X_v = 0$ otherwise. Let $X(v) = \sum_{u \in N(v)} X_u$. It follows:

$$\begin{aligned} r_p(G) &\geq \mathbb{P}\{\forall v \in V_p(G) : d_{V_p(G)}(v) \geq d_{\overline{V_p(G)}}(v)\} \\ &\geq 1 - \sum_{v \in V_p(G)} \mathbb{P}\{X(v) < \frac{d(v)}{2}\} = 1 - \sum_{v \in V_p(G)} \mathbb{P}\{X(v) < (1-\delta)\mathbb{E}(X(v))\} \\ &\geq 1 - \sum_{v \in V_p(G)} e^{-\frac{p\delta^2}{2} d(v)} \geq 1 - ne^{-\frac{p\delta^2}{2} d}. \end{aligned}$$ □

We can apply the previous lemma to graphs for which the degree of every node is high enough. A class of graphs is said to have minimum degree $d(n)$ if the minimum degree of any graph having more than n nodes is at least $d(n)$.

Corollary 3. *Let $\frac{1}{2} < p < 1$ and let $d(n) = \omega(\log(n))$. Then, for every increasing sequence $(G_k)_{k \in \mathbb{N}}$ of graphs with minimum degree $d(n)$, we have $\lim_{k \to \infty} r_p(G_k) = 1$.*

Acknowledgment. The authors would like to thank Martin Loebl for very helpful hints and comments.

References

1. Watts, D.J., Strogatz, S.H.: Collective dynamics of "small-world" networks. Nature 393, 440–442 (1998)
2. Soffer, S.N., Vazquez, A.: Network clustering coefficient without degree-correlation biases. Phys. Rev. E 71(5), 57101 (2005)
3. Abdo, A.H., de Moura, A.P.S.: Measuring the local topology of networks: An extended clustering coefficient, arXiv:physics/0605235 (2006)
4. Caldarelli, G., Pastor-Santorras, R., Vespignani, A.: Cycles structure and local ordering in complex networks. The European Physical Journal B - Condensed Matter 38(2), 183–186 (2004)
5. Buhl, J., Gautrais, J., Solé, R.V., Kuntz, P., Valverde, S., Deneubourg, J., Theraulaz, G.: Efficiency and robustness in ant networks of galleries. The European Physical Journal B - Condensed Matter 42, 123–129 (2004)
6. Schank, T., Wagner, D.: Approximating clustering coefficient and transitivity. Journal of Graph Algorithms and Applications 9(2), 265–275 (2005)
7. Fronczak, A., Holyst, J.A., Jedynak, M., Sienkiewicz, J.: Higher order clustering coefficients in Barabasi-Albert networks. Physica A 316(1), 688–694 (2002)
8. Latora, V., Marchiori, M.: Efficient behavior of small-world networks. Phys. Rev. Lett. 87(19), 198701 (2001)
9. Kristiansen, P., Hedetniemi, S.M., Hedetniemi, S.T.: Alliances in graphs. J. Combin. Math. Combin. Comput. (48), 157–177 (2004)
10. Shafique, K.H.: Partitioning a graph in alliances and its application to data clustering. PhD thesis, School of Comp. Sci., College of Eng. and Comp. Sci., University of Central Florida (2004)

11. Sigarreta, J.M., Rodríguez, J.A.: On defensive alliances and line graphs. Applied Mathematics Letters 12(19), 1345–1350 (2006)
12. Favaron, O., Fricke, G., Goddard, W., Hedetniemi, S.M., Hedetniemi, S.T., Kristiansen, P.: Offensive alliances in graphs. Discuss. Math. Graph Theory 24(2), 263–275 (2004)
13. Haynes, T.W., Hedetniemi, S.T., Henning, M.A.: Global defensive alliances in graphs. Electron. J. Combin (10), 139–146 (2003)
14. Rodríguez, J.A., Sigarreta, J.M.: Global offensive alliances in graphs. Electronic Notes in Discrete Mathematics 25, 157–164 (2006)
15. Brigham, R., Dutton, R., Hedetniemi, S.: A sharp lower bound on the powerful alliance number of $c_m \times c_n$. Congr. Number. 167, 57–63 (2004)
16. Shafique, K.H., Dutton, R.D.: Maximum alliance-free and minimum alliance-cover sets. Congr. Number. 162, 139–146 (2003)
17. Shafique, K.H., Dutton, R.D.: A tight bound on the cardinalities of maximum alliance-free and minimum alliance-cover sets. J. Combin. Math. Combin. Comput. 56, 139–145 (2006)
18. Rodríguez-Velázquez, J.A., Gonzalez-Yero, I., Sigarreta, J.M.: Defensive k-alliances in graphs. eprint arXiv:math/0611180 (2006)
19. Rodríguez-Velázquez, J.A., Sigarreta, J.M.: Global defensive k-alliances in graphs. eprint arXiv:math/0611616 (2006)
20. Flake, G.W., Lawrence, S., Giles, C.L.: Efficient identification of web communities. In: Proc. of the 6th ACM SIGKDD Int. Conference on Knowledge Discovery and Data Mining, pp. 150–160. ACM Press, New York (2000)
21. Flocchini, P., Lodi, E., Luccio, F., Pagli, L., Santoro, N.: Dynamic monopolies in tori. Discrete Appl. Math. 137(2), 197–212 (2004)
22. Bermond, J., Bond, J., Peleg, D., Perennes, S.: Tight bounds on the size of 2-monopolies. In: Proc. 3rd Colloq. on Structural Information and Communication Complexity (1996)
23. Peleg, D.: Local majorities, coalitions and monopolies in graphs: A review. Theor. Comput. Sci. 282(2), 231–257 (2002)
24. Newman, M.E.J.: Finding community structure in networks using the eigenvectors of matrices. Phys. Rev. E 74(3), 36104 (2006)
25. McRae, A., Goddard, W., Hedetniemi, S.M., Hedetniemi, S.T., Kristiansen, P.: The algorithmic complexity of alliances in graphs (Preprint, 2002)
26. Vazirani, V.V.: Approximation Algorithms. Springer, Heidelberg (2004)
27. Sipser, M.: Introduction to the Theory of Computation. International Thomson Publishing (1996)

Appendix: Alliance is *NP*-Complete

It is known that the following problem is NP-complete [27].

HALF-CLIQUE
Instance: Graph G (with n nodes, n even).
Question: Is there any clique in G of size (at least) $\frac{n}{2}$?

Proposition 7. ALLIANCE *is* NP*-complete.*

Proof. Let $G = (V, E)$ be an instance of HALF-CLIQUE (i.e, n is even). In the reduction to ALLIANCE we construct a graph G' of size $4n$ as follows. We first generate graphs $G_1 = (V, E)$, $G_2 = (V, \phi)$, $G_3 = (V, \phi)$ and $G_4 = (V, E)$. The connection between these graphs is made according to the original graph $G = (V, E)$. Here we will abuse the notation, making no distinction between the copies in the four graphs, of a node $v \in V$.

Nodes $t \in G_1$ are connected to those nodes $u \in G_2$ such that $t \in V \setminus (\{t\} \cup N(u))$. Nodes $u \in G_2$ are connected to those nodes $v \in G_3$ such that $u \in N(v)$. Nodes $v \in G_3$ are connected to those nodes $w \in G_4$ such that $v \in (V \setminus (\{v\} \cup N(w))$. For each $i \in \{1, 2, 3, 4\}$, v_i is connected to every vertex of G_i. Notice that G' is n-regular and therefore the smallest alliances are of size $\frac{n}{2}$. The reader should verify that there exists an alliance in G' of size $\frac{n}{2} + 1$ if and only if there exists a clique in G of size $\frac{n}{2}$. □

The Maximum Solution Problem on Graphs

Peter Jonsson*, Gustav Nordh**, and Johan Thapper***

Department of Computer and Information Science
Linköpings universitet
S-581 83 Linköping, Sweden
{petej,gusno,johth}@ida.liu.se

Abstract. We study the complexity of the problem MAX SOL which is a natural optimisation version of the graph homomorphism problem. Given a fixed target graph H with $V(H) \subseteq \mathbb{N}$, and a weight function $w : V(G) \rightarrow \mathbb{Q}^+$, an instance of the problem is a graph G and the goal is to find a homomorphism $f : G \rightarrow H$ which maximises $\sum_{v \in G} f(v) \cdot w(v)$. MAX SOL can be seen as a restriction of the MIN HOM-problem [Gutin et al., Disc. App. Math., 154 (2006), pp. 881-889] and as a natural generalisation of MAX ONES to larger domains. We present new tools with which we classify the complexity of MAX SOL for irreflexive graphs with degree less than or equal to 2 as well as for small graphs ($|V(H)| \leq 4$). We also study an extension of MAX SOL where value lists and arbitrary weights are allowed; somewhat surprisingly, this problem is polynomial-time equivalent to MIN HOM.

Keywords: Constraint satisfaction, homomorphisms, computational complexity, optimisation.

1 Introduction

Throughout this paper, by a graph we mean an undirected graph without multiple edges but possibly with loops. A *homomorphism* from a graph G to a graph H is a mapping f from $V(G)$ to $V(H)$ such that $(f(v), f(v'))$ is an edge of H whenever (v, v') is an edge of G. The homomorphism problem with a fixed target graph H takes a graph G as input and asks whether there is a homomorphism from G to H. Hence, by fixing the graph H we obtain a class of problems, one for each graph H. For example, the graph homomorphism problem with fixed target graph $H = \{(v_0, v_1), (v_1, v_0)\}$, denoted by HOM($H$), is exactly the problem of determining whether the input graph G is bipartite (i.e., the 2-COLORING problem). Similarly, if $H = \{(v_0, v_1), (v_1, v_0), (v_1, v_2), (v_2, v_1), (v_0, v_2), (v_2, v_0)\}$, then HOM($H$) is exactly the 3-COLORING problem. More generally, if H is the clique on k-vertices, then HOM(H) is the k-COLORING problem.

* Partially supported by the *Center for Industrial Information Technology* (CENIIT) under grant 04.01, and by the *Swedish Research Council* (VR) under grant 2006-4532.
** Supported by the *National Graduate School in Computer Science* (CUGS), Sweden.
*** Supported by the *Programme for Interdisciplinary Mathematics*, Department of Mathematics, Linköpings universitet.

L. Kučera and A. Kučera (Eds.): MFCS 2007, LNCS 4708, pp. 228–239, 2007.

Hence, the HOM(H) class of problems contains several well studied problems, some of which are in **P** (e.g, 2-COLORING) and others which are **NP**-complete (e.g., k-COLORING for $k \geq 3$). A celebrated result, due to Hell and Nesetril [9], states that HOM(H) is in **P** if H is bipartite or contains a looped vertex, and that it is **NP**-complete for all other graphs H. For more information on graph homomorphism problems in general and their complexity in particular, we refer the reader to the excellent monograph by Hell and Nesetril [10].

In this paper we study a natural optimisation variant of the HOM(H) problem, i.e., we are not only interested in the existence of a homomorphism but want to find the "best homomorphism". We let the vertices of H be a subset of the natural numbers, $w : V(G) \rightarrow \mathbb{Q}^+$ be a weight function and look for a homomorphism h from G to H that maximise the sum $\sum_{v \in G} w(v) \cdot h(v)$. We call this problem the maximum solution problem (with a fixed target graph H) and denote it by MAX SOL(H).

Just as the HOM(H) problem captures several interesting combinatorial *decision* problems, it is clear that the MAX SOL(H) captures several interesting combinatorial *optimisation* problems. The MAX SOL(H) problem where $H = \{(0,0),(0,1),(1,0)\}$ is exactly the **NP**-hard optimisation problem WEIGHTED MAXIMUM INDEPENDENT SET. MAX SOL can also be seen as a natural generalisation of MAX ONES or, alternatively, as a variation of the integer linear programming problem.

Gutin et al. [7] introduced MIN HOM, another homomorphism optimisation problem motivated by a real-world application in defence logistics. This problem was studied in [5,6] and among other things, a dichotomy was established for (undirected) graphs. In particular, MIN HOM(H) was shown to be tractable whenever H is a *proper interval graph* or a *proper interval bigraph*. When formulated as a minimization problem, MAX SOL is easily seen to be a restriction of MIN HOM.

In [13], MAX SOL was studied as an optimisation variant of the constraint satisfaction problem over arbitrary constraint languages. There, languages defined using a many-valued logic were characterised as being either polynomial time solvable or **APX**-hard. This was accomplished by adopting algebraic techniques from the study of constraint satisfaction problems. In this paper, we continue the study of MAX SOL. We look at languages given by undirected graphs. In particular, we give a complete classification of the tractability of languages given by irreflexive graphs which have degree less than or equal to 2. We also classify the cases when $|V(H)| \leq 3$ and when $V(H) = \{0,1,2,3\}$. An interesting observation in these cases is that for some graphs, the complexity of the problem depends very subtly on the values of the vertices. In particular, applying an order preserving map on the values may change the complexity.

Furthermore, we consider two natural extensions of the MAX SOL-framework. One is to relax the restriction of the weights and allow arbitrary (possibly negative) rational weights on the variables. The other is to attribute a list, $L(v)$, of allowed values to each vertex v in the input instance. The list is a subset of $V(H)$ and any solution must assign v to one of the vertices in $L(v)$. In this paper we focus, apart from the ordinary MAX SOL, on the most general extreme, where we allow both lists and arbitrary weights. This problem, which we call LIST MAX AW SOL, can be seen both as an optimisation version of L-HOM(H), the *list homomorphism problem* (see [3,4]) while it is still a restriction of MIN HOM. We show that for each undirected graph H, LIST MAX AW SOL(H) and MIN HOM(H) are in fact (polynomial time) equivalent.

The paper is organised as follows. In Section 2 we give a formal definition of CSP and the problems MAX SOL and LIST MAX AW SOL. In Section 3 we formalise the algebraic framework for studying MAX SOL. We also give a number of basic results which are used throughout the paper. These results are interesting in their own right, as many of them apply to general constraint languages. The results for MAX SOL are given in Section 4. In Section 5 we show the equivalence of LIST MAX AW SOL and Min Hom for undirected graphs, before concluding in Section 6.

2 Preliminaries

We formally define constraint satisfaction as follows: Let $D \subset \mathbb{N}$ (*the domain*) be a finite set. The set of all n-tuples of elements from D is denoted by D^n. Any subset of D^n is called an n-ary relation on D. The set of all finitary relations over D is denoted by R_D. A constraint language over a finite set, D, is a finite set $\Gamma \subseteq R_D$. Constraint languages are the way in which we specify restrictions on our problems. The constraint satisfaction problem over the constraint language Γ, denoted CSP(Γ), is defined to be the decision problem with instance (V, D, C), where V is a set of variables, D is a finite set of values (the domain), and C is a set of constraints $\{C_1, \ldots, C_q\}$, in which each constraint C_i is a pair (s_i, ϱ_i) with s_i a list of variables of length m_i, called the constraint scope, and ϱ_i an m_i-ary relation over the set D, belonging to Γ, called the constraint relation. The question is whether there exists a solution to (V, D, C) or not, that is, a function from V to D such that, for each constraint in C, the image of the constraint scope is a member of the constraint relation.

List Maximum Solution with Arbitrary Weights over a constraint language Γ, denoted LIST MAX AW SOL(Γ), is the maximization problem with

Instance: Tuple (V, D, C, L, w), where D is a finite subset of $\mathbb{N}$, (V, D, C) is a CSP instance over Γ, $L : V \to 2^D$ is a function from V to subsets of D, and $w : V \to \mathbb{Q}$ is a weight function.
Solution: An assignment $f : V \to D$ to the variables such that all constraints are satisfied and such that $f(v) \in L(v)$ for all $v \in V$.
Measure: $\sum_{v \in V} f(v) \cdot w(v)$

Weighted Maximum Solution over Γ, MAX SOL(Γ), is then defined by restricting w to non-negative rational numbers and letting $L(v) = D$ for all $v \in V$.

Let G be a graph. For a fixed graph H, the *Minimum Cost Homomorphism Problem* [7], MIN HOM(H), is the problem of finding a graph homomorphism f from G to H which minimises $\sum_{v \in V(G)} c_{f(v)}(v)$, where $c_i(v) \in \mathbb{Q}^+$ are costs, for $v \in V(G)$, $i \in V(H)$.

Let G be a graph and H be a subgraph of G. H is a *retract* of G if there exists a graph homomorphism $f : G \to H$ such that $f(v) = v$ for all $v \in V(H)$. The *Retraction Problem*, RET(H), is to determine whether or not H is a retract of G.

Let $F = \{I_1, \ldots, I_k\}$ be a family of intervals on the real line. A graph G with $V(G) = F$ and $(I_i, I_j) \in E(G)$ if and only if $I_i \cap I_j \neq \emptyset$ is called an **interval graph**. If the intervals are chosen to be inclusion-free, G is called a **proper interval graph**.

Let $F_1 = \{I_1, \ldots, I_k\}$ and $F_2 = \{J_1, \ldots, J_l\}$ be two families of intervals on the real line. A graph G with $V(G) = F_1 \cup F_2$ and $(I_i, J_j) \in E(G)$ if and only if $I_i \cap J_j \neq \emptyset$ is called an **interval bigraph**. If the intervals in each family are chosen to be inclusion-free, G is called a **proper interval bigraph**.

Interval graphs are reflexive, while interval bigraphs are irreflexive and bipartite.

3 Methods

3.1 Algebraic Framework

An operation on D is an arbitrary function $f : D^k \to D$. Any operation on D can be extended in a standard way to an operation on tuples over D, as follows: Let f be a k-ary operation on D and let R be an n-ary relation over D. For any collection of k tuples, $\boldsymbol{t_1}, \ldots, \boldsymbol{t_k} \in R$, define $f(\boldsymbol{t_1}, \ldots, \boldsymbol{t_k}) = (f(\boldsymbol{t_1}[1], \ldots, \boldsymbol{t_k}[1]), \ldots, f(\boldsymbol{t_1}[n], \ldots, \boldsymbol{t_k}[n]))$ where $\boldsymbol{t_j}[i]$ is the i-th component in tuple $\boldsymbol{t_j}$. If f is an operation such that for all $\boldsymbol{t_1}, \boldsymbol{t_2}, \ldots, \boldsymbol{t_k} \in R$, we have $f(\boldsymbol{t_1}, \boldsymbol{t_2}, \ldots, \boldsymbol{t_k}) \in R$, then R is said to be invariant under f. If all constraint relations in Γ are invariant under f, then Γ is invariant under f. An operation f such that Γ is invariant under f is called a *polymorphism* of Γ. The set of all polymorphisms of Γ is denoted $Pol(\Gamma)$. Sets of operations of the form Pol(Γ) are known as *clones*, and they are well-studied objects in algebra (cf. [15]).

A first-order formula ϕ over a constraint language Γ is said to be *primitive positive* (we say ϕ is a pp-formula for short) if it is of the form $\exists \mathbf{x} : (P_1(\mathbf{x}_1) \wedge \ldots \wedge P_k(\mathbf{x}_k))$ where $P_1, \ldots, P_k \in \Gamma$ and $\mathbf{x}_1, \ldots, \mathbf{x}_k$ are vectors of variables such that $|P_i| = |\mathbf{x}_i|$ for all i. Note that a pp-formula ϕ with m free variables defines an m-ary relation $R \subseteq D^m$; the relation R is the set of all tuples satisfying the formula ϕ. It follows, for example from the proof of [13, Lemma 4], that there is a polynomial time reduction from MAX SOL($\Gamma \cup \{R\}$) to MAX SOL(Γ).

For any relation R and a unary operation f, let $f(R)$ denote the relation $f(R) = \{f(r) \mid r \in R\}$. Accordingly, let $f(\Gamma)$ denote the constraint language $\{f(R) \mid R \in \Gamma\}$.

Definition 1. *A constraint language Γ is a max-core if and only if there is no non-injective unary operation f in $Pol(\Gamma)$ such that $f(d) \geq d$ for all $d \in D$. A constraint language Γ' is a max-core of Γ if and only if Γ' is a max-core and $\Gamma' = f(\Gamma)$ for some unary operation $f \in Pol(\Gamma)$ such that $f(d) \geq d$ for all $d \in D$.*

We refer to [13] for the proof of the following lemma.

Lemma 1. *If Γ' is a max-core of Γ, then* MAX SOL(Γ) *and* MAX SOL(Γ') *are polynomial time equivalent.*

3.2 Basic Lemmas

We now present a series of lemmas which will prove useful in the coming section. Several of the lemmas are, however, interesting in their own right and can be applied to a wider class of problems. Let Γ denote a constraint language over $D = \{d_1, \ldots, d_k\}$.

Lemma 2. *Arbitrarily choose $D' \subseteq D$, assume without loss of generality that $D' = \{d_1, \ldots, d_l\}$ with $d_1 < d_2 < \ldots < d_l$ and let $F = \{f \in Pol_1(\Gamma) \mid f|_{D'} = \mathbf{id}_{D'}\}$. Assume that there exists constants $a_1, \ldots, a_l > 0$ such that for every $f \in Pol_1(\Gamma) \setminus F$ it holds that $\sum_{i=1}^{l} a_i \cdot d_i > \sum_{i=1}^{l} a_i \cdot f(d_i)$. Then,* MAX SOL$(\Gamma \cup \{\{(d_1)\}, \ldots, \{(d_l)\}\})$ *and* MAX SOL(Γ) *are polynomial time equivalent problems.*

Proof. The non-trivial reduction follows from a construction which uses the concept of an *indicator problem* [11].

Corollary 1. *Let Γ be an arbitrary constraint language and let U be a unary relation on D. If* MAX SOL(Γ) *and* MAX SOL$(\Gamma \cup \{U\})$ *are polynomial time equivalent problems, then so are* MAX SOL(Γ) *and* MAX SOL$(\Gamma \cup \{\{(\max U)\}\})$*. In particular,* MAX SOL(Γ) *and* MAX SOL$(\Gamma \cup \{(d_k)\})$ *are polynomial time equivalent.*

Proof. Use Lemma 2 on $\Gamma \cup \{U\}$, $k = 1, d_1 = m$ and $a_1 > 0$.

Lemma 3. *Let Γ be a constraint language on a finite domain D. Assume that there is a set $F \subseteq Pol_1(\Gamma)$, such that for each $f \in F$,* MAX SOL$(f(\Gamma))$ *is in* **PO** *and such that for each choice of $a_1, \ldots, a_d \in \mathbb{Q}^+$ there is a $f \in F$ for which $\sum_{i=1}^{d} a_i \cdot d_i \leq \sum_{i=1}^{d} a_i \cdot f(d_i)$. Then,* MAX SOL$(\Gamma)$ *is in* **PO**.

Let $\mathcal{H} = \{H_1, \ldots, H_n\}$ be a set of connected graphs and let H be the disjoint union of these graphs. We are interested in the complexity of MAX SOL(H), given the complexities of the individual problems. Let $\mathcal{H}_i = \mathcal{H} \setminus \{H_i\}$. We say that H_i *extends* the set $\mathcal{H}_i$ if there exists an instance $I = (V, D, C, w)$ of MAX SOL(H_i) for which $\text{OPT}(I) > \text{OPT}(I_j)$ where $I_j = (V, D_j, \{xH_jy \mid xH_iy \in C\}, w)$, for all j such that $1 \leq j \neq i \leq n$. We call I a *witness* to the extension.

Assume that for some $1 \leq i \leq n$, it holds that H_i does not extend $\mathcal{H}_i$. It is clear that for any connected instance $I = (V, D, C, w)$ of MAX SOL(H), we have $\text{OPT}(I) = \text{OPT}(I_j)$ for some j, where $I_j = (V, D_j, \{xH_jy \mid xHy \in C\}, w)$. Furthermore, since H_i does not extend $\mathcal{H}_i$, we know that we can choose this $j \neq i$. Let H' be the disjoint union of the graphs in $\mathcal{H}_i$. Then, $\text{OPT}(I) = \text{OPT}(I')$, where $I' = (V, D, \{xH'y \mid xHy \in C\}, w)$ is an instance of MAX SOL(H'). For this reason, we may assume that every $H_i \in \mathcal{H}$ extends every graph in $\mathcal{H}_i$.

Lemma 4. *Let $H_1, \ldots, H_n$ be connected graphs and H their disjoint union. If the problems* MAX SOL(H_i)*, $1 \leq i \leq n$ are all tractable, then* MAX SOL(H) *is tractable. If* MAX SOL(H_i) *is* **NP***-hard and H_i extends the set $\{H_1, \ldots, H_{i-1}, H_{i+1}, \ldots, H_n\}$ for some i, then* MAX SOL(H) *is* **NP***-hard.*

The next lemma can be shown by a reduction from MAXIMUM INDEPENDENT SET.

Lemma 5. *If $a < b$ and $R = \{(a, a), (a, b), (b, a)\}$, then* MAX SOL$(R)$ *is* **NP***-hard.*

4 Results for MAX SOL

Throughout this section, we will assume that all graphs defining constraint languages are max-cores and connected. Due to Lemma 1 and Lemma 4, we can do this without loss of generality. There is a straightforward reduction from MAX SOL to MIN HOM, so polynomiality results for MIN HOM translates directly to MAX SOL. Additionally, the following reduction can sometimes be used to show hardness.

Lemma 6. *Let H be a graph for which the retraction problem is* **NP**-*complete. Then* MAX SOL$(H \cup \{\{(d_1)\}, \ldots, \{(d_k)\}\})$ *is* **NP**-*hard.*

4.1 Irreflexive Graphs with $\deg(v) \leq 2$

There are two types of irreflexive graphs H with $\deg(v) \leq 2$ for all $v \in V(H)$, paths and cycles. Since irreflexive paths are proper interval bigraphs, a reduction to MIN HOM, and [5, Corollary 2.6] shows that:

Proposition 1. *Let H be an irreflexive path. Then* MAX SOL(H) *is in* **PO**.

When H is an odd cycle, we have that CSP(H) is **NP**-complete and therefore MAX SOL(H) is **NP**-hard. It remains to investigate even cycles. Since we do not allow multiple edges, C_2 is a single edge, for which MAX SOL is trivially in **PO**. When $H \cong C_4 \cong K_{2,2}$, there is always an increasing endomorphism from H to one of its edges. Thus no max-core is isomorphic to C_4. For even cycles of length greater or equal to 6, it has been shown in [4] that the retraction problem is **NP**-complete. We will use this with Lemma 6 to prove the **NP**-hard cases. The tractable cases are proven by Lemma 3. We will assume a bipartition $V(H) = \{d_1, \ldots, d_k\} \cup \{d'_1, \ldots, d'_k\}$ of H with $d_1 < d_2 < \cdots < d_k$ and $d'_1 < d'_2 < \cdots < d'_k$ and without loss of generality that $d_k > d'_k$.

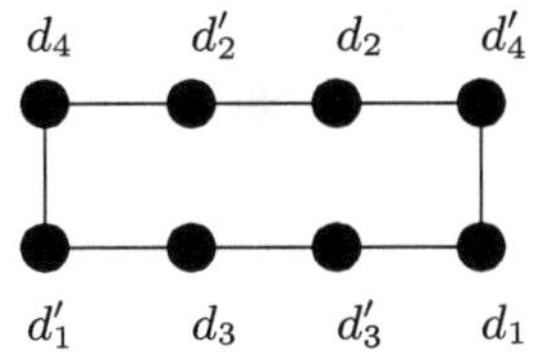

Fig. 1. The graph H in Proposition 3

Proposition 2. *Let H be isomorphic to C_6 and a max-core. Then,* MAX SOL(H) *is* **NP**-*hard.*

Proposition 3. *Let H be isomorphic to C_8 and a max-core. If H is isomorphic to the graph in Figure 1 and $(d_4 - d_3)(d'_4 - d'_3) \geq (d_3 - d_2)(d'_3 - d'_2)$, then* MAX SOL$(H)$ *is in* **PO**. *Otherwise it is* **NP**-*hard.*

In general, for even cycles, the following holds:

Proposition 4. *Let H be a max-core isomorphic to C_{2k}, $k \geq 3$. Then* MAX SOL(H) *and* MAX SOL$(H \cup \{\{(d_k)\}, \{(d'_k)\}\})$ *are polynomial time equivalent problems.*

Assume that there exists non-negative constants $a_1, \ldots, a_{k-1}, a'_1, \ldots, a'_{k-1}$ such that for each $f \in Pol_1(H) \setminus F$, where $F = \{f \in Pol_1(\Gamma) \mid \exists j \neq k : f(d_j) \neq d_j \vee f(d'_j) \neq d'_j\}$, it is true that

$$\sum_{i=1}^{k-1} (a_i \cdot d_i + a'_i \cdot d'_i) > \sum_{i=1}^{k-1} (a_i \cdot f(d_i) + a'_i \cdot f(d'_i)). \quad (1)$$

Then MAX SOL(H) *is* **NP**-*hard, otherwise it is in* **PO**.

4.2 Small Graphs

In this section we determine the complexity of MAX SOL(H) for all graphs $H = (V, E)$ on at most 4 vertices, i.e., $|V(H)| \leq 4$. For $|V(H)| = 4$ we only consider the case where $V = \{0, 1, 2, 3\}$, but for $|V(H)| \leq 3$ we classify the complexity for all $V(H) \subset \mathbb{N}$. In the process we discover a new tractable class for the MAX SOL(H) problem which is closely related to the critical independent set problem [1,17].

We know from Lemma 1 that it is sufficient to consider graphs H that are max-cores. The case $|V(H)| = 1$ is trivial since there are only two such graphs and both are tractable. For $|V(H)| = 2$ there are two graphs that are max-cores, namely, the irreflexive path on two vertices (in **PO** by Proposition 1) and the graph on $V = \{d_1, d_2\}$, $d_1 < d_2$ where d_1 is adjacent to d_2 and d_1 is looped (which is **NP**-hard by Lemma 5).

When $|V(H)| = 3$ we have the following classification.

Theorem 1. *There are six (types of) max-cores over* $\{d_1, d_2, d_3\}$ *where* $d_1 < d_2 < d_3$, *denoted* $H_1, \ldots, H_6$ *and shown in Figure 2.* MAX SOL(H) *is* **NP**-*hard for all of these except* H_5. MAX SOL(H_5) *is in* **PO** *if* $d_3 + d_1 \leq 2d_2$ *and* **NP**-*hard otherwise.*

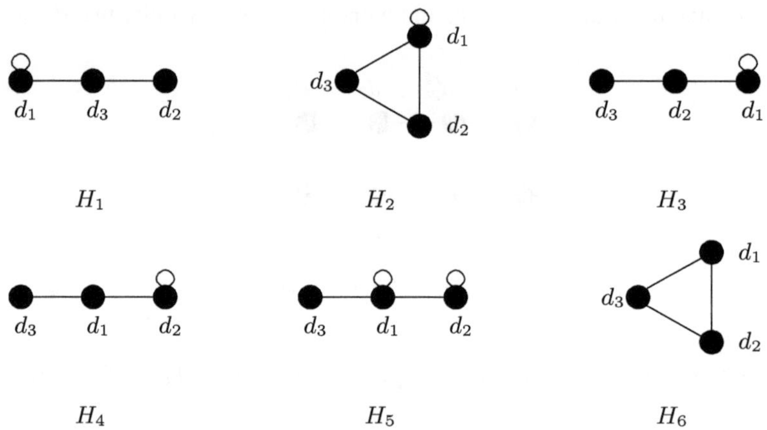

Fig. 2. The graphs H_i

The MAX SOL(H_5) problem is related to the critical independent set problem [1,17] in the following way. An independent set $I_C \subseteq V(G)$ is called critical if

$$|I_C| - |N(I_C)| = \max\{|I| - |N(I)| \mid I \text{ is an independent set in } G\},$$

where $N(I)$ denote the neighborhood of I, i.e., the set of vertices in G that are adjacent to at least one vertex in I. Zhang [17] proved that critical independent sets can be found in polynomial time.

We extend the notion of a critical independent sets to (k, m)-critical independent sets. A (k, m)-critical independent set is an independent set $I_C \subseteq V(G)$ such that

$$k \cdot |I_C| - m \cdot |N(I_C)| = \max\{k \cdot |I| - m \cdot |N(I)| \mid I \text{ is an independent set in } G\}.$$

Note that the maximum independent set problem is exactly the problem of finding a $(1, 0)$-critical independent set. The following proposition shows that that MAX SOL(H_5) is polynomial-time equivalent to the $(d_3 - d_2, d_2 - d_1)$-critical independent set problem.

Proposition 5. *I_C is a $(d_3 - d_2, d_2 - d_1)$-critical independent set in G if and only if the homomorphism h from G to H_5, defined by $h^{-1}(d_3) = I_C$ and $h^{-1}(d_1) = Nbd(I_C)$ is an optimal solution for* MAX SOL*(H_5).*

Proof. Assume that I_C is a $(d_3 - d_2, d_2 - d_1)$-critical independent set in G but h is not an optimal solution to MAX SOL(H_5), i.e., there exists a homomorphism g from G to H_5 such that $m(g) > m(h)$. That is,

$$w(g^{-1}(d_3)) \cdot d_3 + w(g^{-1}(d_1)) \cdot d_1 + w(g^{-1}(d_2)) \cdot d_2 >$$

$$w(h^{-1}(d_3)) \cdot d_3 + w(h^{-1}(d_1)) \cdot d_1 + w(h^{-1}(d_2)) \cdot d_2.$$

Subtracting $w(V(G)) \cdot d_2$ from both sides, we get

$$w(g^{-1}(d_3)) \cdot (d_3 - d_2) - w(g^{-1}(d_1)) \cdot (d_2 - d_1) >$$

$$w(h^{-1}(d_3)) \cdot (d_3 - d_2) - w(h^{-1}(d_1)) \cdot (d_2 - d_1).$$

This contradicts the fact that I_C is a $(d_3 - d_2, d_2 - d_1)$-critical independent set. The proof in the other direction is similar. □

Building upon the results in [1], we are able to completely classify the complexity of the (k, m)-critical independent set problem and, hence, also the complexity of MAX SOL(H_5). More specifically, we prove that the (k, m)-critical independent set problem is in **PO** if $k \leq m$ and that it is **NP**-hard if $k > m$.

Finally, we present the complexity classification of MAX SOL for all graphs $H = (V, E)$ where $V = \{0, 1, 2, 3\}$. Just as in the case where $|V(H)| \leq 3$ we make heavy use of the fact that only graphs that are max-cores need to be classified. Our second tool is the following lemma, stating that we can assume that we have access to all constants.

Lemma 7. *Let H be a max-core over $\{0, 1, 2, 3\}$. Then* MAX SOL(H) *is in* **PO** *(***NP***-hard) if and only if* MAX SOL$(H \cup \{\{(0)\}, \{(1)\}, \{(2)\}, \{(3)\}\})$ *is in* **PO** *(***NP***-hard).*

As an immediate corollary, we get that MAX SOL(H) is **NP**-hard for all max-cores H on $D = \{0, 1, 2, 3\}$ when the retraction problem (RET(H)) is **NP**-complete. Note that the complexity of the retraction problem for all graphs on at most 4 vertices have been classified in [16]. The classification is completed by considering the remaining max-cores (for which RET(H) is in **P**) one by one. Our result is the following.

Theorem 2. *Let H be a max-core on $D = \{0, 1, 2, 3\}$. Then,* MAX SOL(H) *is in* **PO** *if H is an irreflexive path, and otherwise,* MAX SOL(H) *is* **NP***-hard.*

5 Results for LIST MAX AW SOL

The main theorem of this section is stated as follows.

Theorem 3. *Let H be an undirected graph with loops allowed. Then* LIST MAX AW SOL(H) *is solvable in polynomial time if all components of H are proper interval graphs or proper interval bigraphs. Otherwise,* LIST MAX AW SOL(H) *is* **NP**-*hard.*

Corollary 2. *Let H be an undirected graph with loops allowed. Then,* LIST MAX AW SOL(H) *is polynomial time equivalent to* MIN HOM(H).

The reduction from LIST MAX AW SOL(H) is easy. The lists are replaced by weights of ∞ for the appropriate variable-value pairs. Remaining weights are negated and translated so that the smallest weight becomes 0 for MIN HOM. The rest of this section is devoted to proving the other direction. We assume that the input instance is connected. If it is not, then we can solve each component separately and add the solutions.

Lemma 8. *Let H be an undirected graph. Then,* LIST MAX AW SOL(H) *is* **NP**-*hard if there exists a connected component H' of H such that* LIST MAX AW SOL(H') *is* **NP**-*hard. Otherwise, if for each connected component H' of H we have that* LIST MAX AW SOL(H') *is in* **PO**, *then* LIST MAX AW SOL(H) *is in* **PO**.

Lemma 9. *Let H be an undirected graph in which there exists both loop-free vertices and vertices with loops. Then,* LIST MAX AW SOL(H) *is* **NP**-*hard.*

Proof. This is proved by reduction from MAXIMUM INDEPENDENT SET.

Proposition 6. *If H is a connected graph which is a proper interval graph or a proper interval bigraph, then* LIST MAX AW SOL(H) *is polynomial time solvable.*

Proof. This follows from [5, Corollary 2.6] which states that the corresponding MIN HOM(H)-problem is polynomial time solvable. □

Theorem 4 (P. Hell, J. Huang [8]). *A bipartite graph H is a proper interval bigraph if and only if it does not contain an induced cycle of length at least six, or a bipartite claw, or a bipartite net, or a bipartite tent.*

Figure 3 displays the bipartite net and bipartite tent graphs.

Lemma 10. *Let H be a cycle of length at least six. Then* LIST MAX AW SOL(H) *is* **NP**-*hard.*

Proof. The proof is by a simple reduction from the retraction problem on H. This problem is shown to be **NP**-complete in [4]. □

Lemma 11. *Let H be one of the graphs shown in Figure 3. Then* LIST MAX AW SOL(H) *is* **NP**-*hard.*

Proof. The proof follows the same ideas as those in [5]. That is, one reduces from the problem of finding a maximum independent set in a 3-partite graph. The apparent lack of expressive power of the LIST MAX AW SOL-framework, and the dependence on the

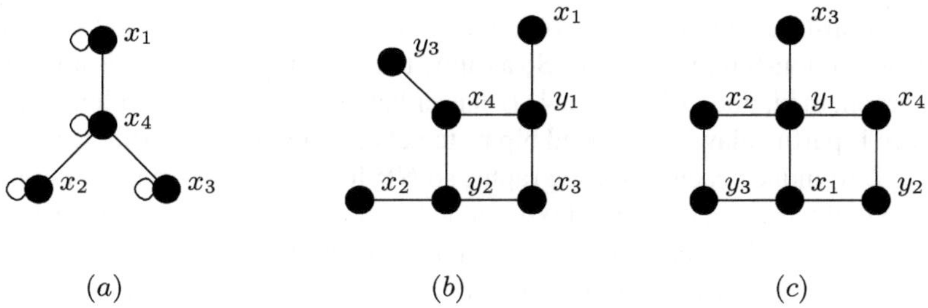

Fig. 3. (a) reflexive claw, (b) bipartite net, (c) bipartite tent

labels of the target graph, are resolved by a precise choice of weights in the constructed instances. We carry out the case in Figure 3(a) in detail.

Let G be a 3-partite graph with partite sets X, Y and Z. We create an instance $I = (V, D, C, L, w)$ of LIST MAX AW SOL(H) as follows. Let $V = V(G)$, $D = V(H) = \{x_1, x_2, x_3, x_4\}$ and create a constraint uHv in C for each $(u, v) \in E(G)$. Now, define the lists and weights as follows.

$$L(u) = \begin{cases} \{x_4, x_1\} & \text{when } u \in X \\ \{x_4, x_2\} & \text{when } u \in Y \\ \{x_4, x_3\} & \text{when } u \in Z. \end{cases}$$

$$w(u) = \begin{cases} 1/(x_1 - x_4) & \text{when } u \in X \\ 1/(x_2 - x_4) & \text{when } u \in Y \\ 1/(x_3 - x_4) & \text{when } u \in Z. \end{cases}$$

Now, if s is a solution to I, let $X_1 = s^{-1}(x_1)$, $X_0 = X \setminus X_1$ and define similarly Y_0, Y_1 and Z_0, Z_1. Note that s defines an independent set $X_1 \cup Y_1 \cup Z_1$ of G. Conversely, it is also clear that any independent set of G yields a solution to I by assigning each variable to x_4 precisely when it is not a part of the independent set. The value of s can be written as

$$\sum_{u \in V} s(u) \cdot w(u) = \sum_{x \in X} s(x) \cdot w(x) + \sum_{y \in Y} s(y) \cdot w(y) + \sum_{z \in Z} s(z) \cdot w(z) =$$

$$\frac{|X_0| \cdot x_4 + |X_1| \cdot x_1}{x_1 - x_4} + \frac{|Y_0| \cdot x_4 + |Y_1| \cdot x_2}{x_2 - x_4} + \frac{|Z_0| \cdot x_4 + |Z_1| \cdot x_3}{x_3 - x_4} =$$

$$\frac{|X| \cdot x_4}{x_1 - x_4} + |X_1| + \frac{|Y| \cdot x_4}{x_2 - x_4} + |Y_1| + \frac{|Z| \cdot x_4}{x_3 - x_4} + |Z_1| = M + |X_1| + |Y_1| + |Z_1|,$$

where M is independent of s and can be calculated in polynomial time from I. Thus, an optimal solution to I gives a maximal independent set in G. □

We now have all the tools necessary to complete the proof of Theorem 3.

Proof of Theorem 3. According to Lemma 8 we can assume that H is connected. Furthermore, due to Lemma 9 we can assume that H is either loop-free or reflexive, or LIST MAX AW SOL(H) is **NP**-hard. Proposition 6 gives the polynomial cases.

If H is loop-free and non-bipartite, we can reduce from HOM(H), which is **NP**-complete for non-bipartite graphs. So assume that H is bipartite. If H is not a proper interval bigraph, then, due to Theorem 4, H has either an induced cycle of length at least 6, an induced bipartite claw, an induced bipartite net or an induced bipartite tent. We can use the lists L to induce each of these graphs, so **NP**-hardness follows from Lemma 10 and 11. Note that hardness for the reflexive claw implies hardness for the bipartite claw.

Finally, if H is reflexive, then it is either not an interval graph, or a non-proper interval graph. If H is not an interval graph, then we can reduce from the list homomorphism problem L-HOMH which is shown to be **NP**-complete for reflexive, non-interval graphs in [3]. In the second case, it has been shown by Roberts [14] that H must contain an induced claw. Lemma 11 shows that this problem is **NP**-hard, which finishes the proof. □

6 Discussion and Future Work

In this paper we have initiated a study of the complexity of the maximum solution problem on graphs. Our results indicate that giving a complete complexity classification of MAX SOL(H) for every fixed graph H is probably harder than first anticipated. In particular, the new tractable class for the MAX SOL problem identified in Section 4.2 depends very subtly on the values of the domain elements and we have not yet been able to characterize this tractable class in terms of polymorphisms. Hence, this tractable class seems to be of a very different flavour compared to the previously identified tractable classes for the MAX SOL problem [13].

On the other hand, we are able to give a complete classification for the complexity of the arbitrary weighted list version of the problem, LIST MAX AW SOL(H). Interestingly, the borderline between tractability and **NP**-hardness for LIST MAX AW SOL(H) coincide exactly with Gutin et al.'s [5] recent complexity classification of MIN HOM(H). This is surprising, since the MIN HOM(H) problem is much more expressive than the LIST MAX AW SOL(H) problem, and hence, we were expecting graphs H such that MIN HOM(H) were **NP**-hard and LIST MAX AW SOL(H) were in **PO**. The obvious question raised by this result is how far can we extend the agreement in complexity between LIST MAX AW SOL(Γ) and MIN HOM(Γ)? To this end, we state the MIN HOM problem for general constraint languages.

Minimum Cost Homomorphism over constraint language Γ, denoted MIN HOM(Γ), is the minimization problem with

Instance: Tuple $(V, D, C, c_i(v))$, where D is a finite subset of $\mathbb{N}$, (V, D, C) is a CSP instance over Γ, $c_i : V \rightarrow Q^+$ are costs for $i \in V(H)$.
Solution: An assignment $f : V \rightarrow D$ to the variables such that all constraints are satisfied.
Measure: $\sum_{v \in V} c_{f(v)}(v)$.

Problem 1. Is it the case that the complexity of LIST MAX AW SOL(Γ) and MIN HOM(Γ) are equal for all constraint languages Γ?

It can be shown, using results from [2,12], that LIST MAX AW SOL(Γ) and MIN HOM(Γ) are polynomial time equivalent when Γ is a boolean constraint language.

References

1. Ageev, A.: On finding critical independent and vertex sets. SIAM J. Discrete Math. 7(2), 293–295 (1994)
2. Cohen, D., Cooper, M., Jeavons, P., Krokhin, A.: The complexity of soft constraint satisfaction. Artif. Intell. 170(11), 983–1016 (2006)
3. Feder, T., Hell, P.: List homomorphisms to reflexive graphs. J. Combin. Th (B) 72, 236–250 (1998)
4. Feder, T., Hell, P., Huang, J.: List Homomorphisms and Circular Arc Graphs. Combinatorica 19, 487–505 (1999)
5. Gutin, G., Hell, P., Rafiey, A., Yeo, A.: A dichotomy for minimum cost graph homomorphisms. European J. Combin. (to appear)
6. Gutin, G., Rafiey, A., Yeo, A.: Minimum cost and list homomorphisms to semicomplete digraphs. Discrete Applied Mathematics 154(6), 890–897 (2006)
7. Gutin, G., Rafiey, A., Yeo, A., Tso, M.: Level of repair analysis and minimum cost homomorphisms of graphs. Discrete Applied Mathematics 154(6), 881–889 (2006)
8. Hell, P., Huang, J.: Interval bigraphs and circular arc graphs. J. Graph Theory 46, 313–327 (2004)
9. Hell, P., Nešetřil, J.: The complexity of H-coloring. J. Combinatorial Theory B 48, 92–110 (1990)
10. Hell, P., Nešetřil, J.: Graphs and Homomorphisms. Oxford Lecture Series in Mathematics and Its Applications. Oxford University Press (2004)
11. Jeavons, P., Cohen, D., Gyssens, M.: A test for tractability. In: Freuder, E.C. (ed.) CP 1996. LNCS, vol. 1118, pp. 267–281. Springer, Heidelberg (1996)
12. Jonsson, P.: Boolean constraint satisfaction: complexity results for optimization problems with arbitrary weights. Theoretical Computer Science 244(1–2), 189–203 (2000)
13. Jonsson, P., Nordh, G.: Generalised integer programming based on logically defined relations. In: Královič, R., Urzyczyn, P. (eds.) MFCS 2006. LNCS, vol. 4162, pp. 549–560. Springer, Heidelberg (2006)
14. Roberts, F.: Indifference graphs. In: Harary, F. (ed.) Proof Techniques in Graph Theory, pp. 139–146. Academic Press, London (1969)
15. Szendrei, Á. (ed.): Clones in Universal Algebra. Seminaires de Mathématiques Supérieures, vol. 99. University of Montreal (1986)
16. Vikas, N.: A complete and equal computational complexity classification of compaction and retraction to all graphs with at most four vertices and some general results. J. Comput. Syst. Sci. 71(4), 406–439 (2005)
17. Zhang, C.-Q.: Finding critical independent sets and critical vertex subsets are polynomial problems. SIAM J. Discrete Math. 3(3), 431–438 (1990)

What Are Iteration Theories?

Jiří Adámek[1], Stefan Milius[1], and Jiří Velebil[2,⋆]

[1] Institute of Theoretical Computer Science, TU Braunschweig, Germany
{adamek,milius}@iti.cs.tu-bs.de
[2] Department of Mathematics, ČVUT Prague, Czech Republic
velebil@math.feld.cvut.cz

Abstract. We prove that iteration theories can be introduced as algebras for the monad $\mathbb{R}\mathrm{at}$ on the category of signatures assigning to every signature Σ the rational-Σ-tree signature. This supports the result that iteration theories axiomatize precisely the equational properties of least fixed points in domain theory: $\mathbb{R}\mathrm{at}$ is the monad of free rational theories and every free rational theory has a continuous completion.

"In the setting of algebraic theories enriched with an external fixed-point operation, the notion of an iteration theory seems to axiomatize the equational properties of all the computationally interesting structures of this kind."
S. L. Bloom and Z. Ésik (1996), see [4]

1 Introduction

In domain theory recursive equations have a clear semantics given by the least solution. The function assigning to every system of recursive equations e the least solution $e^\dagger$ has a number of equational properties. One answer to the question in the title is given by a semantic characterization: iteration theories are those Lawvere theories in which recursive equations have solutions subject to all equational laws that the least-solution-map $e \mapsto e^\dagger$ obeys in domain theory. The same question also has a precise answer given by a list of all the equational axioms involved, see the fundamental monograph [3] of Stephen Bloom and Zoltan Ésik, or Definition 3.1 below. The aim of the present paper is to offer a short and precise syntactic answer:

> Iteration theories are precisely the algebras for the rational-tree monad $\mathbb{R}\mathrm{at}$ on the category of signatures.

To be more specific: let **Sgn** be the category of signatures, that is, the slice category **Set**/$\mathbb{N}$. Every signature Σ generates a free rational theory in the sense of the ADJ-group: it is the theory $\mathbf{RT}_{\Sigma_\perp}$ of all *rational trees* (which means trees having, up to isomorphism, only finitely many subtrees) on the signature

⋆ The support of the grant 201/06/664 of the Grant Agency of the Czech Republic is gratefully acknowledged.

L. Kučera and A. Kučera (Eds.): MFCS 2007, LNCS 4708, pp. 240–252, 2007.

$\Sigma_\perp = \Sigma + \{\perp\}$, for a new nullary symbol $\perp$. This follows from results in [7] and [11], and it yields a free-rational-theory monad

$$\mathbb{R}\mathrm{at}, \qquad \Sigma \mapsto \mathbf{RT}_{\Sigma_\perp}$$

on the category **Sgn** of signatures.

The main result of our paper is the following

Theorem. *The category of iteration theories is isomorphic to the category of Eilenberg-Moore algebras for the rational-tree monad* $\mathbb{R}\mathrm{at}$ *on* **Sgn**.

This theorem gives at first a new semantic characterization of iteration theories as a category of algebras. In addition, it is well-known that a characterization of a category as a category of Eilenberg-Moore algebras for a finitary monad on a locally finitely presentable category means that the category can be presented in the form of identities over the base category, see [8]. Hence, our result bears aspects of a syntactic characterization too.

It is well-known that classical varieties (of one-sorted, finitary algebras) are precisely the finitary, monadic categories over **Set**. We can generalize this to **Sgn** and speak about varieties of theories as categories that are finitary, monadic over **Sgn**. In this sense our main result proves that the iteration theories form a variety which is generated by all rational theories. Is this new? We believe it is—not in its spirit, but in its formal proof. In the monograph [3] the same result is formulated, see Theorem 8.2.5, however, the concept of a variety of theories is not introduced there. In later work, see, e.g., [4], [5], [6], varieties of preiteration theories are defined as classes of theories which satisfy given identities—however, the concept of identity is not defined. We have no doubt that it is possible to introduce the concept of "identity over **Sgn**" and then use it to provide a formal proof, but we decided for a different route using Beck's Theorem to prove the result above. The first (and very crucial) part of such a proof is explicitly contained in the monograph [3]: it is the fact that a free iteration theory on a signature Σ is the rational-tree theory $\mathbf{RT}_{\Sigma_\perp}$ above.

In our previous work we have dealt with iterative theories, e.g., by characterizing the Eilenberg-Moore algebras for the free iterative theories [1]. The present paper uses an analogous method but applied in a second-order setting: whereas Beck's Theorem is used over the given base category in [1], here we apply it to the category of signatures over the base category.

2 Rational Theories

The aim of the present section is to introduce the rational-tree monad $\mathbb{R}\mathrm{at}$ without using the result of [3] that $\mathbb{R}\mathrm{at}$ is the monad of free iteration theories: instead we use the free rational theories of [11].

2.1 Remark. Throughout the paper an (algebraic) *theory* is a category whose objects are natural numbers and every object n is a coproduct $n = 1+1+\cdots+1$

(n copies) with chosen coproduct injections. *Theory morphisms* are the functors which are identity maps on objects and which preserve finite coproducts (on the nose). The resulting category of theories is denoted by

$$\mathbf{Th}.$$

2.2 Examples. (1) Every signature $\Sigma = (\Sigma_n)_{n\in\mathbb{N}}$ defines the *Σ-tree theory*

$$\mathbf{CT}_\Sigma$$

as follows.

By a *Σ-tree* on a set X of generators is meant a tree[1] labelled so that every leaf is labelled in $X + \Sigma_0$ and every node with $n > 0$ children is labelled in Σ_n. The theory $\mathbf{CT}_\Sigma$ has as morphisms $1 \to k$ all Σ-trees on k generators, thus:

$$\mathbf{CT}_\Sigma(n,k) = \text{all } n\text{-tuples of } \Sigma\text{-trees on } k \text{ generators.}$$

Composition is given by tree substitution.

(2) Analogously, the *finite-Σ-tree theory* $\mathbf{FT}_\Sigma$ is given by

$$\mathbf{FT}_\Sigma(n,k) = \text{all } n\text{-tuples of } \textit{finite } \Sigma\text{-trees on } k \text{ generators.}$$

(3) The theory

$$\mathscr{N}$$

which is the full subcategory of **Set** on natural numbers is initial: for every theory $\mathscr{T}$ we have a unique theory morphism $u_{\mathscr{T}} : \mathscr{N} \to \mathscr{T}$; we call the morphisms in $u_{\mathscr{T}}(\mathscr{N})$ *basic*.

2.3 Notation. We denote by

$$U\colon \mathbf{Th} \to \mathbf{Sgn}$$

the forgetful functor assigning to every theory $\mathscr{T}$ the signature $U(\mathscr{T})$ whose n-ary symbols are $\mathscr{T}(1,n)$ for all $n \in \mathbb{N}$.

2.4 Remark. (i) U has a left adjoint

$$\mathbf{FT}\colon \mathbf{Sgn} \to \mathbf{Th}$$

assigning to every signature Σ the finite-tree theory $\mathbf{FT}_\Sigma$. This gives us a monad on the category of signatures:

$$\mathbb{F}\mathrm{in}\colon \Sigma \mapsto U(\mathbf{FT}_\Sigma).$$

Recall that Jean Bénabou [2] proved that U is *monadic*; that means that **Th** is isomorphic to the category of algebras for the monad $\mathbb{F}\mathrm{in}$.

---

[1] Trees are understood to be rooted, ordered, labelled possibly infinite trees that one considers up to isomorphism.

(ii) We denote by

$$\omega\mathbf{CPO}$$

the category whose objects are posets with joins of ω-chains (a least element is not required; if an object has it we speak about a *strict* CPO). Morphisms are the *continuous* functions; they are monotone functions preserving joins of ω-chains. Morphisms between strict CPO's preserving the least element are called *strict continuous maps.*

2.5 Definition (see [11]). (1) *Theories enriched over* **Pos**, *the category of posets and monotone functions, are called* ***ordered theories****. These are the theories with ordered hom-sets such that both composition and cotupling are monotone.*

(2) *An ordered theory is called* ***pointed*** *provided that every hom-set $\mathscr{T}(n,k)$ has a least element (notation: $\perp_{nk}$ or just $\perp$) and composition is left-strict, i.e., $\perp_{kr} \cdot f = \perp_{nr}$ for all $f \in \mathscr{T}(n,k)$.*

(3) ***A continuous theory*** *is a pointed theory enriched over the category* ω**CPO**, *which means that both composition and cotupling preserve joins of ω-chains (but there is no condition on cotupling concerning $\perp$).*

2.6 Remark. (1) In an algebraic theory $\mathscr{T}$ *equation morphisms* are morphisms of the form

$$e\colon n \to n+p.$$

For example, if $\mathscr{T} = \mathbf{FT}_\Sigma$ then e represents a recursive system of n equations

$$\begin{aligned} x_1 &\approx t_1(x_1,\dots,x_n,z_1,\dots,z_p)\\ &\vdots \\ x_n &\approx t_n(x_1,\dots,x_n,z_1,\dots,z_p)\end{aligned} \tag{2.1}$$

where the right-hand sides are terms in $X+Z$ for $X=\{x_1,\dots,x_n\}$ and $Z=\{z_1,\dots,z_p\}$.

(2) A *solution* of an equation morphism $e\colon n \to n+p$ is a morphism $e^\dagger\colon n \to p$ such that the triangle

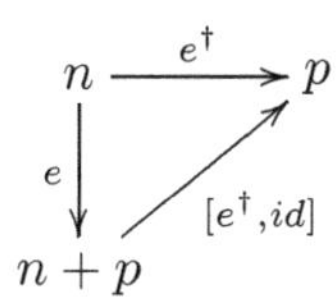

commutes. In case of (2.1) this is a substitution of terms $e^\dagger(x_i)$ for the given variables x_i such that the formal equations of (2.1) become identities in $\mathbf{FT}_\Sigma$. It is obvious that many systems (2.1) fail to have a solution in $\mathbf{FT}_\Sigma$ (because the obvious tree expansions are not finite).

(3) In contrast, in continuous theories all equation morphisms have a solution. In fact, the *least* solution $e^\dagger$ always exists because the endofunction

$$x \mapsto [x, id]\cdot e \qquad \text{of } \mathscr{T}(n,p)$$

is continuous. By Kleene's Fixed-Point Theorem, $e^\dagger$ is the join of the following ω-chain of approximations:

$$e^\dagger = \bigsqcup_{i\in\mathbb{N}} e_i^\dagger : n \to p$$

where $e_0^\dagger = \bot$ and given $e_i^\dagger$ then $e_{i+1}^\dagger$ is the morphism

$$\begin{array}{ccc} n & \xrightarrow{\;e_{i+1}^\dagger\;} & p \\ {\scriptstyle e}\big\downarrow & \nearrow_{[e_i^\dagger, id]} & \\ n+p & & \end{array} \tag{2.2}$$

(4) Observe that the left-hand coproduct injection e in $\mathscr{T}(n, n+p)$ has the solution $e^\dagger = \bot$ in every continuous theory.

2.7 Example: the continuous theory $\mathbf{CT}_{\Sigma_\bot}$. Given a signature Σ we denote by

$$\Sigma_\bot$$

the extension of Σ by a nullary symbol $\bot \notin \Sigma$ (no ordering assumed a priori!). The theory $\mathbf{CT}_{\Sigma_\bot}$ of $\Sigma_\bot$-trees, see 2.2, carries a very "natural" ordering: given trees t and t' in $\mathbf{CT}_{\Sigma_\bot}(1,k)$ then $t \sqsubseteq t'$ holds iff t can be obtained from t' by cutting away some subtrees and labelling the new leaves by $\bot$. And the ordering of $\mathbf{CT}_{\Sigma_\bot}(n,k)$ is componentwise.

2.8 Notation. We denote by

$$\mathbf{CTh}$$

the category of all continuous theories and strict, continuous theory morphisms. Its forgetful functor $\mathbf{CTh} \to \mathbf{Sgn}$ is the domain restriction of U from Notation 2.3.

2.9 Theorem [11]. *For every signature Σ a free continuous theory on Σ is $\mathbf{CT}_{\Sigma_\bot}$. That is, the forgetful functor $\mathbf{CTh} \to \mathbf{Sgn}$ has a left adjoint given by $\Sigma \mapsto \mathbf{CT}_{\Sigma_\bot}$.*

2.10 Remark. Let $\mathscr{T}$ be a pointed ordered theory. For every equation morphism $e \colon n \to n+p$ we can form the morphisms $e_i^\dagger \colon n \to p$ as in 2.2 and we clearly obtain an ω-chain

$$e_0^\dagger \sqsubseteq e_1^\dagger \sqsubseteq e_2^\dagger \ldots \qquad \text{in } \mathscr{T}(n,p).$$

We call these chains admissible and extend this to composites $e_i^\dagger \cdot v$ for morphisms $v \colon m \to n$:

2.11 Definition [11]. *In a pointed ordered theory $\mathscr{T}$ an ω-chain in $\mathscr{T}(m,p)$ is called* ***admissible*** *if it has the form $e_i^\dagger \cdot v$ $(i \in \mathbb{N})$ for some morphisms $e\colon n \to n+p$ and $v\colon m \to n$. The theory $\mathscr{T}$ is called* ***rational*** *if it has joins of all admissible ω-chains and if cotupling preserves these joins.*

2.12 Examples. (1) Every continuous theory is rational.

(2) The free continuous theory $\mathbf{CT}_{\Sigma_\perp}$ has a nice rational subtheory: the theory

$$\mathbf{RT}_{\Sigma_\perp} \quad \text{of all rational } \Sigma_\perp\text{-trees.}$$

Recall that a $\Sigma_\perp$-tree is called *rational* if it has up to isomorphism only finitely many subtrees, see [7]. It is easy to see that $\mathbf{RT}_{\Sigma_\perp}$ is a pointed subtheory of $\mathbf{CT}_{\Sigma_\perp}$.

2.13 Notation. We denote by

$$\mathbf{RTh}$$

the category of rational theories and order-enriched strict theory morphisms preserving least solutions. That is, given rational theories $\mathscr{T}$ and $\mathscr{R}$, a morphism is a theory morphism $\varphi\colon \mathscr{T} \to \mathscr{R}$ which (i) is monotone and strict on hom-sets and (ii) fulfils $\varphi(e^\dagger) = \varphi(e)^\dagger$ for all $e \in \mathscr{T}(n, n+p)$.

2.14 Proposition [11]. *A free rational theory on a signature Σ is the rational-tree theory $\mathbf{RT}_{\Sigma_\perp}$. More precisely, the forgetful functor*

$$W\colon \mathbf{RTh} \to \mathbf{Sig}$$

(a domain restriction of U in 2.3*) has a left adjoint*

$$\Sigma \mapsto \mathbf{RT}_{\Sigma_\perp}.$$

2.15 Corollary. *The monad*

$$\mathbb{R}\mathrm{at}$$

of free rational theories on the category **Sgn** *is defined by*

$$\Sigma \mapsto W(\mathbf{RT}_{\Sigma_\perp}).$$

More precisely, to every signature Σ this monad assigns the signature whose n-ary operation symbols are the rational $\Sigma_\perp$-trees on n generators.

We call $\mathbb{R}$at *the* ***rational-tree*** *monad.*

3 Iteration Theories

In this section we first recall the definition of an iteration theory from [3] and then prove the main result: iteration theories are algebras for the rational-tree monad $\mathbb{R}$at.

Our proof uses Beck's theorem characterizing categories of $\mathbb{T}$-algebras for a monad $\mathbb{T} = (T, \eta, \mu)$ on a category $\mathscr{A}$. Recall that a $\mathbb{T}$-algebra is an object A of $\mathscr{A}$

together with a morphism $\alpha\colon TA \to A$ such that $\alpha \cdot \eta_A = id_A$ and $\alpha \cdot \mu_A = \alpha \cdot T\alpha$. The category $\mathscr{A}^{\mathbb{T}}$ of $\mathbb{T}$-algebras and homomorphisms (defined via an obvious commutative square in $\mathscr{A}$) is equipped with the forgetful functor $I\colon \mathscr{A}^{\mathbb{T}} \to \mathscr{A}$ given by $(A, \alpha) \mapsto A$. This functor is a right adjoint, and it *creates coequalizers of I-split pairs.* The latter means that for every parallel pair $u, v\colon T \to S$ in the category $\mathscr{A}^{\mathbb{T}}$ and every diagram

$$IT \underset{v}{\overset{u}{\underset{\xleftarrow{\;t\;}}{\rightrightarrows}}} IS \underset{s}{\overset{c}{\rightleftarrows}} R$$

in $\mathscr{A}$ whose mappings satisfy the equations

$$cu = cv \tag{i}$$

$$cs = id \tag{ii}$$

$$ut = id \tag{iii}$$

$$vt = sc \tag{iv}$$

there exists a unique morphism $\bar{c}\colon S \to \overline{R}$ in $\mathscr{A}^{\mathbb{T}}$ with $I\bar{c} = c$, and moreover $\bar{c}$ is a coequalizer of u and v. Beck's theorem states that monadic algebras are characterized by the above two properties of the forgetful functor. More precisely, whenever a functor $I\colon \mathscr{B} \to \mathscr{A}$ is a right adjoint creating I-split coequalizers, then $\mathscr{B}$ is isomorphic to $\mathscr{A}^{\mathbb{T}}$ for the monad $\mathbb{T}$ given by the adjoint situation of I. See [9] for a proof.

3.1 Definition. (See [3], 6.8.1.) *An iteration theory is a theory $\mathscr{T}$ together with a function $\dagger$ assigning to every ("equation") morphism $e\colon n \to n+p$ a morphism $e^\dagger\colon n \to p$ in such a way that the following five axioms hold:*

(1) ***Fixed Point Identity.*** *This states that $e^\dagger$ is a solution of e, i.e., a fixed point of $[-, id_p] \cdot e$:*

$$\begin{array}{ccc} n & \xrightarrow{\;e^\dagger\;} & p \\ {\scriptstyle e}\downarrow & \nearrow_{[e^\dagger, id]} & \\ n+p & & \end{array} \tag{3.1}$$

(2) ***Parameter Identity.*** *We use the following notation: given an equation morphism $e\colon n \to n+p$, then every morphism $h\colon p \to q$ yields a new equation morphism*

$$h \bullet e \equiv n \xrightarrow{\;e\;} n+p \xrightarrow{\;id+h\;} n+q. \tag{3.2}$$

The parameter identity tells us how the solutions of e and $h \bullet e$ are related: the triangle

$$\begin{array}{ccc} n & \xrightarrow{\;e^\dagger\;} & p \\ & {\scriptstyle (h\bullet e)^\dagger}\searrow & \downarrow{\scriptstyle h} \\ & & q \end{array} \tag{3.3}$$

commutes.

(3) ***Simplified Composition Identiy.*** *We use the following notation: given morphisms*

$$m \xrightarrow{g} n+p \qquad \text{and} \qquad n \xrightarrow{f} m$$

we obtain an equation morphism

$$f \circ g \equiv m \xrightarrow{g} n+p \xrightarrow{f+id} m+p. \tag{3.4}$$

The simplified composition identity states that the triangle

$$\begin{array}{ccc} n & \xrightarrow{(g\cdot f)^\dagger} & p \\ {\scriptstyle f}\downarrow & \nearrow_{(f\circ g)^\dagger} & \\ m & & \end{array} \tag{3.5}$$

commutes.

(4) ***Double Dagger Identity.*** *This is a statement about morphisms of the form*

$$e\colon n \to n+n+p.$$

A solution yields $e^\dagger\colon n \to n+p$ *which we can solve again and get* $(e^\dagger)^\dagger\colon n \to p$. *On the other hand, the codiagonal* $\nabla\colon n+n \to n$ *yields an equation morphism* $\nabla \circ e\colon n \to n+p$. *The double-dagger identity states*

$$(\nabla \circ e)^\dagger = \left(e^\dagger\right)^\dagger\colon n \to p. \tag{3.6}$$

(5) ***Commutative identity.*** *This is in fact an infinite set of identities: one for every m-tuple of basic endomorphisms of m:*

$$\varrho_0, \ldots, \varrho_{m-1} \in \mathscr{N}(m,m)$$

and for every decomposition $m = n{\cdot}k$ *such that the corresponding codiagonal* $\nabla\colon \coprod_k n \to n$ *in* $\mathscr{N}$ *fulfils*

$$\nabla{\cdot}\varrho_j = \nabla \qquad \text{for } j = 0, \ldots, m-1.$$

The commutative identity concerns an arbitrary morphism

$$f\colon n \to m+p \quad \text{in } \mathscr{T}.$$

We can form two equation morphisms: $\nabla \circ f\colon n \to n+p$ *(see (3.4)) and*

$$\hat{f}\colon m \to m+p$$

defined by the individual components $\hat{f}{\cdot}\mathrm{in}_j\colon 1 \to m+p$ *for* $j = 0, \ldots, m-1$ *as follows:*

$$\hat{f}{\cdot}\mathrm{in}_j \equiv 1 \xrightarrow{\mathrm{in}_j} m \xrightarrow{\nabla} n \xrightarrow{f} m+p \xrightarrow{\varrho_j + id} m+p. \tag{3.7}$$

The conclusion is that the triangle

$$\begin{array}{ccc} m & \xrightarrow{\hat{f}^{\dagger}} & p \\ {\scriptstyle \nabla}\big\downarrow & \nearrow_{(\nabla\circ f)^{\dagger}} & \\ n & & \end{array} \tag{3.8}$$

commutes. (Remark: the notation in [3] *for* $\hat{f}$ *is* $\nabla{\cdot}f\|(\varrho_0,\ldots,\varrho_{m-1})$ *and instead of* ∇ *a general surjective base morphism is assumed. The simplification working with* ∇ *was proved in* [6].*)*

3.2 Definition. *Let* $(\mathscr{T},\dagger)$ *and* $(\mathscr{S},\ddagger)$ *be iteration theories. A theory morphism* $\varphi\colon \mathscr{T}\to\mathscr{S}$ *is said to* ***preserve solutions*** *if for every morphism* $e\in\mathscr{T}(n,n{+}p)$ *we have* $\varphi(e)^{\ddagger}=\varphi(e^{\dagger})$. *The category of iteration theories and solution-preserving morphisms is denoted by*

$$\mathbf{ITh}.$$

We denote by $V\colon \mathbf{ITh}\to\mathbf{Sgn}$ *the canonical forgetful functor (a restriction of* U *in* 2.3*).*

3.3 Example. The rational-tree theory $\mathbf{RT}_{\Sigma_\perp}$ is an iteration theory (for the choice $e^{\dagger}=$ the least solution of e). In fact, as proved in [3], Theorem 6.5.2, this is a free iteration theory on Σ. In other words:

3.4 Theorem. (See [3].) *The forgetful functor* $V\colon \mathbf{ITh}\to\mathbf{Sgn}$ *is a right adjoint and the corresponding monad is the rational-tree monad* $\mathbb{R}\mathrm{at}$.

3.5 Theorem. *The forgetful functor* $V\colon \mathbf{ITh}\to\mathbf{Sgn}$ *is monadic; that means that the category of iteration theories and solution preserving theory morphisms is isomorphic to the category of algebras for the rational-tree monad* $\mathbb{R}\mathrm{at}$ *on* $\mathbf{Sgn}$.

Proof. We are going to use Beck's theorem; due to Theorem 3.4 it is sufficient to verify that $V\colon \mathbf{ITh}\to\mathbf{Sgn}$ creates coequalizers of V-split pairs. From the result of Bénabou mentioned in Remark 2.4(i) we know that $U\colon \mathbf{Th}\to\mathbf{Sgn}$ is monadic, thus, it creates coequalizers of U-split pairs. Consequently, our task is as follows: given a parallel pair of solution-preserving morphisms

$$u,v\colon(\mathscr{T},\ddagger)\to(\mathscr{S},\dagger)\qquad\text{in }\mathbf{ITh},$$

and given a split coequalizer

$$\mathscr{T}\;\underset{Iv}{\overset{Iu}{\rightrightarrows}}\;\overset{t}{\leftarrow}\;\mathscr{S}\;\underset{s}{\overset{c}{\rightleftarrows}}\;\mathscr{R}\qquad\text{in }\mathbf{Sgn},$$

where c is a coequalizer of u and v in $\mathbf{Th}$ and the above equations (i)–(iv) hold, then there exists a unique function $*\colon \mathscr{R}(n,n+p)\to\mathscr{R}(n,p)$ (for all $n,p\in\mathbb{N}$) such that

(a) c is solution preserving:

$$cf^\dagger = (cf)^* \qquad \text{for all } f \in \mathscr{S}(n, n+p), \tag{3.9}$$

(b) the axioms of Definition 3.1 hold for $*$, and
(c) c is a coequalizer of u and v in **ITh**.
In fact, (a) determines $*$ as follows:

$$e^* = c(se)^\dagger \qquad \text{for all } e \in \mathscr{R}(n, n+p). \tag{3.10}$$

To see this, put $f = se$, then $cf = e$ due to (ii), thus

$$e^* = (cf)^* = c(f^\dagger) = c(se)^\dagger.$$

Conversely, by using (3.10) we get (3.9) for every morphism $f \in \mathscr{S}(n, n+p)$

$$\begin{aligned}
(cf)^* &= c(scf)^\dagger && (3.10)\\
&= c(vtf)^\dagger && \text{(iv)}\\
&= cv(tf)^\ddagger && v \text{ in } \mathbf{ITh}\\
&= cu(tf)^\ddagger && \text{(i)}\\
&= c(utf)^\dagger && u \text{ in } \mathbf{ITh}\\
&= cf^\dagger && \text{(iii)}
\end{aligned}$$

We now prove the axioms of iteration theories for $*$ and then we will get immediately (c):

Suppose that $\bar{c}\colon (\mathscr{S}, \dagger) \to (\overline{\mathscr{R}}, @)$ is a morphism of **ITh** with $\bar{c}u = \bar{c}v$. We have a unique theory morphism $r\colon \mathscr{R} \to \overline{\mathscr{R}}$ with $\bar{c} = rc$ in **Th** and we only need to prove that

$$r(e^*) = (re)^@ \qquad \text{for all } e \in \mathscr{R}(n, n+p). \tag{3.11}$$

In fact, since $\bar{c}$ is a morphism of **ITh** we have by (3.10)

$$r(e^*) = rc(se)^\dagger = \bar{c}(se)^\dagger = \big(\bar{c}s(e)\big)^@$$

and it remains to verify $\bar{c}s = r$. This follows from c being an epimorphism since from (iv) and (iii) we get

$$\bar{c}sc = \bar{c}vt = \bar{c}ut = \bar{c} = rc.$$

Consequently, the theorem will be proved by verifying the individual axioms of iteration theories for the function from (3.10) above. Observe that since c is a theory morphism, it preserves $\circ$, see (3.4):

$$c(f \circ g) = (cf) \circ (cg). \tag{3.12}$$

(1) **Fixed Point Identity.** Given $e \in \mathscr{R}(n, n+p)$, we have, since c preserves coproducts:

$$\begin{aligned} e^* &= c(se)^\dagger && (3.10) \\ &= c\Big([(se)^\dagger, id]\cdot se\Big) && (3.1) \\ &= \big[c(se)^\dagger, id\big]\cdot e && \text{(ii)} \\ &= [e^*, id]\cdot e && (3.10) \end{aligned}$$

(2) **Parameter Identity.** Given $e \in \mathscr{R}(n, n+p)$ and $h \in \mathscr{R}(p, q)$, then (ii) implies, since c preserves finite coproducts, the equality

$$h \bullet e = (id + h)\cdot e = c(sh \bullet se). \tag{3.13}$$

Therefore

$$\begin{aligned} (h \bullet e)^* &= c\big[s\cdot c(sh \bullet se)\big]^\dagger && (3.10) \text{ and } (3.13) \\ &= \big[csc(sh \bullet se)\big]^* && (3.9) \\ &= \big[c(sh \bullet se)\big]^* && \text{(ii)} \\ &= c(sh \bullet se)^\dagger && (3.9) \\ &= c\big(sh\cdot(se)^\dagger\big) && (3.3) \\ &= h\cdot c(se)^\dagger && \text{(ii)} \\ &= h\cdot e^* && (3.10). \end{aligned}$$

(3) **Simplified Composition Identity.** Given morphisms $g \in \mathscr{R}(m, n+p)$ and $f \in \mathscr{R}(n, m)$, we have

$$\begin{aligned} (f \circ g)^*\cdot f &= \Big[c\big((sf) \circ (sg)\big)\Big]^*\cdot f && \text{(ii) and } (3.12) \\ &= c\big[(sf) \circ (sg)\big]^\dagger\cdot f && (3.9) \\ &= c\Big(\big[(sf) \circ (sg)\big]^\dagger\cdot sf\Big) && \text{(ii)} \\ &= c(sg\cdot sf)^\dagger && (3.5) \\ &= \Big[c\big((sg)\cdot(sf)\big)\Big]^* && (3.9) \\ &= (g\cdot f)^* && \text{(ii)}. \end{aligned}$$

(4) **Double Dagger Identity.** Given $e \in \mathscr{R}(n, n+n+p)$, since $c(\triangledown) = \triangledown$ (recall that c preserves finite coproducts), we have

$$\begin{aligned} e^{**} &= \big(c(se)^\dagger\big)^* && (3.10) \\ &= c(se)^{\dagger\dagger} && (3.9) \\ &= c(\triangledown \circ se)^\dagger && (3.6) \\ &= (c\triangledown \circ cse)^* && (3.9) \text{ and } (3.12) \\ &= (\triangledown \circ e)^* && \text{(ii) and } c\triangledown = \triangledown. \end{aligned}$$

(5) **Commutative Identity.** Given $\varrho_i \in \mathscr{N}(m,m)$ and $f \in \mathscr{R}(n,n+p)$ then first observe

$$c(\widehat{sf}) = \hat{f}. \tag{3.14}$$

In fact, c preserves coproducts and thus it maps base morphisms (in_j, ∇, ϱ_j etc.) of $\mathscr{S}$ to the corresponding base morphisms of $\mathscr{R}$. Thus (3.14) follows from (3.7). Consequently:

$$\begin{aligned}
\hat{f}^* &= \left[c(\widehat{sf})\right]^* && (3.14)\\
&= c(\widehat{sf})^\dagger && (3.9)\\
&= c\big((\nabla \circ sf)^\dagger \cdot \nabla\big) && (3.8)\\
&= c(\nabla \circ sf)^\dagger \cdot \nabla && c\nabla = \nabla\\
&= \left[c(\nabla \circ sf)\right]^* \cdot \nabla && (3.9)\\
&= (\nabla \circ f)^* \cdot \nabla && \text{(ii) and (3.12)}
\end{aligned}$$

This completes the proof. □

4 Conclusions and Future Research

The goal of our paper was to prove that iteration theories of Stephen Bloom and Zoltan Ésik are monadic over the category **Sgn** of signatures. This provides the possibility of using the corresponding monad $\mathbb{R}\mathrm{at}$ (of rational tree signatures) as a means for defining iteration theories. More important is the way our result supports the result that iteration theories precisely sum up the "equational properties" that the dagger function, assigning to every equation morphism e its least solution $e^\dagger$, satisfies in all continuous theories. In fact, since $\mathbb{R}\mathrm{at}$ is the monad of free rational theories, see [11], and every free rational theory has a solution-preserving completion to a continuous theory, it is obvious that all continuous theories and all rational theories satisfy precisely the same equational laws for †. To make such statements precise, one can either define the concept of "equation over **Sgn**", or use instead finitary monads on **Sgn** (in analogy to the classical varieties over **Set**). We decided for the latter.

In the future we intend to study the analogous question where the base category is, in lieu of **Sgn**, the category of all finitary endofunctors of **Set**. We hope that the corresponding monadic algebras will turn out to be precisely the iteration theories that are parametrically uniform in the sense of Simpson and Plotkin [10].

References

1. Adámek, J., Milius, S., Velebil, J.: Elgot Algebras. Logical Methods in Computer Science 2(5:4), 1–31 (2006)
2. Bénabou, J.: Structures algébriques dans les catégories, Cah. Topol. Géom. Différ. Catég. 10, 1–126 (1968)

3. Bloom, S.L., Ésik, Z.: Iteration Theories. Springer, Heidelberg (1993)
4. Bloom, S.L., Ésik, Z.: Fixed-point operations on ccc's, Part I. Theoret. Comput. Sci. 155, 1–38 (1996)
5. Bloom, S.L., Ésik, Z.: The equational logic of fixed points. Theoret. Comput. Sci. 179, 1–60 (1997)
6. Ésik, Z.: Axiomatizing iteration categories. Acta Cybernetica 14, 65–82 (1999)
7. Ginali, S.: Regular trees and the free iterative theory. J. Comput. Syst. Sci. 18, 228–242 (1979)
8. Kelly, G.M., Power, J.: Adjunctions whose units are coequalizers and presentations of finitary enriched monads. J. Pure Appl. Algebra 89, 163–179 (1993)
9. MacLane, S.: Categories for the Working Mathematician, 2nd edn. Springer, Heidelberg (1998)
10. Simpson, A., Plotkin, G.: Complete axioms for categorical fixed-point operators. IEEE Symposium Logic in Computer Science, 30–41 (1998)
11. Wright, J.B., Thatcher, J.W., Wagner, E.G., Goguen, J.A.: Rational algebraic theories and fixed-point solutions. In: Proc. 17th IEEE Symposium on Foundations of Computing, Houston, Texas, pp. 147–158. IEEE Computer Society Press, Los Alamitos (1976)

Properties Complementary to Program Self-reference*

John Case and Samuel E. Moelius III

Department of Computer & Information Sciences
University of Delaware
103 Smith Hall
Newark, DE 19716
{case,moelius}@cis.udel.edu

Abstract. In computability theory, program self-reference is formalized by the *not-necessarily-constructive* form of Kleene's Recursion Theorem (**krt**). In a programming system in which **krt** holds, for any preassigned, algorithmic task, there exists a program that, in a sense, creates a copy of itself, and then performs that task on the self-copy. Herein, properties *complementary* to **krt** are considered. Of particular interest are those properties involving the implementation of *control structures*. One main result is that *no* property involving the implementation of *denotational* control structures is complementary to **krt**. This is in contrast to a result of Royer, which showed that implementation of **if-then-else** — a denotational control structure — *is* complementary to the *constructive* form of Kleene's Recursion Theorem. Examples of *non-denotational* control structures whose implementation *is* complementary to **krt** are then given. Some such control structures so nearly resemble denotational control structures that they might be called *quasi-denotational*.

Keywords: Computability Theory, Programming Language Semantics, Self-Reference.

1 Introduction

Let $\mathbb{N} = \{0, 1, 2, \ldots\}$, let $\langle \cdot, \cdot \rangle : \mathbb{N} \times \mathbb{N} \to \mathbb{N}$ be any 1-1, onto, computable function, and let ψ be an *effective programming system* (eps)[1] [24,18]. For all $p \in \mathbb{N}$, let $\psi_p = \psi(\langle p, \cdot \rangle)$. One can think of ψ as a *programming language*, and of ψ_p as the partial function (mapping $\mathbb{N}$ to $\mathbb{N}$) computed by the ψ-program with numerical name p. The *not-necessarily-constructive form of Kleene's Recursion Theorem (***krt***)* [24, page 214, problem 11-4] *holds in* $\psi \stackrel{\text{def}}{\Leftrightarrow}$

$$(\forall p)(\exists e)(\forall x)[\psi_e(x) = \psi_p(\langle e, x \rangle)]. \tag{1}$$

* This paper received support from NSF Grant CCR-0208616.

[1] That is, ψ is a partial computable function mapping $\mathbb{N}$ to $\mathbb{N}$ such that, for every partial computable function α mapping $\mathbb{N}$ to $\mathbb{N}$, $(\exists p \in \mathbb{N})[\psi(\langle p, \cdot \rangle) = \alpha]$.

L. Kučera and A. Kučera (Eds.): MFCS 2007, LNCS 4708, pp. 253–263, 2007.

(1) may be interpreted as follows. ψ-program p represents an arbitrary, preassigned, algorithmic task to perform with a self-copy; e represents a ψ-program that

1. creates a copy of itself, external to itself, and, *then*,
2. runs the preassigned task p on the pair consisting of this self-copy and e's input x.

The 'e' on the right-hand side of the equation in (1) *is* the self-copy of the original 'e' on the left-hand side of this equation. The way in which e uses this self-copy is according to how the preassigned task p says to. Thus, in an important sense, e creates a *usable self-model*; it has usable self-knowledge [8].

The form of *recursion* provided by **krt** is *more general* than that built into most programming languages, i.e., that treated by denotational semantics [20,19,28]. The 'e' on the right-hand side of the equation in (1) is e's own syntactic code-script, wiring/flow diagram, etc. Thus, e's I/O behavior can depend, not *only* upon e's *denotational* characteristics,[2] but upon e's *connotational* characteristics (e.g., e's number of symbols or run-time complexity) as well [25].

Self-referential programs were first introduced by Kleene in [16], where he used such programs to prove properties of ordinal notations. His theorem, and its variants [27,24,7],[3] have since found widespread application. Interesting examples can be found in [3,4,24,8,26,13,5].[4]

Our ultimate goal is to characterize insightfully the power of program self-reference, as formalized by **krt**.[5] In this paper, we examine the subject somewhat indirectly, by studying properties *complementary* to **krt**, as described below.

An eps ψ is *acceptable* $\stackrel{\text{def}}{\Leftrightarrow}$ $(\forall \text{ eps } \xi)(\exists \text{ computable } t : \mathbb{N} \to \mathbb{N})\ (\forall p)[\psi_{t(p)} = \xi_p]$ [23,24,18,21,22,25]. Thus, the acceptable epses are exactly those epses into which every eps can be compiled. Any eps corresponding to a real-world, general purpose programming language (e.g., C++, Java, Haskell) is acceptable. A characterization of the acceptable epses due to the first author[6] is that they are exactly those epses having an implementation of *every* control structure (see Definitions 1 (§3.1) and 4 (§4.1), herein).[7]

[2] Since ψ is, itself, a partial computable function mapping $\mathbb{N}$ to $\mathbb{N}$, there exists a *universal simulator* or *self-interpreter* for ψ, i.e., a u such that $\psi_u = \psi$ [2,15]. e *can* inquire about its own denotational characteristics by running u on e, i.e., for all $x \in \mathbb{N}$, $\psi_u(\langle e, x\rangle) = \psi(\langle e, x\rangle) = \psi_e(x)$.

[3] Rogers' *Fixed-Point Recursion Theorem* (**fprt**) is a variant of **krt** that is often attributed to Kleene [24, page 180]. For an eps ψ, **fprt** *holds in* ψ $\stackrel{\text{def}}{\Leftrightarrow}$ for all computable $f : \mathbb{N} \to \mathbb{N}$, $(\exists e)[\psi_e = \psi_{f(e)}]$. **fprt** and **krt** should *not* be confused, however, as **fprt** is *strictly weaker* than **krt** [21, Theorems 5.1 and 5.3].

[4] In the work of Bongard, *et al.* [6], robots employ self-modeling to recover locomotion after injury. See also [1,12].

[5] For recent work in this direction, see [9].

[6] This result appears in [21,22].

[7] The term *control structure* is given a formal definition in the literature [21,22,25], and includes more than what is indicated in Definitions 1 (§3.1) and 4 (§4.1). The first author's result holds for all control structures covered by this formal definition.

In [25], Royer showed that **if-then-else** (Example 1 in §3.1, herein) and the *constructive* form of Kleene's Recursion Theorem (**KRT**) (equation (6) in §3.2, herein) are *complementary*, in the sense that: for each, there is an eps having that one, and *not* the other; *but*, any eps having *both* is acceptable [25, Theorem 4.1.12] (Theorem 2 in §3.2, herein).[8] Given the first author's characterization of the acceptable epses, one way of interpreting Royer's result is: **if-then-else** and **KRT** are independent notions that, together, yield all control structures.

The proof of Royer's result employs, quite essentially, the *constructivity* of **KRT**. Many other *similar* results concerning **KRT** (not described herein) have the same undesirable quality; that is, the constructivity of **KRT** is *all mixed up* with the program self-modeling.

krt is the focus of the present paper, as it embodies *pure* self-modeling with*out* the *additional* constructivity of **KRT**. Specifically, the interest herein is in properties complementary to **krt**, similar to the way in which implementation of **if-then-else** is complementary to **KRT** (see Definition 2 in §3.3, herein).

One main result (Corollary 1 in §3.3, herein) is that **krt** is *not* complementary to the implementation of *any class of denotational control structures* [25,28] (Definition 1 in §3.1, herein) — a type of control structure that includes **if-then-else**. This says, in part, that the constructivity of **KRT** is *essential* to establishing Royer's result.

Despite this outcome, there *do* exist reasonable *non*-denotational control structures whose implementation *is* complementary to **krt**. We give examples of such control structures in Section 4. Some such control structures so nearly resemble denotational control structures that they might be called *quasi-denotational*.

Section 2, just below, covers notation and preliminaries.

Due to space constraints, nearly all proofs are omitted. Complete proofs of all theorems can be found in [10].

2 Notation and Preliminaries

Computability-theoretic concepts not explained below are treated in [24].

Lowercase Roman letters, with or without decorations, range over elements of $\mathbb{N}$, unless stated otherwise. Uppercase Roman letters, with or without decorations, range over subsets of $\mathbb{N}$, unless stated otherwise.

The pairing function $\langle\cdot,\cdot\rangle$ was introduced in Section 1. For all x, $\langle x\rangle \stackrel{\text{def}}{=} x$. For all $x_1,\ldots,x_n$, where $n > 2$, $\langle x_1,\ldots,x_n\rangle \stackrel{\text{def}}{=} \langle x_1, \langle x_2,\ldots,x_n\rangle\rangle$. For all n, all $i \in \{1,\ldots,n\}$, and all $x_1,\ldots,x_n$, $\pi_i^n(\langle x_1,\ldots,x_n\rangle) \stackrel{\text{def}}{=} x_i$.

$\mathcal{P}$ denotes the collection of all partial functions mapping $\mathbb{N}$ to $\mathbb{N}$. α, β, γ, δ, ξ, and ψ, with or without decorations, range over elements of $\mathcal{P}$. For all α and p, $\alpha_p \stackrel{\text{def}}{=} \alpha(\langle p,\cdot\rangle)$. We use Church's lambda-notation [24] to name partial functions, including total functions and predicates, as is standard in many programming languages.[9]

[8] Note that, in [25], **if-then-else** is called **conditional**.

[9] For example, $\lambda x\centerdot(x+1)$ denotes the successor function.

For all α and x, $\alpha(x)\downarrow$ denotes that $\alpha(x)$ converges; $\alpha(x)\uparrow$ denotes that $\alpha(x)$ diverges.[10] We identify a partial function with its graph, e.g., we identify α with the set $\{(x,y) : \alpha(x) = y\}$. We use $\uparrow$ to denote the value of a divergent computation.

As per footnote 1, ψ is an eps $\overset{\text{def}}{\Leftrightarrow}$ ψ is partial computable, *and*, $(\forall$ partial computable $\alpha)(\exists p)[\psi_p = \alpha]$ [24,18]. $\mathcal{EPS}$ denotes the collection of all epses. φ denotes a fixed, acceptable eps [23,24,18,21,22,25].

Intuitively, a mapping $\Gamma : \mathbb{N}^m \times \mathcal{P}^n \to \mathcal{P}$, where $m + n > 0$, is a *computable operator* iff there exists an algorithm for listing the *graph of* the partial function $\Gamma(x_1,\ldots,x_m,\alpha_1,\ldots,\alpha_n)$ from $x_1,\ldots,x_m$ and listings of the *graphs of* the partial functions $\alpha_1,\ldots,\alpha_n$ — independently of the enumeration order chosen for each of $\alpha_1,\ldots,\alpha_n$ [24, §9.8]. Let $\Gamma_0,\Gamma_1,\ldots$ be any standard, algorithmic enumeration of the computable operators of all types $\mathbb{N}^m \times \mathcal{P}^n \to \mathcal{P}$, where $m+n > 0$.[11] Let $\Theta_0,\Theta_1,\ldots$ be a similar enumeration of the computable operators of all types $\mathbb{N}^m \times \mathcal{P}^{n+1} \to \mathcal{P}$, where $m + n > 0$. The Θ_i are those computable operators of a type suitable to determine a recursive denotational control structure (Definition 1(b) in §3.1, herein). For ease of presentation, we shall use the Θ_i exclusively for the purpose of describing recursive denotational control structures. The Γ_i, being of (more or less) arbitrary type, will be used to describe *non*recursive (denotational and *non*-denotational) control structures (Definition 1(a) in §3.1, and Definition 4 in §4.1, herein, respectively).

Let $\mu : \mathbb{N} \to \mathbb{N}$ be a computable function such that, for all i, $\Gamma_{\mu(i)}$ is the *least fixed point* of Θ_i w.r.t. the last argument of Θ_i [24, §11.5] (see also [20,19,28]). That is, if i, m, and n are such that $\Theta_i : \mathbb{N}^m \times \mathcal{P}^{n+1} \to \mathcal{P}$, then $\Gamma_{\mu(i)} : \mathbb{N}^m \times \mathcal{P}^n \to \mathcal{P}$ is such that, for all $x_1,\ldots,x_m$, $\alpha_1,\ldots,\alpha_n$, and β, $\Gamma_{\mu(i)}(x_1,\ldots,x_m,\alpha_1,\ldots,\alpha_n) = \beta$ implies (i) and (ii) below.

(i) $\Theta_i(x_1,\ldots,x_m,\alpha_1,\ldots,\alpha_n,\beta) = \beta$.
(ii) $(\forall\gamma)[\Theta_i(x_1,\ldots,x_m,\alpha_1,\ldots,\alpha_n,\gamma) = \gamma \Rightarrow \beta \subseteq \gamma]$.

The *non*recursive denotational control structure determined by $\Gamma_{\mu(i)}$ is the recursive denotational control structure determined by Θ_i under *least fixed point semantics* [20,19,25,28] (see Definition 1 in §3.1, herein).

In several places, we make reference to the following result.

Theorem 1 (Machtey, *et al.* [17, Theorem 3.2]). For all epses ψ, ψ is acceptable $\Leftrightarrow$ $(\exists$ computable $f : \mathbb{N} \to \mathbb{N})(\forall a,b)[\psi_{f(\langle a,b\rangle)} = \psi_a \circ \psi_b]$.

3 Denotational Control Structures and krt

In this section, we show that there is *no* class of denotational control structures whose implementation is complementary to **krt** (Corollary 1). We begin

[10] For all α and x, $\alpha(x)$ *converges* iff there exists y such that $\alpha(x) = y$; $\alpha(x)$ *diverges* iff there is *no* y such that $\alpha(x) = y$. If α is partial computable, and x is such that $\alpha(x)$ diverges, then one can imagine that a program associated with α *goes into an infinite loop* on input x.

[11] A formal definition of the computable operators (called *recursive operators* in [24]) as well as a construction of the Γ_i's can be found in [24, §9.8].

with a brief introduction to denotational control structures in the context of epses. Then, we formally state Royer's result (Theorem 2), that implementation of **if-then-else** (Example 1, below) — a denotational control structure — *is* complementary to the *constructive* form of Kleene's Recursion Theorem (**KRT**) (equation (6), below).

3.1. In the context of epses, an instance of a control structure [21,22,25,14,11] is a means of forming a composite program from given constituent programs and/or data. An instance of a *denotational* control structure, more specifically, is one for which the I/O behavior of a composite program can depend *only* upon the I/O behavior of the constituent programs and upon the data. So, for example, the I/O behavior of such a composite program can*not* depend upon the connotational characteristics of its constituent programs, e.g., their number of symbols or run-time complexity.

Recursive denotational control structures differ from *non*recursive ones in that, for the former, the composite program is, in a sense, one of the constituent programs. For such a control structure, the I/O behavior of a composite program can*not* depend upon the connotational characteristics of the composite program, itself, just as it can*not* depend upon those of the other constituent programs.

In the following definition, $x_1, \ldots, x_m$ represent data, $x_{m+1}, \ldots, x_{m+n}$ represent constituent programs, and $f(\langle x_1, \ldots, x_{m+n} \rangle)$ represents the composite program formed from $x_1, \ldots, x_{m+n}$.

Definition 1. For all epses ψ, and all $f : \mathbb{N} \to \mathbb{N}$, (a) and (b) below.

(a) Suppose i, m, and n are such that $\Gamma_i : \mathbb{N}^m \times \mathcal{P}^n \to \mathcal{P}$. Then, *$f$ is effective instance in ψ of the* non*recursive denotational control structure determined by Γ_i* $\Leftrightarrow$ f is computable and, for all $x_1, \ldots, x_{m+n}$, $\psi_{f(\langle x_1, \ldots, x_{m+n} \rangle)} = \Gamma_i(x_1, \ldots, x_m, \psi_{x_{m+1}}, \ldots, \psi_{x_{m+n}})$.

(b) Suppose i, m, and n are such that $\Theta_i : \mathbb{N}^m \times \mathcal{P}^{n+1} \to \mathcal{P}$. Then, *$f$ is an effective instance in ψ of the recursive denotational control structure determined by Θ_i* $\Leftrightarrow$ f is computable and, for all $x_1, \ldots, x_{m+n}$, $\psi_{f(\langle x_1, \ldots, x_{m+n} \rangle)} = \Theta_i(x_1, \ldots, x_m, \psi_{x_{m+1}}, \ldots, \psi_{x_{m+n}}, \psi_{f(\langle x_1, \ldots, x_{m+n} \rangle)})$.

For the remainder of the present subsection (3.1), let ψ be any fixed eps.

Example 1. Choose i_{ite} such that $\Gamma_{i_{\text{ite}}} : \mathcal{P}^3 \to \mathcal{P}$, and, for all α, β, and γ,

$$\Gamma_{i_{\text{ite}}}(\alpha, \beta, \gamma)(x) = \begin{cases} \beta(x), & \text{if } [\alpha(x)\downarrow \wedge \alpha(x) > 0]; \\ \gamma(x), & \text{if } [\alpha(x)\downarrow \wedge \alpha(x) = 0]; \\ \uparrow, & \text{otherwise.}^{12} \end{cases} \tag{2}$$

Then, the *nonrecursive denotational control structure determined by $\Gamma_{i_{\text{ite}}}$* is **if-then-else** [21,22]. Furthermore, an effective instance in ψ of **if-then-else** is any computable function $f : \mathbb{N} \to \mathbb{N}$, such that, for all a, b, c, and x,

$$\psi_{f(\langle a,b,c \rangle)}(x) = \begin{cases} \psi_b(x), & \text{if } [\psi_a(x)\downarrow \wedge \psi_a(x) > 0]; \\ \psi_c(x), & \text{if } [\psi_a(x)\downarrow \wedge \psi_a(x) = 0]; \\ \uparrow, & \text{otherwise.} \end{cases} \tag{3}$$

[12] Note that the choice of i_{ite} is *not* unique.

Example 2. Choose $i'_{\rm ite}$ such that $\Theta_{i'_{\rm ite}} : \mathcal{P}^4 \to \mathcal{P}$, and, for all α, β, γ, and δ,

$$\Theta_{i'_{\rm ite}}(\alpha,\beta,\gamma,\delta)(x) = \begin{cases} \beta(x), \text{ if } & [\alpha(x)\downarrow \wedge \alpha(x) > 0] \\ & \vee [\ \beta(x)\downarrow \wedge \gamma(x)\downarrow \wedge \delta(x)\downarrow \\ & \quad \wedge \beta(x) = \gamma(x) = \delta(x) \quad]; \\ \gamma(x), \text{ if } [\alpha(x)\downarrow \wedge \alpha(x) = 0]; & \\ \uparrow, \quad \text{otherwise.} & \end{cases} \tag{4}$$

Then, an f as in (3) above is an effective instance in ψ of the recursive denotational control structure determined by $\Theta_{i'_{\rm ite}}$. Furthermore, if $g : \mathbb{N} \to \mathbb{N}$ is a computable function such that, for all a, b, c, and x,

$$\psi_{g(\langle a,b,c\rangle)}(x) = \begin{cases} \psi_b(x), \text{ if } & [\psi_a(x)\downarrow \wedge \psi_a(x) > 0] \\ & \vee [\psi_b(x)\downarrow \wedge \psi_c(x)\downarrow \wedge \psi_b(x) = \psi_c(x)]; \\ \psi_c(x), \text{ if } [\psi_a(x)\downarrow \wedge \psi_a(x) = 0]; & \\ \uparrow, \quad \text{otherwise;} & \end{cases} \tag{5}$$

then, g is *also* such an effective instance.[13]

Intuitively, in Example 2, g is an instance of a *hasty* variant of **if-then-else**. That is, $g(\langle a,b,c\rangle)$ runs ψ-programs a, b, and c, on x, in parallel. If both b and c halt on x, and yield the same result, then $g(\langle a,b,c\rangle)$ does not wait to see if a halts on x.

As the preceding examples demonstrate, recursive denotational control structures are *more general* than *non*recursive ones, in that they allow greater versatility among their implementations.

Henceforth, when convenient, we will abbreviate *effective instance*, *nonrecursive denotational control structure*, and *recursive denotational control structure*, by ei, ndcs, and rdcs, respectively.

3.2. For any eps ψ, the *constructive form of Kleene's Recursion Theorem (***KRT***) holds in* $\psi \overset{\rm def}{\Leftrightarrow}$

$$(\exists \text{ computable } r : \mathbb{N} \to \mathbb{N})(\forall p, x)[\psi_{r(p)}(x) = \psi_p(\langle r(p), x\rangle)]. \tag{6}$$

In (6), $r(p)$ plays the role of the self-referential e in (1). Thus, in an eps in which **KRT** holds, self-referential programs can be found *algorithmically* from a program for the preassigned task.

The following is the formal statement of Royer's result, that implementation of **if-then-else** is complementary to **KRT**.

Theorem 2 (Royer [25, Theorem 4.1.12]). For all epses ψ, let $P(\psi) \Leftrightarrow$ there is an ei of **if-then-else** in ψ. Then, (a)-(c) below.

(a) $(\exists$ eps $\psi)[$**KRT** *holds* in $\psi \wedge \neg P(\psi)]$.
(b) $(\exists$ eps $\psi)[$**KRT** does *not* hold in $\psi \wedge P(\psi)]$.
(c) $(\forall$ eps $\psi)\big[[$**KRT** holds in $\psi \wedge P(\psi)] \Leftrightarrow \psi$ is acceptable$\big]$.

---

[13] It can be seen that $\Gamma_{i_{\rm ite}} = \Gamma_{\mu(i'_{\rm ite})}$. Thus, f provides a minimal fixed-point solution of (4); whereas, g provides a *non*-minimal fixed-point solution of (4) [20,19,25,28].

3.3. The following definition makes precise what it means for a *property* of an eps to be complementary to **krt**.[14]

Definition 2. For all $P \subseteq \mathcal{EPS}$, *P is complementary to* **krt** $\Leftrightarrow$ (a)-(c) below.

(a) $(\exists$ eps $\psi)[$**krt** *holds* in $\psi \wedge \neg P(\psi)]$.
(b) $(\exists$ eps $\psi)[$**krt** does *not* hold in $\psi \wedge P(\psi)]$.
(c) $(\forall$ eps $\psi)\big[[$**krt** holds in $\psi \wedge P(\psi)] \Leftrightarrow \psi$ is acceptable$\big]$.

The following definition introduces notions used throughout the remainder of this section.

Definition 3. For all I, (a) and (b) below.

(a) $\{\Gamma_i : i \in I\}$ is *nonrecursively denotationally omnipotent* $\Leftrightarrow$ $(\forall$ eps $\psi)$ $[(\forall i \in I)[$there is an ei in ψ of the ndcs determined by $\Gamma_i] \Rightarrow \psi$ is acceptable$]$.
(b) $\{\Theta_i : i \in I\}$ is *recursively denotationally omnipotent* $\Leftrightarrow$ $(\forall$ eps $\psi)$ $[(\forall i \in I)[$there is an ei in ψ of the rdcs determined by $\Theta_i] \Rightarrow \psi$ is acceptable$]$.

Thus, a class of recursive operators is *non*recursively denotationally omnipotent iff it is *so powerful* that: having an effective instance of each *non*recursive denotational control structure that it determines causes an eps to be acceptable. Recursively denotationally omnipotent can be interpreted similarly.

Example 3. Choose i_{comp} such that $\Theta_{i_{\text{comp}}} : \mathcal{P}^3 \rightarrow \mathcal{P}$, and, for all α, β, and γ, $\Theta_{i_{\text{comp}}}(\alpha, \beta, \gamma) = \alpha \circ \beta$. Then, the recursive denotational control structure determined by $\Theta_{i_{\text{comp}}}$ is ordinary composition, and, by Theorem 1, the class consisting of just $\{\Theta_{i_{\text{comp}}}\}$ is recursively denotationally omnipotent.

Corollary 1, our main result of this section, follows from the next theorem.

Theorem 3. Let I be such that $\{\Theta_i : i \in I\}$ is *not* recursively denotationally omnipotent. Then, there exists an eps ψ such that (a)-(c) below.

(a) **krt** holds in ψ.
(b) For each $i \in I$, there is an ei in ψ of the ndcs determined by $\Gamma_{\mu(i)}$.
(c) ψ is *not* acceptable.

The proof of Theorem 3, a finite injury priority argument, is omitted due to space constraints.[15]

Corollary 1. There is *no* I such that $\lambda\psi \in \mathcal{EPS}.(\forall i \in I)[$there is an ei in ψ of the rdcs determined by $\Theta_i]$ is complementary to **krt**.

---

[14] It is relatively straightforward to show that *no* single property *characterizes* the complement of **krt**, e.g., there exist $P \subseteq \mathcal{EPS}$ and $Q \subseteq \mathcal{EPS}$ such that both P and Q are complementary to **krt**, but $P \not\subseteq Q$ and $Q \not\subseteq P$.

[15] Rogers [24] explains priority arguments. We should also mention that our proof of Theorem 3 makes essential use of Royer's [25, Theorem 4.2.15].

Proof of Corollary. Suppose, by way of contradiction, that such an I exists.
CASE $\{\Theta_i : i \in I\}$ is recursively denotationally omnipotent. Then, clearly, the stated property does *not* satisfy Definition 2(b) — a contradiction.
CASE $\{\Theta_i : i \in I\}$ is *not* recursively denotationally omnipotent. Then, clearly, by Theorem 3, the stated property does *not* satisfy (⇒) of Definition 2(c) — a contradiction. □ (*Corollary 1*)

Remark 1. The statement of Theorem 3 is slightly stronger than needed. To establish Corollary 1, it would suffice that, for each $i \in I$, there exist an ei in ψ of the rdcs determined by Θ_i. As stated, the theorem has the following additional corollary, which is of some independent interest.

Corollary 2. For all I, $\{\Theta_i : i \in I\}$ is recursively denotationally omnipotent ⇔ $\{\Gamma_{\mu(i)} : i \in I\}$ is *non*recursively denotationally omnipotent.

Proof of Corollary. (⇒) Immediate. (⇐) Let I be such that $\{\Theta_i : i \in I\}$ is *not* recursively denotationally omnipotent. Then, by Theorem 3, there exists a *non*-acceptable eps ψ such that, for each $i \in I$, there is an ei in ψ of the ndcs determined by $\Gamma_{\mu(i)}$. Thus, $\{\Gamma_{\mu(i)} : i \in I\}$ is is *not* *non*recursively denotationally omnipotent. □ (*Corollary 2*)

4 Control Structures Complementary to krt

In this section, we give examples of control structures whose implementation *is* complementary to **krt**. Each of our examples is drawn from a class of control structures that we call *coded composition* (**CC**). Although, the control structures in this class can, in general, be *non*-denotational, they are still quite reasonable, in that they look much like control structures with which one could actually program.

Note that Definition 2, in the preceding section, formalized what it means for a property of an eps to be complementary to **krt**.

4.1. The following definition introduces the notion of *non*recursive control structures, generally.

Definition 4. Suppose i and m are such that $\Gamma_i : \mathbb{N}^m \times \mathcal{P} \to \mathcal{P}$. Then, for all epses ψ, and all $f : \mathbb{N} \to \mathbb{N}$, *$f$ is an effective instance in ψ of the* non*recursive control structure determined by Γ_i* ⇔ f is computable and, for all $x_1, \ldots, x_m$, $\psi_{f(\langle x_1,\ldots,x_m \rangle)} = \Gamma_i(x_1, \ldots, x_m, \psi)$.

Henceforth, we will abbreviate *nonrecursive control structure* by ncs.

Definition 5

(a) Suppose that $(f^{\mathrm{L}}, g^{\mathrm{L}}, f^{\mathrm{R}}, g^{\mathrm{R}}) : (\mathbb{N} \to \mathbb{N})^4$ is such that (i)-(iii) below.
  (i) Each of f^{L}, g^{L}, f^{R}, and g^{R} is computable.
  (ii) For all a, f^{L}_a and f^{R}_a are onto.[16]
  (iii) For all a, g^{L}_a and g^{R}_a are 1-1.

[16] Recall that $f_a = f(\langle a, \cdot \rangle)$.

Then, $(f^{\rm L}, g^{\rm L}, f^{\rm R}, g^{\rm R})$-**CC** : $\mathbb{N}^2 \times \mathcal{P} \to \mathcal{P}$ is the computable operator, such that, for all a, b, and ψ,

$$(f^{\rm L}, g^{\rm L}, f^{\rm R}, g^{\rm R})\text{-}\mathbf{CC}(a, b, \psi) = f^{\rm L}_a \circ \psi_a \circ g^{\rm L}_a \circ f^{\rm R}_b \circ \psi_b \circ g^{\rm R}_b. \qquad (7)$$

(b) Suppose that $(f^{\rm L}, g^{\rm L}, f^{\rm R}, g^{\rm R}) : (\mathbb{N} \to \mathbb{N})^4$ is as in (a) above. Then, for all epses ψ, $(f^{\rm L}, g^{\rm L}, f^{\rm R}, g^{\rm R})$-**CC** *holds in* $\psi \Leftrightarrow$ there is an ei in ψ of the ncs determined by $(f^{\rm L}, g^{\rm L}, f^{\rm R}, g^{\rm R})$-**CC**.

(c) For all epses ψ, **CC** *holds in* $\psi \Leftrightarrow$ there exists $(f^{\rm L}, g^{\rm L}, f^{\rm R}, g^{\rm R}) : (\mathbb{N} \to \mathbb{N})^4$ as in (a) above such that $(f^{\rm L}, g^{\rm L}, f^{\rm R}, g^{\rm R})$-**CC** holds in ψ.

(L and R are mnemonic for *left* and *right*, respectively. **CC** is mnemonic for *coded composition*.) Thus, if ψ is an eps, and $\diamond : \mathbb{N} \to \mathbb{N}$ is an ei in ψ of the ncs determined by $(f^{\rm L}, g^{\rm L}, f^{\rm R}, g^{\rm R})$-**CC**, then, for all a and b,

$$\psi_{a \diamond b} = f^{\rm L}_a \circ \psi_a \circ g^{\rm L}_a \circ f^{\rm R}_b \circ \psi_b \circ g^{\rm R}_b, \qquad (8)$$

where, in (8), $\diamond$ is written using infix notation.

CC may be thought of as a *collection of* control structures, one for each choice of $(f^{\rm L}, g^{\rm L}, f^{\rm R}, g^{\rm R})$. As the next theorem shows, the property of having an effective instance of *some* control structure in this collection, is complementary to **krt**.

Theorem 4. $\lambda\psi \in \mathcal{EPS}.$[**CC** holds in ψ] is complementary to **krt**.

The proof of Theorem 4 is omitted due to space constraints. The proof employs a trick similar to that used in the proof of Theorem 1 to show that: if ψ is an eps in which both **krt** and **CC** hold, then ψ is acceptable. Intuitively, if **krt** holds in ψ, then ψ-program b as in (8) can *know* its own program number. Thus, b can *decode* its input as *encoded* by the 1-1 function $g^{\rm R}_b$. Similarly, b can *pre-encode* its output, so that the onto function $f^{\rm R}_b$ sends this output to the value that b would *actually* like to produce. The situation is similar for ψ-program a.[17]

4.2. As mentioned above, **CC** may be thought of as a *collection of* control structures, one for each choice of $(f^{\rm L}, g^{\rm L}, f^{\rm R}, g^{\rm R})$. However, it is *not* the case that, for *each* choice of $(f^{\rm L}, g^{\rm L}, f^{\rm R}, g^{\rm R})$, the property $\lambda\psi \in \mathcal{EPS}.[(f^{\rm L}, g^{\rm L}, f^{\rm R}, g^{\rm R})$-**CC** holds in ψ] is complementary to **krt**.[18] An obvious question is: *which* choices of $(f^{\rm L}, g^{\rm L}, f^{\rm R}, g^{\rm R})$ yield properties complementary to **krt**, and *which* do *not*? As the following, somewhat curious result shows, the answer is intimately tied to the choice of $f^{\rm R}$, specifically.

[17] The details of the proof, however, are rather involved. One complication arises from the fact that, if $\diamond$ is as in (8), then the $f^{\rm L}$, $g^{\rm L}$, $f^{\rm R}$, and $g^{\rm R}$ *stack up* when one tries to iterate $\diamond$. This can be seen, for example, from the underlined terms in the following calculation.

$$\begin{aligned}\psi_{a\diamond(b\diamond c)} &= f^{\rm L}_a \circ \psi_a \circ g^{\rm L}_a \circ f^{\rm R}_{b\diamond c} \circ \psi_{b\diamond c} \circ g^{\rm R}_{b\diamond c}\\ &= f^{\rm L}_a \circ \psi_a \circ g^{\rm L}_a \circ \underline{f^{\rm R}_{b\diamond c}} \circ f^{\rm L}_b \circ \psi_b \circ g^{\rm L}_b \circ f^{\rm R}_c \circ \psi_c \circ g^{\rm R}_c \circ \underline{g^{\rm R}_{b\diamond c}}.\end{aligned} \qquad (9)$$

[18] For example, the control structure determined by $(\pi^2_2, \pi^2_2, \pi^2_2, \pi^2_2)$-**CC** is ordinary composition, which, by Theorem 1, causes an eps to be acceptable.

Theorem 5

(a) There exists a computable $f^{\mathrm{R}} : \mathbb{N} \to \mathbb{N}$ and an eps ψ such that

$$(\forall a)(\exists y)(\forall x)\left[f_a^{\mathrm{R}}(x) = \begin{cases} y, \text{ if } x = 0; \\ 0, \text{ if } x = y; \\ x, \text{ otherwise} \end{cases}\right]; \tag{10}$$

$(\pi_2^2, \pi_2^2, f^{\mathrm{R}}, \pi_2^2)$-**CC** holds in ψ, and ψ *not* acceptable.[19]

(b) Suppose that $(f^{\mathrm{L}}, g^{\mathrm{L}}, f^{\mathrm{R}}, g^{\mathrm{R}}) : (\mathbb{N} \to \mathbb{N})^4$ is as in Definition 5(a), and that $f^{\mathrm{R}} = \pi_2^2$. Then, any eps in which $(f^{\mathrm{L}}, g^{\mathrm{L}}, f^{\mathrm{R}}, g^{\mathrm{R}})$-**CC** holds is acceptable.

The proof of Theorem 5 is omitted due to space constraints.[20]

Note that, in Theorem 5(a), for all a, f_a^{R} is a recursive permutation that acts like the identity on all but at most two values. Thus, the *non*recursive control structure determined by $(\pi_2^2, \pi_2^2, f^{\mathrm{R}}, \pi_2^2)$-**CC** is nearly identical to ordinary composition, i.e., $(\pi_2^2, \pi_2^2, \pi_2^2, \pi_2^2)$-**CC**. Such a control structure so nearly resembles a *denotational* control structure that it might be called *quasi-denotational*.[21]

Acknowledgments. We would like to thank several anonymous referees for their useful comments.

References

1. Adami, C.: What do robots dream of? Science 314, 1093–1094 (2006)
2. Amtoft, T., Nikolajsen, T., Träff, J.L., Jones, N.: Experiments with implementations of two theoretical constructions. In: Meyer, A.R., Taitslin, M.A. (eds.) Logic at Botik 1989. LNCS, vol. 363, pp. 119–133. Springer, Heidelberg (1989)
3. Blum, M.: A machine independent theory of the complexity of recursive functions. Journal of the ACM 14, 322–336 (1967)
4. Blum, M.: On the size of machines. Information and Control 11, 257–265 (1967)
5. Bonfante, G., Kaczmarek, M., Marion, J.-Y.: A classification of viruses through recursion theorems. In: Computation and Logic in the Real World - Third Conference of Computability in Europe (CiE 2007). LNCS, vol. 4497, Springer, Berlin (2007)
6. Bongard, J., Zykov, V., Lipson, H.: Resilient machines through continuous self-modeling. Science 314, 1118–1121 (2006)
7. Case, J.: Periodicity in generations of automata. Mathematical Systems Theory 8, 15–32 (1974)
8. Case, J.: Infinitary self-reference in learning theory. Journal of Experimental and Theoretical Artificial Intelligence 6, 3–16 (1994)

[19] Thus, by Theorem 4 (and (⇒) of Definition 2(c)), **krt** does *not* hold in ψ.

[20] The eps used in the proof of Theorem 5(a) is that constructed in [23, page 333] and in [24, page 42, problem 2-11].

[21] We *do* know of control structures that are *not* forms of coded composition and whose implementation is complementary to **krt**, but *none* so nearly resemble denotational control structures.

9. Case, J., Moelius III, S.E.: Characterizing programming systems allowing program self-reference. In: Computation and Logic in the Real World - Third Conference of Computability in Europe (CiE 2007). LNCS, vol. 4497, pp. 125–134. Springer, Berlin (2007)
10. Case, J., Moelius III, S.E.: Properties complementary to program self-reference (expanded version). Technical report, University of Delaware (2007), Available at `http://www.cis.udel.edu/~moelius/publications`
11. Case, J., Jain, S., Suraj, M.: Control structures in hypothesis spaces: The influence on learning. Theoretical Computer Science 270(1-2), 287–308 (2002)
12. Conduit, R.: To sleep, perchance to dream. Science, A letter, including responses from Adami, C., Lipson, H., Zykov, V., Bongard, J. 315(5816), 1219–1220 (2007)
13. Friedman, H.: [FOM] 305:Proofs of Godel's Second. Communication to the Foundations of Mathematics electronic mailing list (December 21, 2006)
14. Jain, S., Nessel, J.: Some independence results for control structures in complete numberings. Journal of Symbolic Logic 66(1), 357–382 (2001)
15. Jones, N.: Computer implementation and applications of Kleene's s-m-n and recursion theorems. In: Moschovakis, Y. (ed.) Logic From Computer Science. Mathematical Science Research Institute Publications, vol. 21, pp. 243–263. Springer, Heidelberg (1992)
16. Kleene, S.C.: On notation for ordinal numbers. Journal of Symbolic Logic 3, 150–155 (1938)
17. Machtey, M., Winklmann, K., Young, P.: Simple Gödel numberings, isomorphisms, and programming properties. SIAM Journal on Computing 7, 39–60 (1978)
18. Machtey, M., Young, P.: An Introduction to the General Theory of Algorithms. North Holland, New York (1978)
19. Manna, Z.: Mathematical theory of computation. MacGraw-Hill, 1974. Reprinted, Dover (2003)
20. Manna, Z., Vuillemin, J.: Fixpoint approach to the theory of computation. Communications of the ACM 15(7), 528–536 (1972)
21. Riccardi, G.: The Independence of Control Structures in Abstract Programming Systems. PhD thesis, SUNY Buffalo (1980)
22. Riccardi, G.: The independence of control structures in abstract programming systems. Journal of Computer and System Sciences 22, 107–143 (1981)
23. Rogers, H.: Gödel numberings of partial recursive functions. Journal of Symbolic Logic, 23:331–341 (1958)
24. Rogers, H.: Theory of Recursive Functions and Effective Computability. McGraw Hill, New York, 1967. Reprinted, MIT Press, Cambridge (1987)
25. Royer, J.: A Connotational Theory of Program Structure. LNCS, vol. 273. Springer, Heidelberg (1987)
26. Royer, J., Case, J.: Subrecursive Programming Systems: Complexity and Succinctness. Progress in Theoretical Computer Science. Birkhäuser, Boston (1994)
27. Smullyan, R.: Theory of formal systems. Annals of Mathematics Studies (1961)
28. Winskel, G.: The Formal Semantics of Programming Languages: An Introduction. Foundations of Computing Series. MIT Press, Cambridge (1993)

Dobrushin Conditions for Systematic Scan with Block Dynamics*

Kasper Pedersen

Department of Computer Science, University of Liverpool, UK
k.pedersen@csc.liv.ac.uk

Abstract. We study the mixing time of systematic scan Markov chains on finite spin systems. It is known that, in a single site setting, the mixing time of systematic scan can be bounded in terms of the influences sites have on each other. We generalise this technique for bounding the mixing time of systematic scan to block dynamics, a setting in which a set of sites are updated simultaneously. In particular we present a parameter α, representing the maximum influence on any site, and show that if $\alpha < 1$ then the corresponding systematic scan Markov chain mixes rapidly. We use this method to prove $O(\log n)$ mixing of a systematic scan for proper q-colourings of a general graph with maximum vertex-degree Δ whenever $q \geq 2\Delta$. We also apply the method to improve the number of colours required in order to obtain mixing in $O(\log n)$ scans for a systematic scan colouring of trees.

1 Introduction

This paper is concerned with the study of finite *spin systems.* A spin system is composed of a set of sites and a set of spins, both of which will be finite throughout this paper. The interconnection between the sites is determined by an underlying graph. A configuration of the spin system is an assignment of a spin to each site. If there are n sites and q available spins then this gives rise to q^n configurations of the system, however some configurations may be illegal. The specification of the system determines how the spins interact with each other at a local level, such that different local configurations on a subset of the graph may have different relative likelihoods. This interaction hence specifies a probability distribution, π, on the set of configurations. One class of configurations that receive much attention in theoretical computer science is *proper q-colourings* of graphs. A proper colouring is a configuration where no two adjacent sites are assigned the same colour. One important example of a spin system is when the set of legal configurations is the set of all proper q-colourings of the underlying graph and π is the uniform distribution on this set. In statistical physics the spin system corresponding to proper q-colourings is known as the q-state anti-ferromagnetic Potts model at zero temperature.

* This work was partly funded by EPSRC projects GR/T07343/02 and GR/S76168/01. A longer version, with all proofs included, is available at http://www.csc.liv.ac.uk/~kasper

L. Kučera and A. Kučera (Eds.): MFCS 2007, LNCS 4708, pp. 264–275, 2007.

Sampling from π is a computationally challenging task. It is, however, an important one and is often carried out by simulating some suitable random *dynamics* on the set of configurations. Such a dynamics must have the following two properties

1. the dynamics eventually converges to π, and
2. the rate of convergence (*mixing time*) is polynomial in the number of sites.

It is generally straightforward to ensure that a dynamics converges to π but much harder provide good upper-bounds on the rate of convergence, which is what we will be concerned with in this paper.

Arguably the simplest dynamics is the heat-bath Glauber dynamics which, at each step, selects a site uniformly at random and updates the spin assigned to that site by drawing a new spin from the distribution on the spin of the selected site induced by π. This procedure is repeated until the distribution of the Markov chain is sufficiently close to π using some suitable measure of closeness between probability distributions. This dynamics falls under a family of Markov chains that we call *random update Markov chains*. We say that a Markov chain is a random update Markov chain if the sites are updated in a random order. This type of Markov chain has been frequently studied in theoretical computer science and much is known about the mixing time of various random update Markov chains.

An alternative to random update Markov chains is to construct a Markov chain that cycles through and updates the sites (or subsets of sites) in a deterministic order. We call this a *systematic scan Markov chain* (or systematic scan for short). Although systematic scan updates the sites in a deterministic order it remains a random process since the procedure used to update the spin assigned to a site is randomised, as specified by the appropriate induced distribution. Systematic scan may be more intuitively appealing that random update in terms of implementation, however until recently little was know about the convergence rates of this type of dynamics. It remains important to know how many steps one needs to simulate a systematic scan for in order to for it to become sufficiently close to its stationary distribution and recently there has been an interest among computer scientists in investigating various approaches for analysing the mixing time of systematic scan Markov chains, see e.g. Dyer, Goldberg and Jerrum [1,2] and Bordewich, Dyer and Karpinski [3]. In this paper we present a new method for analysing the mixing time of systematic scan Markov chains, which is applicable to any spin system. As applications of this method we improve the known parameters required for rapid mixing of systematic scan on (1) proper colourings of general graphs and (2) proper colourings of trees.

A key ingredient in our method for proving mixing of systematic scan is to work with a *block dynamics*. A block dynamics is a dynamics in which we allow a set of sites to be updated simultaneously as opposed to updating one site at a time as in the description of the Glauber dynamics above. Block dynamics is not a new concept and it was used in the mid 1980s by Dobrushin and Shlosman [4] in their study of conditions that imply uniqueness of the Gibbs measure of a spin system, a topic closely related to studying the mixing time of Markov chains (see for example Weitz's PhD thesis [5]). More recently, a block dynamics has been used by Weitz [6] when, in a generalisation of the work of Dobrushin and Shlosman, studying the relationship between various *influence parameters* (also in the context of Gibbs measures) within spin systems and using the influence parameters to establish conditions that imply mixing. Using an influence parameter to establish a condition which implies mixing of systematic scan is a key aspect of the method presented in this paper as we will discuss below. Dyer, Sinclair, Vigoda and Weitz [7] have also used a block dynamics in the context of analysing the mixing time of a Markov chain for proper colourings of the square lattice. Both of these papers consider a random update Markov chain, however several of ideas and techniques carry over to systematic scan as we shall see.

We will bound the mixing time of systematic scan by studying the *influence* that the sites of the graph have on each other. This technique is well-known and the influence parameters generalised by Weitz [6]: "the influence *on* a site is small" (originally attributed to Dobrushin [8]) and "the influence *of* a site is small" (originally Dobrushin and Shlosman [4]) both imply mixing of the corresponding random update Markov chain. It is worth pointing out that a condition of the form "if the influence *on* a site is small then the corresponding dynamics converges to π quickly" is known as a Dobrushin condition. In the context of systematic scan, Dyer et al. [1] point out that, in a single site setting, the condition "the influence *on* a site is small" implies rapid mixing of systematic scan. Our method for proving rapid mixing of systematic scan is a generalisation of this influence parameter to block dynamics.

We now formalise the concepts above and state our results. Let $C = \{1, \ldots, q\}$ be the set of spins and $G = (V, E)$ be the underlying graph of the spin system where $V = \{1, \ldots, n\}$ is the set of sites. We associate with each site $i \in V$ a positive weight w_i. Let Ω^+ be the set of all configurations of the spin system and $\Omega \subseteq \Omega^+$ be the set of all legal configurations. Then let π be a probability distribution on Ω^+ whose support is Ω i.e., $\{x \in \Omega^+ \mid \pi(x) > 0\} = \Omega$. If $x \in \Omega^+$ is a configuration and $j \in V$ is a site then x_j denotes the spin assigned to site j in configuration x. For each site $j \in V$, let S_j denote the set of pairs $(x, y) \in \Omega^+ \times \Omega^+$ of configurations that only differ on the spin assigned to site j, that is $x_i = y_i$ for all $i \neq j$.

We will use Weitz's [6] notation for block dynamics, although we only consider a finite collection of blocks. Define a collection of m blocks $\Theta = \{\Theta_k\}_{k=1,\ldots,m}$ such that each block $\Theta_k \subseteq V$ and Θ covers V, where we say that Θ covers V if $\bigcup_{k=1}^{m} \Theta_k = V$. One site may be contained in several blocks and the size of each block is not required to be the same, we do however require that the size of each block is bounded independently of n. For any block Θ_k and a pair of configurations $x, y \in \Omega^+$ we write "$x = y$ on Θ_k" if $x_i = y_i$ for each $i \in \Theta_k$ and similarly "$x = y$ off Θ_k" if $x_i = y_i$ for each $i \in V \setminus \Theta_k$. We also let $\partial\Theta_k$ denote the set of sites adjacent to but not included in Θ_k; we will refer to $\partial\Theta_k$ as the *boundary* of Θ_k.

With each block Θ_k, we associate a transition matrix $P^{[k]}$ on state space Ω^+ satisfying the following two properties:

1. If $P^{[k]}(x, y) > 0$ then $x = y$ off Θ_k, and also
2. π is invariant with respect to $P^{[k]}$.

Property 1 ensures that an application of $P^{[k]}$ moves the state of the system from from one configuration to another by only updating the sites contained in the block Θ_k and Property 2 ensures that any dynamics composed solely of transitions defined by $P^{[k]}$ converges to π. While the requirements of Property 1 are clear we take a moment to discuss what we mean by Property 2. Consider the following two step process in which some configuration x is initially drawn from π and then a configuration y is drawn from $P^{[k]}(x)$ where $P^{[k]}(x)$ is the distribution on configurations resulting from applying $P^{[k]}$ to a configuration x. We than say that π is invariant with respect to $P^{[k]}$ if for each configuration $\sigma \in \Omega^+$ we have $\Pr(x = \sigma) = \Pr(y = \sigma)$. That is the distribution on configurations generated by the two-step process is the same as if only the first step was executed. In terms of our dynamics this means that once the distribution of the dynamics reaches π, π will continue be the distribution of the dynamics even after applying $P^{[k]}$ to the state of the dynamics. Our main result (Theorem 2) holds for any choice of update rule $P^{[k]}$ provided that it satisfies these two properties.

The distribution $P^{[k]}(x)$, which specifies how the dynamics updates block Θ_k, clearly depends on the specific update rule implemented as $P^{[k]}$. In order to make this idea more clear we describe one particular update rule, known as the *heat-bath* update rule. This example serves a dual purpose as it is a simple way to implement $P^{[k]}$ and we will make use of heat-bath updates in Sects. 3 and 4 when applying our condition to specific spin systems. A heat-bath move on a block Θ_k given a configuration x is performed by drawing a new configuration from the distribution induced by π and consistent with the assignment of spins on the boundary of Θ_k. The two properties of $P^{[k]}$ hold for heat-bath updates since (1) only the assignment of the spin to the sites in Θ_k are changed and (2) the new configuration is drawn from an appropriate distribution induced by π. If the spin system corresponds to proper colourings of graphs then the distribution used in a heat-bath move is the uniform distribution on the set of configurations that agree with x off Θ_k and where no edge containing a site in Θ_k is monochromatic.

With these definitions in mind we are ready to formally define a systematic scan Markov chain.

Definition 1. *We let $\mathcal{M}_{\rightarrow}$ be a systematic scan Markov chain with state space Ω^+ and transition matrix $P_{\rightarrow} = \prod_{k=1}^{m} P^{[k]}$.*

The stationary distribution of $\mathcal{M}_{\rightarrow}$ is π as discussed above, and it is worth pointing out that the definition of $\mathcal{M}_{\rightarrow}$ holds for *any* order on the set of blocks. We will refer to one application of $P_{\rightarrow}$ (that is updating each block once) as one *scan* of $\mathcal{M}_{\rightarrow}$. One scan takes $\sum_k |\Theta_k|$ *updates* and it is generally straight forward to ensure, via the construction of the set of blocks, that this sum is of order $O(n)$.

We will be concerned with analysing the mixing time of systematic scan Markov chains, and consider the case when $\mathcal{M}_{\rightarrow}$ is ergodic. Let $\mathcal{M}$ be any ergodic Markov chain with state space Ω^+ and transition matrix P. By classical theory (see e.g. Aldous [9]) $\mathcal{M}$ has a unique stationary distribution, which we will denote π. The mixing time from an initial configuration $x \in \Omega^+$ is the number of steps, that is applications of P, required for $\mathcal{M}$ to become sufficiently close to π. Formally the mixing time of $\mathcal{M}$ from an initial configuration $x \in \Omega^+$ is defined, as a function of the deviation ε from stationarity, by

$$\mathrm{Mix}_x(\mathcal{M}, \varepsilon) = \min\{t > 0 : \mathrm{d_{TV}}(P^t(x, \cdot), \pi(\cdot)) \leq \varepsilon\}$$

where $\mathrm{d_{TV}}(\cdot, \cdot)$ is the total variation distance between two distributions. The mixing time of $\mathcal{M}$ is then obtained my maximising over all possible initial configurations as as follows $\mathrm{Mix}(\mathcal{M}, \varepsilon) = \max_{x \in \Omega^+} \mathrm{Mix}_x(\mathcal{M}, \varepsilon)$. We say that $\mathcal{M}$ is *rapidly mixing* if the mixing time of $\mathcal{M}$ is polynomial in n and $\log(\varepsilon^{-1})$.

We will now formalise the notion of "the influence *on* a site" in order to state our condition for rapid mixing of systematic scan. For any pair of configurations (x, y) let $\Psi_k(x, y)$ be a coupling of the distributions $P^{[k]}(x)$ and $P^{[k]}(y)$ which we will refer to as "updating block Θ_k". Recall that a coupling $\Psi_k(x, y)$ of $P^{[k]}(x)$ and $P^{[k]}(y)$ is a joint distribution on $\Omega^+ \times \Omega^+$ whose marginal distributions are $P^{[k]}(x)$ and $P^{[k]}(y)$. We write $(x', y') \in \Psi_k(x, y)$ when the pair of configurations (x', y') is drawn from $\Psi_k(x, y)$. Weitz [6] states his conditions for general metrics whereas we will use Hamming distance, which is also how the corresponding condition is defined in Dyer et al. [1]. This choice of metric allows us to define the influence of a site i on a site j under a block Θ_k, which we will denote $\rho^k_{i,j}$, as the maximum probability that two coupled Markov chains differ at the spin of site j following an update of Θ_k starting from two configurations that only differ at the spin on site i. That is

$$\rho^k_{i,j} = \max_{(x,y) \in S_i} \{\Pr_{(x',y') \in \Psi_k(x,y)}(x'_j \neq y'_j)\}.$$

Then let α be the total (weighted) influence *on* any site in the graph site defined by

$$\alpha = \max_k \max_{j \in \Theta_k} \sum_i \frac{w_i}{w_j} \rho^k_{i,j}.$$

We point out that our definition of $\rho^k_{i,j}$ is not the standard definition of ρ used in the literature (see for example Simon [10] or Dyer et al. [1]) since the coupling $\Psi_k(x, y)$ is explicitly included. In the block setting it is, however, necessary to include the coupling directly in the definition of ρ as we discuss in the full version of this paper [11]. In the full version we also show that the condition $\alpha < 1$ is a generalisation of the corresponding condition in Dyer et al. [1] in the sense that if each block contains exactly one site and the coupling minimises the Hamming distance then the conditions coincide.

Our main theorem, which is proved in Sect. 2, states that if the influence on a site is sufficiently small then the systematic scan Markov chain $\mathcal{M}_{\rightarrow}$ mixes in $O(\log n)$ scans.

Theorem 2. *Suppose* $\alpha < 1$. *Then* $\mathrm{Mix}(\mathcal{M}_{\rightarrow}, \varepsilon) \leq \frac{\log(n\varepsilon^{-1})}{1-\alpha}$.

As previously stated we will apply Theorem 2 to two spin systems corresponding to proper q-colourings of graphs in order to improve the parameters for which systematic scan mixes rapidly. In both applications we restrict the state space of the Markov chains to the set of proper colourings, Ω. Firstly we allow the underlying graph to be any finite graph with maximum vertex-degree Δ. Previously, the least number of colours for which systematic scan was known to mix in $O(\log n)$ scans was $q > 2\Delta$ and when $q = 2\Delta$ the best known bound on the mixing time was $O(n^2 \log n)$ scans due to Dyer et al. [1]. For completeness we pause to mention that the least number of colours required for rapid mixing of a random update Markov chain is $q > (11/6)\Delta$ due to Vigoda [12]. We consider the following Markov chain, *edge scan* denoted $\mathcal{M}_{\text{edge}}$, updating the endpoints of an edge during each update. Let $\Theta = \{\Theta_k\}_{k=1,\dots,m}$ be a set of edges in G such that Θ covers V. Using the above notation, $P^{[k]}$ is the transition matrix for performing a heat-bath move on the endpoints of the edge Θ_k and the transition matrix of $\mathcal{M}_{\text{edge}}$ is $\prod_{k=1}^{m} P^{[k]}$. In the full version of this paper we prove the following theorem, which improves the mixing time of systematic scan by a factor of n^2 for proper colourings of general graphs when $q = 2\Delta$ and matches the existing bound when $q > 2\Delta$. An outline of the proof is given in Sect. 3.

Theorem 3. *Let G be a graph with maximum vertex-degree Δ. If $q \geq 2\Delta$ then*

$$\mathrm{Mix}(\mathcal{M}_{\text{edge}}, \varepsilon) \leq \Delta^2 \log(n\varepsilon^{-1}).$$

Next we restrict the class of graphs to trees. It is known that single site systematic scan mixes in $O(\log n)$ scans when $q > \Delta + 2\sqrt{\Delta - 1}$ and in $O(n^2 \log n)$ scans when $q = \Delta + 2\sqrt{\Delta - 1}$ is an integer; see e.g. Hayes [13] or Dyer, Goldberg and Jerrum [14]. More generally it is known that systematic scan for proper colourings of bipartite graphs mixes in $O(\log n)$ scans whenever $q > f(\Delta)$ where $f(\Delta) \rightarrow \beta\Delta$ as $\Delta \rightarrow \infty$ and $\beta \approx 1.76$ due to Bordewich et al. [3]. Again, for completeness, we mention that the mixing time of a random update Markov chain for proper colourings on a tree mixes in $O(n \log n)$ updates when $q \geq \Delta + 2$, a result due to Martinelli, Sinclair and Weitz [15], improving a similar result by Kenyon, Mossel and Peres [16]. We will use a block approach to improve the number of colours required for mixing of systematic scan on trees. We construct the following set of blocks where the height h of the blocks is defined in Table 1. Let a block Θ_k contain a site r along with all sites below r in the tree that are at most $h - 1$ edges away from r. The set of blocks Θ covers the sites of the tree and we construct Θ such that no block has height less than h. Again $P^{[k]}$ is the transition matrix for performing a heat-bath move on block Θ_k and the transition matrix of the Markov chain $\mathcal{M}_{\text{tree}}$ is $\prod_{k=1}^{m} P^{[k]}$ where m is the number of blocks. We outline a proof of the following theorem in Sect. 4

Theorem 4. *Let G be a tree with maximum vertex-degree Δ. If $q \geq f(\Delta)$ where $f(\Delta)$ is specified in Table 1 for small Δ then*

$$\mathrm{Mix}(\mathcal{M}_{\text{tree}}, \varepsilon) = O(\log(n\varepsilon^{-1})).$$

Table 1. Optimising the number of colours using blocks

Δ	h	ξ	$f(\Delta)$	$\lceil \Delta + 2\sqrt{\Delta - 1} \rceil$
3	15	$\frac{4}{7}$	5	6
4	3	$\frac{5}{11}$	7	8
5	12	$\frac{5}{11}$	8	9
6	3	$\frac{1}{2}$	10	11
7	7	$\frac{10}{23}$	11	12
8	13	$\frac{1}{3}$	12	14
9	85	$\frac{5}{19}$	13	15
10	5	$\frac{5}{19}$	15	16

2 Bounding the Mixing Time of Systematic Scan

In this section we will outline a proof of Theorem 2. The proof follows the structure of the proof from the single-site setting in Dyer et al. [1], which follows Föllmer's [17] account of Dobrushin's proof presented in Simon's book [10].

We will make use the following definitions. For any function $f : \Omega^+ \to \mathrm{IR}_{\geq 0}$ let $\delta_i(f) = \max_{(x,y) \in S_i} |f(x) - f(y)|$ and $\Delta(f) = \sum_{i \in V} w_i \delta_i(f)$. Also for any transition matrix P define (Pf) as the function from Ω^+ to $\mathrm{IR}_{\geq 0}$ given by $(Pf)(x) = \sum_{x'} P(x, x') f(x')$. Finally let $\mathbf{1}_{i \notin \Theta_k}$ be the function given by $\mathbf{1}_{i \notin \Theta_k} = 1$ if $i \notin \Theta_k$ and $\mathbf{1}_{i \notin \Theta_k} = 0$ otherwise.

We can think of $\delta_i(f)$ as the deviation from constancy of f at site i and $\Delta(f)$ as the aggregated deviation from constancy of f. Now, Pf is a function where $(Pf)(x)$ gives the expected value of f after making a transition starting from x. Intuitively, if t transitions are sufficient for mixing then $P^t f$ is a very smooth function. An application of $P^{[k]}$ fixes the non-constancy of f at the sites within Θ_k although possibly at the cost of increasing the non-constancy at sites on the boundary of Θ_k. Our aim is then to show that one application of $P_{\rightarrow}$ will on aggregate make f smoother i.e., decrease $\Delta(f)$.

We will establish the following lemma, which corresponds to Corollary 12 in Dyer et al. [1], from which Sect. 3.3 of [1] implies Theorem 2.

Lemma 5. *If* $\alpha < 1$ *then* $\Delta(P_{\rightarrow} f) \leq \alpha \Delta(f)$.

We begin by bounding the effect on f from one application of $P^{[k]}$. The following lemma is a block-move generalisation of Proposition V.1.7 from Simon [10] and Lemma 10 from Dyer et al. [1].

Lemma 6

$$\delta_i(P^{[k]} f) \leq \mathbf{1}_{i \notin \Theta_k} \delta_i(f) + \sum_{j \in \Theta_k} \rho_{i,j}^k \delta_j(f).$$

Proof. Take $\mathbf{E}_{(x',y') \in \Psi_k(x,y))} [f(x')]$ to be the the expected value of $f(x')$ when a pair of configurations (x', y') are drawn from $\Psi_k(x, y)$. Since $\Psi_k(x, y)$ is a coupling of the distributions $P^{[k]}(x)$ and $P^{[k]}(y)$, the distribution $P^{[k]}(x)$ and the first component of $\Psi_k(x, y)$ are the same and hence

$$\mathbf{E}_{(x',y') \in \Psi_k(x,y)} [f(x')] = \mathbf{E}_{x' \in P^{[k]}(x)} [f(x')] \,. \tag{1}$$

Using (1) and linearity of expectation we have

$$\begin{aligned}\delta_i(P^{[k]}f) &= \max_{(x,y)\in S_i}\left|\sum_{x'} P^{[k]}(x,x')f(x') - \sum_{y'} P^{[k]}(y,y')f(y')\right| \\ &= \max_{(x,y)\in S_i}\left|\mathbf{E}_{x'\in P^{[k]}(x)}\left[f(x')\right] - \mathbf{E}_{y'\in P^{[k]}(y)}\left[f(y')\right]\right| \\ &\leq \max_{(x,y)\in S_i} \mathbf{E}_{(x',y')\in\Psi_k(x,y)}\left[|f(x')-f(y')|\right] \\ &\leq \max_{(x,y)\in S_i}\sum_{j\in V}\mathbf{E}_{(x',y')\in\Psi_k(x,y)}\left[\left|f(Z^{(j)})-f(Z^{(j-1)})\right|\right]\end{aligned}$$

where $Z^{(j)}$ is the configuration $(x'_1 \dots x'_j y'_{j+1} \dots y'_n)$.

Now suppose that $j \in \Theta_k$. By definition of $\rho^k_{i,j}$ the coupling $\psi_k(x,y)$ will yield $x'_j \neq y'_j$ with probability at most $\rho^k_{i,j}$ and so

$$\begin{aligned}\mathbf{E}_{(x',y')\in\Psi_k(x,y)}\left[\left|f(Z^{(j)})-f(Z^{(j-1)})\right|\right] &\leq \rho^k_{i,j}\max_{(\sigma,\tau)\in S_j}\{|f(\sigma)-f(\tau)|\} \\ &= \rho^k_{i,j}\delta_j(f)\ .\end{aligned}$$

Otherwise ($j \notin \Theta_k$) we observe that $x_j = x'_j$ and $y_j = y'_j$ since $x = x'$ off Θ_k and $y = y'$ off Θ_k. Hence we can only have $x'_j \neq y'_j$ when $i = j$ which gives

$$\mathbf{E}_{(x',y')\in\Psi_k(x,y)}\left[\left|f(Z^{(j)})-f(Z^{(j-1)})\right|\right] \leq \mathbf{1}_{i=j}\delta_i(f).$$

Adding up the expectations we get the statement of the lemma. □

We will use Lemma 6 to sketch a proof of the following lemma which is similar to (V.1.16) in Simon [10]. It is important to note at this point that the result in Simon is presented for single site heat-bath updates whereas the following lemma applies to any block dynamics (satisfying the stated assumptions) and weighted sites. This lemma is also a block generalisation of Lemma 11 in Dyer et al. [1].

Lemma 7. *Let $\Gamma(k) = \bigcup_{l=1}^{k}\Theta_l$ then for any $k \in \{1,\dots,m\}$, if $\alpha < 1$ then*

$$\Delta(P^{[1]}\cdots P^{[k]}f) \leq \alpha\sum_{i\in\Gamma(k)} w_i\delta_i(f) + \sum_{i\in V\setminus\Gamma(k)} w_i\delta_i(f).$$

Proof. Induction on k. Taking $k = 0$ we get the definition of Δ.

Now assuming that the statement holds for $k-1$ and using Lemma 6 we have

$$\begin{aligned}&\Delta(P^{[1]}\cdots P^{[k]}f) \\ &\quad\leq \alpha\sum_{i\in\Gamma(k-1)} w_i\delta_i(P^{[k]}f) + \sum_{i\in V\setminus\Gamma(k-1)} w_i\delta_i(P^{[k]}f) \\ &\quad\leq \alpha\sum_{i\in\Gamma(k-1)\setminus\Theta_k} w_i\delta_i(f) + \sum_{i\in V\setminus\Gamma(k)} w_i\delta_i(f) + \sum_{j\in\Theta_k}\delta_j(f)\sum_{i\in V} w_i\rho^k_{i,j} \qquad (2)\end{aligned}$$

since $\alpha < 1$. The lemma then follows from (2) by the definition of α. □

Lemma 5 is now a simple consequence of Lemma 7 since

$$\Delta(P_{\rightarrow} f) = \Delta(P^{[1]} \cdots P^{[m]} f) \leq \alpha \sum_{i \in V} w_i \delta_i(f) = \alpha \Delta(f)$$

and Theorem 2 follows as discussed above.

3 Application: Edge Scan on an Arbitrary Graph

In this section we outline a proof of Theorem 3 which states that systematic scan mixes in $O(\log n)$ scans on general graphs whenever $q \geq 2\Delta$. In order to apply Theorem 2 we need to construct a coupling $\Psi_k(x, y)$ of the distributions $P^{[k]}(x)$ and $P^{[k]}(y)$ for each pair of configurations $(x, y) \in S_i$ that differ only at the colour assigned to site i. In the full version of this paper [11] we consider a set of exhaustive cases for this coupling construction, however here we only state the derived bounds on $\rho^k_{i,j}$. The following lemma allows us to establish a bound on the influence on a site, which we use to prove Theorem 3.

Lemma 8. *Let j and j' be the endpoints of an edge Θ_k. If $\{i, j\} \in E$ and $\{i, j'\} \notin E$ then*

$$\rho^k_{i,j} \leq \frac{1}{q - \Delta} \text{ and } \rho^k_{i,j'} \leq \frac{1}{(q - \Delta)^2}.$$

If $\{i, j\} \in E$ and $\{i, j'\} \in E$ then

$$\rho^k_{i,j} \leq \frac{1}{q - \Delta} + \frac{1}{(q - \Delta)^2} \text{ and } \rho^k_{i,j'} \leq \frac{1}{q - \Delta} + \frac{1}{(q - \Delta)^2}.$$

Otherwise $\rho^k_{i,j} = \rho^k_{i,j'} = 0$.

Proof of Theorem 3. We will use Theorem 2 and let $w_i = 1$ for all $i \in V$ so we omit all weights. Let j and j' be the endpoints of an edge represented by a block Θ_k. Let $\alpha_j = \sum_i \rho^k_{i,j}$ be the influence on site j and $\alpha_{j'} = \sum_i \rho^k_{i,j'}$ then influence on j'. Then $\alpha = \max(\alpha_j, \alpha_{j'})$. Suppose that Θ_k is adjacent to t triangles, that is there are t sites $i_1, \ldots, i_t$ such that $\{i, j\} \in E$ and $\{i, j'\} \in E$ for each $i \in \{i_1, \ldots, i_t\}$. Note that $0 \leq t \leq \Delta - 1$. There are at most $\Delta - 1 - t$ sites adjacent to j that are not adjacent to j' and at most $\Delta - 1 - t$ sites adjacent to j' that are not adjacent to j. From Lemma 8 a site adjacent only to j will emit an influence of at most $\frac{1}{q-\Delta}$ on site j and Lemma 8 also guarantees that a site only adjacent to j' can emit an influence at most $\frac{1}{(q-\Delta)^2}$ on site j. Finally Lemma 8 says that a site adjacent to both j and j' can emit an influence of at most $\frac{1}{q-\Delta} + \frac{1}{(q-\Delta)^2}$ on site j and hence

$$\begin{aligned} \alpha_j &\leq t \left(\frac{1}{q - \Delta} + \frac{1}{(q - \Delta)^2} \right) + (\Delta - 1 - t) \left(\frac{1}{q - \Delta} + \frac{1}{(q - \Delta)^2} \right) \\ &= \frac{\Delta - 1}{q - \Delta} + \frac{\Delta - 1}{(q - \Delta)^2}. \end{aligned}$$

We obtain a similar bound on $\alpha_{j'}$ by considering the influence on site j'. Thus whenever $q \geq 2\Delta$ we have

$$\alpha = \max(\alpha_j, \alpha_{j'}) \leq \frac{\Delta-1}{q-\Delta} + \frac{\Delta-1}{(q-\Delta)^2} \leq \frac{\Delta-1}{\Delta} + \frac{\Delta-1}{\Delta^2} = \frac{\Delta^2-1}{\Delta^2} = 1 - \frac{1}{\Delta^2} < 1$$

and we obtain the stated bound on the mixing time by applying Theorem 2. □

4 Application: Colouring a Tree

In this section we sketch a proof of Theorem 4, which improves the least number of colours required for mixing of systematic scan on a tree for individual values of Δ. We will use standard terminology when discussing the structure of the tree. In particular will say that a site i is a *descendant* of a site j (or j is a *predecessor* of i) if j is on the simple path from the root of the tree to i. We will call a site j a *child* of a site i if i and j are adjacent and j is a descendant of i. Finally $N_k(j) = \{i \in \partial\Theta_k \mid i \text{ is a descendant of } j\}$ is the set of descendants of j on the boundary of Θ_k.

It is possible to construct the required couplings recursively for blocks of size h on a tree. For details of the coupling construction as well as the bounds on the disagreement probabilities see the full version of this paper. The $\rho_{i,j}^k$ values are summarised in the following lemma which we use in order to sketch a proof of Theorem 4. This proof will demonstrate the use of weights when bounding the influence on a site.

Lemma 9. *Let $d(i,j)$ denote the number of edges between i and j. Suppose that $j \in \Theta_k$ and $i \in \partial\Theta_k$ then $\rho_{i,j}^k \leq (q-\Delta)^{-d(i,j)}$. Otherwise $\rho_{i,j}^k = 0$.*

Proof of Theorem 4. We will use Theorem 2 and assign a weight to each site i such that $w_i = \xi^{d_i}$ where d_i is the edge distance from i to the root and ξ is defined in Table 1 for each Δ. For a block Θ_k and $j \in \Theta_k$ we let

$$\alpha_{k,j} = \frac{\sum_i w_i \rho_{i,j}^k}{w_j}$$

denote the total weighted influence on site j when updating block Θ_k. For each block Θ_k and each site $j \in \Theta_k$ we will upper-bound $\alpha_{k,j}$ and hence obtain an upper-bound on $\alpha = \max_k \max_{j \in \Theta_k} \alpha_{k,j}$.

We will consider a block Θ_k that does not contain the root. The following labels refer to Fig. 1 in which a solid line is an edge and a dotted line denotes the existence of a simple path between two sites. Let $p \in \partial\Theta_k$ be the predecessor of all sites in Θ_k and $d_r - 1$ be the distance from p to the root of the tree i.e., $w_p = \xi^{d_r-1}$. The site $r \in \Theta_k$ is a child of p. Now consider a site $j \in \Theta_k$ which has distance d to r, hence $w_j = \xi^{d+d_r}$ and $d(j,p) = d+1$. From Lemma 9 it follows that the weighted influence of p on j when updating Θ_k is at most

$$\rho_{p,j}^k \frac{w_p}{w_j} \leq \frac{1}{(q-\Delta)^{d(j,p)}} \frac{\xi^{d_r-1}}{\xi^{d_r+d}} = \frac{1}{(q-\Delta)^{d+1}} \frac{1}{\xi^{d+1}}.$$

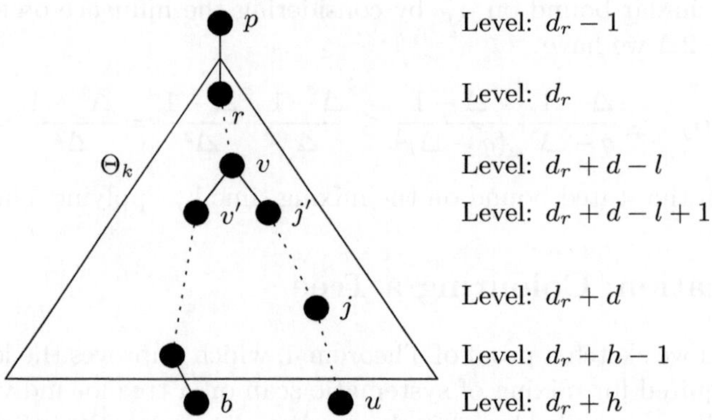

Fig. 1. A block in the tree. A solid line indicates an edge and a dotted line the existence of a path.

Now consider some site $u \in N_k(j)$ which is on the boundary of Θ_k. Since $u \in N_k(j)$ it has weight $w_u = \xi^{d_r+h}$ and so $d(j,u) = h - d$. Hence Lemma 9 says that the weighted influence of u on j is at most

$$\rho_{u,j}^k \frac{w_u}{w_j} \leq \frac{1}{(q-\Delta)^{d(j,u)}} \frac{\xi^{d_r+h}}{\xi^{d_r+d}} = \frac{1}{(q-\Delta)^{h-d}} \xi^{h-d}.$$

Every site in Θ_k has at most $\Delta - 1$ children so the number of sites in $N_k(j)$ is at most $|N_k(j)| \leq (\Delta-1)^{h-d}$ and so, summing over all sites $u \in N_k(j)$, the total weighted influence on j from sites in $N_k(j)$ when updating Θ_k is at most

$$\sum_{u \in N_k(j)} \rho_{u,j}^k \frac{w_u}{w_j} \leq \sum_{u \in N_k(j)} \frac{1}{(q-\Delta)^{h-d}} \xi^{h-d} \leq \frac{(\Delta-1)^{h-d}}{(q-\Delta)^{h-d}} \xi^{h-d}.$$

In the full version of the paper we also consider the influence on j from sites in $\partial\Theta_k \setminus (N_k(j) \cup \{p\})$ and obtain an upper-bounds on both $\alpha_{k,j}$ and $\alpha_{0,j}$ for each $j \in \Theta_k$ where Θ_0 is the block containing the root. We require $\alpha < 1$ which we obtain by satisfying the system of inequalities given by setting

$$\alpha_{k,j} < 1 \tag{3}$$

for all blocks Θ_k and sites $j \in \Theta_k$. In particular we need to find an assignment to ξ and h that satisfies (3) given Δ and q. Table 1 shows the least number of colours $f(\Delta)$ required for mixing for small Δ along with a weight, ξ, that satisfies the system of equations and the required height of the blocks, h. These values were verified by checking the resulting $2h$ inequalities for each Δ using Mathematica. The least number of colours required for mixing in the single site setting is also included in the table for comparison. □

Acknowledgment. I am grateful to Leslie Goldberg for several useful discussions regarding technical issues and for providing detailed and helpful comments on a draft of the full version of this article. I would also like to thank Paul Goldberg for useful comments during the early stages of this work.

References

1. Dyer, M., Goldberg, L.A., Jerrum, M.: Dobrushin conditions and systematic scan. In: Díaz, J., Jansen, K., Rolim, J.D.P., Zwick, U. (eds.) APPROX 2006 and RANDOM 2006. LNCS, vol. 4110, pp. 327–338. Springer, Heidelberg (2006)
2. Dyer, M., Goldberg, L.A., Jerrum, M.: Systematic scan and sampling colourings. Annals of Applied Probability 16(1), 185–230 (2006)
3. Bordewich, M., Dyer, M., Karpinski, M.: Stopping times, metrics and approximate counting. In: Bugliesi, M., Preneel, B., Sassone, V., Wegener, I. (eds.) ICALP 2006. LNCS, vol. 4051, pp. 108–119. Springer, Heidelberg (2006)
4. Dobrushin, R.L., Shlosman, S.B.: Constructive criterion for the uniqueness of Gibbs field. In: Statistical mechanics and dynamical systems. Progress in Physics, vol. 10, pp. 371–403. Birkhäuser, Boston (1985)
5. Weitz, D.: Mixing in Time and Space for Discrete Spin Systems. PhD thesis, University of California, Berkley (2004)
6. Weitz, D.: Combinatorial criteria for uniqueness of Gibbs measures. Random Structures and Algorithms 27(4), 445–475 (2005)
7. Dyer, M., Sinclair, A., Vigoda, E., Weitz, D.: Mixing in time and space for lattice spin systems: A combinatorial view. Random Structures and Algorithms 24(4), 461–479 (2004)
8. Dobrushin, R.L.: Prescribing a system of random variables by conditional distributions. Theory Prob. and its Appl. 15, 458–486 (1970)
9. Aldous, D.J: Random walks on finite groups and rapidly mixing markov chains. In: Séminaire de probabilités XVII, pp. 243–297. Springer, Heidelberg (1983)
10. Simon, B.: The Statistical Mechanics of Lattice Gases. Princeton University Press (1993)
11. Pedersen, K.: Dobrushin conditions for systematic scan with block dynamics. arXiv:math.PR/0703461 (2007)
12. Vigoda, E.: Improved bounds for sampling colourings. J. Math. Phys (2000)
13. Hayes, T.P.: A simple condition implying rapid mixing of single-site dynamics on spin systems. In: Proc. 47th Annual IEEE Symposium on Foundations of Computer Science, pp. 39–46. IEEE Computer Society Press, Los Alamitos (2006)
14. Dyer, M., Goldberg, L.A., Jerrum, M.: Matrix norms and rapid mixing for spin systems. ArXiv math.PR/0702744 (2006)
15. Martinelli, F., Sinclair, A., Weitz, D.: Glauber dynamics on trees: Boundary conditions and mixing time. Communications in Mathematical Physics 250(2), 301–334 (2004)
16. Kenyon, C., Mossel, E., Peres, Y.: Glauber dynamics on trees and hyperbolic graphs. In: Proc. 42nd Annual IEEE Symposium on Foundations of Computer Science, pp. 568–578. IEEE Computer Society Press, Los Alamitos (2001)
17. Föllmer, H.: A covariance estimate for Gibbs measures. J. Funct. Analys. 46, 387–395 (1982)

On the Complexity of Computing Treelength

Daniel Lokshtanov

Department of Informatics, University of Bergen,
N-5020 Bergen, Norway
daniel.lokshtanov@uib.no

Abstract. We resolve the computational complexity of determining the *treelength* of a graph, thereby solving an open problem of Dourisboure and Gavoille, who introduced this parameter, and asked to determine the complexity of recognizing graphs of bounded treelength [6]. While recognizing graphs with treelength 1 is easily seen as equivalent to recognizing chordal graphs, which can be done in linear time, the computational complexity of recognizing graphs with treelength 2 was unknown until this result. We show that the problem of determining whether a given graph has treelength at most k is NP-complete for every fixed $k \geq 2$, and use this result to show that treelength in weighted graphs is hard to approximate within a factor smaller than $\frac{3}{2}$. Additionally, we show that treelength can be computed in time $O^*(1.8899^n)$ by giving an exact exponential time algorithm for the Chordal Sandwich problem and showing how this algorithm can be used to compute the treelength of a graph.

1 Introduction

Treelength is a graph parameter proposed by Dourisboure and Gavoille [6] that measures how close a graph is to being chordal. The treelength of G is defined using tree decompositions of G. Graphs of treelength k are the graphs that have a tree decomposition where the distance in G between any pair of nodes that appear in the same bag of the tree decomposition is at most k. As chordal graphs are exactly those graphs that have a tree decomposition where every bag is a clique [14], [3], [10], we can see that treelength generalizes this characterization.

There are several reasons for why it is interesting to study this parameter. For example, Dourisboure et. al. show that graphs with bounded treelength have sparse additive spanners [5]. Dourisboure also shows that graphs of bounded treelength admit compact routing schemes [4]. One should also note that many graph classes with unbounded treewidth have bounded treelength, such as chordal, interval, split, AT-free, and permutation graphs [6].

In this paper, we show that recognizing graphs with treelength bounded by a fixed constant $k \geq 2$ is NP-complete. The problem of settling the complexity of recognizing graphs of bounded treelength was first posed as an open problem by Doursiboure and Gavoille, and remained open until this result [6]. Our result is somewhat surprising, because by bounding the treelength of G we put heavy restrictions on the distance matrix of G. Another indication that this problem

L. Kučera and A. Kučera (Eds.): MFCS 2007, LNCS 4708, pp. 276–287, 2007.

might be polynomial for fixed k was that the treelength of a graph is fairly easy to approximate within a factor of 3 [6]. In comparison, the best known approximation algorithm for treewidth has an approximation factor of $O(\sqrt{\log k})$ [7]. As Bodlaender showed, recognizing graphs with treewidth bounded by a constant can be done in linear time [1]. Since the above observation about approximately might indicate that determining treelength is "easier" than determining treewidth, one could arrive at the conclusion that recognizing graphs with treelength bounded by a constant should be polynomial. However, there are also strong arguments against this intuition. For instance, graphs of bounded treelength are just bounded diameter graphs that have been glued together in a certain way. Thus, when trying to show that a graph indeed has small treelength one would have to decompose the graph into components of small diameter and show how these components are glued together to form the graph. As the class of bounded diameter graphs is very rich, one would have a myriad of candidates to be such components, making it hard to pick out the optimal ones. This intuition is confirmed when we prove the hardness of recognizing graphs of bounded treelength because the instances we reduce to all have bounded diameter.

In the next section we will give some notation and preliminary results. Next, we present a proof that determining whether the treelength of a weighted graph is less than or equal to k is NP-hard for every fixed $k \geq 2$. Following this, we reduce the problem of recognizing weighted graphs with treelength bounded by k to the problem of recognizing unweighted graphs with the treelength bounded by the same constant k, thereby completing the hardness proof. Finally we also consider the complexity of approximating treelength, and propose a fast exact algorithm to determine the parameter by solving the Chordal Sandwich problem.

2 Notation, Terminology and Preliminaries

For a graph $G = (V, E)$ let $w : E \to \mathbb{N}$ be a *weight function* on the edges. The *length* of a path with respect to a weight function w is the sum of the weights of its edges. The *distance* $d_w(u, v)$ between two vertices is the length of the shortest path with respect to w. Whenever no weight function is specified the unit weight function $w(e) = 1$ for all $e \in E$ is used. G to the *power* of k with respect to the weight function w is $G_w^k = (V, \{uv : d_w(u, v) \leq k\})$. A weight function w is *metric* if it satisfies a generalization of the triangle inequality, that is, if $w((u, v)) = d_w(u, v)$ for every edge (u, v).

A *tree decomposition* of a graph $G = (V, E)$ is a pair (S, T) consisting of a set $S = \{X_i : i \in I\}$ of *bags* and a tree $T = (I, M)$ so that each bag $X_i \in S$ is a subset of V and the following conditions hold:

- $\bigcup_{i \in I} X_i = V$
- For every edge (u, v) in E, there is a bag X_i in S so that $u \in X_i$ and $v \in X_i$
- For every vertex v in V, the set $\{i \in I : v \in X_i\}$ induces a connected subtree of T

The *length* of a bag is the maximum distance in G between any pair of vertices in the bag. The *length* of a tree-decomposition is the maximum length of any bag. The *treelength* of G with weight function w is the minimum length of a tree-decomposition of G, and is denoted by $tl_w(G)$. When no weight function is given on E, then the treelength is denoted by $tl(G)$. A *shortest* tree decomposition is a tree decomposition having minimum length. We will always assume that all weight functions are metric. This can be justified by the fact that if w is not metric, we easily can make a new metric weight function w' by letting $w'((u,v)) = d_w(u,v)$ for every edge (u,v) and observe that $tl_{w'}(G) = tl_w(G)$.

The *neighbourhood* of a vertex v is denoted $N(v)$ and is the vertex set $\{u : (u,v) \in E\}$. When S is a subset of V, $G[S] = (S, E \cap \{(u,v) : u \in S, v \in S\})$ is the subgraph *induced* by S. We will use $G \setminus v$ to denote the graph $G[V \setminus \{v\}]$. G is *complete* if (u,v) is an edge of G for every pair $\{u,v\}$ of distinct vertices in G. A *clique* in G is a set S of vertices in G so that $G[S]$ is complete.

For two graphs $G = (V,E)$ and $G' = (V,E')$, $G \subseteq G'$ means that $E \subseteq E'$. For a graph class Π, G is a *Π-graph* if $G \in \Pi$. G' is a *Π-sandwich* between G and G'' if G' is a Π-graph and $G \subseteq G' \subseteq G''$ [11]. A graph class Π is *hereditary* if every induced subgraph of a Π-graph is a Π-graph.

A graph is *chordal* if it contains no induced cycle of length at least 4. Thus the class of chordal graphs is hereditary. A vertex v is *simplicial* if the neighbourhood of v is a clique. A vertex v is *universal* if $V = \{v\} \cup N(v)$. An ordering of the vertices of G into $\{v_1, v_2, \ldots, v_n\}$ is a *perfect elimination ordering* if for every i, v_i is simplicial in $G[\{v_j : j \geq i\}]$. A *clique tree* of G is a tree decomposition of G where every bag is a maximal clique of G (see e.g., [12] for details).

Theorem 1. *The following are equivalent:*
- *G is chordal.*
- *G has a clique tree. [14], [3], [10]*
- *G has a perfect elimination ordering. [9]*

For more characterizations of chordal graphs and the history of this graph class, refer to the survey by Heggernes [12]. Following Theorem 1 it is easy to see that if v is simplicial then G is chordal if and only if $G \setminus v$ is chordal. Universal vertices share this property, as has been observed by several authors before.

Observation 1. *If v is universal in G then G is chordal if and only if $G \setminus v$ is chordal.*

Proof. If G is chordal then $G \setminus v$ is chordal because the class of chordal graphs is hereditary. Now suppose $G \setminus v$ is chordal. Consider a perfect elimination ordering of $G \setminus v$ appended by v. This is clearly a perfect elimination ordering of G, hence G is chordal.

We now define the problem that we are going to show is NP-complete. In the problem statement below, k is an integer greater than or equal to 2.

k-TREELENGTH
INSTANCE: A graph G
QUESTION: Is $tl(G) \leq k$?

Finally, we define the problem we will reduce from.

CHORDAL SANDWICH [11]
INSTANCE: Two graphs G_1 and G_2 with $G_1 \subseteq G_2$
QUESTION: Is there a chordal sandwich between G_1 and G_2?

3 Weighted k-Treelength is NP-Complete

In this section we are going to show that determining whether the treelength of a given weighted graph is at most k is NP-complete for every fixed $k \geq 2$. In the next section we will conclude the hardness proof for unweighted graphs by showing how one from a weighted graph G in polynomial time can construct an unweighted graph G' with the property that $tl_w(G) \leq k$ if and only if $tl(G') \leq k$.

WEIGHTED k-TREELENGTH
INSTANCE: A graph G with weight function w
QUESTION: Is $tl_w(G) \leq k$?

Observation 2. *For a graph $G = (V, E)$, $tl_w(G) \leq k$ if and only if there exists a chordal sandwich G' between G and G^k_w.*

Proof. Suppose $tl_w(G) \leq k$. Consider a shortest tree decomposition (S, T) of G, and construct the graph $G' = (V, \{(u, v) : \exists i \ u \in X_i, v \in X_i\})$. $G \subseteq G'$ is trivial, $G' \subseteq G^k_w$ holds because the length of the tree decomposition is at most k, and G' is chordal because (S, T) is a clique tree of G'. In the other direction, let G' be a chordal sandwich between G and G^k_w. Consider a clique tree (S, T) of G'. This is a tree decomposition of G, and the length of this decomposition is at most k, as $u \in X_i$ and $v \in X_i$ implies $(u, v) \in E(G') \subseteq E(G^k_w)$.

Corollary 1. *For any graph G, $tl(G) = 1$ if and only if G is chordal.*

From Observation 2, it follows that determining the treelength of a given graph in fact is a special case of the *Chordal Sandwich* problem defined above. In a study of sandwich problems [11], Golumbic et. al. point out that as a consequence of the hardness of Triangulating Colored Graphs, the Chordal Sandwich problem is NP-Complete. Thus, in order to prove that Weighted k-Treelength is indeed hard, we only need to reduce the Chordal Sandwich problem to a special case of itself, namely the one where $G_2 = {G_1}^k_w$ for some weight function w.

We will reduce in the following way. On input $G_1 = (V_1, E_1)$, $G_2 = (V_2, E_2)$ with $G_1 \subseteq G_2$ to the Chordal Sandwich problem, let $E_D = E_2 \setminus E_1$. We construct a new graph G by taking a copy of G_1, adding a new vertex c_{uv} for every edge (u, v) in E_D and making this vertex adjacent to all other vertices of G. We denote the set of added vertices by C, as C is a clique of universal vertices. The weight function is simple, $w(c_{uv}, u) = w(c_{uv}, v) = \lfloor k/2 \rfloor$ for every c_{uv} and $w(e) = k$ for all other edges.

Lemma 3. *Let G, G_1 and G_2 be as described above. Then $tl_w(G) \leq k$ if and only if there is a chordal sandwich G' between G_1 and G_2.*

Proof. Observe that any supergraph G' of G on the same vertex set ($G' \supseteq G$) is chordal if and only if $G'[V_1]$ is chordal since every vertex in C is universal. Also, notice that for every pair u,v of vertices in V_1, $d_w(u,v) \leq k$ if and only if (u,v) is an edge of G_2. Thus it follows that $G_2 = G^k_w[V_1]$. Hence, by Observation 2, $tl_w(G) \leq k$ if and only if there is a chordal sandwich G' between G and G^k_w. By the discussion above, this is true if and only if there is a chordal sandwich between $G[V_1] = G_1$ and $G_2 = G^k_w[V_1]$.

Corollary 2. *Weighted k-Treelength is NP-complete for every $k \geq 2$.*

Proof. By Lemma 3 determining whether a given weighted graph G has $tl_w(G) \leq k$ is NP-hard for every $k \geq 2$. By Observation 2 this problem is polynomial time reducible to the Chordal Sandwich problem, thus it is in NP.

4 k-Treelength is NP-Complete

We will now show how one from a weighted graph G in polynomial time can construct an unweighted graph G'' with the property that $tl_w(G) \leq k$ if and only if $tl(G'') \leq k$. We do this in two steps. First we show how to construct a graph G' and weight function w' from G and w so that $tl_w(G) \leq k$ if and only if $tl_{w'}(G') \leq k$ and $w'(e) = 1$ or $w'(e) = k$ for every edge e in G'. In the second step we show how G'' can be constructed from G' and w'. Both steps are done in an inductive way. Obviously, if G has an edge of weight larger than k then $tl_w(G) > k$. We will therefore assume that $w(e) \leq k$ for all edges e. For an edge (u,v), let $G(u,v) = (V \cup \{r,q\}, (E \setminus (u,v)) \cup \{(u,r),(r,v),(u,q),(q,v)\})$. That is, we build $G(u,v)$ from G by removing the edge (u,v), adding two new vertices r and q and making both of them adjacent to u and v. Let $w_{u,v,k}$ be a weight function of $G(u,v)$ so that $w_{(u,v,k)}(e) = w(e)$ if $e \in E(G) \cap E(G(u,v))$, $w_{(u,v,k)}((u,r)) = w_{(u,v,k)}((r,v)) = k$, $w_{(u,v,k)}((u,q)) = w(u,v) - 1$, and $w_{(u,v,k)}((q,v)) = 1$. Observe that if $w((u,v)) > 1$ then $w_{(u,v,k)}$ is properly defined.

Lemma 4. *Given a graph G, an edge (u,v), and a weight function w with $w((u,v)) > 1$, there is a chordal sandwich between G and G^k_w if and only if there is a chordal sandwich between $G(u,v)$ and $G(u,v)^k_{w_{(u,v,k)}}$.*

Proof. Suppose there is a chordal sandwich $\hat{G}_{(u,v)}$ between $G(u,v)$ and $G(u,v)^k_{w_{(u,v,k)}}$. Then the edge (u,v) must be in $E(\hat{G}_{(u,v)})$ and thus $\hat{G}_{(u,v)} \setminus \{r,q\}$, where r and q are the vertices that were added to G (u,v) to obtain $G(u,v)$, is a chordal sandwich between G and G^k_w. In the other direction, suppose there is a chordal sandwich $\hat{G}$ between G and and G^k_w. Then $\hat{G}' = (V(\hat{G}) \cup \{r,q\}, E(\hat{G}) \cup \{(u,r),(r,v),(u,q),(q,v)\})$ is a chordal sandwich between $G(u,v)$ and $G(u,v)^k_{w_{(u,v,k)}}$ because the r and q are simplicial nodes in $\hat{G}'$.

Now, the idea is that the graph $G(u,v)$ with weight function $w_{(u,v,k)}$ is somewhat closer to not having any edges with weight between 2 and $k-1$. With an appropriate choice of measure, it is easy to show that this is indeed the case. The measure we will use will essentially be the sum of the weights of all edges that have edge weights between 2 and $k-1$. In the following discussion, let $W_w(G) = \sum_{e \in E, w(e)<k}(w(e)-1)$. Observe that if $1 < w(u,v) < k$ then $W_{w_{(u,v,k)}}(G(u,v)) = W_w(G) - 1$, and that if $W_w(G) = 0$ then $w(e) = 1$ or $w(e) = k$ for every edge $e \in E$.

Lemma 5. *For a graph G with weight function w, we can construct in polynomial time a graph G' with weight function w' so that $|V(G')| = |V(G)| + 2W_w(G)$, and $tl_w(G) \leq k$ if and only if $tl_{w'}(G') \leq k$.*

Proof. We prove by induction on $W_w(G)$. If $W_w(G) = 0$ we know that $w(e) = 1$ or $w(e) = k$ for every edge e. Now, suppose the statement of the lemma holds for all graphs with $W_w(G) < t$ for some t and consider a graph G with weight function w so that $W_w(G) = t > 0$. Then, let (u,v) be an edge so that $1 < w((u,v)) < k$. By Lemma 4, $tl_w(G) \leq k$ if and only if $tl_{w_{(u,v,k)}}(G(u,v)) \leq k$. Now, $W_{w_{(u,v,k)}}(G(u,v)) = W_w(G) - 1$. Thus, by the induction assumption, we can in polynomial time construct a graph G' with weight function w' that satisfies $tl_w(G) \leq k \iff tl_{w_{(u,v,k)}}(G(u,v)) \leq k \iff tl_{w'}(G') \leq k$ with $|V(G')| = |V(G(u,v))| + 2W_{w_{(u,v,k)}}(G(u,v)) = |V(G)| + 2 + 2(W_w(G) - 1) = |V(G)| + 2W_w(G)$.

The idea of the above proof is that we can use edges of weight 1 and k to emulate the behaviour of edges with other weights. The method we now will use to prove the hardness of unweighted treelength will be similar - we are going to show that weight k edges can be emulated using only edges with weight 1. In order to do this, we are going to use the following lemma by Dourisboure et. al. concerning the treelength of cycles.

Lemma 6. *[6] The treelength of a cycle on k vertices is $\lceil \frac{k}{3} \rceil$.*

For an edge $(u,v) \in E$, we construct the graph $G[u,v,k]$ in the following way: We replace the edge (u,v) by three paths on $2k-1$, $2k-1$ and $k-1$ vertices respectively. Construct these paths $P_a = \{a_1, a_2, \ldots, a_{2k-1}\}$, $P_b = \{b_1, b_2, \ldots, b_{2k-1}\}$ and $P_c = \{c_1, c_2, \ldots, c_{k-1}\}$ using new vertices. Take a copy of G, remove the edge (u,v) and add edges from u to a_1, b_1 and c_1, and from v to a_{2k-1}, b_{2k-1} and c_{k-1}. For a weight function w of G, $w_{[u,v,k]}$ will be a weight function of $G[u,v,k]$ so that $w_{[u,v,k]}(e) = w(e)$ if $e \in E(G)$ and $w_{[u,v,k]} = 1$ otherwise.

Lemma 7. *Given G, weight function w and an edge $(u,v) \in E$ with $w(u,v) = k$, $tl_w(G) \leq k$ if and only if $tl_{w_{[u,v,k]}}(G[u,v,k]) \leq k$*

Proof. Suppose there is a chordal sandwich $\hat{G}$ between G and G_w^k. We build $\hat{G}'$ from G by taking a copy of $\hat{G}$, adding three new paths $P_a = \{a_1, a_2, \ldots, a_{2k-1}\}$, $P_b = \{b_1, b_2, \ldots, b_{2k-1}\}$ and $P_c = \{c_1, c_2, \ldots, c_{k-1}\}$ and the edge sets $\{(u, a_i) : i \leq k\}$, $\{(u, b_i) : i \leq k\}$, $\{(u, c_i) : i \leq \lfloor \frac{k}{2} \rfloor\}$, $\{(v, a_i) : i \geq k\}$, $\{(v, b_i) : i \geq k\}$, $\{(v, c_i) : i \geq \lfloor \frac{k}{2} \rfloor\}$. We see that $\hat{G}'$ is chordal because $\{a_1, a_2 \ldots a_{k-1}$, $a_{2k-1}, a_{2k-2}, \ldots a_k, b_1, b_2 \ldots b_{k-1}$, b_{2k-1}, b_{2k-2}, $\ldots b_k$, c_1, c_2, $\ldots c_{\lfloor \frac{k}{2} \rfloor - 1}$, c_{k-1},

$c_{k-2}, \ldots c_{\lfloor \frac{k}{2} \rfloor}\}$ followed by a perfect elimination ordering of $\hat{G}$ is a perfect elimination ordering of $\hat{G}'$. Also, $\hat{G}' \subseteq G[u,v,k]^k_{w[u,v,k]}$. Thus $\hat{G}'$ is a chordal sandwich between $G[u,v,k]$ and $G[u,v,k]^k_{w[u,v,k]}$. In the other direction, let $\hat{G}_{[u,v]}$ be a chordal sandwich between $G[u,v,k]$ and $G[u,v,k]^k_{w[u,v,k]}$. It is sufficient to show that $(u,v) \in E(\hat{G}_{[u,v]})$ because then $\hat{G}_{[u,v]}[V(G)]$ is a chordal sandwich between G and G^k_w. Consider the set $V_s = \{u,v\} \cup V(P_a) \cup V(P_b)$, and let C be the subgraph of $G[u,v,k]$ induced by V_s. Now, observe that $E(G[u,v,k]^k_{w[u,v,k]}[S]) = E(C^k) \cup \{(u,v)\}$. Suppose for contradiction that (u,v) is not an edge of $\hat{G}_{[u,v]}$. Then we know that $\hat{G}_{[u,v]}[V_s]$ is a chordal sandwich between C and C^k implying that $tl(C) \leq k$. This contradicts Lemma 6 because C is a cycle on $4k$ vertices.

Lemma 7 gives us a way to emulate edges of weight k using only edges of weight 1. For a graph G with weight function w, let $W_w[G] = |\{e \in E(G) : w(e) = k\}|$. Notice that if $w((u,v)) = k$ then $W_w[G] = W_{w[u,v,k]}[G[u,v,k]] + 1$.

Lemma 8. *For every graph G with weight function w satisfying $w(e) = 1$ or $w(e) = k$ for every edge, we can construct in polynomial time a graph G' with $W_w[G](5k-3) + |V(G)|$ vertices and satisfying $tl_w(G) \leq k$ if and only if $tl(G') \leq k$.*

Proof. We use induction in $W_w[G]$. If $W_w[G] = 0$ the lemma follows immediately. Now, assume the result holds for $W_w[G] < t$ for some $t > 0$. Consider a graph G with weight function w so that $W_w[G] = t$. By Lemma 7 $tl_w(G) \leq k$ if and only if $tl_{w[u,v,k]}(G[u,v,k]) \leq k$. By the inductive hypothesis we can construct in polynomial time a graph G' with $W_{w[u,v,k]}[G[u,v,k]](5k-3) + |V(G[u,v,k])| + 5k - 3 = W_w[G](5k-3) + |V(G)|$ vertices and satisfying $tl(G') \leq k \iff tl_{w[u,v,k]}(G[u,v,k]) \leq k \iff tl_w(G) \leq k$

Corollary 3. *For a graph G and weight function w, we can in polynomial time construct a graph G'' so that $tl_w(G) \leq k$ if and only if $tl(G'') \leq k$.*

Proof. By Lemma 5 we can in polynomial time construct a graph G' with weight function w' so that $tl_{w'}(G') \leq k \iff tl_w(G) \leq k$ and so that $w'(e) = 1$ or $w'(e) = k$ for every edge e in $E(G')$. By Lemma 8 we can from such a G' and w' construct in polynomial time a G'' so that $tl(G'') \leq k \iff tl_{w'}(G') \leq k \iff tl_w(G) \leq k$.

Theorem 2. *Determining whether $tl(G) \leq k$ for a given graph G is NP-complete for every fixed $k \geq 2$.*

Proof. By Corollary 3, k-Treelength is NP-hard. As it is a special case of Weighted k-Treelength it is also NP-complete.

5 Treelength Is Hard to Approximate

Having established that treelength is hard to compute, it is natural to ask how well this parameter can be approximated. We say that a polynomial time algorithm that computes a tree-decomposition of G is a c-approximation algorithm for treelength if there is an integer k so that on any input graph G, the length l of the tree-decomposition returned by the algorithm satisfies the inequality $l \leq c \cdot tl(G) + k$. Dourisboure and Gavoille have already given a 3-approximation algorithm for treelength [6], and have conjectured that the parameter is approximable within a factor 2. In this section we show that as a consequence of the results in the above section, treelength in weighted graphs can not be approximated within a factor $c < \frac{3}{2}$ unless $P = NP$. For the treelength of unweighted graphs we give a weaker inapproximability result, and conjecture that there is no c-approximation algorithm for treelength with $c < \frac{3}{2}$ unless $P = NP$.

Lemma 9. *If $P \neq NP$ then, for any $c < \frac{3}{2}$, there is no polynomial time algorithm that on an input graph G returns a tree-decomposition of G with length $l \leq c \cdot tl(G)$.*

Proof. Suppose there is such an algorithm ALG. We give a polynomial time algorithm for 2-treelength, thereby showing that $P = NP$. On input G, run ALG on G, and let l be the length of the tree-decomposition of G returned by ALG. Answer "$tl(G) \leq 2$" if $l < 3$ and "$tl(G) > 2$" otherwise. We now need to show that $tl(G) \leq 2$ if and only if $l < 3$. Assume $l < 3$. Then $tl(G) \leq l \leq 2$ as l is an integer. In the other direction, assume $tl(G) \leq 2$. In this case $l \leq c \cdot tl(G) < 3$.

Unfortunately, Lemma 9 is not sufficient to prove that it is hard to approximate treelength within a factor $c < \frac{3}{2}$. The reason for this is that an algorithm that guarantees that $l \leq \frac{4}{3} tl(G) + 1$ can not be used to recognize graphs with treelength at most 2 in the above manner. However, we can show that there can be no c-approximation algorithms for the treelength of weighted graphs by using the weights on the edges to "scale up" the gap between 2 and 3.

Theorem 3. *If $P \neq NP$ then there is no polynomial time c-approximation algorithm for weighted treelength for any $c < \frac{3}{2}$.*

Proof. The proof is similar to the one for Lemma 9. Suppose there is a polynomial time c-approximation algorithm ALG for weighted treelength of G, with $c < \frac{3}{2}$. Let k be a non-negative integer so that on any graph G with weight function w, ALG computes a tree-decomposition of G with length $l < c \cdot tl(G) + k$. Now, choose t to be the smallest positive integer so that $(\frac{3}{2} - c) \cdot t \geq k + 1$. Let w be a weight function on the edges of G so that for every edge (u, v), $w((u, v)) = t$. Observe that $tl_w(G) = tl(G) \cdot t$. Run ALG on input (G, w) and let l be the length with respect to w of the tree-decomposition returned by ALG. Answer "$tl(G) \leq 2$" if $l < 3t$ and "$tl(G) > 2$" otherwise. We now need to show that $tl(G) \leq 2$ if and only if $l < 3t$. Assume $l < 3t$. Now, $tl(G) \cdot t = tl_w(G) \leq l < 3t$. Dividing both sides by t yields $tl(G) < 3$ implying $tl(G) \leq 2$ as $tl(G)$ is an integer. In the other direction, assume $tl(G) \leq 2$. In this case $l \leq c \cdot tl_w(G) + k = c \cdot tl(G) \cdot t + k = \frac{3}{2} \cdot tl(G) \cdot t - (\frac{3}{2} - c) \cdot tl(G) \cdot t + k \leq 3t - (k + 1) + k < 3t$. This implies that the described algorithm is a polynomial time algorithm for 2-Treelength implying $P = NP$

In fact, it does not seem that treelength should be significantly harder to compute on weighted than unweighted graphs. The hardness proof for k-Treelength is a reduction directly from weighted k-Treelength. Also, the exact algorithm given in the next section works as well for computing the treelength in weighted as in unweighted graphs. We feel that together with Lemma 9 and Theorem 3 this is strong evidence to suggest that unless $P = NP$, treelength is inapproximable within a factor $c < \frac{3}{2}$, also in unweighted graphs. We state this in the following conjecture.

Conjecture 1. If $P \neq NP$ then there is no polynomial time c-approximation algorithm for treelength for any $c < \frac{3}{2}$.

6 An Exact Algorithm the Chordal Sandwich Problem

In this section we give an exact algorithm that solves the Chordal Sandwich problem. The running time of this algorithm is $O^*(1.8899^n)$. In fact, the algorithm can be obtained by a quite simple modification of an exact algorithm to compute treewidth and minimum fill in given by Fomin et. al [8]. Together with Observation 2 this gives a $O^*(1.8899^n)$ algorithm to compute the treelength of a graph. The algorithm applies dynamic programming using a list of the input graph's minimal separators and potential maximal cliques.

In order to state and prove the results in this section, we need to introduce some notation and terminology. Given two vertices u and v of G, a *minimal u-v-separator* is an inclusion minimal set $S \subseteq V$ so that u and v belong to distinct components of $G \setminus S$. A *minimal separator* is a vertex set S that is a minimal u-v-separator for some vertices u and v. We call a chordal supergraph H of G for a *minimal triangulation* of G if the only chordal sandwich between G and H is H itself. If $C \subseteq V$ is a maximal clique in some minimal triangulation of G, we say that C is a *potential maximal clique* of G. The set of all minimal separators of G is denoted by $\Delta(G)$ and the set of all potential maximal cliques is denoted by $\Pi(G)$. By $\mathcal{C}_G(S)$ we will denote the family of the vertex sets of the connected components of $G \setminus S$. Thus, if the connected components of $G \setminus S$ are $G[C_1]$ and $G[C_2]$, $\mathcal{C}_G(S) = \{C_1, C_2\}$. A *block* is a pair (S, C) where $S \in \Delta(G)$ and $C \in \mathcal{C}(S)$. A block is called *full* if $S = N(C)$. For a block (S, C) the realization of that block is denoted by $R(S, C)$ and is the graph obtained from $G[S \cup C]$ by making S into a clique.

The proof of correctness for algorithm FCS is omitted due to space restrictions.

Theorem 4. *Algorithm FCS returns TRUE if and only if there is a chordal sandwich between G_1 and G_2.*

Theorem 5. *Algorithm FCS terminates in $O^*(|\Pi_1|)$ time.*

Proof. Computing Δ_1 from Π_1 can be done in $O^*(|\Pi_1|)$ time by looping over each potential maximal clique $\Omega \in \Pi_1$ and inserting $N(C)$ into Δ_1 unless already present for every connected component C of $G \setminus \Omega$. $\mathcal{F}_1$ can be computed similarly and then sorted in $O^*(|\Pi_1|)$ time. While building Δ_1 and $\mathcal{F}_1$ we can store a

```
Algorithm: Find Chordal Sandwich – FCS (G1, G2)
Input: Two graphs G1 = (V1, E1) and G2 = (V2, E2) so that G1 ⊆ G2,
       together with a list Π1 of all potential maximal cliques of G1 that
       induce cliques in G2.
Output: TRUE if there is a chordal sandwich between G1 and G2, FALSE
        otherwise.

Δ1 := {S ∈ Δ(G1) : There is an Ω ∈ Π1 so that S1 ⊂ Ω};
F1 := the set of all full blocks (S, C) so that S ∈ Δ1, sorted by |S ∪ C|;
Cs(R(S, C)) := FALSE for every pair of vertex sets S and C;
foreach full block (S, C) in F1 taken in ascending order do
    foreach potential maximal clique Ω ∈ Π1 so that S ⊂ Ω ⊆ S ∪ C do
        ok := TRUE;
        foreach full block (Si, Ci) where Ci ∈ C_G1(Ω) and Si = N(Ci) do
            if Cs(R(Si, Ci)) = FALSE then
                ok := FALSE;
        Cs(R(S, C)) := R(S, C) ∨ ok;
if G1 is a clique then
    RETURN TRUE;
else
    RETURN ⋁_{S∈Δ1} ⋀_{C∈C(S)} Cs(R(S, C));
```

pointer from every full block $(S, C) \in \mathcal{F}$ to all potential maximal cliques $\Omega \in \Pi_1$ satisfying $S \subset \Omega \subseteq S \cup C$. Using these pointers, in each iteration of the second **foreach** loop we can find the next potential maximal clique Ω to consider in constant time. Furthermore, it is easy to see that each iteration of the second **foreach** loop runs in polynomial time. Thus, the total running time is bounded by $O^*(\sum_{(S,C)\in\mathcal{F}_1} |\{\Omega \in \Pi_1 : S \subset \Omega \subseteq S \cup C\}|) = O^*(\sum_{\Omega\in\Pi_1} |\{(S,C) \in \mathcal{F}_1 : S \subset \Omega \subseteq S \cup C\}|)$. But as $|\{(S,C) \in \mathcal{F}_1 : S \subset \Omega \subseteq S \cup C\}| \leq n$ for every potential maximal clique Ω, it follows that the algorithm runs in time $O^*(\sum_{\Omega\in\Pi_1} |\{(S,C) \in \mathcal{F}_1 : S \subset \Omega \subseteq S \cup C\}|) = O^*(|\Pi_1|)$.

Theorem 6. *[8] $\Pi(G)$ can be listed in $O^*(1.8899^n)$ time. Thus $|\Pi(G)| = O^*(1.8899^n)$.*

Corollary 4. *There is an algorithm that solves the Chordal Sandwich problem in time $O^*(1.8899^n)$.*

Proof. Compute $\Pi(G)$. By Theorem 6 this can be done in $O(1.8899^n)$ time. Now, for every $\Omega \in \Pi(G)$ we can test in $O(n^2)$ time whether it is a clique in G_2. If it is, insert Ω into Π_1. We can now call algorithm FCS on G_1, G_2 and Π_1, and return the same answer as algorithm FCS. By Theorem 5 algorithm FCS terminates in time $O^*(|\Pi_1|) = O^*(1.8899^n)$ completing the proof.

Corollary 5. *There is an algorithm that solves the Chordal Sandwich problem in time $O^*(2^{tw(G_2)})$ where $tw(G_2)$ is the treewidth of G_2.*

Proof. For any tree-decomposition of G_2, every clique of G_2 is contained in some bag in this tree-decomposition [12]. Thus, G_2 has at most $O^*(2^{tw(G_2)})$ cliques. We can list all cliques of a graph with a polynomial delay [13]. For every clique Ω in G_2 we can test whether it is a potential maximal clique of G_1 in polynomial time [2]. If it is, we insert Ω into Π_1. Thus $|\Pi_1| = O^*(2^{tw(G_2)})$. Finally, call algorithm FCS on G_1, G_2 and Π_1, and return the same answer as algorithm FCS. By Theorem 5 algorithm FCS terminates in time $O^*(|Pi_1|) = O^*(2^{tw(G_2)})$ completing the proof.

Corollary 6. *There is an algorithm for Weighted k-Treelength that runs in time $O^*(1.8899^n)$.*

Proof. By Observation 2 $tl_w(G) \leq k$ if and only if there is a chordal sandwich between G and G_w^k. By Corollary 4 we can check this in time $O^*(1.8899^n)$.

7 Conclusion

We have proved that it is NP-complete to recognize graphs with treelength bounded by a constant $k \geq 2$. In addition we have proved that unless $P = NP$ there can be no approximation algorithm for the treelength of weighted graphs with approximation factor better than $\frac{3}{2}$ and conjectured that a similar result holds for unweighted graphs. Finally we gave a $O^*(1.8899^n)$ algorithm to solve the Chordal Sandwich problem and showed how it can be used to determine the treelength of a graph within the same time bound. Dourisboure and Gavoille provide two 3-approximation algorithms for treelength in [6], and propose a heuristic that they conjecture is a 2-approximation algorithm. Thus there are currently two unresolved conjectures about the approximability of treelength, and resolving any of these would be of interest.

References

1. Bodlaender, H.L.: A linear time algorithm for finding tree-decompositions of small treewidth. SIAM Journal on Computing 25, 1305–1317 (1996)
2. Bouchitte, V., Todinca, I.: Treewidth and minimum fill-in: Grouping the minimal separators. SIAM Journal on Computing 31(1), 212–232 (2001)
3. Buneman, P.: A characterization of rigid circuit graphs. Discrete Mathematics 9, 205–212 (1974)
4. Dourisboure, Y.: Compact routing schemes for bounded tree-length graphs and for k-chordal graphs. In: Guerraoui, R. (ed.) DISC 2004. LNCS, vol. 3274, pp. 365–378. Springer, Heidelberg (2004)
5. Dourisboure, Y., Dragan, F.F., Gavoille, C., Yan, C.: Sparse additive spanners for bounded tree-length graphs. Theoretical Computer Science (to appear)
6. Dourisboure, Y., Gavoille, C.: Tree-decompositions with bags of small diameter. Discrete Mathematics (to appear)

7. Feige, U., Hajiaghayi, M.T., Lee, J.R.: Improved approximation algorithms for minimum-weight vertex separators. In: 37th Annual ACM Symposium on Theory of Computing (STOC), ACM, New York (2005)
8. Fomin, F.V., Kratsch, D., Todinca, I., Villanger, Y.: Exact algorithms for treewidth and minimum fill-in. SIAM Journal on Computing (submitted) (first appearance at ICALP (2004)
9. Fulkerson, D.R., Gross, O.A.: Incidence matrices and interval graphs. Pacific Journal of Mathematics 15, 835–855 (1965)
10. Gavril, F.: The intersection graphs of subtrees in trees are exactly the chordal graps. Journal of Combinatorial Theory B 16, 47–56 (1974)
11. Golumbic, M.C., Kaplan, H., Shamir, R.: Graph sandwich problems. J. Algorithms 19(3), 449–473 (1995)
12. Heggernes, P.: Minimal triangulations of graphs: A survey. Discrete Mathematics 306(3), 297–317 (2006)
13. Makino, K., Uno, T.: New algorithms for enumerating all maximal cliques, pp. 260–272 (2004)
14. Walter, J.: Representation of rigid cycle graphs. PhD thesis (1972)

On Time Lookahead Algorithms for the Online Data Acknowledgement Problem*

Csanád Imreh[1] and Tamás Németh[2]

[1] Department of Informatics, University of Szeged, Árpád tér 2, H-6720 Szeged, Hungary
cimreh@inf.u-szeged.hu
[2] Móra Ferenc Highschool, Kálvária sgt. 84, H-6724, Szeged, Hungary
tom@moraisk.hu

Abstract. In this work we investigate such online algorithms for the data acknowledgement problem, which have extra information about the arrival time of the packets in the following time interval of length c. We present an algorithm with the smallest possible competitive ratio for the maximum of delay type objective function. In the case of the sum of delay type objective function we present an $1+O(1/c)$-competitive algorithm. Moreover we show that no algorithm may have smaller competitive ratio than $1+\Omega(1/c^2)$ in the case of that objective function.

Keywords: Online algorithms, lookahead, data acknowledgement.

1 Introduction

In the communication of a computer network the information is sent by packets. If the communication channel is not completely safe then the arrival of the packets must be acknowledged. The TCP implementations are also using acknowledgement of the packets (see [10]). In the data acknowledgement problem we try to determine the time of sending acknowledgements. One acknowledgement can acknowledge many packets but waiting for long time can cause the resending of the packets and that results the congestion of the network. On the other hand sending immediately an acknowledgement about the arrival of each packet would cause again the congestion of the network.

The first online optimization model for determining the sending time of the acknowledgements was developed in [4]. In the model each packet has an arrival time, and at any time the algorithm has to make a decision about the acknowledgement of the arrived packets without any information about the further packets. Two objective functions are investigated. Both of them are the convex combination of the number of acknowledgements and the total latency cost assigned to the acknowledgements (with the coefficients $\gamma, 1-\gamma$). The difference is in the definition of the latency cost assigned to an acknowledgement.

* This research has been supported by the Hungarian National Foundation for Scientific Research, Grant F048587.

L. Kučera and A. Kučera (Eds.): MFCS 2007, LNCS 4708, pp. 288–297, 2007.

In the case of function $f_{\max}$ this is the maximum of the delays which the packets have, in the case of f_{sum} it is the sum of the delays of the packets acknowledged.

Optimal 2-competitive online algorithms are presented in both cases. Semi-online algorithms with some lookahead properties are also considered. In the real application the algorithms have to be online, they do not have lookahead information about the further packets. On the other hand investigating lookahead algorithms is also important for this problem (see [4]). Such algorithms can be used to understand how useful some learning algorithm to estimate the further arrivals of the packets can be. In the case of the $f_{\max}$ objective function it is enough to know the arrival time of the next packet to achieve a 1-competitive algorithm. On the other hand, in the case of the f_{sum} function the knowledge of the next k arrivals is not enough to have better competitive ratio than 2 for any constant k.

In this paper we investigate another version of lookahead property. Instead of giving the knowledge of the arrival times of the next k packets we allow the algorithm to see the arrival times of the packets in a time interval of length c. This type of lookahead is called time lookahead property and it is investigated in [2] for online vehicle routing problems. This new type of lookahead property gives completely different results than the lookahead property investigated in [4]. In the case of the function $f_{\max}$ we can obtain a 1-competitive algorithm if $c \geq \gamma/(1-\gamma)$ and we can define a $2-c(1-\gamma)/\gamma$-competitive algorithm for the smaller values of c. We also prove that smaller competitive ratio can not be achieved. In the case of the function f_{sum} the new lookahead definition allows to achieve smaller competitive ratio than 2. We present an algorithm with a competitive ratio tending 1 in order of magnitude $1 + O(1/c)$ as c is increasing. We also show that no 1-competitive algorithm may exist for constant size of lookahead intervals, we prove a $1 + \Omega(1/c^2)$ lower bound on the possible competitive ratio. We also present the lower bound $2\gamma/(c(1-\gamma)+\gamma)$ for the smaller values of c.

There are some further results on the area of data acknowledgement. The offline version of the problem with the function f_{sum} is further investigated in [8] where a faster, linear time algorithm is developed for its solution. Randomized online algorithms are considered in [7] and [9]. In [7] an $e/(e-1)$-competitive algorithm is given for the solution of the problem. In [9] it is shown that no online randomized algorithm can have smaller competitive ratio than $e/(e-1)$.

Some further objective functions are investigated in [1] and [6]. In [1] the objective function is the sum of the number of acknowledgements and the maximum delay of the packets. A generalized version of the function where the maximum delay is on the power p is also investigated. In both cases optimal online algorithms are presented, in the case of $p = 1$ the algorithm has the competitive ratio $\pi^2/6$, in the general case the competitive ratio can be given by Riemann's zeta function, it tends to 1.5 as p tends to ∞. The paper contains also lower bounds on the competitive ratio of randomized algorithms. In [6] a further cost function is investigated, which can be considered as an extension of $f_{\max}$, and it is also required that the difference between the times of the acknowledgements is at least one time unit. In that paper optimal $(1+\sqrt{5})/2$-competitive deterministic and an optimal $(\sqrt{3}+1)/2$-competitive randomized algorithm are presented. Moreover a class of algorithms which are allowed to use only a limited number of random bits is also investigated. A more general problem than the data acknowledgement problem is investigated in [5]. In that paper an online problem motivated by service call management is considered. The model in the case of two customers can be considered as a generalization of the data acknowledgement problem.

Semi-online algorithms with lookahead are investigated for several online optimization problems. We do not collect all of these results here, we just mention here the most recent paper on online vehicle routing ([2]), one can find further examples in its list of references.

2 Preliminaries

We use the mathematical model which was defined in [4]. In the model the input is the list of the arrival times $a_1, \ldots, a_n$ of the packets. We also denote the packets by their arrival time. The decision maker has to determine when to send acknowledgements, these times are denoted by $t_1, \ldots, t_k$. We consider the objective function

$$\gamma k + (1-\gamma) \sum_{j=1}^{k} L_j,$$

where k is the number of the sent acknowledgements and L_j is the extra latency belonging to the acknowledgement t_j and $\gamma \in (0,1)$ is a constant. We consider two different cases. We obtain the objective function $f_{\max}$ if $L_j = \max_{t_{j-1} < a_i \leq t_j} (t_j - a_i)$, the maximal delay collected by j-th acknowledgement. We obtain the objective function f_{sum} if $L_j = \sum_{t_{j-1} < a_i \leq t_j} (t_j - a_i)$, the sum of the delays collected by the j-th acknowledgement.

In the on-line problem at time t the decision maker only knows the arrival times of the packets already arrived and has no information about the further packets. We consider a semi-online model with time lookahead c, where at time t the decision maker knows the arrival times of the packets already arrived and also knows the arrival times of the packets arriving in the time interval $(t, t+c]$. We denote the set of the unacknowledged packets at the arrival time a_i by σ_i. For an arbitrary list L of packets and an algorithm A, we denote by $A(L)$ the total cost of the acknowledgements sent by algorithm A on list L. The total cost of sending optimally the acknowledgements is denoted by $OPT(L)$. We use the competitive analysis to evaluate the algorithms, as it is usual in the case of online and semi-online algorithms ([3]). An algorithm is d-competitive if $A(I) \leq d \cdot OPT(I)$ is valid for every input I. The competitive ratio of an algorithm is the smallest number d for which the algorithm is d-competitive.

We will examine time lookahead extensions of the online alarming algorithms defined in [4], thus we recall here the definition of these algorithms. An alarming algorithm works as follows. At the arrival time a_j an alarm is set for time $a_j + e_j$. If no packet arrives before time $a_j + e_j$, then an acknowledgement is sent at time $a_j + e_j$ which acknowledges all of the unacknowledged packets. Otherwise at the arrival of the next packet at time a_{j+1} the alarm is reset for time $a_{j+1} + e_{j+1}$.

These algorithms are defined and analysed in [4]. In the case of function $f_{\max}$ the alarming algorithm which uses the value $e_j = \gamma/(1-\gamma)$ for each j is an optimal 2-competitive algorithm. In the case of function f_{sum} the alarming algorithm which uses the value $e_j = (\gamma/(1-\gamma) - \sum_{a_i \in \sigma_j}(a_j - a_i))/|\sigma_j|$ for each j is an optimal 2-competitive algorithm. (It is worth noting that in this case e_j is chosen to balance the two types of cost, if no further packet arrives than sending an acknowledgement at time $a_j + e_j$ has latency cost γ).

3 The $f_{\max}$ Objective Function

3.1 Algorithm

In [4] it is shown that the optimal offline solution has a very simple structure in the case of function $f_{\max}$. The following statement is valid.

Proposition 1. *Under $f_{\max}$ there exists an optimal solution S that places an acknowledgement at a_j if and only if $a_{j+1} - a_j \geq \gamma/(1-\gamma)$.*

We consider two cases depending on the value of c. When $c \geq \gamma/(1-\gamma)$, then we obtain a 1-competitive algorithm easily. The size of the lookahead interval is large enough to determine whether $a_{j+1} - a_j > \gamma/(1-\gamma)$ is valid or not, thus an algorithm with lookahead c can find the optimal solution described in Proposition 1.

The case when $c < \gamma/(1-\gamma)$ is more interesting. We define for this case an extended version of the alarming algorithm developed in [4]. This *time lookahead alarming algorithm* (TLA in short) works as follows. At the arrival time a_j an alarm is set for time $a_j + \gamma/(1-\gamma) - c$. If the packet a_{j+1} arrives before the alarm or we can see a_{j+1} at time $a_j + \gamma/(1-\gamma) - c$ in the lookahead interval $(a_j + \gamma/(1-\gamma) - c, a_j + \gamma/(1-\gamma)]$ then we postpone the alarm to the time $a_{j+1} + \gamma/(1-\gamma) - c$. In the opposite case (no packet arrives in the time interval $(a_j, a_j + \gamma/(1-\gamma)]$) an acknowledgement is sent at time $a_j + \gamma/(1-\gamma) - c$ which acknowledges all of the unacknowledged packets.

Theorem 1. *TLA is* $\max\{1, 2 - \frac{1-\gamma}{\gamma}c\}$*-competitive.*

Proof. First we show that the algorithm is $2 - \frac{1-\gamma}{\gamma}c$-competitive. Consider an arbitrary input sequence $a_1, \dots, a_n$. Partition the sequence into phases. Let $S_1 = \{a_1, \dots, a_{k(1)}\}$ where $k(1)$ is the first index with the property $a_{k(1)+1} - a_{k(1)} \geq \gamma/(1-\gamma)$. The other phases are defined in the same way $S_{j+1} = \{a_{k(j)+1}, \dots, a_{k(j+1)}\}$ where $k(j+1)$ is the first index after $k(j)$ with the property $a_{k(j+1)+1} - a_{k(j+1)} \geq \gamma/(1-\gamma)$. The last phase is ended by the last packet. We will also use the value $k(0) = 0$.

Then an optimal offline algorithm sends an acknowledgement at the last packet of each phase. Therefore it has the total cost $OPT(S_j) = \gamma + (1-\gamma)(a_{k(j)} - a_{k(j-1)+1})$ for the acknowledgement of the j-th phase. On the other hand TLA sends the acknowledgement for the phase at time $a_{k(j)} + \gamma/(1-\gamma) - c$, thus it has a total cost $TLA(S_j) = \gamma + (1-\gamma)(a_{k(j)} + \gamma/(1-\gamma) - c - a_{k(j-1)+1})$ for the acknowledgement of the j-th phase. We have $TLA(S_j)/OPT(S_j) > 1$, thus decreasing both values by the same constant increases the ratio. Therefore we obtain that

$$\frac{TLA(S_j)}{OPT(S_j)} \leq \frac{\gamma + (1-\gamma)(\gamma/(1-\gamma) - c)}{\gamma} = 2 - \frac{1-\gamma}{\gamma}c.$$

Since the total cost is the sum of the costs of the phases, thus we obtain that TLA is $2 - \frac{1-\gamma}{\gamma}c$-competitive.

3.2 Lower Bound

TLA has the smallest possible competitive ratio, as the following statement shows.

Theorem 2. *No semi-online algorithm with lookahead c may have smaller competitive ratio than* $\max\{1, 2 - \frac{1-\gamma}{\gamma}c\}$.

Proof. If $c \geq \gamma/(1-\gamma)$ then the statement is obviously true, thus we can assume that $c < \gamma/(1-\gamma)$. Consider an arbitrary online algorithm, denote it by A. Define the following input sequence denoted by I_n. The arrival time of the first packet is $a_1 = 0$, then the i-th packet arrives ($i = 2, \ldots, n$) c time units after the acknowledgement of the $i-1$-th packet ($a_i = t_{i-1} + c$).

We partition the input sequence into phases in the same way as in the proof of Theorem 1. Denote the phases by $S_1, \ldots, S_j$. Consider the phase S_i. To simplify the notation denote the value $k(i) - k(i-1)$ by r_i. Algorithm A sends an acknowledgement for each packet thus we obtain that

$$A(S_i) = \gamma r_i + (1-\gamma) \sum_{p=k(i-1)+1}^{k(i)} (t_p - a_p).$$

The optimal solution sends only one acknowledgement at time $a_{k(i)}$ thus

$$OPT(S_i) = \gamma + (1-\gamma)\Big(\sum_{p=k(i-1)+1}^{k(i)-1} (t_p - a_p) + c(r_i - 1)\Big).$$

Now suppose that $i < j$. If we calculate $A(S_i) - (2 - \frac{1-\gamma}{\gamma}c)OPT(S_i)$ and we use that $t_p - a_p \leq \gamma/(1-\gamma) - c$ for $p = k(i-1)+1, \ldots, k(i)-1$ and $t_{k(i)} - a_{k(i)} > \gamma/(1-\gamma) - c$ then we obtain that

$$A(S_i) - (2 - \frac{1-\gamma}{\gamma}c)OPT(S_i) \geq \gamma(r_i - 2 + \frac{1-\gamma}{\gamma}c) + (1-\gamma)\Big((\frac{1-\gamma}{\gamma}c - 2)(r_i - 1)c +$$

$$(r_i - 1)(\frac{1-\gamma}{\gamma}c - 1)(\frac{\gamma}{1-\gamma} - c) + \frac{\gamma}{1-\gamma} - c\Big) = \gamma(r_i - 2 + \frac{1-\gamma}{\gamma}c) -$$

$$(1-\gamma)(r_i - 1)c + (r_i - 1)((1-\gamma)c - \gamma) + \gamma - c(1-\gamma) = 0.$$

In the second equality we simplified the formula by eliminating $(r_i-1)(\frac{1-\gamma}{\gamma}c-1)c$ which can be found in the formula with the coefficients $+1$ and -1.

Therefore we proved that $A(S_i) \geq (2 - \frac{1-\gamma}{\gamma}c)OPT(S_i)$ if $i < j$. In the case of S_j there is only one difference, we cannot state that the inequality $t_{k(j)} - a_{k(j)} > \gamma/(1-\gamma) - c$ is valid. But we can prove in the same way as above that $A(S_j) + \gamma/(1-\gamma) - c \geq (2 - \frac{1-\gamma}{\gamma}c)OPT(S_j)$.

Since the total cost is the sum of the costs of the phases, thus we obtain that $A(I_n) + \gamma/(1-\gamma) - c \geq (2 - \frac{1-\gamma}{\gamma}c)OPT(I_n)$. On the other hand as n tends to ∞ the value of $OPT(I_n)$ also tends to ∞, thus the above inequality shows that $A(I_n)/OPT(I_n)$ can be arbitrarily close to $2 - \frac{1-\gamma}{\gamma}c$ thus the competitive ratio of A can not be smaller than $2 - \frac{1-\gamma}{\gamma}c$.

4 The Sum Objective Function

4.1 Algorithms

It is a straightforward idea to also use the time lookahead extension of the alarming algorithm from [4] in this case. We can define the TLA extension of the alarming algorithm for this case as follows. At the arrival time a_j an alarm is set for time $a_j + e_j$ where $e_j = (\gamma/(1-\gamma) - \sum_{a_i \in \sigma_j}(a_j - a_i))/|\sigma_j|$. If the packet a_{j+1} arrives before the time $\max\{a_j, a_j+e_j-c\}$ or we can see a_{j+1} at this time in the lookahead interval, then we move to a_{j+1} and reset the alarm. In the opposite case (no packet arrives in the time interval $(a_j, a_j + e_j])$ an acknowledgement is sent at time $\max\{a_j, a_j + e_j - c\}$ which acknowledges all of the unacknowledged packets. Unfortunately this lookahead extension of the algorithm does not make it possible to achieve a smaller competitive ratio than 2.

Theorem 3. *The competitive ratio of TLA is* 2 *for arbitrary* c.

Proof. The cost of this algorithm is at most the cost of the online alarming algorithm from [4] on any input, thus it follows immediately that TLA is 2-competitive from the result that the online alarming algorithm is 2-competitive. TLA has no better performance as the following input sequence shows. Let $I_n = \{a_1, \ldots, a_{2n+1}\}$ where $a_1 = 0$ and $a_{2i} = i\gamma/(1-\gamma) + (i-1)\varepsilon$, $a_{2i+1} = i\gamma/(1-\gamma) + i\varepsilon$ for $i = 1, \ldots, n$. Then TLA sends the acknowledgements at $a_2, \ldots, a_{2n}, \max\{a_{2n+1}, a_{2n+1} + \gamma/(1-\gamma) - c\}$. Thus $TLA(I_n) = (n+1)\gamma + n\gamma + (1-\gamma)\max\{0, \gamma/(1-\gamma) - c\}$. An optimal offline algorithm sends an acknowledgement at a_1, a_3, a_{2n+1} and it has the cost $OPT(I_n) = (n+1)\gamma + (1-\gamma)n\varepsilon$. The ratio of the costs tends to 2 as ε tends to 0 and n tends to ∞, and this yields that the competitive ratio is at least 2.

In the case when $c > \gamma/(1-\gamma)$ we can achieve smaller competitive ratio than 2 by the following algorithm. We present the Lookahead Interval Planning Algorithm (LIP in short). The algorithm partitions the packets into blocks and for each block determines the acknowledgments based on an offline optimal solution. The block always starts at the first unacknowledged packet. First the algorithm examines whether there exist packets a_i, a_{i+1} in the c length lookahead interval with the property $a_{i+1} - a_i > \gamma/(1-\gamma)$. If there exists such pair, then the block is ended at the first such a_i, otherwise the block has length c. Then LIP calculates the optimal solution of the offline acknowledgement problem for the packets in the block, it can use one of the algorithms which solves the offline problem (such algorithms are presented in [4] and [8]) and sends the acknowledgements according to this solution and considers the next block.

Theorem 4. *LIP is* $1 + \frac{\gamma}{(1-\gamma)c}$*-competitive.*

Proof. To prove that LIP is $1 + \frac{\gamma}{(1-\gamma)c}$-competitive consider an arbitrary input I. Divide the input into phases in the same way as in the proof of Theorem 1. Let us observe that there exists an offline optimal algorithm which sends an acknowledgement at the last packet of each phase. (Because, if last packet of the phase is delayed, it incurs a delay cost of more than $(1-\gamma)(\gamma/(1-\gamma)) = \gamma$, whereas it incurs a communication cost of exactly γ if it is acknowledged. Furthermore let us observe that the last packet of a phase is always a last packet of some block, thus LIP also sends an acknowledgement at the last packet of the phase.

Consider an arbitrary phase S_i. Denote by r the number of blocks in the phase. Consider an optimal solution of the phase. If we extend it with $r-1$ further acknowledgements on the end of the first $r-1$ blocks, then we obtain a solution which acknowledges the blocks separately. But LIP gives the best such solution therefore we obtain that $(r-1)\gamma + OPT(S_i) \geq LIP(S_i)$ which yields that $LIP(S_i)/OPT(S_i) \leq 1 + (r-1)\gamma/OPT(S_i)$.

Consider the value of $OPT(S_i)$. Since each block is in the same phase, thus the length of each of the first $r-1$ blocks is c, therefore the length of the phase is at least $(r-1)c$. Suppose that the optimal offline algorithm sends k acknowledgements in this phase. Then after each of the first $k-1$ acknowledgement there is an at most $\gamma/(1-\gamma)$ length interval without packet. This yields that in this case the total latency cost of OPT is at least $(1-\gamma)((r-1)c-(k-1)\gamma/(1-\gamma))$. Therefore $OPT(S_i) \geq k\gamma + (1-\gamma)((r-1)c - (k-1)\gamma/(1-\gamma)) = (1-\gamma)(r-1)c + \gamma$. On the other hand if we use this bound we obtain that $LIP(S_i)/OPT(S_i) \leq 1 + \frac{\gamma}{(1-\gamma)c}$.

4.2 Lower Bounds

First we give a lower bound on the order of magnitude of the possible competitive ratio. This bound is useful for large lookahead intervals, it shows that no constant size lookahead is enough to achieve 1-competitiveness in the case of the f_{sum} function.

Theorem 5. *No online algorithm may have smaller competitive ratio than* $1 + \Omega(1/c^2)$.

Proof. To simplify the calculation suppose that $c = \gamma(k - 1/4)/(1-\gamma)$, where $k \geq 1$ is an integer. We can assume that without loss of generality since allowing larger lookahead can not increase the competitive ratio and we are proving lower bound on the order of magnitude of the possible competitive ratio. Consider an arbitrary algorithm A with lookahead c. Let $x = \frac{\gamma(2k+1)}{4k(1-\gamma)}$ and $y = \gamma/2(1-\gamma)$. Define the following sequences for each $j = 1, \dots, k$.

- $S_{xjx} = \{a_1, a_2, \dots, a_{2j}\}$ where $a_{2i-1} = (i-1)x+(i-1)y$ and $a_{2i} = ix+(i-1)y$ for $i = 1, \dots, j$. Let us note that $a_{2k} = c$ in S_{xkx}.
- $S_{xjy} = \{a_1, a_2, \dots, a_{2j+1}\}$ where $a_{2i-1} = (i-1)x+(i-1)y$ for $i = 1, \dots, j+1$ and $a_{2i} = ix + (i-1)y$ for $i = 1, \dots, j$.
- $S_{yjy} = \{a_1, a_2, \dots, a_{2j}\}$ where $a_{2i-1} = (i-1)y+(i-1)x$ and $a_{2i} = iy+(i-1)x$ for $i = 1, \dots, j$.
- $S_{yjx} = \{a_1, a_2, \dots, a_{2j+1}\}$ where $a_{2i-1} = (i-1)y+(i-1)x$ for $i = 1, \dots, j+1$ and $a_{2i} = iy + (i-1)x$ for $i = 1, \dots, j$.

Since $\gamma + (1-\gamma)(2y+x) = 2\gamma + (1-\gamma)x$, thus we obtain that there exist optimal solutions for the above defined sequences which never acknowledge more than 2 packets with one acknowledgement. Using this observation the following lemma can be easily proven by induction.

Lemma 1. *For each $1 \le j \le k$ we have $OPT(S_{xjy}) = OPT(S_{yjx}) = \gamma(j+1) + (1-\gamma)jy$, $OPT(S_{xjx}) = \gamma j + (1-\gamma)jx$, $OPT(S_{yjy}) = \gamma j + (1-\gamma)jy$.*

Give S_{xkx} as the first part of the input to A and wait till time y. Suppose that the algorithm sends an acknowledgement at time $z \le y$. Then it acknowledges the packet a_1 and it has to handle the remaining part which is $S_{y(k-1)x}$. Therefore by Lemma 1 we obtain that $A(S_{xkx}) \ge \gamma + (1-\gamma)z + k\gamma + (1-\gamma)(k-1)y$. Therefore we obtain that

$$\frac{A(S_{xkx})}{OPT(S_{xkx})} \ge \frac{(k+1)\gamma + (1-\gamma)(k-1)y}{k\gamma + k(1-\gamma)x} = 1 + \frac{1}{6k+1}.$$

Now suppose that A does not send an acknowledgement before time y. Then at time $y + c$ a further packet arrives, thus the input is S_{xky}. The algorithm observes this packet at time y. If it acknowledges the first packet before time x then $A(S_{xky}) \ge \gamma + (1-\gamma)y + OPT(S_{yky})$. Therefore by Lemma 1 we obtain that the following inequality for this case:

$$\frac{A(S_{xky})}{OPT(S_{xky})} \ge \frac{(k+1)\gamma + (1-\gamma)(k+1)y}{(k+1)\gamma + (1-\gamma)ky} = 1 + \frac{1}{3k+2}$$

Finally, suppose that A does not send an acknowledgement before time x. If it sends its first acknowledgement later than x then its total cost is increasing, therefore we can assume that the first acknowledgement is sent at time x. Then the algorithm acknowledges the first two packets and the remaining part is $S_{x(k-1)y}$, thus we obtain $A(S_{xky}) \ge \gamma + (1-\gamma)x + OPT(S_{x(k-1)y})$. Therefore by Lemma 1 we obtain that the following inequality for this case:

$$\frac{A(S_{xky})}{OPT(S_{xky})} \ge \frac{(k+1)\gamma + (1-\gamma)(x + (k-1)y)}{(k+1)\gamma + (1-\gamma)ky} \ge 1 + \frac{1}{6k^2+4k}$$

Since we examined all of the possible cases, we proved the statement of the theorem.

The above bound does not give better value than 1 in the case when the algorithm has only a small lookahead interval. We also show a lower bound for $c \leq \gamma/(1-\gamma)$ by extending the technique which is used to prove a lower bound in [4].

Theorem 6. *No online algorithm with lookahead c can have smaller competitive ratio than* $2\gamma/(c(1-\gamma)+\gamma)$.

Proof. Consider an arbitrary on-line algorithm denote it by A. Analyze the following input. Consider a long sequence of packets where the packets always arrive c time units after the time when A sends an acknowledgement ($t_j + c = a_{j+1}$). Then the on-line cost of a sequence containing $2n+1$ packets is $A(I_{2n+1}) = \gamma(2n+1) + (1-\gamma)\sum_{i=1}^{2n+1}(t_i - a_i)$. Consider the following two offline algorithms. ODD sends the acknowledgements after the odd numbered packets and after the last packet, and EV sends the acknowledgements after the even numbered packets.

Then the costs achieved by these algorithms are

$$\mathrm{EV}(I_{2n+1}) = (n+1)\gamma + (1-\gamma)\sum_{i=1}^{n}(a_{2i+1} - a_{2i}) = (n+1)\gamma + (1-\gamma)(nc + \sum_{i=1}^{n}(t_{2i} - a_{2i}))$$

and

$$\mathrm{ODD} = (n+1)\gamma + (1-\gamma)\sum_{i=1}^{n}(a_{2i} - a_{2i-1}) = (n+1)\gamma + (1-\gamma)(nc + \sum_{i=1}^{n}(t_{2i-1} - a_{2i-1})).$$

On the other hand none of the costs achieved by ODD and EV is smaller than the optimal offline cost, thus $OPT(I_{2n+1}) \leq \min\{EV(I_{2n+1}), \mathrm{ODD}(I_{2n+1})\} \leq (EV(I_{2n+1}) + ODD(I_{2n+1}))/2$. Therefore we obtain that

$$\frac{A(I_{2n+1})}{OPT(I_{2n+1})} \geq \frac{2(\gamma(2n+1) + (1-\gamma)\sum_{i=1}^{2n+1}(t_i - a_i))}{\gamma(2n+2) + (1-\gamma)(\sum_{i=1}^{2n}(t_i - a_i) + 2nc)} \geq 2 - \frac{2\gamma + 4nc(1-\gamma)}{\gamma(2n+2) + 2nc(1-\gamma)}.$$

The ratio which we obtained as a lower bound on $A(I_{2n+1})/OPT(I_{2n+1})$ tends to $2\gamma/(c(1-\gamma)+\gamma)$ as n tends to ∞, and this proves the theorem.

References

1. Albers, S., Bals, H.: Dynamic TCP acknowledgement: Penalizing long delays. SIAM J. Discrete Math. 19(4), 938–951 (2005)
2. Allulli, L., Ausiello, G., Laura, L.: On the power of lookahead in on-line vehicle routing problems. In: Wang, L. (ed.) COCOON 2005. LNCS, vol. 3595, pp. 728–736. Springer, Heidelberg (2005)
3. Borodin, A., El-Yaniv, R.: Online Computation and Competitive Analysis, Cambridge University Press (1998)
4. Dooly, D.R., Goldman, S.A., Scott, S.D.: On-line analysis of the TCP acknowledgment delay problem. J. ACM 48(2), 243–273 (2001)
5. Epstein, L., Kesselman, A.: On the remote server problem or more about TCP acknowledgments. Theoretical Computer Science 369(1-3), 285–299 (2006)

6. Frederiksen, J.S., Larsen, K.S., Noga, J., Uthaisombut, P.: Dynamic TCP acknowledgment in the LogP model. Journal of Algorithms 48(2), 407–428 (2003)
7. Karlin, A.R., Kenyon, C., Randall, D.: Dynamic TCP acknowledgement and other stories about e/(e-1). Algorithmica 36, 209–224 (2003)
8. Noga, J., Seiden, S.S., Woeginger, G.J.: A faster off-line algorithm for the TCP acknowledgement problem. Information Processing Letters 81(2), 71–73 (2002)
9. Seiden, S.S.: A guessing game and randomized online algorithms. In: Proceedings of STOC, pp. 592–601 (2000)
10. Stevens, W.R.: TCP/IP Illustrated, Volume I. The protocols, Addison Wesley, Reading (1994)

Real Time Language Recognition on 2D Cellular Automata: Dealing with Non-convex Neighborhoods

Martin Delacourt and Victor Poupet

LIP (UMR 5668 — CNRS, ENS Lyon, UCB Lyon, INRIA), ENS Lyon,
46 allée d'Italie, 69364 LYON cedex 07, France
martin.delacourt@ens-lyon.fr, victor.poupet@ens-lyon.fr

Abstract. In this paper we study language recognition by two-dimensional cellular automata on different possible neighborhoods. Since it is known that all complete neighborhoods are linearly equivalent we focus on a natural sub-linear complexity class: the real time.

We show that any complete neighborhood is sufficient to recognize in real time any language that can be recognized in real-time by a cellular automaton working on the convex hull of V.

1 Introduction

Cellular automata are a widely studied computing model, very well suited for studying parallel computing (as opposed to most other models such as Turing machines or RAM machines). It is made of infinitely many elementary machines of finite memory (the cells) that evolve synchronously at discrete times according to the states of their neighbors. All cells have the same transition rule and can only see their neighbors. Because of the parallel behavior, it is easy to consider cellular automata in any dimension $d \in \mathbb{N}$ (the cells are arranged on $\mathbb{Z}^d$). It is known that cellular automata are Turing universal [1,8].

The neighborhood of a cellular automaton (the set of cells whose states a given cell can see before changing its own) defines the possible communication between all the cells, and therefore the "geography" of the machine: the neighborhood of a cell is the set of cells from which it can get information in one time step, the neighborhood of the neighborhood is the set from which it can receive information in two time steps, and so on. In that way, the neighborhood defines the shortest paths to exchange information from one point to the other. As such, it can have a great impact on the possible computations that are held on an automaton.

An important result concerning computations on different neighborhoods is due to S. Cole [2] and states that two neighborhoods are either linearly equivalent (any computation that can be done in time T on one can be done in time $k \cdot T$ on the other for some constant k) or that there exists a cell $c \in \mathbb{Z}^d$ such that information can go from c to the origin in one of the neighborhoods but not in the other. If we consider neighborhoods that allow communications between any

L. Kučera and A. Kučera (Eds.): MFCS 2007, LNCS 4708, pp. 298–309, 2007.

two cells (which are the most interesting because they can perform all possible computations), we will want to consider sub-linear complexity classes in order to distinguish them. The *real time* is especially well suited for this study: it corresponds to the shortest possible time so that any letter of the input word could have an impact on the acceptance of the word (we will only deal with language recognition here). This real time depends on the chosen neighborhood.

In one dimension (when the cells are arranged on $\mathbb{Z}$) most studies were done on the *standard* neighborhood $\{-1, 0, 1\}$ (each cell can see its own state and that of its left and right neighbor) and the *one-way* neighborhood $\{0, 1\}$ (cells can only see their own state and their right neighbor's). It has been shown that these two neighborhoods are different [4,6,11](mainly because information can only go in one direction on the *one-way* neighborhood) and many algorithmic results are known [3,5,9,10]. If we only consider neighborhoods that are "complete enough" to perform language recognition (all letters of the input word can affect the outcome of the computation), we have shown in 2005 a stronger version of Cole's equivalence: all neighborhoods are real-time equivalent to either the one-way or standard neighborhood [7]. This was done by showing that it was possible to recognize the same languages in real time on non-convex neighborhoods (neighborhoods that had "holes", for example when a cell c can see $(c+2)$ but not $(c+1)$) than on convex ones.

In two dimensions, the situation is more complicated. Even by only considering complete neighborhoods it is known that the two more studied neighborhoods (the von Neumann and the Moore neighborhoods) are not real time equivalent [12].

In this article we will generalize the work that we had previously done in one-dimension and show that any language that is recognized in real time by a cellular automaton working on the convex hull of a complete neighborhood V can be recognized in real time by a cellular automaton working on V.

To alleviate the notations, we will only consider two-dimensional cellular automata in this article. All results can however easily be generalized to higher dimensions. The only result that might seem complicated to generalize would be Theorem 1, but it can be stated and proved similarly, by considering periodic volumes, surfaces, *etc.* and finite sets in the vincinity of the vertices. Theorem 1 is itself a two-dimensional generalization of Theorem 2.1 from [7].

2 Language Recognition by Cellular Automata

2.1 Cellular Automata

Definition 1. *A* two-dimensional cellular automaton *(2DCA) is a triple* $\mathcal{A} = (\mathcal{Q}, V, f)$ *where*

- $\mathcal{Q}$ *is a finite set called* set of states *containing a special* quiescent *state* $\#$*;*
- $V = \{v_1, \dots, v_{|V|}\} \subseteq \mathbb{Z}^2$ *is a finite set called* neighborhood *that contains 0.*
- $f : \mathcal{Q}^{|V|} \to \mathcal{Q}$ *is the* transition function*. We have* $f(\#, \dots, \#) = \#$.

For a given automaton $\mathcal{A}$, we call *configuration* of $\mathcal{A}$ any function $\mathfrak{C}$ from $\mathbb{Z}^2$ into $\mathcal{Q}$. The set of all configurations is therefore $\mathcal{Q}^{\mathbb{Z}^2}$. From the local function f we can define a global function F

$$\begin{aligned} F : \mathcal{Q}^{\mathbb{Z}^2} &\to \mathcal{Q}^{\mathbb{Z}^2} \\ \mathfrak{C} &\mapsto \mathfrak{C}' \quad \mid \quad \forall x \in \mathbb{Z}^2, \mathfrak{C}'(x) = f(\mathfrak{C}(x+v_1), \dots, \mathfrak{C}(x+v_{|V|})) \end{aligned}$$

Elements of $\mathbb{Z}^2$ are called *cells*. Given a configuration $\mathfrak{C}$, we'll say that a cell c is in state q if $\mathfrak{C}(c) = q$.

If at time $t \in \mathbb{N}$ the 2DCA is in a configuration $\mathfrak{C}$, we'll consider that at time $(t+1)$ it is in the configuration $F(\mathfrak{C})$. We can therefore define the *evolution* of a 2DCA from a configuration. This evolution is completely determined by $\mathfrak{C}$.

2.2 Two-Dimensional Language Recognition

Definition 2. *Given a finite alphabet Σ and two integers n_1 and n_2, we define the set of two-dimensional words of size (n_1, n_2) over the alphabet Σ as:*

$$\Sigma^{(n_1,n_2)} = \Sigma^{[\![0,n_1-1]\!] \times [\![0,n_2-1]\!]}$$

The set of all two-dimensional words over Σ is defined as:

$$\Sigma^{**} = \bigcup_{n_1 \in \mathbb{N}, n_2 \in \mathbb{N}} \Sigma^{(n_1,n_2)}$$

Two-dimensional words over Σ can be seen as rectangular grids of size $n_1 \times n_2$ containing letters of Σ.

Definition 3. *A language over an alphabet Σ is a subset of Σ^{**}.*

Definition 4. *We consider a 2DCA $\mathcal{A} = (\mathcal{Q}, V, f)$ and a set $\mathcal{Q}_{acc} \subseteq \mathcal{Q}$ of* accepting states. *Let $w \in \Sigma^{(n_1,n_2)}$ be a word over a finite alphabet $\Sigma \subseteq \mathcal{Q}$. We define the configuration $\mathfrak{C}_w$ as follows.*

$$\begin{aligned} \mathfrak{C}_w : \quad & \mathbb{Z}^2 \to \mathcal{Q} \\ & \begin{cases} (x,y) \mapsto w(x,y) & \textit{if } (x,y) \in [\![0, n_1 - 1]\!] \times [\![0, n_2 - 1]\!] \\ (x,y) \mapsto \# & \textit{otherwise} \end{cases} \end{aligned}$$

We'll say that the 2DCA $\mathcal{A}$ recognizes *the word w with accepting states $\mathcal{Q}_{acc}$ in time t_w if, starting from the configuration $\mathfrak{C}_w$ at time 0, the cell 0 is in a state in $\mathcal{Q}_{acc}$ at time t_w.*

Definition 5. *Let $\mathcal{A} = (\mathcal{Q}, V, f)$ be a 2DCA and $L \subseteq \Sigma^{**}$ a language on the alphabet $\Sigma \subseteq \mathcal{Q}$. For a given function $T : \mathbb{N}^2 \to \mathbb{N}$, we'll say that the language L is* recognized by $\mathcal{A}$ in time T *if there exists a set $\mathcal{Q}_{acc} \subseteq \mathcal{Q}$ such that, for all words w of size (n_1, n_2) in Σ^{**}, the 2DCA $\mathcal{A}$ recognizes w with accepting states $\mathcal{Q}_{acc}$ in time $T(n_1, n_2)$ if and only if $w \in L$.*

3 Iterated Neighborhoods

Definition 6. *Given two neighborhoods $V_1, V_2 \subseteq \mathbb{Z}^2$, we define*

$$V_1 + V_2 = \{v_1 + v_2 \mid v_1 \in V_1 \text{ and } v_2 \in V_2\}$$

Given a neighborhood V, we define its iterations as $V^0 = \{0\}$ and for all $k \in \mathbb{N}$, $V^{k+1} = V^k + V$ and its multiples as $kV = \{k \cdot v \mid v \in V\}$.

Definition 7. *A neighborhood $V \in \mathbb{Z}^2$ is said to be* complete *if $\bigcup_{k\in\mathbb{N}} V^k = \mathbb{Z}^2$.*

Definition 8. *The* continuous convex hull *of a neighborhood V, denoted* CCH(V)*, is the smallest convex polygon (in $\mathbb{R}^2$) that contains V. The (discrete)* convex hull *of V, denoted* CH(V)*, is the set of all points of $\mathbb{Z}^2$ that are in the continuous convex hull of V.*

Definition 9. *For a given neighborhood V, the* vertices *of the polygon* CCH(V) *are all elements of V (and therefore elements of $\mathbb{Z}^2$). We will call them the vertices of V.*

When considering the set $\{s_1, \dots, s_p\}$ of vertices of a neighborhood, we will always order them as they appear when going clockwise around CCH(V). We will also consider the indexes modulo p (the number of vertices), meaning that $s_0 = s_p$ and $s_{p+1} = s_1$.

3.1 General Form of Iterated Complete Neighborhoods

In this whole subsection V is a complete neighborhood. We will study the shape of the successive iterations of V. First of all, we define the integer $t_c = \min\{t \in \mathbb{N} \mid \mathrm{CH}(V)^2 \subseteq V^{t_c+2}\}$.

We know that t_c is correctly defined because V is complete so there exists an integer t such that $\mathrm{CH}(V)^2 \subseteq V^t$. We have the following proposition:

Proposition 1. *For all integers $t \geq 2$,*

$$\mathrm{CH}(V)^t \subseteq V^{t_c+t} \subseteq \mathrm{CH}(V)^{t_c+t}$$

Proof. The rightmost inclusion is immediate because $V \subseteq \mathrm{CH}(V)$. The other inclusion can be shown by induction using the fact that $V + \mathrm{CH}(V)^2 = \mathrm{CH}(V)^3$.

We have the following theorem:

Theorem 1. *For any two-dimensional complete neighborhood V, if we denote by $(s_1, s_2, \dots, s_p)$ its vertices, there exists an integer t_s such that:*

- *for all $i \in [\![1, p]\!]$, there is a set $A_i \subseteq V^{t_s+t_c} \setminus V^{t_s}$,*
- *for all integer $i \in [\![1, p]\!]$, there is a set B_i included in the trapezoid of sides $h_i = (s_{i+1} - s_i)$, $t_c \cdot s_{i+1}$, $-t_c \cdot h_i$ and $-t_c \cdot s_i$.*
- *for any integer $t \in \mathbb{N}$, the neighborhood $V^{t_c+t_s+t}$ is exactly the union of the following sets:*

- $\mathrm{CH}(V)^{t_s+t}$,
- $(A_i + t \cdot s_i)$ *for all* $i \in [\![1, p]\!]$,
- *copies of* B_i *arranged regularly (translation of vector* h_i*) on the outer strip of the cone* (s_i, s_{i+1}) *to cover the area that isn't covered by the* A_i.

The general form of $V^{t_c+t_s+t}$ (as described by Theorem 1) is illustrated by Figure 1 for two different values of t.

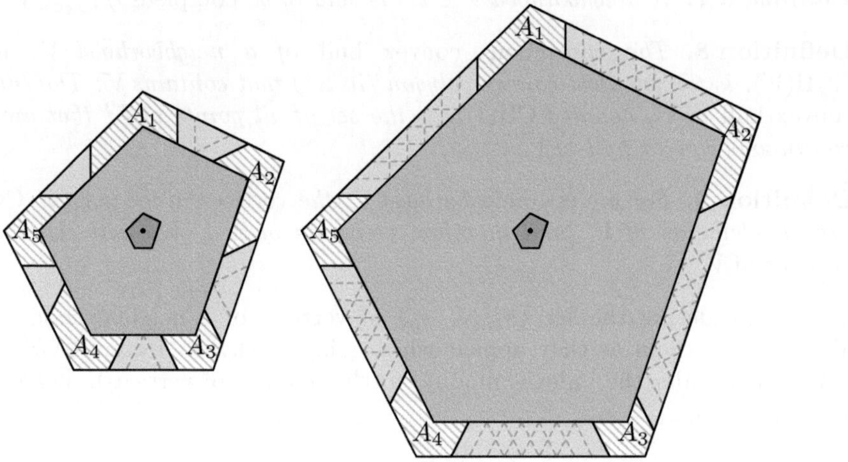

Fig. 1. General form of $V^{t_c+t_s+t}$ (the fillings of the dashed trapezoids are all identical on a given strip)

Even though it is hard to state clearly, Theorem 1 is very important because it shows that no matter how irregular V is, it becomes "regular" after a certain number of iterations.

4 Main Theorem

This whole section will be dedicated to the proof of the following theorem

Theorem 2. *Given a complete neighborhood* V *in* d *dimensions* ($d \in \mathbb{N}$)*, any language that can be recognized in real time by a 2DCA working on the convex hull of* V *can be recognized in real time by a 2DCA working on* V.

To prove this theorem, we'll consider a complete neighborhood V and language L recognized in real time by a 2DCA $\mathcal{A}$ working on $\mathrm{CH}(V)$. We will then describe the behavior of a 2DCA $\mathcal{A}'$ working on the neighborhood V that recognizes L in real time. We define t_c as previously.

4.1 General Behavior of $\mathcal{A}'$

To describe the behavior of $\mathcal{A}'$, we will consider a two-dimensional word w and describe the evolution of $\mathcal{A}'$ on this input. Since the evolution of $\mathcal{A}'$ will mimick that of $\mathcal{A}$, it will be convenient to denote by $\langle c\rangle_t$ the state in which the cell c is at time t in the evolution of $\mathcal{A}$ starting from the initial configuration corresponding to the word w (for instance $\langle 0\rangle_0$ is the lowest and leftmost letter of w).

For some large enough integer t_0 depending on V (we'll explain later how to choose t_0) the automaton $\mathcal{A}'$ will spend the first t_0 steps gathering all possible information on each cell.

After t_0 generations, any cell c knows therefore all states $\{\langle c+x\rangle_0 \mid x \in V^{t_0}\}$. If V^{t_0} is different from $\mathrm{CH}(V)^{t_0}$ there are some states in $\mathrm{CH}(V)^{t_0}(c)$ that c doesn't know. All cells will however assume that the states that they don't know in their neighborhood $\mathrm{CH}(V)^{t_0}$ are #. Obviously, many of these assumptions are false at time t_0, but for cells close enough to the borders of the input word some of these assumptions are true.

The cells of $\mathcal{A}'$ will now apply the transition rule of $\mathcal{A}$ to all the states they hold (including the ones they *assume*). Hence, at time (t_0+t) each cell c of $\mathcal{A}'$ holds a set of states that it assumes to be the states $\{\langle c+x\rangle_t \mid x \in \mathrm{CH}(V)^{t_0}\}$.

4.2 Propagation of Correct Assumptions

As previously, we denote by $\{s_1,\dots,s_p\}$ the set of all vertices of V (ordered clockwise). For all $i \in [\![1,p]\!]$, we separate the cone (s_i,s_{i+1}) of the neighborhood $\mathrm{CH}(V)^{t_0}$ in four parts:

- the inside triangle C_i of sides $(t_0-t_c)s_i$ and $(t_0-t_c)s_{i+1}$ that we know is totally included in V^{t_0};
- a trapezoidal area T_i included in the remaining strip, whose parallel sides lay on the inner and outter borders of the strip and whose two other sides are parallel to the segments $[s_{i-1},s_i]$ and $[s_{i+1},s_{i+2}]$ of the convex hull of V;
- the two parts S_i^d and S_{i+1}^g that are left on each side of the central trapezoidal area.

We also define $S_i = S_i^d \cup S_i^g$.

We choose t_0 large enough so that V^{t_0} is of the "stabilized form" described by Theorem 1 (meaning that $t_0 \geq t_c+t_s$) and also that for all i the trapezoid T_i doesn't extend beyond the central periodic area of the outter strip of the cone (s_i,s_{i+1}) on V^{t_0} (when t grows, the central periodic area becomes arbitrarily large so there is a time t_0 such that we can choose T_i entirely inside of it).

Figure 2 illustrates the general form of such a splitting of $\mathrm{CH}(V)^{t_0}$.

If we consider a cell c of $\mathcal{A}'$ at time (t_0+t), it knows correctly all states $\{\langle c+x\rangle_t \mid x \in C_i\}$ for all i but not necessarily all states in the other regions.

For all i, we will say that a cell is (s_i,s_{i+1})-*correct* if all the assumptions it makes in the area $(c+T_i)$ are correct. We will say that it is s_i-*correct* if it is (s_{i-1},s_i)-correct, (s_i,s_{i+1})-correct and that all the assumptions it makes in the area $(c+S_i)$ are also correct. Figure 3 illustrates these definitions.

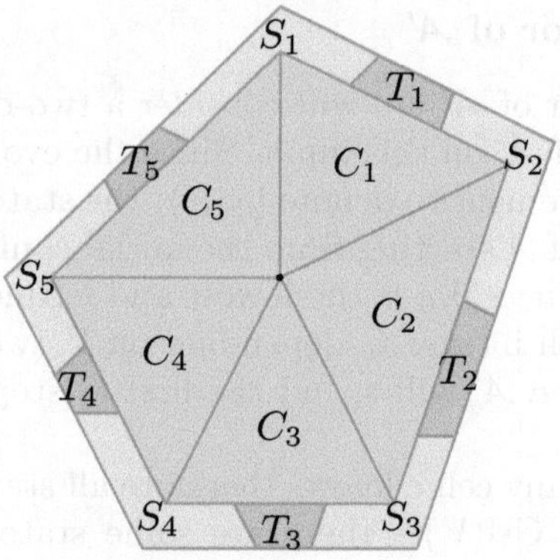

Fig. 2. Splitting of $\mathrm{CH}(V)^{t_0}$

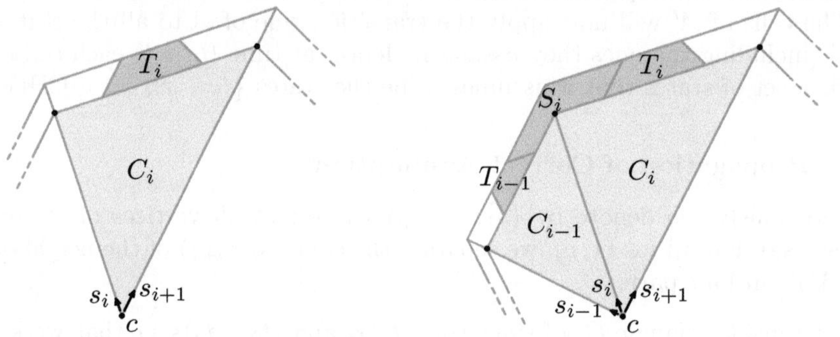

Fig. 3. Correct hypothesis of a (s_i, s_{i+1})-correct (left) and a s_i-correct (right) cell

We can now prove the two following lemmas:

Lemma 1. *If at time* $(t_0 + t)$ *the cells* $(c + s_i)$ *and* $(c + s_{i+1})$ *are both* (s_i, s_{i+1})-*correct then at time* $(t_0 + t + 1)$ *the cell* c *is* (s_i, s_{i+1})-*correct.*

Lemma 2. *If at time* $(t_0 + t)$ *the cell* $(c + s_i)$ *is* s_i-*correct and that both cells* $(c + s_{i-1})$ *and* $(c + s_{i+1})$ *are* (s_{i-1}, s_i)-*correct and* (s_i, s_{i+1})-*correct respectively then at time* $(t_0 + t + 1)$ *the cell* c *is* s_i-*correct.*

Figures 4 and 5 illustrate the proofs of these two lemmas. In both cases we have represented on the left side the area on which the cell must have correct information to be correct at the next step and on the right side the areas on which it can see correct information according to the hypothesis of the lemma.

We see that in both cases the cell has enough information at time $(t_0 + t)$ to compute correct states at time $(t_0 + t + 1)$ (the slope of the central trapezoid has been chosen so that everything works correctly, and we use the fact that $\mathrm{CH}(V)^{t_0}$ is convex). We also use the fact that if there is a conflict between the information

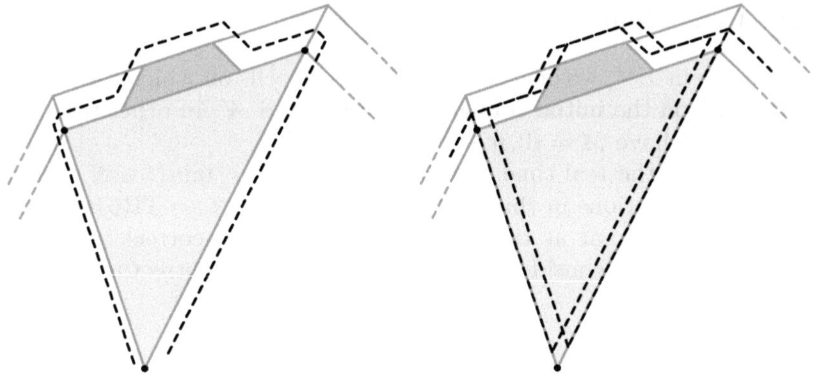

Fig. 4. States that the cell c must know to be (s_i, s_{i+1})-correct at the next time (left) and the correct information held by its neighbors (right)

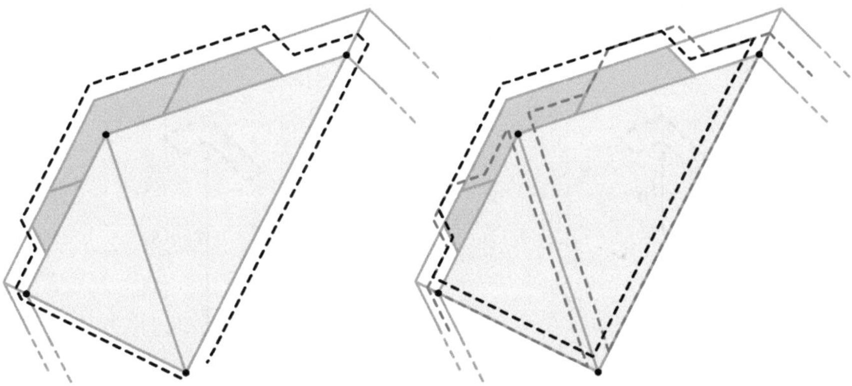

Fig. 5. States that the cell c must know to be s_i-correct at the next time (left) and the correct information held by its neighbors (right)

held by a cell and its neighbors, the priority is given to the information held by the neighbor that is the closest to the disagreeing point.

We know that at time t_0 all cells of $\mathcal{A}'$ that are "close enough" to the border of the word w are correct in the direction pointing outside of the word. Lemmas 1 and 2 show that the the correctness of thes cells "propagates" to their neighbors towards the origin along the vectors s_i until eventually at some time $(t_0 + t_f)$ the origin is correct in all possible directions. At this time, the origin knows correctly all the states in $\{\langle x \rangle_{t_f} \mid x \in \mathrm{CH}(V)^{t_0}\}$ and can hence anticipate the simulation of $\mathcal{A}$ of t_0 steps. At time $(t_0 + t_f)$ the origin is therefore capable of knowing the state $\langle 0 \rangle_{t_0+t_f}$ in which the origin of $\mathcal{A}$ would be at time $(t_0 + t_f)$.

The 2DCA $\mathcal{A}'$ can therefore compensate for the initial t_0 steps that were "lost" at the beginning. Now we have to show that $(t_0 + t_f)$ is exactly the real time corresponding to the input word w.

4.3 The Real Time

Given a word w in Σ^{**}, we denote by M the set of cells on which the word spans when "written" on the initial configuration of $\mathcal{A}$ and $\mathcal{A}'$. In other terms, if w is of size (n, m), we have $M = [\![0, n-1]\!] \times [\![0, m-1]\!]$.

By definition of the real time, we have $\mathrm{TR}_V(n, m) = \min\{t \in \mathbb{N} \mid M \subseteq V^t\}$. To alleviate the notations in this section, we will define $t_r = \mathrm{TR}_V(n, m)$.

We want to show that at time t_r the origin of $\mathcal{A}'$ is correct in all possible directions. We have to consider both cases of angles (s_i-correctness) and cones ((s_i, s_{i+1})-correctness).

Lemma 3. *If $t_r \geq t_0$ then for any vertex s_i of V the cell $(t_r - t_0)s_i$ is s_i-correct at time t_0.*

Proof. According to the definition of real time, we have $M \subseteq V^{t_r}$. If $t_r \geq t_0$, the neighborhood V^{t_r} is of the "stabilized" form described by Theorem 1. Moreover, since V^{t_0} is also "stabilized", we know that the area corresponding to the vertex s_i in both neighborhoods V^{t_0} and V^{t_r} is identical (see Figure 6).

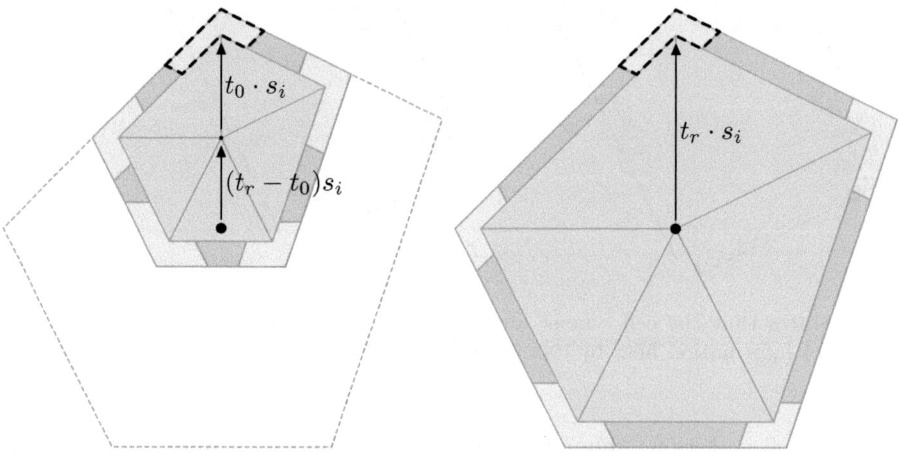

Fig. 6. The sets $V^{t_0} + (t_r - t_0)s_i$ (left) and V^{t_r} (right) coincide on the black dashed area

Since M is included in V^{t_r}, there is no point of M in the black dashed area that isn't in V^{t_r}. By Theorem 1 we know that all points in that area are also in $V^{t_0} + (t_r - t_0)s_i$. Thus the cell $(t_r - t_0)s_i$ makes only correct assumptions in that area at time t_0 when it considers that all the states it doesn't see are #.

Since we have seen that (s_i, s_{i+1})-correctness propagates from $(c + s_i)$ and $(c + s_{i+1})$ to c and that the only cell we are really interested in is the origin, it is sufficient to work on the cells of the form $(a \cdot s_i + b \cdot s_{i+1})$ for $a, b \in \mathbb{N}$.

Lemma 4. *If $t_r \geq t_0$ then for any $i \in [\![1, p]\!]$, all cells of the form $(a \cdot s_i + b \cdot s_{i+1})$ with $a, b \in \mathbb{N}$ and $a + b = t_r - t_0$ are (s_i, s_{i+1})-correct at time t_0 (these cells are all on the segment $[(t_r - t_0)s_i, (t_r - t_0)s_{i+1}]$ as shown by Figure 7).*

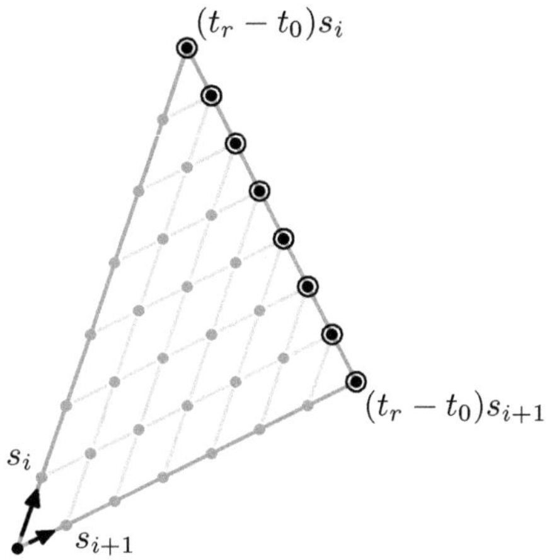

Fig. 7. The cells $(a \cdot s_i + b \cdot s_{i+1})$ where $a, b \in \mathbb{N}$ and $a + b = t_r - t_0$ (represented by black circled dots)

Proof. As previously, if $t_r \geq t_0$ the neighborhood V^{t_r} is of the form described by Theorem 1, as is V^{t_0}. This means that the central trapezoids in the strip of width t_c on both neighborhoods are superpositions of identical trapezoidal fillings periodically translated by a vector $(s_{i+1} - s_i)$.

This means that for any cell $c = (a \cdot s_i + b \cdot s_{i+1})$ with $a, b \in \mathbb{N}$ and $a + b = t_r - t_0$ the filling of $V^{t_0}(c)$ on the trapezoidal area corresponding to (s_i, s_{i+1}) coincides with the neighborhood V^{t_r} (see Figure 8).

Since M in included in V^{t_r}, all letters of the word that are in the area that the cell c has to know in order to be (s_i, s_{i+1})-correct are visible to the cell. It only makes true assumptions in this area when is assumes that all states it cannot see are #. All cells $(a \cdot s_i + b \cdot s_{i+1})$ for $a, b \in \mathbb{N}$ and $a + b = t_r - t_0$ are therefore (s_i, s_{i+1})-correct at time t_0.

4.4 End of the Proof

Using Lemmas 1, 2, 3 and 4 we can now show by induction the following results:

Lemma 5. *If $t_r \geq t_0$, for any vertex s_i of V and any $t \in \mathbb{N}$, the cell $(t_r - t_0 - t)s_i$ is s_i-correct at time $(t_0 + t)$.*

Lemma 6. *If $t_r \geq t_0$, for any vertex s_i of V and any $t \in \mathbb{N}$, all cells $(a \cdot s_i + b \cdot s_{i+1})$ where $a, b \in \mathbb{N}$ and $a + b = t_r - t_0 - t$ are (s_i, s_{i+1})-correct at time $(t_0 + t)$.*

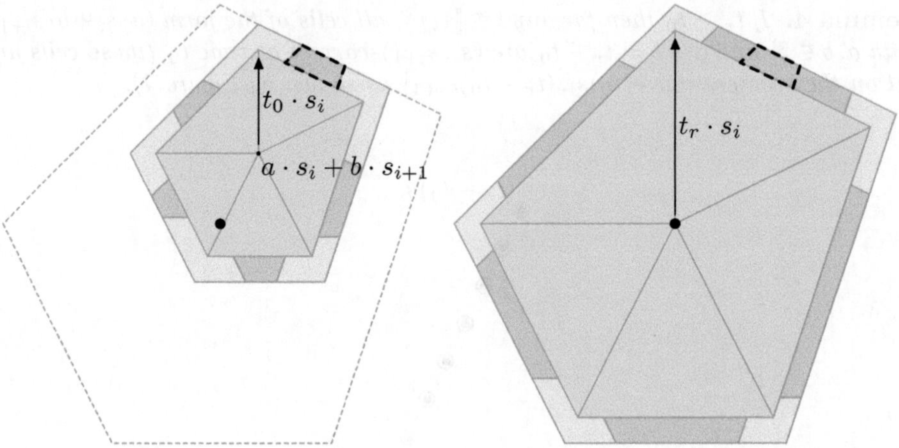

Fig. 8. The sets $V^{t_0}(c)$ (left) and V^{t_r} (right) coincide on the black dashed area

We finally conclude by saying that if $t_r \geq t_0$, the origin is correct in all possible directions at time $t_0 + (t_r - t_0) = t_r$. Hence, on an input w of size (n, m), the 2DCA $\mathcal{A}'$ working on the neighborhood V can compute at time $\mathrm{TR}_V(n, m)$ the state in which the origin of $\mathcal{A}$ would be at the same time from the same input.

Since there are only a finite number of words $w \in \Sigma^{**}$ such that the real time corresponding to these words is smaller than t_0, we can modify the automaton so that it recognizes also correctly in real time these words that are too small. This ends the proof of Theorem 2.

5 Conclusion

Understanding the impact of the choice of the neighborhood in language recognition on cellular automata is a key point to understanding communication in parallel computations. What we have shown here is that, although it might seem important, the neighborhood needs not be convex since the same computation can be done on non-convex neighborhoods than on convex ones, and there is no other loss of time than the obvious one due to the fact that the neighborhood is smaller. Not only it shows that convexity is never fully used by any parallel algorithm, but it also simplifies our study considerably since we will now be able to consider only convex neighborhoods when proving algorithmic results (speed-up theorems for example).

An interesting thing to notice is that we don't have the converse of Theorem 2. Although it might seem unlikely, there might exist languages that can be recognized in real time on a certain neighborhood V but not on its convex hull. Of course, any computation that can be done on V can be done in the same time on $\mathrm{CH}(V)$, but since the real time on $\mathrm{CH}(V)$ can be smaller than the one on V, it is not easy to show that we can go faster on $\mathrm{CH}(V)$. This comes from the fact that we are unable to prove constant time acceleration theorems on some convex neighborhoods (such as the von Neumann one).

Also it might be interesting to determine precisely which neighborhoods are real time equivalent but very little is known in that direction. If we consider complete neighborhoods, we know that there is a language that is recognized in real time on the von Neumann neighborhood but not on the Moore neighborhood [12], but it's the only example (and the converse is still unknown).

References

1. Albert, J., Čulik II, K.: A simple universal cellular automaton and its one-way and totalistic version. Complex Systems 1, 1–16 (1987)
2. Cole, S.N.: Real-time computation by n-dimensional iterative arrays of finite-state machines. IEEE Transactions on Computers C-18, 349–365 (1969)
3. Čulik, K., Hurd, L.P., Yu, S.: Computation theoretic aspects of cellular automata. Phys. D 45, 357–378 (1990)
4. Dyer, C.R.: One-way bounded cellular automata. Information and Control 44, 261–281 (1980)
5. Ibarra, O., Jiang, I.: Relating the power of cellular arrays to their closure properties. Theoretical Computer Science 57, 225–238 (1988)
6. Ibarra, O.: In: Cellular Automata: a Parallel Model. Mathematics and its applications edn., pp. 181–197. Kluwer, Dordrecht (1999)
7. Poupet, V.: Cellular automata: Real-time equivalence between one-dimensional neighborhoods. In: Diekert, V., Durand, B. (eds.) STACS 2005. LNCS, vol. 3404, pp. 133–144. Springer, Heidelberg (2005)
8. Smith III, A.R.: Simple computation-universal cellular spaces. J. ACM 18, 339–353 (1971)
9. Smith III, A.R.: Real-time language recognition by one-dimensional cellular automata. Journal of the Assoc. Comput. Mach. 6, 233–253 (1972)
10. Terrier, V.: Language recognizable in real time by cellular automata. Complex Systems 8, 325–336 (1994)
11. Terrier, V.: Language not recognizable in real time by one-way cellular automata. Theoretical Computer Science 156, 281–287 (1996)
12. Terrier, V.: Two-dimensional cellular automata recognizer. Theor. Comput. Sci. 218, 325–346 (1999)

Towards a Rice Theorem on Traces of Cellular Automata*

Julien Cervelle and Pierre Guillon

Institut Gaspard Monge, Université de Marne la Vallée
77454 Marne la Vallée Cedex 2, France
{julien.cervelle,pierre.guillon}@univ-mlv.fr

Abstract. The trace subshift of a cellular automaton is the subshift of all possible columns that may appear in a space-time diagram. We prove the undecidability of a rather large class of problems over trace subshifts of cellular automata.

Keywords: Discrete-time dynamical systems, cellular automata, symbolic dynamics, formal languages, computability, decidability.

1 Introduction

Cellular automata are well-known formal models for complex systems. They are used in a huge variety of different scientific fields including mathematics, physics and computer science.

A *cellular automaton* (CA for short) consists in an infinite number of identical *cells* arranged on a regular lattice. All cells evolve synchronously according to their own state and those of their neighbors.

Despite the apparent simplicity of the definition of the system, one can observe very complex behaviors. Many attempts have been made to classify them: strictly in topological terms [1], emphasizing limit sets [2], considering computability [3], or yet studying the language appearing in cell evolution [4]. Most behaviors have been proved undecidable: concerning the attractors (nilpotency in [5], other properties in [6]), the equicontinuity classification [7], or universality [8].

Our approach concerns the languages that can be associated to a CA and especially the trace subshift which is the set of all possible sequences of the states a particular cell takes during the evolution. Motivations come both from classical symbolic dynamics but also from physics. Indeed, when observing natural phenomena due to physical constraints, one can keep trace only of a finite number of measurements. This set of measurements takes into account only a minor part of the parameters ruling the phenomenon under investigation. Hence, to some extent, what is observed is the "trace" of the phenomenon left on the instruments.

* This work has been supported by the Interlink/MIUR project "Cellular Automata: Topological Properties, Chaos and Associated Formal Languages", by the ANR Blanc "Projet Sycomore".

L. Kučera and A. Kučera (Eds.): MFCS 2007, LNCS 4708, pp. 310–319, 2007.

We try to prove a "Rice Theorem" for trace subshifts, similarly to what is done in [6,9] for limits sets and tilings, that is to say the undecidability of non-trivial properties over the trace subshift given a CA.

The paper is organized as follows. Section 2 to 4 are devoted to definitions and main concepts of the paper. Section 5 study the case of a Rice theorem for a simpler definition of trace called "subtrace". Section 6 extends the result to the "trace" case.

2 Definitions

Let $\mathbb{N}^* = \mathbb{N}\setminus\{0\}$. For $i, j \in \mathbb{N}$ with $i \leq j$, $[i, j]$ denotes the set of integers between i and j inclusive. For any function F from $A^{\mathbb{Z}}$ into itself, F^n denotes the n-fold composition of F with itself.

Languages. Let A be a finite *alphabet* with at least two *letters*. A *word* is a finite sequence of letters $w = w_0 \dots w_{|w|-1} \in A^*$. A *factor* of a word $w = w_0 \dots w_{|w|-1} \in A^*$ is a word $w_{[i,j]} = w_i \dots w_j$, for $0 \leq i \leq j < |w|$. We note $w_{[i,j]} \sqsubset w$. A *language* on alphabet A is a set of words on alphabet A. Given two languages $C, D \subset A^*$, CD denotes their concatenation, When no confusion is possible, given a word w, we also note w the language $\{w\}$.

Configurations. A *configuration* is a bi-infinite sequence of letters $x \in A^{\mathbb{Z}}$. The set $A^{\mathbb{Z}}$ of configurations is the *phase space*. The definition of *factor* can be naturally extended to configurations: for $x \in A^{\mathbb{Z}}$ and $i \leq j$, $x_{[i,j]} = x_i \dots x_j \sqsubset x$. If $u \in A^*$, then u^ω is the infinite word of $A^{\mathbb{N}}$ consisting in periodic repetitions of u and ${}^\omega u^\omega$ is the configuration of $A^{\mathbb{Z}}$ consisting in periodic repetitions of u.

Topology. We endow the phase space with the *Cantor topology*. For $j, k \in \mathbb{N}$ and a finite set W of words of length j, we note $[W]_k$ the set $\left\{ x \in A^{\mathbb{Z}} \middle| \, x_{[k,k+j-1]} \in W \right\}$. Such a set is called a *cylinder*. We note $[W]_k^C$ the complement of the cylinder $[W]_k$. Cylinders form a base of open sets for Cantor topology.

Cellular automata. A (one-dimensional two-sided) *cellular automaton* (CA for short) is a parallel synchronous computation model consisting in cells distributed over the regular lattice $\mathbb{Z}$. Each cell $i \in \mathbb{Z}$ of the configuration $x \in A^{\mathbb{Z}}$ has a state x_i in the finite alphabet A, which evolves depending on the state of their

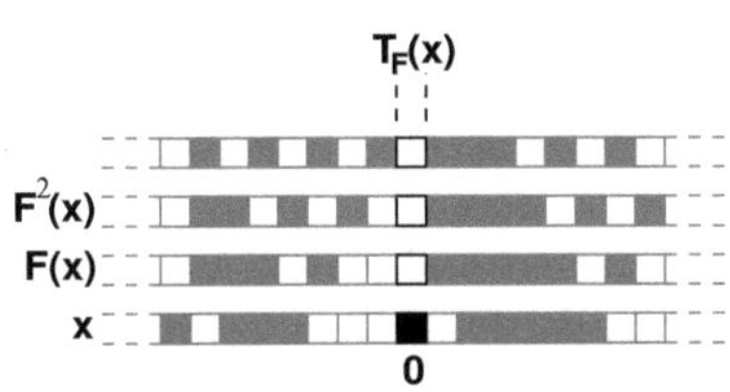

Fig. 1. The trace seen on the space-time diagram

neighbors: $F(x)_i = f(x_{[i-m,i-m+d-1]}))$, where $f : A^d \to A$ is the *local rule*, $m \in \mathbb{Z}$ the *anchor* and $d \in \mathbb{N}^*$ the *diameter* of the CA. By abuse of notation, we use the cellular automaton as its *global function* $F : A^\mathbb{Z} \to A^\mathbb{Z}$. The *orbit* of initial configuration $x \in A^\mathbb{Z}$ is the sequence of the configurations $(F^j(x))_{j\in\mathbb{N}}$. Usually they are graphically represented by a two-dimensional *space-time diagram*, like in Figure 1.

The *shift* $\sigma : A^\mathbb{Z} \to A^\mathbb{Z}$ is a particular CA defined by $\sigma(x)_i = x_{i+1}$ for every $x \in A^\mathbb{Z}$ and $i \in \mathbb{Z}$ which shifts configurations to the left. If the alphabet contains a special letter 0, the *null CA* null : $A^\mathbb{Z} \to A^\mathbb{Z}$ is another particular CA defined by $\text{null}(x)_i = 0$. Its image set is $\{{}^\omega 0^\omega\}$.

Any local rule f of a CA can be extended naturally to an application on words $f(w) = (f(w_{[i,i+d-1]}))_{0\le i\le |w|-d}$, for all $w \in A^*A^d$.

Finally, if a CA has anchor 0, it is called an *one-way cellular automata* (OCA for short), since the new state of a cell only depends on its right neighbors. Hence information can only flow from right to left.

Subshifts. The *one-sided shift* (or simply shift when no confusion is possible), also noted σ by abuse of notation is the self-map of $A^\mathbb{N}$ such that $\sigma(z)_i = z_{i+1}$, for every $z \in A^\mathbb{N}$ and $i \in \mathbb{N}$. A *one-sided subshift* $\Sigma \subset A^\mathbb{N}$ is a σ-stable closed set of infinite words. The language of Σ is $\mathcal{L}(\Sigma) = \{w \in A^* \,|\, \exists z \in \Sigma, w \sqsubset z\}$. It characterizes Σ since $\Sigma = \left\{ z \in A^\mathbb{N} \middle| \forall w \sqsubset z, w \in \mathcal{L}(\Sigma)\right\}$.

A subshift Σ is *nilpotent* if there is some $k \in \mathbb{N}$ such that $\sigma^k(\Sigma) = \{0^\omega\}$. Equivalently, $\Sigma \subset A^k 0^\omega$ for some k. In particular, it is finite.

3 Trace

Trace. In this section we define the main notion introduced in the paper, namely the trace of a CA.

Definition 1 (Trace). *Given a CA F, the* trace *of F with initial condition $x \in A^\mathbb{Z}$ is $T_F(x) = (F^j(x)_0)_{j\in\mathbb{N}}$. In other words, it is the central column of the space-time diagram of initial configuration x (see Figure 1). The* trace subshift *of F is $\tau(F) = T_F(A^\mathbb{Z})$. Similarly, we define $T_F^k(x) = (F^j(x)_{[0,k-1]})_{j\in\mathbb{N}}$ which is the sequence of the words at positions in $[0, k-1]$ in the space-time diagram of initial configuration x.*

For instance, the trace subshift of the shift is $(0 + 1)^\omega$, it is $0^\omega + 1^\omega$ for the identity and $(0 + 1)0^\omega$ for the null CA (see [10] for more examples).

Note that $T_F F = \sigma T_F$ and, as T_F is continuous, it is a *factorization* between the CA and the one-sided full shift which means that their two dynamics are very closely related. For more about factorization, see [11].

Subtrace. If Σ is a subshift on an alphabet $B \subset A^k$, and $i \in [0, k-1]$, then for all $z = ((z_j)_0 \dots (z_j)_{k-1})_{j\in\mathbb{N}} \in \Sigma$, we define its ith projection as $\pi_i(z) = ((z_j)_i)_{j\in\mathbb{N}} \in A^\mathbb{N}$, and $\pi_{[n,m]}(z) = ((z_j)_n, \dots, (z_j)_m)_{j\in\mathbb{N}}$. We also note $\pi(\Sigma) = \bigcup_{0\le i<k} \pi_i(\Sigma)$, which is a subshift on A.

Definition 2. *Given a CA F on the alphabet $B \subset A^k$, the* subtrace subshift *is defined by* $\mathring{\tau}(F) = \bigcup_{0 \le i < k} \left\{ ((F^j(x)_0)_i)_{j \in \mathbb{N}} \middle| x \in B^{\mathbb{Z}} \right\} = \pi(\tau(F))$.

4 The Nilpotency Problem

Definition 3. *Consider a CA F of local rule $f : A^d \to A$ and $q \in A$. A state $q \in A$ is* spreading *if for all word $u \in A^d$ containing letter q, we have $f(u) = q$. A CA on alphabet A is q-*nilpotent *(or simply nilpotent) if there is some integer j for which $\forall x \in A^{\mathbb{Z}}, F^j(x) = {}^{\omega}q^{\omega}$.*

Note that a CA is nilpotent if and only if its trace is nilpotent.

Theorem 1 (Kari[5]). *The problem*

> **Instance:** *a CA F with a spreading state q*
> **Question:** *is F q-nilpotent?*

is undecidable.

Note that it is still undecidable if we restrict to OCA since two CA with the same diameter and same local rule are either both nilpotent or both non nilpotent whatever are their anchor. If the first CA F has anchor m and the second CA F' has anchor m' and if there is a k such that for all configuration x, $F^k(x) = {}^{\omega}q^{\omega}$ then $F'^k(x) = \sigma^{k(m-m')}(F^k(x)) = {}^{\omega}q^{\omega}$.

Proposition 1. *An OCA with a spreading state q is q-nilpotent if and only if q appears in all its space-time diagrams.*

Proof. Consider a CA F. If its diameter is 1 (each cell uses only its state for the local rule), then the equivalence is trivial since for all letter a, the space-time diagram for ${}^{\omega}a^{\omega}$ contains a q after k_a steps and hence for all configuration x, $F^{\max_{a \in A} k_a}(x) = {}^{\omega}q^{\omega}$. Otherwise, for every configuration $x \in A^{\mathbb{N}}$, there are some integers i and j such that $F^j(x)_i = q$. Note that, by spreadingness, $F^{j + \lceil \frac{i}{d} \rceil}(x)_0 = q$. Hence $\bigcup_{j \in \mathbb{N}} F^{-j}([q])$ is a covering by open subsets of the compact $A^{\mathbb{N}}$, so we can find some $k \in \mathbb{N}$ such that $A^{\mathbb{N}} = \bigcup_{j \le k} F^{-j}([q]) = F^{-k}([q])$. By shift-invariance, $F^k(A^{\mathbb{N}}) = \{{}^{\omega}q^{\omega}\}$. □

Proposition 2. *Let F a CA on alphabet A, of diameter d, and with a spreading state $q \in A$. We can build a CA F' on alphabet $\{0,1\}^k$ for some $k \in \mathbb{N}^*$, such that F' is $(0,\ldots,0)$-nilpotent if and only if F is q-nilpotent.*

Proof. Let $k = \lceil \log_2 |A| \rceil$ and $\phi : A \to \{0,1\}^k$ an injection such that $\phi(q) = (0,\ldots,0)$. F' can be defined by the same diameter and anchor as F, and local rule:

$$f' : \begin{array}{l} (\{0,1\}^k)^d \to \{0,1\}^k \\ \qquad u \mapsto \left| \begin{array}{ll} \phi(f(a)) & \text{if } u = \phi(a) \\ (0,\ldots,0) & \text{otherwise} \end{array} \right. \end{array} .$$

□

5 Subtrace Problems

A subtrace is somewhat the trace of a "CA" which does not apply the same evolution rule on each cell. We will first prove, by reduction from the nilpotency problem, the undecidability of a certain class of properties over subtrace subshifts, that is, non-trivial nilpotent-stable properties. In the next section, we will reduce properties over trace subshifts.

5.1 The Full Subtrace Problem

This section is devoted to reducing the nilpotency problem to the problem FULL:

Instance: an OCA on alphabet A^{k+1}
Question: is its subtrace equal to the full shift $A^{\mathbb{N}}$?

Our method presents similarities with [8].

Let F a OCA on A of local rule f, diameter d, and with a spreading state $q \in A$. From Proposition 2, we can assume without loss of generality that $A = \{0,1\}^k$ for some $k \in \mathbb{N}^*$, and $q = (0,\ldots,0)$. We can build the OCA $\sigma \circledcirc F$ on alphabet $\{0,1\}^{1+k}$ with the local rule:

$$g: \begin{array}{l} (\{0,1\} \times \{0,1\}^k)^d \to \{0,1\} \times \{0,1\}^k \\ (((a_0,x_0)\ldots(a_{d-1},x_{d-1}))) \mapsto \left| \begin{array}{ll} (a_1, f(x_0 \ldots x_{d-1})) & \text{if } x_0 \neq q \\ (0,q) & \text{otherwise} \end{array} \right. \end{array}$$

This rule works almost like the Cartesian product of the shift on $\{0,1\}$ and F except that it applies the shift only if the state of the cell for F is not q and else goes to 0. It behaves as if F is supplying power to the shift CA and q means "no power".

Lemma 1. *If F is q-nilpotent, then $\mathring{\tau}\ (\sigma \circledcirc F)$ too. Otherwise, $\mathring{\tau}\ (\sigma \circledcirc F) = A^{\mathbb{N}}$.*

Proof. We can see that the last k projections exactly represent the application of the CA F. In particular, if F is q-nilpotent, then there is some $j \in \mathbb{N}$ for which $\forall x \in A^{\mathbb{Z}}, \pi_{[1,k]}(T_{\sigma\circledcirc F}(x))_{[j,+\infty)} = q^\omega$, and from the rule we get $\pi_0(T_{\sigma\circledcirc F(x)})_{[j+1,+\infty)} = 0^\omega$, so $\mathring{\tau}\ (\sigma \circledcirc F)$ is $(0,q)$-nilpotent.
On the other hand, if F is not q-nilpotent, then from Proposition 1, one can find a configuration $x \in A^{\mathbb{Z}}$ such that $\forall j \in \mathbb{N}, \forall i \in \mathbb{Z}, F^j(x)_i \neq q$. For every $z \in A^{\mathbb{N}}$, define $y \in (\{0,1\}^{1+k})^{\mathbb{Z}}$ such that $\pi_{[1,k]}(y) = x$ and $\pi_0(y)_{[0,+\infty)} = z$. The second case of the rule will never be applied in the orbit of y; hence $\pi_0(T_{\sigma\circledcirc F(y)}) = z$. □

Theorem 2. *Problem* FULL *is undecidable.*

Proof. $\sigma\circledcirc F$ is clearly computable from F, so if we could decide whether $\mathring{\tau}\ (\sigma\circledcirc F)$ is $A^{\mathbb{N}}$, then from Lemma 1, we could decide whether F is nilpotent. □

5.2 Other Problems

In this section, we use the previously-built CA in order to reduce the nilpotency problem to a large class of problems over the subtrace of a CA on $\{0,1\}^{2+k}$.

Definition 4. *A property P on subshifts is* nilpotent-stable *if for every subshift Σ and every nilpotent subshift H, $\Sigma \in P \Leftrightarrow \Sigma \cup H \in P$.*

By *non-trivial* property, we mean that some OCA traces respect it and some do not.

Theorem 3. *For any non-trivial nilpotent-stable property P, the problem*

> **Instance:** *a CA F on alphabet $\{0,1\}^k$, with $k \in \mathbb{N}^*$*
> **Question:** *has the subtrace subshift $\mathring{\tau}(F)$ property P?*

is undecidable.

Proof. Let P be a non-trivial nilpotent-stable property of the subshifts of $A^{\mathbb{N}}$. Should we take its complementary, we can assume P is false for the full shift $A^{\mathbb{N}}$. Let H be an OCA on $\{0,1\}$ such that $\tau(H) \in P$, h its local rule and d its diameter. Let F be an OCA on A^k of local rule f, diameter d and with spreading state q. Without loss of generality, $A = \{0,1\}$ and $q = (0,\dots,0)$. We can build the OCA $G = (\sigma \circledcirc F) \times H$ on alphabet $\{0,1\}^{2+k}$ with the local rule:

$$g: \begin{array}{l} (\{0,1\} \times \{0,1\}^k \times \{0,1\})^d \to \{0,1\} \times \{0,1\}^k \times \{0,1\} \\ ((a_0,x_0,b_0)\dots(a_{d-1},x_{d-1},b_{d-1})) \mapsto \left| \begin{array}{l} a_1, f(x_0\dots x_{d-1}), h(b_0\dots b_{d-1}) \\ \qquad\qquad\qquad\qquad \text{if } x_0 \neq q \\ (0, q, h(b_0\dots b_{d-1})) \text{ otherwise} \end{array} \right. \end{array}$$

The first $1+k$ projections are the previously built automaton $\sigma \circledcirc F$, and the last one represents H. Hence, $\mathring{\tau}(G) = \mathring{\tau}(\sigma \circledcirc F) \cup \tau(H)$, and we can use Lemma 1. If F is nilpotent, then $\mathring{\tau}(\sigma \circledcirc F)$ too, and $\mathring{\tau}(G)$ respects P since $\tau(H)$ does, and P is nilpotent-stable. Otherwise, $\mathring{\tau}(\sigma \circledcirc F) = A^{\mathbb{N}}$ does not respect P. Hence, $\mathring{\tau}(G)$ respects P if and only if F is q-nilpotent. G is clearly computable from F, so if we could decide P, then we could decide whether F is nilpotent. □

6 From Subtrace to Trace

In this section, we assume G is an OCA on A^k, with $A = \{0,1\}$, and we want to build a CA on A that simulates it in a certain way, thus transforming subtrace into trace, provided a little restriction. This construction is similar to that done in [10]. Remark that it is well known that G can be simulated by a CA F on A, as soon as its diameter is wide enough. Each cell can see its neighborhood as words of A^k and evolve accordingly. The problem is that all cells must have the same local rule, so they have to find from the neighborhood which “column” of the A^k simulation they are representing. This can be achieved by delimiting meaningful words with a *border word* 10^k. That simulation allows to "see" the evolution of G as an evolution of F. We want a little more: all evolutions of F must not be far from an evolution of F, especially those that do not correspond to alternating meaningful words and borders words.

We have two *execution modes*: a *simulation mode* will simulate properly the execution of the OCA G on alphabet A^k, and a *default mode* will be applied if the neighborhood contains invalid information. All modes must have an evolution that locally "looks like" some evolution of G, and in particular we have to ensure that when a mode is applied to a cell, the same mode keeps being applied there in the following generations, because a change of mode would produce an invalid trace.

Macrocells. A concatenation of a word of A^k and of the border word 10^k will be called a *macrocell.* Let $B = A^k 10^k \subset A^h$ the set of macrocells, where $h = 2k+1$. By construction, it has the interesting property that it cannot overlap itself on more that half of its length.

Property 1. *If $0 < i \leq k$, then the intersection $BA^i \cap A^i B$ is empty (equivalently $[B] \cap [B]_i = \emptyset$).*

Simulation mode. We will assume in the rest of the section that G has a diameter 2 (this can be easily generalized), and local rule g.

The simulation mode will be applied to macrocells that are not overlapped by another macrocell on their right. For this purpose, they have to "look" on their right (in the local rule). Let $\Theta = BA^h \setminus \bigcup_{0<i<h} A^i BA^{h-i}$ the set of macrocells followed by something that in no way could be understood as an overlapping macrocell. The first thing to notice is that this set is not empty, since it contains successions of two macrocells thanks to Property 1.

Lemma 2. *$BB \subset \Theta$.*

Proof. If $0 < i \leq k$, then $BB \cap A^i BA^{h-i} \subset (BA^i \cap A^i B)A^{h-i} = \emptyset$. If $k < i < h$, then $BB \cap A^i BA^{h-i} \subset A^i(A^{h-i}B \cap BA^{h-i}) = \emptyset$. □

We can now define a rule on the set Θ that represents the evolution of such a macrocell that sees on its right either a macrocell abut, or at least no overlapping macrocell:

$$\Delta : \begin{array}{l} \Theta A^h \to B \\ u \mapsto \left| \begin{array}{ll} g(u_{[0,k-1]}, u_{[h,h+k-1]})10^k & \text{if } u \in B\Theta \\ g(u_{[0,k-1]}, u_{[0,k-1]})10^m & \text{otherwise} \end{array} \right. \end{array} .$$

Default mode. In the case where the neighborhood is not understandable as a valid word, we have to apply a default rule: the null CA, which has the interesting property that it erases the border words that could potentially be in an invalid place.

Combining simulation mode and default mode, we are now able to turn the function Δ into a local rule on $\{0,1\}$. Indeed, we can define, for anchor $m = h-1$ and diameter $d = 4h-1$:

$$f: \begin{array}{l} \{0,1\}^d \to \{0,1\} \\ w \mapsto \left| \begin{array}{ll} \Delta(u)_i & \text{if } w \in A^{m-i}uA^i, \text{ for some } u \in \Theta A^h, i \in [0, h-1] \\ 0 & \text{otherwise} \end{array} \right. \end{array},$$

since such an integer i and such a word u would be unique (from the construction of Θ). This local rule is such that $f(A^m u A^m) = \Delta(u)$ for every $u \in \Theta A^h$, which is what we wanted: it can simulate in one step the behavior of our CA G. Let F the corresponding rule. You can also note that $F([\Theta]) = [\Delta(\Theta A^h)] \subset [B]$ by definition, and that $F^{-1}([1]) \subset \bigcup_{0 \le j \le m} [\Theta]_{-j}$ since the default mode only produces zeros.

Stability. The following lemmas guarantee that no cell changes its evolution mode.

Lemma 3. $F^{-1}([B]) = [\Theta]$.

Proof. By construction, $F^{-1}([B]) \subset F^{-1}([1]_m) \subset \bigcup_{0 \le j \le m} [\Theta]_{-j}$. Now let $x \in F^{-1}([B])$. Then, for some $j \in [0, m]$, $F(x) \in [B] \cap F([\Theta]_{-j}) \subset [B] \cap [B]_{-j}$. By Property 1, we conclude that $j = 0$, *i.e.* $F^{-1}([B]) \subset [\Theta]$. Conversely, we have already seen that $[\Theta] \subset F^{-1}([B])$. □

Lemma 4. $F([\Theta]) \subset [\Theta]$.

Proof. By definition of $[\Theta]$, it is included in $\bigcap_{0<i<h} [\Theta]_i^C$. Moreover, $F([\Theta]^C) \subset [B]^C$ by Lemma 3. Combining the two, we get $F([\Theta]) \subset \bigcap_{0<i<h} [B]_i^C$. Hence, $F([\Theta]) \subset F([B]) \setminus \bigcup_{0<i<h} [B]_i = [\Theta]$. □

Trace. The fact the evolution modes are stable helps us conclude the simulation result.

Lemma 5. *If* $y \in (A^m)^{\mathbb{N}}$ *and* $x \in A^{\mathbb{N}}$ *such that* $\forall i \in \mathbb{N}, x_{[ih, ih+m-1]} = y_i \in A^m$ *and* $x_{[ih+m,(i+1)h-1]} = 10^m$, *then* $T_F^m(x) = T_G(y)$.

Proof. We can prove by induction on the generation $j \in \mathbb{N}$, that for any $i \in \mathbb{N}$, $F^j(x)_{[ih,(i+1)h-1]} = G^j(y)_i 10^m$. This property holds for $j = 0$. Now suppose it is true at generation $j \in \mathbb{N}$, and let us prove it for generation $j+1$. Let $i \in \mathbb{N}$. $F^j(x) \in [BB]_{ih} \subset \Theta$ from the induction hypothesis and Lemma 2. Therefore, we are in execution mode between cells ih and $(i+1)h$:

$$\begin{aligned} F^{j+1}(x)_{[ih,(i+1)h-1]} &= \Delta(F^j(x)_{[ih,(i+1)h-1]}, F^j(x)_{[(i+1)h,(i+2)h-1]}) \\ &= \Delta(F^j(y)_i, F^j(y)_{i+1}) = F^{j+1}(y)_i \ . \end{aligned}$$

In particular, $T_F^m(x) = (F^j(y)_{[0,m-1]})_{j \in \mathbb{N}} = T_G(y)$. □

Lemma 6. *If $p \in \mathbb{N}$ and $x \in A^{\mathbb{N}}$ are such that $\forall i < p, x_{[ih,(i+2)h-1]} \in \Theta$ and $x_{[ph,(p+2)h-1]} \notin \Theta$; let $y \in (A^m)^{\mathbb{N}}$ the ultimately uniform infinite word such that $\forall i < p, y_i = x_{[ih,ih+m-1]}$ and $\forall i \geq p, y_i = x_{[(p-1)h,(p-1)h+m-1]}$. Then $T_F^m(x) = T_G(y)$.*

Proof. Similarly to the proof of Lemma 5, it can easily be proved by induction on the generation $j \in \mathbb{N}$ that for any $i < p$, $F^j(x)_{[ih,(i+1)h-1]} = G^j(y)_i 10^m$, since execution is always applied. Hence, $T_F^m(x) = T_G(y)$. □

Lemma 7. *The trace of F is $\overset{\circ}{\tau}(G) \cup \Sigma$, where $\Sigma \subset \{0^\omega, 10^\omega, 1^\omega\}$.*

Proof. Let $x \in A^{\mathbb{Z}}$.

- If $x \in [\Theta]_{-q}$ for some $q \in [0, m-1]$, and $\forall i \in \mathbb{N}, x_{[ih,(i+2)h-1]} \in \Theta$, and, from Lemma 5, $T_F(x) = \pi_i(T_G(y))$, where $y_i = x_{[ih,ih+m-1]}$.
- If $x \in [\Theta]_{-q}$ for some $q \in [0, m-1]$, and there is some integer p such that $\forall i < p, x_{[ih,(i+2)h-1]} \in \Theta$ and $x_{[ph,(p+2)h-1]} \notin \Theta$, then from Lemma 6, $T_F(x) = \pi_i(T_G(y))$, for some $y \in (A^m)^{\mathbb{N}}$.
- If $x \in [\Theta]_{-q}$ for some $q \in [m, h-1]$, then by induction and Lemma 4, we can see that $\forall j \in \mathbb{N}, F^j(x) \in [\Theta]_{-q}$. Thus, $\forall j \in \mathbb{N}, F^j(x)_0 = 1$ if $q = m$ and 0 if $m < q < h$.
- Lastly, if $x \notin [\Theta]_{-q}$ for any $q \in [0, h-1]$, then default mode is applied, and, according to Lemma 3, will always be: $\forall j > 0, F^j(x)_0 = 0$.

Putting things together, the first two cases give $\pi(\tau(G))$ and the last two give $\{0^\omega, 10^\omega, 1^\omega\}$. □

We are thereby able to "simulate" a OCA G on alphabet A^m by a CA F on alphabet A. This simulation transforms the subtrace of G into the trace of F, provided a little restriction - adding three obviously nilpotent infinite words.

Theorem 4. *For any non-trivial nilpotent-stable property P, the problem*

> **Instance:** *a CA on alphabet $\{0, 1\}$*
> **Question:** *has its trace subshift property P*

is undecidable.

Proof. Let P a non-trivial nilpotent-stable property. From any OCA G on alphabet $\{0,1\}^m$, we can, using previous construction, computably build a CA F on alphabet $\{0,1\}$ such that $\tau(F) = \overset{\circ}{\tau}(G) \cup \Sigma$, where $\Sigma = \{0^\omega, 10^\omega, 1^\omega\}$ are nilpotent subshifts. In particular, we can notice that $\overset{\circ}{\tau}(G)$ respects P if and only if $\tau(F)$ does. So if we could decide, from the entry F, whether $\tau(F)$ respects P, then we could decide from the entry G whether $\overset{\circ}{\tau}(G)$ does. □

7 Conclusion

We have proved the undecidability of a class of problems over CA traces, which could lead us towards a Rice-like Theorem. This class is very comprehensive: fullness, finiteness, ultimate periodicity, soficness, finite type, inclusion of a particular word as a factor... are all nilpotent-stable non-trivial properties. On the other hand, note that context-sensitivity of the language is trivial since true for all trace subshifts according to [12].

The perspectives now would be to decrease the amount of problems that are not concerned. Finite properties (which depend only on the kth prefix of the subshift, for some $k \in \mathbb{N}$) are clearly decidable. Are they the only ones?

Our result can be easily adpated to traces of any fixed width of a CA on any given alphabet, but with arbitrarily wide neighborhoods. Another open problem would be to widen our result to study properties of the canonical factor (of width d), or at least with fixed diameter (and arbitrary alphabet). That would, in particular, comprehend the undecidability of the language classification, proved in [13]. Nonetheless, our construction can hardly be generalized for that matter, since it is based on increasing the diameter to put more information.

References

1. Gilman, R.H.: Classes of linear automata. Erg. Th. & Dyn. Sys. 7, 105–118 (1988)
2. Hurley, M.: Attractors in cellular automata. Erg. Th. & Dyn. Sys. 10, 131–140 (1990)
3. Mazoyer, J., Rapaport, I.: Inducing an order on cellular automata by a grouping operation. In: Meinel, C., Morvan, M. (eds.) STACS 98. LNCS, vol. 1373, pp. 116–127. Springer, Heidelberg (1998)
4. Kůrka, P.: Languages, equicontinuity and attractors in cellular automata. Erg. Th. & Dyn. Sys. 17, 417–433 (1997)
5. Kari, J.: The nilpotency problem of one-dimensional cellular automata. SIAM J. on Computing 21(3), 571–586 (1992)
6. Kari, J.: Rice's theorem for the limit sets of cellular automata. Th. Comp. Sci. 127(2), 229–254 (1994)
7. Durand, B., Formenti, E., Varouchas, G.: On undecidability of equicontinuity classification for cellular automata. In: Morvan, M., Rémila, E. (eds.) DMTCS 2003. DMTCS Proc., Disc. Math. and Th. Comp. Sci, vol. AB, pp. 117–128 (2003)
8. Ollinger, N.: The intrinsic universality problem of one-dimensional cellular automata. In: Alt, H., Habib, M. (eds.) STACS 2003. LNCS, vol. 2607, pp. 632–641. Springer, Heidelberg (2003)
9. Cervelle, J., Durand, B.: Tilings: recursivity and regularity. Th. Comp. Sci. 310(1-3), 469–477 (2004)
10. Cervelle, J., Formenti, E., Guillon, P.: Sofic trace of a cellular automaton. In: CiE, Siena, Italy, LNCS (June 2007)
11. Kůrka, P.: Topological and symbolic dynamics. In: Société Mathématique de France (2003)
12. Gilman, R.H.: Notes on cellular automata (Manuscript 1988)
13. di Lena, P.: Decidable and Computational properties of Cellular Automata. PhD thesis, Università di Bologna e Padova (December 2006)

Progresses in the Analysis of Stochastic 2D Cellular Automata: A Study of Asynchronous 2D Minority

Damien Regnault[1,2], Nicolas Schabanel[1,2], and Éric Thierry[1]

[1] IXXI-LIP, École Normale Supérieure de Lyon, 46 allée d'Italie, 69364 Lyon Cedex 07, France
`http://perso.ens-lyon.fr/{damien.regnault,eric.thierry}`

[2] CNRS, Centro de Modelamiento Matemático, Universidad de Chile, Blanco Encalada 2120 Piso 7, Santiago de Chile
`http://www.cmm.uchile.cl/~schabanel`

Abstract. Cellular automata are often used to model systems in physics, social sciences, biology that are inherently asynchronous. Over the past 20 years, studies have demonstrated that the behavior of cellular automata drastically changed under asynchronous updates. Still, the few mathematical analyses of asynchronism focus on one-dimensional probabilistic cellular automata, either on single examples or on specific classes. As for other classic dynamical systems in physics, extending known methods from one- to two-dimensional systems is a long lasting challenging problem.

In this paper, we address the problem of analysing an apparently simple 2D asynchronous cellular automaton: 2D **Minority** where each cell, when fired, updates to the minority state of its neighborhood. Our experiments reveal that in spite of its simplicity, the minority rule exhibits a quite complex response to asynchronism. By focusing on the fully asynchronous regime, we are however able to describe completely the asymptotic behavior of this dynamics as long as the initial configuration satisfies some natural constraints. Besides these technical results, we have strong reasons to believe that our techniques relying on defining an energy function from the transition table of the automaton may be extended to the wider class of threshold automata.

Due to space constraint, we refer the reader to [16] for the missing proofs.

1 Introduction

In the literature, cellular automata have been both studied as a model of computation presenting massive parallelism, and used to model phenomena in physics, social sciences, biology... Cellular automata have been mainly studied under synchronous dynamics (at each time step, all the cells update simultaneously). But real systems rarely fulfill this assumption and the cell updates rather occur in an asynchronous mode often described by stochastic processes. Over the past 20

L. Kučera and A. Kučera (Eds.): MFCS 2007, LNCS 4708, pp. 320–332, 2007.

years, many empirical studies [2,4,5,13,18] have been carried out showing that the behavior of a cellular automaton may widely vary when introducing asynchronism, thus strengthening the need for theoretical framework to understand the influence of asynchronism. Still, the few mathematical analyses of the effects of asynchronism focus on one-dimensional probabilistic cellular automata, either on single examples like [8,9,15] or on specific classes like [6,7]. As for other classic dynamical systems in physics, such as spin systems or lattice gas, extending known methods from one- to two-dimensional systems is a long lasting challenging problem. For example, understanding how a configuration all-up of spins within a down-oriented external field evolves to the stable configuration all-down has only recently been solved mathematically and only for the limit when the temperature goes to 0, *i.e.*, when only one transition can occur at time (see [3]). Similarly, the resolution of the study of one particular 2D automaton under a given asynchronism regime is already a challenge.

Our Contribution. In this paper, we address the problem of understanding the asynchronous behavior of an apparently simple 2D stochastic cellular automaton: 2D **Minority** where each cell, when fired, updates to the minority state of its neighborhood. We show experimentally in Section 2 that in spite of its simplicity the minority rule exhibits a quite complex response to asynchronism. We are however able to show in Section 3 that this dynamics almost surely converges to a stable configuration (listed in Proposition 3) and that if the initial configuration satisfies some natural constraints, this convergence occurs in polynomial time (and thus is observable) when only one random cell is updated at a time. Our main result (Theorems 1 and 2) rely on extending the techniques based on one-dimensional random walks developed in [6,7] to the study of the two-dimensional random walks followed by the boundaries of the main components of the configurations under asynchronous updates. We have strong reasons to believe that our techniques relying on defining an energy function from the transition table of the automaton may be extended to the wider class of threshold automata.

Our results are of particular interest for modeling regulation network in biology. Indeed, 2D **Minority** cellular automaton represents an extreme simplification of a biological model where the biological cells are organized as a 2D grid and where the regulation network involves only two genes (the two states) which tend to inhibit each other (see [1]). The goal is thus to understand how the concentrations of each gene evolve over time within the biological cells, and in particular, which gene ends up dominating the other in each cell, *i.e.*, in which state ends up each cell. Understanding this simple rule is thus a key step in the understanding of more complex biological systems.

2 Experimental Results

This section is voluntarily informal because it presents experimental observations whose formalizations are already challenging open questions. The next section will present in a proper theoretical framework our progresses in the understanding of

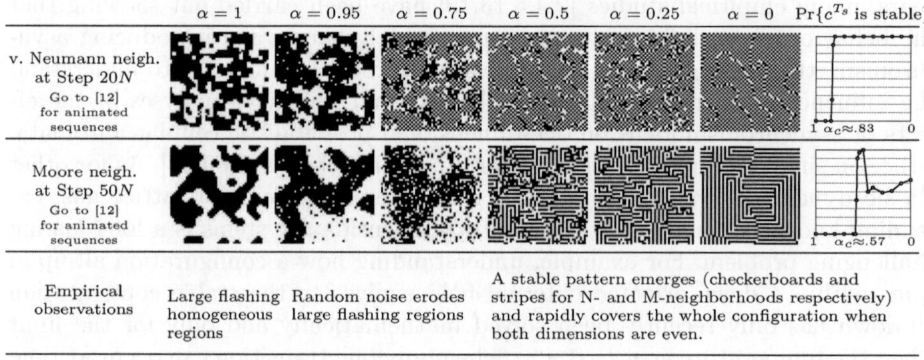

Fig. 1. 2D **Minority** under different α-asynchronous dynamics with $N_{50} = 50 \times 50$ cells. The last column gives, for $\alpha \in [0, 1]$, the empirical probability that an initial random configuration converges to a stable configuration before time step $T_s \cdot N_{50}$ where $T_s = 1000$ and $T_s = 2000$ for von Neumann and Moore neighborhood respectively.

these phenomena. The configurations studied here consist in a set of cells organized as a $n \times m$ torus (n and m are even) in which each cell can take two possible states: 0 (white) or 1 (black). The asynchronous behavior of 2D minority automaton turns out to be surprisingly complex for both of the studied neighborhoods:

- *von Neumann* (N-neighborhood for short), where each selected cell updates to the minority state within itself and its neighbors N, S, E, and W; and
- *Moore* (M-neighborhood for short), where each selected cell updates to the minority state among itself and its 8 closest neighbors N, S, E, W, NE, NW, SE, and SW.

In this section, we present a report on extensive experiments conducted on 2D Minority for both N- and M-neighborhood.

In this section, we consider the *α-asynchronous 2D Minority* dynamics in which at each time step, each cell updates to the minority state in its own neighborhood independently with probability α. We denote by $\alpha = 0$ the *fully asynchronous 2D Minority* dynamics in which at each time step, a daemon selects uniformly at random one cell and updates it to the minority state in its neighborhood.

The Synchronous Regime ($\alpha = 1$) of 2D **Minority** has been thoroughly studied in [10] where it is proved that it converges to cycles of length 1 or 2. Experimentally, from a random configuration, the synchronous dynamics in both neighborhoods converges to sets of large flashing white or black regions.

As Soon As a Little Bit of Asynchronism is introduced, the behavior changes drastically for both neighborhoods (see Fig. 1 and open our website [12] for animated sequences). Due to the asynchronism at each step, some random cells do not update and this creates a *noise* that progressively erodes the flashing homogenous large regions that were stable in the synchronous regime. After few steps, the configuration seems to converge rapidly to a homogeneous flashing background perturbed by random noise.

Experiments Provide Evidences that There Exists a Threshold α_c, $\alpha_c \approx .83$ and $\alpha_c \approx .57$ for the N- and M-neighborhoods respectively, such that if $\alpha \leqslant \alpha_c$, then stable patterns arise (*checkerboards* and *stripes* for N- and M-neighborhood respectively). As it may be observed in [12], above the threshold, when $\alpha > \alpha_c$, these patterns are unstable, but below and possibly at α_c, these patterns are sufficiently stable to extend and ultimately cover the whole configuration.

Convergence in Asynchronous Regimes. The last column of Fig. 1 shows that experimentally, when $\alpha \leqslant \alpha_c$, the asynchronous dynamics appears to converge at least with constant probability, rapidly to very particular stable configurations tiled by simple patterns known to be stable for the dynamics. Above the threshold, when $\alpha_c < \alpha < 1$, the asynchronous dynamics appears experimentally to be stuck into randomly evolving configurations in which no structure seems to emerge.

We will show in Theorem 1 that if at least one of the dimensions is even, the dynamics will almost surely reach a stable configuration, for all $0 \leqslant \alpha < 1$, but after at most an exponential number of steps. We conjecture that below the threshold α_c this convergence occurs in polynomial time on expectation if both dimensions are even (the threshold $T_s = 2000$ is probably too low for the M-neighborhood in Fig.1). We will prove this result in Theorem 2 for the fully asynchronous regime under the N-neighborhood under certain natural constraint on the initial configuration. Similar results to the ones to be presented below have been obtained in [17] for the M-neighborhood by extending of the techniques presented here.

3 Analysis of Fully Asynchronous 2D Minority

We consider now the fully asynchronous dynamics of 2D **Minority** with von Neumann neighborhood. Let n and m be two positive integers and $\mathbb{T} = \mathbb{Z}_n \times \mathbb{Z}_m$ the $n \times m$-torus. A $n \times m$-*configuration* c is a function $c : \mathbb{T} \to \{0,1\}$ that assigns to each *cell* $(i,j) \in \mathbb{T}$ its *state* $c_{ij} \in \{0,1\}$ (0 is white and 1 is black in the figures). We consider here the *von Neumann neighborhood*: the *neighbors* of each cell (i,j) are the four cells $(i \pm 1, j)$ and $(i, j \pm 1)$ (indices are computed modulo n and m, we thus consider periodic boundary conditions). We denote by $N = nm$, the total number of cells.

Definition 1 (Stochastic 2D Minority). *We consider the following dynamics δ that associates to each configuration c a random configuration c' obtained as follows: a cell $(i,j) \in \mathbb{T}$ is selected uniformly at random and its state is updated to the minority state in its neighborhood (we say that cell (i,j) is* fired*), all the other cells remain in their current state: $c'_{ij} = 1$ if $c_{ij} + c_{i-1,j} + c_{i+1,j} + c_{i,j-1} + c_{i,j+1} \leqslant 2$, and $c'_{ij} = 0$ otherwise; and $c'_{kl} = c_{kl}$ for all $(k,l) \neq (i,j)$. We say that a cell is* active *if its neighborhood is such that its state changes when the cell is fired.*

Definition 2 (Convergence). *We denote by c^t the random variable for the configuration obtained from a configuration c after t steps of the dynamics: $c^t = \delta^t(c)$; $c^0 = c$ is the* initial configuration.

We say that the dynamics δ converges *almost surely from an initial configuration c^0 to a configuration $\bar{c}$ if the random variable $T = \min\{t : c^t = \bar{c}\}$ is finite with probability* 1. *We say that the convergence occurs in polynomial (resp., linear, exponential) time on expectation if $\mathbb{E}[T] \leqslant p(N)$ for some polynomial (resp., linear, exponential) function p.*

As seen in Section 2, any configuration tend to converge under this dynamics towards a *stable configuration, i.e.*, towards a configuration where all cells are in the minority state of their neighborhood, *i.e.*, inactive.

Checkerboard Patterns. We say that a subset of cells $R \subseteq \mathbb{T}$ is *connected* if R is connected for the neighborhood relationship. We say that R is *checkerboard-tiled* if all adjacent cells in R are in opposite states. A *horizontal* (resp., *vertical*) *band* of width w is a set of cells $R = \{(i,j) : k \leqslant i < k+w\}$ for some k (resp., $R = \{(i,j) : k \leqslant j < k+w\}$).

3.1 Energy of a Configuration

The following natural parameters measure the stability of a configuration, *i.e.*, how far the cells of the configuration are from the minority state in their neighborhood. Following the seminal work of Tarjan in amortized analysis [19], we define a local potential that measures the amount of local unstability in the configuration. We proceed by analogy with the spin systems in statistical physics (Ising Model [3]): we assign to each cell a potential equal to the benefit of switching its state; this potential is naturally defined as the number of its adjacent cells to which it is opposed (*i.e.*, here, the number of cells which are in the same state as itself); summing the potentials over all the cells defines the total energy of the configuration at that time. As we consider arbitrary initial configuration, the system evolves *out-of-equilibrium* until it (possibly) reaches a stable configuration, thus its energy will vary over time; in particular, as will be seen in Proposition 1, its energy will strictly decrease each time an irreversible transition is performed (*i.e.*, each time a cell of potential $\geqslant 3$ is fired). It turns out that this energy function plays a central role in defining, in Section 3.4, the variant that will be used to prove the convergence of the system. We will see in particular that as observed experimentally in Section 2, the system tends to reach configurations of minimal energy as one would expect in a real physical system.

Definition 3 (Energy). *The* potential v_{ij} *of cell (i,j) is the number of its four adjacent cells that are in the same state as itself. The* energy *of a configuration c is defined as the sum of the potentials of the cells: $E(c) = \sum_{i,j} v_{ij}$.*

Definition 4 (Borders and Homogeneous regions). *We say that there is a* border *between two neighboring cells if they are in the same state. An* alternating path *is a sequence of neighboring cells that does not go through a border, i.e., of* *alternating states. This defines an equivalence relationship «being connected by an alternating path», the equivalence classes of this relationship are called the* homogenous regions *of the configuration.*

By definition, each homogeneous region is connected and tiled by one of the two checkerboard patterns, either or . The boundary of each homogeneous region is exactly the set of borders touching its cells. Note that the potential of a cell is the number of borders among its sides. The energy of a configuration is thus twice the number of borders and a cell is active if and only if at least two of its sides are borders. It follows that: if both dimensions n and m have the same parity, $(\forall c)\, E(c) \in 4\mathbb{N}$; and $(\forall c)\, E(c) \in 2 + 4\mathbb{N}$ otherwise.

There are two configurations of maximum energy $4N$: *all-black* and *all-white*. If n and m are even, there are two configurations of energy zero: the two *checkerboards*. If n is even and m is odd, the minimum energy of a configuration is $2n$ and such a configuration consists in a checkerboard pattern wrapped around the odd dimension creating a vertical band of width 2 tiled with pattern .

Energy of Stable Configurations. A cell is inactive if and only if its potential is $\leqslant 1$. It follows that the energy of any stable configuration belongs to $\{0, 2, \ldots, N\}$. Stable configurations are thus as expected of lower energy. If n and m are even and at least one of them is a multiple of 4, there are stable configurations of maximum energy N, tiled by the "fat"-checkerboard or .

Under the fully asynchronous dynamics δ, the overall variation of the energy of the configuration when the state of a cell of potential v is flipped is $8-4v \leqslant 0$, and since active cells have potential $\geqslant 2$:

Proposition 1 (Energy is non-increasing). *From any initial configuration c, the random variables $E(c^t)$ form a non-increasing sequence and $E(c^t)$ decreases by at least 4 each time a cell of potential $\geqslant 3$ is fired.*

Initial Energy Drop. Furthermore, after a polynomial number of steps and from any *arbitrary* initial configuration, the energy falls rapidly below $5N/3$, which is observed experimentally through the rapid emergence of checkerboard patterns in the very first steps of the evolution. Observing that for any configuration of energy at least $5N/3$, there exists a sequence of at most two updates that decreases strictly the energy, one can show that:

Proposition 2 (Initial energy drop, *proof omitted***).** *The random variable $T = \min\{t : E(c^t) < 5N/3\}$ is almost surely finite and $\mathbb{E}[T] = O(N^2)$.*

Every inactive cell touches at most one border. Thus, the boundaries of homogeneous regions in a stable configuration form straight lines at least 2 cells apart from each other. Thus,

Proposition 3 (Stable configurations). *Stable configurations are the configurations composed of parallel checkerboard-tiled bands of width at least 2.In particular, if n and m are odd, no stable configuration exists.*

It follows that if n and m are odd, the dynamics δ never reaches a stable configuration.

3.2 Coupling with Outer-Totalistic 976

From now on up to the end of section 3, we assume that n and m are even (with the only exception of Corollary 1). We denote by $\boxplus$ the checkerboard configuration of energy 0 defined as follows: $\boxplus_{ij} = (i + j) \bmod 2$. Given two configurations c and c', we denote by $c \oplus c'$ the XOR configuration c'' such that $c''_{ij} = (c_{ij} + c'_{ij}) \bmod 2$.

Dual Configurations. As observed above, the fully asynchronous dynamics c^t tends to converge from any initial configuration c^0 to configurations tiled by large checkerboard regions. It is thus convenient to consider instead, the sequence of *dual configurations* $(\hat{c}^t)$ defined by $\hat{c}^t = \boxplus \oplus c^t$, in which the large checkerboard regions of c^t appear as large homogeneous black or white regions. Clearly, the dual sequence $\hat{c}^t$ evolves according to the dynamics $\hat{\delta}(.) = \boxplus \oplus \delta(\boxplus \oplus .)$, indeed for all t, $\hat{c}^{t+1} = \boxplus \oplus c^{t+1} = \boxplus \oplus \delta(c^t) = \boxplus \oplus \delta(\boxplus \oplus \hat{c}^t) = \hat{\delta}(\hat{c}^t)$.

By construction, the two dual random sequences (c^t) and $(\hat{c}^t)$ as well as their corresponding dynamics δ and $\hat{\delta}$ are *coupled probabilistically* (see [14]): the *same* random cell is fired in both configurations at each time step. A simple calculation shows that the dual dynamics $\hat{\delta}$ associates to each dual configuration $\hat{c}$, a dual configuration $\hat{c}'$ as follows: select uniformly at random a cell (i, j) (the same cell (i, j) as δ fires on the primal configuration c), let $\Sigma = \hat{c}_{i-1,j}+\hat{c}_{i+1,j}+\hat{c}_{i,j-1}+\hat{c}_{i,j+1}$ and set: $\hat{c}'_{ij} = 1$ if $\Sigma \geqslant 3$; $\hat{c}'_{ij} = 1 - \hat{c}_{ij}$ if $\Sigma = 2$; and $\hat{c}'_{ij} = 0$ otherwise; and $\hat{c}'_{kl} = \hat{c}_{kl}$ for all $(k, l) \neq (i, j)$. It turns out that this rule corresponds to the asynchronous dynamics of the cellular automaton Outer-Totalistic **976** [11]. The corresponding transitions are given in Fig. 2.

Stable Configurations of Outer-Totalistic 976. We define the energy of the dual configuration $\hat{c}$ and the potentials of each of its cells (i, j) as the corresponding quantities, $E(c)$ and v_{ij}, in the primal configuration c. By Proposition 3, the stable dual configurations under the dual dynamics $\hat{\delta}$ are the dual configurations composed of homogeneous black or white bands of widths $\geqslant 2$. The two dual configurations of minimum energy 0 are all-white and all-black.

Experimentally, any dual configuration under the fully asynchronous dynamics $\hat{\delta}$ evolves towards large homogeneous black or white regions (corresponding to the checkerboard patterns in the primal configuration). Informally, these regions evolve as follows (see Fig. 2): isolated points tend to disappear as well as peninsulas; borders and surrounded points are stable; large regions are eroded in a random manner from the corners or bridges that can be flipped reversibly and their boundaries follow some kind of 2D random walks until large bands without corners ultimately survive (see Fig. 3 or [12]).

3.3 Convergence from an Arbitrary Initial Configuration

In this section, we consider *arbitrary* initial configurations c^0 and show that indeed the dynamics δ converges to a stable configuration almost surely and after at most an exponential number of steps on expectation.

Von Neumann neighborhoods	**Isolated**	**Peninsula**	**Corner**	**Bridge**	**Border**	**Surrounded**
Minority $\delta(c)$	Inactive	Inactive	Active Reversible $\Delta E(c) = 0$	Active Reversible $\Delta E(c) = 0$	Active Irreversible $\Delta E(c) = -4$	Active Irreversible $\Delta E(c) = -8$
Outer-totalistic 976 $\hat{\delta}(\hat{c}) = \boxplus \oplus \delta(\boxplus \oplus \hat{c})$	Active Irreversible $\Delta E(c) = -8$	Active Irreversible $\Delta E(c) = -4$	Active Reversible $\Delta E(c) = 0$	Active Reversible $\Delta E(c) = 0$	Inactive	Inactive

Fig. 2. Neighborhood's names and transition tables of **Minority** δ and its counterpart Outer-Totalistic **976** $\hat{\delta}$ (see section 3.2): only active cells switch their states when fired

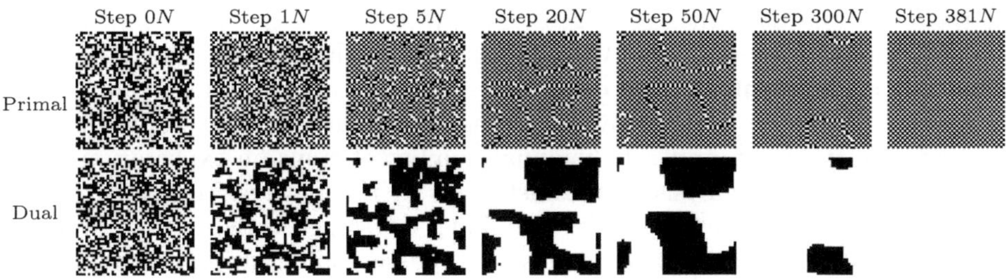

Fig. 3. The coupled evolutions of **Minority** δ on the primal configurations (c^t) (above) and its counterparts Outer-Totalistic **976** $\hat{\delta}$ on dual configurations $(\hat{c}^t)$ (below). Note that from step $50N$ on, (c^t) an $(\hat{c}^t)$ are bounded configurations.

Theorem 1. *From any initial configuration c^0, the dynamics δ convergences to a stable configuration after at most $2N^{2N+1}$ steps on expectation.*

Proof. According to the coupling above, it is equivalent to prove this statement for the dual dynamics. The following sequence of $\hat{\delta}$-updates transforms any dual configuration $\hat{c}$ into a dual stable configuration : *I)* as long as there are active white cells, choose one of them and switch its state to black; *II)* as long as there are active black cells, choose one of them and switch its state to white.

During phase 1, the black regions expand until they fill their surrounding bands or surrounding rectangles. Clearly according to the transition table Fig. 2, after phase 1 of the algorithm, every white cell is inactive and thus is either a border or surrounded. In particular, no white band of width 1 survived. During phase 2, the black cells enclosed in rectangles or in bands of width 1 are eroded progressively and ultimately disapear. Finally, only black bands of width $\geqslant 2$ survive at the end of phase 2 and the configuration is stable since it is composed of homogeneous white or black bands of width $\geqslant 2$ (see Proposition 3). During each phase, at most N cells change their state. We conclude that, from any configuration $\hat{c}$, there exists a path of length at most $2N$ to a stable configuration. Now, splits the sequence (c^t) into segments $(c^{2Nk+1}, ..., c^{2N(k+1)})$ of length $2N$. The sequence of updates in each of these segments has a probability $1/N^{2N}$ to be the sequence of at most $2N$ updates given above that tranforms configuration

c^{2Nk} into a stable configuration. Since these events are independent, this occurs after N^{2N} trials on expectation. We conclude that the dynamics $\hat{\delta}$ and thus δ converge to a stable configuration after at most $2N \cdot N^{2N}$ steps on expectation. □

Corollary 1. *(Proof omitted) From any initial $n \times m$-configuration c^0, where n is even and m is odd, the dynamics δ convergences to a stable configuration after at most $3N^{3N+1}$ steps on expectation.*

3.4 Convergence from a Bounded Configuration

We consider again that n and m are even. We observe experimentally that most of the time, the dynamics converges rapidly to one of the two checkerboard configurations of energy zero. We demonstrate in this section that if the dynamics reaches a configuration composed of an arbitrary region surrounded by a checkerboard, then it will converge to the corresponding checkerboard configuration almost surely after a polynomial number of steps on expectation. This corresponds to the analysis of the last steps of the behavior observed in experimentation. We believe that the techniques developed here may be extended to prove that the dynamics converges to a stable configuration in polynomial expected time from any initial configuration (see discussions in section 4).

Definition 5 (Bounded configuration). *We say that a configuration c is* bounded *if there exists a $(n-2)\times(m-2)$ rectangle such that the states in c of the cells outside this rectangle are equal to the corresponding states in one of the two checkerboard configurations. W.l.o.g., we assume that the upper-left corner of the rectangle is $(1,1)$ and that the checkerboard is ▩, i.e., a configuration c is bounded if $c_{ij} = (i+j) \bmod 2$ for all $(i,j) \in \{(i,j) : (-1 \leqslant i \leqslant 0)$ or $(-1 \leqslant j \leqslant 0)\}$.*

Each cell outside the surrounding rectangle has 3 neighbors in an opposite state as itself, and is thus inactive. It follows that if c is a bounded configuration, $\delta(c)$ is also bounded within the same surrounding rectangle. A bounded configuration is thus equivalent to a finite perturbation of an infinite planar configuration in $\mathbb{Z}^2$ tiled with the ▩ pattern. Since the dual of ▩ is the configuration all-white, the dual of a bounded configuration is thus equivalent to a *finite number of black cells*, included into a $(n-2)\times(m-2)$ rectangle within an *infinite white planar configuration in* $\mathbb{Z}^2$. We shall now consider this setting.

Definition 6 (Convexity). *We say that a set of cells $R \subseteq \mathbb{Z}^2$ is* convex *if for any pair of cells (i,j) and $(i+k,j)$ (resp., $(i,j+k)$) in R, the cells $(i+\ell,j)$ (resp., $(i,j+\ell)$) for $0 \leqslant \ell \leqslant k$ belong to R. We say that R is an* island *if R is connected and convex.*

Our proof of the convergence of the dynamics in polynomial time for bounded configurations relies on the definition of a variant which decreases on expectation over time. It turns out that in order to define the variant, we do not need to consider the exact internal structure of the bounded configuration, but only the structure of the convex hull of its black cells.

Definition 7 (Convex hull of a configuration). *For any finite set of cells $R \in \mathbb{Z}^2$, we denote by* $\mathsf{hull}(R)$ *the convex hull of the cells in* R, *i.e.*, $\mathsf{hull}(R) = \cap\{S \subseteq \mathbb{Z}^2 : S$ *is convex and* $S \supseteq R\}$. *Given a bounded dual configuration* $\hat{c}$, *we define the* convex hull of $\hat{c}$, $\mathsf{hull}(\hat{c})$, *as the dual configuration whose black cells are the cells in the* convex hull *of the black cells of* $\hat{c}$, *i.e., if* $R = \{(i,j) : \hat{c}_{ij} = 1\}$, $\mathsf{hull}(\hat{c})_{ij} = 1$ *if and only if* $(i,j) \in \mathsf{hull}(R)$. *We say that a configuration* c *is* convex *if* $\hat{c} = \mathsf{hull}(\hat{c})$.

We say that $\hat{c} \leqslant \hat{c}'$ *if for all* (i,j), $c_{ij} \leqslant c'_{ij}$. *Let* $\hat{c}$ *be a convex dual bounded configuration. We define for each black cell* (i,j) *in* $\hat{c}$, *the* island of $\hat{c}$ that contains cell (i,j), *as the maximum connected and convex configuration* $\hat{c}'$ *such that* $\hat{c}'_{ij} = 1$ *and* $\hat{c}' \leqslant \hat{c}$. *This defines a unique* decomposition into black islands *of the convex bounded configuration* $\hat{c}$.

The Variant. We now consider the following *variant*: $\Phi(\hat{c}) = E(\mathsf{hull}(\hat{c}))/4 + |\mathsf{hull}(\hat{c}))|$, where $|\mathsf{hull}(\hat{c}))|$ is the number of black cells in the convex hull configuration $\mathsf{hull}(\hat{c})$. We will show that from any initial configuration c^0, $\Phi(c^t)$ decreases by at least $1/N$ on expectation at each time step until it reaches the value 0, *i.e.*, until the primal and dual configurations c^t and $\hat{c}^t$ converge to the infinite checkerboard and the infinite all-white configurations respectively. In order to prove that $\Phi(c^t)$ decreases on expectation, we need to study the evolution of the convex hull of $\hat{c}^t$; for this purpose, we introduce a modified coupled dual dynamics $\bar{\delta}$ that preserves the convexity of a dual configuration. Given a dual configuration $\hat{c}$, we denote by $\bar{\delta}(\hat{c})$ the random configuration $\hat{c}'$ such that: $\hat{c}' = \hat{\delta}(\hat{c})$ if the cell updated by $\hat{\delta}$ is *not* a black bridge, and $\hat{c}' = \hat{c}$ otherwise. Since only firing a black bridge can break the convexity of a black region, then:

Lemma 1. *If* $\hat{c}$ *is a convex bounded configuration,* $\bar{\delta}(\hat{c})$ *is a convex bounded configuration.*

The energy of a convex region is twice the number of borders, *i.e.*, twice the sum of the perimeters of the islands that compose it, so:

Lemma 2. *For all convex bounded configurations* $\hat{c}$ *and* $\hat{c}'$, *if* $\hat{c} \leqslant \hat{c}'$, *then* $E(c) \leqslant E(c')$.

The construction of $\bar{\delta}$ guarantees that the image of the convex hull of $\hat{c}$ by the dynamics $\bar{\delta}$ bounds from above the convex hull of the image of $\hat{c}$ by the dynamics $\hat{\delta}$.

Lemma 3. *(Proof omitted) For all bounded configuration* $\hat{c}$, $\hat{\delta}(\hat{c}) \leqslant \bar{\delta}(\mathsf{hull}(\hat{c}))$.

Let $\Delta\Phi_\lambda(\hat{c})$ be the random variable for the variation of the variant after one step of a dynamics λ from a configuration c, *i.e.*, $\Delta\Phi_\lambda(\hat{c}) = \Phi(\lambda(\hat{c})) - \Phi(\hat{c})$.

Corollary 2. *For all bounded configuration* $\hat{c}$, $\Delta\Phi_{\hat{\delta}}(\hat{c}) \leqslant \Delta\Phi_{\bar{\delta}}(\mathsf{hull}(\hat{c}))$.

Proof. By definition, $\Delta\Phi_{\bar{\delta}}(\mathsf{hull}(\hat{c})) - \Delta\Phi_{\hat{\delta}}(\hat{c}) = \big(|\bar{\delta}(\mathsf{hull}(\hat{c}))| - |\mathsf{hull}(\hat{\delta}(\hat{c}))|\big) + \big(E(\bar{\delta}(\mathsf{hull}(\hat{c}))) - E(\mathsf{hull}(\hat{\delta}(\hat{c})))\big)$. According to lemma 3, $\mathsf{hull}(\hat{\delta}(\hat{c})) \leqslant \bar{\delta}(\mathsf{hull}(\hat{c}))$

and thus $|\mathsf{hull}(\hat{\delta}(\hat{c}))| \leqslant |\bar{\delta}(\mathsf{hull}(\hat{c}))|$. And by Lemma 2, since both configurations are convex, $E(\mathsf{hull}(\hat{\delta}(\hat{c}))) \leqslant E(\bar{\delta}(\mathsf{hull}(\hat{c})))$. □

Lemma 4. *For all bounded configuration $\hat{c}$ that consists of a unique black island, $-4/N \leqslant \mathbb{E}[\Delta\Phi_{\bar{\delta}}(\hat{c})] \leqslant -3/N$.*

Proof. Each active cell is fired with probability $1/N$. According to the dynamics of $\bar{\delta}$ (the same as the dynamics of $\hat{\delta}$, Fig. 2, except that black bridges are inactive), if $\hat{c}$ consists of an island of size $\geqslant 2$, $\mathbb{E}[\Delta\Phi_{\bar{\delta}}(\hat{c})] = -\frac{1}{N}(\#\{\text{black corners}\} + 2\,\#\{\text{black peninsulas}\}) + \frac{1}{N}\#\{\text{white corners}\} = -\frac{1}{N}\#\{\text{salient angles}\} + \frac{1}{N}\#\{\text{reflex angles}\} = -\frac{4}{N}$, since $\#\{\text{salient angles}\} - \#\{\text{reflex angles}\} = 4$ for all convex rectilinear polygon. Finally, if $\hat{c}$ consists of a unique (isolated) black cell, $\Delta\Phi_{\bar{\delta}}(\hat{c}) = -3/N$. □

Lemma 5. *For any bounded not-all-white configuration $\hat{c}$, $\mathbb{E}[\Delta\Phi_{\hat{\delta}}(\hat{c})] \leqslant -\ell/N$, where ℓ is the number of islands that compose $\mathsf{hull}(\hat{c})$.*

Proof. By Corollary 2, $\mathbb{E}[\Delta\Phi_{\hat{\delta}}(\hat{c})] \leqslant \mathbb{E}[\Delta\Phi_{\bar{\delta}}(\mathsf{hull}(\hat{c}))]$. By convexity of $\mathsf{hull}(\hat{c})$, the sets of rows and columns touched by the islands that compose $\mathsf{hull}(\hat{c})$ are pairwise disjoint. Thus, one can index the islands from 1 to ℓ from left to right, and the contacts between islands can only occur between two consecutive islands at the corners of their surrounding rectangles. Each contact creates at most two new active white cells that contribute for $+1/N$ each to $\mathbb{E}[\Delta\Phi_{\bar{\delta}}(\mathsf{hull}(\hat{c}))]$. The contribution of each island to $\mathbb{E}[\Delta\Phi_{\bar{\delta}}(\mathsf{hull}(\hat{c}))]$ is at most $-3/N$ according to Lemma 4. It follows that $\mathbb{E}[\Delta\Phi_{\bar{\delta}}(\mathsf{hull}(\hat{c}))] \leqslant -3\ell/N + 2(\ell-1)/N \leqslant -\ell/N$. □

Theorem 2. *The fully asynchronous minority dynamics δ converges almost surely from any initial bounded configuration c to the stable configuration of minimum energy, ▩, and the expected convergence time is $O(AN)$ where A is the area of surrounding rectangle of the black cells in $\hat{c}$.*

Proof. Initially and for all time $t \geqslant 0$, $\Phi(\hat{c}^t) \leqslant 2(n-2+m-2) + A \leqslant 2N + A$. As long as $\hat{c}^t \not\equiv 0$, $\Phi(\hat{c}^t) > 0$ and according to Lemma 5, $\mathbb{E}[\Delta\Phi_{\hat{\delta}}(\hat{c}^t)] \leqslant -1/N$. It follows that the random variable $T = \min\{t : \Phi(\hat{c}^t) \leqslant 0\}$ is almost surely finite and $\mathbb{E}[T] = O(nA)$ (by applying for example Lemma 2 in [6]); and at that time, $\hat{c}^T$ and c^T are the stable configurations all-white and ▩, respectively. □

Example 1 (Worst case configurations). Consider the initial dual bounded $n \times n$-configuration $\hat{c}$ consisting of a black $2 \times (n-2)$ rectangle. The expected time needed to erase one complete line of the rectangle is at least $\Omega(nN) = \Omega(AN)$.

4 Concluding Remarks

This paper proposes an extension to 2D cellular automata of the techniques based on random walks developped in [6,7] to study 1D asynchronous elementary cellular automata. Our techniques apply as well with some important new ingredients, to the Moore neighborhood where the cell fired updates to the minority state within its height closest neighbors [17]. We believe that these techniques may extend to the wide class of threshold automata, which are of particular interest, in neural networks for instance. We are currently investigating refinements of the tools developed here, based on the study of the boundaries between arbitrary checkerboard regions in order to try to prove that every *arbitrary* $n \times m$-configuration converges to a stable configuration in a polynomial number of steps when n and m are both even (we conjecture a convergence in time $O(N^3)$ for non-bounded toric configurations of even dimensions). This result would conclude the study of this automaton under fully asynchronous dynamics. The experiments lead in Section 2 exhibit an impressive richness of behavior for this yet apparently simple transition rule. An extension of our results to arbitrary α-asynchronous regime is yet a challenging goal, especially if one considers that most of the results concerning spin systems or lattice gas (at the equilibrium) apply only to the limit when the temperature tends to 0, *i.e.*, when only one transition occurs at a time.

Acknowledgements. We would like to thank C. Moore, R. D'Souza and J. Crutchfield for their useful suggestions on the physics related aspects of our work.

References

1. Aracena, J., Lamine, S.B., Mermet, M.-A., Cohen, O., Demongeot, J.: Mathematical modeling in genetic networks: relationships between the genetic expression and both chromosomic breakage and positive circuits. IEEE Trans. on Systems, Man, and Cybernetics Part B 33(5), 825–834 (2003)
2. Bersini, H., Detours, V.: Asynchrony induces stability in cellular automata based models. In: Proceedings of Artificial Life IV, pp. 382–387. MIT Press, Cambridge (1994)
3. Bovier, A., Manzo, F.: Metastability in glauber dynamics in the low temperature limit: Beyond exponential asymptotics. J. Statist. Phys. 107, 757–779 (2002)
4. Buvel, R.L., Ingerson, T.E.: Structure in asynchronous cellular automata. Physica D 1, 59–68 (1984)
5. Fatès, N., Morvan, M.: An experimental study of robustness to asynchronism for elementary cellular automata. Complex Systems 16(1), 1–27 (2005)
6. Fatès, N., Morvan, M., Schabanel, N., Thierry, É.: Fully asynchronous behaviour of double-quiescent elementary cellular automata. Theoretical Computer Science 362, 1–16 (2006) (An extended abstract was also published in Proc. of MFCS'2005)
7. Fatès, N., Regnault, D., Schabanel, N., Thierry, É.: Asynchronous behaviour of double-quiescent elementary cellular automata. In: LATIN 2006. LNCS, vol. 3887, Springer, Heidelberg (2006)
8. Fukś, H.: Non-deterministic density classification with diffusive probabilistic cellular automata. Phys. Rev. E 66(2) (2002)
9. Fukś, H.: Probabilistic cellular automata with conserved quantities. Nonlinearity 17(1), 159–173 (2004)
10. Goles, E., Martinez, S.: Neural and automata networks, dynamical behavior and applications. Maths and Applications, vol. 58. Kluwer Academic Publishers, Dordrecht (1990)

11. http://mathworld.wolfram.com/Outer-TotalisticCellularAutomaton.html
12. http://www.cmm.uchile.cl/~schabanel/2DMINORITY
13. Lumer, E.D., Nicolis, G.: Synchronous versus asynchronous dynamics in spatially distributed systems. Physica D 71, 440–452 (1994)
14. Randall, D.: Mixing. In: Proc. of the Symp. on Foundations of Computer Science (FOCS), pp. 4–15 (2003)
15. Regnault, D.: Abrupt behaviour changes in cellular automata under asynchronous dynamics. In: Proceedings of 2nd European Conference on Complex Systems (ECCS), Oxford, UK (2006) (to appear)
16. Regnault, D., Schabanel, N., Thierry, É.: Progresses in the analysis of stochastic 2D cellular automata: a study of asynchronous 2D minority (Full text). Preprint arXiv:0706.2479 [cs.DM] (2007)
17. Regnault, D., Schabanel, N., Thierry, É.: A study of stochastic 2D Minority CA: Would wearing stripes be a fatality for snob people? Research Report N○ENSL-00140883, École Normale Supérieure de Lyon, 2007.
18. Schönfisch, B., de Roos, A.: Synchronous and asynchronous updating in cellular automata. BioSystems 51, 123–143 (1999)
19. Tarjan, R.E.: Amortized computational complexity. SIAM Journal of Algebraic and Discrete Methods 6(2), 306–318 (1985)

Public Key Identification Based on the Equivalence of Quadratic Forms

Rupert J. Hartung and Claus-Peter Schnorr

Johann Wolfgang Goethe Universität Frankfurt a. M.
Postfach 11 19 32; Fach 238
60054 Frankfurt a. M., Germany
{schnorr,hartung}@mi.informatik.uni-frankfurt.de

Abstract. The computational equivalence problem for quadratic forms is shown to be NP-hard under randomized reductions, in particular for indefinite, ternary quadratic forms with integer coefficients. This result is conditional on a variant of the Cohen-Lenstra heuristics on class numbers. Our identification scheme proves knowledge of an equivalence transform.

1 Introduction

The arithmetic theory of quadratic forms has a long history. Algorithmic problems on lattices and quadratic forms, however, have long been neglected; their study has been significantly pushed by the LLL-algorithm for lattice basis reduction [19]. Recently definite forms, or lattices, gave rise to cryptographic protocols related to the NP-hard problems of finding a shortest, respectively, closest lattice vector; see [22], [20] for hardness results and [2], [14], [15], [13], [12] for the applications. Cryptographic protocols based on NP-hard problems seem to withstand attacks by quantum computers. However, lattice cryptography requires lattices of high dimension. This yields long cryptographic keys and slow protocols.

By contrast, we show that quadratic form cryptography is possible in dimension three. We prove conditional NP-hardness of the equivalence and representation problems of indefinite, ternary forms over the integers using randomized reductions. We build on the work of Adleman and Manders [21] who proved NP-hardness of deciding solvability of inhomogeneous binary quadratic equations over the integers.

In Sect. 3, we present an identification scheme that proves knowledge of an equivalence transform of quadratic forms. This scheme is statistical zero-knowledge under reasonable heuristics. It allows short keys and performs merely one LLL-reduction and a few arithmetic steps per round, but its security requires many independent rounds.

2 The Equivalence Problem for Quadratic Forms

Quadratic Forms. An n-ary *quadratic form* (or simply *form*) f over $\mathbb{Z}$ is a homogeneous quadratic polynomial $f = \sum_{i,j=1}^{n} a_{i,j} x_i x_j = \boldsymbol{x}^t A \boldsymbol{x}$ with integer

L. Kučera and A. Kučera (Eds.): MFCS 2007, LNCS 4708, pp. 333–345, 2007.

coefficients $a_{i,j} = a_{j,i} \in \mathbb{Z}$, $A = (a_{i,j})$ and $\boldsymbol{x} = (x_1, ..., x_n)^t$. Then f has *determinant* $\det A$ and *dimension* n.

Equivalence Classes. Let $f = \boldsymbol{x}^t A \boldsymbol{x}$ be an n-ary form. For $T \in \mathbb{Z}^{n \times n}$ let fT denote the form $\boldsymbol{x}^t T^t A T \boldsymbol{x}$. The forms f, fT are called *equivalent* if $T \in \mathrm{GL}_n(\mathbb{Z})$, i.e., if $|\det T| = 1$, notation $f \sim fT$. The equivalence class of f is simply called the *class* of f. Obviously $\det(fT) = (\det T)^2 \det f = \det f$.

Analogously, we define the equivalence of integral forms over $\mathbb{Z}_p$, the ring of p-adic integers. Two forms f, g over $\mathbb{Z}$ belong to the same *genus* if they are equivalent over $\mathbb{Z}_p$ for all primes p, and over $\mathbb{Z}_\infty = \mathbb{R}$. Clearly, every genus is a disjoint union of classes.

We study the equivalence problem of forms f having certain properties $\mathcal{P}$. Relevant properties are: f is *regular* if $\det f \neq 0$; f is *indefinite* if $f(x)$ takes both positive and negative values, otherwise f is *definite* (definite forms correspond to the Gram matrices $A = B^t B$ of lattice bases B); $f = \boldsymbol{x}^t A \boldsymbol{x}$ is *properly primitive* if $\gcd(a_{ii}, 2a_{ij} \mid i \neq j) = 1$; f is *isotropic* if $f(\boldsymbol{u}) = 0$ holds for some nonzero $\boldsymbol{u} \in \mathbb{Z}^n$, otherwise f is *anisotropic*. Every regular isotropic form is necessarily indefinite.

The Computational Equivalence Problem, CEP

INPUT: equivalent forms f, g satisfying certain properties $\mathcal{P}$.

OUTPUT: $T \in \mathrm{GL}_n(\mathbb{Z})$ such that $g = fT$.

The concept of LLL-reduction [19] extends in a natural way from lattice bases and definite forms to anisotropic indefinite forms. LLL-forms $f = \boldsymbol{x}^t A \boldsymbol{x}, A = (a_{i,j})$ satisfy $a_{1,1}^2 \leq 2^{2/n} \det A^{2/n}$. There is a polynomial time LLL-algorithm that transforms f into an LLL-form fT with $T \in \mathrm{GL}_n(\mathbb{Z})$, see [26], [18], and [25].

Outline. Section 3 presents an identification scheme based on the equivalence problem. Section 4 proves a variant of the equivalence problem to be NP-hard under randomized reductions. This is shown for indefinite, anisotropic forms of dimension $n = 3$ using a number theoretic assumption that guarantees class number 1 in real quadratic fields.

3 Identification Based on the Equivalence Problem

Key Generation. Pick a random, indefinite, anisotropic, ternary LLL-form f_1 and a random $T \in \mathrm{GL}_3(\mathbb{Z})$ following **CT**. LLL-reduce $f_1 T$ to $f_0 := f_1 T T'$. The *public key* is f_0, f_1, the *secret key* is $S := TT'$. S is uniquely determined by f_0, f_1 up to small automorphisms of f_0, f_1.

In the protocol $(\mathcal{P}, \mathcal{V})$ the prover $\mathcal{P}$ proves to the verifier $\mathcal{V}$ knowledge of S.

Identification scheme, $(\mathcal{P}, \mathcal{V})$

1. $\mathcal{P}$ computes an LLL-form $g := f_0 T$ via **CT**, and sends g,

2. $\mathcal{V}$ sends a random one-bit challenge $b \in_R \{0, 1\}$,

3. $\mathcal{P}$ sends $R := ST$, and $\mathcal{V}$ checks that $f_b R = g$.

CT: *Computation of* $T = (T_{i,j}) \in \mathrm{GL}_3(\mathbb{Z})$. Let $\|T\| = \max_{i,j} |T_{i,j}|$ be the *norm*. Set $r := 2^{100}\|S\|$. Pick the $T_{i,j} \in_R [-r, r]$ at random for $j \neq 1$. Compute $T_{1,1}, ..., T_{3,1} \in \mathbb{Z}$ by applying the extended Euclidean algorithm to $T_{1,1}^{adj}, ..., T_{3,1}^{adj}$ in order to achieve that $\det T = \pm 1$.

Note that $\det T = \sum_{i=1}^{3} \pm T_{i,1} T_{i,1}^{adj}$, where the $T_{i,1}^{adj}$ are values of homogeneous, quadratic polynomials in the $T_{i,j}$ with $j \neq 1$. Make sure that $gcd(T_{1,1}^{adj}, ..., T_{3,1}^{adj}) = 1$ by picking, if necessary, some new $T_{i,j}$. The Euclidean algorithm yields $|T_{i,1}| \leq \max_i |T_{i,1}^{adj}| \leq 4r^2$.

Finally LLL-reduce $f_0 T$ into $f_0 T T'$ and replace $T := TT'$. This balances the large $T_{i,1}$ with the smaller $T_{i,j}$, $j > 1$. The leading 100 bits and the least significant 100 bits of the $T_{i,j}$ are nearly random. Think of $f_0 T$ to be a random LLL-form out of a "sphere" of "radius" $\Theta(r^2)$ centered at f_0.

Completeness. The true prover $\mathcal{P}$ withstands the test $f_b R = g$.

Proof of Knowledge. Consider a fraudulent $\widetilde{\mathcal{P}}$ that sends arbitrary $\bar{g}, \bar{R}$. The trivial $\widetilde{\mathcal{P}}$ guesses b in step 1 with probability $\frac{1}{2}$, sends the LLL-form $\bar{g} := f_b \bar{T}_b$ and the reply $\bar{R}_b := \bar{T}_b$. $\widetilde{\mathcal{P}}$ withstands the verification with probability $\frac{1}{2}$. The probability $\frac{1}{2}$ cannot be increased or else $\widetilde{\mathcal{P}}$ obtains an cryptographically equivalent secret key $S' := \bar{R}_0^{-1} \bar{R}_1$ such that $f_0 = f_1 S'$.

Suppose an arbitrary $\widetilde{P}$ withstands the verification for the same $\bar{g}$ and both challenges $b = 0, 1$ replying $\bar{R}_b$. Then $\bar{R}_b$ transforms f_b into $\bar{g}$ and thus $f_0 \bar{R}_0^{-1} \bar{R}_1 = f_1$.

More precisely, letting $|(\widetilde{\mathcal{P}}, \mathcal{V})|$ bound the number of steps of $(\widetilde{\mathcal{P}}, \mathcal{V})$ we have:

Theorem 1. *A fraudulent prover* $\widetilde{P}$ *that withstands* k *independent executions of* $(\widetilde{\mathcal{P}}, \mathcal{V})$ *with probability* $\varepsilon > 2^{-k}$, *obtains an "equivalent" secret key in expected time* $|(\widetilde{\mathcal{P}}, \mathcal{V})|/(\varepsilon - 2^{-k})$.

Statistical Zeroknowledge. The protocol $(\mathcal{P}, \mathcal{V})$ is by definition *statistical zeroknowledge* if for every probabilistic poly-time verifier $\widetilde{\mathcal{V}}$, there is a probabilistic poly-time simulator $\mathcal{S}$, which produces randomized strings whose distribution is statistical close to the communication of $(\mathcal{P}, \widetilde{\mathcal{V}})$ ($\| \cdot \|_1$-distance $\leq 2^{-100}$ suffices in practice). The simulator $\mathcal{S}$ has resettable black-box access to $\widetilde{\mathcal{V}}$ but does not know the secret key.

The simulator $\mathcal{S}$ mimics $\widetilde{\mathcal{P}}$ replying $\bar{R}_b = \bar{T}_b$ in step 3 whereas the true prover replies $R = S^b T$. The distributions of $S^b T, \bar{T}_b$ must be statistical close for both $b = 0, 1$.

Note that $S^b T$, $\bar{T}_b$ are determined, up to small automorphisms of f_0, f_1, as the isomorphisms from f_b to g, resp., from f_b to $\bar{g}$. Consider $g, \bar{g}$ as random LLL-forms out of spheres of radius $\Theta(r^2)$ centered at f_b, resp., $\bar{f}_b$. Small deviations of r should have a negligible impact. In particular, we assume that ST reveals negligible information about S, and the LLL-reduction reveals merely an upper bound of $\|S\|$.

Theorem 2. *The identification scheme $(\mathcal{P}, \mathcal{V})$ is statistical zeroknowledge under reasonable heuristics.*

The zero-knowledge property extends to independent sequential executions of $(\mathcal{P}, \mathcal{V})$. Since $(\mathcal{P}, \mathcal{V})$ is restricted to one-bit challenges, a security level 2^{100} requires to run 100 independent executions of $(\mathcal{P}, \mathcal{V})$. To allow a poly-time simulator $\mathcal{S}$ these executions must be sequential. It is open to extend $(\mathcal{P}, \mathcal{V})$ to long challenges.

4 NP-Completeness

4.1 Introduction and Results

In this section, we prove randomized NP-completeness of a decisional variant of the **CEP**, as well as the related representation problem. We will use a number-theoretic assumption, the special Cohen-Lenstra Heuristics (sCLH), which is discussed in Sect. 4.2.

A quadratic form f of dimension n is said to *represent* a number $m \in \mathbb{Z}$ if there exists $u \in \mathbb{Z}^n \setminus \{0\}$ such that $f(u) = m$; here u is called a *representation* of m by f. The representation is said to be *primitive* if $\gcd(u_1, \ldots, u_n) = 1$. It is natural to ask for an algorithm which, given f and m, computes such a vector u. We will consider the following version of this problem.

Decisional Interval Representation Problem, DIRepr
PARAMETERS: Set $\mathcal{P}$ of properties of quadratic forms.
INPUT: $n \in \mathbb{N}$, quadratic form f of dimension n satifying all properties from $\mathcal{P}$, integer m, vectors $v, w \in (\mathbb{Z} \cup \{\pm\infty\})^n$, factorization of $\det f$.
DECIDE: Whether there is $x \in \mathbb{Z}^n$, $v_i \leq x_i \leq w_i$ for all i s. t. $f(x) = m$.

We can also define the computational problem **IRepr** with the same parameters and inputs as **DIRepr**, but where a representation in the given interval is to be computed. But then by a straightforward divide-and-conquer algorithm, **IRepr**$(\mathcal{P})$ and **DIRepr**$(\mathcal{P})$ are polynomial-time equivalent.

Note that the symbols $\pm\infty$ do not mean greater generality here because there are polynomial bounds on possible representations anyway by [8]. We might have restricted the intervals in the definitions to be origin-symmetric, and to restrict at most one component, as will turn out from the proofs.

We denote by $\operatorname{gen} f$ the genus of a form f, by $\operatorname{cls} f$ its class, and by $\operatorname{cls}^+ f$ its *proper class*, i. e. the set of forms fU with $\det U = +1$.

Theorem 3. *Let $M \in \mathbb{N}$. Let $\mathcal{P}'_M$ consist of the properties*

$$\dim f = 3, \qquad f \textit{ indefinite anisotropic}, \qquad \operatorname{gen} f = \operatorname{cls}^+ f,$$
$$f \textit{ properly primitive}, \qquad \textit{and} \qquad (\det f, M) = 1$$

for a quadratic form f. If the sCLH holds true, then **DIRepr**$(\mathcal{P}'_M)$ *is NP-complete under randomized reductions with one-sided error; more precisely:*

$$NP \subseteq RP^{\mathbf{DIRepr}(\mathcal{P}'_M)}.$$

For the security of our identification scheme, it is important that the extraction of the secret key from public parameters is not feasible; here that means that given equivalent forms f, g, the computation of an equivalence transformation $S \in \mathrm{GL}_n(\mathbb{Z})$ is hard. Again we consider interval constraints on the desired solution.

Decisional Interval Equivalence Problem, DIEP
PARAMETERS: Set $\mathcal{P}$ of properties of quadratic forms.
INPUT: $n \in \mathbb{N}$, n-ary quadratic forms f, g satifying all properties from $\mathcal{P}$, matrices $A, B \in (\mathbb{Z} \cup \{\pm\infty\})^{n \times n}$, factorization of $\det f$.
DECIDE: Whether there exists $T \in \mathrm{GL}_n(\mathbb{Z})$, $A_{ij} \le T_{ij} \le B_{ij}$ for all i, j s. t. $f T = g$.

As with representations, the problem of computing such a transformation is polynomial-time equivalent to **DIEP**, and both are NP-hard:

Theorem 4. *Let $M \in \mathbb{N}$. Let $\mathcal{P}'_M$ consist of the properties*

$$\dim f = 3, \qquad f \textit{ indefinite anisotropic}, \qquad \operatorname{gen} f = \operatorname{cls}^+ f,$$
$$f \textit{ properly primitive}, \qquad \textit{and} \qquad (\det f, M) = 1$$

for a quadratic form f. If the sCLH holds true, then $\mathbf{DIEP}(\mathcal{P}'_M)$ *is NP-complete under randomized reductions with one-sided error; precisely:*

$$NP \subseteq RP^{\mathbf{DIEP}(\mathcal{P}'_M)}.$$

We now explain our assumption in detail and then give proof sketches, while some of its details are elaborated in the appendix.

4.2 The Special Cohen-Lenstra Heuristics

In [5], Cohen and Lenstra suggested a very general heuristic framework for the prediction of the average behavior of the class group of a number field K. Based on the thought experiment that, roughly speaking, all properties of class groups which are not determined a priori (e. g. by the factorization of the discriminant), develop according to a certain random model, they obtain a corresponding very comprehensive conjecture on the distribution of such properties on large sets of discriminants. The CLH intends to give a convincing link between several seemingly independent observations from calculations of class groups. Though still unproven, it has enjoyed a vivid reception, and is thought of as a realistic way of thinking about long-run development of class numbers and groups.

One famous special case will be of central interest to us: Namely, the class numbers of real quadratic fields (cf. [6]). The empirical findings that large class numbers, in particular class numbers with odd part larger than one are rare, have been one of the central motivations to formulate these heuristics. We are going to use this empirically noticeable trait of class numbers. However, we cannot draw upon proven statements here as the conjecture is still wide open; in particular, our variant would imply that there are infinitely many real quadratic fields with class number 1, which is still unknown for number fields in general.

To explain our assumption, we first analyze different parts of the class group and the class number. Let d be an odd squarefree positive integer (for simplicity). Recall that the set $\mathfrak{F}(d)$ of proper classes of integral binary quadratic forms of determinant d forms a group under Gauß composition. The unit element is given by the *principal class*, i. e. the unique class of forms f_0 representing 1.

Let $\mathfrak{Cl}(d)$ be the ideal class group of the real quadratic number field $\mathbb{Q}[\sqrt{d}]$. Then $\mathfrak{Cl}(d) \cong \mathfrak{F}(d)/I$, where I is the subgroup of order 1 or 2 generated by the unique class which represents -1 (see [4, sec. 5.2]). Gauß [11] showed that $\mathfrak{F}(d)^2$ equals the genus of f_0, whence the 2-rank of $\mathfrak{F}(d)$ equals the number of genera, which is $2^{\omega(d)}$. This power of two constitutes the 'deterministic part' of the class number: It is determined by the prime factorization of d.

Beyond that part determined by genus theory, class numbers seem to behave 'randomly'; and essentially, Cohen's and Lenstra's idea was to formulate this impression explicitly and give a probabilistic model to describe the effects observed. In their original paper, however, they consider only the odd part $h^{\not 2}$ of the class number to avoid interference with the genus structure. But the total exclusion of the prime 2 now seems to be overly careful since only the index of $\mathfrak{Cl}^2(d)$ is linked to the genus structure, but not the 2-part of $|\mathfrak{Cl}^2(d)|$. It was conjectured in [16], [17] and, in contrast to the large remaining part of the heuristics, it was proven in [10, thm. 3] that the 2-part of $|\mathfrak{Cl}^2(d)|$ behaves as random as conjectured for the odd part of the class number.

We now want to assume, first put informally, that the Cohen-Lenstra Heuristics is compatible with the theorem on the 2-part still if restricted to primes of certain residue classes, and that this convergence is not too slow. Precisely, we state:

Special Cohen Lenstra Heuristics (sCLH) 4.1. *There are $c, e > 0$ and a polynomial F s. t. the following holds:*

Let $B > 0$ and primes $p_1, \ldots, p_k$ be given, where $k \le e \log B$. Then

$$\#\{q \le F(B) \,|\, q \text{ prime}, \left(\frac{q}{p_i}\right) = -1 \;\forall i; \quad |\mathfrak{Cl}(D(q))^2| = 1\} \ge c\,\frac{F(B)}{B \log F(B)}.$$

Here $D(q)$ denotes the fundamental discriminant corresponding to q.

It should be noted that the restriction to primes, and further to primes in specific residue classes, which is well prepared by tables as [27], already pops up in the original publication (see [5, §9, II. C12]) and is explicitly encouraged in [7, sec. 3].

4.3 Sketch of Proofs

To bridge between the classical NP-complete problems and quadratic forms, we use the following problem on binary Diophantine equations.

MS Narrow Modular Square Problem
PARAMETER: $M \in \mathbb{N}$.
INPUT: Integers $a, b, c \in \mathbb{Z}$ with $c > 0$, a odd, squarefree, s. t. $(ab, M) = 1$ and there is an odd prime p s. t. if $u^2 | b$, then u is a power of p; factorization of b.
DECIDE: Whether there is $x \in \mathbb{Z}$, $|x| \leq c$ s. t. $x^2 \equiv a \bmod b$.

Proposition 1 is similar to the result by Adleman and Manders [21]; they proved it for arbitrary integers a, b in the problem instance.

We denote a deterministic Karp reduction by $\preccurlyeq$, whereas a probabilistic Karp reduction with one-sided error is depicted by $\preccurlyeq_r$.

Proposition 1. *Let $M \in \mathbb{N}$ be arbitrary. Then $3SAT \preccurlyeq_r \mathbf{MS}(M)$.*

This will be proven in the appendix.- Abbreviate the form $a_1 x_1^2 + \ldots + a_n x_n^2$ by $\langle a_1, \ldots, a_n \rangle$. From genus theory, the following can be derived:

Lemma 1. *Let $p \equiv 1 \bmod 4$ be a prime satisfying $\operatorname{cls}\langle 1, -p\rangle = \operatorname{gen}\langle 1, -p\rangle$. Let $m \in \mathbb{Z}$ be odd and satisfy $\left(\frac{m}{p}\right) = 1$ and $\left(\frac{q}{p}\right) = -1 \quad \forall q$ prime, $q|m$. (In particular, $p \not| m$.) Then $\langle 1, -p\rangle$ represents m primitively.*

By the theory of the spinor norm (see [24]) the following can be proven:

Proposition 2. *Let b be odd, p an odd prime, and $p \not| b$. Then the form $f := \langle 2, b, -pb\rangle$ satisfies $\operatorname{gen} f = \operatorname{cls}^+ f$.*

Proof sketch of theorem 3: Membership in NP is covered by [8]. Let Φ' be an instance of 3SAT, i. e. a boolean formula in 3-CNF. Denote by $\varphi := |\Phi'|$ the binary length of Φ. Then by Proposition 1, Φ is randomly mapped to an instance of $\mathbf{MS}'(M)$. For the resulting problem instance proceed as follows:

input: $\mathbf{RMS}'$-instance (a, b, c).
answer := `false`;
repeat polynomially many times
  select random $k \in [0, b]$;
  $a' := a + kb$;
  repeat polynomially many times
    select random prime $p \equiv 1 \bmod 4$, $p > \max\left(\left\lceil \frac{c+|2a'+b|}{|b|} \right\rceil, |b|\right)$,
      and $\left(\frac{-2b}{p}\right) = -1$;

$$\text{ask oracle } \left(2a' + b, \begin{pmatrix} -c \\ -\infty \\ -\infty \end{pmatrix}, \begin{pmatrix} c \\ \infty \\ \infty \end{pmatrix}, \langle 2, b, -bp\rangle\right)$$

    answer := answer $\vee$ (oracle answer)
return answer.

Obviously this establishes a polynomial-time oracle algorithm. Let us examine its correctness for solving $\mathbf{MS}'(M)$. At first, note that if it returns `true` then there are $|x| \le c$, $z_1, z_2 \in \mathbb{Z}$ s.t. $2x^2 + bz_1^2 - bpz_2^2 = 2a' + b$, hence, putting $y := z_1^2 - pz_2^2$, we have, in particular, that there are x, y s.t. $2x^2 + by = 2a' + b$ and thus $x^2 \equiv a' \equiv a \bmod b$ since 2 is invertible modulo the odd integer b. Thus, the $\mathbf{MS}'(M)$ instance has a solution (x, y) and so is a 'yes'-instance.

Conversely, if the algorithm returns `false`, but nevertheless (a, b, c) is a 'yes'-instance, then there is $|x| \le c$ s.t. $x^2 \equiv a \bmod b$; and thus there is $y \in \mathbb{Z}$, necessarily odd, s.t. $2x^2 + by = 2a' + b$, but y is not represented by any of the binary quadratic forms $\langle 1, -p \rangle$. For each of these forms, one of two things may have happened: Either y is represented by the genus of $\langle 1, -p \rangle$, but this genus consists of several classes; or y is not even represented by the genus of $\langle 1, -p \rangle$.

First, the sCLH 4.1 gives us an upper bound on the probability that the first case applies if the second does not. The second case, however, implies that

$$\forall\, (x, y) \in \mathbb{Z}^2,\ |x| \le c,\ x^2 + by = a, \quad \exists\, q | y \text{ prime: } \left(\frac{q}{p}\right) \neq -1$$

by Lemma 45. As the q are odd, the symbol $\left(\frac{q}{p}\right)$ takes the values $1, -1$ according to the uniform distribution and independently for different q as p is randomly chosen; hence if (x, y) is any solution of the $\mathbf{RMS}'$ instance, the probability that the second case applies is bounded by $1 - 2^{-\omega(y)}$ (where $\omega(y)$ counts the number of distinct prime factors of y). We now have to show that if we start with a 'yes'-instance of $\mathbf{MS}'(M)$, then with high probability, in some iteration we obtain an instance of (a', b, c) which has solution (x, y) with y decomposing into only logarithmically many prime factors in the input length. Observe that for all solutions (x, y), $|y|$ is bounded from above by $2(b + 1)$. Assume that (a, b, c) is a 'yes'-instance with some solution (x_0, y_0). Then, for $k = 0, \ldots, b$, the problem instance $(a' = a + kb, b, c)$ necessarily has a solution, namely $(x_0, y_0 + k)$. The range over which y varies thus is an interval $[y_0, y_0 + b] \cap \mathbb{Z}$, where $y_0 < 2b$. As follows directly from a result of Erdős and Nicolas [9, prop. 3], it holds that for $B > 0$,

$$\#\{Y \le B \,|\, \omega(Y) > 2 \ln\ln B\} < \frac{6}{\pi^{5/2}} \frac{B}{(\ln B)^{2\ln(2)-1}\sqrt{\ln\ln B}} \left(1 + \mathcal{O}\left(\frac{1}{\ln\ln B}\right)\right). \tag{1}$$

Combining these insights, we conclude that the innermost *repeat* loop produces at most

$$\mathcal{O}\left(\frac{b}{(\ln b)^{2\ln(2)-1}\sqrt{\ln\ln b}}\right)$$

different a' for which there exists *no* solution (x, y) with y having less than $2 \ln\ln y$ prime factors. This implies that after $\log b$ iterations, we have seen at least one instance with a solution of few prime divisors with exponentially large probability.

Now that we have established the occurrence of at least one solution in which y_0 has few prime divisors with high probability, we may conclude that for every choice of p, the probability of failure according to case two is in each iteration independently bounded from above by

$$1 - 2^{-\omega(y_0)} \leq 1 - 2^{-2\ln\ln y_0} \leq 1 - \frac{1}{\ln^2 \left\lceil \frac{c^2+|a|}{|b|} \right\rceil},$$

which after special treatment of finitely many instances is bounded away from 1. Together with the sCLH in the first case, we have bounded the error probability away from 1, and hence this is a one-sided error probabilistic reduction.

It remains to be shown that the forms constructed here satisfy all the properties entailed on them. Obviously, all forms constructed here are indefinite, of dimension 3, and of determinant prime to M. Next consider anisotropy: By [3, sec. 4.2] and the Hasse principle, a ternary quadratic form f over $\mathbb{Z}$ is isotropic if and only if $c_q f = 1$ for all symbols q (see [3, ch. 4] for the definition of c_p). But we have chosen $\left(\frac{-2b}{p}\right) = -1$, hence it can be computed that $c_p\langle 2, b, -pb\rangle = -1$, so that our forms are anisotropic. Finally, we have to establish the one-proper-class property for all forms constructed above. But this follows directly from Proposition 2. □

Proof Sketch of theorem 4: Membership in NP again follows from [8]. Let f, c, a be taken from an instance of $\mathbf{IRepr}(\mathcal{P}')$ produced in the proof of theorem 3 above. Construct a form g in the genus of f satisfying $g((1,0,0)^t) = a$. This can be accomplished essentially by following the proof of the existence of genera from [3]; the main steps are the following: First write down p-adic forms g_p with $g_p((1,0,0)^t) = a$, for all $p|2d\infty$. From values of the g_p, construct an integer t which is primitively represented over $\mathbb{Z}_p$ by all g_p; to this end, we have to select a prime probabilistically from an arithmetic progression. Then we can compute $U_p \in \mathrm{GL}_3(\mathbb{Z}_p)$ such that $t\, f_p\, U_p = t^2 x_1^2 + f_p^*(x_2, x_3)$. Then the algorithm calls itself recursively and returns a form f^* which is $\mathbb{Z}_p$-equivalent to each f_p^*. Now we obtain the desired f by an application of the Chinese Remainder Theorem to the U_p.

If this fails then we know we have started from a 'no'-instance. Otherwise, a matrix T with $fT = g$ and $|T_{11}| \leq c$ exists if and only if there are $|x| \leq c$, z_1, z_2 s. t. $f((x, z_1, z_2)) = a$. Hence $\mathbf{DIRepr}(\mathcal{P}') \preccurlyeq_r \mathbf{DIEP}(\mathcal{P}')$.

References

1. Agrawal, M., Kayal, N., Saxena, N.: PRIMES is in P. Annals of Mathematics 1602, 781–793 (2004)
2. Ajtai, M., Dwork, C.: A public-key cryptosystem with worst-case/average-case equivalence. In: Proceedings of the 29th annual ACM symposium on theory of computing, El Paso, TX, USA, May 4-6 (New York), pp. 284–293 (Association for Computing Machinery 1997)
3. Cassels, J.W.S.: Rational quadratic forms. London Mathematical Society Monographs, vol. 13. Academic Press, London (1978)
4. Cohen, H.: Course in computational algebraic number theory. Graduate Texts in Mathematics, vol. 138. Springer, Heidelberg (1993)
5. Cohen, H., Lenstra, H.W. jun.: Heuristics on class groups of number fields, Number Theory. In: Queinnec, C., Halstead Jr., R.H., Ito, T. (eds.) PSLS 1995. LNCS, vol. 1068, Springer, Heidelberg (1996)

6. Cohen, H., Martinet, J.: Class groups of number fields: Numerical heuristics. Mathematics of Computation 48(177), 123–137 (1987)
7. Heuristics on class groups: Some good primes are no too good, Mathematics of Computation 63, no. 207, 329–334 (1994)
8. Dietmann, R.: Small solutions of quadratic Diophantine equations. Proceedings of the London Mathematical Society, III. Ser. 86(3), 545–582 (2003)
9. Erdős, P., Nicolas, J.-L.: Sur la fonction: Nombre de facteurs premiers de n. Ens-Math2 27, 3–27 (1981)
10. Fouvry, É., Klüners, J.: On the 4-rank of class groups of quadratic number fields (2006) (preprint)
11. Gauß, C.F.: Untersuchungen über höhere Arithmetik (Disquisitiones Arithmeticae). Springer, Heidelberg (1889) (German translation by H. Maser)
12. Goldreich, O., Goldwasser, S., Halevi, S.: Public-key cryptosystems from lattice reduction problems. In: Kaliski Jr., B.S. (ed.) CRYPTO 1997. LNCS, vol. 1294, pp. 112–131. Springer, Heidelberg (1997)
13. Hoffstein, J., Howgrave-Graham, N., Pipher, J., Silverman, J.H., Whyte, W.: NTRUSign: Digital signatures using the NTRU lattice, Topics in cryptology – CT-RSA 2003. In: Joye, M. (ed.) CT-RSA 2003. LNCS, vol. 2612, pp. 122–140. Springer, Heidelberg (2003)
14. Hoffstein, J., Pipher, J., Silverman, J.H.: A ring-based public key cryptosystem. In: Buhler, J.P. (ed.) Algorithmic number theory. 3rd international symposium, ANTS-III, LNCS, vol. 1423, pp. 267–288. Springer, Heidelberg (1998)
15. NSS: an NTRU lattice-based signature scheme, Advances in cryptology - EUROCRYPT 2001. 20th international conference on theory and application of cryptographic techniques, Innsbruck, Austria, May 6-10, 2001 (Birgit Pfitzmann, ed.), Lecture Notes in Computer Science, vol. 2045, Springer-Verlag, pp. 211–228 (2001)
16. Gerth III, F.: *The 4-class ranks of quadratic fields.* Inventiones Mathematicae 77(3), 489–515 (1984)
17. Gerth III, F.: Extension of conjectures of Cohen and Lenstra. Expositiones Mathematicae 5(2), 181–184 (1987)
18. Ivanyos, G., Szánto, Á.: Lattice basis reduction for indefinite forms and an application. Journal on Discrete Mathematics 153(1–3), 177–188 (1996)
19. Lenstra, H.W. jun., Lenstra, A.K., Lovász, L.: Factoring polynomials with rational coefficients. Mathematische Annalen 261, 515–534 (1982)
20. Khot, S.: Hardness of approximating the shortest vector problem in lattices. Journal of the ACM 52(5), 789–808 (2005)
21. Manders, K.L., Adleman, L.M.: NP-complete decision problems for binary quadratics. Journal of Computer and System Sciences 16, 168–184 (1978)
22. Micciancio, D., Goldwasser, S.: Complexity of lattice problems: a cryptographic perspective. In: The Kluwer International Series in Engineering and Computer Science, Boston, Massachusetts, March 2002, vol. 671, Kluwer Academic Publishers, Dordrecht (2002)
23. Mitrinović, D.S., Sándor, J., Crstici, B. (eds.): Handbook of number theory. Mathematics and Its Applications, vol. 351. Kluwer Academic Publishers, Dordrecht (1996)
24. O'Meara, O.T.: Introduction to quadratic forms. Grundlehren der mathematischen Wissenschaften, vol. 117. Springer, Heidelberg (1963) (reprinted in 2000)
25. Schnorr, C.-P.: Progress on LLL and lattice reduction. In: Proceedings LLL+25, June 29–July 1, 2007, Caen, France (to appear, 2007)
26. Simon, D.: Solving quadratic equations using reduced unimodular quadratic forms. Mathematics of Computation 74(251), 1531–1543 (2005)

27. Tennenhouse, M., Williams, H.C.: A note on the class-number one in certain real quadratic and pure cubic fields. Mathematics of Computation 46(173), 333–336 (1986)

A Proof of Proposition 44

Let Φ be a Boolean formula in 3-CNF which contains each possible clause at most once, and no clause of Φ contains any variable both complemented and uncomplemented. Let ℓ be the number of variables in Φ. Choose an enumeration $\sigma_1, \ldots, \sigma_m$ of all clauses in the variables $x_1, \ldots, x_\ell$ with exactly three literals containing no variable both complemented and uncomplemented, such that both the bijection $i \mapsto \sigma_i$ and its inverse are polynomial-time (e. g. a suitable lexicographic enumeration). Denote by $\sigma \in \Phi$ the assertion that clause σ occurs in Φ, and by $x_j \in \sigma$ ($\bar{x}_j \in \sigma$) that the j-th variable occurs uncomplemented (complemented) in clause σ. Let $n = 2m + \ell$.

For a fixed assignment to the boolean variables x_i, we define $r_i = 1$ if $x_i =$`true` and $r_i = 0$ otherwise. Moreover, for a clause σ, define

$$W(\sigma, r) = \sum_{i:x_i \in \sigma} r_i + \sum_{i:\bar{x}_i \in \sigma} (1 - r_i). \tag{2}$$

For $k = 1, \ldots, m$, let furthermore

$$R_k := \begin{cases} y_k - W(\sigma_k, r) + 1 & \text{if } \sigma_k \in \Phi, \\ y_k - W(\sigma_k, r) & \text{if } \sigma_k \notin \Phi, \end{cases} \tag{3}$$

where y_k are new variables, for $k = 1, \ldots, m$. Since Φ is in 3-CNF, we have $W(\sigma_k, r) = 0$ if assignment r does not render clause σ true, and $1 \leq W(\sigma_k, r) \leq 3$ otherwise. Hence the equation system

$$R_k = 0, \qquad k = 1, \ldots, m \tag{4}$$

has a solution with

$$r \in \{0, 1\}^\ell, \qquad y \in \{0, 1, 2, 3\}^m \tag{5}$$

if and only if Φ is satisfiable. Now choose a prime $p \geq 11$ not dividing the M from the statement of the theorem. As $-3 \leq R_k \leq 4$ for all choices (5) of the variables, (4) is equivalent with

$$\sum_{k=1}^{m} R_k p^k = 0. \tag{6}$$

We may estimate $\left|\sum_{k=1}^m R_k p^k\right| \leq 4 \sum_{k=1}^m p^k < p^{m+1} - 2$ as $p \geq 11$; hence (6) is equivalent with $\sum_{k=1}^m R_k p^k \equiv 0 \bmod p^{m+1}$, or, equivalently, as p is odd, with

$$\sum_{k=1}^{m} (2\, R_k) p^k \equiv 0 \bmod p^{m+1}. \tag{7}$$

Now replace the y_k, $k = 1, \ldots, m$ and the r_i, $i = \ell$, by new variables α_i, $i = 1, \ldots, n$, each ranging independently over $\{1, -1\}$, by the formula

$$\begin{aligned} y_k &= \frac{1}{2}((1-\alpha_{2k-1}) + 2((1-\alpha_{2k})), \\ r_i &= \frac{1}{2}(1-\alpha_{2m+i}), \end{aligned} \tag{8}$$

which obviously induces a bijection between the sets over which the two sequences of variables range.

After this change of variables the left hand side of (7) is still integral, and thus the congruence notation makes sense. Collecting terms, (7) can be rephrased as

$$\sum_{j=1}^{n} c_j \alpha_j \equiv \tau' \bmod p^{m+1} \tag{9}$$

for some $c_j, \tau' \in \mathbb{Z}$; explicitly, we have

$$\begin{aligned} -\tau' &= \sum_{k=1}^{m} (5 - \sum_{i: x_i \in \sigma_k} 1 + \mathbf{1}_{\sigma_k \in \Phi}) p^k, \\ c_{2k-1} &= -p^k, \\ c_{2k} &= -4p^k, \\ c_{2m+i} &= \sum_{k=1}^{m} (\mathbf{1}_{x_i \in \sigma_k} - \mathbf{1}_{\bar{x}_i \in \sigma_k}) p^k, \end{aligned} \tag{10}$$

where $k = 1, \ldots, m$, $i = 1, \ldots, \ell$, and $\mathbf{1}_\Psi = 1$ if Ψ is true and $\mathbf{1}_\Psi = 0$ otherwise.

Without affecting solvability or the number of solutions, we may as well introduce an extra variable α_0, define $c_0 := 1$ and $\tau := \tau' + 1$, and write

$$\sum_{j=0}^{n} c_j \alpha_j \equiv \tau \bmod p^{m+1}. \tag{11}$$

Thus we have learnt that Φ was satisfiable if and only if (11) is solvable for $\alpha \in \{-1, 1\}^{n+1}$. For later use, we verify that $p \not| \tau$: Indeed, the constant term (i.e. independent of the α_i) of $W(\sigma_k, r)$, with the r_i replaced according to (8), equals $w_k := \sum_{x_i \in \sigma_k} 1 + \sum_{\bar{x}_i \in \sigma_k} 1$. Now τ' is obviously divisible by p; and thus

$$\tau = \tau' + 1 \equiv 1 \bmod p. \tag{12}$$

Now define p_0 to be some prime exceeding $4 \cdot p^{m+1}(n+1)$, and let p_j be some prime exceeding p_{j-1}, for $j = 1, \ldots, n$, both of polynomial size in p^m. They can be found by sampling integers uniformly at random in intervals of the shape $[N, 2N(\ln N)^2]$ and testing them for primality [1].

Choose θ_j, for $j = 1, \ldots, n$, as the smallest positive integer satisfying

$$\theta_j \begin{cases} \equiv c_j & \bmod \ p^{m+1}, \\ \equiv 0 & \bmod \ \prod_{i \neq j} p_i, \\ \not\equiv 0 & \bmod \ p_j. \end{cases} \tag{13}$$

Finally, set $K := \prod_{j=0}^{n} p_j$ and $c := \sum_{j=0}^{n} \theta_j$. Then we can reformulate (11) as follows: Φ is satisfiable if and only if there is $\alpha \in \{1, -1\}^{n+1}$ s. t.

$$\sum_{j=0}^{n} \theta_j \alpha_j \equiv \tau \bmod p^{m+1}, \tag{14}$$

and the number of solutions still has not changed.

Now we claim: For $x \in \mathbb{Z}$, $|x| \leq c$ and $c^2 \equiv x^2 \bmod K$ hold if and only if

$$x = \sum_{j=0}^{n} \theta_j \alpha_j \tag{15}$$

for some $\alpha \in \{1, -1\}^{n+1}$.

The proof of this claim is analogous to a lemma in [21] and therefore omitted.

Combining (15) and (14), we obtain that the $3SAT$ formula Φ has a satisfying truth assignments if and only if there is a number $x \in \mathbb{Z}$, $|x| \leq c$ s. t.

$$\begin{aligned} c^2 - x^2 &\equiv 0 \bmod K, \\ x &\equiv \tau \bmod p^{m+1}. \end{aligned} \tag{16}$$

It is easily seen that $(\tau - \xi)(\tau + \xi) = \tau^2 - \xi^2 \equiv 0 \bmod p^{m+1}$ is equivalent with $\xi \equiv \tau \bmod p^{m+1}$ or $\xi \equiv -\tau \bmod p^{m+1}$. So Φ has a satisfying truth assignment if and only if there is an integer x with $|x| \leq c$ s. t.

$$\begin{aligned} c^2 - x^2 &\equiv 0 \quad \bmod K, \\ \tau^2 - x^2 &\equiv 0 \quad \bmod p^{m+1}. \end{aligned} \tag{17}$$

By the Chinese Remainder Theorem, the equations (17) are jointly equivalent to the equation $p^{m+1}(c^2 - x^2) + K(\tau^2 - x^2) \equiv 0 \bmod p^{m+1}K$. But as K is prime to p by the construction of the p_j, and $p^{m+1} + K$ is prime to K, we finally reach the equation

$$x^2 \equiv a \bmod b \tag{18}$$

where

$$a \equiv (p^{m+1} + K)^{-1}(K\tau^2 + p^{m+1}c^2) \bmod p^{m+1}K \tag{19}$$

and $b = p^{m+1}K$. Then (18) has a solution $x \in \mathbb{Z}$ with $|x| \leq c$ if and only if Φ had a satisfying truth assignment. Now by construction, K is odd and squarefree, and a is odd and coprime to b. Now it suffices to note that the arithmetic progression (19) contains sufficiently many squarefree numbers so that one of them can be selected randomly in expected polynomial time. By the Page-Siegel-Walfisz theorem (see [23, §VIII.6]), polynomially many random selections suffice to find a prime a in the arithmetic progression (19), and primes can be efficiently tested as above. Of course, a is then squarefree as well. By construction, it is no problem to output the prime factorization as well.

Reachability Problems in Quaternion Matrix and Rotation Semigroups

Paul Bell and Igor Potapov

Department of Computer Science,
University of Liverpool, Ashton Building,
Ashton St, Liverpool L69 3BX, UK
{pbell,igor}@csc.liv.ac.uk

Abstract. We examine computational problems on quaternion matrix and rotation semigroups. It is shown that in the ultimate case of quaternion matrices, in which multiplication is still associative, most of the decision problems for matrix semigroups are undecidable in dimension two. The geometric interpretation of matrix problems over quaternions is presented in terms of rotation problems for the 2 and 3-sphere. In particular, we show that the reachability of the rotation problem is undecidable on the 3-sphere and other rotation problems can be formulated as matrix problems over complex and hypercomplex numbers.

1 Introduction

Quaternions have long been used in many fields including computer graphics, robotics, global navigation and quantum physics as a useful mathematical tool for formulating the composition of arbitrary spatial rotations and establishing the correctness of algorithms founded upon such compositions.

Many natural questions about quaternions are quite difficult and correspond to fundamental theoretical problems in mathematics, physics and computational theory. Unit quaternions actually form a *double cover* of the rotation group SO_3, meaning each element of SO_3 corresponds to two unit quaternions. This makes them expedient for studying rotation and angular momentum and they are particularly useful in quantum mechanics. The group of unit quaternions form the group SU_2 which is the special unitary group. The large number of applications has renewed interest in quaternions and quaternion matrices ([1], [8], [15], [18], [19]).

Quaternions do not commute and this leads to many problems with their analysis. In particular, defining the determinant and finding the eigenvalues and the inverse of a quaternion matrix are unexpectedly difficult problems [19]. In this paper we study decision questions about semigroups of quaternions, quaternion matrices and rotations, such as several reachability questions, membership problems, freeness problems, etc. There are two major points of this work that we would like to highlight.

First, we investigated classical matrix decision problems for low-dimensional quaternion matrices. The results for matrices over $\mathbb{Z}, \mathbb{Q}, \mathbb{C}$ are not easily transferable to the case of quaternions and thus there are no results on *computational*

L. Kučera and A. Kučera (Eds.): MFCS 2007, LNCS 4708, pp. 346–358, 2007.

problems for quaternions and quaternion matrices. Most of the problems for 2×2 matrices were open for any number field. In this paper, we show that all standard reachability problems are undecidable for 2×2 quaternion matrix semigroups. Moreover, our construction uses unitary quaternions that have a special interest in terms of rotations. After the quaternions, the hypercomplex numbers lose the associativity property and thus no longer form a semigroup. Due to this fact we think that our current research on quaternion matrices gives a more complete picture of decision problems for matrix semigroups. Then we study these problems for a case of Lipschitz integers and state several open problems.

The second important point of the paper is establishing connections between classical matrix semigroup problems and reachability problems for semigroups of rotations. In fact, using unit quaternions for encoding computational problems gives us an opportunity to formulate and prove several interesting results in terms of 3 and 4 dimensional rotations defined by quaternions. In particular, we show that the point-to-point rotation problem for the 3-sphere is undecidable. The same problem for the 2-sphere is open and can be formulated as a special case of the scalar reachability problem for matrix semigroups that we show is undecidable in general. As an additional benefit, the results on rotation semigroups give immediate corollaries for a class of orthogonal matrix semigroups.

The paper is organized as follows. In the second section we give all definitions about quaternions and their matrix representation and a mapping between words and quaternions that will be used in our proofs. The third section contains the main results of the paper on undecidable problems (freeness, membership, reachability) in quaternion matrix semigroups. We prove that the membership for 2×2 rational quaternion matrix semigroups is undecidable. We use a novel technique of PCP encoding, allowing us to encode pairs of words by separate matrices and force them to appear in the right order for a specific product. Then we show that the problem of deciding if any diagonal matrix is in a quaternion matrix semigroup, that has its own interest in a context of control theory, is undecidable. Then we study these problems for the case of Lipschitz integers. In the last section, the geometric interpretation of matrix problems over quaternions is presented in terms of rotation problems for the 2 and 3-sphere.

2 Preliminaries

We use the standard denotations $\mathbb{N}, \mathbb{Z}^+, \mathbb{Q}$ to denote the natural numbers, positive integers and rational numbers respectively.

In a similar style to complex numbers, rational quaternions, which are hypercomplex numbers, can be written $\vartheta = a + b\mathbf{i} + c\mathbf{j} + d\mathbf{k}$ where $a, b, c, d \in \mathbb{Q}$. To ease notation let us define the vector: $\mu = (1, \mathbf{i}, \mathbf{j}, \mathbf{k})$ and it is now clear that $\vartheta = (a, b, c, d) \cdot \mu$ where $\cdot$ denotes the inner or 'dot' product. We denote rational quaternions by $\mathbb{H}(\mathbb{Q})$. Quaternions with real part 0 are called *pure quaternions* and denoted by $\mathbb{H}(\mathbb{Q})_0$.

Quaternion addition is simply the componentwise addition of elements. It is well known that quaternion multiplication is not commutative. Multiplication is completely defined by the equations $\mathbf{i}^2 = \mathbf{j}^2 = \mathbf{k}^2 = -1$, $\mathbf{ij} = \mathbf{k} = -\mathbf{ji}$, $\mathbf{jk} = \mathbf{i} = -\mathbf{kj}$ and $\mathbf{ki} = \mathbf{j} = -\mathbf{ik}$. Thus for two quaternions $\vartheta_1 = (a_1, b_1, c_1, d_1)\mu$ and $\vartheta_2 = (a_2, b_2, c_2, d_2)\mu$, we can define their product as $\vartheta_1\vartheta_2 = (a_1a_2 - b_1b_2 - c_1c_2 - d_1d_2) + (a_1b_2 + b_1a_2 + c_1d_2 - d_1c_2)\mathbf{i} + (a_1c_2 - b_1d_2 + c_1a_2 + d_1b_2)\mathbf{j} + (a_1d_2 + b_1c_2 - c_1b_2 + d_1a_2)\mathbf{k}$.

In a similar way to complex numbers, we define the conjugate of $\vartheta = (a, b, c, d) \cdot \mu$ by $\overline{\vartheta} = (a, -b, -c, -d) \cdot \mu$. We can now define a norm on the quaternions by $||\vartheta|| = \sqrt{\vartheta\overline{\vartheta}} = \sqrt{a^2 + b^2 + c^2 + d^2}$. Any non zero quaternion has a multiplicative (and obviously an additive) inverse [11]. The other properties of being a division ring can be easily checked.

A *unit* quaternion has norm 1 and corresponds to a rotation in three dimensional space. Given a unit vector $\boldsymbol{r} = (r_1, r_2, r_3)$ and a rotation angle $0 \leq \theta < 2\pi$, we would like to find a quaternion transformation to represent a rotation of θ radians of a point $P' = (x, y, z) \in \mathbb{Q}^3$ about the $\boldsymbol{r}$ axis. To facilitate this, we require an encoding of P' as a pure quaternion P, namely $P = (0, x, y, z) \cdot \mu \in \mathbb{H}(\mathbb{Q})_0$.

Let us define a function $\psi_q : \mathbb{H}(\mathbb{Q}) \mapsto \mathbb{H}(\mathbb{Q})$ by $\psi_q(P) = qPq^{-1}$ where $q, P \in \mathbb{H}(\mathbb{Q})$ and $||q|| = 1$. If q is correctly chosen to represent a rotation of θ about a unit axis r, then this function will return a pure quaternion of the form $(0, x', y', z') \cdot \mu$ where $(x', y', z') \in \mathbb{Q}^3$ is the correctly rotated point.

It is well known (see, for example, [11]) that: $\vartheta = \left(\cos\frac{\theta}{2}, \boldsymbol{r}\sin\frac{\theta}{2}\right) \cdot \mu$ represents a rotation of angle θ about the $\boldsymbol{r}$ axis. Therefore using $\psi_\vartheta(P)$ as just described rotates P as required. This will be used in the next section.

All possible unit quaternions correspond to points on the 3-sphere. Any pair of unit quaternions p, q represent a four-dimensional rotation. Given a point $x \in \mathbb{H}(\mathbb{Q})$, we define a rotation of x, by $px\overline{q}$ [17]. Also we use the notation SU_2 to denote the special unitary group, the double cover of the rotation group SO_3.

The length of quaternions is multiplicative and the semigroup of Lipschitz integers with multiplication is closed. The fact that $||q_1q_2|| = ||q_1|| \cdot ||q_2||$ follows since the determinant of the matrix representation of a quaternion we define in Section 2.2 corresponds to the modulus and is multiplicative. This result will be required later.

2.1 Word Morphisms

Let $\Sigma = \{a, b\}$ be a binary alphabet, $\boldsymbol{u} = (1, 0, 0)$ and $\boldsymbol{v} = (0, 1, 0)$. We define the homomorphism $\varphi : \Sigma^* \times \mathbb{Q} \mapsto \mathbb{H}(\mathbb{Q})$ by:

$$\varphi(a, \theta) = (\cos(\tfrac{\theta}{2}), \boldsymbol{u}\sin(\tfrac{\theta}{2})) \cdot \mu \text{ and } \varphi(b, \theta) = (\cos(\tfrac{\theta}{2}), \boldsymbol{v}\sin(\tfrac{\theta}{2})) \cdot \mu$$

where $\theta \in \mathbb{Q} \in [0, 2\pi)$, i.e. $\varphi(a, \theta)$ is a rotation of angle θ about the $\boldsymbol{u}$ axis and $\varphi(b, \theta)$ is a rotation of angle θ about the $\boldsymbol{v}$ axis. $\varphi(\varepsilon, \theta) = \vartheta_I$ is the multiplicative identity element of the division ring of rational quaternions. Note that $\boldsymbol{u} \cdot \boldsymbol{v} = 0$ and $||\boldsymbol{u}|| = ||\boldsymbol{v}|| = 1$, thus these two vectors are orthonormal.

Let us define a specific instance of this morphism. Let $\alpha = 2\arccos(\frac{3}{5}) \in \mathbb{R}$. Now we define $\gamma : \Sigma^* \mapsto \mathbb{H}(\mathbb{Q})$ where $\gamma(a) = \varphi(a, \alpha)$, $\gamma(b) = \varphi(b, \alpha)$ and $\gamma(\varepsilon) = (1, 0, 0, 0)\mu = \vartheta_I$. This gives the homomorphism:

$$\gamma(a) = (\cos(\arccos(\tfrac{3}{5})), \boldsymbol{u}\sin(\arccos(\tfrac{3}{5}))) \cdot \mu = (\tfrac{3}{5}, \tfrac{2}{5}, 0, 0) \cdot \mu \quad (1)$$

$$\gamma(b) = (\cos(\arccos(\tfrac{3}{5})), \boldsymbol{v}\sin(\arccos(\tfrac{3}{5}))) \cdot \mu = (\tfrac{3}{5}, 0, \tfrac{2}{5}, 0) \cdot \mu \quad (2)$$

which follows from the identity $\cos^2\theta + \sin^2\theta = 1$ since $\sqrt{1-(\frac{3}{5})^2} = \frac{2}{5}$.

We can see that the quaternions in the image of γ are unit, i.e. $\forall w \in \Sigma^*$, $||\gamma(w)|| = 1$ since quaternion length is multiplicative ($||q_1 q_2|| = ||q_1|| \cdot ||q_2||$, which we proved in Section 2) and $\gamma(a), \gamma(b)$ have unit length.

Lemma 1. *The mapping $\gamma : \Sigma^* \mapsto \mathbb{H}(\mathbb{Q})$ is a monomorphism.*

Proof. It was proven in [16] that if $\cos(\theta) \in \mathbb{Q}$ then the subgroup of $SO_3(\mathbb{R})$ generated by rotations of angle θ about two perpendicular axes is free iff $\cos(\theta) \neq 0, \pm\frac{1}{2}, \pm 1$. We note that in the definition of γ we use a rotation about two orthonormal axes $\boldsymbol{u}, \boldsymbol{v}$. We use a rotation of $\alpha = 2\arccos\frac{3}{5}$. From basic trigonometry, $\cos(2\arccos(\frac{3}{5})) = -\frac{7}{25}$ and $\sin(2\arccos(\frac{3}{5})) = \frac{24}{25}$ thus the cosine and sine of both angles are rational and not equal to $0, \pm\frac{1}{2}, \pm 1$ (we only require this of the cosine) as required. We showed that all elements of the quaternions are rational, thus we have a free subgroup of $SO_3(\mathbb{Q})$ generated by $\gamma(a), \gamma(b) \in \mathbb{H}(\mathbb{Q})$. Note that the conditions mentioned are guaranteed to give a free group but are not necessary for freeness, see [8]. □

Post's Correspondence Problem (PCP) - Given two (finite) alphabets Γ, Σ and two morphisms $h, g : \Gamma^* \mapsto \Sigma^*$, it is undecidable in general whether there exists a solution $w \in \Gamma^+$ such that $h(w) = g(w)$. We can assume without loss of generality that Σ is binary by using a straightforward encoding. It was shown that the problem is undecidable when the *instance size* $|\Gamma| \geq 7$ in [13]. We denote by n_p the smallest instance size for which PCP is undecidable (thus, $n_p \leq 7$).

2.2 Matrix Representations

It is possible to represent a quaternion $\mathbb{H}(\mathbb{Q})$ by a matrix $M \in \mathbb{C}^{2\times 2}$. For a general quaternion $\vartheta = (a, b, c, d) \cdot \mu$ we define the matrix $M = \begin{pmatrix} a + b\mathbf{i} & c + d\mathbf{i} \\ -c + d\mathbf{i} & a - b\mathbf{i} \end{pmatrix}$.

Corollary 1. *There is a class of 2×2 complex unitary matrices forming a free group.*

Proof. We can define a morphism similar to γ which instead maps to two dimensional complex matrices:

$$\zeta(a) = \begin{pmatrix} \frac{3}{5} + \frac{4}{5}\mathbf{i} & 0 \\ 0 & \frac{3}{5} - \frac{4}{5}\mathbf{i} \end{pmatrix}, \zeta(b) = \begin{pmatrix} \frac{3}{5} & \frac{4}{5} \\ -\frac{4}{5} & \frac{3}{5} \end{pmatrix}, \zeta(\varepsilon) = \begin{pmatrix} 1 & 0 \\ 0 & 1 \end{pmatrix}.$$

Note that these matrices are unitary, therefore let $\zeta(a^{-1}) = \zeta(a)^{-1} = \zeta(a)^*$ and $\zeta(b^{-1}) = \zeta(b)^{-1} = \zeta(b)^*$ where * denotes the Hermitian transpose.

Thus we have an injective morphism $\zeta : (\Sigma \cup \overline{\Sigma})^* \mapsto \mathbb{C}^{2\times 2}$. Since γ forms a free group of quaternions, ζ forms a free group over $\mathbb{C}^{2\times 2}$. □

Also note that we can define such matrices for any three orthonormal vectors where the rotation angle θ satisfies $\cos(\theta) \in \mathbb{Q}$ and $\cos(\theta) \neq 0, \pm\frac{1}{2}, \pm 1$.

3 Quaternion Matrix Semigroups

We will now show an undecidability result similar to one considered by A. Markov, where he showed undecidability for two sets of unimodular 2×2 integral matrices, see [12] and [9].

Theorem 1. *Given two sets $A = \{a_1, a_2, \ldots, a_n\}$ and $B = \{b_1, b_2, \ldots, b_n\}$, where $A, B \subset \mathbb{H}(\mathbb{Q})$, it is undecidable whether there exists a non-empty sequence of indices $(r_1, r_2, \ldots, r_m)$ such that $a_{r_1} a_{r_2} \cdots a_{r_m} = b_{r_1} b_{r_2} \cdots b_{r_m}$, this holds for $n = n_p$.*

Proof. We use a reduction of Post's correspondence problem and the morphism γ defined in Section 2. Given two alphabets Γ, Σ, such that Σ is binary, and an instance of the PCP, $(h, g) : \Gamma^* \mapsto \Sigma^*$. We proved in Lemma 1 that γ is a monomorphism between Σ^* and $\mathbb{H}(\mathbb{Q})$. Thus let us define a new pair of morphisms (ρ, τ) to map $\Gamma^+ \times \Gamma^+$ directly into $\mathbb{H}(\mathbb{Q}) \times \mathbb{H}(\mathbb{Q})$ (we can think of this as $SU_2 \times SU_2$ since each of these unit quaternions represents an element of S^3 (the 3-sphere)). For any $w \in \Gamma^+$, let $\rho(w) = \gamma(h(w))$ and $\tau(w) = \gamma(g(w))$.

Thus for an instance of PCP, $\Gamma = \{a_1, a_2, \ldots, a_m\}$, (h, g), we instead use the pair of morphisms (ρ, τ). Define two semigroups S_1, S_2 which are generated respectively by $\{\rho(a_1), \rho(a_2), \ldots, \rho(a_m)\}$ and $\{\tau(a_1), \tau(a_2), \ldots, \tau(a_m)\}$. We see their exists a solution to the given instance of PCP iff $\exists w \in \Gamma^+$ such that $\rho(w) = \tau(w)$. □

We now move to an extension of the Theorem 1 where it is no longer necessary to consider the index sequence. Markov obtained a similar result by extending the dimension of the integral matrices to 4×4 [12]. See also [3,9], where the authors improve Markov's results to 3×3 integral matrices.

Theorem 2. *Given two semigroups S, T, generated by A, B respectively, such that $A = \{A_1, A_2, \ldots, A_n\}$ and $B = \{B_1, B_2\}$ where $A, B \subset \mathbb{H}(\mathbb{Q})^{2\times 2}$, it is undecidable if $S \cap T = \varnothing$. Furthermore, all matrices in A, B can be taken to be diagonal.*

Proof. Given an instance of PCP, (h, g) where $h, g : \Gamma^* \mapsto \Sigma^*$. We use the monomorphisms $\rho, \tau : \Gamma^* \mapsto \mathbb{H}(\mathbb{Q})$ introduced in Theorem 1. For each $a \in \Gamma$ we define:

$$A_a = \begin{pmatrix} \rho(a) & 0 \\ 0 & \tau(a) \end{pmatrix}$$

and these matrices form the generator for the semigroup S. For the second semigroup, T, we simply wish to encode each symbol from Σ in the $[1, 1]$ and $[2, 2]$ elements using the morphism $\gamma : \Sigma^* \mapsto \mathbb{H}(\mathbb{Q})$ which was shown to be injective in Lemma 1:

$$B_1 = \begin{pmatrix} \gamma(a) & 0 \\ 0 & \gamma(a) \end{pmatrix}, \qquad B_2 = \begin{pmatrix} \gamma(b) & 0 \\ 0 & \gamma(b) \end{pmatrix}.$$

We see that there exists $M \in A$ such that $M_{[1,1]} = M_{[2,2]}$ iff there exists a solution $w \in \Gamma^+$ to the instance of the PCP. This follows since element $[1, 1]$

of M encodes $h(w)$ and element $[2,2]$ encodes $g(w)$. Clearly any such matrix M is also in T since every matrix in T corresponds to an encoding of all words over Σ^+ in the top left and bottom right elements. Note that all matrices are diagonal and unitary. □

The previous two theorems used two separate semigroups. It is more natural to ask whether a particular element is contained within a *single* semigroup. For example, the mortality problem asks if the zero matrix is contained in an integral matrix semigroup and was shown to be undecidable in dimension 3 (see [14]). We showed that in dimension 4 the membership for any k-scalar matrix in an integral (resp. rational) matrix semigroup is undecidable where $k \in \mathbb{Z} \setminus \{0, \pm 1\}$ (resp. $k \in \mathbb{Q} \setminus \{0, \pm 1\}$), (see [4]).

Now we show that the membership problem in 2×2 unitary quaternion matrix semigroups in undecidable. The proof uses our new approach of encoding PCP proposed in [4]. The main idea is to store all words of the PCP separately and use an index coding to ensure they are multiplied in the correct way.

Theorem 3. *Given a unitary quaternion matrix semigroup S which is generated by $X = \{X_1, X_2, \ldots, X_n\} \subseteq \mathbb{H}(\mathbb{Q})^{2\times 2}$, it is undecidable for a matrix Y whether $Y \in S$.*

Proof. Given an instance of the PCP (h, g) where $h, g : \Gamma^* \mapsto \Sigma^*$. Then $w \in \Gamma^+$ is a solution to the PCP iff $h(w) = g(w)$. Assume now that $\forall x \in \Gamma^*$, $g(x)$ has an inverse, $g(x)^{-1}$. In terms of words over Σ, this means that if $g(x) = y$ for some $y \in \Sigma^*$ then $g(x)^{-1} = y^{-1}$ where $y^{-1} \in \overline{\Sigma}^*$ which is a new alphabet containing the inverse of each element of Σ. Formally we say $a \in \Sigma \Leftrightarrow a^{-1} \in \overline{\Sigma}$.

For example, if $g(w) = aabab$ where $w \in \Gamma^+$ and $aabab \in \Sigma^*$ then $g(w)^{-1} = (aabab)^{-1} = b^{-1}a^{-1}b^{-1}a^{-1}a^{-1} \in \overline{\Sigma}^*$.

If there exists a solution to the PCP, $w \in \Gamma^+$, such that $h(w) = g(w)$ then it can be observed that $h(w) \cdot g(w)^{-1} = \varepsilon$. We shall give an example of this simple fact. Let $w = w_1 w_2 \ldots w_k \in \Gamma^+$ be a solution to the PCP. Then $h(w) = g(w) = u$ for some $u = u_1 u_2 \ldots u_m \in \Sigma^+$. It is now clear that $h(w) \cdot g^{-1}(w) = (u_1 u_2 \ldots u_m) \cdot (u_m^{-1} u_{m-1}^{-1} \cdots u_1^{-1}) = \varepsilon$.

We call this type of word an *inverse palindrome*. This allows us to calculate the solution to the PCP instead as a single word. For each new symbol $a \in \Gamma$ we wish to add to the existing word $w \in \Gamma^*$, we concatenate $h(a)$ to the left and $g(a)^{-1}$ to the right of the current word $v \in \Sigma^*$, i.e. $v' = h(a) \cdot v \cdot g(a)^{-1}$. A solution then exists iff $v' = \varepsilon$ after a positive number of steps.

Within a semigroup this constraint is difficult to impose; we cannot say "multiply to the left by U_i and the right by V_i". Such a constraint *is* possible however by encoding *two* words simultaneously. In the first word we store the main word corresponding to the PCP itself such as described above. In the second word, we store the index of the word or its inverse.

Given some $a_i \in \Gamma$, we define two matrices in the semigroup generator Y_{i1}, Y_{i2} corresponding to this symbol. In Y_{i1} we store the two words $h(a_i)$ and $\sigma(i)$ where σ is an injective morphism for each $i \in \mathbb{Z}^+$, $\sigma(i) = a^i b$ where $a, b \in \Sigma$. In Y_{i2}, we store the two words $g(a_i)^{-1}$ and $\mu(i)$ where $\mu(i) = \overline{a}^i\overline{b}$ ($\overline{a} = a^{-1}$, $\overline{b} = b^{-1}$ and γ is the injective group morphism).

We need to store two words separately in one matrix. Let $\Gamma = \{a_1, a_2, \ldots, a_m\}$ and (h, g) be an instance of the PCP. Then for each $1 \leq i \leq m$, define

$$Y_{i1} = \begin{pmatrix} \gamma(h(a_i)) & 0 \\ 0 & \gamma(\sigma(i)) \end{pmatrix}, Y_{i2} = \begin{pmatrix} \gamma(g(a_i))^{-1} & 0 \\ 0 & \gamma(\mu(i)) \end{pmatrix}$$

Note that all quaternions used are unit. Now define two special matrices:

$$M = \begin{pmatrix} \gamma(h(a_1)) & 0 \\ 0 & \gamma(b) \end{pmatrix}, N = \begin{pmatrix} \gamma(g(a_1))^{-1} & 0 \\ 0 & \gamma(b)^{-1} \end{pmatrix}$$

We store the mapping of symbol a_1 in M, N, using the modified PCP to ensure that if there is a solution then there exists a solution using this symbol first. This avoids the pathological case of a product with only M and N in it.

Note that if matrix N appears once in a product equal to I_2 then matrix M appears once also due to the above construction (For the bottom right element to equal 1, $\gamma(b)$ must multiply with $\gamma(b)^{-1}$ at some point, see also [10]). Thus if we consider a semigroup, S, generated by $\{Y_{i1}, Y_{i2}, M\}$ where $1 \leq i \leq m$, then $N^{-1} \in S$ iff the instance of PCP has a solution, thus membership is undecidable. All matrices are diagonal and unitary quaternion matrices which are equivalent to *double quaternions.* Thus the membership for a semigroup of double quaternions is undecidable. □

Corollary 2. *The vector reachability problem for a semigroup of* 2×2 *quaternion matrices is undecidable.*

Proof. The vector reachability question for quaternions is defined as: "Given two vectors $a, b \in \mathbb{H}(\mathbb{Q})^n$ and a semigroup of matrices $S \subset \mathbb{H}(\mathbb{Q})^{n\times n}$, does there exist some $M \in S$ such that $Ma = b$?". The undecidability is straightforward from the Theorem 3. Let $x, y \in \mathbb{H}(\mathbb{Q})^2$ and $x = (1, 1)^T, y = N^{-1}(1, 1)^T$. Then, for some $R \in S$, it is clear that $Rx = y$ iff $R = N^{-1} = Y$ since we use only diagonal matrices. Since determining if $Y \in S$ is undecidable, the vector reachability problem is undecidable. □

The next problem was given as an open problem over matrices of natural numbers $\mathbb{N}$ in any dimension [5]. We show it is undecidable over $\mathbb{H}(\mathbb{Q})^{2\times 2}$.

Theorem 4. *It is undecidable for a finitely generated semigroup* $S \subseteq \mathbb{H}(\mathbb{Q})^{2\times 2}$ *whether there exists* any *diagonal matrix* $D \in S$.

Proof. As before, let $h, g : \Gamma^* \mapsto \Sigma^*$ be an instance of the PCP where $|\Sigma| = 2$. We use the morphisms $\rho, \tau : \Gamma^* \mapsto \mathbb{H}(\mathbb{Q})$ defined for any $w \in \Gamma^*$ as $\rho(w) = \gamma(h(w))$ and $\tau(w) = \gamma(g(w))$. Thus $u, v \in \Gamma^*$, $\rho(u) = \tau(v)$ iff $u = v$. For any two quaternions $q, r \in \mathbb{H}(\mathbb{Q})$ we define $\Psi(q, r) = \frac{1}{2}\begin{pmatrix} q + r & q - r \\ q - r & q + r \end{pmatrix}$.

It is clear that this is still homomorphic [6], since $\Psi(q_1, r_1) \cdot \Psi(q_2, r_2) = \Psi(q_1 q_2, r_1 r_2)$ which is verified easily via:

$$\frac{1}{2}\begin{pmatrix} q_1 + r_1 & q_1 - r_1 \\ q_1 - r_1 & q_1 + r_1 \end{pmatrix} \cdot \frac{1}{2}\begin{pmatrix} q_2 + r_2 & q_2 - r_2 \\ q_2 - r_2 & q_2 + r_2 \end{pmatrix} = \frac{1}{2}\begin{pmatrix} q_1q_2 + r_1r_2 & q_1q_2 - r_1r_2 \\ q_1q_2 - r_1r_2 & q_1q_2 + r_1r_2 \end{pmatrix}$$

It is now obvious that $\Psi(u, v)$ is diagonal iff $u = v$ since the top right and bottom left elements of the matrix equal 0 only if the two quaternions are equal.

Thus we can create one such matrix for each pair of images of letters from Γ using τ and ρ. S contains a diagonal matrix iff a PCP solution exists.

Unfortunately this does not hold when we convert the matrices to four dimensional rational matrices since we only get a *block diagonal* matrix. Thus the decidability for whether any matrix in a semigroup is diagonal remains open for integers, rationals and complex rational numbers. □

Another problem which can be stated is that of *freeness* of quaternion matrix semigroups. We use an almost identical proof to that in [7] to show undecidability, but we obtain the result for matrices over $\mathbb{H}(\mathbb{Q})^{2\times 2}$ rather than $(\mathbb{Z}^+)^{3\times 3}$:

Theorem 5. *Given a semigroup S, finitely generated by $M = \{M_1, \ldots, M_n\} \subset \mathbb{H}(\mathbb{Q})^{2\times 2}$, deciding whether S is free is algorithmically undecidable.*

Proof. Since we can store two words within a matrix $M_i \in \mathbb{H}(\mathbb{Q})^{2\times 2}$ we can use an almost identical proof that was used in [7]. We will give very brief sketch of the proof and refer to [7] for details.

The mixed modification PCP (or MMPCP) is a variant of the standard Post correspondence problem. As in the original PCP, we are given two (finite) alphabets Σ, Δ and two morphisms $h, g : \Sigma^+ \to \Delta^+$. The MMPCP asks whether there exists a word $w = w_1 w_2 \cdots w_m \in \Sigma^+$ such that:

$$h_1(w_1)h_2(w_2)\cdots h_m(w_m) = g_1(w_1)g_2(w_2)\cdots g_m(w_m)$$

where each $h_i, g_i \in \{h, g\}$ and $h_j \neq g_j$ for some $1 \leq j \leq m$. Now, define the set of 2×2 quaternion matrices:

$$M = \left\{ \begin{pmatrix} \gamma(a) & 0 \\ 0 & h(a) \end{pmatrix}, \begin{pmatrix} \gamma(a) & 0 \\ 0 & g(a) \end{pmatrix} ; \quad a \in \Sigma \right\}$$

and it can be seen that if S is not free then there is a word $w = w_1 w_2 \cdots w_n \in \Sigma^+$ such that $h_1(w_1)h_2(w_2)\cdots h_m(w_m) = g_1(w_1)g_2(w_2)\cdots g_m(w_m)$ since any equal matrix product in S must have the same word w in the top left element and the same element in the bottom right which was generated by *different* matrices. Thus the problem of freeness for 2×2 rational quaternion matrix semigroups is undecidable. See [7] for fuller details of the proof method.

Note that an alphabet size of $|\Sigma| = 7$ was required for the undecidability of MMPCP (see [10]), thus the problem is undecidable for 7 matrices. □

We now consider a problem which is *decidable* over complex numbers, but *undecidable* over rational quaternions. This gives a bound between the computational power of complex numbers and quaternions. We must first state the following lemma.

Lemma 2. *[2] Given a semigroup S of commutative matrices of any dimension, then the membership problem for S is decidable.*

Corollary 3. *The problems for diagonal matrices stated in Theorems 1, 2 and 3 are decidable when taken instead over any field up to the complex numbers.*

Proof. In Theorem 1, for each $1 \leq k \leq n$ let us define $M_k = \begin{pmatrix} q_{ik} & 0 \\ 0 & q_{jk} \end{pmatrix} \in \mathbb{C}^{2\times 2}$, and define a semigroup S generated by $\{M_1, M_2, \ldots, M_n\}$. The problem thus becomes "Does there exist a matrix X in S such that $X_{[1,1]} = X_{[2,2]}$?". This is decidable since the matrices commute by Lemma 2.

Theorem 2 concerns the emptiness testing of the intersection of two semigroups A, B. However, B is just the set of matrices with equal elements on the diagonal generated by $\gamma(a)$ and $\gamma(b)$. Thus the problem when taken for complex numbers is simply: "Does there exist some matrix, $X \in A$ with $X_{[1,1]} = X_{[2,2]}$" as in the previous paragraph. Again, since the matrices are diagonal and complex, they commute and the problem is decidable.

For Theorem 3, all matrices in the semigroup commute since they are diagonal with complex entries. By Lemma 2 we can decide if any Y is in the semigroup (in polynomial time) thus concluding the proof. □

3.1 Computational Problems in Lipschitz Integers

We also consider decision problems on the *Lipschitz integers* denoted by $\mathbb{H}(\mathbb{Z})$ which are quaternions with integral parts.

Corollary 4. *The problems stated in Theorems 1 and 2 are undecidable when taken instead over the Lipschitz integers $\mathbb{H}(\mathbb{Z})$.*

Proof. Note that in Lemma 1 we showed γ is injective and in Section 2.2 we showed an isomorphism between quaternions and a subgroup of the 2 dimensional complex matrices, $\mathbb{H}(\mathbb{Q}) \cong \mathbb{C}^{2\times 2}$. If we examine the definition of ζ in 2.2 we see that all elements have 5 as their denominator thus we can multiply $\zeta(a), \zeta(b)$ by the scalar matrix with element 5 thus giving 2 dimensional matrices over the Gaussian integers. This will still be free and is equivalent to the (non-unit) quaternions $q_1 = 5(\frac{3}{5}, \frac{4}{5}, 0, 0) \cdot \mu = (3, 4, 0, 0) \cdot \mu$ and $q_2 = 5(\frac{3}{5}, 0, \frac{4}{5}, 0) \cdot \mu = (3, 0, 4, 0) \cdot \mu$ which will now form a free semigroup. We therefore define $\lambda : \Sigma^* \mapsto \mathbb{H}(\mathbb{Z})$ by

$$\lambda(x) = \begin{cases} 5 \cdot \gamma(x) \text{ if } x \neq \varepsilon \\ \gamma(x) \text{ if } x = \varepsilon \end{cases}$$

Thus in Theorems 1 and 2 we can replace the definitions of ρ, τ to use λ and this will give a free morphism over the Lipschitz integers $\mathbb{H}(\mathbb{Z})$. This cannot be extended to Theorem 3 since the inverse of a non-identity Lipschitz integer is not itself a Lipschitz integer (obviously it must have rational coefficients). □

Theorem 6. *Given a set of Lipschitz integers $S \in \mathbb{H}(\mathbb{Z})$ forming a semigroup $\langle S, \rangle$, the problem of deciding for an arbitrary $L \in \mathbb{H}(\mathbb{Z})$ if $L \in \langle S, \cdot \rangle$ is decidable.*

Proof. Note that all non-zero quaternions have modulus $d \in \mathbb{R}^+$. Furthermore, it is obvious that for any non-zero Lipschitz integer $L \in \mathbb{H}(\mathbb{Z})$, that $d \geq 1$, with equality iff $L \in \Phi = \{(\pm 1,0,0,0)\cdot\mu, (0,\pm 1,0,0)\cdot\mu, (0,0,\pm 1,0)\cdot\mu, (0,0,0,\pm 1)\cdot\mu\}$. We have named this set for ease of explanation.

We see that $\forall q \in \Phi$ that q is of unit length i.e. $||q|| = q\overline{q} = \sqrt{a^2+b^2+c^2+d^2} = 1$. It can be also noted that their fourth powers are all equal to the identity element: $\forall q \in \Phi, q^4 = \vartheta_I = (1,0,0,0)\cdot\mu$ which is easily checked.

For a given L whose membership in S we wish to check, it will have a magnitude $||L|| = m \in \mathbb{R}$. If $m < 1$ then L cannot be a product a Lipschitz integers since the modulus must be at least 1 by definition of the quaternion modulus. If $m = 1$ then L can only be a product of elements from Φ and membership is trivial. Otherwise, $m > 1$. Let $S' = S \setminus \Phi$ (i.e. the generator set S minus any elements of Φ). We can see that there exists only a finite number of products to check since $m > 1$ and for all $x \in [S']$ we have that $||x|| > 1$.

Thus, excluding Φ we have a *finite* set of products of *finite* length to check. However if a (non-identity) element of Φ is in the generator, we must include these in the products. For each product $P = p_1 p_2 \cdots p_n \in S'$ whose magnitude equals m, i.e. $||P|| = m$, we define the set of products:

$$\left\{ P = \left(\prod_{t=1}^{n} r_t p_t \right) r_{n+1} \,|r_t, p_t \in \mathbb{H}(\mathbb{Z}) \right\},$$

where each r_t varies over all elements of $[(\Phi \cap S) \cup \vartheta_I]$ for $1 \leq t \leq n+1$. We must simply prove that $[\Phi]$ (the semigroup over elements of Φ) is finite. This is true since the only Lipschitz integers with moduli 1 are in Φ, the quaternion moduli is multiplicative and the product of two Lipschitz integers is a Lipschitz integer, all of which are very easy to prove. Thus $[\Phi]$ is a finite semigroup and there exists a finite set of products to check for equality to $L \in \mathbb{H}(\mathbb{Q})$ and thus this is a decidable problem. □

4 Geometric Interpretations

In this section, we will move from algebraic point of view to geometric interpretations of quaternion matrix semigroup problems. This leads to an interesting set of problems which we shall now outline.

Problem 1. - Point Rotation Problem (PRP(n)) - *Given points $x, y \in \mathbb{Q}^n$ on the unit length $(n-1)$-sphere and a semigroup S of n-dimensional rotations. Does there exist $M \in S$ such that M rotates x to y?*

In general, we can consider PRP(n) with a semigroup of n-dimensional rotation matrices (i.e. orthogonal matrices with determinant 1). In 3-dimensions, we may take S to be a semigroup of quaternions and define the rotation problem to be $qx'q^{-1} = y'$ where $q \in S$ and $x', y' \in \mathbb{H}(\mathbb{Q})_0$ are pure quaternions with imaginary components corresponding to the vectors x, y.

We shall show that this problem is *decidable* for 2-dimensions. Further, it is *undecidable* in 4-dimensions, and its decidability status is open in 3-dimensions.

Theorem 7. *The Point Rotation Problem, PRP(2) is decidable.*

Proof. Since the rotation of two-dimensional points is commutative, we can represent the problem as a vector reachability problem with a semigroup $S \subset \mathbb{Q}^{2\times 2}$. Since S is commutative, there exists a polynomial time algorithm to solve the membership problem [2]. □

Problem 2. - Quaternion Scalar Reachability Problem (QSRP(n)) - *Given vectors $u, v \in \mathbb{H}(\mathbb{Q})^n$ a scalar $r \in \mathbb{H}(\mathbb{Q})$ and a semigroup of matrices $S' \subset \mathbb{H}(\mathbb{Q})^{n\times n}$. Does there exist $M \in S'$ such that $u^T M v = r$?*

Theorem 8. *The Point Rotation Problem PRP(3) is reducible to the Quaternion Scalar Reachability Problem QSRP(2).*

Proof. Since we are using 3-dimensional rotations, we can convert all elements of the PRP(3) instance to quaternions. We define $x', y' \in \mathbb{H}(\mathbb{Q})_0$ to be pure quaternions with imaginary parts corresponding to x, y vectors respectively. We convert each 3D rotation, R in S to an equivalent unit quaternion q i.e. such that the imaginary vector in $qx'q^{-1}$ is equivalent to Rx for example.

Each quaternion q in the PRP is unit length it is invertible, thus if $qxq^{-1} = y$ we may write $qx = yq$. Let $G = \{q_0, q_1, \ldots, q_m\} = S' \setminus S'^2$ be the generator of S'. Define $\alpha = (y, 1)$ and $\beta = (-1, x)^T$ and let $G' = \{M_0, M_1, \ldots, M_m\}$ where $M_i = \begin{pmatrix} q_i & 0 \\ 0 & q_i \end{pmatrix}$ and let $T = \langle G', \cdot \rangle$ be a new semigroup. Then there exists $M \in T$ such that $\alpha M \beta = 0$ iff $\exists q \in S$ such that $qxq^{-1} = y$. To see this, note that $\alpha M \beta = qx - qy$ where $M = \begin{pmatrix} q & 0 \\ 0 & q \end{pmatrix}$ and $qx - yq = 0 \Rightarrow qx = yq \Rightarrow qxq^{-1} = y$ as required. □

In fact we know that QSRP(2) is undecidable in general:

Theorem 9. *Quaternion Scalar Reachability Problem is undecidable for a semigroup S generated by 5 two-dimensional diagonal quaternion matrices.*

Proof. Let $\gamma : \Sigma^* \mapsto \mathbb{H}(\mathbb{Q})$ be defined as previously and $\{(u_1, v_1), (u_2, v_2), \ldots, (u_n, v_n)\}$ be a Claus instance of PCP. Then we see that if $M_i = \begin{pmatrix} \gamma(u_i) & 0 \\ 0 & \gamma(v_i) \end{pmatrix}$ for each $2 \leq i \leq n-1$ and $\alpha = (\gamma(u_1), \gamma(v_1))$, $\beta = (\gamma(u_n), -\gamma(v_n))^T$ and $r = 0$ then:

$$\alpha M_w \beta = \gamma(u_1 u_w u_n) - \gamma(v_1 v_w v_n) = 0 \Leftrightarrow u_1 u_w u_n = v_1 v_w v_n$$

where $M_w = M_{w_1} M_{w_2} \cdots M_{w_k}$ and $1 \leq w_i \leq n-1$ for each $1 \leq i \leq k$. Since there exists a Claus instance of PCP which is undecidable for $n = 7$ [10], the problem is undecidable for 5 matrices (putting the first and last elements inside α, β). □

But the decidability status of PRP(3) remains open (since the reduction is one way). We next show that PRP(4) is undecidable.

Theorem 10. *The four-dimensional Point Rotation Problem is undecidable.*

Proof. The set of all unit quaternions forms a 3-dimensional sphere (3-sphere) and any pair of unit quaternions a and b can represent a rotation in $4D$ space. We can rotate a point $x = (x_1, x_2, x_3, x_4)$ on the 3-sphere, represented by a quaternion $q_x = (x_1, x_2, x_3, x_4)$, in the following way: aq_xb^{-1}.

Given a finite set of rotations, $\{(a_1, b_1), \ldots, (a_n, b_n)\}$, represented by pairs of quaternions. The question of whether a point x on the 3-sphere can be mapped to itself by the above set of rotations is equivalent to the problem whether there exists a non-empty sequence of indices $(r_1, \ldots, r_m)$ such that $a_{r_1} \cdots a_{r_m} q_x b_{r_m}^{-1} \cdots b_{r_1}^{-1} = q_x$.

If x is a point represented by quaternion $(1, 0, 0, 0) \cdot \mu$, then the above equation only holds when $a_{r_1} a_{r_2} \cdots a_{r_m} = b_{r_1} b_{r_2} \cdots b_{r_m}$. According to Theorem 1 we have that the four-dimensional Point Rotation Problem is undecidable for 7 rotations. Moreover it is easy to see that PRP(4) is undecidable even for 5 rotations using the idea of Claus instances of PCP [10] where two of the rotations (the first and the last one) can be fixed and used only once. □

Corollary 5. *The vector reachability problem for $n \times n$ rational orthogonal matrix semigroups is decidable when $n \leq 2$ and undecidable for $n \geq 4$ with at least 5 matrices in the semigroup generator.*

Open Problem 1. *Given a semigroup of rational quaternions, S, generated by a finite set $Q \subset \mathbb{H}(\mathbb{Q})$, is membership decidable for S? I.e. can we decide if $x \in S$ for any $x \in \mathbb{H}(\mathbb{Q})$?. Also, is the freeness of semigroup S decidable?*

References

1. Au-Yeung, Y.H.: On the Eigenvalues and Numerical Range of a Quaternionic Matrix (1994) (Preprint)
2. Babai, L., Beals, R., Cai, J., Ivanyos, G., Luks, E.M.: Multiplicative Equations over Commuting Matrices. In: Proc. 7th ACM-SIAM Symp. on Discrete Algorithms, pp. 498–507. ACM, New York (1996)
3. Bell, P.: A Note on the Emptiness of Semigroup Intersections. Fundamenta Informaticae 79, 1–4 (2007)
4. Bell, P., Potapov, I.: On the Membership of Invertible Diagonal and Scalar Matrices. Theoretical Computer Science, 37–45 (2007)
5. Blondel, V., Megretski, A.: Unsolved problems in Mathematical Systems and Control Theory. Princeton University Press, Princeton, NJ (2004)
6. Blondel, V., Jeandel, E., Koiran, P., Portier, N.: Decidable and undecidable problems about quantum automata. SIAM Journal on Computing 34(6), 1464–1473 (2005)
7. Cassaigne, J., Harju, T., Karhumäki, J.: On the Undecidability of Freeness of Matrix Semigroups. Intern. J. Alg. & Comp. 9, 295–305 (1999)

8. D'Alessandro, F.: Free Groups of Quaternions. Intern. J. of Alg. and Comp. (IJAC) 14(1) (February 2004)
9. Halava, V., Harju, T.: On Markov's Undecidability Theorem for Integer Matrices, TUCS Technical Report Number 758 (2006)
10. Halava, V., Harju, T., Hirvensalo, M.: Undecidability Bounds for Integer Matrices using Claus Instances, TUCS Technical Report 766 (2006)
11. Lengyel, E.: Mathematics for 3D Game Programming & Computer Graphics, Charles River Media (2004)
12. Markov, A.: On Certain Insoluble Problems Concerning Matrices. Doklady Akad. Nauk SSSR, 539-542 (1947)
13. Matiyasevich, Y., Senizergues, G.: Decision Problems for Semi-Thue Systems with a Few Rules. Theoretical Computer Science 330(1), 145–169 (2005)
14. Paterson, M.: Unsolvability in 3 × 3 Matrices. Studies in Applied Mathematics 49 (1970)
15. So, W., Thomson, R.C., Zhang, F.: Numerical Ranges of Matrices with Quaternion Entries. Linear and Multilinear Algebra 37, 175–195 (1994)
16. Swierczkowski, S.: A Class of Free Rotation Groups. Indag. Math. 5(2), 221–226 (1994)
17. Velichova, D., Zacharias, S.: Projection from 4D to 3D. Journal for Geometry and Graphics 4(1), 55–69 (2000)
18. Wiegmann, N.A.: Some Theorems on Matrices with Real Quaternion Elements. Can. Jour. Math. 7 (1955)
19. Zhang, F.: Quaternions and Matrices of Quaternions. Linear Algebra Appl. 251, 21–57 (1997)

VPSPACE and a Transfer Theorem over the Complex Field

Pascal Koiran and Sylvain Perifel

LIP*, École Normale Supérieure de Lyon
{Pascal.Koiran,Sylvain.Perifel}@ens-lyon.fr

Abstract. We extend the transfer theorem of [15] to the complex field. That is, we investigate the links between the class VPSPACE of families of polynomials and the Blum-Shub-Smale model of computation over $\mathbb{C}$. Roughly speaking, a family of polynomials is in VPSPACE if its coefficients can be computed in polynomial space. Our main result is that if (uniform, constant-free) VPSPACE families can be evaluated efficiently then the class $\mathsf{PAR}_{\mathbb{C}}$ of decision problems that can be solved in parallel polynomial time over the complex field collapses to $\mathsf{P}_{\mathbb{C}}$. As a result, one must first be able to show that there are VPSPACE families which are hard to evaluate in order to separate $\mathsf{P}_{\mathbb{C}}$ from $\mathsf{NP}_{\mathbb{C}}$, or even from $\mathsf{PAR}_{\mathbb{C}}$.

1 Introduction

In algebraic complexity theory, two main categories of problems are studied: evaluation and decision problems. The evaluation of the permanent of a matrix is a typical example of an evaluation problem, and it is well known that the permanent family is complete for the class VNP of "easily definable" polynomial families [19]. Deciding whether a system of polynomial equations has a solution over $\mathbb{C}$ is a typical example of a decision problem. This problem is NP-complete in the Blum-Shub-Smale model of computation over the complex field [1,2].

The main purpose of this paper is to provide a transfer theorem connecting the complexity of evaluation and decision problems. This paper is therefore in the same spirit as [13] and [15] (see also [4]). In the present paper we work with the class of polynomial families VPSPACE introduced in [15]. Roughly speaking, a family of polynomials (of possibly exponential degree) is in VPSPACE if its coefficients can be evaluated in polynomial space. For instance, it is shown in [15] that resultants of systems of multivariate polynomial equations form a VPSPACE family. The main result in [15] was that if (uniform, constant-free) VPSPACE families can be evaluated efficiently then the class $\mathsf{PAR}_{\mathbb{R}}$ of decision problems that can be solved in parallel polynomial time over the real numbers collapses to $\mathsf{P}_{\mathbb{R}}$.

Here we extend this result to the complex field $\mathbb{C}$. At first glance the result seems easier because the order $\leq$ over the reals does not have to be taken into account. The result of [15] indeed makes use of a clever combinatorial lemma of [10] on the existence of a vector orthogonal to roughly half a collection of

* UMR 5668 ENS Lyon, CNRS, UCBL, INRIA.

L. Kučera and A. Kučera (Eds.): MFCS 2007, LNCS 4708, pp. 359–370, 2007.

vectors. More precisely, it relies on the constructive version of this lemma [6]. On the complex field, we do not need this construction.

But the lack of an order over $\mathbb{C}$ makes another part of the proof more difficult. Indeed, over $\mathbb{R}$ testing whether a point belongs to a real variety is done by testing whether the sum of the squares of the polynomials is zero, a trick that cannot be used over the complex field. Hence one of the main technical developments of this paper is to explain how to decide with a small number of tests whether a point is in the complex variety defined by an exponential number of polynomials. This enables us to follow the nonconstructive proof of [12] for our transfer theorem.

Therefore, the main result of the present paper is that if (uniform, constant-free) VPSPACE families can be evaluated efficiently then the class $\mathsf{PAR}_{\mathbb{C}}$ of decision problems that can be solved in parallel polynomial time over the complex field collapses to $\mathsf{P}_{\mathbb{C}}$ (this is precisely stated in Theorem 2). The class $\mathsf{PAR}_{\mathbb{C}}$ plays roughly the same role in the theory of computation over the complex field as PSPACE in discrete complexity theory. In particular, it contains $\mathsf{NP}_{\mathbb{C}}$ [1] (but the proof of this inclusion is much more involved than in the discrete case). It follows from our main result that in order to separate $\mathsf{P}_{\mathbb{C}}$ from $\mathsf{NP}_{\mathbb{C}}$, or even from $\mathsf{PAR}_{\mathbb{C}}$, one must first be able to show that there are VPSPACE families which are hard to evaluate. This seems to be a very challenging lower bound problem, but it is still presumably easier than showing that the permanent is hard to evaluate.

Organization of the Paper. We first recall in Section 2 some usual notions and notations concerning algebraic complexity (Valiant's model, the Blum-Shub-Smale model) and quantifier elimination. The class VPSPACE is defined in Section 3 and some properties proved in [15] are given. Section 4 explains how to decide with a polynomial number of VPSPACE tests whether a point belongs to a variety. The main difficulty here is that the variety is given as a union of an exponential number of varieties, each defined by an exponential number of polynomials. Finally, Section 5 is devoted to the proof of the transfer theorem. Sign conditions are the main tool in this section. We show that $\mathsf{PAR}_{\mathbb{C}}$ problems are decided in polynomial time if we allow Uniform $\mathsf{VPSPACE}^0$ tests. The transfer theorem follows as a corollary.

2 Notations and Preliminaries

2.1 The Blum-Shub-Smale Model

In contrast with boolean complexity, algebraic complexity deals with other structures than $\{0,1\}$. In this paper we will focus on the complex field $(\mathbb{C},+,-,\times,=)$. Although the original definitions of Blum, Shub and Smale [2,1] are in terms of uniform machines, we will follow [18] by using families of algebraic circuits to recognize languages over $\mathbb{C}$, that is, subsets of $\mathbb{C}^\infty = \bigcup_{n\geq 0}\mathbb{C}^n$.

An algebraic circuit is a directed acyclic graph whose vertices, called gates, have indegree 0, 1 or 2. An input gate is a vertex of indegree 0. An output gate is a gate of outdegree 0. We assume that there is only one such gate in the circuit. Gates of indegree 2 are labelled by a symbol from the set $\{+,-,\times\}$. Gates of indegree 1, called test gates, are labelled "$= 0?$". The size of a circuit C, in symbols $|C|$, is the number of vertices of the graph. A circuit with n input gates computes a function from $\mathbb{C}^n$ to $\mathbb{C}$. On input $\bar{u} \in \mathbb{C}^n$ the value returned by the circuit is by definition equal to the value of its output gate. The value of a gate is defined inductively in the usual way: the value of input gate number i is the i-th input u_i; a $+$, $-$ or $\times$ gate respectively adds, subtracts or multiplies its inputs; the value taken by a test gate is 0 if the value of its entry is $\neq 0$ and 1 otherwise. Since we are interested in decision problems, we assume that the output is a test gate: the value returned by the circuit is therefore 0 or 1.

The class $\mathsf{P}_{\mathbb{C}}$ is the set of languages $L \subseteq \mathbb{C}^{\infty}$ such that there exists a tuple $\bar{a} \in \mathbb{C}^p$ and a P-uniform family of polynomial-size circuits (C_n) satisfying the following condition: C_n has exactly $n+p$ inputs, and for any $\bar{x} \in \mathbb{C}^n$, $\bar{x} \in L \Leftrightarrow C_n(\bar{x}, \bar{a}) = 1$. The P-uniformity condition means that C_n can be built in time polynomial in n by an ordinary (discrete) Turing machine. Note that $\bar{a}$ plays the role of the machine constants of [1,2].

As in [5], we define the class $\mathsf{PAR}_{\mathbb{C}}$ as the set of languages over $\mathbb{C}$ recognized by a PSPACE-uniform (or equivalently P-uniform) family of algebraic circuits of polynomial depth (and possibly exponential size), with constants $\bar{a}$ as for $\mathsf{P}_{\mathbb{C}}$. Note at last that we could also define similar classes without constants $\bar{a}$. We will use the superscript 0 to denote these constant-free classes, for instance $\mathsf{P}^0_{\mathbb{C}}$ and $\mathsf{PAR}^0_{\mathbb{C}}$.

We end this section with a theorem on the first-order theory of the complex numbers: quantifiers can be eliminated without much increase of the coefficients and degree of the polynomials. We give a weak version of the result of [9]: in particular, we do not need efficient elimination algorithms. Note that the only allowed constants in our formulae are 0 and 1 (in particular, only integer coefficients can appear). For notational consistency with the remainding of the paper, we denote by 2^s, 2^d and 2^{2^M} the number of polynomials, their degree and the absolute value of their coefficients respectively. This will simplify the calculations and emphasize that s, d and M will be polynomial. Note furthermore that the polynomial $p(n, s, d)$ in the theorem is independent of the formula ϕ.

Theorem 1. *Let ϕ be a first-order formula over $(\mathbb{C}, 0, 1, +, -, \times, =)$ of the form $\forall \bar{x} \psi(\bar{x})$, where $\bar{x}$ is a tuple of n variables and ψ a quantifier-free formula where 2^s polynomials occur. Suppose that their degrees are bounded by 2^d and their coefficients by 2^{2^M} in absolute value.*

There exists a polynomial $p(n, s, d)$, independent of ϕ, such that the formula ϕ is equivalent to a quantifier-free formula ψ in which all polynomials have degree less than $D(n, s, d) = 2^{p(n,s,d)}$, and their coefficients are integers strictly bounded in absolute value by $2^{2^M D(n,s,d)}$.

2.2 Valiant's Model

In Valiant's model, one computes polynomials instead of recognizing languages. We thus use arithmetic circuits instead of algebraic circuits. A book-length treatment of this topic can be found in [3].

An arithmetic circuit is the same as an algebraic circuit but test gates are not allowed. That is to say we have indeterminates $x_1, \ldots, x_{u(n)}$ as input together with arbitrary constants of $\mathbb{C}$; there are $+$, $-$ and $\times$-gates, and we therefore compute multivariate polynomials.

The polynomial computed by an arithmetic circuit is defined in the usual way by the polynomial computed by its output gate. Thus a family (C_n) of arithmetic circuits computes a family (f_n) of polynomials, $f_n \in \mathbb{C}[x_1, \ldots, x_{u(n)}]$. The class $\mathsf{VP_{nb}}$ defined in [16] is the set of families (f_n) of polynomials computed by a family (C_n) of polynomial-size arithmetic circuits, i.e., C_n computes f_n and there exists a polynomial $p(n)$ such that $|C_n| \leq p(n)$ for all n. We will assume without loss of generality that the number $u(n)$ of variables is bounded by a polynomial function of n. The subscript nb indicates that there is no bound on the degree of the polynomial, in contrast with the original class VP of Valiant where a polynomial bound on the degree of the polynomial computed by the circuit is required. Note that these definitions are nonuniform. The class Uniform $\mathsf{VP_{nb}}$ is obtained by adding a condition of polynomial-time uniformity on the circuit family, as in Section 2.1.

We can also forbid constants from our arithmetic circuits in unbounded-degree classes, and define constant-free classes. The only constant allowed is 1 (in order to allow the computation of constant polynomials). As for classes of decision problems, we will use the superscript 0 to indicate the absence of constant: for instance, we will write $\mathsf{VP^0_{nb}}$ (for bounded-degree classes, we are to be more careful: the "formal degree" of the circuits comes into play, see [16,17]).

3 The Class VPSPACE

The class VPSPACE was introduced in [15]. Some of its properties are given there and a natural example of a VPSPACE family coming from algebraic geometry, namely the resultant of a system of polynomial equations, is provided. In this section, after the definition we give some properties without proof and refer to [15] for further details.

3.1 Definition

We fix an arbitrary field K. The definition of VPSPACE will be stated in terms of *coefficient function*. A monomial $x_1^{\alpha_1} \cdots x_n^{\alpha_n}$ is encoded in binary by $\alpha = (\alpha_1, \ldots, \alpha_n)$ and will be written $\bar{x}^\alpha$.

Definition 1. *Let (f_n) be a family of multivariate polynomials with integer coefficients. The coefficient function of (f_n) is the function a whose value on input (n, α, i) is the i-th bit $a(n, \alpha, i)$ of the coefficient of the monomial $\bar{x}^\alpha$ in f_n. Furthermore, $a(n, \alpha, 0)$ is the sign of the coefficient of the monomial $\bar{x}^\alpha$. Thus f_n can be written as*

$$f_n(\bar{x}) = \sum_{\alpha} \Big((-1)^{a(n,\alpha,0)} \sum_{i \geq 1} a(n, \alpha, i) 2^{i-1} \bar{x}^\alpha \Big).$$

The coefficient function is a function $a : \{0,1\}^* \to \{0,1\}$ and can therefore be viewed as a language. This allows us to speak of the complexity of the coefficient function. Note that if K is of characteristic $p > 0$, then the coefficients of our polynomials will be integers modulo p (hence with a constant number of bits). In this paper, we will focus only on the field $\mathbb{C}$ (which is of characteristic 0).

Definition 2. *The class* Uniform VPSPACE^0 *is the set of all families* (f_n) *of multivariate polynomials* $f_n \in K[x_1, \ldots, x_{u(n)}]$ *satisfying the following requirements:*

1. *the number* $u(n)$ *of variables is polynomially bounded;*
2. *the polynomials* f_n *have integer coefficients;*
3. *the size of the coefficients of* f_n *is bounded by* $2^{p(n)}$ *for some polynomial* p*;*
4. *the degree of* f_n *is bounded by* $2^{p(n)}$ *for some polynomial* p*;*
5. *the coefficient function of* (f_n) *is in* PSPACE.

We have chosen to present only Uniform VPSPACE^0, a uniform class without constants, because this is the main object of study in this paper. In keeping with the tradition set by Valiant, however, the class VPSPACE is nonuniform and allows for arbitrary constants. See [15] for a precise definition.

3.2 An Alternative Characterization and Some Properties

Let Uniform VPAR^0 be the class of families of polynomials computed by a PSPACE-uniform family of constant-free arithmetic circuits of polynomial depth (and possibly exponential size). This in fact characterizes Uniform VPSPACE^0. The proof is given in [15].

Proposition 1. *The two classes* Uniform VPSPACE^0 *and* Uniform VPAR^0 *are equal.*

We see here the similarity with PAR$_\mathbb{C}$, which by definition are those languages recognized by uniform algebraic circuits of polynomial depth. But of course there is no test gate in the arithmetic circuits of Uniform VPAR^0.

We now turn to some properties of VPSPACE. The following two propositions come from [15]. They stress the unlikeliness of the hypothesis that VPSPACE has polynomial-size circuits.

Proposition 2. *Assuming the generalized Riemann hypothesis (GRH),* VP$_{\mathsf{nb}}$ = VPSPACE *if and only if* [P/poly = PSPACE/poly *and* VP = VNP].

Proposition 3. Uniform VPSPACE^0 = Uniform VP$^0_{\mathsf{nb}}$ $\Longrightarrow$ PSPACE = P-uniform NC.

Remark 1. To the authors' knowledge, the separation "PSPACE $\neq$ P-uniform NC" is not known to hold (by contrast, PSPACE can be separated from logspace-uniform NC thanks to the space hierarchy theorem).

Let us now state the main result of this paper.

Theorem 2 (main theorem). *If* Uniform VPSPACE^0 = Uniform VP$^0_{\mathsf{nb}}$ *then* PAR$^0_\mathbb{C}$ = P$^0_\mathbb{C}$.

Note that the collapse of the constant-free class $\mathsf{PAR}^0_{\mathbb{C}}$ to $\mathsf{P}^0_{\mathbb{C}}$ implies $\mathsf{PAR}_{\mathbb{C}} = \mathsf{P}_{\mathbb{C}}$: just replace constants by new variables so as to transform a $\mathsf{PAR}_{\mathbb{C}}$ problem into a $\mathsf{PAR}^0_{\mathbb{C}}$ problem, and then replace these variables by their original values so as to transform a $\mathsf{P}^0_{\mathbb{C}}$ problem into a $\mathsf{P}_{\mathbb{C}}$ problem.

The next section is devoted to the problem of testing whether a point belongs to a variety. This problem is useful for the proof of the theorem: indeed, following [12], several tests of membership to a variety will be made; the point here is to make them constructive and efficient. The main difficulty is that the variety can be defined by an exponential number of polynomials.

4 Testing Membership to a Union of Varieties

In this section we explain how to perform in Uniform $\mathsf{VPSPACE}^0$ membership tests of the form "$\bar{x} \in V$", where $V \subseteq \mathbb{C}^n$ is a variety. We begin in Section 4.1 by the case where V is given by s polynomials. In that case, we determine after some precomputation whether $\bar{x} \in V$ in $n+1$ tests. We first need two lemmas given below in order to reduce the number of polynomials and to replace transcendental elements by integers.

Then, in Section 4.2, we deal with the case where V is given as a union of an exponential number of such varieties, as in the actual tests of the algorithm of Section 5. Determining whether $\bar{x} \in V$ still requires $n+1$ tests, but the precomputation is slightly heavier.

Let us first state two useful lemmas. Suppose a variety V is defined by $f_1, \dots, f_s$, where $f_i \in \mathbb{Z}[x_1, \dots, x_n]$. We are to determine whether $\bar{x} \in V$ with only $n+1$ tests, however big s might be. In a nonconstructive manner, this is possible and relies on the following classical lemma already used (and proved) in [12]: any $n+1$ "generic" linear combinations of the f_i also define V (the result holds over any infinite field but here we need it only over $\mathbb{C}$). We state this lemma explicitly since we will also need it in our constructive proof.

Lemma 1. *Let $f_1, \dots, f_s \in \mathbb{Z}[x_1, \dots, x_n]$ be polynomials and V be the variety of $\mathbb{C}^n$ they define. Then for all coefficients $(\alpha_{i,j})_{i=1..s, j=1..n+1} \in \mathbb{C}^{s(n+1)}$ algebraically independent over $\mathbb{Q}$, the $n+1$ linear combinations $g_j = \sum_{i=1}^{s} \alpha_{i,j} f_i$ (for j from 1 to $n+1$) also define V.*

Unfortunately, in our case we cannot use transcendental numbers and must replace them by integers. The following lemma from [11] asserts that integers growing sufficiently fast will do. Once again, this is a weaker version adapted to our purpose.

Lemma 2. *Let $\phi(\alpha_1, \dots, \alpha_r)$ be a quantifier-free first-order formula over the structure $(\mathbb{C}, 0, 1, +, -, \times, =)$, containing only polynomials of degree less than D and whose coefficients are integers of absolute value strictly bounded by C. Assume furthermore that $\phi(\bar{\alpha})$ holds for all coefficients $\bar{\alpha} = (\alpha_1, \dots, \alpha_r) \in \mathbb{C}^r$ algebraically independent over $\mathbb{Q}$.*

Then $\phi(\bar{\beta})$ holds for any sequence $(\beta_1, \dots, \beta_r)$ of integers satisfying $\beta_1 \geq C$ and $\beta_{j+1} \geq C D^j \beta_j^D$ (for $1 \leq j \leq r-1$).

The proof can be found in [11, Lemma 5.4] and relies on the lack of big integer roots of multivariate polynomials.

Let us sketch a first attempt to prove a constructive version of Lemma 1, namely that $n+1$ polynomials with integer coefficients are enough for defining V (this first try will not work but gives the idea of the proof of the next section). The idea is to use Lemma 2 with the formula $\phi(\bar{\alpha})$ that tells us that the $n+1$ linear combinations of the f_i with $\alpha_{i,j}$ as coefficients define the same variety as $f_1, \ldots, f_s$. At first this formula is not quantifier-free, but over $\mathbb{C}$ we can eliminate quantifiers while keeping degree and coefficients reasonably small thanks to Theorem 1. Lemma 1 asserts that $\phi(\bar{\alpha})$ holds as soon as the $\alpha_{i,j}$ are algebraically independent. Then Lemma 2 tells us that $\phi(\bar{\beta})$ holds for integers $\beta_{i,j}$ growing fast enough. Thus V is now defined by $n+1$ linear combinations of the f_i with integer coefficients.

In fact, this strategy fails to work for our purpose because the coefficients involved are growing too fast to be computed in polynomial space. That is why we will proceed by stages in the proofs below: we adopt a divide-and-conquer approach and use induction. Proofs of this section are omitted due to space. They can be found in the long version [14].

4.1 Tests of Membership

The base case of our induction is the following lemma, whose proof is sketched in the end of the preceding section. We only consider here a small number of polynomials, therefore avoiding the problem of too big coefficients mentioned in the preceding section. Then by induction, Proposition 4 follows.

Lemma 3. *There exists a polynomial $q(n,d)$ such that, if $V \subseteq \mathbb{C}^n$ is a variety defined by $2(n+1)$ polynomials $f_1, \ldots, f_{2(n+1)} \in \mathbb{Z}[x_1, \ldots, x_n]$ of degree $\leq 2^d$ and of coefficients bounded by 2^{2^M} in absolute value, then:*

1. *the variety V is defined by $n+1$ polynomials $g_1, \ldots, g_{n+1} \in \mathbb{Z}[x_1, \ldots, x_n]$ of degree $\leq 2^d$ and of coefficients bounded by $2^{2^{M+q(n,d)}}$ in absolute value;*
2. *furthermore, the coefficients of the g_i are bitwise computable from those of the f_j in working space $Mq(n,d)$.*

Proposition 4. *There exists a polynomial $p(n,s,d)$ such that, if V is a variety defined by 2^s polynomials $f_1, \ldots, f_{2^s} \in \mathbb{Z}[x_1, \ldots, x_n]$ of degree $\leq 2^d$ and of coefficients bounded by 2^{2^M} in absolute value, then:*

1. *the variety V is defined by $n+1$ polynomials $g_1, \ldots, g_{n+1} \in \mathbb{Z}[x_1, \ldots, x_n]$ of degree $\leq 2^d$ and of coefficients bounded by $2^{2^{M+p(n,s,d)}}$ in absolute value;*
2. *moreover, the coefficients of the g_i are bitwise computable from those of the f_j in working space $Mp(n,s,d)$.*

4.2 Union of Varieties

In our case, however, the tests made by the algorithm of Section 5 are not exactly of the form studied in the previous section: instead of a single variety given by s polynomials, we have to decide "$x \in W$?" when $W \subseteq \mathbb{C}^n$ is the union of k

varieties. Of course, since the union is finite W is also a variety, but the encoding is not the same as above: now, k sets of s polynomials are given.

A first naive approach is to define $W = \cup_i V_i$ by the different products of the polynomials defining the V_i, but it turns out that there are too many products to be dealt with. Instead, we will adopt a divide-and-conquer scheme as previously.

Lemma 4. *There exists a polynomial $q(n, d)$ such that, if V_1 and V_2 are two varieties of $\mathbb{C}^n$, each defined by $n+1$ polynomials in $\mathbb{Z}[x_1, \ldots, x_n]$, respectively $f_1, \ldots, f_{n+1}$ and $g_1, \ldots, g_{n+1}$, of degree $\le 2^d$ and of coefficients bounded by 2^{2^M} in absolute value, then:*

1. *the variety $V = V_1 \cup V_2$ is defined by $n+1$ polynomials $h_1, \ldots, h_{n+1}$ in $\mathbb{Z}[x_1, \ldots, x_n]$ of degree $\le 2^{d+1}$ and of coefficients bounded by $2^{2^{M+q(n,d)}}$ in absolute value;*
2. *the coefficients of the h_i are bitwise computable from those of the f_j and g_k in space $Mq(n, d)$.*

The next proposition now follows by induction.

Proposition 5. *There exists a polynomial $r(n, s, k, d)$ such that, if $V_1, \ldots, V_{2^k} \subseteq \mathbb{C}^n$ are 2^k varieties, V_i being defined by 2^s polynomials $f_1^{(i)}, \ldots, f_{2^s}^{(i)} \in \mathbb{Z}[x_1, \ldots, x_n]$ of degree $\le 2^d$ and of coefficients bounded by 2^{2^M} in absolute value, then:*

1. *the variety $V = \cup_{i=1}^{2^k} V_i$ is defined by $n+1$ polynomials $g_1, \ldots, g_{n+1}$ in $\mathbb{Z}[x_1, \ldots, x_n]$ of degree $\le 2^{d+k}$ and whose coefficients are bounded in absolute value by $2^{2^{M+r(n,s,k,d)}}$;*
2. *moreover, the coefficients of the g_i are bitwise computable from those of the $f_{j'}^{(j)}$ in space $Mr(n, s, k, d)$.*

Here is the main consequence on membership tests to a union of varieties.

Corollary 1. *Let $p(n)$ and $q(n)$ be two polynomials. Suppose $(f_n(\bar{x}, \bar{y}, \bar{z}))$ is a* Uniform $\mathsf{VPSPACE}^0$ *family with $|\bar{x}| = n$, $|\bar{y}| = p(n)$ and $|\bar{z}| = q(n)$. For an integer $0 \le i < 2^{p(n)}$, call $V_i^{(n)} \subseteq \mathbb{C}^n$ the variety defined by the polynomials $f_n(\bar{x}, i, j)$ for $0 \le j < 2^{q(n)}$ (in this notation, i and j are encoded in binary).*

Then there exists a Uniform $\mathsf{VPSPACE}^0$ *family $g_n(\bar{x}, \bar{y}, \bar{z})$, where $|\bar{x}| = n$, $|\bar{y}| = p(n)$ and $|\bar{z}| = \lceil \log(n+1) \rceil$, such that*

$$\forall \bar{x} \in \mathbb{C}^n, \quad \forall k < 2^{p(n)}, \quad \left(\bar{x} \in \bigcup_{i=0}^{k} V_i^{(n)} \iff \bigwedge_{j=0}^{n} g_n(\bar{x}, k, j) = 0 \right).$$

5 Proof of the Main Theorem

Sign conditions are the main ingredient of the proof. Over $\mathbb{C}$, we define the "sign" of $a \in \mathbb{C}$ by 0 if $a = 0$ and 1 otherwise. Let us fix a family of polynomials $f_1, \ldots, f_s \in \mathbb{Z}[x_1, \ldots, x_n]$. A *sign condition* is an element $S \in \{0, 1\}^s$. Hence there are 2^s sign conditions. Intuitively, the i-th component of a sign condition determines the sign of the polynomial f_i.

5.1 Satisfiable Sign Conditions

The sign condition of a point $\bar{x} \in \mathbb{C}^n$ is the tuple $S^{\bar{x}} \in \{0,1\}^s$ defined by $S_i^{\bar{x}} = 0 \iff f_i(\bar{x}) = 0$. We say that a sign condition is *satisfiable* if it is the sign condition of some $\bar{x} \in \mathbb{C}^n$. As 0-1 tuples, sign conditions can be viewed as subsets of $\{1,\ldots,s\}$. Using a fast parallel sorting algorithm (e.g. Cole's, [7]), we can sort satisfiable sign conditions in polylogarithmic parallel time in a way compatible with set inclusion (e.g. the lexicographic order). We now fix such a compatible linear order on sign conditions and consider our satisfiable sign conditions $S^{(1)} < S^{(2)} < \ldots < S^{(N)}$ sorted accordingly.

The key point resides in the following theorem, coming from the algorithm of [9]: there is a "small" number of satisfiable sign conditions and enumerating them is "easy".

Theorem 3. *Let $f_1,\ldots,f_s \in \mathbb{Z}[x_1,\ldots,x_n]$ and d be their maximal degree. Then the number of satisfiable sign conditions is $N = (sd)^{O(n)}$, and there is a uniform algorithm working in space $\big(n\log(sd)\big)^{O(1)}$ which, on boolean input $f_1,\ldots,f_s$ (in dense representation) and (i,j) in binary, returns the j-th component of the i-th satisfiable sign condition.*

When $\log(sd)$ is polynomial in n, as will be the case, this yields a PSPACE algorithm. If furthermore the coefficients of f_i are computable in polynomial space, we will then be able to use the satisfiable sign conditions in the coefficients of VPSPACE families, as in Lemma 5 below.

Let us explain why we are interested in sign conditions. An arithmetic circuit performs tests of the form $f(\bar{x}) = 0$ on input $\bar{x} \in \mathbb{C}^n$, where f is a polynomial. Suppose $f_1,\ldots,f_s$ is the list of all polynomials that can be tested in *any possible* computation. Then two elements of $\mathbb{C}^n$ with the same sign condition are simultaneously accepted or rejected by the circuit: the results of the tests are indeed always the same for both elements.

Thus, instead of finding out whether $\bar{x} \in \mathbb{C}^n$ is accepted by the circuit, it is enough to find out whether the sign condition of $\bar{x}$ is accepted. The advantage resides in handling only boolean tuples (the sign conditions) instead of complex numbers (the input $\bar{x}$). But we have to be able to find the sign condition of the input $\bar{x}$. This requires first the enumeration of all the polynomials possibly tested in any computation of the circuit.

5.2 Enumerating All Possibly Tested Polynomials

In the execution of an algebraic circuit, the values of some polynomials at the input $\bar{x}$ are tested to zero. In order to find the sign condition of the input $\bar{x}$, we have to be able to enumerate in polynomial space all the polynomials that can ever be tested to zero in the computations of an algebraic circuit. This is done level by level as in [8, Th. 3] and [15].

Proposition 6. *Let C be a constant-free algebraic circuit with n variables and of depth d.*

1. *The number of different polynomials possibly tested to zero in the computations of* C *is* $2^{d^2 O(n)}$.
2. *There exists an algorithm using work space* $(nd)^{O(1)}$ *which, on input* C *and integers* (i, j) *in binary, outputs the* j*-th bit of the representation of the* i*-th polynomial.*

Together with Theorem 3, this enables us to prove the following result which will be useful in the proof of Proposition 7: in Uniform $\mathsf{VPSPACE}^0$ we can enumerate the polynomials as well as the satisfiable sign conditions.

Lemma 5. *Let* (C_n) *be a uniform family of polynomial-depth algebraic circuits with polynomially many inputs. Call* $d(n)$ *the depth of* C_n *and* $i(n)$ *the number of inputs. Let* $f_1^{(n)}, \ldots, f_s^{(n)}$ *be all the polynomials possibly tested to zero by* C_n *as in Proposition 6, where* $s = 2^{O(nd(n)^2)}$. *There are therefore* $N = 2^{O(n^2 d(n)^2)}$ *satisfiable sign conditions* $S^{(1)}, \ldots, S^{(N)}$ *by Theorem 3.*

Then there exists a Uniform $\mathsf{VPSPACE}^0$ *family* $(g_n(\bar{x}, \bar{y}, \bar{z}))$, *where* $|\bar{x}| = i(n)$, $|\bar{y}| = O(n^2 d(n)^2)$ *and* $|\bar{z}| = O(nd(n)^2)$, *such that for all* $1 \le i \le N$ *and* $1 \le j \le s$, *we have:*

$$g_n(\bar{x}, i, j) = \begin{cases} 0 & \text{if } S_j^{(i)} = 1 \\ f_j^{(n)}(\bar{x}) & \text{otherwise.} \end{cases}$$

5.3 Finding the Sign Condition of the Input

In order to find the sign condition $S^{\bar{x}}$ of the input $\bar{x} \in \mathbb{C}^n$, we will give a polynomial-time algorithm which tests some $\mathsf{VPSPACE}$ family for zero. Here is the formalized notion of a polynomial-time algorithm with $\mathsf{VPSPACE}$ tests.

Definition 3. *A polynomial-time algorithm with* Uniform $\mathsf{VPSPACE}^0$ *tests is a* Uniform $\mathsf{VPSPACE}^0$ *family* $(f_n(x_1, \ldots, x_{u(n)}))$ *together with a uniform family* (C_n) *of constant-free polynomial-size algebraic circuits endowed with special test gates of indegree* $u(n)$, *whose value is* 1 *on input* $(a_1, \ldots, a_{u(n)})$ *if* $f_n(a_1, \ldots, a_{u(n)}) = 0$ *and* 0 *otherwise.*

Observe that a constant number of Uniform $\mathsf{VPSPACE}^0$ families can be used in the preceding definition instead of only one: it is enough to combine them all in one by using "selection variables".

The precise result we show now is the following. By the "rank" of a satisfiable sign condition, we merely mean its index in the fixed order on satisfiable sign conditions.

Proposition 7. *Let* (C_n) *be a uniform family of algebraic circuits of polynomial depth and with a polynomial number* $i(n)$ *of inputs. There exists a polynomial-time algorithm with* Uniform $\mathsf{VPSPACE}^0$ *tests which, on input* $\bar{x} \in \mathbb{C}^{i(n)}$, *returns the rank* i *of the sign condition* $S^{(i)}$ *of* $\bar{x}$ *with respect to the polynomials* $g_1, \ldots, g_s$ *tested to zero by* C_n *given by Proposition 6.*

Proof. Take the Uniform $\mathsf{VPSPACE}^0$ family $(g_n(\bar{x}, \bar{y}, \bar{z}))$ as in Lemma 5: in essence, g_n enumerates all the polynomials $f_1, \ldots, f_s$ possibly tested to zero in C_n and enumerates the N satisfiable sign conditions $S^{(1)} < \ldots < S^{(N)}$. The idea now is to perform a binary search in order to find the rank i of the sign condition of the input $\bar{x}$.

Let $S^{(j)} \in \{0,1\}^s$ be a satisfiable sign condition. We say that $S^{(j)}$ is a *candidate* whenever $\forall m \leq s$, $S^{(j)}_m = 0 \Rightarrow f_m(\bar{x}) = 0$. Remark that the sign condition of $\bar{x}$ is the smallest candidate. Call V_j the variety defined by the polynomials $\{f_m | S^{(j)}_m = 0\}$: by definition of g_n, V_j is also defined by the polynomials $g_n(\bar{x}, j, k)$ for $k = 1$ to s. Note that $S^{(j)}$ is a candidate if and only if $\bar{x} \in V_j$.

Corollary 1 combined with Lemma 5 asserts that tests of the form $\bar{x} \in \cup_{k \leq j} V_k$ are in Uniform $\mathsf{VPSPACE}^0$. They are used to perform a binary search by making j vary. In a number of steps logarithmic in N (i.e. polynomial in n), we find the rank i of the sign condition of $\bar{x}$. □

5.4 A Polynomial-Time Algorithm for $\mathrm{PAR}_{\mathbb{C}}$ Problems

Lemma 6. *Let (C_n) be a uniform family of constant-free polynomial-depth algebraic circuits. There is a (boolean) algorithm using work space polynomial in n which, on input i, decides whether the elements of the i-th satisfiable sign condition $S^{(i)}$ are accepted by the circuit C_n.*

Proof. We follow the circuit C_n level by level. For test gates, we compute the polynomial f to be tested. Then we enumerate the polynomials $f_1, \ldots, f_s$ as in Proposition 6 for the circuit C_n and we find the index j of f in this list. By consulting the j-th bit of the i-th satisfiable sign condition with respect to $f_1, \ldots, f_s$ (which is done by the polynomial-space algorithm of Theorem 3), we therefore know the result of the test and can go on like this until the output gate. □

Theorem 4. *Let $A \in \mathsf{PAR}^0_{\mathbb{C}}$. There exists a polynomial-time algorithm with* Uniform $\mathsf{VPSPACE}^0$ *tests that decides A.*

Proof. A is decided by a uniform family (C_n) of constant-free polynomial-depth algebraic circuits. On input $\bar{x}$, thanks to Proposition 7 we first find the rank i of the sign condition of $\bar{x}$ with respect to the polynomials $f_1, \ldots, f_s$ of Proposition 6. Then we conclude by a last Uniform $\mathsf{VPSPACE}^0$ test simulating the polynomial-space algorithm of Lemma 6 on input i. □

Theorem 2 follows immediately from this result. One could obtain other versions of these two results by changing the uniformity conditions or the role of constants.

References

1. Blum, L., Cucker, F., Shub, M., Smale, S.: Complexity and Real Computation. Springer, Heidelberg (1998)
2. Blum, L., Shub, M., Smale, S.: On a theory of computation and complexity over the real numbers: NP-completeness, recursive functions and universal machines. Bulletin of the American Mathematical Society 21(1), 1–46 (1989)

3. Bürgisser, P.: Completeness and Reduction in Algebraic Complexity Theory. Algorithms and Computation in Mathematics, vol. 7. Springer, Heidelberg (2000)
4. Bürgisser, P.: On implications between P-NP-hypotheses: Decision versus computation in algebraic complexity. In: Sgall, J., Pultr, A., Kolman, P. (eds.) MFCS 2001. LNCS, vol. 2136, pp. 3–17. Springer, Heidelberg (2001)
5. Chapuis, O., Koiran, P.: Saturation and stability in the theory of computation over the reals. Annals of Pure and Applied Logic 99, 1–49 (1999)
6. Charbit, P., Jeandel, E., Koiran, P., Perifel, S., Thomassé, S.: Finding a vector orthogonal to roughly half a collection of vectors. Accepted in Journal of Complexity (2006) available from `http://prunel.ccsd.cnrs.fr/ensl-00153736`
7. Cole, R.: Parallel merge sort. SIAM J. Comput. 17(4), 770–785 (1988)
8. Cucker, F., Grigoriev, D.: On the power of real Turing machines over binary inputs. SIAM J. Comput. 26(1), 243–254 (1997)
9. Fitchas, N., Galligo, A., Morgenstern, J.: Precise sequential and parallel complexity bounds for quantifier elimination over algebraically closed fields. Journal of Pure and Applied Algebra 67, 1–14 (1990)
10. Grigoriev, D.: Topological complexity of the range searching. Journal of Complexity 16, 50–53 (2000)
11. Koiran, P.: Randomized and deterministic algorithms for the dimension of algebraic varieties. In: Proc. 38th FOCS, pp. 36–45 (1997)
12. Koiran, P.: Circuits versus trees in algebraic complexity. In: Reichel, H., Tison, S. (eds.) STACS 2000. LNCS, vol. 1770, pp. 35–52. Springer, Heidelberg (2000)
13. Koiran, P., Perifel, S.: Valiant's model: from exponential sums to exponential products. In: Královič, R., Urzyczyn, P. (eds.) MFCS 2006. LNCS, vol. 4162, pp. 596–607. Springer, Heidelberg (2006)
14. Koiran, P., Perifel, S.: VPSPACE and a transfer theorem over the complex field. Technical report, LIP, ENS Lyon (2007) Available from `http://prunel.ccsd.cnrs.fr/ensl-00153701`
15. Koiran, P., Perifel, S.: VPSPACE and a transfer theorem over the reals. In: Thomas, W., Weil, P. (eds.) STACS 2007. LNCS, vol. 4393, pp. 417–428. Springer, Heidelberg (2007). Long version available from `http://prunel.ccsd.cnrs.fr/ensl-00103018`
16. Malod, G.: Polynômes et coefficients. PhD thesis, Université Claude Bernard Lyon 1 (July, 2003), available from `http://tel.archives-ouvertes.fr/tel-00087399`
17. Malod, G., Portier, N.: Characterizing Valiant's algebraic complexity classes. In: Královič, R., Urzyczyn, P. (eds.) MFCS 2006. LNCS, vol. 4162, pp. 704–716. Springer, Heidelberg (2006)
18. Poizat, B.: Les petits cailloux. Aléas (1995)
19. Valiant, L.G.: Completeness classes in algebra. In: Proc. 11th ACM Symposium on Theory of Computing, pp. 249–261. ACM Press, New York (1979)

Efficient Provably-Secure Hierarchical Key Assignment Schemes

Alfredo De Santis, Anna Lisa Ferrara, and Barbara Masucci

Dipartimento di Informatica ed Applicazioni, Università di Salerno
84084 Fisciano (SA), Italy

Abstract. A *hierarchical key assignment scheme* is a method to assign some private information and encryption keys to a set of classes in a partially ordered hierarchy, in such a way that the private information of a higher class can be used to derive the keys of all classes lower down in the hierarchy.

In this paper we design and analyze hierarchical key assignment schemes which are provably-secure and support dynamic updates to the hierarchy with local changes to the public information and without requiring any private information to be re-distributed.

- We first show an encryption based construction which is provably secure with respect to key indistinguishability, requires a single computational assumption and improves on previous proposals.
- Then, we show how to reduce key derivation time at the expense of an increment of the amount of public information, by improving a previous result.
- Finally, we show a construction using as a building block a public-key broadcast encryption scheme. In particular, one of our constructions provides constant private information and public information linear in the number of classes in the hierarchy.

1 Introduction

The *hierarchical access control problem* is defined in a scenario where the users of a computer system are organized in a hierarchy formed by a certain number of disjoint *security classes*. Hierarchical structures arise from the fact that some users have more access rights than others, and are widely employed in many different application areas, including database management systems, computer networks, operating systems, military, and government communications.

In 1983, Akl and Taylor [1] suggested the use of cryptographic techniques to enforce access control in hierarchical structures. In particular, they designed a *hierarchical key assignment scheme* where each class is assigned an encryption key that can be used, along with some public parameters, to compute the key assigned to all classes lower down in the hierarchy. This assignment is carried out by a Trusted Authority (TA), which is active only at the distribution phase. A recent work by Crampton et al. [10] provides a detailed classification of many schemes proposed in the last twenty years and evaluates their merits. Atallah

L. Kučera and A. Kučera (Eds.): MFCS 2007, LNCS 4708, pp. 371–382, 2007.

et al. [3,4] first addressed the problem of formalizing security requirements for hierarchical key assignment schemes. They proposed a first construction based on pseudorandom functions and a second one requiring the use of a symmetric encryption scheme secure against chosen-ciphertext attacks.

In this paper we design and analyze hierarchical key assignment schemes which are provably-secure and efficient. We consider security with respect to *key indistinguishability*, which corresponds to the requirement that an adversary is not able *to learn any information* about a key that it should not have access to. We propose two constructions for hierarchical key assignment schemes. Both constructions support updates to the access hierarchy with local changes to the public information and without requiring any private information to be re-distributed. The first construction, which is based on symmetric encryption schemes, is simpler than the one proposed by Atallah et al. [4], requires a single computational assumption, and offers more efficient procedures for key derivation and key updates. We also focus on improving efficiency of key derivation in hierarchical key assignment schemes. Such a problem has been recently considered by Atallah et al. [4,5], who proposed two different techniques requiring an increment of public information. We show how to further reduce key derivation time by improving one of their techniques. Finally, we show how to construct a hierarchical key assignment scheme by using only a public-key broadcast encryption scheme. In particular, by plugging in the scheme proposed by Boneh et al. [8] we obtain a hierarchical key assignment scheme offering constant private information and public information linear in the number of classes.

The full version of this paper, including a complete analysis and proofs, can be found in [11].

2 Hierarchical Key Assignment Schemes

Consider a set of users divided into a number of disjoint classes, called *security classes.* A binary relation $\preceq$ that partially orders the set of classes V is defined in accordance with authority, position, or power of each class in V. The poset $(V, \preceq)$ is called a *partially ordered hierarchy.* For any two classes u and v, the notation $u \preceq v$ is used to indicate that the users in v can access u's data. Clearly, since v can access its own data, it holds that $v \preceq v$, for any $v \in V$. We denote by A_v the set $\{u \in V : u \preceq v\}$, for any $v \in V$. The partially ordered hierarchy $(V, \preceq)$ can be represented by the directed graph $G^* = (V, E^*)$, where each class corresponds to a vertex in the graph and there is an edge from class v to class u if and only if $u \preceq v$. We denote by $G = (V, E)$ the *minimal representation* of the graph G^*, that is, the directed acyclic graph corresponding to the *transitive and reflexive reduction* of the graph $G^* = (V, E^*)$. Such a graph G has the same transitive and reflexive closure of G^*, i.e., there is a path (of length greater than or equal to zero) from v to u in G if and only if there is the edge (v, u) in E^*.

A hierarchical key assignment scheme for a family Γ of graphs, corresponding to partially ordered hierarchies, is defined as follows.

Definition 1. *A* hierarchical key assignment scheme *for Γ is a pair (Gen, Der) of algorithms satisfying the following conditions:*

1. *The* information generation algorithm *Gen is probabilistic polynomial-time. It takes as inputs the security parameter 1^τ and a graph $G = (V, E)$ in Γ, and produces as outputs*
   (a) a private information s_u, for any class $u \in V$;
   (b) a key k_u, for any class $u \in V$;
   (c) a public information pub.
   We denote by (s, k, pub) the output of the algorithm Gen on inputs 1^τ and G, where s and k denote the sequences of private information and of keys, respectively.
2. *The* key derivation algorithm *Der is deterministic polynomial-time. It takes as inputs the security parameter 1^τ, a graph $G = (V, E)$ in Γ, two classes $u \in V$ and $v \in A_u$, the private information s_u assigned to class u and the public information pub, and produces as output the key k_v assigned to class v.*
   We require that for each class $u \in V$, each class $v \in A_u$, each private information s_u, each key k_v, each public information pub which can be computed by Gen on inputs 1^τ and G, it holds that $Der(1^\tau, G, u, v, s_u, pub) = k_v$.

A hierarchical key assignment scheme is evaluated according to several parameters, such as the amount of private information held by each user, the amount of public data, the complexity of key derivation, and the resistance to collusive attacks. In order to evaluate the security of the scheme, we consider a *static adversary* which wants to attack a class $u \in V$ and which is able to corrupt *all* users not allowed to compute the key k_u. We define an algorithm $Corrupt_u$ which, on input the private information s generated by the algorithm *Gen*, extracts the secret values s_v associated to all classes in the set $F_u = \{v \in V : u \notin A_v\}$. We denote by *corr* the sequence output by $Corrupt_u(s)$. Two experiments are considered. In the first one, the adversary is given the key k_u, whereas, in the second one, it is given a random string ρ having the same length as k_u. It is the adversary's job to determine whether the received challenge corresponds to k_u or to a random string. We require that the adversary will succeed with probability only negligibly different from $1/2$.

If $A(\cdot, \cdot, \ldots)$ is any probabilistic algorithm then $a \leftarrow A(x, y, \ldots)$ denotes the experiment of running A on inputs $x, y, \ldots$ and letting a be the outcome, the probability being over the coins of A. Similarly, if X is a set then $x \leftarrow X$ denotes the experiment of selecting an element uniformly from X and assigning x this value. If w is neither an algorithm nor a set then $x \leftarrow w$ is a simple assignment statement. A function $\epsilon : N \to R$ is *negligible* if for every constant $c > 0$ there exists an integer n_c such that $\epsilon(n) < n^{-c}$ for all $n \geq n_c$.

Definition 2. [IND-ST] *Let Γ be a family of graphs corresponding to partially ordered hierarchies, let $G = (V, E)$ be a graph in Γ, let (Gen, Der) be a hierarchical key assignment scheme for Γ and let $\mathtt{STAT}_u$ be a static adversary which attacks a class u. Consider the following two experiments:*

Experiment $\mathbf{Exp}^{\mathtt{IND-1}}_{\mathtt{STAT}_u}(1^\tau, G)$	Experiment $\mathbf{Exp}^{\mathtt{IND-0}}_{\mathtt{STAT}_u}(1^\tau, G)$
$(s, k, pub) \leftarrow Gen(1^\tau, G)$	$(s, k, pub) \leftarrow Gen(1^\tau, G)$
$corr \leftarrow Corrupt_u(s)$	$corr \leftarrow Corrupt_u(s)$
$d \leftarrow \mathtt{STAT}_u(1^\tau, G, pub, corr, k_u)$	$\rho \leftarrow \{0,1\}^{length(k_u)}$
return d	$d \leftarrow \mathtt{STAT}_u(1^\tau, G, pub, corr, \rho)$
	return d

The advantage of $\mathtt{STAT}_u$ *is defined as*

$$\mathbf{Adv}^{\mathtt{IND}}_{\mathtt{STAT}_u}(1^\tau, G) = |Pr[\mathbf{Exp}^{\mathtt{IND-1}}_{\mathtt{STAT}_u}(1^\tau, G) = 1] - Pr[\mathbf{Exp}^{\mathtt{IND-0}}_{\mathtt{STAT}_u}(1^\tau, G) = 1]|.$$

The scheme is said to be secure in the sense of IND-ST *if, for each graph* $G = (V, E)$ *in* Γ *and each* $u \in V$*, the function* $\mathbf{Adv}^{\mathtt{IND}}_{\mathtt{STAT}_u}(1^\tau, G)$ *is negligible, for each static adversary* $\mathtt{STAT}_u$ *whose time complexity is polynomial in* τ.

In Definition 2 we have considered a static adversary attacking a class. A different kind of adversary, the *adaptive* one, could also be considered. In [6] it has been proven that security against adaptive adversaries is (polynomially) equivalent to security against static adversaries. Hence, in this paper we will only consider static adversaries.

3 An Encryption Based Construction

In this section we consider the problem of constructing a hierarchical key assignment scheme by using as a building block a symmetric encryption scheme. A simple way to realize an encryption based scheme would be to assign a key k_u to each class $u \in V$ and a public information $p_{(u,v)}$, for each edge $(u, v) \in E$, corresponding to the encryption of k_v with the key k_u. This would allow any user in a class u to compute the key k_v held by any class v lower down in the hierarchy, by performing $dist_G(u, v)$ decryptions, where $dist_G(u, v)$ denotes the length of the shortest path from u to v in G.

Unfortunately, the simple solution described above is not secure with respect to *key indistinguishability*. Indeed, consider an adversary attacking a class u and corrupting a class v such that $(u, v) \in E$. The adversary, on input a challenge ρ, corresponding either to the key k_u or to a random value, is able to tell if ρ corresponds to the encryption key k_u simply by checking whether the decryption of the public value $p_{(u,v)}$ with key ρ corresponds to the key k_v held by class v. In order to avoid the above attack, we propose a new construction, described in Figure 1 and referred to as the *Encryption Based Construction (EBC)*, where the key assigned to a class is never used to encrypt the keys assigned to other classes. In particular, in the EBC each class $u \in V$ is assigned a private information s_u, an encryption key k_u, and a public information $\pi_{(u,u)}$, which is the encryption of the key k_u with the private information s_u; moreover, for each edge $(u, v) \in E$, there is a public value $p_{(u,v)}$, which allows class u to compute the private information s_v held by class v. Indeed, $p_{(u,v)}$ consists of the encryption of the private information s_v with the private information s_u. This allows any user in a class u to compute the key k_v held by any class v lower down in the hierarchy, by performing

Let Γ be a family of graphs corresponding to partially ordered hierarchies. Let $G = (V, E) \in \Gamma$ and let $\Pi = (\mathcal{K}, \mathcal{E}, \mathcal{D})$ be a symmetric encryption scheme.

Algorithm $Gen(1^\tau, G)$

1. For any class $u \in V$, let $s_u \leftarrow \mathcal{K}(1^\tau)$ and $k_u \leftarrow \{0,1\}^\tau$;
2. Let s and k be the sequences of private information and keys, respectively, computed in the previous step;
3. For any two classes $u, v \in V$ such that $(u, v) \in E$, compute the public information $p_{(u,v)} = \mathcal{E}_{s_u}(s_v)$;
4. For any class u in V, compute the public information $\pi_{(u,u)} = \mathcal{E}_{s_u}(k_u)$;
5. Let pub be the sequence of public information computed in the previous two steps;
6. Output (s, k, pub).

Algorithm $Der(1^\tau, G, u, v, s_u, pub)$

1. Consider a path $(w_0, w_1), \ldots, (w_{m-1}, w_m) \in E$, from $u = w_0$ to $v = w_m$. For any $i = 1, \ldots, m$, extract the public value $p_{(w_{i-1}, w_i)}$ from pub and compute the private information $s_{w_i} = \mathcal{D}_{s_{w_{i-1}}}(p_{(w_{i-1}, w_i)})$;
2. Extract the public value $\pi_{(v,v)}$ from pub and output the key $k_v = \mathcal{D}_{s_v}(\pi_{(v,v)})$.

Fig. 1. The Encryption Based Construction (EBC)

$dist_G(u, v) + 1$ decryptions. We will show that an adversary attacking a class u is not able to distinguish the key k_u from a random string of the same length unless it is able to break the underlying encryption scheme. We first recall the definition of a symmetric encryption scheme.

Definition 3. *A* symmetric encryption scheme *is a triple* $\Pi = (\mathcal{K}, \mathcal{E}, \mathcal{D})$ *of algorithms satisfying the following conditions:*

1. *The* key-generation algorithm $\mathcal{K}$ *is probabilistic polynomial-time. It takes as input the security parameter* 1^τ *and produces as output a string key.*
2. *The* encryption algorithm $\mathcal{E}$ *is probabilistic polynomial-time. It takes as inputs* 1^τ*, a string key produced by* $\mathcal{K}(1^\tau)$*, and a message* $m \in \{0,1\}^*$*, and produces as output the ciphertext* y*.*
3. *The* decryption algorithm $\mathcal{D}$ *is deterministic polynomial-time. It takes as inputs* 1^τ*, a string key produced by* $\mathcal{K}(1^\tau)$*, and a ciphertext* y*, and produces as output a message* m*. We require that for any string key which can be output by* $\mathcal{K}(1^\tau)$*, for any message* $m \in \{0,1\}^*$*, and for all* y *that can be output by* $\mathcal{E}(1^\tau, key, m)$*, we have that* $\mathcal{D}(1^\tau, key, y) = m$*.*

Before analyzing the security of the EBC we first need to define what we mean by a *secure* symmetric encryption scheme. We formalize security with respect to *plaintext indistinguishability*, which is an adaption of the notion of *polynomial security* as given in [13]. We consider an adversary $A = (A_1, A_2)$ running in two stages. In advance of the adversary's execution, a random key *key* is chosen and kept hidden from the adversary. During the first stage, the adversary A_1 outputs a triple

$(x_0, x_1, state)$, where x_0 and x_1 are two messages of the same length, and *state* is some state information which could be useful later. One message between x_0 and x_1 is chosen at random and encrypted to give the challenge ciphertext y. In the second stage, the adversary A_2 is given y and *state* and has to determine whether y is the encryption of x_0 or x_1. Informally, the encryption scheme is said to be secure with respect to a non-adaptive chosen plaintext attack (IND-P1-C0), if every polynomial-time adversary A, which has access to the encryption oracle only during the first stage of the attack and has never access to the decryption oracle, succeeds in determining whether y is the encryption of x_0 or x_1 with probability only negligibly different from 1/2. The proof of the next theorem can be found in [11].

Theorem 1. *If the encryption scheme* $\Pi = (\mathcal{K}, \mathcal{D}, \mathcal{E})$ *is secure in the sense of* IND-P1-C0, *then the EBC is secure in the sense of* IND-ST.

The EBC requires $|E| + |V|$ public values; on the other hand, each class has to store a single secret value, corresponding to its private information. As for key derivation, a class $u \in V$ which wants to compute the key held by a class $v \in A_u$ is required to perform $dist_G(u, v) + 1$ decryption operations. The EBC supports insertions, but not deletions, of classes and edges in the hierarchy without re-distributing private information to the classes affected by such changes. However, in the full version of this paper [11] a simple modification of the scheme, which avoids such a re-distribution, has been proposed. The modified scheme, referred to as the *Dynamic Encryption Based Construction (DEBC)*, requires $|E| + 2|V|$ public values, whereas, each class has to store a single secret value. Moreover, the number of decryptions needed by class u to compute the key of class $v \in A_u$ is $dist_G(u, v) + 2$.

4 Improving Key Derivation Time

In the EBC, as well as in the schemes proposed by Atallah et al. [3,4], the number of steps that a class u has to perform, in order to compute the key of another class v lower down in the hierarchy, is proportional to the length of the shortest path from u to v. Atallah et al. [3,4,5] analyzed the problem of reducing key derivation time by modifying the graph representing the hierarchy, in order to reduce its diameter, where the *diameter* of a directed graph is defined as the maximum distance between a pair of vertices in the graph connected by a path. To this aim, they proposed some constructions to add additional edges, called *shortcut edges*, as well as *dummy vertices*, to the hierarchy.

4.1 The Shortcutting Technique

The *shortcutting* of a directed graph $G = (V, E)$ consists into inserting *shortcut edges* in E, without changing the transitive closure of G. The goal is to obtain another directed graph, called a *shortcut graph*, having a smaller diameter than G.

The shortcutting technique is quite old, indeed it has been first considered in 1982 by Yao [17]. In particular, Yao considered the problem in a quite different context, where the n elements of V belong to a given semigroup $(S, \circ)$ and one is interested in answering queries of the form *"what is the value of* $v_i \circ v_{i+1} \circ \cdots \circ v_{j-1} \circ v_j$*?"* for any $1 \le i \le j \le n$. In the following we translate to our scenario the main existing results concerning the shortcutting technique when applied to particular kinds of graphs. We start discussing chains, then we analyze trees and finally general graphs.

Chains. By using the techniques proposed by Yao [17] we can add shortcut edges to a chain $(v_1, \ldots, v_n)$ of n vertices. The techniques proposed by Alon and Schieber [2] in 1987 and Bodlaender et al. [7] in 1994 are essentially the same as the ones proposed by Yao, but their description is easier to illustrate. The details of the constructions, translated to our scenario, can be found in [11]. The parameters of such constructions are summarized in Figure 2, where $\log^* n$, is the *iterated logarithmic function.*

Trees. In 1987 Chazelle [9], as well as Alon and Schieber [2], considered the problem of reducing the diameter of free trees, i.e., indirected connected acyclic graphs, by adding shortcut edges. Their results, which are summarized in Figure 2, were also shown to hold for directed trees [16].

Diameter ℓ	Minimal number of shortcut edges
1	$\Theta(n^2)$
2	$\Theta(n \cdot \log n)$
3	$\Theta(n \cdot \log \log n)$
4	$\Theta(n \cdot \log^* n)$
$O(\log^* n)$	$\Theta(n)$

Fig. 2. Minimal number of shortcut edges to be added to chains and trees with n vertices in order to obtain a shortcut graph with diameter ℓ

General Graphs. Thorup [15] conjectured that for any directed graph $G = (V, E)$ one can obtain a shortcut graph with diameter polylogarithmic in $|V|$, i.e., $(\log |V|)^{O(1)}$, by adding at most $|E|$ shortcut edges. He also showed his conjecture to be true for planar directed graphs [16]. However, Hesse [14] gave a counterexample to Thorup's conjecture. He showed how to construct a direct graph requiring the addition of $\Omega(|E| \cdot |V|^{1/17})$ shortcut edges to reduce its diameter below $\Theta(|V|^{1/17})$. By extending his construction to higher dimensions, it is possible to obtain graphs with $|V|^{1+\epsilon}$ edges that require the addition of $\Omega(|V|^{2-\epsilon})$ shortcut edges to reduce their diameter.

All constructions described in this section can be used to reduce key derivation time in hierarchical key assignment schemes. However, the result by Hesse [14] implies that key derivation time cannot be reduced *essentially* below $\Omega(|V|^2)$ for some kinds of graphs by adding only shortcut edges.

4.2 The Shortcutting and Point-Inserting Technique

Atallah et al. [5] also proposed a different technique to reduce the diameter of an access hierarchy. Such a technique consists of the addition of *dummy vertices*, as well as new edges, to the hierarchy. The idea is to obtain a new hierarchy such that there exists a path between two classes u and v in the old hierarchy if and only if there exists a path between u and v in the new one. The addition of dummy vertices results in a smaller number of new edges to be added to the hierarchy.

The technique makes use of the concept of *dimension* of a poset, originally defined by Dushnik and Miller [12]. The dimension of a poset $(V, \preceq)$ is the minimum number of total orders on V whose intersection is $(V, \preceq)$. It can also be seen as the smallest nonnegative integer d for which each $u \in V$ can be represented by a d-vector $(x_{u,1}, \ldots, x_{u,d})$ of integers such that $u \preceq v$ if and only if $x_{u,i} \leq x_{v,i}$, for any $i = 1, \ldots, d$, and any $u, v \in V$. There are efficient algorithms to test if a poset has dimension 1 or 2, but the problem of determining if a poset has dimension 3 is NP-complete. A poset has dimension one if and only if it is a total order.

When applied to a hierarchy with n classes and dimension d, the technique allows to reduce key derivation time to $2d + 1$, by adding $O(n \cdot d \cdot (\log n)^{d-1} \cdot \log\log n)$ dummy classes and new edges. In the following we show how to further reduce key derivation time. Our technique performs a further shortcutting of the graph obtained by Atallah et al.'s technique and allows key derivation time to be independent on d.

4.3 The Improved Shortcutting and Point-Inserting Technique

In this section we consider the problem of reducing the diameter of the graph obtained by the shortcutting and point-inserting technique, on input a poset $(V, \preceq)$ with dimension d. Our construction, which we refer in the following as the *Improved Shortcutting and Point-Inserting Technique (ISPIT)* is recursive, and for the base case $d = 1$ reduces to the construction proposed by Yao [17]. The construction for the case $d \geq 2$ is described in Figure 3. The input is a set of n d-dimensional points corresponding to the vertices in V; for each vertex $v \in V$, let $P_v^{(d)}$ be the corresponding point and let $V^{(d)} = \{P_v^{(d)} : v \in V\}$.

The number $DP(n, d)$ of dummy points added by the ISPIT is $DP(n, d) = 2 \cdot DP(\lceil n/2 \rceil, d) + DP(n, d-1) + \Theta(n)$, where $DP(n, 1) = 0$ and $DP(1, d) = 0$. Indeed, in order to construct $G^{(d)}$, the algorithm adds n dummy points, corresponding to the projections of the points in $V^{(d)}$ on the $(d-1)$-dimensional hyperplane M, plus $DP(n, d-1)$ dummy points for the construction of $G^{(d-1)}$, and then is recursively called on the two sets $V_1^{(d)}$ and $V_2^{(d)}$. The solution of the above recurrence is $DP(n) = \Theta(n \cdot d \cdot (\log n)^{d-1})$. On the other hand, the number $T(n)$ of new edges added by the ISPIT is $T(n, d) \leq 2 \cdot T(\lceil n/2 \rceil, d) + 3 \cdot T(n, d-1) + \Theta(n)$, where $T(n, 1)$ denotes the number of shortcut edges added by Yao's construction [17] for the case $d = 1$ in order to obtain a shortcut graph having a certain diameter, whereas, $T(1, d) = 0$. Indeed, at most $3 \cdot |E^{(d-1)}| + n$ new edges are added in steps 7. and 8. and then the algorithm is recursively called on the two sets $V_1^{(d)}$ and $V_2^{(d)}$. Clearly, the solution of $T(n, d)$, as well as the diameter of the graph

Let $(V, \preceq)$ be a poset with dimension $d \geq 2$, let $V^{(d)}$ be the set of points in the vectorial representation of the Hasse diagram associated to $(V, \preceq)$ and based on its d total orders, and let $\ell \geq 1$.

1. If $|V^{(d)}| = 1$, then output $V^{(d)}$.
2. If $|V^{(d)}| \geq 2$, compute a $(d-1)$-dimensional hyperplane M perpendicular to the d-th dimension that partitions the set of points in $V^{(d)}$ into two sets $V_1^{(d)}$ and $V_2^{(d)}$ of $\lfloor n/2 \rfloor$ and $\lceil n/2 \rceil$ points, respectively, where $V_1^{(d)}$ is the set on the smaller side of the hyperplane (according to the d-th coordinate). Such points are projected on M. Denote by $P_v^{(d-1)}$ the projection of $P_v^{(d)}$ on M. Let $V_1^{(d-1)}$ and $V_2^{(d-1)}$ be the projections of $V_1^{(d)}$ and $V_2^{(d)}$.
3. If $d = 2$, use Yao's construction on the chain whose vertices are the points in the set $V^{(1)}$, in order to obtain a shortcut graph $G^{(1)} = (V^{(1)}, E^{(1)})$, having diameter at most ℓ. The set of dummy points added by the algorithm is $D^{(1)} = \emptyset$ (no dummy points are added).
4. If $d \geq 3$, recursively call the algorithm on the set of points in $V^{(d-1)} = V_1^{(d-1)} \cup V_2^{(d-1)}$, corresponding to a $(d-1)$-dimensional hyperplane; let $G^{(d-1)} = (V^{(d-1)} \cup D^{(d-1)}, E^{(d-1)})$ be the corresponding output.
5. Let $D^{(d)} = V^{(d-1)} \cup D^{(d-1)}$.
6. Let $E^{(d)} = E^{(d-1)}$.
7. Add edges between points in $V^{(d)}$ and corresponding projections:
   (a) For each point $P_v^{(d)} \in V_1^{(d)}$, add an edge $(P_v^{(d-1)}, P_v^{(d)})$ to $E^{(d)}$.
   (b) For each point $P_v^{(d)} \in V_2^{(d)}$, add an edge $(P_v^{(d)}, P_v^{(d-1)})$ to $E^{(d)}$.
8. Add shortcut edges between points in $V^{(d)}$ and dummy points:
   (a) For each edge $(P_u^{(d-1)}, P_v^{(j)}) \in E^{(d-1)}$, add an edge $(P_u^{(d)}, P_v^{(j)})$ to $E^{(d)}$.
   (b) For each edge $(P_u^{(j)}, P_v^{(d-1)}) \in E^{(d-1)}$, add an edge $(P_u^{(j)}, P_v^{(d)})$ to $E^{(d)}$.
9. Recursively call the algorithm on the two sets of points in $V_1^{(d)}$ and $V_2^{(d)}$.
10. Output the graph $G^{(d)} = (V^{(d)} \cup D^{(d)}, E^{(d)})$.

Fig. 3. The Improved Shortcutting and Point-Inserting Technique (ISPIT)

$G^{(d)}$, depends on the the number $T(n, 1)$ of shortcut edges added by Yao's construction. If $T(n, 1) = \Theta(n)$, then $T(n, d) = O(n \cdot d \cdot (3 \log n)^{d-1})$. On the other hand, if $T(n, 1) = \Theta(n \cdot \log \log n)$, then $T(n, d) = O(n \cdot d \cdot (3 \log n)^{d-1} \cdot \log \log n)$ and the diameter of the graph $G^{(d)}$ is three, i.e., it is independent on d. It is easy to see that, for any two vertices $u, v \in V$ such that $u \preceq v$, there exists a path from $P_v^{(d)}$ to $P_u^{(d)}$ in $G^{(d)}$ which has length at most the diameter of the graph $G^{(1)}$ obtained by solving the 1-dimensional problem on $V^{(1)}$.

Compared to the technique in [5], the ISPIT allows a further reduction of the diameter, but in each recursive call, it adds at most three times the number of new edges added by that algorithm. In the following we show a trade-off between the number of edges added by the ISPIT and the diameter of the resulting graph. The idea behind the construction is the following: Assume the 1-dimensional problem is solved by adding $\Theta(n \log \log n)$ shortcut edges. For each $j = 2, \ldots, d$, the j-dimensional problem could be solved either with the technique in [5] or with ours. Let $1 \leq t \leq d$ and assume, for example, that for $j = 2, \ldots, t$, the technique in [5] is used to solve the j-dimensional problem, whereas, our technique is used to solve the problems with dimensions from $t+1$ to d. It is easy to see that the graph resulting by the above construction has diameter $2t+1$. Moreover, the number of new edges added by the modified algorithm is $O(3^{d-t} \cdot n \cdot t \cdot (\log n)^{d-1} \cdot \log \log n)$.

5 A Broadcast Encryption Based Construction

In this section we show how to construct a hierarchical key assignment scheme using as a building block a broadcast encryption scheme. A broadcast encryption scheme allows a sender to broadcast an encrypted message to a set of users in such a way that only legitimate users can decrypt it. Broadcast encryption schemes can be either public key or symmetric key based. In the symmetric key setting, only a trusted authority can broadcast data to the receivers. In contrast, in the public key setting a public key published by a trusted authority allows anybody to broadcast a message. We first recall the definition of a public-key broadcast encryption scheme [8].

Definition 4. *A* public-key broadcast encryption scheme *for a set $\mathcal{U}$ of users is a triple of algorithms (Set, Enc, Dec) satisfying the following conditions:*

1. *The* setup algorithm *Set is probabilistic polynomial-time. It takes as input a security parameter 1^τ and the set of users $\mathcal{U}$ and produces as output a private key sk_u, for each user $u \in \mathcal{U}$, and a public key pk.*
2. *The* encryption algorithm *Enc is probabilistic polynomial-time. It takes as inputs 1^τ, a subset $X \subseteq \mathcal{U}$, and the public key pk, and produces as output a pair (Hdr, k), where Hdr is called the broadcast header and k is a encryption key. Let m be a message to be broadcast in such a way that only users in X are allowed to obtain m and let y be the encryption of m under the symmetric key k. The broadcast message consists of (X, Hdr, y), where the pair (X, Hdr) is called the full broadcast header and y is called the broadcast body.*
3. *The* decryption algorithm *Dec is deterministic polynomial-time. It takes as inputs 1^τ, a subset $X \subseteq \mathcal{U}$, a user $u \in X$ and its private key sk_u, a broadcast header Hdr, and the public key pk, and produces as output the key k. Such a key can be used to decrypt the broadcast body y in order to obtain m.*

We require that for all subsets $X \subseteq \mathcal{U}$, all users $u \in X$, all public keys and private keys which can be output by $Set(1^\tau, \mathcal{U})$, all pairs (Hdr, k), which can be output by $Enc(1^\tau, X, pk)$, we have that $Dec(1^\tau, X, u, sk_u, Hdr, pk) = k$.

The idea behind our construction, referred in the following as the *Broadcast Encryption Based Construction (BEBC)*, is to compute the private and public information by using the broadcast encryption scheme; more precisely, the public information will contain a broadcast header Hdr_u, which corresponds to an encryption of the key k_u, for each class $u \in V$. Such a broadcast header can be decrypted by all classes in the set $I_u = \{v \in V : \text{ there is a path from } v \text{ to } u \text{ in } G\}$, allowing them to compute the key k_u. The Broadcast Encryption Based Construction is described in Figure 4.

Let Γ be a family of graphs corresponding to partially ordered hierarchies. Let $G = (V, E) \in \Gamma$ and let (Set, Enc, Dec) be a public-key broadcast encryption scheme for users in V.

Algorithm $Gen(1^\tau, G,)$

1. Run $Set(1^\tau, V)$ to generate a public key pk and a secret key sk_u for any $u \in V$;
2. For each class $u \in V$, let $s_u = sk_u$;
3. For each class $u \in V$, run $Enc(1^\tau, I_u, pk)$ to obtain the pair (Hdr_u, k_u);
4. Let s and k be the sequences of private information and keys computed in the previous two steps;
5. Let pub be the sequence constituted by the public key pk along with the header Hdr_u, for any $u \in V$;
6. Output (s, k, pub).

Algorithm $Der(1^\tau, G, u, v, s_u, pub)$

1. Extract the public key pk and the header Hdr_v from pub.
2. Output $k_v = Dec(1^\tau, I_v, u, s_u, Hdr_v, pk)$.

Fig. 4. The Broadcast Encryption Based Construction

Before analyzing the security of the BEBC we first need to define what we mean by a *secure* public-key broadcast encryption scheme. The security of a public-key broadcast encryption scheme is defined through a game between an adversary A and a challenger. According to the capabilities of the adversary and the security goal, several types of security notions for public-key broadcast can be defined. We consider the definition of *semantic security* given by Boneh et al. [8], where the adversary is not allowed to issue decryption queries to the challenger. We consider the following game: First, algorithm A outputs a set $X \subseteq \mathcal{U}$ of receivers that it wants to attack. Then, the challenger first runs $Set(1^\tau, \mathcal{U})$ to obtain a private key sk_u for each user $u \in \mathcal{U}$ and a public key pk. Afterwards, it gives the public key pk and all private keys sk_v for which $v \notin X$ to A. The challenger runs $Enc(1^\tau, X, pk)$ to obtain (Hdr, k). Then, it picks a random bit $b \in \{0, 1\}$, sets $k_b = k$ and chooses $k_{\overline{b}}$ as a random key. The challenge (Hdr, k_0, k_1) is given to A. Algorithm A outputs its guess $b' \in \{0, 1\}$ for b and wins the game if $b = b'$. The advantage of the adversary A is defined as $\mathbf{Adv}_{A,\mathcal{U}}(1^\tau) = |Pr[b' = b] - 1/2|$. The scheme is said to be *semantically secure* if the function $\mathbf{Adv}_{A,\mathcal{U}}(1^\tau)$ is negligible, for any adversary A whose time complexity is polynomial in τ. The proof of the next theorem can be found in [11].

Theorem 2. *If the public-key broadcast encryption scheme* (Set, Enc, Dec) *is semantically secure, then the BEBC is secure in the sense of* `IND-ST`.

Boneh et al. [8] showed how to construct a semantically secure public-key broadcast encryption scheme for a set of n users, assuming the intractability of the n*-Bilinear Decisional Diffie-Hellman Exponent* (n-`BDDHE`). The use of such a broadcast encryption scheme allows us to obtain a hierarchical key assignment scheme where the public information consists of $4|V|+1$ group elements, whereas, the private information has constant size. Moreover, key derivation requires a single (complex) decryption operation, which involves at most $|V|-2$ group operations. The above scheme supports dynamic changes to the hierarchy without requiring re-distribution of private information to the classes affected by such changes. Details of the construction can be found in [11].

References

1. Akl, S.G., Taylor, P.D.: Cryptographic Solution to a Problem of Access Control in a Hierarchy. ACM Trans. on Comput. Syst. 1(3), 239–248 (1983)
2. Alon, N., Schieber, B.: Optimal Preprocessing for Answering On-line Product Queries, Tech. Rep, TR 71/87, Inst. of Comput. Sci., Tel-Aviv Univ. (1987)
3. Atallah, M.J., Frikken, K.B., Blanton, M.: Dynamic and Efficient Key Management for Access Hierarchies. In: Proc. of ACM CCS 2005, pp. 190–201(2005)
4. Atallah, M.J., Blanton, M., Fazio, N., Frikken, K.B.: Dynamic and Efficient Key Management for Access Hierarchies, CERIAS Tech. Rep. TR 2006-09, Purdue Univ. (2006)
5. Atallah, M.J., Blanton, M., Frikken, K.B.: Key Management for Non-Tree Access Hierarchies. In: Proc. of ACM SACMAT 2006, pp. 11–18, Full version avail at `http://www.cs.purdue.edu/homes/mbykova/papers/key-derivation.pdf`
6. Ateniese, G., De Santis, A., Ferrara, A.L., Masucci, B.: Provably-Secure Time-Bound Hierarchical Key Assignment Schemes. In: Proc. of ACM CCS 2006, pp. 288–297. Full version avail. as Rep. 2006/225 at the IACR Cryptology ePrint Archive (2006)
7. Bodlaender, H.L., Tel, G., Santoro, N.: Trade-offs in Non-reversing Diameter. Nordic J. on Comput. 1, 111–134 (1994)
8. Boneh, D., Gentry, C., Waters, B.: Collusion Resistant Broadcast Encryption with Short Ciphertexts and Private Keys. In: Shoup, V. (ed.) CRYPTO 2005. LNCS, vol. 3621, pp. 258–275. Springer, Heidelberg (2005)
9. Chazelle, B.: Computing on a Free Tree via Complexity-Preserving Mappings. Algorithmica 2, 337–361 (1987)
10. Crampton, J., Martin, K., Wild, P.: On Key Assignment for Hierarchical Access Control. In: Proc. of IEEE CSFW, pp. 98–111 (2006)
11. De Santis, A., Ferrara, A.L., Masucci, B.: Efficient Provably-Secure Hierarchical Key Assignment Schemes, avail. as Rep. 2006/479 at the IACR Cryptology ePrint Archive (2006)
12. Dushnik, B., Miller, E.W.: Partially Ordered Sets. American Journal of Mathematics 63, 600–610 (1941)
13. Goldwasser, S., Micali, S.: Probabilistic Encryption. Journal of Comp. and System Sci. 28, 270–299 (1984)
14. Hesse, W.: Directed Graphs Requiring Large Number of Shortcuts. In: Proc. of ACM-SIAM SODA 2003, pp. 665–669 (2003)
15. Thorup, M.: On Shortcutting Digraphs. In: Mayr, E.W. (ed.) WG 1992. LNCS, vol. 657, pp. 205–211. Springer, Heidelberg (1993)
16. Thorup, M.: Shortcutting Planar Digraphs. Combinatorics, Probability & Comput. 4, 287–315 (1995)
17. Yao, A.C.: Space-Time Tradeoff for Answering Range Queries. In: Proc. of ACM STOC 1982, pp. 128–136 (1982)

Nearly Private Information Retrieval*

Amit Chakrabarti and Anna Shubina

Department of Computer Science, Dartmouth College
Hanover, NH 03755, USA
{ac,ashubina}@cs.dartmouth.edu

Abstract. A private information retrieval scheme is a protocol whereby a client obtains a record from a database without the database operators learning anything about which record the client requested. This concept is well studied in the theoretical computer science literature. Here, we study a generalization of this idea where we allow a small amount of information about the client's intent to be leaked.

Despite having relaxed the privacy requirement, we are able to prove three fairly strong lower bounds on such schemes, for various parameter settings. These bounds extend previously known lower bounds in the traditional setting of perfect privacy and, in one case, improve upon the previous best result that handled imperfect privacy.

1 Introduction

Private information retrieval (PIR) schemes have been a substantial focus of theoretical research in computer science, beginning with the highly influential work of Chor, Goldreich, Kushilevitz, and Sudan [4]. In that paper, as in most subsequent work, a *PIR scheme* means a communication protocol that specifies an interaction between a *client* or *user* and one or more database *servers*. The user wishes to obtain a record from the database without the servers learning anything about which record the user seeks.

A clean and concrete version of this problem, as proposed by Chor et al., is as follows: the database y is a string of n bits. The client has an index $j \in \{1, 2, \ldots, n\}$ and wishes to obtain the jth bit of y, without the servers obtaining any information about j. As shown by Chor et al., this strong privacy requirement means that if there is only one server that holds the database, the trivial protocol in which the client simply downloads the entire database is optimal in terms of the number of bits communicated. However, as shown in the same paper, if one allows the database to be replicated and copies held by two or more servers that do not talk to each other, the problem can be solved using sublinear communication.

Almost all past work on PIR schemes has required that the servers learn *zero* information about the client's index j. Here, we ask the question: what happens if we allow the protocol to leak a small amount of information about j? To the best of our knowledge, the only other work to have considered this question is that

* This work was supported in part by an NSF CAREER Award CCF-0448277 and startup funds from Dartmouth College.

L. Kučera and A. Kučera (Eds.): MFCS 2007, LNCS 4708, pp. 383–393, 2007.

of Goldreich, Karloff, Schulman, and Trevisan [5]. It is *a priori* conceivable that relaxing the privacy requirement might decrease the communication required in PIR protocols. However, in this work, we prove three lower bounds that show that previously known lower bounds for traditional (perfect privacy) PIR protocols extend to this relaxed setting, up to constant factor losses. One of our bounds improves an earlier result of Goldreich et al. from the aforementioned paper. We also show that another of our bounds is essentially optimal by exhibiting an appropriate upper bound.

To explain our results in detail and compare them to previously known results, we begin with some necessarily definitions.

1.1 Preliminaries

We begin by introducing some notation. For an integer n, we let $[n]$ denote the set $\{1, 2, \ldots, n\}$. For random variables X and Y that take values in the same set S, we write $X \approx_\delta Y$ to denote the fact that the L_1 distance (i.e., twice the variational distance) between the distributions of X and Y is at most δ. To be precise,

$$\sum_{a \in S} |\Pr[X = a] - \Pr[Y = a]| \leq \delta .$$

Definition 1.1. *Let s, ℓ_q, ℓ_a be positive integers and ε, δ be reals in $[0, 1]$. An s-server $(\ell_q, \ell_a; \varepsilon, \delta)$-PIR protocol $\mathcal{P}$ is a communication protocol between a single client, who holds an index $j \in [n]$, and s servers, each of whom holds a string $y \in \{0, 1\}^n$. Formally, $\mathcal{P}$ is specified by a triple (Q, A, Rec) of functions, where $Q : [s] \times [n] \times \{0, 1\}^\rho \to \{0, 1\}^{\ell_q}$, $A : [s] \times \{0, 1\}^n \times \{0, 1\}^{\ell_q} \to \{0, 1\}^{\ell_a}$, and $\mathrm{Rec} : [n] \times \{0, 1\}^\rho \times (\{0, 1\}^{\ell_a})^s \to \{0, 1\}$ for some positive integer ρ. For compactness of notation, we shall write $Q_i(j, R)$ instead of $Q(i, j, R)$ and $A_i(y, z)$ instead of $A(i, y, z)$. Also, we shall drop the subscript on Q_i and A_i altogether when $s = 1$. The protocol operates as follows: the client generates a random string R distributed uniformly in $\{0, 1\}^\rho$ and, for each $i \in [s]$, sends a* query *string $Q_i(j, R)$ to server i. Upon receiving a query string z, server i sends an* answer *string $A_i(y, z)$ to the client. The client then outputs a* recovered bit

$$\mathrm{Out}(j, y, R) \ := \ \mathrm{Rec}(j, R, A_1(y, Q_1(j, R)), \ldots, A_s(y, Q_s(j, R))) ,$$

which is her guess at the value of y_j. The protocol must satisfy the following two conditions.

Correctness: $\forall j \in [n], y \in \{0, 1\}^n : \ \Pr_R[\mathrm{Out}(j, y, R) = y_j] \geq 1 - \varepsilon$.
Privacy: $\forall i \in [s], j, k \in [n] : \ Q_i(j, R) \approx_\delta Q_i(k, R)$.

The parameter ℓ_q is called the query length, *ℓ_a the* answer length, *ε the* recovery error, *and δ the* privacy parameter *of the protocol $\mathcal{P}$. The communication cost of $\mathcal{P}$ is $\mathrm{cost}(\mathcal{P}) = s(\ell_q + \ell_a)$, the total number of bits communicated.*

The goal in designing PIR protocols is to simultaneously reduce ε, δ, and $\mathrm{cost}(\mathcal{P})$. We shall require that all servers receive queries of the same length and return answers of the same length. Since we only deal with constant values of s, this requirement causes no asymptotic loss.

When $\varepsilon = 0$, the protocol is said to have *perfect recovery*, and when $\delta = 0$, it is said to have *perfect privacy*. The bulk of theoretical research on PIR has focused on the case $\varepsilon = \delta = 0$. The work of Goldreich et al. [5] and that of Kerenidis and de Wolf [6] did consider the $\varepsilon > 0$ case. But relatively little attention has been paid to the $\delta > 0$ case, except for one result of Goldreich et al. mentioned below.

1.2 Our Results and Previous Work

We prove three lower bounds that allow $\delta > 0$. Let P be a 1-server $(\ell_q, \ell_a; \varepsilon, \delta)$-PIR protocol. With the privacy requirement relaxed, even the 1-server case becomes nontrivial and it is not *a priori* clear that sublinear communication PIR is not possible. However, we show that for $\varepsilon = 0$, we must have $\mathrm{cost}(P) \geq (1 - \delta/2)\, n = \Omega(n)$. We also show, via an upper bound, that this dependence to δ is essentially tight, up to terms quadratic in δ.

We also consider the more general case when both ε and δ can be nonzero. In this case, we show that $\mathrm{cost}(P) \geq (1 - H(\varepsilon + \delta/2))\, n$ for sufficiently small ε and δ. Here H is the binary entropy function given by $H(x) := -x \lg x - (1-x)\lg(1-x)$; "lg" denotes logarithm to the base 2.

Finally, we consider 2-server schemes for nearly private information retrieval. It is known that, using two servers, $O(n^{1/3})$ communication can be achieved, even with $\varepsilon = \delta = 0$, via a number of different schemes; see, e.g., Chor et al. [4], Beimel, Ishai, and Malkin [3], and Woodruff and Yekhanin [11]. No strong general lower bound is known that comes close to matching this upper bound. However, a recent result of Razborov and Yekhanin [10] provides an $\Omega(n^{1/3})$ bound for protocols whose computation is restricted in a certain way. With arbitrary computations allowed, there *are* strong lower bounds known provided the answer length ℓ_a is short. The cleanest of these results are for the $\ell_a = 1$ case. In this case, Kerenidis and de Wolf [6] prove a lower bound of $(1 - H(11/14 - 4\varepsilon/7))\, n - 2$ on the communication cost when $\delta = 0$. Beigel, Fortnow, and Gasarch [2] prove a tight $n - 2$ lower bound when $\varepsilon = \delta = 0$.

Here, we prove a lower bound of $(1 - H(3/4 + 2\delta/3 - \sqrt{2\delta} - \varepsilon))\, n - 2$ when $\ell_a = 1$, for sufficiently small positive ε and δ. A lower bound handling positive ε and δ was proven by Goldreich et al. [5]. Their bound, for $\ell_a = 1$, is $(1 - 2\varepsilon - \delta)\, n/24 - 4$. (Note that our use of ε and δ is different from theirs; we have translated their bound into our notation.) To see that our bound is an improvement, consider the limiting case $\varepsilon \to 0, \delta \to 0$: our lower bound then approaches $0.19n - 2$, whereas the bound of [5] approaches $0.04n - 4$.

It is worth noting that the issue of lower bounds for PIR schemes with 3 or more servers has recently been largely settled, in a most dramatic way, by Yekhanin [12]: surprisingly low *upper* bounds hold.

2 Simple Upper Bounds

Here, we show simple improvements to the known upper bounds on the communication cost of PIR schemes by allowing imperfect privacy. As we shall see later, the 1-server upper bound we obtain below is essentially optimal in the perfect recovery case.

Theorem 2.1. *For any $\delta > 0$, there is a PIR protocol with perfect recovery, privacy parameter δ, and communication cost at most $\lceil \lg n \rceil + \lceil (1 - \delta/(2+\delta))\, n \rceil = (1 - \delta/2 + O(\delta^2))\, n + O(\log n)$.*

Proof. Let $\delta' = \delta/(2+\delta)$. For each integer $j \in [n]$, define the sets

$$S_j := \{k \in [n] : 0 \le (k - j) \bmod n \le (1 - \delta')n\}\,,$$
$$T_j := \{k \in [n] : 0 \le (j - k) \bmod n \le (1 - \delta')n\}\,.$$

It is important to keep in mind that $[n]$ denotes the set $\{1, 2, \ldots, n\}$ whereas $x \bmod n$ takes values in $\{0, 1, \ldots, n-1\}$.

Design the function Q so that, when R is a uniform random string, $Q(j, R)$ is uniformly distributed on S_j. For $k \in [n]$ and $y \in \{0,1\}^n$, let $A(y, k)$ return the concatenation, in some canonical order, of all y_j such that $j \in T_k$. It is easy to see that $k \in S_j \Leftrightarrow j \in T_k$; therefore $A(y, Q(j, R))$ is guaranteed to contain the desired bit y_j and we can design Rec so as to recover y_j from $Q(j, R)$ and $A(y, Q(j, R))$. Clearly, the PIR protocol given by (Q, A, Rec) has perfect recovery and communication cost at most $\lceil \lg n \rceil + |T_k| \le \lceil \lg n \rceil + \lceil (1 - \delta')n \rceil$.

For all $j \in [n]$, we have $|S_j| \ge (1 - \delta')n$ and for $i \ne j$, we have $|S_i \setminus S_j| + |S_j \setminus S_i| \le 2 \cdot |[n] \setminus S_j| \le 2\delta' n$. Therefore, we can bound the protocol's privacy parameter as follows:

$$Q(i, R) \approx_{\delta''} Q(j, R), \quad \text{where } \delta'' \le \frac{2\delta' n}{(1 - \delta')n} = \delta\,.$$

Thus, the protocol has all the desired properties.

A 2-Server Upper Bound. In a similar manner to the 1-server case, it is possible to add a δ-dependent coefficient to the $O(n^{1/3})$ upper bound for 2-server PIR. The idea is to suitably modify the covering codes scheme of Chor et al. [4]. The details are straightforward and hence omitted from this version.

3 1-Server Lower Bounds

3.1 Perfect Privacy and Recovery

Chor et al. [4] prove that, in the 1-server case with perfect privacy, n bits must be exchanged. Their argument goes as follows. A communication C (the string of exchanged bits) is said to be *possible* for (y, j) if there is a positive probability for C to happen when the database is y, and the user tries to obtain the jth bit. C is said to be *possible* for j if it is possible for some pair (y, j). Let us fix a j and assume that the number of possible communications for j is less than 2^n. Then there exist different databases y, y' and C such that C is possible for both (y, j) and (y', j). But by the privacy requirement, for every $k \in [n]$, C must also be possible for (y, k) and (y', k), since the queries are distributed equally, and the responses are determined by the queries. Pick an index j such that $y_j \neq y'_j$. We know that C is possible for both (y, j) and (y', j), but C determines the output of the protocol, thus the protocol must yield the same bit, and we get a contradiction.

This argument fails in the almost secure case, since there is no requirement that the same communication be possible for all indices if it is possible for one. However, we can still obtain strong lower bounds, as we now show.

3.2 Nearly Private Schemes

Theorem 3.1. *Let $\mathcal{P}$ be a 1-server $(\ell_q, \ell_a; 0, \delta)$-PIR protocol, where $\delta > 0$. Then $\ell_a \geq (1 - \delta/2)n$. In particular,* $\mathrm{cost}(\mathcal{P}) \geq (1 - \delta/2)n$.

Proof. For $j \in [n]$ and $z \in \{0,1\}^{\ell_q}$, let $p_{jz} = \Pr_R[Q(j, R) = z]$. Let $J_z = \{j : p_{jz} > 0\}$. It is easy to verify that

$$|J_z| p_{1z} \;\geq\; \sum_{j=1}^{n} \min\{p_{1z}, p_{jz}\} \;=\; \sum_{j=1}^{n} \left(\frac{p_{1z} + p_{jz}}{2} - \frac{|p_{1z} - p_{jz}|}{2} \right).$$

This implies

$$\begin{aligned} \sum_{z \in \{0,1\}^{\ell_q}} |J_z| p_{1z} \;&\geq\; \sum_{j=1}^{n} \sum_{z \in \{0,1\}^{\ell_q}} \left(\frac{p_{1z} + p_{jz}}{2} - \frac{|p_{1z} - p_{jz}|}{2} \right) \\ &\geq\; \sum_{j=1}^{n} \left(\frac{1+1}{2} - \frac{\delta}{2} \right) \;=\; (1 - \delta/2)n\,, \end{aligned}$$

where the final inequality follows from the privacy guarantee of $\mathcal{P}$. Since we have $\sum_{z \in \{0,1\}^{\ell_q}} p_{1z} = 1$, there must exist a $z \in \{0,1\}^{\ell_q}$ such that $|J_z| \geq (1 - \delta/2)n$. Fix such a z.

Suppose $\ell_a < |J_z|$. Let $Y := \{y \in \{0,1\}^n : y_j = 0 \text{ for } j \notin J_z\}$. Then $|Y| = 2^{|J_z|}$. Meanwhile, the string $A(y, z)$ has length ℓ_a, so it lies in a set of size $2^{\ell_a} < 2^{|J_z|}$. By the pigeonhole principle, there exist distinct strings $y, y' \in Y$ such that $A(y, z) = A(y', z)$. Let j be an index such that $y_j \neq y'_j$. Then $j \in J_z$. Therefore, $p_{jz} > 0$, i.e., there exists an R such that $Q(j, R) = z$. Since $\mathcal{P}$ has perfect recovery, for this R we must have

$$y_j \;=\; \mathrm{Rec}(j, R, A(y, z)) \;=\; \mathrm{Rec}(j, R, A(y', z)) \;=\; y'_j\,,$$

which is a contradiction. This proves that $\ell_a \geq |J_z| \geq (1 - \delta/2)n$.

3.3 Nearly Private Schemes with Imperfect Recovery

We now turn to the imperfect recovery case. We prove our lower bound for this case by a reduction from a communication problem with a well known lower bound. Later, we use a much more sophisticated version of the same idea for a 2-server lower bound.

The problem INDEX_n is a communication problem involving two players: Alice, who holds an n-bit string $x = x_1 x_2 \ldots x_n$ (with each $x_i \in \{0, 1\}$), and Bob, who holds an index $i \in [n]$. A one-way communication protocol for this problem operates as follows: Alice sends Bob a message based on x after which Bob outputs his guess at the bit x_i. Both players may use a public random string in making their decisions, i.e., the protocol is allowed to be public coin. Ablayev [1] proved the following sharp lower bound on the communication cost of such a protocol.

Fact 3.2. *Any public coin one-way communication protocol for* INDEX_n *with error at most* ε *must communicate at least* $(1 - H(\varepsilon))\, n$ *bits.*

Theorem 3.3. *Let* ε *and* δ *be positive reals with* $\varepsilon + \delta/2 < 1/2$. *Then any* 1-*server* $(\ell_q, \ell_a; \varepsilon, \delta)$-*PIR protocol has* $\ell_a \geq (1 - H(\varepsilon + \delta/2))\, n$. *In particular, the communication cost of such a protocol is at least* $(1 - H(\varepsilon + \delta/2))\, n$.

Proof. Suppose P is a 1-server $(\ell_q, \ell_a; \varepsilon, \delta)$-PIR protocol that uses ρ bits of randomness. Let $\mathcal{D}_{jz}$ denote the conditional distribution of R given that $Q(j, R) = z$ and let $\text{Gen} : [n] \times \{0,1\}^{\ell_q} \times \{0,1\}^{\rho'} \to \{0,1\}^{\rho}$ be such that $\text{Gen}(j, z, R')$ is distributed according to $\mathcal{D}_{jz}$ when R' is distributed uniformly in $\{0,1\}^{\rho'}$. Further, define $f : [n] \times \{0,1\}^n \times \{0,1\}^{\ell_q} \times \{0,1\}^{\rho'} \to \{0,1\}$ as follows.

$$f(j, y, z, r') := \begin{cases} 0, & \text{if } \text{Rec}(j, \text{Gen}(j, z, r'), A(y, z)) = y_j\,, \\ 1, & \text{otherwise}. \end{cases}$$

The correctness condition for P implies

$$\begin{aligned} \mathrm{E}_{R,R'}[f(j, y, Q(j, R), R')] &= \Pr_{R,R'}[\text{Rec}(j, \text{Gen}(j, Q(j, R), R'), A(y, Q(j, R))) \neq y_j] \\ &= \Pr_{R}[\text{Rec}(j, R, A(y, Q(j, R))) \neq y_j] \\ &\leq \varepsilon\,. \end{aligned}$$

Now, using the privacy condition $Q(j, R) \approx_\delta Q(1, R)$ and the fact that R and R' are independent, we have

$$\mathrm{E}_{R,R'}[f(j, y, Q(1, R), R')] \;\leq\; \varepsilon + \frac{\delta}{2}\,.$$

In other words, the following is a public coin one-way communication protocol for the problem INDEX_n, with error at most $\varepsilon + \delta/2$. Alice and Bob share a pair of random strings (R, R') distributed uniformly in $\{0,1\}^{\rho} \times \{0,1\}^{\rho'}$. Alice, upon receiving y, sends Bob the message $\mu := A(y, Q(1, R))$. Bob, upon receiving j and μ, outputs $\text{Rec}(j, \text{Gen}(j, Q(1, R), R'), \mu)$ as his guess at y_j. Clearly, this protocol has cost at most ℓ_a. By Fact 3.2, we have $\ell_a \geq (1 - H(\varepsilon + \delta/2))\, n$, which completes the proof.

4 2-Server Lower Bounds

We now turn to the case of 2-server PIR protocols. As mentioned earlier, much less is known about lower bounds for such protocols. In particular, the only strong lower bounds known for protocols that may make arbitrary computations are when the answer size is restricted to be quite small. In particular, there are strong results known for the case of one-bit answers. Here, we prove an asymptotically optimal lower bound for the case of one-bit answers, with imperfect privacy allowed.

Our proof uses a *quantum computation* framework first used by Kerenidis and de Wolf [6]. Below, we quickly review the basics of quantum computation and communication and the Kerenidis - de Wolf framework and argument. We then show how to extend the framework to allow imperfect privacy. For an in-depth explanation of quantum computation we refer the reader to the textbooks by Nielsen and Chuang [9] and by Kitaev, Shen and Vyalyi [7].

4.1 Quantum Communication

For our purposes, a quantum state is to be thought of as analogous to the classical notion of a probability distribution over fixed-length bit strings. A distribution over n-bit strings can be thought of a vector in $[0,1]^{2^n}$ with unit ℓ_1-norm. Analogously, an n-qubit state is a vector in $\mathbb{C}^{2^n} = (\mathbb{C}^2)^{\otimes n}$ with unit ℓ_2-norm. We fix an orthonormal basis for the Hilbert space $(\mathbb{C}^2)^{\otimes n}$ and label the 2^n basis vectors (called basis states) by the 2^n n-bit strings: it is customary to use Dirac notation and denote the vector labeled by the string a as $|a\rangle$. It is also customary to write, e.g., $|5\rangle$ for the 3-qubit state $|101\rangle$ because "101" is the binary representation of 5.

An n-qubit quantum state can evolve by the application of a unitary transformation in $U(2^n)$. It can also be measured in a variety of ways whose details need not concern us here. For our purposes, we need only consider the following type of measurement. Suppose we have a decomposition $(\mathbb{C}^2)^{\otimes n} = \mathcal{W}_1 \oplus \mathcal{W}_2 \oplus \cdots \oplus \mathcal{W}_k$, and suppose W_j denotes the projection onto $\mathcal{W}_j$. Then we can measure an n-qubit state $|\phi\rangle$ according to this decomposition: we will obtain a random outcome in the set $[k]$, with the probability of outcome j being $\|W_j|\phi\rangle\|_2^2 = \langle\phi|W_j|\phi\rangle$.

A quantum communication protocol is like a (classical) communication protocol except that the communicating parties may send qubits (i.e., quantum states) to each other. The communication cost of a protocol is the number of qubits sent.

4.2 Perfect Privacy

Kerenidis and de Wolf prove a number of communication lower bounds for 2-server PIR schemes. However, their arguments only handle the perfect privacy case, although they do handle imperfect recovery. Their arguments are cast in a quantum communication framework whose key observation can be expressed thus: "a single quantum query can simulate two classical queries." Using this observation, they build a 1-server "quantum PIR scheme" and then prove lower bounds on its communication in a way analogous to our 1-server lower bounds. In particular, the appropriate quantum analog of Ablayev's lower bound (Fact 3.2) turns out to be a lower bound for quantum random access codes, due to Nayak [8].

We now outline Kerenidis and de Wolf's argument, using our own terminology. We find it convenient to remove the intermediate steps of a quantum PIR scheme and a quantum random access code; instead, we show that a 2-server PIR scheme with good enough parameters implies a one-way *quantum* communication protocol for INDEX_n with low communication cost. The desired PIR lower bound then follows from the aforementioned result of Nayak [8], which can be restated thus.

Fact 4.1. *A one-way quantum communication protocol for* INDEX_n *with error probability* ε *must communicate at least* $(1 - H(\varepsilon))\, n$ *qubits.*

We now fill in some details. Suppose P is a 2-server $(\ell_q, 1; \varepsilon, \delta)$-PIR protocol, given by (Q, A, Rec), that uses ρ bits of randomness. We associate with P a certain collection $\{|\phi_{jy}\rangle\}$ of $(\rho + 4 + \ell_a)$-qubit quantum states. To define these, we use the basis states $\{|r, i, i, z\rangle : r \in \{0,1\}^\rho, i \in \{0,1,2\}, z \in \{0,1\}^{\ell_q}\}$. We set $c := 1/\sqrt{3 \cdot 2^\rho}$ and, for notational convenience, we define $Q_0(j, r) = 0^{\ell_q}$ and $A_0(y, z) = 0$ for all $j \in [n], r \in \{0,1\}^\rho, y \in \{0,1\}^n$ and $z \in \{0,1\}^{\ell_q}$. Also, for $(i, j, z) \in \{0,1,2\} \times [n] \times \{0,1\}^{\ell_q}$, we define the set $S_{ijz} := \{r \in \{0,1\}^\rho : Q_i(j, r) = z\}$. Finally, we define $|\phi_{jy}\rangle$ as follows:

$$|\phi_{jy}\rangle := \sum_{r \in \{0,1\}^\rho} c\,|r\rangle \Big(|0, 0, 0^{\ell_q}\rangle + (-1)^{A_1(y, Q_1(j,r))} |1, 1, Q_1(j, r)\rangle + (-1)^{A_2(y, Q_2(j,r))} |2, 2, Q_2(j, r)\rangle \Big) .$$

The significance of this quantum state is brought out by the following fact, implicit in the work of Kerenidis and de Wolf.

Fact 4.2 (Kerenidis and de Wolf [6]). *By measuring* $|\phi_{jy}\rangle$ *appropriately, one can obtain a random 2-bit outcome* (β_1, β_2) *such that*

$$\Pr\left[(\beta_1, \beta_2) = (A_1(y, Q_1(j, r)), A_2(y, Q_2(j, r)))\right] \;\geq\; 3/4 .$$

Therefore, by applying the function Rec *to the measured outcome, one can obtain a bit that equals* y_j *with probability at least* $3/4 - \varepsilon$. *In fact, the probability of correctly recovering* y_j *can be further improved to* $11/14 - 4\varepsilon/7$ *by using a (classical) postprocessing trick.*

To see how this fact can be used to obtain the desired communication protocol, note that

$$\begin{aligned}
|\phi_{jy}\rangle &= \sum_{r \in \{0,1\}^\rho} \sum_{i=0}^{2} (-1)^{A_i(y, Q_i(j,r))} c\,|r, i, i, Q_i(j, r)\rangle \\
&= \sum_{i=0}^{2} \sum_{z \in \{0,1\}^{\ell_q}} \sum_{r \in S_{ijz}} (-1)^{A_i(y,z)} c\,|r, i, i, z\rangle \\
&= \sum_{i=0}^{2} \sum_{z \in \{0,1\}^{\ell_q}} |\chi_{ijz}\rangle \cdot (-1)^{A_i(y,z)} c \sqrt{|S_{ijz}|}\; |i, z\rangle ,
\end{aligned}$$

where $|\chi_{ijz}\rangle := |S_{ijz}|^{-1/2} \sum_{r \in S_{ijz}} |r, i\rangle$. Let U_j be a unitary transformation that maps $|0^\rho, 0, i, z\rangle$ to $|\chi_{ijz}\rangle|i, z\rangle$. The protocol for INDEX_n works as follows. Alice, on input y, prepares the quantum state

$$|\psi_{jy}\rangle := \sum_{i=0}^{2} \sum_{z \in \{0,1\}^{\ell_q}} (-1)^{A_i(y,z)} c \sqrt{|S_{ijz}|}\, |i, z\rangle \tag{1}$$

and sends it to Bob. Although it seems at first glance that $|\psi_{jy}\rangle$ depends on j, it in fact doesn't, because the perfect privacy guarantee of P implies that for $j, k \in [n]$,

$$\frac{|S_{ijz}|}{2^\rho} = \Pr_R[Q_i(j, R) = z] = \Pr_R[Q_i(k, R) = z] = \frac{|S_{ikz}|}{2^\rho}. \tag{2}$$

Bob, upon receiving $|\psi_{jy}\rangle$, constructs the state $|0^\rho, 0\rangle|\psi_{jy}\rangle$ using ρ qubits of his own and applies U_j to it. By definition of U_j, the state that Bob obtains is $|\phi_{jy}\rangle$. He then uses the procedure implied by Fact 4.2 to compute his output bit, which is correct with probability at least $11/14 - 4\varepsilon/7$. Since $|\psi_{jy}\rangle$ is a $(2 + \ell_q)$-qubit state, the communication cost of this protocol is $2 + \ell_q$. Fact 4.1 now implies that $\ell_q \geq (1 - H(11/14 - 4\varepsilon/7))\, n - 2$, giving us a lower bound on $\text{cost}(P)$.

4.3 The Nearly Private Case

Without perfect privacy, the argument above does not work. This is because Eq. (2) no longer holds, which makes the above quantum communication protocol ill-defined: Alice can no longer prepare the state $|\psi_{jy}\rangle$ because it *might* depend on j, which Alice does not know. However, we shall show that Alice can get away with sending Bob the state $|\psi_{1y}\rangle$, provided a sufficiently strong privacy guarantee holds.

Theorem 4.3. *Let ε and δ be sufficiently small positive reals. Then any 2-server $(\ell_q, 1; \varepsilon, \delta)$-PIR protocol has $\ell_q \geq (1 - H(3/4 + 2\delta/3 - \sqrt{2\delta} - \varepsilon))\, n - 2$. In particular, the communication cost of such a protocol is at least $\Omega_{\varepsilon,\delta}(n)$.*

Proof. We use the framework and notation of Section 4.2. Suppose P is a 2-server $(\ell_q, 1; \varepsilon, \delta)$-PIR protocol. Consider the following one-way communication protocol for INDEX_n: Alice, on input y, sends Bob the $(2 + \ell_q)$-qubit quantum state $|\psi_{1y}\rangle$. Bob, upon receiving it, constructs the state $|0^\rho\rangle|\psi_{1y}\rangle$ defined by Eq. (1) and applies the unitary transformation U_j to it. He then measures the resulting state $|\phi'_{jy}\rangle$ as mentioned in Fact 4.2.

Let us eschew the additional "11/14 trick" referred to in Fact 4.2 and instead consider the probability p that Bob obtains the "correct" outcome — i.e., the pair of bits $(A_1(y, Q_1(j, r)), A_2(y, Q_2(j, r)))$ — when he uses the same measurement on the state $|\phi'_{jy}\rangle$. Let W be the projection operator corresponding to the desired outcome, so that $\|W|\phi_{jy}\rangle\|_2^2 \geq 3/4$ and $p = \|W|\phi'_{jy}\rangle\|_2^2$. Then

$$\|W|\phi'_{jy}\rangle\|_2 \geq \|W|\phi_{jy}\rangle\|_2 - \|W(|\phi_{jy}\rangle - |\phi'_{jy}\rangle)\|_2 \geq \frac{\sqrt{3}}{2} - \||\phi_{jy}\rangle - |\phi'_{jy}\rangle\|_2. \tag{3}$$

Now,

$$\begin{aligned}
\||\phi_{jy}\rangle - |\phi'_{jy}\rangle\|_2^2 &= \||0^\rho,0\rangle|\phi_{jy}\rangle - |0^\rho,0\rangle|\phi'_{jy}\rangle\|_2^2 \\
&= \||\psi_{jy}\rangle - |\psi_{1y}\rangle\|_2^2 \qquad (4)\\
&= \sum_{i=0}^{2} \sum_{z\in\{0,1\}^{\ell_q}} c^2 \left(\sqrt{|S_{ijz}|} - \sqrt{|S_{i1z}|}\right)^2 \qquad (5)\\
&\leq \sum_{i=0}^{2} \sum_{z\in\{0,1\}^{\ell_q}} c^2 ||S_{ijz}| - |S_{i1z}||
\end{aligned}$$

where Eq. (4) holds because U_j is unitary, and Eq. (5) is obtained by invoking Eq. (1). Since P has privacy parameter δ, for $i \in \{1,2\}$ we have $\sum_{z\in\{0,1\}^\rho} ||S_{ijz}| - |S_{i1z}|| \leq 2^\rho\delta$. Also, by design, $S_{0jz} = S_{01z}$ for all z. Putting these facts together and using Eq. (3) gives

$$p \;=\; \|W|\phi'_{jy}\rangle\|_2^2 \;\geq\; \left(\frac{\sqrt{3}}{2} - \sqrt{2c^2 2^\rho\delta}\right)^2 \;=\; \frac{3}{4} + \frac{2\delta}{3} - \sqrt{2\delta}\,.$$

Since Bob eschews the classical postprocessing (the "11/14 trick"), the probability that he correctly outputs y_j is at least the above quantity minus the probability that the PIR scheme errs, i.e., at least $3/4 + 2\delta/3 - \sqrt{2\delta} - \varepsilon$. The theorem follows.

5 Conclusion

We have found that, in the 1-server case and in the binary 2-server case, relaxing the privacy requirements on a private information retrieval (PIR) scheme by allowing it to leak a small amount of information about the client's index does not allow more than a constant factor improvement in the communication cost. The question of whether improvements can be obtained for the general 2-server case remains open.

References

1. Ablayev, F.: Lower bounds for one-way probabilistic communication complexity and their application to space complexity. Theoretical Computer Science 175(2), 139–159 (1996)
2. Beigel, R., Fortnow, L., Gasarch, W.: A tight lower bound for restricted PIR protocols. Comput. Complexity 15(1), 82–91 (2006)
3. Beimel, A., Ishai, Y., Malkin, T.: Reducing the servers computation in private information retrieval: PIR with preprocessing. In: Bellare, M. (ed.) CRYPTO 2000. LNCS, vol. 1880, pp. 56–74. Springer, Heidelberg (2000)
4. Chor, B., Goldreich, O., Kushilevitz, E., Sudan, M.: Private information retrieval. J. ACM 45(6), 965–982 (1998)

5. Goldreich, O., Karloff, H., Schulman, L., Trevisan, L.: Lower bounds for linear locally decodable codes and private information retrieval. In: Proc. 17th Annual IEEE Conference on Computational Complexity, pp. 175–183. IEEE Computer Society Press, Los Alamitos (2002)
6. Kerenidis, I., de Wolf, R.: Exponential lower bound for 2-query locally decodable codes. J. Comput. Syst. Sci. 69(3), 395–420 (2004) (Preliminary version in Proc. 35th Annual ACM Symposium on the Theory of Computing, pp.106–115 (2003))
7. Kitaev, A.Y., Shen, A.H., Vyalyi, M.N.: Classical and Quantum Computation. American Mathematical Society (2002)
8. Nayak, A.: Optimal lower bounds for quantum automata and random access codes. In: Proc. 40th Annual IEEE Symposium on Foundations of Computer Science, pp. 124–133. IEEE Computer Society Press, Los Alamitos (1999)
9. Nielsen, M.A., Chuang, I.L.: Quantum Computation and Quantum Information. Cambridge University Press, Cambridge (2000)
10. Razborov, A., Yekhanin, S.: An $\Omega(n^{1/3})$ lower bound for bilinear group based private information retrieval. In: Proc. 47th Annual IEEE Symposium on Foundations of Computer Science, pp. 739–748. IEEE Computer Society Press, Los Alamitos (2006)
11. Woodruff, D., Yekhanin, S.: Towards 3-query locally decodable codes of subexponential length. In: Proc. 20th Annual IEEE Conference on Computational Complexity, 2005, pp. 275–284. IEEE Computer Society Press, Los Alamitos (2005)
12. Yekhanin, S.: Towards 3-query locally decodable codes of subexponential length. In: Proc. 39th Annual ACM Symposium on the Theory of Computing, 2007, ACM Press, New York (to appear, 2007)

Packing and Squeezing Subgraphs into Planar Graphs

Fabrizio Frati[1], Markus Geyer[2], and Michael Kaufmann[2]

[1] Dipartimento di Informatica e Automazione – Università Roma Tre, Italy
[2] Wilhelm-Schickard-Institut für Informatik – Universität Tübingen, Germany
frati@dia.uniroma3.it, {geyer,mk}@informatik.uni-tuebingen.de

Abstract. We consider the following problem: Given a set S of graphs, each of n vertices, construct an n-vertex planar graph G containing all the graphs of S as subgraphs. We distinguish the variant in which any two graphs of S are required to have disjoint edges in G (known as 'packing') from the variant in which distinct graphs of S can share edges in G (called 'squeezing'). About the packing variant we show that an arbitrary tree and an arbitrary spider tree can always be packed in a planar graph, improving in this way partial results recently given on this problem. Concerning the squeezing variant, we consider some important classes of graphs, like paths, trees, outerplanar graphs, etc. and establish positive and negative results.

1 Introduction and Motivation

A number of graph algorithms require to find subgraphs satisfying certain properties in a larger graph. Moreover, some of the most studied and attracting topics in graph theory are strictly related to the problem of determining relationships between a graph and its subgraphs. The *subgraph isomorphism* problem asks for finding a subgraph H in a graph G [15,7,3]. The *graph thickness* problem asks for the minimum number of planar subgraphs in which the edges of a graph can be partitioned [12]. The *arboricity* problem asks for determining the minimum number of forests in which a graph can be partitioned [2]. Every planar graph (maximal planar graph) can be partitioned in at most three forests (in three edge-disjoint trees [14]) and it has been recently proved [9] that every planar graph can be partitioned in two edge-disjoint outerplanar graphs.

The study of the relationships between a graph and its subgraphs can be also tackled from the opposite side: Given the n-vertex graphs $G_1, \dots, G_k$, the requirement is to find a graph G satisfying certain properties and containing all the G_i's as subgraphs. This topic occurs with different flavors in the computational geometry and graph drawing literature, motivated by visualization aims, like the display of evolving networks and the simultaneous visualization of relationships involving the same entities. In the *simultaneous embedding* problem [1,8,4] G is given and the goal is to draw it so that the drawing of each G_i is planar. The *simultaneous embedding without mapping* problem [1] is to find a graph G such that: (i) G contains all the G_i's as subgraphs, and (ii) G can be drawn with

L. Kučera and A. Kučera (Eds.): MFCS 2007, LNCS 4708, pp. 394–405, 2007.

(a) (b)

Fig. 1. (a) A caterpillar. (b) A spider tree.

straight-line edges so that the drawing of each G_i is planar. The *packing* problem is the one of finding a graph G containing $G_1, \ldots, G_k$ as edge-disjoint subgraphs. Hedetniemi [10] showed that any two trees with diameter greater than 2, that is with more than three nodes in their longest paths, can be packed in a subgraph of K_n and Maheo et al. [11] gave a characterization of which triples of trees can be packed in K_n.

The *planar packing* problem is the variant of the packing problem in which G is required to be planar. García et al. in [6] conjectured that there exists a planar packing of any two non-star trees, that is of any two trees with diameter greater than 2. Notice that the hypothesis that each tree is different from a star is necessary, since any mapping between the vertices of a star and the vertices of an arbitrary tree leads to at least one common edge. García et al. proved the conjecture for the cases (1) if the trees are isomorphic and (2) if one of the trees is a path respectively. Recently it has been shown in [13] that (3) there exists a planar packing of any two trees if one of them is a caterpillar. In [13] it was also shown the conjecture (4) if one of the trees is a spider with diameter at most 4. A *caterpillar* is a tree which becomes a path when all its leaves are deleted (see Fig. 1.a) and a *spider* is a tree with at most one vertex of degree greater than 2 (see Fig. 1.b).

In this paper we contribute to the state of the art on the planar packing problem, by extending some of the results in [6] and [13]. Namely, in Section 3 we show that there exists a planar packing of any two trees of diameter greater than 2 if one of them is a spider tree. Notice that this result implies results (2) and (4) cited above. The study of the possibility of obtaining a planar packing of a spider tree and an arbitrary tree is motivated by the observation that a spider tree is a subdivision of a star, and hence spider trees are natural candidates for finding counter-examples of the above cited conjecture.

In Section 4 we consider the relaxed version of the planar packing problem in which the subgraphs are not required to be edge-disjoint in the graph containing them. We call such a problem the *planar squeezing* problem and we formally define it as follows: Given the n-vertex graphs $G_1, \ldots, G_k$, find an n-vertex planar graph G containing all the G_i's as subgraphs. We consider some classes of graphs most commonly investigated in the computational geometry and planar graph drawing literature, and we fully determine which ones of them can be generally squeezed in a planar graph. Namely, we show that: (i) there exist a planar graph and a path (a planar graph and a star) that cannot be squeezed in a planar graph; (ii) every two outerplanar graphs (every two trees) can be squeezed in a planar graph; (iii) there exist three caterpillars (three trees) that cannot be squeezed in a planar graph; (iv) there exist two trees that cannot be squeezed in an outerplanar graph; and (v) any number of paths, stars and cycles can be squeezed in an outerplanar graph. Finally, in Section 5 we conclude and suggest some open problems.

2 Definitions

A *drawing* of a graph is a mapping of each vertex to a distinct point in the plane and of each edge to a Jordan curve between the endpoints of the edge. A drawing is *planar* if no two edges intersect but possibly at common endpoints. A *planar graph* is a graph that admits a planar drawing. Two planar drawings of a graph G are *equivalent* if the corresponding circular ordering of the edges incident to each vertex of G is the same for both drawings. An *embedding* of a planar graph G is an equivalence class of planar drawings. An *outerplanar graph* is a planar graph that admits a planar drawing with all its vertices on the same face. An embedding is *outerplanar* if all the vertices lie on the same face. The diameter of a tree is the length of the longest path in the tree. A *star* is a tree with diameter 2, that is a tree where every vertex, but for one, is a leaf, which is a vertex of degree one. A *caterpillar* is a tree such that the graph obtained by deleting its leaves is a path. A *spider* is a tree with at most one vertex, called *root*, of degree greater than 2. The paths starting at the root are called *legs* of the spider. Observe that by definition a star is also a spider and a caterpillar, a path is also a spider and a caterpillar, a caterpillar is also a tree, and a tree is also an outerplanar graph.

Given the n-vertex planar graphs $G_1, \dots, G_k$, a *planar packing* of $G_1, \dots, G_k$ is an n-vertex planar graph containing all the G_i's as edge-disjoint subgraphs (see also [6]). Given the n-vertex planar graphs $G_1, \dots, G_k$, a *planar squeezing* of $G_1, \dots, G_k$ is an n-vertex planar graph containing all the G_i's as subgraphs. In the following, unless otherwise specified, *packing* and *squeezing* will always stand for planar packing and planar squeezing, respectively.

3 Packing Trees in Planar Graphs

In this section we give an algorithm to pack any n-vertex non-star spider tree S and non-star tree T in a planar graph. Observe that we can suppose w.l.o.g. that the diameter of T is greater or equal than 4. In fact, since T is not a star its diameter is greater than 2 and if the diameter of T is 3 then T is a caterpillar, implying that there is a planar packing of T and S [13].

The algorithm we present consists of a *Preprocessing step* and of an *Embedding step* that we sketch here and detail in the following. In the Preprocessing step we root the trees and we fix their embeddings. We also assign a level to each vertex of T. In the Embedding step we embed S on T to obtain a packing of the two trees. After having mapped the root of S to a vertex of T, the legs of S are embedded one at a time sorted by increasing length. For each leg its vertices are embedded one at a time in the order they appear on the leg starting from the nearest to the root and ending with the leaf of the leg. Let v_{cur} denote the vertex of S that has to be embedded. We call the vertex a of S that comes before v_{cur} in the leg of v_{cur} *active vertex*. By the order in which the vertices of S are embedded, a has been already mapped to a vertex of T when v_{cur} is embedded. At every step v_{cur} is mapped to an *'unchosen vertex'*, that is a vertex of T to which no vertex of S has been yet mapped. We analogously call a vertex of T to which a vertex of S has been already mapped *'chosen vertex'*. At every step of the algorithm T and the already embedded edges of S form an embedded graph $\mathcal{E}$. We call the border of the outer face of $\mathcal{E}$ *active border* $\mathcal{F}$. The same vertex of T can have several occurrences in $\mathcal{F}$, since $\mathcal{E}$ is generally a single-connected graph. We denote by $\mathcal{F}(\rightarrow, b, c)$ (by $\mathcal{F}(\leftarrow, b, c)$) the sequence of vertices occurrences that are encountered walking clockwise (resp. counter-clockwise) on $\mathcal{F}$ from an occurrence of a vertex b to an occurrence of a vertex c. When v_{cur} is embedded, edge (a, v_{cur}) is drawn inside the outer face of $\mathcal{E}$.

Preprocessing Step. Pick a leaf l of T such that all the neighbors of the unique neighbor p of l are leaves, but for exactly one vertex r_1. Note that such l always exists since T is different from a star. Let T' denote the tree obtained from T by deleting p and its adjacent leaves. We choose r_1 to be the root of T and the root of T' as well. The root r_2 of S is chosen as usually (see Section 2). Assign a level $l(v)$ to each vertex v of T' so that the root is assigned level 0, all its children are assigned level 1, and so on. Embed T' so that for each vertex v the children of v are in clockwise order $v_1, v_2, \ldots, v_k$ such that $v_i < v_j$ implies that the subtree rooted at v_j contains a vertex w with $l(w) \geq l(u)$, for every vertex u in the subtree rooted at v_i. In the following we will suppose that the children of each node of T' are ordered in clockwise direction. Augment the embedding of T' into an embedding of T by inserting p before the first child of r_1 in T', and by ordering the neighbors of p in clockwise direction so that l is the first vertex and r_1 is the second one. Map r_2 to l. Let r_2 be the first active vertex a.

Embedding Step. This step is repeated until all the vertices and edges of S are embedded on the embedding of T constructed in the Preprocessing step. The legs of S are embedded one at a time sorted by increasing length. For each leg its vertices are embedded one at a time in the order they appear on the leg starting from the nearest to r_2 and ending with the leaf of the leg. Let $p(v)$ denote the parent of a vertex v in T' and $T(v)$ denote the subtree of T' rooted at v. While p has unchosen neighbors, the algorithm will map v_{cur} to the first unchosen vertex in the counter-clockwise order of the neighbors of p starting at r_2. Hence, when v_{cur} is set equal to r_1, all the other neighbors of p will be chosen vertices. Every time v_{cur} has to be embedded, do the following: (i) map v_{cur} to an unchosen vertex u of T; (ii) draw the edge between a and v_{cur} into the outer face of $\mathcal{E}$, and (iii) choose a new vertex of T to be the new active vertex. The choice of the new active vertex a is always done in the following way: If the next v_{cur} is on the same leg of the just embedded v_{cur}, then $a = u$, otherwise $a = r_2$. The choice of the vertex u to which v_{cur} is mapped and the drawing of edge (a, v_{cur}) vary according to several cases:

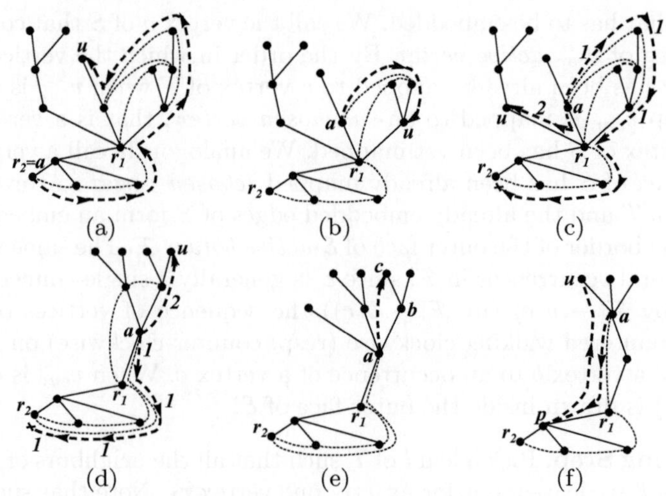

Fig. 2. Illustrations for the different cases of the Embedding step. The dashed edges with arrows represent the searches for unchosen vertices that are done in $\mathcal{F}$ and the drawings of the edges (a, u), where u is the unchosen vertex for which it is set $u = v_{cur}$. (a) Case 1. (b) Case 2 (i). (c) Case 2 (ii): The search labelled by 1 corresponds to the clockwise search for unchosen vertices in $\mathcal{F}$, that does not succeed. The search labelled by 2 corresponds to the counter-clockwise search for unchosen vertices in $\mathcal{F}$, not considering the vertices in $T(a)$. Edge (a, u) will be drawn as the dashed edge labelled by 2. (d) Case 3.1 (ii): The search labelled by 1 corresponds to the clockwise search for unchosen vertices in $\mathcal{F}$, that does not succeed. The search labelled by 2 corresponds to the counter-clockwise search for unchosen vertices in $\mathcal{F}$, considering also the vertices in $T(a)$. Edge (a, u) will be drawn as the dashed edge labelled by 2. (e) Case 3.2, where the dashed edge represents the drawing of (a, c). (f) Case 3.3, where the dashed edges represent the drawing of (a, p) and the drawing of (p, u). Notice that the second edge is drawn only if p is the active vertex after setting $v_{cur} = p$.

Case 1: (refer to Figure 2.a) If a coincides with r_2 or with any other neighbor of p not in T', then walk counter-clockwise on $\mathcal{F}$, starting from the only occurrence of r_2, until an unchosen vertex u is found. Map v_{cur} to u. Draw edge (a, v_{cur}) following the counter-clockwise walk done on $\mathcal{F}$.

Case 2: (refer to Figures 2.b and 2.c) If a does not coincide with r_2 and there is at least one unchosen vertex in the tree $T' \setminus T(a)$ that does not belong to the path from r_1 to a in T', then walk on $\mathcal{F}(\rightarrow, a, r_2)$ not considering the vertices in $T(a)$.

(i) If an unchosen vertex $u \neq p(a)$ has been encountered then map v_{cur} to u and draw the edge (a, v_{cur}) following the clockwise walk done on $\mathcal{F}$.
(ii) If the last occurence of $p(a)$ in $\mathcal{F}(\rightarrow, a, r_2)$ has been encountered and $p(a)$ is not yet chosen or if no unchosen vertex has been found in $\mathcal{F}(\rightarrow, a, r_2)$, then reverse the search direction and map v_{cur} to the first unchosen vertex u in $\mathcal{F}(\leftarrow, a, p)$, not considering the vertices in $T(a)$. Draw the edge (a, u) following the counter-clockwise walk done on $\mathcal{F}$.

Case 3: If a does not coincide with r_2, and if there are no unchosen vertices in $T' \setminus T(a)$, but eventually for those in the path from r_1 to a in T', we distinguish three subcases:

Case 3.1: (refer to Figure 2.d) If no unchosen child of a exists and if $T(a)$ contains unchosen vertices of level $l(a) + 2$ or higher, then search in $\mathcal{F}(\rightarrow, a, r_2)$ not considering the vertices in $T(a)$.

(i) If an unchosen vertex $u \neq p(a)$ has been reached then map v_{cur} to u and draw edge (a, v_{cur}) following the clockwise walk done on $\mathcal{F}$.
(ii) If the last occurence of $p(a)$ in $\mathcal{F}(\rightarrow, a, r_2)$ has been reached and $p(a)$ is not yet chosen or if no unchosen vertex has been found in $\mathcal{F}(\rightarrow, a, r_2)$, then reverse the search direction and map v_{cur} to the first unchosen vertex u in $\mathcal{F}(\leftarrow, a, p)$ starting from the first occurrence of a in $\mathcal{F}(\leftarrow, r_2, p)$. In this case the vertices of $T(a)$ are considered first. Draw edge (a, v_{cur}) following the counter-clockwise walk done on $\mathcal{F}$.

Case 3.2: (refer to Figure 2.e) If there are unchosen children of a and if $T(a)$ contains unchosen vertices of level $l(a)+2$ or higher, then consider the last child b of a in clockwise order. Select the clockwise first child c of b. We will prove later that c is an unchosen vertex. Map v_{cur} to c and draw edge (a, v_{cur}) passing just before edge (a, b) in the clockwise order of the children of a.

Case 3.3: (refer to Figure 2.f) If $T(a)$ does not contain unchosen vertices of level $l(a)+2$ or higher, then we are in the final phase of our algorithm. Notice that the only unchosen vertices in T', but for p, are either at distance one from a or lie on the path from a to r_1. We will prove later that all the unchosen vertices on such a path are pairwise non-adjacent. Map v_{cur} to p draw edge (a, v_{cur}) by walking counter-clockwise on $\mathcal{F}$ starting from the first occurrence of a in $\mathcal{F}(\rightarrow, p, r_2)$. After that, if $a = p$ then search in $\mathcal{F}(\rightarrow, a, r_2)$ until an unchosen vertex u is found. Map v_{cur} to u and draw edge (a, v_{cur}) following the clockwise walk done on $\mathcal{F}$. At this point, or if it was $a = r_2$, only Cases 1, 2, and 3.1 will be applied, until all the remaining vertices of S are mapped to unchosen vertices of T. Notice that Case 3.3 is applied exactly once in one application of the algorithm.

In the following we give some lemmas that will be helpful to prove that the described algorithm constructs a planar packing of S and T. The proofs of such lemmas are in the full version of the paper [5].

Lemma 1. *Let v and $p(v)$ be unchosen vertices in T'. Then all vertices in $T(v)$ are unchosen.*

Corollary 1. *Let $v \in T'$ be a chosen vertex and let $\mathcal{P} = (r_1 = v_1, v_2, \dots, v_{l-1}, v_l = v)$ be the path connecting r_1 and v in T', with $l \geq 2$. There exist no two consecutive unchosen vertices v_i and v_{i+1} in $\mathcal{P}$.*

Lemma 2. *If v is the j-th child of $p(v)$ in T', if $p(v)$ is unchosen, and if v_{cur} has been mapped to v in the current step of the algorithm, then during the last j steps of the application of the algorithm Case 3.2 was applied once to draw an edge from $p(p(v))$ to the first child f of $p(v)$ and Case 2 (i) was applied in the following $j-1$ steps to draw $j-1$ edges connecting the first j children of $p(v)$.*

Corollary 2. *Let a be an active vertex in T' and let $p(a)$ be unchosen. There exists no occurrence of $p(a)$ in $\mathcal{F}(\leftarrow, a, p)$.*

Lemma 3. *Let $v \in T'$ be an occurrence of a vertex in $\mathcal{F}$. In $\mathcal{F}(\rightarrow, v, r_2)$ there exists at least one occurrence of every unchosen vertex belonging to the path connecting r_1 and v in T'. Moreover, all the unchosen vertices of T appear at least once in $\mathcal{F}$.*

Theorem 1. *There exists an algorithm that in polynomial time constructs a planar packing of any n-vertex non-star spider tree S and any n-vertex non-star tree T.*

Proof. Apply the algorithm described in this section to T and S. First, notice that the algorithm can be easily implemented to run in polynomial time. We claim that the constructed embedding $\mathcal{E}$ is a planar packing of T and S. More precisely, we will prove that: (1) $\mathcal{E}$ is planar, (2) every two vertices of S are mapped to distinct vertices of T, (3) there are no common edges between S and T, and finally (4) all the vertices of S are mapped to vertices of T.

(1): The planarity of $\mathcal{E}$ follows from the fact that at every step all the unchosen vertices are incident to the outer face (by Lemma 3) and that by construction every inserted edge is placed inside the outer face of $\mathcal{E}$.

(2): When one of the Cases 1, 2, 3.1, and 3.3 of the Embedding step is applied, by the description of the algorithm v_{cur} is mapped to an unchosen vertex of T. Hence, we have only to show that when Case 3.2 has to be applied in a step s^* of the algorithm the first child c of the last child b of a is unchosen. If before s^* vertex b is chosen, then by Lemma 2 all the children of a were chosen when it was set $v_{cur} = b$, hence Case 3.2 would not be applied in s^*. Otherwise, prior to the choice of a before step $s*$, both b and a were unchosen and so, by Lemma 1, all the vertices in $T(b)$, including c, are unchosen at the beginning of s^*.

(3): Consider the different cases of the Embedding step. In Case 1 a common edge is inserted only if it is set $v_{cur} = p$. If $a \neq r_2$ then setting $v_{cur} = p$ would imply that T is a star, contradicting the hypoteses. Notice that vertices of S are mapped to all the neighbors of p by applications of Case 1 before any other case of the Embedding step is applied. If $a = r_2$ then, since p is the last vertex in $\mathcal{F}(\leftarrow, r_2, p)$, setting $v_{cur} = p$ implies that no other vertex of $\mathcal{F}$ is unchosen and, by Lemma 3, that no other vertex of T is unchosen. Hence, the current leg of S is the last one. Since this leg should have length 1 and since the legs of S are ordered by increasing length, S would be a star, contradicting the hypoteses. In Case 2, the only neighbor of a in T' that belongs to $T' \setminus T(a)$ is $p(a)$. However, in Case 2 (i) it is clear that $v_{cur} = u$ is chosen for a vertex $u \neq p(a)$. In Case 2 (ii) the algorithm chooses for v_{cur} the first unchosen vertex u in $\mathcal{F}(\leftarrow, a, p)$. By Corollary 2 there exists no occurrence of $p(a)$ in such a visit and so $u \neq p(a)$. In Case 3.1 (i) the same considerations done for Case 2 (i) hold. In Case 3.1 (ii) for v_{cur} is a vertex u chosen, that belongs to $T(a)$. Since all the children of a are already chosen, then no common edge is inserted. In Case 3.2 a and c are not neighbors in T. Finally, consider Case 3.3. Since a belongs to T', then a and p are neighbors only if $a = r_1$. However, if $a = r_1$ and there are no unchosen vertices in T', but for the children of a, then the diameter of T would be at most 2, contradicting the hypoteses. Concerning edge (p, u) all the neighbors of p are already chosen before applying Case 3.3, so u cannot be a neighbor of p.

(4): We have to prove that while there are unchosen vertices in T the algorithm applies one of the cases in the Embedding step to map v_{cur} to a vertex of T. All the neighbors of p are chosen at the beginning of the Embedding step by applications of Case 1. After that phase only p and the vertices in $T' \backslash r_1$ are still unchosen. Now let a be the current active vertex. Suppose Case 1 has to be applied. By Lemma 3 at every step of the algorithm all the unchosen vertices are on $\mathcal{F}$, so Case 1 finds an unchosen vertex u to set $v_{cur} = u$. Suppose Case 2 has to be applied. If there are occurrences of unchosen vertices in $\mathcal{F}(\leftarrow, a, r_2)$ not belonging to $T(a)$ or to the path connecting r_1 and a in T', then even if Case 2 (i) fails, then Case 2 (ii) would find such occurrences. Otherwise, suppose that the only unchosen vertices not belonging to $T(a)$ or to the path connecting a and r_1 in T' appear before a in $\mathcal{F}(\leftarrow, r_2, p)$. If $p(a)$ is already chosen, then Case 2 (i) would always succeed. If $p(a)$ is unchosen and if a is the j-th child of $p(a)$, then $T(p(p(a)))$ contains the only unchosen vertices remaining, but for p and for the vertices in the path from r_1 to $p(p(a))$ in T', since Case 3.2 was applied j steps before the current one when $p(p(a))$ was the active vertex (by Lemma 2). Since $p(a)$ is the last child of $p(p(a))$, then the only vertices that can have occurrences before a in $\mathcal{F}(\leftarrow, r_2, p)$ are the vertices in $T(p(a))$. Such occurrences are clearly encountered before the last occurrence of $p(a)$ in $\mathcal{F}(\rightarrow, a, p)$, hence Case 2 (i) finds them and succeeds. Suppose Case 3.1 has to be applied. Then either Case 3.1 (i) succeeds, or Case 3.1 (ii) finds an unchosen vertex in $T(a)$. Such vertex exists by the hypoteses of Case 3.1. Suppose Case 3.2 has to be applied. We have already shown in part (2) of the proof, that if vertex c exists, then it is unchosen. Now we only have to prove the existence of such a vertex. By the construction of the embedding of T' the children of a are clockwise ordered by increasing depth of the subtrees rooted at them; observing that in $T(a)$ there are vertices of level $l(a) + 2$ or higher, then vertex c exists. Finally if Case 3.3 has to be applied, then no problem arises, since p is unchosen and it is on $\mathcal{F}$ before the only application of Case 3.3. Notice that by Corollary 1 and Lemma 3 after Case 3.3 is applied all the remaining unchosen vertices of T are disconnected and are on $\mathcal{F}$. Therefore, Cases 1, 2, and 3.1 can be applied until all the vertices of S are mapped to vertices of T. □

4 Squeezing Planar Graphs in Planar Graphs

When dealing with the planar packing problem, it can be easily observed that two sufficiently dense planar graphs cannot be packed in the same planar graph. For instance, two maximal outerplanar graphs have $2n - 3$ edges each, and a packing of them contains $4n - 6$ edges, that are more than the ones that a planar graph can have.

If you want to obtain planar squeezings of planar graphs, edges of different graphs can overlap, and so edge-counting arguments do not work. However, the following two results are just slightly more than trivial:

Lemma 4. *There exist a planar graph G and a path P that cannot be squeezed in a planar graph.*

Proof. Let G be an n-vertex triangulated planar graph that does not contain any Hamiltonian path, and let P be an n-vertex path. Observe that since G is maximal no edge can be added to it without violating its planarity. However, when squeezing G and P, at least one edge of P is not common to an edge of G, otherwise G would contain an Hamiltonian path. □

Lemma 5. *There exist a planar graph G and a star S that cannot be squeezed in a planar graph.*

Proof. Let G be an n-vertex triangulated planar graph that does not contain a vertex of degree greater than $n-2$, and let S be an n-vertex star. Since G is maximal no edge can be added to it without violating its planarity. However, when squeezing G and S, at least one edge of S is not common to an edge of G, otherwise G would contain a vertex of degree $n-1$. □

Turning the attention from planar to outerplanar graphs, we have:

Lemma 6. *Any two outerplanar graphs can be squeezed in a planar graph.*

Proof. Let O_1 and O_2 be two outerplanar graphs. Assume w.l.o.g. that both O_1 and O_2 are biconnected. Hence O_1 and O_2 contain Hamiltonian cycles, say C_1 and C_2, respectively. Now map the vertices of O_1 and O_2 so that C_1 and C_2 are coincident. Furthermore, embed the edges of O_1 that do not belong to C_1 inside C_1, and embed the edges of O_2 that do not belong to C_2 and that are not common to edges of O_1 outside C_1. By the outerplanarity of O_1 (of O_2) there are no intersections between edges inside C_1 (resp. outside C_1). Further, there are no intersections between edges inside C_1 and edges outside C_1, since they are separated by C_1. □

Corollary 3. *Any two trees can be squeezed in a planar graph.*

Corollary 3 shows that the problem of determining whether for any two trees there exists a planar graph containing them as subgraphs, that has been tackled in [6], in [13] and in Section 3, is easily solvable if common edges are allowed.

However, if one augments the number of trees that must be squeezed, then a planar squeezing is not generally possible. Namely, in the following we provide three caterpillars that cannot be squeezed in the same planar graph.

Theorem 2. *There exist three caterpillars that cannot be squeezed in the same planar graph.*

Proof. Let C_1 be a star with center u and $n-1$ leaves (see Fig. 3.a), let C_2 be a caterpillar with two adjacent vertices v_1 and v_2 of degree $n/2$ and $n-2$ leaves (see Fig. 3.b), and let C_3 be a caterpillar with five vertices $w_1, \ldots, w_5$ of degree at most $n/5+1$ forming a path and with $n-5$ leaves. Each vertex w_i has $n/5-1$ adjacent leaves (see Fig. 3.c).

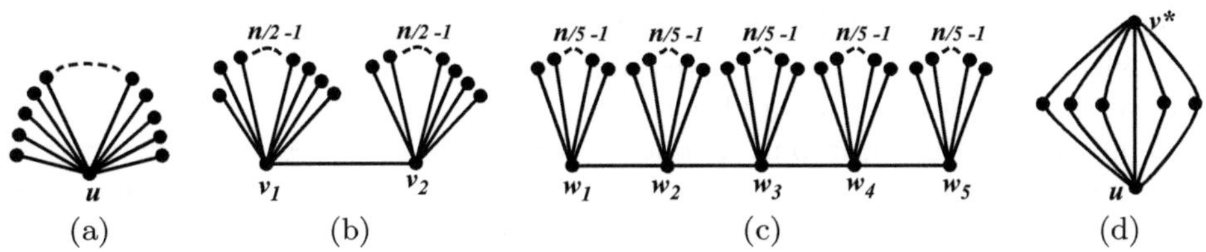

Fig. 3. (a) Star C_1. (b) Caterpillar C_2. (c) Caterpillar C_3. (d) Embedding $\mathcal{E}_A$.

We will try to construct a planar embedding that contains embeddings of C_1, C_2, and C_3 and we will show that this goal is not achievable. Observe that C_1 has a unique embedding $\mathcal{E}_1$ up to a relabelling of its leaves. First, construct a planar embedding $\mathcal{E}_2$ by embedding C_2 on $\mathcal{E}_1$ in any way. Let G_2 denote the planar graph obtained by such a squeezing. Notice that there exists one out of v_1 and v_2, say v^*, that has not been mapped to u and that shares with u exactly $n/2-1$ common neighbors. In fact, if vertex v_1 (vertex v_2) has been mapped to u, then v_2 (resp. v_1) has been mapped to a leaf of C_1 and all the $n/2-1$ leaves adjacent to v_2 (resp. to v_1) have been mapped to leaves of C_1, that are neighbors of u. Otherwise, if both vertices v_1 and v_2 have been mapped to leaves of C_1, then v_1 (or v_2) has exactly $n/2-2$ adjacent leaves that have been mapped to leaves of C_1, that are neighbors of u, and v_2 (resp. v_1) is a neighbor of both u and v_1 (resp. of both u and v_2). Consider the set A of vertices that are neighbors of both u and v^*. Vertex u, vertex v^*, and the vertices in A induce an embedded subgraph $\mathcal{E}_A$ of $\mathcal{E}_2$ that is done by at least one and at most two nested triangles sequences, all sharing edge (u, v^*) (see Fig. 3.d).

Now consider any embedding $\mathcal{E}_3$ of C_3 on $\mathcal{E}_2$. Let us discuss how many vertices of C_3 can be mapped to vertices in A, while preserving the planarity of $\mathcal{E}_3$. Since the degree of each vertex w_i is at most $n/5+1$, at most $2n/5+2$ vertices of A could be neighbors of u and v^* in C_3. Vertices w_1, ..., w_5 of C_3 that are not mapped to u and v^* can have at most two vertices of A as adjacent leaves. In fact, if vertex w_i is mapped to a vertex of A, then it is incident to two adjacent faces of $\mathcal{E}_A$ that have at most two vertices distinct from u, from v^*, and from w_i itself. If vertex w_i is mapped to a vertex not in A and inside any face of $\mathcal{E}_A$, then it can be a neighbor of the at most two vertices of that face that are in A. Hence, for every vertex w_i three vertices internal to $\mathcal{E}_A$ can have a mapping, two with leaves adjacent to w_i and one with w_i itself. Hence less than $2n/5+2+3\cdot 5 = 2n/5+17$ vertices of A can have a mapping with a vertex of C_3 while preserving the planarity of $\mathcal{E}_3$. Choosing $|A| = n/2-1 > 2n/5+17$ (i.e. choosing $n > 180$) implies that the vertices of C_3 cannot be mapped to all the vertices in A while preserving the planarity of $\mathcal{E}_3$ and hence that there is no planar squeezing of C_1, C_2, and C_3 □

Corollary 4. *There exist three trees that cannot be squeezed in the same planar graph.*

If one wants to squeeze trees in trees, trees in outerplanar graphs, or outerplanar graphs in outerplanar graphs, then very few is allowed. Namely, we show two caterpillars that cannot be squeezed in any outerplanar graph. Let C_1 be a star with center u and seven leaves and let C_2 be a caterpillar consisting of two vertices of degree four and six leaves. We claim that C_1 and C_2 cannot be squeezed in the same outerplanar graph. This is proved by showing that any planar embedding of an 8-vertex planar graph that contains embeddings of C_1 and C_2 cannot be an outerplanar embedding. First, observe that C_1 has a unique embedding up to a relabelling of its leaves. So consider it as embedded. Now embed C_2. Since there is just one non-leaf vertex in C_1, at least one of the vertices of C_2 with degree 4 must be mapped to a leaf of C_1. Let v be such a vertex. Again, since there is one non-leaf vertex in C_1, at least three of the neighbors of v must be mapped in leaves of C_1. This implies that in any embedding containing embeddings of C_1 and C_2 there is a cycle formed by v, u and a neighbor of v enclosing a neighbor of v. Hence there exists no outerplanar embedding containing embeddings of C_1 and C_2 and so there exists no outerplanar graph containing C_1 and C_2.

Theorem 3. *There exist two caterpillars that cannot be squeezed in the same outerplanar graph.*

Corollary 5. *There exist two trees that cannot be squeezed in the same outerplanar graph.*

We conclude this section observing that any number of paths, cycles and stars can be squeezed in an outerplanar graph having one vertex of degree $n - 1$.

5 Conclusions and Open Problems

We have considered the problem of packing and squeezing subgraphs in planar graphs. Concerning the planar packing problem, the previous works on this topic [6,13] contain algorithms that construct embeddings of the trees by observing the 'separation principle', i.e. by separating the edges of the two trees in two different portions of the embedding plane, established in advance. This allows to mind only to the presence of common edges for obtaining a planar packing. As far as we know, our algorithm is the first one that does not bind the embeddings of the trees to be separated as described. Also the tree embeddings produced by our algorithm could not be separated in different parts of the plane, since there are vertices with a sequence $[T_1, T_2, T_1, T_2]$ of consecutive edges, where T_1 (T_2) indicates an edge belonging to the first (resp. the second) tree.

Problem 1. Does a planar packing of any two non-star trees with the further constraint of having an embedding where the two trees can be separated by a simple line that intersects the embedding only at vertices of the graph exist?

For the squeezing problem, we considered combinations of important classes of planar graphs like paths, caterpillars, trees, outerplanar graphs and established which combinations can generally be squeezed and which cannot. However, the following open problem is worth of interest:

Problem 2. Which is the time complexity of determining if two planar graphs can be squeezed in a planar graph?

The last question seems to be strictly related to some of the most important problems in graph theory, like *graph isomorphism* and *subgraph isomorphism*.

References

1. Braß, P., Cenek, E., Duncan, C.A., Efrat, A., Erten, C., Ismailescu, D., Kobourov, S.G., Lubiw, A., Mitchell, J.S.B.: On simultaneous planar graph embeddings. Comput. Geom. 36(2), 117–130 (2007)
2. Eppstein, D.: Arboricity and bipartite subgraph listing algorithms. Information Processing Letters 51(4), 207–211 (1994)
3. Eppstein, D.: Subgraph isomorphism in planar graphs and related problems. J. Graph Algorithms & Applications 3(3), 1–27 (1999)
4. Frati, F.: Embedding graphs simultaneously with fixed edges. Graph Drawing, 108–113 (2006)
5. Frati, F., Geyer, M., Kaufmann, M.: Packing and squeezing subgraphs into planar graphs. Tech. Report RT-DIA-114-2007, Dept. of Computer Science and Automation, University of Roma Tre (2007)
6. García, A., Hernando, C., Hurtado, F., Noy, M., Tejel, J.: Packing trees into planar graphs. J. Graph Theory, 172–181 (2002)
7. Garey, M.R., Johnson, D.S.: Computers and Intractability, A Guide to the Theory of NP-Completeness. W.H. Freeman and Company, New York (1979)
8. Geyer, M., Kaufmann, M., Vrtó, I.: Two trees which are self-intersecting when drawn simultaneously. In: Graph Drawing, pp. 201–210 (2005)
9. Gonçalves, D.: Edge partition of planar graphs into two outerplanar graphs. In: STOC, pp. 504–512 (2005)
10. Hedetniemi, S.M., Hedetniemi, S.T., Slater, P.J.: A note on packing two trees into K_N. Ars Combin. 11, 149–153 (1981)
11. Maheo, M., Saclé, J.-F., Woźniak, M.: Edge-disjoint placement of three trees. European J. Combin. 17(6), 543–563 (1996)
12. Mutzel, P., Odenthal, T., Scharbrodt, M.: The thickness of graphs: A survey. Graphs and Combinatorics 14(1), 59–73 (1998)
13. Oda, Y., Ota, K.: Tight planar packings of two trees. In: European Workshop on Computational Geometry, pp. 215–216 (2006)
14. Schnyder, W.: Planar graphs and poset dimension. Order 5, 323–343 (1989)
15. Ullmann, J.R.: An algorithm for subgraph isomorphism. J. ACM 23(1), 31–42 (1976)

Dynamic Matchings in Convex Bipartite Graphs

Gerth Stølting Brodal[1,⋆], Loukas Georgiadis[2], Kristoffer Arnsfelt Hansen[3], and Irit Katriel[4,⋆⋆]

[1] University of Aarhus, Århus, Denmark
gerth@daimi.au.dk
[2] Hewlett-Packard Laboratories, Palo Alto, CA, USA
loukas.georgiadis@hp.com
[3] University of Chicago, Chicago, IL, USA
arnsfelt@cs.uchicago.edu
[4] Brown University, Providence, RI, USA
irit@cs.brown.edu

Abstract. We consider the problem of maintaining a maximum matching in a convex bipartite graph $G = (V, E)$ under a set of update operations which includes insertions and deletions of vertices and edges. It is not hard to show that it is impossible to maintain an explicit representation of a maximum matching in sub-linear time per operation, even in the amortized sense. Despite this difficulty, we develop a data structure which maintains the set of vertices that participate in a maximum matching in $O(\log^2 |V|)$ amortized time per update and reports the status of a vertex (matched or unmatched) in constant worst-case time. Our structure can report the mate of a matched vertex in the maximum matching in worst-case $O(\min\{k \log^2 |V| + \log |V|, |V| \log |V|\})$ time, where k is the number of update operations since the last query for the same pair of vertices was made. In addition, we give an $O(\sqrt{|V|} \log^2 |V|)$-time amortized bound for this pair query.

1 Introduction

Let $G = (V, E)$ be a finite undirected graph. A subset $M \subseteq E$ of the edges is a *matching* if each vertex is incident with at most one edge in M. A matching of maximum cardinality is called a *maximum matching*. In this paper we study the problem of maintaining a *maximum matching* in a *convex bipartite* graph; G is bipartite if V can be partitioned to subsets X and Y such that $E \subseteq X \times Y$. A bipartite graph $G = (X \cup Y, E)$ is convex if there is a linear arrangement $y_1, \ldots, y_m$ of the nodes of Y such that the neighborhood of every node in X is an interval of the nodes of Y. Given a linear order for Y we can represent G in $O(|V|) = O(n+m)$ space (instead of $O(|V| + |E|)$), where $n = |X|$ and $m = |Y|$, since it suffices to specify the smallest and largest neighbor in Y for each $x \in X$.

⋆ MADALGO - Center for Massive Data Algorithmics, a Center of the Danish National Research Foundation.

⋆⋆ Supported in part by NSF award DMI-0600384 and ONR Award N000140610607.

L. Kučera and A. Kučera (Eds.): MFCS 2007, LNCS 4708, pp. 406–417, 2007.

Matchings in convex bipartite graphs correspond to a basic variant of the ubiquitous job scheduling problem. Namely, the case of unit-length tasks that need to be scheduled on a single disjunctive resource (i.e., one at a time) and where we are given, for each job, a release time and a deadline. Another application for convex matching is in constraint programming, where filtering algorithms for the *AllDifferent* constraint [21] essentially identify edges in a bipartite graph which do not belong to any maximum matching. The vertices of X then represent variables of the program while the vertices of Y represent the values in their domains and the edges connect every variable with every value in its domain. In some cases, the domains are intervals and the graph is convex. In other cases, the constraint solver *assumes* that they are intervals as a heuristic that speeds up computations. Motivated by this application, and to avoid confusion, in the rest of the paper we call the vertices of X *variables* and the vertices of Y *values*. Furthermore, we sometimes refer to the interval of neighbors of a variable as its *domain*.

The problem of finding a maximum matching has a rich history [5]. Currently, the best upper bound for both bipartite and general matchings is randomized $O(|V|^{\omega})$ [13], where ω is the exponent of matrix multiplication ($\omega < 2.376$ [2]). A simpler randomized algorithm with the same bound was presented recently [9]. The best deterministic bound is $O(|E|\sqrt{|V|})$ both for bipartite graphs [10] and general graphs [12].

Finding maximum matchings in convex bipartite graphs has also been extensively studied. Glover [7] showed that a maximum matching in such a graph can be found by traversing the values in increasing order and matching each value $y \in Y$ with the variable whose domain ends earliest, among those that have y in their domain and were not matched with smaller values, breaking ties according to a predetermined lexicographic order of the variables. Such a matching is called a *greedy matching*. This algorithm runs in $O(m+n \log\log n)$ time with the use of a fast priority queue [20]. Lipski and Preparata [11] presented an $O(n+m\alpha(m))$-time algorithm. (Here α is a functional inverse of Ackermann's function.) This running time was reduced to $O(n+m)$ in [4]. Further research [6,18] aimed at decreasing the dependence on m, finally leading to an $O(n)$-time algorithm by Steiner and Yeomans [19].

In this paper we study the problem of maintaining a maximum matching in a dynamic convex bipartite graph. In the job scheduling view, this problem amounts to maintaining an optimal schedule for a dynamic set of unit-length jobs. Notice that Glover's algorithm can easily be adapted to operate in a dynamic setting where a job may be deleted or its release time and deadline may be available before the job is actually released. But still we cannot infer the service time of a given job (or even whether the job will eventually be serviced) before it is either actually dispatched or its deadline elapses.

Intuitively it is clear that maintaining a maximum matching dynamically must be hard; even a single edge insertion can change all the edges in the matching. Apparently, due to this difficulty, this problem has not received much attention. To the best of our knowledge, the only related previous work is presented in [16,17]. In [16], Sankowski considers dynamic graphs, modified by edge insertions

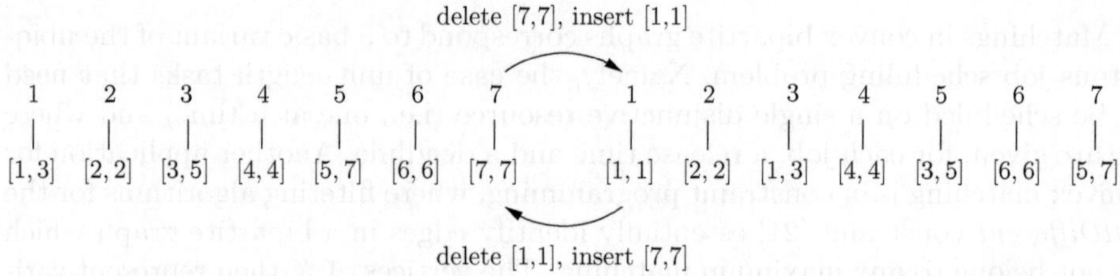

Fig. 1. Sequence of operations inducing linear number of changes in any explicit representation of a convex matching. Only edges in the maximum matching are shown.

and deletions, that maintain the invariant that a *perfect matching* exists, i.e., all vertices are matched. He shows how to update the graph in $O(n^{1.495})$ time per operation and support queries asking whether a given edge is contained in any perfect matching. The same asymptotic update bound is achieved in [17] for maintaining the size of a maximum matching in general dynamic graphs (modified by edge insertions and deletions). The running time in both results depends on the value of the the exponent ω of matrix multiplication.

The situation is similar in the case of convex bipartite graphs; it is not hard to construct examples where a single update changes all the matching edges. Still, due to the special structure of the convex graphs it is less obvious that *any* representation of a maximum matching can change dramatically due to a small change in the input. For example, since the values in Y are ordered, a matching can be described by giving a corresponding order for X (together with a dummy variable ϵ for each unmatched value). To measure the difference between two convex matchings more accurately we can count the number of inversions between pairs of successive variables in the two corresponding orderings of X. Figure 1 shows that even with respect to this measure, the difference between two matchings after $O(1)$ updates can be $\Theta(|V|)$, also in the amortized sense. Such problematic cases rule out the possibility of maintaining any explicit representation of a maximum matching in sub-linear time, even for convex bipartite graphs. On the other hand, it is easy to verify that the sets of matched vertices in X (and Y) can change only by one element per update. (This fact holds also for general graphs modified by edge insertions and deletions, since a matching M is maximum iff there is no augmenting path relative to M [1].) The structure we develop is based on the previous observation and on a parallel algorithm of Dekel and Sahni for convex bipartite matchings [3]. We review their algorithm in Section 2.1.

2 Dynamic Convex Matchings

Recall that Y is totally ordered and the input graph $G = (X \cup Y, E)$ is represented succinctly by giving an interval $[s(x), t(x)]$, for each $x \in X$; $s(x)$ ($t(x)$) is the first (last) vertex in Y adjacent to x. For convenience, we identify the vertices in Y by their rank in the underlying linear arrangement, and henceforth assume

$Y = \{1, \ldots, m\}$. We allow G to be modified by a collection $\mathcal{U}$ composed of the following update operations:

- Insert a vertex x into X together with an interval $[s(x), t(x)]$; make x adjacent to every $y \in Y$ such that $s(x) \leq y \leq t(x)$.
- Delete a vertex x from X; remove all edges incident to x.
- Insert a vertex y into Y; make y adjacent to any $x \in X$ such that $s(x) < y < t(x)$.
- Delete a vertex y from Y if there is no $x \in X$ such that $s(x) = y$ or $t(x) = y$; remove all edges adjacent to y.

Clearly, the above operations maintain the convexity of G, and also maintain the given order of Y. As a result, we do not need to test if G remains convex after each operation (in fact we are not aware of any efficient dynamic algorithm for this test), but this also implies that we cannot support arbitrary edge updates. Restricted edge updates are possible by deleting a variable and re-inserting it with a different interval. Given a dynamic convex bipartite graph G that is modified subject to $\mathcal{U}$, we wish to support a collection $\mathcal{Q}$ of the following query operations:

- Given a vertex $x \in X$ or $y \in Y$, return the status *matched* or *unmatched* of this vertex.
- Given a matched vertex u ($u \in X \cup Y$), return its mate w in the maximum matching; the pair $\{u, w\}$ is an edge of the maximum matching.

Our results are summarized in the following theorems.

Theorem 1. *We can maintain a maximum matching of a convex bipartite graph $G = (X \cup Y, E)$, under the update operations in $\mathcal{U}$, in $O(\log^2 |X|)$ amortized time per update and $O(|Y| + |X| \log |X|)$ space. A status query can be answered in constant worst-case time. A pair query for a vertex u can be answered in $O(\min\{k \log^2 |X| + \log |X|, |X| \log |X|\})$ worst-case time, where k is the number of update operations that occurred since the last pair query for u or its mate.*

Theorem 2. *The amortized running time for a pair query is $O(\sqrt{|X|} \log^2 |X|)$. Also, there is an update sequence which forces our structure to spend $\Omega(\sqrt{|X|})$ amortized time for each pair query.*

2.1 The Dekel-Sahni Algorithm

Let $S = \{s_1, \ldots, s_k\}$ be the set of start-points of the n variable domains ($s_i = s(x_i)$). The base structure in the algorithm of Dekel and Sahni [3] is a balanced binary search tree $\mathcal{T}$ over S. Each leaf s_i is associated with the interval of values $[s_i, s_{i+1} - 1]$; each internal node is associated with the interval of values formed by the union of the intervals of its children in $\mathcal{T}$. We denote the interval of values associated with a node P in $\mathcal{T}$ by *values*(P); we let $s(P)$ and $t(P)$ be, respectively, the smallest and largest value in *values*(P). We let *variables*(P) denote the set of variables $x \in X$ such that $s(x) \in$ *values*(P). Also, define *matched*(P) to be the set of variables that belong to the greedy matching of the subgraph induced by *variables*(P) $\cup$ *values*(P). Let *transferred*(P) be the variables x in *variables*$(P) \setminus$ *matched*(P) such that $t(x) > t(P)$, i.e., their end-points extend beyond *values*(P); these variables, although not matched at P, may become matched at an ancestor of P in $\mathcal{T}$ and thus participate in the final (global) matching, computed at the root of $\mathcal{T}$. The remaining variables in *variables*(P) form a set *infeasible*(P) and are discarded from later computations.

Now let P be an internal node of $\mathcal{T}$ with left child L and right child R. Assume that *matched*(L), *transferred*(L), *matched*(R) and *transferred*(R) were already computed. Dekel and Sahni show that *matched*(P) and *transferred*(P) can be computed as follows: Let *variables'*(R) = *transferred*$(L) \cup$ *matched*(R). First compute a greedy matching in the subgraph G'_R induced by *variables'*$(R) \cup$ *values*(R). Let *matched'*(R) be the subset of *variables'*(R) in that matching, and let *transferred'*$(R) = \{x \mid x \in$ *variables'*$(R) \setminus$ *matched'*(R) and $t(x) > t(R)\}$. Then, *matched*(P) = *matched*$(L) \cup$ *matched'*(R) and *transferred*(P) = *transferred*$(R) \cup$ *transferred'*(R). The key theorem of Dekel and Sahni [3, Theorem 2.1], shows that there is a simple way to compute *matched'*(R): It is safe to assume that the start-point of *every* interval in *variables'*(R) is $s(R)$. Although the matching produced this way may not be feasible in G'_R, the set of matched variables that are identified is correct, i.e., there is a greedy matching in this graph that matches exactly this set of variables.

After identifying the matched vertices in X, the algorithm of Dekel and Sahni performs a top-down phase that constructs a feasible matching. For each node P of $\mathcal{T}$, it constructs the set *matching*(P), containing the variables that are matched with a value in *values*(P). Although *matched*(P) is a subset of *variables*(P), this is not true, in general, for *matching*(P). Clearly, if P is the root of $\mathcal{T}$, *matching*(P) = *matched*(P). The *matching* sets for the children L and R of P are constructed by partitioning *matching*(P) as follows. Let W be the subset of *matching*(P) consisting of variables x such that $s(x) < s(L)$ or $x \in$ *matched*(P). (Note that W does not contain any variable in *transferred*(L).) If $|W| \leq |$*values*$(L)|$, then set *matching*$(L) = W$. Otherwise, pick $|$*values*$(L)|$ variables $x \in W$ with the smallest $t(x)$ and insert them in *matching*(L). The remaining variables are inserted into *matching*(R). Finally, we notice that for any leaf F, each $x \in$ *matching*(F) satisfies $s(x) \leq s(F)$. (Because we cannot have $s(F) < s(x) \leq t(F)$ for any $x \in X$.) Hence, sorting *matching*(F) by increasing right endpoint, gives us the matching at F; the variable with rank i in the sorted order is matched to the ith value.

2.2 Overview of the Dynamic Algorithm

We achieve an efficient dynamization of the Dekel-Sahni algorithm by showing (1) how to perform logarithmic-time local updates of the sets *matched*(P) and *transferred*(P), for each node $P \in \mathcal{T}$, and (2) that each operation in $\mathcal{U}$ requires a logarithmic number of local updates. These updates will allow us to report the status of any vertex in $O(1)$ worst-case time. The construction of this data structure is described in Section 3. In order to report the mate of a matched vertex we also need to update the *matching* sets. These updates are performed just-in-time, as required by a pair query. We analyze the performance of this method in Section 4.

3 Data Structure Supporting Status Queries

Here we present a structure supporting the update operations in $\mathcal{U}$ in $O(\log^2 n)$ amortized time, and answers status queries in $O(1)$ worst-case time. Due to the limited space, we only consider the case of the insertion of a new variable. The remaining update operations are performed in a similar manner. Our first building block is a structure which allows us to maintain a greedy matching in $O(\log n)$ time per update, for the special case where all intervals have the same starting point. We describe this in Section 3.1. Then, in Section 3.2, we show how an update can propagate bottom-up in T and give the new set of matched vertices in X. Given this set, we can update the set of matched vertices in Y in $O(\log n)$ time, using the same ideas as in Section 3.1. (Due to limited space we omit this last part.)

3.1 The Equal Start-Points Case

Suppose we wish to maintain a greedy matching when all intervals have the same start-point. That is, the input graph $G = (X \cup Y, E)$ is such that $s(x) = 1$ for all $x \in X$, and $Y = \{1, \ldots, m\}$. We say that an $x \in X$ has *higher priority* than an $x' \in X$ if $t(x) < t(x')$. Define n_i as the number of variables whose interval is $[1, i]$. Also define a_i to be the number of matched variables $x \in X$, such that $t(x) \leq i$. These quantities satisfy the recurrence $a_i = \min\{a_{i-1} + n_i, i\}$, for $1 \leq i \leq m$ and $a_0 = 0$. Inserting (deleting) a variable x with domain $[1, k]$ amounts to incrementing (decrementing) n_k. Let n'_i (a'_i) be the value of n_i (a_i) after the update (insertion or deletion of $[1, k]$).

Consider the insertion of the variable $[1, k]$. For the given k we have $n'_k = n_k + 1$ and for $i \neq k$, $n'_i = n_i$. Let j be the smallest integer such that $j \geq k$ and $a_j = j$, if such a j exists, and let $j = m + 1$ otherwise. If $j = k$ then all the first k values are matched with variables that have the same priority as x or higher, and nothing changes. Otherwise, $a_k < k$ and the new variable will be matched; this happens because we always maintain a greedy matching and $[1, k]$ has higher priority than any interval that ends after k. Thus, $a'_i = a_i + 1$ if $k \leq i < j$, and $a'_i = a_i$ otherwise. The remaining update operations are analyzed similarly. In every update operation, we are interested in the quantities $b_j = j - a_j - n_{j+1}$ for all $j \in Y$. When we insert a variable with domain $[1, k]$, we need to locate the first $j \geq k$ with $a_j = j$. Then, by the definition of a_j we have $a_{j-1} + n_j \geq j$, so $j' - a_{j'} - n_{j'+1} \leq -1$ for $j' = j - 1$. Similar inequalities are derived for the remaining update operations.

We construct a data structure that maintains the b_i quantities implicitly. This implicit representation is necessary because a single update can change many b_i's. The data structure consists of a balanced binary search tree T (e.g., a red-black tree [8]) over Y with each leaf corresponding to a value. Leaf i stores n_i and a number $\hat{b}_i$; an internal node v stores the numbers $add(v)$ and $min(v)$. We define these quantities later. Given a node or leaf u and an ancestor v of u in T, define $sum(u, v)$ to be the sum of $add(w)$ for all nodes w on the tree path from u to v, excluding v. Given a node or leaf u, define $sum(u)$ to be $sum(u, root)$, where *root* is the root of T. Let $leaves(v)$ be the set of leaves contained in the subtree rooted at v. The numbers stored in the tree will satisfy the invariants: (a) $b_j = \hat{b}_j + sum(j)$, and (b) $min(v) = \min\{\hat{b}_j + sum(j,v) \mid j \in leaves(v)\}$. Combining (a) and (b) we get $\min\{b_j \mid j \in leaves(v)\} = \min\{\hat{b}_j + sum(j,v) \mid j \in leaves(v)\} + sum(v) = min(v) + sum(v)$. Initially $add(v) = 0$ for all nodes $v \in T$. As T is modified after an update operation the values $add(v)$ keep track of the changes that have to be applied to the affected b_j values. By its definition $add(v)$ affects the b_j values of all the leaves $j \in leaves(v)$. Also, $\hat{b}_j$ is simply the value of b_j before the update. We note that the numbers stored at the internal nodes are easy to maintain in constant time for each restructuring operation of the balanced tree (e.g., rotation).

We sketch how T is updated when we insert a variable with domain $[1, k]$. After locating leaf k we increment n_k. Since this affects b_{k-1}, we increment $\hat{b}_{k-1}$ and update $min(v)$ for the ancestors v of $k-1$ in T bottom-up. As we ascend the path toward the root we calculate $sum(k-1, v)$ at each node v, which takes constant time per node. Next we locate the first leaf $j \geq k$ such that $b_j < 0$. To that end we perform a top-down search on the appropriate subtree of T. Note that as we descend the tree we can compute $sum(v)$ at each node v that we visit, and thus compute $min(v) + sum(v)$ (in constant time per node). So, let $P = (v_0 = root, v_1, \ldots, k)$ be the search path for k in T. Let $P' = (u_1, u_2, \ldots)$ be the subsequence of P composed of the nodes v_i such that v_{i+1} is the left child of v_i; let $r_{i'}$ be the right child of $u_{i'}$. Then, j is located at the subtree rooted at r_{i^*} where i^* is the largest index i' that satisfies $min(r_{i'}) + sum(r_{i'}) < 0$. We can locate r_{i^*} in $O(\log m)$ time. Then the search for j proceeds top-down in the subtree of r_{i^*} as follows: Let v be the current node, and let u be its left child. If $min(u) + sum(u) < 0$ then the search continues in the subtree of u. Otherwise, we move to the right child of v. The last step is to subtract one from b_i, for $k \leq i \leq j$. Since the involved leaves may be too many, we perform the decrements implicitly using the *add* quantities. Let η be the nearest common ancestor of k and j in T; we can locate η easily in $O(\log m)$ time since T is balanced. Let P_k be the path from η to k, excluding the end-points, and let P_j be the path from η to j excluding the end-points. Let v be a node on P_k such that its right child u is not on P_k. If u is a leaf then decrement $\hat{b}_u$; otherwise decrement $add(u)$. Perform the symmetric steps for P_j (visiting the left children of nodes on this path). Finally, decrement $\hat{b}_k$ and $\hat{b}_j$ and update *min* for all $O(\log m)$ nodes on P_k, P_j and on the path from *root* to η; each update takes constant time if we process the paths bottom-up.

Notice that by storing at each leaf i the variables x with $t(x) = i$, we can locate a variable that is replaced in the maximum matching by the newly inserted variable. This is a variable stored in the leaf j that satisfies $\alpha_j = j$ and $j \geq k$. Similarly, after a deletion we can locate a variable that replaces the deleted one in the maximum matching. We will use this capability in the next section. Also, we note that by grouping successive empty leaves (i.e., leaves j with $n_j = 0$) we can reduce the number of leaves in T to $\min\{m, n\}$. Thus, we get:

Lemma 1. *Updates of variables and values in the equal start-points case take worst-case $O(\log n)$ time.*

3.2 The Complete Data Structure

The complete data structure maintains the *matched* and *transferred* sets of all nodes of $\mathcal{T}$. Its building block is the special-case structure of Section 3.1 which maintains these sets, assuming that all start-points of *variables* are equal. By [3, Theorem 2.1], such a data structure can be used to maintain the matching *matched'*(R) for every internal node and the complete matching for every leaf. Hence, each node P in $\mathcal{T}$ is associated with such a local structure T_P, as follows. First, the interval $[s(R), t(R)]$ of available values for matching with *variables'*(R) is translated to $[1, s(R)-t(R)+1]$, and every variable in *variables'*(R) is assumed to have start-point equal to one. Also, each such variable with end-point $t(x) > t(R)$ is stored in the leaf of T_P that corresponds to value $s(R)-t(R)+1$. When we have an update in *variables'*(R) we use the structure T_P to update *matched'*(R) accordingly. As noted in Section 3.1, this process will also return the variables that enter or leave *matched'*(R), so we can propagate the changes towards the root of $\mathcal{T}$.

For simplicity, we assume that the base tree $\mathcal{T}$ is fixed (i.e., we do not insert and delete leaves). Given this, we show that the global data structure can be updated upon the insertion or deletion of a variable or value in $O(\log^2 n)$ time, by proving the next lemma.

Lemma 2. *Each operation in $\mathcal{U}$ causes $O(\log n)$ operations on local data structures of the nodes of $\mathcal{T}$.*

The fully-dynamic data structure can be obtained by applying standard techniques, e.g., maintaining $\mathcal{T}$ as a (weight-balanced) BB[α]-tree [14,15]. As a result, however, our update bounds become amortized.

Insertion of a Variable. Assume that we insert a new variable x with domain $[s, t]$ $(s, t \in Y)$. Then s determines the leaf F of $\mathcal{T}$ where the variable is inserted. After inserting x into T_F, the sets *matched*(F) and *transferred*(F) may change. The changes to these sets need to be propagated to the parent of F in $\mathcal{T}$, in turn causing new changes to the parent of the parent of F, and so on; potentially, changes occur at all nodes on the path from F to the root of $\mathcal{T}$. We will show that at each node, only a constant number of update operations is necessary.

Consider the two possible results of inserting x to the leaf F. The first is that x entered *matched*(F) and, possibly, some other variable x' left *matched*(F). The insert operation on T_F will report if x has entered *matched*(F) and will return x' if it exists. If this x' exists, it either entered *transferred*(F) or not, depending on its end-point, which can be tested in constant time. The second case is that x did not enter *matched*(F), and depending on its end-point either entered *transferred*(F) or not. In both cases, the information that needs to be propagated to the parent of F in $\mathcal{T}$ is a triplet of the form (a, b, c), where each of a and b is either a variable or the symbol ϵ which is a place-holder that represents nothing, and c is either *transferred* or *infeasible*. Such a triplet is interpreted to mean "a is inserted into *matched*(F), b is deleted from *matched*(F) and inserted into c". Note that a and b are not necessarily distinct. For instance, the case in which the inserted variable x is not matched and is inserted into *transferred*(F) is encoded as (x, x, *transferred*).

We now show by induction on the height of the nodes in $\mathcal{T}$, that the information that needs to be propagated from an internal node to its parent can also be represented as a triplet of this form. To that end, let P be a node of $\mathcal{T}$ with left child L and right child R. By the induction hypothesis, P received a triplet (a, b, c) from one of its children, indicating the changes that have occurred at this child. We analyze the two possible cases separately.

First, assume that it was the left child who sent the triplet (a, b, c). This implies a was inserted into *matched*(L) and b was removed from *matched*(L) and inserted into c. The effects of these changes on *matched*(P) and *transferred*(P) are as follows. The first part of *matched*(P) is simply *matched*(L), so it changes in the same manner. The second half is the matching obtained from *variables'*(R). If c is equal to *infeasible*, neither of these sets has changed and the changes that P needs to report to its parent are (a, b, *infeasible*). If c is equal to *transferred*, then b may belong to the second part of *matched*(P), possibly replacing some other variable b' which was matched there before. Again, notice that b' is returned after the update operation on T_P. Now, the important point to note is that P does not need to report anything about b to its parent; b was and remains in *matched*(P). The changes that P needs to report are summarized in the triplet (a, b', c'), where c' is *transferred* or *infeasible*, depending on the end-point of b'.

Now, assume that the triplet (a, b, c) was reported to P from its right child. In this case a was inserted into *matched*(R) and may or may not need to be inserted into *matched*(P). If not, then b was not in *matched*(P) before and will not be in *matched*(P) afterwards; since a replaced b in *matched*(R), we know that its end-point is smaller than b's, hence, it cannot be that $b \in$ *matched*(P) if a was not inserted. Therefore, P does not report any change to its parent. On the other hand, if a was inserted into *matched*(P), then if b was in *matched*(P) before the update it becomes unmatched, and if b was not in *matched*(P) before the update then some other variable b' may have become unmatched to make space for a. In the first case, the change reported by P to its parent is the triplet (a, b, c') where c' depends on the end-point of b. In the second case, P sends the triplet (a, b', c') where c' depends on the end-point of b', and b' is found during the update in T_R. It is important to note that even if b was inserted into *transferred*(R) after the update, it is not inserted into *transferred*(P). This is implied by the fact that b is not in *matched*(P): If b is in *transferred*(P) after the update then b must have already been in *transferred*(P) before the update.

4 Pair Queries

Now we augment the structure of Section 3 in order to support the pair query. As Figure 1 suggests, we cannot afford to change all the affected *matching* sets after an update operation. (In this example we would need to change $O(n)$ such sets.) However, we can get a more efficient solution, based on the following observation:

Observation 3. *An update operation in $\mathcal{U}$ can change only one member of a matching(P) set, for any $P \in \mathcal{T}$.*

A pair query will need to visit the nodes of a root-to-leaf path $\mathcal{T}_q$ of $\mathcal{T}$, and our plan is to update the *matching*(P) sets of the nodes P on this path. If the query vertex is $y \in Y$ then $\mathcal{T}_q$ is simply the path from the root to the leaf corresponding to y. Otherwise, the query vertex is $x \in X$ and $\mathcal{T}_q$ is discovered as we decide at each node P if x belongs to *matching*(L) or *matching*(R), where L and R are the children (left and right respectively) of P. To make this decision fast, we store at node P a list $\ell(P)$ representing the set W defined in Section 2.1. This list is sorted by the end-points $t(x)$ of the variables $x \in W$. A variable $x \in$ *matching*(P) belongs to *matching*(L) iff it appears in $\ell(P)$ and its rank in W is at most $|$*values*(L)$|$. Otherwise, $x \in$ *matching*(R). Hence the decision of whether to follow the left or the right child of P can be made in $O(\log n)$ time if the lists *matching*(P) and $\ell(P)$ are updated. To handle the updates in *matching*(P) (and $\ell(P)$), node P also maintains a list of variables *update*(P), which keeps track of the operations that need to be performed to bring *matching*(P) back to date. This list is arranged in the order the variables were inserted. The meaning of $x \in$ *update*(P) is that x has to be inserted into *matching*(P) if it does not already appear there, otherwise x has to be deleted. Notice that we can construct *matching*(P) in $O(|$*values*(P)$|)$ time, using the lists maintained at its parent in $\mathcal{T}$. Therefore, in our analysis we can assume that the size of *update*(P) is $O(|$*values*(P)$|)$. We use this assumption in the amortized analysis in Section 4.1.

Now consider what happens after an update operation. Observation 3 implies that at most one variable has to be inserted into *matching*(P) and at most one variable has to be deleted. If P is the root of $\mathcal{T}$ then these are exactly the changes in *matched*(P). Hence, the updates can be performed immediately at the root. Let L and R be, respectively, the left and right child of P. Using $\ell(P)$ we can find the variables that change in *matching*(L) and *matching*(R) and insert them into *update*(L) and *update*(R), all in $O(\log n)$ time. (At most two variables are inserted in each of these lists.) To update *matching*(P) for any non-root node of $\mathcal{T}$, we process the variables in *update*(P) one by one. Each variable can be processed in the same way an update for the root of $\mathcal{T}$ was performed. After each update we insert the appropriate variables in the *update* lists of the children of P in $\mathcal{T}$, and so the changes propagate downwards the tree. As noted earlier, the time spent per variable in *update*(P) is $O(\log n)$, and every *update* list contains a number of variables at most twice the number of update operations in $\mathcal{U}$. To get a bound on the pair query that is always $O(n \log n)$, we note that we can rebuilt the whole structure for $\mathcal{T}$ in $O(n \log n)$ time. If we do that after $O(n/\log n)$ total updates, each *update* list will have $O(n/\log n)$ variables at any time. Rebuilding $\mathcal{T}$ results to an additional $O(\log^2 n)$ amortized cost for the updates in $\mathcal{U}$, so the asymptotic update bound is not affected and the bounds stated in Theorem 1 follow. Next we sketch the proof of Theorem 2.

4.1 Amortized Cost of Pair Queries

Lower Bound. We begin with $\mathcal{T}$ being a full binary tree with n leaves, where each leaf stores a unique variable: The ith leaf stores a variable with domain $[i, n]$. For simplicity we take $m = n$. Next we insert $\sqrt{n}$ variables with domains $[1, j]$, $j = 1, \ldots, \sqrt{n}$. These variables will shift the matching by $\sqrt{n}$ positions, and the last $\sqrt{n}$ variables will become infeasible. Consider the $\sqrt{n}$ nodes of $\mathcal{T}$ at height $\log \sqrt{n}$. For each such node P_i we make a pair query for a leaf in P_i's subtree. The corresponding query paths intersect only above the P_i's. To get the claimed bound it suffices to count the work done at each P_i. When the pair query algorithm visits P_i, *update*(P_i) will have $\Theta(\sqrt{n})$ variables, so it will spend $\Omega(\sqrt{n})$ time to process these variables. Hence the total work for all nodes at height $\log \sqrt{n}$ is $\Omega(n)$. By deleting the variables $[1, j]$ and reinserting them, we can repeat this process $\sqrt{n}$ times. This implies an amortized cost of $\Omega(\sqrt{n})$ for a pair query.

Upper Bound. For simplicity, we consider $\mathcal{T}$ to be a full binary tree with n leaves, each storing a single variable. Informally, having more variables in some leaves cannot increase the asymptotic cost bound because both the height of the tree decreases and also each pair query updates the matching for more than one variable. We analyze the worst case running time of a sequence of operations composed of ν updates, followed by μ pair queries. This will give us the worst-case scenario, because the total cost for querying a set of paths is maximized after all updates are performed. At any node P, processing *update*(P) takes $O(2^i \log n)$ if the height i of P is at most $\log \nu$, and $O(\nu \log n)$ for $i > \log \nu$. We calculate separately the cost for processing the nodes at height above $\log \nu$ (top subtree) and below $\log \nu$ (bottom subtrees). First, for the top subtree of $\mathcal{T}$ we observe that each pair query visits $\log n/\nu$ of its nodes and at each node it spends at most $O(\nu \log n)$ time. Hence, if $\mu \leq n/\nu$, the total cost is $O(\mu\nu \log n/\nu \log n) = O(\mu\nu \log^2 n)$, so the amortized cost is $O(\frac{\mu\nu}{\nu+\mu} \log^2 n) = O(\sqrt{n} \log^2 n)$. For $\mu > n/\nu$, the total cost is $O(n \log n + \mu \log n)$, because the top subtree has $O(n/\nu)$ nodes. So, in this case, the amortized cost is $O(\frac{n}{\nu+\mu} \log n + \frac{\mu}{\nu+\mu} \log n) = O(\frac{n\nu}{\nu^2+n} \log n + \log n)$, which is $O(\sqrt{n} \log n)$. We now turn to the bottom subtrees. The contribution of a path inside a bottom subtree is $O(\nu \log n)$. Hence, if $\mu \leq n/\nu$, the total cost is $O(\mu\nu \log n)$, which gives $O(\sqrt{n} \log n)$ amortized cost. The total contribution of a bottom subtree to a pair query is $O(\nu \log n)$, because the total number of variables in all *update* lists is $O(\nu)$. Since we have $O(n/\nu)$ bottom subtrees, for $\mu > n/\nu$ the total cost is $O(n \log n + \mu \log n)$, i.e., $O(\sqrt{n} \log n)$ amortized cost.

Acknowledgement. We thank Willem-Jan van Hoeve for useful references.

References

1. Berge, C.: Two theorems in graph theory. Proc. Nat. Acad. Sci. 43, 842–844 (1957)
2. Coppersmith, D., Winograd, S.: Matrix multiplication via arithmetic progressions. Journal of Symbolic Computation 9(3), 251–280 (1990)

3. Dekel, E., Sahni, S.: A parallel matching algorithm for convex bipartite graphs and applications to scheduling. Journal of Parallel and Distributed Computing 1, 185–205 (1984)
4. Gabow, H.N., Tarjan, R.E.: A linear-time algorithm for a special case of disjoint set union. Journal of Computer and System Sciences 30(2), 209–221 (1985)
5. Galil, Z.: Efficient algorithms for finding maximum matching in graphs. ACM Comput. Surv. 18(1), 23–38 (1986)
6. Gallo, G.: An $O(n \log n)$ algorithm for the convex bipartite matching problem. Operations Research Letters 3(1), 31–34 (1984)
7. Glover, F.: Maximum matching in convex bipartite graphs. Naval Research Logistic Quarterly 14, 313–316 (1967)
8. Guibas, L., Sedgewick, R.: A dichromatic framework for balanced trees. In: Proc. 19th IEEE Symp. on Foundations of Computer Science, pp. 8–21 (1978)
9. Harvey, N.J.A.: Algebraic structures and algorithms for matching and matroid problems. In: Proc. 47th IEEE Symp. on Foundations of Computer Science, pp. 531–542 (2006)
10. Hopcroft, J.E., Karp, R.M.: An $n^{5/2}$ algorithm for maximum matchings in bipartite graphs. SIAM Journal on Computing 2(4), 225–231 (1973)
11. Lipski, W., Preparata, F.P.: Efficient algorithms for finding maximum matchings in convex bipartite graphs and related problems. Acta Informatica 15, 329–346 (1981)
12. Micali, S., Vazirani, V.: An $O(\sqrt{|V|} \cdot |E|)$ algorithm for finding maximal matching in general graphs. In: Proc. 21st IEEE Symp. on Foundations of Computer Science, pp. 17–27 (1980)
13. Mucha, M., Sankowski, P.: Maximum matchings via gaussian elimination. In: Proc. 45th IEEE Symp. on Foundations of Computer Science, pp. 248–255 (2004)
14. Nievergelt, J., Reingold, E.M.: Binary search trees of bounded balance. In: Proc. 4th ACM Symp. on Theory of Computing, pp. 137–142 (1972)
15. Nievergelt, J., Wong, C.K.: Upper bounds for the total path length of binary trees. Journal of the ACM 20(1), 1–6 (1973)
16. Sankowski, P.: Dynamic transitive closure via dynamic matrix inverse. In: Proc. 45th IEEE Symp. on Foundations of Computer Science, pp. 509–517 (2004)
17. Sankowski, P.: Faster dynamic matchings and vertex connectivity. In: Proc. 18th ACM-SIAM Symp. on Discrete Algorithms, pp. 118–126 (2007)
18. Scutellà, M.G., Scevola, G.: A modification of Lipski-Preparata's algorithm for the maximum matching problem on bipartite convex graphs. Ricerca Operativa 46, 63–77 (1988)
19. Steiner, G., Yeomans, J.S.: A linear time algorithm for determining maximum matchings in convex, bipartite graphs. Computers and Mathematics with Applications 31(12), 91–96 (1996)
20. van Emde Boas, P.: Preserving order in a forest in less than logarithmic time and linear space. Information Processing Letters 6(3), 80–82 (1977)
21. van Hoeve, W.-J.: The AllDifferent Constraint: A Survey. In: Proceedings of the Sixth Annual Workshop of the ERCIM Working Group on Constraints (2001)

Communication in Networks with Random Dependent Faults

Evangelos Kranakis[1], Michel Paquette[1], and Andrzej Pelc[2]

[1] School of Computer Science, Carleton University,
Ottawa, Ontario, K1S 5B6, Canada
kranakis@scs.carleton.ca, michel.paquette@polymtl.ca
[2] Département d'informatique et d'ingénierie, Université du Québec en Outaouais.
Gatineau, Québec, J8X 3X7, Canada
pelc@uqo.ca

Abstract. The aim of this paper is to study communication in networks where nodes fail in a random dependent way. In order to capture fault dependencies, we introduce the *neighborhood fault* model, where damaging events, called *spots*, occur randomly and *independently* with probability p at nodes of a network, and cause faults in the given node and all of its neighbors. Faults at distance at most 2 become dependent in this model and are positively correlated. We investigate the impact of spot probability on feasibility and time of communication in the fault-free part of the network. We show a network which supports fast communication with high probability, if $p \leq 1/c \log n$. We also show that communication is not feasible with high probability in most classes of networks, for constant spot probabilities. For smaller spot probabilities, high probability communication is supported even by bounded degree networks. It is shown that the torus supports communication with high probability when p decreases faster than $1/n^{1/2}$, and does not when $p \in 1/O(n^{1/2})$. Furthermore, a network built of tori is designed, with the same fault-tolerance properties and additionally supporting fast communication. We show, however, that networks of degree bounded by a constant d do not support communication with high probability, if $p \in 1/O(n^{1/d})$. While communication in networks with independent faults was widely studied, this is the first analytic paper which investigates network communication for random dependent faults.

Keywords: Fault-tolerance, dependent faults, communication, crash faults, network connectivity.

1 Introduction

As interconnection networks grow in size and complexity, they become increasingly vulnerable to component failures. Links and nodes of a network may fail, and these failures often result in delaying, blocking, or even distorting transmitted messages. It becomes important to design networks in such a way that the desired communication task be accomplished efficiently in spite of these faults,

L. Kučera and A. Kučera (Eds.): MFCS 2007, LNCS 4708, pp. 418–429, 2007.

usually without knowing their location ahead of time. Such networks are called fault-tolerant.

The fundamental questions of network reliability have received much attention in past research under the assumption that components fail randomly and independently (cf., e.g. [1,2,3,4] and the survey [5]). On the other hand, empirical work has shown that positive correlation of faults is a more reasonable assumption for networks [6,7,8]. In [8], the authors provide empirical evidence that data packets losses are spatially correlated in networks, and in [7], the authors use the assumption of failure spatial correlation to enhance network traffic management. Furthermore, in [6], the authors simulate failures in a sensor network using a model much like that of the present paper; according to these authors, the environment provides many spatially correlated phenomena resulting in such fault patterns. Physical and logical phenomena generally affect physical components, causing failures in a positively correlated way. E.g., on August 14, 2003, faults cascaded on the power distribution network and deprived part of North America of electricity. Logical phenomena, like computer viruses and worms, also cause dependent faults. Lightning strikes hitting one node of an electric network cause power outages in entire city blocks.

As our society is increasingly dependent on information networks, it becomes essential to study questions relating to tolerance of dependent positively correlated faults. However, no analytic work has been done for communication networks under this assumption about faults.

In this paper, we consider the problem of feasibility and time of communication in networks with dependent positively correlated faults. To the best of our knowledge, this is the first analytic paper which provides this type of results for network communication.

1.1 Model and Problem Definition

A communication network is modeled as an undirected graph $G = (V, E)$ with a set of nodes V connected by a set of undirected links E. We say that two nodes are *adjacent* (or *neighbors*) if they share a link. The distance between nodes $u, v \in V$ is the minimum number of links which must be traversed from u to reach v; it is denoted by $dist(u, v)$. In a network, $\Gamma(u)$ is the set of nodes adjacent to u; $\Gamma_i(u)$ is the set of nodes $v \in V$ whose distance from u is i; we also denote by $\Gamma_{\leq i}(u)$ the set of nodes $v \in V$ whose distance from u is at most i. A node is said to be *functional*, or *fault-free*, when it executes only its predefined algorithm without any deviation, and doing so, transmits all messages correctly, in a timely manner and without any loss; a node which is not functional is said to be *faulty*. Faults can be of different types: at opposite ends of the spectrum are *crash* and *Byzantine* faults. Faults of the crash type cause faulty components to stop all communication; these components can neither send, receive nor relay any message. Faulty nodes of the Byzantine type may behave arbitrarily (even maliciously) as transmitters. We say that faults are permanent when they affect the nodes for the entire duration of a communication process; otherwise, the faults are said to be transient. In this paper, we assume that faults are permanent and of crash type. Throughout the paper, log means logarithm with base 2 and ln means the natural logarithm.

We consider communication in the fault-free part of the network, where all nodes exchange messages with each other. Communication among functional nodes is feasible if the fault-free part of the network is connected and contains at least two nodes. We measure communication time under the all-port message passing model, where nodes can communicate with all their neighbors in one round, and under the 1-port model, in which every node can send a message to at most one neighbor in one round. Under the all-port model, communication can be completed in time D if the fault-free part of the network has diameter D. Hence, we study the connectivity and diameter of the fault-free part of the network. Moreover, we seek networks of low maximum degree Δ. Since in the 1-port model communication can be completed in time $D\Delta$, networks of low maximum degree and low diameter of the fault-free part support fast communication also in the 1-port model.

In order to capture fault dependencies, we introduce the *neighborhood fault* model, where damaging events, called *spots*, occur randomly and *independently* at nodes of a network, with probability p, and cause permanent crash faults in the given node and all of its neighbors. Faults at distance at most 2 become dependent in this model and are positively correlated. We investigate the impact of spot probability on feasibility and time of communication in the fault-free part of the network.

We design general networks and bounded degree networks which support fast and highly reliable communication despite relatively high spot probabilities. We also prove bounds on spot probability such that highly reliable communication is not supported.

We focus attention on the problem of feasibility and time of communication, guaranteed with *high probability*, i.e., with probability converging to 1 as the size of the network grows. Under the all-port model, in which time of communication is proportional to the diameter, this problem reduces to the question for what spot probability the fault-free part of the network is connected and when it has diameter at most D, with high probability. Under the 1-port model, the same reduction is valid for networks of a given degree.

1.2 Related Work

Dependent fault models were introduced in the study of integrated circuit manufacturing yields. This research models defects as the result of impurities, positioned randomly and independently, affecting nearby circuit components in a dependent way. Results were proposed mainly according to the *quadrat-based* and *center-satellite* approaches. In [9], the author proposed a coarse approach to analyzing production yields based on the assumption that faults occurred in clusters inside a defined grid pattern on Very Large Scale Integration (VLSI) wafers; this quadrat-based model offered provably good results and ease of use required by the industry. Then, in [10], the authors introduced a detailed model of manufacturing defects in VLSI wafers based on the *center-satellite concept* for ecological sampling [11]. Later on, in [12], the authors proposed a simplified center-satellite model of manufacturing defects on VLSI wafers for the study of the memory array reconfiguration problem. In fact, both the center-satellite and quadrat-based approaches are still in use for System on Chip (SoC) (cf., e.g., [13]) and VLSI (cf., e.g., [14,15]) applications. Throughout this field of literature, the consensus is that results originating from the center-satellite approach, as opposed to quadrat-based approaches, are more difficult to apply but provide better prediction quality.

The above approach should be contrasted with the literature on fault-tolerant communication in networks. Many results concerned random link and/or node failures (cf., e.g. [1,2,3,4] and the survey [5]) but, to the best of our knowledge, in all cases faults were assumed to be independent. In [1], the author shows the existence of networks in which $O(\log n)$-time broadcast can be done, under the 1-port model, with high probability, despite links which fail randomly and independently with positive constant probability. In [2], the authors design a network of logarithmic degree which can support high probability communication in time $O(\log n)$ when faults occur randomly and independently on links and nodes with any constant probabilities smaller than 1. In [4], the authors design a similar network which can support communication with high probability in time $O(\log^2 n)$ with Byzantine faults.

Our present research focuses on communication network failures which occur in a dependent way. We consider networks modeled by arbitrary graphs, hence the geometry-dependent, quadrat-based approach to fault dependencies is not appropriate. Our neighborhood fault model, more appropriate for general graphs, is a simplified version of the center-satellite approach.

1.3 Our Results

All our results address the general problem for which spot probabilities p there exist networks supporting communication with high probability, and if so, if this communication is fast in the all-port and 1-port models. Hence we ask for which spot probabilities the fault-free part of the network is connected of size larger than 1, and if so, does it have a small diameter. Moreover, in our positive results we seek networks of low maximum degree.

In Section 2, we address the questions regarding general networks. We first show that there exists a constant c, such that for spot probability $p \leq 1/c \log n$, there exists an n-node graph whose fault-free part has logarithmic diameter and logarithmic degree, with high probability. Hence it supports high probability communication in time $O(\log n)$ in the all-port model and in time $O(\log^2 n)$ in the 1-port model. On the negative side, we show that for constant spot probability p, there exist constants c_1 and c_2 such that: if all degrees in a graph are at most $c_1 \log n$ then the graph is disconnected with high probability; if all degrees in a graph are at least $c_2 \log n$ then the graph has all nodes faulty with high probability. In either case, highly reliable communication is not possible. This leaves some very particular networks undecided. For example, this negative result does not cover the important case of the hypercube, for some constant spot probabilities. Therefore, we study the hypercube separately and prove that, for any constant spot probability $0 < p \leq 1$, this network does not support high probability communication. The above should be contrasted with the results from [2,3] showing that, for *independent* faults, fast highly reliable communication is possible for arbitrary constant fault probabilities in some graphs and for small constant fault probability, even in the hypercube.

In Section 3, we investigate communication in bounded degree networks. We show that the torus supports communication with high probability when $p \in 1/\omega(n^{1/2})$. (As usual, $\omega(f)$ denotes the set of functions g such that $g/f \to \infty$.) However, the diameter of an n-node torus is at least $\Theta(\sqrt{n})$ and the fault-free part has the same large diameter. Hence we seek networks with the same fault-tolerance properties, but with small diameter. We construct a bounded degree network built of tori, whose fault-free part has diameter $O(\log n)$ whenever $p \in 1/\omega(n^{1/2})$. Hence this network supports high probability communication in logarithmic time, both in the all-port and in the 1-port models. On the negative side, we show that neither the torus nor the above network can support highly reliable communication when $p \in 1/O(n^{1/2})$. Finally, we prove that networks of degree bounded by a constant d cannot support communication with high probability when $p \in 1/O(n^{1/d})$. Due to lack of space, many proofs are deferred to the journal version of this paper.

2 General Networks

In this section, we focus on general networks. We first design a network which supports communication with high probability when the spot probability is at most $1/c\log n$, for some positive constant c. We then establish two bounds on node degrees showing that a large class of networks cannot support communication with high probability when spot probability is a positive constant.

2.1 Upper Bounds

This section is dedicated to proving the following result.

Theorem 1. *There exists an n-node graph whose fault-free part has diameter $O(\log n)$ and logarithmic degree, with high probability, for spot probability $p \leq 1/c\log n$, where c is some positive constant.*

The network construction is based on a binary tree structure where each node of the tree represents a group of nodes and each link of the tree represents a random set of links between nodes in adjacent groups. To be more precise, for a fixed m, we define a random n-node graph $G(n,m)$. Let $x = \lceil n/m \rceil$. Partition the set of all nodes into subsets $S_1, \ldots, S_x$, of size m, (S_x of size at most m) called *supernodes*. Let $\mathcal{S} = \{S_1, \ldots, S_x\}$ be the set of all supernodes.

Let $L = \lfloor \log x \rfloor$. Arrange all supernodes into a binary tree T with $L+1$ levels $0, 1, \ldots, L$, placing each supernode S_i on level $\lfloor \log i \rfloor$. Level 0 contains the root and levels $L-1$ and L contain leaves of T. The supernode S_1, is the root of T.

For every $1 \le i \le \lfloor x/2 \rfloor$, S_{2i} is the left child of S_i and S_{2i+1} is the right child of S_i in T (S_{2i+1} exists if $x \ge 2i+1$). For every $1 < i \le x$, supernode $S_{\lfloor i/2 \rfloor}$ is the parent of S_i. If a supernode is a parent or a child of another supernode, we say that these supernodes are adjacent in T.

The set of edges of $G(n, m)$ is defined as follows. If supernodes S_i and S_j are adjacent in T, then there is an edge in $G(n, m)$ between any node in S_i and any node in S_j with probability p_l. Moreover, supernodes have no interior links. The graph $G(n, m)$ is called a Random Binary Thick Tree ($RBTT$).

In the remainder of this section, we analyze $RBTT$ and show that, if $p \le 1/c \log n$, for some constant $c > 0$ to be defined below, then it supports communication with high probability in time $O(\log n)$. We consider the n-node $RBTT$ with link probability $p_l = 1/18 \ln n$ and $m = \lceil 1152 \ln^2 n \rceil$ nodes per supernode, when spot probability is $p \le 1/(768 \ln n)$. Hence, we take $c = 768/\ln 2$.

Let C_1 be the event that all supernodes in $RBTT$ contain less than $6 \ln n + 1$ spots.

Lemma 1. *The event C_1 occurs with probability at least $1 - 1/n$.*

For a given constant $0 < \epsilon \le 1$, let C_2 be the event that all supernodes in $RBTT$ have more than $288(1-\epsilon)\ln^2 n$ functional nodes.

Lemma 2. *The event C_2 occurs with probability at least $1 - 1/n^{d \log n}$, for some positive constant d.*

Using the previous results, we now present two connectivity lemmas in preparation for the proof of the main theorem of this section.

Lemma 3. *All functional nodes are connected to at least one functional node in each supernode adjacent to their own, with probability exceeding $1 - 1/n^{13}$.*

Proof. Fix a node u. Let $N(u)$ denote the set of supernodes adjacent to the supernode containing u. Consider the event γ_{u,S_k} that u has a link to at least one functional node in a given supernode $S_k \in N(u)$. The event γ_{u,S_k} occurs unless all links from u to functional nodes in S_k do not exist. From Lemma 2, we get for any constants $0 < \epsilon', \epsilon'' \le 1$

$$
\begin{aligned}
\Pr[\gamma_{u,S_k}] &\ge \Pr[\gamma_{u,S_k} \wedge C_2] = \Pr[C_2] \Pr[\gamma_{u,S_k} \mid C_2] \\
&\ge \Pr[C_2] \left(1 - (1 - 1/18 \ln n)^{288(1-\epsilon')(1-\epsilon'') \ln^2 n}\right) \\
&\ge \left(1 - n^{-d \ln n}\right) \left(1 - n^{-16(1-\epsilon')(1-\epsilon'')}\right),
\end{aligned}
$$

and hence $\Pr[\gamma_{u,S_k}] \ge 1 - n^{-15}$. Furthermore, since the graph contains at most n functional nodes, which should be connected to at least one functional node in at most 3 supernodes, the estimated probability is at least

$$
\begin{aligned}
\Pr[(\forall u \in V\ (\forall S_k \in N(u)))\ \gamma_{u,S_k}] &\ge 1 - \sum_{u \in V} \sum_{S_k \in N(u)} \Pr[\neg \gamma_{u,S_k}] \\
&\ge 1 - 3nn^{-15} > 1 - n^{-13}.
\end{aligned}
$$

□

Lemma 4. *All functional node pairs in supernodes at distance 3 are connected by a fault-free path with probability at least* $1 - 1/n^{1.9}$.

Proof. This lemma is proven in steps, defining connection probabilities and lower bounds on the number of connected nodes at distances 1, 2, and 3.

Fix 4 supernodes, S_u, S_i, S_j, S_k, which form a simple path in RBTT. I.e., S_u is adjacent to S_i, which is adjacent to S_j, which is adjacent to S_k.

Fix a node u in S_u. Let X_i be the random variable which counts the number of functional nodes $i \in \Gamma(u)$ located in S_i. From Lemma 2, each supernode contains more than $288(1-\epsilon)\ln^2 n$ fault-free nodes, for any $0 < \epsilon \leq 1$ with probability $1 - 1/n^{d \log n}$, for some positive constant d. Since the link probability is $p_l = 1/18 \ln n$,

$$E[X_i] \geq \Pr[C_2] \frac{288(1-\epsilon)\ln^2 n}{18 \ln n} = (1 - n^{-d \log n}) 16(1-\epsilon)\ln n \geq 16(1-\epsilon')\ln n,$$

with some $1 > \epsilon' > \epsilon$. We also have that a fixed functional node has at most $16(1 - \sqrt{3/8(1-\epsilon')})(1-\epsilon')\ln n$ such neighbors with probability

$$\Pr\left[X_i \leq 16\left(1 - \sqrt{\frac{3}{8(1-\epsilon')}}\right)(1-\epsilon')\ln n\right] \leq e^{-\left(\sqrt{\frac{3}{8(1-\epsilon')}}\right)^2 \frac{16(1-\epsilon')\ln n}{2}} = n^{-3}.$$

Let A be the event that node u has at least $16(1 - \sqrt{3/(8(1-\epsilon'))})(1-\epsilon')\ln n$ functional neighbors in S_i.

Assume event A occurs. Now, fix a node x in S_j. Fix a subset $S \subseteq \Gamma(u) \cap S_i$ of functional nodes, with size $16(1 - \sqrt{3/(8(1-\epsilon'))})(1-\epsilon')\ln n$. Denote by P_{Sx} the event that there exists a link between the node x and any node from S. This event occurs unless x has no link to some node in S. Hence,

$$\begin{aligned}\Pr[P_{Sx}|A] &= 1 - (1 - 1/18\ln n)^{16\left(1-\sqrt{3/(8(1-\epsilon'))}\right)(1-\epsilon')\ln n} \\ &\geq 1 - e^{-8\left(1-\sqrt{3/(8(1-\epsilon'))}\right)(1-\epsilon')/9} \geq 1/4\end{aligned}$$

for some small ϵ'.

Let X_j be the random variable which counts the number of functional nodes $j \in S_j$ which are adjacent to some node in S. We have that $E[X_j] \geq (1/4)\cdot 288(1-\epsilon')\ln^2 n$, assuming that A holds. Let B be the event that $X_j \geq 72(1-\epsilon'')\ln^2 n$, for some small $\epsilon'' > \epsilon'$. Since all events P_{Sx}, for fixed S and varying x, are independent, we use a Chernoff bound to show that, if event A occurs, event B occurs with probability $1 - 1/n^{k' \log n}$, for some positive constant k'.

Assume event $A \cap B$. Fix a functional node k in S_k. Fix a subset $S' \subseteq S_j$ of functional nodes, each of which is a neighbor of some element of S, with size $72(1-\epsilon'')\ln^2 n$. Denote by $P_{S'k}$ the event that there exists a link between node k and some node in S'. This event occurs unless k has no link to any node in S'. Hence,

$$\begin{aligned}\Pr[P_{S'k}|B \cap A] &= \left(1 - (1 - 1/18\ln n)^{72(1-\epsilon'')\ln^2 n}\right) \\ &\geq 1 - e^{-72(1-\epsilon'')\ln^2 n/(18\ln n)} \geq 1 - n^{-4(1-\epsilon'')}.\end{aligned}$$

Consider the event P_{uijk} that there exists a fault-free path of the form $uijk$ from a fixed node u to a fixed node k. Clearly, P_{uijk} is a subset of the event detailed in the above argument. Hence,

$$\begin{aligned}\Pr[P_{uijk}] &\geq \Pr[P_{S'k} \cap B \cap A] = \Pr[A]\Pr[B|A]\Pr[P_{S'k}|B \cap A] \\ &\geq (1-n^{-3})\left(1-1/n^{k'\log n}\right)\left(1-n^{-4(1-\epsilon'')}\right) \geq \left(1-n^{-3+\epsilon'''}\right),\end{aligned}$$

for some $0 < \epsilon''' < 0.1$.

There are at most n functional nodes in $RBTT$, each with $O(\log^2 n)$ other functional nodes in supernodes at distance 3. Hence, there are $O(n\log^2 n)$ functional node pairs in supernodes at distance 3. It follows that all node pairs in supernodes at distance 3 are connected with probability at least $1 - n^{-1.9}$. □

Combining the previous lemmas, we are now ready to prove Theorem 1.

Proof of Theorem 1. The $RBTT$ contains $O(n/\log^2 n)$ supernodes connected in a binary-tree structure of diameter $D \in O(\log n)$. It follows from the construction that the maximum degree of the $RBTT$ is $O(\log n)$, with high probability. By Lemma 4, all functional node pairs in supernodes at distance 3 are connected by at least one fault-free path of length 3 with probability greater than $1 - 1/n^{1.9}$. Therefore, all functional nodes in the subgraph $RBTT'$ composed of the root supernode S_1 and of all supernodes at distances multiple of 3 from S_1 are connected with this probability. Clearly, functional nodes not in $RBTT'$ are in supernodes adjacent to supernodes in $RBTT'$. Thus, by Lemma 3, all these functional nodes are also connected to at least one functional node in $RBTT'$ with probability exceeding $1-1/n^{13}$. Hence, with probability exceeding $1 - 1/n^{1.8}$, the fault-free part of $RBTT$ is connected.

We now investigate the diameter of the fault-free part of $RBTT$. From the above argument, we observe that 1) nodes in supernodes at distances multiple of 3 are connected with high probability by a path of length equal to the distance of the supernodes; 2) functional nodes in all other supernodes are connected with high probability by a path of length at most 2 longer than the distance of the supernodes. This leads to the conclusion that the diameter of the fault-free part of $RBTT$ is also in $O(\log n)$, with high probability. □

2.2 Lower Bounds

We have shown that it is possible to build a logarithmic-degree graph which supports communication with high probability in spite of spot probabilities $p \leq 1/c\log n$, for some positive constant c. The natural question then is whether it is possible to build arbitrarily large networks which can support communication with high probability despite larger spot probabilities. In what follows, we show that for constant spot probabilities, most networks do not have this property. More formally, the following theorem holds.

Theorem 2. *For any constant spot probability $p > 0$, there exist constants c_1 and c_2 such that: if all degrees in a graph are at most $c_1 \log n$ then the fault-free* *part of the graph is disconnected with high probability; if all degrees in a graph are at least* $c_2 \log n$ *then the graph has all nodes faulty with high probability. In either case, highly reliable communication is not possible.*

The preceding theorem leads to the conclusion that high probability communication is not possible, for a large class of graphs, when spot probability is a positive constant. However, the bounds $c_1 \log n$ and $c_2 \log n$ do not coincide. Since $c_1 < \frac{1}{\log(1/(p(1-p)))}$ and $c_2 > \frac{1}{\log(1/(1-p))}$, we have $c_1 < c_2$ for all positive values of p. It remains open whether or not there exists an arbitrarily large graph which supports reliable communication despite constant spot probabilities.

We will now attempt to provide insight into the question of what happens when node degrees lie between these bounds. For example, when $p = 1/2$, we have $c_1 < 1/2$ and $c_2 > 1$. Thus, with degree $\log n$, the important case of the n-node hypercube is not covered by Theorem 2. We will investigate this case in the following section.

2.3 Communication in the Hypercube

The hypercube H_k of dimension k is a 2^k-node graph with the set of nodes with identifiers from $\{0,1\}^k$ and the set of links between nodes whose identifiers have a Hamming distance of 1. Hence the n-node hypercube H_k has dimension $\log n$.

Theorem 3. *The n-node hypercube H_k does not support high probability communication for any constant spot probability* $0 < p \leq 1$.

We first show that for constant $0 < p < 1/2$, the fault-free part of the graph is disconnected with high probability. We then show that for $1/2 < p \leq 1$, the graph has all nodes faulty with high probability, and that for $p = 1/2$, the graph has all nodes faulty with constant probability. This will prove Theorem 3.

3 Bounded Degree Networks

The RBTT presented in Section 2 remained connected despite relatively high spot probabilities. However, its degree is unbounded. For certain applications, smaller-degree networks may be preferred as they are easier to implement and give shorter communication time in the 1-port model. Therefore, it is natural to ask if bounded-degree networks can also support high-probability communication with comparable spot probabilities.

In this section we construct bounded-degree networks which tolerate inverse polynomial spot probabilities and which support high-probability communication with optimal time complexity. Furthermore, we prove that bounded-degree networks can tolerate at most inverse polynomial spot probabilities.

3.1 Upper Bounds

We now study the properties of two networks: the torus and a torus-based tree-like network that we call the *toroidal tree*. We show that the torus supports high-probability communication for spot probability in $1/\omega(n^{1/2})$. However, the diameter of the torus is quite large, which prohibits fast communication. Thus we design a tree-like structure based on the torus which provides the same fault-tolerance properties and supports communication in time $O(\log n)$, even in the 1-port model.

The Torus. In this section, we show an upper bound on the spot probability such that the fault-free part of the torus remains connected. Denote by $\mathcal{T}_{m\times k}$ the $m \times k$ torus with $m, k \geq 4$. The torus has the set of nodes $\{u = (u_x, u_y) : u_x \in \{0, 1, \ldots, m-1\}, u_y \in \{0, 1, \ldots, k-1\}\}$ and the set of links $\{(u, v) : |u_x - v_x| \bmod m + |u_y - v_y| \bmod k = 1\}$.

Theorem 4. *The fault-free part of the n-node torus $\mathcal{T}_{m\times k}$ is connected with high probability for $p \in 1/\omega(n^{1/2})$.*

The Toroidal Tree. We now design a network which provides the same fault-tolerance as the torus, while also providing optimal-order communication time for bounded-degree graphs. Since the diameter of a bounded-degree graph is at least logarithmic, our aim is to construct a network whose fault-free part has logarithmic diameter. Such a network supports highly reliable communication in optimal time $O(\log n)$, even in the 1-port model. The network construction is based on two binary trees, T and T', connected by a link between their root nodes. Each node of T, T' represents a group of nodes, and groups adjacent in T, T' have a subset of nodes in common. More precisely, for constant $k \geq 4$, we define a n-node graph $G(n, k)$. Assume that the set of nodes can be partitioned exactly as described below; this is easy to modify in the general case, by adding nodes to a leaf group.

Let the sets $\mathcal{T}_1, \ldots, \mathcal{T}_x$ and $\mathcal{T}'_1 \ldots, \mathcal{T}'_{x'}$ be tori with $2k$ rows $\{0, 1, \ldots 2k-1\}$ and k columns $\{0, 1, \ldots k-1\}$; $|x - x'| \leq 1$. We describe the construction for the tree T; the same construction is applied for the tree T'. Arrange all $\mathcal{T}_i$ as the nodes of T, with $L+1$ levels $0, 1, 2, \ldots, L$, placing each $\mathcal{T}_i$ on level $\lfloor \log i \rfloor$ of T. Level 0 contains the root of T and levels $L-1$ and L contain the leaves. For every $1 \leq i \leq \lfloor x/2 \rfloor$, $\mathcal{T}_{2i}$ is the left child of $\mathcal{T}_i$ in T and $\mathcal{T}_{2i+1}$ is the right child of $\mathcal{T}_i$ in T ($\mathcal{T}_{2i+1}$ exists if $x \geq 2i+1$). For every $1 < i \leq x$, $\mathcal{T}_{\lfloor i/2 \rfloor}$ is the parent of $\mathcal{T}_i$. Use row 0 of each child torus to connect it to its parent in T. Use row k of each parent torus to connect it to both its children in T. Use row 0 of both roots in T, T' to connect them together. Connections between tori adjacent in T, T' are done by identifying the respective rows.

It follows from the above description that $x + x' = \lfloor (n - 2k^2)/((2k-1)k) \rfloor + 1$ tori are located on $L = \lfloor \log(x + x' + 1) \rfloor$ levels in $G(n, k)$. The graph $G(n, k)$ is called a Toroidal Tree. It has bounded maximal degree.

Theorem 5. *For $p \in 1/\omega(n^{1/2})$, the n-node Toroidal Tree supports high probability communication in time $O(\log n)$.*

3.2 Lower Bounds

In this section, we show that bounded-degree graphs do not support high probability communication even for relatively small spot probabilities. We first show that the bounds on spot probability provided in Theorem 4 and Theorem 5 are tight for tori and toroidal trees. We then show that for general bounded-degree networks, if spot probability is the inverse of some polynomial, then high probability communication is not supported.

The Torus and Toroidal Tree. The following lower bounds match the upper bounds from Theorem 4 and Theorem 5, thus showing that the results are tight.

Theorem 6. *For spot probability* $p \in 1/O(n^{1/2})$, *the n-node torus* $\mathcal{T}_{m \times k}$ *does not support high probability communication.*

Theorem 7. *For spot probability* $p \in 1/O(n^{1/2})$, *the n-node Toroidal Tree does not support high probability communication.*

General Bounded Degree Graphs. We showed in the preceding section that in the case of the torus and the Toroidal Tree, spot probabilities at most $1/\omega(n^{1/2})$ can be tolerated if these graphs support high-probability communication. In the following theorem, we show that a similar phenomenon occurs for all graphs whose degree is bounded by a constant.

Theorem 8. *For spot probability* $p \in 1/O(n^{1/d})$, *no n-node graph with degree bounded above by* $d \in \Theta(1)$ *supports high probability communication.*

4 Conclusion

We provided what is, to the best of our knowledge, the first analytic results on fault-tolerance of networks in the presence of dependent, positively correlated faults. To do so, we introduced the *neighborhood fault* model where damaging events, called *spots*, occur randomly and independently at nodes of a network with probability p, and cause faults in the affected node and its neighbors.

We addressed questions regarding the connectivity and diameter of the fault-free part of networks in this fault model, as these characteristics of the network are responsible for the feasibility of communication and for its time. Our results show clear differences between the assumption of independent faults and that of the neighborhood fault model. For example, while under independent faults with small constant fault probability $p > 0$ the fault-free part of the hypercube remains connected with high probability [3], this is not the case under the neighborhood fault model with any positive constant spot probability. Likewise, the fault-free part of the torus is connected with high probability for fault probability $p \in 1/\Omega(n^{1/4})$ when faults are independent, but this is not the case for such spot probabilities under the neighborhood fault model.

It remains open whether or not there exists a network, which, under the neighborhood fault model, has the fault-free part connected with high probability despite constant spot probabilities. We conjecture that this is not the case.

The neighborhood fault model is the first step in modeling dependent positively correlated faults in networks. It would be interesting to analyze more precise center-satellite based models in which independent spots yield faults in nodes with probability decreasing with the distance of the node from the spot.

Acknowledgements. Evangelos Kranakis and Michel Paquette were supported by NSERC and MITACS. Andrzej Pelc was supported by NSERC and the Research Chair in Distributed Computing of the Université du Québec en Outaouais.

References

1. Bienstock, D.: Broadcasting with random faults. Discr. Appl. Math. 20, 1–7 (1988)
2. Chlebus, B.S., Diks, K., Pelc, A.: Sparse networks supporting efficient reliable broadcasting. Nordic Journal of Computing 1, 332–345 (1994)
3. Chlebus, B.S., Diks, K., Pelc, A.: Reliable broadcasting in hypercubes with random link and node failures. Comb., Prob. and Computing 5, 337–350 (1996)
4. Paquette, M., Pelc, A.: Fast broadcasting with byzantine faults. International Journal of Foundations of Computer Science 17(6), 1423–1439 (2006)
5. Pelc, A.: Fault-tolerant broadcasting and gossiping in communication networks. Networks 28(6), 143–156 (1996)
6. Ganesan, D., Govindan, R., Shenker, S., Estrin, D.: Highly-resilient, energy-efficient multipath routing in wireless sensor networks. ACM SIGMOBILE Mobile Computing and Communications Review 5(4), 11–25 (2001)
7. Thottan, M., Ji, C.: Using network fault predictions to enable IP traffic management. J. Network Syst. Manage 9(3), 327–346 (2001)
8. Yajnik, M., Kurose, J., Towsley, D.: Packet loss correlation in the MBone multicast network. Proceedings of IEEE Global Internet (May 27, 1996)
9. Stapper, C.H.: On yield, fault distributions and clustering of particles. IBM Journal of Research and Development 30(3), 326–338 (1986)
10. Meyer, F.J., Pradhan, D.K.: Modeling defect spatial distribution. IEEE Trans. Computers 38(4), 538–546 (1989)
11. Warren, W.: The center-satellite concept as a basis for ecological sampling. Stat. Ecol. 2, 87–118 (1971)
12. Blough, D.M., Pelc, A.: A clustered failure model for the memory array reconfiguration problem. IEEE Trans. Computers 42(5), 518–528 (1993)
13. Meyer, F.J., Park, N.: Predicting defect-tolerant yield in the embedded core context. IEEE Trans. Computers 52(11), 1470–1479 (2003)
14. Choi, A., Park, N., Meyer, F.J., Lombardi, F., Piuri, V.: Reliability measurement of fault-tolerant onboard memory system under fault clustering. In: Proceedings of 19th Instrumentation and Measurement Technology Conference, 2002. IMTC, vol. 2, pp. 1161–1166. IEEE Computer Society Press, Los Alamitos (2002)
15. Yu, F., Tsai, C.H., Huang, Y.W., Lee, D.T., Lin, H.Y., Kuo, S.Y.: Efficient exact spare allocation via boolean satisfiability. In: 20th IEEE Int. Symp. on Defect and Fault Tolerance in VLSI Systems (DFT'05), pp. 361–370. IEEE Computer Society Press, Los Alamitos (2005)

Optimal Gossiping in Directed Geometric Radio Networks in Presence of Dynamical Faults*
(Extended Abstract)

Andrea E.F. Clementi[1], Angelo Monti[2], Francesco Pasquale[1,**], and Riccardo Silvestri[2]

[1] Dipartimento di Matematica, Università di Roma "Tor Vergata"
{clementi,pasquale}@mat.uniroma2.it
[2] Dipartimento di Informatica, Università di Roma "La Sapienza"
{monti,silvestri}@di.uniroma1.it

Abstract. We study deterministic fault-tolerant gossiping protocols in *directed Geometric Radio Networks* (in short, directed GRN). Unpredictable node and link faults may happen during every time slot of the protocol's execution.

We first consider the *single-message* model where every node can send at most one message per time slot. We provide a protocol that, in any directed GRN G of n nodes, completes gossiping in $O(n\Delta)$ time (where Δ is the maximal in-degree of G) and has message complexity $O(n^2)$. Both bounds are then shown to be optimal.

As for the *combined-message* model, we give a protocol working in optimal completion time $O(D\Delta)$ (where D is the maximal source eccentricity) and message complexity $O(Dn)$. Finally, our protocol performs the (single) broadcast operation within the same optimal time and optimal message complexity $O(n)$.

1 Introduction

In a *radio* network, every node (station) can directly transmit to some subset of the nodes depending on the power of its transmitter and on the topological characteristics of the surrounding region. When a node u can directly transmit to a node v, we say that there is a (wireless) directed link (u, v). The set of nodes together with the set of these links form a directed communication graph that represents the radio network. In the radio network model [BGI92, CGR02, CGGPR00, CR06], the communication is assumed to be synchronous: this allows to focus on the impact of the *interference* phenomenon on the network performance. When a node sends a message, the latter is sent in parallel on all outgoing links. However, since a single radio frequence is used (see [ABLP89, BGI92, CGGPR00]), when two or more neighbors of a node transmit at the same time slot, a *collision* occurs (due to interference) and the message is lost. So, a node can recover a message from one of its incoming links if and only if this link is the only one bringing in a message. The broadcast task consists of sending a *source message* from a given *source* node to all nodes of the network. The *completion time* of a broadcast protocol is the number of time slots required by the protocol to inform all (reachable) nodes. A node is *informed* if it has received the source message.

* Research partially supported by the EU under the EU/IST Project 15964 AEOLUS.
** Corresponding author.

L. Kučera and A. Kučera (Eds.): MFCS 2007, LNCS 4708, pp. 430–441, 2007.

Another important task in radio networks is *gossiping*, i.e., n simultaneous and independent broadcast operations, each one from a different node [CGR02, CMS03, GPX05]. The completion time of a gossiping protocol is the number of time slots the protocol requires so that every source message m is received by all nodes reachable from the source of m. We will consider two transmission models: the *single-message* model [BII93] and the *combined-message* one [CGR02]: in the former every node can transmit and receive at most one source message per time-slot while, in the latter, source messages can be arbitrarily combined and sent/received in one time slot [CGR02, GPX05]. Broadcasting and gossiping are fundamental communication tasks in radio networks and they are the subject of several research works in both algorithmic and networking areas [BGI92, CGR02, CGGPR00, PR97, R96]. It is reasonable to claim that almost all major theoretical questions related to such tasks can be considered closed as far as *static* networks are considered: the network never changes during the entire protocol's execution (see Subsection 1.1).

However, radio networks are typically adopted in scenarios where *unpredictable* node and link faults happen very frequently. Node failures happen when some hardware or software component of a station does not work, while link failures are due to the presence of a new (artificial or natural) hurdle that does not allow the communication along that link. In ad-hoc networking, while it is sometimes reasonable to assume that nodes (thus the protocol) know the *initial* topology, nothing is known about the duration and the location of faults. Such faults may clearly happen *even during the execution of a protocol.* In the sequel, such kind of faults will be called *dynamical faults* or, simply, faults.

Theoretical results on broadcast and gossiping protocols in any scenario where the network topology may change during the protocol's execution are very few (see Subsection 1.1).

The Model of Faulty Networks. We follow a high-level approach by considering *adversarial networks* [AS98, ABBS01, CMS04, P02, S01]. Arbitrary dynamical faults are decided by a deterministic adaptive *adversary*. We analyze the completion time and the message complexity (i.e. maximum number of transmitted messages) of broadcast and gossiping protocols with respect to *worst-case* adversary's strategies.

The (worst-case) completion time of a *Fault-tolerant Broadcast* (in short, *FB*) protocol on a network G is defined as the maximal number (with respect to any possible adversarial strategy) of time slots required to *inform* all nodes reachable from the source in the *unpredictable* fault-free part of the network. More precisely, according to the fault-tolerance model adopted in [KKP98, P02, CMS04], a *fault pattern* F is a function (managed by the adaptive adversary) that maps every time-slot t to the subset $F(t)$ of nodes and links that are faulty during time slot t. The residual subgraph G^F is the graph obtained from G by removing *all* those nodes and links that belong to $F(t)$, *for some* time-slot t during the protocol's execution. Then, a FB protocol for a graph G is a broadcast protocol that, for any source s, and for any *fault pattern* F, guarantees that every node, which is reachable from s in the *residual subgraph* G^F, will receive the source message. The *residual eccentricity* of a node v is its eccentricity in the residual graph. The eccentricity of v is the maximal oriented distance (i.e. number of hops) from v to a reachable node.

The above definitions can be easily extended to *Fault-tolerant Gossiping* (in short *FG*) protocols: For any source s, message m_s must be received by every node reachable from s in G^F, for any choice of fault pattern F.

It is important to remark that if a node v is not reachable from a source in the residual subgraph, then the arrival of m_s to v is not considered in the analysis of the completion time. This assumption might be considered too strong but it is *necessary*. Indeed, it is easy to see that any attempt to consider larger residual subgraphs makes the worst-case completion time of *any* deterministic FG protocol *infinite*. This is well-explained by the following simple game. Consider k informed nodes that are in the in-neighborhood of a non informed node w. It is easy to see that *any* deterministic protocol, trying to inform w, fails *forever* against the following simple adversary's strategy: if at least two of the k in-neighbors transmit then the adversary leaves all edges on, while if there is exactly one of them transmitting, then the adversary makes only this link faulty. Observe that w is always *connected* to the informed part of the network but it will never receive the message (w is indeed not in the *residual* graph).

On the other hand, broadcasting and gossiping (and their analysis) in the residual graph is much harder than the same operation in fault-free radio networks. This is mainly due to the presence of unknown collisions that the adversary can produce at any time-slot *on the residual graph too*. As a matter of fact, while the completion time of broadcast on general fault-free radio networks of source eccentricity D is $O(D+\log^3 n)$ [GPX05], it turns out that there is a class of radio networks of *constant* source eccentricity where the same operation, in the above fault model, requires $\Theta(n\sqrt{n})$ time slots [CMS04]. So, in general graphs of "small" source eccentricity, the completion time gap may be exponential. The lower bound $\Omega(n\sqrt{n})$ in [CMS04] provides also a strong evidence of the significant difference between *dynamical* faults (on the residual graph) and *permanent* faults: in the latter network scenario, worst-case broadcasting time is $O(n \log^2 n)$ [CGR02].

Our Results. We investigate *directed Geometric Radio Networks*, in short directed GRN [ENW00, KKKP00, CKOZ03, CCPRV01, DP07]. A directed GRN $G(V, E)$ is constructed by arbitrarily placing n nodes on the Euclidean plane; then, to each node v a transmission range $r_v \geq 0$ is assigned. These transmission ranges uniquely determine the set E of *directed* links: $(u, v) \in E$ iff $d(u, v) \leq r_u$, where $d(u, v)$ denotes the Euclidean distance between u and v. When all nodes have the same transmission range, the resulting graph is symmetric: this restriction is denoted as *symmetric* GRN.

We provide the first optimal bounds on the completion time and message complexity of FG protocols (and FB ones) in directed GRN for both single-message and combined-message models. More precisely, for the first model, given any directed GRN G of n nodes and maximal in-degree Δ, our FG protocol works in $O(n\Delta)$ time-slots and it has message complexity $O(n^2)$. Such bounds are then shown to be optimal.

Then, we consider the combined-message model and provide an FG protocol that works in optimal $O(D\Delta)$ time-slots (D denotes the maximal residual source eccentricity) and it has message complexity $O(n^2)$. We emphasize that this is the first FG protocol whose completion-time does not (explicitly) depend on n. Furthermore, the protocol can be easily analyzed for the (single) broadcast task: in this case, the completion time is still $O(D\Delta)$ while the message complexity reduces to $O(n)$. Both upper bounds are again optimal and, as for time complexity, it improves over the best (polynomial-time constructible) FB upper bound for general graphs by an $O(\log^3 n)$ factor ([CMS04] - see Subsection 1.1).

Adopted Techniques. Since the fault pattern is unpredictable, an FG protocol must have the following "connectivity" property: it must consider all possible paths from a source to any node reachable from that source. To this aim, our protocols make an iterative use of collision-free families. A *collision-free family* is a set family (defined on the out-neighborhoods of the input graph - see Definition 2.1) that induces a transmission scheduling that somewhat guarantees the above connectivity property and yields *no* collision. So, when a node is scheduled as transmitter, its message is safely received by *all* its out-neighbors in the residual graph. This important fact is one of the key ingredients to get optimal message complexity (and thus energy efficiency) of our protocols. On the other hand, the size of the collision-free family is a linear factor in the completion time of our FG protocols. A crucial step in our protocol design is thus the efficient construction of a collision-free family for the input graphs. We indeed provide an algorithm that constructs an optimal-size collision-free family for any directed GRN working in time $O(n^2)$.

We observe that, given access to a collision-free family for the input graph, our protocols run in a fully-distributed fashion. However, in order to construct such optimal collision-free family it is necessary to know the *initial* graph topology. In Section 3, we also provide an efficient distributed construction of collision-free families under a much weaker knowledge condition: each node construct its own scheduling (so, "its" component of the collision-free family) by assuming that it only knows its position and a good approximation of the minimal distance among nodes. We then prove that if the (unknown) initial topology is *well spread* [CPS04], the returned collision-free family has optimal size, thus yielding the same protocol's performance given by the centralized construction. Well spread instances (see Definition 3.8) are a natural and broad generalization of *grid* networks. Due to lack of space, some proofs are omitted and they are available in the full version [CMPS07].

1.1 Related Works

Permanent Faults. In [KKP98], the authors consider the broadcast operation in presence of permanent unknown node faults for two restricted classes of networks. They derive a $\Theta(D+\log\min\{\Delta, t\})$ bound where D is the source eccentricity in the residual graph and t is the number of faults. More recently, the issue of permanent-fault-tolerant broadcasting in general networks has been studied in [CGGPR00, CGR02, CMS03]. In these papers, several lower and upper bounds on the completion time of broadcasting are obtained in the *unknown* fault-free network model. We observe that the results obtained in unknown networks apply to general networks with permanent faults. In particular, in [CMS03], an $\Omega(n \log D)$ lower bound for the broadcast completion time is proved. The best general upper bound is $O(n \log^2 n)$ [CGR02]. In [CMS03], the authors provide a protocol having $O(D\Delta \log^2 n)$ completion time.

In [GL02], a gossiping protocol for unknown networks is given that works in $O(n^{1.5} \log^2 n)$ time. [CMS03] provides a permanent-fault tolerant gossiping protocol having $O(D\Delta^2 \log^2 n)$ completion time. The above results work for the combined-message model. As for the single-message model, in [CMS03], a deterministic gossiping protocol is given that has $O(n\Delta^2 \log^3 n)$ completion time. We also mention the protocol for unknown directed GRN working in $O(n)$ time given in [DP07], even though it does not work for faulty networks.

Dynamical Faults. We emphasize that *all* the above protocols *do not work* in presence of dynamical faults. As mentioned before, this is mainly due to the collisions yielded by any unpredictable wake-up of a faulty node/link during the protocol execution. Our dynamical fault model has been studied in [CMS04] where the *round robin* strategy is proved to be optimal for general graphs. Then, they show the existence of a deterministic FG protocol having $O(D\Delta \log n)$ completion time. The protocol is based on a probabilistic construction of *ad-hoc strongly-selective families* [CMS03, I02] for general graphs. Such families have a weaker property than collision-free ones: this weakness yields a not efficient message complexity. By adopting the efficient construction of such families in [I97], they can efficiently construct a FG protocol having $O(D\Delta \log^3 n)$ completion time. These protocols only hold for the combined-message model. In [PP05] an initial graph is given and, at each time slot, every node is faulty with probability p, where p is a fixed positive *constant* such that $0 < p < 1$. They prove an $O(\mathit{opt} \log n)$ bound for the broadcast completion time where *opt* is the optimal completion time in the fault-free case. They also prove that it is impossible to achieve $O(\mathit{opt} + \log n)$ completion time.

It is not hard to see that, when the graph is *symmetric*, any *distance-2 coloring* [C06] of size k yields a collision-free family of size k and viceversa. For some classes of undirected graphs, there are efficient constant-factor approximation algorithms that find a distance-2 coloring. In particular, for *unit disk graphs* [C06, CCJ90, SM97] a 7-approximation algorithm is presented in [SM97]. Since symmetric GRN in the plane are equivalent to unit disk graphs, the latter algorithm can be used to construct a collision-free family for this class of symmetric radio networks. However, this coloring algorithm *does not work* for *directed* GRN.

2 Collision-Free Families and Fault-Tolerant Gossiping

In this section we introduce collision-free families and we show how to exploit them to design fault-tolerant gossiping protocols.

Definition 2.1 (Collision-free families). *Let $G(V,E)$ be a directed graph and let V' be the set of nodes that have at least one out-neighbor. A* collision-free family *$\mathcal{S}$ for G is a partition $\mathcal{S} = \{S_1, \ldots, S_k\}$ of V', such that, for each $S \in \mathcal{S}$ and for each $x, y \in S$ with $x \neq y$, $N^{out}(x) \cap N^{out}(y) = \emptyset$.*

In the sequel, we assume that, given any directed graph $G(V,E)$, we have at hand a collision-free family $\mathcal{S} = \{S_1, S_2, \ldots, S_k\}$ for G. In Section 3 we will then show how to construct collision-free families of small size.

Single-Message Model. In this model every transmission can contain only one of the source messages. We assume that each message contains the unique ID number of its source so that different messages have different ID's. The following FG protocol makes use of message IDs to define a *priority* queue in every node.

Protocol PRIO-SELECT($\mathcal{S}$) consists of a sequence of consecutive *phases*. Each *phase* consists of $k = |\mathcal{S}|$ time-slots. At the very beginning, the priority queue of every node u contains only m_u. At the beginning of every phase, every node v extracts (if any) the message $\hat{m}$ of *highest priority* (i.e. the maximal ID number) from its priority queue. Then, at time-slot j of a phase, node v acts according to the following rules

- If $v \in S_j$ and $\hat{m}$ exists then v transmits $\hat{m}$.
- In all other cases, v acts as receiver. If v receives a message m for the *first time* then m is enqueued, otherwise it is discarded.

Theorem 2.2. *Given a collision-free family $\mathcal{S}$ of size k for a directed graph G,* PRIO-SELECT($\mathcal{S}$) *completes fault-tolerant gossiping in G within $O(nk)$ time slots and message complexity $O(n^2)$.*

Combined-Message Model. In this model, source messages can be arbitrarily combined and sent in one transmission.

Protocol MULTI-SELECT($\mathcal{S}$). Each node v keeps the set $M_{old}(v)$ of the messages already sent by node v and the set $M_{new}(v)$ of the messages that node v has to send. At the beginning of the protocol, $M_{new}(v)$ contains only the source message of node v and the set $M_{old}(v)$ is empty. The protocol consists of a sequence of consecutive *phases*. Each *phase* consists of $k = |\mathcal{S}|$ time-slots. All phases are identical. At time slot j of a phase, node v acts according to the following rules

- If $v \in S_j$ and $M_{new}(v)$ is not empty then v transmits all the messages in $M_{new}(v)$ and moves all these messages to the set $M_{old}(v)$;
- In all other cases, v acts as receiver. When v receives a message m, if it is not in $M_{old}(v)$ then it is added to $M_{new}(v)$. Otherwise m is discarded.

Theorem 2.3. *Given a collision-free family $\mathcal{S}$ of size k for a directed graph G,* MULTI-SELECT($\mathcal{S}$) *completes fault-tolerant gossiping in G within $O(Dk)$ time-slots and message complexity $O(Dn)$, where D is the maximal residual source eccentricity. Moreover, an easy adaptation of* MULTI-SELECT($\mathcal{S}$) *for the broadcast operation works with the same completion time while the message complexity reduces to $O(n)$.*

3 Explicit Constructions of Collision-Free Families

Centralized Construction. Given a set V of points (i.e. nodes) in $\mathbb{R}^2$ and a range assignment $r : V \rightarrow \mathbb{R}^+$, the directed GRN is uniquely determined and it will be denoted as $G_r(V)$. Indeed, for each node $v \in V$, let $B(v)$ be the closed disk of center v and radius $r(v)$, i.e., $B(v) = \{x \in \mathbb{R}^2 \ : \ d(v,x) \leqslant r(v)\}$. We define the in-neighborhood of a node $v \in V$ as the set $N^{in}(v) = \{w \in V \ : \ v \in B(w)\}$. We define $\Delta(v) = |N^{in}(v)|$ and the maximal in-degree of $G_r(V)$ as $\Delta = \max_{v \in V} \Delta(v)$.

We will show that, given any directed GRN $G_r(V)$ as input, the following algorithm CFF returns a collision-free family $\mathcal{S}$ for $G_r(V)$ of size $O(\Delta)$. Since $\Omega(\Delta)$ is a trivial lower bound for such families, the one returned by CFF is asymptotically optimal.

The algorithm constructs every set of $\mathcal{S}$ by inserting nodes whose range disks are pairwise disjoint. Nodes are inserted in a non increasing order w.r.t. their ranges. This set construction is repeated until no node of V' is left outside $\mathcal{S}$.

Algorithm CFF (a finite set $V \subseteq \mathbb{R}^2$, a function $r : V \rightarrow \mathbb{R}^+$)

1 Let $X := V' = \{v \in V : N^{out}(v) \neq \emptyset\}$; $\mathcal{S} := \emptyset$; $i := 0$;

2 **while** $X \neq \emptyset$ **do**

3 $\quad i := i + 1$; $S_i := \emptyset$; $U := \emptyset$; $Y := X$;

4 $\quad$ **while** $Y \neq \emptyset$ **do**

5 $\quad\quad$ Choose $v \in Y$ such that $r(v)$ is maximum;

6 $\quad\quad$ **if** $U \cap B(v) = \emptyset$ **then**

7 $\quad\quad\quad S_i := S_i \cup \{v\}$; $U := U \cup B(v)$;

8 $\quad\quad Y := Y - \{v\}$;

9 $\quad \mathcal{S} := \mathcal{S} \cup \{S_i\}$; $X := X - S_i$;

10 **return** $\mathcal{S}$.

It is easy to see that, by using standard data structures, the algorithm works in $O(n^2)$ time. Moreover, family $\mathcal{S}$ returned by the algorithm is collision free by construction. We now provide a preliminary bound on the size of $\mathcal{S}$. For every $v \in V'$, we define the set $I(v)$ of all nodes of V' that could interfere with v and that have range not smaller than the range of v, i.e., $I(v) = \{w \in V' : B(v) \cap B(w) \neq \emptyset \text{ and } r(w) \geqslant r(v)\}$.

Lemma 3.1. *Family $\mathcal{S}$ has size at most* $\max_{v \in V'} |I(v)|$.

Proof. At every iteration of the external loop (line 2), a new set of $\mathcal{S}$ is constructed. Consider the i-th iteration and let $v \in V'$ be any node not yet inserted in any of sets $S_1, S_2, \ldots, S_{i-1}$ constructed in the previous iterations. For every $j = 1, 2, \ldots, i-1$, S_j must contain at least one node in $I(v)$. Indeed, assume by contradiction that there exists $j \leqslant i - 1$ such that $S_j \cap I(v) = \emptyset$. Then, for every $w \in S_j$ with $r(w) \geqslant r(v)$, it holds that $B(w) \cap B(v) = \emptyset$. When the algorithm selects v in line 5, the condition at line 6 is true, so v should be inserted in S_j: a contradiction. Since the sets of $\mathcal{S}$ are pairwise disjoints, the number of iterations of the external loop does not exceed $\max_{v \in V'} |I(v)|$. □

Our next goal is to prove that $\max_{v \in V'} |I(v)| \in O(\Delta)$. To this aim, we will show that, for every $v \in V'$, we can partition $\mathbb{R}^2$ into a constant number of *regions* so that each region contains at most Δ nodes of $I(v)$.

Lemma 3.2. *For every $v \in V'$, it holds that $|B(v) \cap I(v)| \leqslant \Delta$.*

Proof. Nodes in $I(v)$ have range at least $r(v)$. Hence, all nodes of $I(v)$ in $B(v)$ are points of $N^{in}(v)$, i.e., $I(v) \cap B(v) \subseteq N^{in}(v)$. □

We now consider the region outside disk $B(v)$ and define the circular crown

$$C_\lambda(v) = \{y \in \mathbb{R}^2 \ : \ r(v) < d(v, y) \leqslant \lambda r(v)\}, \text{ where } \lambda > 1.$$

Lemma 3.3. *Let $1 < \lambda < 2$ and let $k \in \mathbb{N}$ be large enough such that $\cos \frac{2\pi}{k} \geqslant \lambda/2$. Then, for any $v \in V'$, $C_\lambda(v)$ contains at most $k\Delta$ nodes of $I(v)$.*

Proof. Consider a polar coordinate system centered in v and consider the partition of $C_\lambda(v)$ defined by the regions $]r(v), \lambda r(v)] \times [\vartheta_i, \vartheta_{i+1}[$ where $\vartheta_i = \frac{2\pi i}{k}$ for $i = 0, 1, \ldots, k-1$. Then, since $\cos\frac{2\pi}{k} \geqslant \lambda/2$, it is easy to see that the square of the maximal distance between two points in the same region is $r(v)^2 + \lambda^2 r(v)^2 - 2\lambda r(v)^2 \cos\frac{2\pi}{k} \leqslant r(v)^2$. For any $w \in I(v)$, it holds that $r(w) \geqslant r(v)$, so w is in the in-neighborhood of all points in the same region of w. So, in every region there are at most Δ points of $I(v)$ and, since there are k regions in $C_\lambda(v)$, the thesis follows. □

Consider the function $g(\lambda) = \frac{\lambda^2+2\lambda-1}{2\lambda^2}$ and observe that $1/2 < g(\lambda) < 1$, for any $\lambda > 1$. It is possible to prove that, if k is such that $\cos\frac{2\pi}{k} \geqslant g(\lambda)$, then for any $a \geqslant b \geqslant \lambda$ it holds that

$$a^2 + b^2 - 2ab\cos\frac{2\pi}{k} \leqslant (a-1)^2. \tag{1}$$

We will use this fact in proving next Lemma.

Lemma 3.4. *Let $\lambda > 1$ and let $k \in \mathbb{N}$ be large enough such that $\cos\frac{2\pi}{k} \geqslant g(\lambda)$. Then, for any $v \in V'$, there are at most $k\Delta$ nodes of $I(v)$ outside $B(v) \cup C_\lambda(v)$.*

Proof. Consider a polar coordinate system centered in v, and define a partition of the space outside $B(v) \cup C_\lambda(v)$ in the regions $[\lambda r(v), +\infty[\times [\vartheta_i, \vartheta_{i+1}[$ where $\vartheta_i = \frac{2\pi i}{k}$ for $i = 0, 1, \ldots, k-1$. Let $x = (\varrho_x, \varphi_x)$ and $y = (\varrho_y, \varphi_y)$ two nodes of $I(v)$ that lie in the same region and suppose wlog that $\varrho_x \geqslant \varrho_y$. Then, two constants $a, b \in \mathbb{R}$ exist with $a \geqslant b \geqslant \lambda$ such that $\varrho_x = a \cdot r(v)$ and $\varrho_y = b \cdot r(v)$. We thus get

$$d(x,y)^2 = \varrho_x^2 + \varrho_y^2 - 2\varrho_x\varrho_y\cos(\varphi_x - \varphi_y) \leqslant r(v)^2\left(a^2 + b^2 - 2ab\cos\frac{2\pi}{k}\right)$$

where in the inequality we used the fact that x and y lie in the same region. From (1), we get $d(x,y)^2 \leqslant r(v)^2(a-1)^2 = (a \cdot r(v) - r(v))^2 = (\varrho_x - r(v))^2$. Since $x \in I(v)$, it must hold that $B(x) \cap B(v) \neq \emptyset$, so $\varrho_x - r(v) \leqslant r(x)$, and $d(x,y)^2 \leqslant r(x)^2$. Therefore, y lies in $B(x)$ and, thus, $x \in N^{in}(y)$. It follows that, for every region T, if $y \in T \cap I(v)$ is a node with minimum distance from v, i.e, a node with minimum ϱ_y, then $T \cap I(v) \subseteq N^{in}(y)$. This implies that in every region there are at most Δ points of $I(v)$: since the regions are k, the thesis follows. □

Lemma 3.5. *Let $1 < \lambda < 2$ and let $k \in \mathbb{N}$ be such that $\cos\frac{2\pi}{k} \geqslant \max\left\{\frac{\lambda}{2}, g(\lambda)\right\}$. Then, for any $v \in V'$, it holds that $|I(v)| \leqslant (1+2k)\Delta$.*

Proof. Consider the partition of $\mathbb{R}^2$ into the three sets: $B(v)$, $C_\lambda(v)$, and the complement of $B(v) \cup C_\lambda(v)$. By combining Lemmas 3.2, 3.3, and 3.4, we get $|I(v)| \leqslant (1+k+k)\Delta$. □

Theorem 3.6. *Algorithm* CFF *returns a collision-free family $\mathcal{S}$ for $G_r(V)$ of size at most $c\Delta$, where $c \leqslant 33$.*

Proof. Family $\mathcal{S}$ is collision-free for $G_r(V)$ by construction. Let λ be such that $1 < \lambda < 2$. From Lemmas 3.1 and 3.5, we obtain $|\mathcal{S}| \leqslant \max_{v \in V'} |I(v)| \leqslant (1+2k)\Delta$, with $k \in \mathbb{N}$ such that $\cos(2\pi/k) \geqslant \max\{\lambda/2, g(\lambda)\}$. Then, in order to minimize k, we choose λ such that $\lambda/2 = (\lambda^2+2\lambda-1)/(2\lambda^2)$. Consider the function $f(\lambda) = \lambda^3-\lambda^2-2\lambda+1$. Then $f(1) = -1$ and $f(2) = 1$, so there exists a solution between 1 and 2. By numerical arguments, we can set $\lambda \approx 1.8$ and get

$$\cos\frac{2\pi}{k} \geqslant \max\left\{\frac{\lambda}{2}, \frac{\lambda^2+2\lambda-1}{2\lambda^2}\right\}, \quad \text{for any } k \geqslant 16. \qquad \square$$

Distributed Construction. Let us consider GRN $G_r(V)$ where $r(v) = R$ for each $v \in V$ (so $G_r(V)$ is symmetric). Directed GRN will be discussed at the end of this section. Our distributed construction of a collision-free family for $G_r(V)$ is based on the following idea. Consider a partition of $\mathbb{R}^2$ into squares small enough to guarantee that in each square there is at most one node of V. Then we partition the set of such small squares so that the distance between two squares in the same set of the partition is at least $2R$. Finally, consider the subsets of V obtained by collecting all nodes in the same set of squares.

Let $\gamma = \min\{d(u,v) \ : \ u,v \in V, u \neq v\}$. For any $x \in \mathbb{R}$ we define $[x]$ as the nearest integer to x. We now assume that each node knows its own position, the transmission range R and the minimum distance γ. In the following protocol, $\varepsilon > 0$ is an arbitrary small constant: we need it in order to have strict inequalities.

PROTOCOL FOR NODE u (position (x_u, y_u), transmission range R, min distance γ)

---

1 **Define** $\lambda = \gamma/\sqrt{2} - \varepsilon$;
2 **Define** $k = \lceil (2R+\varepsilon)/\lambda \rceil + 1$;
3 **Define** $\hat{x}_u = [x_u/\lambda]$ and $\hat{y}_u = [y_u/\lambda]$;
4 **Return** $f(u) = (\hat{x}_u \bmod k,\ \hat{y}_u \bmod k)$;

Let us consider the family $\mathcal{S} = \{S_{i,j}\}_{i,j=0,1,\ldots,k-1}$ where $S_{i,j} = \{u \in V \ : \ f(u) = (i,j)\}$. We now show that $\mathcal{S}$ is a collision-free family.

Theorem 3.7. *Family $\mathcal{S}$ is collision-free family of size $O(R^2/\gamma^2)$ for $G_r(V)$.*

Sketch of the Proof. By definition of k (line 2 of Protocol) we have that $|\mathcal{S}| = k^2 \in O(R^2/\gamma^2)$. Let $u,v \in S_{i,j}$ with $u \neq v$. Assume, by contradiction, that $N^{out}(u) \cap N^{out}(v) \neq \emptyset$ and let $w \in N^{out}(u) \cap N^{out}(v)$. So $d(u,w) \leqslant R$ and $d(v,w) \leqslant R$. By triangular inequality we get $d(u,v) \leqslant 2R$, but this is a contradiction, because since u, v are two different nodes in the same set of the family, by construction it must be $d(u,v) > 2R$. $\square$

We now show that when nodes are *well spread*, the size of the family is asymptotically optimal.

Definition 3.8 (Well spread instances). *Let $V \subseteq \mathbb{R}^2$ be a set of n points in the Euclidean plane. Let γ and Γ be respectively the minimal and the maximal distance between two points in V. Let c be any positive constant, set V is said* c-well spread *if $\Gamma/\gamma \leqslant c\sqrt{n}$.*

Observe that square-grid networks are the most regular case of c-well spread instances where $c = \sqrt{2}$ [CPS04].

Theorem 3.9. *If $V \subseteq \mathbb{R}^2$ is a c-well spread instance, then $R^2/\gamma^2 \in O(c^2\Delta)$, where Δ is the maximal degree of $G_r(V)$.*

Proof. There exists a disk of radius Γ that contains all the n nodes. That disk can be covered with $O(\Gamma^2/R^2)$ disks of radius R. Then there exists a disk U with radius R such that it contains $\Omega\left(\frac{nR^2}{\Gamma^2}\right)$ nodes. Since V is c-well spread, $n/\Gamma^2 \in \Omega(1/c^2\gamma^2)$ and so disk U contains $\Omega(R^2/c^2\gamma^2)$ nodes. It follows that $R^2/\gamma^2 \in O(c^2\Delta)$. □

Our distributed construction also works for directed GRN where parameter R is replaced by the maximal node range $R_{\max}$. Theorem 3.7 holds with $R_{\max}$ in place of R and Theorem 3.9 holds with $R_{\min}$ in place of R, where $R_{\min}$ is the minimal node range. Thus if $R_{\max} \in O(R_{\min})$ the construction is still optimal.

4 Optimal Bounds

The results obtained in the previous two sections allow us to get optimal bounds for fault-tolerant protocols.

Single-Message Model

Corollary 4.1. *Given a directed GRN $G_r(V)$, there exists an explicit FG protocol having completion time $O(n\Delta)$ and message complexity $O(n^2)$, where Δ is the maximal in-degree of $G_r(V)$.*

There exists a distributed FG protocol that, on any c-well spread symmetric GRN $G_r(V)$, completes gossiping in $O(nc^2\Delta)$ time slots and has message complexity $O(n^2)$. The protocol requires the knowledge of the minimal distance γ.

Theorem 4.2. *For any sufficiently large n and Δ, such that $n - \Delta \in \Omega(n)$, there exists a GRN $G_r(V)$ of n nodes and maximal in-degree Δ such that, for any FG protocol for $G_r(V)$, an adversary's fault-pattern F exists such that the protocol is forced to execute $\Omega(n\Delta)$ time-slots and to have message complexity $\Omega(n^2)$.*

Combined-Message Model

Corollary 4.3. *Given a directed GRN $G_r(V)$, there exists an explicit FG protocol having completion time $O(D\Delta)$ and message complexity $O(Dn)$, where D is the maximal residual source eccentricity.*

There exists a distributed FG protocol that, on any c-well spread symmetric GRN $G_r(V)$, completes gossiping in $O(Dc^2\Delta)$ time slots and has message complexity $O(Dn)$. The protocol requires the knowledge of the minimal distance γ.

As for the (single) broadcast operation, the same protocols work in the same completion time while the message complexity reduces to $O(n)$ that is optimal.

Theorem 4.4. *For any n, Δ and D such that $D\Delta \leqslant n$, there exists a GRN $G_r(V)$ of n nodes and maximal in-degree Δ such that, for any FB protocol for $G_r(V)$, there are a source $s \in V$ and an adversary's fault-pattern F, yielding source eccentricity D, such that the protocol is forced to execute $\Omega(D\Delta)$ time-slots and to have message complexity $\Omega(n)$.*

As for the case $D \cdot \Delta > n$, we observe that a lower bound $\Omega(n\sqrt{n})$ holds for FB protocols on directed GRN of unbounded maximal in-degree and residual source eccentricity $D = \Theta(\sqrt{n})$. This result is an easy consequence of the lower bound for arbitrary graphs proved in [CMS04]: The graph yielding such lower bound is indeed a GRN of maximal in-degree $\Delta = \Theta(n)$.

5 Open Questions

It is an open question whether the $O(Dn)$ bound for the FG message complexity is optimal. Another future work is that of extending our distributed construction of collision-free families to other important classes of radio networks. Finally, an interesting issue is that of designing *randomized* FG protocols. Such protocols may yield a much better completion time on the residual graph and, more importantly, they might have good performances *outside* the residual graph too.

References

[AS98] Adler, M., Scheideler, C.: Efficient communication strategies for ad hoc wireless networks. Proc. of 10th ACM SPAA, pp. 259–268 (1998)

[ABLP89] Alon, N., Bar-Noy, A., Linial, N., Peleg, D.: A lower bound for radio broadcast. Journal of Computer and System Science 43, 290–298 (1991)

[ABBS01] Awerbuch, B., Berenbrink, P., Brinkmann, A., Scheideler, C.: Simple routing strategies for adversarial systems. In: Proc. of 42th IEEE FOCS, pp. 158–167 (2001)

[BGI92] Bar-Yehuda, R., Goldreich, O., Itai, A.: On the time-complexity of broadcast in multi-hop radio networks: An exponential gap between determinism and randomization. Journal of Computer and System Science 45, 104–126 (1992)

[BII93] Bar-Yehuda, R., Israeli, A., Itai, A.: Multiple communication in multi-hop radio networks. SICOMP 22(4), 875–887 (1993)

[C06] Calamoneri, T.: The $L(h, k)$-Labeling problem: Survey and Annotated Bibliography. The Computer Journal 49, 585–608 (2006)

[CKOZ03] Calinescu, G., Kapoor, S., Olshevsky, A., Zelikovsky, A.: Network Life Time and Power Assignment in ad hoc Wireless Networks. In: Di Battista, G., Zwick, U. (eds.) ESA 2003. LNCS, vol. 2832, pp. 114–126. Springer, Heidelberg (2003)

[CGGPR00] Chlebus, B., Gasieniec, L., Gibbons, A., Pelc, A., Rytter, W.: Deterministic broadcasting in unknown radio networks. In: Proc. of 11th ACM-SIAM SODA, pp. 861–870 (2000)

[CGR02] Chrobak, M., Gasieniec, L., Rytter, W.: Fast Broadcasting and Gossiping in Radio Networks. Journal of Algorithms 43(2), 177–189 (2002)

[CCJ90] Clark, B.N., Colbourn, C.J., Johnson, D.S.: Unit disk graphs. Discrete Mathematics 86, 165–177 (1990)

[CCPRV01] Clementi, A.E.F., Crescenzi, P., Penna, P., Vocca, P.: On the complexity of minimum energy consumption broadcast subgraphs. In: Proc of 11th STACS, pp. 121–131 (2001)

[CMS03] Clementi, A.E.F., Monti, A., Silvestri, R.: Distributed broadcast in radio networks of unknown topology. Theoretical Computer Science 302, 337–364 (2003)

[CMS04] Clementi, A.E.F., Monti, A., Silvestri, R.: Round Robin is optimal for fault-tolerant broadcasting on wireless networks. Journal of Parallel and Distributed Computing 64(1), 89–96 (2004)

[CMPS07] Clementi, A.E.F., Monti, A., Pasquale, F., Silvestri, R.: Optimal gossiping in directed geometric radio networks in presence of dynamical faults (2007), available at www.mat.uniroma2.it/~pasquale

[CPS04] Clementi, A.E.F., Penna, P., Silvestri, R.: On the power assignment problem in radio networks. MONET (9), 125–140 (2004)

[CR06] Czumaj, A., Rytter, W.: Broadcasting algorithms in radio networks with unknown topology. Journal of Algorithms 60(2), 115–143 (2006)

[DP07] Dessmark, A., Pelc, A.: Broadcasting in geometric radio networks. Journal of Discrete Algorithms 5, 187–201 (2007)

[ENW00] Ephremides, A., Nguyen, G., Wieselthier, J.: On the construction of energy-efficient broadcast and multi-cast trees in wireless networks. In: Proc. of 19th IEEE INFOCOM, pp. 585–594 (2000)

[GL02] Gasieniec, L., Lingas, A.: On adaptive deterministic gossiping in ad-hoc radio networks. Information Processing Letters 2, 89–93 (2002)

[GPX05] Gasieniec, L., Peleg, D., Xin, Q.: Faster Communication in Known Topology Radio Networks. In: Proc. of 24th ACM PODC, pp. 129–137 (2005)

[I97] Indyk, P.: Deterministic Superimposed Coding with Application to Pattern Matching. In: Proc. of 38th IEEE FOCS, pp. 127–136 (1997)

[I02] Indyk, P.: Explicit constructions of selectors and related combinatorial structures, with applications. In: Proc of 13th ACM-SIAM SODA, pp. 697–704 (2002)

[KKKP00] Kirousis, L., Kranakis, E., Krizanc, D., Pelc, A.: Power consumption in packet radio networks. Theoretical Computer Science 243, 289–305 (2000)

[KKP98] Kranakis, E., Krizanc, D., Pelc, A.: Fault-Tolerant Broadcasting in Radio Networks. In: Bilardi, G., Pietracaprina, A., Italiano, G.F., Pucci, G. (eds.) ESA 1998. LNCS, vol. 1461, pp. 283–294. Springer, Heidelberg (1998)

[P02] Pelc, A.: Broadcasting in radio networks. In: Handbook of Wireless Networks and Mobile Computing, pp. 509–528. John Wiley and Sons, Inc., Chichester (2002)

[PP05] Pelc, A., Peleg, D.: Feasibility and complexity of broadcasting with random transmission failures. In: Proc. of 24th ACM PODC, pp. 334–341 (2005)

[PR97] Pagani, E., Rossi, G.: Reliable Broadcast in Mobile Multihop Packet Networks. In: Proc. of 3rd ACM-IEEE MOBICOM, pp. 34–42 (1997)

[R96] Rappaport, T.S.: Wireless Communications: Principles and Practice. Prentice-Hall, Englewood Cliffs (1996)

[SM97] Sen, A., Malesinska, E.: On Approximation algorithms for packet radio scheduling. In: Proc. of 35th Allerton Conference on Communication, Control and Computing, Allerton, pp. 573–582 (1997)

[S01] Scheideler, C.: Models and Techniques for Communication in Dynamic Networks. In: Proc. of 18th STACS, pp. 27–49 (2001)

A Linear Time Algorithm for the *k* Maximal Sums Problem

Gerth Stølting Brodal and Allan Grønlund Jørgensen*

BRICS**, MADALGO***, Department of Computer Science,
University of Aarhus, Denmark
{gerth,jallan}@daimi.au.dk

Abstract. Finding the sub-vector with the largest sum in a sequence of n numbers is known as the maximum sum problem. Finding the k sub-vectors with the largest sums is a natural extension of this, and is known as the k maximal sums problem. In this paper we design an optimal $O(n+k)$ time algorithm for the k maximal sums problem. We use this algorithm to obtain algorithms solving the two-dimensional k maximal sums problem in $O(m^2 \cdot n+k)$ time, where the input is an $m \times n$ matrix with $m \leq n$. We generalize this algorithm to solve the d-dimensional problem in $O(n^{2d-1}+k)$ time. The space usage of all the algorithms can be reduced to $O(n^{d-1}+k)$. This leads to the first algorithm for the k maximal sums problem in one dimension using $O(n+k)$ time and $O(k)$ space.

1 Introduction

To solve the maximum sum problem one must locate the maximal sum sub-vector of an array A of n numbers. The maximal sub-vector of A is the sub-vector $A[i, \ldots, j]$ maximizing $\sum_{s=i}^{j} A[s]$. The problem originates from Ulf Grenander who defined the problem in the setting of pattern recognition [1]. Solutions to the problem also have applications in areas such as Data Mining [2] and Bioinformatics [3].

The problem, and an optimal linear time algorithm credited to Jay Kadane, are described by Bentley [1] and Gries [4]. The algorithm they describe is a scanning algorithm which remembers the best solution, $\max_{1 \leq i \leq j \leq t} \sum_{s=i}^{j} A[s]$, and the best suffix solution, $\max_{1 \leq i \leq t} \sum_{s=i}^{t} A[s]$, in the part of the input array, $A[1, \ldots, t]$, scanned so far. Both values are updated in $O(1)$ time in each step yielding a linear time algorithm using $O(1)$ space.

The problem can be extended to any number of dimensions. In two dimensions the input is an $m \times n$ matrix of numbers and the task is to find the connected

* Supported in part by an Ole Roemer Scholarship from the Danish National Science Research Council.

** Basic Research in Computer Science, research school.

*** Center for Massive Data Algorithmics, a Center of the Danish National Research Foundation.

L. Kučera and A. Kučera (Eds.): MFCS 2007, LNCS 4708, pp. 442–453, 2007.

Table 1. Previous and new results for the k maximal sums problem

Paper	Time complexity
Bae & Takaoka [8]	$O(n \cdot k)$
Bengtson & Chen [9]	$O(\min\{k + n \log^2 n, n\sqrt{k}\})$
Bae & Takaoka [10]	$O(n \log k + k^2)$
Bae & Takaoka [11]	$O((n + k) \log k)$
Lie & Lin [12]	$O(n \log n + k)$ expected
Cheng *et al.* [13]	$O(n + k \log k)$
Liu & Chao [14][1]	$O(n + k)$
This paper	$O(n + k)$

sub-matrix with the largest aggregate. The two-dimensional version was the original problem, introduced as a method for maximum likelihood estimations of patterns in digitized images [1].

With $m \leq n$, the problem can be solved by a reduction to $\binom{m}{2} + m$ one-dimensional problems resulting in an $O(m^2 \cdot n)$ time algorithm. The same reduction technique can be applied iteratively to solve the problem in any dimension. But unlike the one dimensional case these algorithms are not optimal. In [5] and [6] asymptotically faster algorithms for the two-dimensional problem are described. In [6] Takaoka designed an $O(m^2 n \sqrt{\log\log m / \log m})$ time algorithm by a reduction to (min,+) matrix multiplication [7].

A simple extension of the maximum sum problem is to compute the k largest sub-vectors for $1 \leq k \leq \binom{n}{2} + n$. The sub-vectors are allowed to overlap, and the output is k triples of the form (i, j, sum) where $sum = \sum_{s=i}^{j} A[s]$. This extension was introduced in [8]. The solution for $k = 1$ described above does not seem to be extendable in any simple manner to obtain a linear algorithm for any k. Therefore, different solutions to this extended problem has emerged over the past few years. These results are summarized in Table 1.

A lower bound for the k maximal sums problem is $\Omega(n+k)$, since an adversary can force any algorithm to look at each of the n input elements and the output size is $\Omega(k)$.

1.1 Results

In this paper we close the gap between upper and lower bounds for the k maximal sums problem. We design an algorithm computing the k sub-vectors with the largest sums in an array of size n in $O(n + k)$ time. We also describe algorithms solving the problem extended to any dimension. We begin by solving the two-dimensional problem where we obtain an $O(m^2 \cdot n + k)$ time algorithm for an $m \times n$ input matrix with $m \leq n$. This improves the previous best result [13], which was

[1] The k maximal sums problem can also be solved in $O(n + k)$ time by a reduction to Eppstein's solution for the k shortest paths problem [15] which also makes essential use of Fredericksons heap selection algorithm. This reduction was observed independently by Hsiao-Fei Liu and Kun-Mao Chao [14].

an $O(m^2 \cdot n + k \log k)$ time algorithm. This solution is then generalized to solve the d dimensional problem in $O(n^{2d-1} + k)$ time, assuming for simplicity that all sides of the d-dimensional input matrix are equally long. Furthermore we describe how to minimize the additional space usage of our algorithms. The additional space usage of the one dimensional algorithm is reduced from $O(n + k)$ to $O(k)$. The input array is considered to be read only. The additional space usage for the algorithm solving the two-dimensional problem is reduced from $O(m^2 \cdot n + k)$ to $O(n + k)$ and for the general algorithm solving the d dimensional problem the space is reduced from $O(n^{2(d-1)} + k)$ to $O(n^{d-1} + k)$.

The main contribution of this paper is the first algorithm solving the k maximal sums problem using $O(n + k)$ time and $O(k)$ space. The result is achieved by generating a binary heap that implicitly contains the $\binom{n}{2} + n$ sums in $O(n)$ time. The k largest sums from the heap are then selected in $O(n + k)$ time using the heap selection algorithm of Frederickson [16]. The heap is build using partial persistence [17]. The space is reduced by only processing k elements at time. The resulting algorithm can be viewed as a natural extension of Kadane's linear time algorithm for solving the maximum sum problem introduced earlier.

1.2 Outline of Paper

The remainder of the paper is structured as follows. In Section 2 the overall structure of our solution is explained. Descriptions and details regarding the algorithms and data structures used to achieve the result are presented in Sections 3, 4 and 5. In Section 6 we combine the different algorithms and data structures completing our algorithm. This is followed by Section 7 where we show how to use our algorithm to solve the problem in d dimensions. Finally in Section 8 we explain how to reduce the additional space usage of the algorithms without penalizing the asymptotic time bounds.

2 Basic Idea and Algorithm

In this paper the term heap denotes a max-heap ordered binary tree. The basic idea of our algorithm is to build a heap storing the sums of all $\binom{n}{2} + n$ sub-vectors and then use Fredericksons binary heap selection algorithm to find the k largest elements in the heap.

In the following we describe how to construct a heap that implicitly stores all the $\binom{n}{2} + n$ sums in $O(n)$ time. The triples induced by the $\binom{n}{2} + n$ sums in the input array are grouped by their end index. The *suffix set* of triples corresponding to all sub-vectors ending at position j we denote Q^j_{suf}, and this is the set $\{(i, j, sum) \mid 1 \leq i \leq j \;\wedge sum = \sum_{s=i}^{j} A[s]\}$. The Q^j_{suf} sets can be incrementally defined as follows:

$$Q^j_{\text{suf}} = \{(j, j, A[j])\} \cup \{(i, j, s + A[j]) \mid (i, j-1, s) \in Q^{j-1}_{\text{suf}}\}. \quad (1)$$

As stated in equation (1) the suffix set Q^j_{suf} consists of all suffix sums in Q^{j-1}_{suf} with the element $A[j]$ added as well as the single element suffix sum $A[j]$.

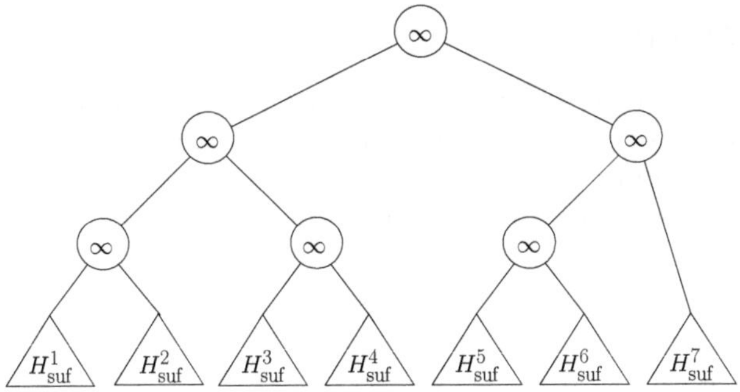

Fig. 1. Example of a complete heap H constructed on top of the H^j_{suf} heaps. The input size is 7.

Using this definition, the set of triples corresponding to all $\binom{n}{2}+n$ sums in the input array is the union of the n disjoint Q^j_{suf} sets. We represent the Q^j_{suf} sets as heaps and denote them H^j_{suf}. Assuming that for each suffix set Q^j_{suf}, a heap H^j_{suf} representing it has been build, we can construct a heap H containing all possible triples by constructing a complete binary heap on top of these heaps. The keys for the $n-1$ top elements is set to ∞ (see Figure 1). To find the k largest elements, we extract the $n-1+k$ largest elements in H using the binary heap selection algorithm of Frederickson [16] and discard the $n-1$ elements equal to ∞.

Since the suffix sets contain $\Theta(n^2)$ elements the time and space required is still $\Theta(n^2)$ if they are represented explicitly. We obtain a linear time construction of the heap by constructing an implicit representation of a heap that contains all the sums. We make essential use of a heap data structure to represent the Q^j_{suf} sets that supports insertions in amortized constant time.

Priority queues represented as heap ordered binary trees supporting insertions in constant time already exist. One such data structure is the self-adjusting binary heaps of Tarjan and Sleator described in [18] called Skew Heaps. The Skew heap is a data structure reminiscent of Leftist heaps [19,20]. Even though the Skew heap would suffice for our algorithm it is able to do much more than we require. Therefore, we design a simpler heap which we will name *Iheap.* The essential properties of the Iheap are that it is represented as a heap ordered binary tree and that insertions are supported in amortized constant time.

We build H^{j+1}_{suf} from H^j_{suf} in $O(1)$ time amortized without destroying H^j_{suf} by using the partial persistence technique of [17] on the Iheap. This basically means that the H^j_{suf} heaps become different versions of the same Iheap. To make our Iheap partially persistent we use the node copying technique [17]. The cost of applying this technique is linear in the number of changes in an update. Since only the insertion procedure is used on the Iheap, the extra cost of using partial persistence is the time for copying amortized $O(1)$ nodes per insert operation. The overhead of traversing a previous version of the data structure is $O(1)$ per data/pointer access.

3 Binary Heaps

The main data structure of our algorithm is a heap supporting constant time insertions in the amortized sense. The heap is not required to support operations like deletions of the minimum or an arbitrary element. All we do is insert elements and traverse the structure top down during heap selection. We design a simple binary heap data structure Iheap by reusing the idea behind the Skew heap and perform all insertions along the rightmost path of the tree starting from the rightmost leaf.

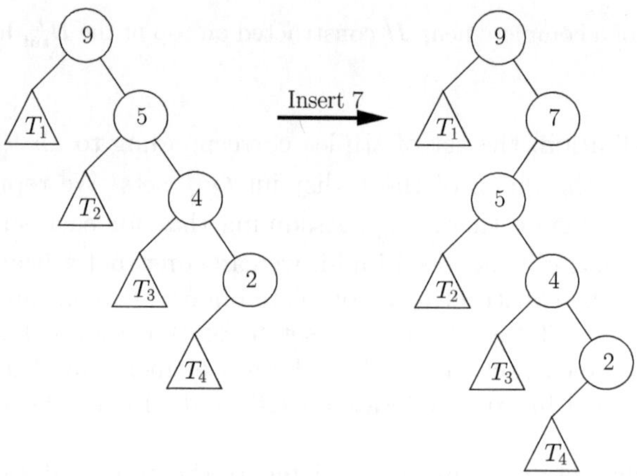

Fig. 2. An example of an insertion in the Iheap. The element 7 is compared to 2,4 and 5 in that order, and these elements are then removed from the rightmost path.

A new element is inserted into the Iheap by placing it in the first position on the rightmost path where it satisfies the heap order. This is performed by traversing the rightmost path bottom up until a larger element is found or the root is passed. The element is then inserted as a right child of the larger element found (or as the new root). The element it is replacing as a right child (or as root) becomes the left child of the inserted element. An insertion in an Iheap is illustrated in Figure 2. If $O(\ell)$ time is used to perform an insertion operation because ℓ elements are traversed, the rightmost path of the heap becomes $\ell - 1$ elements shorter. Using a potential function on the length of the rightmost path of the tree we get amortized constant time insertions for the Iheap. Each element is passed on the rightmost path only once, since it is then placed on the left-hand side of element passing it, and never returns to the rightmost path.

Lemma 1. *The Iheap supports insertion in amortized constant time.*

4 Partial Persistence and H^j_{suf} Construction

As mentioned in Section 2 the H^j_{suf} heaps are build based on equation (1) using the partial persistence technique of [17] on an Iheap.

Data structures are usually *ephemeral*, meaning that an update to the data structure destroys the old version, leaving only the new version available for use. An update changes a pointer or a field value in a node. Persistent data structures allow access to any version old or new. Partially persistent data structures allow updates to the newest version, whereas fully persistent data structures allow updates to any version. With the partial persistence technique known as node copying, linked ephemeral data structures, with the restriction that for any node the number of other nodes pointing to it is $O(1)$, can be made partially persistent [17]. The Iheap is a binary tree and therefore trivially satisfies the above condition. The amortized cost of using the node copying technique is bounded by the cost of copying and storing $O(1)$ nodes from the ephemeral structure per update.

The basic idea of applying node copying to the Iheap is the following (see [17] for further details). Each persistent node contains one version of each information field in an original node, but it is able to contain several versions of each pointer (link to other node) differentiated by time stamps (version numbers). However, there are only a constant number of versions of any pointer, why each partially persistent Iheap node only uses constant space. Accessing relatives of a node in a given version is performed by finding the pointer associated with the correct time stamp. This is performed in constant time making the access time in the partially persistent Iheap asymptotically equal to the access time in an ephemeral Iheap.

According to equation (1), the set Q^{j+1}_{suf} can be constructed from Q^{j}_{suf} by adding $A[i+1]$ to all elements in Q^{j}_{suf} and then inserting an element representing $A[i+1]$. To avoid adding $A[i+1]$ to each element in Q^{j}_{suf}, we represent each Q^{j}_{suf} set as a pair $\langle \delta_j, H^{j}_{\text{suf}} \rangle$, where H^{j}_{suf} is a version of a partial persistent Iheap containing all sums of Q^{j}_{suf} and δ_j is an element that must be added to all elements. With this representation a constant can be added to all elements in a heap implicitly by setting the corresponding δ. Similar to the way the Q^{j}_{suf} sets were defined by equation (1) we get the following incremental construction of the pair $\langle \delta_{j+1}, H^{j+1}_{\text{suf}} \rangle$:

$$\langle \delta_0, H^{0}_{\text{suf}} \rangle = \langle 0, \emptyset \rangle \,, \tag{2}$$

$$\langle \delta_{j+1}, H^{j+1}_{\text{suf}} \rangle = \langle \delta_j + A[i+1], H^{j}_{\text{suf}} \cup \{-\delta_j\} \rangle \,. \tag{3}$$

Let $\langle \delta_j, H^{j}_{\text{suf}} \rangle$ be the latest pair built. To construct $\langle \delta_{j+1}, H^{j+1}_{\text{suf}} \rangle$ from this pair, an element with $-\delta_j$ as key is inserted into H^{j}_{suf}. We insert this value, since δ_j has not been added to any element in H^{j}_{suf} explicitly, and because the sum $A[i+1]$ that the new element are to represent must be added to all elements in H^{j}_{suf} to obtain H^{j+1}_{suf}. Since we apply partial persistence on the heap, H^{j}_{suf} is still intact after the insertion, and a new version of the Iheap with the inserted element included has been constructed. H^{j+1}_{suf} is this new version and δ_{j+1} is set to $\delta_j + A[i+1]$. Therefore, the newly inserted element represents the sum $-\delta_j + \delta_j + A[i+1] = A[i+1]$. This ends the construction of the new pair $\langle \delta_{j+1}, H^{j+1}_{\text{suf}} \rangle$. Since all sums from H^j_{suf} gets $A[i+1]$ added because of the increase of δ_{j+1} compared to δ_j and the new element represents $A[i+1]$ we conclude that $\langle \delta_{j+1}, H^{j+1}_{\text{suf}} \rangle$ represents the set Q^{j+1}_{suf}. The time needed for constructing H^{j+1}_{suf} is the time for inserting an element into a partial persistent Iheap. Since the size of an Iheap node is $O(1)$, by Lemma 1 and the node copying technique, this is amortized constant time

Lemma 2. *The time for constructing the n pairs $\langle \delta_j, H^j_{suf} \rangle$ is $O(n)$.*

5 Fredericksons Heap Selection Algorithm

The last algorithm used by our algorithm is the heap selection algorithm of Frederickson, which extracts the k largest[2] elements in a heap in $O(k)$ time. Input to this algorithm is an infinite heap ordered binary tree. The infinite part is used to remove special cases concerning the leafs of the tree, and is implemented by implicitly appending nodes with keys of $-\infty$ to the leafs of a finite tree. The algorithm starts at the root, and otherwise only explores a node if the parent already has been explored.

The main part of the algorithm is a method for locating an element, e, with $k \leq rank(e) \leq ck$ for some constant c[3]. After this element is found the input heap is traversed and all elements larger than e are extracted. Standard selection [21] is then used to obtain the k largest elements from the $O(k)$ extracted elements. To find e, elements in the heap are organized into appropriately sized groups named clans. Clans are represented by their smallest element, and these are managed in classic binary heaps [22].

By fixing the size of clans to $\log k$ one can obtain an $O(k \log\log k)$ time algorithm as follows. Construct the first clan by locating the $\lfloor \log k \rfloor$ largest elements and initialize a *clan-heap* with the representative of this clan. The children of the elements in this clan are associated with it and denoted its *offspring*.

A new clan is constructed from a set of $\log k$ nodes in $O(\log k \log\log k)$ time using a heap. However, not all the elements in an offspring set are necessarily put into such a new clan. The leftover elements are then associated to the newly created clan and are denoted the *poor relatives* of the clan.

Now repeatedly delete the maximum clan from the clan heap, construct two new clans from the offspring and poor relatives, and insert their representatives into the clan-heap. After $\lceil k/\lfloor \log k \rfloor \rceil$ iterations an element of rank at least k is found, since the representative of the last clan deleted, is the smallest of $\lceil k/\lfloor \log k \rfloor \rceil$ representatives. Since $2\lceil k/\lfloor \log k \rfloor \rceil + 1$ clans are created at most, each using time $O(\log k \log\log k)$, the total time becomes $O(k \log\log k)$.

By applying this idea recursively and then bootstrapping it Frederickson obtains a linear time algorithm.

Theorem 1 ([16]). *The k largest elements in a heap can be found in $O(k)$ time.*

---

[2] Actually Frederickson [16] considers min-heaps.

[3] The largest element has rank one.

6 Combining the Ideas

The heap constructed by our algorithm is actually a graph because the H_{suf}^j heaps are different versions of the same partially persistent Iheap. Also, the roots of the H_{suf}^j heaps include additive constants δ_j to be added to all of their descendants. However, if we focus on any one version, it will form an Iheap. This Iheap we can construct explicitly in a top down traversal starting from the root of this version, by incrementally expanding it as the partial persistent nodes are encountered during the traversal. Since the size of a partially persistent Iheap node is $O(1)$, the explicit representation of an Iheap node in a given version can be constructed in constant time.

However, the entire partially persistent Iheap does not need to be expanded into explicit heaps, only the parts actually visited by the selection algorithm. Therefore, we adjust the heap selection algorithm to build the visited parts of the heap explicitly during the traversal. This means that before any node in a H_{suf}^j heap is visited by the selection algorithm, it is build explicitly, and the newly built node is visited instead. We remark that the two children of an explicitly constructed Iheap node, can be nodes from the partially persistent Iheap.

The additive constants associated with the roots of the H_{suf}^j are also moved to the expanding heaps, and they are propagated downwards whenever they are encountered. An additive constant is pushed downwards from node v by adding the value to the sum stored in v, removing it from v, and instead inserting it into the children of v. Since nodes are visited top down by Fredericksons selection algorithm, it is possible to propagate the additive constants downwards in this manner while building the visited parts of the partially persistent Iheap. Therefore, when a node is visited by Fredericksons algorithm the key it contains is equal to the actual sum it represents.

Lemma 3. *Explicitly constructing t connected nodes in any fixed version of a partially persistent Iheap while propagating additive values downwards can be done in $O(t)$ time.*

Theorem 2. *The algorithm described in Section 2 is an $O(n+k)$ time algorithm for the k maximal sums problem.*

Proof. Constructing the pairs $\langle \delta_j, H_{\text{suf}}^j \rangle$ for $i = 1, \ldots, n$ takes $O(n)$ time by Lemma 2. Building a complete heap on top of these n pairs, see Figure 1, takes $O(n)$ time. By Lemma 3 and Theorem 1 the result follows.

7 Extension to Higher Dimensions

In this section we use the optimal algorithm for the one-dimensional k maximum sums problem to design algorithms solving the problem in d dimensions for any natural number d. We start by designing an algorithm for the k maximal sums problem in two dimensions, which is then extended to an algorithm solving the problem in d dimensions for any d.

Theorem 3. *There exists an algorithm for the two-dimensional k maximal sums problem, where the input is an $m \times n$ matrix, using $O(m^2 \cdot n + k)$ time and space with $m \leq n$.*

Proof. Without loss of generality assume that m is the number of rows and n the number of columns. This algorithm uses the reduction to the one-dimensional case mentioned in Section 1 by constructing $\binom{m}{2} + m$ one-dimensional problems. For all i, j with $1 \leq i \leq j \leq m$ we take the sub-matrix consisting of the rows from i to j and sum each column into a single entrance of an array. The array containing the rows from i to j can be constructed in $O(n)$ time from the array containing the rows from i to $j-1$. Therefore, we for each $i = 1, \ldots, m$ construct the arrays containing rows from i to j for $j = i, \ldots, m$ in this order.

For each one-dimensional instance we construct the n heaps H^j_{suf}. These heaps are then merged into one big heap by adding nodes with ∞ keys, by the same construction used in the one-dimensional algorithm, and use the heap selection algorithm to extract the result. This gives $(\binom{m}{2} + m) \cdot (n-1) + \binom{m}{2} + m - 1$ extra values equal to ∞.

It takes $O(n)$ time to build the H^j_{suf} heaps for each of the $\binom{m}{2} + m$ one-dimensional instances and $O(m^2 \cdot n + k)$ time to do the final selection. □

The above algorithm is naturally extended to an algorithm for the d-dimensional k maximum sums problem, for any constant d. The input is a d-dimensional vector A of size $n_1 \times n_2 \times \cdots \times n_d$.

Theorem 4. *There exists an algorithm solving the d-dimensional k maximal sums problem using $O(n_1 \cdot \prod_{i=2}^{d} n_i{}^2)$ time and space.*

Proof. The dimension reduction works for any dimension d, *i.e.* we can reduce an d-dimensional instance to $\binom{n_d}{2} + n_d$ instances of dimension $d-1$. We iteratively use this dimension reduction, reducing the problem to one-dimensional instances. Let $A^{i,j}$ be the $d-1$-dimensional matrix, with size $n_1 \times n_2 \times \cdots \times n_{d-1}$ and $A^{i,j}[i_1] \cdots [i_{d-1}] = \sum_{s=i}^{j} A[i_1] \cdots [i_{d-1}][s]$.

We obtain the following incremental construction of a $d-1$-dimensional instance in the dimension reduction, $A^{i,j} = A^{i,j-1} + A^{j,j}$. Therefore, we can build each of the $\binom{n_d}{2} + n_d$ instances of dimension $d-1$ by adding $\prod_{i=1}^{d-1} n_i$ values to the previous instance. The time for constructing all these instances is bounded by:

$$T(1) = 1$$

$$T(d) = \left(\binom{n_d}{2} + n_d\right) \cdot \left(T(d-1) + \prod_{i=1}^{d-1} n_i\right),$$

which solves to $O(n_1 \cdot \prod_{i=2}^{d} n_i{}^2)$ for $n_i \geq 2$ and $i = 1, \ldots, d$. This adds up to $\prod_{i=2}^{d}(\binom{n_i}{2} + n_i) = O(\prod_{i=2}^{d} n_i{}^2)$ one-dimensional instances in total. For each one-dimensional instance the n_1 heaps, H^j_{suf}, are constructed. All heaps are assembled into one complete heap using $n_1 \cdot \prod_{i=2}^{d} \left(\binom{n_i}{2} + n_i\right) - 1$ infinity keys (∞) and heap selection is used to find the k largest sums. □

8 Space Reduction

In this section we explain how to reduce the space usage of our linear time algorithm from $O(n+k)$ to $O(k)$. This bound is optimal in the sense that at least k values must be stored as output.

Theorem 5. *There exists an algorithm solving the k maximal sums problem using $O(n+k)$ time and $O(k)$ space.*

Proof. The original algorithm uses $O(n+k)$ space. Therefore, we only need to consider the case where $k \leq n$. Instead of building all n heaps at once, only k heaps are built at a time. We start by building the k first heaps, $H_{\text{suf}}^1, \dots, H_{\text{suf}}^k$, and find the k largest sums from these heaps using heap selection as in the original algorithm. These elements are then inserted into an *applicant set*. Then all the heaps except the last one are deleted. This is because the last heap is needed to build the next k heaps. Remember the incremental construction of H_{suf}^{j+1} from H_{suf}^j defined in equation (3) based on a partial persistent Iheap.

We then build the next k heaps and find the k largest elements as before. These elements are merged with the applicant set and the k smallest are deleted using selection [21]. This is repeated until all H_{suf}^j heaps have been processed. The space usage of the last heap grows by $O(k)$ in each iteration, ruining the space bound if it is reused. To remedy this, we after each iteration find the k largest elements in the last heap and build a new Iheap with these elements using repeated insertion. The old heap is then discarded. Only the k largest elements in the last heap can be of interest for the suffix sums not yet constructed, thus the algorithm remains correct.

At any time during this algorithm we store an applicant set with k elements and k heaps which in total contains $O(k)$ elements. The time bound remains the same since there are $O(\frac{n}{k})$ iterations each performed in $O(k)$ time. □

In the case where $k = 1$, it is worth noticing the resemblance between the algorithm just described and the optimal algorithm of Jay Kadane described in the introduction. At all times we remember the best sub-vector seen so far. This is the single element residing in the applicant set. In each iteration we scan one entrance more of the input array and find the best suffix of the currently scanned part of the input array. Because of the rebuilding only two suffixes are constructed in each iteration and only the best suffix is kept for the next iteration. We then update the best sub-vector seen so far by updating the applicant set. In these terms with $k = 1$ our algorithm and the algorithm of Kadane are the same and for $k > 1$ our algorithm can be seen as a natural extension of it.

The original algorithm solving the two-dimensional version of the problem requires $O(m^2 \cdot n + k)$ space. Using the same ideas as above, we design an algorithm for the two-dimensional k maximal sums problem using $O(m^2 \cdot n + k)$ time and $O(n+k)$ space.

Theorem 6. *There exists an algorithm for the k maximal sums problem in two dimensions using $O(m^2 \cdot n + k)$ time where $m \leq n$ and $O(n+k)$ additional space.*

Proof. Using $O(n)$ space for a single array we iterate through all $\binom{m}{2}+m$ one-dimensional instances in the standard reduction creating each new instance from the last one in $O(n)$ time. We only store in memory $\lceil\frac{k}{n}\rceil$ instances at a time.

We start by finding the k largest sub-vectors from the first $\lceil\frac{k}{n}\rceil$ instances by concatenating them into a single one-dimensional instance separated by $-\infty$ values and use our one-dimensional algorithm. No returned sum will contain values from different instances because that would imply that the sum also included a $-\infty$ value. The k largest sums are saved in the applicant set. We then repeatedly find the k largest from the next $\lceil\frac{k}{n}\rceil$ instances in the same way and update the applicant set in $O(k)$ time using selection. When all instances have been processed the applicant set is returned.

If $k \leq n$ we only consider one instance in each iteration. The k largest sums from this instance is found and the applicant set is updated. This can all be done in $O(n+k) = O(n)$ time using the linear algorithm for the one-dimensional problem and standard selection. There are $\binom{m}{2}+m$ iterations resulting in an $O(m^2 \cdot n) = O(m^2 \cdot n + k)$ time algorithm.

If $k > n$ each iteration considers $\lceil\frac{k}{n}\rceil \geq 2$ instances. These instances are concatenated using $\lceil\frac{k}{n}\rceil - 1$ extra space for the ∞ values. The k largest sums from these instances are found from the concatenated instance using the linear one-dimensional algorithm in $O((\lceil\frac{k}{n}\rceil \cdot n) + k) = O(k)$ time. The number of iterations is $(\binom{m}{2}+m)/\lceil\frac{k}{n}\rceil \leq (\binom{m}{2}+m)\cdot\frac{n}{k}$, leading to an $O(m^2 \cdot n + k)$ time algorithm.

For both cases the additional space usage is at most $O(n+k)$ at any point during the iteration since only the applicant set, $\lceil\frac{k}{n}\rceil$ instances, and $\lceil\frac{k}{n}\rceil - 1$ dummy values are stored in memory at any one time. □

The above algorithm is extended naturally to solve the problem for d dimensional inputs of size $n_1 \times n_2 \times \cdots \times n_d$.

Theorem 7. *There exists an algorithm solving the d-dimensional k maximal sums problem using* $O(n_1 \cdot \prod_{i=2}^{d} {n_i}^2)$ *time and* $O(\prod_{i=1}^{d-1} n_i + k)$ *additional space.*

Proof. As in Theorem 4, we apply the dimension reduction repeatedly, using $d-1$ vectors of dimension $1, 2, \ldots, d-1$ respectively, to iteratively construct each of the $\prod_{i=2}^{d}(\binom{n_i}{2} + n_i) = O(\prod_{i=2}^{d} {n_i}^2)$ one-dimensional instances. Every time a $d-1$-dimensional instance is created we recursively solve it. Again only $\lceil\frac{k}{n_1}\rceil$ one-dimensional instances and the applicant set is kept in memory at any one time and the algorithm proceeds as in the two-dimensional case. The space required for the arrays is $\sum_{i=1}^{d-1}\prod_{j=1}^{i} n_j = O(\prod_{i=1}^{d-1} n_i)$ with $n_i \geq 2$ for all i. □

References

1. Bentley, J.: Programming pearls: algorithm design techniques. Commun. ACM 27(9), 865–873 (1984)
2. Fukuda, T., Morimoto, Y., Morishita, S., Tokuyama, T.: Data mining with optimized two-dimensional association rules. ACM Trans. Database Syst. 26(2), 179–213 (2001)

3. Allison, L.: Longest biased interval and longest non-negative sum interval. Bioinformatics 19(10), 1294–1295 (2003)
4. Gries, D.: A note on a standard strategy for developing loop invariants and loops. Sci. Comput. Program. 2(3), 207–214 (1982)
5. Tamaki, H., Tokuyama, T.: Algorithms for the maximum subarray problem based on matrix multiplication. In: Proceedings of the ninth annual ACM-SIAM symposium on Discrete algorithms, pp. 446–452. Society for Industrial and Applied Mathematics, Philadelphia, PA, USA (1998)
6. Takaoka, T.: Efficient algorithms for the maximum subarray problem by distance matrix multiplication. Electr. Notes Theor. Comput. Sci. 61 (2002)
7. Takaoka, T.: A new upper bound on the complexity of the all pairs shortest path problem. Inf. Process. Lett. 43(4), 195–199 (1992)
8. Bae, S.E., Takaoka, T.: Algorithms for the problem of k maximum sums and a vlsi algorithm for the k maximum subarrays problem. In: 7th International Symposium on Parallel Architectures, Algorithms, and Networks (I-SPAN 2004), Hong Kong, SAR, China, 10-12 May 2004, pp. 247–253. IEEE Computer Society, Los Alamitos (2004)
9. Bengtsson, F., Chen, J.: Efficient algorithms for k maximum sums. In: Fleischer, R., Trippen, G. (eds.) ISAAC 2004. LNCS, vol. 3341, pp. 137–148. Springer, Heidelberg (2004)
10. Bae, S.E., Takaoka, T.: Improved algorithms for the k-maximum subarray problem for small k. In: Wang, L. (ed.) COCOON 2005. LNCS, vol. 3595, pp. 621–631. Springer, Heidelberg (2005)
11. Bae, S.E., Takaoka, T.: Improved algorithms for the k-maximum subarray problem. Comput. J. 49(3), 358–374 (2006)
12. Lin, T.-C., Lee, D.T.: Randomized algorithm for the sum selection problem. In: Deng, X., Du, D.-Z. (eds.) ISAAC 2005. LNCS, vol. 3827, pp. 515–523. Springer, Heidelberg (2005)
13. Cheng, C.-H., Chen, K.-Y., Tien, W.-C., Chao, K.-M.: Improved algorithms for the k maximum-sums problems. Theoretical Computer Science 362(1-3), 162–170 (2006)
14. Chao, K.M., Liu, H.F.: Personal communication (2007)
15. Eppstein, D.: Finding the k shortest paths. SIAM J. Comput. 28(2), 652–673 (1999)
16. Frederickson, G.N.: An optimal algorithm for selection in a min-heap. Inf. Comput. 104(2), 197–214 (1993)
17. Driscoll, J.R., Sarnak, N., Sleator, D.D., Tarjan, R.E.: Making data structures persistent. Journal of Computer and System Sciences 38(1), 86–124 (1989)
18. Sleator, D.D., Tarjan, R.E.: Self adjusting heaps. SIAM J. Comput. 15(1), 52–69 (1986)
19. Crane, C.A.: Linear lists and priority queues as balanced binary trees. Technical Report STAN-CS-72-259, Dept. of Computer Science, Stanford University (1972)
20. Knuth, D.E.: The art of computer programming, sorting and searching, 2nd ed., vol. 3 Addison Wesley Longman Publishing Co., Inc., Redwood City, CA, USA (1998)
21. Blum, M., Floyd, R.W., Pratt, V.R., Rivest, R.L., Tarjan, R.E.: Time bounds for selection. J. Comput. Syst. Sci. 7(4), 448–461 (1973)
22. Williams, J.W.J.: Algorithm 232: Heapsort. Communications of the ACM 7(6), 347–348 (1964)

A Lower Bound of $1 + \phi$ for Truthful Scheduling Mechanisms*

Elias Koutsoupias and Angelina Vidali

Department of Informatics, University of Athens
{elias,avidali}@di.uoa.gr

Abstract. We give an improved lower bound for the approximation ratio of truthful mechanisms for the unrelated machines scheduling problem. The mechanism design version of the problem which was proposed and studied in a seminal paper of Nisan and Ronen is at the core of the emerging area of Algorithmic Game Theory. The new lower bound $1+\phi \approx 2.618$ is a step towards the final resolution of this important problem.

1 Introduction

We study the classical scheduling problem on unrelated machines [15,21,16] from the mechanism-design point of view. There are n machines and m tasks each with different execution times on each machine. The objective of the mechanism is to schedule the tasks on the machines to minimize the makespan, i.e. to minimize the time we have to wait until all tasks are executed. In the mechanism-design version of the problem, the machines are selfish players that want to minimize the execution time of the tasks assigned to them. To overcome their "laziness" the mechanism pays them. With the payments, the objective of each player/machine is to minimize the time of its own tasks minus the payment. A loose interpretation of the payments is that they are given to machines as an incentive to tell the truth. A mechanism is called *truthful* when telling the truth is a dominant strategy for each player: for all declarations of the other players, an optimal strategy of the player is to tell the truth. A classical result in mechanism design, the Revelation Principle, states that for every mechanism, in which each player has a dominant strategy, there is a truthful mechanism which achieves the same objective. The Revelation Principle allows us to concentrate on truthful mechanisms (at least for the class of centralized mechanisms).

A central question in the area of Algorithmic Mechanism Design is to determine the best approximation ratio of mechanisms. This question was raised by Nisan and Ronen in their seminal work [23] and remains wide open today. The current work improves the lower bound on the approximation to $1 + \phi \approx 2.618$, where ϕ is the golden ratio.

A lower bound on the approximation ratio can be of either computational or information-theoretic nature. A lower bound is computational when it is based on some assumption about the computational resources of the algorithm, most

* Supported in part by IST-15964 (AEOLUS) and the Greek GSRT.

L. Kučera and A. Kučera (Eds.): MFCS 2007, LNCS 4708, pp. 454–464, 2007.

commonly that the algorithm is polynomial-time. It is of information-theoretic nature when the source of difficulty is not computational but is imposed by the restrictions of the mechanism framework and more specifically by the truthfulness condition. Our lower bound is entirely information-theoretic: No (truthful) mechanism—including exponential and even recursive algorithms—can achieve approximation ratio better than 2.618.

When we consider the approximation ratio of a mechanism, we ignore the payments and care only about the allocation part of the mechanism. A natural question then arises: Which scheduling (allocation) algorithms are part of truthful mechanisms? There is an elegant and seemingly simple characterization of these mechanisms: Monotone Algorithms. The characterizing property of these algorithms, the Monotonicity Property, implies that when we increase the time of some tasks on a specific machine, the machine will not get any new tasks. Similarly when we decrease the time of some tasks of a specific machine the machine can only get more tasks. In a loose sense, the Monotonicity Property is a combination of these two facts and can be expressed very succinctly. Despite its simplicity, we do not know how to take full advantage of the Monotonicity Property and in this work (as in all previous works on this problem) we apply a very restricted variant of it (Lemma 1).

1.1 Related Work

The scheduling problem on unrelated machines is one of the most fundamental scheduling problems [15,16]. The problem is NP-complete, it can be approximated within a factor of 2, and it is known that no polynomial-time algorithm can have approximation ratio better than 3/2, unless P=NP [21].

Here we study its mechanism-design version and we improve the results of [8], where we (together with George Christodoulou) gave a lower bound of $1 + \sqrt{2}$. The current work can be considered a continuation of it as we use similar tools, in particular Lemma 1; this is essentially the only known tool (used also in the seminal work of Nisan and Ronen [23]), and from this perspective it is unavoidable that we use it again here. However, our techniques are completely different and much more sophisticated than the ones used in [8], where we used simple instances of 3 players and 5 tasks. In this work on the other hand, we use arbitrarily many players and tasks and we obtain the bound of $1 + \phi$ only as the limit when the number of players tends to infinity.

Surprisingly, we are not using anything about the geometrical structure of the mechanism, even though it seemed as if the use of a geometric lemma was a crucial part of the proof in [8]. The main connection between the proof of the 2.61 and the 2.41 lower bound is the use of the second part of Lemma 1 which, albeit being a very simple observation seems very powerful.

Nisan and Ronen [23,24], who introduced the problem and initiated the algorithmic theory of Mechanism Design gave a truthful n-approximate (polynomial-time) algorithm; they also showed that no mechanism (polynomial-time or not) can achieve approximation ratio better than 2. They conjectured that there is no deterministic mechanism with approximation ratio less than n.

Recent work by Lavi and Swamy [20] improves the upper bound for a special case of the same problem—namely when the processing times have only two possible values low or high—and devise a deterministic 2-approximation truthful mechanism.

Archer and Tardos [4] considered the variant of the problem for related machines. In this case, for each machine there is a single value (instead of a vector), its speed. They provided a characterization of all truthful algorithms in this class, in terms of a monotonicity condition. Using this characterization, they showed that there is an optimal algorithm which is truthful (albeit exponential-time). They also gave a polynomial-time randomized 3-approximation mechanism, which was later improved to a 2-approximation, in [2]. This mechanism is truthful in expectation. Andelman, Azar and Sorani [1] gave a 5-approximation deterministic truthful mechanism, in the same framework, which was then improved by Kovacs [18] to 3-approximation deterministic truthful mechanism, while finally the ratio was reduced to 2.8.

For randomized mechanisms, Nisan and Ronen [23] gave a randomized truthful mechanism for two players, that achieves an approximation ratio of 7/4. Recently, Mu'alem and Schapira [22] extended this result and showed a $7n/8$ upper bound. They also proved a lower bound of $2-\frac{1}{n}$ for any randomized truthful mechanism for n machines.

Very recently Koutsoupias, Christodoulou and Kovacs [12] gave a more general result. They considered the fractional variant of the same problem and showed a lower bound of $2-\frac{1}{n}$ (which naturally extends to randomized algorithms). They also gave a $\frac{n+1}{2}$ fractional approximation algorithm.

Related is also some work which has been done in the context of combinatorial auctions which is a generalization of the scheduling problem (see for example [3,6,7,10,5,11] and the references within). Saks and Yu [25] proved that for convex domains the Monotonicity Property characterizes the class of social choice functions implementable by truthful mechanisms, generalizing results of [14,19].

2 Problem Definition

We recall here the definitions of the scheduling problem, of the concept of mechanisms, as well as some fundamental properties of them (see [8] for more details, references, and proofs).

Definition 1 (The unrelated machines scheduling problem). *The input to the scheduling problem is a nonnegative matrix t of n rows, one for each machine-player, and m columns, one for each task. The entry t_{ij} (of the i-th row and j-th column) is the time it takes for machine i to execute task j. Let t_i denote the times for machine i, which is the vector of the i-th row. The output is an allocation $x = x(t)$, which partitions the tasks into the n machines. We describe the partition using indicator values $x_{ij} \in \{0,1\}$: $x_{ij} = 1$ iff task j is allocated to machine i. We should allocate each task to exactly one machine, or more formally $\sum_{j=1}^{m} x_{ij} = 1$.*

In the mechanism-design version of the problem we consider direct-revelation mechanisms. That is, we consider mechanisms that work according to the following protocol:

- Each player i declares the values in row t_i, which is known only to player i.
- The mechanism, based on the declared values, decides how to allocate the tasks to the players.
- The mechanism, based on the declared values, and the allocation of the previous step, decides how much to pay each player.

The mechanism consists of two algorithms, an allocation algorithm and a payment algorithm. The cost of a player (machine) is the sum of the times of the tasks allocated to it minus the payment. One way to think of it is as if the players are lazy and don't want to execute tasks, and the mechanism pays them enough to induce them to execute the tasks. On the other hand, the players know both the allocation and the payment algorithm and may have an incentive to lie in the first step. The class of mechanisms for which the players have no incentive to lie are called truthful mechanisms. Here we consider the strictest version of truthfulness which is the class of dominant truthful mechanisms: In these mechanism truth telling is a dominant strategy, i.e., for every possible declaration of the other players, an optimal strategy of a player is to reveal its true values. This restricts significantly the set of possible algorithms.

On the other hand every mechanism can be turned into an equivalent truthful mechanism. This fact, known in the literature as the Revelation Principle, allows as to concentrate only on truthful mechanisms.

Here we care only about the approximation ratio of the allocation part of the mechanisms. So when we refer to the approximation ratio of a mechanism, we mean the approximation ratio of its allocation part. Since payments are of no importance in this consideration, it would be helpful if we could find a necessary and sufficient condition which characterizes which allocations algorithms are ingredients of truthful mechanisms. Fortunately such condition exists:

Definition 2 (Monotonicity Property). *An allocation algorithm is called monotone if it satisfies the following property: for every two sets of tasks t and t' which differ only on some machine i (i.e., on the i-the row) the associated allocations x and x' satisfy $\sum_{j=1}^{m}(x_{ij} - x'_{ij})(t_{ij} - t'_{ij}) \leq 0$, which can be written more succinctly as a dot product:*

$$(x_i - x'_i) \cdot (t_i - t'_i) \leq 0.$$

The Monotonicity Property characterizes the allocation part of truthful mechanisms. The fact that is necessary and sufficient was shown in [23] and [25] respectively. Although this is a complete characterization, it is not easy to use it and we don't know how to take complete advantage of it.

One fundamental open problem is to find a useful characterization of truthful mechanisms for the scheduling problem. For the much more general problem of mechanism design in arbitrary domains, there is simple characterization by

Roberts [17]: The only truthful mechanisms are generalized VCG mechanisms [26,9,13]. The scheduling problem is at the other end of the spectrum, where the domain is restricted yet general enough to allow for interesting mechanisms.

Not only we lack such a nice characterization as Roberts Characterization for the domain of the scheduling problem, but we also employ a very specific way to apply the Monotonicity Property. In particular, the only known way to take advantage of the Monotonicity Property is the following lemma [8], which will be the main ingredient of our proof. For completeness we also include the proof of this lemma.

Lemma 1. *Let t be a matrix of processing times and let $x = x(t)$ be the allocation produced by a truthful mechanism. Suppose that we change only the tasks of machine i and in such a way that $t'_{ij} > t_{ij}$ when $x_{ij} = 0$, and $t'_{ij} < t_{ij}$ when $x_{ij} = 1$. The mechanism does not change the allocation to machine i, i.e., $x_i(t') = x_i(t)$. (However, it may change the allocation of other machines).*

Moreover for mechanisms with bounded approximation ratio, suppose that there exists a task with ∞ processing time in all machines except machine i. Suppose further that we change the processing time of this task on machine i to some bounded value and the processing times of the remaining tasks on machine i as above. Then again the allocation of machine i is not affected.

Proof. By the Monotonicity Property, we have

$$\sum_{j=1}^{m}(t_{ij} - t'_{ij})(x_{ij}(t) - x_{ij}(t')) \leq 0.$$

If a task can only be processed by machine i then we must have $x_{ij}(t) = x_{ij}(t') = 1$ and consequently the corresponding term is 0.

For a task j with $x_{ij}(t) = 0$ we have $x_{ij}(t) - x'_{ij}(t) \leq 0$ (whichever the allocation $x_{ij}(t') \in \{0, 1\}$ is). Raising the value of such a tasks means $t_{ij} - t'_{ij} \leq 0$. On the other hand if $x_{ij}(t) = 1$ we have $x_{ij}(t) - x'_{ij}(t) \geq 0$. Lowering the value of such a task we get $t_{ij} - t'_{ij} \geq 0$.

In either case the corresponding product satisfies $(x_{ij}(t) - x'_{ij}(t))(t_{ij} - t'_{ij}) \geq 0$. Every term of this sum is nonnegative and consequently the only way to satisfy the inequality is with equality, by setting $x_{ij}(t) = x_{ij}(t')$ for all j.

To simplify the presentation, when we apply Lemma 1, we will increase or decrease only some values of a machine, not all its values. The understanding will be that *the rest of the values increase or decrease appropriately by a tiny amount which we omit* to keep the expressions simple.

3 A Lower Bound of $1 + \phi$ for $n \to \infty$ Machines

The main result of this work is

Theorem 1. *There is no deterministic mechanism for the scheduling problem with $n \to \infty$ machines with approximation ratio less than $1 + \phi$.*

We shall build the proof of the theorem around the instance

$$\begin{pmatrix} 0 & \infty & \cdots & \infty & \infty & 1 & a & \cdots & a^{n-2} \\ \infty & 0 & \cdots & \infty & \infty & a & a^2 & \cdots & a^{n-1} \\ & & \ddots & & & & & & \\ \infty & \infty & \cdots & 0 & \infty & a^{n-2} & a^{n-1} & \cdots & a^{2n-4} \\ \infty & \infty & \cdots & \infty & 0 & a^{n-1} & a^{n} & \cdots & a^{2n-3} \end{pmatrix},$$

where $a \geq 1$ is a parameter and ∞ denotes an arbitrarily high value. Eventually, we will set $a = \phi$ when $n \to \infty$. We let however a to be a parameter for clarity and for obtaining better bounds for finite n.

The lower bound will follow from the fact (which we will eventually prove) that every truthful mechanism with approximation ratio less than $1 + a$ must allocate all $n - 1$ rightmost tasks to the first player. The proof of this fact is by induction. However, the induction needs a stronger induction hypothesis which involves instances of the form

$$T(i_1, \ldots, i_k) = \begin{pmatrix} 0 & \infty & \cdots & \infty & a^{i_1} & a^{i_2} & \cdots & a^{i_k} \\ \infty & 0 & \cdots & \infty & a^{i_1+1} & a^{i_2+1} & \cdots & a^{i_k+1} \\ \vdots & & \ddots & & \vdots & & \ddots & \vdots \\ \infty & \infty & \cdots & 0 & a^{i_1+n-1} & a^{i_2+n-1} & \cdots & a^{i_k+n-1} \end{pmatrix},$$

where $0 \leq i_1 < i_2 < \ldots < i_k$ are natural numbers and $k \leq n - 1$. We allow these instances to have additional tasks for which some value is 0, i.e., additional columns with at least one 0 entry in each one. This is only for technical reasons and will play no significant role in the proof (and it definitely does not affect the optimal cost).

We will call the first n tasks *dummy*. Observe that every mechanism with bounded approximation ratio must allocate the i-th dummy task to player i.

Remark 1. Notice that the optimal allocation has cost a^{i_k}. Furthermore, if $i_1, i_2, \ldots, i_k$ are all successive natural numbers, then the optimal allocation is unique and coincides with the diagonal assignment. Otherwise there are more than one allocations with optimal cost. For example the allocations indicated by stars:

$$\begin{pmatrix} 0* & \infty & \infty & \infty & \infty & 1 & a & a^3* \\ \infty & 0* & \infty & \infty & \infty & a & a^2* & a^4 \\ \infty & \infty & 0* & \infty & \infty & a^2* & a^3 & a^5 \\ \infty & \infty & \infty & 0* & \infty & a^3 & a^4 & a^6 \\ \infty & \infty & \infty & \infty & 0* & a^4 & a^5 & a^7 \end{pmatrix}, \quad \begin{pmatrix} 0* & \infty & \infty & \infty & \infty & 1 & a & a^3* \\ \infty & 0* & \infty & \infty & \infty & a & a^2 & a^4 \\ \infty & \infty & 0* & \infty & \infty & a^2 & a^3* & a^5 \\ \infty & \infty & \infty & 0* & \infty & a^3* & a^4 & a^6 \\ \infty & \infty & \infty & \infty & 0* & a^4 & a^5 & a^7 \end{pmatrix}$$

both have the optimal cost a^3.

We will now show the main technical lemma of the proof.

Lemma 2. *Suppose that a truthful mechanism on $T(i_1, \ldots, i_k)$, does not allocate all non-dummy tasks to the first player. Then we can find another instance for which the approximation ratio is at least $1 + a$.*

Proof. Fix a truthful mechanism and suppose that the first player does not get all non-dummy tasks. In the first part, we manipulate the tasks (by changing their values and using the tools from the previous section) in such a way that we obtain an instance with equal or fewer tasks for which in the allocation of the mechanism

- the first player gets no non-dummy task, and
- every other player gets at most one non-dummy task.

In the second part, we show that instances which satisfy the above two conditions, can be changed to obtain an instance with approximation ratio at least $1 + a$.

1st Part: Suppose that the first of the above conditions is not satisfied. That is, suppose that the first player gets some non-dummy task. We can then decrease its value (for the first player) to 0. By the Monotonicity Property and in particular by Lemma 1, the same set of tasks will be allocated to the first player, so he still does not get all non-dummy tasks.

Suppose that the second condition is not satisfied, i.e., there is a player in $\{2, \dots, n\}$ who gets at least two tasks. We can then lower all the non-zero values allocated to this player to 0 except for one. By the Monotonicity Property and in particular by Lemma 1, the same tasks will be allocated to the player. This guarantees that the first player still does not get all non-dummy tasks.

By repeating the above operations, we decrease the number of non-dummy tasks. We will end up with an instance in which the first player gets no non-dummy task and every other player will get at most one non-dummy task. This process will definitely stop when there is only one non-dummy task left.

Notice that the tasks whose value was changed to 0 remain part of the instance but they will play no particular role in the induction. This is exactly the reason for which we allowed $T(i_1, \dots, i_k)$ to have additional tasks with at least one 0 entry.

2nd Part: We can now assume that there is some $T(i_1, \dots, i_k)$ for which the above two conditions are satisfied, i.e, the mechanism allocates no non-dummy task to the first player and at most one non-dummy task to each of the other players.

The optimal cost is a^{i_k}. Our aim is to find a task which is allocated to some player j with value at least a^{i_k+1}; we will then increase player j's dummy 0 value to a^{i_k}. Then by Lemma 1, player j will get both tasks with total value at least $a^{i_k+1} + a^{i_k}$. If the optimal value is still a^{i_k}, then the approximation ratio is at least $1 + a$. However, when we raise the dummy 0 to a^{i_k} we may increase the optimal value. The crux of the proof is that there is always an allocated value of at least a^{i_k+1} for which this bad case does not occur. To find such a value we consider two cases:

Case 1: The algorithm assigns a task with value at least a^{i_k+1} to one of the last $n - k$ players. This is the easy case, because we increase the dummy 0 value of this player to a^{i_k} and the optimum is not affected. The reason is that we can allocate the non-dummy tasks to the first k players with cost a^{i_k}.

Example 1. Consider the following instance with $n = 5$ and $k = 3$. Suppose that the mechanism has the allocation indicated by the stars.

$$\begin{pmatrix} 0* & \infty & \infty & \infty & \infty & 1 & a & a^3 \\ \infty & 0* & \infty & \infty & \infty & a & a^2 & a^4* \\ \infty & \infty & 0* & \infty & \infty & a^2* & a^3 & a^5 \\ \infty & \infty & \infty & 0* & \infty & a^3 & a^4* & a^6 \\ \infty & \infty & \infty & \infty & 0* & a^4 & a^5 & a^7 \end{pmatrix}$$

Then we can raise the dummy 0 of the 4-th player to a^3. This does not affect the optimum (which is a^3) but raises the cost of the 4-th player to $a^4 + a^3$.

Case 2: The value of all tasks assigned to the last $n - k$ players is at most a^{i_k}. Consequently the indexes i_ℓs are not successive integers (Remark 1). Let q be the length of the last block of successive indexes, i.e., $k-q$ is the maximum index where there is a gap in the i_ℓ's. More precisely, let $k - q$ be the maximum index for which $i_{k-q} + 1 < i_{k-q+1}$. Since player 1 gets no non-dummy task, there is a player $p \in \{q+1, \ldots, n\}$ such that some of the last q tasks is allocated to p. We raise the dummy 0 value of player p to a^{i_k}.

We have to show two properties: First that the allocated value to p was at least a^{i_k+1} and that the optimum is not affected. Indeed, the first property follows from the fact that $p > q$ (and by the observation that all values of the last q tasks for the players in $\{q + 1, \ldots, n\}$ are at least a^{i_k+1}). To show that the optimal solution is not affected consider the allocation which assigns

- the ℓ-th from the end non-dummy task to the ℓ-player, for $\ell < p$
- the ℓ-th from the end non-dummy task to the $(\ell + 1)$-player, for $\ell \geq p$

Notice that this allocation assigns no non-dummy task to the p-th player, as it should. The p-th player is allocated the dummy task which was raised from 0 to a^{i_k}. Also, since there is a gap at the $k - q$ position, all allocated values are at most a^{i_k}.

Example 2. Consider the following instance with $n = 5$, $k = 3$, and $q = 2$. Suppose that the mechanism has the allocation indicated by the stars.

$$\begin{pmatrix} 0* & \infty & \infty & \infty & \infty & 1 & a^2 & a^3 \\ \infty & 0* & \infty & \infty & \infty & a & a^3* & a^4 \\ \infty & \infty & 0* & \infty & \infty & a^2 & a^4 & a^5* \\ \infty & \infty & \infty & 0* & \infty & a^3* & a^5 & a^6 \\ \infty & \infty & \infty & \infty & 0* & a^4 & a^6 & a^7 \end{pmatrix}$$

Then $p = 3$, and we can raise the dummy 0 of the 3-rd player to a^3. This does not affect the optimum (which allocates the a^3 values), but raises the cost of the 4-th player to $a^5 + a^3 \geq a^4 + a^3$.

With the above lemma, we can easily prove the main result:

Proof (Proof of Theorem 1). Consider the instance

$$\begin{pmatrix} 0 & \infty & \cdots & \infty & \infty & 1 & a & \cdots & a^{n-2} \\ \infty & 0 & \cdots & \infty & \infty & a & a^2 & \cdots & a^{n-1} \\ & & \ddots & & & & & & \\ \infty & \infty & \cdots & 0 & \infty & a^{n-2} & a^{n-1} & \cdots & a^{2n-4} \\ \infty & \infty & \cdots & \infty & 0 & a^{n-1} & a^n & \cdots & a^{2n-3} \end{pmatrix}.$$

By the previous lemma, either the approximation ratio is at least $1+a$ or all non-dummy tasks are allocated to the first player. In the latter case, we raise the dummy 0 of the 1-st player to a^{n-1}. The optimal cost becomes a^n while the cost of the first player is $1+a+a^2+\ldots+a^{n-1}$.

The approximation ratio is at least

$$\min\{1+\frac{1}{a}+\frac{1}{a^2}+\ldots+\frac{1}{a^{n-1}}, a+1\}.$$

We select a so that

$$1+\frac{1}{a}+\frac{1}{a^2}+\ldots+\frac{1}{a^{n-1}}=1+a. \tag{1}$$

For $n\to\infty$, this gives $\dfrac{1}{1-\dfrac{1}{a}}=1+a$.

Thus $a^2=1+a$, and the solution to this equation is $a=\phi$. So the approximation ratio of any mechanism is at least $1+\phi$. For fixed number of players n, the solution of Equation 1 determines a lower bound for the approximation ratio. For small values of n, the approximation ratio is less than $1+\phi$ but it converges to it rapidly, as shown in Table 1.

Table 1. The lower bound given by Theorem 1 for a small number of machines

n	2	3	4	5	6	7	8	...	∞
$1+a$	2	2.324	2.465	2.534	2.570	2.590	2.601	...	$1+\phi$

References

1. Andelman, N., Azar, Y., Sorani, M.: Truthful approximation mechanisms for scheduling selfish related machines. In: 22nd Annual Symposium on Theoretical Aspects of Computer Science (STACS), pp. 69–82 (2005)
2. Archer, A.: Mechanisms for Discrete Optimization with Rational Agents. PhD thesis, Cornell University (January 2004)

3. Archer, A., Papadimitriou, C.H., Talwar, K., Tardos, É.: An approximate truthful mechanism for combinatorial auctions with single parameter agents. In: Proceedings of the Fourteenth Annual ACM-SIAM Symposium on Discrete Algorithms (SODA), pp. 205–214. ACM Press, New York (2003)
4. Archer, A., Tardos, É.: Truthful mechanisms for one-parameter agents. In: 42nd Annual Symposium on Foundations of Computer Science (FOCS), pp. 482–491 (2001)
5. Babaioff, M., Lavi, R., Pavlov, E.: Mechanism design for single-value domains. In: Proceedings, The Twentieth National Conference on Artificial Intelligence and the Seventeenth Innovative Applications of Artificial Intelligence Conference (AAAI), pp. 241–247 (2005)
6. Bartal, Y., Gonen, R., Nisan, N.: Incentive compatible multi unit combinatorial auctions. In: Proceedings of the 9th Conference on Theoretical Aspects of Rationality and Knowledge (TARK), pp. 72–87 (2003)
7. Briest, P., Krysta, P., Vöcking, B.: Approximation techniques for utilitarian mechanism design. In: Proceedings of the 37th Annual ACM Symposium on Theory of Computing (STOC), pp. 39–48. ACM Press, New York (2005)
8. Christodoulou, G., Koutsoupias, E., Vidali, A.: A lower bound for scheduling mechanisms. In: Proceedings of the Eighteenth Annual ACM-SIAM Symposium on Discrete Algorithms (SODA), pp. 1163–1169. ACM Press, New York (2007)
9. Clarke, E.: Multipart pricing of public goods. Public Choice 8, 17–33 (1971)
10. Dobzinski, S., Nisan, N., Schapira, M.: Approximation algorithms for combinatorial auctions with complement-free bidders. In: Proceedings of the 37th Annual ACM Symposium on Theory of Computing (STOC), pp. 610–618. ACM Press, New York (2005)
11. Dobzinski, S., Nisan, N., Schapira, M.: Truthful randomized mechanisms for combinatorial auctions. In: Proceedings of the 38th Annual ACM Symposium on Theory of Computing (STOC), pp. 644–652. ACM Press, New York (2006)
12. Christodoulou, E.K.G., Kovacs, A.: Mechanism design for fractional scheduling on unrelated machines. In: ICALP'07 (to appear, 2007)
13. Groves, T.: Incentives in teams. Econometrica 41, 617–631 (1973)
14. Gui, H., Müller, R., Vohra, R.V.: Dominant strategy mechanisms with multidimensional types. In: Computing and Markets (2005)
15. Hochbaum, D.S.: Approximation algorithms for NP-hard problems. PWS Publishing Co. Boston, MA, USA (1996)
16. Horowitz, E., Sahni, S.: Exact and approximate algorithms for scheduling nonidentical processors. J. ACM 23(2), 317–327 (1976)
17. Kevin, R.: The characterization of implementable choice rules. Aggregation and Revelation of Preferences, 321–348 (1979)
18. Kovacs, A.: Fast monotone 3-approximation algorithm for scheduling related machines. In: Brodal, G.S., Leonardi, S. (eds.) ESA 2005. LNCS, vol. 3669, pp. 616–627. Springer, Heidelberg (2005)
19. Lavi, R., Mu'alem, A., Nisan, N.: Towards a characterization of truthful combinatorial auctions. In: 44th Symposium on Foundations of Computer Science (FOCS), pp. 574–583 (2003)
20. Lavi, R., Swamy, C.: Truthful mechanism design for multi-dimensional scheduling via cycle-monotonicity. In: Proceedings 8th ACM Conference on Electronic Commerce (EC), ACM Press, New York (2007)
21. Lenstra, J.K., Shmoys, D.B., Tardos, É.: Approximation algorithms for scheduling unrelated parallel machines. Mathematical Programming 46(1), 259–271 (1990)

22. Mu'alem, A., Schapira, M.: Setting lower bounds on truthfulness. In: Proceedings of the Eighteenth Annual ACM-SIAM Symposium on Discrete Algorithms (SODA), pp. 1143–1152. ACM Press, New York (2007)
23. Nisan, N., Ronen, A.: Algorithmic mechanism design (extended abstract). In: Proceedings of the Thirty-First Annual ACM Symposium on Theory of Computing (STOC), pp. 129–140. ACM Press, New York (1999)
24. Nisan, N., Ronen, A.: Algorithmic mechanism design. Games and Economic Behavior 35, 166–196 (2001)
25. Saks, M.E., Yu, L.: Weak monotonicity suffices for truthfulness on convex domains. In: Proceedings 6th ACM Conference on Electronic Commerce (EC), pp. 286–293. ACM Press, New York (2005)
26. Vickrey, W.: Counterspeculations, auctions and competitive sealed tenders. Journal of Finance 16, 8–37 (1961)

Analysis of Maximal Repetitions in Strings*

Maxime Crochemore[1,**] and Lucian Ilie[2,***,†]

[1] Institut Gaspard-Monge, Université de Marne-la-Vallée
77454 Marne-la-Vallée, Cedex 2, France
mac@univ-mlv.fr

[2] Department of Computer Science, University of Western Ontario
N6A 5B7, London, Ontario, Canada
ilie@csd.uwo.ca

Abstract. The cornerstone of any algorithm computing all repetitions in strings of length n in $\mathcal{O}(n)$ time is the fact that the number of maximal repetitions (runs) is linear. Therefore, the most important part of the analysis of the running time of such algorithms is counting the number of runs. Kolpakov and Kucherov [FOCS'99] proved it to be cn but could not provide any value for c. Recently, Rytter [STACS'06] proved that $c \leq 5$. His analysis has been improved by Puglisi et al. to obtain 3.48 and by Rytter to 3.44 (both submitted). The conjecture of Kolpakov and Kucherov, supported by computations, is that $c = 1$. Here we improve dramatically the previous results by proving that $c \leq 1.6$ and show how it could be improved by computer verification down to 1.18 or less. While the conjecture may be very difficult to prove, we believe that our work provides a good approximation for all practical purposes.

For the stronger result concerning the linearity of the sum of exponents, we give the first explicit bound: $5.6n$. Kolpakov and Kucherov did not have any and Rytter considered "unsatisfactory" the bound that could be deduced from his proof. Our bound could be as well improved by computer verification down to $2.9n$ or less.

Keywords: Combinatorics on words, repetitions in strings, runs, maximal repetitions, maximal periodicities, sum of exponents.

1 Introduction

Repetitions in strings constitute one of the most fundamental areas of string combinatorics with very important applications to text algorithms, data compression, or analysis of biological sequences. They have been studied already in the papers of Axel Thue [20], considered as having founded stringology. While Thue was interested in finding long sequences with few repetitions, one of the most important problems from the algorithmic point of view was finding all repetitions

* This work has been done during the second author's stay at Institut Gaspard-Monge.
** Research supported in part by CNRS.
*** Corresponding author.
† Research supported in part by CNRS and NSERC.

L. Kučera and A. Kučera (Eds.): MFCS 2007, LNCS 4708, pp. 465–476, 2007.

fast. A major obstacle for a linear-time algorithm was finding a way to encode all repetitions in linear space. The problem was studied first by Crochemore [2] where maximal (non-extendable) integer powers were introduced and an (optimal) $\mathcal{O}(n \log n)$ algorithm for finding them all was given. Moreover, the bound was shown to be optimal as it is reached by the Fibonacci strings.

Occurrences of fractional repetitions were considered next, of right-maximal (non-extendable to the right) repetitions by Apostolico and Preparata [1] and then of maximal (non-extendable both ways, called *runs* for the rest of the paper) repetitions by Main [15] who gave a linear-time algorithm for finding all leftmost occurrences of runs. Iliopoulos et al. [9] showed that for Fibonacci strings the number of maximal repetitions is linear. Even if their result applies to a particular class of strings, it is important since the Fibonacci strings were known to contain many repetitions. The breakthrough came in the paper of Kolpakov and Kucherov [12], where it was finally proved that encoding all occurrences of repetitions into runs was the right way to obtain a linear-sized output. They modified Main's algorithm to compute all runs in linear time; see [11]. For more details on various algorithms computing repetitions, see Chapter 8 of [14].

Kolpakov and Kucherov [12] showed that the number of runs in a string of length n is at most cn but their proof could not provide any value for the constant c. Another breakthrough came very recently, when Rytter [17] proved that $c \leq 5$. Puglisi et al. [16] improved Rytter's analysis to bring the constant down to 3.48 and then Rytter [18] produced his own version of the improved analysis with a constant factor of 3.44. The fact that the two bounds are so close may show that the ideas in Rytter's initial paper have been well squeezed.

The conjecture in [12] is that $c = 1$ for binary alphabets, as supported by computations for string lengths up to 31. A stronger conjecture is proposed in [6] where a family of strings is given with the number of runs equal to $\frac{3}{2\phi} = 0.927\ldots$ (ϕ is the golden ratio), thus proving $c \geq 0.927\ldots$. The authors of [6] conjectured that this bound is optimal. Some reasons which might indicate that the optimal bound may be less than n are discussed in Section 7. The proof of [12] is extremely long and complex and the one of [17] is still very intricate. (The improvements in [16] and [18] only make a careful analysis of the ideas of [17].) A simple proof for the linearity is given by the authors in [3] where an improvement of the notion of neighbors of [17] is introduced.

It is interesting to note that the number of runs having the same starting point is logarithmic, due to the three-square lemma of [4], that is, they are not uniformly distributed, which makes proving linearity hard. The situation is even worse for centers (beginning of the second period of the run – see next section for details) where linearly many runs can share the same center. However, while Rytter [17] counted runs by their beginnings, we count them by centers and obtain much better results. A more detailed comparison of the two approaches is included in Section 3.

In this paper we improve significantly the previous results by proving the bound $1.6n$ on the number of runs in a string of length n. This bound can be lowered a bit by extra effort for the runs with short periods, but we show also how it could be improved by computer verification down to $1.18n$ or even further. Note that this bound has an important direct impact on the running time of all algorithms computing all repetitions since it says how many runs we expect the algorithm to output. It is important as well from mathematical point of view, that is, to find the best upper bound on the number of runs, and from algorithm-design point of view, as it may lead to simpler algorithms for finding all repetitions (the algorithm of [11] uses relatively complicated data structures such as suffix trees). While the conjecture may be very difficult to solve, we believe that our work provides a good approximation for all practical purposes.

The approach in [3] is used also to give a simple proof for the stronger result concerning the linearity of the sum of exponents. This result has been proved by Kolpakov and Kucherov [10]. (It follows also from Rytter's construction in [17].) It has applications to the analysis of various algorithms, such as computing branching tandem repeats: the linearity of the sum of exponents solves a conjecture of [19] concerning the linearity of the number of maximal tandem repeats and implies that all can be found in linear time. For other applications, see [10].

But the proof of [10] is very complex and could not provide a constant. A bound can be derived from the proof of Rytter [17] but he mentioned only that the bound that he obtains is "unsatisfactory." It seems to be $25n$. The improved analysis in [18] does not mention the sum of exponents at all. We provide here the first explicit bound, which is $5.6n$. As with the other bound, extra effort for the runs with short periods can lower the bound, but we show how it could be improved by computer verification down to $2.9n$ or further. As mentioned in [10], computations seem to indicate a $2n$ bound.

The paper is organized as follows. In the next section we give the basic definitions needed in the paper. The new bound is given in the Section 3 which contains also a comparison between our approach and the one of Rytter [17,18]. Our approach is more accurate for both long and short periods. The analyses needed for the new bound are presented in Section 4, for runs with arbitrarily long periods, and Section 5, for runs with periods 9 or less. The sum of exponents in discussed in Section 6. Some comments on both bounds, as well as ways to improve them further by computer verification are included in Section 7. We conclude with a discussion on several other related problems. Some proofs are omitted due to lack of space.

2 Definitions

Let A be an alphabet and A^* the set of all finite strings over A. We denote by $|w|$ the length of a string w, its ith letter by $w[i]$ and the factor $w[i]w[i+1]\dots w[j]$ by $w[i\,..\,j]$. We say that w has period p iff $w[i] = w[i+p]$, for all i, $1 \le i \le |w| - p$. The shortest period of w is called *the period* of w. The ratio between the length and the period of w is called the *exponent* of w.

For a positive integer n, the nth power of w is defined inductively by $w^1 = w$, $w^n = w^{n-1}w$. A string is *primitive* if it cannot be written as a proper (two or more) integer power of another string. Any string can be uniquely written as an integer power of a primitive string, called its *primitive root*. The following well-known *synchronization* property will be useful for us: if w is primitive, then w appears as a factor of ww only as a prefix and as a suffix (not in-between). Another property we use is *Fine and Wilf's periodicity lemma*: If w has periods p and q and $|w| \geq p+q$, then w has also period $\gcd(p,q)$. (This is a bit weaker than the original lemma which works as soon as $|w| \geq p+q-\gcd(p,q)$, but it is good enough for our purpose.) We refer the reader to [13,14] for further information on all concepts used here.

For a string w, a *run*[1] (or maximal repetition) is an interval $[i\,..\,j]$ such that both (i) the factor $w[i\,..\,j]$ has its shortest period at most $\frac{j-i+1}{2}$ and (ii) $w[i-1\,..\,j]$ and $w[i\,..\,j+1]$, if defined, have a strictly higher shortest period. As an example, consider $w = \mathsf{abbababbbaba}$; $[3\,..\,7]$ is a run with period 2 and exponent 2.5; we have $w[3\,..\,7] = \mathsf{babab} = (\mathsf{ba})^{2.5}$. Other runs are $[2\,..\,3], [7\,..\,8], [8\,..\,11], [5\,..\,10]$ and $[1\,..\,11]$.

By definition, a run is a maximal occurrence of a repetition of exponent at least two. Therefore, it starts with a square and continues with the same period. But the square is the only part of the run we can count on. Therefore, for a run starting at i and having period $|x| = p$, we shall call $w[i\,..\,i+2p-1] = x^2$ the *square* of the run. Note that x is primitive and the square of a run is not left-extendable (with the same period) but may be extendable to the right. The *center* of the run is the position $c = i + p$. We shall denote the *beginning* of the run by $i_x = i$, the *end* of its square by $j_x = i_x + 2p - 1$, and its center by $c_x = c$.

3 The Bound

The idea is to group together the runs having close centers and similar periods and then prove that we have only one on the average in each group. For any $\delta > 0$, two runs having squares x^2 and y^2 are *δ-close* if both (i) $|c_x - c_y| \leq \delta$ and (ii) $2\delta \leq |x|, |y| \leq 3\delta$. Abusing the language, we shall sometimes say that the squares, instead of the runs, are δ-close. Another notion that we shall use frequently is that of runs with the periods between 2δ and 3δ; we shall call those *δ-runs*.

We prove (Section 4) that the number of runs is highest when, for all δ, any interval of length δ contains only one center of a δ-run. That means that the number of δ-runs in a string of length n is at most $\frac{n}{\delta}$. We could then sum up for a sequence $\delta_1, \delta_2, \ldots$ of values of δ such that the corresponding intervals $[2\delta_i\,..\,3\delta_i]$ collectively cover the set of all possible periods but we make one further improvement. Since any bound for arbitrarily long runs performs poorly on runs with short periods, we bound separately the number of runs with short periods; precisely we prove that there are at most n runs with period at most 9 in any string of length n (Section 5).

Summing up the above for all values $\delta_i = \frac{10}{2}\left(\frac{3}{2}\right)^i$, $i \geq 0$, to cover all periods greater than 9, we obtain the following upper bound for the number of runs in a string of length n, which is the main result of our paper:

[1] Runs were introduced in [15] under the name *maximal periodicities*; they are called *m-repetitions* in [12] and *runs* in [9].

$$n + \sum_{i=0}^{\infty} \frac{n}{\delta_i} = n + \left(\frac{2}{10} \sum_{i=0}^{\infty} \left(\frac{2}{3}\right)^i \right) n = 1.6n \ . \tag{1}$$

Theorem 1. *The number of runs in a string of length n is less than* $1.6n$.

Our approach differs from the one of Rytter [17] in several respects. First, our notion of δ-closeness is different from his notion of *neighbors*. We consider the case when the centers of the runs are close to each other as opposed to beginnings as this gives us a better overlap between the runs. Thus we can choose a better interval length for the periods. Second, we make a combinatorially finer analysis of the close runs which enables us to count all runs together; [17] splits them into weekly and highly periodic. Doing so, the proof becomes conceptually simpler. For runs with long periods we can say that our approach is about ten times better than Rytter's. He explicitly states that the number of runs with periods larger than 87 is at most $0.67n$. With our approach, this number is about ten times smaller: $\left(\frac{2}{87}\sum_{i=0}^{\infty}\left(\frac{2}{3}\right)^i\right)n \leq 0.06897n$. Third, our approach for runs with short periods is different from the one of [17]. We essentially verify that the conjecture is true up to a certain threshold for the periods of the runs. Due to the complexity of the analysis, we restricted this threshold to 9 but it can be checked automatically for higher thresholds, every time improving the bound. More on this is in Section 7.

4 Runs with Close Centers

In this section we show that, for a given δ, each interval of positions of length δ contains at most 1 center of a δ-run on the average. The result is used for runs having a period greater than 9 in the sum (1).

We investigate what happens when two (or, sometimes, three) runs in a string w are δ-close. Denote their squares by x^2, y^2, z^2, their root lengths by $|x| = p$, $|y| = q$, $|z| = r$, and assume $p \leq q \leq r$.

We discuss below all ways in which occurrences of x^2 and y^2 can be positioned relative to each other and see that long factors of both runs have short periods. When we have only two δ-close runs, synchronization properties show that the next (to the right) interval of length δ (as counted in (1)) does not contain any center of a δ-run.

When we have three δ-close runs, z^2 has to synchronize the short periods mentioned above, which restricts the beginning of z^2 to only one choice as otherwise some run would be left extendable (which is not possible). Stronger periodicity properties are implied by the existence of the third run and we can find an interval of length at least 4δ which contains no other center of δ-runs. Such an interval covers at least three intervals of length δ no matter how the decomposition of $[1\mathinner{..}n]$ into such intervals is done. Thus, less runs than in (1) are obtained.

It is also possible to have arbitrarily many δ-close runs, that is, when they all have the same center; case (i). A similar global counting approach is performed in this case. The existence of such runs implies strong periodicity properties of

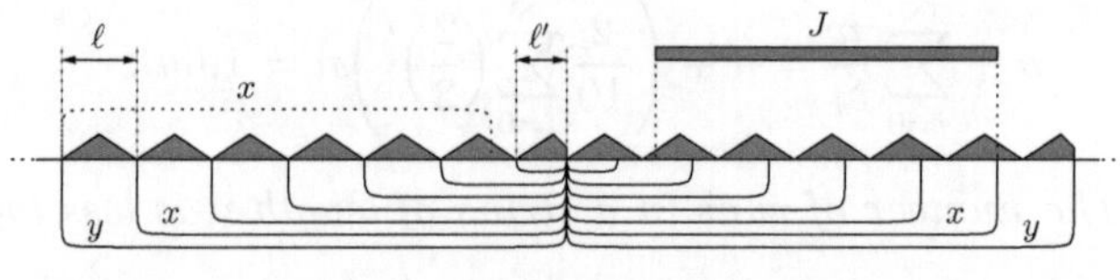

Fig. 1. The runs with the same center in case (i)

a factor of w and we exhibit a long interval without any center of runs with certain periods. In total, less runs than in (1) are obtained.

There can exist several δ-close runs such that some of them only share the same center. Therefore, we shall discuss the case of many runs having the same center first. It helps solving some situations in the other cases.

(i) $c_x = c_y$. First, both x and y have the same small period $\ell = q-p$; see Fig. 1. If we denote $c = c_y$ then we have $h = \lceil \frac{q}{q-p} \rceil$ runs $x_j^{\alpha_j}$, $2 \le \alpha_j \in \mathbb{Q}$, for $1 \le j \le h$, having period $|x_j| = (j-1)\ell + \ell'$ and beginning at $i_{x_j} = c - ((j-1)\ell + \ell')$. If we set $x_j = u^{j-1}u'$, with $|u| = \ell, |u'| = \ell'$, then the last letters of u and u' are different, as otherwise x would be left-extendable. As an example, take $w = (\mathsf{ab})^6\mathsf{aa}(\mathsf{ba})^6$, where $c = 14, h = 7, \ell = 2, \ell' = 1, u = \mathsf{ab}$, and $u' = \mathsf{a}$.

We show that for $h \ge 6$ we have less runs than in (1). Note that only for $h \ge 7$ we can have three of the $x_j^{\alpha_j}$s mutually δ-close. Therefore, we may assume for cases ((ii)–(v)) that there are no three δ-close runs with the same center.

There exists δ_{i_0} such that $\frac{\ell}{2} \le \delta_{i_0} \le \frac{3\ell}{4}$, that is, this δ_{i_0} is considered in (1). The periods corresponding to δ_{i_0} are between ℓ and $\frac{9}{4}\ell$.

We claim that there is no run in w with period between ℓ and $\frac{9}{4}\ell$ and center inside the interval $J = [c+\ell+1\,..\,c+(h-2)\ell+\ell']$. Indeed, assume there is a run with the initial square t^2, $c_t \in J$. If $i_t \ge c$, then the prefixes of length ℓ of the first and second occurrences of t, respectively, must synchronize. If $i_t > c$, then t^2 is left-extendable, a contradiction. If $i_t = c$, then ℓ divides $|t|$ and hence t is not primitive, a contradiction. If $i_t < c$, then synchronization is obtained (either the prefixes or the suffixes of length ℓ of the two occurrences of t synchronize) and we get that the last letters of u and u' are the same, a contradiction.

Then, the length of J is larger than $(h-3)\ell$ which in turn is larger than $(h-2)\delta_{i_0}$ (since $3\ell \ge 4\delta_{i_0}$ and $h \ge 6$). Thus J covers at least $h-3$ intervals of length δ_{i_0} that would contain, if considered in (1), $h-3$ runs. This is enough as we need to account, for each δ from (1), for the extra runs, that is, all but one. At least three δs are needed for all our h runs, so we need to account for at most $h-3$ runs, which we already did.

We need also mention that these h intervals of length δ_{i_0} are not reused by a different center with multiple runs since such centers cannot be close to each other. Indeed, assume we have two centers c_j with the above parameters h_j, ℓ_j, $j = 1, 2$. Then the periods satisfy $\frac{\ell_j}{2} \le \delta_{i_0} \le \frac{3\ell_j}{4}$, $j = 1, 2$, and so $\ell_j \le \frac{3}{2}\ell_{3-j}$, $j = 1, 2$. As soon as the longest runs with centers at c_1 and c_2, respectively, overlap over $\ell_1 + \ell_2$ positions, we have $\ell_1 = \ell_2$, due to Fine and Wilf's lemma. Thus, the closest positions of J_1 and J_2 cannot be closer than $\ell_1 = \ell_2 \ge \delta_{i_0}$ as otherwise some of the runs become non-primitive, a contradiction.

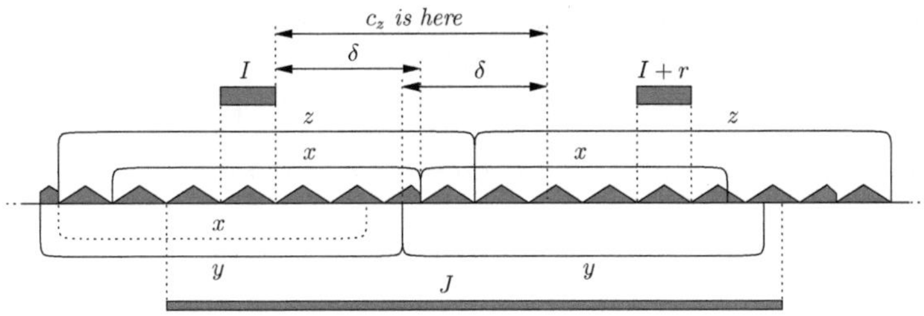

Fig. 2. Relative position of x^2, y^2 and z^2 in case (ii)

(ii) $(i_y < i_x) < c_y \leq c_x < e_x \leq e_y$. Then x and the suffix of length $c_y - i_x + (q-p)$ of y have period $q-p$; see Fig. 2. (Only the periods to the left of e_y are implied by x^2 and y^2, the ones to the right of e_y are obtained in the case when a third run, z^2, exists – see below.) We may assume this period is a primitive string as otherwise we can make the same reasoning with its primitive root.

It is not difficult to see that no δ-run can have its center in the interval $[c_y + \delta \mathinner{.\,.} c_y + 2\delta]$ as it would be left extendable. This argument is sufficient for the case when no third δ-close run exists.

If there is a third run, z^2, then we need a stronger argument to account for the three centers in the same interval of length δ. Since z^2 is δ-close to both x^2 and y^2, it must be that $c_z \in [c_x - \delta \mathinner{.\,.} c_y + \delta]$. Consider the interval of length $q-p$ that ends at the leftmost possible position for c_z, that is, $I = [c_x - \delta - (q-p) \mathinner{.\,.} c_x - \delta - 1]$; see Fig. 2. The following arguments show that it is included in the first period of z^2, that is, $[i_z \mathinner{.\,.} c_z - 1]$, and in $[i_x \mathinner{.\,.} c_y - 1]$. It is clear that I ends before both $c_z - 1$ and $c_y - 1$. The other two inequalities are proved as follows. First $i_z = c_z - r \leq c_y + \delta - q \leq c_x + p - \delta - q$. Then $i_x = c_x - p \leq c_x - \delta - q + p$. Subsequently, all such inequalities can be proved similarly and are left as exercises.

Thus $w[I]$ is primitive and equal, due to z^2, to $w[I+r]$ which is contained in $[c_x \mathinner{.\,.} e_y]$. Therefore, the periods inside the former must synchronize with the ones in the latter. It follows, in the case $i_z > i_x - (q-p)$, that $w[i_z - 1] = w[c_z - 1]$, that is, z^2 is left extendable, a contradiction. If $i_z < i_x - (q-p)$, then $w[c_x - 1] = w[i_x - (q-p) - 1] = w[i_x - 1]$, that is, x^2 is left extendable, a contradiction. The only possibility is that $i_z = i_x - (q-p)$ and r equals q plus a multiple of $q-p$. Here is an example that this is indeed possible: $w =$ baababababaababababaab, $x^2 = w[5 \mathinner{.\,.} 14]$, $y^2 = w[1 \mathinner{.\,.} 14]$, and $z^2 = w[3 \mathinner{.\,.} 20]$.

The existence of z^2 implies that the period of the second occurrence of y extends past e_y, as seen in Fig. 2. Consider the interval $J = [i_z + 2(q-p) \mathinner{.\,.} e_z - 2(q-p)]$. The existence of a fourth δ-run with center inside J but different from c_x would imply that either x^2 or z^2 is left extendable, a contradiction. (Note that we allowed the length of two $(q-p)$-periods at either end of J so that the hiccup of the period ending at c_x does not cause any problems.) On the other hand, such a run can have c_x as its center. If so, then all such runs are accounted for by case (i) since we have at least three periods at c_x between 2δ and 3δ: q, $q-(q-p)$, and $q-2(q-p)$. The length of J is $2r - 4(q-p) \geq 2(q+q-p) - 4(q-p) = 2p \geq 4\delta$

and therefore it covers at least three intervals of length δ. In total, we have at most the number of runs as counted in (1).

The three remaining cases are similar with (ii). We proved

Proposition 1. *There is at most 1 center of a δ-run on average in each interval of length δ.*

5 Microruns

We prove in this section that the number of runs with periods at most 9, which we call *microruns*,[2] in a string of length n is at most n. All runs we are talking about in this proof have periods at most 9.

The idea of the proof is as follows. We pick an arbitrary position i and consider all possible microruns with center at i. Then we show that the number of microruns with centers in an interval $[i-j\,..\,i]$ is at most j where j can vary but is always less than 5. Put otherwise, any position with two or more centers of microruns is amortized within up to 4 previous positions to its left.

For the analysis, the number of possible subsets of periods is very high ($2^9 = 512$) but we have the following lemma to help. It not only reduces significantly the number of cases but it helps with the analysis of each case as well.

Assume we have a string w. We shall say that w *has p at i* if there is a run with period p and center at position i in w. Denote also $C(i) = \{p \mid p \text{ at } i\}$. (This set depends also on w but we shall consider w implicit and omit it.)

Note that there are two differences between (a run of period) p (and center) at i and a *period p centered at i*: the former is not left-extendable and its root is primitive whereas the latter may have none of the two properties.

Lemma 1. *Consider a string w and the periods p and $p-\ell$, $0<\ell<p$. Let h be the smallest integer such that $h\ell \geq p$ ($h = \lceil p/\ell \rceil$).*

(i) (periods) If w has the period $p-\ell$ at i and the period p at $i+j$ or $i-j$ with $j \leq \ell$, then w has the also periods $p-k\ell$, $2 \leq k \leq h-1$, at i.

(ii) (runs) If w has $p-\ell$ at i and either (a) p at $i+j$ with $j \leq \ell-1$, or (b) p at $i-j$ with $j \leq \ell$, then w has $p-k\ell$ at i, for $2 \leq k \leq h-3$ (that is, all but the shortest two).

Another useful remark, with an obvious proof, is next.

Remark 1. If p at i, then there is not p at j for any $j, i-p \leq j \leq i+p$, $j \neq i$.

It is also obvious how Remark 1 is used. As far as Lemma 1 is concerned, it can be used in many ways. The first is, as already announced, to reduce as much as possible sets of periods of microruns with the same center. For instance, if we do not have periods 1,2,3 at i but do have 5, then we cannot have anything else: having 4 would imply having 1,2,3; 6 implies 1,2,3,4; 7 implies 1,3; 8 implies 2; 9 implies 1. This way our potential 512 cases are reduced to 24.

[2] By analogy with the *microsatellites* in bioinformatics; these correspond to the concatenation of short DNA sequences (1 to 4 nucleotides) that are similar.

The lemma helps also with the analysis. For example, consider the case when $C(i) = \{1, 3\}$. We shall use many times the lemma without mentioning. What we know so far about w is that $w[i-4\,..\,i+2] = \bar{\mathsf{a}}\mathsf{abaaba}$, where $\mathsf{a} \neq \mathsf{b}$ and $\bar{\mathsf{a}}$ means any letter different from a. The smallest element of $C(i-1)$ is 5. If we have 5 at $i-1$, then $w[i-7\,..\,i+3] = \bar{\mathsf{b}}\mathsf{aababaabab}$. Thus, $C(i-1) = \{5\}$ and $C(i-2) = \emptyset$, which means that the two centers at i are amortized, in this case, within the previous two positions, since the total number of centers inside the interval $[i-2\,..\,i]$ is 3. If there is not 5 at $i-1$, then the next that can be is 7 and the reasoning is identical. If there is not 7 at $i-1$, the next can be 8. If so, then $C(i-1) = \{8\}$ and the only possible (but not necessary) candidate at $i-2$ is 2. If there is 2 at $i-2$ then $C(i-2) = \{2\}$, $C(i-3) = \emptyset$, and in this case the two centers at i are amortized within the previous three positions.

The reasoning continues like this until all possibilities are completely analyzed. Actually the case $C(i) = \{1, 3\}$ has the longest analysis and there are very few comparable ones. This is the reason why we proved the result for periods up to 9. For higher numbers it gets quickly too complicated. It proves the result we claimed at the beginning of the section:

Lemma 2. *The number of runs with periods at most 9 in a string of length n is bounded by n.*

6 Sum of Exponents

We give in this section our bound on the sum of exponents which relies heavily on the results we have proved so far. The strategy is similar. We show that the sum of exponents of runs with periods four or less is at most $2n$.

Lemma 3. *The sum of exponents of runs with periods at most 4 in a string of length n is bounded by $2n$.*

For runs with periods higher than 4, we shall use the discussion in Section 4 and Fine and Wilf's lemma. The lemma can be rephrased as follows: For two primitive strings x and y, any powers x^α and y^β, $\alpha \geq 2$ and $\beta \geq 2$, cannot have a common factor longer than $|x| + |y|$ as such a factor would have also period $\gcd(|x|, |y|)$, contradicting the primitivity of x and y.

Next consider a fixed δ and two δ-runs, x^α and y^β, $\alpha, \beta \in \mathbb{Q}$, and denote their periods $|x| = p$ and $|y| = q$. The strings x^α and y^β cannot overlap more than $2.5\min(p, q)$ as otherwise Fine and Wilf's lemma would imply that x and y are not primitive, a contradiction. Therefore, their suffixes $x^{\alpha-2.5}$ and $y^{\beta-2.5}$ (assuming the exponents large enough) cannot overlap at all. Therefore, the sum of exponents of δ-runs is at most 2.5 times the number of runs plus whatever exponent is left of each run after removing the prefix of exponent 2.5. For x^α, that means $\alpha - 2.5 = \frac{|x^{\alpha-2.5}|}{|x|} \leq \frac{|x^{\alpha-2.5}|}{2\delta}$ and when summing up all these, as they cannot overlap, we obtain $\frac{n}{2\delta}$.

Assuming that the number of runs as above is at most $\frac{n}{\delta}$ and using Lemma 3, we obtain the following bound on the sum of exponents, where $\delta_i = \frac{5}{2}\left(\frac{3}{2}\right)^i$, $i \geq 0$:

$$2n + \sum_{i=2}^{\infty}\left(2.5\frac{n}{\delta_i} + \frac{n}{2\delta_i}\right) = 2n + \left(3\frac{2}{5}\sum_{i=2}^{\infty}\left(\frac{2}{3}\right)^i\right)n = 5.6n\,. \quad (2)$$

Note however, that in our analysis from Section 4, for the case (i) of many runs with the same center, we accounted for some of the runs using other runs with periods belonging to a different δ-class. That means the number of runs for each δ need not be $\frac{n}{\delta}$. Still our bound is exact because the runs we account for in case (i) have very small exponents. Recall that we need to account, for each δ, for all runs but one. Using the notation from case (i), any run $x_j^{\alpha_j}$, $2 \le j \le h-1$, cannot extend its period $|x_j|$ by more than ℓ positions to the right past the end of the initial square, and therefore has $\alpha_j \le 2 + \frac{1}{j} \le 2.5$. The runs with the shortest and the longest periods, $x_1^{\alpha_1}$ and $x_h^{\alpha_h}$, respectively, may have arbitrarily large exponents but we need not account for either one. The bound in (2) therefore holds and we proved

Theorem 2. *The sum of exponents in a string of length n is less than $5.6n$.*

7 Comments

A small improvement of our main result in Theorem 1 can be rather easily obtained as follows. We can make a better choice of the δ_is to cover all periods:

$\delta_0 = \frac{10}{2}$ – covers the periods between 10 and 15,
$\delta_1 = \frac{16}{2}$ – covers the periods between 16 and 24,
$\delta_i = \frac{25}{2}\left(\frac{3}{2}\right)^{i-2}$, for all $i \ge 2$ – cover all periods larger than or equal to 25.

The bound becomes: $n + \left(\frac{2}{10} + \frac{2}{16} + \frac{2}{25}\sum_{i=2}^{\infty}\left(\frac{2}{3}\right)^{i-2}\right)n = 1.565n$. The method of choosing values can be extended in the same manner to all δ_is but the improvements to the final bound are less and less significant. One would have to modify the proof of Theorem 1 to accommodate these changes, which is not difficult to do, but we preferred to keep the proof simpler.

The proof technique in Section 5 can be automatized so that larger periods for microruns can be considered. If one can prove it, for instance, for microruns with periods up to 32, then the bound improves to $1.18n$. A similar computer-aided approach can be applied to the bound on the sum of exponents which could be improved down to $2.9n$, assuming one can verify that the result in Lemma 3 holds for runs with periods up to 20.

Actually solving the conjecture using the above approach may be possible but certainly not easy. For instance, one could attempt to verify by computer that the number of runs with periods less than 40 is at most $0.85n$ (the remaining ones are less than $0.15n$ by our reasoning) but this seems very difficult.

We need to comment a bit more here. Our approach essentially approximates the number of runs, as it is very difficult, if not impossible, to account for all runs in this way. Therefore, the fact that we can attempt solving the conjecture shows, on the one hand, that our approach must be very close to reality, that is, the approximation we obtain is very good, yet, on the other, we believe that the bound n is not optimal as we seem to be able to get too close to it. Recall however, that it has to be more than $0.92n$, according to the lower bound of [6], that means, not too far away.

A promising approach is extending Lemma 1 to cover all periods. Of course, removing the bound on the length of periods of microruns in Lemma 1 makes it identical to the conjecture but we believe that the proof supporting the result can be extended. Precisely, we conjecture that each position with two or more centers can be amortized within at most half of the length of the longest possible period of a run, that is, at most a quarter of the length of the string.

8 Further Research

We discuss here several related problems. The first one is old but the others are proposed here.

Squares. As the number of all square occurrences in a string may be quadratic and that of primitively rooted square occurrences can still be superlinear, as already mentioned in the Introduction, it is natural to count squares, that is, each square is counted only once no matter how many occurrences it has. As proved by Fraenkel and Simpson [5], there are at most $2n$ squares in a string of length n. A simple proof has been given by Ilie [7]. Based on the numerical evidence, it has been conjectured that this number is actually less than n; see also Chapter 8 of [14]. The best bound to date is $2n - \Theta(\log n)$ due to Ilie [8].

Runs. In this paper we counted in fact occurrences of repetitions because the runs are defined as intervals. Inspired by the square problem, we may look at their associated strings and count only the number of runs associated with different strings. Note that the number of nonequivalent runs and that of squares do not seem to be obviously related to each other. The same run may contain several distinct squares (e.g., ababa contains the squares abab and baba) but we can have also different runs corresponding to a single squares (e.g., aa and aaa can be different runs but only the square aa is involved).

$(2+\varepsilon)^+$*-Repetitions.* A way to weaken the conjecture on the number of squares is to increase the exponent of the repetition. Given a non-negative ε, one could count only the number of repetitions of exponent $2+\varepsilon$ or higher. We need first to make it precise what we are talking about. We count primitively rooted repetitions of exponent at least $2+\varepsilon$ and having distinct roots. That is, x^α and y^β, x and y primitive, $\alpha \geq 2+\varepsilon$, $\beta \geq 2+\varepsilon$, are different if and only if $x \neq y$.

This conjecture might be easier to prove. At least for $2+\varepsilon = 1+\phi$, where ϕ is the golden ratio, we can prove it immediately. We count each square at the position where its rightmost occurrence starts and show that no two distinct squares can have the same rightmost starting position. Assume $x^{1+\phi}$ is a prefix of $y^{1+\phi}$ and denote $|x| = p < q = |y|$. Then necessarily $|x^{1+\phi}| = (1+\phi)p > \phi q = |y^\phi|$ as otherwise $x^{1+\phi}$ would have another occurrence to the right. That means $\phi^2 p = (1+\phi)p > \phi q$, or $\phi p > q$. Therefore, the overlap between the two runs has the length $|x^{1+\phi}| = (1+\phi)p = p + \phi p > p + q$. Fine and Wilf's lemma implies that x and y are powers of the same string, thus not primitive, a contradiction.

References

1. Apostolico, A., Preparata, F.: Optimal off-line detection of repetitions in a string. Theoret. Comput. Sci. 22(3), 297–315 (1983)
2. Crochemore, M.: An optimal algorithm for computing the repetitions in a word. Inform. Proc. Letters 12, 244–250 (1981)
3. Crochemore, M., Ilie, L.: A simple proof that the number of runs in a word is linear (manuscript, 2006)
4. Crochemore, M., Rytter, W.: Squares, cubes, and time-space efficient string searching. Algorithmica 13, 405–425 (1995)
5. Fraenkel, A.S., Simpson, J.: How many squares can a string contain? J. Combin. Theory, Ser. A 82, 112–120 (1998)
6. Franek, F., Simpson, R.J., Smyth, W.F.: The maximum number of runs in a string. In: Miller, M., Park, K. (eds.) Proc. 14th Australasian Workshop on Combinatorial Algorithms, pp. 26–35 (2003)
7. Ilie, L.: A simple proof that a word of length n has at most $2n$ distinct squares. J. Combin. Theory, Ser. A 112(1), 163–164 (2005)
8. Ilie, L.: A note on the number of squares in a word. Theoret. Comput. Sci. 380(3), 373–376 (2007)
9. Iliopoulos, C.S., Moore, D., Smyth, W.F.: A characterization of the squares in a Fibonacci string. Theoret. Comput. Sci. 172, 281–291 (1997)
10. Kolpakov, R., Kucherov, G.: On the sum of exponents of maximal repetitions in a word, Tech. Report 99-R-034, LORIA (1999)
11. Kolpakov, R., Kucherov, G.: Finding maximal repetitions in a word in linear time. In: Proc. of FOCS'99, pp. 596–604. IEEE Computer Society Press, Los Alamitos (1999)
12. Kolpakov, R., Kucherov, G.: On maximal repetitions in words. J. Discrete Algorithms 1(1), 159–186 (2000)
13. Lothaire, M.: Algebraic Combinatorics on Words. Cambridge Univ. Press, Cambridge (2002)
14. Lothaire, M.: Applied Combinatorics on Words. Cambridge Univ. Press, Cambridge (2005)
15. Main, M.G.: Detecting lefmost maximal periodicities. Discrete Applied Math. 25, 145–153 (1989)
16. Puglisi, S.J., Simpson, J., Smyth, B.: How many runs can a string contain? (submitted, 2006)
17. Rytter, W.: The number of runs in a string: improved analysis of the linear upper bound. In: Durand, B., Thomas, W. (eds.) STACS 2006. LNCS, vol. 3884, pp. 184–195. Springer, Heidelberg (2006)
18. Rytter, W.: The number of runs in a string (submitted, 2006)
19. Stoye, J., Gusfield, D.: Simple and flexible detection of contiguous repeats using a suffix tree. In: Farach-Colton, M. (ed.) CPM 1998. LNCS, vol. 1448, pp. 140–152. Springer, Heidelberg (1998)
20. Thue, A., Zeichenreihen, Ü.u.: Kra. Vidensk. Selsk. Skrifter. I. Mat.-Nat. Kl. Cristiana 7 (1906)

Series-Parallel Languages on Scattered and Countable Posets

Nicolas Bedon and Chloé Rispal

Université de Paris-Est, CNRS
Laboratoire d'informatique de l'Institut Gaspard Monge, UMR CNRS 8049
5, boulevard Descartes, Champs-sur-Marne
77454 Marne-la-Vallée Cedex 2, France
Nicolas.Bedon@univ-mlv.fr, Chloe.Rispal@univ-mlv.fr

Abstract. We initiate a study on automata recognizing labelled posets constructed from scattered and countable linear orderings. More precisely, the class of labelled posets considered in this paper is the smallest containing letters, closed under finite parallel operation and sequential product indexed by all countable and scattered linear orderings. The first result of this paper establishes that those labelled posets are precisely the N-free ones. The second result is a Kleene-like theorem, which establishes that the class of languages of labelled posets accepted by branching automata is exactly the class of rational languages. This generalizes both the finite [9] and ω-labelled posets [2,6] cases, and the Kleene-like theorem on words on linear orderings [3].

1 Introduction

In concurrency theory, actions are partially ordered and the order in which independent actions are executed is irrelevant or unknown. Several theoretical frameworks are used in order to model concurrency. Let us cite, as examples, Petri nets, Mazurkiewicz's traces, or rational posets. In Mazurkiewicz and traces theory, the sequences of actions are modeled as linear words in which some letters commute: if the action a commutes with b, then the word abc is identified with bac. An other approach consists in modeling a sequence of actions with a partially ordered set (poset), where two incomparable elements represent actions that can be concurrently executed. Thus, a sequence of actions which contains a and b with indifferent order and ends with c is modeled indifferently by $(a \parallel b)c$ or $(b \parallel a)c$. The class of labelled posets used in this approach, usually named *series-parallel labelled posets*, is the smallest containing the letters and closed under finite union (also named *parallel product*) and sequential (or *series*) product. This class has also been characterized in other terms, by an examination of the relation order between all 4-tuples of elements: the series-parallel labelled posets are exactly the N-free labelled posets [12,13]. Lodaya and Weil defined *branching* automata for the recognition of languages of such finite labelled posets. They also introduced rational expressions equivalent to automata [9], and studied recognizability by algebra [7,8]. Automata and algebra are not in general equivalent

L. Kučera and A. Kučera (Eds.): MFCS 2007, LNCS 4708, pp. 477–488, 2007.

as in the sequential word case: the two formalisms coincide only for languages of bounded-width. Those results on finite labelled posets have been generalized to ω-labelled posets. The class of ω-series-parallel labelled posets is obtained from the series-parallel finite labelled posets with an ω-product operation. It has been studied in [2] in the special case where the ω-product is the infinite repetition of a given finite labelled posets, and more generally in [6,5]. Again, the ω-series-parallel labelled posets are precisely the N-free ω-labelled posets. Rational expressions, automata and algebra adapted to the ω case have been introduced, and the results from the finite case extended.

In this paper, these notions are generalized again to a wider class of infinite labelled posets. Actually, we are interested in labelled posets constructed from the letters and closed under finite parallel product and J-sequential product, where J is any scattered (nowhere dense) and countable linear ordering. This includes in particular the finite and ω cases. We first prove that the class of such labelled posets coincides again with the N-free labelled posets. Then, we introduce branching automata adapted to such labelled posets, rational expressions, and prove that a language of labelled posets is defined by an automaton if and only if it is rational. The automata and rational expressions used are a mixture between those of Lodaya and Weil for finite labelled posets, and those introduced by Bruyère and Carton for words on linear orderings [3]. This could be a first step in the study of algebraic and logical theories including both those for linear orderings and finite posets.

The paper is organized as follows. Section 2 is devoted to notation and basic definitions. Section 3 contains several equivalent definitions of series-parallel posets. In particular, we generalize the Hausdorff characterization of countable and scattered linear orderings to series-parallel posets, and prove that a poset is series-parallel if and only if it is N-free. This extends the well-known result on finite and ω-posets. Words on posets and rational languages are defined in Section 4, while branching automata are introduced in Section 5. A Kleene-like theorem is proved in Section 6.

2 Notation and Basic Definitions

Let E be a set. We denote by $\mathcal{P}(E)$, $\mathcal{P}^*(E)$ and $\mathcal{P}^+(E)$ respectively the set of subsets of E, the set of non-empty subsets of E and the set of subsets of E with at least two elements. For any integer n, the group of permutations of $\{1,\dots,n\}$ is denoted by S_n. Let J be a set equipped with an order $<$. The ordering J is *linear* if all elements are comparable : for any distinct j and k in J, either $j<k$ or $k<j$. A linear ordering J is *dense* if for any j and k in J such that $j<k$, there exists an element i of J such that $j<i<k$. It is *scattered* if it contains no dense subordering. The ordering ω of natural integers and the ordering ζ of all the integers (negative, 0 and positive) are scattered. More generally, ordinals are scattered orderings. We denote by $\mathcal{N}$ the subclass of finite linear orderings, $\mathcal{O}$ the class of countable ordinals and by $\mathcal{S}$ the class of countable scattered linear orderings. We refer to [11] for more details on linear orderings and ordinals.

3 Posets

A poset is a partially ordered set. A subset P' of a poset P is *independent* if all elements of P' are incomparable. The *width* of P is

$$wd(P) = \max\{|E| : E \text{ is an independent subset of } P\} - 1$$

In the following, we restrict to *countable scattered posets* which are thus partially ordered sets without any dense sub-ordering.

Let $(P, <_P)$ and $(Q, <_Q)$ be two disjoint posets.

The *union* of P and Q is the set $P \cup Q$ equipped with the orderings on P and Q such that the elements of P and Q can not be compared. It is defined as $(P \cup Q, <)$ where $x < y$ if and only if:

- $x, y \in P$ and $x <_P y$ or
- $x, y \in Q$ and $x <_Q y$.

The *sum* of P and Q, denoted by $P + Q$ is the poset $(P \cup Q, <)$ such that $x < y$ if and only if one of the following condition is true:

- $x \in P$, $y \in P$ and $x <_P y$;
- $x \in Q$, $y \in Q$ and $x <_Q y$;
- $x \in P$ and $y \in Q$.

The finite sum of two posets can be generalized to any linearly ordered sequence of posets: if J is a linear ordering and $((P_j, <_j))_{j \in J}$ is a sequence of posets, then $\sum_{j \in J} P_j = (\cup_{j \in J} P_j, <)$ such that $x < y$ if and only if one of the following condition is true:

- $x \in P_j$, $y \in P_j$ and $x <_j y$;
- $x \in P_j$ and $y \in P_k$ and $j < k$.

Definition 1. *The set $SP^\diamond$ is defined as the smallest set containing the posets with zero and one element and closed under finite union and sum indexed by countable scattered linear orderings.*

Note that any poset of $SP^\diamond$ is countable, scattered and of finite width.

Hausdorff characterized the countable and scattered linear orderings using an induction on ordinals: any such ordering can be obtained from the finite linear orderings using finite sums, ω-sums and $-\omega$-sums. We adapt such a characterization to countable and scattered posets:

Theorem 1. *Let S_{sp} be the class of countable and scattered posets defined by $S_{sp} = \underset{\alpha \in \mathcal{O}}{\cup} W_\alpha$ where W_α is inductively defined by*

$$V_0 = \{0, 1\}$$
$$W_0 = \mathcal{C}_{\cup,+}(V_0)$$
$$V_\alpha = \{\sum_{i \in J} p_i : J \in \{\omega, -\omega\} \text{ and } \forall i \in J, p_i \in \underset{\beta<\alpha}{\cup} W_\beta\} \bigcup \bigcup_{\beta<\alpha} W_\beta$$
$$W_\alpha = \mathcal{C}_{\cup,+}(V_\alpha)$$

where 0 *and* 1 *denote respectively the empty set and the poset with one element, and* $\mathcal{C}_{\cup,+}(E)$ *denotes the closure of* E *under finite union and finite sum. Then* $S_{sp} = SP^{\diamond}$.

The *rank* $r(p)$ of a countable and scattered poset $p \in S_{sp}$ is the smallest ordinal α such that $p \in W_{\alpha}$.

Posets can also be thought as oriented graphs, where the nodes are the elements of the posets and the edges represent the ordering relation. It is a well-known result that the finite posets of $SP^{\diamond}$ are precisely the N-free finite posets. A poset P is *N-free* if it does not contain N as a subposet, that is if it does not contain elements $p, q, r, s \in P$ such that the ordering relations between those four elements are *precisely* $p < r$, $q < r$ and $q < s$. The poset N is pictured by Figure 1.

Fig. 1. The poset N

Note that there exist two definitions of N-freeness in the literature. The reader must be aware that the definition used in this paper is not the same as in [10].

The class of all scattered, countable and N-free posets of finite width is denoted by $NF^{\diamond}$. Actually, if a poset P is represented as an oriented graph, the N-free property can be expressed in terms of paths. Let $p, q \in P$ and let n be an integer. A *path* $p \to q$ from p to q of length n is a sequence $(q_i)_{i<n}$ such that $q_0 = p$, $q_{n-1} = q$ and either $q_i < q_{i+1}$ or $q_i > q_{i+1}$ for any $i < n-1$. The poset P is said to be *connected* if for any $p, q \in P$ there exists a path from p to q. Otherwise, P is *disconnected.*

Lemma 1. *Let* P *be a N-free poset, let* p *and* q *be elements of* P *and let* $(q_i)_{i<n}$ *be a shortest path from* p *to* q*. Then* $n < 3$.

The result on finite N-free [12,13] posets also extends to our case:

Theorem 2. $NF^{\diamond} = SP^{\diamond}$.

The proof of this theorem is based on the following property:

Remark 1. Any scattered, countable and N-free poset P of finite width can be written as a sum indexed by a linear ordering of trivial or disconnected posets:

$$P = \sum_{j \in J} P_j \text{ where } J \in \mathcal{S} \text{ and } \forall j \in J$$

$$(|P_j| = 1) \vee \exists m \in \mathbb{N}, \exists (P_{i,j})_{1<i\leq m} \text{ such that } P_j = \bigcup_{1<i\leq m} P_{i,j}$$

Furthermore, the linear ordering J and the posets $(P_j)_{j\in J}$ are unique.

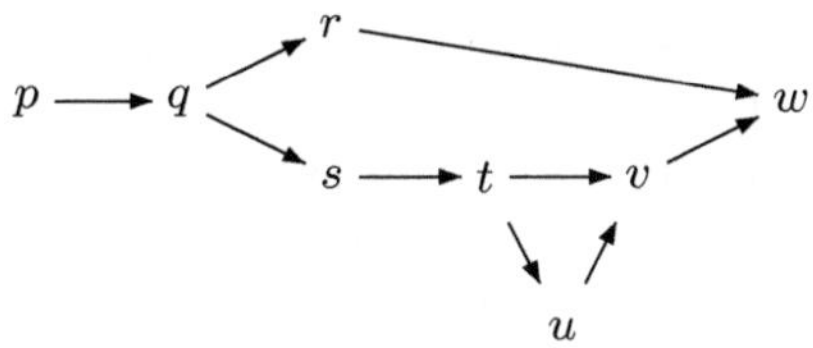

Fig. 2. A N-free poset

Example 1. Let $(P, <)$ be the poset represented by the graph of Figure 2. The edges represent the relation "is an immediate predecessor of" induced by $<$. Then $P = \sum_{j \in \{0,1,2,3\}} P_j$ with $P_0 = \{p\}$, $P_1 = \{q\}$, $P_2 = \{r, s, t, u, v\}$, $P_3 = \{w\}$, and P_2 can be decomposed into $P_2 = P_{1,2} \cup P_{2,2}$ with $P_{1,2} = \{r\}$ and $P_{2,2} = \{s, t, u, v\}$. Observe that $P_{2,2}$ can also be decomposed using the same principle.

Remark 2. The hypothesis "scattered" and "countable" on the elements of $SP^\diamond$ and $NF^\diamond$ are not necessary in the statement of Theorem 2. Thus $NF^\diamond = SP^\diamond$ also holds for classes of posets with dense or uncountable parts, for example.

4 Rational Languages

Let A be a finite alphabet. A poset P *labelled* by A is a map which associates a letter of A to any element of P. This generalizes the definition of words and such a labelled poset is called a *word* of length P over A. The length of u is denoted by $|u| = P$. We denote by $SP^\diamond(A)$ the set of words over A whose lengths belong to $SP^\diamond$.

4.1 Operations on Words

The operations of finite union and sum defined for posets correspond respectively to the parallel and sequential product on words. Let u and v be two words of $SP^\diamond(A)$ of length respectively P and Q. The *parallel product* of u and v denoted by $u \parallel v$ is of length $P \cup Q$. It is defined as the map from the poset $P \cup Q$ into A such that its restriction to P is u and its restriction to Q is v. The *sequential product* of u and v, denoted by $u \cdot v$ is of length $P + Q$. More generally, let J be a linear ordering and let $(u_j)_{j \in J}$ be words of respective length P_j for any $j \in J$. The sequential product $u = \prod_{j \in J} u_j$ obtained by concatenation of the words u_j with respect to the ordering on J has length $P = \sum_{j \in J} P_j$. The sequence $(u_j)_{j \in J}$ of words is called a *J-factorization* of the word $u = \prod_{j \in J} u_j$.

A word $u \in SP^\diamond(A)$ is a *sequential word* if it admits a J-factorization where J contains more than one element. It is a *parallel word* otherwise. Using Remark 1, any word $u \in SP^\diamond(A)$ admits a unique factorization $u = \prod_{j \in J} u_j$ such that $J \in \mathcal{S}$

and $\forall j \in J$, u_j is a parallel word. This factorization is called the *maximal sequential factorization* of u.

A *language* over an alphabet A is a subset of $SP^\diamond(A)$. The sequential and parallel product of words can naturally be extended to languages. If $L_1, L_2 \subseteq SP^\diamond(A)$, then $L_1 \cdot L_2 = \{u \in SP^\diamond(A) : \exists v \in L_1\ \exists w \in L_2 \text{ such that } u = v \cdot w\}$ and $L_1 \parallel L_2 = \{u \in SP^\diamond(A) : \exists v \in L_1\ \exists w \in L_2 \text{ such that } u = v \parallel w\}$.

4.2 Rational Languages

Let A and B be two alphabets and let $u \in SP^\diamond(A)$, $L \subseteq SP^\diamond(B)$ and $\xi \in A$. The word u in which the letter ξ is replaced by the language L is denoted by $L \circ_\xi u$. This substitution is a language over the alphabet $A \cup B$ defined inductively on $wd(|u|)$ by:

- if $wd(|u|) = 0$ and u is a letter, then either $u = \xi$ and $L \circ_\xi u = L$, or u is a letter $a \neq \xi$ and $L \circ_\xi u = \{u\}$. If $wd(|u|) = 0$ and u is not a letter, then there exists J such that $u = \prod_{j \in J} u_j$ and u_j is a letter for every $j \in J$. Then $L \circ_\xi u = \prod_{j \in J}(L \circ_\xi u_j)$;
- if $wd(|u|) > 0$, and there exists n such that $u = \parallel_{i<n} u_i$, then $L \circ_\xi u = \parallel_{i<n} (L \circ_\xi u_i)$. Otherwise, by Remark 1, there exist a linear ordering J and a factorization $u = \prod_{j \in J} u_j$ such that u_j is either a letter or a parallel word. Then $L \circ_\xi u = \prod_{j \in J}(L \circ_\xi u_j)$.

Using the previous definition of substitution on words, we define the substitution and the iterated substitution on languages. By the way the usual rational operations on linear orderings are recalled. Let L and L' be languages of $SP^\diamond(A)$:

$$L \circ_\xi L' = \{v \in SP^\diamond(A) : \exists u \in L' \text{ such that } v \in L \circ_\xi u\}$$

$$L^{*\xi} = \bigcup_{i \in \mathcal{N}} L^{i\xi} \text{ with } L^{0\xi} = \{\xi\} \text{ and } L^{(i+1)\xi} = \bigcup_{j \leq i} L^{j\xi} \circ_\xi L$$

$$L^* = \{\prod_{j \in n} u_j | n \in \mathcal{N}, u_j \in L\}$$

$$L^\omega = \{\prod_{j \in \omega} u_j | u_j \in L\}, \qquad L^{-\omega} = \{\prod_{j \in -\omega} u_j | u_j \in L\}$$

$$L^\natural = \{\prod_{j \in \alpha} u_j | \alpha \in \mathcal{O}, u_j \in L\}, \qquad L^{-\natural} = \{\prod_{j \in -\alpha} u_j | \alpha \in \mathcal{O}, u_j \in L\}$$

For technical reasons, we do not define the $\diamond$ operator as in the original article of Bruyère and Carton [3] on the extension of Kleene's theorem on linear orderings. We use instead an equivalent definition extracted from [1]. Informally speaking, the words of $L_1 \diamond L_2$ are obtained by alternating labelled posets of L_1 and L_2.

Lemma 2. *Let L_1 and L_2 be two languages. Then $u \in L_1 \diamond L_2$ if and only if there exist a scattered and countable linear ordering $K \neq \emptyset$, a sequence $(u_k)_{k \in K}$ of words and a map $f : K \to \{1, 2\}$ such that the following conditions are true:*

1. $u = \prod_{k \in K} u_k$;
2. *if* $f(k) = i$ *and* $k+1 \in K$ *then* $f(k+1) \neq i$;
3. *if* $k \in K$, *k is not the last element of* K, *and k has no successor, then* $f(k) = 2$;
4. *if* $k \in K$, *k is not the first element of* K, *and k has no predecessor, then* $f(k) = 2$;
5. *if k is the first or the last element of* K, *then* $f(k) = 1$;
6. K *is complete;*
7. $f(k) = i$ *implies* $u_k \in L_i$.

The previous rational operators are illustrated by examples in the following section with automata. A language $L \subseteq SP^{\diamond}(A \cup \{\xi\})$ is *rational* if it is obtained from the letters of the alphabet $A \cup \{\xi\}$ using usual rational operators : finite union $\cup$, finite concatenation $\cdot$, and finite iteration $*$, using rational operations on linear words: ω and $-\omega$-products, iteration on ordinals $\natural$ and reverse iteration on ordinals $-\natural$ as well as diamond operator $\diamond$, and using also the rational operators of finite parallel product $\|$, substitution $\circ_\xi$ and iterated substitution $^{*\xi}$, provided that the letter ξ appears only inside parallel factors.

Note that the rational expressions are precisely those of Bruyère and Carton [3] over words on scattered and countable linear orderings, with two additional operators $\|$ and $^{*\xi}$ for parallelism and substitution.

5 Automata

Automata on countable, scattered and N-free posets are a generalization of automata on finite N-free posets [8], N-free ω-posets [6] and automata on linear orderings [3].

A *branching automaton* over an alphabet A is a tuple $\mathcal{A} = (Q, A, E, E_{fork}, E_{join}, I, F)$ where Q is a finite set of states, $I \subseteq Q$ is the set of *initial states*, $F \subseteq Q$ the set of *final states*, and $E \cup E_{fork} \cup E_{join}$ is the set of *transitions* of $\mathcal{A}$. The transitions of $E \subseteq (Q \times A \times Q) \cup (Q \times \mathcal{P}^*(Q)) \cup (\mathcal{P}^*(Q) \times Q)$ are named *sequential*, while $E_{fork} \subseteq Q \times \mathcal{P}^+(Q)$ and $E_{join} \subseteq \mathcal{P}^+(Q) \times Q$ are respectively the sets of *fork* and *join* transitions.

Let p and q be two states of an automaton $\mathcal{A}$ and let $u \in SP^{\diamond}(A)$. The existence of a path from p to q labelled by u of content $C \subseteq Q$ is defined by induction on the rank of $|u| = P$:

- $P \in V_0$: There exists $a \in A$ such that $u = a$, and there exists a path from p to q labelled by u of content $C = \{p, q\}$ if $(p, a, q) \in E$;
- $P \in W_0$: In this case u is a finite word and we suppose that u is not a letter. There are two cases: either u has a parallel factorization, or u has a sequential factorization.
  - If u has a parallel factorization $u = \|_{i<n} u_i$ for some integer $n > 1$ then there exists a partition $\{A_0, \ldots, A_j\}$ of $\{0, \ldots, n-1\}$ and $j+1$ pairs of states $(p_k, q_k)_{k<j+1}$ such that $(p, \{p_0, \ldots, p_j\}) \in E_{fork}$, $(\{q_0, \ldots, q_j\}, q) \in E_{join}$ and there exists a path γ_k from p_k to q_k labelled by $\|_{l \in A_k} u_l$

for each $k < j+1$. In this case the content of the path labelled u is the set $\{p, q\}$. Note that this content does not depend on the contents of the smaller parallel paths γ_k.
- If u has a sequential factorization $u = \prod_{i<n} u_i$ for some integer $n > 1$, then there exist $n+1$ states $(q_i)_{i<n+1}$ such that there exists a path from q_i to q_{i+1} labelled by u_i of content C_i for each $i < n$ and $q_0 = p$, $q_n = q$ and $C = \underset{i<n}{\cup} C_i$.

– $P \in V_\alpha$ for some countable ordinal $\alpha > 0$:
If $P \in \bigcup_{\beta<\alpha} W_\beta$ then the path is already defined by induction.
Suppose that $P = \sum_{i\in\omega} P_i$ where $\forall i \in \omega, P_i \in \bigcup_{\beta<\alpha} W_\beta$. Let $u = \prod_{i\in\omega} u_i$ be the corresponding factorization of u :$\forall i \in \omega, |u_i| = P_i$. There exists a path from p to q labelled u of content C if the following conditions holds:
- there exist states $(p_i)_{i\in\omega}$ with $p = p_0$ such that for any $i \in \omega$, there is a path from p_i to p_{i+1} labelled u_i of content C_i;
- If $P = \{q \in Q : \forall i \in \omega, \exists j > i, q \in C_j\}$ then $\exists q \in Q$ such $(P, q) \in E$ and $C = (\underset{i<n}{\cup} C_i) \cup \{q\}$.

The $-\omega$ case is treated symmetrically.

– $P \in W_\alpha$ for some countable ordinal $\alpha > 0$:
This case is similar to the W_0 case.

A path from p to q labelled by u of content C is denoted by $p \overset{u}{\underset{C}{\Longrightarrow}} q$. The states p and q are respectively the *source* (or *origin*) and *destination* of the path. When the content is not useful, the path is simply denoted by $p \overset{u}{\Longrightarrow} q$, or $p \Longrightarrow q$ when the label is unuseful too. A word is *accepted* by an automaton if it is the label of a successful path leading from an initial state to a final state. The language $L(\mathcal{A})$ is the set of words accepted by the automaton $\mathcal{A}$.

Note that branching automata without fork and join transitions are precisely the automata on scattered and countable linear orderings defined by Bruyère and Carton [3].

Example 2. The automaton of Figure 3 with the initial state 0 and final state 6 accepts the word $((a \parallel b)c)^\omega$. It has one fork transition $0 \to \{1, 2\}$, one join transition $\{3, 4\} \to 5$ and four sequential transitions: the three successor transitions denoted by $1 \xrightarrow{a} 3$, $2 \xrightarrow{b} 4$ and $5 \xrightarrow{c} 0$ and the transition $\{0, 5\} \to 6$. Let $\mathcal{B}$ be the automaton pictured in Figure 3 where the final state 6 is deleted, the new final state is 0 and the limit transition $\{0, 5\} \to 6$ is replaced by $\{0, 5\} \to 0$. Then $\mathcal{B}$ accepts the language $((a \parallel b)c)^\natural$.

Example 3. An automaton accepting the language $(a \parallel b) \diamond c$ is pictured in Figure 4.

In order to simplify the proofs, we use branching automata with the property of *behavedness*:

Definition 2. *A branching automaton is* misbehaved *if one of the following conditions holds:*

- *there exists a fork transition* $(q, \{p_1, \cdots, p_n\})$ *and there exists* $1 \leq i \leq n$ *such that there is a path from* p_i *to some final state;*
- *there exists a join transition* $(\{p_1, \cdots, p_n\}, q)$ *and there exists* $1 \leq i \leq n$ *such that there is a path from some initial state to* p_i.

Note that this definition does not exactly correspond to the definition given in [9]. An automaton which is not misbehaved is said to be *behaved.* The proof of the following Proposition is adapted from the case of finite posets (see [9]):

Proposition 1. *Let* L *be a language of* $SP^\diamond(A)$. *If* L *is accepted by a branching automaton then* L *is accepted by a behaved automaton.*

6 A Kleene Theorem for Scattered Posets

In this section, we give an extension of Kleene's theorem for words indexed by countable scattered and N-free posets. This generalizes both the theorem of Weil and Lodaya [9] for words on finite posets, that of Kuske [6] for ω-posets and that of Bruyère and Carton [3] for words on scattered linear orderings.

Proposition 2. *Any rational language* $L \subseteq SP^\diamond(A)$ *is accepted by a branching automaton.*

Proof. Since the letters of the alphabet A are obviously languages accepted by branching automata, it suffices to prove that the set of languages accepted by a branching automaton is closed under the rational operations. Let L and L' be two languages of $SP^\diamond(A)$ accepted by branching automata. Concerning operators on linear orderings, that is the constructions of automata accepting $L \cup L'$, $L \cdot L'$,

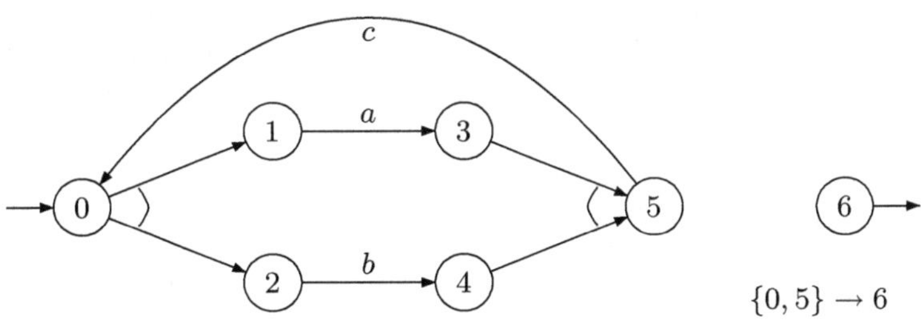

Fig. 3. An automaton accepting $((a \parallel b)c)^\omega$

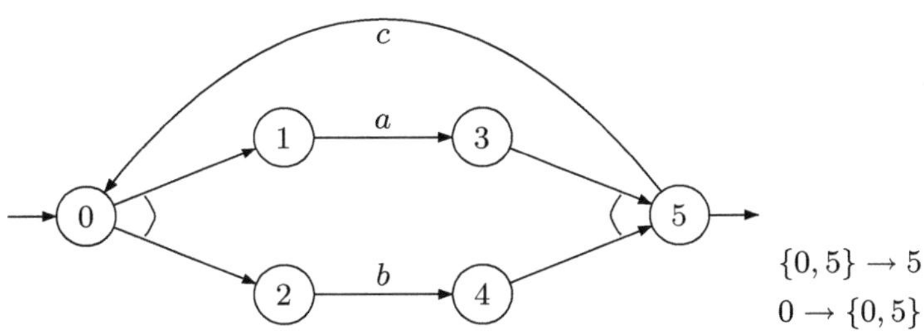

Fig. 4. An automaton accepting $(a \parallel b) \diamond c$

L^*, L^ω, $L^{-\omega}$, $L^\natural$, $L^{-\natural}$ and $L \diamond L'$, we refer the reader to [3]. For the parallel operators, the method used in [9] for the constructions of automata accepting $L \parallel L'$ and $L \circ_\xi L'$ also works but the construction for $L^{*\xi}$ is modified as follows:

Let $\mathcal{A} = (Q, A, E, E_{fork}, E_{join}, I, F)$ be a branching automaton accepting L. According to Proposition 1, $\mathcal{A}$ can be chosen behaved. Let $K = \{(p,q) \in Q^2 : (p, \xi, q) \in E\}$. For each element $(p,q) \in K$ build two copies $\mathcal{A}_{1,p,q}$ and $\mathcal{A}_{2,p,q}$ of $\mathcal{A}$. For each $c \in (\{1,2\} \times K)$, the set of states of $\mathcal{A}_c$ is denoted by Q_c and the copy of a state $s \in Q$ in $\mathcal{A}_c$ is sometime denoted by s_c. The automaton $\mathcal{B}$ is defined as a disjoint union of $\mathcal{A}$ denoted by $\mathcal{A}_0$ and all the copies $(\mathcal{A}_c)_{c\in\{1,2\}\times K}$, with the following additional transitions:

- for each sequential transition (i, a, s) of $\mathcal{A}$ starting from an initial state $i \in I$, for each $(p,q) \in K$ and each copy $c \in \{0\} \cup (\{1,2\} \times K)$, add a transition $(p_c, a, s_{1,p,q})$, except if $c = (1,p,q)$ (i.e. the new transition would loop into the same copy); instead, in this case, add a new transition $(p_c, a, s_{2,p,q})$. Informally speaking, those new transitions make a move to the copy $\mathcal{A}_{j,p,q}$ when a transition (p, ξ, q) is encountered to start a substitution. The copy $\mathcal{A}_{2,p,q}$ can only be reached from $\mathcal{A}_{1,p,q}$, whereas $\mathcal{A}_{1,p,q}$ can be reached from any other copy (except itself). Operate similarly for fork and limit transitions from an initial state of $\mathcal{A}$. The set of all transitions constructed in this step and whose destinations belong to Q_c is denoted by In_c;
- operate symmetrically for sequential, fork and limit transitions going to a final state of $\mathcal{A}$. Those new transitions go back to previous copy when the substitution is complete. Thus, each transition of the form (p, ξ, q) is replaced by a successful path of $\mathcal{A}_{j,p,q}$ for some $j \in \{1,2\}$. The set of all transitions constructed in this step whose sources belong to Q_c is denoted by Out_c;
- in order to take into account a possible replacement of an infinite sequence of transitions labelled by ξ some additional limit transitions are necessary. For instance, consider a path in $\mathcal{A}$ ending with a transition (P, r), and using an infinity of transitions labelled by ξ. Now consider the same path where all transitions labelled by ξ are replaced by successful paths in different copies. These replacements may change infinitely often the content of the path. Thus, the transition (P, r) must be replaced by a transition $(P \cup P', r)$, where P' contains some states visited in the copies. Formally, for any copy $\mathcal{A}_c$ and any transition (P, r) in this copy, if P contains at least one pair of K, that is, if $\exists (p,q) \in K$ such that $p_c, q_c \in P$ then add a transition $(P \cup R, r)$ for any set R of states of other copies which do not contain any pair of K: $R \subseteq Q_\mathcal{B} \setminus (Q_c \cup \{p_{c'}, q_{c'} : (p,q) \in K, c \neq c'\})$ where $Q_\mathcal{B}$ denotes the set of states of $\mathcal{B}$. This last condition is justified by the fact that nested substitutions do not change the content. Symmetrical transitions are added for limit transitions of the form (r, P).
- for each transition of the form (i, a, f) of $\mathcal{A}$, where $i \in I$ and $f \in F$, and for each transition (p, ξ, q) of one of each copy, add a transition (p, a, q). The set of such new transitions is denoted by $InOut$.

The initial and final states of $\mathcal{B}$ are those of $\mathcal{A}$. In order to ensure the acceptance of ξ, add a new initial state i and new final state f with a new transition (i, ξ, f).

It can be proved that $\mathcal{B}$ accepts the language $L^{*\xi}$. Informally speaking, every transition $p \xrightarrow{\xi} q$ can be replaced by $p \overset{u}{\Longrightarrow} q$, where $u \in L$. □

Let us turn to the converse:

Proposition 3. *Any language $L \subseteq SP^{\diamond}(A)$ accepted by a branching automaton is rational.*

Proof. Let $\mathcal{A} = (Q, A, E, E_{fork}, E_{join}, I, F)$ be a branching automaton accepting a language $L \subseteq SP^{\diamond}(A)$. The automaton $\mathcal{A}' = (Q, A', E', \emptyset, \emptyset, I, F)$ on linear ordering is defined such that:

- for every pair $(f, j) \in E_{fork} \times E_{join}$ such that f and j have the same arity, add a new letter $a_{f,j}$ into A', and a transition $(q, a_{f,j}, p)$, where q and p are respectively the source of f and the destination of j;
- A' contains all the letters of A;
- all the fork and join transitions have been removed.

Denote by $L_{p,q}(\mathcal{A})$ the set of labels of paths leading from p to q in $\mathcal{A}$. Then $L_{p,q}(\mathcal{A})$ is the set $L_{p,q}(\mathcal{A}')$ in which any letter $a_{r,s}$ is replaced by the set $M_{r,s}$ of parallel labels of paths leading from r to s.

$$L_{p,q}(\mathcal{A}) = \underset{r,s \in Q}{\circ} (M_{r,s} \circ_{a_{r,s}})(L_{p,q}(\mathcal{A}'))$$

where

$$M_{r,s} = \bigcup_{(r,\{r_1,\ldots,r_n\}) \in E_{fork}} \bigcup_{(\{s_1,\ldots,s_n\},s) \in E_{join}} \bigcup_{\sigma \in S_n} L_{r_1,s_{\sigma(1)}}(\mathcal{A}) \parallel \cdots \parallel L_{r_n,s_{\sigma(n)}}(\mathcal{A})$$

Since $\mathcal{A}'$ is an automaton on linear ordering without any fork or join transition, it coincides with automata studied in [3]. The extended Kleene theorem [3] gives that, for any $p, q \in Q$, $L_{p,q}(\mathcal{A}')$ is rational. Moreover, $L_{p,q}(\mathcal{A})$ is an equation which contains variables $L_{p',q'}(\mathcal{A})$ for some $p', q' \in Q$, which appear only in parallel parts of the equation. The finite system of equations can be solved using substitution and the $^{*\xi}$ operation to eliminate recursivity. A rational expression is obtained for each $L_{p,q}(\mathcal{A})$ once all variables have been eliminated thus L is rational. □

Propositions 2 and 3 prove the extended Kleene theorem:

Theorem 3. *A language $L \subseteq SP^{\diamond}(A)$ is accepted by a branching automaton if and only if it is rational.*

7 Conclusion

In this paper, we proved that a countable and scattered poset of finite width is series-parallel if and only if it is N-free and we have established a Kleene-like theorem for those posets. It would be interesting to investigate the algebraic and logic counterparts of branching automata. For the finite posets case, Lodaya and Weil have established [8,7] that automata and recognition by finite semigroups extended with a commutative operation for parallelism are not equivalent in general. However, the equivalence holds when the width of the posets recognized by the automaton is bounded. In this case, the equivalence between monadic second-order logic, bounded-width automata and adapted algebra also holds. Those equivalence results were obtained also for ω-posets (obtained like the finite posets but with an infinite serial product allowed) (see [6,5] or [2]). There remains to explore the case of $SP^{\diamond}$.

References

1. Bedon, N., Rispal, C.: Recognizable languages and decidability of the monadic second-order logic on countable scattered linear orderings. In: Logical Methods in Computer Science (Submitted, 2006)
2. Bloom, S.L., Ésik, Z.: Shuffle binoids. Theoretical Informatics and Applications 32(4–5–6), 175–198 (1998)
3. Bruyère, V., Carton, O.: Automata on linear orderings. In: Ito, M., Toyama, M. (eds.) DLT 2002. LNCS, vol. 2450, pp. 103–115. Springer, Heidelberg (2003)
4. Ésik, Z., Németh, Z.L.: Automata on series-parallel biposets. In: Kuich, W., Rozenberg, G., Salomaa, A. (eds.) DLT 2001. LNCS, vol. 2295, pp. 217–227. Springer, Heidelberg (2002)
5. Kuske, D.: Infinite series-parallel posets: logic and languages. In: Welzl, E., Montanari, U., Rolim, J.D.P. (eds.) ICALP 2000. LNCS, vol. 1853, pp. 648–662. Springer, Heidelberg (2000)
6. Kuske, D.: Towards a language theory for infinite N-free pomsets. Theoret. Comput. Sci. 299, 347–386 (2003)
7. Lodaya, K., Weil, P.: Series-parallel posets: algebra, automata and languages. In: Meinel, C., Morvan, M. (eds.) STACS 98. LNCS, vol. 1373, pp. 555–565. Springer, Heidelberg (1998)
8. Lodaya, K., Weil, P.: Series-parallel languages and the bounded-width property. Theoret. Comput. Sci. 237(1–2), 347–380 (2000)
9. Lodaya, K., Weil, P.: Rationality in algebras with a series operation. Information and Computation, 269–293 (2001)
10. Rival, I.: Optimal linear extension by interchanging chains. Proc. AMS 89(3), 387–394 (1983)
11. Rosenstein, J.G.: Linear Orderings. Academic Press, London (1982)
12. Valdes, J.: Parsing flowcharts and series-parallel graphs. Technical Report STAN-CS-78-682, Computer science departement of the Stanford University, Standford, Ca. (1978)
13. Valdes, J., Tarjan, R.E., Lawler, E.L.: The recognition of series parallel digraphs. SIAM J. Comput. 11, 298–313 (1982)

Traces of Term-Automatic Graphs

Antoine Meyer

LIAFA – Université Paris Diderot – Paris 7
Case 7014, 2 place Jussieu, 75251 Paris Cedex 05, France
ameyer@liafa.jussieu.fr

Abstract. In formal language theory, many families of languages are defined using grammars or finite acceptors like pushdown automata and Turing machines. For instance, context-sensitive languages are the languages generated by growing grammars, or equivalently those accepted by Turing machines whose work tape's size is proportional to that of their input. A few years ago, a new characterisation of context-sensitive languages as the sets of traces, or path labels, of rational graphs (infinite graphs defined by sets of finite-state transducers) was established.

We investigate a similar characterisation in the more general framework of graphs defined by term transducers. In particular, we show that the languages of term-automatic graphs between regular sets of vertices coincide with the languages accepted by alternating linearly bounded Turing machines. As a technical tool, we also introduce an arborescent variant of tiling systems, which provides yet another characterisation of these languages.

Introduction

In classical language theory, context-sensitive languages, one of the families of the Chomsky hierarchy [Cho59], are defined as the languages generated by growing grammars. They were later characterised as the languages accepted by linearly space-bounded Turing machines [Kur64], i.e. Turing machines whose runs on any input word of length n use at most $k \cdot n$ work tape cells, for some constant k. In [LS97], it was shown that context-sensitive languages also coincide with the languages accepted by bounded tiling systems.

In 2001, [MS01] provided yet another characterisation of this family as the set of path languages of rational graphs [Mor00], i.e. infinite graphs whose vertices are words and whose sets of edges are defined by finite transducers. This result was later extended in [Ris02] to the more restricted family of automatic graphs (Cf. [KN95]), and even to synchronous rational graphs when an infinite number of initial and final vertices are considered (see also [MR05]). In a way, this provides a "forward", automata-based characterisation of context-sensitive languages, as opposed to linearly bounded machines which are essentially a two-way mechanism. To prove the inclusion of context-sensitive languages in the set of path languages of these families of graphs, these papers use a normal form for growing grammars, due to Penttonen [Pen74]. In [CM06], these results were

L. Kučera and A. Kučera (Eds.): MFCS 2007, LNCS 4708, pp. 489–500, 2007.

reformulated using simpler proof techniques based on tiling systems. This also allowed to investigate interesting sub-cases, in particular concerning deterministic context-sensitive languages or various sub-classes of rational graphs.

The aim of this work is to extend the results of [LS97] and [CM06] to the more general family ETIME of languages accepted by deterministic Turing machines working in time less than $2^{O(n)}$, or equivalently by alternating linearly bounded machines. This family lies between context-sensitive and recursively-enumerable languages in the Chomsky hierarchy. We obtain two new characterisations of ETIME, first as the languages accepted by arborescent tiling systems and second as the traces of infinite graphs defined by various classes of term transducers, namely term-synchronous and term-automatic (or tree-automatic) graphs [BG00].

After recalling definitions and notations in Section 1, we introduce the notion of arborescent tiling systems in Section 2 and prove that they characterise ETIME. Finally, we extend previously mentioned proofs over rational graphs to the family of term-automatic graphs in Section 3.

1 Notations

1.1 Words, Terms and Trees

A word u over alphabet Σ can be seen as a tuple $(a_1, \ldots, a_n)$ of elements of Σ, usually written $a_1 \ldots a_n$. Its i-th letter is denoted by $u(i) = a_i$. The set of all words over Σ is written Σ^*. The number of letters occurring in u is its length, written $|u|$ (here $|u| = n$). The empty word is written ε. The concatenation of two words $u = a_1 \ldots a_n$ and $v = b_1 \ldots b_m$ is the word $uv = a_1 \ldots a_n b_1 \ldots b_m$. The concatenation operation extends to sets of words: for all $A, B \subseteq \Sigma^*$, AB stands for the set $\{uv \mid u \in A \text{ and } v \in B\}$.

Let $F = \bigcup_{n \geq 0} F_n$ be a finite ranked alphabet, each F_n being the set of symbols of F of arity n, and X be a finite set of *variables* disjoint from F (all sets F_n are also disjoint). We denote the arity of a symbol $f \in F$ by $\mathrm{a}(f)$. Variables are considered of arity 0. The set of finite first-order terms on F with variables in X, written $T(F, X)$, is the smallest set including X such that $f \in F_n \wedge t_1, \ldots, t_n \in T(F, X) \Rightarrow f t_1 \ldots t_n \in T(F, X)$. Words can be seen as terms over a ranked alphabet whose symbols have arity exactly 1 and whose last symbol is a variable or a special constant. To improve readability, $f t_1 \ldots t_n$ will sometimes be written $f(t_1, \ldots, t_n)$.

A finite ordered tree t over a set of labels Σ is a mapping from a prefix-closed set $\mathrm{dom}(t) \subseteq \mathbb{N}^*$ into Σ. Elements of $\mathrm{dom}(t)$ are called positions, and for every $p \in \mathrm{dom}(t)$, $t(p)$ is the label of the node at position p. The node at position ε is called the root of the tree, nodes at maximal positions (i.e. positions x such that $\nexists y \neq \varepsilon,\ xy \in \mathrm{dom}(t)$) are called leaves, other nodes are called internal.

Any term t over a ranked alphabet F and set of variables X can be represented as a finite ordered ranked tree, whose leaves are labelled with constants in F_0 or variables in X and whose internal nodes are labelled with symbols of arity equal to the number of children of that node. In that case, the domain of t, additionally to being prefix-closed, also has the following properties:

1. $\forall p \in \mathrm{dom}(t),\ t(p) \in F_{n\geq 1} \implies \{j \mid pj \in \mathrm{dom}(t)\} = [1, n]$,
2. $\forall p \in \mathrm{dom}(t),\ t(p) \in F_0 \cup X \implies \{j \mid pj \in \mathrm{dom}(t)\} = \emptyset$.

In such a tree, position pi with $i \in \mathbb{N}$ always denotes the i-th child of node p. Conversely, any finite ordered tree t labelled over Σ can be represented as a ranked tree t', and hence as a term, by mapping each node label a to a set of symbols (a, n) in $\Sigma \times \mathbb{N}$, with $\mathrm{a}(a, n) = n$, and by renumbering all positions such that $\mathrm{dom}(t')$ verifies the above properties. This will usually be left implicit.

A finite tree (or term) automaton is a tuple $A = \langle Q, F, q_0, \delta \rangle$, where Q is a set of control states, F a ranked alphabet, q_0 the initial set and δ the set of transition rules of A of the form $(q, f, q_1, \ldots, q_n)$ with $\mathrm{a}(f) = n$. A run of A over a tree t is a mapping ρ from $\mathrm{dom}(t)$ to Q such that $\rho(\varepsilon) = q_0$ and for all node $u \in \mathrm{dom}(t)$ of arity $\mathrm{a}(u) = n$, $(\rho(u), t(u), \rho(u1), \ldots, \rho(un)) \in \delta$. If A has a valid run on t, we say that t is accepted by A. The set of trees accepted by a finite automaton is called its language, and all such languages are said to be *regular*.

1.2 Graphs

A labelled, directed and simple *graph* is a set $G \subseteq V \times \Sigma \times V$ where Σ is a finite set of labels and V an arbitrary countable set. An element (s, a, t) of G is an *edge* of *source* s, *target* t and *label* a, and is written $s \xrightarrow[G]{a} t$ or simply $s \xrightarrow{a} t$ if G is understood. An edge with the same source and target is called a *loop*. The set of all sources and targets of a graph form its *support* V_G, its elements are called *vertices*. A sequence of edges $(s_1 \xrightarrow{a_1} t_1, \ldots, s_k \xrightarrow{a_k} t_k)$ with $\forall i \in [2, k],\ s_i = t_{i-1}$ is called a *path*. It is written $s_1 \xrightarrow{u} t_k$, where $u = a_1 \ldots a_k$ is the corresponding *path label*. Vertex s_1 is called the origin of the path, t_k its destination. A path is called a *cycle* if its origin and destination are the same vertex. The language, or set of traces of a labelled graph between two sets I and F of vertices is the set of all words w such that there exists a path labelled by w whose origin is in I and destination in F.

1.3 Turing Machines

A Turing machine is a tuple $M = \langle \Gamma, \Sigma, Q, q_0, F, \delta \rangle$ where Σ is the input alphabet, Γ the tape or work alphabet (with $\Sigma \subseteq \Gamma$), Q is a set of states among which q_0 is an initial state and F is a set of final states, and δ is a set of transition rules of the form $pA \to qB\epsilon$ where $p, q \in Q$, $A, B \in \Gamma \cup \{\square\}$ ($\square$ being a blank symbol not in Γ) and $\epsilon \in \{+, -\}$.

Configurations of M are denoted as words upv, where uv is the content of the work tape (where prefix and suffix blank symbols are omitted), p is the current control state and the head scans the cell containing the first letter of v. A transition $d = pA \to qB\epsilon$ is enabled on any configuration c of the form $upAv$, and yields a new configuration $d(c) = uBqv'$ (with $v' = v$ if $v \neq \varepsilon$, or $\square$

otherwise) if $\epsilon = +$ and $u'qCBv$ (with $u'C = u$ if $u \neq \varepsilon$ or $u' = \varepsilon$ and $C = \Box$ otherwise) if $\epsilon = -$. If d is not enabled on c, then $d(c)$ is left undefined.

An *alternating* Turing machine M is defined similarly, with the exception that rules are of the form $d = pA \to \bigwedge_{i \in [1,n]} q_i B_i \epsilon_i$. The alternation degree n of d is written $\mathrm{a}(d)$, by analogy with the notion of arity. For all $i \leq \mathrm{a}(d)$, we write d_i the non-alternating transition $pA \to q_i B_i \epsilon_i$. A run of M on input word w is a tree whose root is labelled by configuration $q_0 w$, and such that the children of any node labelled by configuration c are labelled by $c_1, \ldots, c_n$ if and only if there exists a transition $d \in \delta$ enabled on c such that $\mathrm{a}(d) = n$ and $\forall i \in [1, n]$, $c_i = d_i(c)$. Such a run is successful if all its leaves are labelled by configurations whose control state is in F.

A Turing machine is *linearly bounded* if on every run the total work tape space it uses is at most proportional to the length of its input word. By standard coding techniques, it is sufficient to consider machines whose tape is limited to the cells initially containing the input word. This may be enforced by forbidding transition rules to rewrite the blank symbol $\Box$. The languages of non-alternating linearly bounded machines form the complexity class SPACE($O(n)$), which is equivalent to context-sensitive languages [Kur64]. Adding alternation, one obtains the more general class ASPACE($O(n)$). By classical complexity results [CKS81], it is also equivalent to the class DTIME($2^{O(n)}$), also called ETIME.

2 Arborescent Tiling Systems

To facilitate the proofs of our main results, this section provides an important technical tool, which was also central to some versions of the corresponding proofs on rational graphs and context-sensitive languages (Cf. [CM06]).

Tiling systems were originally defined to recognise or specify picture languages, i.e. sets of two-dimensional words on finite alphabets [GR96], called local picture languages. However, by only looking at the words contained in the first row of each picture of a local picture language, one obtains a context-sensitive language, and the converse is true : for any context-sensitive language there exists a local picture language (and a tiling system accepting it) whose set of upper frontiers is that language [LS97].

In this section, we extend this result to an arborescent extension of tiling systems, and prove that this new formalism characterises precisely the class ETIME.

2.1 Definitions

Instead of planar pictures, we consider so-called arborescent pictures, which are to ordinary pictures what terms are to words.

Definition 1 (Arborescent picture). *Let Γ be a finite alphabet, an arborescent picture p over Γ is a mapping from the set $X \times [1, m]$ to Γ, where $X \subseteq \mathbb{N}_+{}^*$ is a finite, prefix-closed set of sequences of positive integers (called positions in the framework of trees) and m is a positive integer called the width of p. The set*

$\mathrm{dom}(p) = X \times [1, m]$ *is the domain of* p. *The set of arborescent pictures over* $X \times [1, m]$ *is written* $\mathrm{AP}(X, m)$.

Like in the case of trees, we assume that X is not only prefix-closed but also left-closed, i.e. $\forall i > 0,\ ui \in X \implies \forall j < i,\ uj \in X$. For a given picture $p \in \mathrm{AP}(X, m)$, we write $\mathrm{fr}(p)$ the word $w \in \Gamma^m$ such that $w(i) = p(\varepsilon, i)$, which we call the (upper) frontier of p.

Arborescent pictures of domain $X \times [1, m]$ are isomorphic to ordered trees of domain X with nodes labelled over the set Γ^m. As such, if $m = 1$ they are isomorphic to Γ-labelled ordered trees. One can observe that any branch of an arborescent picture seen as a Γ^m-labelled tree, as well as any arborescent picture whose set of positions X is a subset of 1^*, is an ordinary, planar picture.

Definition 2 (Sub-picture). *For any arborescent picture* $p \in \mathrm{AP}(X, m)$, *the sub-picture* $p' = p|_{x,i,Y,n}$ *of* p *at offset* $o = (x, i)$ *with* $x \in X$ *and* $i \in [0, m-1]$ *is the arborescent picture of domain* $Y \times [1, n]$ *such that* Y *is prefix- and left-closed and* $\forall (y, j) \in Y \times [1, n]$, $(xy, i+j) \in X \times [1, m]$ *and* $p'(y, j) = p(xy, i+j)$.

We can now define arborescent tiling systems, which allow the specification of sets of arborescent pictures. Similarly to planar tiling systems, in order to be able to recognise meaningful sets of pictures, we first add a border or frame to each picture using a new symbol #.

Definition 3 (Framed picture). *Let* p *be an arborescent picture of domain* $X \times [1, m]$ *over* Γ *and* $\# \notin \Gamma$ *a new symbol, we define the #-framed picture* $p_\#$ *as the picture of domain* $X' \times [1, m+2]$ *with* $X' = \{\varepsilon\} \cup \{1\}X \cup X''$ *and* $X'' = \{1x1 \mid x \in X \wedge \nexists y \in \mathbb{N}, xy \in X\}$ *such that*

$$\begin{aligned}
&p_\#(\varepsilon, i) = \# && \text{for all } i \in [1, m+2],\\
&p_\#(1x, 1) = \# \text{ and } p_\#(1x, m+2) = \# && \text{for all } x \in X,\\
&p_\#(x, i) = \# && \text{for all } x \in X'', i \in [1, m+2],\\
&p_\#(1x, i+1) = p(x, i) && \text{for all } x \in X, i \in [1, m].
\end{aligned}$$

An arborescent tiling system is then defined as a set of tiling elements of width and height 2, which can then be combined to form larger framed pictures.

Definition 4 (Arborescent tiling system). *An arborescent tiling system (or ATS)* S *is a triple* $(\Gamma, \#, \Delta)$, *where* Γ *is a finite alphabet,* $\# \notin \Gamma$ *a frame symbol and* Δ *is a set of arborescent tiling elements (tiles) in* $\{\bar{\Gamma} \times \bar{\Gamma} \times \bar{\Gamma}^n \times \bar{\Gamma}^n \mid n > 0\}$ *with* $\bar{\Gamma} = \Gamma \cup \{\#\}$.

Each tiling element $d \in \Delta$ is of the form $d = (A, B, \bar{C}, \bar{D})$ with $A, B \in \bar{\Gamma}$ and $\bar{C}, \bar{D} \in \bar{\Gamma}^n$ for some positive integer n. We define additional notations to conveniently manipulate tiling elements. Let $d = (A, B, \bar{C}, \bar{D})$ with $\bar{C} = C_1 \dots C_n$ and $\bar{D} = D_1 \dots D_n$, we write $\mathrm{a}(d) = n$ to denote the arity of d, and d_i with $i \in [1, \mathrm{a}(d)]$ to denote the (planar) tile (A, B, C_i, D_i).

Note that any tiling element $d = (A, B, \bar{C}, \bar{D})$ of arity n is isomorphic to an arborescent picture p_d of domain $X \times [1, 2]$, where $X = \{\varepsilon, 1, \dots, n\}$ and $p_d(\varepsilon, 1)$,

$p_d(\varepsilon, 2)$, $p_d(i, 1)$ and $p_d(i, 2)$ are respectively equal to A, B, C_i and D_i (for all $i \in [1, n]$). In general we do not distinguish p_d from d and write simply d.

Well-formed tiling systems should obey a certain number of restrictions over their set of tiles, regarding in particular the occurrences of the frame symbol # inside tiles. For all $d = (A, B, \bar{C}, \bar{D})$,

1. $(A, B) = (\#, \#) \implies \mathrm{a}(d) = 1 \wedge (C_1, D_1) \neq (\#, \#)$,
2. $\exists i, (C_i, D_i) = (\#, \#) \implies \mathrm{a}(d) = 1 \wedge (A, B) \neq (\#, \#)$,
3. $\exists i, C_i = \# \wedge D_i \neq \# \implies A = \# \wedge \forall i, C_i = \#$,
4. $\exists i, D_i = \# \wedge C_i \neq \# \implies B = \# \wedge \forall i, D_i = \#$,

Before defining the set of pictures and the word language accepted by an arborescent tiling system, we define for any arborescent picture p of domain $X \times [1, m]$ over Γ the set $\mathrm{T}(p)$ of tiling elements of p as the set of all sub-pictures $p|_{x,j,X',2}$ of p such that x is an internal position in X, $j \in [1, m-1]$ and $X' = \{\varepsilon\} \cup \{i' > 0 \mid xi' \in X\}$.

Definition 5 (Language of a tiling system). *The set of arborescent pictures accepted by an arborescent tiling system $S = (\Gamma, \#, \Delta)$ is the set $\mathrm{P}(S) = \{p \in \mathrm{AP} \mid \mathrm{T}(p_{\#}) \subseteq \Delta\}$. The (word) language accepted by S is the set $\mathrm{L}(S) = \{w \in \Gamma^* \mid \exists p \in \mathrm{P}(S), w = \mathrm{fr}(p)\}$ of all upper frontiers of pictures of $\mathrm{P}(S)$.*

As previously, note that arborescent tiling systems are a syntactical generalisation of planar tiling systems : framed pictures with a domain $X \subseteq 1^*$ or branches of framed arborescent pictures are planar framed pictures, and arborescent tiling systems whose elements all have arity 1 are ordinary, planar tiling systems.

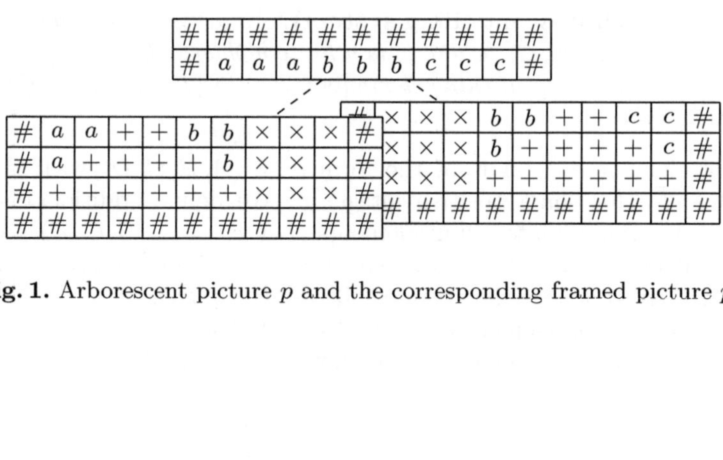

Fig. 1. Arborescent picture p and the corresponding framed picture $p_{\#}$

Example 1. Figure 2.1 represents an arborescent picture p whose frontier is the word $a^3b^3c^3$, as well as the corresponding framed picture. For the sake of clarity, the tree-structure of p is denoted by dashed lines only where the arity is greater than 1. By considering all sub-pictures of height and width 2 of that framed picture, one obtains a set of tiling elements Δ, which contains, among others, tiling elements $(\#, \#, a, b)$, $(a, +, +, +)$ and $(+, +, \#, \#)$ of arity 1 and $(b, c, b+, \times+)$ of arity 2, but not $(b, c, b+, c+)$ or $(\#, +, \#, +)$ for instance.

One can see that the tiling system $S = (\{a, b, c, +, \times\}, \#, \Delta)$ accepts all arborescent pictures similar to p whose frontiers are words of the form $a^n b^n c^n$ with $n \geq 2$: the left branch of each such picture ensures that the number of a's and b's is equal by replacing at each successive row one occurrence of a and one occurrence of b by some symbol $+$. Occurrences of c are irrelevant and are replaced with symbol $\times$. A lower frame borders can only be generated once all occurrences of a and b have been replaced. A similar check is performed by the right branch for symbols b and c.

Note that S does not accept the word abc, since accepting a similar picture with frontier abc would require some additional tiling elements, like for instance $(a, b, +\times, ++)$ and $(b, c, ++, \times+)$. Consequently, the language $\mathrm{L}(S)$ is $\{a^n b^n c^n \mid n \geq 2\}$.

2.2 Languages of Arborescent Tiling Systems

In this section, we prove that arborescent tiling systems and alternating linearly bounded machines define the same family of languages, namely ASPACE($O(n)$), also equal as previously mentioned to DTIME($2^{O(n)}$) = ETIME.

Proposition 1. *For every arborescent tiling system S, there exists an alternating linearly bounded machine M such that* $\mathrm{L}(M) = \mathrm{L}(S)$.

Proof sketch. Let $S = (\Gamma, \#, \Delta)$ be an arborescent tiling system. We build an alternating linearly bounded machine $M = (\Gamma, \Gamma', Q, q_\#, f, \delta)$ accepting $\mathrm{L}(S)$. Its work alphabet Γ' is the union of all $\bar{\Gamma}^k$ for $k \in [1, \mathrm{a}(S)]$, where $\bar{\Gamma} = \Gamma \cup \{\#\}$ and $\mathrm{a}(S) = \max\{\mathrm{a}(d) \mid d \in \Delta\}$. We informally describe M's behaviour as follows:

1. M starts in configuration $[q_\# w]$, where $w \in \Gamma^*$ is the input word. In a first sweep, it checks that w is a possible frontier of a picture accepted by S.
2. In the next sweep, M generates a n-tuple of possible next rows based on the current configuration and the tiles in Δ. M then uses universal branching to evaluate the sub-pictures whose upper frontiers are each of these rows.
3. The last generated row consists in a sequence of frame symbols $\#$ if and only if the last written symbol is $\#$. If this is the case on the current computation branch, reach accepting state f. Otherwise, repeat the previous step.

Steps 2 and 3 are repeated until all computation branches have reached the accepting state f. □

Proposition 2. *For every alternating linearly bounded machine M, there exists an arborescent tiling system S such that* $\mathrm{L}(S) = \mathrm{L}(M)$.

Proof sketch. Let $M = (\Sigma, \Gamma, Q, q_0, F, \delta)$ be an alternating linearly bounded machine. We build an arborescent tiling system $S = (\Gamma', \#, \Delta)$ such that $\mathrm{L}(S) = [\mathrm{L}(M)]$, where [and] are two new symbols. The set of tiling elements Δ is built in order to conform to the following informal specification.

S first needs to set an input word w as the upper frontier of any picture it accepts. It then encodes the initial configuration of M on w as the second row. Subsequent tiles simulate the application of a transition of M on the configuration encoded by the current row, and check that the previous transition was correctly simulated. This requires additional information, in particular about the position of the head and the index of the last simulated transition, to be added to the picture alphabet. Arity n tiling elements are used when the simulated rule is of alternation degree n. This process goes on until an accepting state is reached by M on a given execution branch. In that case, a bottom border is generated by S on the corresponding picture branch. □

From Propositions 1 and 2, we deduce the announced theorem.

Theorem 1. *The languages of arborescent tiling systems form the complexity class* ETIME.

Note that the language accepted by the tiling system of Example 1 is a context-sensitive language, which could also be accepted by a non-arborescent tiling system.

3 Traces of Term-Automatic Graphs

We now turn to the main result of this paper, which is the study of languages of graphs characterised by automata-defined binary relations over terms, and in particular term-automatic graphs. We define these relations and the graphs they generate, then present a two-steps proof that the languages of term-automatic graphs indeed coincide with ASPACE($O(n)$). First, we establish this result for the simpler term-synchronous graphs in Section 3.2, then generalise it to term-automatic graphs in Section 3.3.

3.1 Definitions

Let $s = f(s_1 \dots s_m)$ and $t = g(t_1 \dots t_n)$ be two terms over some ranked alphabet F. We define the overlap $[st]$ of s and t as a term over domain $\mathrm{dom}(s) \cup \mathrm{dom}(t)$ and extended alphabet $(F \cup \{\bot\})^2$ (each element (f, g) of this alphabet being written simply fg), such that $\forall p \in \mathrm{dom}(s) \cup \mathrm{dom}(t)$, $[st](p) = fg$ with $f = s(p)$ if $p \in \mathrm{dom}(s)$ or $\bot$ otherwise, and $g = t(p)$ if $p \in \mathrm{dom}(t)$ or $\bot$ otherwise. This notation is extended to sets in the natural way.

We can now define term-automatic and term-synchronous relations. We say a binary relation R is *term-* (or *tree-*)*automatic* if the term language $[R] = \{[st] \mid (s,t) \in R\}$ is regular. If furthermore for all $(s,t) \in R$, $\mathrm{dom}(s) = \mathrm{dom}(t)$, it is called *synchronous*. In other words, a synchronous relation is an automatic relation which only associates terms with the same domain. Both families of relations are closed under relational composition. Term-automatic and term-synchronous relations are syntactical extensions of the corresponding families of relations over words. As such, they also define extended families of graphs.

Definition 6. *A Σ-graph G is term-automatic (resp. term-synchronous) if it is isomorphic to a graph $\{u \xrightarrow{a} v \mid a \in \Sigma,\ (u,v) \in R_a\}$, where $(R_a)_{a\in\Sigma}$ is a family of term-automatic (resp. term-synchronous) relations.*

3.2 Term-Synchronous Graphs

This section presents direct simulations of alternating tiling systems by synchronous graphs and conversely, showing that the languages of term-synchronous graphs between regular sets of vertices form the class ETIME.

Proposition 3. *For every term-synchronous graph G and regular sets I and F there exists an arborescent tiling system S such that* $\mathrm{L}(S) = \mathrm{L}(G, I, F)$.

Proof sketch. Let $G = (R_a)_{a\in\Sigma}$ be a synchronous graph, and I, F two regular sets of vertices of G. We build a tiling system $S = (\Gamma, \#, \Delta)$ such that $\mathrm{L}(S) = \mathrm{L}(G, I, S)$.

For all $a \in \Sigma$, let A_a be a finite top-down term automaton accepting the language $[R_a]$ (as defined in Section 3.1), and A_I, A_F similar automata for I and F respectively. For every $a \in \Sigma$, we also define relations $R_{I\circ a} = Id_I \circ R_a$ and $R_{a\circ F} = R_a \circ Id_F$, where Id_L denotes the identity relation over some set L. Let also $A_{I\circ a}$ and $A_{a\circ F}$ be two automata accepting the languages $[R_{I\circ a}]$ and $[R_{a\circ F}]$ respectively. The control state sets of all these automata are supposed disjoint.

The idea of this construction is that, for every path $t_0 \xrightarrow{a_1} t_1 \ldots \xrightarrow{a_n} t_n$ in G with $t_0 \in I$, $t_n \in F$ and $\forall i,\ \mathrm{dom}(t_i) = X$, S should accept an arborescent picture p whose upper frontier is w and whose successive vertical "slices" correspond to encodings of runs of $A_{I\circ a_1}$, A_{a_2}, ..., $A_{a_{n-1}}$ and $A_{a_n\circ F}$ respectively. Conversely, S should only accept all such pictures which correspond to paths in G between I and F. These conditions are sufficient for $L(S)$ to be equal to $\mathrm{L}(G, I, F)$. To ensure they indeed hold, we define Δ in order to be able to check that the i-th and $(i + 1)$-th "slices" are indeed compatible. □

Proposition 4. *For every arborescent tiling system S, there exists a term-synchronous graph G and regular sets I and F such that* $\mathrm{L}(G, I, F) = \mathrm{L}(S)$.

Proof sketch. Let $S = (\Gamma, \#, \Delta)$ be an arborescent tiling system. We build a term-synchronous graph G such that $\mathrm{L}(S) = \mathrm{L}(G, I, F)$ for some regular sets I and F. In the following, symbol $\#$ is overloaded to make the notation less cumbersome, and represents functional symbols of varying arities, which can be deduced from the context. In particular, we write $\#_X$ for a given prefix-closed set X the term of domain X whose nodes are all labelled with $\#$.

Let R_a, $a \in \Sigma$, be the binary relation between all terms $\#(s)$ and $\#(t)$ (i.e. s and t with an additional unary $\#$ at the root) such that a labels the root of t and for a given $p \in P(S)$, either $s = p|_{\varepsilon,i,X,1}$ and $t = p|_{\varepsilon,i+1,X,1}$ for some

$i > 0$ or $s = \#_X$ and $t = p|_{\varepsilon,0,X,1}$. Let G be the graph defined by $(R_a)_{a\in\Sigma}$, we show that G is term-synchronous by constructing automata $(A_a)_{a\in\Sigma}$ such that $\mathrm{L}(A_a) = [R_a] = \{[st] \mid (s,t) \in R_a\}$. For all a, A_a has transitions:

$$\begin{array}{ll} q_0\#\# \to q_{AB,1} & \text{if } (\#,\#,A,B) \in \Delta, \\ q_{\bar{A}\bar{B},i}AB \to q_{\bar{C}\bar{D},1} \cdots q_{\bar{C}\bar{D},k} & \text{if } d = (A_i, B_i, \bar{C}, \bar{D}) \in \Delta,\ k = \mathrm{a}(d), \\ & \quad A_i = \bar{A}(i) \text{ and } B_i = \bar{B}(i), \\ q_{\bar{A}\bar{B},i}A_iB_i \to \varepsilon & \text{if } (A_i, B_i, \#, \#) \in \Delta, \\ & \quad A_i = \bar{A}(i) \text{ and } B_i = \bar{B}(i). \end{array}$$

We define I as the regular set of all terms labelled over $\{\#\}$, and F as the set of all possible rightmost columns of pictures accepted by S. This set of terms is accepted by an automaton A_F whose construction is straightforward.

By construction of I, A_F and each of the A_a, there is a path in G labelled by a word w between a vertex in I and a vertex in F iff the vertices along that path are the successive columns of a picture in $P(S)$ with frontier w. □

3.3 Term-Automatic Graphs

In this section, we show that the more general family of term-automatic graphs defines the same family of languages as their synchronous counterparts.

Proposition 5. *For every term-automatic graph G and regular sets of terms I and F, there exists a term-synchronous graph G' and regular sets I' and F' such that* $\mathrm{L}(G', I', F') = \mathrm{L}(G, I, F)$.

Proof sketch. Let G be a term-automatic graph defined by a family $(R_a)_{a\in\Sigma}$ of automatic relations and I, F be two regular languages, each $[R_a]$ being accepted by an automaton A_a, I by A_I and F by A_F. We define a synchronous graph $G' = (R'_a)_{a\in\Sigma}$ and two regular sets I' and F' such that $\mathrm{L}(G, I, F) = \mathrm{L}(G', I', F')$.

Recall that term-automatic relations are defined using a notion of overlap between terms (Cf. Section 3.1). Two terms s and t with different domains belong to a term-automatic relation R defined by automaton A if the overlap $[st]$ of s and t is accepted by A. This notion of overlap consists in "unifying" the domains of s and t, and padding undefined positions with a special symbol $\perp$.

We wish to reuse this idea, but instead of unifying the domains of two terms only, we have to unify the domains of all vertices along a given path. Indeed, in a term-synchronous graph, edges can only exist between terms with precisely the same domain. For every term s standing for a vertex in G, we will thus have to consider an infinite set of "versions" of s in G', one for each possible term domain larger than that of s.

Let Γ be a ranked alphabet, we define alphabet Γ' as $\Gamma' = \Gamma_0 \cup \Gamma_n$ with $\Gamma'_0 = \#_0$ and $\Gamma'_n = \Gamma \cup \#_n$, where n is the maximal arity of symbols in Γ. Let ϕ be a mapping from $T(\Gamma)$ to $2^{T(\Gamma')}$ such that for any term $t \in T(\Gamma)$,

$$\begin{aligned} \phi(t) \ = \ \{t' \in T(\Gamma') \mid \mathrm{dom}(t) \subset \mathrm{dom}(t'),\ \forall p \in \mathrm{dom}(t), t'(p) = t(p) \\ \text{and } \forall p \in \mathrm{dom}(t') \setminus \mathrm{dom}(t), t'(p) \in \{\#_0, \#_n\}\}. \end{aligned}$$

In other words, to any term t, ϕ associates the set of all terms obtained by "padding" t with silent symbols $\#_0$ and $\#_n$. This mapping is extended to sets of terms in the natural way. Note that, given any $t' \in F(\Gamma')$, there exists at most one term $t \in T(\Gamma)$ such that $t' \in \phi(t)$.

We now define, for every $a \in \Sigma$, relation R'_a as $\{(s', t') \mid (s, t) \in R_a,\ s' \in \phi(s),\ t' \in \phi(t) \text{ and } \mathrm{dom}(s') = \mathrm{dom}(t')\}$. This synchronous relation can be characterised by a finite tree automaton A'_a defined from A_a. We also let $I' = \phi(I)$ and $F' = \phi(F)$, for which automata can be similarly defined from A_I and A_F.

Let G' be the term-synchronous graph defined by $(R'_a)_{a \in \Sigma}$. One can show that for every path labelled w in G' between some $i' \in I'$ and $f' \in F'$, there exists a unique path between I and F in G with the same label, and that conversely for every w-path between I and F in G there must exist at least one corresponding path in G' between I' and F'. This ensures that $L(G, I, F)$ and $L(G', I', F')$ are indeed equal. □

Remark 1. Note that for every term-automatic graph G and regular sets I and F, there exists a term-automatic graph G' and finite sets I' and F' such that $\mathrm{L}(G', I', F') = \mathrm{L}(G, I, F)$. Indeed, for any regular I and F and finite I' and F' the relations $I' \times I$ and $F \times F'$ are automatic. Since term-automatic relations are closed under union and composition, this can be used to build G' from G.

This, however, does not hold in the term-synchronous case. Indeed, since each connected component of a term-synchronous graph is finite, the language of any such graph from a finite set of initial vertices is regular.

Combining Theorem 1, Propositions 3, 4 and 5, as well as Remark 1, we obtain the following result concerning the family of languages accepted by term-synchronous and term-automatic graphs.

Theorem 2. *The languages of term-synchronous graphs between regular sets of vertices and of term-automatic graphs between regular or finite sets of vertices form the complexity class* ETIME.

4 Conclusion

We have proved that the class of languages accepted by alternating linearly bounded machines (ETIME) can also be characterised as the sets of first rows of pictures accepted by arborescent tiling systems, as well as the sets of path labels of term-automatic graphs between regular or finite sets of initial and final vertices.

A natural extension of this work would be to generalise Theorem 2 to graphs defined by more expressive classes of tree transducers, in order to fully extend the existing results on rational graphs. In practice, this would require extending the construction for Proposition 5 to more general padding techniques.

Further points of interest concern the extension of other results from [CM06] to term-automatic graphs, in particular regarding structural restrictions of these graphs, like finite or bounded degree, or the restriction to a single initial vertex, as well as a similar study of related complexity classes or families of languages.

References

[BG00] Blumensath, A., Grädel, E.: Automatic structures. In: Proceedings of the 15th IEEE Symposium on Logic in Computer Science (LICS 2000), pp. 51–62. IEEE, Los Alamitos (2000)

[Cho59] Chomsky, N.: On certain formal properties of grammars. Information and Control 2, 137–167 (1959)

[CKS81] Chandra, A., Kozen, D., Stockmeyer, L.: Alternation. Journal of the ACM 28(1), 114–133 (1981)

[CM06] Carayol, A., Meyer, A.: Context-sensitive languages, rational graphs and determinism. Logical Methods in Computer Science 2(2) (2006)

[GR96] Giammarresi, D., Restivo, A.: Handbook of Formal Languages. In: Two-dimensional languages, vol. 3, Springer, Heidelberg (1996)

[KN95] Khoussainov, B., Nerode, A.: Automatic presentations of structures. In: Leivant, D. (ed.) LCC 1994. LNCS, vol. 960, pp. 367–392. Springer, Heidelberg (1995)

[Kur64] Kuroda, S.: Classes of languages and linear-bounded automata. Information and Control 7(2), 207–223 (1964)

[LS97] Latteux, M., Simplot, D.: Context-sensitive string languages and recognizable picture languages. Information and Computation 138(2), 160–169 (1997)

[Mor00] Morvan, C.: On rational graphs. In: Tiuryn, J. (ed.) ETAPS 2000 and FOSSACS 2000. LNCS, vol. 1784, pp. 252–266. Springer, Heidelberg (2000)

[MR05] Morvan, C., Rispal, C.: Families of automata characterizing context-sensitive languages. Acta Informatica 41(4-5), 293–314 (2005)

[MS01] Morvan, C., Stirling, C.: Rational graphs trace context-sensitive languages. In: Sgall, J., Pultr, A., Kolman, P. (eds.) MFCS 2001. LNCS, vol. 2136, pp. 548–559. Springer, Heidelberg (2001)

[Pen74] Penttonen, M.: One-sided and two-sided context in formal grammars. Information and Control 25(4), 371–392 (1974)

[Ris02] Rispal, C.: The synchronized graphs trace the context-sensitive languages. In: Proceedings of the 4th International Workshop on Verification of Infinite-State Systems (INFINITY 2002). Electronic Notes in Theoretical Computer Science, vol. 68 (2002)

State Complexity of Basic Operations on Suffix-Free Regular Languages

Yo-Sub Han[1,⋆] and Kai Salomaa[2,⋆⋆]

[1] Intelligence and Interaction Research Center,
Korea Institute of Science and Technology
P.O.BOX 131, Cheongryang, Seoul, Korea
emmous@kist.re.kr

[2] School of Computing, Queen's University
Kingston, Ontario K7L 3N6, Canada
ksalomaa@cs.queensu.ca

Abstract. We investigate the state complexity of basic operations for suffix-free regular languages. The state complexity of an operation for regular languages is the number of states that are necessary and sufficient in the worst-case for the minimal deterministic finite-state automaton that accepts the language obtained from the operation. We establish the precise state complexity of catenation, Kleene star, reversal and the Boolean operations for suffix-free regular languages.

1 Introduction

Codes are useful in many areas such as information processing, data compression, cryptography and information transmission [16]. Some of well-known codes are prefix codes, suffix codes, bifix codes and infix codes. People use different codes for different applications based on the characteristic of each code [1,16]. Since codes are sets of strings over an alphabet, they are closely related to formal languages: a code is a language. Thus, the condition defining a class of codes defines a corresponding subfamily of each language family. For regular languages, for example, suffix-freeness defines suffix-free regular languages, which constitute a subfamily of regular languages.

There are different ways to define the complexity of a regular language L. One classical definition is the total number of states in the minimal deterministic finite-state automaton (DFA) for L since the minimal DFA for L is unique (up to isomorphism) [13,21]. Based on this definition, Yu and his co-authors [24] defined the state complexity of an operation for regular languages to be the number of states that are necessary and sufficient in the worst-case for the minimal DFA that accepts the language obtained from the operation. Yu [23] gave a comprehensive survey of the state complexity of regular languages. Salomaa et al. [20] studied classes of languages for which the reversal operation reaches the

⋆ Han was supported by the KIST Research Grants 2E20050 and 2Z03050.

⋆⋆ Salomaa was supported by the Natural Sciences and Engineering Research Council of Canada Grant OGP0147224.

L. Kučera and A. Kučera (Eds.): MFCS 2007, LNCS 4708, pp. 501–512, 2007.

exponential upper bound. As special cases of the state complexity, researchers examined the state complexity of finite languages [3,8], the state complexity of unary language operations [19] and the nondeterministic descriptional complexity of regular languages [11,12]. There are several other results with respect to the state complexity of different operations [4,5,6,14,15,18].

Recently, Han et al. [9] examined the state complexity of prefix-free regular languages. They tackled the problem based on the structural property of prefix-free DFAs: A prefix-free DFA must be non-exiting assuming all states are useful [10]. It turns out that the state complexity for the prefix-free case is strictly less than the corresponding state complexity for regular languages over some basic operations. We know that if a language L is prefix-free, then its reversal L^R is suffix-free by definition. Moreover, if L is regular and non-empty, then the start state of a DFA for L^R should not have any in-transitions. However, this condition is necessary but not sufficient. Due to this fact, the state complexity of suffix-free regular languages is not symmetric to the prefix-free case. This leads us to investigate the state complexity of basic operations on suffix-free regular languages. Interestingly, the results for catenation and Kleene star turn out to be of a totally different order than in the case of prefix-free regular languages.

In Section 2, we define some basic notions. In Section 3, we examine the state complexity of Kleene star and reversal of suffix-free regular languages. We then look at the catenation of two suffix-free minimal DFAs in Section 4. Next, we investigate the state complexity of intersection and union of suffix-free regular languages based on the Cartesian product of states in Section 5. We present the comparison table of the state complexity on different types of regular languages and conclude the paper in Section 6.

2 Preliminaries

Let Σ denote a finite alphabet of characters and Σ^* denote the set of all strings over Σ. The size $|\Sigma|$ of Σ is the number of characters in Σ. A language over Σ is any subset of Σ^*. Given a set X, 2^X denotes the power set of X. For a string $x \in \Sigma^*$ and a character a, $|x|_a$ denotes the number of symbol a occurrences in x. We say that a string x is a *suffix* of a string y if $y = ux$ for some string u. We define a set X of strings to be a *suffix-free set* if a string from X is not a suffix of any other string in X. Given a string x from a set X, let x^R be the reversal of x, in which case $X^R = \{x^R \mid x \in X\}$.

The symbol $\emptyset$ denotes the empty language and the character λ denotes the null string. A finite-state automaton (FA) A is specified by a tuple $(Q, \Sigma, \delta, s, F)$, where Q is a finite set of states, Σ is an input alphabet, $\delta : Q \times \Sigma \rightarrow 2^Q$ is a transition function, $s \in Q$ is the start state and $F \subseteq Q$ is a set of final states. If F consists of a single state f, we use f instead of $\{f\}$ for simplicity. $|Q|$ denotes the number of states in Q. We define a state d to be a *sink state* if d is reachable from s of A and, for any $a \in \Sigma$, $\delta(d, a) = d$ and $d \notin F$. Since all sink states are always equivalent, we can assume that A has a unique sink state. For a transition $\delta(p, a) = q$ in A, we say that p has an *out-transition* and q has an *in-transition*. Furthermore, p is a *source state* of q and q is a *target state* of p. The transition function δ can be extended to a function $Q \times \Sigma^* \rightarrow 2^Q$ that reflects sequences of inputs. A string x over Σ is accepted by A if there is a labeled path from s to a state in F such that this path spells out the string x. Namely, $\delta(s, x) \cap F \neq \emptyset$. The language $L(A)$ of an FA A is the set of all strings that are spelled out by paths from s to a final state in F. We say that A is *non-returning* if the start state of A does not have any in-transitions and A is *non-exiting* if all out-transitions of every final state in A go to the sink state.

Given an FA $A = (Q, \Sigma, \delta, s, F)$, we define the *right language* L_q of a state q to be the set of strings that are spelled out by some path from q to a final state in A; namely, L_q is the language accepted by the FA obtained from A by changing the start state to q. We say that two states p and q are *equivalent* if $L_p = L_q$.

We define an FA A to be a DFA if the number of target states for each pair of a state q and a character $a \in \Sigma$ is one: namely, $|\delta(q, a)| = 1$. Given a DFA A, we assume that A is complete; namely, each state has $|\Sigma|$ out-transitions. If A has m states, then we say that A is an m-state DFA.

We define a (regular) language L to be suffix-free if L is a suffix-free set. A regular expression E is suffix-free if $L(E)$ is suffix-free. Similarly, an FA A is suffix-free if $L(A)$ is suffix-free. Moreover, if $L(A)$ is suffix-free and non-empty, then A must be non-returning. Similarly, we can define prefix-free regular expressions and languages. Note that if a language L is suffix-free, then L^R is prefix-free.

For complete background knowledge in automata theory, the reader may refer to textbooks [13,21].

Due to the limit on the number of pages, we omit all the proofs in the following sections. The proofs can be found in a full version [7].

3 Kleene Star and Reversal

Before examining the state complexity of various operations, we establish that any suffix-free (complete) DFA must always have a sink state. Recall that the state complexity of a regular language L is the number of states in its minimal DFA. If L is a regular language, its minimal DFA does not necessarily have have a sink state. However, if L is prefix-free, then its minimal DFA A must have a sink state since A is non-exiting. Therefore, we have to verify the existence of the sink state in a suffix-free minimal DFA before investigating the state complexity for each operation. This is crucial for computing the correct state complexity.

Lemma 1. *Let $A = (Q, \Sigma, \delta, s, F)$ be a minimal DFA for a suffix-free language and $k = |Q|$. Then, A has a sink state $d \in Q$ and for every string $w \in \Sigma^+$, $\delta(s, w^k) = d$.*

Lemma 1 shows that we must always consider the sink state for computing the state complexity of suffix-free regular languages. From now, we assume that a suffix-free minimal DFA has the unique sink state.

3.1 Kleene Star of Suffix-Free Regular languages

We first start with the Kleene star operation.

Lemma 2. *Given an m-state suffix-free minimal DFA $A = (Q, \Sigma, \delta, s, F)$, $2^{m-2}+1$ states are sufficient for $L(A)^*$.*

We now define a DFA A such that $L(A)$ is suffix-free and the state complexity of $L(A)^*$ reaches the upper bound in Lemma 2. Let $A = (Q, \Sigma, \delta, s, F)$, where $Q = \{0, 1, \ldots, m-1\}$, for $m \geq 4$, $\Sigma = \{a, b, c, d\}$, $s = m-2$, $F = \{0\}$ and δ is defined as follows:

(i) $\delta(m-2, c) = 0$,
(ii) $\delta(i, a) = i+1$, for $0 \leq i \leq m-4$, and $\delta(m-3, a) = 0$,
(iii) $\delta(i, d) = i$, for $1 \leq i \leq m-3$,
(iv) $\delta(m-2, b) = 1$, $\delta(0, b) = 0$, $\delta(i, b) = i$ for $2 \leq i \leq m-3$,
(v) all transitions not defined above go to the sink state $m-1$.

Fig. 1 depicts the DFA A. The figure omits the sink state $m-1$.

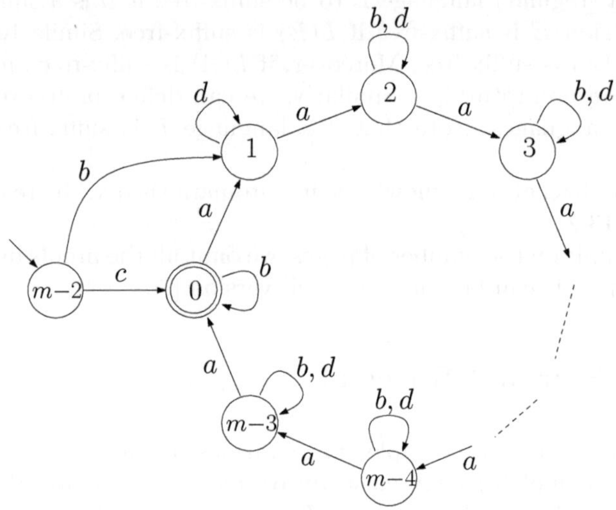

Fig. 1. The DFA A for the worst-case lower bound for the Kleene star of $L(A)$, for $m \geq 4$. Note that we omit the sink state $m-1$.

Lemma 3. *Let A be the DFA in Fig. 1 for $m \geq 4$.*

1. *The language $L(A)$ is suffix-free.*
2. *The state complexity of $L(A)^*$ is $2^{m-2}+1$.*

Combining Lemma 2 and Lemma 3, we have the following result.

Theorem 1. *Given an m-state suffix-free minimal DFA A, $2^{m-2}+1$ states are necessary and sufficient in the worst-case for the minimal DFA of $L(A)^*$.*

The proof of Lemma 3 uses a four character alphabet. It remains an open question whether the bound of Theorem 1 can be reached using an alphabet of size 2 or 3.

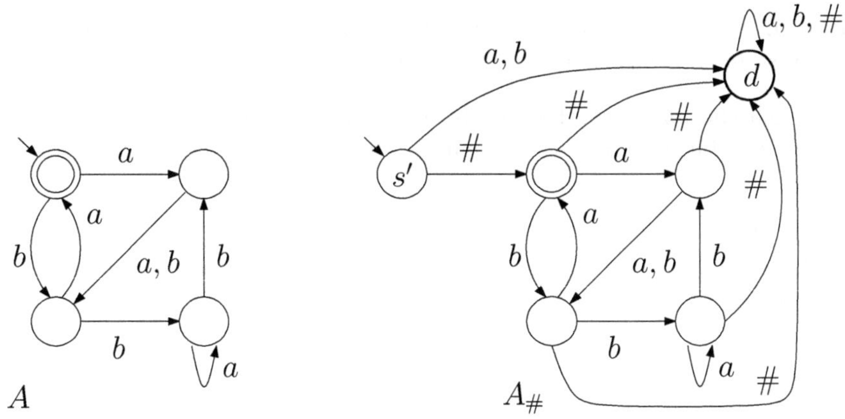

Fig. 2. An example of a minimal DFA A in Proposition 1. Note that $A_\#$ is also a minimal DFA and $L(A_\#)$ is suffix-free.

3.2 Reversal of Suffix-Free Regular Languages

We examine the reversal operation of suffix-free regular languages. First, we recall the state complexity of reversal on regular languages. If a regular language L is accepted by an m-state minimal DFA, then its reversal L^R is accepted by an m-state NFA. By the well-known subset argument, we can conclude that the state complexity of L^R is at most 2^m.

Proposition 1 (Leiss [17] and Salomaa et al. [20]). *There are classes of regular languages for which 2^m states are necessary and sufficient for the reversal of an m-state minimal DFA. Note that such an m-state minimal DFA does not have the sink state.*

Given a suffix-free minimal DFA $A = (Q, \Sigma, \delta, s, F)$, we flip all transition directions in A and obtain a new FA A^R for $L(A)^R$. If we apply the subset construction on A^R, then the resulting DFA is the minimal DFA for $L(A^R)$ [2,22].

Lemma 4. *Given an m-state suffix-free minimal DFA A, $2^{m-2}+1$ states are sufficient in the worst-case for the minimal DFA of $L(A)^R$.*

Next, we show that $2^{m-2}+1$ states are necessary for the reversal of a suffix-free minimal DFA. Given a (regular) language L over Σ, $\#L$ is suffix-free if the character $\#$ is not in Σ.

We construct a suffix-free minimal DFA that has m states as follows: Let $A = (Q, \Sigma, \delta, s, F)$ be a minimal DFA as in Proposition 1 over Σ, which is not suffix-free in general. We introduce a new start state s' and a new transition $\delta(s', \#) = s$. We also introduce a sink state d. Note that a minimal DFA for a regular language in Proposition 1 does not have a sink state. Consequently, d is not equivalent with any of the states of A. Then, the new FA $A_\#$ is deterministic and minimal by construction. Furthermore, $L(A_\#)$ is suffix-free. Thus, if A has $m-2$ states, then $A_\#$ has m states. See Fig. 2 for an example.

Lemma 5. *Given an m-state suffix-free minimal DFA $A_{\#}$ as shown in Fig. 2, $2^{m-2}+1$ states are necessary for the minimal DFA of $L(A_{\#})^R$, where $\# \notin \Sigma$.*

We establish the following theorem from Lemmas 4 and 5. Note that Salomaa et al. [20] established that the result of Proposition 1 holds also for binary alphabets.

Theorem 2. *Given an m-state suffix-free minimal DFA A over Σ, $2^{m-2}+1$ states are necessary and sufficient in the worst-case for the minimal DFA of $L(A)^R$, where $|\Sigma| \geq 3$.*

4 Catenation

We investigate the state complexity of the catenation of two suffix-free regular languages. We first compute the upper bound and after that present a matching lower bound example.

Lemma 6. *Given two suffix-free minimal DFAs $A = (Q_1, \Sigma, \delta_1, s_1, F_1)$ and $B = (Q_2, \Sigma, \delta_2, s_2, F_2)$, $(m-1)2^{n-2}+1$ states are sufficient for the minimal DFA of $L(A) \cdot L(B)$, where $m = |Q_1|$ and $n = |Q_2|$.*

We present two suffix-free minimal DFAs A and B such that the state complexity of $L(A)L(B)$ reaches the upper bound in Lemma 6. In the following, let $\Sigma = \{a, b, c, d\}$. We define

$$A = (Q_1, \Sigma, \delta_1, s_1, F_1), \tag{1}$$

where $Q_1 = \{0, 1, \ldots, m-1\}$, $m \geq 3$, $s_1 = 0$, $F_1 = \{1\}$ and δ_1 is defined as follows:

(i) $\delta_1(0, c) = 1$,
(ii) $\delta_1(i, a) = i + 1$, $1 \leq i \leq m-3$, $\delta_1(m-2, a) = 1$,
(iii) $\delta_1(i, b) = i$, $1 \leq i \leq m-2$,
(iv) $\delta_1(1, d) = 1$,
(v) all transitions not defined above go to the sink state $m-1$.

The DFA A is depicted in Fig. 3. The figure does not show the sink state $m-1$ or the transitions into the sink state.

Next we define

$$B = (Q_2, \Sigma, \delta_2, s_2, F_2), \tag{2}$$

where $Q_2 = \{0, 1, \ldots, n-1\}$, $n \geq 3$, $s_2 = 0$, $F_2 = \{1\}$, and δ_2 is defined by the following:

1. $\delta_2(0, d) = 1$,
2. $\delta_2(i, b) = i + 1$, $1 \leq i \leq n-3$, $\delta_2(n-2, b) = 1$,
3. $\delta_2(i, a) = \delta_2(i, c) = i$, $1 \leq i \leq n-2$,
4. $\delta_2(i, d) = i$, $2 \leq i \leq n-2$,
5. all transitions not defined above go to the sink state $n-1$.

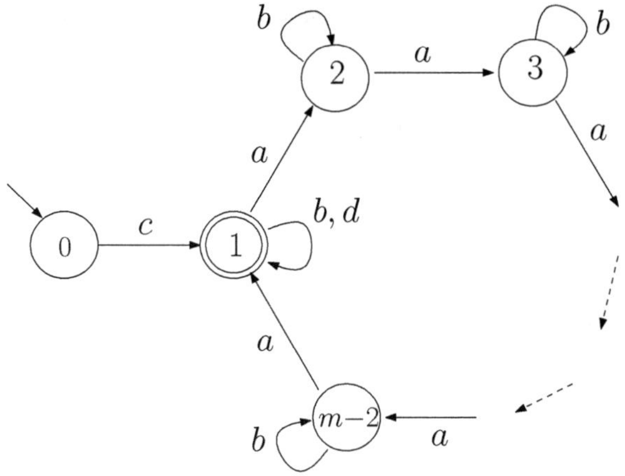

Fig. 3. The DFA A for the worst-case lower bound for catenation

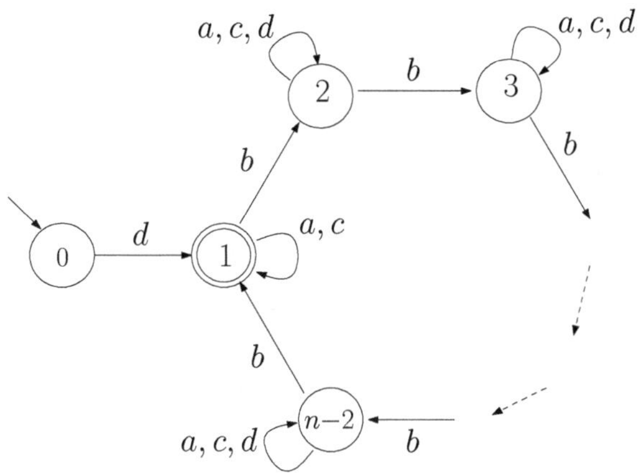

Fig. 4. The DFA B for the worst-case lower bound for catenation

The DFA B is depicted in Fig. 4. Again the figure does not show the sink state $n-1$.

Lemma 7. *Let A be as in (1) and B as in (2), for $m, n \geq 3$.*

1. *The languages $L(A)$ and $L(B)$ are suffix-free.*
2. *The state-complexity of $L(A) \cdot L(B)$ is $(m-1)2^{n-2}+1$.*

Lemma 7 shows that the upper bound in Lemma 6 is tight when $|\Sigma| \geq 4$.

Theorem 3. *For arbitrary $m, n \geq 3$, $(m-1)2^{n-2}+1$ states are necessary and sufficient in the worst-case for the catenation of, respectively, an m-state and an n-state suffix-free minimal DFAs.*

The worst-case example in Lemma 7 uses an alphabet with 4 characters. We do not know whether the upper bound can be reached using an alphabet of size 2 or 3.

5 Intersection and Union

Note that for the complement operation of an m-state suffix-free DFA, it is easy to verify that m states are necessary and sufficient. In the following, we consider the operations of intersection and union.

5.1 Intersection of Suffix-Free Regular Languages

Given two DFAs A and B, we can construct a DFA for the intersection of $L(A)$ and $L(B)$ based on the Cartesian product of states. For details on the Cartesian product construction, refer to Hopcroft and Ullman [13].

Proposition 2. *Given two DFAs* $A = (Q_1, \Sigma, \delta_1, s_1, F_1)$ *and* $B = (Q_2, \Sigma, \delta_2, s_2, F_2)$, *let* $M = (Q_1 \times Q_2, \Sigma, \delta, (s_1, s_2), F_1 \times F_2)$, *where for all* $p \in Q_1$ *and* $q \in Q_2$ *and* $a \in \Sigma$,

$$\delta((p, q), a) = (\delta_1(p, a), \delta_2(q, a)).$$

Then, $L(M) = L(A) \cap L(B)$.

Since the automaton M constructed in Proposition 2 is deterministic, it follows that mn states are sufficient for the intersection of $L(A)$ and $L(B)$, where $|A| = m$ and $|B| = n$. Note that mn is a tight bound for the intersection of two regular languages [24].

We assign a unique number for each state from 1 to m in A and from 1 to n in B, where $|A| = m$ and $|B| = n$. Assume that the mth state and the nth state are the sink states in A and B, respectively. Let $A \cap_c B$ denote the resulting intersection automaton that we compute using the Cartesian product of states. By the construction, $A \cap_c B$ is deterministic since A and B are deterministic. Therefore, we obtain a DFA for $L(A) \cap L(B)$. Next, we minimize $A \cap_c B$ by merging all equivalent states and removing unreachable states from the start state.

Proposition 3 (Han et al. [9]). *For a state* (i, j) *in* $A \cap_c B$, *the right language* $L_{(i,j)}$ *of* (i, j) *is the intersection of* L_i *in* A *and* L_j *in* B.

Since a suffix-free DFA A has the sink state as proved in Lemma 1, $L_{(m,i)} = \emptyset$, for $1 \le i \le n$, by Proposition 3, where m is the sink state of A. Therefore, we can merge all these states. Similarly, all states (j, n), for $1 \le j \le m$, of $A \cap_c B$ are equivalent and, therefore, can be merged.

Observation 1. *Given suffix-free minimal DFAs* A *and* B, *all states* (m, i) *for* $1 \le i \le n$ *and all states* (j, n) *for* $1 \le j \le m$ *of* $A \cap_c B$ *are equivalent.*

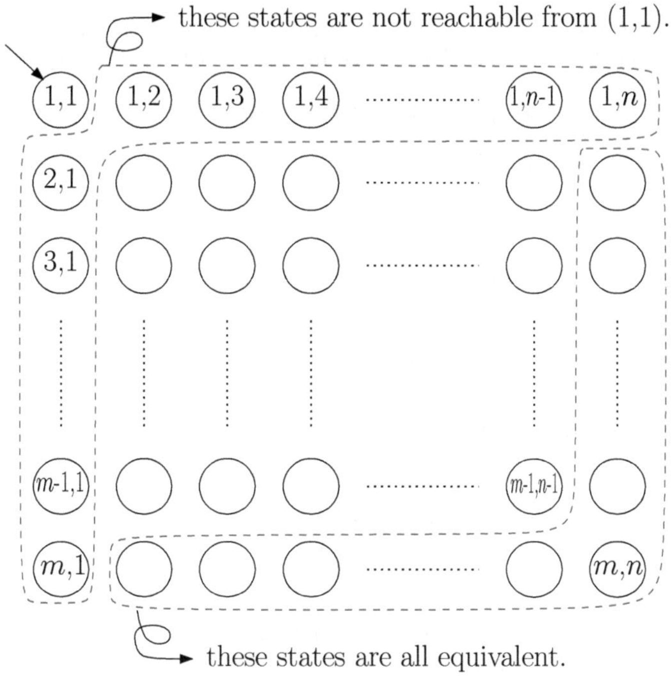

Fig. 5. The figure depicts the intersection automaton $A \cap_c B$ constructed for two suffix-free minimal DFAs A and B. Note that, by Observation 1, all states in the last row and in the last column are equivalent. Similarly, by Observation 2, all states, except for the start state (1,1), in the first row and in the first column are unreachable from (1,1).

Consider all states $(1, i)$, for $1 < i \le n$, of $A \cap_c B$. Since $L(A)$ is suffix-free, the start state of A has no in-transitions. It implies that $(1, i)$ is not reachable from $(1, 1)$ in $A \cap_c B$ and, therefore, these states are useless as shown in Fig. 5. We can establish a similar result for the the states $(j, 1)$, for $1 < j \le m$.

Observation 2. *Given suffix-free minimal DFAs A and B, all states* $(1, i)$, *for* $1 < i \le m$, *and all states* $(j, 1)$, *for* $1 < j \le n$, *are useless in* $A \cap_c B$.

Once we minimize $A \cap_c B$ based on Observations 1 and 2, the resulting minimal DFA has $mn - 2(m + n) + 6$ states.

Theorem 4. *Given two suffix-free minimal DFAs A and B,* $mn - 2(m+n) + 6$ *states are necessary and sufficient in the worst-case for the minimal DFA of* $L(A) \cap L(B)$, *where* $|\Sigma| \ge 3$.

5.2 Union of Suffix-Free Regular Languages

We now investigate the union of two suffix-free regular languages. We compute the union DFA for $L(A)$ and $L(B)$ using the Cartesian product of states. Given two suffix-free minimal DFAs $A = (Q_1, \Sigma, \delta_1, s_1, F_1)$ and $B = (Q_2, \Sigma, \delta_2, s_2, F_2)$, let $M = (Q_1 \times Q_2, \Sigma, \delta, (s_1, s_2), F)$, where for all $p \in Q_1$ and $q \in Q_2$ and $a \in \Sigma$,

$$\delta((p, q), a) = (\delta(p, a), \delta(q, a))$$

and $F = (F_1 \times Q_2) \cup (Q_1 \times F_2)$. Then, $L(M) = L(A) \cup L(B)$ and M is deterministic. Let $A \cup_c B$ denote M. Consider the right language of a state (i, j) in $A \cup_c B$.

Proposition 4 (Han et al. [9]). *For a state (i, j) in $A \cup_c B$, the right language $L_{(i,j)}$ of (i, j) is the union of L_i in A and L_j in B.*

Note that the two constructions for $A \cap_c B$ and $A \cup_c B$ are different. This implies that we may not be able to apply the same approach that we used for $A \cap_c B$ for computing the upper bound for $L(A) \cup L(B)$. For example, since $L_{(n,j)} = L_n \cup L_j \neq \emptyset$ by Proposition 4, all states (m, i) and (j, n) for $1 \leq i \leq n$ and $1 \leq j \leq m$, in $A \cup_c B$ are not necessarily equivalent. Thus, these states cannot be merged. On the other hand, we observe that all states $(1, i)$ and $(j, 1)$, for $1 < i \leq n$ and $1 < j \leq m$, are useless since $L(A)$ and $L(B)$ are suffix-free. Therefore, we minimize $A \cup_c B$ by removing these $m + n - 2$ states.

Theorem 5. *Given two suffix-free minimal DFAs A and B, $mn - (m + n) + 2$ states are necessary and sufficient in the worst-case for the minimal DFA of $L(A) \cup L(B)$, where $|\Sigma| \geq 5$.*

6 Conclusion

The state complexity of an operation for regular languages is the number of states that are necessary and sufficient for the minimal DFA that accepts the language obtained from the operation. Yu et al. [24] studied the operational state complexity of general regular languages and Han et al. [9] examined the state complexity of basic operations on prefix-free regular languages. Since suffix-freeness is reversal of prefix-freeness, it was a natural problem to examine the state complexity of basic operations on suffix-free regular languages.

operation	regular languages	prefix-free case	suffix-free case
L_1^*	$2^{m-1} + 2^{m-2}$	m	$2^{m-2} + 1$
L_1^R	2^m	$2^{m-2} + 1$	$2^{m-2} + 1$
$L_1 \cdot L_2$	$(2m - 1)2^{n-1}$	$m + n - 2$	$(m - 1)2^{n-2} + 1$
$L_1 \cap L_2$	mn	$mn - 2(m + n) + 6$	$mn - 2(m + n) + 6$
$L_1 \cup L_2$	mn	$mn - 2$	$mn - (m + n) + 2$

Fig. 6. Operational state complexity of general, prefix-free and suffix-free regular languages

Based on the structural property that a suffix-free minimal DFA must be non-returning, we have tackled Kleene star, reversal, catenation, intersection and union cases and obtained the tight bound for each operation.

Fig. 6 shows the comparison table of the state complexity on regular languages, prefix-free regular languages and suffix-free regular languages. We have established the tight state complexity bounds for each of the operations using languages over a fixed alphabet. However, the constructions usually require an alphabet of size 3 or 4 and, then, for most operations, it is open whether or not the upper bound for the state complexity of each operation can be reached using a small size alphabet such as $|\Sigma| = 2$ or 3.

References

1. Berstel, J., Perrin, D.: Theory of Codes. Academic Press, Inc., London (1985)
2. Brzozowski, J.: A survey of regular expressions and their applications. IEEE Transactions on Electronic Computers 11, 324–335 (1962)
3. Câmpeanu, C., Culik II, K., Salomaa, K., Yu, S.: State complexity of basic operations on finite languages. In: Boldt, O., Jürgensen, H. (eds.) WIA 1999. LNCS, vol. 2214, pp. 60–70. Springer, Heidelberg (2001)
4. Câmpeanu, C., Salomaa, K., Yu, S.: Tight lower bound for the state complexity of shuffle of regular languages. Journal of Automata, Languages and Combinatorics 7(3), 303–310 (2002)
5. Domaratzki, M.: State complexity of proportional removals. Journal of Automata, Languages and Combinatorics 7(4), 455–468 (2002)
6. Domaratzki, M., Salomaa, K.: State complexity of shuffle on trajectories. Journal of Automata, Languages and Combinatorics 9(2-3), 217–232 (2004)
7. Han, Y.-S., Salomaa, K.: State complexity of basic operations on suffix-free regular languages. Technical Report Technical Report 2007-534, Queen's University (2007) `http://www.cs.queensu.ca/TechReports/Reports/2007-534.pdf`
8. Han, Y.-S., Salomaa, K.: State complexity of union and intersection of finite languages. In: Proceedings of DLT'07. LNCS, vol. 4588, pp. 217–228 (2007)
9. Han, Y.-S., Salomaa, K., Wood, D.: State complexity of prefix-free regular languages. In: Proceedings of DCFS'06, pp. 165–176 (2006) (Full version is submitted for publication)
10. Han, Y.-S., Wood, D.: The generalization of generalized automata: Expression automata. International Journal of Foundations of Computer Science 16(3), 499–510 (2005)
11. Holzer, M., Kutrib, M.: Unary language operations and their nondeterministic state complexity. In: Ito, M., Toyama, M. (eds.) DLT 2002. LNCS, vol. 2450, pp. 162–172. Springer, Heidelberg (2003)
12. Holzer, M., Kutrib, M.: Nondeterministic descriptional complexity of regular languages. International Journal of Foundations of Computer Science 14(6), 1087–1102 (2003)
13. Hopcroft, J., Ullman, J.: Introduction to Automata Theory, Languages, and Computation, 2nd edn. Addison-Wesley, Reading, MA (1979)
14. Hricko, M., Jirásková, G., Szabari, A.: Union and intersection of regular languages and descriptional complexity. In: Proceedings of DCFS'05, pp. 170–181 (2005)

15. Jirásek, J., Jirásková, G., Szabari, A.: State complexity of concatenation and complementation of regular languages. In: Domaratzki, M., Okhotin, A., Salomaa, K., Yu, S. (eds.) CIAA 2004. LNCS, vol. 3317, pp. 178–189. Springer, Heidelberg (2005)
16. Jürgensen, H., Konstantinidis, S.: Codes. In: Rozenberg, G., Salomaa, A. (eds.) Word, Language, Grammar. Handbook of Formal Languages, vol. 1, pp. 511–607. Springer, Heidelberg (1997)
17. Leiss, E.L.: Succint representation of regular languages by boolean automata. Theoretical Computer Science 13, 323–330 (1981)
18. Nicaud, C.: Average state complexity of operations on unary automata. In: Kutyłowski, M., Wierzbicki, T., Pacholski, L. (eds.) MFCS 1999. LNCS, vol. 1672, pp. 231–240. Springer, Heidelberg (1999)
19. Pighizzini, G., Shallit, J.: Unary language operations, state complexity and Jacobsthal's function. International Journal of Foundations of Computer Science 13(1), 145–159 (2002)
20. Salomaa, A., Wood, D., Yu, S.: On the state complexity of reversals of regular languages. Theoretical Computer Science 320(2-3), 315–329 (2004)
21. Wood, D.: Theory of Computation. John Wiley & Sons, Inc., New York, NY (1987)
22. Yu, S.: Regular languages. In: Rozenberg, G., Salomaa, A. (eds.) Word, Language, Grammar. Handbook of Formal Languages, vol. 1, pp. 41–110. Springer, Heidelberg (1997)
23. Yu, S.: State complexity of regular languages. Journal of Automata, Languages and Combinatorics 6(2), 221–234 (2001)
24. Yu, S., Zhuang, Q., Salomaa, K.: The state complexities of some basic operations on regular languages. Theoretical Computer Science 125(2), 315–328 (1994)

Exact Algorithms for $L(2,1)$-Labeling of Graphs

Jan Kratochvíl[1], Dieter Kratsch[2], and Mathieu Liedloff[2]

[1] Department of Applied Mathematics, and Institute for Theoretical Computer Science*, Charles University, Malostranské nám. 25, 118 00 Praha 1, Czech Republic
honza@kam.ms.mff.cuni.cz

[2] Laboratoire d'Informatique Théorique et Appliquée, Université Paul Verlaine - Metz, 57045 Metz Cedex 01, France
{kratsch,liedloff}@univ-metz.fr

Abstract. The notion of distance constrained graph labelings, motivated by the Frequency Assignment Problem, reads as follows: A mapping from the vertex set of a graph $G = (V, E)$ into an interval of integers $[0..k]$ is an $L(2,1)$-labeling of G of span k if any two adjacent vertices are mapped onto integers that are at least 2 apart, and every two vertices with a common neighbor are mapped onto distinct integers. It is known that for any fixed $k \geq 4$, deciding the existence of such a labeling is an NP-complete problem. We present exact exponential time algorithms that are faster than the naive $O((k+1)^n)$ algorithm that would try all possible mappings. The improvement is best seen in the first NP-complete case of $k = 4$ – here the running time of our algorithm is $O(1.3161^n)$.

1 Introduction

The Frequency Assignment Problem asks for assigning frequencies to transmitters in a broadcasting network with the aim of avoiding undesired interference. One of the graph theoretical models of FAP which is well elaborated is the notion of *distance constrained labeling* of graphs. An $L(2,1)$-labeling of a graph G is a mapping from the vertex set of G into nonnegative integers such that the labels assigned to adjacent vertices differ by at least 2, and labels assigned to vertices of distance 2 are different. The *span* of such a labeling is the maximum label used. In this model, the vertices of G represent the transmitters and the edges of G express which pairs of transmitters are too close to each other so that an undesired interference may occur, even if the frequencies assigned to them differ by 1. This model was introduced by Roberts [15] and since then the concept has been intensively studied. Undoubtedly, distance constrained graph labelings provide a graph invariant of significant theoretical interest. Let us mention a few of the known results and open problems: Griggs and Yeh [11] proved that determining the minimum possible span of G – denoted by $L_{2,1}(G)$ – is an NP-hard problem.

* Supported by Research grant 1M0021620808 of the Czech Ministry of Education.

L. Kučera and A. Kučera (Eds.): MFCS 2007, LNCS 4708, pp. 513–524, 2007.

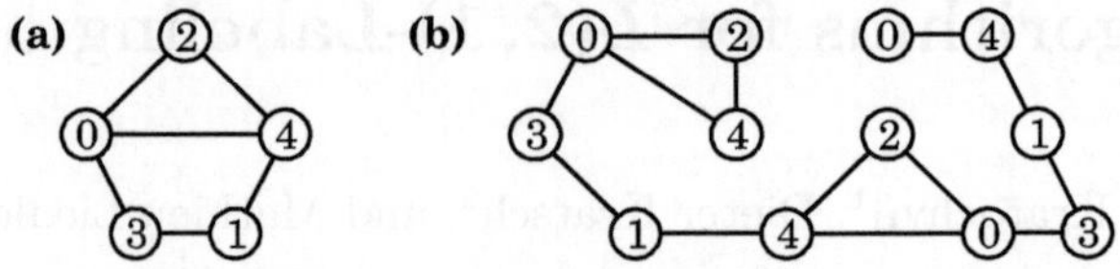

Fig. 1. (a) The graph $H = \overline{P_5}$. (b) A graph G with an $L(2,1)$-labeling of span 4 as a locally injective homomorphism into H.

Fiala et al. [4] later proved that deciding $L_{2,1}(G) \leq k$ remains NP-complete for every fixed $k \geq 4$, while Bodlaender et al. [1] proved NP-hardness for planar inputs for $k = 8$. (For $4 \leq k \leq 7$ and planar inputs, the complexity is still open.) When the span k is part of the input, the problem is nontrivial even for trees – though a polynomial time algorithm based on bipartite matching was presented in [2], existence of a linear time algorithm for trees is still open. Moreover, somewhat surprisingly, the problem becomes NP-complete for series-parallel graphs [3], and thus the $L(2,1)$-labeling problem belongs to a handful of problems known to separate graphs of tree-width 1 and 2 by P/NP-completeness dichotomy. From the structural point of view, Griggs and Yeh [11] conjectured that every graph of maximum degree Δ satisfies $L_{2,1}(G) \leq \Delta^2$. The so far best published upper bound $L_{2,1}(G) \leq \Delta^2 + \Delta - 2$ was proved by Gonçalves [10]. However, for $\Delta > 2$, the Moore graphs are the only graphs known to require span Δ^2, and it is an open problem if there are infinitely many graphs satisfying $L_{2,1}(G) > \Delta^2 - o(\Delta)$.

Generalizations have been considered, both in the direction of taking into account larger distances and in the direction of allowing a more complicated structure of the frequency space. Circular metric was considered by Leese et al. [13] and Liu et al. [14], showing that in certain sense the circular metric is easier than the linear one (e.g., the circular span of a tree is uniquely determined by its maximum degree and can thus be determined in linear time). Fiala and Kratochvíl consider in [5] the utmost generalization for the case when the metric in the frequency space can be described by a graph, say H. They define the notion of an $H(2,1)$-labeling of G, which is a mapping from the vertex set of G into the vertex set of H such that vertices adjacent in G are mapped onto nonadjacent (distinct) vertices of H, and vertices with a common neighbor (in G) are mapped onto distinct vertices of H. They also show that $H(2,1)$-labelings are exactly locally injective homomorphisms from G to $\overline{H}$, the complement of H. In particular, an $L(2,1)$-labeling of span k is a locally injective homomorphism into the complement of the path of length k. (The complement of the path of length 4 is depicted in Figure 1.) The complexity of locally injective homomorphisms was considered in [5,6,7] where a number of NP-complete cases were identified, but the complete characterization is still open.

The goal of this paper is to explore exact exponential time algorithms for the $L(2,1)$-labeling problem of fixed span. Since one cannot hope for polynomial time algorithms (unless P = NP), our aim is to design algorithms with running time

$O^*(c^n)$ and minimizing the constant c.[1] First we show that it is not difficult to beat the trivial bound $c \leq k+1$ (which follows from merely checking all possible mappings from $V(G)$ into $\{0, 1, \ldots, k\}$) by presenting an algorithm of running time $O^*((k-1)^n)$ which can also be generalized to the $H(2,1)$-labeling problem. Then we refine the branching algorithm for the case of span 4 to achieve an algorithm of running time $O^*(1.3161^n)$ (which beats not only $c = k - 1 = 3$, but also $k - 2 = 2$). Finally, a running time of $O^*(1.3006^n)$ of the same algorithm is established using a refined analysis.

Throughout the paper we consider finite undirected graphs without multiple edges or loops. The vertex set (edge set) of a graph G is denoted by $V(G)$ ($E(G)$, respectively). The open neighborhood of a vertex u in G is denoted by $N_G(u)$. The symbol n is reserved for the number of vertices of the input graph, which will always be denoted by G.

2 Exact Algorithm for Locally Injective Homomorphisms

A *graph homomorphism* is an edge preserving vertex mapping between two graphs. More formally, a mapping $f : V(G) \to V(H)$ is a homomorphism from G to H if $f(u)f(v) \in E(H)$ whenever $uv \in V(G)$. Such a mapping is sometimes referred to as an H-*coloring* of G since homomorphisms provide a generalization of the concept of graph colorings – k-colorings of G are exactly homomorphisms from G to the complete graph K_k. Hell and Nešetřil proved that from the computational complexity point of view, homomorphisms admit a complete dichotomy – deciding existence of a homomorphism into a fixed graph H is polynomial when H is bipartite and NP-complete otherwise [12]. The study of exact algorithms for graph homomorphisms was initiated in [9].

A homomorphism $f : G \to H$ is called *locally injective* if for every vertex $u \in V(G)$, its neighborhood is mapped injectively into the neighborhood of $f(u)$ in H, i.e., if every two vertices with a common neighbor in G are mapped onto distinct vertices in H. When deciding the existence of a locally injective homomorphism, one might try to utilize the known algorithms that list all possible homomorphisms and then check if any of them is locally injective. It is not surprising that using the local injectivity, one can often do much better. The known algorithms for H-homomorphism relate the base of the exponential function that expresses the running time to the number of vertices of H [9], while we prove in Theorem 1 that H-locally-injective-homomorphism can be solved in time $O^*((\Delta(H) - 1)^n)$, where $\Delta(H)$ is the maximum degree of a vertex of H. This, in most cases considerable, speed-up is achieved when we label the vertices consecutively using the fact that a neighbor of an already labeled vertex has only a limited number of candidate labels.

Without loss of generality we may assume that G is a connected graph, since otherwise we solve the problem on each connected component of G separately.

[1] Here we use the so called O^* notation: $f(n) = O^*(g(n))$ if $f(n) \leq p(n) \cdot g(n)$ for some polynomial $p(n)$.

In the algorithm, f denotes a partial labeling of the vertices of G by vertices of H which is a candidate for a locally injective homomorphism from G to H.

Algorithm-H-LIH(G)

```
if ∃v ∈ V(G) s.t. v is unlabeled and v has at least one neighbor u which
was already labeled then
    foreach c ∈ N_H(f(u)) \ f(N_G(u)) do
        set f(v) = c
        Algorithm-H-LIH(G)
else
    if ∃u ∈ V(G) s.t. u is unlabeled then
        foreach c ∈ V(H) do
            set f(u) = c
            Algorithm-H-LIH(G)
    else
        if the labeling f is a locally injective homomorphism from G to H
        then
            return the labeling
```

Theorem 1. *The H-Locally-Injective-Homomorphism problem is solved in time $O^*((\Delta(H)-1)^n)$ by Algorithm-H-LIH.*

Proof. In the first step the algorithm picks an unlabeled vertex, say u, and labels it in $|V(H)|$ ways. In the second step, the first rule is used and a neighbor v of u is labeled in $\deg_H(f(u)) \leq \Delta(H)$ ways. From this time on, the algorithm branches each time into at most $\Delta(H)-1$ ways. To see this, let $T = (V(G), E(T))$ be an auxiliary graph which contains the edges uv from the application of the first rule. The loop invariant of the algorithm is that T is an acyclic graph consisting of one connected component – containing the so far labeled vertices – and remaining isolated (and unlabeled) vertices. Also, $f(u)f(v) \in E(H)$ for every edge $uv \in E(T)$. From the third round on, the first rule is always applied, and one edge uv is added to T. And since u had another neighbor w in T, $f(w) \in N_H(f(u))$ and so $N_H(f(u)) \setminus f(N_G(u))$ has at most $\Delta(H)-1$ available labels for v. □

Corollary 1. *The $L(2,1)$-labeling problem of span k can be decided in time $O^*((k-1)^n)$. Hence, $L(2,1)$-labeling of span 4 can be solved in time $O^*(3^n)$.*

Proof. The maximum degree of a vertex in the complement of the path of length k is $\Delta(\overline{P_{k+1}}) = k-1$. □

3 A Branching Algorithm for $L(2,1)$-Labeling of Span 4

In this section we present a significantly faster algorithm for the case of $L(2,1)$-labeling of span 4. The main idea is the same as for Algorithm-H-LIH – in the first two steps we label two adjacent vertices in all possible (i.e., at most 12) ways. Then we keep labeling the vertices one by one (and branching into several possibilities when necessary) so that the so far labeled part of the input graph G remains connected. It follows that every newly labeled vertex has (at least) one labeled neighbor, and this labeled neighbor has another (at least one) labeled neighbor. The key idea of the speed-up is two-fold. First, we list several rules and apply them in order of their preferences, thus aiming at reducing the number of branching steps. Secondly, we often label several vertices at a time which leads to a more convenient recurrence for the upper bound of the running time.

Before we describe the details of the algorithm, we introduce a technical definition and a lemma. Throughout this section, we assume that we are in the middle of a run of our algorithm and that $f : X \to \{0, 1, 2, 3, 4\}$ is a partial $L(2,1)$-labeling of G such that the labeled vertices $X \subseteq V(G)$ induce a connected subgraph. Note that in order to have a chance to admit a valid labeling, G must have maximum degree at most 3. It is also clear that every vertex of degree 3 must be labeled by 0 or 4, and we will keep checking that this condition is satisfied by each candidate labeling f. To avoid trivial cases we assume that G has at least one vertex of degree 3.

Definition 1. *A path in G is called an* extension path *if all inner vertices are unlabeled and of degree 2, at least one endpoint is labeled and the unlabeled endpoint (if there is one) has either degree 1 or 3. (With a slight abuse of notation we allow that the endpoints are the same vertex, so such an extension path is in fact a cycle and the endpoint is labeled).*

Lemma 1. *Let $P = v_0 v_1 \dots v_k$ be an extension path such that v_0 is labeled and v_k has degree 1 (and is unlabeled). Let $G' = G[V(G) \setminus \{v_1, \dots, v_k\}]$ be the subgraph obtained by deleting the path P and let $f' : V(G') \to \{0, 1, 2, 3, 4\}$ be a valid extension of f to an $L(2,1)$-labeling of G'. Then f' can be extended to an $L(2,1)$-labeling of the entire graph G.*

Proof. Since $\deg(v_0) \leq 3$ and $f'(v_0) \in \{0, 4\}$ if $deg(v_0) = 3$, there is always a label, say ℓ, available for v_1. The edge $f'(v_0)\ell$ belongs to a cycle in $\overline{P_5}$, and we label the path P wrapping around this cycle. □

Now we describe the rules and discuss their effect on the running time. When a rule is applied to a partially labeled graph it has at least two labeled vertices and its labeled vertices induce a connected subgraph. Note that only Rules 4 and 5 use branchings.

Rule 1 - Forced Extensions

- If u is an unlabeled vertex whose labeled neighbor v has two labeled neighbors, then the possible label of u is uniquely determined by the labels of v and its neighbors;
- if u is an unlabeled vertex with a neighbor v labeled by 1, 2 or 3, then, since v has another labeled neighbor, the label of u is uniquely determined by the labels of v and its neighbor;

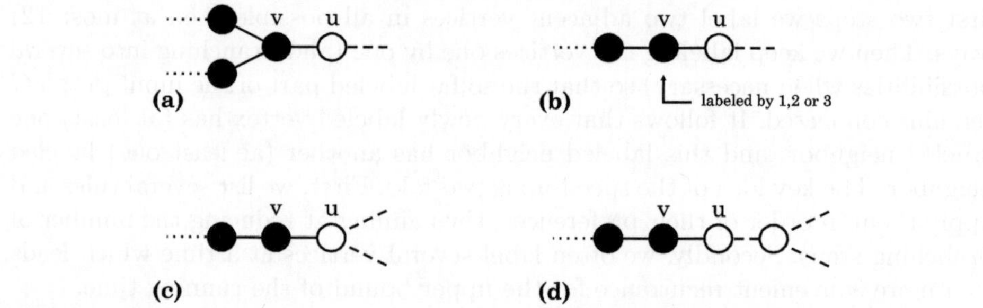

Fig. 2. Forced extensions. (a) An unlabeled vertex u whose labeled neighbor v has two labeled neighbors. (b) An unlabeled vertex u with a neighbor v having label 1,2 or 3. (c) An unlabeled vertex u of degree 3 with a labeled neighbor v. (d) An unlabeled vertex u of degree 2 with one labeled neighbor v and one (possibly unlabeled) neighbor of degree 3.

- if u is an unlabeled vertex of degree 3 with a labeled neighbor v, then the label of u is either 0 or 4 and is uniquely determined by the label of v and its other labeled neighbor;
- if u is an unlabeled vertex of degree 2 such that one of its neighbors is labeled and the other one is a (possibly unlabeled) degree 3 vertex, then the label of u is uniquely determined by the labels of its neighbor(s).

Note that if Rule 1 cannot be applied, every unlabeled vertex that is adjacent to a labeled one has degree at most 2 and each of its adjacent labeled vertices is labeled by 0 or 4.

Rule 2 - Easy Extension

- If P is an extension path with one endpoint of degree 1, Lemma 1 says that the unlabeled vertices of P are irrelevant – we delete them from G and continue with the reduced graph.

If neither Rule 1 nor Rule 2 can be applied, every unlabeled vertex that is adjacent to a labeled one has degree 2.

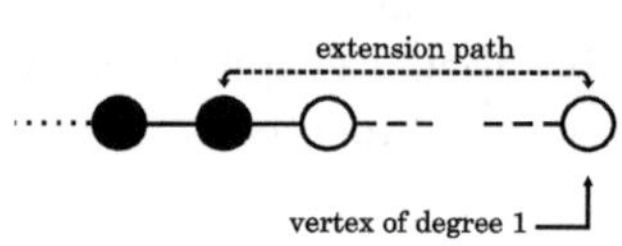

Fig. 3. Easy extension. An extension path with one unlabeled endpoint of degree 1.

Rule 3 - Cheap Extensions

- If P is an extension path with both endpoints labeled and of degree 2, we can decide by dynamic programming whether P has an $L(2,1)$-labeling

compatible with the labeling of the labeled neighbors of its endpoints. In the affirmative case we just delete the unlabeled vertices and continue with the reduced graph, otherwise we reject the current f as allowing no extension.
- If P is an extension path with identical endpoints, we again decide by dynamic programming if the path has an $L(2,1)$-labeling compatible with the label of the endpoint and its labeled neighbor. And we either reduce G or reject f, depending on the outcome.

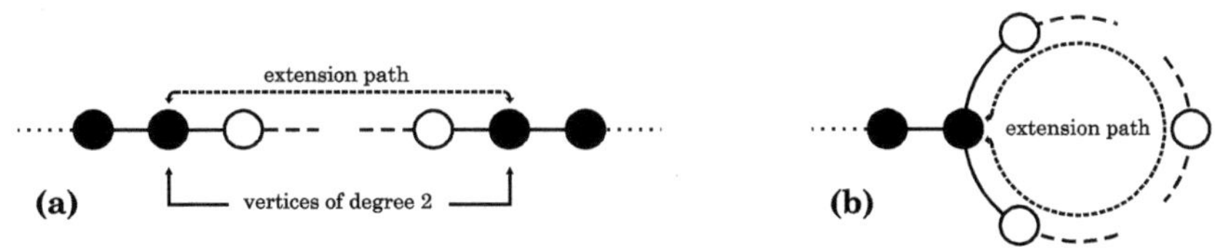

Fig. 4. Cheap extensions. (a) An extension path with both endpoints labeled and of degree 2. (b) An extension path with indentical endpoints.

The dynamic programming consumes only time polynomial in the length of the path, and so does not affect the base of the exponential function bounding the running time. (To be honest, it only consumes constant time – it can be shown by case analysis that if the path is long enough, then any combination of labelings of its terminal edges is feasible, and so the dynamic programming is only applied to short paths of constant length).

Rule 4 - Extensions with Strong Constraints

- Let P be an extension path with both endpoints labeled, each with 0 or 4, such that each endpoint has only one labeled neighbor and at least one of them has another unlabeled neighbor that does not belong to P. In this case we branch along possible labelings of the (at most 4) unlabeled neighbors of the endpoints of P, while extending each of these labelings to entire P (by dynamic programming approach).

Now we discuss the details of this branching rule and the consequences for the running time. The illustrative Fig. 5 (a) will be helpful. Let b and c be the labeled endpoints of P, b of degree 3 and c of degree 2 or 3, and let a and d, respectively,

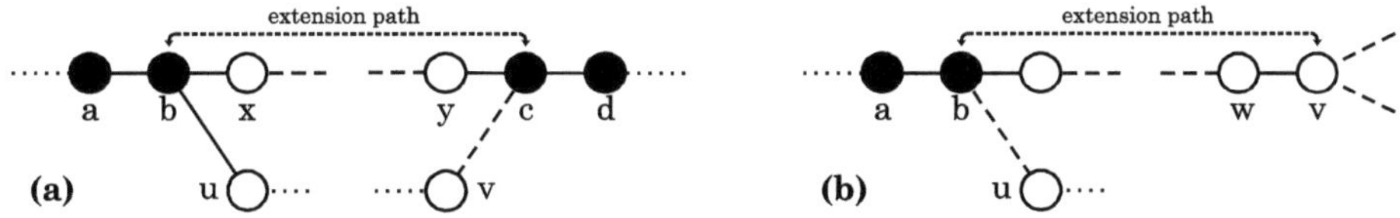

Fig. 5. Extensions with strong (a) and weak (b) constraints

be their labeled neighbors. Let further x and y be the unlabeled neighbors of b and c, respectively, on the path P, and let $u \neq x$ be the other unlabeled neighbor of b; and let $v \neq y$ be the other unlabeled neighbor of c, if it exists.

Length 1. If the length of P (measured by the number of unlabeled vertices) is 1, then the label of $x = y$ is uniquely determined (in fact, it has to be 2), and we do not really branch.

Length 2. If the length of P is 2, we have the following possible labelings of $ab..cd$ and their extensions to $abxycd$ (up to the symmetric labeling $f' = 4 - f$):

$$\begin{array}{lll} 40xy40 \rightarrow 403140 & 40xy42 \rightarrow 403142 & 40xy02 \rightarrow 402402 \\ 40xy03 \rightarrow 402403 & 20xy03 \rightarrow 204203 & 20xy04 \rightarrow 204204 \\ 20xy40 \rightarrow 203140 & 20xy42 \rightarrow 203142 & 30xy02 \rightarrow 302402 \\ 30xy04 \rightarrow 304204 & & 30xy03 \rightarrow 302403, 304203. \end{array}$$

We see that most cases allow only one extension of the labeling to P, except for the last case, where branching into two cases occurs. If this happens, we gain at least three newly labeled vertices (x, y, u and possibly also v).

To analyze the running time of our algorithm we determine an upper bound on the maximum number $T(n)$ of leaves in the search tree corresponding to an execution of the algorithm on an input with n unlabeled vertices. The overall running time will then be $O^*(T(n))$ since the application of every rule takes only polynomial time and reduces the number of unlabeled vertices by at least one.

From our above analysis for Rule 4, we obtain two recurrences: $T(n) = 2T(n-3)$ (if c does not have another unlabeled neighbor v or if $v = u$) and $T(n) = 2T(n-4)$ (if the neighbor v exists and is distinct from u). The solution of a recurrence $T(n) = \alpha T(n-\beta)$ is of the form $\Theta(c^n)$ for $c = \sqrt[\beta]{\alpha}$. Here we obtain $c = \sqrt[3]{2}$ for the first recurrence and $c = \sqrt[4]{2}$ for the second one.

Length 3. The maximum number of possible extensions of the labelings of an extension path P of length 3 can be established from exhaustive search trees giving all $L(2,1)$-labelings.

Table 1 summarizes the numbers of extensions and corresponding upper bounds for $T(n)$ for extension paths of lengths 1 to 3.

Length at least 4. If the path is longer, we have two possible extensions of the labeling to the vertices x and u, and two extensions to y and v. For each of these 4 cases we check (in polynomial time, by dynamic programming) if it extends to a labeling of P. If v exists and $v \neq u$, we gain length$(P) + 2$ newly labeled vertices (the unlabeled vertices of P plus u and v), which leads to the recurrence $T(n) = 4T(n - \text{length}(P) - 2)$ and $T(n) = O(4^{\frac{n}{6}}) = O(1.2600^n)$.

If v does not exist, it may seem to mind that we only gain length$(P) + 1$ newly labeled vertices at the same cost of branching. However, in this case we only consider two possible labelings of the pair x and u, and for each of them we only check if it extends to a labeling of P or not. The actual label of y is irrelevant since c has degree 2 in this case. This leads to the recurrence $T(n) = 2T(n - \text{length}(P) - 1)$ and $T(n) = O(2^{\frac{n}{5}}) = O(1.1487^n)$.

If $v = u$ then u would be treated by Rule 1 since u is adjacent to c and in that case c has degree 3. Thus $v = u$ is not possible when applying Rule 4.

Comparing all cases we see that the worst case is achieved when $\deg(c) = 2$ and $\text{length}(P) = 2$. Thus using any branching of Rule 4 leads to $T(n) = O(1.2600^n)$.

Table 1. Branching on extension paths of length ≤ 3 with strong constraints

length l of the path P	maximum number of branchings t_1 if $\deg(c) = 2$	solution of the recurrence $T(n) = t_1T(n-l-1)$	maximum number of branchings t_2 if $\deg(c) = 3$	solution of the recurrence $T(n) = t_2T(n-l-2)$
1	1	no branching	1	no branching
2	2	$O(2^{\frac{n}{3}}) = O(1.2600^n)$	2	$O(2^{\frac{n}{4}}) = O(1.1893^n)$
3	2	$O(2^{\frac{n}{4}}) = O(1.1893^n)$	2	$O(2^{\frac{n}{5}}) = O(1.1487^n)$

If none of Rules 1-4 can be applied, then every unlabeled vertex that is adjacent to a labeled one belongs to an extension path with one unlabeled endpoint of degree 3. This is treated by the last branching rule.

Rule 5 - Extensions with Weak Constraints

- Let P be an extension path with one unlabeled endpoint v of degree 3. Let w be the neighbor of v in P, let the labeled endpoint of P be b, let its labeled neighbor be a and let u be (if it exists) the unlabeled neighbor of b not belonging to P (see Fig. 5 (b)). In this case we branch along possible labelings of v, w and (possibly) u, while extending each of these labelings to entire P (by dynamic programming).

The table below summarizes the numbers of branchings for paths of length at most 8 (these numbers can be established from exhaustive search trees giving all such possible $L(2,1)$-labelings) Again, when $\deg(b) = 2$, we only count the number of labelings of v and w that extend to a labeling of P compatible with the labeling of a and b, since the actual label used on the neighbor of b in P is irrelevant. On the other hand, when $\deg(b) = 3$, we count the number of labelings of v, w and u that allow an extension to a labeling of P compatible with the labels of a and b. Note that in either case both v and b may only receive labels 0 or 4.

In the case of a longer path, we have at most 6 possible labelings of v and w, yielding the recurrence $T(n) = 6T(n - \text{length}(P))$ if $\deg(b) = 2$. If $\deg(b) = 3$, we have at most 12 possible labelings of v, w and u, yielding the recurrence $T(n) = 12T(n - \text{length}(P) - 1)$.

Since $\sqrt[8]{6} < \sqrt[4]{3}$ and $\sqrt[9]{12} < \sqrt[4]{3}$, the overall worst case for Rule 5 is achieved when b is a degree 2 vertex and the extension path has length 4. Hence $T(n) = O(1.3161^n)$.

Summarizing the analysis of the algorithm, we obtain

Theorem 2. *The existence of an $L(2,1)$-labeling of span 4 can be decided in time $O^*(1.3161^n)$. If such a labeling exists it can be computed within the same time.*

Table 2. Branching on extension paths of lengths ≤ 8 with weak constraints

length l of the path P	maximum number of branchings t_1 if $\deg(b) = 2$	solution of the recurrence $T(n) = t_1 T(n-l)$	maximum number of branchings t_2 if $\deg(b) = 3$	solution of the recurrence $T(n) = t_2 T(n-l-1)$
1	1	no branching	1	no branching
2	1	no branching	1	no branching
3	2	$O(2^{\frac{n}{3}}) = O(1.2600^n)$	2	$O(2^{\frac{n}{4}}) = O(1.1893^n)$
4	3	$O(3^{\frac{n}{4}}) = O(1.3161^n)$	3	$O(3^{\frac{n}{5}}) = O(1.2458^n)$
5	3	$O(3^{\frac{n}{5}}) = O(1.2458^n)$	3	$O(3^{\frac{n}{6}}) = O(1.2010^n)$
6	5	$O(5^{\frac{n}{6}}) = O(1.3077^n)$	6	$O(6^{\frac{n}{7}}) = O(1.2918^n)$
7	5	$O(5^{\frac{n}{7}}) = O(1.2585^n)$	6	$O(6^{\frac{n}{8}}) = O(1.2511^n)$
8	5	$O(5^{\frac{n}{8}}) = O(1.2229^n)$	7	$O(7^{\frac{n}{9}}) = O(1.2414^n)$

4 A Refined Time Analysis

In this section we report on an attempt to improve upon the upper bound of $O^*(1.3161^n)$ for the running time of our algorithm. To do this we use a Measure & Conquer approach (see e.g. [8]). To each graph G with a partial labeling f we assign the following measure

$$\mu = \mu(G, f) = \tilde{n} + \epsilon \hat{n}$$

where $\tilde{n}$ is the number of unlabeled vertices with no labeled neighbor and $\hat{n}$ is the number of unlabled vertices having a labeled neighbor. Furthermore, ϵ is a constant to be chosen later such that $0 \leq \epsilon \leq 1$.

This means that the weight of a vertex is 0 if it is already labeled, it is ϵ if it is unlabeled with a labeled neighbor, and it is 1 otherwise. Note that $\mu(G, f) \leq n$, where n is the number of vertices of G.

Theorem 3. *The existence of an $L(2,1)$-labeling of span 4 can be decided in time $O^*(1.3006^n)$.*

Proof. Let $G = (V, E)$ be a graph with a partial labeling f. We consider the measure $\mu = \mu(G, f)$ for analysing the running time of the algorithm presented in Section 3. The analysis of the running time is quite similar to the one provided in Section 3, but since the measure involves the use of different weights (i.e., 1 or ϵ) depending on the status –labeled or unlabeled– of the vertices and their neighborhood, we obtain new recurrences. To simplify the notation, given a vertex v we denote by $w(v)$ the weight of v. Namely $w(v) = 1$ for each unlabeled vertex v with no labeled neighbor, $w(v) = \epsilon$ for each unlabled vertex v with a labeled neighbor and $w(v) = 0$ for each labeled vertex v. Thus summing over all vertices of G the equality $\mu(G, f) = \sum_{v \in V} w(v)$ holds.

Rule 1, 2 and 3. The application of rules 1,2 and 3 needs only a polynomial time and cannot increase the measure μ.

Rule 4 - Extensions with strong constraints. We consider an extension path P with both endpoints labeled and we branch on the possible labelings of the unlabeled neighbors of the endpoints of P (see Section 3 and Figure 6 (a)).

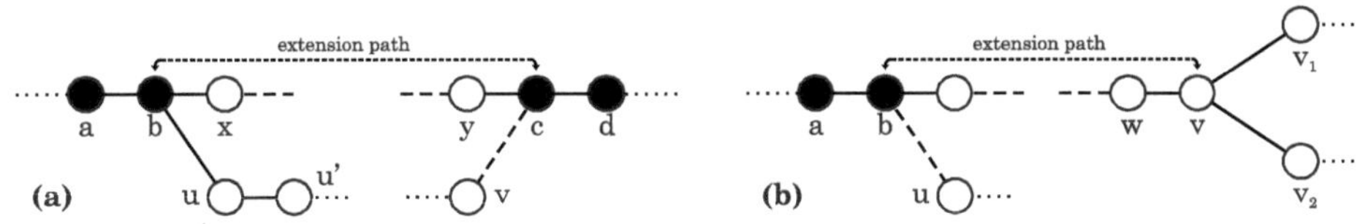

Fig. 6. Extensions with strong (a) and weak (b) constraints

Note that by application of Rule 1 and Rule 2, the degrees of u and v (if it exists) are precisely 2. Let u' be the unlabeled neighbor of u. The weight $w(u')$ can be either equal to 1 or equal to ϵ. We distinguish two cases:

- If $w(u') = 1$, labeling u would decrease the weight of u' to ϵ.
- If $w(u') = \epsilon$ then denote by u'' a labeled neighbor of u'. Due to Rule 1, it follows that u' has degree 2 and thus labeling u would create an extension path $P' = uu'u''$ of length one (u' is the unlabeled vertex of P') that can be labeled without any branching by Rule 4 (see analysis of Rule 4 in Section 3). Thus, labeling u would decrease the weight of u' to 0.

If v exists, we can assume that u and v are different since otherwise if $u = v$, Rule 1 would label u. However, in the case that v exists it is possible that $u' = v$. Consequently, labeling the path P and the vertices u and v would decrease the measure by at least $(2\epsilon + (\text{length}(P) - 2) + 2\epsilon)$ if v exists, and by at least $(2\epsilon + (\text{length}(P) - 2) + \epsilon + \min(1 - \epsilon, \epsilon))$ otherwise. We recall that length(P) is the number of unlabeled vertices of P.

Rule 5 - Extensions with weak constraints. We consider an extension path P with one unlabeled endpoint of degree 3 and we branch on the possible labelings of v, w and (possibly) u (see Section 3 and Figure 6 (b)).

Let v_1 and v_2 be the two neighbors of v which do not belong to P. Note that due to Rule 1, neither v_1 nor v_2 are labeled or adjacent to a labeled vertex.

If u exists, we can assume that u and v_i, $i \in \{1, 2\}$, are different since otherwise if $u = v_i$, Rule 1 would label u. Thus, labeling the path P and the vertex u would decrease the measure by at least $(\epsilon + (\text{length}(P) - 1) + 2 - 2\epsilon + \epsilon)$ if u exists, and by at least $(\epsilon + (\text{length}(P) - 1) + 2 - 2\epsilon)$ otherwise.

Setting $\epsilon = 0.8190$ and solving the corresponding recurrences establish a running time bounded by $O(1.3006^n)$. □

It is an interesting question whether a more clever choice of the measure can help to prove a more significant improvement of the running time of the algorithm.

References

1. Bodlaender, H.L., Kloks, T., Tan, R.B., van Leeuwen, J.: Approximations for lambda-Colorings of Graphs. Computer Journal 47, 193–204 (2004)
2. Chang, G.J., Kuo, D.: The $L(2,1)$-labeling problem on graphs. SIAM Journal of Discrete Mathematics 9, 309–316 (1996)
3. Fiala, J., Golovach, P., Kratochvíl, J.: Distance Constrained Labelings of Graphs of Bounded Treewidth. In: Caires, L., Italiano, G.F., Monteiro, L., Palamidessi, C., Yung, M. (eds.) ICALP 2005. LNCS, vol. 3580, pp. 360–372. Springer, Heidelberg (2005)
4. Fiala, J., Kloks, T., Kratochvíl, J.: Fixed-parameter complexity of λ-labelings. Discrete Applied Mathematics 113, 59–72 (2001)
5. Fiala, J., Kratochvíl, J.: Complexity of partial covers of graphs. In: Eades, P., Takaoka, T. (eds.) ISAAC 2001. LNCS, vol. 2223, pp. 537–549. Springer, Heidelberg (2001)
6. Fiala, J., Kratochvíl, J.: Partial covers of graphs. Mathematicae Graph Theory 22, 89–99 (2002)
7. Fiala, J., Kratochvíl, J., Pór, A.: On the computational complexity of partial covers of theta graphs. Electronic Notes in Discrete Mathematics 19, 79–85 (2005)
8. Fomin, F., Grandoni, F., Kratsch, D.: Measure and conquer: Domination - A case study. In: Caires, L., Italiano, G.F., Monteiro, L., Palamidessi, C., Yung, M. (eds.) ICALP 2005. LNCS, vol. 3580, pp. 192–203. Springer, Heidelberg (2005)
9. Fomin, F., Heggernes, P., Kratsch, D.: Exact algorithms for graph homomorphisms. In: Liśkiewicz, M., Reischuk, R. (eds.) FCT 2005. LNCS, vol. 3623, pp. 161–171. Springer, Heidelberg (2005)
10. Gonçalves, D.: On the $L(p,1)$-labelling of graphs. In: DMTCS Proceedings, vol. AE, pp. 81–86
11. Griggs, J.R., Yeh, R.K.: Labelling graphs with a condition at distance 2. SIAM Journal of Discrete Mathematics 5, 586–595 (1992)
12. Hell, P., Nešetřil, J.: On the complexity of H-colouring. Journal of Combinatorial Theory Series B 48, 92–110 (1990)
13. Leese, R.A., Noble, S.D.: Cyclic labellings with constraints at two distances. Electronic Journal of Combinatorics, Research paper 16, 11, (2004)
14. Liu, D., Zhu, X.: Circular Distance Two Labelings and Circular Chromatic Numbers. Ars Combinatoria 69, 177–183 (2003)
15. Roberts, F.S.: Private communication to J. Griggs

On (k, ℓ)-Leaf Powers

Andreas Brandstädt and Peter Wagner*

Institut für Informatik, Universität Rostock, D-18051 Rostock, Germany
{ab,Peter.Wagner}@informatik.uni-rostock.de

Abstract. We say that, for $k \geq 2$ and $\ell > k$, a tree T is a *(k, ℓ)-leaf root of a graph* $G = (V_G, E_G)$ if V_G is the set of leaves of T, for all edges $xy \in E_G$, the distance $d_T(x, y)$ in T is at most k and, for all non-edges $xy \notin E_G$, $d_T(x, y)$ is at least ℓ. A graph G is a *(k, ℓ)-leaf power* if it has a (k, ℓ)-leaf root. This new notion modifies the concept of k-leaf power which was introduced and studied by Nishimura, Ragde and Thilikos motivated by the search for underlying phylogenetic trees. Recently, a lot of work has been done on k-leaf powers and roots as well as on their variants phylogenetic roots and Steiner roots. For $k = 3$ and $k = 4$, structural characterisations and linear time recognition algorithms of k-leaf powers are known, and, recently, a polynomial time recognition of 5-leaf powers was given. For larger k, the recognition problem is open.

We give structural characterisations of (k, ℓ)-leaf powers, for some k and ℓ, which also imply an efficient recognition of these classes, and in this way we also improve and extend a recent paper by Kennedy, Lin and Yan on strictly chordal graphs and leaf powers.

Keywords: (k, ℓ)-leaf powers, leaf powers, leaf roots, strictly chordal graphs, linear time algorithms.

1 Introduction

Nishimura, Ragde and Thilikos [20] introduced the notion of k-leaf power and k-leaf root, motivated by the following: "... a fundamental problem in computational biology is the reconstruction of the *phylogeny*, or evolutionary history, of a set of species or genes, typically represented as a *phylogenetic tree* ...". The species occur as leaves of the phylogenetic tree.

Let $G = (V_G, E_G)$ be a finite undirected graph. For $k \geq 2$, a tree T is a *k-leaf root* of G if V_G is the leaf set of T and two vertices $x, y \in V_G$ are adjacent in G if and only if their distance $d_T(x, y)$ in T is at most k, i.e., $xy \in E_G \iff d_T(x, y) \leq k$. The graph G is a *k-leaf power* if it has a k-leaf root. Obviously, a graph is a 2-leaf power if and only if it is the disjoint union of cliques, i.e., it contains no induced P_3. See [17] for the related notions of phylogenetic root and Steiner root and [2,3,4,6,9,11,14,15,16,21] for recent work on leaf powers (including characterisations of 3- and 4-leaf powers) and their variants. It is

* Supported by DFG research grant BR 2479/7-1.

L. Kučera and A. Kučera (Eds.): MFCS 2007, LNCS 4708, pp. 525–535, 2007.

well known that every k-leaf power is a strongly chordal graph. For $k \geq 6$, no characterisation of k-leaf powers and no efficient recognition is known.

In this paper, we modify the notion of k-leaf power and k-leaf root in the following natural way: For $k \geq 2$ and $\ell > k$, a tree T is a *(k, ℓ)-leaf root of a graph* $G = (V_G, E_G)$ if V_G is the set of leaves of T, for all edges $xy \in E_G$, $d_T(x, y) \leq k$ and, for all non-edges $xy \notin E_G$, $d_T(x, y) \geq \ell$. A graph G is a *(k, ℓ)-leaf power* if it has a (k, ℓ)-leaf root. Thus, every k-leaf power is a $(k, k+1)$-leaf power, and every (k, ℓ)-leaf power is an (i, j)-leaf power, for all i and j with $k \leq i < j \leq \ell$. In particular, every (k, ℓ)-leaf power is simultaneously a k'-leaf power, for all k' with $k \leq k' \leq \ell - 1$. In a similar way, Steiner roots and powers can be generalised.

In [16], Kennedy, Lin and Yan study so-called strictly chordal graphs which were originally defined via (rather complicated) hypergraph properties but finally turn out to be exactly the (dart,gem)-free chordal graphs [14]. It is not hard to see that a graph is (dart,gem)-free chordal if and only if it results from substituting cliques into the vertices of a block graph (i.e., a graph whose blocks are cliques). We will show that these graphs are exactly the $(4, 6)$-leaf powers which explains various of their properties, and the same class appears as (k, ℓ)-leaf powers for infinitely many other values of k and ℓ, e.g., as $(6, 10)$-leaf powers and in general as $(4 + 2i, 6 + 4i)$-leaf powers, for $i \geq 0$. Moreover, it is known from [2,11,21] that 3-leaf powers (i.e., $(3, 4)$-leaf powers) are exactly the (bull,dart,gem)-free chordal graphs which in turn result from substituting cliques into the vertices of a tree. By a simple argument, every class of $(3 + 2i, 4 + 4i)$-leaf powers, for $i \geq 0$, is also exactly the same class of (bull,dart,gem)-free chordal graphs.

We give structural characterisations of (k, ℓ)-leaf powers, for some k and ℓ (and in particular for (8,11)-leaf powers) which also imply efficient recognition of these classes. Most of the proofs are omitted due to space constraints of this extended abstract.

2 Basic Notions and Results

Throughout this paper, let $G = (V, E)$ be a finite undirected graph without loops and multiple edges and with vertex set V and edge set E, and let $|V| = n$, $|E| = m$. For a vertex $v \in V$, let $N_G(v) = N(v) = \{u \mid uv \in E\}$ denote the *(open) neighbourhood* of v in G, and let $N_G[v] = N[v] = \{v\} \cup \{u \mid uv \in E\}$ denote the *closed neighbourhood* of v in G. A *clique* is a set of vertices which are mutually adjacent. A *stable set* is a set of vertices which are mutually non-adjacent.

A vertex subset $U \subseteq V$ is a *module* in G if, for all $v \in V \setminus U$, either v is adjacent to all vertices of U or v is adjacent to none of them. A *clique module* in G is a module which induces a clique in G. A vertex $z \in V$ *distinguishes* $x, y \in V$ if $zx \in E$ and $zy \notin E$. Two vertices $x, y \in V$ are *true twins* in G if they have the same neighbors in G and are adjacent to each other.

Let $d_G(x, y)$ (or $d(x, y)$ for short if G is understood) be the length, i.e., number of edges, of a shortest path in G between x and y. Let $N_G^k(x) = \{y \mid d_G(x, y) = k\}$ and let $G^k = (V, E^k)$ with $xy \in E^k$ if and only if $d_G(x, y) \leq k$ denote the *k-th power of G*.

For $U \subseteq V$, let $G[U]$ denote the subgraph of G induced by U. Throughout this paper, all subgraphs are understood to be induced subgraphs. Let $\mathcal{F}$ denote a set of graphs. A graph G is *$\mathcal{F}$-free* if none of its induced subgraphs is in $\mathcal{F}$.

For $k \geq 1$, let P_k denote a chordless path with k vertices and $k-1$ edges, and, for $k \geq 3$, let C_k denote a chordless cycle with k vertices and k edges. A *diamond* (or $K_4 - e$, see Figure 1) consists of four vertices a, b, c, d and five edges ab, ac, bc, bd and cd.

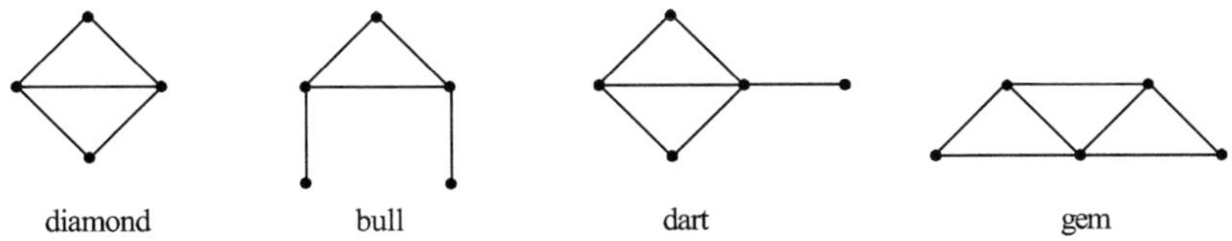

Fig. 1. Diamond, bull, dart and gem

For $k \geq 2$, let S_k denote the (*complete*) *sun* with $2k$ vertices $u_1, \ldots, u_k$ and $w_1, \ldots, w_k$ such that $u_1, \ldots, u_k$ is a clique, $w_1, \ldots, w_k$ is a stable set and, for $i \in \{1, \ldots, k\}$, w_i is adjacent to u_i and u_{i+1} (index arithmetic modulo k).

A graph is *chordal* if it contains no induced C_k, $k \geq 4$. A graph is *strongly chordal* if it is chordal and sun-free - see e.g. [5] for various characterisations of chordal and strongly chordal graphs. In particular, in a chordal graph G, the maximal cliques of G can be arranged in a tree T_G (a so-called *clique tree* of G) such that for every vertex v, the maximal cliques containing v form a subtree of T_G.

In [10,18,22], it is shown that the class of strongly chordal graphs is closed under powers:

Proposition 1 ([10,18,22]). *If G is strongly chordal then, for every $k \geq 1$, G^k is strongly chordal.*

A graph is a *block graph* if its 2-connected components (which are also called *blocks*) are cliques. It is well known that the following holds:

Proposition 2. *A graph G is a block graph if and only if G is diamond-free and chordal.*

In [17], the notion of k-th Steiner root of an undirected graph G, $k \geq 1$, is defined as follows: A tree $T = (V_T, E_T)$ is a *k-th Steiner root* of the graph $G = (V_G, E_G)$ if $V_G \subseteq V_T$ and $xy \in E_G$ if and only if $d_T(x, y) \leq k$. In this case, G is a *k-th Steiner power*. The vertices in $V_T \setminus V_G$ are called *Steiner nodes* in T.

In [6], we say that a graph G is a *basic k-leaf power* if G has a k-leaf root T such that no two leaves of T are attached to the same parent vertex in T (a so-called *basic k-leaf root*). Obviously, for $k \geq 2$, the set of leaves having the same parent node in T form a clique, and G is a k-leaf power if and only if G results from a basic k-leaf power by substituting cliques into its vertices. If T is a basic k-leaf root of G then T minus its leaves is a $(k-2)$-th Steiner root of G. Summarising, the following obvious equivalence holds:

Proposition 3. *For a graph G, the following conditions are equivalent for all $k \geq 2$:*

(i) *G has a k-th Steiner root.*
(ii) *G is an induced subgraph of the k-th power of a tree.*
(iii) *G is a basic $(k+2)$-leaf power.*

Similar to basic k-leaf roots, we say that a (k,l)-leaf root T is *basic* if no two leaves x and y of T have a distance satisfying $2 \leq d_T(x,y) \leq l-k+1$.

3 Some Basic Facts on (k,ℓ)-Leaf Powers

The following facts are well known for k-leaf powers (see, e.g., [2]) and can easily be shown for (k,l)-leaf powers.

Proposition 4

(i) *Every induced subgraph of a (k,l)-leaf power is a (k,l)-leaf power.*
(ii) *A graph is a (k,l)-leaf power if and only if each of its connected components is a (k,l)-leaf power.*

Let T be a k-leaf root of a graph G. Then, by definition, G is isomorphic to the subgraph of T^k induced by the leaves of T. Since trees are strongly chordal and induced subgraphs of strongly chordal graphs are strongly chordal, Proposition 1 implies:

Proposition 5. *For every $k \geq 1$, k-leaf powers are strongly chordal.*

This strengthens the fact that k-leaf powers are chordal, which is observed in some previous papers dealing with k-leaf powers, and also implies that (k,ℓ)-leaf powers are strongly chordal. The converse implication is not true: In [1], based on [7], an example of a strongly chordal graph is given which is no k-leaf power, for any $k \geq 2$.

The following simple facts are helpful:

Proposition 6

(i) *For $k \leq k' < \ell$, if G is a (k,ℓ)-leaf power then it is a (k',ℓ)-leaf power.*
(ii) *For $k < \ell' \leq \ell$, if G is a (k,ℓ)-leaf power then it is a (k,ℓ')-leaf power.*
(iii) *If G is a (k,ℓ)-leaf power then it is a $(k+2i,\ell+2i)$-leaf power, for all $i \geq 1$.*
(iv) *If G is a (k,ℓ)-leaf power then it is a $(k+i(k-2),\ell+i(\ell-2))$-leaf power, for all $i \geq 1$.*

Proof. (i) and (ii) are obviously true, by definition. (iii) is shown by subdividing each edge of a (k,l)-leaf root T of G containing a leaf of T. (iv) is shown by subdividing each edge of T not containing a leaf of T. □

Thus, obviously every k-leaf power is also a $(k+2)$-leaf power but it is not known whether every k-leaf power is also a $(k+1)$-leaf power. For 3-leaf powers, however, it is noted in [2]:

Proposition 7. *Every 3-leaf power is a k-leaf power, for all $k \geq 3$.*

By the proof of Proposition 6 (*iv*), every 3-leaf power is a (4,6)-leaf power. By Proposition 6 (*iii*), we get the next proposition which also implies Proposition 7:

Proposition 8. *Every $(4, 6)$-leaf power is a k-leaf power, for all $k \geq 4$, and, in general, every $(k, k + 2)$-leaf power is an ℓ-leaf power, for all $k \leq \ell$.*

A graph $H = (V_H, E_H)$ results from a graph $G = (V_G, E_G)$ by *substituting* a clique Q into a vertex $v \in V_G$ (or *substituting a vertex v by a clique Q*), if V_H is the union of $V_G \setminus \{v\}$ and the vertices in Q, and E_H results from E_G by removing all edges containing v, adding all clique edges in Q and adding all edges between vertices in Q and in $N_G(v)$.

Proposition 9. *For every graph G and for every $k \geq 2$ and $\ell > k$, G is a (k, ℓ)-leaf power if and only if every graph resulting from G by substituting its vertices by cliques is a (k, ℓ)-leaf power.*

Proof. If T is a (k, ℓ)-leaf root for the (k, ℓ)-leaf power $G = (V, E)$, and G' is the result of substituting a clique Q into a vertex $u \in V$, then attach all vertices in Q at the same parent in T as u and skip u; the resulting tree T' is a (k, ℓ)-leaf root for G'. The converse direction obviously holds. □

4 Metric Properties of (k, ℓ)-Leaf Powers

Obviously, a graph is P_3-free if and only if it is the disjoint union of cliques, and G is a 2-leaf power if and only if it is P_3-free.

Proposition 10. *Let G be a (k, ℓ)-leaf power.*

(i) *If $\ell > 2k - 2$ then G is P_3-free.*
(ii) *If $\ell = 2k - 2$ then P_3 has a unique (k, ℓ)-leaf root.*

Proof. Let T be a (k, ℓ)-leaf root of G, and suppose that G contains a P_3 with vertices a, b, c and edges ab and bc. Then $d_T(a, b) \leq k$ and $d_T(b, c) \leq k$, i.e., $d_T(a, c) \leq 2k-2$. On the other hand, $d_T(a, c) \geq \ell$ since $ac \notin E$. Thus, $\ell \leq 2k-2$. If $\ell = 2k - 2$ then a (k, ℓ)-leaf root of the P_3 has distance exactly ℓ between a and c, and the leaf b is attached to the central vertex of the path between a and c. This is obviously the only (k, ℓ)-leaf root of the P_3 abc. □

A well known fact for distances in trees found by Buneman [8] (respectively, for block graphs found by Howorka [13]) is the following characterisation in terms of a four-point condition:

Theorem 1. *Let $G = (V, E)$ be a connected graph.*

(i) *G is a tree if and only if G contains no triangles and G satisfies the following four-point condition: For all $u, v, x, y \in V$,*
$(*)$ $d_G(u, v) + d_G(x, y) \leq \max\{d_G(u, x) + d_G(v, y), d_G(u, y) + d_G(v, x)\}$.
(ii) *G is a block graph if and only if G satisfies $(*)$, for all $u, v, x, y \in V$.*

From now on, let G be a (k,ℓ)-leaf power with (k,ℓ)-leaf root T. We apply this four-point condition to various induced subgraphs of G such as P_4 as well as diamond, dart and gem (see Figure 1).

Proposition 11. *Let the four vertices a,b,c,d with non-edge ad induce a diamond in G. Then $d_T(b,c) \le 2k-\ell$.*

Proof. According to condition $(*)$, $d_T(a,d)+d_T(b,c) \le \max\{d_T(a,b)+d_T(c,d), d_T(a,c)+d_T(b,d)\} \le 2k$ holds since $ab, cd, ac, bd \in E$. Since $ad \notin E$, we have $d_T(a,d) \ge \ell$. Thus $d_T(b,c) \le 2k-\ell$. □

Proposition 12. *Let the four vertices a,b,c,d with edges ab, ac, bc, bd and cd induce a diamond in G such that b and c can be distinguished in G. Then $2\ell \le 3k-2$.*

Proof. According to Proposition 11, $d_T(b,c) \le 2k-\ell$. Let z be a vertex which distinguishes b and c, say $bz \in E$ and $cz \notin E$. Then $d_T(b,z) \le k$ and $d_T(c,z) \ge \ell$. The T-paths P_{bz} between b and z and P_{bc} between b and c have at least two vertices in common, namely b and its parent, say b'. Let x be the last common vertex of P_{bz} and P_{bc}. Then $d_T(x,b)+d_T(x,c) \le 2k-\ell$, by Proposition 11, and $d_T(x,b)+d_T(x,z) \le k$, which implies that $d_T(x,c)+d_T(x,z)+2d_T(x,b) \le 3k-\ell$. On the other hand, $2d_T(x,b) \ge 2$ and $d_T(x,c)+d_T(x,z) \ge \ell$, which implies $\ell+2 \le 3k-\ell$, i.e., $2\ell \le 3k-2$. □

Corollary 1. *If dart or gem is a (k,ℓ)-leaf power then $2\ell \le 3k-2$.*

Proposition 13. *Let the four vertices a,b,c,d with edges ab, bc and cd induce a P_4 in G. Then $d_T(a,d) \ge 2\ell-k$.*

Proof. According to condition $(*)$, $d_T(a,c)+d_T(b,d) \le \max\{d_T(a,b)+d_T(c,d), d_T(a,d)+d_T(b,c)\}$ holds. Since $ac \notin E$ and $bd \notin E$ and T is a (k,ℓ)-leaf root of G, we have $d_T(a,c)+d_T(b,d) \ge 2\ell$. On the other hand, since $ab \in E$ and $cd \in E$, we have $d_T(a,b)+d_T(c,d) \le 2k$, and this sum cannot be the maximum of the two sums on the right hand side of inequality $(*)$. Thus $d_T(a,c)+d_T(b,d) \le d_T(a,d)+d_T(b,c)$ holds, which implies that $d_T(a,d) \ge 2\ell-k$. □

Proposition 14. *If $2\ell \le 3k-2$ then dart and gem are (k,ℓ)-leaf powers.*

5 Characterisations of $(4,6)$-Leaf Powers

The characterisation of $(4,6)$-leaf powers given in this section is very similar to the following one for 3-leaf (i.e., $(3,4)$-leaf) powers:

Theorem 2 ([2,11,21]). *The following conditions are equivalent:*

(i) *G is a 3-leaf power.*
(ii) *G is (bull, dart, gem)-free chordal.*
(iii) *G results from substituting cliques into the vertices of a tree.*

Now we consider the class of $(4,6)$-leaf powers. Recall that every $(4,6)$-leaf power is a k-leaf power, for all $k \ge 4$. In [16], the authors study so-called strictly chordal graphs which are defined via (rather complicated) hypergraph properties but finally turn out to be exactly the (dart,gem)-free chordal graphs as Corollary 2.2.2. in [14] says:

Proposition 15 ([14]). *G is strictly chordal if and only if it is (dart, gem)-free chordal.*

The next theorem has been our motivation for defining and investigating the notion of (k, l)-leaf powers:

Theorem 3. *The following conditions are equivalent:*

(i) *G is a* $(4, 6)$*-leaf power.*
(ii) *G is (dart, gem)-free chordal (i.e., strictly chordal).*
(iii) *G results from substituting cliques into the vertices of a block graph.*

For the proof of Theorem 3 (which is omitted due to space constraints) we use:

Proposition 16. *Every block graph is a (basic)* $(4, 6)$*-leaf power, and a (basic)* $(4, 6)$*-leaf root of a given block graph can be determined in linear time.*

Now Theorem 3 together with Proposition 16 implies:

Corollary 2. *Strictly chordal graphs are k-leaf powers for all* $k \geq 4$*, and a k-leaf root of G can be determined in linear time.*

Corollary 2 is one of the main results (namely Theorem 4.1) in [16]. It has also been mentioned in [16] that strictly chordal graphs can be recognised in linear time. We give a simpler proof for it.

Corollary 3. $(4, 6)$*-leaf powers (and thus also strictly chordal graphs) can be recognised in linear time.*

Proof. By Theorem 3, we know that a graph G is a $(4, 6)$-leaf power if and only if G results from substituting cliques into the vertices of a block graph, and we check the last condition in the following way. For a given graph G, first check whether G is chordal. If not then G is not a $(4, 6)$-leaf power, else determine a clique tree of G (which can be done in linear time, see, e.g., [23]). It is well known (see, e.g., [19]) that the minimal separators of G are given as the intersections of cliques which are adjacent in the clique tree.

In a block graph, the minimal separators are the cut vertices. If G results from substituting cliques into the vertices of a block graph, then the cliques which replace cut vertices are pairwise disjoint minimal separators (which are also clique modules).

Thus, determine the minimal separators in G, check whether they are pairwise disjoint (if not, G is no $(4, 6)$-leaf power), shrink them to one vertex, respectively, and check whether the resulting graph G' is a block graph. If yes, G results from substituting cliques into the (cut) vertices of the block graph G', otherwise G is no $(4, 6)$-leaf power. □

6 The Main Results

Very similar to Proposition 16, we obtain:

Proposition 17. *Every block graph is a* $(5,7)$*-leaf power, and a* $(5,7)$*-leaf root of a given block graph can be determined in linear time.*

Theorem 4

(i) *For all* k, ℓ *with* $k \geq 2$ *and* $\ell > 2k-2$*, the class of* (k,ℓ)*-leaf powers is the class of* P_3*-free graphs, i.e., disjoint unions of cliques.*

(ii) *For all* k, ℓ *with odd* $k = 2i+1$*,* $i \geq 1$*, and* $\ell = 4i$*, the class of* (k,ℓ)*-leaf powers is the class of 3-leaf powers, i.e., graphs obtained from substituting cliques into trees.*

(iii) *For all* k, ℓ *with* $k \geq 2$ *and* $2\ell > 3k-2$ *but* $\ell \leq 2k-2$ *and not the situation of (ii), the class of* (k,ℓ)*-leaf powers is the class of* $(4,6)$*-leaf powers, i.e., graphs obtained from substituting cliques into block graphs.*

Proof.
(i): This follows from Proposition 10 and the obvious fact that disjoint unions of cliques have (k,ℓ)-leaf roots with $k \geq 2$ and $\ell > 2k-2$ by simply connecting the central vertices of stars realising cliques by paths which are long enough.

(ii): The case $k = 3$ and thus $i = 1$ is the case of 3-leaf powers, and we can refer to Theorem 2. For larger odd k, we first have to show that every (3,4)-leaf power is a $(2i+1,4i)$-leaf power; to show this we subdivide the internal edges of a (3,4)-leaf root T, more precisely, we replace every internal edge of a (3,4)-leaf root T by a P_4 (P_6, respectively) and obtain a (5,8)-leaf root ((7,12)-leaf root, respectively), and we perform in the same way for larger k. Conversely, $(2i+1,4i)$-leaf powers are dart- and gem-free, by Corollary 1, since, for $k = 2i+1$ and $\ell = 4i$, the inequality $2\ell \leq 3k-2$ in Corollary 1 is not fulfilled. We claim that $(2i+1,4i)$-leaf powers are also bull-free: By Proposition 10, every P_3 has a unique $(2i+1,4i)$-leaf root since $\ell = 4i = 2(2i+1)-2 = 2k-2$. Let a, b, c, d, e induce a bull with the P_4 $abcd$ with edges ab, bc, cd and vertex e adjacent to b and c. Then the unique $(2i+1,4i)$-leaf root for the P_4 is a path of length $6i-1$ between a and d, and b and c are attached as leaves to this path such that $d_T(a,b) = 2i+1$, $d_T(b,c) = 2i+1$ and $d_T(c,d) = 2i+1$. Now a, b, e induce a P_3, and e, c, d induce a P_3, whose roots are unique and require that $d_T(b,e) = 2i+1$ and $d_T(e,c) = 2i+1$, which leads to a contradiction. Thus, $(2i+1,4i)$-leaf powers are bull-, dart- and gem-free chordal, and, by Theorem 2, they are 3-leaf powers.

(iii): For $k = 2$ and $k = 3$ there is nothing to prove. As $2\ell > 3k-2$, by Corollary 1, (k,l)-leaf powers in this case are dart- and gem-free. As they are also chordal, by Theorem 3, they are $(4,6)$-leaf powers. For the other direction, again by Theorem 3, it suffices to show that every block graph has a basic (k,l)-leaf root. And, by Proposition 6, it suffices to show this for the largest possible l, for every k, i.e., for the (k,l)-pairs $(4,6), (5,7), (6,10), (7,11)$ and so on, i.e.,

for the (k,l)-pairs $(4+2i, 6+4i), (5+2i, 7+4i)$, for all $i \geq 0$. Proposition 16 and Proposition 17 together with their proofs deal with the case $i = 0$. In the case $i = 0$, we start with block roots which are stars whose edges are subdivided exactly once. For a general $i \geq 0$, we use the same construction with stars whose edges are subdivided exactly $i+1$ times. □

For the graphs $H_1, \ldots, H_8$ in Theorem 5 see Figure 2. Rautenbach [21] has shown that a graph without true twins is a 4-leaf power if and only if it is $(H_1, \ldots, H_8)$-free chordal. In [6], the following more detailed characterisation is shown:

Theorem 5. *G is a basic 4-leaf power if and only if G is $(H_1, \ldots, H_8)$-free chordal. In particular, G is the square of a tree if and only if G is 2-connected $(H_1, \ldots, H_5)$-free chordal.*

In fact, the forbidden subgraphs $H_1, \ldots, H_5$ are responsible for the blocks of a basic 4-leaf power, and H_6, H_7, H_8 represent the gluing conditions of blocks.

In Theorem 6 characterising the $(8,11)$-leaf powers, we additionally need graph H_9 which is given in Figure 2 and replaces the role of H_8 as a gluing condition.

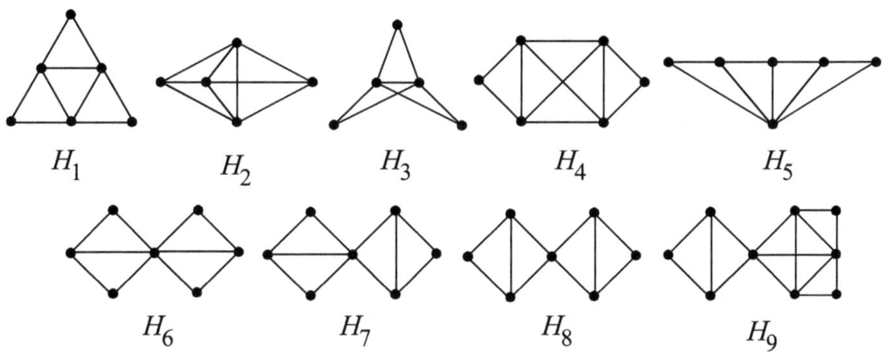

Fig. 2. Some forbidden subgraphs; the graphs $H_1, \ldots, H_8$ characterise basic 4-leaf powers.

Theorem 6. *The $(8,11)$-leaf powers are exactly the graphs obtained from substituting cliques into $(H_1, \ldots, H_7, H_9)$-free chordal graphs, i.e., the basic $(8,11)$-leaf powers are exactly the $(H_1, \ldots, H_7, H_9)$-free chordal graphs.*

Corollary 4. *$(8,11)$-leaf powers can be recognised in polynomial time.*

7 Conclusion

In this paper, we gave structural characterisations of (k,ℓ)-leaf powers, for some k and ℓ which imply efficient recognition of these classes, and in this way we improve and extend a recent paper [16] by Kennedy, Lin and Yan on strictly chordal graphs and leaf powers. Our main results are presented in Theorem 4 and 6. Other characterisations can be expected for "limit" classes (i.e., for every k, the largest l which is not yet covered by Theorem 4) such as (12,17)-leaf powers. We expect that our new notion of (k, ℓ)-leaf powers will shed new light on the open problem of characterising and recognising k-leaf powers for $k \geq 6$. We also have a characterisation of $(6, 8)$-leaf powers in terms of induced subgraphs of squares of block graphs as well as in terms of forbidden induced subgraphs which will be described in a forthcoming paper.

References

1. Bibelnieks, E., Dearing, P.M.: Neighborhood subtree tolerance graphs. Discrete Applied Math. 98, 133–138 (2006)
2. Brandstädt, A., Le, V.B.: Structure and linear time recognition of 3-leaf powers. Information Processing Letters 98, 133–138 (2006)
3. Brandstädt, A., Le, V.B.: Dieter Rautenbach. Exact leaf powers, manuscript (submitted, 2006)
4. Brandstädt, A., Le Dieter, V.B.: Rautenbach, Distance-hereditary 5-leaf powers, manuscript (submitted, 2006)
5. Brandstädt, A., Le, V.B., Spinrad, J.P.: Graph Classes: A Survey. SIAM Monographs on Discrete Mathematics and Applications 3 (1999)
6. Brandstädt, A., Le, V.B., Sritharan, R.: Structure and linear time recognition of 4-leaf powers, manuscript (submitted, 2006)
7. Broin, M.W., Lowe, T.J.: A dynamic programming algorithm for covering problems with (greedy) totally balanced constraint matrices. SIAM J. Alg. Disc. Meth. 7, 348–357 (1986)
8. Buneman, P.: A note on the metric properties of trees. J. Combin. Th. 1(B), 48–50 (1974)
9. Chang, M.-S., Ko, T.: The 3-Steiner Root Problem. In: Proceedings WG 2007. LNCS, manuscript, as extended abstract (to appear, 2007)
10. Dahlhaus, E., Duchet, P.: On strongly chordal graphs. Ars Combinatoria 24 B, 23–30 (1987)
11. Dom, M., Guo, J., Hüffner, F., Niedermeier, R.: Error compensation in leaf root problems. In: Fleischer, R., Trippen, G. (eds.) ISAAC 2004. LNCS, vol. 3341, pp. 363–381. Springer, Heidelberg (2004), Algorithmica 44(4), 363–381(2006)
12. Dom, M., Guo, J., Hüffner, F., Niedermeier, R.: Extending the tractability border for closest leaf powers. In: Kratsch, D. (ed.) WG 2005. LNCS, vol. 3787, Springer, Heidelberg (2005)
13. Howorka, E.: On metric properties of certain clique graphs. J. Combin. Th. 27(B), 67–74 (1979)
14. Kennedy, W.: Strictly chordal graphs and phylogenetic roots, Master Thesis, University of Alberta (2005)
15. Kennedy, W., Lin, G.: 5-th phylogenetic root construction for strictly chordal graphs, extended abstract. In: Deng, X., Du, D.-Z. (eds.) ISAAC 2005. LNCS, vol. 3827, pp. 738–747. Springer, Heidelberg (2005)
16. Kennedy, W., Lin, G., Yan, G.: Strictly chordal graphs are leaf powers. J. Discrete Algorithms 4, 511–525 (2006)
17. Lin, G.-H., Jiang, T., Kearney, P.E.: Phylogenetic k-root and Steiner k-root, Extended abstract. In: Lee, D.T., Teng, S.-H. (eds.) ISAAC 2000. LNCS, vol. 1969, pp. 539–551. Springer, Heidelberg (2000)

18. Lubiw, A.: Γ-free matrices, Master Thesis, Dept. of Combinatorics and Optimization, University of Waterloo, Canada (1982)
19. McKee, T.A., McMorris, F.R.: Topics in Intersection Graph Theory. SIAM Monographs on Discrete Mathematics and Applications 2 (1999)
20. Nishimura, N., Ragde, P., Thilikos, D.M.: On graph powers for leaf-labeled trees. J. Algorithms 42, 69–108 (2002)
21. Rautenbach, D.: Some remarks about leaf roots. Discrete Math. 306(13), 1456–1461 (2006)
22. Raychaudhuri, A.: On powers of strongly chordal and circular arc graphs. Ars Combinatoria 34, 147–160 (1992)
23. Spinrad, J.P.: Efficient Graph Representations, Fields Institute Monographs, American Mathematical Society, Providence, Rhode Island (2003)

An Improved Claw Finding Algorithm Using Quantum Walk

Seiichiro Tani[1,2]

[1] Quantum Computation and Information Project, ERATO-SORST, JST
[2] NTT Communication Science Laboratories, NTT Corporation
tani@theory.brl.ntt.co.jp

Abstract. The claw finding problem has been studied in terms of query complexity as one of the problems closely connected to cryptography. For given two functions, f and g, as an oracle which have domains of size N and M ($N \leq M$), respectively, and the same range, the goal of the problem is to find x and y such that $f(x) = g(y)$. This paper describes a quantum-walk-based algorithm that solves this problem; it improves the previous upper bounds. Our algorithm can be generalized to find a claw of k functions for any constant integer $k > 1$, where the domains of the functions may have different size.

Keywords: Quantum computing, query complexity, oracle computation.

1 Introduction

The most significant discovery in quantum computation would be Shor's polynomial-time quantum algorithms for factoring integers and computing discrete logarithms [14], both of which are believed to be hard to solve in classical settings and are thus used in arguments for the security of the widely used cryptosystems. Another significant discovery is Grover's quantum algorithm for the problem of searching an unstructured set [10]; it has yielded a variety of generalization approaches [3,11]. Grover's algorithm and its generalizations assume the *oracle computation model*, in which a problem instance is given as a black box (called an oracle) and any algorithm needs to issue queries to the black box to get sufficient information on the instance.

One of the earliest applications of Grover's algorithm was the bounded-error algorithm of Brassard, Høyer and Tapp [4]; it addressed the *collision* problem in a cryptographic context, i.e., finding pair (x, y) such that $f(x) = f(y)$, in a given 2-to-1 function f of domain size N. Their quantum algorithm requires $O(N^{1/3})$ queries, whereas any bounded-error classical algorithm needs $\Theta(N^{1/2})$ queries. Subsequently, Aaronson and Shi [1] proved the matching lower bound. Brassard et al. [4] considered two more related problems: the *element distinctness* problem and the *claw finding* problem. These problems are also important in a cryptographic sense. Furthermore, studying these problems has deepened our understanding of the power of quantum computation.

The element distinctness problem is to decide whether or not given N integers are all distinct. Buhrman et al. [7] gave a bounded-error algorithm for the problem, which makes $O(N^{3/4})$ queries (strictly, they assumed a comparison oracle, which returns just

L. Kučera and A. Kučera (Eds.): MFCS 2007, LNCS 4708, pp. 536–547, 2007.

the result of comparing the function values for two specified inputs, and, in this case, the number of required queries, i.e., the query complexity, is $O(N^{3/4}\log N)$). Subsequently, Ambainis [2] gave an improved upper bound $O(N^{2/3})$ by introducing a new framework of quantum walk (his quantum walk algorithm was reviewed from a slightly more general point of view in [13,9], and a much more general framework was given by Szegedy [15]). This upper bound matches the lower bound proved by Aaronson and Shi [1].

The *claw finding problem* is defined as follows. Given two functions $f : X \to Z$ and $g : Y \to Z$ as an oracle, *find* a pair $(x, y) \in X \times Y$, called a *claw*, such that $f(x) = g(y)$ if it exists, where X and Y are domains of size N and M ($N \leq M$), respectively. By $\mathbf{claw}_{\text{finding}}(N, M)$, we mean this problem. Unlike the collision problem and the element distinctness problem, there is still a big gap between the upper and lower bounds of the query complexity for the claw finding problem.

After Brassard et al. [4] considered a special case of the claw finding problem, Buhrman et al. [6] gave a quantum algorithm that requires $O(N^{1/2}M^{1/4})$ queries for $N \leq M < N^2$ and $O(M^{1/2})$ queries for $M \geq N^2$ (strictly, they assumed a comparison oracle, and, in this case, the query complexity is multiplied by $\log N$). They also proved that any algorithm requires $\Omega(M^{1/2})$ queries by reducing the search problem over an unstructured set to the claw finding problem. Thus, while their bounds of the query complexity are tight when $M \geq N^2$, there is still a big gap when $N \leq M < N^2$. Furthermore, they gave a quantum algorithm that finds a claw for k functions (called the k-claw finding problem) of the common domain of size N with $O(N^{1-1/2^k})$ queries. In the case of three functions, Iwama and Kawachi [12] gave a quantum algorithm that is more efficient than that of Buhrman et al. [6], if the number of claws of two functions among the three is at most $N^{7/8}$. In fact, it is shown in [13] that the algorithm in [2] for the element distinctness problem is so general that it can be applied to the claw finding problem with slight modification; it yields $\tilde{O}((N + M)^{2/3}) = \tilde{O}(M^{2/3})$ queries for the two-function case. This upper bound is better than the previous upper bounds only when $M = O(N^{6/5}/poly(\log N))$.

This paper gives a bounded-error quantum algorithm for the claw finding problem that achieves a better bound when $N \leq M < N^2$ than the previous best bound (even when $M \geq N^2$, our algorithm can work well with the optimal query complexity $O(M^{1/2})$). More precisely,

$$Q_2(\mathbf{claw}_{\text{finding}}(N, M)) = \begin{cases} O\left((NM)^{1/3}\right) & (N \leq M < N^2) \\ O\left(M^{1/2}\right) & (M \geq N^2), \end{cases}$$

where $Q_2(P)$ means the number of queries required to solve problem P with bounded error. Our algorithm can easily be modified to solve the more general problem of finding a tuple $(x_1, \ldots, x_p, y_1, \ldots, y_q) \in X^p \times Y^q$ such that $x_i \neq x_j$ and $y_i \neq y_j$ for any $i \neq j$, and $(f(x_1), \ldots, f(x_p), g(y_1), \ldots, g(y_q)) \in R$, for given $R \subseteq Z^{p+q}$, where p and q are positive constant integers; the query complexity is $O((N^p M^q)^{1/(p+q+1)})$ for $N \leq M < N^{1+1/q}$ and $O(M^{q/(1+q)})$ for $M \geq N^{1+1/q}$.

Our algorithm first finds subsets $X'' \subseteq X$ and $Y'' \subseteq Y$ of size $O(1)$ such that there is a claw in $X'' \times Y''$, by using binary and 4-ary searches over X and Y; in order to decide which branch we should proceed at each visited node in the search trees, we use a subroutine that *detects* the existence of a claw of two functions f and g with bounded error. The algorithm then classically searches $X'' \times Y''$ for a claw. If we use the bounded-error subroutine in a straight-forward manner in the search trees, a "log" factor would be multiplied to the total query complexity to guarantee bounded error as a whole. Instead, at the sth node of the search trees, we repeat the subroutine $O(s)$ times to amplify success probability. This achieves bounded error as a whole, while pushing up the query complexity by just a constant multiplicative factor. (Høyer et al. [11] introduced an error reduction technique with a similar flavor; however, their technique is used in an algorithmic context different from ours: their error reduction is performed at each recursion level while ours is sequentially used at each step of the search tree).

The subroutine is developed around the Szegedy's quantum walk framework [15] over a Markov chain on the graph categorical product of two Johnson graphs, which correspond to the two functions (with an idea similar to the one used in [8]).

We also prove a lower bound, which is better than the known lower bound $\Omega(M^{1/2})$ given in [6] when $M = o(N^{3/2})$, by a simple reduction from the element distinctness problem: $Q_2(\mathbf{claw}_{\text{finding}}(N, M)) = \Omega(N^{1/2}M^{1/6})$. This bound together with the previous bound $\Omega(M^{1/2})$ implies that our algorithm is optimal when $M = \Theta(N)$ or $M = \Omega(N^2)$. As a by-product, an optimal quantum algorithm of $O(N^{2/3})$ queries for the element distinctness problem can be obtained by combining the above reduction with our claw finding algorithm (note that the first $O(N^{2/3})$-query algorithm was presented in [2]).

Our algorithm can be generalized to the k-claw finding problem. For any constant integer $k > 1$, let $\mathbf{k\text{-}claw}_{\text{finding}}(N_1, \ldots, N_k)$ denote the k-claw finding problem for k functions $f_i : X_i := [N_i] \to Z$ $(i = 1, \ldots, k)$ given as an oracle, where $N_i \leq N_j$ for $i < j$,

$$Q_2(\mathbf{k\text{-}claw}_{\text{finding}}(N_1, \ldots, N_k)) = \begin{cases} O\left(\left(\prod_{i=1}^{k} N_i\right)^{\frac{1}{k+1}}\right) & \text{if } \prod_{i=2}^{k} N_i = O(N_1^k), \\ O\left(\sqrt{\prod_{i=2}^{k} N_i / N_1^{k-2}}\right) & \text{otherwise.} \end{cases}$$

No nontrivial bounds were known in the case where the domain sizes of the given functions are not linearly correlated.

Our algorithms can work with slight modification even against a comparison oracle (i.e., against an oracle that, for a given pair of inputs $(x_i, x_j) \in X_i \times X_j$, only decides which is the larger of two function values $f_i(x_i)$ and $f_j(x_j)$); the query complexity increases by a multiplicative factor of $\log N_1$ for the k-function case ($\log N$ for the two-function case).

2 Preliminaries

This section defines problems and introduces some useful techniques. We denote the set of positive integers by $\mathbb{Z}^*$, the set of $\{i \mid j \leq i \leq k \text{ for } i, j, k \in \mathbb{Z}^*\}$ by $[j.k]$, and $[1.k]$ by $[k]$ for short.

Problem 1 (Claw Finding Problem). Given two functions $f : X := [N] \to Z$ and $g : Y := [M] \to Z$ as an oracle for $N \leq M$, where $Z = [|Z|]$, find a pair $(x, y) \in X \times Y$ such that $f(x) = g(y)$ if such a pair exists.

In a quantum setting, the two functions are given as quantum oracle $O_{f,g}$ which is defined as $O_{f,g} : |p,z,w\rangle \longrightarrow |p, z \oplus P(p) \pmod{|Z|}, w\rangle$, where $p \in X \cup Y$, $z \in Z$, w is work space, $P(p)$ is defined as $f(p)$ if $p \in X$ and $g(p)$ if $p \in Y$ (note that it easy to know whether p is in X or Y by using one more bit to represent p). This kind of oracle, which returns the value of the function(s), is called a *standard oracle*.

Another type of oracle is called the *comparison oracle*, which, for given two inputs, only decides which is the larger of the two function values corresponding to the inputs. More formally, comparison oracle $O_{f,g}$ is defined as $O_{f,g} : |p,q,b,w\rangle \longrightarrow |p,q,b \oplus [P(p) \leq Q(q)], w\rangle$, where $p,q \in X \cup Y$, $b \in \{0,1\}$, w and P are defined as in the standard oracle, Q is defined in the same way as P, and $[P(p) \leq Q(q)]$ is the predicate such that its value is 1 if and only if $P(p) \leq Q(q)$.

It is obvious that, if we are given a standard oracle, we can realize a comparison oracle by issuing $O(1)$ queries to the standard oracle. Thus, upper bounds for a comparison oracle are those for a standard oracle, and lower bounds for a standard oracle are those for a comparison oracle, if we ignore constant multiplicative factors.

Buhrman et al. [6] generalized the claw finding problem to a k-function case.

Problem 2 (k-Claw Finding Problem). Given k functions $f_i : X_i := [N_i] \rightarrow Z$ $(i \in [k])$ as an oracle, where $N_i \leq N_j$ if $i < j$, and $Z := [|Z|]$, find a k-claw, i.e., a k-tuple $(x_1, \ldots, x_k) \in X_1 \times \cdots \times X_k$ such that $f_i(x_i) = f_j(x_j)$ for any $i, j \in [k]$, if it exists.

Standard and comparison oracles are defined in almost the same way as in the two-function case, except that inputs p and q belong to one of X_i's, respectively, for $i \in [k]$.

The next theorem describes Szegedy's framework, which we use to prove our upper bounds.

Theorem 1 ([15]). *Let $\mathcal{M}$ be a symmetric Markov chain with state set V and transition matrix P and let $\delta_{\mathcal{M}}$ be the spectral gap of P, i.e., $1 - \max_i |\lambda_i|$ for the eigenvalues λ_i's of P. For a certain subset $V' \subseteq V$ with the promise that $|V'|$ is either 0 or at least $\epsilon|V|$ for $0 < \epsilon < 1$, any element in V' is marked. For $T = O(1/\sqrt{\epsilon\delta_{\mathcal{M}}})$, the next quantum algorithm decides whether $|V'|$ is 0 ("false") or at least $\epsilon|V|$ ("true") with one-sided bounded error with cost $O(C_U + (C_F + C_W)/\sqrt{\delta_{\mathcal{M}}\epsilon})$, where $C = \sum_i |c_i\rangle\langle c_i|$ for $|c_i\rangle = \sum_j \sqrt{P_{i,j}}|i\rangle|j\rangle$ and $R = \sum_j |r_j\rangle\langle r_j|$ for $|r_j\rangle = \sum_i \sqrt{P_{j,i}}|i\rangle|j\rangle$:*

1. *Prepare $|0\rangle$ in a one-qubit register $\mathbf{R}_0$, and prepare a uniform superposition $|\phi_0\rangle := \frac{1}{\sqrt{r|V|}} \sum_{i,j \in V, P_{i,j} \neq 0} |i\rangle|j\rangle$ in a register $\mathbf{R}_1$ with cost at most C_U, where r is the number of adjacent states (of any state) in $\mathcal{M}$.*
2. *Apply the Hadamard operator to $\mathbf{R}_0$.*
3. *For randomly and uniformly chosen $1 \leq t \leq T$, apply the next operation W t times to $\mathbf{R}_1$ if the content of $\mathbf{R}_0$ is "1."*
   (a) *To any $|i\rangle|j\rangle$, perform the next steps: (i) Check if $i \in V'$ with cost at most C_F, (ii) If $i \notin V'$, apply diffusion operator $2C - I$ with cost at most C_W.*
   (b) *To any $|i\rangle|j\rangle$, perform the next steps: (i) Check if $j \in V'$ with cost at most C_F, (ii) If $j \notin V'$, apply diffusion operator $2R - I$ with cost at most C_W.*
4. *Apply the Hadamard operator to $\mathbf{R}_0$, and measure registers $\mathbf{R}_0$ and $\mathbf{R}_1$ with respect to the computational basis.*
5. *If the result of measuring $\mathbf{R}_0$ is 1 or a marked element is found by measuring $\mathbf{R}_1$, output "true"; otherwise output "false."*

3 Claw Detection

In this section, we describe "claw-detection" algorithms that detect the existence of a claw. The claw-detection algorithms will be used as subroutines in the "claw-search" algorithms presented in the next section that find a claw.

Before presenting the claw-detection algorithm, we introduce some notions. The *Johnson graph* $J(n,k)$ is a connected regular graph with $\binom{n}{k}$ vertices such that every vertex is a subset of size k of $[n]$; two vertices are adjacent if and only if the symmetric difference of their corresponding subsets has size 2. The *graph categorical product* $G = (V_G, E_G)$ of two graphs $G_1 = (V_{G_1}, E_{G_1})$ and $G_2 = (V_{G_2}, E_{G_2})$, denoted by $G = G_1 \times G_2$, is a graph having vertex set $V_G = V_{G_1} \times V_{G_2}$ such that $((v_1, v_2), (v_1', v_2')) \in E_G$ if and only if $(v_1, v_1') \in E_{G_1}$ and $(v_2, v_2') \in E_{G_2}$.

The next two propositions are useful in analyzing the claw-detection algorithms we will describe.

Proposition 1. *For Markov chains $\mathcal{M}, \mathcal{M}_1, \ldots, \mathcal{M}_k$, the spectral gap δ of $\mathcal{M}$ is the minimum of those $\delta_1, \ldots, \delta_k$ of $\mathcal{M}_1, \ldots, \mathcal{M}_k$, i.e., $\delta = \min_i\{\delta_i\}$, if the underlying graph of $\mathcal{M}$ is the graph categorical product of those of $\mathcal{M}_1, \ldots, \mathcal{M}_k$.*

The eigenvalues of the Markov chain on $J(n,k)$ are $\frac{(k-j)(n-k-j)-j}{k(n-k)}$ for $j \in [0.k]$ [5, pages 255–256], from which the next proposition follows.

Proposition 2. *The Markov chain on Johnson graph $J(n,k)$ has spectral gap $\delta = \Omega(\frac{1}{k})$, if $2 \le k \le n/2$.*

We will first describe a claw-detection algorithm against a comparison oracle, from which we can almost trivially obtain a claw-detection algorithm against a standard oracle. Let Claw_Detect denote the algorithm. To construct Claw_Detect, we apply Theorem 1 on the graph categorical product of two Johnson graphs $J_f = J(|X|, l)$ and $J_g = J(|Y|, m)$ for the domains X and Y of functions f and g, respectively, where l and m ($l \le m$) are integers fixed later.

More precisely, let F and G be any vertices of J_f and J_g, respectively, i.e., any l-element subset and m-element subset of X and Y, respectively. Then (F, G) is a vertex in $J_f \times J_g$. Similarly, for any edges (F, F') and (G, G') of J_f and J_g, respectively, $((F,G),(F',G'))$ is an edge connecting two vertices (F,G) and (F',G') in $J_f \times J_g$. We next define "marked vertices" as follows. Vertex (F,G) is marked if there is a pair of $(x,y) \in F \times G$ such that $f(x) = g(y)$. To check if (F,G) is marked or not, we just sort all elements in $F \cup G$ on their function values. Although we have to sort all elements in the initial vertex, we have only to change a small part of the sorted list we have already had when moving to an adjacent vertex. For every vertex (F,G), we maintain a representation $L_{F,G}$ of the sorted list of all elements in $F \cup G$ on their function values, and we identify $(F, G, L_{F,G})$ as a vertex of $J_f \times J_g$. Here, we want to guarantee that $L_{F,G}$ is uniquely determined for any pair (F,G) in order to avoid undesirable quantum interference; we have just to introduce some appropriate rules that break ties, i.e., the situation where there are multiple elements in $F \cup G$ that have the same function value.

As the state $|\phi_0\rangle$ in Theorem 1, we prepare

$$|\phi_0\rangle = \frac{1}{\sqrt{\binom{N}{l}\binom{M}{m}l(N-l)m(M-m)}} \bigotimes_{\substack{|F\triangle F'|=|G\triangle G'|=2\\ F,F'\subseteq X,|F|=|F'|=l\\ G,G'\subseteq Y,|G|=|G'|=m}} |F,G,L_{F,G}\rangle|F',G',L_{F',G'}\rangle,$$

in register $\mathbf{R}_1$. The number $1 \le t \le \frac{c}{\sqrt{\delta\epsilon}}$ of repeating W is chosen randomly and uniformly for some constant c, $\delta := \Omega(1/m)$ and $\epsilon := lm/(NM)$.

We next describe the implementation of operation W. Since diffusion operator $2C-I$ depends on $L_{F,G}$'s, it cannot be performed without queries to the oracle. We thus divide operator $2C-I$ into a few steps. For every unmarked vertex $(F,G,L_{F,G})$, we first transform $|F,G,L_{F,G}\rangle|F',G',L_{F',G'}\rangle$ into $|F,G,L_{F,G}\rangle|F',G',L_{F,G}\rangle$ with queries to the oracle. We then perform a diffusion operator on the registers where the contents "F,G" and "F',G'" are stored, to obtain a superposition of $|F,G,L_{F,G}\rangle|F'',G'',L_{F,G}\rangle$ over all (F'',G'') adjacent to (F,G). Finally, we transform $|F,G,L_{F,G}\rangle|F'',G'',L_{F,G}\rangle$ into $|F,G,L_{F,G}\rangle|F'',G'',L_{F'',G''}\rangle$. Operator $2R-I$ can be implemented in a similar way.

Lemma 1. *Let $Q_2(\mathbf{claw}_{\text{detect}}(N,M))$ be the number of queries needed to decide whether there is a claw or not for functions $f : X := [N] \to Z$ and $g : Y := [M] \to Z$ given as a comparison oracle. Then,*

$$Q_2(\mathbf{claw}_{\text{detect}}(N,M)) = \begin{cases} O((NM)^{1/3}\log N) & (N \le M < N^2) \\ O(M^{1/2}\log N) & (M \ge N^2). \end{cases}$$

Proof. We will estimate C_U, C_F and C_W for Claw_Detect, and then apply Theorem 1.

To generate $|\phi_0\rangle$, we first prepare the uniform superposition of $|F,G\rangle|F',G'\rangle$ over all F,F',G,G' such that (F,F') and (G,G') are edges of J_f and J_g, respectively. Obviously, this requires no queries. We then compute $L_{F,G}$ and $L_{F',G'}$ for each basis state by issuing $O((l+m)\log(l+m))$ queries to oracle $O_{f,g}$. Thus, $C_U = O((l+m)\log(l+m))$.

We can check if there is a pair of $(x,y) \in F \times G$ such that $f(x) = g(y)$ by looking through $L_{F,G}$ (without any queries). Thus, $C_F = 0$.

For every unmarked $(F,G,L_{F,G})$, step (a).ii of operation W transforms $|F,G,L_{F,G}\rangle|F',G',L_{F',G'}\rangle$ into a superposition over all $|F,G,L_{F,G}\rangle|F'',G'',L_{F'',G''}\rangle$ such that $|F\triangle F''| = |G\triangle G''| = 2$. This can be realized by insertion and deletion of $O(1)$ elements to/from the sorted list of $O(l+m)$ elements, and diffusion operators without queries. Each insertion or deletion can be performed with $O(\log(l+m))$ queries by using binary search. Similarly, step (b).ii of operation W needs $O(\log(l+m))$ queries. Thus, we have $C_W = O(\log(l+m))$.

We set ϵ to $\frac{l}{N} \times \frac{m}{M}$, since the probability that a state is marked is minimized when only one claw exists for f and g, in which case the probability is $\frac{l}{N} \times \frac{m}{M}$. Since, from Proposition 2, the spectral gaps of the Markov chains on $J(N,l)$ and $J(M,m)$ are $\Omega(\frac{1}{l})$ and $\Omega(\frac{1}{m})$, respectively, the spectral gap of the Markov chain on $J(N,l) \times J(M,m)$ is $\Omega(\min\{\frac{1}{l},\frac{1}{m}\}) = \Omega(\frac{1}{m})$ due to $l \le m$ and Proposition 1.

From Theorem 1, the total number of queries is $Q_2(\mathbf{claw}_{\text{detect}}(N,M)) = O((l+m)\log(l+m) + \log(l+m)\sqrt{m(NM/(lm))}) = O((l+m)\log(l+m) + \sqrt{NM/l}\log(l+m))$.

When $N \leq M < N^2$, we set $l = m = \Theta((NM)^{1/3})$, which satisfies condition $l \leq N$. The total number of queries is $Q_2(\mathbf{claw}_{\text{detect}}(N, M)) = O((NM)^{1/3} \log N)$. When $M \geq N^2$, we set $l = m = N$, implying that $Q_2(\mathbf{claw}_{\text{detect}}(N, M)) = O(M^{1/2} \log N)$. □

The standard oracle case can be handled by using almost the same approach.

Corollary 1. *Let $Q_2(\mathbf{claw}_{\text{detect}}(N, M))$ be the number of queries needed to decide whether there is a claw or not for functions $f : X = [N] \to Z$ and $g : Y = [M] \to Z$ given as a standard oracle. Then,*

$$Q_2(\mathbf{claw}_{\text{detect}}(N, M)) = \begin{cases} O((NM)^{1/3}) & (N \leq M < N^2) \\ O(M^{1/2}) & (M > N^2). \end{cases}$$

The claw-detection algorithm against a standard oracle can easily be modified in order to solve the more general problem of *detecting* a tuple $(x_1, \ldots, x_p, y_1, \ldots, y_q) \in X^p \times Y^q$ such that $x_i \neq x_j$ and $y_i \neq y_j$ for any $i \neq j$, and $(f(x_1), \ldots, f(x_p), g(y_1), \ldots, g(y_q)) \in R$, for given $R \subseteq Z^{p+q}$, where p and q are any constant positive integers. A modification is made to the part of the algorithm that decides whether a vertex of the underlying graph is marked or not; the modification can be made without changing the number of queries. The query complexity can be analyzed by using almost the same approach as used in claw detection with $\epsilon = \binom{N-p}{l-p}/\binom{N}{l} \times \binom{M-q}{m-q}/\binom{M}{m} \geq l^p m^q (1 - o(1))/(N^p M^q)$; the query complexity is $O((N^p M^q)^{1/(p+q+1)})$ for $N \leq M < N^{1+1/q}$ and $O(M^{q/(1+q)})$ for $M \geq N^{1+1/q}$. The problem of *finding* such a tuple can also be solved with the same order of complexity as above by using the algorithm for *detecting* it as a subroutine.

Our algorithm for detecting a claw can easily be generalized to the case of k functions of domains of size $N_1, \ldots, N_k$, respectively. More concretely, we apply Theorem 1 to the Markov chain on the graph categorical product of the k Johnson graphs, each of which corresponds to one of the k functions. We denote this "k-claw detection" algorithm by k-Claw_Detect in the next section.

Lemma 2. *For any positive integer $k > 1$, let $Q_2(\mathbf{k\text{-}claw}_{\text{detect}}(N_1, \ldots, N_k))$ be the number of queries needed to decide whether there is a k-claw or not for functions $f_i : X_i := [N_i] \to Z$ $(i \in [k])$ given as a comparison oracle, where $N_i \leq N_j$ if $i < j$. If k is constant,*

$$Q_2(\mathbf{k\text{-}claw}_{\text{detect}}(N_1, \ldots, N_k)) = \begin{cases} O\left(\left(\prod_{i=1}^k N_i\right)^{\frac{1}{k+1}} \log N_1\right) & \text{if } \prod_{i=2}^k N_i = O(N_1^k), \\ O\left(\sqrt{\prod_{i=2}^k N_i / N_1^{k-2}} \log N_1\right) & \text{otherwise.} \end{cases}$$

Proof (Sketch). In a way similar to the case of two functions, we apply Theorem 1 on the graph categorical product of k Johnson graphs $J_{f_i} := J(|X_i|, l_i)$ $(i \in [k])$ for the domains X_i's of functions f_i's, where l_i's are integers fixed later such that $l_i \leq l_j$ for $i < j$.

To generate $|\phi_0\rangle$, we first prepare the uniform superposition of $|F_1, \ldots, F_k\rangle |F_1', \ldots, F_k'\rangle$ over all F_i and F_i' such that (F_i, F_i') is an edge of J_{f_i} for every i. This requires no queries. As in the case of two functions, define $L_{F_1,\ldots,F_k}$ for any $F_1, \ldots, F_k$ as a representation of the sorted list of all elements in $\bigcup_{i=1}^k F_i$ so that it can be uniquely determined for each tuple $(F_1, \ldots, F_k)$. We then compute $L_{F_1,\ldots,F_k}$ and $L_{F_1',\ldots,F_k'}$

for each basis state by issuing $O((l_1 + \cdots + l_k)\log(l_1 + \cdots + l_k))$ queries to the oracle. Thus, $C_U = O((l_1 + \cdots + l_k)\log(l_1 + \cdots + l_k))$. C_F and C_W can be estimated as 0 and $O(\log(l_1 + \cdots + l_k))$, respectively, in a way similar to the case of two functions. We set ϵ to $\prod_{i=1}^{k} l_i/N_i$ and δ to $\min_i\{1/l_i\} = 1/l_k$.

When $\prod_{i=2}^{k} N_i = O(N_1^k)$, we set $l_i := \Theta\left(\left(\prod_{i=1}^{k} N_i\right)^{\frac{1}{k+1}}\right)$ for every $i \in [k]$, which satisfies condition $l_i \leq N_1 \leq N_i$ for every $i \in [k]$. When $\prod_{i=2}^{k} N_i = \Omega(N_1^k)$, we set $l_i := \Theta(N_1)$ for every $i \in [k]$. □

Against a standard oracle, we obtain a similar result.

Corollary 2. *For any positive integer $k > 1$, let $Q_2(\textbf{k-claw}_{\text{detect}}(N_1, \ldots, N_k))$ be the number of queries needed to decide whether there is a k-claw or not for functions $f_i : X_i := [N_i] \to Z$ $(i \in [k])$ given as a standard oracle, where $N_i \leq N_j$ if $i < j$. If k is constant,*

$$Q_2(\textbf{k-claw}_{\text{detect}}(N_1, \ldots, N_k)) = \begin{cases} O\left(\left(\prod_{i=1}^{k} N_i\right)^{\frac{1}{k+1}}\right) & \text{if } \prod_{i=2}^{k} N_i = O(N_1^k), \\ O\left(\sqrt{\prod_{i=2}^{k} N_i/N_1^{k-2}}\right) & \text{otherwise.} \end{cases}$$

4 Claw Finding

We now describe an algorithm, Claw_Search, that finds a claw. The algorithm consists of three stages. In the first stage, we find an $O(N)$-sized subset Y' of Y such that there is a claw in $X \times Y'$, by performing binary search over Y with Claw_Detect. In the second stage, we perform 4-ary search over X and Y' with Claw_Detect to find $O(1)$-sized subsets X'' and Y'' of X and Y', respectively, such that there is a claw in $X'' \times Y''$. In the final stage, we search $X'' \times Y''$ for a claw by issuing classical queries. To keep the error rate moderate, say, at most 1/3, Claw_Detect is repeated $O(s)$ times against the same pair of domains at the sth node of the search tree at each stage. This pushes up the query complexity by only a constant multiplicative factor.

Figure 1 precisely describes Claw_Search. Steps 2, 3 and 4 in the figure correspond to the first, second and final stages, respectively.

Theorem 2. *Let $Q_2(\textbf{claw}_{\text{finding}}(N, M))$ be the number of queries needed to locate a claw if it exists for functions $f : X = [N] \to Z$ and $g : Y = [M] \to Z$ given as a comparison oracle. Then,*

$$Q_2(\textbf{claw}_{\text{finding}}(N, M)) = \begin{cases} O\left((NM)^{1/3}\log N\right) & N \leq M < N^2 \\ O(M^{1/2}\log N) & M \geq N^2. \end{cases}$$

Proof. We will analyze Claw_Search in Fig. 1.

When there is no claw, Claw_Search always outputs the correct answer. Suppose that there is a claw. The algorithm may output a wrong answer if at least one of the following two cases happens. In case (1), one of $O(\log M/N)$ runs of step 2.(b) errs; in case (2), one of $O(\log N)$ runs of step 3.(b) errs.

Algorithm Claw_Search

Input: Integers M and N such that $M \geq N$; Comparison oracle $O_{f,g}$ for functions $f : X \to Z$ and $g : Y \to Z$, respectively, such that $X := [N]$ and $Y := [M]$.
Output: Claw pair $(x, y) \in X \times Y$ such that $f(x) = g(y)$ if such a pair exists; otherwise $(-1, -1)$.

1. Set $\tilde{X} := X$ and $\tilde{Y} := Y$.
2. Set $s := 1$, and repeat the next steps until $u_{\tilde{Y}} - l_{\tilde{Y}} \leq |\tilde{X}|$, where $u_{\tilde{Y}}$ and $l_{\tilde{Y}}$ are the largest and smallest values, respectively, in $\tilde{Y}$.
   (a) Set $\Xi_Y := \{[l_{\tilde{Y}}.m_{\tilde{Y}} - 1], [m_{\tilde{Y}}.u_{\tilde{Y}}]\}$, where $m_{\tilde{Y}} := \lceil (l_{\tilde{Y}} + u_{\tilde{Y}})/2 \rceil$.
   (b) For every $\tilde{Y}' \in \Xi_Y$, do the following.
   If all $\tilde{Y}' \in \Xi_Y$ are examined, output $(-1, -1)$ and halt.
      i. Apply Claw_Detect $(s + 2)$ times to f and g restricted to domains $\tilde{X}$ and $\tilde{Y}$, respectively.
      ii. If at least one of the $(s + 2)$ results is "true," set $\tilde{Y} := \tilde{Y}'$, and break (leave (b)).
   (c) Set $s := s + 1$.
3. Set $s := 1$, and repeat the next steps until $u_D - l_D \leq c$ for every $D \in \{\tilde{X}, \tilde{Y}\}$ and some constant c, say, 100, where u_D and l_D are the largest and smallest values, respectively, in D.
   (a) For every $D \in \{\tilde{X}, \tilde{Y}\}$, set $\Xi_D := \{[l_D.u_D]\}$ if $u_D - l_D \leq c$, and otherwise, set $\Xi_D := \{[l_D.m_D - 1], [m_D.u_D]\}$ where $m_D := \lceil (l_D + u_D)/2 \rceil$.
   (b) For every pair $(\tilde{X}', \tilde{Y}') \in \Xi_{\tilde{X}} \times \Xi_{\tilde{Y}}$, do the following.
   If all the pairs are examined, output $(-1, -1)$ and halt.
      i. Apply Claw_Detect $(s + 3)$ times to f and g restricted to domains $\tilde{X}'$ and $\tilde{Y}'$, respectively.
      ii. If at least one of the $(s + 3)$ results is "true," set $\tilde{X} := \tilde{X}'$ and $\tilde{Y} := \tilde{Y}'$, and break (leave (b)).
   (c) Set $s := s + 1$.
4. Classically search $\tilde{X} \times \tilde{Y}$ for a claw.
5. Output claw $(x, y) \in \tilde{X} \times \tilde{Y}$ if it exists; otherwise output $(-1, -1)$.

Fig. 1. Algorithm Claw_Search

Without loss of generality, the error probability of Claw_Detect can be assumed to be at most 1/3. The error probability of each single run of step 2.(b).i is at most $\frac{1}{3^{s+2}}$. The error probability of each run of step 2.(b) is at most $\frac{2}{3^{s+2}} < \frac{1}{3^{s+1}}$. The error probability of case (1) is thus at most $\sum_{s=1}^{\lceil \log M/N \rceil} \frac{1}{3^{s+1}} < \frac{1}{6}$. The error probability of case (2) is also at most $\sum_{s=1}^{\lceil \log N_1 \rceil} \frac{1}{3^{s+1}} < \frac{1}{6}$ by similar calculation. Therefore, the overall error probability is at most 1/6+1/6=1/3.

We next estimate the number of queries. If $N \leq M < N^2$, the size of $\tilde{Y}$ is always at most quadratically different from that of $\tilde{X}$. Thus, the sth repetition of step 2 requires $O(s(NM/2^s)^{1/3} \log N)$ queries by Lemma 1. Similarly, the sth repetition of step 3 requires $O(s(N/2^s)^{2/3} \log N)$ queries by Lemma 1.

The total number of queries is

$$O\left(\sum_{s=1}^{\lceil \log(M/N) \rceil} \left(s \left(N \frac{M}{2^s} \right)^{1/3} \log N \right) + \sum_{s=1}^{\lceil \log N \rceil} \left(s(N/2^s)^{2/3} \log N \right) \right) = O\left((NM)^{1/3} \log N \right).$$

If $M \geq N^2$, the sth repetition of step 2 requires $O(s((NM/2^s)^{1/3} + (M/2^s)^{1/2}) \log N)$ by Lemma 1. Thus, similar calculation gives $O(M^{1/2} \log N)$ queries. □

We can easily obtain the standard-oracle version of the above theorem by using Corollary 1 instead of Lemma 1.

Corollary 3. *Let $Q_2(\mathbf{claw}_{\text{finding}}(N, M))$ be the number of queries needed to locate a claw if it exists for functions $f : X := [N] \to Z$ and $g : Y := [M] \to Z$ given as a standard oracle. Then,*

$$Q_2(\mathbf{claw}_{\text{finding}}(N, M)) = \begin{cases} O\left((NM)^{1/3}\right) & N \leq M < N^2 \\ O(M^{1/2}) & M \geq N^2. \end{cases}$$

Similarly, we can find a k-claw by using k-Claw_Detect as a subroutine. First, we find $O(N_1)$-sized subset X_i' of X_i for every $i \in [2.k]$ such that there is a k-claw in $X_1 \times X_2' \times \cdots \times X_k'$, by performing 2^{k-1}-ary search over X_i''s for all $i \in [2.k]$ with k-Claw_Detect. Let $X_1' := X_1$. We then perform 2^k-ary search over X_i's for all $i \in [k]$ with k-Claw_Detect to find $O(1)$-sized subset X_i'' of X_i' for every $i \in [k]$ such that there is a k-claw in $X_1'' \times \cdots \times X_k''$. Finally, we search $X_1'' \times \cdots \times X_k''$ for a k-claw by issuing classical queries. A more precise description of the algorithm, k-Claw_Search, is given in Fig. 2.

Theorem 3. *For any positive integer $k > 1$, let $Q_2(\mathbf{k\text{-}claw}_{\text{finding}}(N_1, \ldots, N_k))$ be the number of queries needed to locate a k-claw if it exists for k functions $f_i : X_i := [N_i] \to Z$ $(i \in [k])$ given as a comparison oracle, where $N_i \leq N_j$ if $i < j$. If k is constant,*

$$Q_2(\mathbf{k\text{-}claw}_{\text{finding}}(N_1, \cdots, N_k)) = \begin{cases} O\left(\left(\prod_{i=1}^k N_i\right)^{\frac{1}{k+1}} \log N_1\right) & \text{if } \prod_{i=2}^k N_i = O(N_1^k), \\ O\left(\sqrt{\prod_{i=2}^k N_i / N_1^{k-2}} \log N_1\right) & \text{otherwise.} \end{cases}$$

We can easily obtain the standard-oracle version of the above theorem by using Corollary 2 instead of Lemma 2.

Corollary 4. *For any positive integer $k > 1$, let $Q_2(\mathbf{k\text{-}claw}_{\text{finding}}(N_1, \ldots, N_k))$ be the number of queries needed to locate a k-claw if it exists for k functions $f_i : X_i := [N_i] \to Z$ $(i \in [k])$ given as a standard oracle, where $N_i \leq N_j$ if $i < j$. If k is constant,*

$$Q_2(\mathbf{k\text{-}claw}_{\text{finding}}(N_1, \cdots, N_k)) = \begin{cases} O\left(\left(\prod_{i=1}^k N_i\right)^{\frac{1}{k+1}}\right) & \text{if } \prod_{i=2}^k N_i = O(N_1^k), \\ O\left(\sqrt{\prod_{i=2}^k N_i / N_1^{k-2}}\right) & \text{otherwise.} \end{cases}$$

5 Lower Bound

We reduce the element distinctness problem ED(N) to the claw finding problem to obtain a lower bound. We define ED(N) as follows.

Problem 3 (Element Distinctness Problem (ED(N))). Given function $h : V = [N] \to Z$ as an oracle, is there a pair $(x, y) \in V \times V$ such that $h(x) = h(y)$?

Algorithm k-Claw_Search

Input: k integers $N_1, \ldots, N_k$ such that $N_i \leq N_j$ if $i < j$.
Comparison oracle $O_{f_1,\ldots,f_k}$ for functions $f_i : X_i \to Z$ such that $X_i := [N_i]$ for every $i \in [k]$.

Output: k-claw $(x_1, \ldots, x_k) \in X_1 \times \cdots \times X_k$ such that $f_i(x_i) = f_j(x_j)$ for every $i, j \in [k]$ if it exists; otherwise $(-1, \ldots, -1)$.

1. Set $\tilde{X}_i := X_i$ for every $i \in [k]$.
2. Set $s := 1$, and repeat the next steps until $u_i - l_i \leq |\tilde{X}_1|$ for all $i \in [2.k]$, where u_i and l_i are the largest and smallest values, respectively, in $\tilde{X}_i$.
   (a) For every $i \in [2.k]$, set $\Xi_i := \{[l_i.u_i]\}$ if $u_i - l_i \leq |\tilde{X}_1|$, and otherwise, set $\Xi_i := \{[l_i.m_i - 1], [m_i.u_i]\}$ where $m_i := \lceil (l_i + u_i)/2 \rceil$.
   (b) For every tuple $(\tilde{X}'_1, \tilde{X}'_2, \ldots, \tilde{X}'_k) \in \{\tilde{X}_1\} \times \Xi_2 \times \cdots \times \Xi_k$, do the following. If all the tuples are examined, output $(-1, \ldots, -1)$ and halt.
      i. Apply k-Claw_Detect $(s + 1) + \lceil \log_3 2^{k-1} \rceil$ times to the k functions f_i restricted to domains $\tilde{X}'_i$, respectively, for every $i \in [k]$.
      ii. If at least one of the $(s + 1) + \lceil \log_3 2^{k-1} \rceil$ results is "true," set $\tilde{X}_i := \tilde{X}'_i$ for every $i \in [2.k]$, and break (leave (b)).
   (c) Set $s := s + 1$.
3. Set $s := 1$, and repeat the next steps until $u_i - l_i \leq c$ for all $i \in [k]$ and some constant c, say, 100, where u_i and l_i are the largest and smallest values, respectively, in $\tilde{X}_i$.
   (a) For every $i \in [k]$, set $\Xi_i := \{[l_i.u_i]\}$ if $u_i - l_i \leq c$, and otherwise, set $\Xi_i := \{[l_i.m_i - 1], [m_i.u_i]\}$ where $m_i = \lceil (l_i + u_i)/2 \rceil$.
   (b) For every tuple $(\tilde{X}'_1, \tilde{X}'_2, \ldots, \tilde{X}'_k) \in \Xi_1 \times \cdots \times \Xi_k$, do the following. If all the tuples are examined, output $(-1, \ldots, -1)$ and halt.
      i. Apply k-Claw_Detect $(s + 1) + \lceil \log_3 2^k \rceil$ times to the k functions f_i restricted to domains $\tilde{X}'_i$ for every $i \in [k]$.
      ii. If at least one of the $(s+1)+\lceil \log_3 2^k \rceil$ results is "true," set $\tilde{X}_i := \tilde{X}'_i$ for every $i \in [k]$, and break (leave (b)).
   (c) Set $s := s + 1$.
4. Classically search $\tilde{X}_1 \times \cdots \times \tilde{X}_k$ for a k-claw.
5. Output k-claw $(x_1, \ldots, x_k) \in X'_1 \times \cdots \times X'_k$ if it exists; otherwise output $(-1, \ldots, -1)$.

Fig. 2. Algorithm k-Claw_Search

For problem ED(N), Aaronson and Shi [1] proved the tight lower bound of the bounded error quantum query complexity against a standard oracle: $\Omega(N^{2/3})$. This lower bound holds for a comparison oracle up to a constant multiplicative factor. Ambainis [2] gave an algorithm that needs $O(N^{2/3})$ queries for a search version of ED(N). Thus, ED(N) has query complexity $\Theta(N^{2/3})$ for both the search and decision versions.

The reduction is as follows. Suppose that we are given an instance of ED($N + M$) as comparison oracle $O_{h,h} : |p, q, b, w\rangle \longrightarrow |p, q, b \oplus [h(p) \leq h(q)], w\rangle$, where $p, q \in V$, $b \in \{0, 1\}$, $[h(p) \leq h(q)]$ is the predicate such that it is 1 if and only if $h(p) \leq h(q)$, and w is work space. Let A be any bounded-error quantum algorithm for the claw finding problem for functions f and g of domains X and Y, respectively, such that $|X| = N$ and $|Y| = M$. We will construct an algorithm to solve ED($N + M$) by using A. First, we randomly divide the domain of ED($N + M$) into two disjoint subdomains of sizes N and M, respectively, and we then apply A to the sub-domains. If A finds a claw, the answer to ED($N+M$) is *no*; otherwise the answer is *yes*. Finally, we amplify the success probability of this procedure by using the quantum amplitude amplification [3]. Then we have the next lemma.

Lemma 3. *$Q_2(\mathbf{claw}_{\text{finding}}(N, M))$ is $\Omega(N^{1/2}M^{1/6})$.*

This bound together with the previous bound $\Omega(M^{1/2})$ given in [6] implies that our algorithm for the claw finding problem is optimal when $M = \Theta(N)$ or $M = \Omega(N^2)$.

As a by-product, an optimal quantum algorithm of $O(N^{2/3})$ queries for the element distinctness problem, which is different from the first optimal algorithm in [2], can be obtained by combining the above reduction in the case of $N = M$ with our claw finding algorithm.

References

1. Aaronson, S., Shi, Y.: Quantum lower bounds for the collision and the element distinctness problems. J. ACM 51(4), 595–605 (2004)
2. Ambainis, A.: Quantum walk algorithm for element distinctness. In: Proc. 45th IEEE FOCS, pp. 22–31. IEEE Computer Society Press, Los Alamitos (2004)
3. Brassard, G., Høyer, P., Mosca, M., Tapp, A.: Quantum amplitude amplification and estimation, Quantum Computation and Quantum Information: A Millennium Volume. AMS Contem. Math. 305, 53–74 (2002)
4. Brassard, G., Høyer, P., Tapp, A.: Quantum cryptanalysis of hash and claw-free functions. In: Lucchesi, C.L., Moura, A.V. (eds.) LATIN 1998. LNCS, vol. 1380, pp. 163–169. Springer, Heidelberg (1998)
5. Brouwer, A.E., Cohen, A.M., Neumaier, A.: Distance-Regular Graphs. In: Brouwer, A.E., Cohen, A.M., Neumaier, A. (eds.) A series of Modern Surveys in Mathematics, Springer, Heidelberg (1989)
6. Buhrman, H., Dürr, C., Heiligman, M., Høyer, P., Magniez, F., Santha, M., de Wolf, R.: Quantum algorithms for element distinctness. In: Proc. 16th IEEE Conference on Computational Complexity, pp. 131–137. IEEE Computer Society Press, Los Alamitos (2001)
7. Buhrman, H., Dürr, C., Heiligman, M., Høyer, P., Magniez, F., Santha, M., de Wolf, R.: Quantum algorithms for element distinctness. SIAM J. Comput. 34(6), 1324–1330 (2005)
8. Buhrman, H., Špalek, R.: Quantum verification of matrix products. In: Proc. 17th ACM/SIAM SODA, pp. 880–889 (2006)
9. Childs, A.M., Eisenberg, J.M.: Quantum algorithms for subset finding. Quantum Information and Computation 5(7), 593–604 (2005)
10. Grover, L.K.: A fast quantum mechanical algorithm for database search. In: Proc. of 28th ACM STOC, pp. 212–219. ACM Press, New York (1996)
11. Høyer, P., Mosca, M., de Wolf, R.: Quantum search on bounded-error inputs. In: Baeten, J.C.M., Lenstra, J.K., Parrow, J., Woeginger, G.J. (eds.) ICALP 2003. LNCS, vol. 2719, pp. 291–299. Springer, Heidelberg (2003)
12. Iwama, K., Kawachi, A.: A new quantum claw-finding algorithm for three functions. New Generation Computing 21(4), 319–327 (2003)
13. Magniez, F., Santha, M., Szegedy, M.: Quantum algorithms for the triangle problem. In: Proc. 16th ACM/SIAM SODA, pp. 1109–1117 (2005)
14. Shor, P.W.: Polynomial-time algorithms for prime factorization and discrete logarithms on a quantum computer. SIAM J. Comput. 26(5), 1484–1509 (1997)
15. Szegedy, M.: Quantum speed-up of markov chain based algorithms. In: Proc. 45th IEEE FOCS, pp. 32–41. IEEE Computer Society Press, Los Alamitos (2004)

Complexity Upper Bounds for Classical Locally Random Reductions Using a Quantum Computational Argument*

Rahul Tripathi

Department of Computer Science and Engineering, University of South Florida,
Tampa, FL 33620, USA
tripathi@cse.usf.edu

Abstract. We use a *quantum computational* argument to prove, for any integer $k \geq 2$, a complexity upper bound for nonadaptive k-query classical locally random reductions (LRRs) that allow bounded-errors. Extending and improving a recent result of Pavan and Vinodchandran [PV], we prove that if a set L has a nonadaptive 2-query classical LRR to functions g and h, where both g and h can output $O(\log n)$ bits, such that the reduction succeeds with probability at least $1/2 + 1/\text{poly}(n)$, then $L \in \text{PP}^{\text{NP}}/\text{poly}$. Previous complexity upper bound for nonadaptive 2-query classical LRRs was known only for much restricted LRRs: LRRs in which the target functions can only take values in $\{0, 1, 2\}$ and the error probability is zero [PV]. For $k > 2$, we prove that if a set L has a nonadaptive k-query classical LRR to *boolean* functions $g_1, g_2, \ldots, g_k$ such that the reduction succeeds with probability at least $2/3$ and the distribution on $(k/2+\sqrt{k})$-element subsets of queries depends only on the input length, then $L \in \text{PP}^{\text{NP}}/\text{poly}$. Previously, for no constant $k > 2$, a complexity upper bound for nonadaptive k-query classical LRRs was known even for LRRs that do not make errors.

Our proofs follow a two stage argument: (1) simulate a nonadaptive k-query classical LRR by a 1-query quantum *weak* LRR, and (2) upper bound the complexity of this quantum weak LRR. To carry out the two stages, we formally define nonadaptive quantum weak LRRs, and prove that if a set L has a 1-query quantum weak LRR to a function g, where g can output *polynomial* number of bits, such that the reduction succeeds with probability at least $1/2 + 1/\text{poly}(n)$, then $L \in \text{PP}^{\text{NP}}/\text{poly}$.

1 Introduction

1.1 Background

A *locally random reduction* (LRR) of a set L to a database f is an efficient computational procedure that allows to determine the membership of any instance x in L by using random queries to the database. The concept of LRR is motivated from the standpoint of cryptographic security and can be understood from the following example. Suppose Alice holds an object (encoded as a binary string) and wants to efficiently

* Research supported by the New Researcher Grant of University of South Florida.

L. Kučera and A. Kučera (Eds.): MFCS 2007, LNCS 4708, pp. 548–558, 2007.

retrieve some information about the object by using queries to Bob (database f). However for security reasons, Alice cannot reveal her object to Bob. Therefore, she makes random queries to Bob so that Bob gets no clue about the object from the queries. An LRR is an efficient computational procedure that allows Alice to retrieve the information without leaking Bob anything more than the size of the object. (See Section 2.4 for a more general definition of LRR.) An LRR where the set L and the database f are the same, i.e., f is the characteristic function of L, is called a *random self-reduction* (RSR).

In general, one can define an LRR of a set L to *several* databases f_1, f_2, ..., f_k such that the reduction (1) is an efficient randomized procedure, (2) allows determining the membership of any instance x in L with bounded-error probability by using random queries α_1, α_2, ..., α_k to f_1, f_2, ..., f_k, respectively, and (3) leaks no detail more than $|x|$ to an adversary even if the adversary is revealed any subset of at most t queries, for some fixed $t \geq 1$. For the special case $f_1 = f_2 = \ldots = f_k = f$, Beaver et al. [BFKR97] called this reduction a (t, k)-locally random reduction of L to f. This general notion of LRR subsumes earlier studied notions of random reductions known as *single-oracle* [AFK89] and *multioracle* [Riv86] instance-hiding schemes. Here "leaking no detail more than $|x|$ to an adversary even if the adversary is revealed any subset of t queries" has the following interpretation: If instances x and y are such that $|x| = |y|$, then for any $i_1, i_2, \ldots, i_t$, the distribution on the t queries $\langle \alpha_{i_1}, \alpha_{i_2}, \ldots, \alpha_{i_t} \rangle$ induced by the randomized reduction on input x is identical to the distribution induced by the randomized reduction on input y. Thus, from the viewpoint of an adversary who has access to some subset of t queries, any instance y of length $|x|$ is equally likely to be the input of the reduction.

Both LRR and RSR have proved to be useful at several places in computational complexity theory. They have found implicit or explicit applications in worst-case to average-case reductions [Lip91], random oracle separations [Bab87], interactive proof systems [BFL91, LFKN92, Sha92], program checkers and self-testing/correcting pairs [BK95, Lip91, BLR93], probabilistically checkable proof systems [AS98, ALM+98, FGL+96], cryptography [GM84, BM84], instance hiding schemes [Riv86, AFK89, BF90, BFKR97], zero knowledge proofs on committed bits [BFKR97], private information retrieval [BFG06], and locally decodable codes [PV].

1.2 Related Work

A direct consequence of a result of Beaver and Feigenbaum [BF90] is that for every set L, there is a function f such that L is $(n + 1)$-query locally random reducible to f. Beaver, Feigenbaum, Kilian, and Rogaway [BFKR97] extended and improved this result: For any constant $c > 0$ and any function $t : \mathbb{N} \to \mathbb{N}$, every set L is $t\lfloor n/c \log n \rfloor$-query locally random reducible to some function f, where the distribution on t-element subsets of queries depends only on n and where the lengths of answers from f could be $\Theta(\log n + \log t)$.

There have been work on understanding the complexity of functions that can be locally random reduced via k queries to some function f, for constants $k \geq 1$. Abadi, Feigenbaum, and Kilian [AFK89] proved that if a set L is 1-query locally random reducible to some function, then L is in NP/poly $\cap$ coNP/poly. Yao [Yao90] proved that if a set L is 2-query locally random reducible to some *boolean* function, then L is in PSPACE/poly. Fortnow and Szegedy [FS92] improved upon Yao's result and showed that any such set in fact belongs to NP/poly $\cap$ coNP/poly. Pavan and Vinodchandran [PV] addressed the question whether the results of Yao, and Fortnow and Szegedy can be extended for LRRs where the reductions are to functions other than the boolean functions. Building on the work of Yao [Yao90] and Fortnow and Szegedy [FS92], they proved that if a set L is 2-query locally random reducible to functions g and h that take values in $\{0, 1, 2\}$, then L is in PSPACE/poly. The LRRs considered in the last three papers, i.e., in [Yao90, FS92, PV] do not allow errors.

1.3 Our Results

A comparison of our results with previously known results is summarized in Table 1. The notations "(t, k, ℓ, ϵ)-clr" and "(t, k, ℓ, ϵ)-qwlr," used in Table 1, capture generalizations of previously studied notions of classical LRRs. They are defined as follows: (a) A nonadaptive (t, k, ℓ, ϵ)-clr is an LRR of a set L to some functions $g_1, g_2, \ldots, g_k$ such that (1) the reduction makes k nonadaptive queries $\alpha_1, \alpha_2, \ldots, \alpha_k$ to $g_1, g_2, \ldots, g_k$, respectively, (2) the distribution on t-element subsets of queries $\langle \alpha_{i_1}, \alpha_{i_2}, \ldots, \alpha_{i_t} \rangle$ is dependent only on the input length n to the reduction, (3) each g_i returns $\ell(n)$ bits on input length n to the reduction, and (4) the reduction succeeds with probability at least $1/2 + \epsilon(n)$, and (b) A nonadaptive (t, k, ℓ, ϵ)-qwlr is a *quantum* analog of (t, k, ℓ, ϵ)-clr, where the reduction can be a bounded-error quantum polynomial-time algorithm that can make quantum queries. (See Definition 3 and Definition 4 for a formal definition of these notions.) For notational convenience, if the answers returned by the target functions in LRRs can only take values in $\{0, 1, 2\}$, then for such reductions we define the number of answer bits ℓ to be 3/2 (see, for instance, the second column of Table 1 in the entry corresponding to [PV]).

Note that our proofs for classical LRRs use quantum computational arguments. The application of quantum arguments in proving results related to classical computing is a surprising phenomenon witnessed only in the last few years. See, for instance, the

Table 1. Summary of results showing complexity upper bounds for various nonadaptive LRRs

Papers	Type of nonadaptive LRRs (t, k, ℓ, ϵ)-clr / (t, k, ℓ, ϵ)-qwlr	Complexity Upper Bound
[AFK89]	$(1, 1, \mathrm{poly}(n), 1/\mathrm{poly}(n))$-clr	NP/poly $\cap$ coNP/poly
[Yao90]	$(1, 2, 1, 1/2)$-clr	PSPACE/poly
[FS92]	$(1, 2, 1, 1/2)$-clr	NP/poly $\cap$ coNP/poly
[BFKR97]	$(t, t\lfloor n/O(\log n)\rfloor, O(\log n + \log t), 1/2)$-clr	None
[PV]	$(1, 2, 3/2, 1/2)$-clr	PSPACE/poly
This paper	$\mathbf{(1, 2, O(\log n), 1/poly(n))}$**-clr**	$\mathbf{PP^{NP}/poly}$
This paper	$\mathbf{(k/2 + \sqrt{k}, k, 1, 1/6)}$**-clr**	$\mathbf{PP^{NP}/poly}$
This paper	$\mathbf{(1, 1, poly(n), 1/poly(n))}$**-qwlr**	$\mathbf{PP^{NP}/poly}$

papers [KdW04, WdW05, AR03, AR05, Aar05, dW06, Ker05, LLS05, Aar06] where quantum computational arguments have been used to prove classical complexity results.) This paper fits in this growing body of research on proving classical complexity results using quantum computational arguments. Our results also shed light on the role of quantum computational arguments in the understanding of classical computation.

2 Preliminaries

2.1 Notations

Let $\mathbb{N}$ denote the set of all positive integers. Our alphabet is $\Sigma = \{0, 1\}$. For a binary string $b \in \Sigma^\ell$, we use b_i to denote its i'th bit. We identify a binary string $b \in \Sigma^\ell$ alternatively as a bit vector $\boldsymbol{b} = (b_1, b_2, \ldots, b_\ell)$. Given two binary strings a, $b \in \Sigma^\ell$, their *inner product* $\boldsymbol{a} \cdot \boldsymbol{b}$ is the integer $\boldsymbol{a} \cdot \boldsymbol{b} =_{df} \sum_{i=1}^{\ell} a_i \cdot b_i$, and their *xor* is the binary string $a \oplus b$ obtained by taking the xor of the individual bits of a and b, i.e., $a \oplus b =_{df} (a_1 \oplus b_1) \ldots (a_\ell \oplus b_\ell)$. For a string $a \in \Sigma^\ell$, we use $|\boldsymbol{a}|$ to denote the number of 1's in a and for a set A, we use $|A|$ to denote the cardinality of A (which sense is being used for "$|\cdot|$" will be clear from the context). For a binary string $b \in \Sigma^\ell$, let $\text{int}(b) \in \{0, 1, \ldots, 2^\ell - 1\}$ denote the integer representation of b. Let $[n] =_{df} \{1, 2, \ldots, n\}$ for all $n \in \mathbb{N}$.

2.2 Basics of Quantum Computing

Let $\mathcal{H}$ denote a two-dimensional Hilbert space, i.e., a complex vector space equipped with an inner product $\langle \cdot \mid \cdot \rangle$ operation. A qubit $|u\rangle =_{df} (\alpha, \beta)^T$ represent the states $|0\rangle$ and $|1\rangle$, respectively, associated with a qubit. The states $|0\rangle$ and $|1\rangle$ are called the *computational basis* states. We can express the qubit $|u\rangle$ as a linear combination of the computational basis states: $|u\rangle = \alpha|0\rangle + \beta|1\rangle$. Here α and β are complex numbers, called the *amplitudes* of $|u\rangle$. Since $|u\rangle$ is a unit vector, the amplitudes (of $|u\rangle$) must satisfy $|\alpha|^2 + |\beta|^2 = 1$.

An m-qubit is a unit vector in the 2^m-dimensional Hilbert space $\mathcal{H}^{\otimes m} =_{df} \mathcal{H} \otimes \cdots \otimes \mathcal{H}$, the m-fold tensor product of $\mathcal{H}$ with itself. A multiple qubit is an m-qubit for some integer $m > 1$. The computational basis states of $\mathcal{H}^{\otimes m}$ are the m-fold tensor product of the computational basis states of $\mathcal{H}$. That is, they are $|b_1\rangle \otimes \cdots \otimes |b_m\rangle$, where for each $1 \le i \le m$, b_i ranges over 0 and 1. The vector representation of a computational basis state $|b_1\rangle \otimes \cdots \otimes |b_m\rangle$ is the column vector with 2^m rows in which the only row containing 1 is at location $1 + \text{int}(b_1 b_2 \ldots b_m)$ and all other rows contain 0. We sometimes use the standard abbreviation $|a\rangle|b\rangle$ for $|a\rangle \otimes |b\rangle$, where $|a\rangle$ and $|b\rangle$ can be arbitrary multiple qubits. An m-qubit $|u\rangle =_{df} (\alpha_{0^m}, \alpha_{0^{m-1}1}, \ldots, \alpha_{1^m})^T$ can be expressed as a linear combination of the computation basis states of $\mathcal{H}^{\otimes m}$: $|u\rangle = \sum_{i \in \Sigma^m} \alpha_i |i\rangle$. Here, the complex numbers α_is are the amplitudes of $|u\rangle$. These amplitudes satisfy $\sum_{i \in \Sigma^m} |\alpha_i|^2 = 1$ because $|u\rangle$ is a unit vector in $\mathcal{H}^{\otimes m}$.

The conjugate transpose of a vector $|u\rangle$, i.e., $|u\rangle^\dagger$, is denoted by $\langle u|$. The inner product $\langle \cdot|\cdot \rangle$ of vectors $|u\rangle$ and $|v\rangle$ can be expressed as: $\langle u|v\rangle = \langle u| \cdot |v\rangle$, i.e., the matrix product of $\langle u|$ and $|v\rangle$. The vectors $|u\rangle$ and $|v\rangle$ are orthogonal if their inner product $\langle u|v\rangle$ is zero. The norm of $|u\rangle$ is $||u|| =_{df} \sqrt{\langle u|u\rangle}$.

A quantum system that can take one of a number of states $|\psi_i\rangle$ with respective probabilities p_i is said to be in a *mixed* state. A quantum system whose state is known exactly is said to be in *pure* state. (Thus, a pure quantum state is also a mixed quantum state, but not the vice-versa.) A mixed quantum state is described by an ensemble $\{p_i, |\psi_i\rangle\}$ of pure quantum states. We can equivalently describe this mixed quantum state in terms of the density operator ρ: $\rho = \sum_i p_i |\psi_i\rangle\langle\psi_i|$. In particular, for a pure quantum state $|\psi\rangle$, the density operator is $|\psi\rangle\langle\psi|$.

Any operation on a quantum system is either a *unitary* operation or a measurement operation. A unitary operation is described by a linear transformation U, which preserves the ℓ_2 norm: for any state $|\psi\rangle$, $||\psi|| = ||U|\psi\rangle||$. When a unitary operation U is performed on a state $|\psi\rangle$, the resulting state is $U|\psi\rangle$. In the terminology of density operators, U transforms the state ρ into state $U\rho U^\dagger$.

The most general measurement in quantum mechanics is the POVM measurement, which is described by a collection of positive semidefinite measurement operators $E_m = M_m^\dagger M_m$ satisfying $\sum_m E_m = I$. If a measurement described by the measurement operators E_m is performed on a quantum system in state $|\psi\rangle$, then the probability $p(m)$ of getting outcome m is given by $p(m) =_{df} \langle\psi|M_m^\dagger M_m|\psi\rangle = \langle\psi|E_m|\psi\rangle$ and the resulting state is $\frac{M_m|\psi\rangle}{\sqrt{p(m)}}$. In the terminology of density operators ρ, the probability $p(m)$ is given by $p(m) =_{df} \mathrm{Tr}(M_m \rho M_m^\dagger) = \mathrm{Tr}(E_m \rho)$ and the resulting state is $\frac{M_m \rho M_m^\dagger}{p(m)}$. By *measuring in the computational basis* of a 2^m-dimensional Hilbert space $\mathcal{H}^{\otimes m}$, we mean that we perform a measurement whose measurement operators E_m are given by $E_m = M_m = |\psi_m\rangle\langle\psi_m|$, where $|\psi_m\rangle$s are the computational basis states of $\mathcal{H}^{\otimes m}$.

A *bipartite* quantum system consists of two subsystems. Let $\mathcal{H}$ and $\mathcal{K}$ be Hilbert spaces and let ρ be the density operator of a bipartite quantum system over the Hilbert space $\mathcal{H} \otimes \mathcal{K}$. A *partial trace* $\mathrm{Tr}_{\mathcal{K}}$ of ρ over $\mathcal{K}$ is the following mapping: $\mathrm{Tr}_{\mathcal{K}}(\rho) = \sum_{j=1}^{n} (I \otimes \langle e_j|)\rho(I \otimes |e_j\rangle)$, where $\{|e_1\rangle, |e_2\rangle, \ldots, |e_n\rangle\}$ is any orthonormal basis of $\mathcal{K}$. Intuitively, the *partial trace* $\mathrm{Tr}_{\mathcal{K}}(\rho)$ of a mixed state ρ of a bipartite system over the Hilbert space $\mathcal{H} \otimes \mathcal{K}$ is the density operator of the first part (i.e., $\mathcal{H}$) of the system obtained by discarding the second part (i.e., $\mathcal{K}$) of the system.

We will consider quantum queries whose answers are ℓ bits long, for some $\ell \geq 1$. A quantum query to an oracle $\mathcal{O} : \Sigma^m \to \Sigma^\ell$ is the unitary transformation given by $|s\rangle|z\rangle \longrightarrow |s\rangle|z \oplus \mathcal{O}(s)\rangle$, where $z \in \Sigma^\ell$ is called the target register. For convenience, we store the query answer in the phase of the quantum state instead of storing it in the target register. To store the answer in the phase, define for any $R \in \Sigma^\ell$, the quantum state $|z_R\rangle = \frac{1}{\sqrt{2^\ell}} \bigotimes_{i=1}^{\ell} (|0\rangle + (-1)^{R_i}|1\rangle)$. A quantum query to an oracle $\mathcal{O} : \Sigma^m \longrightarrow \Sigma^\ell$ then, for any $R \in \Sigma^\ell$, maps $|s\rangle|z_R\rangle$ to $(-1)^{\boldsymbol{R}\cdot\boldsymbol{\mathcal{O}}(\boldsymbol{s})}|s\rangle|z_R\rangle$.

Finally, we refer the reader to the excellent textbook [NC00] for any relevant concept in quantum computing that is not explained here.

2.3 Complexity Classes

The quantum complexity class BQP/qpoly is the class of all sets decidable by a polynomial-time quantum computer when given a polynomial-size quantum advice state, which depends only on the input length.

Definition 1. $\mathrm{BQP/qpoly}$ *is the class of all sets* L *for which there exist a polynomial-size quantum circuit family* $\{C_n\}_{n\in\mathbb{N}}$ *and a polynomial-size family of quantum states* $\{|\Psi_n\rangle\}_{n\in\mathbb{N}}$ *such that for all* $n \in \mathbb{N}$ *and* $x \in \Sigma^n$,

1. *if* $x \in L$, *then* C_n *accepts* $|x\rangle|\Psi_n\rangle$ *with probability at least* $2/3$, *and*
2. *if* $x \notin L$, *then* C_n *accepts* $|x\rangle|\Psi_n\rangle$ *with probability at most* $1/3$.

We will require the following result of Aaronson [Aar04] on the power of $\mathrm{BQP/qpoly}$. This result holds in every relativized world.

Theorem 2. [Aar04] $\mathrm{BQP/qpoly} \subseteq \mathrm{PP/poly}$.

2.4 Locally Random Reduction

Beaver et al. [BFKR97] formally introduced the notion of LRRs. They defined LRRs that reduce a function f to a *single* function g using k random queries, succeed with probability at least $3/4$, and leak no detail more than $|x|$ even if any t-element subset of queries is revealed.

We present a more general definition of (nonadaptive) LRRs in Definition 3. This new definition also takes into account the number of answer bits returned by the target functions and the success probability of the reduction.

Definition 3 (Nonadaptive Classical Locally Random Reduction). *Let* $t, k \in \mathbb{N}$, $\ell : \mathbb{N} \to \mathbb{N}$, *and* $\epsilon : \mathbb{N} \to (0, 1/2]$. *A set* L *is nonadaptively* (t, k, ℓ, ϵ)*-classically-locally-random (or, "nonadaptively* (t, k, ℓ, ϵ)*-clr" in short) reducible to functions* g_1, g_2, ..., g_k, *where each* g_i *outputs* $\ell(n)$ *bits on inputs of length* n *to the reduction, if there exist a classical bounded-error probabilistic polynomial-time algorithm* A, *a polynomial-time function* σ, *and a polynomial* $p(\cdot)$ *such that:*

1. **[Query Reduction]** *For all* $n \in \mathbb{N}$, $x \in \Sigma^n$, *and* $r \in \Sigma^{p(n)}$, A *makes* k *nonadaptive queries* $\sigma(1, x, r)$, $\sigma(2, x, r)$, ..., $\sigma(k, x, r)$ *to* g_1, g_2, ..., g_k, *respectively.*
2. **[Local Randomness]** *For all* $n \in \mathbb{N}$ *and* $\{i_1, i_2, \ldots, i_t\} \subseteq [k]$, *if* $r \in \Sigma^{p(n)}$ *is chosen uniformly at random, then for any* $x, y \in \Sigma^n$, *the distribution on* $\langle\sigma(i_1, x, r), \sigma(i_2, x, r), \ldots, \sigma(i_t, x, r)\rangle$ *is identical to that on* $\langle\sigma(i_1, y, r), \sigma(i_2, y, r), \ldots, \sigma(i_t, y, r)\rangle$.
3. **[Correctness]** *For all* $n \in \mathbb{N}$ *and* $x \in \Sigma^n$, *it holds that*

$$\mathrm{Prob}_r\left[A(x, r, g_1(\sigma(1, x, r)), \ldots, g_k(\sigma(k, x, r))) = L(x)\right] \geq \frac{1}{2} + \epsilon,$$

where the probability is over the uniform random choice of $r \in \Sigma^{p(n)}$.

If the algorithm A *receives* $h(n)$ *bits of advice on input length* n, *then we say that* L *is nonuniformly nonadaptively* (t, k, ℓ, ϵ)*-clr reducible to functions* g_1, g_2, ..., g_k *with* $h(n)$ *bits of advice.*

Our proofs of complexity upper bounds for nonadaptive classical LRRs use (as a tool) the notion of nonadaptive quantum *weak* LRRs, defined in Definition 4. We mention that our notion may not fully capture the most general notion of nonadaptive quantum LRRs.[1]

Definition 4 (Nonadaptive Quantum Weak Locally Random Reduction). *Let $t, k \in \mathbb{N}$, $\ell : \mathbb{N} \to \mathbb{N}$, and $\epsilon : \mathbb{N} \to (0, 1/2]$. A set L is nonadaptively (t, k, ℓ, ϵ)-quantumly-weakly-locally-random (or, "nonadaptively (t, k, ℓ, ϵ)-qwlr" in short) reducible to functions g_1, g_2, …, g_k, where each g_i outputs $\ell(n)$ bits on inputs of length n to the reduction, if there exist a bounded-error quantum polynomial-time algorithm A, and polynomials $p(\cdot)$ and $q(\cdot)$ such that:*

1. **[Query Reduction]** *For all $n \in \mathbb{N}$, $x \in \Sigma^n$, A uniformly at random selects a string $r \in \Sigma^{p(n)}$, deterministically computes k sets of strings $T^1_{x,r}$, $T^2_{x,r}$, …, $T^k_{x,r} \subseteq \Sigma^{q(n)}$, and makes k quantum queries of the form: $|Q(x, r)\rangle =$*
$$\frac{1}{\sqrt{\prod_{i=1}^{k} |T^i_{x,r}|}} \sum_{s_{j_1} \in T^1_{x,r}} \sum_{s_{j_2} \in T^2_{x,r}} \cdots \sum_{s_{j_k} \in T^k_{x,r}} |s_{j_1}, s_{j_2}, \ldots, s_{j_k}\rangle \otimes |\psi\rangle^{\otimes k},$$
where $|\psi\rangle = \frac{1}{\sqrt{2^\ell}} \sum_{R \in \Sigma^\ell} |z_R\rangle$. Also, each $T^i_{x,r} \subseteq \Sigma^{q(n)}$ depends only on i, x, and r, and $|T^i_{x,r}|$ depends only on i and n.
2. **[Local Randomness]** *For all $n \in \mathbb{N}$ and $\{i_1, i_2, \ldots, i_t\} \subseteq [k]$, and for any $x \in \Sigma^n$, the distribution on strings at locations i_1, i_2, …, i_t induced by measuring the query state $\frac{1}{\sqrt{2^{p(n)}}} \sum_{r \in \Sigma^{p(n)}} |r\rangle |Q(x, r)\rangle$ in the* computational basis *is independent of x, but may perhaps depend on n. In other words, if $r \in \Sigma^{p(n)}$ is chosen uniformly at random, then for any $x, y \in \Sigma^n$ and for all s_1, s_2, …, $s_t \in \Sigma^{q(n)}$, it holds that*
$$\text{Prob}_r \left[s_1 \in T^{i_1}_{x,r} \wedge \ldots \wedge s_t \in T^{i_t}_{x,r}\right] = \text{Prob}_r \left[s_1 \in T^{i_1}_{y,r} \wedge \ldots \wedge s_t \in T^{i_t}_{y,r}\right].$$

[1] Definition 4 may not fully capture nonadaptive quantum LRRs in the most general way for the following two reasons: (1) In the *query reduction* property, we require the quantum algorithm A, on input x, to *randomly* select $r \in \Sigma^{p(n)}$ and generate a superposition over strings belonging to the *polynomial-time computable* sets $T^i_{x,r}$, for $1 \leq i \leq k$. Meaning, the quantum state describing the quantum queries just before their answers are received is given by $|\Phi\rangle =_{df} \frac{1}{\sqrt{2^{p(n)}}} \sum_{r \in \Sigma^{p(n)}} |r\rangle |Q(x, r)\rangle$. While this requirement on the form of $|\Phi\rangle$ helps to serve our purpose, which is obtaining complexity upper bounds for nonadaptive classical LRRs, it may not be an essential requirement for the most general definition of quantum LRRs. (2) In the *local randomness* property, we consider measurements only in the *computational basis*. Again, we restrict to only such measurements because they suffice to obtain complexity upper bounds for nonadaptive classical LRRs. If we consider general (POVM) measurements, then the *local randomness* property may be stated as follows:

Let ρ_x denote the density operator describing the state $|\Phi\rangle$ of the quantum queries just before their answers are received, i.e., $\rho_x = |\Phi\rangle\langle\Phi|$.
[Local Randomness] For all $n \in \mathbb{N}$, $\{i_1, i_2, \ldots, i_t\} \subseteq [k]$, and for any $x, y \in \Sigma^n$, it holds that
$$\text{Tr}_{r,[k]-\{i_1,i_2,\ldots,i_t\}}(\rho_x) = \text{Tr}_{r,[k]-\{i_1,i_2,\ldots,i_t\}}(\rho_y),$$
where $\text{Tr}_{r,[k]-\{i_1,i_2,\ldots,i_t\}}(\rho_x)$ denotes the reduced density operator obtained by taking the partial trace of ρ_x over the qubits storing r and the qubits corresponding to the query locations $[k] - \{i_1, i_2, \ldots, i_t\}$.

3. **[Correctness]** *For all* $n \in \mathbb{N}$ *and* $x \in \Sigma^n$, *it holds that*

$$\mathrm{Prob}_{r,A}\left[A\left(x, r, g_1 \circ g_2 \circ \cdots \circ g_k(|Q(x,r)\rangle)\right) = L(x)\right] \geq \frac{1}{2} + \epsilon,$$

where the probability is over the uniform random choice of $r \in \Sigma^{p(n)}$ *and over the inherent randomness of* A. *Here* $g_1 \circ g_2 \circ \cdots \circ g_k(|Q(x,r)\rangle)$ *denotes*

$$\frac{1}{\sqrt{\prod_{i=1}^{k} |T^i_{x,r}|}} \sum_{s_{j_1} \in T^1_{x,r}} \cdots \sum_{s_{j_k} \in T^k_{x,r}} |s_{j_1}, \ldots, s_{j_k}\rangle \otimes \left(\bigotimes_{i=1}^{k} \frac{1}{\sqrt{2^\ell}} \sum_{R \in \Sigma^\ell} (-1)^{R \cdot g_i(s_{j_i})} |z_R\rangle \right),$$

i.e., the outcome of the queries to the functions $g_1, \ldots, g_k$.

If the algorithm A *receives* $h(n)$ *bits (qubits) of advice on input length* n, *then we say that* L *is nonuniformly nonadaptively* (t, k, ℓ, ϵ)*-qwlr reducible to functions* g_1, g_2, ..., g_k *with* $h(n)$ *bits (respectively, qubits) of classical (respectively, quantum) advice.*

3 Results

3.1 The Case of Two Queries

Theorem 5 shows that a nonadaptive 2-query classical LRR with answer length ℓ and success probability at least $1/2 + \epsilon$ can be simulated by a 1-query quantum weak LRR with success probability at least $1/2 + \epsilon/2^\ell$. This simulation of nonadaptive 2-query classical LRRs by 1-query quantum weak LRRs along with Theorem 6, which proves a complexity upper bound for 1-query quantum weak LRRs, allow us to obtain a complexity upper bound for nonadaptive 2-query classical LRRs.

The proofs of Theorem 5 and Theorem 6 are inspired from those in the papers by Kerenidis and de Wolf [KdW04] and Wehner and de Wolf [WdW05]. However, there are at least two technical features that indicate that our proofs are conceptually different from those in [KdW04, WdW05]. First, efficiency of algorithms is an issue in our proofs (since LRRs are required to be efficient algorithms), whereas efficiency is not an issue in the papers [KdW04, WdW05] (since algorithms in these papers are for information-theoretic LDCs and PIRs). Second, we do not require any major result from quantum information theory in our proofs, whereas the proofs in [KdW04, WdW05] use Nayak's [Nay99] linear lower bound on the length of quantum random access codes. Nayak's [Nay99] lower bound proof in turn requires some deep results from quantum information theory.

Theorem 5. *Let* $\ell(n) = \mathrm{poly}(n)$ *and* $\epsilon(n) \in (0, 1/2]$. *If a set* L *is nonadaptively* $(1, 2, \ell, \epsilon)$*-clr reducible to functions* g_1 *and* g_2, *then* L *is* $(1, 1, \ell, \frac{\epsilon}{2^\ell})$*-qwlr reducible to some function* g. *Here, each of* g_1, g_2, *and* g *outputs* $\ell(n)$ *bits on inputs of length* n *to their corresponding reductions.*

Theorem 6 shows that any set that has a 1-query quantum weak LRR in which the target function outputs *polynomial* number of bits and the reduction succeeds with probability at least $1/2 + 1/\mathrm{poly}(n)$ is in $\mathrm{BQP}^{\mathrm{NP}}/\mathrm{qpoly}$.

Theorem 6. *Let* $\ell(n) = \text{poly}(n)$ *and* $\epsilon(n) = 1/\text{poly}(n) \in (0, 1/2]$. *If a set* L *is* $(1, 1, \ell, \epsilon)$*-qwlr reducible to a function* g*, where* g *outputs* $\ell(n)$ *bits on inputs of length* n *to the reduction, then* L *is in* $\text{BQP}^{\text{NP}}/\text{qpoly}$.

Aaronson [Aar04] gave a relativizable proof of the inclusion $\text{BQP}/\text{qpoly} \subseteq \text{PP}/\text{poly}$. As a consequence, we obtain the following corollary of Theorem 5 and Theorem 6.

Corollary 7. *Let* $\ell(n) = O(\log n)$ *and* $\epsilon(n) = 1/\text{poly}(n) \in (0, 1/2]$. *If a set* L *is nonadaptively* $(1, 2, \ell, \epsilon)$*-clr reducible to functions* g_1, g_2*, where each* g_i *outputs* $\ell(n)$ *bits on inputs of length* n *to the reduction, then* $L \in \text{PP}^{\text{NP}}/\text{poly}$.

Note that Theorem 6 holds even if the reduction is nonuniform and requires a polynomial-size quantum advice state. Thus, we get the following strengthening of Theorem 6.

Theorem 8. *Let* $\ell = \text{poly}(n)$ *and* $\epsilon(n) = 1/\text{poly}(n) \in (0, 1/2]$. *If a set* L *is nonuniformly* $(1, 1, \ell, \epsilon)$*-qwlr reducible to a function* g *with a polynomial-size quantum advice, then* L *is in* $\text{BQP}^{\text{NP}}/\text{qpoly}$. *Here* g *outputs* $\ell(n)$ *bits on inputs of length* n *to the reduction.*

3.2 The Case of More Than Two Queries and Binary Answers

Theorem 5 shows that a nonadaptive 2-query classical LRR can be simulated by a 1-query quantum weak LRR. We show in Theorem 9 that, for any constant $k > 2$, a nonadaptive k-query classical LRR can also be simulated by a 1-query quantum weak LRR provided that in the classical LRR, the target functions are boolean and the distribution on sufficiently large subsets of queries depends only on the input length.

Theorem 9. *Let* $k > 2$ *be some fixed integer and let* $\epsilon \in (0, 1/2]$ *be a fixed constant. If a set* L *is nonadaptively* $(k/2 + O(\sqrt{k}), k, 1, \epsilon)$*-clr reducible to boolean functions* g_1, g_2, …, g_k*, then* L *is nonuniformly* $(1, 1, 1, \epsilon/2)$*-qwlr reducible to some boolean function* g *with* k *qubits of quantum advice. (The constant inside the* O*-notation depends only on* ϵ*.)*

In the statement of Theorem 9, the constant inside the O-notation depends only on ϵ. In particular, it can be shown that for any $\epsilon \geq 0.055$ and integer $k > 2$, if a set L is nonadaptively $(k/2 + \sqrt{k}, k, 1, \epsilon)$-clr reducible to boolean functions $g_1, g_2, \ldots, g_k$, then L is nonuniformly $(1, 1, 1, \epsilon/2)$-qwlr reducible to some boolean function g with k qubits of quantum advice.

The following corollary is an immediate consequence of Theorem 8 and Theorem 9.

Corollary 10. *Let* $\epsilon \in (0, 1/2]$ *be a fixed constant. If a set* L *is nonadaptively* $(k/2 + O(\sqrt{k}), k, 1, \epsilon)$*-clr reducible to boolean functions* $g_1, g_2, \ldots, g_k$*, then* $L \in \text{PP}^{\text{NP}}/\text{poly}$.

Acknowledgment. We thank Aduri Pavan and Vinodchandran Variyam for several insightful comments during an early stage of this work.

References

[Aar04] Aaronson, S.: Limitations of quantum advice and one-way communication. In: Proceedings of the 19th Annual IEEE Conference on Computational Complexity, pp. 320–332. IEEE Computer Society Press, Los Alamitos (2004)

[Aar05] Aaronson, S.: Quantum computing, postselection, and probabilistic polynomial-time. Technical Report 05-003, Electronic Colloquium on Computational Complexity (ECCC) (January 2005), `http://www.eccc.uni-trier.de/eccc/`

[Aar06] Aaronson, S.: Lower bounds for local search by quantum arguments. SIAM Journal on Computing 35(4), 804–824 (2006)

[AFK89] Abadi, M., Feigenbaum, J., Kilian, J.: On hiding information from an oracle. Journal of Computer and System Sciences 39(1), 21–50 (1989)

[ALM$^+$98] Arora, S., Lund, C., Motwani, R., Sudan, M., Szegedy, M.: Proof verification and the hardness of approximation problems. Journal of the ACM 45(3), 501–555 (1998)

[AR03] Aharonov, D., Regev, O.: A lattice problem in quantum NP. In: Proceedings of the 44th IEEE Symposium on Foundations of Computer Science, pp. 210–219. IEEE Computer Society Press, Los Alamitos (2003)

[AR05] Aharonov, D., Regev, O.: Lattice problems in NP $\cap$ coNP. Journal of the ACM 52(5), 749–765 (2005)

[AS98] Arora, S., Safra, S.: Probabilistic checking of proofs: A new characterization of NP. Journal of the ACM 45(1), 70–122 (1998)

[Bab87] Babai, L.: A random oracle separates PSPACE from the Polynomial Hierarchy. Information Processing Letters 26(1), 51–53 (1987)

[BF90] Beaver, D., Feigenbaum, J.: Hiding instances in multioracle queries. In: Choffrut, C., Lengauer, T. (eds.) STACS 90. LNCS, vol. 415, pp. 37–48. Springer, Heidelberg (1990)

[BFG06] Beigel, R., Fortnow, L., Gasarch, W.: A tight lower bound for restricted PIR protocols. Computational Complexity 15, 82–91 (2006)

[BFKR97] Beaver, D., Feigenbaum, J., Kilian, J., Rogaway, P.: Locally random reductions: Improvements and applications. Journal of Cryptology 10(1), 17–36 (1997)

[BFL91] Babai, L., Fortnow, L., Lund, C.: Non-deterministic exponential time has two-prover interactive protocols. Computational Complexity 1(1), 3–40 (1991)

[BK95] Blum, M., Kannan, S.: Designing programs that check their work. Journal of the ACM 42(1), 269–291 (1995)

[BLR93] Blum, M., Luby, M., Rubinfeld, R.: Self-testing/correcting with applications to numerical problems. Journal of Computer and System Sciences 47(3), 549–595 (1993)

[BM84] Blum, M., Micali, S.: How to generate cryptographically strong sequences of pseudo-random bits. SIAM Journal on Computing 13(4), 850–864 (1984)

[dW06] de Wolf, R.: Lower bounds on matrix rigidity via a quantum argument. In: Proceedings of the 33rd International Colloquium on Automata, Languages, and Programming, pp. 62–71 (2006)

[FGL$^+$96] Feige, U., Goldwasser, S., Lovász, L., Safra, S., Szegedy, M.: Interactive proofs and the hardness of approximating cliques. Journal of the ACM 43, 268–292 (1996)

[FS92] Fortnow, L., Szegedy, M.: On the power of two-local random reductions. Information Processing Letters 44(6), 303–306 (1992)

[GM84] Goldwasser, S., Micali, S.: Probabilistic encryption. Journal of Computer Security 28, 270–299 (1984)

[KdW04] Kerenidis, I., de Wolf, R.: Exponential lower bound for 2-query locally decodable codes via a quantum argument. Journal of Computer and System Sciences 69(3), 395–420 (2004)

[Ker05] Kerenidis, I.: Quantum multiparty communication complexity and circuit lower bounds. Technical Report quant-ph/0504087, Los Alamos e-Print Quantum Physics Technical Report Archive (April 12, 2005)

[LFKN92] Lund, C., Fortnow, L., Karloff, H., Nisan, N.: Algebraic methods for interactive proof systems. Journal of the ACM 39(4), 859–868 (1992)

[Lip91] Lipton, R.: New directions in testing. In: Feigenbaum, J., Merritt, M. (eds.) Distributed Computing and Cryptography. DIMACS series in Discrete Mathematics and Theoretical Computer Science, pp. 191–202. American Mathematical Society (1991)

[LLS05] Laplante, S., Lee, T., Szegedy, M.: The quantum adversary method and classical formula size lower bounds. In: Proceedings of the 20th Annual IEEE Conference on Computational Complexity, pp. 76–90. IEEE Computer Society Press, Los Alamitos (2005)

[Nay99] Nayak, A.: Optimal lower bounds for quantum automata and random access codes. In: Proceedings of the 40th IEEE Symposium on Foundations of Computer Science, pp. 369–377. IEEE Computer Society Press, Los Alamitos (1999)

[NC00] Nielsen, M., Chuang, I.: Quantum Computation and Quantum Information. Cambridge University Press, Cambridge (2000)

[PV] Pavan, A., Vinodchandran, N.: 2-local random reductions to 3-valued functions. Computational Complexity (to appear)

[Riv86] Rivest, R.: Workshop on communication and computing. MIT, Cambridge (1986)

[Sha92] Shamir, A.: IP=PSPACE. Journal of the ACM 39(4), 869–877 (1992)

[WdW05] Wehner, S., de Wolf, R.: Improved lower bounds for locally decodable codes and private information retrieval. In: Proceedings of the 32nd International Colloquium on Automata, Languages, and Programming. LNCS, pp. 1424–1436. Springer, Heidelberg (2005)

[Yao90] Yao, A.: An application of communication complexity to cryptography. In: Lecture at DIMACS Workshop on Structural Complexity and Cryptography (1990)

On the Complexity of Game Isomorphism*
(Extended Abstract)

Joaquim Gabarró, Alina García, and Maria Serna

LSI Dept. Universitat Politècnica de Catalunya, Barcelona
{gabarro,agarcia,mjserna}@lsi.upc.edu

Abstract. We consider the question of when two games are equivalent and the computational complexity of deciding such a property for strategic games. We introduce three types of isomorphisms depending on which structure of the game is preserved: *strict*, *weak*, and *local*. We show that the computational complexity of the game isomorphism problem depends on the level of succinctness of the description of the input games but it is independent of the way the isomorphism is defined. Utilities or preferences in games can be represented by Turing machines (general form) or tables (explicit form). When the games are given in general form, we show that the game isomorphism problem is equivalent to the circuit isomorphism problem. When the games are given in explicit form, we show that the game isomorphism problem is equivalent to the graph isomorphism problem.

Keywords: Game isomorphism, succinct representations, boolean formulas, computational complexity, boolean isomorphism, graph isomorphism.

1 Introduction

We are interested in the computational aspects of game equivalence. Surprisingly there is not a fully accepted definition of game equivalence in game theory books. In 1951, J. Nash [11] gave a definition of automorphism between strategic games. More recently, B. Peleg, J. Rosemuller and P. Sudhölter [13,16] consider isomorphisms for strategic and extensive games with incomplete information. J.C Harsanyi and R. Selten have introduced other definitions of isomorphism [8]. Equivalence by the way of transformations to a common form have been considered in [4].

We consider the case of strategic games. Our motivation is twofold. First, a strategic game is a special type of combinatorial structure and a natural question is to study the computational effort needed to decide when two such structures are isomorphic. Secondly, we can ask if such a question is interesting to the

* Partially supported by Network of Excellence CoreGRID (IST-2002-004265), FET Proactive IP 001907 (DELIS) and by the spanish project TIN2005-09198-C02-02 (ASCE). The second author is supported by a FPI Spanish grant.

L. Kučera and A. Kučera (Eds.): MFCS 2007, LNCS 4708, pp. 559–571, 2007.

game theory community. In practice strategic games are used as ingredient of more complicated games, but usually there is a way to transform any game into a strategic game. Furthermore, in [4] equivalence between extensive games is defined in terms of strategic games. Therefore strategic games are the first game structure to start analyzing game equivalence.

In defining a concrete equivalence we have to pay attention to the structural properties that are preserved in equivalent games. We consider three versions of isomorphisms. A *strong isomorphism* preserves utilities corresponding to the notion introduced in [11]. A *weak isomorphism* preserves preferences. A *local isomorphism* preserves preferences defined only on "close" strategy profiles. Each of them requires to preserve less information about the relative structure of profiles while preserving part of the structure of the Nash equilibria. More precisely, strong isomorphisms preserve pure and mixed Nash equilibria, while weak and local isomorphisms only preserve pure Nash equilibria.

In this paper we are interested in the computational complexity of deciding whether two games are equivalent. We consider four problems related to isomorphisms. In the IsIso problem, given two games Γ and Γ' and a mapping ψ we have to decide whether ψ is an isomorphism. In the IsAuto problem, given Γ and ψ we have to decide whether ψ is an automorphism. In the Iso problem we decide whether two games are isomorphic. Finally in Auto we ask to decide when a game has an automorphism different from the identity. In order to study the computational aspects of problems on strategic games and isomorphisms, we need first to decide how to represent games and morphisms as inputs.

For games we consider the two levels of succinctness proposed in [2]. When a game is given in *general* form the actions are listed explicitly but utilities and mappings are given by deterministic Turing machines. In the *explicit* case utilities are stored in tables. In both cases morphisms are always represented by tables. This is not a restriction as in polynomial time we can pass a representation by Turing machines into a representation by tables. The main results of the paper are

- The IsIso and the IsAuto problems are coNP-complete, for games given in general form, and NC when games are given in explicit form.
- The Iso and the Auto problems belong to Σ_2^p, for games given in general form, and to NP when games are given in explicit form.
- The Iso problem is equivalent to boolean circuit isomorphism, for games in general form, and to graph isomorphism, for games given in explicit form.

The above results hold independently of the type of isomorphism considered, observe that boolean circuit isomorphism is believed not to be Σ_2^p-hard [1], as well as graph isomorphism is conjectured not to be NP-hard [9]. Therefore the same result applies for the Iso.

In our presentation we consider games defined by utilities, preferences, or local preferences to present our complexity results as general as possible and matching the corresponding definition of morphism. However a game defined by utilities can be defined by preferences or local preferences. All the results hold for games defined through utilities when the isomorphism is required to preserve utilities (strong), the induced preference relation (weak) or the induced local preferences (local).

The paper is organized as follows. In Section 2, we consider games defined through utilities and define the notion of strong morphisms. In this perspective we introduce the different problems on game isomorphism and formalize the different levels of succinctness for the representation of games and morphisms. In Section 3 we develop the complexity results for strong isomorphisms. In Section 4 we study the complexity results for strategic games defined through preferences in the case of weak isomorphisms and we introduce the idea local preferences (well adapted to the notion of Nash equilibrium). In Section 5 we study the structure of the Nash equilibria in the case of two player games, each player with two actions using weak isomorphisms. We conclude in Section 6, with additional comments and open problems.

Due to the lack of space all the proofs are omitted we refer the interested reader to the full version [6] for details. However, before formulating a hardness result we provide a definition of the transformation used in the reduction.

2 Games, Isomorphisms, Problems, and Representations

We start stating the mathematical definition of strategic games given in [12].

Definition 1. *A* strategic game $\Gamma = (N, (A_i)_{i\in N}, (u_i)_{i\in N})$ *is a tuple. The set of players is* $N = \{1, \ldots, n\}$. *Player* $i \in N$ *has a finite set of actions* A_i, *we note* a_i *any action belonging to* A_i. *The elements* $a = (a_1, \ldots, a_n) \in A_1 \times \ldots \times A_n$ *are the strategy profiles. The* utility *(or* payoff*) function* u_i *for each player* $i \in N$ *is a mapping from* $A_1 \times \ldots \times A_n$ *to the rationals.*

First of all we consider game mappings which do not consider utilities. We adapt notations and definitions given in [13,16].

Definition 2. *Given* $\Gamma = (N, (A_i)_{i\in N}, (u_i)_{i\in N})$ *and* $\Gamma' = (N, (A'_i)_{i\in N}, (u'_i)_{i\in N})$, *a* game mapping ψ *from* Γ *to* Γ' *is a tuple* $\psi = (\pi, (\varphi_i)_{i\in N})$ *where* π *is a bijection from* N *to* N, *the* player's bijection, *and for any* $i \in N$, φ_i *is a bijection from* A_i *to* $A'_{\pi(i)}$, *the i-th* actions bijection.

Observe that the player bijection identifies player $i \in N$ with player $\pi(i)$ and the corresponding actions bijection φ_i maps the set of actions of player i to the set of actions of player $\pi(i)$. A game mapping ψ from Γ to Γ' induces, in a natural way, a function from $A_1 \times \cdots \times A_n$ to $A'_1 \times \cdots \times A'_n$ where strategy profile $(a_1, \ldots, a_n)$ is mapped into the strategy profile $(a'_1, \ldots, a'_n)$ defined as $a'_{\pi(i)} = \varphi_i(a_i)$, for all $1 \le i \le n$. We note, overloading the use of ψ, this mapping as $\psi(a_1, \ldots, a_n) = (a'_1, \ldots, a'_n)$. A mixed strategy profile $p = (p_1, \ldots, p_i, \ldots p_n)$ is given by probabilities p_i on A_i (such that $\sum_{a_i \in A_i} p_i(a_i) = 1$) for $1 \le i \le n$. A game mapping ψ also induces a mapping $\psi(p_1, \ldots, p_n) = (p'_1, \ldots, p'_n)$ such that $p'_{\pi(i)}$ is a probability on $A'_{\pi(i)}$ defined by $p'_{\pi(i)}(\varphi_i(a_i)) = p_i(a_i)$. Isomorphisms are game mappings fulfilling some extra restrictions about utilities or preferences. We start defining the stronger version of an isomorphism introduced by J. Nash [11], look also [13,16].

$$\begin{array}{c} \text{Player 2} \\ \text{Player 1} \begin{array}{c|c|c|} & l & r \\ \hline t & 0,0 & 0,1 \\ \hline b & 1,1 & 1,0 \\ \hline \end{array} \\ \Gamma \end{array} \xrightarrow{\psi} \begin{array}{c} \text{Player 2} \\ \text{Player 1} \begin{array}{c|c|c|} & l' & r' \\ \hline t' & 1,0 & 0,1 \\ \hline b' & 0,0 & 1,1 \\ \hline \end{array} \\ \Gamma' \end{array}$$

The strong isomorphism $\psi : \Gamma \to \Gamma'$ is $\psi = (\pi, \varphi_1, \varphi_2)$ where $\pi = (1 \to 2, 2 \to 1)$, $\varphi_1 = (t \to l', b \to r')$ and $\varphi_2 = (l \to b', r \to t')$. This strong isomorphism maps strategy profiles as $\psi(t,l) = (b',l')$, $\psi(t,r) = (t',l')$, $\psi(b,l) = (b',r')$ and $\psi(b,r) = (t',r')$.

$$\begin{array}{c} \text{Player 2} \\ \text{Player 1} \begin{array}{c|c|c|} & l & r \\ \hline t & 0,1 & 1,0 \\ \hline b & 1,0 & 0,1 \\ \hline \end{array} \\ \Gamma \end{array} \xrightarrow{\psi'} \begin{array}{c} \text{Player 2} \\ \text{Player 1} \begin{array}{c|c|c|} & l & r \\ \hline t & 0,1 & 1,0 \\ \hline b & 1,0 & 0,1 \\ \hline \end{array} \\ \Gamma \end{array}$$

The strong automorphism is $\psi' = (\pi', \varphi'_1, \varphi'_2)$ where $\pi' = (1 \to 2, 2 \to 1)$ and the action bijections are $\varphi'_1 = (t \to r, b \to l)$ and $\varphi'_2 = (l \to t, r \to b)$.

Fig. 1. Example of a strong isomorphism ψ and of an automorphism ψ'

Definition 3. *Given* $\Gamma = (N, (A_i)_{i \in N}, (u_i)_{i \in N})$ *and* $\Gamma' = (N, (A'_i))_{i \in N}, (u'_i)_{i \in N})$, *a game mapping* $\psi : \Gamma \to \Gamma'$ *with* $\psi = (\pi, (\varphi_i)_{i \in N})$ *is called a* strong isomorphism $\psi : \Gamma \to \Gamma'$ *when, for any player* $1 \le i \le n$ *and any strategy profile* a*, it holds* $u'_{\pi(i)}(\psi(a)) = u_i(a)$*. In the particular case that* Γ' *is* Γ *a strong isomorphism is called a* strong automorphism.

In Figure 1 we provide an example of strong isomorphism and another of strong automorphism. Given a strong isomorphism ψ between Γ and Γ', observe that a strategy profile a is a pure Nash equilibrium in Γ iff $\psi(a)$ is a pure Nash equilibrium in Γ' and the same holds for mixed Nash equilibria.

We consider the following computational problems related to games and morphisms.

Is Game Isomorphism (IsIso). Given two games Γ, Γ' and a game mapping $\psi : \Gamma \to \Gamma'$, decide whether ψ is a strong isomorphism.

Game Isomorphism (Iso). Given two games Γ, Γ', decide whether there exists a strong isomorphism between Γ and Γ'.

Is Game Automorphism (IsAuto). Given a game Γ and a game mapping $\psi : \Gamma \to \Gamma$, decide whether ψ is a strong automorphism.

Game Automorphism (Auto). Given a game Γ, decide whether Γ has a non trivial (different from the identity) strong automorphism.

In the context of computational complexity it is very important to fix how games and morphisms are represented. We take the representations of games given in [2]. We assume a pre-fixed alphabet. Hence, we can describe the pay-off functions of a game by a tuple $\langle M, 1^t \rangle$ where M is a deterministic TM and t is a natural number bounding its computation time on any input. The idea is that given a strategy profile a and a natural number i, the output of M on input $\langle a, i \rangle$ is the value of the pay-off function of the i-th player on input a. From the three levels of succinctness in the representations introduced in [2] we consider only two.

Strategic games in general form. A game is given by a tuple

$$\Gamma = \langle 1^n, A_1, \ldots, A_n, M, 1^t \rangle.$$

It has n players, and for each player i, where $1 \leq i \leq n$, their set of actions A_i is given by listing all its elements. The description of their pay-off functions is given by $\langle M, 1^t \rangle$.

Strategic games in explicit form. A game is given by a tuple

$$\Gamma = \langle 1^n, A_1, \ldots, A_n, T \rangle,$$

where T is a table such that $u_i(a) = T[a][i]$.

In Definition 1 a game Γ is defined in an abstract way using set theory. When the computational aspects of Γ have to be studied, Γ has to be encoded. Here we consider two encodings with different levels of succinctness, the general form and the explicit form. We use the same symbol Γ to denote both, the abstract game and the encoded version. Observe that in the explicit form, games are described as it is done in elementary books, just giving explicitly the tables of utility functions. In order to describe a game mapping, we consider the less succinct approach.

Game mapping in explicit form. All is given explicitly, actions are given listing all its elements and mappings are given by tables, then

$$\psi = \langle 1^n, A_1, \ldots, A_n, A'_1, \ldots, A'_n, T_\pi, T_{\varphi_1}, \ldots, T_{\varphi_n} \rangle$$

where $T_\pi, T_{\varphi_1}, \ldots, T_{\varphi_n}$ are tables such that $T_{\varphi_i}[a_i] = a'_{T_\pi[i]}$.

We have not considered the description of a mapping by Turing machines,

$$\psi = \langle 1^n, A_1, \ldots, A_n, A'_1, \ldots, A'_n, M_\pi, M_{\varphi_1}, \ldots, M_{\varphi_n}, 1^t \rangle$$

because in such a case we can construct an explicit coding of ψ with size bounded by $2|\psi|$ in time $|\psi|^2$.

3 Complexity Results for Strong Isomorphism

First we consider the IsIso and IsAuto problems and later on the Iso problem. Our coNP hardness results follow from reductions from the following coNP-complete problem [7]:

Validity problem(VALIDITY): Given a boolean formula F decide whether F is satisfiable by all truth assignments.

We also consider the following problems on boolean circuits. Recall that two circuits $C_1(x_1, \ldots, x_n)$ and $C_2(x_1, \ldots, x_n)$ are *isomorphic* if there is a permutation π of $\{1, \ldots, n\}$ such that, for any truth assignment $x \in \{0,1\}^n$, $C_1(x) = C_2(\pi(x))$.

Boolean circuit isomorphism problem (CIRCUITISO): Given two boolean circuits C_1 and C_2 decide whether C_1 and C_2 are isomorphic.

A related problem is based on the notion of congruence. A *congruence* between two circuits on n variables, $C_1(x_1, \ldots, x_n)$ and $C_2(x_1, \ldots, x_n)$ is a mapping $\psi = (\pi, f_1, \ldots, f_n)$, where π is a permutation of $\{1, \ldots, n\}$ and, for any $1 \leq i \leq n$, f_i is a permutation on $\{0, 1\}$ (either the identity or the negation function). As in the case of game morphism, the image $\psi(x)$ is obtained by permuting the positions of the input bits, according to permutation π, and then applying to any bit i the permutation f_i.

Boolean circuit congruence problem (CIRCUITCONG): Given two circuits C_1 and C_2 decide whether C_1 and C_2 are congruent.

The CIRCUITISO problem has been studied by B. Borchert, D. Ranjan and F. Stephan in [3], among many other results they show that CIRCUITISO and CIRCUITCONG are equivalent. It is known that CIRCUITISO $\in \Sigma_2^p$, but it cannot be Σ_2^p-hard unless the polynomial hierarchy collapses [1].

Two graphs are *isomorphic* if there is a one-to-one correspondence between their vertices and there is an edge between two vertices of one graph if and only if there is an edge between the two corresponding vertices in the other graph.

Graph isomorphism (GI): Given two graphs, decide whether they are isomorphic.

It is well known that GI is not expected to be NP-hard [9]. Let us start with the complexity for ISISO problem in the case of strategic games.

Theorem 1. *The* ISISO *and the* ISAUTO *problems for strong morphisms are coNP-complete when the games are given in general form. Both problems belong to NC whenever the games are given in explicit form. The strong isomorphism is given in both cases in explicit form.*

Our next step is to provide upper bounds for the complexity of the ISO and the AUTO problems. Later on we show that the bounds are best possible for the ISO problems.

Theorem 2. *The* ISO *and the* AUTO *problems for strong morphisms belong to* Σ_2^p *when the games are given in general form. Both problems belong to NP when the games are given in explicit form.*

We prove that ISO is equivalent to CIRCUITISO for games in general form. This is done through a series of reductions transforming the game while preserving the existence of strong isomorphism. First, we show how to construct a game in which the set of actions for each player is $\{0, 1\}$, which we call a *binary action game.*

Second, we show how to construct from a binary action game another binary action game in which the utility functions range is $\{0,1\}$, which we call a *binary game*. Finally, we show the equivalence with the Boolean circuit congruence. All the transformations presented in the paper can be computed in polynomial time, thus we avoid to mention this fact all through the paper. Let us start with the first transformation.

Let $\Gamma = (N, (A_i)_{i\in N}, (u_i)_{i\in N})$ be a game and let $\mu_\Gamma = \min\{u_i(a) | a \in A, i \in N\}$ be the smaller payoff obtained by any player, take $\mu < \mu_\Gamma$ and call such value μ a "penalty payoff". Without loss of generality we assume that $N = \{1,\dots,n\}$ and that, for any $i \in N$, $A_i = \{1,\dots,k_i\}$ for suitable values. Given $A_i = \{1,\dots,k_i\}$ we "binify" an action $j \in A_i$ coding it with k_i bits, defined by $\mathsf{binify}(j) = 0^{j-1}10^{k_i-j}$. Moreover we associate with A_i a block B_i of k_i players each one taking care of one bit. In this case we get $k = \sum_{i\in N} k_i$ players partitioned into $B_1,\dots,B_n$ blocks. Given $i \in B_j$ we say that i belongs to block j of players and write $\mathsf{block}(i) = j$. The binify process can be used in a strategy profile to clarify notation, given $a = (a_1,\dots,a_n)$, we write $\mathsf{binify}(a) = (\mathsf{binify}(a_1),\dots,\mathsf{binify}(a_n))$. Often we look at $\mathsf{binify}(a)$ as a k tuple of bits. For instance, given Γ with 3 players $A_1 = A_3 = \{1,2\}$ and $A_2 = \{1,2,3\}$ we have $\mathsf{binify}(1,2,2) = (10,010,01) = (1,0,1,0,0,1)$. Set $A' = \{0,1\}^k$, as for any $a \in A$ it holds $\mathsf{binify}(a) \in A'$, we can define $\mathsf{good}(A') = \{\mathsf{binify}(a) | a \in A\}$ and $\mathsf{bad}(A') = A' \setminus \mathsf{good}(A)$. When $a' \in \mathsf{good}(A)$ we say that a' is good, otherwise is bad. Note that $\mathsf{binify} : A \to \mathsf{good}(A')$ is a bijection, therefore the inverse function is also a bijection, for instance $\mathsf{binify}^{-1}(1,010,01) = (1,2,2)$.

BinaryAct(Γ,μ) is defined as

$$\text{BinaryAct}(\Gamma,\mu) = (N', (A'_i)_{i\in N'}, (u'_i)_{i\in N'})$$

where $N' = \{1,\dots,k\}$ and, for any $i \in N'$, $A'_i = \{0,1\}$ and thus the set of action profiles is $A' = \{0,1\}^k$. The utilities are defined by

$$u'_i(a') = \begin{cases} u_{\mathsf{block}(i)}(\mathsf{binify}^{-1}(a')) & \text{if } a' \in \mathsf{good}(A'), \\ \mu & \text{if } a' \in \mathsf{bad}(A'). \end{cases}$$

Notice that, for $a \in A$, $u'_i(\mathsf{binify}(a)) = u_{\mathsf{block}(i)}(a)$, furthermore, all the players in a given block have the same utility. Each strategy profile a' in BinaryAct(Γ,μ) can be factorized giving the actions taken by the k players as $a' = (a'_1,\dots,a'_k)$ or grouping the actions according to teams $B_1,\dots,B_n$ as $a' = (b_1,\dots,b_n)$ where b_i is a strategy profile for B_i. The big gap in the utility function is used to created a gap that separates the profiles in BinaryAct(Γ,μ) that codify correctly a profile of Γ from those that do not.

Lemma 1. *Let Γ_1, Γ_2 be two games given in general form and set $t = \max\{t_1,t_2\}$, t_i for $1 \le i \le 2$, is the time allowed to the utility* TM *of the game Γ_i. There is a strong isomorphism between Γ_1 and Γ_2 iff there is a strong isomorphism between the games* BinaryAct(Γ_1,μ) *and* BinaryAct(Γ_2,μ) *where $\mu = -2^t$.*

Let us now transform a binary actions game into a binary game. Given a game $\Gamma = (N, (A_i)_{i\in N}, (u_i)_{i\in N})$ in which $A_i = \{0,1\}$, for any $i \in N$, and $N = \{1,\dots,n\}$. Given positive values t and m such that, for any action profile a and any player i, $|u_i(a)| \le t$ and $m \ge \{n,t\}$. We set $k = n + tn + m + 2$.

BINARY(Γ, t, m) is defined as

$$\text{BINARY}(\Gamma, t, m) = (N', (A'_i)_{i\in N'}, (u'_i)_{i\in N'})$$

where $N' = \{1,\dots,k\}$ and, for any $i \in N'$, $A'_i = \{0,1\}$.

Before defining the utilities we need some additional notation. The set N' is partitioned into $n+2$ consecutive intervals $B_0,\dots,B_n,B_{n+1}$ so that the interval B_0 has exactly n players, for $1 \le i \le n$, the block B_i has t players, finally block B_{i+1} has $m+2$ players. Inside the blocks we use relative coordinates to identify the players. In all the blocks coordinates start at 1 except for the last block that starts with 0. In this situation a strategy profile a is usually factorized as $a = x\, b_1 \dots b_n\, z$ where $x = x_1 \dots x_n$, $b_i = b_{i_1} \dots b_{i_t}$ and $z = z_0 \dots z_{m+1}$. We define the utility function by properties of the strategy profile, assume that $a = x\, b_1 \dots b_n\, z$ is a strategy profile of BINARY(Γ, t, m).

- In the case that, for some ℓ, $0 \le \ell \le m+1$, the last ℓ bits of z are 1, all the players except the last ℓ get utility 0. The remaining players get utility 1.
- In the case that, for some j, $1 \le j \le t$, the j-th bit of z is the unique 1 in z, all the players in blocks $B_1,\dots,B_n$ that do not occupy position j in their block get utility 0, all the players in blocks B_0 and B_{n+1} get utility 1, all the remaining players get as utility their action.
- In the case that, the 0-th bit of z is the unique 1 in z, for any i, $1 \le i \le n$, player i and all the players in block B_i get utility 1 when $u_i(x) = b^i$. All the players in block B_{n+1} get utility 0.
- In the remaining cases all the players get utility 1.

Notice that the utilities for all the players are either 0 or 1.

Lemma 2. *Let Γ_1, Γ_2 be two games given in general form, set $t = \max\{t_1, t_2, 3\}$, where t_i is the time allowed to the utility* TM *of game Γ_i, and $m = \max\{t, n_1, n_2\}$, where n_i is the number of players in game Γ_i. There is a strong isomorphism between Γ_1 and Γ_2 iff there is a strong isomorphism between* BINARY(Γ_1, t, m) *and* BINARY(Γ_2, t, m).

Given a binary game $\Gamma = (N, (A_i)_{i\in N}, (u_i)_{i\in N})$ with n players, such that for any $1 \le i \le n$, utility u_i has range $\{0,1\}$ and $A_i = \{0,1\}$. We construct a circuit C_Γ on $4n+2$ variables. Remind that, when $u_i(x)$ is computed by a Turing machine in polynomial time, Ladner's construction [15] gives us a polynomial size circuit.

Circuit C_Γ. The variables in C_Γ are grouped in four blocks, the X-block contains the first n-variables, the Y-block is formed by the variables in positions $n+1$ to $2n$, the C-block contain the following $n+2$ variables, and the D-block the remaining variables. For sake of readability we split the set of variables into four parts $a = (x, y, c, d)$ where $x = (x_1,\dots,x_n)$, $y = (y_1,\dots,y_n)$, $c = (c_1,\dots c_{n+2})$, and $d = (d_1,\dots,d_n)$.

We define C_Γ with the help of $n+2$ following circuits.

$$C_1(x,y,d) = [(x_1 = \overline{d_1}) \wedge \cdots \wedge (x_n = \overline{d_n}) \wedge (u_1(x) = y_1) \wedge \cdots \wedge (u_n(x) = y_n)]$$
$$C_2(y) = [y_1 \vee \cdots \vee y_n]$$
$$C_{i+2}(x_i, y_i, d_i) = [\overline{y_i} \wedge (x_i = \overline{d_i})] \qquad \text{for } 1 \leq i \leq n.$$

Finally

$$C_\Gamma(x,y,c,d) = \begin{cases} 0 & \text{if } \sum_{1\leq i\leq n+2} c_i = 0 \text{ or } \sum_{1\leq i\leq n+2} c_i > 1 \\ C_j & \text{if } \sum_{1\leq i\leq n+2} c_i = 1 \text{ and } c_j = 1 \end{cases}$$

The previous construction is used to reduce the Iso problem to the CircuitCong problem.

Lemma 3. *Let Γ and Γ' be two games in general form with at least two players each, boolean utilities and actions $\{0,1\}$. There is a congruence isomorphism between C_Γ and $C_{\Gamma'}$ iff there is a strong isomorphism between Γ and Γ'.*

Proving NP-completeness in the case of explicit form appears to be a difficult task. Observe, that a game in explicit form can be seen as a graph with edge labels and weights. As the total number of different weights appearing in both games is polynomial the problem can be reduced to the Graph isomorphism (GI) problem [17]. Therefore the NP-hardness of Iso will imply the NP-hardness of GI. Our proof provides a reduction that shows that the opposite direction is true. It is easy to show that CircuitCong is reducible to Iso, just consider a game with as many players as variables in which the utilities for all the players are identical and coincide with the evaluation of the circuit. Taking into account that CircuitCong is equivalent to CircuitIso and putting all together we have:

Theorem 3. *The* Iso *problem for strong isomorphism and games given in general form is equivalent to the circuit isomorphism problem. In the case of games given in extensive form the problem is equivalent to graph isomorphism.*

4 Weak and Local Isomorphism

There are several ways to relax the notion of strong isomorphism while maintaining the structure of Nash equilibria. For instance, Harsanyi and Selten [8] substitute $u_{\pi(i)}(\psi(a)) = u_i(a)$ for $u_{\pi(i)}(\psi(a)) = \alpha_i u_i(a) + \beta_i$. In order to generalize this approach we consider, following [12], games in which utility functions are replaced by preference relations $(\preceq_i)_{i\in N}$. All the preference relations must be total, that is, given any pair a, a' holds $a \preceq_i a'$ or $a' \preceq_i a$. In this case, a game is a tuple $\Gamma = (N, (A_i)_{i\in N}, (\preceq_i)_{i\in N})$. We note *strict preference* as usual, $a \prec_i a'$ iff $a \preceq_i a'$ but not $a' \preceq_i a$. We note *indifference* by $a' \sim_i a'$, as usual indifference occurs when $a \preceq_i a'$ and $a' \preceq_i a$ holds. The definition of isomorphism can be adapted to respect preference relations instead of utility functions.

Definition 4. *A weak isomorphism $\psi : \Gamma \to \Gamma'$ is a mapping $\psi = (\pi, (\varphi_i)_{i \in N})$ such that any triple a, a' and i verifies:*

$$\mathit{Preserve}(a, a', \psi, i) \equiv$$
$$(a \prec_i a' \Rightarrow \psi(a) \prec_{\pi(i)} \psi(a')) \vee (a \sim_i a' \Rightarrow \psi(a) \sim_{\pi(i)} \psi(a')).$$

Preference relations can be defined using utility functions, $a \prec_i a'$ iff $u_i(a) < u_i(a')$ and $a \sim_i a'$ iff $u_i(a) = u_i(a')$.

Weak isomorphisms preserves preferences for any pair of strategy profiles and therefore maintains the structure of pure Nash equilibria. However, if we are interested to maintain this structure through isomorphisms we need to consider only preferences $a \preceq_i a'$ such that $a_{-i} = a'_{-i}$. We call these preferences *local preferences*.

Definition 5. *A local isomorphism $\psi : \Gamma \to \Gamma'$ is a mapping ψ such that for any triple a, a' and i such that $a_{-i} = a'_{-i}$ verifies $\mathit{Preserve}(a, a', \psi, i)$.*

It is easy to see that weak and local isomorphisms preserve pure Nash equilibria.

In order to describe games in general form with total or local preferences we consider the following TM

$$M_{\preceq}(i, a, a'_i) = \begin{cases} i & \text{if } a \sim_i (a_{-i}, a'_i) \text{ (indifference case)} \\ b & \text{if } a \prec_i (a_{-i}, a'_i) \text{ (better case)} \\ w & \text{if } a \succ_i (a_{-i}, a'_i) \text{ (worse case)} \end{cases}$$

Given $\Gamma = \langle 1^n, A_1, \ldots, A_n, M, 1^t \rangle$ such that $M(i, a) = u_i(a)$ we can easily build $\Gamma = \langle 1^n, A_1, \ldots, A_n, M_{\preceq}, 1^t \rangle$ in polynomial time. In order to describe games in explicit form we consider a tuple $\Gamma = \langle 1^n, A_1, \ldots, A_n, L_{\prec}, L_{\sim} \rangle$ where $L_{\prec}$ and and $L_{\sim}$ are two adjacency lists. The list $L_{\prec}[a][i] = (a'_{i_1}, a'_{i_2}, \ldots, a'_{i_\ell})$ stores all the actions $a_{i_j} \in A_i$ such that $a \prec_i (a_{-i}, a'_{i_j})$. and $L_{\sim}$ store the elements a_{i_j} such that $a \sim_i (a_{-i}, a'_{i_j})$. Replacing strict by weak or local isomorphisms does not modify complexity bounds.

Theorem 4. *In the case of weak or local isomorphisms, the* IsIso *is coNP-complete, for games given in general form, and it belongs to NC when the games are given in explicit form. The* Iso *problem belongs to Σ_2^p, when the games are given in general form and it belongs to NP when the games are given in explicit form.*

In the case of Iso we can also prove

Theorem 5. *The* Iso *problem for weak and local isomorphism and games given in general form is equivalent to the circuit isomorphism problem. In the case of games given in explicit form the problem is equivalent to graph isomorphism.*

5 Small Case Study

An interesting problem is to have a classification of strategic games with the same number of players according to the structure of the pure Nash equilibria. This requires the development of acceptable definitions of equivalence between games. A first naive approach is to consider games as equivalent if they have the same number of Nash equilibria. This approach has been undertaken via probabilistic analysis by I. Y. Powers [14]. She studied the limit distributions of the number of pure strategies Nash equilibria for n players strategic games. Further results in [10].

Observe that for 2-players each one with m actions we get a bimatrix with m^2 positions. As numbers $\{0, \ldots, m^2 - 1\}$ are enough to encode preferences, there are m^{4m^2} different bimatrix games. In the particular case of 2 players and 2 actions the games can be grouped, according to the number of Nash equilibria, as follows:

Number of PNE	0	1	2	3	4	total
Number of Games	2592	29376	27936	5376	256	65536

The above table has been obtained by exhaustive computation. For bigger m this analysis becomes quickly intractable.

This simple approach has limitations because games with different structure can have the same number of Nash equilibria. Therefore there is a need to provide a better answer to the question: *When are two games the same?* The obvious mathematical setting to deal with equivalence is through isomorphisms. When games are defined with preferences, we consider weak isomorphisms. Two games are equivalent iff they are weakly isomorphic and we partition the set of 65536 games into equivalence classes.

In the case of games with two players and two actions, the set of strategy profiles is $\{00, 01, 10, 11\}$. This set can be represented as the set of nodes of a square where edges represent the preferences. We note, for instance $00 \leftrightarrow 01$ to mean $00 \sim_2 01$ and $00 \rightarrow 01$ iff $00 \prec_2 01$ (when dealing with local preferences, the

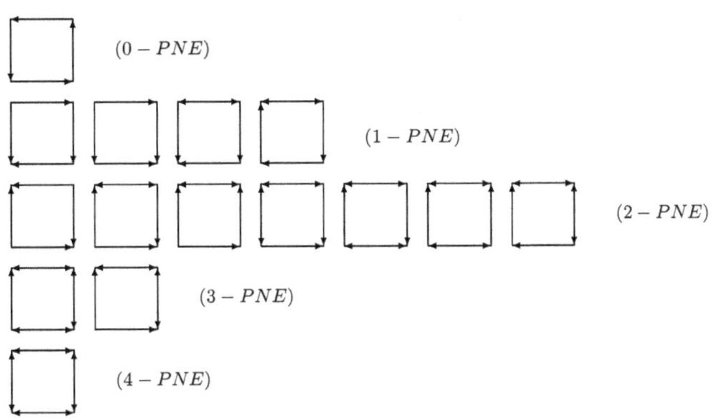

Fig. 2. Different game classes

subindex 2 can be avoided). Therefore, each class of equivalence is represented by a square fulfilling some conditions about preferences. In figure 2 we provide all the equivalence classes, giving information about the possible structure of the Nash equilibria. In fact, as one can see, the number is much smaller than 65536.

Number of PNE	0	1	2	3	4
Number of classes	1	4	7	2	1

Therefore we have more concise information about the structure of the Nash equilibria. The graphs appearing in Figure 2 are close to the Nash dynamics graphs defined in [5], observe that here we are considering double connections between equally likely profiles.

6 Comments and Open Questions

The equivalence betwen the Iso problem for games in general form and boolean circuit equivalence, has been obtained using the simulation of Turing machines by circuits. For families of games whose utility functions are defined by boolean formulas, the Iso problem turns out to be equivalent to boolean formula equivalence. Recall that boolean formula equivalence is not believed to be Σ_2^p-hard [1], and that it is still open whether the formula isomorphism is equivalent to circuit isomorphism.

We are working towards extending the definitions of game isomorphism to extensive games avoiding the use of strategic forms. An interesting open question is defining game isomorphisms for games without perfect information.

References

1. Agrawal, M., Thierauf, T.: The formula isomorphism problem. SIAM Journal on Computing 30(3) (2000)
2. Àlvarez, C., Gabarro, J., Serna, M.: Pure Nash equilibrium in strategic games with a large number of actions. In: Jedrzejowicz, J., Szepietowski, A. (eds.) MFCS 2005. LNCS, vol. 3618, pp. 95–106. Springer, Heidelberg (2005)
3. Borchet, B., Ranjan, D., Stephan, F.: On the computational complexity of some classical equivalence relations on boolean functions. Theory Comput. Systems 31, 679–693 (1998)
4. de Bruin, B.: Game transformations and game equivalence. Technical Report X-1999-01, IIL Technical Note (1999)
5. Fabrikant, A., Papadimitriou, C.H., Talwar, K.: The complexity of pure nash equilibria. In: STOC, pp. 604–612 (2004)
6. Gabarró, J., García, A., Serna, M.: On the complexity of game isomorphism. Technical Report LSI TR-07-19-R, Technical University of Catalunya (2007)
7. Garey, M., Johnson, D.: Computers and Intractability: A Guide to the Theory of NP-completeness. W.H.Freeman and Co. (1979)
8. Harsanyi, J., Selten, R.: A General Theory of Equilibrium Selection in Games. MIT Press, Cambridge, MA (1988)

9. Kobler, J., Schoning, U., Torán, J.: The Graph Isomorphism Problem: Its Structural Complexity. Birkhauser (1993)
10. McLennan, A., Berg, J.: Asymptotic expected number of Nash equilibria of two-player normal form games. Games and Economic Behavior 51, 264–295 (2005)
11. Nash, J.: Non-cooperative games. In: Classics in Game Theory, pp. 14–26 (1997)
12. Osborne, M., Rubinstein, A.: A Course in Game Theory. MIT Press, Cambridge (1994)
13. Peleg, B., Rosenmüller, J., Sudhölder, P.: The canonical extensive form of a game form: Part I symmetries. In: Current Trends in Economics, Advancement of Studies in Economics, pp. 367–387 (1999)
14. Powers, I.: Limiting distribution of the number of pure strategy Nash equilibria in n-person games. International Journal of Game Theory 19(3), 277–286 (1990)
15. Greenlaw, J.J.R., Ruzzo, W.: Limits to Parallel Computation. Oxford (1995)
16. Sudhölter, P., Rosenmüller, J., Peleg, B.: The canonical extensive form of a game form. Part II. Representation. Journal of Mathematical Economics 33(3), 299–338 (2000)
17. Torán, J.: Personal Communication

Hardness Results for Tournament Isomorphism and Automorphism

Fabian Wagner*

Institut für Theoretische Informatik,
Universität Ulm, 89073 Ulm, Germany
fabian.wagner@uni-ulm.de

Abstract. A tournament is a graph in which each pair of distinct vertices is connected by exactly one directed edge. Tournaments are an important graph class, for which isomorphism testing seems to be easier to compute than for the isomorphism problem of general graphs. We show that tournament isomorphism and tournament automorphism is hard under DLOGTIME uniform AC^0 many-one reductions for the complexity classes NL, $C_=L$, PL (probabilistic logarithmic space), for logarithmic space modular counting classes $\mathrm{Mod_kL}$ with odd $k \geq 3$ and for DET, the class of problems, NC^1 reducible to the determinant. These lower bounds have been proven for graph isomorphism, see [21].

1 Introduction

The graph isomorphism problem (GI) consists in determining whether there is a bijection between the vertices of two graphs, preserving the edge-relations. Until today, it is open whether GI is contained in P or complete for NP. A proof of the NP-completeness for GI would cause a collapse of the polynomial time hierarchy to its second level, see [7],[20]. Concerning lower bounds, DET $\leq_m^{AC^0}$ GI [21].

For many graph classes, polynomial time algorithms for isomorphism testing are known, e.g. for graphs of bounded degree [16], or planar graphs [13]. Even fast parallel algorithms for isomorphism testing have been developed, e.g. for planar graphs [18], trees [15], [9] or graphs with bounded color-class size [17].

A tournament is a directed graph with exactly one arc between every pair of distinct vertices. Tournaments comprise a large and important class of directed graphs and can be found in many applications, see e.g. [12]. The tournament isomorphism problem (TI) is GI restricted to tournaments. The best known algorithm for TI takes $n^{log(n)}$ time [6] and for GI takes $\exp(\sqrt{cn\log(n)})$ time (Luks, Zemlyachenko, cf. [6]). Arvind et al. [1] reduced TI onto $\mathrm{Mod_2GA}$ which is an intermediate problem between GA and GI and contains the class of graphs with an even number of automorphisms. Thus TI seems to be an easier problem than GI. Since the relation between GI and TI is not clear, we contribute to analyze the complexity status of tournament isomorphism. We show that TI and the tournament automorphism problem (TA) are hard for NC^1, L, NL, $\mathrm{Mod_kL}$ with

* Supported by DFG grant TO 200/2-1.

L. Kučera and A. Kučera (Eds.): MFCS 2007, LNCS 4708, pp. 572–583, 2007.

$k \geq 3$ odd integer, #L and DET under AC^0 many-one reductions. For proving that GI is hard for Mod_kL, we need a graph gadget with subgraphs, having orbits of size k to encode an integer in $\mathbb{Z}_k$. Since the order of automorphism groups of tournaments always are *odd* [14], we cannot directly encode an integer in $\mathbb{Z}_k$ with even $k \geq 2$. In order to encode Boolean values, we need another graph gadget. We encode value 0 as the identical mapping and value 1 as the switching of two subgraphs. which again leads to orbits of even size.

Since TA $\leq_m^{AC^0}$ TI (see Corollary 1, TA is prefix-TA without prefix) and the converse direction is unknown, proving DET $\leq_m^{AC^0}$ TA is a stronger result.

Due to space reasons some proofs are missing and would be included in the final version. We refer to the authors home page for the complete proofs.

The layout of this paper is as follows: In Section 2 we denote complexity classes and graph isomorphism problems. Section 3 contains hardness results for TI, that is DET $\leq_m^{AC^0}$ TI. In Section 4, we prove these results for TA.

2 Preliminaries

Complexity Classes. We assume familiarity with basic notions of complexity theory such as can be found in standard textbooks. NL is the class of languages accepted by nondeterministic Turing machines using a work tape bounded by logarithmic space. #L [4] is the class of functions $f : \Sigma^* \to \mathbb{N}$ that counts the number of accepting paths of a NL machine on a input. Based on #L we define:

$$\begin{aligned}
\text{PL} &= \{A : \exists p \in Poly, f \in \#\text{L}\ ,\ \forall x \in \Sigma^*\ x \in A \Leftrightarrow f(x) \geq 2^{p(|x|)}\} \ [11]\ [19] \\
\text{C}_=\text{L} &= \{A : \exists p \in Poly, f \in \#\text{L}\ ,\ \forall x \in \Sigma^*\ x \in A \Leftrightarrow f(x) = 2^{p(|x|)}\} \ [3] \\
\text{Mod}_k\text{L} &= \{A : \exists f \in \#\text{L}\ ,\ \forall x \in \Sigma^*\ x \in A \Leftrightarrow f(x) \equiv 1 \mod k\} \ [8]
\end{aligned}$$

Mod_k circuits ($k \geq 2$) are circuits with input variables over $\mathbb{Z}_k$ and gates computing addition in $\mathbb{Z}_k$. The evaluation problem for such circuits (given fixed values for the inputs, testing if the output is 1) is complete for Mod_kL under AC^0 many-one reductions.

DET (also denoted $NC^1(\#L)$) is the class of functions NC^1 Turing reducible to the determinant. That is the class of problems, solvable by NC^1 circuits with additional oracle gates for computing the determinant of an integer matrix.

The known relations among the considered classes are: $\text{Mod}_k\text{L} \subseteq$ DET and NL $\subseteq$ $\text{C}_=\text{L}$ $\subseteq$ PL $\subseteq$ DET. Therefore, the hardness of GI for DET implies hardness with respect to the other classes. We denote AC^0 many-one reductions by $\leq_m^{AC^0}$ and logspace Turing reductions by $\leq_T^L$.

Graph Isomorphism Problems. Let $G = (V, E)$ be a *graph* with a set of *vertices* $V = V(G)$ and *edges* $E = E(G)$. A *directed* edge or *arc* is denoted (v_1, v_2) and an *undirected* edge $\{v_1, v_2\}$. $G[X]$ is a subgraph of G *induced* on vertex set X. Let H be a subgraph of G then $G \setminus H = G[V(G) \setminus V(H)]$. Let $E' \subseteq E(G)$ then $G \setminus E' = (V(G), E(G) \setminus E')$.

For shorter notations we write $[k, n] = \{k, \dots, n\}$ for integers $k < n$ and $[n]$ if $k = 1$. Let $\oplus$ denote the modulo addition in $\mathbb{Z}_n$. The set $Sym(V)$ is the *symmetric group* over a set V and $S_n = Sym([n])$.

An *automorphism* of graph G is a permutation $\phi : V(G) \mapsto V(G)$ preserving adjacency: $(u,v) \in E(G) \Leftrightarrow (\phi(u), \phi(v)) \in E(G)$. The automorphisms except the identity are called *nontrivial.* A *rigid* graph contains no nontrivial automorphisms. The *graph automorphism problem* (GA) decides, whether a graph is rigid or not. The *automorphism group* $Aut(G)$ is the set of automorphisms of G.

An *isomorphism* between graphs G and H is a bijective mapping of vertices in G onto vertices in H that preserves adjacency. Both graphs are *isomorphic* if such an isomorphism exists. The *graph isomorphism problem* (GI) is the problem of deciding, whether two given graphs G, H are isomorphic, write $G \cong H$. Further, define the tuple of graph pairs $PGI = \{ ((G,H)(I,J)) \mid G \cong H \Leftrightarrow I \ncong J \}$ with the promise that exactly one pair is isomorphic [21].

Let H be a subgraph of G. Define $Aut_G(H) \subseteq Aut(H)$ as the set of automorphisms, which can be extended to an automorphism in $Aut(G)$. An automorphism $\phi \in Sym(V)$ *acts cyclically* on a vertex set $V = \{v_0, \ldots, v_{n-1}\}$, if there exists $a \in [0, n-1]$ such that $\phi(v_i) = v_{i \oplus a}$ for all $i \in [0, n-1]$. We further say $\phi \in Sym(V(G))$ acts cyclically on subgraphs $G[V_0], \ldots, G[V_{k-1}]$, if there exists $a \in [0, k-1]$ such that $\phi(v) \in V_{i \oplus a}$ for all $v \in V_i$ and $i \in [0, k-1]$.

Let $S_1, \ldots, S_k \subseteq V(G)$ be a set of distinct vertices of graph G. The set of automorphisms, mapping for all $i \in [k]$ vertices in S_i onto vertices in S_i in any order, are called *setwise stabilizer* of $S_1, \ldots, S_k$. $G_{[S_1,\ldots,S_k]}$ denotes graph G with $S_1, \ldots, S_k$ *setwise stabilized* in automorphism group of G.

Let k be a fixed integer. A *coloring* of a graph G is a function $f : V(G) \mapsto [k]$. For any isomorphism between colored graphs, the color relations have to be preserved. The decision problem is called the *isomorphism problem for colored graphs* (color-GI). Observe that color-GI $\leq_m^{\mathrm{AC}^0}$ GI [14].

Let $\{x_1, \ldots, x_k\}, \{y_1, \ldots, y_k\} \subseteq V(G)$ be vertex sets of graph G. The *prefix automorphism problem* (prefix-GA) as denoted in [14] is to find an automorphism $\phi \in Aut(G)$ such that $\phi(x_i) = y_i \; \forall i \in [k]$. Observe that prefix-GA $\leq_m^{\mathrm{AC}^0}$ GI [21]. If the vertex sets are given as mapping ϕ with $\phi(x_i) = y_i, i \in [k]$ then (G, ϕ) is an instance for prefix-GA.

A *tournament* is a directed graph with one arc between each pair of distinct vertices. A *cyclic tournament* T is a tournament on n vertices $x_0, \ldots, x_{n-1}$ such that $(x_i, x_{i \oplus j}) \in E(T)$ for all $i \in [0, n-1], j \in [1, \lfloor \frac{n}{2} \rfloor]$. The *tournament isomorphism problem* (TI) is the same as GI, if the input-graphs are tournaments. Then we also write color-TI, TA, prefix-TA instead of color-GI, GA, prefix-GA. Arvind et.al. [2] showed that color-TI is polynomial-time many-one reducible to TI without coloring. By verifying the proof, we observe that it is an AC^0 many-one reduction. Adapting the proof of prefix-GA $\leq_m^{\mathrm{AC}^0}$ GI [21], we obtain the following chain of reductions. Thus we prove lower bounds for prefix-TA.

Corollary 1. *prefix-TA* $\leq_m^{\mathrm{AC}^0}$ *color-TI* $\leq_m^{\mathrm{AC}^0}$ *TI.*

In some reductions, we construct graph gadgets $G(C)$ for simulating the evaluation of circuits C. By $G(C_i)$ we denote the simulation of a gate C_i of circuit C with index i. We also denote this way vertex sets or vertices, e.g. $v(C) \in U(C) \subseteq V(G(C))$. If the context is clear then (C) will be omitted.

3 Hardness Results for Tournament Isomorphism

In this section we show that TI is hard for some complexity classes under AC^0 many-one reductions.

3.1 Hardness Results of TI for Modular Counting Classes

GI is hard for the logarithmic space modular counting classes $\mathrm{Mod_kL}$ for all $k \geq 2$ [21]. In the proof, a graph gadget is defined for encoding input values over $\mathbb{Z}_k$ and simulating a circuit gate for modulo addition, see Definition 3.1 in [21]. Since the circuit value problem restricted to modulo addition gates over $\mathbb{Z}_k$ for $k \geq 2$ is complete for $\mathrm{Mod_kL}$ and with Corollary 1, we prove that TI is hard for $\mathrm{Mod_kL}$ with odd $k \geq 3$. Lemma 1 describes automorphism properties of this graph gadget.

Lemma 1. [21] *Fix $k \geq 2$. Then for any $a, b \in [0, k-1]$ there is a unique automorphism $\phi_{ab} \in Aut(G^k)$ with $\phi_{ab}(x_i) = x_{a \oplus i}$ and $\phi_{ab}(y_i) = y_{b \oplus i}$ for $i \in [0, k-1]$ and with $\phi_{ab}(u_i) = u_{a \oplus i, b \oplus j}$ and $\phi_{ab}(z_i) = z_{a \oplus b \oplus i}$.*

With this graph gadget, we construct a graph $G(C)$, simulating a circuit C of wired modulo addition gadgets.

Definition 1. [21] *Let C be a circuit of m modulo addition gates $C_1, \ldots, C_m$ and $k \geq 2$. Define $G(C)$ with*

$$V(G(C)) = \bigcup_{p \in [m]} V(G^k(C_p)),$$
$$E(G(C)) = \bigcup_{p \in [m]} E(G^k(C_p)) \cup \bigcup_{p,q \in [m], p<q} E_{p,q} \text{ with}$$
$$E_{p,q} = \begin{cases} \{z_i(C_p), x_i(C_q)\} & \text{if } C_p \text{ and left input of } C_q \text{ are wired,} \\ \{z_i(C_p), y_i(C_q)\} & \text{if } C_p \text{ and right input of } C_q \text{ are wired,} \\ \emptyset & \text{if } C_p, C_q \text{ are not wired directly.} \end{cases}$$

Furthermore, color vertices in $U(C_j), X(C_j), Y(C_j), Z(C_j)$ with colors (u, j), (x, j), (y, j), (z, j) in this order for all $j \in [m]$.

The reduction of this decision problem for circuit C to prefix-GA is as follows: compute a graph $G(C)$ and define prefixes for input- and output values. Thus $\mathrm{Mod_kL} \leq_m^{AC^0}$ GI. We will show, how far hardness for $\mathrm{Mod_kL}$ also holds for TI.

Theorem 1. $\mathrm{Mod}_k\mathrm{L} \leq_m^{AC^0}$ *TI with $k \geq 3$ an odd integer.*

The main proof idea is to transform the graph gadgets G^k and the circuit $G(C)$ (containing graph gadgets G^k as subgraphs) into tournaments. For this task, we first describe how graphs can be modified without changing the automorphism group.

Lemma 2. *Let H be an induced subgraph of a graph G with $Aut(G)$ setwise stabilizing vertices in H. Let H' be a graph with $V(H') = V(H)$ and let G' be G after replacing the induced subgraph H by subgraph H', setwise stabilizing vertices in H'. If $Aut_G(H) \subseteq Aut(H') \subseteq Aut(H)$ then $Aut(G) = Aut(G')$.*

The next lemma shows, how to connect two subgraphs with arcs, such that each vertex in the first graph is connected to every vertex in the second graph.

Lemma 3. *Let $G[X], G[Y]$ be vertex-disjoint and setwise stabilized subgraphs of G. Suppose that in G all the edges with one endpoint in X and one endpoint in Y point from X to Y. Then G can be transformed with an AC^0 computable function into a graph G' over the same vertex set such that $Aut(G'_{[X,Y]}) = Aut(G_{[X,Y]})$. If $G[X], G[Y]$ are tournaments then G' is a tournament.*

With Lemmas 2 and 3, we can now transform graph gadgets into tournaments and prove that they obey the same automorphism properties as the modulo addition graph gadget. Now we define a tournament with the same automorphism properties as G^k.

Definition 2. *Fix $k \geq 3$ odd integer. The* tournament modulo addition graph gadget *T^k is defined by vertex set $V(T^k) = \{ x_a, y_a, z_a, u_{a,b} \mid a, b \in [0, k-1] \}$. Let $U, X, Y, Z \subseteq V(T^k)$ contain vertices denoted by indexed lower case letters each. Let $U_a = \{ u_{a,b} \mid b \in [0, k-1] \}$ for any $a \in [0, k-1]$. $E(T^k)$ unifies*

1. $\{ (x_a, x_{a \oplus i}), (y_a, y_{a \oplus i}), (z_a, z_{a \oplus i}) \mid a \in [0, k-1],\ i \in [1, \lfloor \frac{k}{2} \rfloor] \}$,
2. $\{ (u_{a,b}, u_{a,b \oplus i}) \mid a, b, \in [0, k-1],\ i \in [1, \lfloor \frac{k}{2} \rfloor] \}$,
3. $\{ (u_{a,b}, u_{a \oplus i, b \oplus b'}) \mid a, b, b' \in [0, k-1],\ i \in [1, \lfloor \frac{k}{2} \rfloor] \}$,
4. $\{ (x_a, u_{a,b}), (u_{i,b}, x_a) \mid a, b \in [0, k-1],\ i \in [0, k-1] \backslash \{a\} \}$,
5. $\{ (y_b, u_{a,b}), (u_{a,i}, y_b) \mid a, b \in [0, k-1],\ i \in [0, k-1] \backslash \{b\} \}$,
6. $\{ (u_{a,b}, z_{a \oplus b}), (z_i, u_{a,b}) \mid a, b \in [0, k-1],\ i \in [0, k-1] \backslash \{a \oplus b\} \}$,
7. $\{ (x_a, y_b), (x_a, z_b), (y_a, z_b) \mid a, b \in [0, k-1] \}$.

Remark that $V(T^k) = V(G^k)$. In item 1 we define cyclic tournaments for induced subgraphs on vertex sets X, Y and Z. In item 2 we define cyclic tournaments $T^k[U_a]$ induced on vertices $U_a = \{ u_{a,b} \mid b \in [0, k-1] \}$ for any fixed

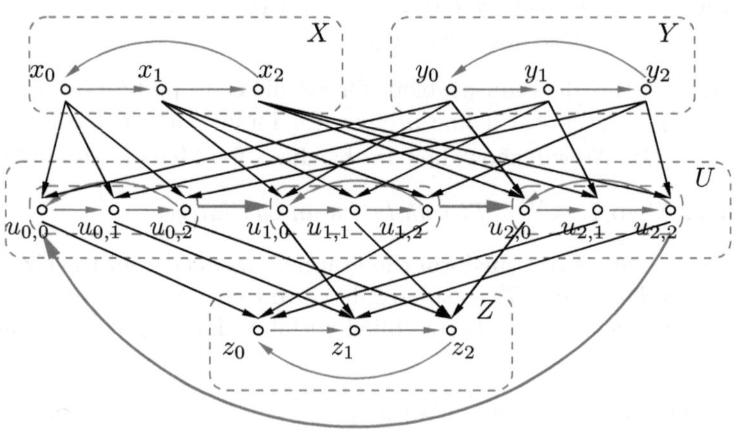

Fig. 1. Sketch of tournament modulo addition graph gadget T^3

$a \in [0, k-1]$. In item 3 we describe the connection between subgraphs $T^k[U_a]$, such that automorphisms in $Aut(T^k)$ act cyclically on subgraphs $T^k[U_a]$ for all $a \in [0, k-1]$. Items 4 to 7 describe complete bipartite edge sets among U, X, Y, Z. Figure 1 shows T^3 and contains edge sets of items 4 to 7 partially.

Lemma 4. *There is an* AC^0 *computable function that transforms* G^k *with odd* $k \geq 3$ *into the tournament modulo addition graph gadget* T^k *containing unique automorphisms as described in Lemma 1.*

Proof. In Lemma 1 the unique automorphisms act cyclically on vertex sets $G^k[X]$,$G^k[Y]$ and $G^k[Z]$. First, regard $G^k[X]$; $Aut(G^k[X]) = Sym(G^k[X])$ and $Aut_{G^k}(G^k[X])$ is generated by permutation $(x_0 \dots x_{k-1})$.

Clearly, $Aut(T^k[X]) \subseteq Aut(G^k[X])$ and because of $T^k[X]$ containing cyclic automorphisms, $Aut_{G^k}(G^k[X]) \subseteq Aut(T^k[X])$. Apply Lemma 2 and replace $G^k[X]$ by $T^k[X]$ in G^k, without changing the automorphism group. The same holds for $G^k[Y]$ and $G^k[Z]$. Second, regard $G^k[U]$. $Aut(G^k[U]) = Sym(G^k[U])$ and $Aut_{G^k}(G^k[U])$ is generated by ϕ, ψ, which are defined by the relation:

$\phi(u_{i,j}) \rightarrow u_{i,j\oplus 1}$ and $\psi(u_{i,j}) \rightarrow u_{i\oplus 1,j}$ for all $i, j \in [0, k-1]$.

It follows that $Aut_{G^k}(G^k[U]) \subseteq Aut(T^k[U])$. Apply Lemma 2 and replace $G^k[U]$ by $T^k[U]$ in G^k, without changing the automorphism group. Third, the edge sets of item 4 to item 7 in definition of $E(T^k)$ can be described as exchanging undirected edges between stabilized vertex sets X, Y, Z, U by arcs. By Lemma 3, this also keeps the automorphism group unchanged. Thus T^k is the union of all these modifications on G^k which can be computed in AC^0. □

With Lemma 4 we can prove that the replacement of gadgets G^k in $G(C)$ by tournament gadgets T^k does not change the automorphism group of $G(C)$. With Lemma 5, we complete the proof of Theorem 1.

Lemma 5. *Let* C *be a circuit of modulo addition gates in* $\mathbb{Z}_k$*, with odd* $k \geq 3$ *and output value* $s \in [0, k-1]$*. Construct under* AC^0 *many-one reductions a tournament* $T(C)$ *containing nontrivial prefix automorphisms,iff* C *outputs* s*.*

Transform $G(C)$ (of Definition 1) into a tournament $T(C)$, such that $Aut(G(C))$ $= Aut(T(C))$ and $T(C)$ contains a nontrivial prefix automorphism, iff $G(C)$ does. Therefore, we apply Lemmas 2 and 4.

3.2 Hardness Results of TI for NL, #L , $\mathrm{C_=L}$ and PL

Now we introduce graph gadgets for simulation of AND- and OR-gates in circuits. We transform them into tournament gadgets, to get the same hardness results for TI which hold for GI. A NC^1 circuit can be simulated by a balanced DLOGTIME uniform family of circuits with fan-out 1, logarithmic depth, polynomial size and alternating layers of and-gates and or-gates [5].

Definition 3. *Let* $((G_\wedge, H_\wedge)(I_\wedge, J_\wedge)) \in PGI$ *be the graph tuple for simulation of conjunction and* $((G_\vee, H_\vee)(I_\vee, J_\vee)) \in PGI$ *of disjunction, containing* $((G_0, H_0)(I_0, J_0))$*,* $((G_1, H_1)(I_1, J_1)) \in PGI$ *as in proof of Theorem 4.3 in [21].*

The graph tuples for conjunction have the following properties: $G_\wedge \cong H_\wedge$ iff $G_0 \cong H_0$ and $G_1 \cong H_1$; $I_\wedge \cong J_\wedge$ iff $G_0 \not\cong H_0$ or $G_1 \not\cong H_1$ (in this case $I_0 \cong J_0$ or $I_1 \cong J_1$). Similarly, the graph tuples for disjunction: exchange $\wedge$ with $\vee$, also exchange 'and' with 'or' and vice versa. For clear notation, we will apply prefixes e.g. PGI-$G_\wedge$, PGI-$G_\vee$. If the context is clear, we omit these prefixes. Now we transform PGI-tuples into tuples of tournaments.

Lemma 6. *There is an* AC^0 *computable function for translation of graph gadgets PGI-$G_\vee$ and PGI-$H_\vee$ into tournament graph gadgets PTI-$G_\vee$, PTI-$H_\vee$ (see Definition 4) having the same isomorphism properties as in Definition 3.*

Definition 4. *A* PTI-graph tuple *is a tuple of rigid tournaments $((G,H),(I,J))$ with $G \cong H$, iff $I \not\cong J$. Let PTI be the set of all such tuples. Define for conjunction $((G_\wedge, H_\wedge)(I_\wedge, J_\wedge)) \in PTI$ (write e.g. PTI-$G_\wedge$) and for disjunction $((G_\vee, H_\vee)(I_\vee, J_\vee)) \in PTI$ (write e.g. PTI-$G_\vee$) as follows:*

First, PTI-$G_\wedge$ contains tournaments G_0, G_1 and a set of arcs, pointing from every vertex in G_0 to every vertex in G_1. Replace G_0, G_1 in PTI-$G_\wedge$ by H_0, H_1 for obtaining PTI-$H_\wedge$, by I_0, I_1 for PTI-$I_\vee$ and by J_0, J_1 for PTI-$J_\vee$.

Second, let $i \in [0,1]$ and $j \in [0,2]$. Let X be a graph as in Figure 2. PTI-$G_\vee$ contains subgraphs $X, G_0, G_0', G_1, G_1', H_0$ and H_1 with G_0', G_1' copies of G_0, G_1. Let $E(PTI\text{-}G_\vee) = E_1 \cup \ldots \cup E_4$. E_1 unifies edges of all subgraphs. E_2 contains edges $(x_{i,0}, v)$ for all $v \in G_i$ (call G_i associated *to $x_{i,0}$), and similar edge sets with $x_{i,1}$ associated to G_i' and $x_{i,2}$ to H_i. E_3 contains the following arcs: If $(x, x') \in E(X)$ then connect every vertex of the subgraph associated to x with arcs, pointing to every vertex of the subgraph associated to x'. E_4 contains (u,v) for all $u \in V(PTI\text{-}G_\vee \setminus X)$, $v \in V(X)$, iff $(v,u) \notin E_2$. Now construct PTI-$H_\vee$ with minor changes. Associate $x_{1,1}$ with H_1 and $x_{1,2}$ with G_1'. The rest of the construction is the same. Now replace subgraphs G_0, G_1, H_0, H_1 in PTI-$G_\vee$ (and PTI-$H_\vee$) by I_0, I_1, J_0, J_1 in this order and obtain PTI-$I_\wedge$ (and PTI-$J_\wedge$).*

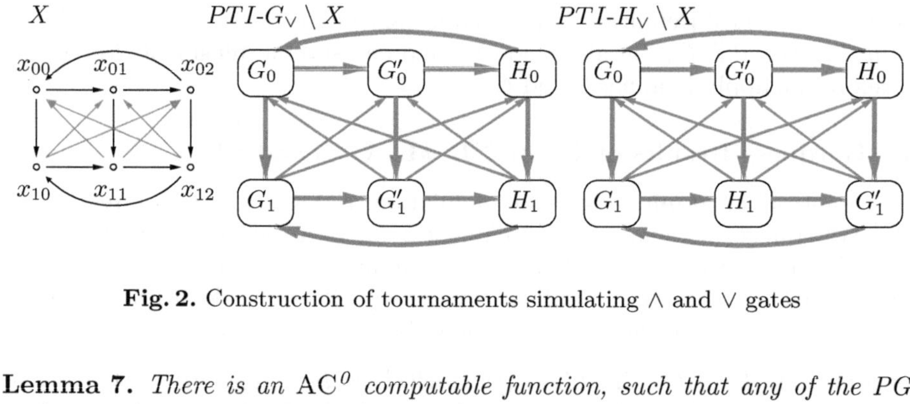

Fig. 2. Construction of tournaments simulating $\wedge$ and $\vee$ gates

Lemma 7. *There is an* AC^0 *computable function, such that any of the PGI-tuples simulating and-gates and or-gates can be transformed into PTI-tuples with the same isomorphism properties.*

With all the graph gadgets (G^k, PGI-$G_\wedge$, PGI-$G_\vee, \dots$) as defined so far, Torán proved that GI is hard for NL, #L , $C_=L$ and PL under AC^0 many-one reductions [21]. We prove that the same lower bounds hold for TI.

Theorem 2. *Tournament isomorphism is hard for* NL, #L , $C_=L$ *and* PL *under* AC^0 *many-one reductions.*

Proof. For proving all the hardness bounds, graph gadgets are needed as described above. Regard Theorems 4.1, 4.4 and Corollaries 4.5, 4.6 in [21] for details. First, MOD_kL circuits are needed to compute the result of a #L function $f(x) \mod k$. These gadgets encode $f(x) \mod k$ for a set of r different primes $k \in \{ k_1, \dots, k_r \mid 3 \leq k_1 < \dots < k_r \}$ (in Chinese remainder representation). The results are inputs to a NC^1 -circuit, which compute bits of $f(x)$. Since tournament modulo addition gadgets for even k are not defined, the prime 2 cannot be chosen. In every step, the graph gadgets serve as subgraphs in new PGI-tuples. Applying Lemma 5 and 7, the graph gadgets can be transformed into tournaments under AC^0 many-one reductions. □

3.3 Hardness Results of TI for DET

Observe that DET $\leq_m^{AC^0}$ GI (Theorem 4.9 in [21]) and that the complexity class DET coincides with NC^1(#L) . We already described, how NC^1 -circuits and #L -functions can be reduced to graph gadgets. For implementing oracle questions, with another graph gadget every #L function f can be transformed in AC^0 into a sequence of PGI-tuples, encoding the bits of $f(x)$. The input $x \in \Sigma^n$ is also encoded as PGI-tuples. For details see proof of Lemma 4.7 in [21].

Definition 5. [21] *The* oracle graph gadget Gad_k *contains subgraphs* G_a, H_a^h, $I_a^i, J_a^{i,j}$ *with* $h \in [1, k-1]$, $i, j \in [0, k-1]$, *which are copies of graphs in* $((G_a, H_a)(I_a, J_a)) \in PGI$, *encoding bit* x_a *of* $f(x) \mod k$. *Let* $W = \{w_0, \dots, w_{k-1}\}$, $Z = \{z_0, \dots, z_{k-1}\} \subseteq V(Gad_k)$. *Henceforth, for simplifying notations, let* $W^0 = G_a$ *and* $W^h = H_a^h$. *We also denote* $Z[i,j] = J_a^{i,j}$ *for* $j \neq i$ *and* $Z[i,i] = I_a^i$ *for* $i, j \in [0, k-1]$. *Let* $Z^i = \bigcup_{j \in [0,k-1]} Z[i,j]$. *We now describe the edge set* $E(Gad_k)$ *as the union of the following edge sets*

1. $E(G_a), E(H_a^h), E(I_a^i), E(J_a^{i,j})$,
2. $\{ \{u, v\} \mid u = z_i,\ v \in Z^i \text{ or } u \in W^i,\ v = w_i \}$,
3. $\{ \{u, v\} \mid u \in Z[i,j],\ v \in W^j \}$,
4. $\{ (u, v) \mid u = w_i, v = w_{i \oplus 1} \text{ or } u = z_i, v = z_{i \oplus 1} \}$.

For properties of Gad_k see Lemma 4.8 in [21]. Observe that any automorphism of Gad_k acts cyclically on W, Z. Therefore, we can introduce arcs as in item 4, to restrict the automorphism group of Gad_k and for simplifying proofs.

We need all graph gadgets described so far to prove that GI is hard for DET. We want to show the same result for TI.

Theorem 3. *DET* $\leq_m^{AC^0}$ *TI.*

More precisely, we show hardness of TI for $\mathrm{NC}^1(\#\mathrm{L})$. Oracle questions will be implemented by using oracle graph gadgets like Gad_k. We transform Gad_k into a tournament and consider its automorphism properties.

Definition 6. *Let $k \geq 3$ be an odd integer and $i, j \in [0, k-1]$. The oracle tournament gadget $TGad_k$ has vertex set $V(TGad_k) = V(Gad_k)$ and edge set $E(TGad_k)$ as the union of the following edge sets (see also Figure 3)*

1. *$E(Gad_k)$ but write '(u,v)' for items 2, 3 in Definition 5 of $E(Gad_k)$,*
2. *$E(W) \cup E(Z) = \{(w_i, w_{i\oplus h}), (z_i, z_{i\oplus h}) \mid h \in [1, \lfloor \frac{k}{2} \rfloor]\}$,*
3. *$\{(u,v) \mid u \in W^i,\ v \in W^j,\ \text{ iff } (w^i, w^j) \in E(W)\}$,*
4. *$\{(u,v) \mid u \in Z^i,\ v \in Z^j,\ \text{ iff } (z_i, z_j) \in E(Z)\}$,*
5. *$\{(u,v) \mid u \in Z[i,j],\ v \in Z[i, j \oplus h] \text{ with } h \in [1, \lfloor \frac{k}{2} \rfloor]\}$,*
6. *$\{(u,v) \mid u \in Z^i,\ v = z_j \text{ with } j \neq i\}$,*
7. *$\{(u,v) \mid u \in W^i,\ v \in Z^j \text{ with } j \neq i\}$,*
8. *$\{(u,v) \mid u = w_i,\ v \in W^j \text{ with } j \neq i\}$.*

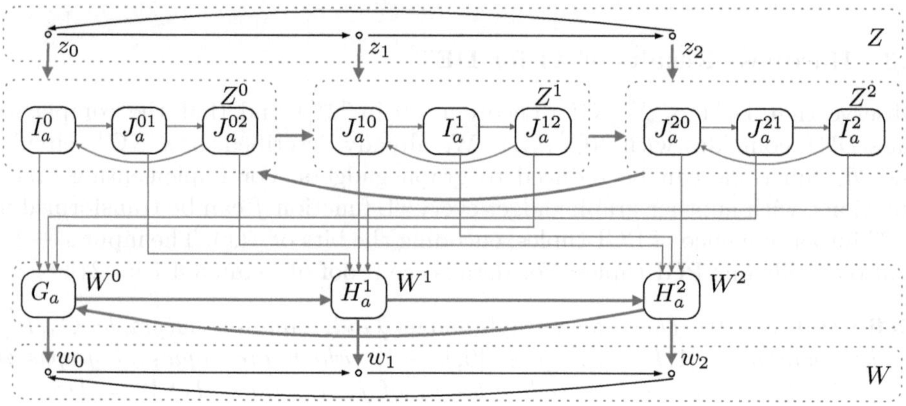

Fig. 3. Graph gadget $TGad_3$ after performing edge sets of items 1 to 5

Lemma 8. *If Gad_k contains tournaments (PTI-tuples) as subgraphs of item 1 in Definition 5, then Gad_k with odd $k \geq 3$ can be transformed into a tournament $TGad_k$ under AC^0 many-one reductions, without changing the automorphism group.*

In the proof we transform Gad_k into $TGad_k$ and apply Lemmas 2 and 3 to construct each subset of edges in $E(TGad^k)$.

Our aim is to construct a graph G^* with prefixes, which contains nontrivial prefix automorphisms, iff an NC^1 circuit with #L oracle questions outputs *true*. The oracle questions thereby must have the same size. This makes it possible to use exactly one type of oracle graph gadget Gad_k (for exactly one value of k). For more details we refer to [21]. Any subgraph like Gad_k inside of G^* is setwise stabilized in $Aut(G^*)$. By Lemma 2 and 8, transform any subgraph in G^* isomorphic to Gad_k into a tournament $TGad_k$ without changing the automorphism group of G^*. Then apply Lemma 3 in order to connect this graph gadget with any other graph gadget. Since G^* contains only graph gadgets as described in this chapter and every graph gadget is setwise stabilized in automorphism group of G^*, replace them all by tournaments and connect them with each other as described in Lemma 3. Thus we can transform G^* into a tournament T^* under AC^0 many-one reductions. It follows that TI is hard for NC^1 (#L) and thus for DET. Corollary 2 follows by the fact that the $\mathrm{Mod_kL}$ hierarchy is logspace Turing reducible to DET.

Corollary 2. $\mathrm{Mod}_k\mathrm{L} \leq_T^L$ *TI for any* $k \geq 2$.

4 Hardness Results for Tournament Automorphism

GA is many-one hard for the $\mathrm{Mod_kL}$ hierarchy (Theorem 5.1 [21]). Transform a circuit C into a rigid graph $G(C)$ as in Definition 1, having a unique automorphism satisfying prefixes, iff the output value of the circuit is 1. Take two copies G_1 and G_2 of graph G and apply colorings $Col(G_1), Col(G_2)$ to the vertices which represent the input and output values of the circuit in order to encode the prefixes the same way as for prefix-GA. Thus the graph $G_1 \cup G_2$ has a nontrivial automorphism, iff the output of the original circuit is 1. We show how this result also holds for TA.

Theorem 4. $\mathrm{Mod}_k\mathrm{L} \leq_m^{\mathrm{AC}^0}$ *TA with* $k \geq 3$ *odd integer.*

Proof. First, transform $G(C)$ into a tournament $T(C)$ as described in proof of Lemma 5. Instead of taking two copies we need three copies T_0, T_1, T_2 of $T(C)$ and apply the colorings $Col(G_1)$ to T_0 and $Col(G_2)$ to T_1 and T_2. Then include complete bipartite edge sets $\{ (u,v) \mid u \in V(T_i), v \in V(T_{i\oplus 1}), i \in [0,2] \}$. Thus $T(C)$ is a tournament and contains two nontrivial automorphisms (that is mapping T_0 onto T_1 or T_2), iff $G(C)$ contains one nontrivial automorphism. □

Now we discuss the counterpart of PGI and PTI tuples, in order to prove lower bounds for TA. Therefore, we define the following graph tuples.

Theorem 5. *DET* $\leq_m^{\mathrm{AC}^0}$ *TA.*

Definition 7. [21] *A* PGA-graph tuple *is a tuple of rigid graphs* $((G,H),(I,J))$ *with* $G \cong H \Leftrightarrow I \ncong J$. *Let PGA be the set of all such tuples. The graph tuple* $(G_\wedge, H_\wedge)(I_\wedge, J_\wedge)) \in PGA$ *(write e.g. PGA-*$G_\wedge$*) for simulating conjunction and* $(G_\vee, H_\vee)(I_\vee, J_\vee)) \in PGA$ *(write e.g. PGA-*$G_\vee$*) for disjunction. For a detailed definition and its properties see proof of Theorem 5.3 in [21].*

If all the subgraphs are rigid then the new graphs forming tuples are rigid as well. We transform now these PGA-tuples into tuples of tournaments.

Lemma 9. *There is an* AC^0 *computable function, such that the PGA graph tuples can be transformed into tournaments, having the same automorphism properties and rigidity properties as in Definition 7.*

Definition 8. *A* PTA-graph tuple *is a tuple of rigid tournaments* $((G,H),(I,J))$ *with* $G \cong H \Leftrightarrow I \ncong J$. *Let PTA be the set of all such tuples.*

The graph tuple $(G_\wedge, H_\wedge)(I_\wedge, J_\wedge)) \in PTA$ *(write e.g. PTA-*$G_\wedge$*) for simulating conjunction and* $(G_\vee, H_\vee)(I_\vee, J_\vee)) \in PTA$ *(write e.g. PTA-*$G_\vee$*) for disjunction is defined as follows:*

The graph PTA-$G_\vee$ *contains subgraphs* $X, G_0, G_0', H_0, A_0, A_1, A_2$. *Let subgraphs* G_0', G_1', I_0' *be copies of* G_0, G_1, I_0. *Let* X *be defined as shown in Figure 4.* A_0 *contains subgraphs* I_0 *and* G_1*, with vertices in* I_0 *pointing to all vertices in* G_1. A_1 *is a copy of* A_0. A_2 *is constructed like* A_0*, containing* J_0 *instead of* I_0 *and* H_1 *instead of* G_1. *Let subgraph* G_0 *be associated to* $x_{0,0}$, G_0' *to* $x_{0,1}$ *and* H_0 *to* $x_{0,2}$ *and let the* A_i *be associated to* $x_{1,i}$ *for* $i \in [0,2]$. *Concerning edge sets, the rest of the construction is similar to that of PTI-*$G_\vee$.

The subgraph $H_\vee$*, is constructed like* $G_\vee$ *but with* A_1 *associated to* $x_{1,2}$ *and* A_2 *to* $x_{1,1}$. *Obtain* $I_\wedge$ *from* $G_\vee$ *and* $J_\wedge$ *from* $H_\vee$*, if subgraphs* G_i, H_i, I_i, J_i *will be replaced by* I_i, J_i, G_i, H_i *for* $i \in [0,1]$ *in this order.*

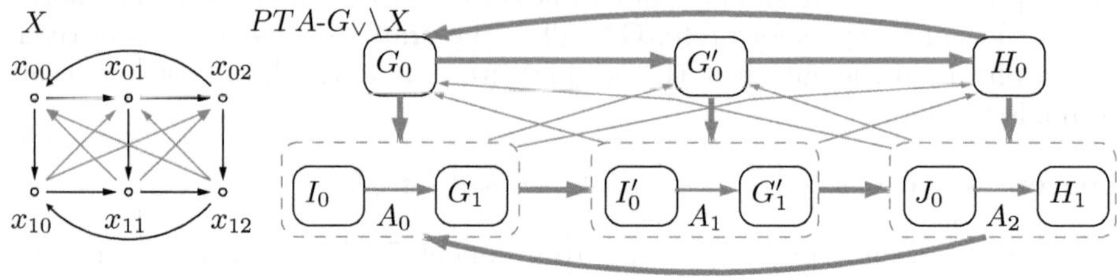

Fig. 4. PTA-tuples simulating ∧ and ∨ functions

The proof is like that of Lemma 7, simulate the alternating layers of AND's and OR's of an NC^1 circuit with certain PTA-tuples. The main difficulty is to preserve the rigidity of the tuple components.

It immediately follows, that TA is hard for NC^1 under AC^0 many-one reductions. By applying Theorem 4, it is possible to prove hardness of TA for complexity class DET. The proof of this result follows (similar to that in [21]) exactly the same lines as that for Theorem 3 taking in consideration that the tournament graph pairs produced in the reduction from Theorem 1 are rigid and that the gadgets in the proof of Theorem 3 also preserve rigidity. Similar to the Corollary 2, the Corollary 3 immediately follows:

Corollary 3. $\text{Mod}_k\text{L} \leq_T^L$ *TA for any* $k \geq 2$.

Acknowledgments. I am grateful to my supervisor Jacobo Torán , Sebastian Dörn, Thanh Minh Hoang for helpful discussion and the anonymous referees for helpful comments and suggestions.

References

1. Arvind, V., Beigel, R., Lozano, A.: The Complexity of Modular Graph Automorphism. Symp. on Theoret. Aspects of Computer Sci., 172–182 (1998)
2. Arvind, V., Das, B., Mukhopadhyay, P.: On Isomorphism and Canonization of Tournaments and Hypertournaments. In: Asano, T. (ed.) ISAAC 2006. LNCS, vol. 4288, pp. 449–459. Springer, Heidelberg (2006)
3. Allender, E., Ogihara, M.: Relationships among PL, #L and the determinant. RAIRO Inform. Theor. Appl. 30, 1–21 (1996)
4. Alvarez, C., Jenner, B.: A very hard logspace counting class. Theoretical Computer Science 107, 3–30 (1993)
5. Barrington, D.A.M., Immerman, N., Straubing, H.: On uniformity within NC^1. J. Comput. System Sci. 41, 274–306 (1990)
6. Babai, L., Luks, E.: Canonical labeling of graphs. Proceedings of the 15th Annual ACM Symposium on Theory of Computing, 171–183 (1983)
7. Boppana, R., Hastad, J., Zachos, S.: Does co-NP have short interactive proofs? Inform. Process. Lett. 25, 27–32 (1987)
8. Buntrock, G., Damm, C., Hertrampf, U., Meinel, C.: Structure and importance of logspace-MOD-classes. Math. System Theory 25, 223–237 (1992)
9. Buss, S.R.: Alogtime algorithms for tree isomorphism, comparison, and canonization. In: Gottlob, G., Leitsch, A., Mundici, D. (eds.) KGC 1997. LNCS, vol. 1289, pp. 18–33. Springer, Heidelberg (1997)
10. Cook, S.A.: A taxonomy of problems with fast parallel algorithms. Information and Control 64, 2–22 (1985)
11. Gill, J.: Computational complexity of probabilistic Turing machines. SIAM J. Comput. 6, 675–695 (1977)
12. Gross, J.L., Yellen, J.: Discrete Mathematics and its Applications - Handbook of Graph Theory. CRC Press LLC (2004)
13. Hopcroft, J.E., Tarjan, R.E.: A V^2 algorithm for determining isomorphism of planar graphs, pp. 32–34 (1971)
14. Köbler, J., Schöning, U., Torán, J.: The Graph Isomorphism Problem - Its Structural Complexity. In: Prog. Theor. Comp.Sci., Birkhaeuser, Boston, MA (1993)
15. Lindell, S.: A logspace algorithm for tree canonization. Proceedings of the 24th Annual ACM Symposium on Theory of Computing, 400–404 (1992)
16. Luks, E.: Isomorphism of bounded valence can be tested in polynomial time. J. Comput. System Sci. 25, 42–65 (1982)
17. Luks, E.: Parallel algorithms for perumtation groups and graph isomorphism. Proc of the 27th IEEE Symp. on Found. of Comp. Sci., 292–302 (1986)
18. Miller, G.L., Reif, J.H.: Parallel tree contraction Part 2: further applications. SIAM Journal on Computing 20(6), 1128–1147 (1991)
19. Ruzzo, W., Simon, J., Tompa, M.: Space bounded hierarchies and probabilistic computations. J. Comput. System Sci. 28, 216–230 (1984)
20. Schöning, U.: Graph isomorphism is in the low hierarchy. J. Comput. System Sci. 37, 312–323 (1988)
21. Torán, J.: On the Hardness of Graph Isomorphism. SIAM J. Comput. 33(5), 1093–1108 (2004)

Relating Complete and Partial Solution for Problems Similar to Graph Automorphism

Takayuki Nagoya[1] and Seinosuke Toda[2]

[1] Department of Mathematical Sciences, Tokyo Denki University
nagoya@r.dendai.ac.jp

[2] Department of Computer Science and System Analysis,
College of Humanities and sciences, Nihon University
toda@cssa.chs.nihon-u.ac.jp

Abstract. It is known that, given a graph G, finding a pair of vertices (v_i, v_j) such that v_i is mapped to v_j by some non-trivial automorphism on G is as hard as computing a non-trivial automorphism. In this paper, we show that, given a graph G, computing even a single vertex that is mapped to a different vertex by a non-trivial automorphism is as hard as computing a non-trivial automorphism. We also show that RightGA has the same property. On the other hand, we show that if PrefixGA has this property then GI $\leq_T^p$ GA.

Keywords: Computational complexity, reducibility, graph automorphism, partial solution.

1 Introduction

Graph Isomorphism problem (GI) is to determine whether two given graphs are isomorphic or not. Closely related to GI is the Graph Automorphism problem (GA) : given a graph G, decide whether its automorphism group contains a non-trivial automorphism. These problems are trivially in NP, but the problems are not known to be in P and not known to be NP-complete either [9], [4]. In spite of their similarity, there seem to be differences between the computational structures of these problems. For example, although GI has a polynomial-time computable and-function [5], it is not known whether there exists such function for GA. Actually, GA seems to be computationally easier than GI. Although GA is polynomial-time many-one reducible to GI [5], it is not known whether GI is reducible to GA or not. Lubiw [6] has left this problem as an open question. There are many problems similar to GA. It is known that the prefix set of GA (PrefixGA) is polynomial-time many-one equivalent to GI. Lozano and Toran [5] have shown that the left set of GA (LeftGA) is polynomial-time many-one equivalent to GI. On the other hand, they also prove that the right set of GA (RightGA) is polynomial-time many-one equivalent to GA. Other discussions between GI and GA are found in e.g. [1], [4], [6], [7], [10].

For several NP problems, Gál et al.[3] proved that it is enough for solving their corresponding search problems to provide an efficient algorithm for computing

L. Kučera and A. Kučera (Eds.): MFCS 2007, LNCS 4708, pp. 584–595, 2007.

only a small part of a solution. Große et al. [2] proved that an isomorphism on given graph can be computed by using any oracle that gives a single pair of vertices that are mapped onto each other by an isomorphism between them. In [8], it has been proved that a non-trivial automorphism on a given graph can be computed by using any oracle that gives a single pair (v_i, v_j) of vertices such that v_i is mapped to v_j by a non-trivial automorphism on the graph.

In this paper, we prove that a non-trivial automorphism on a given graph can be computed in polynomial time by using any oracle that gives a single vertex that is mapped to a different vertex by a non-trivial automorphism. We notice that the result is quite a contrast to known results of GI and GA, because our oracle returns only a single vertex but not a pair. We prove that the same result holds for RightGA. That is, the search version of RightGA can be computed by using any oracle that gives a single-vertex solution. These results indicate that, for GA and RightGA, computing a single-vertex solution is as hard as computing a (complete) solution. On the other hand, we prove that computing a single-vertex solution of PrefixGA is reducible to GA. This indicates an evidence that, for PrefixGA, computing a single-vertex solution is apparently easier than computing a (complete) solution, since PrefixGA is many-one equivalent to GI. This also seems to indicate a structural difference between GA and GI.

2 Computing Graph Automorphism from Single-Vertex Solutions

Throughout this paper, we suppose that all graphs are undirected and simple. For a graph G, its vertex set and its edge set are denoted by $V(G)$ and $E(G)$, respectively. We sometimes identify a graph with its vertex set if there is no fear of confusion. For a vertex $v \in V(G)$, we define $N_G(v) = \{w \in V(G) | (v, w) \in E(G)\}$ and define $G \backslash \{v\}$ to be a graph obtained by removing v and all edges incident to v. For a graph G, M_G is the maximum number of vertices in a connected component of G. For two graphs G and H, $G \cup H$ is the disjoint union of G and H. We denote by id_G the identity automorphism of a graph G. A labeled graph is a graph with labels assigned to its vertices. In this paper, we assume that each label is a set of distinct positive integers. For two functions $f : A \to B$ and $g : C \to D$, if $A \cap C = \phi$ then we define $f \sqcup g$ by, for all $v \in A \cup C$, $f \sqcup g(v) = f(v)$ if $v \in A$ and $f \sqcup g(v) = g(v)$ otherwise. For a function $f : A \to B$ we denote $f|_X$ to be the restriction of f whose domain is $X \subseteq A$. The following theorems was proved in [2] and [8].

Theorem 1. *[2] Let f be any oracle that, given any two graphs $\hat{G}$ and $\hat{H}$, outputs a pair of vertices $(v_i, w_j), v_i \in V(\hat{G}), w_j \in V(\hat{H})$ such that if there exists an isomorphism from $\hat{G}$ to $\hat{H}$, then $\varphi(v_i) = w_j$ for some isomorphism φ from $\hat{G}$ to $\hat{H}$. Then there is an algorithm such that, given any two graphs G and H, it uses the oracle f and correctly determines whether G and H are isomorphic, and if so, it constructs an isomorphism from G to H in polynomial time.*

Theorem 2. *[8] Let f be any oracle that, given any graph $\hat{G}$, outputs a pair of vertices (v_i, v_j) such that if there exists a non-trivial automorphism on $\hat{G}$ then v_i is mapped to v_j by some non-trivial automorphism on $\hat{G}$. Then there is an algorithm such that, given any graph G, it uses the oracle f and correctly determines whether a non-trivial automorphism on G exists, and if so, it constructs a non-trivial automorphism on G in polynomial time.*

These results indicates that computing a pair of vertices that is a piece of a (complete) solution is as hard as computing a (complete) solution. In this section, we prove that computing a single vertex that is moved by some non-trivial automorphism is as hard as computing a non-trivial automorphism. To prove the claim, we first focus our attention on labeled graphs. Then the claim is obtained by using a fact that the labels on the vertices of a labeled graph can be simulated in an unlabeled graph by attaching suitable gadgets to the vertices. We state the detail of the fact at the end of this section.

The next theorem is the result on labeled graphs.

Theorem 3. *Let f^l_{GA} be any oracle that, given any labeled graph $\hat{G}$, outputs a vertex v_i such that if there exists a non-trivial automorphism on $\hat{G}$ then v_i is mapped to a different vertex by some non-trivial automorphism on $\hat{G}$. Then there is an algorithm such that, given any labeled graph G, it uses the oracle f^l_{GA} and correctly determines whether a non-trivial automorphism on G exists, and if so, it constructs a non-trivial automorphism on G in polynomial time.*

Before proving the above theorem, we would like to state some properties of f^l_{GA}. If a queried graph does not have a non-trivial automorphism, then f^l_{GA} returns an arbitrary vertex without revealing that the graph does not have a non-trivial automorphism. If we query f^l_{GA} on the same graph repeatedly, then f^l_{GA} could keep returning the same vertex. We also emphasize that f^l_{GA} does not seem to be powerful enough to solve GI. For, f^l_{GA} is trivially reducible to the search version of GA and GA has self-computable solutions (refer to [4] for the definition of self-computability). So, f^l_{GA} is reducible to (the decision version of) GA. This indicates that f^l_{GA} does not seem to be available to compute GI-complete problems, like AUTOMORPHISM WITH 1 RESTRICTION that was introduced in [6], i.e. given a graph G and a vertex $v \in V(G)$, determining whether there is a non-trivial automorphism that maps v to a different vertex. In spite of the similarity between f^l_{GA} and AUTOMORPHISM WITH 1 RESTRICTION, f^l_{GA} is reducible to GA and AUTOMORPHISM WITH 1 RESTRICTION is GI-complete. So, f^l_{GA} does not seem to be available to solve AUTOMORPHISM WITH 1 RESTRICTION.

As stated above, if a queried graph does not have a non-trivial automorphism, then f^l_{GA} returns an arbitrary vertex without revealing that the graph does not have a non-trivial automorphism. Thus, in general, it seems difficult to determine, by only using information from the oracle, whether a graph has a non-trivial automorphism. However, in some special case, we can take notice of the lack of a non-trivial automorphism by checking simple conditions.

Definition 1. *Let G be a labeled graph and let $v_1, v_2, (v_1 \neq v_2)$ be vertices of G. Let W_1 and W_2 be the connected components of G containing v_1 and v_2 respectively. We say that v_2 is a candidate for v_1 if these vertices have the same label, $|N_{W_1}(v_1)| = |N_{W_2}(v_2)|$, and $W_1 \backslash \{v_1\}$ is disconnected if and only if $W_2 \backslash \{v_2\}$ is disconnected.*

Note that if the oracle returns a vertex that has no candidate vertex, then, by the property of the oracle, there is no non-trivial automorphism on the graph.

Definition 2. *Let $X_1, \ldots, X_a, Y_1, \ldots, Y_a$ be $2a$ connected labeled graphs such that $|X_1| = \ldots = |X_a| = |Y_1| = \ldots = |Y_a|$. We denote by $\langle (X_1, Y_1), \ldots, (X_a, Y_a) \rangle$ the labeled graph constructed as follows: Let $l_1, l_2, \ldots, l_a$ be integers which are not used in the labels of vertices in $X_1 \cup \ldots \cup X_a \cup Y_1 \cup \ldots \cup Y_a$. For each s $(1 \leq s \leq a)$, add l_s to the label of every vertex in $X_s \cup Y_s$. Let $\hat{X}_1, \ldots, \hat{X}_a, \hat{Y}_1, \ldots, \hat{Y}_a$ be the resulting graphs. Then, the labeled graph $\langle (X_1, Y_1), \ldots, (X_a, Y_a) \rangle$ is the disjoint union of $\hat{X}_1, \ldots, \hat{X}_a, \hat{Y}_1, \ldots, \hat{Y}_a$.*

Note that any automorphism on $\langle (X_1, Y_1), \ldots, (X_a, Y_a) \rangle$ maps $X_s \cup Y_s$ to $X_s \cup Y_s$.

Now, we are ready to prove two lemmas that state the detail of the algorithm for Theorem 3 and that guarantee the correctness of Theorem 3.

Lemma 1. *Let G be a connected labeled graph and let $V(G) = \{v_1, v_2, \ldots, v_n\}$, $n \geq 2$. Let v_i be a vertex resulting from a query $f^l_{GA}(G)$. If there exists a candidate for v_i in G then there exists a labeled graph $\hat{G}$ that satisfies the following conditions.*

1. *$M_{\hat{G}} < M_G$ and $|V(\hat{G})| \leq 4M_G^2$.*
2. *$\hat{G}$ can be constructed in polynomial time.*
3. *G has a non-trivial automorphism if and only if $\hat{G}$ has a non-trivial automorphism. Furthermore, if a non-trivial automorphism ψ on $\hat{G}$ is given then a non-trivial automorphism on G can be constructed from ψ in polynomial time.*

Proof. We construct a labeled graph $\hat{G}$ from G as follows: Let l be an integer that is not used in the labels of vertices of G. Let $v_{j_1}, v_{j_2}, \ldots, v_{j_a} \in V(G)$ be the candidates for v_i. Then we create $2a$ copies $X_1, X_2, \ldots, X_a, Y_1, Y_2, \ldots, Y_a$ of G. For each X_s $(1 \leq s \leq a)$, add l to the label of each neighbor of v_i, and then remove the vertex v_i and all edges incident to v_i. For each Y_s $(1 \leq s \leq a)$, add l to the label of each neighbor of v_{j_s}, and then remove the vertex v_{j_s} and all edges incident to v_{j_s} (see Figure 1). Since every v_{j_s} is a candidate for v_i, $G \backslash \{v_i\}$ is disconnected if and only if each constructed graph X_s and Y_s is disconnected. If $G \backslash \{v_i\}$ is disconnected then we consider the complement $\bar{G}$ instead of G, and then construct X_s and Y_s from $\bar{G}$ in the same manner[1]. As a consequence, we can assume that each constructed labeled graph $X_1, X_2, \ldots X_a, Y_1, Y_2, \ldots, Y_a$ is connected. Finally, let $\hat{G}$ be the labeled graph $\langle (X_1, Y_1), \ldots, (X_a, Y_a) \rangle$.

[1] Note that the complement of a disconnected graph is always connected.

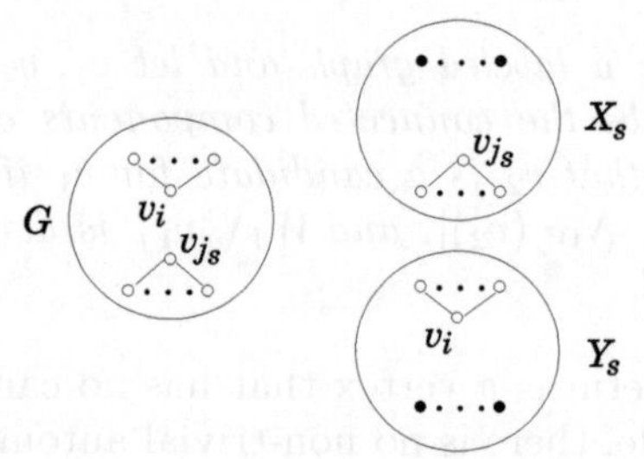

Fig. 1. Construction of X_s and Y_s in Lemma 1. Large black circles are the vertices to which l is added.

We easily see that the first and the second conditions are satisfied. We bellow prove the third condition is satisfied. Assume that there exists a non-trivial automorphism on G. Then, by the property of the oracle, there exists a non-trivial automorphism φ on G that maps v_i to a different vertex. Therefore, there exists a candidate $v_{j_s} \in V(G)$ of v_i such that $\varphi(v_i) = v_{j_s}$. We easily see that, using φ, we can construct an isomorphism ξ from X_s to Y_s. So we have that $\xi \sqcup \xi^{-1}$ is a non-trivial automorphism on $X_s \cup Y_s$. Now we have a non-trivial automorphism ψ on $\hat{G}$ defined as follow:

$$\psi(v) = \begin{cases} \xi \sqcup \xi^{-1}(v) & \text{if } v \in X_s \cup Y_s \\ v & \text{otherwise.} \end{cases} \tag{1}$$

Conversely, we assume that there exists a non-trivial automorphism ψ on $\hat{G}$. By the construction of $\hat{G}$, for each $s, (1 \leq s \leq a)$, ψ must map $X_s \cup Y_s$ to $X_s \cup Y_s$. Since ψ is a non-trivial automorphism on $\hat{G}$, we have that there exists $s, (1 \leq s \leq a)$ such that $\psi|_{X_s \cup Y_s}$ is a non-trivial automorphism on $X_s \cup Y_s$. Now we have the following two cases.

1. $\psi(X_s) = X_s$. In this case, we have that either $\psi|_{X_s}$ is a non-trivial automorphism on X_s or $\psi|_{Y_s}$ is a non-trivial automorphism on Y_s. If $\psi|_{X_s}$ is a non-trivial automorphism on X_s then $\psi|_{X_s}$ can be considered as a non-trivial automorphism on G that fixes v_i. If $\psi|_{Y_s}$ is a non-trivial automorphism on Y_s then $\psi|_{Y_s}$ can be considered as a non-trivial automorphism on G that fixes v_{j_s}.
2. $\psi(X_s) = Y_s$. In this case, $\psi|_{X_s}$ is an isomorphism from X_s to Y_s. So, using $\psi|_{X_s}$, we can construct a non-trivial automorphism on G that maps v_i to v_{j_s}.

We conclude that there exists a non-trivial automorphism on G. Furthermore, such an automorphism on G can be constructed from ψ in polynomial time. This completes the proof of Lemma 1.

Lemma 2. *Let $G =< (X_1, Y_1), \ldots, (X_a, Y_a) >$ be a labeled graph constructed from $2a$ connected labeled graphs $X_1, X_2, \ldots, X_a, Y_1, Y_2, \ldots, Y_a$ such that $|X_1| = \ldots = |X_a| = |Y_1| = \ldots = |Y_a| \geq 2$. Let v_i be a vertex resulting from a query $f^l_{GA}(G)$. If there exists a candidate for v_i in G then there exists a labeled graph $\hat{G}$ that satisfy the following conditions.*

1. $M_{\hat{G}} < M_G$ *and* $|V(\hat{G})| \leq 4M_G^2$.
2. $\hat{G}$ *can be constructed in polynomial time.*
3. G *has a non-trivial automorphism if and only if* $\hat{G}$ *has a non-trivial automorphism. Furthermore, if a non-trivial automorphism* ψ *on* $\hat{G}$ *is given then a non-trivial automorphism on* G *can be constructed from* ψ *in polynomial time.*

Proof. By symmetry, we below assume that some X_s contains v_i. For the case that some Y_s contains v_i, we can prove the claim similarly. We construct a labeled graph $\hat{G}$ (only) from $X_s \cup Y_s$ as follows: Let l be an integer that is not used in the labels of vertices of $X_s \cup Y_s$. Let $v_{j_1}, v_{j_2}, \ldots, v_{j_b} \in X_s$ be the candidates for v_i in X_s. Then we create $2b$ copies $U_1, \ldots, U_b, W_1, \ldots, W_b$ of X_s. For each U_t, $(1 \leq t \leq b)$, add l to the label of each neighbor of v_i, and then remove the vertex v_i and all edges incident to v_i. For each W_t, $(1 \leq t \leq b)$, add l to the label of each neighbor of v_{j_t}, and then remove the vertex v_{j_t} and all edges incident to v_{j_t} (see the left of Figure 2). Let $v_{k_1}, v_{k_2}, \ldots, v_{k_c} \in Y_s$ be the candidates for v_i in Y_s. Then we create c copies $S_1, \ldots, S_c$ of X_s and create c copies $T_1, \ldots, T_c$ of Y_s. For each S_u, $(1 \leq u \leq c)$, add l to the label of each neighbor of v_i and then remove the vertex v_i and all edges incident to v_i. For each T_u, $(1 \leq u \leq c)$, add l to the label of each neighbor of v_{k_u} and then remove the vertex v_{k_u} and all edges incident to v_{k_u} (see the right of Figure 2). By the similar argument of the proof of Lemma 1, we can assume that each constructed labeled graph U_t, W_t, S_u, and T_u is connected. Finally, let $\hat{G}$ be the labeled graph $< (U_1, W_1), \ldots, (U_b, W_b), (S_1, T_1), \ldots, (S_c, T_c) >$.

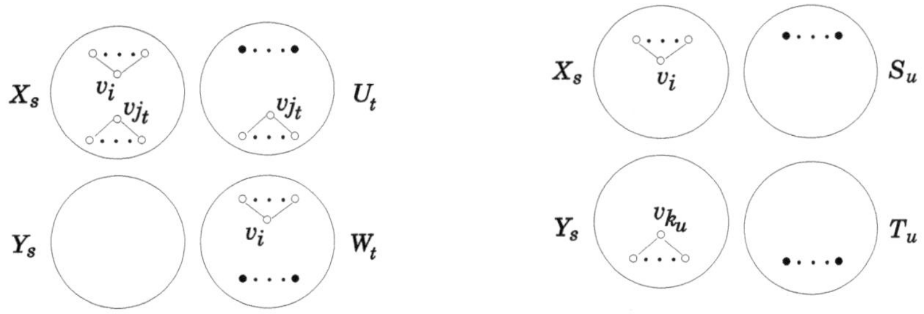

Fig. 2. Construction of the labeled graph U_t, W_t (the left of the figure) and S_t, T_u (the right of the figure) in Lemma 2. Large black circles are the vertices to which l is added.

We easily see that the first and second conditions are satisfied. Note that $\hat{G}$ is constructed from X_s and Y_s, and we do not use other connected components of G to construct $\hat{G}$. Therefore, we see that $|V(\hat{G})| \leq (2b + 2c)M_G \leq 4M_G^2$.

We will bellow prove the third condition is satisfied. Assume that there exists a non-trivial automorphism on G. Then, by the property of the oracle, there exists a non-trivial automorphism φ on G that maps v_i to a different vertex. Therefore, we have that $\varphi|_{X_s \cup Y_s}$ is a non-trivial automorphism on $X_s \cup Y_s$. Now we have the following two cases.

1. $\varphi(X_s) = X_s$. In this case, $\varphi|_{X_s}$ is a non-trivial automorphism on X_s that maps v_i to a different vertex. So there exists a candidate $v_{j_t} \in X_s$ of v_i such that $v_{j_t} = \varphi(v_i)$. Since, using $\varphi|_{X_s}$, we can construct an isomorphism ξ from U_t to W_t, $\xi \sqcup \xi^{-1}$ is a non-trivial automorphism on $U_t \cup W_t$. Now we have a non-trivial automorphism ψ on $\hat{G}$ defined as follows:
$$\psi(v) = \begin{cases} \xi \sqcup \xi^{-1}(v) & \text{if } v \in U_t \cup W_t \\ v & \text{otherwise.} \end{cases} \quad (2)$$
2. $\varphi(X_s) = Y_s$. In this case, $\varphi|_{X_s}$ is an isomorphism from X_s to Y_s. Furthermore, there exists a candidate $v_{k_u} \in Y_s$ for v_i such that $v_{k_u} = \varphi(v_i)$. Since $\varphi|_{X_s}$ can be considered as an isomorphism ξ from S_u to T_u, we have that $\xi \sqcup \xi^{-1}$ is a non-trivial automorphism on $S_u \cup T_u$. Now we have a non-trivial automorphism ψ on $\hat{G}$ defined as follows:
$$\psi(v) = \begin{cases} \xi \sqcup \xi^{-1}(v) & \text{if } v \in S_u \cup T_u \\ v & \text{otherwise.} \end{cases} \quad (3)$$

Conversely, we assume that there exists a non-trivial automorphism ψ on $\hat{G}$. We have the following cases.

1. There exists $t, (1 \le t \le b)$ such that $\psi|_{U_t \cup W_t}$ is a non-trivial automorphism on $U_t \cup W_t$. In this case, we have the following two cases.
   (a) $\psi(U_t) = U_t$. In this case, either $\psi|_{U_t}$ is a non-trivial automorphism on U_t or $\psi|_{W_t}$ is a non-trivial automorphism on W_t. If $\psi|_{U_t}$ is a non-trivial automorphism on U_t then $\psi|_{U_t}$ can be considered as a non-trivial automorphism ξ on X_s that fixes v_i. If $\psi|_{W_t}$ is a non-trivial automorphism on W_t then $\psi|_{W_t}$ can be considered as a non-trivial automorphism ξ on X_s that fixes v_{j_t}. In both cases, using ξ, we can construct a non-trivial automorphism τ on G defined as follows:
$$\tau(v) = \begin{cases} \xi(v) & \text{if } v \in X_s \\ v & \text{otherwise.} \end{cases} \quad (4)$$
   (b) $\psi(U_t) = W_t$. In this case, $\psi|_{U_t}$ is an isomorphism from U_t to W_t and it can be considered as a non-trivial automorphism ξ on X_s that maps v_i to v_{j_t}. So we have a non-trivial automorphism τ on G defined as follows:
$$\tau(v) = \begin{cases} \xi(v) & \text{if } v \in X_s \\ v & \text{otherwise.} \end{cases} \quad (5)$$
2. There exists $u, (1 \le u \le c)$ such that $\psi|_{S_u \cup T_u}$ is a non-trivial automorphism on $S_u \cup T_u$. In this case, we can consider $\psi|_{S_u \cup T_u}$ to be a non-trivial automorphism ξ on $X_s \cup Y_s$ that fixes v_i and v_{k_u} or exchanges v_i and v_{k_u}. Now, we have a non-trivial automorphism τ on G defined as follows:
$$\tau(v) = \begin{cases} \xi(v) & \text{if } v \in X_s \cup Y_s \\ v & \text{otherwise.} \end{cases} \quad (6)$$

We conclude that there exists a non-trivial automorphism on G. Furthermore, such an automorphism on G can be constructed from ψ in polynomial time. This completes the proof of Lemma 2.

The algorithm of Theorem 3 is as follows. The algorithm computes a bijection on G such that if there exists a non-trivial automorphism on G then the bijection is a non-trivial automorphism.

Algorithm 1 (labeled graph G)

1. Query f_{GA}^{l} on G. Let v_i be the returned vertex.
2. If there is no candidate for v_i in G then return an arbitrary bijection on $V(G)$.
3. Otherwise, construct $\hat{G}$ from G and v_i as in Lemma 1.
4. Call **Algorithm 2($\hat{G}$)**. Let ψ be the returned bijection on $V(\hat{G})$.
5. If ψ is a non-trivial automorphism on $\hat{G}$ then compute a non-trivial automorphism φ on G from ψ as in Lemma 1, and return φ as a non-trivial automorphism on G. Otherwise, return an arbitrary bijection on $V(G)$.

Algorithm 2 (labeled graph $G =< (X_1, Y_1), \ldots, (X_a, Y_a) >$)

1. If $|X_1| = \ldots = |X_a| = |Y_1| = \ldots = |Y_a| = 1$ then, look for $s\,(1 \leq s \leq a)$ such that the vertex of X_s and the vertex of Y_s have the same label. If there exists such s then return a non-trivial automorphism on G that exchanges the only vertex of X_s and the only vertex of Y_s but fixes the other vertices. Otherwise, return an arbitrary bijection on $V(G)$.
2. Query f_{GA}^{l} on G. Let v_i be the returned vertex.
3. If there is no candidate for v_i in G then return an arbitrary bijection on $V(G)$.
4. Otherwise, construct $\hat{G}$ from G and v_i as in Lemma 2.
5. Recursively call **Algorithm 2($\hat{G}$)**. Let ψ be the returned bijection on $\hat{G}$.
6. If ψ is a non-trivial automorphism on $\hat{G}$ then compute a non-trivial automorphism φ on G from ψ as in Lemma 2, and return φ as a non-trivial automorphism on G. Otherwise, return an arbitrary bijection on G.

Lemma 3. *The algorithm terminates in polynomial time. Furthermore, if there exists a non-trivial automorphism on a given labeled graph then the bijection that is computed by the algorithm is a non-trivial automorphism on the graph.*

Proof. By the construction of Lemma 1 and Lemma 2, the constructed labeled graph $\hat{G} =< (X_1, Y_1), \ldots, (X_a, Y_a) >$ satisfies the condition $|X_1| = \ldots = |X_a| = |Y_1| = \ldots = |Y_a|$. Furthermore, Lemma 1 and 2 decrease M_G by one. So, we have that the algorithm decreases the number of vertices of each connected component to one. We also have that the number of recursive calls of **Algorithm 2** is $|M_G| - 1$. Furthermore, we see $|V(\hat{G})| \leq 4M_G^2$. We conclude that the algorithm terminates in polynomial time.

Next, we prove that if there exists a non-trivial automorphism on a given labeled graph then the bijection that is computed by the algorithm is a non-trivial automorphism on the graph. Let G be a labeled graph $< (X_1, Y_1), \ldots, (X_a, Y_a) >$ such that $|X_1| = \ldots = |X_a| = |Y_1| = \ldots = |Y_a| = 1$. If G has a non-trivial automorphism then there exists $s\,(1 \leq s \leq a)$ such that both the only vertex of X_s and the only vertex of Y_s have the same label and the bijection on $V(G)$ that exchanges these two vertices but fixes the other vertices is a non-trivial automorphism on G. Together with Lemma 1 and Lemma 2, we have the claim immediately.

Theorem 3 has dealt with labeled graphs. By using a well known gadget technique, we will prove that the oracle of Theorem 3 can be computed by using the unlabeled version of the oracle. This fact leads to the result for unlabeled graphs similar to Theorem 3.

Lemma 4. *Let f_{GA} be any oracle that, given any unlabeled graph $\hat{H}$, outputs a vertex w_i such that if there exists a non-trivial automorphism on $\hat{H}$ then w_i is mapped to a different vertex by some non-trivial automorphism on $\hat{H}$. Then, there is an algorithm that, given any labeled graph G, uses the oracle f_{GA} to compute a vertex $v_i \in G$ such that if there exists a non-trivial automorphism on G then v_i is mapped to a different vertex by some non-trivial automorphism on G in polynomial time.*

Proof. Let G be a labeled graph. Let $L(v_i)$ be the label of vertex v_i. We can assume that every integer that is used in the labels of vertices in G is between 1 and $|V(G)|^c$ for some constant c. Let m be the maximum integer in $\bigcup_{v_i \in V(G)} L(v_i) \cup \{|V(G)|\}$.

Firstly, the algorithm constructs an unlabeled graph H from the labeled graph G as follows. For each vertex $v_i \in V(G)$, construct a gadget of Figure 3. $2m+3$ vertices of the gadget form a chain that is connected to v_i. For each $l_j \in L(v_i)$, a chain of l_j vertices starts from the $(m+2)$-th vertex of the above chain. We easily see that, for every non-trivial automorphism φ on H, $\varphi|_G$ is a non-trivial automorphism on G.

Secondly, the algorithm queries f_{GA} about H. Let w_i be the returned vertex. If w_i is a vertex in $V(G)$, then the algorithm outputs w_i. If w_i is a vertex in a gadget that was attached to some vertex v_j of G, then the algorithm outputs v_j.

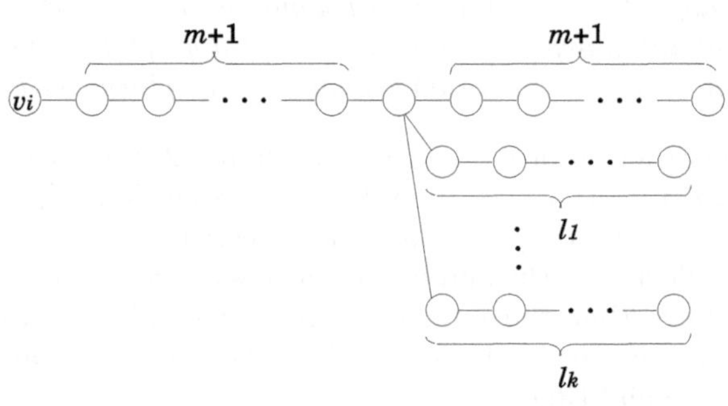

Fig. 3. The gadget attached to v_i whose label is $\{l_1, \ldots, l_k\}$

It is easy to see that the computed vertex is a vertex in $V(G)$ and satisfies that if there exists a non-trivial automorphism on G then the vertex is mapped to a different vertex by some non-trivial automorphism on G. This completes the proof of Theorem 4.

From Theorem 3 and Lemma 4, we have the following theorem.

Theorem 4. *There is an algorithm such that, given any unlabeled graph H, it uses the oracle f_{GA} and correctly determines whether a non-trivial automorphism on H exists, and if so, it constructs a non-trivial automorphism on H in polynomial time.*

3 Other Problems Related to Graph Automorphism

In this section, we consider some problems related to GA. To state the problems, we need the following definitions.

An injective function $\sigma : \{1, \ldots, t\} \mapsto \{1, \ldots, n\}$ is represented by a string $\sigma(1) \cdots \sigma(t)$ over the alphabet $\{1, \ldots, n\}$. In this way, two functions $\sigma, \varphi : \{1, \ldots, t\} \mapsto \{1, \ldots n\}$ represented by the strings $\sigma(1) \cdots \sigma(t)$ and $\varphi(1) \cdots \varphi(t)$ can be compared and we say that σ is smaller than φ if the string $\sigma(1) \cdots \sigma(t)$ is smaller than $\varphi(1) \cdots \varphi(t)$ in lexicographical order.

We consider the following problems related to graph automorphism. The symbol $\leq$ refers to lexicographical order and the symbol φ^i refers to the subsequence $\varphi(1) \cdots \varphi(i)$ of φ.

RightGA $= \{(G, \sigma) : V(G) = \{1, \ldots, n\}, \sigma : \{1, \ldots, t\} \mapsto \{1, \ldots, n\}$ for some $t \leq n$, and there is a non-trivial automorphism φ on G such that $\varphi^t \leq \sigma\}$.

PrefixGA $= \{(G, \sigma) : V(G) = \{1, \ldots, n\}, \sigma : \{1, \ldots, t\} \mapsto \{1, \ldots, n\}$ for some $t \leq n$, and there exists a non-trivial automorphism φ on G such that $\varphi^t = \sigma\}$.

Lozano and Torán [5] proved the following theorem.

Theorem 5. *[5] RightGA $\equiv_m^p$ GA.*

The following proposition is noted in [1].

Proposition 1. *PrefixGA $\equiv_m^p$ GI.*

In the previous section, we have shown that a non-trivial automorphism can be computed by using any oracle that gives us a single-vertex solution. Our interest in this section is whether or not those automorphisms related to the above problems have a similar property. We will show that those in RightGA have the property. On the other hand, those in PrefixGA do not have the property unless GI $\leq_T^p$ GA.

Definition 3. *Let $f_{RightGA}$ be any oracle that, given any pair (G, σ), $V(G) = \{1, \ldots, n\}$, $\sigma : \{1, \ldots, t\} \mapsto \{1, \ldots, n\}$, outputs a vertex v_i such that if $(G, \sigma) \in$ RightGA then v_i is mapped to a different vertex by some non-trivial automorphism φ on G such that $\varphi^t \leq \sigma$.*

Definition 4. *Let $f_{PrefixGA}$ be any oracle that, given any pair (G, σ), $V(G) = \{1, \ldots, n\}$, $\sigma : \{1, \ldots, t\} \mapsto \{1, \ldots, n\}$, outputs a vertex v_i such that if $(G, \sigma) \in$ PrefixGA then v_i is mapped to a different vertex by some non-trivial automorphism φ on G such that $\varphi^t = \sigma$.*

We first prove that the search version of RightGA can be computed by using $f_{RightGA}$.

Theorem 6. *There is an algorithm such that, given any pair (G, σ), it uses the oracle $f_{RightGA}$ and correctly determines whether $(G, \sigma) \in$ RightGA, and if so, it constructs a non-trivial automorphism φ on G such that $\varphi^t \leq \sigma$.*

Proof. We first prove that f_{GA} is reducible to $f_{RightGA}$. Let G be a query for f_{GA} and let $V(G) = \{1, \ldots, n\}$. We consider a pair (G, σ) with $\sigma(1) = n$. Let v_i be the vertex resulting from a query $f_{RightGA}((G, \sigma))$. It is easy to see that if there exists a non-trivial automorphism on G then v_i is mapped to a different vertex by some non-trivial automorphism on G. We have that f_{GA} can be computed by using $f_{RightGA}$. Now, together with Theorem 5 and Theorem 4, we have that (the decision version of) RightGA is reducible to $f_{RightGA}$. Our remaining task is to prove that RightGA has self-computable solutions. However, we omit the proof of this fact because of a lack of the space.

Finally, we prove that PrefixGA is not reducible to $f_{PrefixGA}$ unless GI $\leq_T^p$ GA.

Theorem 7. *If there exists a polynomial time algorithm such that, given any pair (G, σ), it uses the oracle $f_{PrefixGA}$ and correctly determines whether $(G, \sigma) \in$ PrefixGA, then GI $\leq_T^p$ GA.*

Proof. We prove that there exists a polynomial-time algorithm that computes $f_{PrefixGA}$ by using GA. This indicates that, together with the result of Proposition 1, if an algorithm mentioned above exists, then GI is polynomial time Turing reducible to GA. Let (G, σ) be an input for $f_{PrefixGA}$ such that $V(G) = \{1, \ldots, n\}$, $\sigma : \{1, \ldots, t\} \mapsto \{1, \ldots, n\}$ for some $t \leq n$. The algorithm computes $f_{PrefixGA}$ by using GA as follows.

CASE 1: There exists k, $(1 \leq k \leq t)$ such that $\sigma(k) \neq k$. In this case, if there exists a non-trivial automorphism φ on G that is an extension of σ, then φ must map the vertex k to a different vertex. So, the algorithm outputs k.

CASE 2: $\sigma(i) = i$ for every $1 \leq i \leq t$. Let $G_{[1,\ldots,j]}$ denote the graph obtained from G by labeling vertex 1 with label l_1, vertex 2 with label l_2, $\cdots$, vertex j with label l_j. Note that every non-trivial automorphism on $G_{[1,\ldots,t]}$ is a non-trivial automorphism on G that is an extension of σ and vice-versa. The algorithm first queries GA on $G_{[1,\ldots,t]}$. If the answer is "NO", then we have that there is no non-trivial automorphism on G that is an extension of σ. So, the algorithm outputs an arbitrary vertex. Otherwise, for each $i = t+1, t+2, \cdots, n$ in this order, the algorithm asks GA on $G_{[1,\ldots,i]}$. Let $k = \min\{i \mid t+1 \leq i \leq n$ and $G_{[1,\ldots,i]} \notin$ GA$\}$. Such k must exists because $G_{[1,\ldots,n]} \notin$ GA. Now, we have that there exists a non-trivial automorphism on $G_{[1,\ldots,k-1]}$ that maps k to a different vertex. This automorphism is an extension of σ. So, the algorithm outputs k.

It is easy to see that the algorithm computes $f_{PrefixGA}$. Note that the algorithm outputs an arbitrary vertex if there is no non-trivial automorphism on G that is an extension of σ. This completes the proof of Theorem 7.

References

1. Agrawal, M., Arvind, V.: A note on decision versus search for graph automorphism. Information and computation 131, 179–189 (1996)
2. Große, A., Rothe, J., Wechung, G.: Computing complete graph isomorphisms and hamiltonian cycles from partial ones. Theory of Computing Systems 35, 81–93 (2002)
3. Gál, A., Halevi, S., Lipton, R., Petrank, E.: Computing from partial solutions. In: Proceedings of the 14th Annual IEEE Conference on Computational Complexity, May 1999, pp. 34–45. IEEE Computer Society Press, Los Alamitos (1999)
4. Köbler, J., Schöning, U., Torán, J.: The Graph Isomorphism Problem: Its Structural Complexity. Birkhauser (1993)
5. Lozano, A., Torán, J.: On the nonuniform complexity of the graph isomorphism problem. In: Proceedings of the 7th Structure in Complexity Theory Conference, 1992, pp. 118–129 (1992)
6. Lubiw, A.: Some NP-complete problems similar to graph isomorphism. SIAM Journal on Computing 10, 11–21 (1981)
7. Mathon, R.: A note on the graph isomorphism counting problem. Information Processing Letters 8(3), 131–132 (1979)
8. Nagoya, T.: Computing graph automorphism from partial solutions. Theory of Computing Systems (to appear)
9. Read, R.C., Corneil, D.G.: The graph isomorphism disease. Journal of Graph Theory 1, 339–363 (1977)
10. Toran, J.: On the hardness of graph isomorphism. SIAM Journal on Computing 33, 1093–1108 (2004)

Well Supported Approximate Equilibria in Bimatrix Games: A Graph Theoretic Approach*

Spyros C. Kontogiannis[1,2] and Paul G. Spirakis[2]

[1] Dept. of Comp. Sci., Univ. of Ioannina, 45110 Ioannina, Greece
kontog@cs.uoi.gr

[2] Research Academic Computer Technology Institute, N. Kazantzaki Str., Univ. Campus, 26500 Rio-Patra, Greece
{kontog,spirakis}@cti.gr

Abstract. We study the existence and tractability of a notion of approximate equilibria in bimatrix games, called *well supported approximate Nash Equilibria* (SuppNE in short). We prove existence of $\varepsilon-$SuppNE for any constant $\varepsilon \in (0,1)$, with only logarithmic support sizes for both players. Also we propose a polynomial–time construction of SuppNE, both for win lose and for arbitrary (normalized) bimatrix games. The quality of these SuppNE depends on the *girth* of the Nash Dynamics graph in the win lose game, or a (rounded–off) win lose image of the original normalized game. Our constructions are very successful in sparse win lose games (ie, having a constant number of $(0,1)-$elements in the bimatrix) with large girth in the Nash Dynamics graph. The same holds also for normalized games whose win lose image is sparse with large girth.

Finally we prove the simplicity of constructing SuppNE both in random normalized games and in random win lose games. In the former case we prove that the uniform full mix is an $o(1)$ −SuppNE, while in the case of win lose games, we show that (with high probability) there is either a PNE or a 0.5-SuppNE with support sizes only 2.

1 Introduction

One of the most appealing concepts in game theory is the notion of a Nash equilibrium: *A collection of strategies from which no player has an incentive to unilaterally deviate from its own strategy.* The nice thing about Nash equilibria is that they always exist in any finite $k-$person game in normal form [18]. This is one of the most important reasons why Nash equilibria are considered to be the prevailing solution concept for finite games in normal form. The problem with Nash equilibria is that there can be exponentially many of them, of quite different characteristics, even for bimatrix games. Additionally, it is not yet known if one of them can be found in subexponential time. Therefore, $k-$NASH, the problem of computing an arbitrary Nash equilibrium of a finite $k-$person game in normal form, is considered as a fundamental problem in algorithmic game theory [19].

* Partially supported by the Future and Emerging Technologies Unit of EC (IST priority - 6th FP), under contract no. FP6-021235-2 (ARRIVAL) and 001907 (DELIS).

L. Kučera and A. Kučera (Eds.): MFCS 2007, LNCS 4708, pp. 596–608, 2007.

Its complexity has been a long standing open problem, since the introduction of the pioneering (pivoting) algorithm of Lemke and Howson [16]. Unfortunately, it was recently shown by Savani and von Stengel [20] that this algorithm requires an exponential number of steps; it is also known that the smoothed complexity of the algorithm is likely to be superpolynomial [4]. Moreover, many (quite natural) refinements of $k-$NASH are known to be **NP**−complete problems [11,6].

Due to the (recently proved) completeness even of 2−NASH, approximate solutions have attracted the attention of the research community. There are two different notions of approximate Nash equilibria: Those which require that each player gets the maximum possible payoff, within some *additive constant* ε (denoted here by ApproxNE), and those which require that each player is allowed to adopt **wpp**[1] only actions that are *approximate best responses* to the opponent's strategy, within an additive constant ε (denoted here by SuppNE). ApproxNE are the dominant notion of approximate equilibria in the literature, while SuppNE is a rather new notion (eg, see [4,5,8]). As will be explained later, SuppNE seem to be also harder to construct. On the other hand they might be more naturally motivated by the players' *selfish behavior*: Each player is assumed to choose *approximate best responses*, within some *additive* constant, (rather than exact best responses) against the other player's strategy. This is in contrast to the notion of ApproxNE, in which the two players have no restriction in what kind of actions they choose to play **wpp**, so long as their payoffs are close to their best response payoffs. We would like to argue that SuppNE is a quite interesting notion of approximate equilibrium, due to both its mathematical challenge and also its highly intuitive property that the players are assumed to avoid playing actions that are indeed *meaningless* to them. The present paper tries to shed some light on this new notion of approximate equilibria.

Related Work. The most popular algorithm for computing NE in bimatrix games, is the algorithm of Lemke and Howson [16] (LH in short), a special case of Lemke's algorithm for finding solutions (if such exist) for arbitrary instances of the Linear Complementarity Problem. It has been proved [20] that LH may require an exponential number of steps before reaching a NE. Moreover, it is well known that various (quite natural) restrictions of $k-$NASH (eg, uniqueness, bounds on support sizes, etc) lead to **NP**−hard problems [11,6].

A very recent series of research papers within the last two years deal with the complexity of $k-$NASH. Initially [7,12] proved that 4−NASH is **PPAD**−complete. This result was extended to 3−player games [10]. Surprisingly, [3] proved the same complexity result even 2−NASH. In view of all these hardness results for the $k-$NASH, understanding the limitations of the (in)approximability of the problem is quite important. To our knowledge, the first result that provides $\varepsilon-$ApproxNE within *subexponential* time is [17]. In particular, for any *constant* $\varepsilon > 0$, they prove the existence of an $\varepsilon-$ApproxNE for arbitrary $n \times n$ bimatrix games, which additionally is a uniform profile that has supports of size at most $\left\lceil \frac{\log n}{\varepsilon^2} \right\rceil$. This obviously implies a simple subexponential

[1] With positive probability.

algorithm for constructing $\varepsilon-$ApproxNE for $[0,1]-$bimatrix games. This approach still remains the fastest strategy to date, for the general problem of providing $\varepsilon-$ApproxNE for any *constant* $\varepsilon > 0$. [15] shows a similar existential argument of $\varepsilon-$SuppNE with logarithmic support sizes, for any constant $\varepsilon > 0$.

With respect to the tractability of a FPTAS for NE, [4] proved that providing a FPTAS for 2−NASH is also **PPAD**−complete. Namely, they proved that unless **PPAD** $\subseteq$ **P**, there is no algorithm that constructs $\varepsilon-$ApproxNE in time $poly(n, 1/\varepsilon)$, for any $\varepsilon = n^{-\Theta(1)}$. Moreover, they proved that unless **PPAD** $\subseteq$ **RP**, there is no algorithm that constructs a NE in time $poly(n, 1/\sigma)$, where σ is the size of the deviation of the elements of the bimatrix.

Two independent results [8,14] have recently made progress in the direction of providing the first $\varepsilon-$ApproxNE and $\varepsilon-$SuppNE for $[0,1]-$bimatrix games and some *constant* $1 > \varepsilon > 0$. In particular, [8] gave a nice and simple $\frac{1}{2}-$ApproxNE for $[0,1]-$bimatrix games, involving only two strategies per player. Very recently they improved this upper bound to 0.38 [9]. In [8] an algorithm for SuppNE was proposed, which, under a graph theoretic conjecture, would construct in polynomial time a $(2/3)-$SuppNE. The status of this conjecture is still unknown. [8] made also a quite interesting connection of the problem of constructing $\frac{1+\varepsilon}{2}-$SuppNE in an arbitrary $[0,1]-$bimatrix game, to that of constructing $\varepsilon-$SuppNE for a properly chosen win lose game of the same size. As for [14], based on linear programming techniques, they provided a $\frac{3}{4}-$ApproxNE, as well as a parameterized $\frac{2+\lambda}{4}-$ApproxNE for arbitrary $[0,1]-$bimatrix games, where λ is the minimum payoff of a player at a NE.

Concerning random $[0,1]-$bimatrix games, the work of Bárány, Vempala and Vetta [2] considers the case where all the cells of the payoff matrices are (either uniform, or normal) **iid**[2] random variables in $[0,1]$. They analyze a simple Las Vegas algorithm for finding a NE in such a game, by brute force on the support sizes, starting from smaller ones. The running time of their algorithm is $\mathcal{O}(m^2 n \log\log n + n^2 m \log\log m)$, **whp**[3].

Contribution and Roadmap. In [15] we adopted a methodology for constructing SuppNE in bimatrix games, based on Linear Programming (the **LP approach**). In particular, we constructed in polynomial time a $0.5-$SuppNE for win lose games and a $0.658-$SuppNE for $[0,1]-$bimatrix games. Here we start by proving existence of SuppNE with logarithmic support sizes, and then we adopt an alternative, graph theoretic approach (the **GT approach**). We also consider random win lose and random $[0,1]-$bimatrix games. Table 1 summarizes our results for SuppNE in these two papers.

The structure (and contribution) of this paper is the following: Section 2 provides some preliminaries and notation. In Section 3 we prove existence of uniform $\varepsilon-$SuppNE with logarithmic support sizes, for any *constant* $\varepsilon > 0$. In Section 4 we present the GT approach for constructing SuppNE in both win lose and normalized games. We prove (Theorem 3) that every win lose game contains

[2] Independent, identically distributed.

[3] With high probability, ie, with probability $1 - m^{-c}$, for some constant $c > 0$.

Table 1. Synopsis of results for ε−SuppNE

	GT		**LP** [15]	**Random**
Win Lose	$1 - 2/g$		0.5	**whp**, either $\exists$ *PNE* or $\exists$ 0.5−SuppNE with support sizes 2
	λ−sparse with large girth g, ie, $\lambda = o(g)$	o(1)		
Normalized	$1 - 1/g$		0.658	Uniform full mix is **whp** $\mathcal{O}\left(\sqrt{\frac{\log m}{m}}\right)$ −SuppNE
	λ−sparse win lose image with large girth	$\frac{1+o(1)}{2}$		

either a PNE, or a subgame that is a $\frac{g}{2}$−MP game (a generalization of the Matching Pennies game, to be defined later), where $g \geq 4$ is the girth of the Nash Dynamics graph. Based on this, we construct (Theorem 4) in polynomial time a $\left(1 - \frac{2}{g}\right)$−SuppNE for the whole game. Our technique provides a o(1)−SuppNE (Theorem 5) for $\lambda - sparse$ games with girth g for some $\lambda = o(g)$, and an *exact NE* for 1−sparse games. The supports of such a NE can be arbitrary and the rank of the payoff matrices can be arbitrary as well. Thus, an alternative attack (eg, by support enumeration, or exploitation of small ranks) would not necessarily work in such a game. We extend these results to $\left(1 - \frac{1}{g}\right)$−SuppNE for any $[0, 1]$−bimatrix game (Theorem 6), exploiting an observation of [8] that connects SuppNE of a bimatrix game to SuppNE of a properly constructed win lose game. For $[0, 1]$−bimatrix games whose *win lose image* (as defined in [8]) is λ−sparse and girth g such that $\lambda = o(g)$, we get a $\left(\frac{1+o(1)}{2}\right)$−SuppNE.

Finally, we explore the tractability of SuppNE in random bimatrix games (Section 5). We propose two random models, one for $[0, 1]$−bimatrix games and one for random win lose games. For $[0, 1]$−bimatrix games, we propose a random model that is more general than that of [2]. We prove (Theorem 7) that the strict uniform full mix $\left(\mathbf{1}\frac{1}{m}, \mathbf{1}\frac{1}{n}\right)$ is an ε−SuppNE **whp**, for any $\varepsilon = \Omega\left(\sqrt{\log m/m}\right)$. The proposed solution is straightforward to construct, and the analysis of our model is quite simple, only based on Hoeffding bounds. For the case of win lose games, we consider a random model that (at least **whp**) disallows the existence of PNE. We prove (Theorem 8) that, when no PNE appears, a $\frac{1}{2}$−SuppNE with support size 2 for both strategies exists **whp**. We show that inexistence of PNE can happen only for a very restricted family of (sparse) random win lose games.

2 Preliminaries

Mathematical Notation. For any $k \in \mathbb{N}$, let $[k] \equiv \{1, 2, \ldots, k\}$. $M \in F^{m \times n}$ denotes a $m \times n$ matrix (denoted by capital letters) whose elements belong to set F. We call a pair $(A, B) \in (F \times F)^{m \times n}$ (ie, an $m \times n$ matrix whose elements are *ordered pairs* of values from F) a **bimatrix**. A $k \times 1$ matrix is also considered to be an k-**vector**. Vectors are denoted by bold small letters (eg, $\mathbf{x}$). $\mathbf{e_i}$ denotes a vector having a 1 in the i-th position and 0 everywhere else. $\mathbf{1_k}$ ($\mathbf{0_k}$) is the k-vector having 1s (0s) in all its coordinates. The $k \times k$ matrix $E = \mathbf{1_k} \cdot \mathbf{1_k}^T \in \{1\}^{k\times k}$ has value 1 in all its elements. For any $\mathbf{x}, \mathbf{y} \in \mathbb{R}^n$, $\mathbf{x} \geq \mathbf{y}$ implies their component–wise comparison: $\forall i \in [n], x_i \geq y_i$. For any $m \times n$ (bi)matrix M, M_j is its j-th column (as an $m \times 1$ vector), M^i is the i-th row (as a (transposed) $1 \times n$ vector) and $M_{i,j}$ is the (i, j)-th element.

For any integer $k \geq 1$, $\Delta_k = \{\mathbf{z} \in \mathbb{R}^k : \mathbf{z} \geq \mathbf{0};\ (\mathbf{1_k})^T\mathbf{z} = 1\}$ is the $(k-1)$-simplex. For any point $\mathbf{z} \in \Delta_k$, its **support** is the set of coordinates with positive value: $supp(\mathbf{z}) \equiv \{i \in [k] : z_i > 0\}$. For an arbitrary logical expression $\mathcal{E}$, we denote by $\mathbb{P}\{\mathcal{E}\}$ the probability of this expression being true, while $\mathbb{I}_{\{\mathcal{E}\}}$ is the indicator variable of whether $\mathcal{E}$ is true or false. For any random variable x, $\mathbb{E}\{x\}$ is its expected value (wrt some probability measure).

Game Theoretic Definitions and Notation. For $2 \leq m \leq n$, an $m \times n$ **bimatrix game** $\langle A, B\rangle$ is a 2−person game in normal form, determined by the bimatrix $(A, B) \in (\mathbb{R} \times \mathbb{R})^{m\times n}$ as follows: The first player (the **row player**) has an m−element *action set* $[m]$, and the second player (the **column player**) has an n−element *action set* $[n]$. Each row (column) of the bimatrix corresponds to a different action of the row (column) player. The row and the column player's payoffs are determined by the $m \times n$ real matrices A and B respectively. $\langle A, B\rangle$ is a **zero sum** game, if $B = -A$. In that case the game is solvable in polynomial time using linear programming. If both payoff matrices belong to $[0, 1]^{m\times n}$ then we have a $[0, 1]$−**bimatrix** (aka **normalized**) **game**. The special case of bimatrix games in which all elements of the bimatrix belong to $\{0, 1\} \times \{0, 1\}$, is called a $\{0, 1\}$−**bimatrix** (aka **win lose**) **game**. A win lose game having (for integer $\lambda \geq 1$) at most λ $(1, 0)$−elements per row and at most λ number $(0, 1)$−element per column of the bimatrix, is called a λ−**sparse** game. Any point $\mathbf{x} \in \Delta_m$ is a **mixed strategy** for the row player: She determines her action independently from the column player, according to the probability vector $\mathbf{x}$. Similarly, any point $\mathbf{y} \in \Delta_n$ is a mixed strategy for the column player. Each extreme point $\mathbf{e_i} \in \Delta_m$ ($\mathbf{e_j} \in \Delta_n$), that enforces the use of the i-th row (j-th column) by the row (column) player, is a **pure strategy** for her. Any element $(\mathbf{x}, \mathbf{y}) \in \Delta_m \times \Delta_n$ is a (mixed in general) **strategy profile** for the players. The notion of *approximate best responses* will help us simplify the forthcoming definitions:

Definition 1 (Approximate Best Response). *Fix any bimatrix game* $\langle A, B\rangle$ *and* $0 \leq \varepsilon \leq 1$. *The sets of* **approximate (pure) best responses** *of the column (row) player against* $\mathbf{x} \in \Delta_m$ ($\mathbf{y} \in \Delta_n$) *are:* $BR(\varepsilon, B^T, \mathbf{x}) = \{\mathbf{y} \in \Delta_n : \mathbf{y}^T B^T \mathbf{x} \geq \mathbf{z}^T B^T \mathbf{x} - \varepsilon, \forall \mathbf{z} \in \Delta_n\}$, $BR(\varepsilon, A, \mathbf{y}) = \{\mathbf{x} \in \Delta_m : \mathbf{x}^T A\mathbf{y} \geq \mathbf{z}^T A\mathbf{y} - \varepsilon, \forall \mathbf{z} \in \Delta_m\}$, $PBR(\varepsilon, B^T, \mathbf{x}) = \{j \in [n] : B_j^T \mathbf{x} \geq B_s^T \mathbf{x} - \varepsilon, \forall s \in [n]\}$ *and* $PBR(\varepsilon, A, \mathbf{y}) = \{i \in [m] : A^i\mathbf{y} \geq A^r\mathbf{y} - \varepsilon, \forall r \in [m]\}$.

Definition 2 (Approximate Nash Equilibria). *For any bimatrix game* $\langle A, B\rangle$ *and* $0 \leq \varepsilon \leq 1$, $(\mathbf{x}, \mathbf{y}) \in \Delta_m \times \Delta_n$ *is: (1) An* ε−**approximate Nash**

Equilibrium *($\varepsilon-$ApproxNE) iff each player chooses an $\varepsilon-$approximate best response against the opponent:* $[\mathbf{x} \in BR(\varepsilon, A, \mathbf{y})] \wedge [\mathbf{y} \in BR(\varepsilon, B^T, \mathbf{x})]$ *. (2) An $\varepsilon-$***well–supported Nash Equilibrium** *($\varepsilon-$SuppNE) iff each player assigns positive probability only to $\varepsilon-$approximate* pure best responses *against the strategy of the opponent:* $\forall i \in [m], x_i > 0 \Rightarrow i \in PBR(\varepsilon, A, \mathbf{y})$ *and* $\forall j \in [n],\ y_j > 0 \Rightarrow j \in PBR(\varepsilon, B^T, \mathbf{x})$.

To see the difference between the two notions of approximate equilibria, consider the Matching Pennies game with payoffs $(A,B)_{1,1} = (A,B)_{2,2} = (1,0)$ (row player wins) and $(A,B)_{1,2} = (A,B)_{2,1} = (0,1)$ (column player wins). $(\mathbf{e_1}, \frac{1}{2}\cdot(\mathbf{e_1}+\mathbf{e_2}))$ is a $0.5-$ApproxNE, but only a $1-$SuppNE for the game.

Any NE is both a $0-$ApproxNE and a $0-$SuppNE. Moreover, any $\varepsilon-$SuppNE is also an $\varepsilon-$ApproxNE, but *not necessarily vice versa*, as was shown in the previous example. Indeed, the only thing we currently know towards this direction is that from an arbitrary $\frac{\varepsilon^2}{8n}-$ApproxNE one can construct an $\varepsilon-$SuppNE in polynomial time [4]. Note that both notions of approximate equilibria are defined wrt an *additive* error term ε. Although (exact) NE are known not to be affected by any positive scaling of the payoffs, approximate notions of NE are indeed affected. Therefore, from now on we adopt the commonly used assumption in the literature (eg, [17,8,14,3,4]) that, when referring to $\varepsilon-$ApproxNE or $\varepsilon-$SuppNE, we consider a $[0,1]-$bimatrix game.

Definition 3 (Uniform Profiles). $\mathbf{x} \in \Delta_r$ *is a $k-$***uniform strategy** *iff* $\mathbf{x} \in \Delta_r(k) = \Delta_r \cap \{0, \frac{1}{k}, \frac{2}{k}, \ldots, \frac{k-1}{k}, 1\}^r$ *. If* $\mathbf{x} \in \hat{\Delta}_r(k) = \Delta_r \cap \{0, \frac{1}{k}\}^r$*, then we refer to a* **strict** $k-$**uniform strategy**. $(\mathbf{x}, \mathbf{y}) \in \Delta_m \times \Delta_n$ *for which* $\mathbf{x}$ *is a (strict) $k-$uniform strategy and* $\mathbf{y}$ *is a (strict) $\ell-$uniform strategy, is called a* **(strict)** $(k,\ell)-$**uniform** *profile.*

3 Existence of Uniform SuppNE

Existence of uniform $\varepsilon-$ApproxNE with small support sizes is already known from [17]. We prove here a similar result for SuppNE, based solely on the *Approximation Lemma* of Althöfer [1]:

Theorem 1 (Approximation Lemma [1]). *Fix any real matrix* $C \in [0,1]^{m\times n}$ *and any* constant $\varepsilon > 0$*. For any* $\mathbf{p} \in \Delta_m$*, there exists a $k-$uniform strategy* $\hat{\mathbf{p}} \in \Delta_r(k)$ *with* $k = \left\lceil \frac{log(2n)}{2\varepsilon^2} \right\rceil$*, such that* $|\mathbf{p}^T C_j - \hat{\mathbf{p}}^T C_j| \le \varepsilon,\ \forall j \in [n]$.

Proposition 1. *Fix any real matrix* $C \in \mathbb{R}^{m\times n}$ *and any* $\mathbf{p} \in \Delta_m$*. For* $\hat{\mathbf{p}} \in \Delta_m$ *indicated by the Approximation Lemma, it holds that* $supp(\hat{\mathbf{p}}) \subseteq supp(\mathbf{p})$.

Proof. See full version of the paper.

We now demonstrate how the Approximation Lemma guarantees the existence of a $k-$uniform $(2\varepsilon)-$SuppNE where $k = \left\lceil \frac{\log(2n)}{2\varepsilon^2} \right\rceil$, for any *constant* $\varepsilon > 0$:

Theorem 2. *Fix any positive constant $\varepsilon > 0$ and an arbitrary $[0,1]-$bimatrix game $\langle A, B\rangle$. There is at least one $(k,\ell)-$uniform profile which is also a $(2\varepsilon)-$SuppNE for this game, where $k \le \left\lceil \frac{\log(2n)}{2\varepsilon^2} \right\rceil$ and $\ell \le \left\lceil \frac{\log(2m)}{2\varepsilon^2} \right\rceil$.*

Proof. Consider any (exact) NE $(\mathbf{p}, \mathbf{q})$ of $\langle A, B\rangle$, which exists for any finite game in normal form [18]. By the Approximation Lemma, there is some $k-$uniform strategy $\hat{\mathbf{p}} \in \Delta_m$ with $|supp(\hat{\mathbf{p}})| \le k = \left\lceil \frac{\log(2n)}{2\varepsilon^2} \right\rceil$, such that $|\mathbf{p}^T B_j - \hat{\mathbf{p}}^T B_j| \le \varepsilon$, $\forall j \in [n]$. and some $\ell-$uniform strategy $\hat{\mathbf{q}} \in \Delta_n$ with $|supp(\hat{\mathbf{q}})| \le \ell = \left\lceil \frac{\log(2m)}{2\varepsilon^2} \right\rceil$, such that $|A^i\mathbf{q} - A^i\hat{\mathbf{q}}| \le \varepsilon$, $\forall i \in [m]$. Observe that $\hat{\mathbf{p}}^T B - \varepsilon \mathbf{1}^T \le \mathbf{p}^T B \le \hat{\mathbf{p}}^T B + \varepsilon \mathbf{1}^T$ and $A\hat{\mathbf{q}} - \varepsilon\mathbf{1} \le A\mathbf{q} \le A\hat{\mathbf{q}} + \varepsilon\mathbf{1}$. Exploiting $supp(\hat{\mathbf{p}}) \subseteq supp(\mathbf{p})$, the Nash Property of $(\mathbf{p}, \mathbf{q})$ and the Approximation Lemma, we have: $\forall i \in [m]$, $\hat{p}_i > 0 \Rightarrow p_i > 0 \Rightarrow A^i\mathbf{q} \ge A^r\mathbf{q}$, $\forall r \in [m] \Rightarrow A^i\hat{\mathbf{q}} \ge A^r\hat{\mathbf{q}} - 2\varepsilon$, $\forall r \in [m]$. The argument for the column player is identical. We thus conclude that $(\hat{\mathbf{p}}, \hat{\mathbf{q}})$ is a $(k,\ell)-$uniform $(2\varepsilon)-$SuppNE for $\langle A, B\rangle$.

In the following Lemma we show that what we really should look for, is a polynomial–time algorithm for *strict* $(k,\ell)-$uniform $(2\varepsilon)-$SuppNE for $\langle A, B\rangle$:

Lemma 1. *Fix $\varepsilon > 0$, $k = \left\lceil \frac{\log(2n)}{2\varepsilon^2} \right\rceil$ and $\ell = \left\lceil \frac{\log(2m)}{2\varepsilon^2} \right\rceil$. For any $m \times n$ $[0,1]-$bimatrix game $\langle A, B\rangle$, there is some $(km) \times (\ell n)$ $[0,1]-$bimatrix game $\langle A', B'\rangle$ that is polynomial–time equivalent with $\langle A, B\rangle$ with regard to $(k,\ell)-$uniform $\varepsilon-$SuppNE. That is, there are polynomial–time computable maps $F_I : \Delta_m \mapsto \Delta_{km}$ and $F_{II} : \Delta_n \mapsto \Delta_{\ell n}$ such that for any $\varepsilon-$SuppNE $(\mathbf{p}, \mathbf{q}) \in \Delta_m \times \Delta_n$ of $\langle A, B\rangle$ there is a unique $\varepsilon-$SuppNE $(\tilde{\mathbf{p}}, \tilde{\mathbf{q}}) = (F_I(\mathbf{p}), F_{II}(\mathbf{q})) \in \Delta_{km} \times \Delta_{\ell n}$ of $\langle A', B'\rangle$. Conversely, there are polynomial–time computable maps $H_I : \Delta_{km} \mapsto \Delta_m$ and $H_{II} : \Delta_{\ell n} \mapsto \Delta_n$ such that for any $\varepsilon-$SuppNE $(\tilde{\mathbf{p}}, \tilde{\mathbf{q}}) \in \Delta_{km} \times \Delta_{\ell n}$ of $\langle A, B\rangle$, there is a unique $(\hat{\mathbf{p}}, \hat{\mathbf{q}}) = (H_I(\tilde{\mathbf{p}}), H_{II}(\tilde{\mathbf{q}})) \in \Delta_m \times \Delta_n$ $\varepsilon-$SuppNE of $\langle A, B\rangle$. Finally, the proposed mappings assign $(k,\ell)-$uniform profiles of $\langle A, B\rangle$ to strict $(k,\ell)-$uniform profiles of $\langle A', B'\rangle$ and vice versa.*

Proof. See full version of the paper.

4 A Graph Theoretic Approach for Constructing SuppNE

SuppNE in Win Lose Games. We now focus on SuppNE for win lose bimatrix games. This is particularly interesting because of a result of [8] that connects the construction of SuppNE for win lose games to the construction of SuppNE for $[0,1]-$bimatrix games. We distinguish the case in which there is a PNE in the game, because such a PNE can be easily detected in polynomial time. The following lemma characterizes all the win lose bimatrix games having no PNE.

Lemma 2. *An $m \times n$ $\{0,1\}-$bimatrix game $\langle A, B\rangle$ has no PNE if and only if all the following conditions hold:* **(PNE1)** $\forall (i,j) \in [m] \times [n]$, $A_{i,j} +$

$B_{i,j} \leq 1$. **(PNE2)** $\forall i \in [m]$, $\left(\sum_{j\in[n]} B_{i,j} \geq 1\right) \vee \left(\sum_{j\in[n]} A_{i,j} = 0\right)$. **(PNE3)** $\forall j \in [n]$, $\left(\sum_{i\in[m]} A_{i,j} \geq 1\right) \vee \left(\sum_{i\in[m]} B_{i,j} = 0\right)$. **(PNE4)** $\forall (i,j) \in [m] \times [n]$, $\left(\sum_{r\in[m]} A_{r,j} \geq 1\right) \vee \left(\sum_{s\in[n]} B_{i,s} \geq 1\right)$.

Proof. See full paper.

We now show the existence of a special kind of subgame, in any win lose game complying with (PNE1)–(PNE4) of Lemma 2. We first define this subgame:

Definition 4 (Generalized Matching Pennies). *For any integer $k \geq 2$ and any $(R, C) \in (\{0,1\} \times \{0,1\})^{k\times k}$, $\langle R, C\rangle$ is a* **k-Matching Pennies game** *($k-$MP) iff there are permutations $\pi_r : [k] \mapsto [k]$ of the rows and $\pi_c : [k] \mapsto [k]$ of the columns of (R, C), s.t.: (a) $\forall i \in [k]$, $(R, C)_{\pi_r(i),\pi_c(i)}$ is $(1, 0)-$element; (b) $\forall i \in [k]\setminus\{1\}$, $(R, C)_{\pi_r(i-1),\pi_c(i)}$, as well as $(R, C)_{\pi_r(k),\pi_c(1)}$ are $(0, 1)-$elements; (c) all the remaining elements of (R, C) are $(0, 0)-$elements.*

Exploiting the existence of such a subgame, we shall propose a polynomial–time construction of SuppNE for both win lose and $[0, 1]-$bimatrix games. But first we prove the existence of such a subgame, in win lose games without PNE. The next proposition is a simple observation that will help us in our argument:

Proposition 2. *For any $(A, B) \in (\{0, 1\} \times \{0, 1\})^{m\times n}$ that complies with (PNE1)–(PNE4), it holds that: (a) Any row of (A, B) containing a $(1, 0)-$element, must also contain a $(0, 1)-$element. (b) Any column of (A, B) containing a $(0, 1)-$element, must also contain an $(1, 0)-$element.*

Proof. A simple consequence of Lemma 2, see full version of the paper.

Consider now the (directed) graph G that we get from the Nash Dynamics graph of a game $\langle A, B\rangle$ that complies with (PNE1)–(PNE4) when we remove all the vertices corresponding to $(0, 0)-$elements. By Proposition 2, each $(1, 0)-$element of G has an outgoing arc (a **horizontal arc**) towards every $(0, 1)-$element of the *same row* in (A, B). Similarly, each $(0, 1)-$element of (A, B) has an outgoing arc (a **vertical arc**) towards every $(1, 0)-$element of the *same column* in (A, B). Obviously, horizontal arcs represent selfish movements of the column player, while the vertical arcs represent selfish movements of the row player. Observe that each cycle in the Nash Dynamics graph is also a cycle of G, since $(0, 0)-$elements cannot participate in a cycle. The following properties of G are easy to prove:

Lemma 3. *The following statements are true for any $m \times n$ $\{0, 1\}-$bimatrix (A, B) that complies with (PNE1)–(PNE4): (a) Any $(1, 0)-$element of (A, B) has at least one outgoing horizontal arc in G, and any $(0, 1)-$element of (A, B) has at least one outgoing vertical arc in G. (b) Any cycle in G alternates vertical and horizontal arcs and has even length. (c) G contains at least one cycle.*

Proof. See full version of the paper.

Having proved the existence of cycles in G, our next step is to determine in a submatrix of (A, B) that comprises a $k-$MP, for some $2 \le k \le m$.

Theorem 3. *Consider an arbitrary $m \times n$ win lose game $\langle A, B\rangle$ whose bimatrix complies with (PNE1)–(PNE4). There is a $k \times k$ submatrix (A', B') of (A, B) that comprises a k-Matching Pennies subgame, for some $2 \le k \le m$.*

Proof. We prove this by determining in polynomial time such a submatrix of (A, B). First we determine a directed cycle K of G of *minimum total length* (assuming that each arc has length exactly 1). This can be done easily in polynomial time, eg, by determining for each arc (v, u) of G a shortest path γ from u to v (if such paths exist) in the graph $(V, E \setminus \{(v, u)\})$. γ along with (v, u) comprise a shortest–length cycle containing (v, u) in G. By Lemma 3 we already know that there is at least one cycle in G. Then we compare the lengths of all the produced directed cycles and choose a cycle K of minimum total length. Obviously, the girth $g = |K| \ge 4$ of G is an even number.

Proposition 3. *There are permutations of the rows and columns of (A, B) such that, wrt K, all the $(1, 0)-$elements lie in the main diagonal and all the $(0, 1)-$elements lie in the diagonal just above it, plus the element $(A, B)_{g/2,1}$.*

Proof. See full version of the paper.

Given these permutations of the rows and columns of (A, B), we consider the $(g/2) \times (g/2)$ submatrix M of (A, B) consisting of the first $g/2$ rows and the first $g/2$ columns of (A, B). The following proposition ensures that the positions of M not involved in our cycle, are indeed $(0, 0)-$elements.

Proposition 4. *The elements of M not involved in K are all $(0, 0)-$elements.*

Proof. See full version of the paper.

M has the shape of a $(g/2)-$MP and this completes the proof of Theorem 3.

Our next theorem exploits the existence of a $\frac{g}{2}-$MP subgame, in order to construct a $\left(1 - \frac{2}{g}\right)-$SuppNE for the whole win lose game:

Theorem 4. *For any $m \times n$ win lose bimatrix game $\langle A, B\rangle$, it is possible to construct in polynomial time either a PNE, or else a $\left(1 - \frac{2}{g}\right)-$SuppNE, where g the girth of the Nash Dynamics graph.*

Proof. Consider the following algorithm (call it SuppNE4WinLose): Check whether the conditions (PNE1)–(PNE4) are violated, and if this is the case, return a PNE. Otherwise, find a *shortest cycle* K in the Nash Dynamics graph, and return a uniform full mix of the rows and columns involved in this cycle (rows and columns not involved get a zero probability mass).

It is trivial to verify that for $k = \frac{g}{2}$, the $k-$MP subgame induced by the rows and columns of (A, B) involved in K, the equiprobable fully mixed profile $\left(\frac{1}{k}\mathbf{1}, \frac{1}{k}\mathbf{1}\right) \in \Delta_k \times \Delta_k$ is a NE (for the subgame). Additionally, each used row (column) ensures for the row (column) player a payoff of $1/k$, provided that the opponent adopts the proposed strategy. The same holds if we extend the profile with zero probabilities to all the rows and columns not involved in K. Let $(\mathbf{x}, \mathbf{y})$ be the resulting profile from this extension. Any other row r of (A, B) (chosen with zero probability by the row player) may ensure a profit $A^r\mathbf{y} \le 1$. Therefore, no row may ensure more than $1 - \frac{1}{k}$ positive gain for the row player, given that the column player chooses strategy $\mathbf{x}$. The argument is identical for the column player. We conclude that the proposed profile is a $\left(1 - \frac{2}{g}\right)$ −SuppNE.

As for the time complexity, checking the existence of a PNE requires $\mathcal{O}(m \cdot n)$ elementary operations. The exploration of a shortest cycle can be performed in polynomial time: Run an All Pairs Shortest Path algorithm for directed graphs with integer weights. [13] ensures a running time of $\mathcal{O}\big(|E| \cdot |V| + |V|^2 \log\log |V|\big)$ for a graph $G = (V, E)$ with integer arc lengths. Observe that G has at most $mn^2/4$ horizontal arcs and at most $nm^2/4$ vertical arcs. Since $m \le n$, $|E| = \mathcal{O}(n^3)$ and $|V| = m \cdot n = \mathcal{O}(n^2)$. So, this step needs $\mathcal{O}(n^5)$ operations and dominates the time complexity of SuppNE4WinLose.

Unfortunately the quality of the produced SuppNE depends on the girth of the Nash Dynamics graph. Thus, we have no guarantee that our algorithm provides an ε−SuppNE for some $0 \le \varepsilon < 1$ that is far away from 1 (as the number of strategies grows). Nevertheless, the following theorem proves that our algorithm is quite efficient for sparse win lose games with large girth:

Theorem 5. *For any λ−sparse win lose game $\langle A, B\rangle$ complying with (PNE1–PNE4),* SuppNE4WinLose *provides a $\left(\frac{\min\{g,2\lambda\}-2}{g}\right)$ −SuppNE, where g is the girth of the Nash Dynamics graph. If additionally $\lambda = o(g)$, then we get an $o(1)$ −SuppNE. For 1−sparse games* SuppNE4WinLose *returns an exact NE.*

Proof. The uniform profile over the rows and columns involved in the shortest cycle ensure profit of $\frac{2}{g}$ per *used* row and column of the bimatrix, while any unused row or column of the bimatrix may provide (given the strategy of the opponent does not change) a profit of at most $\min\{1, \frac{2\lambda}{g}\}$. Therefore, our algorithm produces in polynomial time a $\left(\frac{\min\{g,2\lambda\}-2}{g}\right)$ −SuppNE. In the special case of 1−sparse game, since there is no row (column) of (A, B) having more than one $(1, 0)$−elements (respectively $(0, 1)$−elements), the uniform full mix on the rows and columns of the $(g/2)$−MP that we compute is indeed an exact NE.

SuppNE for Normalized Games. We now generalize our result for SuppNE in win lose games to SuppNE in arbitrary $[0, 1]$−bimatrix games, by exploiting a recent observation of [8]:

Lemma 4 ([8], Lemma 4.6). *For any $[0, 1]$−bimatrix game $\langle A, B\rangle$, and its* **relaxed win lose image** *$\langle \tilde{A}, \tilde{B}\rangle$ defined by $\tilde{A}_{i,j} = \mathbb{I}_{\{A_{i,j} \ge 1/2\}}$, $\tilde{B}_{i,j} = \mathbb{I}_{\{B_{i,j} \ge 1/2\}}$, $\forall (i, j) \in [m] \times [n]$, if $(\mathbf{x}, \mathbf{y}) \in \Delta_m \times \Delta_n$ is an ε−SuppNE for $\langle \tilde{A}, \tilde{B}\rangle$ then it is also an $\frac{1+\varepsilon}{2}$−SuppNE for $\langle A, B\rangle$.*

Theorem 6. *There exists a polynomial-time algorithm which constructs a* $\left(1-\frac{1}{g}\right)$*−SuppNE for any* $[0,1]-$*bimatrix game* $\langle A,B\rangle$*. If the relaxed win lose image of* $\langle A,B\rangle$ *is* $\lambda-$*sparse, then this algorithm returns a* $\left(\frac{1}{2}+\min\{\frac{1}{2},\frac{\lambda}{g}\}-\frac{1}{g}\right)$*−SuppNE for* $\langle A,B\rangle$*.*

Proof. The algorithm (call it SuppNE4Normalized) is the quite simple: First construct the relaxed win lose game $\langle A',B'\rangle$, and then return the profile that computed by SuppNE4WinLose(A',B'). The proof of its quality is a rather straightforward application of Theorem 4 and Lemma 4. Step (2) returns either a $\frac{1}{2}$−SuppNE for $\langle A,B\rangle$ (if the relaxed win lose image has a PNE), or a $\left(1-\frac{1}{g}\right)$ −SuppNE (if no PNE exist in the win lose image). In case of a $\lambda-$sparse game, step (2) returns a $\left(\frac{1}{2}+\min\{\frac{1}{2},\frac{\lambda}{g}\}-\frac{1}{g}\right)$ −SuppNE for $\langle A,B\rangle$.

5 SuppNE in Random Games

Random Normalized Games. Consider $[0,1]-$bimatrix games where the $m\times n$ bimatrix is constructed randomly: The row player chooses the payoff matrix A so that all its entries are independent (not necessarily **iid**) random variables in $[0,1]$. The column player chooses its payoff matrix B so that all its entries are independent random variables in $[0,1]$. We denote by $\bar{A^i}\equiv\frac{1}{n}\cdot\sum_{j\in[n]}A_{ij}=\frac{A^i\cdot\mathbf{1}}{n}$ (resp. $\bar{B}_j\equiv\frac{1}{m}\cdot\sum_{i\in[m]}B_{ij}=\frac{B_j^T\cdot\mathbf{1}}{m}$) the *average value* of a cell in the ith row of A (j^{th} column of B).

Theorem 7. *Consider a* $[0,1]-$*bimatrix game* $\langle A,B\rangle$*, where the cells of* A *are independent random variables such that* $\forall i\in[m],\mathbb{E}\left\{\bar{A^i}\right\}\in[\alpha-\gamma,\alpha+\gamma]$ *and the cells of* B *are independent random variables such that* $\forall j\in[n],\mathbb{E}\left\{\bar{B}_j\right\}\in[\beta-\gamma,\beta+\gamma]$*, for some sufficiently small deviation parameter* $\gamma\geq 0$*.* $(\mathbf{1}\frac{1}{m},\mathbf{1}\frac{1}{n})$ *is a* $2(\gamma+\varepsilon)-$*SuppNE with probability* $1-m^{-\delta}$ *for any constant* $\delta>0$*, and any* $\varepsilon\geq\sqrt{[\log n+\delta\log m+\log 4]/(2m)}$*.*

Proof. See full version of the paper.

Random Win Lose Games. We finally study the construction of SuppNE for random win lose games. Since the game is trivial when any of (PNE1)–(PNE4) is violated, we are interested only in random models that are unlikely to violate any of these conditions. Additionally, we assume that the random elements are the entries of the bimatrix itself, rather than the payoff matrices of the two players (otherwise the random game would be trivial to solve): For each entry of the bimatrix, there is a probability $p_{\alpha\beta}$ that it becomes an $(\alpha,\beta)-$element, for any $\alpha,\beta\in\{0,1\}$. If $p_{00}=p_{11}=0$ then we get a (solvable in polynomial time) constant sum game. If additionally a $(1,1)-$element appears, this is trivially a PNE. Therefore, we consider the case in which $0>p_{00}>p_{11}=0$.

Theorem 8. *For a random win lose game with* $0 > p_{00} > p_{11} = 0$ *if* $p_{01}p_{10} = o\left(\sqrt{(\log m)/(mn)}\right)$, *then* **whp** *there is a PNE. If* $p_{01}p_{10} = \Omega\left(\sqrt{(\log m)/(mn)}\right)$, *then* **whp** *there is a* $0.5-$*SuppNE with support size* 2 *for both players.*

Proof. See full version of the paper.

Acknowledgements. We wish to thank Panagiota Panagopoulou for contributing to this work an earlier version of Theorem 7.

References

1. Althöfer, I.: On sparse approximations to randomized strategies and convex combinations. Linear Algebra and Applications 199, 339–355 (1994)
2. Bárány, I., Vempala, S., Vetta, A.: Nash equilibria in random games. In: Proc. of the 46th IEEE Symp. on Found. of Comp. Sci (FOCS '05), pp. 123–131. IEEE Computer Society Press, Los Alamitos (2005)
3. Chen, X., Deng, X.: Settling the complexity of 2-player nash equilibrium. In: Proc. of the 47th IEEE Symp. on Found. of Comp. Sci (FOCS '06), pp. 261–272. IEEE Computer Society Press, Los Alamitos (2006)
4. Chen, X., Deng, X., Teng, S.: Computing nash equilibria: Approximation and smoothed complexity. In: Proc. of the 47th IEEE Symp. on Found. of Comp. Sci (FOCS '06), pp. 603–612. IEEE Computer Society Press, Los Alamitos (2006)
5. Chen, X., Deng, X., Teng, S.: Sparse games are hard. In: Spirakis, P.G., Mavronicolas, M., Kontogiannis, S.C. (eds.) WINE 2006. LNCS, vol. 4286, pp. 262–273. Springer, Heidelberg (2006)
6. Conitzer, V., Sandholm, T.: Complexity results about nash equilibria. In: Proc. of the 18th Int. Joint Conf. on Art. Intel (IJCAI '03), pp. 765–771 (2003)
7. Daskalakis, C., Goldberg, P., Papadimitriou, C.: The complexity of computing a nash equilibrium. In: Proc. of the 38th ACM Symp. on Th. of Comp (STOC '06), pp. 71–78. ACM Press, New York (2006)
8. Daskalakis, C., Mehta, A., Papadimitriou, C.: A note on approximate equilibria. In: Spirakis, P.G., Mavronicolas, M., Kontogiannis, S.C. (eds.) WINE 2006. LNCS, vol. 4286, pp. 297–306. Springer, Heidelberg (2006)
9. Daskalakis, C., Mehta, A., Papadimitriou, C.: Progress in approximate nash equilibrium. In: Proc. of the 8th ACM Conf. on El. Comm (EC '07), ACM Press, New York (2007)
10. Daskalakis, C., Papadimitriou, C.: Three player games are hard. Technical Report TR05-139, Electr. Coll. on Comp. Compl (ECCC) (2005)
11. Gilboa, I., Zemel, E.: Nash and correlated equilibria: Some complexity considerations. Games & Econ. Behavior 1, 80–93 (1989)
12. Goldberg, P., Papadimitriou, C.: Reducibility among equilibrium problems. In: Proc. of the 38th ACM Symp. on Th. of Comp (STOC '06), pp. 61–70. ACM Press, New York (2006)
13. Hagerup, T.: Improved shortest paths on the word ram. In: Welzl, E., Montanari, U., Rolim, J.D.P. (eds.) ICALP 2000. LNCS, vol. 1853, pp. 61–72. Springer, Heidelberg (2000)

14. Kontogiannis, S., Panagopoulou, P., Spirakis, P.: Polynomial algorithms for approximating nash equilibria in bimatrix games. In: Spirakis, P.G., Mavronicolas, M., Kontogiannis, S.C. (eds.) WINE 2006. LNCS, vol. 4286, pp. 286–296. Springer, Heidelberg (2006)
15. Kontogiannis, S., Spirakis, P.: Efficient algorithms for constant well supported approximate equilibria in bimatrix games. In: Proc. of the 34th Int. Col. on Aut., Lang. and Progr (ICALP '07). LNCS, vol. 4596, pp. 595–606. Springer, Heidelberg (2007)
16. Lemke, C., Howson, J.T.: Equilibrium points of bimatrix games. Journal of the Society for Industrial and Applied Mathematics 12, 413–423 (1964)
17. Lipton, R., Markakis, E., Mehta, A.: Playing large games using simple strategies. In: Proc. of the 4th ACM Conf. on El. Comm (EC '03), pp. 36–41. ACM Press, New York (2003)
18. Nash, J.: Noncooperative games. Annals of Mathematics 54, 289–295 (1951)
19. Papadimitriou, C.: Algorithms, games and the internet. In: Proc. of the 33rd ACM Symp. on Th. of Comp (STOC '01), pp. 749–753. ACM Press, New York (2001)
20. Savani, R., von Stengel, B.: Hard-to-solve bimatrix games. Econometrica 74(2), 397–429 (2006)

Selfish Load Balancing Under Partial Knowledge*

Elias Koutsoupias[1], Panagiota N. Panagopoulou[2,3], and Paul G. Spirakis[2,3]

[1] Department of Informatics and Telecommunications, University of Athens, Greece
elias@di.uoa.gr
[2] Computer Engineering and Informatics Department, Patras University, Greece
[3] Research Academic Computer Technology Institute, Greece
{panagopp,spirakis}@cti.gr

Abstract. We consider n selfish agents or players, each having a load, who want to place their loads to one of two bins. The agents have an incomplete picture of the world: They know some loads exactly and only a probability distribution for the rest. We study Nash equilibria for this model, we compute the Price of Anarchy for some cases and show that sometimes extra information adversely affects the Divergence Ratio (a kind of subjective Price of Anarchy).

1 Introduction

We consider a simple version of load balancing in which n agents or players, each having a load, want to place their loads to one of two bins. We assume that the players are selfish and each one wants to minimize their own expected load of their own bin. This is a typical problem in the study of the Price of Anarchy. What distinguishes our approach here is that we aim at studying how the information regime affects the Price of Anarchy. We consider local strategies for the players in which a player has *exact* knowledge about the loads only of some players and *statistical* knowledge about the rest. Such a situation can practically arise in distributed situations, especially when there is not sufficient time for the n agents to communicate and coordinate.

There are n selfish players (agents) and $m = 2$ bins. Each player i has a load w_i and has to select a bin in $\{0, 1\}$ to place her load. The cost c_i of player i is the total load of the bin she selects, i.e., $c_i = \sum_{k\,:\,b_k=b_i} w_k$. The cost is apparently influenced by the decision of the other players which suggests that game theory is a proper framework to study this situation. However, before we are able to analyze the situation, we need to specify what kind of information is available to the players. What does player i know about the load of some other player j? Among the many possibilities, we will concentrate on two extreme cases: Player

* Partially supported by the EU within the 6th Framework Programme under contracts 001907 "Dynamically Evolving, Large Scale Information Systems" (DELIS) and 015964 "Algorithmic Principles for Building Efficient Overlay Computers" (AEOLUS), and by the General Secretariat for Research and Technology of the Greek Ministry of Development within the Programme PENED 2003.

L. Kučera and A. Kučera (Eds.): MFCS 2007, LNCS 4708, pp. 609–620, 2007.

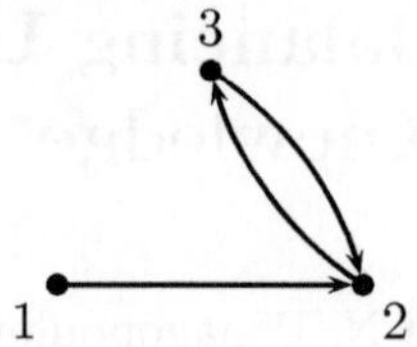

Fig. 1. Some information regime graph for 3 players

i knows either the exact value w_j (a real number in $[0,1]$) or a probability distribution on w_j (the uniform distribution in $[0,1]$). We can represent the information regimes by directed graphs which we will call *information regime graphs* (example in Fig. 1). An edge (i,j) denotes that player i knows the value w_j, while its absence denotes that the player i knows only that the value w_j is a random variable uniformly distributed in $[0,1]$.

At the one extreme, when every player knows all w_j's (which corresponds to a complete graph), the problem has been studied before as selfish task allocation [8]. At the other extreme, when each player knows only the distributions of the other players but not their actual values, the situation is very similar to the model studied by Papadimitriou and Yannakakis [11] for the purpose of quantifying the value of information. There are two main differences between our consideration and the model of [11]: First, we take the game-theoretic approach and assume that the players are selfish and care only about their own objective, while in the model of [11] the players want to optimize the global objective. Second, the cost in [11] was the probability of overflow, but in this work, the cost of a player is the total load of the bin she selects.

The main motivation of our work is to study the effects of information hiding on selfish systems. What happens to the Price of Anarchy in the more general case when players know only some values w_i?

In this work, we consider only pure Nash equilibria. A trivial observation is that when we allow mixed Nash equilibria, there is always the fully-mixed equilibrium in which every player selects uniformly at random a bin. This Nash equilibrium has been studied extensively. For example, it was shown in [8] that the Price of Anarchy is 3/2 in this case even when there are only 2 players.

To be able to compute the Price of Anarchy, we need to decide about the social cost and the optimal cost, that is the numerator and denominator of the Price of Anarchy. We consider the social cost to be either the sum (equivalent to the average) of the cost of all players or the maximum cost among players (which is equal to the makespan). The definition of the optimal cost is clear for the complete information regime, but in the case of incomplete information there are at least two natural choices. Either the optimal algorithm is a centralized one which knows all w_i's, or the optimal algorithm is a distributed algorithm which knows exactly as much as the players of the game. In the latter case, each node knows only the w_i's indicated by the information regime graph. The objective of each node, unlike the selfish objective that contributes to the numerator of the Price of Anarchy, is to minimize the global objective, the social cost. In this work, we will compute the Price of Anarchy using a distributed optimal algorithm.

Related Work. The game of n non-cooperating agents and m parallel links (or bins) was defined in [8] and has an extended literature e.g., [2,3,4,7,9,10]. In particular, the work of Monien et al [6] is closer in spirit to this work since they consider Bayesian games. Their model though is very different than ours.

The load balancing problem studied in this work was originally introduced by Papadimitriou and Yannakakis [11] in an effort to understand the crucial economic value of information [1] as a computational resource in a distributed system (e.g. in the context of Computational Complexity [12]). That work considered only $n = 3$ agents. In order to understand how the optimum solution achieved by the agents varies as a function of the amount of information available to them, Papadimitriou and Yannakakis [11] considered each possible communication pattern and discovered the corresponding optimal decision protocol to be unexpectedly sophisticated. For the special case where no communication is allowed, i.e. when each agent i is aware of only her own load w_i, it was conjectured that the simple decision rule: "if $w_i \leq 1 - \frac{1}{\sqrt{7}}$ then put w_i on bin 0 else put w_i on bin 1" is optimal; the conjecture is still open. Georgiades et al. [7] studied the extension of the load balancing problem to the case of n agents. Their work was focused on the special cases of oblivious decision rules, for which agents do not "look at" their inputs, and non-oblivious decision rules, for which they do. In either case, optimality conditions were derived in the form of combinatorial polynomial equations.

Our Results. We study the Nash equilibria of these games and give some results about the Divergence Ratio and the Price of Anarchy. In particular since the players have incomplete picture of the world, the cost that they compute may differ from the actual cost. To capture this situation, we define the Divergence Ratio, a kind of subjective Price of Anarchy. We show that for the regime of total lack of information, the Divergence Ratio is $(n+1)/n$ for even n and 1 for odd n. For the regime of complete information the Divergence Ratio is 4/3. In contrast, the Divergence Ratio is $\Theta(n)$ for some intermediate information regimes. We also estimate the Price of Anarchy for the total lack of information regime.

2 The Model

Consider a set $N = \{1, 2, \ldots, n\}$ of n selfish agents. Each agent i has a *load* $w_i \in [0, 1]$. Let $\mathbf{w} = (w_i)_{i \in N}$ denote a vector of input loads, one per agent. For any pair of agents $(i, j) \in N \times N$, agent i knows either (a) the exact value of w_j or (b) that w_j is uniformly distributed on $[0, 1]$. For each $i \in N$ let $I_i = \{j \in N :$ agent i knows the exact value of $w_j\}$, and thus for each $j \notin I_i$, agent i knows that w_j is uniformly distributed on $[0, 1]$. Denote by $\mathbf{I} = (I_i)_{i \in N}$ the collection of I_i for all $i \in N$. Let's also denote the cardinality of I_i by γ_i.

Each agent $i \in N$ has to select one of two available bins (bin 0 and bin 1) to put her load. The bin is selected based on the values of w_i's in I_i. Thus, a *strategy* for agent i is a function s_i from $[0, 1]^{\gamma_i}$ to $\{0, 1\}$: $s_i((w_j)_{j \in I_i})$.

Of particular interest are the *single-threshold* strategies where each agent i places her load to the first bin iff w_i is below a threshold value t_i. This threshold value depends on the known loads of the agents, i.e. t_i is a function of $(w_j)_{j \in I_i}$.

Here we study threshold strategies, but *most of our results can be extended easily to all strategies.* The reason is simple: What matters in most of our considerations is the expected load that a player places in one bin. In particular, for a threshold t_i, the expected load that player i places in bin 0 is $\int_0^{t_i} w_i dw_i = t_i^2/2$ and in bin 1 it is $\int_{t_i}^1 w_i dw_i = 1/2 - t_i^2/2$. From the point of view of the other players that do not know the value w_i, only the expected value $t_i^2/2$ matters. But this, in general, can be achieved with many strategies (for example, by the inverse threshold strategy that places the load w_i in bin 0 when its value is *greater* that $\bar{t}_i$ where $\bar{t}_i^2/2 = 1/2 - t_i^2/2$).

A *strategy profile* $\mathbf{s} = (s_1, \ldots, s_n)$ is a collection of strategies, one for each agent. Denote by $(s'_i, \mathbf{s}_{-i})$ the strategy profile that is identical to $\mathbf{s}$ except for agent i, who chooses strategy s'_i instead of s_i. Since we will mainly consider threshold strategies, we will denote such a strategy profile by $\mathbf{t} = (t_1, \ldots, t_n) \in [0,1]^n$.

For weights $\mathbf{w}$ and strategies $\mathbf{s}$, we define two costs of each player i. The *subjective cost* of a player is the expected cost that player computes based on her knowledge of the weights in I_i and assuming that the other weights are uniformly (and independently) distributed in $[0,1]$. The *(objective) cost* is the actual cost which can be computed from the full knowledge of $\mathbf{w}$ and $\mathbf{s}$. The *social cost* is the cost of the system for these weights and strategies and it is either the sum of the cost of all agents or the maximum cost among agents; the latter corresponds to the makespan. In analogy to the costs of the players, we have the subjective social cost and the (objective) social cost. We will denote the subjective cost by $\mathsf{cost}_i(\mathbf{s}; \mathbf{w}; I_i)$.

Definition 1. *The strategy profile* $\mathbf{s} = (s_1, \ldots, s_n)$ *is a Nash equilibrium if and only if, for all* $\mathbf{w}$, $i \in N$, *and* s'_i, $\mathsf{cost}_i(\mathbf{s}; \mathbf{w}; I_i) \leq \mathsf{cost}_i((s'_i, \mathbf{s}_{-i}); \mathbf{w}; I_i)$.

3 The Structure of Nash Equilibria

Lemma 1. *For any threshold strategy profile* $\mathbf{t} \in [0,1]^n$ *and for all* $i \in N$,

$$i \notin I_i \Rightarrow \mathsf{cost}_i(\mathbf{t}; \mathbf{w}; I_i) = t_i \left(\sum_{j \in I_i : w_j \leq t_j} w_j - \sum_{j \in I_i : w_j > t_j} w_j + \sum_{j \notin I_i, j \neq i} t_j^2 - \frac{n - \gamma_i - 1}{2} \right)$$
$$+ \frac{1}{2} + \sum_{j \in I_i : w_j > t_j} w_j + \frac{n - \gamma_i - 1}{2} - \frac{1}{2} \sum_{j \notin I_i, j \neq i} t_j^2 \tag{1}$$

$$i \in I_i \Rightarrow \mathsf{cost}_i(\mathbf{t}; \mathbf{w}; I_i) = \begin{cases} \sum_{j \in I_i : w_j \leq t_j} w_j + \frac{1}{2} \sum_{j \notin I_i} t_j^2 & \text{if } w_i \leq t_i \\ \sum_{j \in I_i : w_j > t_j} w_j + \frac{n - \gamma_i}{2} - \frac{1}{2} \sum_{j \notin I_i} t_j^2 & \text{if } w_i > t_i \end{cases} . \tag{2}$$

Proof. Fix a strategy profile $\mathbf{t} \in [0,1]^n$ and an agent $i \in N$. Assume that $i \notin I_i$. Then i does not know the exact value of her own load, so she expects that her load will be put in bin 0 with probability t_i and in bin 1 with probability $1 - t_i$.

Therefore (i) with probability t_i, the cost for agent i is her own expected load, plus the sum of the loads of all $j \in I_i$ such that $w_j \leq t_j$, plus the expected load that every other agent $j \notin I_i$ puts on bin 0 (that is, $t_j^2/2$), and (ii) with probability $1 - t_i$, the cost for agent i is her own expected load, plus the sum of the loads of all $j \in I_i$ such that $w_j > t_j$, plus the expected load that every other agent $j \notin I_i$ puts on bin 1 (that is, $1/2 - t_j^2/2$).

Agent's i expected load is 1/2 (i.e. $\int_0^1 w_i dw_i$). Therefore, if $i \notin I_i$, then the cost of agent $i \in N$ in the strategy profile $\mathbf{t}$ is

$$\begin{aligned}\mathsf{cost}_i(\mathbf{t}; \mathbf{w}; I_i) &= \frac{1}{2} + t_i \left(\sum_{j \in I_i : w_j \le t_j} w_j + \sum_{j \notin I_i, j \ne i} \frac{t_j^2}{2} \right) \\ &\quad + (1 - t_i) \left(\sum_{j \in I_i : w_j > t_j} w_j + \sum_{j \notin I_i, j \ne i} \frac{1 - t_j^2}{2} \right) \\ &= t_i \left(\sum_{j \in I_i : w_j \le t_j} w_j - \sum_{j \in I_i : w_j > t_j} w_j + \sum_{j \notin I_i, j \ne i} t_j^2 - \frac{n - \gamma_i - 1}{2} \right) \\ &\quad + \frac{1}{2} + \sum_{j \in I_i : w_j > t_j} w_j + \frac{n - \gamma_i - 1}{2} - \frac{1}{2} \sum_{j \notin I_i, j \ne i} t_j^2.\end{aligned}$$

Now, if $i \in I_i$, i.e. i knows the exact value of her load, then

$$\begin{aligned}\mathsf{cost}_i(\mathbf{t}; \mathbf{w}; I_i) &= \begin{cases} \sum_{j \in I_i : w_j \le t_j} w_j + \sum_{j \notin I_i} t_j^2/2 & \text{if } w_i \le t_i \\ \sum_{j \in I_i : w_j > t_j} w_j + \sum_{j \notin I_i} (1/2 - t_j^2/2) & \text{if } w_i > t_i \end{cases} \\ &= \begin{cases} \sum_{j \in I_i : w_j \le t_j} w_j + \frac{1}{2} \sum_{j \notin I_i} t_j^2 & \text{if } w_i \le t_i \\ \sum_{j \in I_i : w_j > t_j} w_j + \frac{n - \gamma_i}{2} - \frac{1}{2} \sum_{j \notin I_i} t_j^2 & \text{if } w_i > t_i \end{cases}.\end{aligned}$$

□

3.1 The Total Lack of Information Case, $I_i = \emptyset$

Assume $I_i = \emptyset$ for all $i \in N$. From Equation 1 of Lemma 1, the cost of any agent $i \in N$ for a strategy profile $\mathbf{t} = (t_1, \ldots, t_n) \in [0, 1]^n$ is

$$\mathsf{cost}_i(\mathbf{t}; \mathbf{w}; I_i) = t_i \left(\sum_{j \ne i} t_j^2 - \frac{n - 1}{2} \right) + \frac{n}{2} - \frac{1}{2} \sum_{j \ne i} t_j^2. \tag{3}$$

Proposition 1. *Consider the case where $I_i = \emptyset$ for all $i \in N$. Then the strategy profile $\mathbf{t} \in [0, 1]^n$ is a Nash equilibrium if and only if, for all $i \in N$,*

$$\begin{aligned} t_i = 0 &\Rightarrow \sum_{j \ne i} t_j^2 \ge \frac{n - 1}{2} \\ t_i = 1 &\Rightarrow \sum_{j \ne i} t_j^2 \le \frac{n - 1}{2} \\ t_i \in (0, 1) &\Rightarrow \sum_{j \ne i} t_j^2 = \frac{n - 1}{2}. \end{aligned}$$

Proof. Fix a strategy profile $\mathbf{t} \in [0,1]^n$. By Definition 1, $\mathbf{t}$ is a Nash equilibrium if and only if, for all $i \in N$, $\mathsf{cost}_i(\mathbf{t};\mathbf{w};I_i) = \min_{t'_i \in [0,1]} \mathsf{cost}_i((t'_i, \mathbf{t}_{-i});\mathbf{w};I_i)$. Now fix an agent $i \in N$. From Equation 3, $\mathsf{cost}_i(\mathbf{t};\mathbf{w};I_i)$ is an affine function of t_i. Therefore, it is minimized at $t_i = 0$ only if it is non-decreasing, i.e. only if $\sum_{j \neq i} t_j^2 \geq \frac{n-1}{2}$. It is minimized at $t_i = 1$ only if it is non-increasing, i.e. only if $\sum_{j \neq i} t_j^2 \leq \frac{n-1}{2}$. Finally, it is minimized at some $t_i \in (0,1)$ only if it is a constant function, i.e. only if $\sum_{j \neq i} t_j^2 = \frac{n-1}{2}$. □

Observe that, in a Nash equilibrium, all $i \in N$ such that $t_i \in (0,1)$ must have equal t_i's. This is because, for all $i \in N$ such that $t_i \in (0,1)$, $\sum_{j \neq i} t_j^2 = \frac{n-1}{2}$ which implies that $t_i^2 = \sum_{j \in N} t_j^2 - \frac{n-1}{2}$.

With this, we can now characterize all the Nash equilibria of the total lack of information case.

Theorem 1. *Consider the case where $I_i = \emptyset$ for all $i \in N$. Then the strategy profile $\mathbf{t} \in [0,1]^n$ is a Nash equilibrium if and only if κ agents choose threshold 1, λ agents choose threshold $t_A \in (0,1)$, $n - \kappa - \lambda$ agents choose threshold 0 and*

(1) $\frac{n-1}{2} - \lambda \leq \kappa \leq \frac{n-1}{2}$, $\lambda > 1$, $t_A^2 = \frac{n-1}{2(\lambda-1)} - \frac{\kappa}{\lambda-1}$ *or*
(2) *n is even,* $\kappa = \frac{n}{2}$, $\lambda = 0$ *or*
(3) *n is odd,* $\kappa = \frac{n+1}{2}$, $\lambda = 0$ *or*
(4) *n is odd,* $\kappa = \frac{n-1}{2}$, $\lambda = 0$ *or*
(5) *n is odd,* $\kappa = \frac{n-1}{2}$, $\lambda = 1$.

Moreover, the maximum, over all Nash equilibria, Social Cost is $\frac{n+1}{4}$.

Proof. In order to find all Nash equilibria we have to find all the possible partitions of the set of agents into three sets A, B and C so that

– For all $i \in A$, $t_i = t_A$ for some $t_A \in (0,1)$ and $\sum_{j \neq i} t_j^2 = \frac{n-1}{2}$, or equivalently

$$(|A|-1) \cdot t_A^2 + |B| \cdot 0 + |C| \cdot 1 = \frac{n-1}{2}$$
$$(|A|-1)t_A^2 + |C| = \frac{n-1}{2}.$$

– For all $i \in B$, $t_i = 0$ and $\sum_{j \neq i} t_j^2 \geq \frac{n-1}{2}$, or equivalently

$$|A|t_A^2 + |C| \geq \frac{n-1}{2}.$$

– For all $i \in C$, $t_i = 1$ and $\sum_{j \neq i} t_j^2 \leq \frac{n-1}{2}$, or equivalently

$$|A|t_A^2 + |C| - 1 \leq \frac{n-1}{2}.$$

We consider the following cases.

1. $|A| = 0$. Then there is a Nash equilibrium if and only if $\frac{n-1}{2} \leq |C| \leq \frac{n+1}{2}$. Since $|B| = n - |C|$, it must also hold that $\frac{n-1}{2} \leq |B| \leq \frac{n+1}{2}$. This is possible if and only if (1) n is even and $|B| = |C| = n/2$, in which case the cost for any agent is $n/4$, or (2) if n is odd, $|B| = \frac{n+1}{2}$ and $|C| = \frac{n-1}{2}$, in which case the cost for any agent that chooses threshold 0 is $\frac{n+1}{4}$ and the cost for any agent that chooses threshold 1 is $\frac{n-1}{4}$ or if (3) n is odd, $|B| = \frac{n-1}{2}$ and $|C| = \frac{n+1}{2}$, in which case the cost for any agent that chooses threshold 0 is $\frac{n-1}{4}$ and the cost for any agent that chooses threshold 1 is $\frac{n+1}{4}$.
2. $|A| = 1$. Then there is a Nash equilibrium if and only if $|C| = \frac{n-1}{2}$ and $0 < t_A < 1$. Then $|B| = n - 1 - \frac{n-1}{2} = \frac{n-1}{2}$. So in this case we have a Nash equilibrium if and only if n is odd, $|B| = |C| = \frac{n-1}{2}$ and $0 < t_A < 1$. Moreover, the cost for any agent that chooses threshold 0 is $\frac{n+1}{4} - \frac{t_A^2}{2}$, the cost for any agent that chooses threshold 1 is $\frac{n-1}{4} + \frac{t_A^2}{2}$, and the cost for the agent that chooses $t_A \in (0,1)$ is $\frac{n+1}{4}$.
3. $|A| > 1$. Then there is a Nash equilibrium if and only if

$$(|A|-1)t_A^2 + |C| = \frac{n-1}{2} \quad \text{and}$$

$$|A|t_A^2 + |C| \geq \frac{n-1}{2} \quad \text{and}$$

$$|A|t_A^2 + |C| - 1 \leq \frac{n-1}{2}.$$

Simple calculations yield that the above are equivalent to $|C| \leq \frac{n-1}{2}$ and $|C| \geq \frac{n+1}{2} - |A|$ and $t_A^2 = \frac{n-1}{2(|A|-1)} - \frac{|C|}{|A|-1}$. Furthermore, the cost for any agent $i \in N$ such that $t_i = 0$ is

$$\mathsf{cost}_i(\mathbf{t}; \mathbf{w}; I_i) = \frac{n}{2} - \frac{1}{2}(\lambda t_A^2 + \kappa) < \frac{n+1}{4},$$

the cost for any agent $i \in N$ such that $t_i = 1$ is

$$\mathsf{cost}_i(\mathbf{t}; \mathbf{w}; I_i) = \kappa - 1 + \lambda t_A^2 - \frac{n-1}{2} + \frac{n}{2} - \frac{1}{2}(\kappa - 1 + \lambda t_A^2) < \frac{n+1}{4},$$

and the cost for any agent $i \in N$ such that $t_i = t_A$ is

$$\mathsf{cost}_i(\mathbf{t}; \mathbf{w}; I_i) = \frac{n}{2} - \frac{1}{2}\left((\lambda - 1)t_A^2 + \kappa\right) = \frac{n+1}{4}.$$

□

4 The Divergence Ratio

In order to study the impact that the information regime plays on the performance of the system, we first consider the *Divergence Ratio*. This is essentially the subjective Price of Anarchy and it is the worst-case ratio of the system cost at a Nash equilibrium over the minimum system cost.

We first define the *social cost* $\mathsf{SC}(\mathbf{t}, \mathbf{I})$ to be the maximum subjective selfish cost over all agents, i.e. $\mathsf{SC}(\mathbf{t}, \mathbf{I}) = \max_{i \in N} \mathsf{cost}_i(\mathbf{t}; \mathbf{w}; I_i)$. Notice that here consider the social cost as the maximum among the costs of all players. We could define it also as the sum of the costs of all players. All results in this section extend easily to this social cost as well, although we omit them from this abstract for lack of space.

The *Players' Optimum* $\mathsf{PO}(\mathbf{I})$ is the minimum, over all possible strategy profiles $\mathbf{t} \in [0,1]^n$, Social Cost: $\mathsf{PO}(\mathbf{I}) = \min_{\mathbf{t}\in[0,1]^n} \mathsf{SC}(\mathbf{t},\mathbf{I})$. That is, the Players' Optimum corresponds to a strategy profile that minimizes the maximum cost seen by the agents.

The *Divergence Ratio* $\mathsf{DR}(\mathbf{I})$ is the worst-case, over all weights $\mathbf{w}$ and Nash equilibria $\mathbf{t}$, of the ratio $\frac{\mathsf{SC}(\mathbf{t},\mathbf{I})}{\mathsf{PO}(\mathbf{I})}$:

$$\mathsf{DR}(\mathbf{I}) = \max_{\mathbf{w}} \max_{\mathbf{t}:\mathbf{t} \text{ N.E.}} \frac{\mathsf{SC}(\mathbf{t},\mathbf{I})}{\mathsf{PO}(\mathbf{I})}.$$

4.1 The Case of $I_i = \emptyset$

We will now show that the Divergence Ratio for the total lack of information regime is small. Recall from Theorem 1 that the Social Cost for this information regime is $(n+1)/4$. We now need to compute the Players' Optimum for this regime.

Lemma 2. *Consider the case where* $I_i = \emptyset$ *for all* $i \in N$. *Then* $\mathsf{PO}(\mathbf{I}) = \frac{n}{4}$ *if* n *is even and* $\mathsf{PO}(\mathbf{I}) = \frac{n+1}{4}$ *if* n *is odd.*

Proof (Sketch). Let $\mathbf{t}^*$ denote the strategy profile that corresponds to the Players' Optimum. We consider the following cases:

- Case 1: $t_i^* \in \{0,1\}$ for all $i \in N$. In this case, the set of agents N is partitioned into 2 subsets N_0 and N_1 such that $t_i^* = 0$ for all $i \in N_0$ and $t_i^* = 1$ for all $i \in N_1$, $|N_0| + |N_1| = n$ and the Social Cost is minimized, i.e. $\mathsf{PO}(\mathbf{I}) = \min_{N_0 \subseteq N} \max\left\{\frac{|N_0|}{2}, \frac{n-|N_0|}{2}\right\}$. If n is even, then $|N_0| = n/2$ and $\mathsf{PO}(\mathbf{I}) = \frac{n}{4}$. If n is odd, then $|N_0| = (n-1)/2$ or $|N_0| = (n+1)/2$ and $\mathsf{PO}(\mathbf{I}) = \frac{n+1}{4}$.
- Case 2: There exists some $i \in N$ such that $t_i^* \in (0,1)$. By contradiction, it can be shown that in this case there must be some agent j such that $\mathsf{cost}_j(\mathbf{t}^*; \mathbf{w}; \mathbf{I}) \geq \frac{n+1}{4}$ if n is odd or some agent j such that $\mathsf{cost}_j(\mathbf{t}^*; \mathbf{w}; \mathbf{I}) \geq \frac{n}{4}$ if n is even. □

Theorem 1 and Lemma 2 immediately yield:

Theorem 2. *Consider the case where* $I_i = \emptyset$ *for all* $i \in N$. *Then* $\mathsf{DR}(\mathbf{I}) = 1 + \frac{1}{n}$ *if* n *is even and* $\mathsf{DR}(\mathbf{I}) = 1$ *if* n *is odd.*

4.2 The Case of $I_i = N$

Assume that $I_i = N$ for all $i \in N$. From Equation 2 of Lemma 1, the cost of any agent $i \in N$ for a strategy profile $\mathbf{t} = (t_1, \ldots, t_n) \in [0,1]^n$ is

$$\mathsf{cost}_i(\mathbf{t}; \mathbf{w}; I_i) = \begin{cases} \sum_{j \in N: w_j \leq t_j} w_j & \text{if } w_i \leq t_i \\ \sum_{j \in N: w_j > t_j} w_j & \text{if } w_i > t_i \end{cases}.$$

An important observation in this case is that, since each agent knows the exact values of the loads of all other agents, it suffices to consider single-threshold strategies of the form $t_i = 0$ or $t_i = 1$, for all $i \in N$. For example, consider a Nash equilibrium $\mathbf{t}$ such that $0 < t_k < 1$ for some $k \in N$. Assume that $w_k \leq t_k$. All agents know that $w_k \leq t_k$, so all agents know that w_k goes on bin 0. Therefore, all strategy profiles $(t'_k, \mathbf{t}_{-k})$ such that $w_k \leq t'_k \leq 1$ are Nash equilibria, which are equivalent (as regards the selfish costs of the agents) to the Nash equilibrium $\mathbf{t}$. Similar arguments apply for the case $w_k > t_k$. Therefore the Nash equilibria in this case correspond to the *pure* Nash equilibria of the KP model [8] with n agents and 2 links. So in this case the Divergence Ratio is identical to the *pure* Price of Anarchy in the KP model with n agents and 2 links. In [5] (Theorem 7.1), it is shown that the pure Price of Anarchy in this setting is at most $\frac{4}{3}$ and this bound is tight. We give an alternative proof of this bound:

Theorem 3. *Consider the case where $I_i = N$ for all $i \in N$. Then* $\mathsf{DR}(\mathbf{I}) = \frac{4}{3}$.

Proof. Consider an arbitrary Nash equilibrium $\mathbf{t}^*$. Then there is an equivalent, with respect to the costs and the Divergence Ratio, Nash equilibrium $\mathbf{t}$ such that for all $i \in N$, $t_i = 0$ or $t_i = 1$. The total load on bin 0 is $B_0(\mathbf{t}) = \sum_{i:t_i=1} w_i$ and the total load on bin 1 is $B_1(\mathbf{t}) = \sum_{i:t_i=1} w_i$. Therefore the cost for agent $i \in N$ is

$$\mathsf{cost}_i(\mathbf{t}; \mathbf{w}; I_i) = \begin{cases} B_0(\mathbf{t}) & \text{if } t_i = 1 \\ B_1(\mathbf{t}) & \text{if } t_i = 0 \end{cases}.$$

Without loss of generality, assume that $B_0(\mathbf{t}) \geq B_1(\mathbf{t})$. Thus $\mathsf{SC}(\mathbf{t}, \mathbf{I}) = B_0(\mathbf{t})$. Moreover, $\mathsf{PO}(\mathbf{I}) \geq \frac{\sum_{i \in N} w_i}{2} = \frac{B_0(\mathbf{t}) + B_1(\mathbf{t})}{2}$. Now if only one agent, say agent i, places her load on bin 0 (i.e. $t_i = 1$ and $t_j = 0$ for all $j \neq i$) then $\mathsf{PO}(\mathbf{I}) = B_0(\mathbf{t})$ and $\mathsf{DR}(\mathbf{I}) = 1$.

Otherwise, there are at least two loads on bin 0. Thus there exists an agent i that chooses bin 0 (i.e. with $t_i = 1$) such that $w_i \leq \frac{B_0(\mathbf{t})}{2}$. Since $\mathbf{t}$ is a Nash equilibrium, it holds that $B_0(\mathbf{t}) \leq B_1(\mathbf{t}) + w_i$, implying that $B_0(\mathbf{t}) \leq B_1(\mathbf{t}) + \frac{B_0(\mathbf{t})}{2}$ and that $B_1(\mathbf{t}) \geq \frac{B_0(\mathbf{t})}{2}$. Therefore, $\mathsf{PO}(\mathbf{I}) \geq \frac{B_0(\mathbf{t}) + B_1(\mathbf{t})}{2} \geq \frac{3B_0(\mathbf{t})}{4}$ and $\mathsf{DR}(\mathbf{I}) = \max_{\mathbf{t}:\mathbf{t} \text{ N.E.}} \frac{\mathsf{SC}(\mathbf{t}, \mathbf{I})}{\mathsf{PO}(\mathbf{I})} \leq \frac{4}{3}$.

Now consider the case where n is even and $n > 2$, $w_1 = w_2 = (n-2)\alpha$ and $w_i = \alpha$ for all $i \neq 1, 2$, for some $\alpha \in \left(0, \frac{1}{n-2}\right]$. Then the strategy profile $\mathbf{t}$ where $t_1 = t_2 = 1$ and $t_i = 0$ for all $i \neq 1, 2$ is a Nash equilibrium which gives a Social Cost equal to $2(n-2)\alpha$. In this case however $\mathsf{PO}(\mathbf{I}) = \frac{3}{2}(n-2)\alpha$ and thus $\mathsf{DR}(\mathbf{I}) \geq \frac{2(n-2)\alpha}{\frac{3}{2}(n-2)\alpha} = \frac{4}{3}$. □

4.3 The Case of Arbitrary I_i

We have shown that, in the case where the agents have no information about the exact value of their weights, then the Divergence Ratio is very close to 1.

The same holds when the agents have complete information. In contrast, we will next show that if $i \in I_i$ and the cardinality of I_i is sufficiently small, then the Divergence Ratio grows significantly.

Theorem 4. *If $\gamma_i = \gamma \leq \frac{n-2}{3}$ and $i \in I_i$ for all $i \in N$, then $\mathsf{DR}(\mathbf{I}) \geq \frac{n+\gamma+2}{4\gamma+4}$.*

Proof. For the proof, we focus on the instance where $w_i = 1$ for all $i \in N$. Our goal is to find (a) a Nash equilibrium $\mathbf{t}$ of low Social Cost, so as to upper bound the Players' Optimum, and (b) a Nash equilibrium $\mathbf{t}'$ of high Social Cost, so as to lower bound the worst possible Social Cost:

(a) Consider the strategy profile $\mathbf{t}$ such that $t_i = 1 - \frac{1}{n-\gamma}$ for all $i \in N$. Then the cost for any agent $i \in N$ is

$$\mathsf{cost}_i(\mathbf{t};\mathbf{w};I_i) = \gamma + \frac{n-\gamma}{2} - \frac{n-\gamma}{2}\left(1 - \frac{1}{n-\gamma}\right)^2 = \gamma + 1 - \frac{1}{2(n-\gamma)}.$$

The profile $\mathbf{t}$ is a Nash equilibrium, since the cost for i if she chose bin 0 would be

$$1 + \frac{n-\gamma}{2}\left(1 - \frac{1}{n-\gamma}\right)^2 \geq \gamma + 1 + \frac{1}{2(n-\gamma)} > \mathsf{cost}_i(\mathbf{t};\mathbf{w};I_i).$$

The Social Cost of $\mathbf{t}$ is $\mathsf{SC}(\mathbf{t},\mathbf{I}) = \max_{i \in N} \mathsf{cost}_i(\mathbf{t};\mathbf{w};\mathbf{I}) \leq \gamma + 1$.

(b) Now consider the profile $\mathbf{t}'$ where $t'_i = \sqrt{\frac{1}{2} + \frac{\gamma-1}{n-\gamma}}$ for all $i \in N$ (since $\gamma \leq \frac{n-2}{3}$, $t'_i \in (0,1)$). Then the cost for any agent $i \in N$ is

$$\mathsf{cost}_i(\mathbf{t}';\mathbf{w};I_i) = \gamma + \frac{n-\gamma}{2} - \frac{n-\gamma}{2}(t'_i)^2 = \frac{n+\gamma+2}{4}.$$

The profile $\mathbf{t}'$ is also a Nash equilibrium, since the cost for i if she chose bin 0 would be

$$1 + \frac{n-\gamma}{2}(t'_i)^2 = \frac{n+\gamma+2}{4} = \mathsf{cost}_i(\mathbf{t}';\mathbf{w};I_i).$$

The Social Cost of $\mathbf{t}'$ is $\mathsf{SC}(\mathbf{t}',\mathbf{I}) = \max_{i \in N} \mathsf{cost}_i(\mathbf{t}';\mathbf{w};\mathbf{I}) = \frac{n+\gamma+2}{4}$.

Thus the Divergence Ratio is

$$\mathsf{DR}(\mathbf{I}) = \max_{\hat{\mathbf{t}}:\hat{\mathbf{t}} \text{ N.E.}} \frac{\mathsf{SC}(\hat{\mathbf{t}},\mathbf{I})}{\mathsf{PO}(\mathbf{I})} \geq \frac{\mathsf{SC}(\mathbf{t}',\mathbf{I})}{\mathsf{SC}(\mathbf{t},\mathbf{I})} \geq \frac{\frac{n+\gamma+2}{4}}{\gamma+1} = \frac{n+\gamma+2}{4\gamma+4}.$$ □

Corollary 1. *If $\gamma_i = \gamma = o(n)$ and $i \in I_i$ for all $i \in N$, then $\lim_{n\to\infty} \mathsf{DR}(\mathbf{I}) = \infty$.*

Interestingly, the above observations hold also for the case of the social cost which is the sum of the costs of all players and it is an easy consequence of the symmetry of the Nash equilibria in this section.

5 The Price of Anarchy

We now consider the objective cost and the Price of Anarchy for the total lack of information regime. In particular, for a given strategy profile $\mathbf{t}$ (or more general strategy profile $\mathbf{s}$), we define the objective cost of player i to be the expected load of the bin selected by i. The expectation is over uniformly distributed $\mathbf{w} \in [0,1]^n$. Accordingly, we define the social cost as the sum of the costs of all players. Notice that the social cost here is the sum of the cost of all players. The case of max social cost is more complicated and we leave it as an interesting open problem.

The Price of Anarchy is the worst-case ratio of the social cost at a Nash equilibrium over the expected optimum. To compute the optimum, we assume that the optimal algorithm is distributed and has the same information as the agents. In the total lack of information regime, this is easy to define when we consider only pure strategies: $\lceil n/2 \rceil$ of the agents select bin 0 and the rest select bin 1. The expected optimal cost of each agent is either $\frac{1}{2}\lceil n/2 \rceil$ or $\frac{1}{2}\lfloor n/2 \rfloor$. The sum of the costs of all agents is

$$\mathsf{OPT} = \begin{cases} \frac{n^2}{4} & \text{for even } n \\ \frac{n^2+1}{4} & \text{for odd } n \end{cases}.$$

Theorem 5. *The Price of Anarchy for the total lack of information regime is*

$$\mathsf{PA} = \begin{cases} \frac{n+1}{n} & \textit{for even } n \\ \frac{n(n+1)}{n^2+1} & \textit{for odd } n \end{cases}.$$

Proof. From the characterization of the Nash equilibria in Theorem 1, we can compute the cost for each agent. For the non-deterministic agents, i.e., the agents that have a threshold in $(0,1)$, the cost is exactly $(n+1)/4$. This follows either by computing the cost explicitly: $1/2$, which is the expected cost of her own load, plus the expected cost of bin 0 due to other agents $\kappa/2 + t_A^2/(\lambda-1) = (n+1)/4$ (the notation is from Theorem 1). A more direct way follows from the definition of the Nash equilibrium: The expected load on the bins due to other agents should be the same which happens when the load is $(n-1)/4$. Adding the expected load $1/2$ of her own, the cost of agent i is $(n+1)/4$.

By similar considerations, for the deterministic agents the cost is at most $(n+1)/4$. The worst case happens when all agents are non-deterministic. The total cost of all players is $n(n+1)/4$. The Price of Anarchy follows by dividing by the optimal OPT. □

References

1. Arrow, K.: The Economics of Information. Harvard University Press (1984)
2. Christodoulou, G., Koutsoupias, E., Nanavati, A.: Coordination Mechanisms. In: Díaz, J., Karhumäki, J., Lepistö, A., Sannella, D. (eds.) ICALP 2004. LNCS, vol. 3142, pp. 345–357. Springer, Heidelberg (2004)

3. Fotakis, D., Kontogiannis, S., Koutsoupias, E., Mavronicolas, M., Spirakis, P.: The Structure and Complexity of Nash Equilibria for a Selfish Routing Game. In: Widmayer, P., Triguero, F., Morales, R., Hennessy, M., Eidenbenz, S., Conejo, R. (eds.) ICALP 2002. LNCS, vol. 2380, pp. 123–134. Springer, Heidelberg (2002)
4. Gairing, M., Lücking, T., Mavronicolas, M., Monien, B., Rode, M.: Nash Equilibria in Discrete Routing Games with Convex Latency Functions. In: Díaz, J., Karhumäki, J., Lepistö, A., Sannella, D. (eds.) ICALP 2004. LNCS, vol. 3142, pp. 645–657. Springer, Heidelberg (2004)
5. Gairing, M., Lücking, T., Mavronicolas, M., Monien, B., Spirakis, P.: Structure and Complexity of Extreme Nash Equilibria. Theoretical Computer Science 343(1-2), 133–157 (2005)
6. Gairing, M., Monien, B., Tiemann, K.: Selfish routing with incomplete information. SPAA 2005 , 203–212 (2005)
7. Georgiades, S., Mavronicolas, M., Spirakis, P.: Optimal, Distributed Decision-Making: The Case of No Communication. In: Proceedings of the 12th International Symposium on Fundamentals of Computation Theory, pp. 293–303 (1999)
8. Koutsoupias, E., Papadimitriou, C.H.: Worst-Case Equilibria. In: Meinel, C., Tison, S. (eds.) STACS 99. LNCS, vol. 1563, pp. 404–413. Springer, Heidelberg (1999)
9. Lücking, T., Mavronicolas, M., Monien, B., Rode, M., Spirakis, P., Vrto, I.: Which is the Worst-case Nash Equilibrium? In: Rovan, B., Vojtáš, P. (eds.) MFCS 2003. LNCS, vol. 2747, pp. 551–561. Springer, Heidelberg (2003)
10. Mavronicolas, M., Spirakis, P.: The Price of Selfish Routing. In: Mavronicolas, M., Spirakis, P. (eds.) Proceedings of the 33rd Annual ACM Symposium on Theory of Computing, July 2001, pp. 510–519 (2001), Also, accepted to Algorithmica
11. Papadimitriou, C.H., Yannakakis, M.: On the Value of Information in Distributed Decision-Making. In: Proceedings of the 10th Annual ACM Symposium on Principles of Distributed Computing, pp. 61–64. ACM Press, New York (1991)
12. Yao, A.C.: Some Complexity Questions Related to Distributive Computing. In: Proceedings of the 11th ACM Symposium on Theory of Computing (STOC 1979), pp. 209–213. ACM Press, New York (1979)

Extending the Notion of Rationality of Selfish Agents: Second Order Nash Equilibria

Vittorio Bilò[1] and Michele Flammini[2]

[1] Dipartimento di Matematica "Ennio De Giorgi", Università del Salento
Provinciale Lecce-Arnesano, P.O. Box 193, 73100 Lecce, Italy
vittorio.bilo@unile.it
[2] Dipartimento di Informatica, Università di L'Aquila
Via Vetoio, Loc. Coppito, 67100 L'Aquila, Italy
flammini@di.univaq.it

Abstract. Motivated by the increasing interest of the Computer Science community in the study and understanding of non-cooperative systems, we present a novel model for formalizing the rational behavior of agents with a more farsighted view of the consequences of their actions. This approach yields a framework creating new equilibria, which we call Second Order equilibria, starting from a ground set of traditional ones. By applying our approach to pure Nash equilibria, we define the set of Second Order Nash equilibria and present their applications to the Prisoner's Dilemma game, to an instance of Braess's Paradox in the $Wardrop$ model and to the KP model with identical machines.

1 Introduction

Central to the theory and study of multiplayer non-cooperative games is the notion of Nash equilibrium [33,34], due to its ability to model the rational behavior of selfish agents. All the agents (players) participating in a game have a set of strategies they can adopt and, for any combination of the strategies adopted by everyone, they obtain a certain payoff. A Nash equilibrium is a particular combination of strategies such that none of the players can improve his payoff by changing his strategy. It is well known that Nash equilibria fail in optimizing the overall satisfaction of the players in several games, the pragmatic example being the Prisoner's Dilemma. One of the reasons for this suboptimality lies in the fact that players always perform deviations from a particular strategy only motivated by a transient improvement on their payoffs, without considering what will be their final payoffs when the game eventually reaches a Nash equilibrium.

This observation naturally yields the following question: *"Can we really consider rational an agent taking decisions only based on what will be their short term consequences, without considering what these decisions will cause tomorrow?"*. In this paper we propose a novel model for formalizing the rational behavior of agents with a more farsighted view of the consequences of their deviations. We achieve this purpose by defining a framework yielding new notions of equilibria which we call Second Order equilibria. They can be based upon different

L. Kučera and A. Kučera (Eds.): MFCS 2007, LNCS 4708, pp. 621–632, 2007.

notions of traditional non-cooperative equilibria. Throughout the paper we will deal with Second Order Nash equilibria leaving the discussion of other extensions to future works.

Games, Equilibria and Optimality. A strategic game $\mathcal{G} = (P, S, \omega)$ is defined as follows. There is a set of n players P. Any player $p_i \in P$ has a set of strategies S_i and the set $S = S_1 \times \ldots \times S_n$ is called the set of all possible states of the game. The payoff function $\omega_i : S \rightarrow \Re$ defines the cost that player p_i has to incur when the game is on state $s \in S$. Usually, each game has an associated global function $\gamma : S \rightarrow \Re$, called the social function, that is required to be optimized.

Let $s = (s_1, \ldots, s_i, \ldots, s_n)$ and $s' = (s_1, \ldots, s'_i, \ldots, s_n)$ be two states of $\mathcal{G}$ such that $\omega_i(s') < \omega_i(s)$. We call the transition of game $\mathcal{G}$ from s to s' an improving step performed by player p_i. A pure Nash equilibrium is a state in which no player possesses an improving step. Unfortunately, pure Nash equilibria are not guaranteed to exist for all games, thus Nash himself generalized this definition by introducing the concept of mixed strategy. A mixed strategy for player p_i is a probability distribution on the set of strategies S_i. The payoff obtained by p_i is now defined as the expected value of the related random variable and the definition of mixed Nash equilibrium is obtained consequently. The property of mixed Nash equilibria, stated by Nash's famous Theorem, is that they exist for any game. However, there are cases in which the use of mixed strategies is unrealistic or unacceptable. Throughout the paper we will only deal with pure Nash equilibria and will refer to them simply as Nash equilibria.

The evolution of a game $\mathcal{G}$ resulting from the interactions among players performing improving steps can be easily captured by a graph $G_{\mathcal{G}} = (N, A)$, called the state graph of $\mathcal{G}$, where $N = S$ and A is such that there exists an edge between s and s' if and only if there exists an improving step from s to s'. An important issue related to the notion of Nash equilibrium is that of convergence towards such an equilibrium point. A game $\mathcal{G}$ is said to be convergent if for any $s \in S$, any sequence of improving steps starting from s ends in a Nash equilibrium, or, analogously, $\mathcal{G}$ does not admit an infinite sequence of improving steps. By using the representation of $\mathcal{G}$ through its state graph, we have that $\mathcal{G}$ is convergent if and only if $G_{\mathcal{G}}$ is acyclic.

Example 1 (Prisoner's Dilemma). Two suspects in a major crime are held in separate cells. There is enough evidence to convict each of them of a minor offense, but not enough evidence to convict either of them of the major crime unless one of them acts as an informer against the other (finks). If they both stay quiet, each will be convicted of the minor offense and spend one year in prison. If one and only one of them finks, he will be freed and used as a witness against the other, who will spend three years in prison. If they both fink, each will spend two years in prison.

This situation can be modeled as a strategic game in which we have two players p_1 and p_2. The set of strategies is the same for both of them and is $S_i = \{Quiet, Fink\}$, for $i \in \{1, 2\}$. Finally, the payoff function is shown in the table depicted in Figure 1.

		Player 2	
		Quiet	Fink
Player 1	Quiet	1,1	3,0
	Fink	0,3	2,2

Fig. 1. The Prisoner's Dilemma Game

As it can be easily seen by inspection, the game admits one Nash equilibrium represented by the state $\{Fink, Fink\}$. This non-cooperative solution contrasts with the state $\{Quiet, Quiet\}$ in which both players enjoy a better payoff.

Two metrics have been introduced in the literature in order to capture the loss of optimality yielded by non-cooperative equilibria, that is the price of anarchy [35] and the price of stability [1]. Let $\mathcal{N}(\mathcal{G})$ be the set of equilibria of game $\mathcal{G}$ and s^* be a state optimizing the social function γ. The price of anarchy of $\mathcal{G}$ is defined as $PA(\mathcal{G}) = \sup_{s \in \mathcal{N}(\mathcal{G})} \frac{\gamma(s)}{\gamma(s^*)}$, while the price of stability of $\mathcal{G}$ is defined as $PS(\mathcal{G}) = \inf_{s \in \mathcal{N}(\mathcal{G})} \frac{\gamma(s)}{\gamma(s^*)}$.

Network Games and Selfish Routing. In recent years a considerable research effort has been devoted to the estimation of the price of anarchy of different network games. The reasons of such an interest in network games come from the affirmation of the Internet and, in general, of huge unregulated networks, where the traffic generated by their users is not controlled by some central authority, but it is rather the outcome resulting from the interaction of the users when routing their traffic selfishly and independently on the network. Two major models have been deeply investigated by researchers: the KP model [28] and the $Wardrop$ model [13,41], both being convergent games.

In the KP model there are n players and m parallel links. Each player owns a certain unsplittable traffic and wants to route it on one of the links. This game can also be interpreted as the non-cooperative version of the problem of scheduling n jobs on m parallel machines. The payoff obtained by a player is the completion time of the chosen machine. The social function is the makespan, that is the maximum completion time of all the machines.

In the $Wardrop$ model there are infinitely many players who want to route their traffic over an arbitrary network. There is a convex latency function associated with each link which is defined in terms of its load. The traffic can be split into arbitrary pieces each being handled by a selfish player, so that unregulated traffic can be modeled as a network flow. The payoff obtained by a player is the sum of the latencies experienced on the edges he uses. The social function is the sum of the products between the payoff of each player and the amount of traffic he owns. The $Wardrop$ model can also be seen as a congestion game [36] where all the players own the same infinitesimal small amount of traffic.

Related Work. The KP model was introduced by [28] and then has been further studied in [11,12,14,16,17,18,21,27,31,32]. The $Wardrop$ model was defined in [41] and then studied in [4,5,9,13]. Recent papers interested in the price of anarchy of this model include [37,38,39]. Several other works have dealt with the study of the price of anarchy of pure and/or mixed Nash equilibria in different network games [2,6,7,8,15,19,20,30].

Other equilibrium concepts have been introduced and studied in the literature. We recall here perfect equilibria [40], sequential equilibria [29], stable equilibria [26], stochastic adjustment models [25], correlated equilibria [3], Bayesian equilibria [24] and, recently, sink equilibria [23]. In particular, correlated and sink equilibria are generalizations of mixed and pure Nash equilibria respectively, while the others are refinements of mixed (and consequently also pure) Nash equilibria.

Our Contribution. Critics and improvements upon the classical notion of pure Nash equilibrium have had different targets such as *existence*, as in the case of correlated and sink equilibria, *stability*, as in the case of stable equilibria and stochastic adjustment models, *irrationality*, as in the case of perfect and sequential equilibria, and *need of complete information*, as in the case of Bayesian equilibria. However, to the best of our knowledge, no effort has been done in order to model the rational behavior of players on a longer range basis.

We thus introduce a new definition of selfish agents by giving them a more farsighted view of their actions. An agent knows he is part of a multiplayer game and also knows that the game will not stop right after he has performed an improving step. In this scenario the agent will be mostly interested in evaluating the fairness of the current state with respect to the equilibrium that the game will reach after he has performed his improving step. When more than just one equilibrium can be reached from a particular state, by following a classical worst-case analysis, we assume that the agent will compare the current state with the equilibrium yielding the worst payoff for him. Such a view point is clearly based upon the definition of a ground set of equilibria the agents will compare a generic state with. According to these comparisons, a possible non empty set of new equilibria may arise and the process may be iterated recursively until a fixed point is reached and the final set of desired equilibria is created. We call such a set the set of Second Order equilibria. Using different definitions of equilibrium for defining the ground set, we can achieve different sets of Second Order equilibria. In this paper we concentrate our attention on the definition and the evaluation of Second Order Nash equilibria, that is with a ground set given by the set of Nash equilibria. In particular, we present applications of these equilibria to the Prisoner's Dilemma, to an instance of Braess's Paradox [9] in the $Wardrop$ model and to the KP model with identical machines. We want to stress here that the approaches pursued so far in the literature always try to prune unwanted solutions from the set of Nash equilibria. Thus, our approach is novel in the sense that it provides an expanded set of equilibria.

Paper Organization. In the next section we give the formal definition of Second Order Nash equilibria, while in Section 3 we present their applications. In Section 4 we discuss other possible models of farsighted selfish agents and finally, in the last section we give conclusive remarks and open questions. Due to space limitations several details have been omitted and will be given in the full version of the paper.

2 Second Order Nash Equilibria

We first propose a generalization of the state graph related to a game $\mathcal{G}$. Given a set of equilibria states $E \subseteq S$, let $G_{\mathcal{G},E} = (N, A)$ be the directed graph in which $N = S$ and A is such that there exists an edge between s and s' if and only if there exists an improving step from s to s' and $s \notin E$. Clearly, $G_{\mathcal{G},\emptyset}$ coincides with the state graph $G_{\mathcal{G}}$. If p_i is the unique player changing his strategy from s to s', we label the arc $\langle s, s' \rangle$ with the index i.

We define $\rho_E(s)_i^k$ as the set of all the states of $\mathcal{G}$ that can be reached starting from s by following a path of length at most k whose first arc is labeled with index i in the graph $G_{\mathcal{G},E}$. The set $\rho_E(s)^k = \bigcup_{i=1}^n \rho_E(s)_i^k$ will denote the set of all states that can be reached from s by following a path of length at most k in the graph $G_{\mathcal{G},E}$. When $E = \emptyset$, we will simply remove the subscript E from the notation. We also define $P(s)$ as the set of players that can perform an improving step starting from state s.

We now give a recursive definition of the new set of equilibria that will be further clarified in the following.

Definition 1. *Let $\mathcal{G}$ be a convergent game. The set $N^k(\mathcal{G}) = \{s \in S : \forall p_i \in P(s)$ and $\forall s' \in \rho(s)_i^1$, $\exists s'' \in N^k(\mathcal{G})$ such that $s'' \in \rho_{N^k(\mathcal{G})}(s')^k$ and $\omega_i(s) < \omega_i(s'')\}$ is the set of all the Second Order k-Nash equilibria of game $\mathcal{G}$, for any integer $k \geq 0$.*

Intuitively, this rather involved definition says that a state s is a Second Order k-Nash equilibrium, for some integer $k \geq 0$, if all the players who can perform an improving step in s would experience, in one of the Second Order k-Nash equilibria resulting from an evolutive process of at most k improving steps taking place after their first defection, a payoff which is worse than the one they get in state s. Such a definition is clearly recursive. However, in the following we show that it is well posed, in the sense that it admits a unique set of solutions or fixed point. First of all, we prove that $N^0(\mathcal{G})$ coincides with the set of the Nash equilibria of $\mathcal{G}$ and that each Nash equilibrium is a Second Order k-Nash equilibrium, for any integer $k \geq 1$.

Lemma 1. *$N^0(\mathcal{G}) = \{s \in S : s$ is a Nash equilibrium$\}$ and $N^0(\mathcal{G}) \subseteq N^k(\mathcal{G})$, for any integer $k \geq 1$.*

We now define an algorithm constructing a set of states $\tilde{N}^k(\mathcal{G})$ that we will after show to coincide with $N^k(\mathcal{G})$. To this aim, we first introduce some necessary notation. Given a directed graph $G = (N, A)$ and a set of vertices $T \subseteq N$, we define $leaves(T)$ as the maximal subset of T such that for any $s \in leaves(T)$ and $s' \in T$ there exists no (s, s')-directed path in G.

Lemma 2. *For any acyclic directed graph $G = (N, A)$ and any non empty subset of vertices $T \subseteq N$, $leaves(T)$ is unique and $leaves(T) \neq \emptyset$.*

Corollary 1. *For any acyclic directed graph $G = (N, A)$, $leaves(T) = \emptyset$ if and only if $T = \emptyset$.*

The algorithm Construct $\tilde{N}^k(\mathcal{G})$ for determining $\tilde{N}^k(\mathcal{G})$ is defined as follows.

Construct $\tilde{N}^k(\mathcal{G})$:

1. $\tilde{N}^k(\mathcal{G}) \leftarrow N^0(\mathcal{G})$
2. $T(k) \leftarrow \{s \in S \setminus \tilde{N}^k(\mathcal{G}) : \forall p_i \in P(s)$ and $\forall s' \in \rho(s)_i^1, \exists s'' \in \tilde{N}^k(\mathcal{G})$ such that $s'' \in \rho_{\tilde{N}^k(\mathcal{G})}(s')^k$ and $\omega_i(s) < \omega_i(s'')\}$
3. if $T(k) \neq \emptyset$
   (a) $\tilde{N}^k(\mathcal{G}) \leftarrow \tilde{N}^k(\mathcal{G}) \cup leaves(T(k))$
   (b) goto 2
4. else return $\tilde{N}^k(\mathcal{G})$

Clearly, since at each step $\tilde{N}^k(\mathcal{G})$ grows, the algorithm terminates. Moreover, because of Lemma 2, we have that $\tilde{N}^k(\mathcal{G})$ is unique. In the following theorem we show that $N^k(\mathcal{G}) = \tilde{N}^k(\mathcal{G})$ for any integer $k \geq 0$, thus proving the uniqueness of $N^k(\mathcal{G})$.

Theorem 1. *Let $\mathcal{G}$ be a convergent game then $N^k(\mathcal{G}) = \tilde{N}^k(\mathcal{G})$ for any integer $k \geq 0$.*

In the following lemma we show that there exists a value k^* for which all the sets $N^k(\mathcal{G})$ become the same for any $k \geq k^*$.

Lemma 3. *Let $\mathcal{G}$ be a convergent game and let k^* be the diameter of $G_\mathcal{G}$ minus 1, then $N^{k^*}(\mathcal{G}) = N^k(\mathcal{G})$ for any integer $k \geq k^*$.*

We can now define the general notion of Second Order Nash equilibrium as follows.

Definition 2. *Given a convergent game $\mathcal{G}$, each state $s \in N^{k^*}(\mathcal{G}) =_{def} N(\mathcal{G})$ is a Second Order Nash equilibrium.*

3 Applications of Second Order Nash Equilibria

The power of Second Order equilibria lies in the fact that they introduce a sort of cooperation among the players in the following particular sense: the players are interested in not hurting each other, that is, in not leading the game to a state that is worse than the current one for each of them. By quoting a classical Italian proverb we can say that they *"do not awake the sleeping dog"*.

Quantitatively speaking, the effects of the introduction of Second Order Nash equilibria are the following.

Proposition 1. *The price of stability of Second Order k-Nash equilibria is not worse than that of Nash equilibria, while the price of anarchy of Nash equilibria is not worse than that of Second Order k-Nash equilibria, for any $k \geq 1$.*

Proof. Since $N^0(\mathcal{G}) \subseteq N^k(\mathcal{G})$ for any $k \geq 1$, the claim holds trivially. □

In particular, a good behavior of Second Order Nash equilibria can be appreciated by considering applications to some well-known paradoxes arising in Game Theory, such as the Prisoner's Dilemma and Braess's paradox in the *Wardrop* model.

3.1 The Prisoner's Dilemma

It can be easily shown that $N(\mathcal{G}) = \{\{Fink, Fink\}, \{Quiet, Quiet\}\}$. In fact we have $\{Fink, Fink\} \in N(\mathcal{G})$ as $\{Fink, Fink\} \in N^0(\mathcal{G})$ and $\{Quiet, Quiet\} \in N(\mathcal{G})$ since $\{Fink, Fink\} \in \rho(\{Quiet, Quiet\})_i^1$ for $i \in \{1,2\}$ and $\omega_i(\{Fink, Fink\}) > \omega_i(\{Quiet, Quiet\})$ for $i \in \{1,2\}$ and no other state belongs to $N(\mathcal{G})$.

Theorem 2. *The price of stability of Second Order Nash equilibria for the Prisoner's Dilemma is 1.*

3.2 Braess's Paradox

Suppose one unit of traffic needs to be routed from s to t in the first network of Figure 2, where each edge is labelled with its latency function of the link congestion x. In the unique flow at Nash equilibrium, which coincides with the optimal flow, half of the traffic takes the upper path and the other half travels along the lower path, and thus all agents are routed on a path of latency $\frac{3}{2}$. Next suppose a fifth edge of latency 0 (independent of the congestion) is added to the network, with the result shown on Figure 2(b). The optimal flow is unaffected by this augmentation (there is no way to use the new link to decrease the total latency) while in the new (unique) flow at Nash equilibrium all traffic follows the path $s \to v \to w \to t$; here, the latency experienced by each agent is 2. Thus, the intuitively helpful (or at least innocuous) action of adding a new zero-latency link may negatively impact on the payoffs all of the agents.

Let Π_1 be the path $s \to v \to t$, Π_2 be the path $s \to w \to t$, Π_3 be the path $s \to v \to w \to t$ and define f_i as the amount of flow routed on path Π_i. We can thus denote a state $s \in S$ as $s = (f_1, f_2, f_3)$ by specifying the amount of flow routed on the three different paths.

Since our model of Second Order equilibria, based on the structure of the state graph, is clearly a discrete one, we consider the instance of the Braess's Paradox

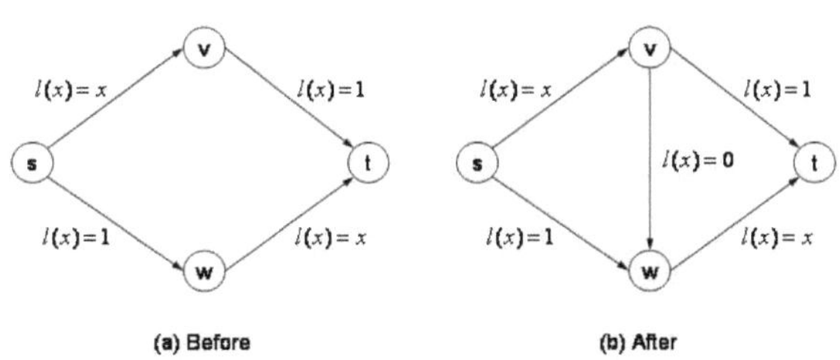

Fig. 2. Braess's Paradox

in which the unitary flow representing the traffic on the network is split into infinitely many atomic pieces all having the same infinitesimal dimension $\epsilon > 0$ and denote this game by $\mathcal{G}_\epsilon$. This assumption does not change the properties of the *Wardrop* model as well as those of Braess's Paradox. In particular, the set of the Nash equilibria becomes $N^0(\mathcal{G}_\epsilon) = \{(0,0,1), (\epsilon, 0, 1-\epsilon), (0, \epsilon, 1-\epsilon), (\epsilon, \epsilon, 1-2\epsilon)\}$. However letting ϵ go to zero, the two games become essentially the same. The set of Nash equilibria, for example, collapses to $N^0(\mathcal{G}) = \{(0,0,1)\}$.

The main result of this section is the characterization of the set of Second Order Nash equilibria for $\mathcal{G}_\epsilon$.

Theorem 3. $N(\mathcal{G}_\epsilon) = N^0(\mathcal{G}_\epsilon) \cup \{((\ell+3j)\epsilon, \ell\epsilon, 1-(2\ell+3j)\epsilon), (\ell\epsilon, (\ell+3j)\epsilon, 1-(2\ell+3j)\epsilon) | j = 0, \ldots, \lfloor \frac{1-4\epsilon}{3\epsilon} \rfloor$ *and* $\ell = 3, \ldots, \lfloor \frac{1-3j\epsilon}{2\epsilon} \rfloor\}$.

Corollary 2. *The price of stability of the Second Order Nash equilibrium for* $\mathcal{G}_\epsilon$ *is 1.*

In this problem, as well as in other non trivial multiplayer games, we have seen that the structure of the state graph can be really intricate thus making the definition of Second Order Nash equilibria a very challenging task. In order to ease the computation one can work on a simplified version of the state graph by considering, for example, the existence of a particular ordering in which improving steps are performed during the evolution of the game. Such a scenario seems perfectly reasonable and different ordering strategies can be considered indeed as some sort of coordination mechanisms [10] that can be adopted during the game in order to lead the players towards a desired behavior.

To this aim consider the following ordering mechanism $\mathcal{M}$: *at each state s, the player* $p_i \in P(s)$ *using the most congestioned path is the one allowed to change his strategy (breaking ties arbitrarily).*

We can state the following result.

Theorem 4. *When mechanism* $\mathcal{M}$ *is adopted* $N(\mathcal{G}_\epsilon) = N^0(\mathcal{G}_\epsilon) \cup \{((2^{i+1}-1)\epsilon, (2^{i+1}-1)\epsilon, 1-2(2^{i+1}-1)\epsilon) | i = 1, \ldots \lfloor \log_2(\frac{1+2\epsilon}{2\epsilon}) \rfloor - 1\}$.

Corollary 3. *When mechanism* $\mathcal{M}$ *is adopted the price of stability of the Second Order Nash equilibrium of* $\mathcal{G}_\epsilon$ *falls in the interval* $[1; \frac{13}{12}]$.

Thus, if from one hand the introduction of the ordering mechanism $\mathcal{M}$ has simplified the set of Second Order Nash equilibria, on the other hand it has worsened its price of stability.

We argue here that even when ϵ goes to zero the results of Corollary 2 and 3 still hold since $\ell\epsilon = \frac{1}{2}$ when $j = 0$ and $\ell = \lfloor \frac{1-3j\epsilon}{2\epsilon} \rfloor$ in the claim of Theorem 3 and $(2^{\lfloor \log_2(\frac{1+2\epsilon}{2\epsilon}) \rfloor} - 1)\epsilon \in [\frac{1}{4}; \frac{1}{2}]$ in the claim of Theorem 4 respectively. Hence our analysis carries over also to the general *Wardrop* model.

3.3 Selfish Load Balancing

In this section we analyze the applications of Second Order Nash equilibria to the load balancing game, the special case of the *KP* model when all the machines have identical speed.

We show that no proper Second Order Nash equilibrium exists for the load balancing game, i.e., $N(\mathcal{G})$ collapses to the set of Nash equilibria.

Theorem 5. *Let $\mathcal{G}$ be any load balancing game, it holds $N(\mathcal{G}) = N^0(\mathcal{G})$.*

4 Other Notions of Selfish Farsighted Behavior

What we have seen so far are agents always comparing the current situation with the worst Second Order equilibrium the game can reach after their improving steps. We will call these agents *prudent agents* because if there is a chance of worsening their payoffs they will stay quiet and will not perform any improving step. Clearly, it is also possible to consider *rash agents* choosing to perform improving steps when there is a chance of reaching a good equilibrium for them, thus comparing with the best Second Order equilibrium. When considering rash agents, the definition of Second Order Nash equilibria becomes the following.

Definition 3. *Let $\mathcal{G}$ be a convergent game. The set $N^k(\mathcal{G}) = \{s \in S : \forall p_i \in P(s)$ and $\forall s' \in \rho(s)_i^1$ and $\forall s'' \in N^k(\mathcal{G})$ such that $s'' \in \rho_{N^k(\mathcal{G})}(s')^k$ it holds $\omega_i(s) \leq \omega_i(s'')\}$ is the set of all the Second Order k-Nash equilibria of game $\mathcal{G}$, for any integer $k \geq 0$.*

The notion of Nash equilibrium and that of improving steps are strictly related. Moreover, improving steps are the fundamental argument needed to understand how good the process of convergence towards Nash equilibria is, by measuring the speed of convergence and the quality of the reached equilibrium. To this aim, we will define in this section the notion of Second Order improving step as well as that of *patient* and *impatient agents*. In particular, according to our definitions of Second Order equilibria, we will have four types of selfish agents (patient prudent agents, patient rash agents, impatient prudent agents and impatient rash agents) characterized by how and when they perform improving steps. We formalize this discussion in the following definition.

Definition 4. *Let $\mathcal{G}$ be a convergent game. An improving step from $s \notin N(\mathcal{G})$ to s' is a Second Order improving step for player $p_i \in P(s)$ if*

- **(Impatient prudent agents)** *$\forall s'' \in N(\mathcal{G})$ such that $s'' \in \rho_{N(\mathcal{G})}(s')^{k^*}$, it holds $\omega_i(s) \geq \omega_i(s'')$.*
- **(Impatient rash agents)** *$\exists s'' \in N(\mathcal{G})$ such that $s'' \in \rho_{N(\mathcal{G})}(s')^{k^*}$ and $\omega_i(s) > \omega_i(s'')$.*
- **(Patient prudent agents)** *$worst(s') \leq worst(s, s')$, where $worst(s') = \max_{s'' \in N(\mathcal{G}) : s'' \in \rho_{N(\mathcal{G})}(s')^{k*}} \{\omega_i(s'')\}$ and $worst(s, s') = \max_{s'' \in N(\mathcal{G}) : s'' \in \rho_{N(\mathcal{G}) \cup \{s'\}}(s)^{k*}} \{\omega_i(s'')\}$.*
- **(Patient rash agents)** *$best(s') < best(s, s')$, where $best(s') = \min_{s'' \in N(\mathcal{G}) : s'' \in \rho_{N(\mathcal{G})}(s')^{k*}} \{\omega_i(s'')\}$ and $best(s, s') = \min_{s'' \in N(\mathcal{G}) : s'' \in \rho_{N(\mathcal{G}) \cup \{s'\}}(s)^{k*}} \{\omega_i(s'')\}$.*

5 Conclusion

In this paper we try to give an impulse towards the definition of a better model for selfish rational agents by exploiting a simple and intuitive observation. If starting from a state s in which a set of players $P(s)$ are unhappy of their payoffs, the game can reach an equilibrium s' in which for any $p_i \in P(s)$ it holds $\omega_i(s) < \omega_i(s')$, then it is rational to consider s as an equilibrium state, since all players who have an incentive in deviating from s discover that such an incentive is just illusory. This view point lead us to Second Order Nash equilibria which, in spite of a simple intuitive nature, required not trivial arguments in order to be captured in a formal definition.

A well studied problem in Game Theory has been that of reducing the set of Nash equilibria of a game by eliminating those which can be considered in some sense "irrational". This process can surely provide an improvement on the price of anarchy of Nash equilibria. However, no benefits can be achieved in all those games in which the price of stability of Nash equilibria is too high. Our definition of Second Order Nash equilibria has the property of expanding the set of Nash equilibria thus being able to potentially improve on the price of stability.

We provided different types of applications of Second Order Nash equilibria to games such as the Prisoner's Dilemma, the *Wardrop* model and the KP model. We believe that our work can open a new window on this young and fascinating research field, by widening the notion of rationality of players.

A lot of open questions are thus introduced by this vision. The first one is certainly that of giving further validation of Second Order equilibria by using them together with other known equilibria and presenting good applications. To this aim, the definition of Second Order Sink equilibria seems to be a promising research direction. Moreover, there is the important issue of understanding the power of different ordering strategies in influencing the performances of Second Order equilibria. An interesting question can be also that of trying to understand if the use of Second Order equilibria can lead sequences of improving steps towards better states. In the paper we have only considered impatient prudent agents. A final open issue is certainly that of analyzing the other three possible definitions for rational agents as well as the Second Order equilibria yielded by rash agents.

Acknowledgements. The authors would like to thank an anonymous referee for pointing out an error on an earlier version of Theorem 3.

References

1. Anshelevich, E., Dasgupta, A., Kleinberg, J., Tardos, E., Wexler, T., Roughgarden, T.: The Price of Stability for Network Design with Fair Cost Allocation. In: Proceedings of FOCS'04, pp. 295–304. IEEE Computer Society Press, Los Alamitos (2004)
2. Anshelevich, E., Dasgupta, A., Tardos, E., Wexler, T.: Near-Optimal Network Design with Selfish Agents. In: Proceedings of STOC'03, pp. 511–520. ACM Press, New York (2003)

3. Aumann, R.J.: Subjectivity and Correlation in Randomized Strategies. Journal of Mathematical Economics 1, 67–96 (1974)
4. Beckmann, M.J.: On the theory of Traffic Flow in Networks. Traffic Quart 21, 109–116 (1967)
5. Beckmann, M.J., McGuire, C.B., Winsten, C.B.: Studies in the Economics of Transportation. Yale University Press (1956)
6. Bilò, V.: On the Packing of Selfish Items. In: Proceedings of IPDPS'06, IEEE Computer Society Press, Los Alamitos (2006)
7. Bilò, V., Flammini, M., Moscardelli, L.: On Nash Equilibria in Non-cooperative All-Optical Networks. In: Diekert, V., Durand, B. (eds.) STACS 2005. LNCS, vol. 3404, pp. 448–459. Springer, Heidelberg (2005)
8. Bilò, V., Flammini, M., Melideo, G., Moscardelli, L.: On Nash Equilibria for Multicast Transmissions in Ad-Hoc Wireless Networks. In: Fleischer, R., Trippen, G. (eds.) ISAAC 2004. LNCS, vol. 3341, pp. 172–183. Springer, Heidelberg (2004)
9. Braess, D.: Uber ein Paradoxon der Verkehrsplanung. Unternehmensforschung 12, 258–268 (1968)
10. Christodoulou, G., Koutsoupias, E., Nanavati, A.: Coordination Mechanisms. In: Díaz, J., Karhumäki, J., Lepistö, A., Sannella, D. (eds.) ICALP 2004. LNCS, vol. 3142, pp. 345–357. Springer, Heidelberg (2004)
11. Czumaj, A., Vocking, B.: Tight Bounds for Worst-Case Equilibria. In: Proceedings of SODA'02, pp. 413-420, ACM-SIAM (2002)
12. Czumaj, A., Krysta, P., Vocking, B.: Selfish Traffic Allocation for Server Farms. In: Proceedings of STOC'02, pp. 287–296. ACM Press, New York (2002)
13. Dafermos, S.C., Sparrow, F.T.: The Traffic Assignment Problem for a General Network. Journal of Research of the National Bureau of Standards - B. Mathematical Sciences 73B(2), 91–118 (1969)
14. Even-Dar, E., Kesselman, A., Mansour, Y.: Convergence Time to Nash Equilibria. In: Baeten, J.C.M., Lenstra, J.K., Parrow, J., Woeginger, G.J. (eds.) ICALP 2003. LNCS, vol. 2719, pp. 502–513. Springer, Heidelberg (2003)
15. Fabrikant, A., Luthra, A., Maneva, E., Papadimitriou, C.H., Shenker, S.: On a Network Creation Game. In: Proceedings of PODC'03, pp. 347–351. ACM Press, New York (2003)
16. Feldmann, R., Gairing, M., Lúcking, T.: Nashification and the Coordination Ratio for a Selfish Routing Game. In: Baeten, J.C.M., Lenstra, J.K., Parrow, J., Woeginger, G.J. (eds.) ICALP 2003. LNCS, vol. 2719, pp. 514–526. Springer, Heidelberg (2003)
17. Fotakis, D., Kontogiannis, S., Koutsoupias, E., Mavronicolas, M., Spirakis, P.: The Structure and Complexity of Nash Equilibria for a Selfish Routing Game. In: Widmayer, P., Triguero, F., Morales, R., Hennessy, M., Eidenbenz, S., Conejo, R. (eds.) ICALP 2002. LNCS, vol. 2380, pp. 123–134. Springer, Heidelberg (2002)
18. Gairing, M., Lucking, T., Mavronicolas, M., Monien, B.: Computing nash equilibria for scheduling on restricted parallel links. In: Proceedings of STOC'04, pp. 613–622. ACM Press, New York (2004)
19. Gairing, M., Lucking, T., Mavronicolas, M., Monien, B.: The price of anarchy for polynomial social cost. In: Fiala, J., Koubek, V., Kratochvíl, J. (eds.) MFCS 2004. LNCS, vol. 3153, pp. 574–585. Springer, Heidelberg (2004)
20. Gairing, M., Lucking, T., Mavronicolas, M., Monien, B., Rode, M.: Nash equilibria in discrete routing games with convex latency functions. In: Díaz, J., Karhumäki, J., Lepistö, A., Sannella, D. (eds.) ICALP 2004. LNCS, vol. 3142, pp. 645–657. Springer, Heidelberg (2004)

21. Gairing, M., Lucking, T., Mavronicolas, M., Monien, B., Spirakis, P.: Extreme Nash Equilibria. In: Blundo, C., Laneve, C. (eds.) ICTCS 2003. LNCS, vol. 2841, pp. 1–20. Springer, Heidelberg (2003)
22. Goemans, M.X., Li, L., Mirrokni, V.S., Thottan, M.: Market Sharing Games Applied to the Content Distribution in Ad-Hoc Networks. In: Proceedings of MobiHoc'04, pp. 55–66. ACM Press, New York (2004)
23. Goemans, M.X., Mirrokni, V.S., Vetta, A.: Sink Equilibria and Convergence. In: Proceedings of FOCS'05, pp. 142–154. IEEE Computer Society Press, Los Alamitos (2005)
24. Harsanyi, J.C.: Games with Incomplete Information Played by 'Bayesian' Players. Management Science, 14:159–182, 320–334, 486–502 (1967)
25. Kandori, M., Mailath, G., Rob, R.: Learning, Mutation and Long-Run Equilibria in Games. Econometrica 61, 29–56 (1993)
26. Kohlberg, M., Mertens, J.: On the Strategic Stability of Equilibria. Econometrica 54(5), 1003–1037 (1986)
27. Koutsoupias, E., Mavronicolas, M., Spirakis, P.: Approximate Equilibria and Ball Fusion. In: Proceedings of SIROCCO'02. Proceedings in Informatics. Carleton Scientific, vol. 13, pp. 223–235 (2002)
28. Koutsoupias, E., Papadimitriou, C.H.: Worst-case Equilibria. In: Meinel, C., Tison, S. (eds.) STACS 99. LNCS, vol. 1563, pp. 404–413. Springer, Heidelberg (1999)
29. Kreps, D., Wilson, R.: Sequential Equilibria. Econometrica 50, 863–894 (1982)
30. Lucking, T., Mavronicolas, M., Monien, B., Rode, M.: A new model for selfish routing. In: Diekert, V., Habib, M. (eds.) STACS 2004. LNCS, vol. 2996, pp. 547–558. Springer, Heidelberg (2004)
31. Lucking, T., Mavronicolas, M., Monien, B., Rode, M., Spirakis, P., Vrto, I.: Which is the Worst-case Nash Equilibrium? In: Rovan, B., Vojtáš, P. (eds.) MFCS 2003. LNCS, vol. 2747, pp. 551–561. Springer, Heidelberg (2003)
32. Mavronicolas, M., Spirakis, P.: The Price of Selfish Routing. In: Proceedings of STOC'01, pp. 510–519. ACM Press, New York (2001)
33. Nash, J.: Equilibrium points in n-person games. Proceedings of the National Academy of Sciences 36, 48–49 (1950)
34. Nash, J.: Non-cooperative Games. Annals of Mathematics 54(2), 286–295 (1951)
35. Papadimitriou, C.H.: Algorithms, Games, and the Internet. In: Proceedings of STOC'01, pp. 749–753. ACM Press, New York (2001)
36. Rosenthal, R.W.: A Class of Games Possessing Pure-Strategy Nash Equilibria. International Journal of Game Theory 2, 65–67 (1973)
37. Roughgarden, T.: The Price of Anarchy is Independent of the Network Topology. In: Proceedings of STOC'02, pp. 428–437. ACM Press, New York (2002)
38. Roughgarden, T.: Selfish Routing. Ph. D. Thesis, Department of Computer Science, Cornell University (May 2002)
39. Roughgarden, T., Tardos, E.: How Bad is Selfish Routing? Journal of ACM 49, 236–259 (2002)
40. Selten, R.: Reexamination of the Perfectness Concept for Equilibrium Points in Extensive Games. International Journal of Game Theory 4, 25–55 (1975)
41. Wardrop, J.G.: Some Theoretical Aspects of Road Traffic Research. In: Proceedings of the Institute of Civil Engineers, Pt. II, vol. 1, pp. 325–378 (1956)

Congestion Games with Player-Specific Constants*

Marios Mavronicolas[1], Igal Milchtaich[2],
Burkhard Monien[3], and Karsten Tiemann[3,**]

[1] Department of Computer Science,
University of Cyprus, 1678 Nicosia, Cyprus
mavronic@cs.ucy.ac.cy

[2] Department of Economics,
Bar-Ilan University, Ramat Gan 52900, Israel
milchti@mail.biu.ac.il

[3] Faculty of Computer Science, Electrical Engineering, and Mathematics,
University of Paderborn, 33102 Paderborn, Germany
{bm, tiemann}@uni-paderborn.de

Abstract. We consider a special case of *weighted congestion games* with *player-specific latency functions* where each player uses for each particular resource a fixed (non-decreasing) *delay function* together with a player-specific *constant*. For each particular resource, the resource-specific delay function and the player-specific constant (for that resource) are composed by means of a *group operation* (such as addition or multiplication) into a player-specific latency function. We assume that the underlying group is a *totally ordered abelian group*. In this way, we obtain the class of *weighted congestion games with player-specific constants*; we observe that this class is contained in the new intuitive class of *dominance weighted congestion games*. We obtain the following results:

Games on parallel links:

- Every unweighted congestion game has a *generalized ordinal potential*.
- There is a weighted congestion game with 3 players on 3 parallel links that does not have the *Finite Best-Improvement Property*.
- There is a particular *best-improvement cycle* for general weighted congestion games with player-specific latency functions and 3 players whose outlaw implies the existence of a pure Nash equilibrium. This cycle is indeed outlawed for dominance weighted congestion games with 3 players – and hence for weighted congestion games with player-specific constants and 3 players.

Network congestion games:
For unweighted *symmetric network congestion games* with player-specific *additive* constants, it is $\mathcal{PLS}$-complete to find a pure Nash equilibrium.

Arbitrary (non-network) congestion games:
Every weighted congestion game with *linear* delay functions and player-specific additive constants has a *weighted potential*.

* This work was partially supported by the IST Program of the European Union under contract number IST-15964 (AEOLUS).

** Supported by the International Graduate School of Dynamic Intelligent Systems (University of Paderborn, Germany).

L. Kučera and A. Kučera (Eds.): MFCS 2007, LNCS 4708, pp. 633–644, 2007.

1 Introduction

Motivation and Framework. Originally introduced by Rosenthal [15], *congestion games* model resource sharing among *(unweighted) players*. Here, the strategy of each player is a set of *resources*. The cost for a player on resource e is given by a *latency function* for e, which depends on the total weight of all players choosing e. In *congestion games with player-specific latency functions*, which were later introduced by Milchtaich [13], players are *weighted* and each player chooses her own latency function for each resource, which determines her own player-specific cost on the resource. These choices reflect different preferences, beliefs or estimates by the players; for example, such differences occur in multiclass networks or in networks with uncertain parameters.

In this work, we introduce a special case of (weighted) congestion games with player-specific latency functions [13], which we call *(weighted) congestion games with player-specific constants*. Here, each player-specific latency function is made up of a resource-specific *delay function* and a player-specific *constant* (for the particular resource); the two are composed by means of a *group* operation. We will be assuming that the underlying group is a *totally ordered abelian group* (see, for example, [9, Chapter 1]). Note that this new model of congestion games restricts Milchtaich's one [13] since player-specific latency functions are no longer completely arbitrary; simultaneously, it generalizes the weighted generalization of Rosenthal's model [15] since it allows composing player-specific constants into each (resource-specific) latency function. For example, *(weighted) congestion games with player-specific additive constants* (resp., *multiplicative constants*) correspond to the case where the group operation is addition (resp., multiplication).

We will sometimes focus on *network congestion games*, where the resources and strategies correspond to edges and paths in a (directed) *network*, respectively; network congestion games offer an appropriate model for some aspects of *routing* problems. In such games, each player has a *source* and a *destination* node and her strategy set is the set of all paths connecting them. In a *symmetric* network congestion game, all players use the same pair of source and destination; else, the network congestion game is *asymmetric*. The simplest symmetric network congestion game is the *parallel links* network with only two nodes.

The *Individual Cost* for a player is the sum of her costs on the resources in her strategy. In a *(pure) Nash equilibrium*, no player can decrease her Individual Cost by unilaterally deviating to a different strategy. We shall study questions of existence of, computational complexity of, and convergence to pure Nash equilibria for (weighted) congestion games with player-specific constants.

For convergence, we shall consider sequences of *improvement* and *best-improvement* steps of players; in such steps, a player *improves* (that is, decreases) and *best-improves* her Individual Cost, respectively. A game has the *Finite Improvement Property* [14] (resp., the *Finite Best-Improvement Property*, also called *Finite Best-Reply Property* [13]) if all *improvement paths* (resp., *best-improvement paths*) are finite. Both properties imply the existence of a pure Nash equilibrium [14]; clearly, the first property implies the second. Also, the existence of a *generalized ordinal potential* is equivalent to the Finite Improvement Property [14] (and hence it implies the Finite Best-Improvement Property and the existence of a pure Nash equilibrium as well). A *weighted potential*

[14] is a particular case of a generalized ordinal potential; an *exact potential* [14] is a particular case of a weighted potential.

We observe that the class of (weighted) congestion games with player-specific constants is contained in the more general, intuitive class of *dominance (weighted) congestion games* that we introduce (Proposition 1). In this more general class, it holds that for any pair of players, the preference of some of the two players with regard to any arbitrary pair of resources necessarily induces an identical preference for the other player (Definition 2).

State-of-the-Art. It is known that every *unweighted* congestion game has a pure Nash equilibrium [15]; Rosenthal's original proof uses an *exact potential* [14]. It is possible to compute a pure Nash equilibrium for an unweighted symmetric network congestion game in polynomial time by reduction to the *min-cost flow problem* [3]. However, the problem becomes $\mathcal{PLS}$-complete for either (arbitrary) symmetric congestion games [3] or asymmetric network congestion games where the edges of the network are either directed [3] or undirected [1]. Weighted asymmetric network congestion games with affine latency functions are known to have a pure Nash equilibrium [6]; in contrast, there are weighted symmetric network congestion games with non-affine latency functions that have no pure Nash equilibrium (even if there are only 2 players) [6,12]. Weighted (network) congestion games on parallel links have the Finite Improvement Property (and hence a pure Nash equilibrium) if all latency functions are non-decreasing; in this setting, [5] proves that a pure Nash equilibrium can be computed in polynomial time by using the classical LPT algorithm due to Graham [10] when latency functions are linear. (This is the well-known setting of *related parallel links*, which is equivalent to using the identity function for all delay functions in a weighted congestion game with multiplicative constants.) In the general case, it is strongly $\mathcal{NP}$-complete to determine whether a given weighted network congestion game has a pure Nash equilibrium [2].

For weighted congestion games with (non-decreasing) player-specific latency functions on parallel links, there is a counterexample to the existence of a pure Nash equilibrium with only 3 players and 3 links [13]. This result is *tight* since such games with 2 players have the Finite Best-Improvement Property [13]. Unweighted congestion games with (non-decreasing) player-specific latency functions have a pure Nash equilibrium but not necessarily the Finite Best-Improvement Property [13].

The special case of (weighted) congestion games with player-specific *linear* latency functions (without a constant term) was studied in [7,8]. Such games have the Finite Improvement Property if players are unweighted [7], while there is a game with 3 weighted players that does not have it [7]. For the case of 3 weighted players, every congestion game with player-specific linear latency functions (without a constant term) has a pure Nash equilibrium but not necessarily an exact potential [8]. For the case of 2 links, there is a polynomial time algorithm to compute a pure Nash equilibrium [8]. A larger class of (incomplete information) unweighted congestion games with player-specific latency functions that have the Finite Improvement Property has been identified in [4]; the special case of our model where the player-specific constants are *additive* is contained in this larger class.

Contribution and Significance. We partition our results on congestion games with player-specific constants according to the structure of the strategy sets in the congestion game:

Games on parallel links:

- Every unweighted congestion game with player-specific constants has a generalized ordinal potential (Theorem 1). (Hence, every such game has the Finite Improvement Property and a pure Nash equilibrium.) The proof employs a potential function involving the group operation; the proof that this function is a generalized ordinal potential explicitly uses the assumption that the underlying group is a totally ordered abelian group. We remark that Theorem 1 does *not* need the assumption that the (resource-specific) delay functions are non-decreasing.

  Theorem 1 simultaneously broadens two corresponding state-of-the-art results for two very special cases: *(i)* each delay function is the identity function and the group operation is *multiplication* [7] and *(ii)* the group operation is addition [4]. We note that, in fact, the potential function we used is a generalization of the potential function used in [4] (for addition) to an arbitrary group operation. However, [4] applies to *all* unweighted congestion games.
- It is *not* possible to generalize Theorem 1 to weighted congestion games (with player-specific constants): there is such a game with 3 players on 3 parallel links that does not have the Finite Best-Improvement Property – hence, neither the Finite Improvement Property (Theorem 2). To prove this, we provide a simple counterexample for the case of player-specific additive constants.
- Note that Theorem 2 does not outlaw the possibility that every weighted congestion game with player-specific constants has a pure Nash equilibrium. Although we do not know the answer for the general case with an arbitrary number of players, we have settled the case with 3 players: every weighted congestion game with player-specific constants and 3 players has a pure Nash equilibrium (Corollary 3). The proof proceeds in two steps.

  First, we establish that there is a particular best-improvement cycle whose outlaw implies the existence of a pure Nash equilibrium (Theorem 3). We remark that an identical cycle had been earlier constructed by Milchtaich for the more general class of weighted congestion games with player-specific latency functions [13, Section 8].

  Second, we establish that this particular best-improvement cycle is indeed outlawed for the more specific class of dominance weighted congestion games (Theorem 4). Since a weighted congestion game with player-specific constants is a dominance weighted congestion game, the cycle is outlawed for weighted congestion games with player-specific constants as well; this implies the existence of a pure Nash equilibrium for them (Corollary 3). This implies, in particular, a separation of this specific class from the general class of congestion games with player-specific latency functions with respect to best-improvement cycles.

  We remark that Corollary 3 broadens the earlier result by Georgiou *et al.* [8, Lemma B.1] for congestion games with player-specific multiplicative constants and identity delay functions.

Network congestion games:
Recall that every unweighted congestion game with player-specific additive constants has a pure Nash equilibrium [4]. Nevertheless, we establish that it is $\mathcal{PLS}$-complete to compute one (Theorem 5) even for a symmetric network congestion game. The proof uses a simple reduction from the $\mathcal{PLS}$-complete problem of computing a pure Nash equilibrium for an unweighted asymmetric network congestion game [3].

Arbitrary (non-network) congestion games:
Note that Theorem 2 outlaws the possibility that every weighted congestion game with player-specific constants has the Finite Best-Improvement Property. Nevertheless, we establish that every weighted congestion game with player-specific constants has a weighted potential for the special case of linear delay functions and player-specific additive constants (Theorem 6). (Hence, every such game has the Finite Improvement Property and a pure Nash equilibrium).

The proof employs a potential function and establishes that it is a weighted potential. For the special case of weighted asymmetric network congestion games with affine latency functions (which are not player-specific), the potential function we used reduces to the potential function introduced in [6] for the corresponding case.

Theorems 1 and 3 suggest that the class of congestion games with player-specific constants provides a vehicle for reaching the limit of the existence of potential functions towards the direction of player-specific costs.

2 Framework and Preliminaries

Totally Ordered Abelian Groups. A *group* $(G, \odot)$ consists of a *ground set* G together with a binary operation $\odot : G \times G \rightarrow G$; $\odot$ is associative and allows for an *identity element* and *inverses*. The group $(G, \odot)$ is *abelian* if $\odot$ is commutative. We will consider *totally ordered abelian groups* with a *total order* on G [9] which satisfies *translation invariance*: for all triples $r, s, t \in G$, if $r \leq s$ then $r \odot t \leq s \odot t$. Examples of totally ordered abelian groups include (i) $(\mathbb{R}_{>0}, \cdot)$ under the usual number-ordering, and (ii) $(\mathbb{R}^2, +)$ under the *lexicographic ordering* on pairs of numbers. We will often focus on the case where G is $\mathbb{R}$ (the set of reals).

Congestion Games. For all integers $k \geq 1$, we denote $[k] = \{1, \ldots, k\}$. A *weighted congestion game with player-specific latency functions* [13] is a tuple $\Gamma = (n, E, (w_i)_{i\in[n]}, (S_i)_{i\in[n]}, (f_{ie})_{i\in[n],e\in E})$. Here, n is the number of *players* and E is a finite set of *resources*. For each player $i \in [n]$, $w_i > 0$ is the *weight* and $S_i \subseteq 2^E$ is the *strategy set* of player i. For each pair of player $i \in [n]$ and resource $e \in E$, $f_{ie} : \mathbb{R}_{>0} \rightarrow \mathbb{R}_{>0}$ is a non-decreasing player-specific *latency function*. In the *unweighted* case, $w_i = 1$ for all players $i \in [n]$.

In a (weighted) *network congestion game* (with player-specific latency functions), resources and strategies correspond to edges and paths in a directed network. In such games, each player has a *source* and a *destination* node, each of her strategies is a path from source to destination and all paths are possible. In a *symmetric* network congestion game, all players use the same pair of source and destination; else, the network congestion game is *asymmetric*. In the *parallel links* network, there are only two nodes; this gives rise to symmetric network congestion games.

Definition 1. *Fix a totally ordered abelian group $(G, \odot)$. A* ***weighted congestion game with player-specific constants*** *is a weighted congestion game Γ with player-specific latency functions such that (i) for each resource $e \in E$, there is a non-decreasing* delay function $g_e : \mathbb{R}_{>0} \rightarrow \mathbb{R}_{>0}$, *and (ii) for each pair of a player $i \in [n]$ and a resource $e \in E$, there is a player-specific constant $c_{ie} > 0$, so that for each player $i \in [n]$ and resource $e \in E$, $f_{ie} = c_{ie} \odot g_e$.*

In a *weighted congestion game with player-specific additive constants* (resp., *player-specific multiplicative constants*), G is $\mathbb{R}$ and $\odot$ is $+$ (resp., G is $\mathbb{R}_{>0}$ and $\odot$ is $\cdot$). The special case of weighted congestion games with player-specific constants where for all players $i \in [n]$ and resources $e \in E$, $c_{ie} = \epsilon$ (the *identity element* of G) yields the *weighted congestion games* generalizing the unweighted congestion games introduced by Rosenthal [15]. So, (weighted) congestion games with player-specific constants fall between the weighted generalization of congestion games [15] and (weighted) congestion games with player-specific latency functions [13].

We now prove that, in fact, congestion games with player-specific constants are contained within a more restricted class of congestion games with player-specific latency functions that we introduce. Fix a weighted congestion game Γ with player-specific latency functions. Consider a pair of (distinct) players $i, j \in [n]$ and a pair of (distinct) resources $e, e' \in E$. Say that i *dominates* j for the ordered pair $\langle e, e' \rangle$ if for every pair of positive numbers $x, y \in \mathbb{R}_{>0}$, $f_{ie}(x) > f_{ie'}(y)$ implies $f_{je}(x) > f_{je'}(y)$. Intuitively, i dominates j for $\langle e, e' \rangle$ if the decision of i to switch her strategy from e to e' always implies a corresponding decision for j; in other words, j always follows the decision of i (to switch or not) for the pair $\langle e, e' \rangle$.

Definition 2. *A weighted congestion game with player-specific latency functions is a* ***dominance (weighted) congestion game*** *if for all pairs of players $i, j \in [n]$, for all pairs of resources $e, e' \in E$, either i dominates j for $\langle e, e' \rangle$ or j dominates i for $\langle e, e' \rangle$.*

We prove:

Proposition 1. *A (weighted) congestion game with player-specific constants is a dominance (weighted) congestion game.*

Proof. Fix a pair of players $i, j \in [n]$ and a pair of resources $e, e' \in E$. We proceed by case analysis. Assume first that $c_{ie} \odot c_{je'} \geq c_{ie'} \odot c_{je}$. We will show that j dominates i for $\langle e, e' \rangle$. Fix a pair of numbers $x, y \in \mathbb{R}_{>0}$. Assume that $f_{je}(x) > f_{je'}(y)$ or $c_{je} \odot g_e(x) > c_{je'} \odot g_{e'}(y)$. By translation-invariance, it follows that $c_{ie} \odot c_{je} \odot g_e(x) > c_{ie} \odot c_{je'} \odot g_{e'}(y)$. The assumption that $c_{ie} \odot c_{je'} \geq c_{ie'} \odot c_{je}$ implies that $c_{ie} \odot c_{je'} \odot g_{e'}(y) \geq c_{ie'} \odot c_{je} \odot g_{e'}(y)$. It follows that $c_{ie} \odot g_e(x) > c_{ie'} \odot g_{e'}(y)$ or $f_{ie}(x) > f_{ie'}(y)$. Hence, j dominates i for $\langle e, e' \rangle$.

Assume now that $c_{ie'} \odot c_{je} > c_{ie} \odot c_{je'}$. We will show that i dominates j for $\langle e, e' \rangle$. Fix a pair of numbers $x, y \in \mathbb{R}_{>0}$. Assume that $f_{ie}(x) > f_{ie'}(y)$ or $c_{ie} \odot g_e(x) > c_{ie'} \odot g_{e'}(y)$. By translation-invariance, it follows that $c_{je} \odot c_{ie} \odot g_e(x) > c_{je} \odot c_{ie'} \odot g_{e'}(y)$. The assumption that $c_{ie'} \odot c_{je} > c_{ie} \odot c_{je'}$ implies that $c_{je} \odot c_{ie'} \odot g_{e'}(y) > c_{je'} \odot c_{ie} \odot g_{e'}(y)$. It follows that $c_{je} \odot g_e(x) > c_{je'} \odot g_{e'}(y)$ or $f_{je}(x) > f_{je'}(y)$. Hence, i dominates j for $\langle e, e' \rangle$. □

Profiles and Individual Cost. A *strategy* for player $i \in [n]$ is some specific $s_i \in S_i$. A *profile* is a tuple $\mathsf{s} = (s_1, \ldots, s_n) \in S_1 \times \ldots \times S_n$. For the profile s, the *load* $\delta_e(\mathsf{s})$ on resource $e \in E$ is given by $\delta_e(\mathsf{s}) = \sum_{i \in [n] \mid s_i \ni e} w_i$. For the profile s, the *Individual Cost* of player $i \in [n]$ is given by $\mathsf{IC}_i(\mathsf{s}) = \sum_{e \in s_i} f_{ie}(\delta_e(\mathsf{s})) = \sum_{e \in s_i} c_{ie} \odot \cdot g_e(\delta_e(\mathsf{s}))$.

Pure Nash Equilibria. Fix a profile s. A player $i \in [n]$ is *satisfied* if she cannot decrease her Individual Cost by unilaterally changing to a different strategy; else, player i is *unsatisfied*. So, an unsatisfied player i can take an *improvement step* to decrease her Individual Cost; if player i is satisfied after the improvement step, the improvement step is called a *best-improvement step*. An *improvement cycle* (resp., *best-improvement cycle*) is a cyclic sequence of improvement steps (resp., best-improvement steps). A game has the *Finite Improvement Property* (resp., *Finite Best-Improvement Property*) if all sequences of improvement steps (resp., best-improvement steps) are finite; clearly, the Finite Improvement Property (resp., the Finite Best-Improvement Property) outlaws improvement cycles (resp., best-improvement cycles). Clearly, the Finite Improvement Property implies the Finite Best-Improvement Property. A profile is a *(pure) Nash equilibrium* if all players are satisfied. Clearly, the Finite Improvement Property implies the existence of a pure Nash equilibrium (as also does the Finite Best-Improvement Property), but not vice versa [14].

A *generalized ordinal potential* for the game Γ [14] is a function $\Phi : S_1 \times \ldots \times S_n \rightarrow \mathbb{R}$ that decreases when a player takes an improvement step. Say that a function $\Phi : S_1 \times \ldots \times S_n \rightarrow \mathbb{R}$ is a *weighted potential* for the game Γ [14] if there is a *weight vector* $\mathsf{b} = (b_i)_{i \in [n]}$ such that for every player $k \in [n]$, for every profile s, and for every strategy $t_k \in S_k$ that transforms s to t, it holds that $\mathsf{IC}_k(\mathsf{s}) - \mathsf{IC}_k(\mathsf{t}) = b_k \cdot (\Phi(\mathsf{s}) - \Phi(\mathsf{t}))$. If this even holds for the vector b with $b_i = 1$ for all $i \in [n]$, the function Φ is an *exact potential* [14]. A game has a generalized ordinal potential if and only if it has the Finite Improvement Property (and hence the Finite Best-Improvement Property and a pure Nash equilibrium) [14].

$\mathcal{PLS}$(-complete) Problems. $\mathcal{PLS}$ [11] includes optimization problems where the goal is to find a *local optimum* for a given instance; this is a feasible solution with no feasible solution of better objective value in its well-determined neighborhood. A problem Π in $\mathcal{PLS}$ has an associated set of instances $\mathcal{I}_\Pi$. There is, for every instance $I \in \mathcal{I}_\Pi$, a set of feasible solutions $\mathcal{F}(I)$. Furthermore, there are three polynomial time algorithms A, B and C. A computes for every instance I a feasible solution $S \in \mathcal{F}(I)$; B computes for a feasible solution $S \in \mathcal{F}(I)$, the objectice value; C determines, for a feasible solution $S \in \mathcal{F}(I)$, whether S is locally optimal and, if not, it outputs a feasible solution in the neighborhood of S with better objective value.

A $\mathcal{PLS}$-problem Π_1 is *$\mathcal{PLS}$-reducible* [11] to a $\mathcal{PLS}$-problem Π_2 if there are two polynomial time computable functions F_1 and F_2 such that F_1 maps instances $I \in \mathcal{I}_{\Pi_1}$ to instances $F_1(I) \in \mathcal{I}_{\Pi_2}$ and F_2 maps every local optimum of the instance $F_1(I)$ to a local optimum of I. A $\mathcal{PLS}$-problem Π is *$\mathcal{PLS}$-complete* [11] if every problem in $\mathcal{PLS}$ is $\mathcal{PLS}$-reducible to Π.

3 Congestion Games on Parallel Links

We now introduce a function $\Phi : S_1 \times \ldots \times S_n \to \mathbb{R}$ with

$$\Phi(\mathsf{s}) = \bigodot_{e \in E} \bigodot_{i=1}^{\delta_e(\mathsf{s})} g_e(i) \odot \bigodot_{i=1}^{n} c_{is_i} .$$

for any profile s. We prove that this function is a generalized ordinal potential:

Theorem 1. *Every unweighted congestion game with player-specific constants on parallel links has a generalized ordinal potential.*

Proof. Fix a profile s. Consider an improvement step of player $k \in [n]$ to strategy t_k, which transforms s to t. Clearly, $\mathsf{IC}_k(\mathsf{s}) > \mathsf{IC}_k(\mathsf{t})$ or

$$g_{s_k}(\delta_{s_k}(\mathsf{s})) \odot c_{ks_k} > g_{t_k}(\delta_{t_k}(\mathsf{t})) \odot c_{kt_k} .$$

Note also that $\delta_{s_k}(\mathsf{t}) = \delta_{s_k}(\mathsf{s}) - 1$ and $\delta_{t_k}(\mathsf{t}) = \delta_{t_k}(\mathsf{s}) + 1$, while $\delta_e(\mathsf{t}) = \delta_e(\mathsf{s})$ for all $e \in E \setminus \{s_k, t_k\}$. Hence,

$$\begin{aligned}
&\Phi(\mathsf{s}) \\
&= \bigodot_{e \in E \setminus \{s_k,t_k\}} \bigodot_{i=1}^{\delta_e(\mathsf{s})} g_e(i) \odot \bigodot_{i \in [n] \setminus \{k\}} c_{is_i} \odot \bigodot_{i=1}^{\delta_{s_k}(\mathsf{s})} g_{s_k}(i) \odot \bigodot_{i=1}^{\delta_{t_k}(\mathsf{s})} g_{t_k}(i) \odot c_{ks_k} \\
&= \bigodot_{\substack{e \in E \setminus \\ \{s_k,t_k\}}} \bigodot_{i=1}^{\delta_e(\mathsf{s})} g_e(i) \odot \bigodot_{\substack{i \in [n] \\ \setminus \{k\}}} c_{is_i} \odot \bigodot_{i=1}^{\delta_{s_k}(\mathsf{s})-1} g_{s_k}(i) \odot \bigodot_{i=1}^{\delta_{t_k}(\mathsf{s})} g_{t_k}(i) \odot g_{s_k}(\delta_{s_k}(\mathsf{s})) \odot c_{ks_k} \\
&> \bigodot_{\substack{e \in E \setminus \\ \{s_k,t_k\}}} \bigodot_{i=1}^{\delta_e(\mathsf{s})} g_e(i) \odot \bigodot_{\substack{i \in [n] \\ \setminus \{k\}}} c_{is_i} \odot \bigodot_{i=1}^{\delta_{s_k}(\mathsf{s})-1} g_{s_k}(i) \odot \bigodot_{i=1}^{\delta_{t_k}(\mathsf{s})} g_{t_k}(i) \odot g_{t_k}(\delta_{t_k}(\mathsf{t})) \odot c_{kt_k} \\
&= \bigodot_{e \in E \setminus \{s_k,t_k\}} \bigodot_{i=1}^{\delta_e(\mathsf{t})} g_e(i) \odot \bigodot_{i \in [n] \setminus \{k\}} c_{is_i} \odot \bigodot_{i=1}^{\delta_{s_k}(\mathsf{t})} g_{s_k}(i) \odot \bigodot_{i=1}^{\delta_{t_k}(\mathsf{t})} g_{t_k}(i) \odot c_{kt_k} \\
&= \Phi(\mathsf{t}),
\end{aligned}$$

so that Φ is a generalized ordinal potential. □

Theorem 1 immediately implies:

Corollary 1. *Every unweighted congestion game with player-specific constants on parallel links has the Finite Improvement Property and a pure Nash equilibrium.*

We continue to prove:

Theorem 2. *There is a weighted congestion game with additive player-specific constants and 3 players on 3 parallel links that does* not *have the Finite Best-Improvement Property.*

Proof. By construction. The weights of the 3 players are $w_1 = 2$, $w_2 = 1$, and $w_3 = 1$. The player-specific constants and resource-specific delay functions are as follows:

c_{ie}	Link 1	Link 2	Link 3
Player 1	0	∞	5
Player 2	0	0	∞
Player 3	∞	0	2

	Link 1	Link 2	Link 3
$g_e(1)$	1	2	1
$g_e(2)$	8	13	2
$g_e(3)$	14	∞	10

Notice that the profiles $\langle 1,2,3\rangle$ and $\langle 3,1,2\rangle$ are both Nash equilibria. Consider now the cycle $\langle 1,1,3\rangle \to \langle 1,1,2\rangle \to \langle 1,2,2\rangle \to \langle 3,2,2\rangle \to \langle 3,2,3\rangle \to \langle 3,1,3\rangle \to \langle 1,1,3\rangle$. The Individual Cost of the deviating player decreases in each of these steps:

	IC_1	IC_2	IC_3
$\langle 1,1,3\rangle$	14		3
$\langle 1,1,2\rangle$		14	2

	IC_1	IC_2	IC_3
$\langle 1,2,2\rangle$	8	13	
$\langle 3,2,2\rangle$	7		13

	IC_1	IC_2	IC_3
$\langle 3,2,3\rangle$		2	12
$\langle 3,1,3\rangle$	15	1	

So, this is an improvement cycle. Furthermore, note that each step in this cycle is a best-improvement step, so this is actually a best-improvement cycle. The claim follows. □

We continue to consider the special case of 3 players but for the general case of weighted congestion games with player-specific constants. We prove:

Theorem 3. *Let Γ be a weighted congestion game with player-specific latency functions and 3 players on parallel links. If Γ does* not *have a best-improvement cycle $\langle l,j,j\rangle \to \langle l,l,j\rangle \to \langle k,l,j\rangle \to \langle k,l,l\rangle \to \langle k,j,l\rangle \to \langle l,j,l\rangle \to \langle l,j,j\rangle$ (where $l \neq j$, $j \neq k$, $l \neq k$ are any three links and $w_1 \geq w_2 \geq w_3$), then Γ has a pure Nash equilibrium.*

We now continue to prove:

Theorem 4. *Every dominance weighted congestion game with 3 players on parallel links does not have an improvement cycle of the form $\langle l,j,j\rangle \to \langle l,l,j\rangle \to \langle k,l,j\rangle \to \langle k,l,l\rangle \to \langle k,j,l\rangle \to \langle l,j,l\rangle \to \langle l,j,j\rangle$ where $l \neq j$, $j \neq k$, $l \neq k$ are any three links and $w_1 \geq w_2 \geq w_3$.*

Proof. Assume, by way of contradiction, that there is a dominance congestion game with such a cycle. Since all steps in the cycle are improvement steps, one gets for player 2 that $f_{2j}(w_2 + w_3) > f_{2l}(w_1 + w_2)$ and $f_{2l}(w_2 + w_3) > f_{2j}(w_2)$. In the same way, one gets for player 3 that $f_{3j}(w_3) > f_{3l}(w_2 + w_3)$ and $f_{3l}(w_1 + w_3) > f_{3j}(w_2 + w_3)$. We proceed by case analysis on whether 2 dominates 3 or 3 dominates 2 for $\langle j,l\rangle$.

Assume first that 2 dominates 3 for $\langle j,l\rangle$. Then, the first inequality for player 2 implies that $f_{3j}(w_2+w_3) > f_{3l}(w_1+w_2) \geq f_{3l}(w_1+w_3)$ (since f_{3l} is non-decreasing and $w_2 \geq w_3$), a contradiction to the second inequality for player 3. Assume now that 3 dominates 2 for $\langle j,l\rangle$. Then, the first inequality for player 3 implies that $f_{2l}(w_2+w_3) < f_{2j}(w_3) \leq f_{2j}(w_2)$ (since f_{2j} is non-decreasing and $w_2 \geq w_3$), a contradiction to the second inequality for player 2. □

Since dominance (weighted) congestion games are a subclass of (weighted) congestion games with player-specific latency functions, Theorems 3 and 4 immediately imply:

Corollary 2. *Every dominance weighted congestion game with 3 players on parallel links has a pure Nash equilibrium.*

By Proposition 1, Corollary 2 immediately implies:

Corollary 3. *Every weighted congestion game with player-specific constants and 3 players on parallel links has a pure Nash equilibrium.*

4 Network Congestion Games

Theorem 5. *It is $\mathcal{PLS}$-complete to compute a pure Nash equilibrium in an unweighted symmetric network congestion game with player-specific additive constants.*

Proof. Clearly, the problem of computing a pure Nash equilibrium in an unweighted symmetric congestion game with player-specific additive constants is a $\mathcal{PLS}$-problem. (The set of feasible solutions is the set of all profiles and the neighborhood of a profile is the set of profiles that differ in the strategy of exactly one player; the objective function is the generalized ordinal potential since a local optimum of this functions is a Nash equilibrium [14].) To prove $\mathcal{PLS}$-hardness, we use a reduction from the $\mathcal{PLS}$-complete problem of computing a pure Nash equilibrium for an unweighted, asymmetric network congestion game [3]. For the reduction, we construct the two functions F_1 and F_2:

$\underline{F_1}$: Given an unweighted, asymmetric network congestion game Γ on a network G, where $(a_i, b_i)_{i\in[n]}$ are the source and destination nodes of the n players and $(f_e)_{e\in E}$ are the latency functions, F_1 constructs a symmetric network congestion game Γ' with n players on a graph G', as follows:

- G' includes G, where for each edge e of G, $g'_e := f_e$ and $c'_{ie} = 0$ for each $i \in [n]$.
- G' contains a new common source a' and a new common destination b'; for each player $i \in [n]$, we add an edge (a', a_i) with $g'_{(a',a_i)}(x) := 0$, $c'_{i(a',a_i)} := 0$, and $c'_{k(a',a_i)} := \infty$ for all $k \neq i$; in addition we add for each player $i \in [n]$ an edge (b_i, b') with $g'_{(b_i,b')}(x) := 0$, $c'_{i(b_i,b')} := 0$, and $c'_{k(b_i,b')} := \infty$ for all $k \neq i$.

$\underline{F_2}$: Consider now a pure Nash equilibrium t for Γ'. The function F_2 maps t to a profile s for Γ (which, we shall prove, is a Nash equilibrium for Γ) as follows:

- Note first that for each player $i \in [n]$, t_i (is a path that) includes both edges (a', a_i) and (b_i, b') (since otherwise $\mathsf{IC}_i(\mathsf{t}) = \infty$). Construct s_i from t_i by eliminating the edges (a', a_i) and (b_i, b').

It remains to prove that $\mathsf{s} = F_2(\mathsf{t})$ is a Nash equilibrium for Γ. By way of contradiction, assume otherwise. Then, there is a player k that can decrease her Individual Cost in Γ by changing her path s_k to s'_k. But then player k can decrease her Individual Cost in Γ' by changing her path $t_k = (a', a_k), s_k, (b_k, b')$ to $t'_k = (a', a_k), s'_k, (b_k, b')$. So, t is not a Nash equilibrium for Γ'. A contradiction. □

We remark that Theorem 5 holds also for unweighted symmetric network congestion games with player-specific additive constants and *undirected* edges since the problem of computing a pure Nash equilibrium for an unweighted, asymmetric network congestion game with undirected edges is also $\mathcal{PLS}$-complete [1].

5 Arbitrary Congestion Games

We now restrict attention to weighted congestion games with player-specific additive constants c_{ie} and linear delay functions $f_e(x) = a_e \cdot x$. This gives rise to *weighted congestion games with player-specific affine latency functions* $f_{ie}(x) = a_e \cdot x + c_{ie}$, where $i \in [n]$ and $e \in E$. For this case, we introduce a function $\Phi : S_1 \times \ldots \times S_n \rightarrow \mathbb{R}$ with $\Phi(\mathsf{s}) = \sum_{i=1}^{n} \sum_{e \in s_i} w_i \cdot (2 \cdot c_{ie} + a_e \cdot (\delta_e(s) + w_i))$, for any profile s. For any pair of player $i \in [n]$ and resource $e \in E$, define $\phi(\mathsf{s}, i, e) = w_i \cdot (2 \cdot c_{ie} + a_e \cdot (\delta_e(\mathsf{s}) + w_i))$, so that $\Phi(\mathsf{s}) = \sum_{i=1}^{n} \sum_{e \in s_i} \phi(\mathsf{s}, i, e)$. We now prove that this function is a weighted potential:

Theorem 6. *Every weighted congestion game with player-specific affine latency functions has a weighted potential.*

Proof. Fix a profile s. Assume that player $k \in [n]$ unilaterally changes to the strategy t_k, which transforms s to t. Clearly,

$$\begin{aligned}
&\Phi(\mathsf{s}) - \Phi(\mathsf{t}) \\
&= \sum_{i \in [n]} \sum_{e \in s_i} \phi(\mathsf{s}, i, e) - \sum_{i \in [n]} \sum_{e \in t_i} \phi(\mathsf{t}, i, e) \\
&= \sum_{e \in s_k} \phi(\mathsf{s}, k, e) - \sum_{e \in t_k} \phi(\mathsf{t}, k, e) + \sum_{i \in [n] \setminus \{k\}} \left(\sum_{e \in s_i} \phi(\mathsf{s}, i, e) - \sum_{e \in t_i} \phi(\mathsf{t}, i, e) \right)
\end{aligned}$$

We treat separately the first and the second part of this expression. On one hand,

$$\begin{aligned}
\sum_{e \in s_k} \phi(\mathsf{s}, k, e) - \sum_{e \in t_k} \phi(\mathsf{t}, k, e) &= \sum_{e \in s_k \setminus t_k} \phi(\mathsf{s}, k, e) - \sum_{e \in t_k \setminus s_k} \phi(\mathsf{t}, k, e) \\
= \sum_{e \in s_k \setminus t_k} w_k (2 \cdot c_{ke} + a_e \cdot (\delta_e(\mathsf{s}) + w_k)) &- \sum_{e \in t_k \setminus s_k} w_k (2 \cdot c_{ke} + a_e \cdot (\delta_e(\mathsf{t}) + w_k)).
\end{aligned}$$

On the other hand,

$$\begin{aligned}
&\sum_{i \in [n] \setminus \{k\}} \left(\sum_{e \in s_i} \phi(\mathsf{s}, i, e) - \sum_{e \in t_i = s_i} \phi(\mathsf{t}, i, e) \right) = \sum_{i \in [n] \setminus \{k\}} \sum_{e \in s_i} (\phi(\mathsf{s}, i, e) - \phi(\mathsf{t}, i, e)) \\
&= \sum_{\substack{i \in \\ [n] \setminus \{k\}}} \left(\sum_{e \in s_i \cap (s_k \setminus t_k)} (\phi(\mathsf{s}, i, e) - \phi(\mathsf{t}, i, e)) + \sum_{e \in s_i \cap (t_k \setminus s_k)} (\phi(\mathsf{s}, i, e) - \phi(\mathsf{t}, i, e)) \right) \\
&= \sum_{e \in s_k \setminus t_k} \sum_{\substack{i \in [n] \setminus \{k\} \\ | e \in s_i}} (\phi(\mathsf{s}, i, e) - \phi(\mathsf{t}, i, e)) + \sum_{e \in t_k \setminus s_k} \sum_{\substack{i \in [n] \setminus \{k\} \\ | e \in s_i}} (\phi(\mathsf{s}, i, e) - \phi(\mathsf{t}, i, e)) \\
&= \sum_{\substack{e \in \\ s_k \setminus t_k}} \sum_{\substack{i \in [n] \setminus \{k\} \\ | e \in s_i}} (w_i \cdot a_e \cdot (\delta_e(\mathsf{s}) - \delta_e(\mathsf{t}))) + \sum_{\substack{e \in \\ t_k \setminus s_k}} \sum_{\substack{i \in [n] \setminus \{k\} \\ | e \in s_i}} (w_i \cdot a_e \cdot (\delta_e(\mathsf{s}) - \delta_e(\mathsf{t}))) \\
&= w_k \cdot \sum_{e \in s_k \setminus t_k} a_e \cdot (\delta_e(\mathsf{s}) - w_k) - w_k \cdot \sum_{e \in t_k \setminus s_k} a_e \cdot (\delta_e(\mathsf{t}) - w_k) .
\end{aligned}$$

Putting these together yields that Φ is a weighted potential with weight vector b having $b_i = \frac{1}{2w_i}$, $i \in [n]$. □

Theorem 6 immediately implies:

Corollary 4. *Every weighted congestion game with player-specific affine latency functions has the Finite Improvement Property and a pure Nash equilibrium.*

References

1. Ackermann, H., Röglin, H., Vöcking, B.: On the Impact of Combinatorial Structure on Congestion Games. In: Proceedings of the 47th Annual IEEE Symposium on Foundations of Computer Science, pp. 613–622. IEEE Computer Society Press, Los Alamitos (2006)
2. Dunkel, J., Schulz, A.: On the Complexity of Pure-Strategy Nash Equilibria in Congestion and Local-Effect Games. In: Spirakis, P.G., Mavronicolas, M., Kontogiannis, S.C. (eds.) WINE 2006. LNCS, vol. 4286, pp. 50–61. Springer, Heidelberg (2006)
3. Fabrikant, A., Papadimitriou, C.H., Talwar, K.: The Complexity of Pure Nash Equilibria. In: Proceedings of the 36th Annual ACM Symposium on Theory of Computing, pp. 604–612. ACM Press, New York (2004)
4. Facchini, G., van Megan, F., Borm, P., Tijs, S.: Congestion Models and Weighted Bayesian Potential Games. Theory and Decision 42, 193–206 (1997)
5. Fotakis, D., Kontogiannis, S., Koutsoupias, E., Mavronicolas, M., Spirakis, P.: The Structure and Complexity of Nash Equilibria for a Selfish Routing Game. In: Widmayer, P., Triguero, F., Morales, R., Hennessy, M., Eidenbenz, S., Conejo, R. (eds.) ICALP 2002. LNCS, vol. 2380, pp. 123–134. Springer, Heidelberg (2002)
6. Fotakis, D., Kontogiannis, S., Spirakis, P.: Selfish Unsplittable Flows. Theoretical Computer Science 348(2–3), 226–239 (2005)
7. Gairing, M., Monien, B., Tiemann, K.: Routing (Un-)Splittable Flow in Games with Player-Specific Linear Latency Functions. In: Bugliesi, M., Preneel, B., Sassone, V., Wegener, I. (eds.) ICALP 2006. LNCS, vol. 4051, pp. 501–512. Springer, Heidelberg (2006)
8. Georgiou, C., Pavlides, T., Philippou, A.: Network Uncertainty in Selfish Routing. In: CD-ROM Proceedings of the 20th IEEE International Parallel & Distributed Processing Symposium, p. 105. IEEE Computer Society Press, Los Alamitos (2006)
9. Goodearl, K.R.: Partially Ordered Abelian Groups with Interpolation. American Mathematical Society (1986)
10. Graham, R.L.: Bounds on Multiprocessing Timing Anomalies. SIAM Journal of Applied Mathematics 17(2), 416–429 (1969)
11. Johnson, D.S., Papadimitriou, C.H., Yannakakis, M.: How Easy is Local Search? Journal of Computer and System Sciences 37(1), 79–100 (1988)
12. Libman, L., Orda, A.: Atomic Resource Sharing in Noncooperative Networks. Telecommunication Systems 17(4), 385–409 (2001)
13. Milchtaich, I.: Congestion Games with Player-Specific Payoff Functions. Games and Economic Behavior 13(1), 111–124 (1996)
14. Monderer, D., Shapley, L.S.: Potential Games. Games and Economic Behavior 14(1), 124–143 (1996)
15. Rosenthal, R.W.: A Class of Games Possessing Pure-Strategy Nash Equilibria. International Journal of Game Theory 2, 65–67 (1973)

Finding Patterns in Given Intervals

Maxime Crochemore[1,2], Costas S. Iliopoulos[1,*], and M. Sohel Rahman[1,**,***]

[1] Algorithm Design Group
Department of Computer Science
King's College London
Strand, London WC2R 2LS, England
Maxime.Crochemore@kcl.ac.uk,
{csi,sohel}@dcs.kcl.ac.uk
http://www.dcs.kcl.ac.uk/adg
[2] Institut Gaspard-Monge
University of Marne-la-Vallée, France

Abstract. In this paper, we study the pattern matching problem in given intervals. Depending on whether the intervals are given a priori for pre-processing, or during the query along with the pattern or, even in both cases, we develop solutions for different variants of this problem. In particular, we present efficient indexing schemes for each of the above variants of the problem.

1 Introduction

The classical pattern matching problem is to find all the occurrences of a given pattern $\mathcal{P} = \mathcal{P}[1..m]$ of length m in a text $\mathcal{T} = \mathcal{T}[1..n]$ of length n, both being sequences of characters drawn from a finite character set Σ. This problem is interesting as a fundamental computer science problem and is a basic need of many applications, such as text retrieval, music retrieval, computational biology, data mining, network security, to name a few. Several of these applications require, however, more sophisticated forms of searching. As a result, most recent works in pattern matching has considered *'inexact matching'*. Many types of differences have been defined and studied in the literature, namely, errors (Hamming distance, LCS [10,17], edit distance [10,20]), wild cards or don't cares [10,11,14,23], rotations [3,7], scaling [4,5], permutations [9] among others.

Contemporary research on pattern matching has taken many other different and interesting directions ranging from position restricted pattern matching [21] to pattern matching with address error [2] and property matching [6]. In this paper, we are interested in pattern matching in given intervals and focus on building an index data structure to handle this problem efficiently. This particular variant of the classic pattern matching problem is motivated by practical

* Supported by EPSRC and Royal Society grants.
** Supported by the Commonwealth Scholarship Commission in the UK under the Commonwealth Scholarship and Fellowship Plan (CSFP).
*** On Leave from Department of CSE, BUET, Dhaka-1000, Bangladesh.

L. Kučera and A. Kučera (Eds.): MFCS 2007, LNCS 4708, pp. 645–656, 2007.

applications depending on different settings. For example, in many text search situations one may want to search only a part of the text collection, e.g. restricting the search to a subset of dynamically chosen documents in a document database, restricting the search to only parts of a long DNA sequence, and so on. In these cases we need to find a pattern in a text interval where the intervals are given with the query pattern. On the other hand, in a different setting, the interval or a set thereof may be supplied with the text for preprocessing. For example, in molecular biology, it has long been a practice to consider special genome areas by their structure. Examples are repetitive genomic structures [18] such as tandem repeats, LINEs (Long Interspersed Nuclear Sequences) and SINEs (Short Interspersed Nuclear Sequences) [19]. In this setting, the task may be to find occurrences of a given pattern in a genome, provided it appears in a SINE, or LINE. Finally a combination of these two settings is also of particular interest: find occurrences of a given pattern in a particular part of a genome, provided it appears in a SINE, or LINE.

Note that, if we consider the 'normal' (non-indexing) pattern matching scenario, the pattern matching in given intervals become straightforward to solve: we solve the classic pattern matching problem and then output only those that belong to the given intervals. However, the indexing version of the problem seems to be much more complex. Depending on whether the intervals are given a priori for pre-processing (Problem PMGI), or during the query along with the pattern (Problem PMQI) or, even in both the cases (Problem PMI), we develop solutions for different variants of this problem. A slightly different variant of Problem PMGI was studied in [6], whereas Problem PMQI was introduced and handled in [21] (See Section 2 for details).

The contribution of this paper is as follows. We first handle the more general problem PMI (Section 3) and present an efficient data structure requiring $O(n \log^3 n)$ time and $O(n \log^2 n)$ space with a query time of $O(m+\log\log n+K)$ per query, where K is the size of the output. We then solve Problem PMGI (Section 4) optimally ($O(m + K)$ query time on a data structure with $O(n)$ time and $O(n \log n)$-bit space complexity). Finally, we improve the query time of [21] for Problem PMQI (Section 5) to optimal i.e. $O(m + K)$ per query. The corresponding data structure, however, requires $O(n^2)$ time due to a costly preprocessing of an intermediate problem, which remains as the bottleneck in the overall running time.

The rest of the paper is organized as follows. In Section 2, we present the preliminary concepts. The contributions of this paper are presented in Section 3 to 5. We conclude briefly in Section 6.

2 Preliminaries

A *text*, also called a *string*, is a sequence of zero or more symbols from an alphabet Σ. A text $\mathcal{T}$ of length n is denoted by $\mathcal{T}[1..n] = \mathcal{T}_1\mathcal{T}_2\ldots\mathcal{T}_n$, where $\mathcal{T}_i \in \Sigma$ for $1 \leq i \leq n$. The *length* of $\mathcal{T}$ is denoted by $|\mathcal{T}| = n$. A string w is a *factor* or *substring* of $\mathcal{T}$ if $\mathcal{T} = uwv$ for $u, v \in \Sigma^*$; in this case, the string w occurs at position $|u|+1$ in $\mathcal{T}$. The factor w is denoted by $\mathcal{T}[|u|+1..|u|+|w|]$. A *prefix* (*suffix*) of $\mathcal{T}$ is a factor $\mathcal{T}[x..y]$ such that $x=1$ ($y=n$), $1 \le y \le n$ ($1 \le x \le n$).

In traditional pattern matching problem, we want to find the occurrences of a given pattern $\mathcal{P}[1..m]$ in a text $\mathcal{T}[1..n]$. The pattern $\mathcal{P}$ is said to occur at position $i \in [1..n]$ of $\mathcal{T}$ if and only if $\mathcal{P} = \mathcal{T}[i..i+m-1]$. We use $Occ^{\mathcal{P}}_{\mathcal{T}}$ to denote the set of occurrences of $\mathcal{P}$ in $\mathcal{T}$.

The problems we handle in this paper can be defined formally as follows.

Problem "PMQI" (Pattern Matching in a Query Interval). *Suppose we are given a text $\mathcal{T}$ of length n. Preprocess $\mathcal{T}$ to answer following form of queries.*

Query: *Given a pattern $\mathcal{P}$ and a query interval $[\ell..r]$, with $1 \le \ell \le r \le n$, construct the set*

$$Occ^{\mathcal{P}}_{\mathcal{T}[\ell..r]} = \{i \mid i \in Occ^{\mathcal{P}}_{\mathcal{T}} \text{ and } i \in [\ell..r]\}.$$

Problem "PMGI" (Pattern Matching in Given Intervals). *Suppose we are given a text $\mathcal{T}$ of length n and a set of disjoint intervals $\pi = \{[s_1..f_1], [s_2..f_2], \ldots, [s_{|\pi|}..f_{|\pi|}]\}$ such that $s_i, f_i \in [1..n]$ and $s_i \le f_i$, for all $1 \le i \le |\pi|$. Preprocess $\mathcal{T}$ to answer following form of queries.*

Query: *Given a pattern $\mathcal{P}$ construct the set*

$$Occ^{\mathcal{P}}_{\mathcal{T},\pi} = \{i \mid i \in Occ^{\mathcal{P}}_{\mathcal{T}} \text{ and } i \in \varpi \text{ for some } \varpi \in \pi\}.$$

Problem "PMI" (Generalized Pattern Matching with Intervals). *Suppose we are given a text $\mathcal{T}$ of length n and a set of intervals $\pi = \{[s_1..f_1], [s_2..f_2], \ldots, [s_{|\pi|}..f_{|\pi|}]\}$ such that $s_i, f_i \in [1..n]$ and $s_i \le f_i$, for all $1 \le i \le |\pi|$. Preprocess $\mathcal{T}$ to answer following form of queries.*

Query: *Given a pattern $\mathcal{P}$ and a query interval $[\ell..r]$ such that $\ell, r \in [1..n]$ and $\ell \le r$, construct the set*

$$Occ^{\mathcal{P}}_{\mathcal{T}[\ell..r],\pi} = \{i \mid i \in Occ^{\mathcal{P}}_{\mathcal{T}} \text{ and } i \in [\ell, r] \bigcap \varpi \text{ for some } \varpi \in \pi\}.$$

Problem PMQI was studied extensively in [21]. The authors in [21] presented a number of algorithms depending on different trade-offs between the time and space complexities. The best query time they achieved was $O(m + \log\log n + |Occ^{\mathcal{P}}_{\mathcal{T}[\ell..r]}|)$ against a data structure exhibiting $O(n \log^{1+\epsilon} n)$ space and time complexity, where $0 \le \epsilon \le 1$. A slightly different version of Problem PMGI was studied in [6][1]. In particular, the difference lies in the fact that the problem handled in [6], looks for the occurrences of the given pattern completely confined in the given set of intervals, π, whereas in Problem PMGI, only the occurreces that starts in π are of interest. Problem PMI, as is evident from the definition, is the combination of Problem PMQI and PMGI and hence is a more general problem in this regard.

[1] In [6] a data structure requiring $O(n \log \Sigma + n \log\log n)$ time was presented to support $O(m+K)$ time query, where K is the output size.

In traditional indexing problem one of the basic data structures used is the suffix tree data structure. In our indexing problem, we make use of this suffix tree data structure. A complete description of a suffix tree is beyond the scope of this paper, and can be found in [22,25] or in any textbook on stringology (e.g. [12,16]). However, for the sake of completeness, we define the suffix tree data structure as follows. Given a string $\mathcal{T}$ of length n over an alphabet Σ, the suffix tree $ST_{\mathcal{T}}$ of $\mathcal{T}$ is the compacted trie of all suffixes of $\mathcal{T}\$$, where $\$ \notin \Sigma$. Each leaf in $ST_{\mathcal{T}}$ represents a suffix $\mathcal{T}[i..n]$ of $\mathcal{T}$ and is labeled with the index i. We refer to the list (in left-to-right order) of indices of the leaves of the subtree rooted at node v as the leaf-list of v; it is denoted by $LL(v)$. Each edge in $ST_{\mathcal{T}}$ is labeled with a nonempty substring of $\mathcal{T}$ such that the path from the root to the leaf labeled with index i spells the suffix $\mathcal{T}[i..n]$. For any node v, we let ℓ_v denote the string obtained by concatenating the substrings labeling the edges on the path from the root to v in the order they appear. Several algorithms exist that can construct the suffix tree $ST_{\mathcal{T}}$ in $O(n \log \Sigma)$ time[2] [22,25,13]. The space requirement of suffix tree is $O(n \log n)$ bits. Given the suffix tree $ST_{\mathcal{T}}$ of a text $\mathcal{T}$ we define the "locus" $\mu^{\mathcal{P}}$ of a pattern $\mathcal{P}$ as the node in $ST_{\mathcal{T}}$ such that $\ell_{\mu^{\mathcal{P}}}$ has the prefix $\mathcal{P}$ and $|\ell_{\mu^{\mathcal{P}}}|$ is the smallest of all such nodes. Note that the locus of $\mathcal{P}$ does not exist, if $\mathcal{P}$ is not a substring of $\mathcal{T}$. Therefore, given $\mathcal{P}$, finding $\mu^{\mathcal{P}}$ suffices to determine whether $\mathcal{P}$ occurs in $\mathcal{T}$. Given a suffix tree of a text $\mathcal{T}$, a pattern $\mathcal{P}$, one can find its locus and hence the fact whether $\mathcal{T}$ has an occurrence of $\mathcal{P}$ in optimal $O(|\mathcal{P}|)$ time.In addition to that, all such occurrences can be reported in constant time per occurrence.

3 Problem PMI

In this section, we handle Problem PMI. Since this is a more general problem than both PMQI and PMGI, any solution to PMI would also be a solution to both PMQI and PMGI. Our basic idea is to build an index data structure that would solve the problem in two steps. First, it will (implicitly) give us the set $Occ^{\mathcal{P}}_{\mathcal{T}}$. Then, the index would 'select' some of the occurrences to provide us with our desired set $Occ^{\mathcal{P}}_{\mathcal{T}[\ell..r],\pi}$.

We describe now the idea we employ. We first construct a suffix tree $ST_{\mathcal{T}}$. According to the definition of suffix tree, each leaf in $ST_{\mathcal{T}}$ is labeled by the starting location of its suffix. We do some preprocessing on $ST_{\mathcal{T}}$ as follows. We maintain a linked list of all leaves in a left-to-right order. In other words, we realize the list $LL(\mathcal{R})$ in the form of a linked list, where $\mathcal{R}$ is the root of the suffix tree. In addition to that, we set pointers $v.left$ and $v.right$ from each tree node v to its leftmost leaf v_ℓ and rightmost leaf v_r (considering the subtree rooted at v) in the linked list. It is easy to realize that, with these set of pointers at our disposal, we can indicate the set of occurrences of a pattern $\mathcal{P}$ by the two leaves $\mu^{\mathcal{P}}_\ell$ and $\mu^{\mathcal{P}}_r$ because all the leaves between and including $\mu^{\mathcal{P}}_\ell$ and $\mu^{\mathcal{P}}_r$ in $LL(\mathcal{R})$ correspond to the occurrences of $\mathcal{P}$ in $\mathcal{T}$. In what follows, we define the terms $\ell_{\mathcal{T}}$ and $r_{\mathcal{T}}$ such that $LL(\mathcal{R})[\ell_{\mathcal{T}}] = \mu^{\mathcal{P}}_\ell$ and $LL(\mathcal{R})[r_{\mathcal{T}}] = \mu^{\mathcal{P}}_r$, where $\mathcal{R}$ is the root of $ST_{\mathcal{T}}$.

[2] For bounded alphabet the running time remains linear, i.e. $O(n)$.

Now recall that our data structure has to be able to somehow "select" and report only those occurrences that lies in the intersection of the query interval and one of the given intervals. To solve this we use the following two interesting problems.

Problem "CRSI" (Colored Range Set Intersection Problem). *Suppose* $V[1..n]$ *and* $W[1..n]$ *are two permutations of* $[1..n]$. *Also, assume that each* $i \in [1..n]$ *is assigned a not necessarily distinct color. Preprocess* V *and* W *to answer the following form of queries.*

Query: *Find the distinct colors of the intersection of the elements of* $V[i..j]$ *and* $W[k..\ell]$, $1 \le i \le j \le n, 1 \le k \le \ell \le n$.

Problem "CRSG" (Colored Range Search Problem on Grid). *Suppose* $A[1..n]$ *is a set of* n *colored points on the grid* $[0..U] \times [0..U]$. *Preprocess* A *to answer the following form of queries.*

Query: *Given a query rectangle* $q \equiv (a,b) \times (c,d)$, *find the set of distinct colors of points contained in* q.

Our idea is to first reduce Problem PMI to Problem CRSI and then to the much more studied Problem CRSG. Recall that, we have an array $LL(\mathcal{R})$ and an interval $[\ell_T..r_T]$, which implicitly gives us the set $Occ_{\mathcal{T}}^{\mathcal{P}}$. Recall also that, our goal is to select those $i \in Occ_{\mathcal{T}}^{\mathcal{P}}$ such that i occurs in one of the intervals of π and also in $[\ell..r]$. We first construct an array $\mathcal{M} = \mathcal{M}[1..n]$, such that for all $k \in [1..n], \mathcal{M}[k] = k$. Also, we construct a 'color array' $\mathcal{C}$ to assign colors to each $k \in [1..n]$ as follows. For each $k \in [1..n]$ we assign $\mathcal{C}[k] = c_k$, if there exists an i such that $s_i \le k \le f_i, [s_i..f_i] \in \pi$; we then say that the color of k is c_k. Any other $k \in [1..n]$ is assigned a fixed different color, say c_{fixed}. In other words, all the positions of the text $\mathcal{T}$, not covered by any of the intervals of π are given a fixed color c_{fixed} and every other position carries a distinct color each. We also realize the inverse relation in the form of the array, $\mathcal{C}^{-1}$, such that $C^{-1}[c_k] = k$, if and only if, $\mathcal{C}[k] = c_k$ and $c_k \ne c_{fixed}$. Note that, there may exist more than one positions having color c_{fixed}. We define $\mathcal{C}^{-1}(c_{fixed}) = \infty$.

Now we can reduce our problem to Problem CRSI as follows. We have two arrays $LL(\mathcal{R})$ and $\mathcal{M}$ and, respectively, two intervals $[\ell_T..r_T]$ and $[\ell..r]$. Also we have color array $\mathcal{C}$, which associates a (not necessary distinct) color to each $i \in [1..n]$. Now it is easy to see that, if we can find the distinct colors in the set of intersections of elements of $LL(\mathcal{R})[\ell_T..r_T]$ and $\mathcal{M}[\ell..r]$, then we are (almost) done. The only additional thing we need to take care of is that if we have the color c_{fixed} in our output, we need to discard it. So, the Problem PMI is reduced to Problem CRSI.

On the other hand, we can see that Problem CRSI is just a different formulation of the Problem CRSG. This can be realized as follows. We set $U = n$. Since V and W in Problem CRSI are permutations of $[1..n]$, every number in $[1..n]$ appears precisely once in each of them. We define the coordinates of every number $i \in [1..n]$ to be (x, y), where $V[x] = W[y] = i$. Thus we get the n colored points (courtesy to $\mathcal{C}$) on the grid $[0..n] \times [0..n]$, i.e. the array A of Problem

CRSG. The query rectangle q is deduced from the two intervals $[i..j]$ and $[k..\ell]$ as follows: $q \equiv (i,k) \times (j,\ell)$. It is straightforward to verify that the above reduction is correct and hence we can solve Problem CRSI using the solution of Problem CRSG.

Algorithm 1. Algorithm to build IDS_PMI

Build a suffix tree $ST_{\mathcal{T}}$ of $\mathcal{T}$. Let the root of $ST_{\mathcal{T}}$ is $\mathcal{R}$.
Label each leaf of $ST_{\mathcal{T}}$ by the starting location of its suffix.
Construct a linked list $\mathcal{L}$ realizing $LL(\mathcal{R})$. Each element in $\mathcal{L}$ is the label of the corresponding leaf in $LL(\mathcal{R})$.
for each node v in $ST_{\mathcal{T}}$ **do**
  Store $v.left = i$ and $v.right = j$ such that $\mathcal{L}[i]$ and $\mathcal{L}[j]$ corresponds to, respectively, (leftmost leaf) v_ℓ and (rightmost leaf) v_r of v.
end for
for $i = 1$ *to* n **do**
  Set $\mathcal{M}[i] = i$
end for
for $i = 1$ *to* n **do**
  Set $\mathcal{C}[i] = c_{fixed}$
end for
for $i = 1$ *to* $|\pi|$ **do**
  for $j = s_i$ *to* f_i **do**
    $\mathcal{C}[j] = c_j$
  end for
end for
for $i = 1$ *to* n **do**
  Set $A[i] = \epsilon$
end for
for $i = 1$ to n **do**
  if there exists (x,y) such that $\mathcal{M}[x] = \mathcal{L}[y] = i$ **then**
    Set $A[i] = A[i] \bigcup (x,y)$
  end if
end for
Preprocess A (and $\mathcal{C}$) for Colored Range Search on a Grid $[0..n] \times [0..n]$.

To solve Problem CRSG, we are going to use the data structure of Agarwal et al. [1][3]. This data structure can answer the query of Problem CRSG in $O(\log\log U + K)$ time, where K is the number of points contained in the query rectangle q. The data structure can be built in $O(n \log n \log^2 U)$ time and requires $O(n \log^2 U)$ space. Algorithm 1 formally states the steps to build our data structure. In the rest of this paper, we refer to this data structure as IDS_PMI. One final remark is that, we can use the suffix array instead of suffix tree as well with some standard modifications in Algorithm 1.

[3] To the best of our knowledge, this is the only data structure that handles the colored range query exploiting the grid property to gain efficiency.

3.1 Analysis

Let us now analyze the cost of building the index data structure IDS_PMI. To build IDS_PMI, we first construct a traditional suffix tree requiring $O(n \log \Sigma)$ time. The preprocessing on the suffix tree can be done in $O(n)$ by traversing $ST_{\mathcal{T}}$ using a breadth first or in order traversal. The color array $\mathcal{C}$ can be setup in $O(n)$ because π is a set of disjoint intervals and it can cover, at most, n points. The construction of the set A of points in the grid $[0..n] \times [0..n]$, on which we will apply the range search, can also be done in $O(n)$ as follows. Assume that $\mathcal{L}$ is the linked list realizing $LL(\mathcal{R})$. Each element in $\mathcal{L}$ is the label of the corresponding leaf in $LL(\mathcal{R})$. We construct $\mathcal{L}^{-1}$ such that $\mathcal{L}^{-1}[\mathcal{L}[i]] = i$. It is easy to see that with $\mathcal{L}^{-1}$ in our hand we can easily construct A in $O(n)$. After A is constructed we build the data structure to solve Problem CRSG which requires $O(n \log^3 n)$ time and $O(n \log^2 n)$ space because $U = n$. Since, we can assume $\Sigma \leq n$, the index IDS_PMI can be constructed in $O(n \log^3 n)$ time.

Algorithm 2. Algorithm for Query Processing

1: Find $\mu^{\mathcal{P}}$ in $ST_{\mathcal{T}}$.
2: Set $i = \mu^{\mathcal{P}}.left$, $j = \mu^{\mathcal{P}}.right$.
3: Compute the set B, where B is the set of distinct colors in the set of points contained in $q \equiv (i, \ell) \times (j, r)$
4: **return** $Occ^{\mathcal{P}}_{\mathcal{T}[\ell..r],\pi} = \{\mathcal{C}^{-1}[x] \mid x \in B \text{ and } x \neq c_{fixed}\}$

3.2 Query Processing

So far we have concentrated on the construction of IDS_PMI. Now we discuss the query processing. Suppose we are given a query pattern $\mathcal{P}$ along with a query interval $[\ell..r]$. We first find the locus $\mu^{\mathcal{P}}$ in $ST_{\mathcal{T}}$. Let $i = \mu^{\mathcal{P}}.left$ and $j = \mu^{\mathcal{P}}.right$. Then we perform a colored range search query on A with the rectangle $q \equiv (i, \ell) \times (j, r)$. Let B is the set of those colors as output by the query. Then it is easy to verify that $Occ^{\mathcal{P}}_{\mathcal{T}[\ell..r],\pi} = \{\mathcal{C}^{-1}[x] \mid x \in B \text{ and } x \neq c_{fixed}\}$. The steps are formally presented in the form of Algorithm 2.

The running time of the query processing is deduced as follows. Finding the locus $\mu^{\mathcal{P}}$ requires $O(m)$ time. The corresponding pointers can be found in constant time. The construction of the set B is done by performing the range query and hence requires $O(\log \log n + |B|)$ time. Note that $|B|$ is either equal to $|Occ^{\mathcal{P}}_{\mathcal{T}[\ell..r],\pi}|$ or just one unit more than that. The latter happens when we have $c_{fixed} \in B$. So, in total the query time is $O(m + \log \log n + |Occ^{\mathcal{P}}_{\mathcal{T}[\ell..r],\pi}| + 1) = O(m + \log \log n + |Occ^{\mathcal{P}}_{\mathcal{T}[\ell..r],\pi}|)$. We state the results of this section in the form of following theorem.

Theorem 1. *For Problem PMI, we can construct the IDS_PMI data structure in $O(n \log^3 n)$ time and $O(n \log^2 n)$ space and we can answer the relevant queries in $O(m + \log \log n + |Occ^{\mathcal{P}}_{\mathcal{T}[\ell..r],\pi}|)$ time per query.*

4 Problem PMGI

In Section 3, we have presented an efficient index data structure, namely IDS_PMI, to solve Problem PMI. In this section, we consider Problem PMGI. Since PMI is a generalized version of PMGI, we can easily use the solution in Section 3 to solve PMGI. We use the same data structure IDS_PMI. During the query, since PMGI doesn't have any query interval, we just need to assume the query interval to be $[1..n]$. So we have the following theorem.

Theorem 2. *For Problem PMGI, we can construct the IDS_PMI data structure in $O(n \log^3 n)$ time and $O(n \log^2 n)$ space and we can answer the relevant queries in $O(m + \log\log n + |Occ^{\mathcal{P}}_{\mathcal{T},\pi}|)$ time per query.*

However, as it turns out, we can achieve better results for Problem PMGI. And in fact, as we show below, we can solve Problem PMGI optimally. We first discuss how we construct the data structure, namely IDS_PMGI, to solve PMGI. As before, we start by constructing a suffix tree (or suffix array) $ST_{\mathcal{T}}$. Then we do all the preprocessing done on $ST_{\mathcal{T}}$ as we did to construct IDS_PMI. We also construct the color array $\mathcal{C}$. This time however, we do a slightly different encoding as follows. For each $k \in [1..n]$, we assign $\mathcal{C}[k] = -1$, if there exists an i such that $s_i \leq k \leq f_i, [s_i..f_i] \in \pi$. For all other $k \in [1..n]$ we assign $\mathcal{C}[k] = 0$. In other words, all the positions of the text $\mathcal{T}$, not covered by any of the intervals of π gets 0 as their color and every other positions gets the color -1. Now we make use of the following interesting problem.

Problem "RMIN" (Range Minima Query Problem). *We are given an array $A[1..n]$ of numbers. We need to preprocess A to answer following form of queries:*

Query: Given an interval $I = [i_s..i_e], 1 \leq i_s \leq i_e \leq n$, the goal is to find the index k (or the value $A[k]$ itself) with minimum (maximum, in the case of Range Maxima Query) value $A[k]$ for $k \in I$.

Problem RMIN has received much attention in the literature and Bender and Farach-Colton showed that we can build a data structure in $O(n)$ time using $O(n \log n)$-bit space and can answer subsequent queries in $O(1)$ time per query [8][4]. Recently, Sadakane [24] presented a succinct data structure which achieves the same time complexity using $O(n)$ bits of space.

Now, we preprocess the array $\mathcal{C}$ to answer the range minima queries (RMQ). Note that, in $\mathcal{C}$, we have only two values. So to define a unique value in case of a tie, we consider the index along with the value. More formally, we define $\mathcal{C}[i] \prec \mathcal{C}[j], 1 \leq i \neq j \leq n$ if and only if $C[i] \leq C[j]$ and $i < j$. And we employ RMQ using the relation $\prec$. This can be easily done in $O(n)$ slightly modifying the preprocessing used in [8][5]. Finally, for each $\mathcal{C}[i] = -1$, we maintain a pointer

[4] The same result was achieved in [15], albeit with a more complex data structure.

[5] In particular, the only modification needed is in the construction of the Cartesian tree.

to $\mathcal{C}[j] = -1$ such that $j > i$ and j is the smallest index with this property; if there doesn't exist any such $\mathcal{C}[j]$, then $\mathcal{C}[i]$ points to 'NULL'. More formally, we maintain another array $\mathcal{D}[1..n]$ such that for all $i \in [1..n]$ with $\mathcal{C}[i] = -1$, we have $\mathcal{D}[i] = j$, if and only if, $\mathcal{C}[j] = -1$ and $\mathcal{C}[k] = 0, i < k < j$. For all other index the $\mathcal{D}$ is given a 'NULL' value. This completes the construction of IDS_PMGI. Note that, the overall running time to construct IDG_PMGI remains dominated by the construction of the suffix tree $ST_{\mathcal{T}}$. As a result, the construction time is $O(n)$ for bounded alphabet and $O(n \log \Sigma)$ otherwise.

Now we discuss how we perform the query on IDS_PMGI. Suppose we are given a query pattern $\mathcal{P}$. We first find the locus $\mu^{\mathcal{P}}$ in $ST_{\mathcal{T}}$. Let $i = \mu^{\mathcal{P}}.left$ and $j = \mu^{\mathcal{P}}.right$. Now we basically have the set $Occ^{\mathcal{P}}_{\mathcal{T}}$ in $\mathcal{L}[i..j]$. Now we perform a range minima query on $\mathcal{C}$ with the query interval $[i..j]$. This gives us, in constant time [8], the first index $k \in [i..j]$ such that $C[k] = -1$. Then we follow the linked list realized by $\mathcal{D}$ to report all the indices in the range $[i..j]$ having color -1. More formally, we construct the set $B = \{k \mid k \in [i..j] \text{ and } \mathcal{C}[k] = -1\}$. With the help of $\mathcal{D}$ this can be done in $O(|B|)$ time. And it is easy to realize that $Occ^{\mathcal{P}}_{\mathcal{T},\pi} = \{\mathcal{L}[i] \mid i \in B\}$. Therefore we can perform the query in optimal $O(m + |Occ^{\mathcal{P}}_{\mathcal{T},\pi}|)$ time. The following theorem present the results achieved in this section.

Theorem 3. *For Problem PMGI, we can construct the IDS_PMGI data structure in $O(n \log \Sigma)$ time and $O(n \log n)$ bits of space and the relevant queries can be answered optimally in $O(m + |Occ^{\mathcal{P}}_{\mathcal{T},\pi}|)$ time per query.*

For bounded alphabets, we have the following result.

Theorem 4. *For Problem PMGI, we can construct the IDS_PMGI data structure in $O(n)$ time and $O(n \log n)$ bits of space and the relevant queries can be answered optimally in $O(m + |Occ^{\mathcal{P}}_{\mathcal{T},\pi}|)$ time per query.*

5 Problem PMQI

This section is devoted to Problem PMQI. As is mentioned above, PMQI was studied extensively in [21]. The best query time achieved in [21] was $O(m + \log\log n + |Occ^{\mathcal{P}}_{\mathcal{T}[\ell..r]}|)$ against a data structure exhibiting $O(n \log^{1+\epsilon} n)$ space and time complexity, where $0 \le \epsilon \le 1$. Note that, we can easily use IDS_PMI to solve PMQI by assuming $\pi = \{[1..n]\}$. So, with a slightly worse data structure construction time, we can achieve the same query time of [21] to solve PMQI using our data structure IDS_PMI to solve a more general problem, namely PMI. However, as pointed out in [21], it would be really interesting to get an optimal query time for this problem. In this section, we attain an optimal query time for PMQI. However, the optimal query time is achieved against a $O(n^2)$ preprocessing time.

The data structure, namely IDS_PMQI, is constructed as follows. As before, we start by constructing a suffix tree (or suffix array) $ST_{\mathcal{T}}$. Then we do all the preprocessing done on $ST_{\mathcal{T}}$ as we did to construct IDS_PMI and IDS_PMGI.

Recall that, with $ST_{\mathcal{T}}$ in our hand, preprocessed as above, we can have the set $Occ^{\mathcal{P}}_{\mathcal{T}}$ in the form of $\mathcal{L}[i..j]$ in $O(m)$ time. To achieve the optimal query time we now must 'select' $k \in \mathcal{L}[i..j]$ such that $k \in [\ell..r]$ without spending more than constant time per selection. To achieve this goal we introduce the following interesting problem.

Problem "RNV" (Range Next Value Query Problem). *We are given an array $A[1..n]$, which is an permutation of $[1..n]$. We need to preprocess A to answer the following form of queries.*

Query: Given an integer $k \in [1..n]$, and an interval $[i..j], 1 \leq i \leq j \leq n$, the goal is to return the index of the immediate higher (or equal) number ('next value') than k from $A[i..j]$ if there exists one. More formally, we need to return ℓ (or $A[\ell]$ as the value itself) such that $i \leq \ell \leq j$ and $A[\ell] = \min\{A[q] \mid A[q] \geq k$ and $i \leq q \leq j\}$

Despite extensive results on various range searching problems we are not aware of any result that directly addresses this problem. Recall that our goal now is to answer the RNV queries in $O(1)$ time per query. We below give a solution where we can preprocess A in $O(n^2)$ time and then can answer the subsequent queries in $O(1)$ time per query. The idea is is as follows. We maintain n arrays $B_i, 1 \leq i \leq n$. Each array, B_i has n elements. So we could view B as a two dimensional array as well. We fill each array B_i depending on A as follows. For each $1 \leq i \leq n$ we store in B_i the difference between i and the corresponding element of A and then replace all negative entries of B_i with ∞. More formally, for each $1 \leq i \leq n$ and for each $1 \leq j \leq n$ we set $B_i[j] = A[j] - i$ if $A[j] \geq i$; otherwise we set $B_i[j] = \infty$. Then we preprocess each $B_i, 1 \leq i \leq n$ for range minima query [8]. This completes the construction of the data structure. It is clear that it will require $O(n^2)$ time. The query processing is as follows. Suppose the query parameters are k and $[i..j]$. Then we simply apply range minima query in B_k for the interval $[i..j]$. So we have the following theorem.

Theorem 5. *For Problem RNV, we can construct a data structure in $O(n^2)$ time and space to answer the relevant queries in $O(1)$ time per query.*

Now we show how we can use the result of Theorem 5 to answer the queries of Problem PMQI optimally. To complete the construction of IDS_PMQI, we preprocess the array $\mathcal{L}$ for Problem RNV. The query processing is as follows. Recall that, we can have the set $Occ^{\mathcal{P}}_{\mathcal{T}}$ in the form of $\mathcal{L}[i..j]$ in $O(m)$ time. Recall also that as part of the PMQI query, we are given an interval $[\ell..r]$. Now we perform an RNV query on $\mathcal{L}$ with the parameters ℓ and $[i..j]$. Suppose the query returns the index q. It is easy to see that if $\mathcal{L}[q] \leq r$, then $\mathcal{L}[q] \in Occ^{\mathcal{P}}_{\mathcal{T}[\ell..r]}$. And then we repeat the RNV query with parameters $\mathcal{L}[q]$ and $[i..j]$. We stop as soon as a query returns an index q such that $\mathcal{L}[q] > r$. So, in this way, given $\mathcal{L}[i..j]$, we can get the set $Occ^{\mathcal{P}}_{\mathcal{T}[\ell..r]}$ in $O(|Occ^{\mathcal{P}}_{\mathcal{T}[\ell..r]}|)$ time. So we have the following theorem.

Theorem 6. *For Problem PMQI, we can construct a data structure, namely IDS_PMQI, in $O(n^2)$ time and space to answer the relevant query in optimal $O(m + |Occ^{\mathcal{P}}_{\mathcal{T}[\ell..r]}|)$ time.*

It is clear that the bottleneck in the construction time lies in the preprocessing of Problem RNV. So any improvement on Theorem 5 would improve Theorem 6 as well. One interesting fact is that, the occurrences in $Occ^{\mathcal{P}}_{\mathcal{T}[\ell..r]}$ as output by our algorithm will always remain sorted according to their position in $\mathcal{T}$, which in many applications may turn out to be useful.

6 Conclusion

In this paper, we have considered the problem of pattern matching in given intervals and focused on building index data structure to handle different versions of this problem efficiently. We first handled the more general problem PMI and presented an efficient data structure requiring $O(n \log^3 n)$ time and $O(n \log^2 n)$ space with a query time of $O(m + \log\log n + |Occ^{\mathcal{P}}_{\mathcal{T}[\ell..r],\pi}|)$ per query. We then solved Problem PMGI optimally ($O(n)$ time and $O(n \log n)$-bits space data structure and $O(m + |Occ^{\mathcal{P}}_{\mathcal{T},\pi}|)$ query time). Finally, we improved the query time of [21] for Problem PMQI to optimal i.e. $O(m + |Occ^{\mathcal{P}}_{\mathcal{T}[\ell..r]}|)$ per query, although, at the expense of a more costly data structure requiring $O(n^2)$ time and space. It would be interesting to improve the preprocessing time of both Problem PMI and PMQI and also the query time of the former. Of particular interest is the improvement of the $O(n^2)$ data structure of PMQI without sacrificing the optimal query time. Furthermore, we believe that Problem RNV is of independent interest and could be investigated further.

References

1. Agarwal, P.K., Govindarajan, S., Muthukrishnan, S.: Range searching in categorical data: Colored range searching on grid. In: Möhring, R.H., Raman, R. (eds.) ESA 2002. LNCS, vol. 2461, pp. 17–28. Springer, Heidelberg (2002)
2. Amir, A., Aumann, Y., Benson, G., Levy, A., Lipsky, O., Porat, E., Skiena, S., Vishne, U.: Pattern matching with address errors: rearrangement distances. In: SODA, pp. 1221–1229. ACM Press, New York (2006)
3. Amir, A., Butman, A., Crochemore, M., Landau, G.M., Schaps, M.: Two-dimensional pattern matching with rotations. Theor. Comput. Sci. 314(1-2), 173–187 (2004)
4. Amir, A., Butman, A., Lewenstein, M.: Real scaled matching. Inf. Process. Lett. 70(4), 185–190 (1999)
5. Amir, A., Chencinski, E.: Faster two dimensional scaled matching. In: Lewenstein, M., Valiente, G. (eds.) CPM 2006. LNCS, vol. 4009, pp. 200–210. Springer, Heidelberg (2006)
6. Amir, A., Chencinski, E., Iliopoulos, C., Kopelowitz, T., Zhang, H.: Property matching and weighted matching. In: CPM, pp. 1–15 (2006)

7. Amir, A., Kapah, O., Tsur, D.: Faster two dimensional pattern matching with rotations. In: Sahinalp, S.C., Muthukrishnan, S.M., Dogrusoz, U. (eds.) CPM 2004. LNCS, vol. 3109, pp. 409–419. Springer, Heidelberg (2004)
8. Bender, M.A., Farach-Colton, M.: The lca problem revisited. In: Latin American Theoretical INformatics (LATIN), pp. 88–94 (2000)
9. Butman, A., Eres, R., Landau, G.M.: Scaled and permuted string matching. Inf. Process. Lett. 92(6), 293–297 (2004)
10. Cole, R., Gottlieb, L.-A., Lewenstein, M.: Dictionary matching and indexing with errors and don't cares. In: Babai, L. (ed.) STOC, pp. 91–100. ACM Press, New York (2004)
11. Cole, R., Hariharan, R.: Verifying candidate matches in sparse and wildcard matching. In: STOC, pp. 592–601 (2002)
12. Crochemore, M., Rytter, W.: Jewels of Stringology. World Scientific (2002)
13. Farach, M.: Optimal suffix tree construction with large alphabets. In: FOCS, pp. 137–143 (1997)
14. Fischer, M., Paterson, M.: String matching and other products. In: Karp, R.M. (ed.) Complexity of Computation. SIAM AMS Proceedings, vol. 7, pp. 113–125 (1974)
15. Gabow, H., Bentley, J., Tarjan, R.: Scaling and related techniques for geometry problems. In: Symposium on the Theory of Computing (STOC), pp. 135–143 (1984)
16. Gusfield, D.: Algorithms on Strings, Trees, and Sequences - Computer Science and Computational Biology. Cambridge University Press, Cambridge (1997)
17. Hirschberg, D.S.: Algorithms for the longest common subsequence problem. J. ACM 24(4), 664–675 (1977)
18. Jurka, J.: Human repetitive elements. In: Meyers, R.A. (ed.) Molecular Biology and Biotechnology
19. Jurka, J.: Origin and evolution of alu repetitive elements. In: Maraia, R. (ed.) The impact of short interspersed elements (SINEs) on the host genome
20. Levenshtein, V.: Binary codes capable of correcting, deletions, insertions and reversals. Soviet Phys. Dokl. 10, 707–710 (1966)
21. Mäkinen, V., Navarro, G.: Position-restricted substring searching. In: LATIN, pp. 1–12 (2006)
22. McCreight, E.M.: A space-economical suffix tree construction algorithm. J. ACM 23(2), 262–272 (1976)
23. Rahman, M.S., Iliopoulos, C., Lee, I., Mohamed, M., Smyth, W.: Finding patterns with variable length gaps or don't cares. In: Chen, D.Z., Lee, D.T. (eds.) COCOON 2006. LNCS, vol. 4112, pp. 146–155. Springer, Heidelberg (2006)
24. Sadakane, K.: Succinct data structures for flexible text retrieval systems. Journal of Discrete Algorithms 5(1), 12–22 (2007)
25. Ukkonen, E.: On-line construction of suffix trees. Algorithmica 14(3), 249–260 (1995)

The Power of Two Prices: Beyond Cross-Monotonicity*

Yvonne Bleischwitz[1,2], Burkhard Monien[1], Florian Schoppmann[1,2], and Karsten Tiemann[1,2]

[1] Faculty of Computer Science, Electrical Engineering and Mathematics, University of Paderborn, Fürstenallee 11, 33102 Paderborn, Germany
{yvonneb,bm,fschopp,tiemann}@uni-paderborn.de
[2] International Graduate School of Dynamic Intelligent Systems

Abstract. Assuming strict consumer sovereignty (CS*), when can cost-sharing mechanisms simultaneously be group-strategyproof (GSP) and β-budget-balanced (β-BB)? Moulin mechanisms are GSP and 1-BB for submodular costs. We overcome the submodularity requirement and instead consider arbitrary—yet symmetric—costs:

- Already for 4 players, we show that symmetry of costs is not sufficient for the existence of a GSP and 1-BB mechanism. However, for only 3 players, we give a GSP and 1-BB mechanism.
- We introduce *two-price cost-sharing forms* (2P-CSFs) that define players' cost shares and present a novel mechanism that is GSP given any such 2P-CSF. For subadditive costs, we give an algorithm to compute 2P-CSFs that are $\frac{\sqrt{17}+1}{4}$-BB (≈ 1.28). This result is then shown to be tight for 2P-CSFs. Yet, this is a significant improvement over 2-BB, which is the best Moulin mechanisms can achieve.
- We give applications to the minimum makespan scheduling problem.

A key feature of all our mechanisms is a preference order on the set of players. Higher cost shares are always payed by least preferred players.

1 Introduction and Model

1.1 Motivation

Consider a computing center with a large cluster of parallel machines that offers (uninterrupted) processing times. First, potential customers submit a maximum payment they would be willing to contribute for having their jobs processed. Then, solely based on these messages, the computing center uses a (commonly known) algorithm to determine both the served customers and their eventual payments. In addition, it computes a schedule for the accepted jobs.

Deciding on prices only after receiving *binding bids* by customers enables the computing center to determine an outcome that both ensures recovery of its own cost as well as competitive prices in that its surplus is always relatively small.

* This work was partially supported by the IST Program of the European Union under contract number IST-15964 (AEOLUS).

L. Kučera and A. Kučera (Eds.): MFCS 2007, LNCS 4708, pp. 657–668, 2007.

In fact, if the computing center is a public institution, this *budget-balance* might even be a legal requirement. Since prices are bid-dependent, selfishness of customers elevates the computing center's scheduling problems to a game-theoretic context: How can the service provider ensure *group-strategyproofness* such that (coalitional) strategic bidding corresponds to submitting true *valuations*?

The above scenario is an example for *cost sharing*: A service provider offers some service which $n \in \mathbb{N}$ selfish players are interested in. This interest is defined by a player's true valuation $v_i \in \mathbb{Q}$ for being served, which is private information. The protocol used for negotiation is simple: The service provider elicits bids $b_i \in \mathbb{Q}$ from the players that are supposed to indicate their (alleged) valuations for being served. It uses these bids b_i as a proxy for the true v_i's and employs a commonly known *cost-sharing mechanism* to determine both the set of served players and their payments, referred to as *cost shares* in the following.

1.2 The Model

Notation. For $n \in \mathbb{N}_0$, let $[n] := \{1, \ldots, n\}$ and $[n]_0 := [n] \cup \{0\}$. For all $A \subseteq [n]$ and all $i \in A$, let $\text{RANK}(i, A) := |\{j \in A \mid j \leq i\}|$. Given $\boldsymbol{x}, \boldsymbol{y} \in \mathbb{Q}^n$ and $A \subseteq [n]$, let $\boldsymbol{x}_A := (x_i)_{i \in A} \in \mathbb{Q}^{|A|}$ and $\boldsymbol{x}_{-A} := \boldsymbol{x}_{[n] \setminus A}$. Let $(\boldsymbol{x}_{-A}, \boldsymbol{y}_A) \in \mathbb{Q}^n$ denote the vector where the components in $\boldsymbol{x}$ for A are replaced by the respective ones from $\boldsymbol{y}$. For $\boldsymbol{z} \in \mathbb{Q}^m$, $(\boldsymbol{x}; \boldsymbol{z}) \in \mathbb{Q}^{n+m}$ is the vector containing the components of both $\boldsymbol{x}$ and $\boldsymbol{z}$. By convention, the vector of the players' true valuations is always $\boldsymbol{v} \in \mathbb{Q}^n$, whereas an actual bid vector is denoted $\boldsymbol{b} \in \mathbb{Q}^n$.

Definition 1. *A* cost-sharing mechanism $M = (Q \times x) : \mathbb{Q}^n \to 2^{[n]} \times \mathbb{Q}^n$ *is a function where* $Q(\boldsymbol{b}) \in 2^{[n]}$ *is the set of players to be served and* $x(\boldsymbol{b}) \in \mathbb{Q}^n$ *is the vector of cost shares.*

All cost-sharing mechanisms are required to fulfill three standard properties:

- *No positive transfers* (NPT): Players never get paid, i.e., $x_i(\boldsymbol{b}) \geq 0$.
- *Voluntary participation* (VP): Players never pay more than they bid and are only charged when served, i.e., if $i \in Q(\boldsymbol{b})$ then $x_i(\boldsymbol{b}) \leq b_i$, else $x_i(\boldsymbol{b}) = 0$.
- *Consumer sovereignty* (CS): For any player $i \in [n]$ there is a threshold bid $b_i^+ \in \mathbb{Q}_{\geq 0}$ such that i is served if bidding at least b_i^+, regardless of the other players' bids; i.e., there is a $b_i^+ \in \mathbb{Q}_{\geq 0}$ such that if $b_i \geq b_i^+$ then $i \in Q(\boldsymbol{b})$.

Note that in our model, VP implies that players may opt to not participate (by submitting a negative bid). This property together with CS is referred to as *strict consumer sovereignty* (CS*).

We assume the utility $u_i : \mathbb{Q}^n \to \mathbb{Q}$ of any player $i \in [n]$ to be *quasi-linear*, i.e., $u_i(\boldsymbol{b}) := v_i - x_i(\boldsymbol{b})$ if $i \in Q(\boldsymbol{b})$ and 0 otherwise. Under this premise, mechanisms should elicit truthful bids ($\boldsymbol{b} = \boldsymbol{v}$) even if players may collude:

Definition 2. *A mechanism is* group-strategyproof *(GSP) if for every true valuation vector* $\boldsymbol{v} \in \mathbb{Q}^n$ *and any coalition* $K \subseteq [n]$ *there is no bid vector* $\boldsymbol{b} \in \mathbb{Q}^n$ *with* $\boldsymbol{b}_{-K} = \boldsymbol{v}_{-K}$ *such that* $u_i(\boldsymbol{b}) \geq u_i(\boldsymbol{v})$ *for all* $i \in K$ *and* $u_i(\boldsymbol{b}) > u_i(\boldsymbol{v})$ *for at least one* $i \in K$.

The provider's cost of serving a set of players $Q \subseteq [n]$ is defined by a *cost function* $C : 2^{[n]} \to \mathbb{Q}_{\geq 0}$ with $C(A) = 0 \iff A = \emptyset$. Of particular interest are:

- *Symmetric costs*: Costs depend only on the cardinality, i.e., for all $A, B \subseteq [n]$ with $|A| = |B|$: $C(A) = C(B)$. In this case, we usually define costs solely by a function of the cardinality, $c : [n]_0 \to \mathbb{Q}_{\geq 0}$, where $C(A) := c(|A|)$.
- *Subadditive costs*: The marginal cost of adding a set of players is never more than the stand-alone cost of this set. That is, for any two sets $A, B \subseteq [n]$: $C(A) + C(B) \geq C(A \cup B)$.
- *Submodular costs*: For all $A, B \subseteq [n]$: $C(A) + C(B) \geq C(A \cup B) + C(A \cap B)$.

A mechanism should be budget-balanced with respect to the incurred cost:

Definition 3. *A mechanism $M = (Q \times x)$ is* β-budget-balanced (β-BB, for $\beta \geq 1$) w.r.t. cost C *if for all $\boldsymbol{b} \in \mathbb{Q}^n$: $C(Q(\boldsymbol{b})) \leq \sum_{i \in Q(\boldsymbol{b})} x_i(\boldsymbol{b}) \leq \beta \cdot C(Q(\boldsymbol{b}))$.*

GSP is a strong property in that cost shares selected by a GSP mechanism only depend on the set of served players and not on the bids [13]. This gives rise to:

Definition 4. *A* cost-sharing method *is a function $\xi : 2^{[n]} \to \mathbb{Q}^n_{\geq 0}$ that maps each set of players to a vector of cost shares.*

Clearly, every GSP mechanism *induces* a unique cost-sharing method ξ, by setting $\xi(A) := x(\boldsymbol{b})$ where $b_i < 0$ if $i \notin A$ and $b_i = b_i^+$ if $i \in A$. Three special properties of a cost-sharing method ξ are of importance in this work:

- *β-Budget-balance (w.r.t. C)*: For all $A \subseteq [n] : C(A) \leq \sum_{i \in A} \xi_i(A) \leq \beta \cdot C(A)$.
- *Cross-monotonicity*: For any player $i \in [n]$, cost shares are non-increasing as the set of players gets larger, i.e., for all $A, B \subseteq [n] : \xi_i(A \cup B) \leq \xi_i(A)$.
- *Preference order* (for symmetric costs): ξ has a succinct representation by vectors $\boldsymbol{\xi}^j \in \mathbb{Q}^j_{\geq 0}$ for all $j \in \mathbb{N}$: $\xi_i(A) := \xi^{|A|}_{\text{RANK}(i,A)}$ if $i \in A$ and 0 otherwise. Example: If $n = 5$, $A = \{2, 4\}$, and $\boldsymbol{\xi}^2 = (2, 1)$, then $\xi(A) = (0, 2, 0, 1, 0)$.

While the first two properties are standard in the realm of cost sharing, the third one is essentially "new" and crucial for our results. A preference order ensures that the cost share of any player i in a set of served players $A \subseteq [n]$ only depends on the rank of i in A and the cardinality of A.

In his seminal work [13], Moulin gave the straightforward mechanism *Moulin*$_\xi$ that is GSP given any cross-monotonic cost-sharing method ξ. *Moulin*$_\xi$ repeatedly rejects players whose bids are below their current cost shares until all remaining players can afford their payments. For any GSP mechanism M with induced cross-monotonic cost shares ξ, *Moulin*$_\xi$ produces the same utility for each player as M. Therefore, we call any GSP mechanism with cross-monotonic cost shares a *Moulin mechanism*.

1.3 Related Work

Moulin [13] completely characterizes the impact of submodular costs on GSP and 1-BB: Any GSP mechanism that is 1-BB w.r.t. submodular costs is a Moulin mechanism. Conversely, for any submodular cost function $C : 2^{[n]} \to \mathbb{Q}_{\geq 0}$, a cross-monotonic 1-BB cost-sharing method always exists. Besides this result, characterizations have also been obtained for so-called *upper-continuous* mechanisms [8,14], which are a superclass of Moulin mechanisms.

Essentially all known GSP mechanisms are Moulin mechanisms; confer, e.g., [1,7,11,2,9,10,3,6,4]. For scheduling, when costs are defined as the optimal makespan, Bleischwitz and Monien [3] give $\frac{2m}{m+1}$-BB cross-monotonic cost-sharing methods in case of identical jobs (m is the number of machines). They show that no Moulin mechanism can perform better. For arbitrary processing times, they give a $2d$-BB cross-monotonic cost-sharing method, where d is the number of different processing times, and show tightness up to a factor of 2.

In the (implausible) case that players have no means of refusing service (technically: when bids are required to be non-negative and CS* is not required), Immorlica et al. [8] present simple 1-BB and GSP mechanisms.

Recently, trading off *social cost* and BB has become an active direction of research. For details and a definition confer [12,4,6] and their references.

1.4 Contribution

Assuming strict consumer sovereignty (CS*), we obtain the following results:

1. In the vein of previous characterization attempts [13,8], we study the impact of symmetric costs on GSP and 1-BB in Section 3: Already for 4 players, we show that symmetry of costs is not sufficient for the existence of a GSP and 1-BB mechanism any more. However, for only 3 players, we give a GSP and 1-BB mechanism (based on the techniques from Section 2).
2. For symmetric costs, we introduce *two-price cost-sharing forms* (2P-CSFs) in Section 2 that define players' cost shares. The attribute 'two-price' is to indicate that for any set of served players, there are at most two different cost shares. For any 2P-CSF F, we present a novel mechanism $\mathit{MechCSF}_F$ that is GSP. This is analogous to Moulin_ξ which is GSP if the cost-sharing method ξ is cross-monotonic. The usefulness of our new technique lies in the fact that 2P-CSFs do not necessarily represent cross-monotonic cost-sharing methods. Hence, for certain classes of cost functions, 2P-CSFs allow for better budget approximations than cross-monotonic cost-sharing methods. In particular, for symmetric and subadditive costs, we give an algorithm to compute 2P-CSFs that are $\frac{\sqrt{17}+1}{4}$-BB. We show that, in general, this is the best 2P-CSFs can yield. Yet, this significantly improves over 2-BB, which is the best possible for cross-monotonic cost shares [3].
3. We apply our technique to the scheduling problem of minimizing makespan on related machines in Section 2.3. We obtain a quadratic-time algorithm for computing GSP and $(\frac{\sqrt{17}+1}{4} \cdot d)$-BB mechanisms, where d is the number of different processing times. This beats the previously best-known BB of $2d$ [3]. We are able to extend our techniques to guarantee GSP and 1-BB for the case of identical jobs and identical machines and a specific (non-symmetric) scheduling setting on 3 identical machines and processing times 1 and 2. Unfortunately, the same approach fails for 4 identical machines.

All our algorithms are based on one basic idea: They compute mechanisms with preference-ordered cost shares. Most players are charged a reasonable lower cost share (at least the minimum average per-player cost over all possible sets of players) while some less preferred players have to reimburse the remaining cost.

Omitted proofs are given in the extended version of this paper.

2 Two Price Cost-Sharing Forms

Before looking at *two* prices, we state what can be achieved with *one* price:

Lemma 1. *For any non-decreasing, symmetric, and subadditive cost function* $c : [n]_0 \to \mathbb{Q}_{>0}$*, there is a GSP and* 2*-BB mechanism that always charges all served players equally. If c is also submodular, this mechanism is even* 1*-BB. However, for any* $\varepsilon > 0$*, there is a non-decreasing, symmetric, and subadditive cost function for which no such GSP and* $(2-\varepsilon)$*-BB mechanism exists.*

In the following, we give GSP mechanisms that perform better with respect to BB. They use at most two different cost shares for any set of served players.

Definition 5. *A* two-price cost-sharing form *(2P-CSF) is a tuple* $F = (n, \boldsymbol{h}, \boldsymbol{l}, \boldsymbol{d})$ *where for each cardinality* $i \in [n]_0$ *of the set of served players*

- $h_i \in \mathbb{Q}_{>0}$ *is the* **h***igher,* $l_i \in \mathbb{Q}_{>0}$ *(with* $l_i < h_i$*) is the* **l***ower cost share, and*
- $d_i \in [i]_0$ *is the number of* **d***isadvantaged players paying* h_i*.*

Note that $d_0 = 0$ by definition and neither h_0 nor l_0 are actually used; cardinality 0 is included only to avoid undesired case analyses later on. A 2P-CSF $(n, \boldsymbol{h}, \boldsymbol{l}, \boldsymbol{d})$ is a succinct representation of vectors $\boldsymbol{\xi}^i \in \mathbb{Q}^i$, $i \in [n]$, which define the preference ordered cost-sharing method $\xi : 2^{[n]} \to \mathbb{Q}^n$ by

$$\boldsymbol{\xi}^i := (\underbrace{h_i, \ldots, h_i}_{d_i}, \underbrace{l_i, \ldots, l_i}_{i-d_i \text{ elements}}).$$

We call a contiguous range $\{s, s+1, \ldots, t\} \subseteq [n]_0$ of cardinalities with $d_s = d_{t+1} = 0$ (let $d_{t+1} := 0$ if $t = n$), and $d_k > 0$ for $k \in \{s+1, \ldots, t\}$ a *segment*. That is, only at the beginning of a segment there is no disadvantaged player paying the higher cost share. Furthermore, a segment is maximal with this property.

Definition 6. *A 2P-CSF* $(n, \boldsymbol{h}, \boldsymbol{l}, \boldsymbol{d})$ *is* valid *if for each cardinality* $i \in [n]$*:*

- *Lower cost shares are non-increasing (in the cardinality) and stay the same within a segment, i.e.,* $l_i \le l_{i-1}$ *and* $(l_i < l_{i-1} \Longrightarrow d_i = 0)$*.*
- *Higher cost shares may only increase at the beginning of a segment, i.e.,* $h_i > h_{i-1} \Longrightarrow d_i = 0$*.*
- *Adding a single player may not increase the number of disadvantaged players by more than 1. Moreover, if the higher cost share decreases, only one disadvantaged player may remain, i.e.,* $d_i \le d_{i-1} + 1$ *and* $(h_i < h_{i-1} \Longrightarrow d_i \le 1)$*.*

We define $\gamma : [n]_0 \to \mathbb{Q}_{\ge 0}$, $\gamma(i) := d_i \cdot h_i + (i - d_i) \cdot l_i$ as the *recovered cost.*

2.1 GSP Mechanisms for Two-Price Cost-Sharing Forms

Algorithm 1 (Computing mechanism $\mathit{MechCSF}_F$**, for a 2P-CSF** F**)**

Input: 2P-CSF $F = (n, \boldsymbol{h}, \boldsymbol{l}, \boldsymbol{d})$, bid vector $\boldsymbol{b} \in \mathbb{Q}^n$

Output: set of players $Q \in 2^{[n]}$, vector of cost shares $\boldsymbol{x} \in \mathbb{Q}^n_{\geq 0}$

1: $k := \max\left\{ i \in [n]_0 \;\middle|\; |\{j \in [n] \mid b_j \geq l_i\}| \geq i \right\}$; $l := l_k$ ▷ Find segment

2: $Q := \{i \in [n] \mid b_i \geq l\}$ ▷ Players still in the game

3: $I := \{i \in [n] \mid b_i = l\}$ ▷ Indifferent players

4: $D := \emptyset$ ▷ Disadvantaged players with higher cost share

5: **loop**

6: $q := \max\{i \in [|Q|]_0 \mid d_i = |D|\}$ ▷ Max # players if players in D pay $> l$

7: **if** $q \geq |Q \setminus I|$ **then** ▷ Possible to serve all non-indifferent players?

8: $Q := Q \setminus \{|Q| - q$ smallest elements of $I\}$ ▷ Remove "excess"

9: **break**

10: $\ell := \min(Q \setminus D)$ ▷ Least preferred player not yet in D

11: **if** $b_\ell \geq h_{|Q|}$ **then** $D := D \cup \{\ell\}$ ▷ Make disadvantaged

12: **else** $Q := Q \setminus \{\ell\}$; $I := I \setminus \{\ell\}$

13: $x_i := h_{|Q|}$ for $i \in D$; $x_i := l$ for $i \in Q \setminus D$; $x_i := 0$ for $i \in [n] \setminus Q$

Informally, Algorithm 1 works as follows.

1. Find largest $k \in \mathbb{N}$ such that there are k players who bid at least the respective lower cost share $l_k =: l$. Note that k is already in the correct segment.
2. Reject all players who do not even bid l.
3. If possible, include all players still in the game with bid $b_i > l$ for price l, by rejecting a suitable subset of the indifferent players ($b_i = l$); then stop.
4. Otherwise, include the least preferred player (with lowest number) for the current higher cost share or, if she bids less than that, reject her. Go to 3.

The intuition is that including the least preferred player for the current higher cost share never harms the other players. Instead, it may even benefit in that more players can be served for l afterwards. Once a player is included for a higher cost share, this cost share remains fixed during the further execution.

Theorem 1. *Let $F = (n, \boldsymbol{h}, \boldsymbol{l}, \boldsymbol{d})$ be a valid 2P-CSF. Then $\mathit{MechCSF}_F$ is GSP and can be computed by Algorithm 1 in time $O(n^2)$.*

Proof (Sketch). For any input $\boldsymbol{b} \in \mathbb{Q}^n$, denote by $k(\boldsymbol{b})$ and $l(\boldsymbol{b})$ the values of k and l in line 1 of Algorithm 1. Moreover, set $s(\boldsymbol{b}) := \max\{j \in [k(\boldsymbol{b})]_0 \mid d_j = 0\}$ to the beginning of the segment that $k(\boldsymbol{b})$ is in. Note that $s(\boldsymbol{b}) \leq |Q(\boldsymbol{b})| \leq k(\boldsymbol{b})$.

Denote by $\boldsymbol{v} \in \mathbb{Q}^n$ the true valuation vector and let $K \subseteq [n]$ be a successful coalition, i.e., there is a bid vector $\boldsymbol{b} \in \mathbb{Q}^n$ with $\boldsymbol{b}_{-K} = \boldsymbol{v}_{-K}$ such that $u_j(\boldsymbol{b}) \geq u_j(\boldsymbol{v})$ for all $j \in K$ and $u_i(\boldsymbol{b}) > u_i(\boldsymbol{v})$ for at least one $i \in K$.

First assume $s(\boldsymbol{b}) > s(\boldsymbol{v})$. Then also $s(\boldsymbol{b}) > k(\boldsymbol{v})$, thus $|Q(\boldsymbol{b})| > k(\boldsymbol{v})$. Hence, there is a player $j \in Q(\boldsymbol{b})$ with $v_j < l(\boldsymbol{b})$ because $k(\boldsymbol{v})$ would not have been maximal in line 1 otherwise. Since $j \in Q(\boldsymbol{b})$, it holds that $b_j \geq x_j(\boldsymbol{b}) \geq l(\boldsymbol{b}) > v_j$, so we have $j \in K$ and $u_j(\boldsymbol{b}) < 0 \leq u_j(\boldsymbol{v})$, a contradiction.

Now consider $s(\boldsymbol{b}) < s(\boldsymbol{v})$. Then also $k(\boldsymbol{b}) < s(\boldsymbol{v})$. Define $L := \{j \in [n] \mid b_j \geq l(\boldsymbol{v})\}$. Clearly, $|L| < s(\boldsymbol{v})$ as $k(\boldsymbol{b})$ would not have been maximal in line 1 otherwise. Now, let M be a set of $s(\boldsymbol{v}) - |L|$ players $j \in [n] \setminus L$ with $l(\boldsymbol{v}) \leq v_j$. Such a set M exists. Define a new bid vector $\boldsymbol{b}' \in \mathbb{Q}^n$ by $b'_j := l(\boldsymbol{v})$ for $j \in M$ and $b'_j := b_j$ otherwise. Then $s(\boldsymbol{b}') = |Q(\boldsymbol{b}')| = k(\boldsymbol{b}') = s(\boldsymbol{v})$ and $u_j(\boldsymbol{b}') \geq u_j(\boldsymbol{b})$ for all players $j \in [n]$. W.l.o.g., we may thus assume $s(\boldsymbol{b}) = s(\boldsymbol{v})$ in the following. For clarity let $l := l(\boldsymbol{b}) = l(\boldsymbol{v})$.

Since $u_i(\boldsymbol{b}) > u_i(\boldsymbol{v})$, we have $i \in Q(\boldsymbol{b})$ and $x_i(\boldsymbol{b}) < h_{|Q(\boldsymbol{v})|}$ and ($i \notin Q(\boldsymbol{v})$ or $x_i(\boldsymbol{v}) = h_{|Q(\boldsymbol{v})|}$). At least one of the following has to be fulfilled:

- The number of disadvantaged players paying the higher cost share increased from $\boldsymbol{v}$ to $\boldsymbol{b}$. That is, $\exists j \in [i-1] : b_j \geq h_{|Q(\boldsymbol{b})|}$ and $h_{|Q(\boldsymbol{v})|} > v_j$.
- A player j that the mechanism prefers to i (and who got the service for l) waived being served. That is, $\exists j \in \{i+1, \dots, n\} : b_j \leq l < v_j$ and $j \notin Q(\boldsymbol{b})$.
- The total number of players served for the lower cost share l increased. That is, $\exists j \in [n] : b_j \geq l > v_j$ and $j \in Q(\boldsymbol{b})$.

In either case, player j is part of the coalition K and $u_j(\boldsymbol{b}) < u_j(\boldsymbol{v})$. □

2.2 $\frac{\sqrt{17}+1}{4}$-BB Two-Price Cost-Sharing Forms for Subadditive Costs

Theorem 2. *Let $\varepsilon > 0$. For any non-decreasing, symmetric, and subadditive cost function $c : [n]_0 \to \mathbb{Q}_{\geq 0}$, there is a $\left(\frac{\sqrt{17}+1}{4} + \varepsilon\right)$-BB 2P-CSF. Moreover (if c is given as an array of n function values), it can be computed in time $O(n)$.*

Proof. Algorithm 2 computes a 2P-CSF for increasing cardinalities. Fix an arbitrary $\beta \geq \frac{\sqrt{17}+1}{4}$. The ε in the formulation of the theorem is solely to account for the fact that we require bids and cost shares to be rational.

Algorithm 2 (Computing a 2P-CSF for subadditive costs)
Input: subadditive cost function $c : [n]_0 \to \mathbb{Q}_{\geq 0}$
Output: valid 2P-CSF $(n, \boldsymbol{h}, \boldsymbol{l}, \boldsymbol{d})$

$l_0 := l_1 := \beta \cdot c(1)$; $h_0 := h_1 := \infty$; $d_0 := d_1 := 0$; $f := 1$
for $i := 2, \dots, n$ **do**
  if $\beta \cdot \frac{c(i)}{i} \leq l_f$ **then** $l_i := \beta \cdot \frac{c(i)}{i}$; $h_i := \infty$; $d_i := 0$; $f := i$
  else
    $l_i := l_{i-1}$; $h_i := \min\{\beta \cdot c(i) - (i-1) \cdot l_i, h_{i-1}\}$
    if $h_i + (i-1) \cdot l_i < c(i)$ **then**
      $d_i := 2$
    else if $h_i + (i-1) \cdot l_i \geq 2 \cdot c(f)$ **then**
      $d_i := 1$
    else if $h_i \geq (\beta^2 - \beta) \cdot c(f)$ **then**
      $d_i := 1$
      if $(\beta^2 - \beta) \cdot c(f) + (i-1) \cdot l_i \geq c(i)$ **then** $h_i := (\beta^2 - \beta) \cdot c(f)$
    else
      $d_i := 0$; $h_i := \infty$

Here, "∞" is a placeholder for any "sufficiently large" value (a value strictly larger than $\beta \cdot c(f)$ is sufficient) to simplify the presentation. Let $(n, \boldsymbol{h}, \boldsymbol{l}, \boldsymbol{d})$ be the output 2P-CSF. Since $d_0 = 0$ and $\gamma(0) = c(0)$, consider an arbitrary cardinality $i \in [n]$. We first show validity. Clearly, $d_i \in \{0, 1, 2\}$.

- $l_i \leq l_{i-1}$ and $(l_i < l_{i-1} \Longrightarrow d_i = 0)$ since $l_i \neq l_{i-1}$ only if l_i was set in line 3.
- Moreover, line 5 ensures $d_i > 0 \Longrightarrow h_i \leq h_{i-1}$.
- $d_i \leq d_{i-1} + 1$ and $(d_i = 2 \Longrightarrow h_i = h_{i-1})$. Clearly, we only need to consider $d_i = 2$: Then $h_i < c(i) - (i-1) \cdot l_i$ because line 6 evaluated to true and thus $h_i = \min\{\beta \cdot c(i) - (i-1) \cdot l_i, h_{i-1}\} = h_{i-1}$. Now assume $d_i > d_{i-1} + 1$, i.e., $d_{i-1} = 0$. Then, $h_i = h_{i-1} = \infty$, a contradiction to $h_i < c(i) - (i-1) \cdot l_i$.

We now prove β-BB. If $d_i = 1$, $c(i) \leq \gamma(i) \leq \beta \cdot c(i)$ due to lines 5 and 6. For $d_i \in \{0, 2\}$, we define $f := \max\{j \in [i] \mid \beta \cdot \frac{c(j)}{j} = l_i\}$ to be the last cardinality previous or equal to i for which the lower cost share was set in lines 1 or 3. Furthermore, let $g := \min(\{j \in \{i+1, i+2, \ldots, n\} \mid \beta \cdot \frac{c(j)}{j} \leq l\} \cup \{n+1\})$ be the next such cardinality after i (or $g = n+1$ if f is the largest such cardinality). It is $f \leq i < g \leq 2f$. Otherwise, f would not be maximal, due to $\frac{c(2f)}{2f} \leq \frac{2 \cdot c(f)}{2f} = \frac{c(f)}{f}$ because of subadditivity. Since c is non-decreasing, $c(i) \leq c(2f) \leq 2 \cdot c(f)$. Set $h_i' := \min\{\beta \cdot c(i) - (i-1) \cdot l_i, h_{i-1}\}$. We will make use of the following property:

$$d_{i-1} = 1 \text{ and } \gamma(i-1) \geq 2 \cdot c(f) \Longrightarrow \forall j \in \{i, i+1, \ldots, g-1\} : d_j = 1 \,. \quad (1)$$

Proof of (1): If $h_i' = h_{i-1}$, then $h_i' + (i-1) \cdot l_i = \gamma(i-1) + l_i$. If $h_i' = \beta \cdot c(i) - (i-1) \cdot l_i$, then $h_i' + (i-1) \cdot l_i = \beta \cdot c(i) \geq \beta \cdot c(i-1) \geq \gamma(i-1)$. In any case, $h_i' + (i-1) \cdot l_i \geq \gamma(i-1) \geq 2 \cdot c(f) \geq c(i)$, so $h_i = h_i'$ and $d_i = 1$ by line 8. Inductively, (1) follows. Consider now $d_i = 2$ and $d_i = 0$:

For $d_i = 2$, we first show $h_i = (\beta^2 - \beta) \cdot c(f)$. Define $i' := \max\{j \in [i] \mid d_j = 1\}$. By validity, $f < i' < i$ and $h_{i'} = h_i$. Since line 8 evaluated to false for cardinality i' because of (1), line 10 must have evaluated to true, implying $h_i = h_i' \geq (\beta^2 - \beta) \cdot c(f)$. Now assume ">". Let $s := \max\{j \in [i] \mid d_i = 0\}$ be the start of the segment that i is in. By validity, $f \leq s \leq i - 2$, $d_{s+1} = 1$, and $h_{s+1} \geq h_i$. Lines 8 and 12 evaluated to false for cardinality $(s+1)$, meaning that $\beta^2 \cdot c(f) \leq (\beta^2 - \beta) \cdot c(f) + s \cdot l_{s+1} < c(s+1)$. Then, however, $h_{s+1} + s \cdot l_{s+1} = \beta \cdot c(s+1) > \beta^3 \cdot c(f) > 2 \cdot c(f)$, a contradiction. Hence,

$$\gamma(i) = 2h_i + (i-2) \cdot \beta \cdot \frac{c(f)}{f} \geq (2\beta^2 - \beta) \cdot c(f) \geq 2 \cdot c(f) \geq c(i) \,.$$

On the other hand, $\beta \cdot c(f) < \beta^2 \cdot c(f) = h_i + \beta \cdot c(f) < h_i + (i-1) \cdot l_i < c(i)$, where the last inequality holds since line 6 evaluated to true. Hence,

$$\gamma(i) = 2h_i + (i-2) \cdot l_i < h_i + c(i) < (\beta - 1) \cdot c(i) + c(i) = \beta \cdot c(i) \,.$$

Now let $d_i = 0$. Since line 10 evaluated to false, $h_i' < (\beta^2 - \beta) \cdot c(f)$. We first show that $h_i' = \beta \cdot c(i) - (i-1) \cdot l_i$. Assume otherwise. Then, $h_i' = h_{i-1}$ and $d_{i-1} = 1$ since $h_{i-1} \notin \{\infty, (\beta^2 - \beta) \cdot c(f)\}$. Yet, line 8 evaluated to false for $(i-1)$ because of (1). Thus, $h_i' = h_{i-1} \geq (\beta^2 - \beta) \cdot c(f)$ by line 10. Contradiction.

Now $\beta \cdot c(i) < (\beta - 1) \cdot \beta \cdot c(f) + (i-1) \cdot l_i$. Furthermore, $\beta \cdot c(f) \leq \gamma(i)$ and $(i-1) \cdot l_i \leq \gamma(i)$. Putting everything together gives

$$c(i) = \frac{\beta \cdot c(i)}{\beta} \leq \frac{(\beta - 1) \cdot \gamma(i) + \gamma(i)}{\beta} = \gamma(i)\,.$$

Moreover, $\gamma(i) = i \cdot l_i \leq \beta \cdot c(i)$ as otherwise $l_f = l_i > \beta \cdot \frac{c(i)}{i}$. □

Theorem 3. *For all $\varepsilon > 0$, there is a non-decreasing, symmetric, and subadditive cost function $c : [n]_0 \to \mathbb{Q}_{\geq 0}$ for which no valid $\left(\frac{\sqrt{17}+1}{4} - \varepsilon\right)$-BB 2P-CSF exists.*

Proof (Sketch). Fix $\alpha := \frac{\sqrt{17}+1}{4}$. W.l.o.g., let $0 < \varepsilon \leq \alpha - 1$ and set $\beta := \alpha - \varepsilon$. Finally, let $k, l \in \mathbb{N}$ with $l > \ln_\alpha \frac{\alpha - 1}{\varepsilon}$ and $k > \frac{(l+1)\cdot\beta}{\varepsilon} = \frac{(l+1)\cdot\alpha}{\varepsilon} - (l+1)$. Set $m := k + l + 1$ and $n := m + 1$ and consider the cost function $c : [n]_0 \to \mathbb{Q}_{\geq 0}$:

i	$1 \cdots k$	$k+1$	$k+2$	$\cdots$	$k+l$	m	n
$c(i)$	$1 \cdots 1$	$\alpha - \frac{\alpha-1}{\alpha^1}$	$\alpha - \frac{\alpha-1}{\alpha^2}$	$\cdots$	$\alpha - \frac{\alpha-1}{\alpha^l}$	α	2

Clearly, c is subadditive. Now assume there is a valid 2P-CSF $(n, \boldsymbol{h}, \boldsymbol{l}, \boldsymbol{d})$ which is β-BB. It can be shown that $d_m \geq 1$ and $d_n = d_m + 1$. Let $s := \max\{j \in [n] \mid d_j = 0\}$ be the start of the segment that cardinality n is in. Clearly, $s < m$ and $h_n \leq h_{s+1}$. By case analysis, $h_{s+1} \leq \beta \cdot c(s+1) - c(s) < \alpha^2 - \alpha$. Thus $\gamma(n) \leq h_n + \beta \cdot \alpha < 2 \cdot \alpha^2 - \alpha = 2 = c(n)$, a contradiction to β-BB. □

2.3 Applications to Scheduling

We apply our technique to the scheduling problem of minimizing makespan on related machines. To keep the service provider's task computationally tractable, we want algorithms to be polynomial-time in the size of the scheduling instance plus the players' bids. An instance is given by a tuple $(n, m, \boldsymbol{p}, \boldsymbol{s})$, where $n \in \mathbb{N}$ is the number of players, $m \in \mathbb{N}$ the number of machines, $\boldsymbol{s} \in \mathbb{Q}^m_{>0}$ a vector containing the machine speeds, and $\boldsymbol{p} \in \mathbb{Q}^n_{>0}$ a vector with the processing times. For identical jobs, computing the (symmetric) optimal makespan cost function $c : [n]_0 \to \mathbb{Q}_{\geq 0}$ in time $O(n \cdot \log m)$ is straightforward using LPT [5]. If jobs are not identical, an optimal schedule is in general NP-hard to compute and costs are not symmetric any more. Let d be the number of different processing times. We treat each processing time separately: First compute the costs $c : [n]_0 \to \mathbb{Q}_{\geq 0}$ of an optimal schedule when all processing times are 1. Second, compute the corresponding 2P-CSF F in time $O(n)$. Finally, for each processing time $\phi \in \bigcup\{p_i\}$, compute $\mathit{MechCSF}_F$ (though cost shares from F multiplied with ϕ) with just the bids of all players i with $p_i = \phi$. Each run with n_j players takes $O(n_j^2)$ time. Return the union of all selected players. Since $\forall (n_1, \ldots, n_d) \in [n]_0^d$ with $\sum n_j = n$: $\sum n_j^2 \leq n^2$, this algorithm is quadratic-time, GSP, and $(\frac{\sqrt{17}+1}{4} \cdot d)$-BB both w.r.t. the actual *and* the optimal scheduling cost:

Lemma 2. *For sharing the makespan cost on related machines, there is always a quadratic-time computable, GSP, and* $\left(\frac{\sqrt{17}+1}{4} \cdot d\right)$*-BB cost-sharing mechanism where* d *is the number of different processing times.*

In particular, dividing all cost shares by $\min_{i\in[n]}\{\frac{\gamma(i)}{c(i)}\}$ leads to:

Lemma 3. *For sharing the makespan cost of identical jobs on identical machines there is always a quadratic-time, GSP, and* 1*-BB mechanism.*

Extending our techniques in order to achieve 1-BB for non-symmetric costs seems to be very challenging. For 3 identical machines and processing times either 1 or 2, we manage to construct 1-BB and GSP mechanisms by generalizing the notion of a preference order and making cost shares dependent on the rank as well as the cardinalities of *both* classes of served players. Unfortunately, this approach is not directly extendable to more than three machines.

Theorem 4. *For sharing the makespan cost on 3 identical machines and jobs with processing times* 1 *or* 2*, there is always a GSP and* 1*-BB mechanism.*

Finally, we find it interesting that the costs used for Theorem 2 are optimal makespan costs for $n = m + 1$ identical jobs and speed $\frac{1}{c(i)}$ for machine $i \in [m]$.

3 The Impact of Symmetric Costs on GSP and 1-BB

Theorem 5. *There is no GSP mechanism that is 1-BB w.r.t.* $c : [4]_0 \to \mathbb{Q}_{\geq 0}$ *with* $c(1) := 1, c(2) := 3, c(3) := 6,$ *and* $c(4) := 7$.

Theorem 6. *For any cost function* $c : [3]_0 \to \mathbb{Q}_{\geq 0}$*, there is a 1-BB and GSP mechanism.*

Proof. In the following, we describe how the techniques from Section 2 can be reused to construct an algorithm for computing GSP and 1-BB mechanisms, given any arbitrary symmetric 3-player cost function $c : [3]_0 \to \mathbb{Q}_{\geq 0}$. It turns out that only the case where marginal costs are strictly increasing needs special treatment, i.e., the case $c(3) - c(2) > c(2) - c(1) > c(1)$.

First, for $i \in [3]$, vectors $\boldsymbol{\xi}^i \in \mathbb{Q}^i_{>0}$ are computed, where $\boldsymbol{\xi}^1 := c(1)$ and

$$\boldsymbol{\xi}^2 := \begin{cases} (\frac{c(2)}{2}, \frac{c(2)}{2}) & \text{if } \frac{c(2)}{2} \leq \xi^1_1 = c(1) \\ (c(2) - c(1); \boldsymbol{\xi}^1) & \text{otherwise} \end{cases}$$

$$\boldsymbol{\xi}^3 := \begin{cases} (\frac{c(3)}{3}, \frac{c(3)}{3}, \frac{c(3)}{3}) & \text{if } \frac{c(3)}{3} \leq \xi^2_2 \\ (c(3) - 2 \cdot \xi^2_2, \xi^2_2, \xi^2_2) & \text{otherwise, if } c(3) - c(2) < \xi^2_2 \\ (\xi^2_1, c(3) - c(2), \xi^2_2) & \text{otherwise, if } c(3) - c(2) < \xi^2_1 \\ (c(3) - c(2); \boldsymbol{\xi}^2) & \text{otherwise.} \end{cases}$$

It is a simple observation that the defined cost-sharing vectors correspond to a valid 2P-CSF if at most two prices are used for each cardinality. Hence, we can call Algorithm 1 as a subroutine. In the following, we consider the only remaining case $\xi^3_1 > \xi^3_2 > \xi^3_3$. We distinguish:

- $\xi_1^3 = \xi_1^2$, i.e., $\xi_1^3 = c(2) - c(1) > \xi_2^3 = c(3) - c(2) > \xi_3^3 = c(1)$:
  Here, we can actually reuse Algorithm 1 as well, by replacing references to the cost-sharing form by their equivalent in terms of cost-sharing vectors:
  - Replace line 1 by "$l := c(1)$"
  - In line 6, replace "$d_i = |D|$" with "$\xi^i_{|D|+1} = l$".
  - In line 11, replace $h_{|Q|}$ with $\xi^{|Q|}_{|D|+1}$.

  With these modifications, the computed mechanism is GSP: Assume true valuations $\boldsymbol{v} \in \mathbb{Q}^3$ and that there is a player $i \in [3]$ who improved for bid vector $\boldsymbol{b} \in \mathbb{Q}^3$. This can only happen if ($i \notin Q(\boldsymbol{v})$ or $x_i(\boldsymbol{v}) > c(1)$) and $\max Q(\boldsymbol{b}) = i$. Hence, there is a more preferred player j (i.e., $j > i$) with $b_j \leq c(1) < v_j$. Thus, j is part of the coalition and $u_j(\boldsymbol{b}) < u_j(\boldsymbol{v})$.
- $\xi_1^3 > \xi_1^2$, i.e., $\xi_1^3 = c(3) - c(2) > \xi_2^3 = c(2) - c(1) > \xi_3^3 = c(1)$:
  Here, including indifferent players never helps. Also, including less preferred players does not help more preferred players. Hence, we can simply go through the players in the order of preference and accept them if this gives them a strictly positive utility and reject them otherwise:

  $Q := \emptyset$
  for $i := 3, 2, 1$ **do**
      if $b_i > \xi_1^{|Q|+1}$ **then** $Q := Q \cup \{i\}$

  This is GSP: Assume true valuations $\boldsymbol{v} \in \mathbb{Q}^3$ and that there is a player $i \in [3]$ who improved for bid vector $\boldsymbol{b} \in \mathbb{Q}^3$. This can only happen if some player $j > i$ bids $b_j < v_i$ such that $j \notin Q(\boldsymbol{b})$ but $j \in Q(\boldsymbol{v})$. Then, however, j is part of the coalition and $u_j(\boldsymbol{b}) < u_j(\boldsymbol{v})$. □

4 Conclusion and Future Work

We regard as the main asset of our work that it is a systematic first step for finding GSP mechanisms that perform better than Moulin mechanisms. Furthermore, we continued the line of characterization efforts by specifically looking at symmetric costs. It came as a surprise that despite their simplicity, these costs do not necessarily allow for GSP and 1-BB mechanisms. While symmetric costs are arguably of limited practical interest, we yet transferred our techniques to the minimum makespan scheduling problem as an application and also to a setting with non-symmetric costs. Clearly, our work has to leave open many issues:

- For symmetric and/or subadditive costs, we still need an exact characterization with respect to the *best* possible BB that GSP mechanisms can achieve.
- Can our techniques be generalized to all/most non-symmetric cost functions? What is the potential of Algorithm 1? What is achievable with more prices? How would more general cost-sharing forms have to be like?
- Finally: Does the better BB (compared to Moulin mechanisms) come at the price of increased social cost (which was not considered in this work)?

Acknowledgment. We would like to thank Marios Mavronicolas for many fruitful discussions. Furthermore, we thank the anonymous referee who provided us with thorough feedback and helpful suggestions.

References

1. Archer, A., Feigenbaum, J., Krishnamurthy, A., Sami, R.: Approximation and collusion in multicast cost sharing. Games and Economic Behaviour 47, 36–71 (2004)
2. Becchetti, L., Könemann, J., Leonardi, S., Pál, M.: Sharing the cost more efficiently: improved approximation for multicommodity rent-or-buy. In: Proceedings of the 16th Annual ACM-SIAM Symposium on Discrete Algorithms, pp. 375–384. ACM Press, New York (2005)
3. Bleischwitz, Y., Monien, B.: Fair cost-sharing methods for scheduling jobs on parallel machines. In: Calamoneri, T., Finocchi, I., Italiano, G.F. (eds.) CIAC 2006. LNCS, vol. 3998, pp. 175–186. Springer, Heidelberg (2006)
4. Brenner, J., Schäfer, G.: Cost sharing methods for makespan and completion time scheduling. In: Thomas, W., Weil, P. (eds.) STACS 2007. LNCS, vol. 4393, Springer, Heidelberg (2007)
5. Graham, R.: Bounds on multiprocessing timing anomalies. SIAM Journal of Applied Mathematics 17(2), 416–429 (1969)
6. Gupta, A., Könemann, J., Leonardi, S., Ravi, R., Schäfer, G.: An efficient cost-sharing mechanism for the prize-collecting Steiner forest problem. In: Proceedings of the 18th Annual ACM-SIAM Symposium on Discrete Algorithms, ACM Press, New York (2007)
7. Gupta, A., Srinivasan, A., Tardos, E.: Cost-sharing mechanisms for network design. In: Jansen, K., Khanna, S., Rolim, J.D.P., Ron, D. (eds.) RANDOM 2004 and APPROX 2004. LNCS, vol. 3122, pp. 139–152. Springer, Heidelberg (2004)
8. Immorlica, N., Mahdian, M., Mirrokni, V.: Limitations of cross-monotonic cost sharing schemes. In: Proceedings of the 16th Annual ACM-SIAM Symposium on Discrete Algorithms, pp. 602–611. ACM Press, New York (2005)
9. Könemann, J., Leonardi, S., Schäfer, G.: A group-strategyproof mechanism for Steiner forests. In: Proceedings of the 16th Annual ACM-SIAM Symposium on Discrete Algorithms, pp. 612–619. ACM Press, New York (2005)
10. Könemann, J., Leonardi, S., Schäfer, G., van Zwam, S.: From primal-dual to cost shares and back: A stronger LP relaxation for the Steiner forest problem. In: Caires, L., Italiano, G.F., Monteiro, L., Palamidessi, C., Yung, M. (eds.) ICALP 2005. LNCS, vol. 3580, pp. 930–942. Springer, Heidelberg (2005)
11. Leonardi, S., Schäfer, G.: Cross-monotonic cost-sharing methods for connected facility location games. In: Proceedings of the ACM Conference on Electronic Commerce, pp. 224–243. ACM Press, New York (2004)
12. Mehta, A., Roughgarden, T., Sundararajan, M.: Beyond Moulin mechanisms. In: The Proceedings of the 8th ACM Conference on Electronic Commerce, ACM Press, New York (to appear, 2007)
13. Moulin, H.: Incremental cost sharing: Characterization by coalition strategy-proofness. Social Choice and Welfare 16(2), 279–320 (1999)
14. Penna, P., Ventre, C.: The algorithmic structure of group strategyproof budget-balanced cost-sharing mechanisms. In: Durand, B., Thomas, W. (eds.) STACS 2006. LNCS, vol. 3884, pp. 337–348. Springer, Heidelberg (2006)

Semisimple Algebras of Almost Minimal Rank over the Reals

Markus Bläser[1] and Andreas Meyer de Voltaire[2]

[1] Computer Science Department, Saarland University
[2] Chair of Information Technology and Education, ETH Zurich

Abstract. A famous lower bound for the bilinear complexity of the multiplication in associative algebras is the Alder–Strassen bound. Algebras for which this bound is tight are called algebras of minimal rank. After 25 years of research, these algebras are now well understood. We here start the investigation of the algebras for which the Alder–Strassen bound is off by one. As a first result, we completely characterize the semisimple algebras over $\mathbb{R}$ whose bilinear complexity is by one larger than the Alder–Strassen bound.

1 Introduction

A central problem in algebraic complexity theory is the question about the costs of multiplication in associative algebras. Let A be a finite dimensional associative k-algebra with unity 1. By fixing a basis of A, say $v_1, \ldots, v_N$, we can define a set of bilinear forms corresponding to the multiplication in A. If $v_\mu v_\nu = \sum_{\kappa=1}^{N} \alpha_{\mu,\nu}^{(\kappa)} v_\kappa$ for $1 \leq \mu, \nu \leq N$ with **structural constants** $\alpha_{\mu,\nu}^{(\kappa)} \in k$, then these constants and the identity

$$\left(\sum_{\mu=1}^{N} X_\mu v_\mu\right)\left(\sum_{\nu=1}^{N} Y_\nu v_\nu\right) = \sum_{\kappa=1}^{N} b_\kappa(X, Y) v_\kappa$$

define the desired bilinear forms $b_1, \ldots, b_N$. The **bilinear complexity** or **rank** of these bilinear forms $b_1, \ldots, b_N$ is the smallest number of essential bilinear multiplications necessary and sufficient to compute $b_1, \ldots, b_N$ from the indeterminates $X_1, \ldots, X_N$ and $Y_1, \ldots, Y_N$. More precisely, the bilinear complexity of $b_1, \ldots, b_N$ is the smallest number r of products $p_\rho = u_\rho(X_i) \cdot v_\rho(Y_j)$ with linear forms u_ρ and v_ρ in the X_i and Y_j, respectively, such that $b_1, \ldots, b_N$ are contained in the linear span of $p_1, \ldots, p_r$. From this characterization, it follows that the bilinear complexity of $b_1, \ldots, b_N$ does not depend on the choice of $v_1, \ldots, v_N$, thus we may speak about the bilinear complexity of (the multiplication in) A. For a modern introduction to this topic and to algebraic complexity theory in general, we recommend [6].

A fundamental lower bound for the rank of an associative algebra A is the so-called **Alder–Strassen bound** [1]. It states that the rank of A is bounded from below by twice the dimension of A minus the number of maximal twosided ideals in A. This bound is sharp in the sense that there are algebras for which equality holds. For instance, for $A = k^{2\times 2}$, we get a lower bound of 7, since $k^{2\times 2}$ is a simple algebra and has only one

L. Kučera and A. Kučera (Eds.): MFCS 2007, LNCS 4708, pp. 669–680, 2007.

maximal twosided ideal (other than $k^{2\times 2}$). 7 is a sharp bound, since we can multiply 2×2–matrices with 7 multiplications by Strassen's algorithm.

An algebra A has **minimal rank** if the Alder–Strassen bound is sharp, that is, the rank of A equals twice the dimension minus the number of maximal two-sided ideals. After 25 years of effort [7,8,5,9], the algebras of minimal rank were finally characterized in terms of their algebraic structure [4]: An algebra over some field k has minimal rank if and only if

$$A \cong C_1 \times \cdots \times C_s \times k^{2\times 2} \times \cdots \times k^{2\times 2} \times A',$$

where $C_1, \ldots, C_s$ are local algebras of minimal rank with $\dim(C_\sigma / \operatorname{rad} C_\sigma) \geq 2$ (as characterized in [5]), $\#k \geq 2 \dim C_\sigma - 2$, and A' is an algebra of minimal rank such that $A' / \operatorname{rad} A' \cong k^t$ for some t. Such an algebra A' has minimal rank if and only if there exist $w_1, \ldots, w_m \in \operatorname{rad} A$ with $w_i w_j = 0$ for $i \neq j$ such that

$$\operatorname{rad} A = \mathsf{L}_A + A w_1 A + \cdots + A w_m A = \mathsf{R}_A + A w_1 A + \cdots + A w_m A$$

and $\#k \geq 2N(A) - 2$. Here L_A and R_A denote the left and right annihilator of $\operatorname{rad} A$, respectively, and $N(A)$ is the largest natural number s such that $(\operatorname{rad} A)^s \neq \{0\}$.

Algebraic preliminaries. In this work, the term **algebra** always means a finite dimensional associative algebra with unity 1 over some field k. The terms **left module** and **right module** always means a finitely generated left module and right module, respectively, over some algebra A. Every A-left module and A-right module is also a finite dimensional k-vector space (by the embedding $k \mapsto k \cdot 1$). If we speak of a basis of an algebra or a module, we always mean a basis of the underlying vector space.

A left ideal I (and in the same way, a right ideal or twosided ideal) is called **nilpotent**, if $I^n = \{0\}$ for some positive integer n. For all finite dimensional algebras A, the sum of all nilpotent left ideals of A is a nilpotent twosided ideal, which contains every nilpotent right ideal of A. This twosided ideal is called the **radical** of A.

We call an algebra A **semisimple**, if $\operatorname{rad} A = \{0\}$. The quotient algebra $A / \operatorname{rad} A$ is semisimple. An algebra A is called **simple**, if there are no twosided ideals in A except the zero ideal and A itself. Wedderburn's theorem states that every semisimple algebra A is isomorphic to a product of simple algebras and every simple algebra is of the form $D^{n\times n}$ for some division algebra D.

Model of computation. In the remainder of this work, we use a coordinate-free definition of rank, which is more appropriate when dealing with algebras of minimal rank, see [6, Chap. 14]. For a vector space V, V^* denotes the dual space of V, that is, the vector space of all linear forms on V. For a set of vectors U, $\langle U \rangle$ denotes the linear span of U, i.e., the smallest vector space that contains U.

Definition 1. *Let k be a field, U, V, and W finite dimensional vector spaces over k, and $\phi : U \times V \to W$ be a bilinear map.*

1. *A sequence $\beta = (f_1, g_1, w_1; \ldots; f_r, g_r, w_r)$ such that $f_\rho \in U^*$, $g_\rho \in V^*$, and $w_\rho \in W$ is called a bilinear computation of length r for ϕ if*

$$\phi(u, v) = \sum_{\rho=1}^{r} f_\rho(u) g_\rho(v) w_\rho \quad \textit{for all } u \in U, v \in V.$$

2. *The length of a shortest bilinear computation for ϕ is called the bilinear complexity or the rank of ϕ and is denoted by $R(\phi)$ or $R_k(\phi)$ if we want to stress the underlying field k.*
3. *If A is a finite dimensional associative k-algebra with unity, then the rank of A is defined as the rank of the multiplication map of A, which is a bilinear map $A \times A \to A$. The rank of A is denoted by $R(A)$ or $R_k(A)$.*

Equivalence of computations. Often, proofs become simpler when we normalize computations. A simple equivalence transformation of computations is the **permutation** of the products.

Trickier is the so-called **sandwiching**. Let $\beta = (f_1, g_1, w_1; \ldots; f_r, g_r, w_r)$ be a computation for an algebra A, i.e.,

$$xy = \sum_{\rho=1}^{r} f_\rho(x) g_\rho(y) w_\rho.$$

Let a, b, c be invertible elements of A. Then

$$xy = a(a^{-1}xb)(b^{-1}yc)c^{-1} = \sum_{\rho=1}^{r} f_\rho(a^{-1}xb) g_\rho(b^{-1}yc) a w_\rho c^{-1}.$$

Thus we can replace each f_ρ by $\hat{f}_\rho$ defined by $\hat{f}_\rho(x) = f_\rho(a^{-1}xb)$, g_ρ by $\hat{g}_\rho$ defined by $\hat{g}_\rho(y) = g_\rho(b^{-1}yc)$, and w_ρ by $\hat{w}_\rho = a w_\rho c^{-1}$.

For the next two equivalence transformations, we assume that A is a simple algebra, that is, $A \cong D^{n\times n}$ for some division algebra A. For an element $x \in A$, x^T denotes the transposed of x. Let $\beta = (f_1, g_1, w_1; \ldots; f_r, g_r, w_r)$ be a computation for an algebra A. Then

$$y^T x^T = (xy)^T = \sum_{\rho=1}^{r} \tilde{g}_\rho(y^T) \tilde{f}_\rho(x^T) w_\rho^T,$$

where $\tilde{g}_\rho(y)$ is defined by $\tilde{g}_\rho(y) := g_\rho(y^T)$ and $\tilde{f}_\rho(x)$ is defined by $\tilde{f}_\rho(x) := f_\rho(x^T)$. So we can change the f's with the g's (at the cost of **transposing** the w's but this will not do any harm since in our proofs, we usually only care about the rank of the w_ρ and other quantities that are invariant under transposing).

Finally, with every matrix $x \in A$, we can associate a linear form, namely, $y \mapsto \langle\langle x, y\rangle\rangle$, where $\langle\langle ., .\rangle\rangle$ denotes the standard inner product. (We here view x and y as vectors in $k^{n^2 \cdot \dim D}$.) In this way, we will often identify f_ρ with an element of A, which we abusively call f_ρ again. For all $x, y \in A$ we have

$$xy = \sum_{\rho=1}^{r} f_\rho(x) g_\rho(y) w_\rho \qquad \text{iff} \qquad \langle\langle xy, z\rangle\rangle = \sum_{\rho=1}^{r} \langle\langle f_\rho, x\rangle\rangle \langle\langle g_\rho, y\rangle\rangle \langle\langle w_\rho, z\rangle\rangle$$

for all $z \in A$. Since $\langle\langle xy, z\rangle\rangle = \langle\langle xyz^T, 1\rangle\rangle = \operatorname{trace}(xyz^T)$ and $\operatorname{trace}(xyz^T) = \operatorname{trace}(z^T xy) = \operatorname{trace}(yz^T x)$, we can **cyclically shift** the f's, g's, and w's in this way. Altogether, the latter two equivalence transformations allow us to permute the f's, g's, and w's in an arbitrary way.

Our results. It is a natural question to ask which are the algebras whose rank is exactly one larger than the minimum.[1] We say that an algebra has **minimal rank plus one** if

$$R(A) = 2\dim A - t + 1,$$

where t is the number of maximal twosided ideals in A. We completely solve this question here for semisimple algebras over $\mathbb{R}$. A semisimple $\mathbb{R}$-algebra has minimal rank plus one iff $A = \mathbb{H} \times B$ where B is a semisimple algebra of minimal rank, that is, $B = \mathbb{C}^{2\times 2} \times \cdots \times \mathbb{C}^{2\times 2} \times \mathbb{C} \times \cdots \times \mathbb{C} \times \mathbb{R} \times \cdots \times \mathbb{R}$. Note that over $\mathbb{R}$, there is only one division algebra of dimension two, namely the complex numbers $\mathbb{C}$ (viewed as an $\mathbb{R}$-algebra), and one division algebra of dimension four, the **Hamiltonian quaternions** $\mathbb{H}$. There are no further nontrivial $\mathbb{R}$-division algebras. $\mathbb{C}$ is also the only commutative division algebra, that is, extension field over $\mathbb{R}$.

Characterization results as the one that we prove in this paper are important, since they link the algebraic structure of an algebra to the complexity. We can read off the complexity of the algebra from its structure or get at least lower bounds by inspecting the algebraic structure.

One result on the way of our characterization is a new lower bound of 17 for $\mathbb{C}^{2\times 2}$ (viewed as an $\mathbb{R}$-algebra). This bound holds for any other extension field of dimension two over arbitrary fields. This new bound improves the current best answer to an open question posed by Strassen [10, Section 12, Problem 3].

Outline of the proof. A semisimple algebra A consists of simple factors of the form $D^{n\times n}$, where D is a division algebra. It follows from results by Alder and Strassen that no factor of A can have rank $\geq 2\dim D^{n\times n} + 1$ and at least one factor has to have rank $2\dim D^{n\times n}$, i.e., has minimal rank plus one. We show that the only simple $\mathbb{R}$-algebra that has minimal rank plus one is $\mathbb{H}$, the Hamiltonian quaternions. In particular, we show that $\mathbb{C}^{2\times 2}$ does not have minimal rank plus one in Section 3. (This is the "hardest case".) Next, we show that A cannot have two factors of the form $\mathbb{H}$ in Section 2. With this, we show the characterization result in Section 4 (Theorem 3).

2 A Lower Bound for $\mathbb{H} \times \mathbb{H}$ over $\mathbb{R}$

In this section, we will prove the the following theorem.

Theorem 1. *We have* $R_{\mathbb{R}}(\mathbb{H} \times \mathbb{H}) = 16$.

Proof. It is well known that $R_{\mathbb{R}}(\mathbb{H}) = 8$, which implies that $R_{\mathbb{R}}(\mathbb{H} \times \mathbb{H}) \leq 16$. To prove the lower bound, we first will show the following claim:

Claim. *If* x, $y \in \mathbb{H}$ *are such that* x, y, *and* 1 *are linearly independent over* $\mathbb{R}$, *then* $\langle 1, x, y, x{\cdot}y\rangle = \mathbb{H}$.

Let x, $y \in \mathbb{H}$ have the above mentioned properties. The inner automorphisms act on $\mathbb{H}$ via rotation in $\mathbb{R}^3$ on the last three coordinates of each quaternion. Hence, we can assume w.l.o.g. that $x = x_1{\cdot}1 + x_2{\cdot}i$ and $y = y_1{\cdot}1 + y_2{\cdot}i + y_3{\cdot}j$, x_ν, $y_\nu \in \mathbb{R}$.

[1] This already characterizes the algebras in terms of complexity. Of course, we seek a characterization in terms of their algebraic structure.

Since 1, x, and y are still linearly independent, we know that $x_2 \neq 0 \neq y_3$ and hence $\langle 1, x, y \rangle = \langle 1, i, j \rangle$. Furthermore, the last coordinate of $x \cdot y$ equals $x_2 y_3$ and is hence not equal to zero, which proves the claim.

Let $\beta = (f_1, g_1, w_1; \ldots; f_r, g_r, w_r)$ be a computation for $\mathbb{H} \times \mathbb{H}$. We can choose two elements $\hat{a} = (a, a')$ and $\hat{b} = (b, b') \in \mathbb{H} \times \mathbb{H}$ such that their span is contained in the intersection of at least six of the kernels of $f_1, \ldots, f_r$ and a and b are linearly independent vectors in $\mathbb{R}^4$. W.l.o.g., assume that $\langle \hat{a}, \hat{b} \rangle \subseteq \ker f_1 \cap \cdots \cap \ker f_6$.

If for all possible choices $a' = 0$ and $b' = 0$, then we can split the computation into two separate computations for $\mathbb{H}$ and get a lower bound of $8 + 8 = 16$. Thus we can assume that $a' \neq 0$. Via sandwiching, we can achieve that $a = 1$ and furthermore, by letting inner automorphisms act, that $b \in \langle 1, i \rangle$. Since $a' \neq 0$, it follows that $g_7, \ldots, g_r$ generate $(\mathbb{H} \times \mathbb{H})^*$. Now, choose a vector $\hat{c} = (c, c')$, $c \neq 0$, that is contained in the intersection of the kernels of at least seven of the vectors $g_7, \ldots, g_r$ and use sandwiching to achieve $c = 1$. W.l.o.g., let $\hat{c}$ be contained in $\ker g_7 \cap \cdots \cap \ker g_{13}$. Finally, we can choose an element $\hat{d} = (d, d')$ in the intersection of the kernels of at least six of $g_7, \ldots, g_{13}$ such that 1, b, and d are linearly independent over $\mathbb{R}$. W.l.o.g., assume that $\langle \hat{c}, \hat{d} \rangle \subseteq \ker g_9 \cap \cdots \cap \ker g_{12}$. The above claim shows that $a \cdot c = 1$, $a \cdot d = d$, $b \cdot c = b$, and $b \cdot d$ span $\mathbb{H}$. In particular, the products $\hat{a} \cdot \hat{c}$, $\hat{a} \cdot \hat{d}$, $\hat{b} \cdot \hat{c}$, and $\hat{b} \cdot \hat{d}$ span a four dimensional vector space over $\mathbb{R}$. On the other hand, we know that by construction, each of these products lies in the span of $\langle w_{13}, \ldots, w_r \rangle$. Hence, r has to be at least 16. □

3 A Lower Bound for $\mathbb{C}^{2\times 2}$ over $\mathbb{R}$

The goal of this section is to prove the following theorem.

Theorem 2. *We have* $R_{\mathbb{R}}(\mathbb{C}^{2\times 2}) \geq 17$.

We will prove this theorem in two steps. We define the following property for computations. A computation $\beta := (f_1, g_1, w_1; \ldots; f_r, g_r, w_r)$ has the property (*) if the following holds:

(*) Let $x \in \mathbb{C}^{2\times 2} \setminus \{0\}$ such that there exist three different indices ν_1, ν_2, and $\nu_3 \in \{1, \ldots, r\}$ such that

$$\langle x, i \cdot x \rangle \subseteq \ker f_{\nu_1} \cap \ker f_{\nu_2} \cap \ker f_{\nu_3} \text{ or}$$
$$\langle x, i \cdot x \rangle \subseteq \ker g_{\nu_1} \cap \ker g_{\nu_2} \cap \ker g_{\nu_3} \text{ or}$$
$$\langle x, i \cdot x \rangle \subseteq \langle w_{\nu_1}, w_{\nu_2}, w_{\nu_3} \rangle^{\perp},$$

where $V^{\perp}$ is the space of all vectors u that fulfill $\langle\langle v, u \rangle\rangle = 0$ for all $v \in V$. Then x is a matrix of rank two.

In Subsection 3.1 we show that a computation for $\mathbb{C}^{2\times 2}$ of length 16 must satisfy (*) and in Subsection 3.2 we show that no such computation exists.

3.1 Computations Not Satisfying Property (*)

For a field k, let $\langle e, h, l \rangle_k$ denote the matrix multiplication tensor of dimensions $e \times h$, $h \times l$, and $e \times l$ having coefficients in k.

Lemma 1. $R_{\mathbb{R}}(\langle 1,1,2\rangle_{\mathbb{C}}) = 6$.

Proof. This tensor has rank at most six, since the complex multiplication has rank three over $\mathbb{R}$. Assume that there exists a computation $(f_1, g_1, w_2; \ldots; f_5, g_5, w_5)$ of length five for $\langle 1,1,2\rangle$. Then we can (possibly after permuting the products) assume that f_1, f_2 are a basis of $\mathbb{C}^*$ and that $g_2, \ldots, g_5$ form a basis of $(\mathbb{C}^{1\times 2})^*$. Let x_1, x_2 and $y_1, \ldots, y_4$ be the bases dual to f_1, f_2 and $g_2, \ldots, g_5$, respectively. Then we can choose an index $\nu \in \{2, \ldots, 4\}$ such that y_1 and y_ν are linearly independent over $\mathbb{C}$, which means that the span of $\langle x_1y_1, x_1y_\nu, x_2y_1, x_2y_\nu\rangle$ is a four dimensional vector space over $\mathbb{R}$. But for $i \in \{1,2\}$ and $j \in \{1,\nu\}$, $x_iy_j \in \langle w_1, w_2, w_\nu\rangle$. Since the latter is a vector space over $\mathbb{R}$ with dimension at most three, we get a contradiction. □

Lemma 2. *Let u, v, and $w \in \mathbb{C}^{2\times 2}$ and assume that there exists a rank one matrix x such that $\langle x, ix\rangle \subset \langle u, v, w\rangle^\perp$ over $\mathbb{R}$. Then we can find invertible matrices a and b such that $(aub)_{11} = (avb)_{11} = (awb)_{11} = 0$, where $(.)_{11}$ denotes the entry in position $(1,1)$.*

Proof. Let $x = (x_{11}, x_{12}, x_{21}, x_{22})$, $x_{\nu\mu} = (x'_{\nu\mu}, x''_{\nu\mu}) \in \mathbb{C}$, be a matrix with the above property. (To save some space, we write matrices occasionally as column vectors.) Let z be any of the vectors u, v, or w. The vectors $-ix$ (for convenience) and x being perpendicular to $z = (z_{11}, z_{12}, z_{21}, z_{22})$, $z_{\nu\mu} = (z'_{\nu\mu}, z''_{\nu\mu}) \in \mathbb{C}$, means that we have

$$\sum_{\nu,\mu=1}^{2} \begin{pmatrix} x''_{\nu\mu} & -x'_{\nu\mu} \\ x'_{\nu\mu} & x''_{\nu\mu} \end{pmatrix} \begin{pmatrix} z'_{\nu\mu} \\ z''_{\nu\mu} \end{pmatrix} = \begin{pmatrix} 0 \\ 0 \end{pmatrix}. \quad \text{Since the matrix} \quad \begin{pmatrix} x''_{\nu\mu} & -x'_{\nu\mu} \\ x'_{\nu\mu} & x''_{\nu\mu} \end{pmatrix}$$

is the left multiplication matrix of $\hat{x}_{\nu\mu} = i\cdot\bar{x}_{\nu\mu}$, we can also write the above sum as $\sum_{\nu,\mu=1}^{2} \hat{x}_{\nu\mu}\cdot z_{\nu\mu} = 0$. Note that the matrix $\hat{x} := (\hat{x}_{11}, \hat{x}_{12}, \hat{x}_{21}, \hat{x}_{22})$ with $\hat{x}_{\nu\mu} := i\cdot\bar{x}_{\nu\mu} = (x''_{\nu\mu}, x'_{\nu\mu})$ has rank one, too. On the other hand, multiplying z from the left by $a = (a_{11}, a_{12}, a_{21}, a_{22})$ an from the right by $b = (b_{11}, b_{12}, b_{21}, b_{22})$ yields $(azb)_{11} = a_{11}b_{11}z_{11} + a_{11}b_{21}z_{12} + a_{12}b_{11}z_{21} + a_{12}b_{21}z_{22}$. Hence, we have to find a_{11}, a_{12}, b_{11}, $b_{21} \in \mathbb{C}$ such that $a_{11}b_{11} = \hat{x}_{11}$, $a_{11}b_{21} = \hat{x}_{12}$, $a_{12}b_{11} = \hat{x}_{21}$, and $a_{12}b_{21} = \hat{x}_{22}$. This is equivalent to finding two 2-dimensional vectors (a_{11}, a_{12}) and (b_{11}, b_{21}) with complex entries such that

$$(a_{11}, a_{12}) \otimes (b_{11}, b_{21}) = \begin{pmatrix} \hat{x}_{11} & \hat{x}_{12} \\ \hat{x}_{21} & \hat{x}_{22} \end{pmatrix}.$$

This is possible if and only if $\hat{x}$ has rank one, which had been one of our assumptions. Furthermore, since $\hat{x} \neq 0$, neither both a_{11} and a_{12} nor both b_{11} and b_{21} can be zero. Hence, we can construct invertible matrices $(a_{11}, a_{12}, a_{21}, a_{22})$ and $(b_{11}, b_{12}, b_{21}, b_{22})$ such that $(azb)_{11} = 0$ for all $z \in \{u, v, w\}$. □

Proposition 1. *Let $\beta := (f_1, g_1, w_1; \ldots; f_r, g_r, w_r)$ be a computation that does not satisfy (*). Then $r \geq 17$.*

Proof. Since β does not satisfy (*), we can find three indices ν_1, ν_2, $\nu_3 \in \{1, \ldots, r\}$ and a rank one matrix x such that $\langle x, i\cdot x\rangle \subseteq \ker f_{\nu_1} \cap \ker f_{\nu_2} \cap \ker f_{\nu_3}$, $\langle x, i\cdot x\rangle \subseteq$

$\ker g_{\nu_1} \cap \ker g_{\nu_2} \cap \ker g_{\nu_3}$, or $\langle x, i \cdot x \rangle \subseteq \langle w_{\nu_1}, w_{\nu_2}, w_{\nu_3} \rangle^\perp$. W.l.o.g., assume that $\nu_1 = 1$, $\nu_2 = 2$, and $\nu_3 = 3$ and that $\langle x, i \cdot x \rangle \subseteq \langle w_1, w_2, w_2 \rangle^\perp$, for otherwise, we could exchange the f's or g's with the w's.[2] Then, by Lemma 2, we can achieve (via sandwiching) that

$$W := \langle w_1, w_2, w_3 \rangle \subseteq \begin{pmatrix} 0 & * \\ * & * \end{pmatrix}.$$

Define the two left and two right ideals L_1, L_2, R_1, and R_2 as follows:

$$L_1 := \begin{pmatrix} * & 0 \\ * & 0 \end{pmatrix}, \quad L_2 := \begin{pmatrix} 0 & * \\ 0 & * \end{pmatrix}, \quad R_1 := \begin{pmatrix} * & * \\ 0 & 0 \end{pmatrix}, \quad \text{and} \quad R_2 := \begin{pmatrix} 0 & 0 \\ * & * \end{pmatrix}$$

Each ideal is a four dimensional vector space over $\mathbb{R}$. For the following claims, define the computation $\beta' := (\tilde{g}_1, \tilde{f}_1, w_1^T; \ldots; \tilde{g}_r, \tilde{f}_r, w_r^T)$, that is obtained by transposing as described in Section 1.

Claim 1. *The triple $(\{0\}, L_2, W)$ is separable (see [6, Notation 17.15]) by β and the triple $(\{0\}, L_2, W^T)$ is separable by β', where $W^T := \{w^T : w \in W\}$.*

Assume $(\{0\}, L_2, W)$ is not separable by β. By the Extension Lemma [6, Lemma 17.18], there exists an element $l \in L_2 \setminus \{0\}$ such that $\mathbb{C}^{2\times 2} \cdot l \subseteq \{0\} \cdot l + W = W$. But $\mathbb{C}^{2\times 2} \cdot l = L_2$ is four dimensional (over $\mathbb{R}$), whereas W has dimension at most three. The second part of the claim is shown in a similar fashion.

Claim 2. *The triple (R_2, L_2, W) is separable by β or the triple (R_2, L_2, W^T) is separable by β'.*

Assume (R_2, L_2, W) is not separable by β. By the Extension Lemma, there exists an element $r \in R_2 \setminus \{0\}$ such that $r \cdot \mathbb{C}^{2\times 2} \subseteq R_2 \cdot L_2 + W$. Now, $r \cdot \mathbb{C}^{2\times 2} = R_2$ and $R_2 \cdot L_2$ contains exactly all matrices with a nonzero entry only in the lower right corner. We distinguish three different cases:

(i) $\dim(W + R_2) \geq 6$: Then the image of the projection

$$\pi_{12} : W \to \mathbb{C}, \quad \begin{pmatrix} 0 & b \\ c & d \end{pmatrix} \mapsto b,$$

is two dimensional and hence, the space $W \cap R_2$ is at most one dimensional. Furthermore, $R_2 \cdot L_2$ is two dimensional. Thus the four dimensional space R_2 cannot be contained in $R_2 \cdot L_2 + W$.

(ii) $\dim(W + L_2) \geq 6$: In this case, we can use the computation β'. But then from $\dim(W + L_2) \geq 6$ it follows that $\dim(W^T + R_2) \geq 6$ and hence, by case (i), (R_2, L_2, W^T) is separable by β'.

(iii) $\dim(W + L_2) \leq 5$: Then the image of the projection

$$\pi_{21} : W \to \mathbb{C}, \quad \begin{pmatrix} 0 & b \\ c & d \end{pmatrix} \mapsto c,$$

is at most one dimensional, which shows that the whole ideal R_2 cannot lie in the space $R_2 \cdot L_2 + W$. This proves Claim 2.

[2] Strictly speaking, we can only exchange the adjoints of the f's and g's with the w's, see Section 1. But since "having rank one" is invariant under transposing, this does not matter.

W.l.o.g., assume that (R_2, L_2, W) is separable by β and define the projection

$$\pi : \mathbb{C}^{2\times 2} \longrightarrow \mathbb{C}^{2\times 2}, \begin{pmatrix} a & b \\ c & d \end{pmatrix} \mapsto \begin{pmatrix} a & 0 \\ 0 & 0 \end{pmatrix}.$$

Let ϕ be the multiplication of $\mathbb{C}^{2\times 2}$. Since $W \subseteq \ker \pi$, it follows that

$$R(\pi \circ \phi / R_2 \times L_2) + \dim(R_2 \times L_2) + \dim W \leq r$$

by [6, Lemma 17.17]. Hence $R(\pi \circ \phi / R_2 \times L_2) + 11 \leq r$. Now, the bilinear map $\pi \circ \phi / R_2 \times L_2$ is a map

$$\pi \circ \phi / R_2 \times L_2 : \mathbb{C}^{2\times 2}/R_2 \times \mathbb{C}^{2\times 2}/L_2 \to \mathbb{C}^{2\times 2}/\ker \pi.$$

But $\mathbb{C}^{2\times 2}/R_2 = R_1$, $\mathbb{C}^{2\times 2}/L_2 = L_1$, and $\mathbb{C}^{2\times 2}/\ker \pi = \begin{pmatrix} * & 0 \\ 0 & 0 \end{pmatrix}$. It follows that $\pi \circ \phi / R_2 \times L_2 \cong \langle 1, 2, 1 \rangle$, the complex matrix multiplication tensor $\langle 1, 2, 1 \rangle$ over $\mathbb{R}$. By Lemma 1, the tensor $\langle 1, 1, 2 \rangle$ has rank six. Since this tensor is isomorphic to the tensor $\langle 1, 2, 1 \rangle$, we get $r \geq R(\pi \circ \phi / R_2 \times L_2) + 11 = 6 + 11 = 17$. □

3.2 Computations Satisfying Property (*)

Lemma 3. *Let $\beta := (f_1, g_1, w_1; \ldots; f_{16}, g_{16}, w_{16})$ satisfy (*). Then we can achieve (possibly after permutation), that $f_1, \ldots, f_8$ and $w_9, \ldots, w_{16}$ are bases of $\mathbb{R}^8$.*

Proof. We can assume that $f_1, \ldots, f_8$ is a basis. We can also assume that $g_9, \ldots, g_{16}$ and $w_9, \ldots, w_{16}$ are linearly dependent (otherwise, after probably exchanging the g's and w's, we are finished). Then the following claim holds:

Claim. *$g_1, \ldots, g_8$ and $w_1, \ldots, w_8$ are bases of $\mathbb{C}^{2\times 2}$ and for all $\nu \in \{1, \ldots, 8\}$, $\dim\langle g_9, \ldots, g_{16}, g_\nu \rangle = \dim\langle w_9, \ldots, w_{16}, w_\nu \rangle = 8$.*

Exchanging the f's and w's (again we can skip the adjoints here) gives a computation $\beta' := (w_1, g_1, f_1; \ldots; w_{16}, g_{16}, f_{16})$ for the same tensor. Assume that a nonzero matrix $y \in \ker g_9 \cap \cdots \cap \ker g_{16}$ has rank one. We know that there is a rank one matrix x such that $x \cdot y = 0 = ix \cdot y$. But this means that $x \cdot y = \sum_{\nu=1}^{8} w_\nu(x) g_\nu(y) f_\nu = 0$ and $ix \cdot y = \sum_{\nu=1}^{8} w_\nu(ix) g_\nu(y) f_\nu = 0$. Since $f_1, \ldots, f_8$ are linearly independent, we get $w_\nu(x) g_\nu(y) = w_\nu(ix) g_\nu(y) = 0$ for $\nu \in \{1, \ldots, 8\}$. Now, the image of R_y, the right multiplication with y, is four dimensional, hence, at least four of $g_\nu(y)$, $\nu \leq 8$, are nonzero. But then at least for four indices $\nu \leq 8$ we have $w_\nu(x) = w_\nu(ix) = 0$, which is a contradiction to property (*). This means that the matrix y has rank two and thus the image of $R_y = \sum_{\nu=1}^{8} g_\nu(y) w_\nu \otimes f_\nu$ is full dimensional. On the one hand, this implies that $w_1, \ldots, w_8$ has to be a basis. On the other hand, we see that $g_\nu(y)$ has to be nonzero for all $\nu \leq 8$, which proves the second part of the claim. (Note that $\dim\langle g_9, \ldots, g_{16} \rangle \geq 7$, since otherwise, we could find an invertible matrix in $\ker g_8 \cap \cdots \cap \ker g_{16}$ with the same arguments as above, which is a contradiction.) Similarly, after exchanging the g's and w's, one can conclude the same assertions for the g's. This proves the claim.

Showing that there exists a partition $I, J \subseteq \{1, \ldots, 16\}$ such that $|I| = |J| = 8$ and $\{f_i : i \in I\}$ and $\{w_j : j \in J\}$ are both bases would prove the lemma.

Now, the claim above shows that if we choose an index set $J' \subset \{9, \ldots, 16\}$, $|J'| = 7$, such that $\{g_j : j \in J'\}$ are linearly independent, every g_ν, $\nu \leq 8$, would lead to a basis $\{g_j : j \in J_\nu\}$, where $J_\nu := J' \cup \{\nu\}$. Let μ be such that $\{\mu\} = \{9, \ldots, 16\} - J'$. Then, by Steinitz exchange, there has to be a $\nu \in \{1, \ldots, 8\}$ such that

$$\{w_j : j \in (\{1, \ldots, 8\} - \{\nu\}) \cup \{\mu\}\}$$

is a basis. Exchanging the g's and the f's and setting $I := (\{1, \ldots, 8\} - \{\nu\}) \cup \{\mu\}$ and $J := J_\nu$ gives a partition with the desired properties. □

Lemma 4. *Let $x_1, \ldots, x_5 \in \mathbb{C}^{2\times 2}$ be five matrices that are linearly independent over $\mathbb{R}$. Then $\langle x_1, \ldots, x_5 \rangle$ contains a matrix of rank two.*

Proof. Omitted. □

Lemma 5. *Let $U \subseteq \mathbb{C}^{2\times 2}$ be a three dimensional subspace of rank one matrices. Then there exists an $x \in U$ such that $ix \in U$.*

Proof. Omitted. □

Proposition 2. *There does not exist any computation for $\mathbb{C}^{2\times 2}$ over $\mathbb{R}$ of length 16 that satisfies (*).*

Proof. Assume there exists such a computation $\beta := (f_1, g_1, w_1; \ldots; f_{16}, g_{16}, w_{16})$ that satisfies (*). By Lemma 3, we can assume that $f_1, \ldots, f_8$ and $w_9, \ldots, w_{16}$ are bases. Let $x_1, \ldots, x_8$ be the basis dual to $f_1, \ldots, f_8$.

Claim. *For each $j \leq 8$, the rank of x_j is two.*

Assume that the rank of x_j is one. Since the rank of L_{x_j}, the 8×8-matrix induced by the left multiplication with x_j, is four, there are four matrices $y_1, \ldots, y_4$ that are linearly independent over $\mathbb{R}$ such that $x_j \cdot y_k = 0$ for all $k \leq 4$. Define the subspace $U := \langle y_1, \ldots, y_4 \rangle \cap \ker g_j$. For each $y \in U$ we then have

$$x_j \cdot y = \sum_{\nu=1}^{16} f_\nu(x_j) g_\nu(y) w_\nu = \sum_{\nu=9}^{16} f_\nu(x_j) g_\nu(y) w_\nu = 0.$$

But $w_9, \ldots, w_{16}$ is a basis. So $(f_9(x_j)g_9(y), \ldots, f_{16}(x_j)g_{16}(y))$ must be the zero vector. Since the rank of L_{x_j} is four, at least three of the $f_\nu(x_j)$, $\nu \geq 9$, are nonzero. This means that at least for three indices $\nu \geq 9$ we have $g_\nu(y) = 0$ for every $y \in U$. But U is at least three dimensional and contains only rank one matrices. Hence, Lemma 5 tells us that we can find a vector $x \in U$ such that $ix \in U$. Now x has rank one and $\langle x, ix \rangle$ is contained in intersection of at least three $\ker g_\nu$, which contradicts property (*) and hence proves the claim.

This shows that via sandwiching we can achieve that $x_1 = I_2$ is the unit matrix and x_2 is in Jordan normal form. We consider three different cases depending on the Jordan normal form of x_2.

(i) x_2 has two different eigenvalues $\lambda_1, \lambda_2 \in \mathbb{C}$:
In this case, we use [2, Lemma 3.8]. For this, note that since

$$\begin{pmatrix}\lambda_1 & 0\\ 0 & \lambda_2\end{pmatrix}\begin{pmatrix}x_{11} & x_{12}\\ x_{21} & x_{22}\end{pmatrix}-\begin{pmatrix}x_{11} & x_{12}\\ x_{21} & x_{22}\end{pmatrix}\begin{pmatrix}\lambda_1 & 0\\ 0 & \lambda_2\end{pmatrix}=\begin{pmatrix}0 & (\lambda_1-\lambda_2)x_{12}\\ (\lambda_2-\lambda_1)x_{21} & 0\end{pmatrix},$$

$[x_2, x]$ is invertible if $x_{12} \neq 0 \neq x_{21}$.

Claim. *There is an index* $\nu \in \{3, \dots, 8\}$ *such that* $[x_2, x_\nu]$ *is invertible.*
Assume that none of the matrices $x_3, \dots, x_8$ fulfills this property, i.e., that either $(x_\nu)_{12}$ or $(x_\nu)_{21}$ is zero. Then we can find at least three matrices $x_{\nu_1}, x_{\nu_2}, x_{\nu_3}$, $\nu_j \geq 3$, such that $(x_{\nu_1})_{12} = (x_{\nu_2})_{12} = (x_{\nu_3})_{12} = 0$ or $(x_{\nu_1})_{21} = (x_{\nu_2})_{21} = (x_{\nu_3})_{21} = 0$. W.l.o.g., assume that we are in the first case and that $\nu_1 = 3$, $\nu_2 = 4$, and $\nu_3 = 5$. Then consider the space U defined by

$$U := \langle x_1, \dots, x_5\rangle \cap \langle \begin{pmatrix}(1,0) & (0,0)\\ (0,0) & (0,0)\end{pmatrix}, \begin{pmatrix}(0,1) & (0,0)\\ (0,0) & (0,0)\end{pmatrix}\rangle^{\perp}.$$

Since $\langle x_1, \dots, x_5\rangle$ is five dimensional, the dimension of U is at least three. Furthermore, U contains only matrices where the entries in the first row are zero, i.e., only matrices of rank one. By Lemma 5, U contains a rank one matrix x such that $ix \in U$. But, by construction, x and ix are in $\ker f_6 \cap \ker f_7 \cap \ker f_8$, which is a contradiction to property (*).

W.l.o.g., let x_3 be such that $[x_2, x_3]$ is invertible. Then, choosing $m = 8{-}3 = 5$ in [2, Lemma 3.8], we get that the length of the computation is at least

$$m + 8 + \frac{1}{2}\dim([x_2, x_3]\mathbb{C}^{2\times 2}) = 5 + 8 + 4 = 17.$$

(ii) x_2 has twice the same eigenvalue λ and a nilpotent part:
This means, x_2 is of the form $x_2 = \lambda I_2 + n$, where n is the matrix that has a one in the upper right corner and zeros elsewhere. But for any matrix x we then have

$$[x_2, x] = [n, x] = \begin{pmatrix}0 & 1\\ 0 & 0\end{pmatrix}\begin{pmatrix}x_{11} & x_{12}\\ x_{21} & x_{22}\end{pmatrix}-\begin{pmatrix}x_{11} & x_{12}\\ x_{21} & x_{22}\end{pmatrix}\begin{pmatrix}0 & 1\\ 0 & 0\end{pmatrix}=\begin{pmatrix}x_{21} & x_{22}-x_{11}\\ 0 & -x_{21}\end{pmatrix},$$

which is an invertible matrix if $x_{21} \neq 0$. Since $x_1, \dots, x_8$ is a basis, we can find an index $\nu \in \{3, \dots, 8\}$, such that $(x_\nu)_{21} \neq 0$. W.l.o.g., let $\nu = 3$ be such an index. Then $[x_2, x_3]$ is invertible and by [2, Lemma 3.8], we get that the computation must have length at least 17 (as in case (i)).

(iii) x_2 has twice the same eigenvalue without a nilpotent part:
Then, since x_2 is also invertible and linearly independent from x_1, we know that $\langle x_1, x_2\rangle = \langle I_2, i{\cdot}I_2\rangle$. Since L_{x_1} is invertible, we know that $g_1, g_9, \dots, g_{16}$ generate $\mathbb{C}^{2\times 2}$ as an $\mathbb{R}$ vector space. Hence, we can choose indices $\nu_1 \dots, \nu_8 \in \{1, 9, \dots, 16\}$ such that $g_{\nu_1}, \dots, g_{\nu_8}$ is a basis. Let $y_1, \dots, y_n$ be the corresponding dual basis. W.l.o.g. we can assume that $y_1, \dots, y_4$ generate $\mathbb{C}^{2\times 2}$ as a $\mathbb{C}$-vector space. This means that

$$\langle x_i y_j : 1 \leq i \leq 2, 1 \leq j \leq 4\rangle = \mathbb{C}^{2\times 2} \qquad (1)$$

over $\mathbb{R}$. On the other hand, we have

$$x_i y_j = g_i(y_j)w_i + \sum_{\mu=1}^{4} f_{\nu_\mu}(x_i)g_{\nu_\mu}(y_j)w_{\nu_\mu} + f_l(x_i)g_l(y_j)w_l,$$

where $l = \{1, 9, \ldots, 16\} - \{\nu_1, \ldots, \nu_8\}$, and hence

$$x_i y_j \in \langle w_1, w_2, w_{\nu_1}, w_{\nu_2}, w_{\nu_3}, w_{\nu_4}, w_l \rangle$$

for $1 \leq i \leq 2$ and $1 \leq j \leq 4$. But the latter is a vector space of dimension at most seven, which is a contradiction to (1). □

4 Semisimple Algebras of Minimal Rank Plus One

Theorem 3. *Let A be a semisimple algebra over $\mathbb{R}$ of rank $2\dim A - t + 1$, where t is number of maximal twosided ideals of A. Then A is of the form*

$$A \cong \mathbb{H} \times \mathbb{R}^{2\times 2} \times \cdots \times \mathbb{R}^{2\times 2} \times \mathbb{C} \times \cdots \times \mathbb{C} \times \mathbb{R} \times \cdots \times \mathbb{R}.$$

Proof. Let A be a semisimple $\mathbb{R}$-algebra. By Wedderburn's Theorem, we know that A is isomorphic to an algebra $A_1 \times \cdots \times A_t$, A_τ simple, i.e., $A_\tau \cong D_\tau^{n_\tau \times n_\tau}$, D_τ a division algebra over $\mathbb{R}$. By [6, Lemma 17.23] and using induction, we obtain

$$R(A) \geq 2\dim A - t - (2\dim A_\tau - 1) + R(A_\tau).$$

Since A is supposed to have rank $2\dim A - t + 1$, we see that

$$R(A_\tau) \leq R(A) - 2\dim A + t + (2\dim A_\tau - 1) = 2\dim A_\tau. \tag{2}$$

Hence, by [2, Theorem 1], no factor can be a matrix algebra of the form $D^{n\times n}$, $n \geq 3$ and $\dim D \geq 2$. If $\dim D = 1$, i.e., $D = \mathbb{R}$, then this follows from the lower bound for matrix multiplication in [3]. Consider an algebra $B = D^{2\times 2}$, D a finite dimensional $\mathbb{R}$-division algebra such that $\dim D \geq 4$. Then [2, Theorem 1] tells us that

$$R(B) \geq \frac{5}{2}\dim B - 6 = 10\dim_{\mathbb{R}}(D) - 6,$$

which is greater than $2\dim B$, since $\dim D \geq 4$. Because of (2), this also excludes algebras of the above form from being a factor of A. Furthermore, there is no real division algebra of dimension three and Theorem 2 shows that also $\mathbb{C}^{2\times 2}$ cannot be one of the factors.

This shows that the only factors can be $\mathbb{R}$, $\mathbb{C}$, $\mathbb{R}^{2\times 2}$, and $\mathbb{H}$. From these factors, only the latter one is an algebra that is not of minimal rank, hence it must be contained in A at least once. On the other hand, from Theorem 1 it follows that

$$R_{\mathbb{R}}(\mathbb{H} \times \mathbb{H}) = 16 > 2\dim(\mathbb{H} \times \mathbb{H}) - 1,$$

which shows that $\mathbb{H} \times \mathbb{H}$ cannot be a factor of A. □

5 Conclusion

Two natural questions arise. First, can we extend our results to other fields than $\mathbb{R}$? And second, can we extend our results to arbitrary algebras (with radical)?

Over $\mathbb{R}$, there are only two nontrivial division algebras, $\mathbb{C}$ and $\mathbb{H}$. We used this fact several times in our proofs. Over $\mathbb{Q}$, there are more division algebras. The key question to solve the problem over $\mathbb{Q}$ is the following. For any numbers a, b, we can define quaternion algebras $H(a,b)$. Over $\mathbb{R}$, they are all either isomorphic to $\mathbb{R}^{2\times 2}$ or $\mathbb{H}$. Over $\mathbb{Q}$, the situation is more complicated. Question: What is $R_{\mathbb{Q}}(H(a,b))$ (in dependence on a, b)? If $H(a,b)$ is a division algebra, then it is clear that its rank is ≥ 8, since it is not a division algebra of minimal rank. The questions is whether 8 bilinear products are also sufficient.

To the second question, we have the following partial answer: If A is an algebra of minimal rank plus one and $A/\operatorname{rad} A$ contains one factor $\mathbb{H}$, then $A = \mathbb{H} \times B$ where B is an algebra of minimal rank. If $A/\operatorname{rad} A$ does not contain the factor $\mathbb{H}$, then $A = \mathbb{R}^{2\times 2} \times \cdots \times \mathbb{R}^{2\times 2} \times B$ where B is a superbasic algebra of minimal rank plus one. So far, we do not have a complete characterization of the superbasic algebras of minimal rank plus one.

Acknowledgements. We would like to thank the anonymous referees for their helpful comments.

References

1. Alder, A., Strassen, V.: On the algorithmic complexity of associative algebras. Theoret. Comput. Sci. 15, 201–211 (1981)
2. Bläser, M.: Lower bounds for the bilinear complexity of associative algebras. Comput. Complexity 9, 73–112 (2000)
3. Bläser, M.: On the complexity of the multiplication of matrices of small formats. J. Complexity 19, 43–60 (2003)
4. Bläser, M.: A complete characterization of the algebras of minimal bilinear complexity. SIAM J. Comput. 34(2), 277–298 (2004)
5. Büchi, W., Clausen, M.: On a class of primary algebras of minimal rank. Lin. Alg. Appl. 69, 249–268 (1985)
6. Bürgisser, P., Clausen, M., Shokrollahi, M.: Algebraic Complexity Theory. Springer, Heidelberg (1997)
7. de Groote, H.F.: Characterization of division algebras of minimal rank and the structure of their algorithm varieties. SIAM J. Comput. 12, 101–117 (1983)
8. de Groote, H.F., Heintz, J.: Commutative algebras of minimal rank. Lin. Alg. Appl. 55, 37–68 (1983)
9. Heintz, J., Morgenstern, J.: On associative algebras of minimal rank. In: Poli, A. (ed.) Applied Algebra, Algorithmics and Error-Correcting Codes. LNCS, vol. 228, pp. 1–24. Springer, Heidelberg (1986)
10. Strassen, V.: Algebraic complexity theory. In: van Leeuwen, J. (ed.) Handbook of Theoretical Computer Science, vol. A, pp. 634–672. Elsevier Science Publishers B.V, Amsterdam (1990)

Structural Analysis of Gapped Motifs of a String

Esko Ukkonen*

Helsinki Institute for Information Technology
Helsinki University of Technology and University of Helsinki
P.O. Box 68 (Gustaf Hällströmin katu 2b)
FIN-00014 University of Helsinki
Esko.Ukkonen@cs.helsinki.fi

Abstract. We investigate the structure of the set of gapped motifs (repeated patterns with don't cares) of a given string of symbols. A natural equivalence classification is introduced for the motifs, based on their pattern of occurrences, and another classification for the occurrence patterns, based on the induced motifs. Quadratic–time algorithms are given for finding a maximal representative for an equivalence class while the problems of finding a minimal representative are shown NP–complete. Maximal gapped motifs are shown to be composed of blocks that are maximal non–gapped motifs. These can be found using suffix–tree techniques. This leads to a bound on the number of gapped motifs that have a fixed number of non–gapped blocks.

1 Introduction

A gapped pattern is a string of symbols consisting of regular alphabet symbols and of joker ('don't care') symbols. A regular symbol a matches only a itself while the joker symbol matches any alphabet symbol. Given a string S of regular symbols, we are interested in finding gapped patterns, called the motifs of S, that occur in S repeatedly. Such questions are of interest from the point of view of better understanding of various forms of repetitions in strings. The applications in the analysis of DNA and other molecular biological sequences are also important. For example, motifs can be used for modeling different putative gene regulatory elements in DNA.

The number of different gapped motifs can be large. For example, string $a^m b a^m$ has $\Omega(2^m)$ different gapped motifs [10]. If the joker is not allowed, the situation becomes much simpler. Then we are dealing with non–gapped motifs that are simply contiguous substrings of S, and hence their number is always at most quadratic in the length of S. The suffix–tree of S is a powerful tool that can be used for finding and completely enumerating and indexing the substring motifs. This can even be accomplished in time that is linear in the length of S.

In this paper we continue the recent work on gapped motifs, e.g., in [1, 2, 3, 8, 9, 10]. We contribute to the problems of classifying and discovering gapped motifs for S. We introduce equivalence classifications for motifs and for their

* Supported by the Academy of Finland under grant 21196 (From Data to Knowledge).

L. Kučera and A. Kučera (Eds.): MFCS 2007, LNCS 4708, pp. 681–690, 2007.

occurrence patterns that seem natural and helpful in clarifying the conceptual basis of this very subtle domain. Finding maximal and minimal representatives for an equivalence class are shown polynomial–time solvable and, respectively, NP–complete. We also exhibit a connection between gapped and non–gapped motifs, showing that a maximal gapped motif is composed of blocks, that by themselves are maximal substring motifs and correspond to some internal nodes of the suffix–tree.

We assume that the reader has basic familiarity with the core string matching techniques (see, e.g., [4, 6]) such as the Aho–Corasick algorithm, suffix-tree techniques as well as the Fast Fourier Transform based algorithm for string matching with don't cares in $O(n \log n)$ time [5]. It should be emphasized that the conceptual framework introduced here is a major part of our contribution, also simplifying the proofs of some earlier results.

2 Patterns and Motifs

A *gapped pattern* P is a string of symbols consisting of regular alphabet symbols and of *joker* (also called *don't care*) symbols. Denoting the regular alphabet by Σ and the joker by ?, ? $\notin \Sigma$, a gapped pattern is a string in alphabet $\Sigma \cup \{?\}$. We additionally require that a pattern is either empty or its first and last symbols are from Σ. Hence a pattern P is a string in $\Sigma^* \cup \Sigma^+(\{?\}^+\Sigma^+)^+$. Any regular symbol $a \in \Sigma$ of P matches only a itself while ? matches any symbol.

Pattern $P = p_1p_2 \ldots p_m$ *occurs* at location i in a string $S = s_1s_2 \ldots s_n$ over Σ, if $p_j = s_{(i-1)+j}$ for all $j = 1, \ldots, m$ such that $p_j \neq$?. The list of all occurrence locations of P in S is denoted $L(P) = (l_1, \ldots, l_k)$ where $l_1 < \ldots < l_k$ are the occurrence locations. Pattern P is called a *motif* of S if it occurs in S repeatedly. We use throughout the mildest possible repetitiveness constraint by requiring that to be a motif, P has to occur at least *two times* in S. Hence pattern P is a motif if and only if $|L(P)| \geq 2$.

Given pattern P and string S of length n, a basic task is to find all occurrences of P in S. Here we can utilize the structure of P that is composed of alternating blocks of regular symbols and joker symbols ?. Any nonempty P can uniquely be decomposed as

$$P = A_1J_2A_2 \cdots J_kA_k$$

where each *solid block* A_i is in Σ^+ and each *joker block* J_i is in $\{?\}^+$. For example, pattern `AAB?A??BB` has solid blocks `AAB`, `A`, and `BB`.

For completeness, we sketch three alternative algorithms for finding the occurrences of P, all based on well–known techniques. The basic on–line algorithm builds an Aho–Corasick automaton for the solid blocks A_i and finds their occurrences in S with the automaton. A subsequent scan over the occurences finds the locations where $A_i, A_2, \ldots, A_k$ occur at the distances as given by P. All this takes time $O(kn)$ where k is the number of solid blocks.

Another possibility is to build a suffix–tree or a suffix–array for S, to find for each A_i the list $L(A_i)$ of its occurrences in S, and to build the occurrence list for P by constructing the intersection $\cap_i(L(A_i) - d_i)$ where $(L(A_i) - d_i)$ is the list $L(A_i)$ shifted to the left by the amount d_i where d_i is the distance between the first symbol of A_1 and the first symbol of A_i in P. The intersection can be constructed in time $O(\sum_i |L(A_i)|)$ which is always $O(kn)$ but can in the best case be much less.

The third possibility is to find the occurrences of P in time $O(n \log n)$ using the FFT algorithm for pattern matching with don't cares.

Unfortunately the number of different motifs of some S is very high. A well–known example from Pisanti et al [10] is $S = \mathtt{A}^m\mathtt{B}\mathtt{A}^m$. Any pattern $P \in \mathtt{A}(\mathtt{A} \cup \mathtt{?})^{m-2}\mathtt{A}$ is a motif of this S. There are 2^{m-2} such motifs.

3 Equivalence of Motifs with Similar Occurrences

The equivalence concept to be presented in this section captures what is invariant of motifs that occur similarly.

Definition 1. *Any two motifs A and B of S are* occurrence–equivalent, *denoted $A \sim_S B$, if the pattern of their occurrences in S is the same. That is, $A \sim_S B$ if $L(A) = L(B) + d$ for some constant d where $L(A)$ and $L(B)$ are the sets of the occurrence locations of A and B in S.*

Clearly, $\sim_S$ is really an equivalence relation (i.e., the relation is reflexive, symmetric, and transitive).

For example, for string HATTIVATTI the motifs T?I and A?T are occurrence–equivalent as $L(\mathtt{T?I}) = \{3, 8\}$ and $L(\mathtt{A?T}) = \{2, 7\} = L(\mathtt{T?I}) - 1$.

Let us define a *translation representation* $\tau(L(P))$ for the list $L(P)$ of the occurrences of motif P as follows. Assume that $L(P) = (l_1, l_2, \ldots, l_k)$ where the locations l_i are given in increasing order $l_i < l_{i+1}$. We define $\tau(L(P)) = (l_2 - l_1, l_3 - l_1, \ldots, l_k - l_1)$. Then it should be obvious that $A \sim_S B$ iff $\tau(L(A)) = \tau(L(B))$. Translation representation is translation–invariant, i.e., it does not change if the locations l_i are translated as $l_i + d$ by some constant d.

Theorem 1. *Given gapped patterns A and B, one can test whether or not $A \sim_S B$ in time $O(kn)$ or $O(n \log n)$ where k is the total number of solid blocks of A and B.*

Proof. Using the methods of the previous section, find the lists $L(A)$ and $L(B)$, in sorted order. This takes time $O(kn)$ or $O(n \log n)$. Then test in time $O(n)$ if the corresponding elements of the two lists differ by the same constant (or construct $\tau(L(A))$ and $\tau(L(B))$ and test if $\tau(L(A)) = \tau(L(B))$). □

The *content* of a pattern or a motif P is defined as the number of its non–joker symbols. For example, AA??A?A has content 4.

Definition 2. *Motif P is a* maximal motif *of S if P has the largest content among the motifs that belong to the equivalence class of P in the occurrence–equivalence $\sim_S$.*

Motif P is a minimal motif *of S if P has the smallest content among the motifs that belong to the equivalence class of P in the occurrence–equivalence $\sim_S$.*

For string HATTIVATTIHHT, we have ATTI??T $\sim_S$ ATTI $\sim_S$ I $\sim_S$ A where ATTI??T is maximal and I and A are minimal motifs.

One easily sees that an occurrence–equivalence class has a unique maximal motif; otherwise one could join two separate maximal motifs to get a motif with larger content. The above example shows, however, that an equivalence class can have several minimal motifs.

It is not difficult to see that our concept of a maximal motif coincides with that of [10] although the definitions differ technically. Equivalence idea similar to ours was used in [2, 8] to introduce maximal motifs.

We describe the construction of the maximal motif for a given motif using an illustrative technique based on *multiple self–alignments* of S. A self–alignment of S with respect to translations $I = (d_1, \ldots, d_k)$, $d_1 < \ldots < d_k$, is formed by taking $k+1$ copies of S and translating them such that the locations $1, 1+d_1, \ldots, 1+d_k$ of copies $1, 2, \ldots, k+1$, respectively, become on top of each other. With respect to the first copy, the second copy is translated d_1 units to the left and so on, such that the last copy is translated d_k units to the left. The aligned symbols of different copies are called the *columns* of the self–alignment. For example, the column of the first symbol s_1 of the first copy contains $s_1, s_{1+d_1}, \ldots, s_{1+d_k}$. The motif construction inspects the content of the *full* columns (the columns for symbols $s_1, s_2, \ldots, s_{n-d_k}$ of the first copy) that intersect all copies.

A quadratic–time algorithm for constructing the maximal occurrence–equivalent motif $M(P)$ for a given pattern P builds a self–alignment of $S = s_1 s_2 \ldots s_n$ with respect to all occurrences of P, and then reads the maximal equivalent motif by expanding P whenever the aligned strings agree on full columns. Note that the occurrences of P become aligned and therefore the resulting motif contains P. The operation is essentially the same as the extended *Merge* of [10].

The construction proceeds as follows.

1. Find the occurrences $L(P)$ of P in S. If $|L(P)| < 2$, stop as P is not a motif of S.
2. Let $\tau(L(P)) = (d_1, d_2, \ldots, d_k)$ be the translation invariant presentation of $L(P)$.
3. Self–align S with respect to $\tau(L(P))$ and build from the full columns of the self–alignment a *consensus motif* $C = c_1 \cdots c_{n-d_k}$ where for $i = 1, \ldots, n-d_k$, the symbol c_i is defined as

$$c_i = \begin{cases} s_i \text{ if } s_i = s_{i+d_1} = \ldots = s_{i+d_k} \\ ? \text{ otherwise} \end{cases}$$

4. The maximal motif $M(P)$, $M(P) \sim_S P$, is the motif obtained from C by removing the leading and trailing joker symbols ?, to get a syntactically correct motif.

By the construction, $\tau(L(M(P))) = \tau(L(P))$ and $M(P)$ has largest possible content, hence $M(P) \sim_S P$. Step 1 takes time $O(kn) = O(n^2)$. As $|L(P)| = O(n)$, Step 2 takes time $O(n)$, and Step 3 time $O(n^2)$. Step 4 needs $O(n)$, and the total time hence becomes $O(n^2)$.

Summarized, we have obtained the following; c.f., Lemma 4 of [2].

Theorem 2. *Given a gapped pattern P, the equivalent maximal motif $M(P)$ of S can be constructed in time $O(n^2)$.*

To illustrate the above construction of $M(P)$, let us consider a string S = `HATTIVAHATTIVHTTTIVAT` and a pattern P = `T??V`. Then $L(P) = \{3, 10, 16\}$ and hence $\tau(L(P)) = (7, 13)$. The self–alignment becomes as follows, with the consensus C of the full columns shown below the horizontal line:

```
                    HATTIVAHATTIVHTTTIVAT
             HATTIVAHATTIVHTTTIVAT
      HATTIVAHATTIVHTTTIVAT
      ----------------------------------
                    H?TTIV??
```

Removing the leading and trailing jokers from the consensus gives $M($`T??V`$)$ = `H?TTIV`.

Finding an equivalent motif with smallest content is hard:

Theorem 3. *The problem of finding an occurrence–equivalent minimal gapped motif for a given gapped motif of a binary string S is NP–complete.*

Proof. We reduce the NP–hard Set Cover problem to the following decision version of our problem.

Minimal Gapped Motif: Given a string S, a gapped motif P of S and an integer κ, has S a motif P', $P' \sim_S P$, such that the content of P' is $\leq \kappa$?

Let $(U, \{C_1, \ldots, C_k\}, K)$ be an instance of Set Cover where each C_j is a subset of the basic set $U = \{1, \ldots, t\}$ and K is an integer. The problem is to decide whether or not there are K sets among $\{C_1, \ldots, C_k\}$ such that their union equals U.

This instance is transformed into an instance of Minimal Gapped Motif as follows. String S is of the form $S = B_0 B_1 \ldots B_t B_{t+1}$ where the $t+2$ blocks B_i are strings of length k. The block $B_i = b_{i1} \cdots b_{ik}$, $1 \leq i \leq t$, tells which sets C_j cover the element $i \in U$:

$$b_{ij} = \begin{cases} 0 \text{ if } i \in C_j \\ j \text{ if } i \in U \setminus C_j \end{cases}$$

Moreover, the extra blocks B_0 and B_{t+1} are defined as $B_0 = B_{t+1} = 12 \cdots k$. We may assume that each $i \in U$ belongs to some C_j, hence blocks B_i, $1 \leq i \leq t$, differ from blocks $B_0 = B_{t+1}$.

Claim: Collection $\{C_{j_1}, \ldots, C_{j_s}\}$ where $j_1 < \ldots < j_s$ is a cover of U if and only if motif $Q = j_1 ?^{j_2 - j_1} j_2 ?^{j_3 - j_2} \ldots ?^{j_s - j_{s-1}} j_s$ is occurrence–equivalent with motif $P = 12 \cdots k$ of S.

Proof of Claim. Only if: First note that P and Q are really motifs of S: they occur in B_0 and B_{t+1}, P occurs at location 1 and Q at location j_1 of the two blocks. Motif P does not occur elsewhere. To show that $P \sim_S Q$, it suffices to prove that Q does not occur somewhere else in S. To derive a contradiction, assume Q has such an extra occurrence. Then Q must occur at location j_1 of some B_i, $i \neq 0$, $i \neq t+1$. Hence $b_{ij} = j$ for $j = j_1, \ldots, j_s$. This means, by the construction of the block B_i, that the element i of U does not belong to C_j for $j = j_1, \ldots, j_s$. But then the collection $\{C_{j_1}, \ldots, C_{j_s}\}$ does not cover i, hence not U, contradicting our assumption.

If: Assume $P \sim_S Q$. Then no block B_i, $1 \leq i \leq t$, has an occurrence of Q meaning that none of these blocks contains all of $j_1, \ldots, j_s$. Hence $b_{ij} = 0$ for some $j = j_1, \ldots, j_s$, which means by the construction of S that $i \in C_j$ for some $j = j_1, \ldots, j_s$, that is, $\{C_{j_1}, \ldots, C_{j_s}\}$ is a cover of U. This completes the proof of *Claim*.

NP–hardness of MINIMAL GAPPED MOTIF now follows as by *Claim*, there is a cover of size $\leq K$ if and only if there is Q, $Q \sim_S P$, with content $\leq K$. As MINIMAL GAPPED MOTIF clearly is in NP, Theorem 3 follows for strings S in unlimited alphabet.

However, the reduction to MINIMAL GAPPED MOTIF can be modified into binary alphabet. To this end encode in the construction of S, P, and Q each symbol "0" by "00", each symbol $\neq$ "0" by "01", and add at the end of each block a separator string "011". Add the separator "011" also at the end of motif P. In motif Q, also replace each "?" by "0?" and add to the end the string "$(0?)^{k-j_s}011$" in which "0?" is repeated $k - j_s$ times. Checking that the required properties are valid in the binary case is left to the reader. □

While the equivalence classification of $\sim_S$ reduces the number of essentially different motifs, their number can still be high. The bad example $\mathtt{A}^m\mathtt{B}\mathtt{A}^m$ is still bad as the number of equivalence classes (or equivalently, the number of different maximal motifs) is $> 2^{m-2}$.

4 Equivalence of Self–alignments

In Section 3 we constructed maximal and minimal motifs under equivalence $\sim_S$ starting from a given motif P. The multiple self–alignment of S that aligns all occurrences of P was used as a technical tool in the algorithm of Theorem 2. Here we start from a given self–alignment. We define an equivalence for self–alignments such that the classes of this equivalence will be in one–to–one correspondence with the classes of $\sim_S$.

Let $I = (d_1, \ldots, d_k)$, $d_1 < \ldots < d_k$, be the translation representation of some locations of S. The *consensus* defined by the multiple self–alignment of S with respect to I is – as already defined in the previous section – string $C = c_1 \ldots c_{n-d_k}$ where for $i = 1, \ldots, n - d_k$:

$$c_i = \begin{cases} s_i \text{ if } s_i = s_{i+d_1} = \ldots = s_{i+d_k} \\ ? \text{ otherwise} \end{cases}$$

By removing from the consensus C the possible ?'s before the first and after the last regular symbol, we obtain a motif (which may also be the empty string). We denote this motif by $\mu(I)$. It is easy to see that a non–empty $\mu(I)$ is a maximal motif of equivalence $\sim_S$ (as also stated in Lemma 5 of [10]) but it may have more occurrences than just those that give I.

Definition 3. *Translation representations I and I' are* motif–equivalent *in S, denoted $I \simeq_S I'$, if $\mu(I) = \mu(I')$.*

Again, $\simeq_S$ is an equivalence relation.

As an example, for string `BABAACACCACA` and for representations $I = (3)$ and $I' = (8)$ we have $I \simeq_S I'$ as $\mu(I) = \mu(I') =$ `A?A`.

Note that the classes of the equivalences $\sim_S$ and $\simeq_S$ correspond to each other one–to–one via the maximal motifs: each maximal motif P has a unique occurrence–equivalence class $\{P' \mid P' \sim_S P\}$ and a unique motif–equivalence class $\{I \mid \mu(I) = P\}$.

It is immediate from the construction of $\mu(I)$, that if $\mu(I) = \mu(I')$ then also $\mu(I \cup I') = \mu(I)$. This gives the following theorem. Here $I \cup I'$ is the merge of the two lists.

Theorem 4. *If $I \simeq_S I'$ then $I \simeq_S (I \cup I')$.*

Definition 4. *A translation representation I is* maximal *if it is a largest element of its class in the motif–equivalence $\simeq_S$, and I is* minimal *if it is a smallest element in its class.*

By Theorem 4, the maximal element of an motif–equivalence class is the union (merge) of all members in the class. Therefore the maximal element is unique. The example after Definition 3 shows that there can be several different minimal elements.

Theorem 5. *a) A translation representation has a unique motif–equivalent maximal representation, and this representation can be found in time $O(n^2)$;*
b) The problem of finding a minimal motif–equivalent translation representation for a given translation representation is NP–complete for binary strings S.

Proof. a) Construct in time $O(n^2)$ the motif $\mu(I)$. Then find $L(\mu(I))$ in time $O(n^2)$, and finally construct the translation invariant presentation $\tau(L(\mu(I)))$. Obviously $\tau(L(\mu(I)))$ is maximal, otherwise $L(\mu(I))$ should be larger.

b) The membership in NP is clear. The hardness is shown again by reducing Set Cover to the following decision version of our problem: Given a string S, a translation representation I of some locations of S, and a constant κ, is there a translation representation I' such that $I' \simeq_S I$ and $|I'| \leq \kappa$.

Let $(U, \{C_1, \ldots, C_k\}, K)$ be an instance of Set Cover where $U = \{1, 2, \ldots, k\}$, each $C_j \subseteq U$, and K is an integer. Build a string $S = \#B_0\#B_1\# \cdots \#B_k\#B_{k+1}\#$ where each block $B_i = b_{i1} \cdots b_{i,t+2}$ is a string of length $t + 2$ and $\#$ is a separator. Block B_0 is a special block defined as $b_{0j} = 0$ for all j, and B_{k+1} is a special block defined as $b_{k+1,j} = 0$ for $j = 1, \ldots, t$ and

$b_{k+1,t+1} = b_{k+1,t+2} = 1$. The remaining blocks B_i are bit–vector presentations of sets C_i, defined as

$$b_{ij} = \begin{cases} 1 \text{ if } j \in C_i \\ 0 \text{ if } j \in U \setminus C_i \end{cases}$$

Moreover, $b_{i,t+1} = 1$ and $b_{i,t+2} = 0$. The translation representation I is $I = (t+2, 2(t+2), \ldots, (k+1)(t+2))$ which aligns all blocks B_i on top of each other. Then $\mu(I)$ is the motif $\#?^{t+2}\#$ that contains $t+2$ joker symbols in the middle.

Now the NP–hardness is shown simply by verifying that the set cover instance has a solution of size $\leq K$ if and only if there is a translation representation I', $I' \simeq_S I$, such that $|I'| \leq K+1$. Here representation I' will contain elements $i(t+2)$ of I such that C_i belongs to the set cover. Moreover, I' must always contain $(k+1)(t+2)$ as this is the only possibility to get the last two jokers of the motif $\mu(I) = \mu(I') = \#?^{t+2}\#$. This forcing in the construction is necessary to guarantee that $\mu(I')$ will not become longer than one block (plus the two separators).

This proves NP–hardness when S is in alphabet of size 3. By replacing the symbol # by binary string $01^{t+3}0$ we obtain a binary variant for which the above proof is still valid. □

Testing whether or not S has motifs with large content and many occurrences is known to be hard [7]. However, if we do not pose constraints on the number of occurrences, then the motif with largest content can be found fast. In fact, let $I \subseteq I'$ be translation invariant representations. Then the motif $\mu(I')$ is a submotif of $\mu(I)$, and the content of $\mu(I')$ is $\leq$ the content of $\mu(I)$. To find a motif with largest content it therefore suffices to consider representations I with only one element, i.e., consider $I_d = (d)$ for $d = 1, \ldots, n-1$.

The motif $\mu(I_d)$ has a solid symbol on locations where the two copies of S, when shifted by d with respect to each other, match. Hence to find a motif with largest content we have to count the matching symbols for each possible d, and take the maximum. Let

$$\mathcal{C}(d) = |\{\, i \mid s_i = s_{i+d} \text{ and } 1 \leq i \leq n-d\}|$$

be the *match count* of I_d.

Theorem 6. *The motif of S with largest content has* $\max_d \mathcal{C}(d)$ *solid symbols, and it can be constructed in time $O(n \log n)$.*

Proof. The construction directly from the definition of $\mathcal{C}(d)$ would take $O(n^2)$ time. However, match counts $\mathcal{C}(d)$ can also be evaluated in time $O(n \log n)$ using FFT (see e.g. [6]). □

5 Representation by Maximal Non–gapped Motifs

In this section we give a representation for gapped motifs in terms of non–gapped ones. This will also give an improved upper bound for the number of gapped motifs, parametrized with the number blocks of the motif.

A *non–gapped pattern* P, also called a *substring pattern*, is a pattern that does not contain any jokers, i.e., P is a string in Σ^*. The occurrence–equivalence $\sim_S$ generalizes immediately for non–gapped motifs. A non–gapped P is maximal if it has largest content in its occurrence–equivalence class of non–gapped motifs. Note that an equivalence class may have more than one maximal non–gapped motif.

For example, the maximal gapped motifs of string `HATTIVATTIAA` are `ATTI?A`, `TT`, and `A` while its maximal non–gapped motifs are `ATTI`, `TT`, and `A`.

Given a non–gapped pattern P it is of interest to find the maximal non–gapped pattern P' that contains P and is occurrence–equivalent to P. This can be solved fast using the suffix–tree $T(S)$ of S; recall that $T(S)$ is the compacted trie that represents all suffixes of S [6,4]. The so–called *suffix links* are important tools in the linear–time suffix–tree construction as well as in many applications of suffix–trees: if Y is a node of $T(S)$ for string ax, $a \in \Sigma$, and X is the node for string x, then there is a suffix link from Y to X.

First construct $T(S)$ and find the internal node $V(P)$ of $T(S)$ such that

(i) P is a prefix of the string $\overline{V(P)}$ that spells out the path from the root to $V(P)$, and
(ii) $\overline{V(P)}$ is shortest possible.

Now, $\overline{V(P)}$ is a right–maximal substring pattern that contains P, that is, $\overline{V(P)}$ cannot be expanded to the right without loosing some occurrence of P. To make it also left–maximal, we must check whether or not it is possible to add something to the beginning of $\overline{V(P)}$. If $V(P)$ is on the path that spells out the entire S or if there are at least two suffix links that point to $V(P)$, we are done. In that case an expansion to the left is not possible as there are at least two alternatives. If, however, exactly one suffix link points to $V(P)$, then we go to the source node of the link and repeat this until a node W is found such that W is on the path for the entire S or at least two suffix links enter W. Then the string $\overline{W}$ from root to W is the maximal non–gapped motif that contains P.

This takes time $O(n)$, and we have obtained the following.

Theorem 7. *The maximal non–gapped motif of S that contains a given non–gapped motif P and is occurrence–equivalent to P can be found in time $O(n)$.*

It is well-known that S has only $\Theta(n)$ right–maximal non–gapped motifs: each internal node of $T(S)$ gives one such motif. Maximality in both directions still reduces the motifs:

Theorem 8. *String S has $\leq n$ maximal non–gapped motifs; they can be found in time $O(n)$.*

Proof. Construct $T(S)$ in time $O(n)$ and find the internal nodes W of $T(S)$ that correspond to the maximal non–gapped motifs using the criteria that either W is on the path for S or the number of suffix links entering W is at least 2. This can be accomplished by traversing the tree and its $O(n)$ suffix links, taking time $O(n)$. The number of such nodes W is $\leq n$ as $T(S)$ has $\leq n$ internal nodes. □

The next representation theorem follows quite directly from our definitions.

Theorem 9. *If B is a non–gapped block of a maximal gapped motif of S, then B is a maximal non–gapped motif of S.*

This leads to the following upper bound for the number of different gapped motifs.

Corollary 1. *String S has $\leq n^{2k-1}$ different maximal gapped motifs that have k non–gapped blocks.*

Proof. By Theorem 8, there are $< n$ maximal non–gapped motifs, and each of the $k-1$ gaps between the k blocks must have length $< n$. □

References

1. Apostolico, A., Parida, L.: Incremental Paradigms of Motif Discovery. J. Computational Biology 11, 15–25 (2004)
2. Arimura, H., Uno, T.: A polynomial space and polynomial delay algorithm for enumeration of maximal motifs in a sequence. In: Deng, X., Du, D.-Z. (eds.) ISAAC 2005. LNCS, vol. 3827, pp. 724–737. Springer, Heidelberg (2005)
3. Crochemore, M., Iliopoulos, C.S., Mohamed, M., Sagot, M.-F.: Longest repeats with a block of k don't cares. In: Theoretical Computer Science, vol. 362, pp. 248–254 (2006)
4. Crochemore, M., Rytter, W.: Jewels of Stringology, World Scientific (2002)
5. Fischer, M.J., Paterson, M.S.: String matching and other products. In: Complexity of Computation. SIAM–AMS Proc, pp. 113–125 (1974)
6. Gusfield, D.: Algorithms on Strings, Trees, and Sequences. Cambridge University Press, Cambridge (1997)
7. Michael, M., Nicolas, F., Ukkonen, E.: On the complexity of finding gapped motifs. Submitted manuscript (2007)
8. Pelfrêne, J., Abdeddaïm, S., Alexandre, J.: Extracting approximate patterns. J. Discrete Algorithms 3, 293–320 (2005)
9. Pisanti, N., Crochemore, M., Grossi, R., Sagot, M.F.: A basis of tiling motifs for generating repeated patterns and its complexity for higher quorum. In: Rovan, B., Vojtáš, P. (eds.) MFCS 2003. LNCS, vol. 2747, pp. 622–631. Springer, Heidelberg (2003)
10. Pisanti, N., Crochemore, M., Grossi, R., Sagot, M.F.: Bases of motifs for generating repeated patterns with wild cards. IEEE/ACM Transactions on Computational Biology and Bioinformatics 2, 40–50 (2005)

Online and Offline Access to Short Lists

Torben Hagerup

Institut für Informatik, Universität Augsburg, 86135 Augsburg, Germany
hagerup@informatik.uni-augsburg.de

Abstract. We consider the list-update problem introduced by Sleator and Tarjan, specializing it to the case of accesses only and focusing on short lists. We describe a new optimal offline algorithm, faster than the best previous algorithm when the number of accesses is sufficiently large relative to the number ℓ of items. We also give a simple optimal offline algorithm for $\ell = 3$. Taking c_ℓ to denote the best competitive ratio of a randomized online algorithm for the list-access problem with ℓ items, we demonstrate that $c_3 = 6/5$ and give new upper and lower bounds on c_4. Finally we prove a strengthened lower bound for general ℓ.

1 Introduction

The List-Access Problem

In the *list-access* or *LA* problem, a list L containing the items in a finite set I is to be maintained during a sequence of accesses to items in I. The set I is fixed, but its items may be reordered in L in the course of the accesses to improve the performance. An access is assumed to have a cost equal to the position in L, just before the access, of the item x accessed, i.e., to one more than the number of items preceding x in L at that time. As part of the access to an item x, x can be moved at no cost to an earlier position in L, with the order in L of the other items remaining unchanged. This is considered to happen in a number of *free exchanges*, each of which interchanges x with the item currently preceding it in L. In addition, at all times except during an access, it is possible to interchange two arbitrary adjacent items in L with a *paid exchange* at a cost of 1. We formalize the list L as a permutation of I, namely as the function from I to $\{1, \ldots, |I|\}$ that maps each item to its position in L. An instance of the LA problem is given by a pair (π_0, σ), where, for some finite set I, $\pi_0 : I \to \{1, \ldots, |I|\}$ is the initial permutation and $\sigma \in I^*$, called the *request sequence*, is the sequence of items to be accessed. A solution to the instance (π_0, σ) specifies, for each access requested in σ, the free and paid exchanges to be carried out as part of the access and just before it, respectively, starting from the initial permutation π_0. The cost of the solution is the total cost of the accesses plus the number of paid exchanges. A solution to (π_0, σ) whose cost is minimal, over those of all solutions to (π_0, σ), is said to be *optimal*, and we denote its cost by $Opt(\pi_0, \sigma)$. Similarly, the cost of the solution produced by an algorithm $\mathcal{A}$ for the LA problem on the input (π_0, σ) is denoted by $\mathcal{A}(\pi_0, \sigma)$. The algorithm $\mathcal{A}$ may be *offline*, in which case it can base each of its decisions on knowledge of the entire sequence σ, or *online*,

L. Kučera and A. Kučera (Eds.): MFCS 2007, LNCS 4708, pp. 691–702, 2007.

in which case, when an item to be accessed is revealed to it, it must decide on the paid and free exchanges to be carried out before that access and during it, respectively, before it learns about the following item to be accessed, if any.

Previous Work

The more general (dynamic) list-update problem was formalized by Sleator and Tarjan [8], who proposed to judge the quality of online algorithms by comparing them with optimal offline algorithms, i.e., offline algorithms that always compute an optimal solution. For $c \geq 1$, a deterministic online algorithm $\mathcal{A}$ for the LA problem is *c-competitive* if there is a constant α such that for every instance (π_0, σ) of the LA problem, $\mathcal{A}(\pi_0, \sigma) \leq c \cdot \mathit{Opt}(\pi_0, \sigma) + \alpha$. If $\mathcal{A}$ is randomized, in the definition $\mathcal{A}(\pi_0, \sigma)$ is replaced by $E(\mathcal{A}(\pi_0, \sigma))$, the expected cost incurred by $\mathcal{A}$ on the instance (π_0, σ). If the constant α can be taken to be zero, $\mathcal{A}$ is *strictly c-competitive.* The *competitive ratio* of $\mathcal{A}$ is the infimum of the set of those c for which $\mathcal{A}$ is c-competitive (taken to be ∞ if this set is empty).

Sleator and Tarjan showed that the simple *Move-To-Front* or *MTF* algorithm, which carries out all possible free exchanges and no paid exchanges, is 2-competitive, and that no deterministic online algorithm for the LA problem has a competitive ratio smaller than 2. Later authors specialized the LA problem to the LA_ℓ problem, for every $\ell \in \mathbb{N} = \{1, 2, \ldots\}$, by fixing the number $|I|$ of items to be ℓ. Irani [5] established that for all $\ell \in \mathbb{N}$, the smallest possible competitive ratio of a deterministic online algorithm for the LA_ℓ problem (defined in the obvious way) is exactly $2 - 2/(\ell + 1)$.

In the case of randomized algorithms, our knowledge is less complete. For all $\ell \in \mathbb{N}$, denote by c_ℓ the infimum of the set of competitive ratios of (randomized) online algorithms for the LA_ℓ problem, and take $c_\infty = \sup_{\ell \in \mathbb{N}} c_\ell$. The best currently known randomized online algorithm for the LA problem, the COMB algorithm of Albers, von Stengel and Werchner [1], has the competitive ratio 8/5 and therefore proves that $c_\infty \leq 8/5$. On the other hand, Teia [9] establishes the lower bound $c_\ell \geq 3/2 - 5/(\ell + 5)$ for all $\ell \in \mathbb{N}$ and therefore $c_\infty \geq 3/2$. A number of results for small values of ℓ are mentioned in the literature. The lower bound $c_2 \geq 9/8$ is attributed to Karp and Raghavan [4,7], and a matching upper bound $c_2 \leq 9/8$ was obtained by Chrobak and Larmore [4]. The bounds $c_3, c_4 \leq 3/2$ are stated in [1], and Reingold, Westbrook and Sleator [7] report the bounds $c_3 \geq 1.1998$ and $c_4 \geq 1.2467$, obtained by iterative numerical methods. A parenthetical remark in [7] also mentions the bounds $c_3 \geq 1.2$ and $c_4 \geq 1.25$, but gives no proof or indication of how they were obtained. In addition to the standard so-called i or *full cost model* defined above, in which accessing a list item in position i costs i, researchers have studied an $i-1$ or *partial cost model*, in which an access to the item in position i is assumed to cost only $i - 1$. This model is easier to analyze. If we define $\hat{c}_\ell$ as c_ℓ for all $\ell \in \mathbb{N}$, but with respect to the partial cost model, Albers, von Stengel and Werchner [2] argue that $\hat{c}_2 = \hat{c}_3 = \hat{c}_4 = 3/2$.

In the following, we will consistently use ℓ to denote the number $|I|$ of items and n to denote the number $|\sigma|$ of requests. The following simple optimal offline algorithm for $\ell = 2$ was described by Reingold and Westbrook [6]: In every access to the item x at the back of the list L, move x to the front of L exactly if the next request exists and is also to x; do no paid exchanges. For $\ell \geq 3$, no simple optimal offline algorithm was known prior to the work reported here, but Reingold and Westbrook propose an algorithm based on dynamic programming that runs in $O(2^\ell(\ell-1)!n)$ time. The space needed by the algorithm is $\Theta(\ell!n)$ or $\Theta(\ell!)$, depending on whether the solution itself or only its cost is needed.

New Results

The present paper deals mostly with the case of small values of ℓ; specifically, $\ell \in \{2,3,4\}$. Clearly not too much practical interest is tied up with maintaining lists of up to four items. However, since the associated problems are already theoretically interesting and challenging, they seem a natural starting point for learning more about the general list-access problem. In detail, our results are the following.

A new optimal offline algorithm. Building on the work of Reingold and Westbrook [6], we propose a new optimal offline algorithm with a running time of $O(2^\ell \ell! f(\ell)+n+r)$, where $r = O(\ell n)$ is the size of the output and $f : \mathbb{N} \to \mathbb{N}$ is a certain function. The space needed is $O((\ell+1)!f(\ell)+n)$ or $O(\ell! f(\ell))$, depending on whether the solution itself or only its cost is needed. It is easy to establish that $f(\ell) \leq \ell! 3^{\ell!}$ for all $\ell \in \mathbb{N}$, but f is likely to be significantly smaller. Formally speaking, the new algorithm is incomparable with the algorithm of Reingold and Westbrook, being faster if n is sufficiently large relative to ℓ and slower in the opposite case. From a practical point of view, the new algorithm is probably the best algorithm known for $\ell = 4$, but it loses out rapidly for larger values of ℓ due to the growth of $f(\ell)$.

A simple optimal offline algorithm for $\ell = 3$. The algorithm of Reingold and Westbrook [6] works in $O(n)$ time for each fixed value of ℓ, but we describe a new $O(n)$-time algorithm for $\ell = 3$ that is simple, intuitive and easy to execute by hand, properties not shared by the dynamic-programming algorithm.

Improved bounds on c_ℓ. We determine that $c_3 = 6/5$. This shows the lower bound of 1.2 obtained by Reingold, Westbrook and Sleator [7] to be tight. We also prove a bound of the form $5/4 - \epsilon \leq c_4 \leq 4/3$, where $\epsilon \approx 3 \cdot 10^{-5}$. Reingold, Westbrook and Sleator already indicated a lower bound of 1.25, but without giving any details. Finally, we modify Teia's proof [9] to obtain a slightly strengthened lower bound of $3/2 - 2/(\ell+3)$ for general values of ℓ. The previous and new bounds on c_ℓ are summarized in the following table.

ℓ	Old lower bound	New lower bound	New upper bound	Old upper bound
2	9/8 (Karp and Raghavan)	—	—	9/8 [4]
3	1.2 [7]	6/5	6/5	3/2 [5]
4	1.25 [7]	$5/4-\epsilon$	4/3	3/2 [1]
5	1.2728 [7]	5/4	—	8/5 [1]
6	23/22 [9], 1.268 [7]	23/18	—	8/5 [1]
⋮	⋮	⋮	⋮	⋮
ℓ	$3/2-5/(\ell+5)$ [9]	$3/2-2/(\ell+3)$	—	8/5 [1]

2 A New Optimal Offline Algorithm

Let an instance (π_0, σ) of the LA problem be given, let I be the domain of π_0 and take $\ell = |I|$ and $n = |\sigma|$. In the interest of simplicity, let us consider only the problem of computing $Opt(\pi_0, \sigma)$. Write $\sigma = r_1 \cdots r_n$, with $r_i \in I$ for $i = 1, \ldots, n$. Let $Perm(I)$ be the set of permutations of I and, for $\pi_1, \pi_2 \in Perm(I)$, denote by $dist(\pi_1, \pi_2)$ the number of interchanges of adjacent items needed to transform π_1 into π_2. Call π_1 and π_2 *adjacent* if $dist(\pi_1, \pi_2) = 1$. Let $\mathbb{N}_0 = \{0, 1, \ldots\}$, let S be the semiring $(\mathbb{N}_0 \cup \{\infty\}, \min, +, \infty, 0)$, let $\mathcal{V}$ be the set of column vectors with entries in S and indexed by permutations in $Perm(I)$, and let $\mathcal{M}$ be the set of square matrices with entries in S and with rows and columns indexed by permutations in $Perm(I)$. Reingold and Westbrook [6] observe that

$$Opt(\pi_0, \sigma) = \mathbf{0}^{\mathrm{T}} A_{r_n} A_{r_{n-1}} \cdots A_{r_1} v_0, \tag{1}$$

where $A_x \in \mathcal{M}$ for all $x \in I$, $\mathbf{0}^{\mathrm{T}}$ is the transpose of the all-zero vector $\mathbf{0} \in \mathcal{V}$, and $v_0 \in \mathcal{V}$ maps each $\pi \in Perm(I)$ to $dist(\pi_0, \pi)$ (i.e., the entry of π is $dist(\pi_0, \pi)$). The main thrust of their proof is to show that for all $x \in I$, A_x can be chosen to have $O(2^\ell(\ell-1)!)$ non-∞ entries that can be computed in $O(2^\ell(\ell-1)!)$ time. Because of this, $Opt(\pi_0, \sigma)$ can be found via (1) in $O(2^\ell(\ell-1)!n)$ time.

Suppose that the product (1) is evaluated from left to right. For $i = n+1, n, \ldots, 1$, $w_i^{\mathrm{T}} = \mathbf{0}^{\mathrm{T}} A_{r_n} \cdots A_{r_i}$ is a row vector that maps each $\pi \in Perm(I)$ to $Opt(\pi, r_i \cdots r_n)$. Let us call $w_{n+1}^{\mathrm{T}}, \ldots, w_1^{\mathrm{T}}$ *cost vectors.* The general step in the computation is to multiply a cost vector with one of the matrices $A_1, \ldots, A_\ell$ to obtain another cost vector. The $n+1$ cost vectors arising in the computation are all distinct, but some of them differ in just an additive constant, which is what we will exploit. Let ρ be the function that maps each $w \in \mathcal{V}$ to the pair $(\min w, w - \min w)$, where $\min w$ is the smallest entry in w and $w - \min w$ is the vector obtained from w by subtracting $\min w$ from each of its entries. It is easy to see that if $A \in \mathcal{M}$, $w \in \mathcal{V}$ and $\rho(w) = (m, z)$, then $w^{\mathrm{T}} A = m + z^{\mathrm{T}} A$ and $w^{\mathrm{T}} v_0 = m + z^{\mathrm{T}} v_0$. Because of this, we can compute $Opt(\pi_0, \sigma)$ by initializing $z_{n+1} := \mathbf{0}$ and $m_{n+1} := 0$ and then, for $i = n, \ldots, 1$, if $\rho(z_{i+1}^{\mathrm{T}} A_{r_i}) = (m, z)$, taking $z_i := z$ and $m_i := m_{i+1} + m$, with $m_1 + z_1^{\mathrm{T}} v_0$ being the final answer. This computation can be viewed as the processing of the input $r_n \cdots r_1$ by a slightly embellished deterministic finite automaton (DFA) G_I that has elements of $\mathcal{V}$ as its states and an edge (a transition) labeled $(x, m) \in I \times \mathbb{N}_0$ from $z \in \mathcal{V}$ to $z' \in \mathcal{V}$ if $\rho(z^{\mathrm{T}} A_x) = (m, z')$. As concerns the finiteness of G_I, it suffices for G_I to have the state $\mathbf{0}$ and all states reachable from it via edges defined as above. But every such state, viewed as an element of $\mathcal{V}$, has at least one entry equal to zero, and the entries of adjacent permutations cannot differ by more than one. Therefore the number of states in G_I, which is what we call $f(\ell)$, is bounded by $\ell! 3^{\ell!}$. G_I can be constructed in $O(2^\ell(\ell-1)!)$ time per edge, after which the request sequence can be processed in constant time per request. This leads to the following result.

Theorem 1. *Given an instance (π_0, σ) of the LA problem, $Opt(\pi_0, \sigma)$ can be computed in $O(2^\ell \ell! f(\ell) + n)$ time using $O(\ell! f(\ell))$ space, where ℓ is the number of items and $n = |\sigma|$ is the number of requests.*

By way of example, take $I = \{x, y\}$. We represent $\pi \in Perm(I)$ by the sequence $\pi^{-1}(1) \cdots \pi^{-1}(\ell)$. For $\pi \in Perm(I)$ and $v \in \mathcal{V}$, let $v[\pi]$ be the entry of π in v and represent v by the sequence $v[xy]v[yx]$. Then G_I takes the form shown in Fig. 1.

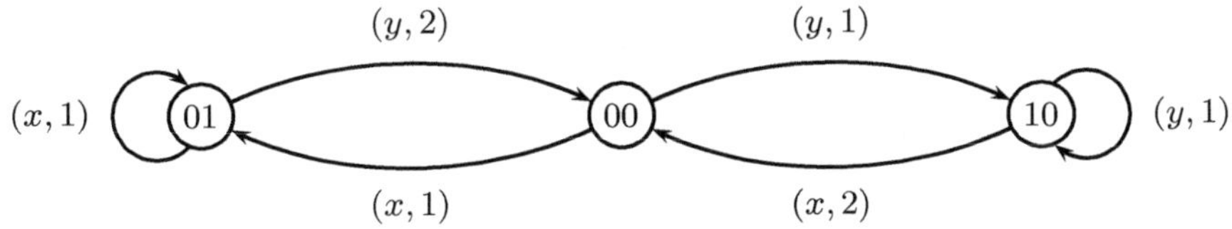

Fig. 1. The DFA G_{xy} for computing the optimal offline cost of sequences in $\{x, y\}^*$

To determine the optimal offline cost of processing the request sequence $yxyyxx$, e.g., one can trace out the path taken by the DFA on the reverse sequence, starting from the state $\mathbf{0} = 00$, and sum the second components of the edges followed. The path ends in the state 10, and the sum of the second components of the edge labels is 8. If the initial permutation is $\pi_0 = xy$, a final multiplication with the vector $v_0 = 01$, which simply picks out the entry of π_0, yields the total cost of 9.

Let $z_1, \ldots, z_{n+1}$ be defined as above, i.e., z_i is the state of G_I after the processing of $r_n \cdots r_i$, for $i = 1, \ldots, n + 1$. One can prove the following:

Lemma 2. *For $\ell = 2$, a solution of (π_0, σ) is optimal if and only if it carries out no paid exchanges after the last access and adheres to the following rule: For $i = 1, \ldots, n$,*

(1) do not carry out a paid exchange, changing the current permutation from π to π', just before the access to r_i unless $z_i[\pi'] < z_i[\pi]$.
(2) carry out a free exchange, changing the current permutation from π to π', as part of the access to r_i if $z_{i+1}[\pi'] < z_{i+1}[\pi]$, but not if $z_{i+1}[\pi'] > z_{i+1}[\pi]$.

3 A Simple Optimal Offline Algorithm for $\ell = 3$

Given a request sequence $\sigma \in I^*$ and two distinct items $x, y \in I$, the *projection* of σ on $\{x, y\}$ is the subsequence σ_{xy} of σ obtained from σ by deleting all occurrences of items other than x and y. Similarly, a permutation π_0 of I and a solution to the instance (π_0, σ) can be projected on $\{x, y\}$, simply by ignoring all items other than x and y and all interchanges involving such items. The use of projections in the analysis of algorithms for the LA problem is a standard technique that goes back to Bentley and McGeoch [3] and Irani [5]. The cost model of the LA problem can be thought of as assigning a cost of 1 to each item that is in front of an item y when y is accessed and a cost of 1 to the access itself. Because of this, if a solution to an instance with $\ell = 3$ is projected on each of the three subsets of I of size 2, the resulting total cost of the three projections will exceed the cost of the original solution by exactly n, the original number of requests: Each "in front of" cost unit is counted in exactly one projection, whereas each "access itself" cost unit is counted in exactly two projections, which results in an "overcount" of n. In particular, this means that if each of the three projections is optimal, the original solution is also optimal. Our goal, therefore, is to show that at each point during the solution of an instance with $\ell = 3$ (in the "forward" direction, but with "lookahead"), one can proceed in a way that does not violate optimality in any of the three projections on pairs of items.

Suppose that the current list is xyz. It is easy to see that the only troublesome case is that of an access to z in which the projection on $\{x, z\}$ requires z to move in front of x and the projection on $\{y, z\}$ requires z to stay behind y. With Lemma 2, this can be seen to imply that after the processing of the outstanding requests in reverse order, not including the request to z, the DFA G_{xz} as in Fig. 1 is in the state 10, and G_{yz} is in the state 01. But forming the cross product of the three DFAs G_{xy}, G_{xz} and G_{yz}, after first extending them to do nothing upon receipt of the "foreign" symbol, shows that then G_{xy} must be in the state 10 before and after the processing of the request to z. By Lemma 2, this state allows a paid exchange of x and y, after which z can move in front of x but stay behind y, as required.

A formulation of the optimal offline algorithm for $\ell = 3$ that is imprecise, but easy to remember, goes as follows: At an access to an item z, use the optimal offline algorithm for $\ell = 2$ for z and each item preceding z to decide whether to interchange z with that item. If a conflict arises, resolve it by interchanging the two items in front of z and then move z between them.

The principle behind the algorithm above does not extend to $\ell = 4$. The reason is that for $\ell = 4$, there are request sequences whose optimal solutions do not have projections on pairs of items that are also optimal. A shortest example, for the initial permutation *abcd*, is the request sequence *dbccdad*.

4 Randomized Lower Bounds for Small ℓ

Fix $\ell \in \mathbb{N}$ and, for definiteness, take $I = \{1, \ldots, \ell\}$. Suppose that for some $n \in \mathbb{N}$, Σ is a random sequence drawn from the uniform distribution over I^n. Then for every (deterministic or randomized) online algorithm $\mathcal{A}$, $E(\mathcal{A}(\pi_0, \Sigma)) \geq n(\ell+1)/2$ for every initial permutation $\pi_0 \in \mathit{Perm}(I)$ (with equality if the algorithm carries out no paid exchanges). The reason is that no matter how the algorithm orders its list before an access, the access will be to an item whose position in the list is uniformly distributed over $\{1, \ldots, \ell\}$. But with $d_\ell = \liminf_{n\to\infty} E(\mathit{Opt}(\pi_0, \Sigma))/n$ (the choice of π_0 is immaterial), this means that $c_\ell \geq (\ell+1)/(2d_\ell)$. A similar observation was used to obtain lower bounds on c_ℓ by Reingold, Westbrook and Sleator [7], who attribute it to Karp. The difference to the work reported here is that whereas Reingold, Westbrook and Sleator bound d_ℓ from above by means of an offline algorithm with bounded lookahead found by an extensive search, here we compute the exact value of d_ℓ by analyzing the optimal offline algorithm of Section 2.

Indeed, for $\ell = 2$, the random request sequence Σ sends the DFA of Fig. 1 on a random walk of exactly n steps, each step of which chooses the next edge to follow from the uniform distribution over the edges that leave the current state.

The corresponding cost is the sum of the second components of the labels of the edges traversed plus a quantity bounded by $\binom{\ell}{2}$, namely what was called $z_1^{\mathrm{T}} v_0$ in Section 2.

The graph of Fig. 1, considered as a Markov chain, is irreducible and aperiodic and so has a unique stationary distribution, which, as is easy to verify, assigns probability 1/3 to each of the three states. Because of this, d_2 is the average of the second components of the edges of the graph, i.e., 4/3. Together with the formula $c_2 \geq 3/(2d_2)$, this reproves the result $c_2 \geq 9/8$ of Karp and Raghavan.

For general $\ell \in \mathbb{N}$, we can consider $G_\ell := G_I$ as a Markov chain with some state set V_ℓ and a probability of $1/\ell$ associated with each edge (note, however, that G_ℓ in general contains parallel edges and loops). For each $v \in V_\ell$, let r_v be the average of the second components of the labels of the edges that leave v. One can prove that G_ℓ is irreducible and aperiodic. If its stationary distribution assigns probability p_v to v for each $v \in V_\ell$, then $d_\ell = \sum_{v \in V_I} p_v r_v$. Thus d_ℓ can be computed automatically.

In practice, for the present purposes there is no need to distinguish between states, such as 01 and 10 in Fig. 1, that can be transformed into each other by permuting the elements of I (in the example, by interchanging x and y), so each group of states that are equivalent in this sense can be merged into a single state. Moreover, the label on each edge can be simplified to its second component. For $\ell = 3$, this leads to the graph of Fig. 2. For a verification, it is helpful to know that the vertex labels use the order $123, 213, 231, 321, 312, 132$ of the permutations of $I = \{1, 2, 3\}$.

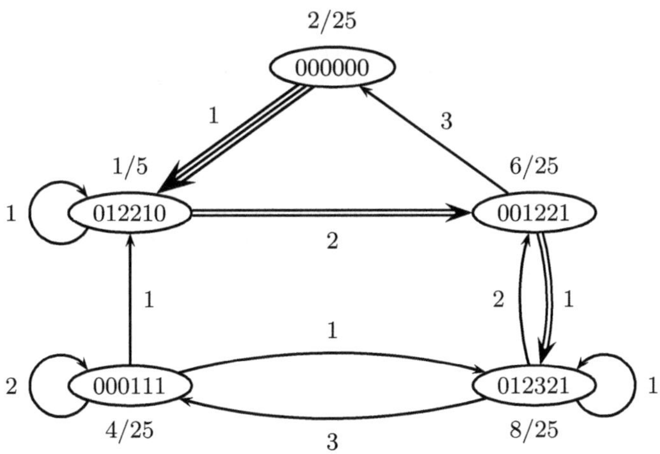

Fig. 2. The graph G_3 after the merging of equivalent states

Each vertex is shown with its probability in the stationary distribution, and an easy calculation demonstrates that $d_3 = 5/3$ and therefore that $c_3 \geq 6/5$. The corresponding Markov chain for $\ell = 4$ has 80 states. Computing its stationary distribution proves that $c_4 \geq p/q$, where $p = 4575402276067795174675434$55 and $q = 3660396535859744114839327$64, i.e., $c_4 > 5/4 - \epsilon$ with $\epsilon \approx 2.5515 \cdot 10^{-5}$.

5 An Optimal Randomized Online Algorithm for $\ell = 2$

This section introduces the randomized online *odd-even* algorithm for $\ell = 2$ and proves that it is optimal, i.e., has the smallest competitive ratio of any online algorithm (whereas an offline algorithm is optimal if it always produces an optimal solution). An optimal randomized online algorithm for $\ell = 2$ was given previously by Chrobak and Larmore [4].

Let y_0 be the item at the back of the initial list. The odd-even or OE algorithm never carries out a paid exchange. At the first access to y_0, it moves y_0 to the front of the list with probability 1/2 and initializes $prev := y_0$ and $even := true$. At every subsequent access, say to an item x, it does the following, where MTF(p), for $0 \le p \le 1$, denotes the action of moving the item accessed to the front of the list (if it is not already there) with probability p.

```
if even then MTF(3/4);
if prev = x then even := false else even := ¬even;
prev := x;
if even then MTF(3/7);
```

The key to analyzing the rather unintuitive algorithm and especially extensions of it is to focus on what we will call the *state* of the execution. For $\ell = 2$, the state is the probability that the previous access was to the item at the front of the algorithm's list. It is easy to see that immediately after the first access to y_0, OE's state is 1/2 and $even = true$. Between accesses from then on, the state is 1/2 if $even = true$ and 7/8 if $even = false$. To see this, suppose that $even = true$ before an access. Then the access changes *even* to *false* and executes MTF(3/4), which changes the state from 1/2 to $(1/2)\cdot(3/4)+(1/2)\cdot 1 = 7/8$. If $even = false$ before an access, on the other hand, either (if the condition $prev = x$ is satisfied) the access changes neither *even* nor the state, or it changes *even* to *true* and executes MTF(3/7), which changes the state from 7/8 to $(7/8)\cdot(3/7)+(1/8)\cdot 1 = 1/2$.

Suppose now that OE and an arbitrary offline algorithm $\mathcal{A}$ process a request sequence σ in an interleaved fashion, starting from a common initial list π_0 with the item y_0 at the back. We analyze the process using the potential function Φ defined as follows: Until the first access to y_0, $\Phi = 0$. Between accesses after that, $\Phi = 1/4$ if $even = true$. If $even = false$, $\Phi = -1/8$ if *prev* is at the front of $\mathcal{A}$'s list, and $\Phi = 1$ otherwise. We will show that the inequality

$$E(\mathrm{OE}(r)) + \Delta\Phi(r) \le \frac{9}{8}\mathcal{A}(r) \qquad (2)$$

holds for each request r in σ, where $\Delta\Phi(r)$ denotes the increase in Φ caused by the processing of r by both OE and $\mathcal{A}$. A paid exchange by $\mathcal{A}$ costs 1 to $\mathcal{A}$ and increases Φ by at most 9/8, so we can disregard paid exchanges in the following. The first access to y_0, if there is one, costs 2 to both algorithms and increases Φ by 1/4, in accordance with (2), and any earlier accesses cost the same to both algorithms and leave Φ unchanged. Now let r be a request to an item x after the first request to y_0 and consider three cases, distinguished by the effect on *even* of the processing of r by OE:

Case 1: *even* changes from *true* to *false*. Because its state is $1/2$, the expected cost to OE of serving r is $3/2$. If $\mathcal{A}$ has x at the back of its list after serving r, it incurs a cost of 2 in doing so, and the potential increases by $3/4$, namely from $1/4$ to 1. Otherwise $\mathcal{A}$ incurs a cost of at least 1, and Φ decreases by $3/8$, namely from $1/4$ to $-1/8$. In either case, (2) holds.

Case 2: *even* changes from *false* to *true*. The expected cost to OE is at most $15/8$, and the previous request was to the item y other than x. Either the potential drops by $3/4$, namely from 1 to $1/4$, and the cost to $\mathcal{A}$ is 1, or the potential increases from $-1/8$ to $1/4$, i.e., by $3/8$, and the cost to $\mathcal{A}$ is 2. In each case, (2) holds.

Case 3: *even* is *false* before and after the processing of r. The previous request was also to x, so the expected cost to OE is $9/8$, and the potential cannot increase. Once again, (2) holds.

Summing the inequality (2) over all requests r in σ and noting that the overall decrease in Φ is bounded by $1/8$ shows that

$$E(\text{OE}(\pi_0, \sigma)) \leq \frac{9}{8}\mathcal{A}(\pi_0, \sigma) + \frac{1}{8}.$$

Recalling that $c_2 \geq 9/8$, we obtain the following.

Theorem 3. OE *is* $(9/8)$*-competitive, and this is optimal.*

The proof leaves open the possibility that OE may not be strictly $(9/8)$-competitive. This is indeed the case, as evidenced by the request sequence $y_0 y_0$. This effect can be viewed as caused by the impossibility of starting in the state $7/8$. The strict competitive ratio of OE is $7/6$, and this is the best possible for any online algorithm for LA_2. To see the latter, consider the initial list xy and the two request sequences yx and yy.

6 An Optimal Randomized Online Algorithm for $\ell = 3$

Searching for a good randomized online algorithm $\mathcal{R}$ for $\ell \geq 3$ whose competitive ratio is demonstrated similarly as in the previous section can be formulated as solving a linear program $\mathcal{L}_\ell$. Before the constraints of $\mathcal{L}_\ell$ can be specified, we need additional terminology.

Take $I = \{1, \ldots, \ell\}$ and let $\mathcal{P}$ be the set of unordered pairs of items in I. Let $\mathcal{F}$ be the set of functions that map $\mathcal{P}$ to $\{0, 1\}$. A function $f \in \mathcal{F}$ represents the values of the variable *even* in all $\binom{\ell}{2}$ projections of $\mathcal{R}$ on elements of $\mathcal{P}$. More precisely, suppose that a permutation $\tau \in \mathit{Perm}(I)$, initially the identity permutation $\mathit{id} \in \mathit{Perm}(I)$, is maintained during the processing of a sequence of requests to I such that each access moves the item x accessed to the front, i.e., replaces τ by $\mathit{MTF}(\tau, x)$, where $\mathit{MTF}(\tau, x)$ is the permutation τ' of I with

$$\tau'(y) = \begin{cases} 1, & \text{if } y = x, \\ \tau(y) + 1, & \text{if } \tau(y) < \tau(x), \\ \tau(y), & \text{if } \tau(y) > \tau(x). \end{cases}$$

We call τ the *most-recently-accessed* or *MRA* permutation. For all $\{i,j\} \in \mathcal{P}$, $f(\{i,j\}) = 0$ means that $even = true$ in the projection of $\mathcal{R}$ on $\{\tau^{-1}(i), \tau^{-1}(j)\}$, and $f(\{i,j\}) = 1$ means that $even = false$ in that projection. For $f, g \in \mathcal{F}$ and $i \in I$, we say that i *changes* f *to* g, written $f \xrightarrow{i} g$, if, informally, a request to $\tau^{-1}(i)$ changes the values of $even$ in all projections of $\mathcal{R}$ in a way that corresponds to stepping from f to g. More precisely, with $\pi' = MTF(id, i)$, $f \xrightarrow{i} g$ if $g(\{1, \pi'(j)\}) = 1 - f(\{i,j\})$ for all $j \in I$ with $j < i$, $g(\{1, \pi'(j)\}) = 1$ for all $j \in I$ with $j > i$, and $g(\{\pi'(j), \pi'(k)\}) = f(\{j,k\})$ for all $\{j,k\} \in \mathcal{P}$ with $i \notin \{j,k\}$. For $f, h \in \mathcal{F}$, h is *reachable* from f if $f = h$ or there is $g \in \mathcal{F}$ and $i \in I$ such that $f \xrightarrow{i} g$ and h is reachable from g. Let $\mathcal{F}'$ be the subset of $\mathcal{F}$ of those functions reachable from the function f_0 in $\mathcal{F}$ that maps every element of $\mathcal{P}$ to 1. We call the elements of $\mathcal{F}'$ *parity vectors*.

Some of the ideas expressed above were anticipated independently by Albers, von Stengel and Werchner [2]. In particular, for every request sequence σ, they defined a partial order $\langle\sigma]$ on I that contains largely the same information as conveyed here in the pair (τ, f) of the current MRA permutation and parity vector after the processing of σ, starting from $(\tau, f) = (id, f_0)$.

For all $\pi \in Perm(I)$ and all $x \in I$, let $ML(\pi, x)$ be the set of those permutations $\pi' \in Perm(I)$ that can be obtained from π by moving x part or all of the way to the front as in a number of free exchanges, i.e., the set of those π' with $\pi'(x) \le \pi(x)$ such that $\pi'(y) = \pi(y) + 1$ for all $y \in I$ with $\pi'(x) \le \pi(y) < \pi(x)$ and $\pi'(y) = \pi(y)$ for all $y \in I$ with $\pi(y) < \pi'(x)$ or $\pi(y) > \pi(x)$.

For every parity vector f and every permutation $\pi \in Perm(I)$, $\mathcal{L}_\ell$ has a nonnegative variable $p_{f,\pi}$, intended to represent the probability with which $\pi \circ \tau$ is the list of $\mathcal{R}$ after the processing of a request sequence that leads to the MRA permutation τ and the parity vector f, and an unrestricted variable $q_{f,\pi}$, intended to represent a potential associated with a situation in which the parity vector is f and an adversary algorithm $\mathcal{A}$ has its list ordered as $\pi \circ \tau$, where τ is the current MRA permutation. Finally, it has a variable c, which represents the competitive ratio to be demonstrated, and the objective is to minimize c.

$\mathcal{L}_\ell$ has the following linear constraints: For all $f \in \mathcal{F}'$,

$$\sum_{\pi \in Perm(I)} p_{f,\pi} = 1, \tag{3}$$

which ensures that for all parity vectors f, $\{p_{f,\pi}\}_{\pi \in Perm(I)}$ is a probability distribution on $Perm(I)$, called the *state* of f.

For all $f, g \in \mathcal{F}'$, all $i \in I$ such that $f \xrightarrow{i} g$ and all $\pi \in Perm(I)$,

$$\sum_{\mu \in ML(\pi, i)} p_{f,\mu} \le \sum_{\mu \in ML(\pi \circ (MTF(id,i))^{-1}, 1)} p_{g,\mu}, \tag{4}$$

which ensures that for all parity vectors f and g and all items x, it is possible to go from the state of f to that of g using only free exchanges when a request to x changes f to g (seeing this takes some effort).

For all $f, g \in \mathcal{F}'$, all $i \in I$ such that $f \xrightarrow{i} g$ and all $\pi_1, \pi_2 \in Perm(I)$ such that π_2 can be obtained from π_1 by moving a subset of the items before the item i to just after i without additional interchanges (see [6, Theorem 2]),

$$\sum_{\pi \in Perm(I)} \pi(i) p_{f,\pi} + q_{g,\pi_2 \circ (MTF(id,i))^{-1}} - q_{f,\pi_1} \leq c(dist(\pi_1, \pi_2) + \pi_2(i)). \quad (5)$$

The inequalities (5) are best understood by comparing them with the inequality (2). The term $\sum_{\pi \in Perm(I)} \pi(i) p_{f,\pi}$ is the expected cost to the online algorithm of serving a request to $\tau^{-1}(i)$ when the parity vector is f and the MRA permutation is τ, i.e., it corresponds to the term $E(\text{OE}(r))$ in (2). $q_{g,\pi_2 \circ (MTF(id,i))^{-1}} - q_{f,\pi_1}$ is the potential increase when the offline algorithm $\mathcal{A}$ changes its list from $\pi_1 \circ \tau$ to $\pi_2 \circ \tau$, while the parity vector changes from f to g and the MRA permutation changes from τ to $MTF(\tau, \tau^{-1}(i))$, i.e., it corresponds to the term $\Delta\Phi(r)$ in (2). Finally, $dist(\pi_1, \pi_2) + \pi_2(i)$ is the cost to $\mathcal{A}$ of changing its list from $\pi_1 \circ \tau$ to $\pi_2 \circ \tau$ and then serving a request to $\tau^{-1}(i)$ without further interchanges of items, i.e., it corresponds to the quantity $\mathcal{A}(r)$ in (2).

This concludes the description of the linear program $\mathcal{L}_\ell$. $\mathcal{L}_3$ and $\mathcal{L}_4$ have the optimal values 6/5 and 4/3, respectively. Combining this with what was shown in previous sections, we obtain the following.

Theorem 4. *$c_3 = 6/5$ and $5/4 - \epsilon < c_4 \leq 4/3$ for $\epsilon = 2.5515 \cdot 10^{-5}$.*

7 A Strengthened Lower Bound for General ℓ

In this section we show that the method of Teia [9] can yield the bound $c_\ell \geq 3/2 - 2/(\ell + 3)$, which is slightly stronger than the bound $c_\ell \geq 3/2 - 5/(\ell + 5)$ claimed by Teia.

Fix $\ell \in \mathbb{N}$ and let π_0 be a permutation of a set of ℓ items. For an integer $r \in \mathbb{N}$ that can be chosen arbitrarily, Teia considers a random request sequence Σ_r constructed by the following process that uses a list L, initially ordered as π_0: Move a pointer through L r times. As each item x is encountered, generate either one or three successive requests to x, choosing each possibility with probability 1/2. If three requests to x are generated, move x to the front of L (after passing the pointer to the next item). Teia observes that for every request sequence σ that can be generated by this process, $Opt(\pi_0, \sigma) \leq r\ell(\ell + 5)/2$. Here we instead use that $E(Opt(\pi_0, \Sigma_r)) \leq r\ell(\ell + 3)/2$. To see this, note that by keeping its list an exact copy of L and moving each item requested three times in a row to the front of the list during the first access to the item, an offline algorithm can serve the requests generated during one pass over L at an average expected cost per item equal to the average position in L of an item at the time just before its processing, i.e., $(\ell + 1)/2$, plus the average number of accesses to an item x after the first access to x, i.e., 1.

Teia proves that for every deterministic online algorithm $\mathcal{A}$,

$$E(\mathcal{A}(\pi_0, \Sigma_r)) \geq r\left(\frac{3\ell(\ell - 1)}{4} + 2\ell\right) - \binom{\ell}{2}, \quad (6)$$

where the term $\binom{\ell}{2}$ bounds the increase of a potential function. Because a randomized algorithm can be viewed as a probability distribution over a set of deterministic algorithms, (6) in fact holds for every randomized online algorithm $\mathcal{A}$, where the expectation is now taken both over the distribution of Σ_r and over the random choices made by $\mathcal{A}$. Combining (6) with the upper bound on $E(Opt(\pi_0, \Sigma_r))$ stated above, one obtains that for $c \geq 1$, $E(\mathcal{A}(\pi_0, \Sigma_r) - c \cdot Opt(\pi_0, \Sigma_r)) \geq \frac{r}{4}(3\ell(\ell-1) + 8\ell - 2c\ell(\ell+3)) - \binom{\ell}{2}$. If $\mathcal{A}$ is c-competitive, the left-hand side must remain bounded as $r \to \infty$. This is possible only if $3\ell(\ell-1) + 8\ell - 2c\ell(\ell+3) \leq 0$, i.e., if $c \geq 3/2 - 2/(\ell+3)$.

Theorem 5. *For all $\ell \in \mathbb{N}$, $c_\ell \geq 3/2 - 2/(\ell+3)$.*

References

1. Albers, S., von Stengel, B., Werchner, R.: A combined BIT and TIMESTAMP algorithm for the list update problem. Inform. Process. Lett. 56, 135–139 (1995)
2. Albers, S., von Stengel, B., Werchner, R.: List update posets, Manuscript (1996)
3. Bentley, J.L., McGeoch, C.C.: Amortized analyses of self-organizing sequential search heuristics. Comm. Assoc. Comput. Mach. 28, 404–411 (1985)
4. Chrobak, M., Larmore, L.L.: The server problem and on-line games. DIMACS Series in Disc. Math. and Theoret. Comput. Sci. 7, 11–64 (1992)
5. Irani, S.: Two results on the list update problem. Inform. Process. Lett. 38, 301–306 (1991)
6. Reingold, N., Westbrook, J.: Off-line algorithms for the list update problem. Inform. Process. Lett. 60, 75–80 (1996)
7. Reingold, N., Westbrook, J., Sleator, D.D.: Randomized competitive algorithms for the list update problem. Algorithmica 11, 15–32 (1994)
8. Sleator, D.D., Tarjan, R.E.: Amortized efficiency of list update and paging rules. Comm. Assoc. Comput. Mach. 28, 202–208 (1985)
9. Teia, B.: A lower bound for randomized list update algorithms. Inform. Process. Lett. 47, 5–9 (1993)

Optimal Randomized Comparison Based Algorithms for Collision

Riko Jacob*

Computer Science Department, Technische Universität München
jacob@in.tum.de

Abstract. We consider the well known problem of finding two identical elements, called a collision, in a list of n numbers. Here, the (very fast) comparison based algorithms are randomized and will only report existing collisions, and do this with (small) probability p, the success probability. We find a trade-off between p and the running time t, and show that this trade-off is optimal up to a constant factor. For worst-case running time t, the optimal success probability is $p = \Theta\left(\min\{t/n, 1\}t/(n \log t)\right)$. For expected running time t, the success probability is $p = \Theta\left(t/(n \log n)\right)$.

1 Introduction

We consider the problem of finding two equal elements, called a collision, in a list of numbers. This is a fundamental problem in computational complexity theory and among the first decision problems for which non-trivial lower bounds were known. It is usually formulated as the decision if all elements of a list are different, and is hence called **"element uniqueness"** or **"element distinctness."** One way to guarantee that no element is repeated is to provide a linear sequence of strict inequalities. Actually, in many models of computation where numbers are treated as atoms, the lower bound of $\Omega(n \log n)$ for sorting carries over to this decision problem. The problem has been studied in all kinds of machine models, for example in the stronger algebraic decision tree model [1], in a general decision tree setting [2], or on a quantum computer [3,4].

Motivation. The investigation in this paper was triggered by a question in cryptography, namely the precise complexity of computing discrete logarithms in a cyclic group of roughly $n \approx 2^h$ numbers, where the input can be specified using $O(h)$ bits. Here, cryptographic primitives are build on the assumption that exponentiation is a good one-way function, i.e., it is easy to compute $x = a^z \bmod n$, but it is difficult to invert this function, namely to compute z from x, the discrete logarithm of x to the base a.

One classical algorithm for the discrete logarithm is the so called baby-step giant-step algorithm that produces two lists A, B of $\sqrt{n}$ elements and determines the discrete log from a collision between the lists, i.e., elements $a \in A$ and $b \in B$ with $a = b$, a generalization of the collision problem considered here. If we allow

* Work done at ETH Zurich, Institute of Theoretical Computer Science, Switzerland.

L. Kučera and A. Kučera (Eds.): MFCS 2007, LNCS 4708, pp. 703–714, 2007.

the algorithm to use a hash table and assume that the algebraic operations have unit cost, this algorithm takes $O(\sqrt{n})$ time, whereas a comparison based algorithm, where a hash table is not allowed and is replaced by a balanced search tree, takes $O(\sqrt{n}\log n)$ time. Now, in the cryptographic setting, of course lower bounds for the discrete log computation are even more interesting than upper bounds. Today, no good general lower bounds are known, which led to the investigation of restricted classes algorithms. One such restriction are abstract models of computation [5], that treat the group as a black box that allows only the group operations and a comparison of elements. Now, one can show a $\sqrt{n}$ lower bound to compute the discrete log with constant probability if the group order n is prime [5], matching the performance of the hash-table based version of the baby-step giant-step algorithm. Here, we consider the situation of allowing only comparisons, which makes a hash table impossible. We show that this strengthens the lower bound by the logarithmic factor the algorithm is slowed down, and that this remains the case for small success probabilities. This setting also motivates to consider worst case running times which translate into the assumption that a code-breaking algorithm can run only for a limited time, which might be in contrast to its expected running time.

Randomized Algorithms. Here, we consider randomized algorithms (in the Monte-Carlo sense) with one-sided error. Such an algorithm must announce an existing collision with probability p, but never claim a collision for an input consisting of distinct elements. We are interested in how the running time depends on the success probability p. More precisely, we consider both the success probability p and the running time t as functions of n, and are mainly interested in the asymptotic behavior for large n.

Randomized Input. Intuitively, it is clear that a completely random input will make it particularly difficult to find a collision, and, because the input is random, even a deterministic algorithm has access to randomness. This idea directly suggests two particular uniform distributions that are discussed in detail in Section 1.2. First, there is the uniform distribution $\mathbf{q}$ of all inputs with precisely one collision. This kind of input reveals the success probability of the algorithm. Secondly, there is the uniform distribution $\mathbf{p}$ of all inputs without collision. This is useful to analyze the worst case running time of the algorithm. We analyze the asymptotic worst-case and expected running time of deterministic algorithms achieving success probability p. Using Yao's Minimax-principle [6,7, p. 35], we can transfer the obtained lower bound to randomized algorithms.

Output and allowed errors. For a comparison based algorithm, it is actually equivalent to find a collision or to state the existence of a collision without making an error. Certainly, if the algorithm needs to show a collision to have success, it can as well just state that there is a collision. But also if a comparison based algorithm announces the existence of a collision only if it is actually present in the input, then it must have compared two identical elements, and hence is in the position to report this collision.

It is also meaningful to consider only the decision problem and to allow the algorithm to announce a non-existing collision with a certain probability $q < p$. For a comparison based algorithm, this means that in some situations where no comparison showed equality, the algorithm still announces a collision. By eliminating this behavior one gets an algorithm that announces only existing collisions and has success probability at least $p - q$.

New Results. We parametrize the algorithms by a goal running time function $t(n)$. The deterministic Algorithm 1 achieves in worst-case $O(t(n))$ time a success probability $p_t(n) = \min\{\frac{t}{n}, 1\}\frac{t}{n \log t}$ when input is drawn from distribution $\mathbf{q}$. Here, and throughout the paper, log stands for the binary logarithm. In Section 2.3, this algorithm is randomized, resulting in an algorithm with the same worst-case performance on arbitrary input. Additionally, we analyze the above situation for expected running times (instead of worst-case). There we find that success probability $p(n)$ requires running time $t_p(n) = \Theta(p(n) \cdot n \log n)$, or put the other way around, that expected running time $t(n)$ allows for success probability $p_t(n) = \Theta\left(\frac{t \log t}{n \log n}\right)$.

Observe that in both situations, for any positive constant c we have $p_{c \cdot t}(n) = \Theta(p_t(n))$. Vice versa, to change the success probability by a constant factor c, it is always possible to choose a $t'(n) = \Theta(t(n))$ with $p_{t'} = c \cdot p_t$. Hence, the trade-off between running time and success probability can be formulated either way, and it is meaningful to state that the mentioned algorithm is optimal up to a constant factor.

On a side note, we also analyze the maximal number of random bits needed by an algorithm to be $\Theta(-\log p)$.

As is usual, the lower bounds are valid in a strong, non-uniform model of computation (comparison trees), whereas the algorithms are quite general and can be implemented in many uniform models of computation.

1.1 Related Work

The inverse setting, where a randomized comparison based algorithm solves **"element uniqueness"** (and not **"collision"**), has been studied by Snir [8]. There, the algorithm needs to recognize that all elements are different with probability p, but is not allowed to misclassify an input with collision. The result is a lower bound of $\lambda n \log n + \lambda \log(p(1-\lambda))$ for all $0 < \lambda < 1$, which yields, for example, for $p = 1/\log n$ (with $\lambda = 1/2$) an $\Omega(n \log n)$ lower bound. This shows an important difference to collision, where this success probability can be achieved in $O(n)$ time. Grigoriev, Karpinski, Meyer auf der Heide, and Smolensky [9] show that if the algorithm is allowed to misclassify with probability $p < 1/2$ in both directions (two-sided errors), even for algebraic decision trees of constant degree, there is an $\Omega(n \log n)$ lower bound.

For the problem **"collision"**, or **"element non-uniqueness"** as they call it, Manber and Tompa [10,11] give a lower bound of $\Omega(n \log n)$ for comparison based algorithms that are not allowed to announce a collision if there is none, and need to announce an existing collision with probability 1/2. This is the special case of our result with $p = 1/2$. Similar to our main technical Lemma 3, they bound the number of linear extensions of a graph, given that at least half of the edges must be used, an idea going back to [12]. In contrast to the results presented here, their estimate heavily depends on the success probability being 1/2, and does not give a lower bound for smaller probabilities like $1/\log n$.

Our Lemma 3 is an observation about the structure of the directed acyclic graph G_c describing the outcomes of comparisons at a (leaf) node c of the comparison tree. The observation is that a high success probability can only be achieved if the number of linear extensions is small. The permutations ending at c describe linear extensions of G_c, and the success probability of one permutation/linear extension π is given by the number of successor relations of π that coincide with edges of G. This number is also known as π, the number of steps in G [13], and often analyzed as the number of jumps $n - 1 - \pi$.

1.2 Preliminaries

The input to **Collision**$_n$ consists of a list of n numbers $x_1, \ldots, x_n$. The answer is YES if there exist $i \neq j$ such that $x_i = x_j$, and otherwise NO. The numbers are assumed to be atomic to the algorithms, and the only operation on numbers is the comparison $x \leq y$. A set of inputs that are indistinguishable by such comparisons are called an **order type**, and for **Collision**$_n$ there is no loss of generality in assuming that all $x_i \in \{1, \ldots, n\}$. An input without collision is then a permutation π understood as $x_i = \pi(i)$. The variable j with $x_j = \pi(j) = \pi(i)+1$ is called the successor of i. The distribution $\mathbf{p}$ is defined to be the uniform distribution over these $n!$ different inputs.

Choosing uniformly from the inputs with precisely one collision is called the distribution $\mathbf{q}$. There are two interesting procedures to draw from this distribution.

The **collision first** procedure starts by choosing uniformly a pair $i \neq j$ of indices. Set $x_i = x_j$, and then choose a random ordering (permutation) of the values of the variables $x_1, \ldots, x_{j-1}, x_{j+1}, \ldots, x_n$. In this way, the $(n-1)!n(n-1)/2$ different order types with precisely one collision are created with equal probability.

Alternatively, the **permutation first** procedure chooses a permutation π and a pivot $o \in \{1, \ldots, n-1\}$ of the collision, both with uniform probability. The input is created by changing the value o to the value $o+1$, creating two elements with value $o+1$. More precisely, the input is given by $x_i = \pi(i)$ if $\pi(i) \neq o$, and $x_i = o + 1 = \pi(i) + 1$ if $\pi(i) = o$. Consider the permutation π' that is identical to π, only the values of o and $o+1$ are exchanged. Then the input (π, o) is indistinguishable from the input (π', o). Further, these two representations are the only way to create this order-type. Again, the $n!(n-1)/2$ different inputs are created with equal probability.

We denote the uniform distribution of inputs without collision when the defining permutation is restricted to belong to S as $\mathbf{p}|_S$. With $\mathbf{q}|_S$ we denote the restriction of $\mathbf{q}$ where in the permutation first procedure π is restricted to the set S. Note that $\mathbf{q}|_S$ is not a uniform distribution because inputs can be created from one or two permutations of S.

The focus of our considerations is the number of comparisons performed by an algorithm. Hence, we model the algorithm as a family of comparison trees of the following type: For every input size n (number of variables), there is a comparison tree T, a rooted tree where every internal node c has a left and a right child and is annotated by a pair (i, j). Every input $\mathbf{x} = x_1, \ldots, x_n$ (with or without collision) defines a path $P(\mathbf{x})$ in T, the next node is given by the outcome of the comparison, for $x_i < x_j$ the left child, for $x_i > x_j$ the right child. If the comparison yields $x_i = x_j$, the path stops at the corresponding node. Here, we assume that all nodes of the tree are reached for some input. This is a non-uniform model of computation, the tree for n elements need not be similar in any way to the trees for other n.

The success probability p_T of T is the probability for an input $\mathbf{x}$ (with collision) drawn from distribution $\mathbf{q}$ that the last node of $P(\mathbf{x})$ is an internal node, and hence the collision is indeed detected. More precisely, we define the random variable $S(\mathbf{x})$ to be 1 if $P(\mathbf{x})$ ends at an internal node, and 0 otherwise, i.e., $P(\mathbf{x})$ ends at a leaf of T. For a given probability distribution on the input, we define the success probability as $E[S(\mathbf{x})]$.

This modeling reflects that we require our algorithms to have found the collision if they answer YES. Indeed, any well defined comparison based algorithm with this property can be described by a comparison tree T, with the same success probability and performing not more comparisons than the original algorithm.

By $|P(\mathbf{x})|$ we denote the number of internal nodes on $P(\mathbf{x})$, which is for randomly chosen $\mathbf{x}$ a random variable, representing the number of comparisons or running time on input $\mathbf{x}$. We are interested in the expected running time $C_{\mathbf{q}} = E_{\mathbf{q}}[|P(\mathbf{x})|]$ and $C_{\mathbf{p}} = E_{\mathbf{p}}[|P(\mathbf{x})|]$. We are also interested in the **worst-case running time** of T which is the maximal running time (number of comparisons) for a possible input, which is given by the height of T.

2 The Algorithms

We start with the slightly simpler deterministic algorithms that rely upon random input, then we also consider randomized algorithms. The connection between element uniqueness and sorting is well known for comparison based models of computation. Not surprisingly, all algorithms presented here are based on sorting some subset of the input values. We use that the rank k element of a list can be found in linear time [14], and that sorting k values takes $O(k \log k)$ comparisons.

2.1 Deterministic or Worst-Case Time

Consider the following Algorithm 1 that is designed to run in worst-case $O(t(n))$ time. For $t(n) = n \log n$ time, this algorithm sorts the complete input and hence finds the collision with probability 1. For $t(n) = O(1)$ the success probability is $O(\frac{1}{n^2})$, comparable to testing a single edge. Note that for the interesting case $t(n) \leq n \log n$ we always have $k = \frac{t(n)}{\log t(n)}$.

Algorithm 1: Deterministic collision find

1 Determine $r = \min\{n, t(n)\}$ and $k = \min\{r, t(n)/\log t(n)\}$;
2 Select the k-smallest element x_j of $R = \{x_1, \ldots, x_r\}$;
  Determine $S := \{x \in R \mid x \le x_j\}$ /* $|S| = k$ */
3 Sort S;
  return the collision if two elements of S are equal;

Lemma 1. *For a function t that can be computed in $O(t(n))$ time, Algorithm 1 runs in worst-case time $O(t(n))$ and computes* **Collision**$_n$ *on input drawn from distribution* $\mathbf{q}$ *with success probability at least* $p = \min\left\{\frac{t(n)}{n}, 1\right\} \frac{t(n)}{n \log t}$.

Proof. The selection in Line 2 can be achieved in worst-case $O(t(n))$ time [14]. Sorting k elements can be achieved in $O(k \log k) = O\left(\frac{t}{\log t} \log \frac{t}{\log t}\right) = O(t)$ worst-case time, for example with heap sort.

To compute the success probability of Algorithm 1, consider the "collision first" procedure to draw an element from distribution $\mathbf{q}$. The probability that the collision is in the first variables ($i < j \le r$) is $\frac{r}{n}\frac{r-1}{n-1}$. The rank of the value $x_i = x_j$ within $x_1, \ldots, x_r$ is uniform between 1 and $r - 1$, hence the probability for it to be $\le k$ is $\frac{k}{r-1}$. Hence, the success probability of Algorithm 1 is $\frac{r}{n}\frac{r-1}{n-1}\frac{k}{r-1} > \frac{kr}{n^2}$. If $t(n) > n \log n$, we have $r = k = n$ and hence $p = 1$, as stated in the lemma. For $n \le t(n) \le n \log n$ we have $r = n$ and hence $p = \frac{k}{n}$. Finally, for $t(n) < n$, we have $r = t(n)$ and hence $\frac{r}{n} < 1$, leading to $p = \frac{r}{n} kn$. □

In Section 3.4 Lemma 8 shows that the trade-off between success probability and worst-case running time is asymptotically optimal if $t(n) < n$. Otherwise, this follows from Section 3.3, Lemma 6 because the worst case running time is an upper bound on the expected running time.

2.2 Expected Time

In comparison to the worst-case time, it is easier to achieve good expected running times because on some inputs the algorithm may be slow if it is fast on others. More precisely, if a fraction p of the inputs is sorted completely, the success probability is p, and the expected running time is $O(pn \log n)$, as long as the expected running time of a non-successful input is $O(1)$. If $p \le 1/n^2$, comparing x_1 and x_2 suffices.

To achieve this, we use that in distribution $\mathbf{q}$ and $\mathbf{p}$, the outcomes of the *i-th* **canonical test**, comparing x_{2i} with x_{2i+1} are independent for different $i \in \{1, \ldots, \lfloor n/2 \rfloor\}$, and will be used to emulate random bits. Choose the integer k such that $2^{-k} \ge p > 2^{-k-1}$. For $p > 1/n^2$, we have $k \le 2 \log n < \lfloor n/2 \rfloor$ if $n > 7$. The algorithm performs the canonical tests in the natural order. As soon as one of the tests fails, the algorithm stops. Once test k succeeds, the algorithm sorts the input and hence finds all collisions. Hence, the success probability is at least $2^{-k} \ge p$. The expected running time until a failing test is reached is bounded by $\sum_{i=1}^{k} i2^{-i} = O(1)$. Hence, the expected running time is $O(1 + pn \log n)$.

This running time is asymptotically optimal, for distribution $\mathbf{p}$ this is shown in Section 3.3 (Lemma 6), for distribution $\mathbf{q}$ in Section 3.5 (Lemma 9).

2.3 Randomized Algorithms

Expected time. Again, if only a good expected running time should be achieved, the algorithm is very simple if we allow to toss an arbitrarily biased coin. With probability p, we solve the problem deterministically in $O(n \log n)$ time, otherwise we do nothing and declare that we find no collision. This algorithm has expected success probability p and expected running time $O(pn \log n)$ on any input. If only unbiased binary coins are allowed, p should be overestimated as 2^{-k}, leading to the same asymptotic performance, which is optimal by Theorem 1.

Worst-case time. Now consider the case where the randomized algorithm should never exceed a certain running time, and still find a collision with reasonably high probability. The idea here is to use few random bits to "simulate" distribution $\mathbf{q}$ in Algorithm 1.

Let $t = t(n) \leq n \log n$ be the goal for a asymptotic worst-case running time. We design an algorithm with running time $O(t(n))$ and high success probability. For the case $t < n/2$ the variables are divided equally into $k = \lfloor n/t \rfloor$ classes, such that the size is $\lfloor n/k \rfloor$ or $\lceil n/k \rceil$. Now, choose uniformly at random two different such classes and call the resulting set of variables R and define $r = |R|$. Observe that $\lceil n/k \rceil \leq n/k + 1 \leq n/(n/t - 1) + 1 = t/(1 - t/n) + 1 \leq 2t + 1$, and hence $r = O(t(n))$. Divide the set $[r]$ equally into ranges of length at least $t/\log t$ and at most $2t/\log t$. Choose one such range $[a, b]$, determine the rank-a element of R and the rank-b element, such that another scan over R yields the set S of elements whose rank is in the range $[a, b]$, and sort this set S. By a similar calculation as for Algorithm 1, the worst case running time of this algorithm is $O(t(n))$.

To have success, the algorithm needs to randomly choose the two classes where the variables of the collision are located, and it must randomly choose the rank of this collision within the set R.

For the case $t < n$, there are $k \leq n/t$ classes, and at most $\log t$ ranges, such that this success probability is at least $p = (t/n)^2(1/\log t)$. Otherwise, only the choice of the range is random, the range with the collision is chosen with probability at least $\frac{t}{\log t}\frac{1}{n}$.

We summarize the above discussion in the following Lemma.

Lemma 2. *Let t be a function that can be computed in $O(t(n))$ time, then there is a randomized algorithm that runs in worst-case time $O(t(n))$ and computes* ***Collision***$_n$ *with success probability at least* $p = \min\left\{\frac{t}{n}, 1\right\} \frac{t}{n \log t}$.

The performance of the described algorithm is asymptotically optimal as discussed in Section 3.

3 The Lower Bound

The purpose of this section is to show that the four algorithms introduced in Section 2 achieve an asymptotically optimal trade-off between running time and success probability, for all functions t.

By Yao's minimax principle, this task reduces to showing lower bounds for deterministic algorithms working on random input, i.e., the two algorithms of Section 2.1 and Section 2.2. It then follows that also the randomized algorithms cannot be faster. For the sake of completeness, we summarize the results of this section (in particular Lemma 8) in the following theorem. It also states a lower bound on the amount of randomness required.

Theorem 1. *Assume a randomized algorithm $\mathcal{A}$ solves* **Collision$_n$** *for all inputs in time t and with positive success probability. Then, with $r = \min\{t, n\}$ and $p \le p_t = \frac{8r^2}{(n-1)^2 \log(2t)}$, the success probability of $\mathcal{A}$ is at most p and there exists an input where it uses at least $-\log p_t$ random bits.*

At the heart of the lower bound is the consideration about a single leaf of the comparison tree that relates the fraction of the input ending at this leaf to the success probability induced by this leaf. This basic trade-off is formulated between success for input from $\mathbf{q}$ versus running time (fraction of input reaching the leaf) of $\mathbf{p}$, which is the worst-case running time of the tree. Transforming this into bounds on the expected running time for $\mathbf{q}$, and taking into account that sublinear algorithms cannot access all the variables, requires some extra work.

3.1 High-Probability DAGs

For the purposes of analyzing T, we annotate every node c of T by the directed graph G_c that reflects the already performed comparisons. The vertices of G'_c are the variables, and there is a directed arc from x_i to x_j if the variables were compared, and $x_i < x_j$. By this definition, G'_c is a directed acyclic graph. Since the order relation is transitive, and because we are interested in single collisions we consider the irreducible core G_c of G'_c, i.e., the graph with the fewest edges and the same transitive closure as G'_c.

This leads to an alternative way of computing the success probability of T following the "permutation first" procedure to draw an input with collision. Choose uniformly a permutation π. The corresponding input vector defines a path to a leaf c of T, and π can be understood as a linear extension of G_c. Actually, the node c is the only leaf of T with this property. Now, uniformly choose the pivot $o \in \{1, \ldots, n-1\}$ (identifying the value o and $o+1$). This collision is detected if and only if there is an arc between the vertex i with $\pi(i) = o$ of G_c and its successor in π, which is called the **success probability of the permutation** π in T, and hence in G_c.

Define the success probability p_c of G_c by the probability that a uniformly chosen linear extension of G_c and a uniformly chosen collision is detected by G_c. Let u_c be the number of linear extensions of G_c, and define $f_c = u/n!$. Then, the probability of reaching (ending at) c with a uniformly chosen permutation is f_c, and we can express the success probability of T as the weighted sums of the success probabilities of the leaves:

$$p_T = \sum_{c \text{ is leaf of T}} f_c \cdot p_c \,.$$

The following information theoretic consideration gives a precise relation between the success probability p_c and the number of linear extensions f_c.

Lemma 3. *Let $G = (V, E)$ be a directed acyclic graph with n vertices, $V = \{1, \ldots, n\}$. Then, the number of linear extensions of G with at least k arcs in G is at most*

$$\binom{n}{k} \cdot \frac{n!}{k!}$$

Proof. Any linear extension with at least k arcs in G can be described by the set A of additional arcs that need to be inserted into G to yield a directed path (thinking of G as an order-relation, the additional comparisons that are necessary to make all elements comparable). Since at least k arcs of G are used we have $|A| \leq n - k - 1 < n - k$. Define $T_A \subseteq V$ to be the starting points of the arcs in A. Now, the arcs form an injective mapping $\gamma\colon T_A \to \{1, \ldots, n\}$. There are at most $\binom{n}{n-k} = \binom{n}{k}$ possibilities for T_A, and at most $\frac{n!}{k!}$ possibilities for γ. This leads to the claimed bound. □

By Stirling's formula, we get the following lemma.

Lemma 4. *Given a graph G on n vertices and a set S of permutations on $[n]$. Assume that the success probability of G when drawing uniformly permutations from S that are linear extensions of G and uniformly the collision is at least $p \geq n^{-\frac{1}{6}}$. Then the number u of permutations in S that are linear extensions of G is bounded by $-\log \frac{u}{|S|} \geq \frac{pn}{6} \log n - 3pn - \frac{\log n}{6} - 1 + \log \frac{|S|}{n!}$.*

Proof omitted due to space limitations.

3.2 Easy Low-Probability Bound

If an algorithm runs in sublinear time, it cannot access all the input elements, leading to a fairly small success probability. We make this precise as a minimum number of accessed variables that is required to achieve a certain success probability.

Lemma 5. *Assume a leaf node c of a decision tree T for* **Collision$_n$** *has success probability p_c for* **q**. *Then the depth d_c of c is at least $d_c \geq \frac{n-1}{2}\sqrt{p_c}$. For $p_c \leq n^{-\frac{1}{6}}$ this implies $d_c \geq \frac{p_c n \log n}{6} - 3p_c n - \frac{\log n}{6} - 1$*

Proof omitted due to space limitations.

3.3 Expected Running Time Without Collisions

We want to show a lower bound of $\Omega(pn \log n)$ for the expected running time. As a first step, we consider the running time implied by input without collision (drawn from $\mathbf{p}$). This certainly implies the corresponding lower bound on the worst-case running time, the depth of the tree.

Lemma 6. *Let T be a comparison tree solving* **Collision**$_n$*, and S a set of permutations. Assume an input drawn from $\mathbf{q}|_S$ (uniform permutation in S and uniform collision) has success probability at least p and expected running time D for input from $\mathbf{p}|_S$, i.e., a uniformly chosen permutation from S without collision. Then $D \geq \frac{pn}{6} \log n - 3pn - \frac{\log n}{6} - 1 + \log \frac{|S|}{n!}$*

Proof. Every leaf c of T has success probability p_c, a depth d_c, and a fraction f_c of the permutations from S that end at c. Now, by definition, $D = \sum f_c \cdot d_c$, and $p = \sum f_c \cdot p_c$.

We define two classes of nodes, the high-probability nodes $H = \{c \mid p_c > n^{-\frac{1}{6}}\}$, and the remaining nodes L. Define further for these two classes of permutations in S the split $f_H = \sum_{c \in H} f_c$, and similarly $f_L = \sum_{c \in L} f_c$, such that $f_H + f_L = 1$. The restricted probabilities p_H and p_L, and the restricted expected running times D_L and D_H are defined by $f_H \cdot p_H = \sum_{c \in H} f_c \cdot p_c$, $f_H \cdot D_H = \sum_{c \in H} f_c \cdot d_c$, $f_L \cdot p_L = \sum_{c \in L} f_c \cdot p_c$, $f_L \cdot D_L = \sum_{c \in L} f_c \cdot d_c$. These values satisfy $p = f_H \cdot p_H + f_L \cdot p_L$, and $D = f_H \cdot D_H + f_L \cdot D_L$.

Define for $c \in H$ the relative reach-probability by $f'_c = f_c / f_H$. Note that the f'_c sum to 1, i.e., they form a probability distribution. Define $a_c = 2^{-d_c}$, $A_H = \sum_{c \in H} a_c$, such that the values $a'_c = a_c / a_H$ sum to 1 and form a probability distribution.

With this, we get

$$D_H = -\sum_{c \in H} f'_c \log a_c = -\log a_H - \sum_{c \in H} f'_c \log a'_c \tag{1}$$

$$\geq -\log a_H - \sum_{c \in H} f'_c \log f'_c \tag{2}$$

$$= -\log f_H - \log a_H - \sum_{c \in H} f'_c \log f_c \tag{3}$$

$$\geq \log -f_H + \sum_{c \in H} f'_c \left(\frac{p_c n \log n}{6} - 3p_c n - \frac{\log n}{6} - 1 + \log \frac{|S|}{n!} \right) \tag{4}$$

$$= -\log f_H + \frac{pn}{6} \log n - 3pn - \frac{\log n}{6} - 1 + \log \frac{|S|}{n!}. \tag{5}$$

Where the inequality (2) is Gibbs' inequality and the inequality (4) is the statement of Lemma 4, together with the fact that $-\log a_H \geq 0$. Now, consider a node $c \in L$ of T, i.e., with low probability $p_c \leq n^{-\frac{1}{6}}$. By Lemma 5 we have the depth-bound $d_c \geq \frac{p_c n}{6} \log n - 3pn - \frac{\log n}{6} - 1$. This inequality yields $D_L = \sum_{c \in L} f'_c \cdot d_c \geq \sum_{c \in L} f'_c \cdot \left(\frac{p_c n}{6} \log n - 3p_c n - \frac{\log n}{6} - 1 \right) \frac{p_L n}{6} \log n - 3p_L n - \frac{\log n}{6} - 1$. Now, the lemma follows by $D = f_H \cdot D_H + f_L \cdot D_L \geq (f_H p_H + f_L p_L)\left(\frac{n}{6}(\log n - 18)\right) - \frac{\log n}{6} - 1 + f_H \log \frac{f_H|S|}{n!} \geq \frac{pn}{6}\log n - 3pn - \frac{\log n}{6} - 1 + \log \frac{|S|}{n!}$. □

3.4 Strong Low Probability Bound for Worst-Case Time

Certainly, the lower bound on the expected running time is also lower bound on the worst-case running time. Still, for sub-linear time algorithms the success probability is significantly lowered by the impossibility to touch all vertices.

Lemma 7. *Let T be a comparison tree for* **Collision$_n$** *with maximal depth $r < n/2$, and that input is drawn from* $\mathbf{p}$ *or* $\mathbf{q}$. *Then there is a comparison tree T' with the same success probability, expected and worst-case running time as T, and T' uses only the variables $x_1, \ldots, x_{2r}$.*

Lemma 8. *Any comparison tree T with worst-case running time $t \leq n$ has success probability $p \leq p_t = \frac{16t^2}{(n-1)^2 \log(2t)}$ when input is drawn from* $\mathbf{q}$.

Proof. With $r = 2t$, by Lemma 7 w.l.o.g. the comparison tree T solves **Collision$_r$** in worst-case time t and success probability q. From Lemma 6 follows $t \geq \frac{qr}{6}\log r - 3pr - \frac{\log r}{6} - 1$, which yields $q \leq 6(r/2 + \frac{\log r}{6} + 1)/r(\log r - 18) < (3r + \log r + 1)/r \log r < 4/\log r = 4/\log(2t)$. Now, by the argument of Lemma 5 $p \leq \frac{r(r-1)}{n(n-1)} q \leq \frac{16t^2}{(n-1)^2 \log(2t)}$. □

3.5 Expected Time for Random Input with Collision

Finally, we can also conclude an asymptotic lower bound of $\Omega(pn \log n)$ for the expected running time when input is drawn according to $\mathbf{q}$. By Yao's Minimax Principle, the same lower bound holds for the expected running time of randomized algorithms on input with collision.

Lemma 9. *Let T be a linear decision tree solving* **Collision$_n$**. *Assume that the success probability for input drawn according to distribution* $\mathbf{q}$ *is p, and the expected running time is $C_\mathbf{q}$. Then, $C_\mathbf{q} \geq \frac{pn}{48}(\log n - 18) - \frac{\log n}{12} - 4$.*

Proof. Let S be the set of permutations that have success probability $> 1/2$. With $f_S := |S|/n!$ we can express running time and probability as $C = f_S C_S + (1 - f_S) C_{\bar{S}}$, where C_S is the expected running time for permutations in S, and $C_{\bar{S}}$ the expected running time for permutations not in S. Similarly, we can write the success probability as $p = f_S p_S + (1 - f_S) p_{\bar{S}}$.

For permutations not in S, half the contribution to the average running times stems from undetected collision. Hence, it can be estimated using $C_\mathbf{p}$, by Lemma 6, we have $C_{\bar{S}} \geq \left(\frac{np_{\bar{S}}}{6}(\log n - 18) - \frac{\log n}{6} - 1 + \log(1 - f_S)\right)/2$.

By a Markov inequality, at least half of the inputs in $\mathbf{q}|_S$ stop at times before $2C_S$. Cut T at depth (time) $2C_S$, leading to T' with success probability $p'_S \geq 1/4 \geq p_S/4$. Now, because the expected running time of T' is less than the worst-case running time $2C_S$ of T', Lemma 6 yields $2C_S \geq \frac{p_S n}{4 \cdot 6}(\log n - 18)$

$-\frac{\log n}{6} - 1 + \log f_S$. It remains to take the weighted sums of the bounds on $2C_S$ and $2C_{\bar{S}}$, yielding $2C \geq \frac{pn}{4 \cdot 6}(\log n - 18) - \frac{\log n}{6} - 2$. Here, the last term stems from $f_S \log f_S + (1 - f_S) \log(f_S - 1) > -1$. □

Acknowledgment

I would like to thank Ueli Maurer and Dominik Raub for introducing me to the problem and for several fruitful discussions, and an anonymous referee for suggestions improving the introduction.

References

1. Ben-Or, M.: Lower bounds for algebraic computation trees. In: Proc. 15th Annual ACM Symposium on Theory of Computing, pp. 80–86. ACM Press, New York (1983)
2. Boppana, R.B.: The decision-tree complexity of element distinctness. Inf. Process. Lett. 52, 329–331 (1994)
3. Buhrman, H., Dürr, C., Heiligman, M., Høyer, P., Magniez, F., Santha, M., de Wolf, R.: Quantum algorithms for element distinctness. SIAM J. Comput. 34, 1324–1330 (2005)
4. Aaronson, S., Shi, Y.: Quantum lower bounds for the collision and the element distinctness problems. J. ACM 51, 595–605 (2004)
5. Maurer, U.M.: Abstract models of computation in cryptography. In: Smart, N.P. (ed.) Cryptography and Coding. LNCS, vol. 3796, pp. 1–12. Springer, Heidelberg (2005)
6. Yao, A.C.C.: Probabilistic computations: towards a unified measure of complexity. In: Proc. 18th FOCS, IEEE 1977, pp. 222–227. IEEE Computer Society Press, Los Alamitos (1977)
7. Motwani, R., Raghavan, P.: Randomized Algorithms. Cambridge University Press, Cambridge (1995)
8. Snir, M.: Lower bounds on probabilistic linear decision trees. Theoret. Comput. Sci. 38, 69–82 (1985)
9. Grigoriev, D., Karpinski, M., Meyer auf der Heide, F.: A lower bound for randomized algebraic decision trees. Comput. Complexity 6, 357–375 (1996/1997)
10. Manber, U., Tompa, M.: Probabilistic, nondeterministic, and alternating decision trees (preliminary version). In: STOC '82: Proceedings of the fourteenth annual ACM symposium on Theory of computing, New York, NY, USA, pp. 234–244. ACM Press, New York (1982)
11. Manber, U., Tompa, M.: The complexity of problems on probabilistic, nondeterministic, and alternating decision trees. J. Assoc. Comput. Mach. 32, 720–732 (1985)
12. Manber, U., Tompa, M.: The effect of number of Hamiltonian paths on the complexity of a vertex-coloring problem. SIAM J. Comput. 13, 109–115 (1984)
13. Chein, M., Habib, M.: The jump number of DAGs and posets: An introduction. Annals of Discrete Mathematics 9, 189–194 (1980)
14. Blum, M., Floyd, R., Pratt, V., Rivest, R., Tarjan, R.: Time bounds for selection. Journal of Computer and System Sciences 7, 448–461 (1973)

Randomized and Approximation Algorithms for Blue-Red Matching

Christos Nomikos[1], Aris Pagourtzis[2,⋆], and Stathis Zachos[2,3,⋆]

[1] Department of Computer Science, University of Ioannina
cnomikos@cs.uoi.gr
[2] School of Electrical and Computer Engineering,
National Technical University of Athens
{pagour,zachos}@cs.ntua.gr
[3] CIS Department, Brooklyn College, Cuny

Abstract. We introduce the Blue-Red Matching problem: given a graph with red and blue edges, and a bound w, find a maximum matching consisting of at most w edges of each color. We show that Blue-Red Matching is at least as hard as the problem Exact Matching (Papadimitriou and Yannakakis, 1982), for which it is still open whether it can be solved in polynomial time. We present an RNC algorithm for this problem as well as two fast approximation algorithms. We finally show the applicability of our results to the problem of routing and assigning wavelengths to a maximum number of requests in all-optical rings.

1 Introduction

We define and study a matching problem on graphs with blue and red edges; we call the new problem Blue-Red Matching (BRM for short). The goal is to find a maximum matching under the constraint that the number of edges of each color in the matching does not exceed a given bound w.

We are motivated for this study by a problem that arises in all-optical networks, namely DirMaxRWA [12]. In particular, it was implicit in [12] that solving BRM exactly would imply an improved approximation ratio for DirMaxRWA in rings. Moreover, BRM can capture several interesting scenarios such as the following: Consider a team of friends that would like to play chess or backgammon. Some pairs prefer to play chess, while other pairs prefer backgammon. There could even exist pairs that would like to play either game. Now, if the number of available boards for each game is limited we need to solve BRM if we want to maximize the number of pairs that will manage to play the game of their preference.

In this work we first show that BRM is at least as hard as Exact Matching, a problem defined by Papadimitriou and Yannakakis [13], for which it is still an open question whether it can be solved in polynomial time. Therefore, an

⋆ Research supported in part by the General Secretariat of Research and Technology of Greece, through ΠΕΝΕΔ 2003 program, grant nr. ΕΔ285.

L. Kučera and A. Kučera (Eds.): MFCS 2007, LNCS 4708, pp. 715–725, 2007.

exact polynomial time algorithm for BRM would answer that question in the affirmative.

Our main result is that BRM can be solved exactly by a polynomial time randomized (in fact RNC^2) algorithm, which uses ideas from [10]. Since the sequential version of the randomized algorithm is quite slow, we also present two approximation algorithms for BRM; the first is a fast and simple greedy algorithm that achieves a $\frac{1}{2}$-approximation ratio, the second is a more involved algorithm that achieves an asymptotic $\frac{3}{4}$-approximation ratio.

We finally demonstrate the relation between BRM and DirMaxRWA in rings, by showing that an algorithm for BRM with (asymptotic) approximation ratio a results in an algorithm for DirMaxRWA in rings with (asymptotic) approximation ratio $\frac{a+1}{a+2}$. Combining all the above results we obtain as a corollary that DirMaxRWA in rings admits a randomized approximation algorithm with ratio $\frac{2}{3}$ and a (much faster) deterministic approximation algorithm with asymptotic ratio $\frac{7}{11}$.

As far as we know BRM has not been studied before. As mentioned earlier, a related problem is Exact Matching [13] which admits an RNC algorithm due to Mulmuley, Vazirani and Vazirani [10]. Polynomial time algorithms for Exact Matching are known only for special classes of graphs, e.g. complete graphs [7] and complete bipartite graphs [7][17].

2 Problem Definition and Hardness

Let $G = (V, E_{blue} \cup E_{red})$ be a graph in which each edge is colored either blue or red; E_{blue} is the set of blue edges and E_{red} the set of red edges. A matching M in G is called *w-blue-red matching* if $M \cap E_{blue} \leq w$ and $M \cap E_{red} \leq w$, that is, if it contains at most w edges of each color.

The notion of w-blue-red matching can be extended for multigraphs that may contain edges of both colors between two vertices. It is easy to see that in fact we do not have to use multigraphs; it suffices to specify a third set of initially uncolored (white) edges as follows. Let $G = (V, E_{blue} \cup E_{red} \cup E_{white})$ be a graph in which E_{blue}, E_{red}, and E_{white} are sets of blue, red, and white edges respectively. A matching M in G is called w-blue-red matching if there exists a partition $\{E_{wb}, E_{wr}\}$ of E_{white} such that $M \cap (E_{blue} \cup E_{wb}) \leq w$ and $M \cap (E_{red} \cup E_{wr}) \leq w$. In other words M is a w-blue-red matching if we can choose a color for each white edge in G so that M contains at most w edges of each color.

We define BRM to be the following optimization problem: given a graph $G = (V, E_{blue} \cup E_{red} \cup E_{white})$ and a positive integer w, find a w-blue-red matching of maximum cardinality. In the decision version of this problem, denoted by BRM(D), a bound B is also given and the question is whether G has a w-blue-red matching of cardinality at least B.

It turns out that BRM(D) is closely related to a well known problem, namely Exact Matching, defined in [13]. In this problem, the input is a graph $G = (V, E)$, a set of red edges $E' \subseteq E$ and a positive integer k and the question is whether G contains a perfect matching involving exactly k edges in E'. The next theorem shows that BRM(D) is at least as hard as EXACT MATCHING.

Theorem 1. *There is a logarithmic space reduction from* EXACT MATCHING *to* BRM(D).

Proof. Consider a graph $G = (V, E)$, a set of red edges $E' \subseteq E$ and a positive integer k.

If $|V|$ is an odd number or $k > \frac{|V|}{2}$, then G does not contain a perfect matching involving exactly k edges in E'. In that case we construct a 'no' instance of BRM(D) (for example, any instance with $2w < B$).

Otherwise, let $w = \max(k, \frac{|V|}{2} - k)$ and $r = w - \min(k, \frac{|V|}{2} - k)$. Graph G^* is obtained from G by adding $2r$ new vertices $u_1, \ldots, u_r$, $v_1, \ldots, v_r$ and r edges $\{u_1, v_1\}$, ..., $\{u_r, v_r\}$. The additional edges are colored blue if $k > \frac{|V|}{2} - k$, otherwise they are colored red. Furthermore, edges in $E - E'$ are colored blue and edges in E' remain red in G^*. Let $B = 2w$.

The above construction requires logarithmic space. It is not hard to check that G contains a perfect matching involving exactly k edges in E' if and only if G^* contains a w-blue-red matching of cardinality B. □

The above theorem indicates that it is probably not a trivial task to find a polynomial time (deterministic) algorithm for BRM, since this would imply polynomial time solvability for EXACT MATCHING as well. Therefore, we will restrict our attention to approximation and randomized algorithms.

3 Approximation Algorithms for Blue-Red Matching

We first observe that there exists a simple approximation algorithm for BRM, which requires time linear in the number of edges. The algorithm, which we call Greedy-BRM, constructs a w-blue-red matching M in a greedy manner: edges are examined in an arbitrary order; an edge e is added to M if both endpoints of e are unmatched and M contains fewer than w edges of the same color as e (or M contains fewer than w edges of any color if e is white). It is not hard to prove the following:

Theorem 2. *Algorithm* Greedy-BRM *returns a solution with at least* $\frac{1}{2} \cdot \mu_{OPT}$ *edges, where* μ_{OPT} *is the cardinality of an optimal solution.*

In the remaining of this section we present an approximation algorithm for BRM, which achieves asymptotic approximation ratio $\frac{3}{4}$.

The algorithm first computes a maximum cardinality matching M (Step 1). In Step 2, a color is assigned to each white edge of the graph. If after Step 2 M contains more than w edges of one color and fewer than w edges of the other color then Step 3 is executed in order to produce a more balanced matching. Finally, Step 4 eliminates superfluous edges of any of the two colors.

Algorithm Balance-BRM

Input: graph $G = (V, E_{blue} \cup E_{red} \cup E_{white})$, integer w.
Output: a w-blue-red matching of G.

1. find a maximum matching M in G
2. **for** every white edge e **do**
    color e with the color which is currently less used in M, breaking ties arbitrarily
  let E'_{red}, E'_{blue} be the sets of red and blue edges after coloring the white edges
3. **if** M contains $> w$ edges of one color and $< w$ edges of the other color **then**
  (Assume w.l.o.g. that the majority color in M is blue—the other case is symmetric)
  (a) find a maximum matching M_{red} in graph $G_{red} = (V, E'_{red})$
  (b) let G' be the graph resulting by superimposing M and M_{red}
  (c) let S be the set of all connected components in G, in which the number of edges that belong to M_{red} is greater than the number of red edges that belong to M
  (d) **while** M contains more than $w+1$ blue edges and fewer than w red edges and S is not empty **do**
    (i) choose (arbitrarily) a connected component F in S. Let b_M, b_F be the number of blue edges in M, F respectively
    (ii) **if** $b_M - w < b_F$ **then** pick a chain F' of edges in F containing exactly $b_M - w$ blue edges, such that F' begins and ends with a blue edge
    else let $F' = F$
    (iii) delete from M all edges that belong to F'; add to M all edges in F' that belong to M_{red}
    (iv) delete F from S
4. **if** M contains more than w blue (red) edges **then**
    choose arbitrarily w of them and eliminate the rest
5. **return** M

We will next prove that algorithm Balance-BRM achieves an asymptotic $\frac{3}{4}$-approximation ratio. Let us first note that if after the first two steps there are either at most w edges of each color in matching M or at least w edges of each color in M, then M (after removing surplus edges, in the latter case, in Step 4) is an optimal solution. Therefore, it remains to examine the case in which there are more than w edges of one color and fewer than w edges of the other after Step 2. W.l.o.g. we assume that the majority color is blue. We will first give two lemmata concerning Step 3.

Each substitution in Step 3 increases the number of red edges in M. However, it may decrease the number of blue edges. In the extreme case one red edge replaces two blue edges. Therefore, we have:

Lemma 1. *If Step 3 of algorithm* Balance-BRM *decreases the number of blue edges by δ, then it increases the number of red edges by at least $\delta/2$.*

If M contains more than $w+1$ blue edges, then Step 3 can always perform a substitution, unless the number of red edges has reached its maximum possible value. Therefore we have:

Lemma 2. *If after Step 3, M contains more than $w+1$ blue edges, then algorithm* Balance-BRM *returns an optimal solution.*

We are now ready to state the main theorem of this section.

Theorem 3. *Algorithm* Balance-BRM *returns a solution of cardinality at least $\frac{3}{4}\mu_{OPT} - \frac{1}{2}$, where μ_{OPT} is the cardinality of an optimal solution.*

Proof. We prove the claim for the case in which the number of blue edges is greater than the number of red edges. The other case is symmetric.

Let μ_{SOL} be the number of edges in the solution returned by Balance-BRM, μ_r, μ_b be the number of blue and red edges respectively contained in M after Step 2, and μ_{red} be the size of M_{red}. For convenience, let us also define $z = \min(\mu_{red}, w, \mu_b + \mu_r - w)$.

All red edges in an optimal matching belong to $E_{red} \cup E_{white}$ which is equal to E'_{red} in the case in which blue is the majority color. Therefore μ_{red} is an upper bound for the number of red edges in an optimal matching, which implies $\mu_{OPT} \leq w + \mu_{red}$. Since $\mu_b + \mu_r$ is the size of the maximum cardinality matching M, it also holds $\mu_{OPT} \leq \mu_b + \mu_r = w + (\mu_b + \mu_r - w)$. Moreover, by definition $\mu_{OPT} \leq 2w$. Combining the above inequalities we obtain:

$$\mu_{OPT} \leq w + z \tag{1}$$

Lemma 2 implies that in any non-optimal solution, M contains at most $w+1$ blue edges after Step 3. Hence, in Step 3 the number of blue edges decreases by at least $\mu_b - w - 1$. By Lemma 1, the number of additional red edges is at least $\frac{(\mu_b - w - 1)}{2}$. Since after Step 3 the number of blue edges in M is at least w, we get $\mu_{SOL} \geq w + \mu_r + \frac{(\mu_b - w - 1)}{2}$. Using the fact that, by definition, $z \leq \mu_b + \mu_r - w$, it turns out that:

$$\mu_{SOL} \geq w + \frac{\mu_b + \mu_r - w}{2} + \frac{\mu_r - 1}{2} \geq w + \frac{z}{2} - \frac{1}{2} + \frac{\mu_r}{2} \geq w + \frac{z}{2} - \frac{1}{2} \tag{2}$$

From (2) and the fact that, by definition, $z \leq w$ we obtain:

$$\mu_{SOL} \geq \frac{3z}{2} - \frac{1}{2} \tag{3}$$

From (1) and (2) we get that:

$$\mu_{OPT} \leq \mu_{SOL} + \frac{z}{2} + \frac{1}{2} \tag{4}$$

Finally, from (3) and (4) it follows that $\mu_{OPT} \leq \mu_{SOL} + \frac{1}{3}(\mu_{SOL} + \frac{1}{2}) + \frac{1}{2} = \frac{4}{3}(\mu_{SOL} + \frac{1}{2})$, which is equivalent to $\mu_{SOL} \geq \frac{3}{4}\mu_{OPT} - \frac{1}{2}$. □

It can be shown that the above asymptotic approximation ratio is tight. The complexity of the algorithm is $O(n^{2.5})$: Steps 1 and 3 require $O(n^{2.5})$ time to construct M and M_{red} and all the remaining tasks require time that is linear in the number of edges, which is at most $O(n^2)$.

4 A Randomized Algorithm for Blue-Red Matching

In this section we present a randomized polynomial time algorithm, called Random-BRM that finds an optimal solution for BRM with high probability. This algorithm makes use of some ideas proposed in [10].

Algorithm Random-BRM operates as follows: First, it augments G to a complete graph G^* by adding edges of a new color (say black). Then it assigns a random weight to each edge of G^* and constructs a variation of the Tutte matrix of G^*, in which each indeterminate is replaced by a constant value or by a monomial, depending on the weight and the color of the corresponding edge. In particular the indeterminate that corresponds to a blue (red) edge e_{ij} of weight w_{ij} is replaced by the monomial $x2^{w_{ij}}$ (resp. $y2^{w_{ij}}$). Then, the algorithm computes the Pfaffian of this matrix, which in this case is a polynomial in the variables x, y. Finally, it uses the coefficients of this polynomial in order to find a specific matching in G^*, from which it obtains an optimal solution for BRM.

The detailed algorithm is given at the end of the section; its correctness is based on a series of lemmata which are stated below, together with some necessary definitions.

Consider a graph $G = (V, E)$, where $V = \{v_1, v_2, \ldots, v_n\}$ and $E = E_{blue} \cup E_{red} \cup E_{white}$, and an positive integer w. Without loss of generality we may assume that n is even (otherwise an isolated vertex can be added to G). Let G^* be the complete graph with set of vertices V. We denote the edge $\{v_i, v_j\}$ by e_{ij}. We assume that edges not in E are colored black, i.e. $G^* = (V, E \cup E_{black})$. A perfect matching of G^* with exactly p blue and q red edges is called (p, q)-perfect matching. We denote by $\mathcal{M}$ (resp. $\mathcal{M}_{pq}$) the set of all perfect matchings (resp. (p, q)-perfect matchings) of G^*.

Perfect matchings in G^* can be used in order to obtain w-blue-red matchings in G. For fixed w, let us define a function $\mathrm{sol}_w(p, q, t) = \min(2w, \min(w, p) + \min(w, q) + t)$. The lemma below explains the use of function sol_w:

Lemma 3. *Let M be a (p, q)-perfect matching of G^* with t white edges. Then there exists a w-blue-red matching $M_w \subseteq M$ of G of cardinality $\mathrm{sol}_w(p, q, t)$.*

Proof. We can construct M_w as follows: we first select arbitrarily $\min(w, p)$ blue edges and $\min(w, q)$ red edges from M and add them to M_w; then we repeatedly select a white edge from M, color it with the color which is currently less used by edges in M_w, and add it to M_w, until the cardinality of M_w becomes $2w$ or we run out of white edges. In the latter case $M_w = \min(w, p) + \min(w, q) + t$. □

Suppose that a number s_{ij} is selected at random from $\{1, 2, \ldots, n^4\}$, for each (i, j), $1 \le i < j \le n$, and define the weight w_{ij} of e_{ij} as follows:

$$w_{ij} = \begin{cases} s_{ij} & \text{if } e_{ij} \in E \\ n^5 + s_{ij} & \text{if } e_{ij} \in E_{black} \end{cases}$$

The weight of a perfect matching M is $W_M = \sum_{e_{ij} \in M} w_{ij}$. We denote by W_{pq} the minimum weight of a matching among all matchings in $\mathcal{M}_{pq}$. The number of white edges in a (p, q)-perfect matching with weight W_{pq} can be easily computed, using the next lemma:

Lemma 4. *Let p, q be integers, with $0 \leq p, q \leq \frac{n}{2}$ and let M_{pq} be a minimum weight (p, q)-perfect matching of G^*. Then the number of white edges in M_{pq} is $\frac{n}{2} - p - q - \lfloor \frac{W_{pq}}{n^5} \rfloor$.*

The following lemma can be used to compute the number of edges in an optimal w-blue-red matching:

Lemma 5. *The number of edges in an optimal w-blue-red matching of graph G is*

$$C = \max_{(p,q):\mathcal{M}_{pq} \neq \emptyset} \mathrm{sol}_w(p, q, \frac{n}{2} - p - q - \lfloor \frac{W_{pq}}{n^5} \rfloor).$$

The Tutte matrix A of G^* is defined as follows:[1]

$$a_{ij} = \begin{cases} 0 & \text{if } i = j \\ 2^{w_{ij}} & \text{if } i < j \text{ and } e_{ij} \in E_{white} \cup E_{black} \\ x2^{w_{ij}} & \text{if } i < j \text{ and } e_{ij} \in E_{blue} \\ y2^{w_{ij}} & \text{if } i < j \text{ and } e_{ij} \in E_{red} \\ -a_{ji} & \text{if } i > j \end{cases}$$

The *canonical permutation* for a perfect matching $M \in \mathcal{M}$, denoted by π_M, is the unique permutation of $\{1, 2, \ldots, n\}$ that satisfies the following properties:

- $\{v_{\pi_M(2i-1)}, v_{\pi_M(2i)}\} \in M$, for every i, $1 \leq i \leq \frac{n}{2}$
- $\pi_M(2i-1) < \pi_M(2i)$, for every i, $1 \leq i \leq \frac{n}{2}$
- $\pi_M(2i-1) < \pi_M(2i+1)$, for every i, $1 \leq i \leq \frac{n}{2} - 1$

For every matching M, let $sign(\pi_M) = (-1)^{|\{(i,j):\ i<j,\ \pi_M(i) > \pi_M(j)\}|}$ and $value(\pi_M) = \prod_{i=1}^{n/2} a_{\pi_M(2i-1),\pi_M(2i)}$.

The *Pfaffian* of A is defined as follows: $\mathcal{PF}(A) = \sum_{M \in \mathcal{M}} sign(\pi_M) \cdot value(\pi_M)$. The Pfaffian of A is a polynomial of the form: $\mathcal{PF}(A) = \sum_{p=0}^{n/2} \sum_{q=0}^{n/2} c_{pq} x^p y^q$ and it can be computed by interpolation (see [6]), using an algorithm that computes arithmetic Pfaffians [4,8] as a subroutine.

The term $x^p y^q$ of $\mathcal{PF}(A)$ corresponds to the (p, q)-perfect matchings of G^*. Therefore, if c_{pq} is nonzero, then a (p, q)-perfect matching exists in G^*. The converse does not necessarily hold: it is possible that the coefficient of c_{pq} is zero although G^* contains (p, q)-perfect matchings, in the case where the terms corresponding to these matchings are mutually cancelled. The following lemma gives a sufficient condition so that the coefficient of c_{pq} is nonzero.

Lemma 6. *Let p, q be integers, with $0 \leq p, q \leq \frac{n}{2}$ and suppose that there exists a unique minimum weight (p, q)-perfect matching M_{pq} of G^*. Then the coefficient c_{pq} of $\mathcal{PF}(A)$ is nonzero. Furthermore, W_{pq} is the maximum power of 2 that divides c_{pq}.*

[1] Strictly speaking, A is a special form of the Tutte matrix of G^*, where each indeterminate has been replaced either by a specific value or by an indeterminate multiplied by a specific value.

Proof. We have: $c_{pq} = sign(\pi_{M_{pq}}) \cdot 2^{W_{pq}} + \sum_{M \in \mathcal{M}_{pq}-\{M_{pq}\}} sign(\pi_M) \cdot 2^{W_M}$ Since M_{pq} is a unique minimum weight (p,q)-perfect matching, $W_{pq} < W_M$ for every $M \in \mathcal{M}_{pq} - \{M_{pq}\}$. Therefore $c_{pq} \bmod 2^{W_{pq}} = 0$ and $c_{pq} \bmod 2^{W_{pq}+1} = 1$, which imply that $c_{pq} \neq 0$ and that W is the maximum power 2 that divides c_{pq}. □

For every edge e_{ij} of G^* we define

$$Z_{ij} = \sum_{M \in \mathcal{M}: e_{ij} \in M} sign(\pi_M) \cdot value(\pi_M)$$

(that is, Z_{ij} is the part of $\mathcal{PF}(A)$ that involves e_{ij}). The following lemma shows how Z_{ij} can be computed up to sign.

Lemma 7. *For every edge e_{ij} of G^*, $Z_{ij} = \sigma \cdot a_{ij} \cdot \mathcal{PF}(A_{ij})$ where $\sigma \in \{-1, 1\}$ and A_{ij} is the matrix obtained from A by removing the i-th and j-th row and the the i-th and j-th column.*

If G^* has a unique minimum weight (p,q)-perfect matching M_{pq}, we can decide whether an edge belongs to M_{pq}, using the following lemma:

Lemma 8. *Suppose that G^* has a unique minimum weight (p,q)-perfect matching M_{pq}. Let c be the coefficient of $x^p y^q$ in Z_{ij}. Then $e_{ij} \in M_{pq}$ if and only if $2^{W_{pq}+1}$ does not divide c.*

In order to bound from below the probability that the algorithm returns an optimal solution, we make use of the following strong version of the Isolating Lemma:

Lemma 9. (**Isolating Lemma** [10]) *Let $B = \{b_1, b_2, \ldots, b_k\}$ be a set of elements, let $\mathcal{S} = \{S_1, S_2, \ldots, S_\ell\}$ be a collection of subsets of B and let $a_1, a_2, \ldots, a_\ell$ be integers. If we choose integer weights $w_1, w_2, \ldots w_k$ for the elements of B at random from the set $\{1, 2, \ldots, m\}$, and define the weight of set S_j to be $a_j + \sum_{b_i \in S_j} w_i$ then the probability that the minimum weight subset in $\mathcal{S}$ is not unique is at most $\frac{k}{m}$.*

Theorem 4. *Algorithm* Random-BRM *returns an optimal solution with probability at least $\frac{1}{2}$.*

Proof. If there exists a unique minimum weight element M_{pq} in every non-empty set $\mathcal{M}_{pq}$, $0 \leq p, q \leq \frac{n}{2}$ then it follows from Lemmata 3, 5, 6, 7 and 8 that the above algorithm returns an optimal w-blue-red matching.

The probability that $\mathcal{M}_{pq}$ contains at least two minimum weight elements for fixed values p and q is at most $\frac{1}{2n^2}$ by the Isolating Lemma. Thus, the probability that there exist values p, q such that $\mathcal{M}_{pq}$ contains at least two minimum weight elements is at most $(\frac{n}{2}+1)^2 \cdot \frac{1}{2n^2} \leq \frac{1}{2}$. Therefore the algorithm returns a correct solution with probability at least $\frac{1}{2}$. □

Algorithm Random-BRM

Input: graph $G = (V, E_{blue} \cup E_{red} \cup E_{white})$ with even number of vertices n, positive integer w.
Output: maximum w-blue-red matching (with probability $\geq \frac{1}{2}$).

1. augment G to a complete graph G^* by adding a set E_{black} of black edges
2. **for** every $e_{ij} \in G^*$ **do**
  – choose at random a number s_{ij} from $\{1, 2, \ldots, n^4\}$
  – **if** $e_{ij} \in E_{black}$ **then** $w_{ij} := n^5 + s_{ij}$
   else $w_{ij} := s_{ij}$
3. construct the Tutte matrix A of G^*
4. compute $\mathcal{PF}(A)$; let c_{pq} be the coefficient of $x^p y^q$ in in $\mathcal{PF}(A)$, $0 \leq p, q, \leq \frac{n}{2}$
5. **for** every $(p, q) \in \{0, 1, \ldots \frac{n}{2}\}^2$ **do**
  – **if** $c_{pq} \neq 0$ **then** let W_{pq} be the maximum power of 2 that divides c_{pq}
   else $W_{pq} := \infty$
  find $(p, q) \in \{0, 1, \ldots \frac{n}{2}\}^2$ such that $\text{sol}_w(p, q, \frac{n}{2} - p - q - \lfloor \frac{W_{pq}}{n^5} \rfloor)$ is maximum
6. $M_{pq} := \emptyset$
  for every $e_{ij} \in G^*$ **do**
  – compute $|Z_{ij}| := |a_{ij} \cdot \mathcal{PF}(A_{ij})|$
  – let c be the coefficient of $x^p y^q$ in $|Z_{ij}|$
  – **if** $c \bmod 2^{W_{pq}+1} \neq 0$ **then** $M_{pq} := M_{pq} \cup \{e_{ij}\}$
7. compute a w-blue-red matching M from M_{pq}, by a greedy coloring of white edges
8. **return** M

Complexity. Sequentially, the algorithm requires $O(n^7)$ time, since the computation of the symbolic Pfaffian requires $O(n^5)$ time, using the algorithm in [4] which computes arithmetic Pfaffians in $O(n^3)$ time and Step 6 requires the computation of $O(n^2)$ minor Pfaffians. However all steps can be parallelized resulting in a RNC algorithm (in the parallel version, the algorithm from [8] is used to compute arithmetic Pfaffians). In fact, it can be shown by careful analysis that Random-BRM is an RNC^2 algorithm.

5 Application to Optical Networking

In this section we show how solving BRM can help in approximately solving the Directed Maximum Routing and Wavelength Assignment (DirMaxRWA) problem in rings.

DirMaxRWA is defined as follows [12]: Given are a directed symmetric graph G, a set of requests (pairs of nodes) R on G, and an integer w (bound on the number of available wavelengths). The goal is to find a routing and wavelength assignment to an as large as possible set of requests $R' \subseteq R$ such that any two requests routed via edge-intersecting paths receive different wavelengths and only wavelengths from $\{1, \ldots, w\}$ are used.

It can be shown that the algorithm for DirMaxRWA in rings proposed in [12] can be modified to make an explicit call to an algorithm for solving BRM (instead of implicitly solving it, as was the case originally).

The following theorem relates the approximation ratio of the modified algorithm to the approximation ratio achieved by the algorithm for BRM that is employed. The proof is an adaptation of the analysis of the algorithm presented in [12] and will appear in the full version.

Theorem 5. *An algorithm for* BRM *which returns a w-blue-red matching containing at least* $a \cdot \mu_{OPT} - b$ *edges, where* μ_{OPT} *is the size of an optimal solution and* $a > 0, b \geq 0$ *are constants, results in an algorithm for* DirMaxRWA *in rings that satisfies at least* $\frac{a+1}{a+2} \cdot OPT - \frac{b}{a+2}$ *requests, where* OPT *is the size of an optimal solution for* DirMaxRWA.

Therefore, by using the algorithms for BRM proposed in the previous sections (Random-BRM and Balance-BRM) we obtain the following:

Corollary 1. DirMaxRWA *in rings admits a randomized approximation algorithm with ratio* $\frac{2}{3}$ *and a deterministic approximation algorithm with asymptotic ratio* $\frac{7}{11}$.

Note that the $\frac{2}{3}$ approximation ratio is tight, as can be shown by appropriate examples.[2] The deterministic algorithm is slightly worse in terms of approximation ratio, but is considerably faster. An even faster deterministic approximation algorithm with ratio $\frac{3}{5}$ is obtained if we use algorithm Greedy-BRM as a subroutine. As regards time requirements, it can be shown that the complexity of the algorithm for DirMaxRWA is dominated by the complexity of the algorithm for BRM that is employed; therefore it is $O(n^7)$ if we use Random-BRM for solving BRM, while it is $O(n^{2.5})$ if we use Balance-BRM and $O(n^2)$ if we use Greedy-BRM.

References

1. Caragiannis, I.: Wavelength Management in WDM Rings to Maximize the Number of Connections. In: Thomas, W., Weil, P. (eds.) STACS 2007. LNCS, vol. 4393, pp. 61–72. Springer, Heidelberg (2007)
2. Carlisle, M.C., Lloyd, E.L.: On the k-Coloring of Intervals. Discrete Applied Mathematics 59, 225–235 (1995)
3. Erlebach, T., Jansen, K.: The Complexity of Path Coloring and Call Scheduling. Theoretical Computer Science 255(1-2), 33–50 (2001)
4. Galbiati, G., Maffioli, F.: On the Computation of Pfaffians. Discrete Applied Mathematics 51(3), 269–275 (1994)
5. Garey, M., Johnson, D., Miller, G., Papadimitriou, C.: The Complexity of Coloring Circular Arcs and Chords. SIAM Journal on Algebraic Discrete Methods 1(2), 216–227 (1980)
6. Horowitz, E., Sahni, S.: On Computing the Exact Determinant of Matrices with Polynomial Entries. Journal of the ACM 22(1), 38–50 (1975)

[2] The $\frac{2}{3}$ approximation ratio improves upon the ratio obtained in [12] and has been the best known so far for DirMaxRWA in rings until very recently, when Caragiannis [1] gave a 0.708-approximation algorithm for the problem.

7. Karzanov, A.V.: Maximum Matching of Given Weight in Complete and Complete Bipartite Graphs. Kibernetika 23(1), 7–11 (1987) (English translation in CYBNAW 23(1), 8–13 (1987))
8. Mahajan, M., Subramanya, P.R., Vinay, V.: The Combinatorial Approach Yields an NC Algorithm for Computing Pfaffians. Discrete Applied Mathematics 143(1-3), 1–16 (2004)
9. Micali, S., Vazirani, V.V.: An $O(n^{2.5})$ Algorithm for Maximum Matching in General Graphs. In: Proceedings Twenty-first Annual Symposium on the Foundations of Computer Science, pp. 17–27 (1980)
10. Mulmuley, K., Vazirani, U.V., Vazirani, V.V.: Matching is as Easy as Matrix Inversion. Combinatorica 7(1), 105–113 (1987)
11. Nomikos, C., Pagourtzis, A., Zachos, S.: Satisfying a Maximum Number of Pre-Routed Requests in All-Optical Rings. Computer Networks 42(1), 55–63 (2003)
12. Nomikos, C., Pagourtzis, A., Zachos, S.: Minimizing Request Blocking in All-Optical Rings. In: Proceedings INFOCOM 2003, pp. 1771–1780 (2003)
13. Papadimitriou, C.H., Yannakakis, M.: The Complexity of Restricted Spanning Tree Problems. Journal of the ACM 29(2), 285–309 (1982)
14. Raghavan, P., Upfal, E.: Efficient Routing in All-Optical Networks. In: Proceedings of the 26th Annual ACM Symposium on the Theory of Computing STOC 1994, pp. 134–143. ACM Press, New York (1994)
15. Stamoulis, G.: Maximum Matching Problems with Constraints (in Greek). Diploma Thesis, Department of Computer Science, University of Ioannina (2006)
16. Wan, P.-J., Liu, L.: Maximal Throughput in Wavelength-Routed Optical Networks. DIMACS Series in Discrete Mathematics and Theoretical Computer Science 46, 15–26 (1998)
17. Yi, T., Murty, K.G., Spera, C.: Matchings in Colored Bipartite Networks. Discrete Applied Mathematics 121(1-3), 261–277 (2002)

Real Computational Universality: The Word Problem for a Class of Groups with Infinite Presentation
(Extended Abstract)

Klaus Meer[1,*] and Martin Ziegler[2,**]

[1] IMADA, Syddansk Universitet, Campusvej 55,
5230 Odense M, Denmark
meer@imada.sdu.dk
[2] University of Paderborn
ziegler@upb.de

Abstract. The word problem for discrete groups is well-known to be undecidable by a Turing Machine; more precisely, it is reducible both to and from and thus equivalent to the discrete Halting Problem.

The present work introduces and studies a real extension of the word problem for a certain class of groups which are presented as quotient groups of a free group and a normal subgroup. As main difference with discrete groups, these groups may be generated by *un*countably many generators with index running over certain sets of real numbers. This includes a variety of groups which are not captured by the finite framework of the classical word problem.

Our contribution extends computational group theory from the discrete to the Blum-Shub-Smale (BSS) model of real number computation. It provides a step towards applying BSS theory, in addition to semi-algebraic geometry, also to further areas of mathematics.

The main result establishes the word problem for such groups to be not only semi-decidable (and thus reducible *to*) but also reducible *from* the Halting Problem for such machines. It thus gives the first non-trivial example of a problem *complete*, that is, computationally universal for this model.

1 Introduction

In 1936, Alan M. Turing introduced the now so-called Turing Machine and proved the associated Halting Problem H, that is the question of termination of a given such machine M, to be undecidable. On the other hand simulating a machine M on a Universal Turing Machine establishes H to be semi-decidable. In

* Partially supported by the IST Programme of the European Community, under the PASCAL Network of Excellence, IST-2002-506778 and by the Danish Agency for Science, Technology and Innovation FNU. This publication only reflects the author's views.

** Supported by DFG (project Zi1009/1-1) and by JSPS (ID PE 05501).

L. Kučera and A. Kučera (Eds.): MFCS 2007, LNCS 4708, pp. 726–737, 2007.

the sequel, several other problems P were also revealed semi-, yet un-decidable. Two of them, Hilbert's Tenth and the Word Problem for groups, became particularly famous, not least because they arise and are stated in purely mathematical terms whose relation to computer science turned out considerable a surprise. The according undecidability proofs both proceed by constructing from a given Turing Machine M an instance x_M of the problem P under consideration such that $x_M \in P$ iff M terminates; in other words, a reduction from H to P. As P is easily seen to be semi-decidable this establishes, conversely, reducibility to H and thus Turing-completeness of P.

1.1 Real Computability

Turing Machines are still nowadays, 70 years after their introduction, considered the appropriate model of computation for discrete problems, that is, over bits and integers. For real number problems of Scientific Computation as for example in Numerics, Computer Algebra, and Computational Geometry on the other hand, several independent previous formalizations were in 1989 subsumed in a real counterpart to the classical Turing Machines called the Blum-Shub-Smale, for short BSS model [BSS89, BCSS98]. Essentially a (real) BSS-machine can be considered as a Random Access Machine over $\mathbb{R}$ which is able to perform the basic arithmetic operations at unit cost and whose registers can hold arbitrary real numbers; its inputs are thus finite sequences over $\mathbb{R}$ of possibly unbounded length. This model bears many structural similarities to the discrete setting like for example the existence of a Universal Machine, the notion of (e.g. $\mathcal{NP}$–) completeness, or undecidable problems:

Definition 1.1. *The* real Halting Problem $\mathbb{H}$ *is the following decision problem. Given the code* $\langle \mathbb{M} \rangle \in \mathbb{R}^\infty$ *of a BSS machine* $\mathbb{M}$*, does* M *terminate its computation (on empty input) ?*

Both the existence of a coding $\langle \cdot \rangle$ for BSS machines and the undecidability of $\mathbb{H}$ in the BSS model were shown in [BSS89]. Concerning BSS-complete problems $\mathbb{P}$ however, not many are known so far. The Turing-complete ones for example and, more generally, any discrete problem becomes decidable over the reals [BSS89, EXAMPLE §1.6]; and *extending* an undecidable discrete problem to the reals generally does not work either:

Example 1.2. Hilbert's Tenth Problem (over R) is the task of deciding, given a multivariate polynomial equation over R, whether it has a solution in R. For integers $R = \mathbb{Z}$, this problem has been proven (Turing-) undecidable [Mati70]. For reals $R = \mathbb{R}$ however, it *is* (BSS-)decidable by virtue of TARSKI's Quantifier Elimination [BCSS98, top of p.97]. □

1.2 Subsumption of Our Results

Provably undecidable problems over the reals, such as the Mandelbrot Set or the rationals $\mathbb{Q}$ are supposedly (concerning the first) or, concerning the latter, have actually been established [MeZi05] *not* reducible from, and thus strictly easier than, $\mathbb{H}$. In fact the only BSS-complete $\mathbb{P}$ essentially differing from $\mathbb{H}$ we are aware of is a certain countable existential theory in the language of ordered fields [Cuck92, THEOREM 2.13].

The present work closes this structural gap by presenting a real generalization of the word problem for groups and proving it to be reducible both from and to the real Halting Problem. (Such a result had been envisioned in Section 4 of [MeZi06].) On the way to that, we significantly extend notions from classical and computational (discrete, i.e.) combinatorial group theory to the continuous setting of BSS-computability. Several examples reveal these new notions as mathematically natural and rich. They bear some resemblance to certain recent presentations of continuous fundamental groups from topology [CaCo00] where, too, the set of generators ('alphabet') is allowed to be infinite and in fact of continuum cardinality. There however words generally have transfinite length whereas we require them to consist of only finitely many symbols.

1.3 Further Related Work

We find our synthesis of computational group theory and real number computability to also differ significantly from the usual problems studied in the BSS model which typically stem from semi-algebraic geometry. Indeed, the papers dealing with groups G in the BSS setting [Bour01, Gass01, Prun02] treat such G as underlying structure of the computational model, that is, not over the reals $\mathbb{R}$ and its arithmetic. [Tuck80] considers the question of computational realizing G and its operation, not of deciding properties of (elements of) G. An exception, [DJK05] does consider BSS-decidability (and complexity) of properties of a real group; however without completeness results. There also the group is not fixed nor presented but given by some matrix generators.

1.4 Overview

Section 2 starts with a review of the classical word problem in *finitely* presented groups. Then we introduce real counterparts called algebraically presented groups, the core objects of our interest. (Guiding examples of mathematical groups that fit into this framework can be found in the full version...) The word problem for these groups is defined and shown to be semi-decidable in the BSS model of computation over the reals. Section 3 proves our main result: The real Halting Problem can be reduced to the word problem of algebraically presented real groups.

2 Word-Problem for Groups

Groups occur ubiquitously in mathematics, and having calculations with and in them handled by computers constitutes an important tool both in their theoretical investigation and in practical applications as revealed by the flourishing field of *Computational Group Theory* [FiKa91, FiKa95, HEoB05]. Unfortunately already the simplest question, namely equality '$a = b$' of two elements $a, b \in G$

is in general undecidable for groups G reasonably presentable to a digital computer, that is, in a finite way — the celebrated result obtained in the 1950ies independently by NOVIKOV [Novi59] and BOONE [Boon58]. In the BSS model of real number decidability[1] on the other hand, *every* discrete problem $L \subseteq \Sigma^*$ is solvable [BSS89, EXAMPLE §1.6], rendering the word problem for finitely presented groups trivial.

However whenever we deal with computational questions involving groups of real or complex numbers, the Turing model seems not appropriate anyway. As an example take the unit circle in $\mathbb{R}^2$ equipped with complex multiplication. There is a clear mathematical intuition how to compute in this group; such computations can be formalized in the BSS model. We thus aim at a continuous counterpart to the discrete class of finitely presented groups for which the word problem is universal for the BSS model.

2.1 The Classical Setting

Here, the setting for the classical word problem is briefly recalled. A review of the main algebraic concepts needed in our proofs is postponed to Section 3.

Definition 2.1. *a) Let X be a set. The* free group generated by X, *denoted by $F = (\langle X\rangle, \circ)$ or more briefly $\langle X\rangle$, is the set $(X \cup X^{-1})^*$ of all finite sequences $\bar{w} = x_1^{\epsilon_1} \cdots x_n^{\epsilon_n}$ with $n \in \mathbb{N}$, $x_i \in X$, $\epsilon_i \in \{-1, +1\}$, equipped with concatenation $\circ$ as group operation subject to the rules*

$$x \circ x^{-1} \quad = \quad 1 \quad = \quad x^{-1} \circ x \qquad \forall x \in X \tag{1}$$

where $x^1 := x$ and where 1 denotes the empty word, that is, the unit element.

b) For a group H and $W \subseteq H$, denote by

$$\langle W\rangle_H \;:=\; \{w_1^{\epsilon_1} \cdots w_n^{\epsilon_n} : n \in \mathbb{N}, w_i \in W, \epsilon_i = \pm 1\}$$

the subgroup of H generated by W. The normal *subgroup of H generated by W is $\langle W\rangle_{H\mathrm{n}} := \langle\{h \cdot w \cdot h^{-1} : h \in H, w \in W\}\rangle_H$. For $h \in H$, we write h/W for its W–coset $\{h \cdot w : w \in \langle W\rangle_{H\mathrm{n}}\}$ of all $g \in H$ with $g \equiv_W h$.*

c) Fix sets X and $R \subseteq \langle X\rangle$ and consider the quotient group $G := \langle X\rangle / \langle R\rangle_{\mathrm{n}}$, denoted by $\langle X|R\rangle$, of all $\langle R\rangle$–cosets of $\langle X\rangle$.

If both X and R are finite, the tuple (X, R) will be called a finite presentation *of G; if X is finite and R recursively enumerable (by a Turing machine, that is in the discrete sense; equivalently: semi-decidable), it is a* recursive[2] presentation*; if X is finite and R arbitrary, G is* finitely generated.

[1] We remark that in the other major and complementary model of *real* (as opposed to rational or algebraic [KMPSY04]) number computation, the concept of decidability is inapplicable because, there, discontinuous functions are generally uncomputable due to the so-called Main Theorem of Recursive Analysis [Weih00, THEOREM 4.3.1].

[2] This notion seems misleading as R is in general *not* recursive; nevertheless it has become established in literature.

Intuitively, R induces further rules "$\bar{w} = 1$" ($\bar{w} \in R$) in addition to Equation (1); put differently, distinct words $\bar{u}, \bar{v} \in \langle X \rangle$ might satisfy $\bar{u} = \bar{v}$ *in* G, that is, by virtue of R. Observe that the rule "$w_1^{\epsilon_1} \cdots w_n^{\epsilon_n} = 1$" induced by an element $\bar{w} = (w_1^{\epsilon_1} \cdots w_n^{\epsilon_n}) \in R$ can also be applied as "$w_1^{\epsilon_1} \cdots w_k^{\epsilon_k} = w_n^{-\epsilon_n} \cdots w_{k+1}^{-\epsilon_{k+1}}$".

Definition 2.1 (continued)

d) *The* word problem *for* $\langle X|R \rangle$ *is the task of deciding, given* $\bar{w} \in \langle X \rangle$, *whether* $\bar{w} = 1$ *holds in* $\langle X|R \rangle$.

The famous work of Novikov and, independently, Boone establishes the word problem for finitely presented groups to be Turing-complete:

Fact 2.2. *a) For any finitely presented group* $\langle X|R \rangle$, *its associated word problem is semi-decidable (by a Turing machine).*

b) There exists a finitely presented group $\langle X|R \rangle$ *whose associated word problem is many-one reducible from the discrete Halting Problem* H. □

For the nontrivial Claim b), see e.g. one of [Boon58, Novi59, LySc77, Rotm95].

Example 2.3. $\mathcal{H} := \langle \{a, b, c, d\} \,|\, \{a^{-i}ba^i = c^{-i}dc^i : i \in H\} \rangle$ is a *recursively* presented group with word problem reducible from H; compare the proof of [LySc77, Theorem §IV.7.2]. □

Fact 2.2b) requires the group to be *finitely* presented group. This step is provided by the remarkable

Fact 2.4 (Higman Embedding Theorem). *Every recursively presented group can be embedded in a finitely generated one.*

Proof. See, e.g., [LySc77, Section §IV.7] or [Rotm95, Theorem 12.18]. □

Fact 2.4 asserts the word problem from Example 2.3 to be in turn reducible to that of the finitely presented group $\mathcal{H}$ is embedded into, because any such embedding is automatically effective:

Observation 2.5. *Let* $G = \langle X \rangle / \langle R \rangle_{\mathrm{n}}$ *and* $H = \langle Y \rangle / \langle S \rangle_{\mathrm{n}}$ *denote finitely generated groups and* $\psi : G \to H$ *a homomorphism. Then,* ψ *is (Turing-) computable in the sense that there exists a computable homomorphism* $\psi' : \langle X \rangle \to \langle Y \rangle$ *such that* $\psi'(\bar{x}) \in \langle S \rangle_{\mathrm{n}}$ *whenever* $\bar{x} \in \langle R \rangle_{\mathrm{n}}$*; that is,* ψ' *maps* R*-cosets to* S*-cosets and makes Diagram (2) commute.*

$$\begin{array}{ccc} \langle X \rangle & \xrightarrow[\psi']{} & \langle Y \rangle \\ \downarrow & & \downarrow \\ \langle X \rangle / \langle R \rangle_{\mathrm{n}} & \xrightarrow{\psi} & \langle Y \rangle / \langle S \rangle_{\mathrm{n}} \end{array} \qquad (2)$$

Indeed, due the homomorphism property, ψ is uniquely determined by its values on the finitely many generators $x_i \in X$ of G, that is, by $\psi(x_i) = \bar{w}_i / \langle S \rangle_{\mathrm{n}}$ where $\bar{w}_i \in \langle Y \rangle$. Setting (and storing in the machine) $\psi'(x_i) := \bar{w}_i$ yields the claim.

2.2 Presenting Real Groups

Regarding that the BSS-machine is the natural extension of the Turing machine from the discrete to the reals, the following is equally natural a generalization of Definition 2.1c+d):

Definition 2.6. *Let $X \subseteq \mathbb{R}^\infty$ and $R \subseteq \langle X \rangle \subseteq^3 \mathbb{R}^\infty$. The tuple (X, R) is called a* presentation *of the* real group $G = \langle X|R \rangle$. *This presentation is* algebraically generated *if X is BSS-decidable and $X \subseteq \mathbb{R}^N$ for some $N \in \mathbb{N}$. G is termed* algebraically enumerated *if R is in addition BSS semi-decidable; if R is even BSS-decidable, call G* algebraically presented. *The* word problem *for the presented real group $G = \langle X|R \rangle$ is the task of BSS-deciding, given $\bar{w} \in \langle X \rangle$, whether $\bar{w} = 1$ holds in G.*

CLASSICAL DISCRETE AND OUR NEW REAL NOTIONS:

Turing	BSS
finitely generated	algebraically generated
recursively presented	algebraically enumerated
finitely presented	algebraically presented

Remark 2.7. a) Although X inherits from $\mathbb{R}$ algebraic structure such as addition $+$ and multiplication $\times$, the Definition 2.1a) of the free group $G = (\langle X \rangle, \circ)$ considers X as a plain set only. In particular, (group-) inversion in G must not be confused with (multiplicative) inversion: $5 \circ \frac{1}{5} \neq 1 = 5 \circ 5^{-1}$ for $X = \mathbb{R}$. This difference may be stressed notationally by writing 'abstract' generators $x_{\bar{a}}$ indexed with real vectors $\bar{a}$: $x_5^{-1} \neq x_{1/5}$.

b) Isomorphic (that is, essentially identical) groups $\langle X|R \rangle \cong \langle X'|R' \rangle$ may have different presentations (X, R) and (X', R'). Even when $R = R'$, X need not be unique! Nevertheless we adopt from literature such as [LySc77] the convention of speaking of "the group $\langle X|R \rangle$", meaning a group with presentation (X, R).

This however requires some care, for instance when $\bar{w}$ is considered (as in Definition 2.1d) both an element of $\langle X \rangle$ and of $\langle X|R \rangle$! For that reason we prefer to write $\langle W \rangle_H$ rather than, e.g., $\mathrm{Gp}(W)$: to indicate in which group we consider a subgroup to be generated.

For a BSS-machine to read or write a word $\bar{w} \in \langle X \rangle = (X \cup X^{-1})^*$ of course means to input or output a vector $(w_1, \epsilon_1, \dots, w_n, \epsilon_n) \in (\mathbb{R}^N \times \mathbb{N})^n$. In this sense, the Rules (1) implicit in the free group are obviously decidable and may w.l.o.g. be included in R.

We first show that, parallel to Fact 2.2a), the word problem for any algebraically enumerated real group is not harder than the BSS Halting Problem:

Theorem 2.8. *Let $G = \langle X|R \rangle$ denote a algebraically enumerated real group. Then the associated word problem is BSS semi-decidable.*

This and all further proofs will be available in the full version.

[3] R is a set of *vectors of vectors* of varying lengths. By suitably encoding delimiters we shall regard R as effectively embedded into *single* vectors of varying lengths.

3 Reduction *from* the Real Halting Problem

This section proves the main result of the paper and continuous counterpart to Fact 2.2b): The word problem for algebraically presented real groups is in general not only undecidable (cmp. [MeZi05]) in the BSS model but in fact as hard as the real Halting Problem.

Theorem 3.1. *There exists an algebraically presented real group $\mathcal{H} = \langle X|R\rangle$ such that $\mathbb{H}$ is BSS-reducible to the word problem in $\mathcal{H}$.*

Our proof has been guided by, and borrows concepts from, that of the discrete case [LySc77, SECTION §IV.7]. However a simple transfer fails because many properties heavily exploited in the discrete case (e.g., that the homeomorphic image of a finitely generated group is again finitely generated) are not immediately clear how to carry over to the reals (Section 3.2). For instance, a proof for the classical result may exploit MATIYASEVICH's famous solution of Hilbert's Tenth Problem, namely a Diophantine formulation of H [Mati70], which is infeasible for $\mathbb{H}$ (recall Example 1.2).

3.1 Basics from Group Theory and Their Presentations

This subsection briefly recalls some constructions from group theory and their properties which will heavily be used later on. For a more detailed exposition as well as proofs of the cited results we refer to the two textbooks [LySc77, Rotm95].

Here, no (e.g. effectivity) assumptions are made concerning the set of generators nor relations presenting a group. To start with and just for the records, let us briefly extend the standard notions of a subgroup and a homomorphism to the setting of *presented* groups:

Definition 3.2. *A* subgroup *U of the presented group $G = \langle X|R\rangle$ is a tuple (V, S) with $V \subseteq \langle X\rangle$ and $S = R \cap \langle V\rangle$. This will be denoted by $U = \langle V|R_V\rangle$ or, more relaxed, $U = \langle V|R\rangle$.*

A realization *of a homomorphism $\psi : G \to H$ between presented groups $G = \langle X|R\rangle$ and $H = \langle Y|S\rangle$ is a mapping $\psi' : X \to \langle Y\rangle$ whose unique extension to a homomorphism on $\langle X\rangle$ maps R-cosets to S-cosets, that is, makes Equation (2) commute.*

A realization of an isomorphism ϕ is a realization of ϕ as a homomorphism.

In the above notation, $\langle \psi'(X)\big|S\rangle$ is a presentation of the subgroup $\psi(G)$ of H. For an embedding ψ, G is classically isomorphic to $\psi(G)$; Lemma 3.14 below contains a computable variation of this fact.

Remark 3.3. The intersection $A \cap B$ of two subgroups A, B of G is again a subgroup of G. For presented sub-groups $A = \langle U|R\rangle$ and $B = \langle V|R\rangle$ of $G = \langle X|R\rangle$ however, $\langle U \cap V|R\rangle$ is in general *not* a presentation of $A \cap B$.

Definition 3.4 (Free Product). *Consider two presented groups $G = \langle X|R\rangle$ and $H = \langle Y|S\rangle$ with disjoint generators $X \cap Y = \emptyset$ — e.g. by proceeding to*

$X' := X \times \{1\}, Y' := Y \times \{2\}$, $R' := R \times \{1\}$, $S' := S \times \{2\}$. *The* free product *of G and H is the presented group* $G * H := \langle X \cup Y \mid R \cup S \rangle$.
Similarly for the free product $\ast_{i \in I} G_i$ *with* $G_i = \langle X_i | R_i \rangle$, *$I$ an arbitary index set.*

In many situations one wants to identify certain elements of a free product of groups. These are provided by two basic constructions: *amalgamation* and *Higman-Neumann-Neumann* (or shortly HNN) extension, see [LySc77, Rotm95] and in particular the nice illustration [Rotm95, FIGURE 11.9].

Definition 3.5 (Amalgamation). *Let* $G = \langle X|R \rangle$, $H = \langle Y|S \rangle$ *with* $X \cap Y = \emptyset$. *Let* $A = \langle V|R \rangle$ *and* $B = \langle W|S \rangle$ *be respective subgroups and* $\phi' : \langle V \rangle \to \langle W \rangle$ *realization of an isomorphism* $\phi : A \to B$. *The free product of G and H* amalgamating *the subgroups A and B via ϕ is the presented group*

$$\langle G * H \mid \phi(a) = a \forall a \in A \rangle \quad := \quad \big\langle X \cup Y \mid R \cup S \cup \{\phi'(\bar{v})\bar{v}^{-1} : \bar{v} \in V\} \big\rangle \ . \quad (3)$$

Definition 3.6 (HNN Extension). *Let* $G = \langle X|R \rangle$, $A = \langle V|R \rangle, B = \langle W|R \rangle$ *subgroups of G, and ϕ' a realization of an isomorphism between A and B. The* Higman-Neumann-Neumann *(HNN) extension of G relative to A, B and ϕ is the presented group*

$$\langle G; t \mid ta = \phi(a)t \forall a \in A \rangle \quad := \quad \big\langle X \cup \{t\} \mid R \cup \{\phi'(\bar{v})t\bar{v}^{-1}t^{-1} : \bar{v} \in V\} \big\rangle \ .$$

G is the base *of the HNN extension, $t \notin X$ is a new generator called the* stable letter, *and A and B are the* associated subgroups *of the extension. Similarly for the HNN extension* $\langle G; (t_i)_{i \in I} \mid t_i a = \phi_i(a) t_i \forall a \in A_i \forall i \in I \rangle$ *with respect to a family of isomorphisms* $\phi_i : A_i \to B_i$ *and subgroups* $A_i, B_i \subseteq G$, $i \in I$.

Both HNN extensions and free products with amalgamation admit simple and intuitive characterizations for a word to be, in the resulting group, equivalent to 1. These results are connected to some very famous names in group theory. Proofs can be found, e.g., in [LySc77, CHAPTER IV] or [Rotm95, CHAPTER 11].

Fact 3.7 (Higman-Neumann-Neumann). *Let* $G^* := \langle G; t | ta = \phi(a)t \forall a \in A \rangle$ *be a HNN extension of G. Then,* $\mathrm{id} : g \mapsto g$ *is an embedding of G into G^*.* □

Fact 3.8 (Britton's Lemma). *Let* $G^* := \langle G; t | ta = \phi(a)t \forall a \in A \rangle$ *be an HNN extension of G. Consider a sequence* $(g_0, t^{\epsilon_1}, g_1, \ldots, t^{\epsilon_n}, g_n)$ *with* $n \in \mathbb{N}$, $g_i \in G$, $\epsilon_i \in \{-1, 1\}$. *If it contains no consecutive subsequence* (t^{-1}, g_i, t) *with* $g_i \in A$ *nor* (t, g_j, t^{-1}) *with* $g_j \in B$, *then it holds* $g_0 \cdot t^{\epsilon_1} \cdot g_1 \cdots t^{\epsilon_n} \cdot g_n \neq 1$ *in G^*.* □

Fact 3.9 (Normal Form). *Let* $P = \langle G * H | \phi(a) = a \forall A \rangle$ *denote a free product with amalgamation. Consider* $c_1, \ldots, c_n \in G * H$, $2 \leq n \in \mathbb{N}$, *such that i) each c_i is either in G or in H; ii) consecutive c_i, c_{i+1} come from different factors; iii) no c_i is in A nor B. Then* $c_1 \cdots c_n \neq 1$ *in P.* □

3.2 First Effectivity Considerations

Regarding finitely generated groups, the cardinalities of the sets of generators (that is their *ranks*) add under free products [LySc77, COROLLARY §IV.1.9]. Consequently, they can straight forwardly be bounded under both HNN extensions and free products with amalgamation. Similarly for real groups, we have easy control over the *dimension* N of set of generators according to Definition 2.6:

Observation 3.10. *For groups $G_i = \langle X_i | R_i \rangle$ with $X_i \subseteq \mathbb{R}^N$ for all $i \in I \subseteq \mathbb{R}$, the free product $\ast_{i \in I} G_i = \langle \bigcup_{i \in I}(X \times \{i\}) \mid \bigcup_{i \in I}(R \times \{i\}) \rangle$ is of dimension at most $N+1$. In the countable case $I \subseteq \mathbb{N}$, the dimension can even be achieved to not grow at all: by means of a bicomputable bijection $\mathbb{R} \times \mathbb{N} \to \mathbb{R}$ like $(x,n) \mapsto \langle \lfloor x \rfloor, n \rangle + (x - \lfloor x \rfloor)$.*

Similarly for free products with amalgamation and for HNN extensions...

Moreover, free products, HNN extensions, and amalgamations of algebraically generated/enumerated/presented groups are, under reasonable presumptions, again algebraically generated/enumerated/presented:

Lemma 3.11. *a) Let $G_i = \langle X_i | R_i \rangle$ for all $i \in I \subseteq \mathbb{N}$. If I is finite and each G_i algebraically generated/enumerated/presented, then so is $\ast_{i \in I} G_i$. Same for $I = \mathbb{N}$, provided that G_i is algebraically generated/enumerated/ presented* uniformly in i.

b) Let $G = \langle X | R \rangle$ and consider the HNN extension $G^* := \langle G; (t_i)_{i \in I} \mid t_i a = \phi_i(a) t_i \forall a \in A_i \forall i \in I \rangle$ w.r.t. a family of isomorphisms $\phi_i : A_i \to B_i$ between subgroups $A_i = \langle V_i | R \rangle, B_i = \langle W_i | R \rangle$ for $V_i, W_i \subseteq \langle X \rangle$, $i \in I$.

Suppose that I is finite, each G_i is algebraically enumerated/presented, $V_i \subseteq \mathbb{R}^\infty$ is semi-/decidable, and finally each ϕ_i is effective as a homomorphism; then G^* is algebraically enumerated/presented as well. Same for $I = \mathbb{N}$, provided that the V_i are uniformly *semi-/decidable and effectivity of the ϕ_i holds* uniformly.

c) Let $G = \langle X | R \rangle$ and $H = \langle Y | S \rangle$; let $A = \langle V | R \rangle \subseteq G$ and $B = \langle W | S \rangle \subseteq H$ be subgroups with $V \subseteq \langle X \rangle$, $W \subseteq \langle Y \rangle$, $V \subseteq \mathbb{R}^\infty$ semi-/decidable, and $\phi : A \to B$ an isomorphism and effective homomorphism. Then the free product with amalgamation (3) is algebraically enumerated/presented whenever G and H are.

Remark 3.12. Uniform (semi-) decidability of a family $V_i \subseteq \mathbb{R}^\infty$ of course means that every V_i is (semi-)decidable not only by a corresponding BSS-machine $\mathbb{M}_i$, but all V_i by one common machine $\mathbb{M}$; similarly for *uniform* computability of a family of mappings. By virtue of (the proof of) [Cuck92, THEOREM 2.4], a both necessary and sufficient condition for such uniformity is that the real constants employed by the $\mathbb{M}_i$ can be chosen to all belong to one common finite field extension $\mathbb{Q}(c_1, \ldots, c_k)$ over the rationals. □

Recall (Observation 2.5) that a homomorphism between finitely generated groups is automatically effective and, if injective, has decidable range and effective inverse. For real groups however, in order to make sense out of the prerequisites in Lemma 3.11b+c), we explicitly have to specify the following

Definition 3.13. *An homomorphism $\psi : \langle X|R\rangle \to \langle Y|S\rangle$ of presented real groups is called an* effective homomorphism *if it admits a BSS-computable realization $\psi' : X \to \langle Y\rangle$ in the sense of Definition 3.2.*

For ψ to be called an effective embedding, *it must not only be an effective homomorphism and injective; but ψ' is also required to be injective and have decidable image $\psi'(X)$ plus a BSS-computable inverse $\chi' : \psi'(X) \subseteq \langle Y\rangle \to X$.*

Effective embeddings arise in Lemmas 3.14 and 3.17. For an injective effective homomorphism ϕ as in Lemma 3.11c) on the other hand, a realization needs not to be injective; for instance, ϕ' might map two equivalent (w.r.t. the relations R) yet distinct elements to the same image word.

Lemma 3.14. *Let $\psi : G = \langle X|R\rangle \to \langle Y|S\rangle = K$ denote an effective embedding.*

a) There is an effective embedding $\chi : \psi(G) \to G$ (i.e. we have an effective isomorphism).
b) If $V \subseteq \langle X\rangle$ is decidable, then the restriction $\psi|_H$ to $H = \langle V|R\rangle \subseteq G$ is an effective embedding again.
c) If G is algebraically generated and K algebraically presented then $\psi(G)$ is algebraically presented.

3.3 Benign Embeddings

The requirement in Lemma 3.11b+c) that the subgroup(s) A be recursively enumerable or even decidable, is of course central but unfortunately violated in many cases. This suggests the notion of *benign* subgroups, in the classical case (below, Item a). Recall that there, effectivity of an embedding drops off automatically.

Definition 3.15. *a) Let X be finite, $V \subseteq \langle X\rangle$. The subgroup $A = \langle V|R\rangle$ of $G = \langle X|R\rangle$ is (classically)* benign *in G if the HNN extension $\langle X; t \,|\, ta = at \forall a \in A\rangle$ can be embedded into some finitely presented group $K = \langle Y|S\rangle$.*
b) Let $X \subseteq \mathbb{R}^\infty$, $V \subseteq \langle X\rangle$. The subgroup $A = \langle V|R\rangle$ of $G = \langle X|R\rangle$ is effectively benign *in G if the HNN extension $\langle G; t \,|\, ta = at \forall a \in A\rangle$ admits an effective embedding into some algebraically presented group $K = \langle Y|S\rangle$.*
c) Let $I \subseteq \mathbb{N}$. A family $(A_i)_{i\in I}$ of subgroups of G is uniformly effectively benign *in G if, in the sense of Remark 3.12, there are groups K_i uniformly algebraically presented and uniformly effective embeddings $\phi_i : \langle G; t_i|t_i a_i = a_i t_i \forall a_i \in A_i\rangle \to K_i$.*

The benefit of benignity is revealed in the following

Remark 3.16. In the notation of Definition 3.15b), if A is effectively benign in G then the word problem for A is reducible to that for K: Fact 3.7.
Moreover in this case, the *membership problem* for A in G — that is the question whether given $\bar{x} \in \langle X\rangle$ is equivalent (w.r.t. R) to an element of A — is also reducible to the word problem for K: According to Fact 3.8, $a := \bar{x}/R$ satisfies $t \cdot a \cdot t^{-1} \cdot a^{-1} = 1 \Leftrightarrow a \in A$. □

The following fundamental properties extend corresponding results from the finite framework. Specifically, Lemma 3.17b) generalizes [LySc77, LEMMA §IV.7.7(i)] and Claims d+e) generalize [LySc77, LEMMA §IV.7.7(ii)].

Lemma 3.17. *a) Let $A = \langle V|R\rangle \subseteq H = \langle W|R\rangle \subseteq G = \langle X|R\rangle$ denote a chain of sub-/groups with $V \subseteq \langle W\rangle$ and $W \subseteq \langle X\rangle$. If W is decidable and A effectively benign in G, then it is also effectively benign in H.*

b) If $G = \langle X|R\rangle$ is algebraically presented and subgroup $A = \langle V|R\rangle$ has decidable generators $V \subseteq \langle X\rangle$, then A is effectively benign in G.

c) If A is effectively benign in G and $\phi : G \to H$ an effective embedding, then $\phi(A)$ is effectively benign in $\phi(G)$.

d) Let A and B be effectively benign in algebraically presented G. Then $A \cap B$ admits a presentation effectively benign in G.

e) Let A, B, G as in d); then $\langle A \cup B\rangle_G$ admits a presentation (possibly different from $\langle V \cup W|R\rangle$) effectively benign in G.

f) Let $(A_i)_{i\in I}$ be uniformly effectively benign in G (Definition 3.15c). Then $\langle \bigcup_{i\in I} A_i\rangle$ admits a presentation effectively benign in G.

The above claims hold uniformly, corresponding effective embeddings do not introduce new real constants.

3.4 Putting It All Together

Let $\mathbb{H} \subseteq \mathbb{R}^\infty$ denote the real Halting Problem, semi-decided by some (constant-free) universal BSS Machine $\mathbb{M}$. Denote by $n \mapsto \gamma_n$ an effective enumeration of all computational paths of $\mathbb{M}$, $\mathbb{A}_n \subseteq \mathbb{H} \cap \mathbb{R}^{d(n)}$ the set of inputs accepted at path γ_n. Let $X := \{x_r : r \in \mathbb{R}\} \uplus \{k_n : n \in \mathbb{N}\} \uplus \{s\}$, $G := \langle X\rangle$, and $U := \langle \bar{k}_n^{-1} \cdot w_{\vec{r}} \cdot k_n : n \in \mathbb{N}, \vec{r} \in \mathbb{A}_n\rangle$ where $\bar{w}_{r_1,\ldots,r_d} := x_{r_d}^{-1} \cdots x_{r_1}^{-1} \cdot s \cdot x_{r_1} \cdots x_{r_d}$. Finally let $V := \langle U; (k_n)\rangle \cap \langle s; x_r : r \in \mathbb{R}\rangle$. Based on Lemma 3.17 we can show

Proposition 3.18. *a) U is decidable. b) U and V are effectively benign in G. c) The words $k_n^{-1} \cdot w_{\vec{r}} \cdot k_n$ freely generate U. d) $V = \langle \bar{w}_{\vec{r}} : \exists n \in \mathbb{N} : \vec{r} \in \mathbb{A}_n\rangle$.*

Proof (Theorem 3.1). Let K denote the algebraically presented group which the HNN extension $\langle G; t|tv = vt \forall v \in V\rangle$ effectively embeds into (Proposition 3.18b) by some effective embedding ψ. Then, according to Fact 3.8 and by Proposition 3.18d), $\bar{v} := \psi\big(t \cdot \bar{w}_{\vec{r}} \cdot t^{-1} \cdot \bar{w}_{\vec{r}}^{-1}\big)$ equals 1 in K iff $\vec{r} \in \mathbb{H} = \bigcup_n \mathbb{A}_n$. □

References

[BCSS98] Blum, L., Cucker, F., Shub, M., Smale, S.: Complexity and Real Computation. Springer, Heidelberg (1998)

[BSS89] Blum, L., Shub, M., Smale, S.: On a Theory of Computation and Complexity over the Real Numbers: $\mathcal{NP}$-Completeness, Recursive Functions, and Universal Machines. Bulletin of the American Mathematical Society (AMS Bulletin), vol. 21, pp. 1–46 (1989)

[Boon58] Boone, W.W.: The word problem. Proc. Nat. Acad. Sci. 44, 265–269 (1958)

[Bour01] Bourgade, M.: Séparations et transferts dans la hiérarchie polynomiale des groupes abéliens infinis. Mathematical Logic Quarterly 47 (4), 493–502 (2001)

[CaCo00] Cannon, J.W., Conner, G.R.: The combinatorial structure of the Hawaiian earring group. Topology and its Applications 106, 225–271 (2000)

[Cuck92] Cucker, F.: The arithmetical hierarchy over the reals. Journal of Logic and Computation 2(3), 375–395 (1992)

[DJK05] Derksen, H., Jeandel, E., Koiran, P.: Quantum automata and algebraic groups. J. Symbolic Computation 39, 357–371 (2005)

[FiKa91] Finkelstein, L., Kantor, W.M. (eds.): Groups and Computation. The DIMACS Series in Discrete Mathematics and Theoretical Computer Science, AMS, vol. 11. AMS, Providence, RI (1991)

[FiKa95] Finkelstein, L., Kantor, W.M. (eds.): Groups and Computation II. The DIMACS Series in Discrete Mathematics and Theoretical Computer Science, AMS, vol. 28. AMS, Providence, RI (1995)

[Gass01] Gassner, C.: The $\mathcal{P} = \mathcal{DNP}$ problem for infinite abelian groups. Journal of Complexity 17, 574–583 (2001)

[HEoB05] Holt, D.F., Eick, B., O'Brien, E.: Handbook of Computational Group Theory. Chapman&Hall/CRC (2005)

[KMPSY04] Kettner, L., Mehlhorn, K., Pion, S., Schirra, S., Yap, C.K.: Classroom Examples of Robustness Problems in Geometric Computations. In: Albers, S., Radzik, T. (eds.) ESA 2004. LNCS, vol. 3221, pp. 702–713. Springer, Heidelberg (2004)

[LySc77] Lyndon, R.C., Schupp, P.E.: Combinatorial Group Theory. Springer, Heidelberg (1977)

[Mati70] Matiyasevich, Y.: Enumerable sets are Diophantine. Soviet Mathematics. Doklady 11(2), 354–358 (1970)

[MeZi05] Meer, K., Ziegler, M.: An Explicit Solution to Post's Problem over the Reals. In: Liśkiewicz, M., Reischuk, R. (eds.) FCT 2005. LNCS, vol. 3623, pp. 456–467. Springer, Heidelberg (2005), full version to appear in the journal of complexity, see also `arXiv:cs.LO/0603071`

[MeZi06] Meer, K., Ziegler, M.: Uncomputability Below the Real Halting Problem. In: Beckmann, A., Berger, U., Löwe, B., Tucker, J.V. (eds.) CiE 2006. LNCS, vol. 3988, pp. 368–377. Springer, Heidelberg (2006)

[Novi59] Novikov, P.S.: On the algorithmic unsolvability of the word problem in group theory. Trudy Mat. Inst. Steklov 44, 1–143 (1959)

[Prun02] Prunescu, M.: A model-theoretic proof for $\mathcal{P} \neq \mathcal{NP}$ over all infinite abelian groups. The Journal of Symbolic Logic 67, 235–238 (2002)

[Rotm95] Rotman, J.J.: An Introduction to the Theory of Groups 4th Edition. Springer, Heidelberg (1995)

[Tuck80] Tucker, J.V.: Computability and the algebra of fields. J. Symbolic Logic 45, 103–120 (1980)

[Turi36] Turing, A.M.: On Computable Numbers, with an Application to the Entscheidungsproblem. Proc. London Math. Soc. 42(2), 230–265 (1936)

[Weih00] Weihrauch, K.: Computable Analysis. Springer, Heidelberg (2000)

Finding Paths Between Graph Colourings: PSPACE-Completeness and Superpolynomial Distances

Paul Bonsma[1,*] and Luis Cereceda[2]

[1] Institut für Mathematik, Sekr. MA 6-1, Technische Universität Berlin,
Straße des 17. Juni 136, 10623 Berlin, Germany
bonsma@math.tu-berlin.de
[2] Centre for Discrete and Applicable Mathematics
Department of Mathematics, London School of Economics
Houghton Street, London WC2A 2AE, U.K.
luis@maths.lse.ac.uk

Abstract. Suppose we are given a graph G together with two proper vertex k-colourings of G, α and β. How easily can we decide whether it is possible to transform α into β by recolouring vertices of G one at a time, making sure we always have a proper k-colouring of G? This decision problem is trivial for $k = 2$, and decidable in polynomial time for $k = 3$. Here we prove it is PSPACE-complete for all $k \geq 4$, even for bipartite graphs, as well as for: (i) planar graphs and $4 \leq k \leq 6$, and (ii) bipartite planar graphs and $k = 4$. The values of k in (i) and (ii) are tight.

We also exhibit, for every $k \geq 4$, a class of graphs $\{G_{N,k} : N \in \mathbb{N}^*\}$, together with two k-colourings for each $G_{N,k}$, such that the minimum number of recolouring steps required to transform the first colouring into the second is superpolynomial in the size of the graph. This is in stark contrast to the $k = 3$ case, where it is known that the minimum number of recolouring steps is at most quadratic in the number of vertices. The graphs $G_{N,k}$ can also be taken to be bipartite, as well as (i) planar for $4 \leq k \leq 6$, and (ii) planar and bipartite for $k = 4$. This provides a remarkable correspondence between the tractability of the problem and its underlying structure.

Keywords: vertex-recolouring, colour graph, PSPACE-complete, superpolynomial distance.

1 Introduction

We deal exclusively with finite, simple, loopless graphs. Most of our terminology and notation is standard and can be found in any textbook on graph theory such as, for example, [1]. A standard reference for complexity theory is [2].

* Supported by the Graduate School "Methods for Discrete Structures" in Berlin, DFG grant GRK 1408.

L. Kučera and A. Kučera (Eds.): MFCS 2007, LNCS 4708, pp. 738–749, 2007.

We regard a k-colouring of a graph $G = (V, E)$ as proper; that is, as a function $\alpha : V \to \{1, 2, \ldots, k\}$ such that $\alpha(u) \neq \alpha(v)$ for all $uv \in E$. For a positive integer k and a graph G, we define the *k-colour graph of G*, denoted $\mathcal{C}_k(G)$, as the graph that has the k-colourings of G as its node set, with two k-colourings joined by an edge in $\mathcal{C}_k(G)$ if they differ in colour on just one vertex of G. We assume throughout that $k \geq \chi(G) \geq 2$, where $\chi(G)$ is the chromatic number of G. Having defined the colourings as nodes of $\mathcal{C}_k(G)$, the meaning of a path between two colourings should be clear. In addition, other graph-theoretical notions such as distance and adjacency can now be used for colourings. If $\mathcal{C}_k(G)$ is connected, we say that G is *k-mixing*. We use the term *frozen* for a k-colouring of a graph G that forms an isolated node in $\mathcal{C}_k(G)$.

In [3], [4], some preliminary investigations into the connectedness of the k-colour graph are made. In particular, [4] settles the computational complexity of the following decision problem: given a 3-colourable graph G, is G 3-mixing? This problem is proved to be coNP-complete for bipartite graphs but polynomial-time solvable for bipartite planar graphs. For G a 3-chromatic graph, the answer is always in the negative.

The question of when the k-colour graph is connected is not new: it has been addressed by researchers when studying the Glauber dynamics for sampling k-colourings of a given graph. This is a Markov chain used to obtain efficient algorithms for approximately counting or almost uniformly sampling k-colourings of a graph, and the connectedness of the k-colour graph is a necessary condition for such a Markov chain to be rapidly mixing. For full details, see, for example, [5] and references therein.

A related problem is that of recognising when two given k-colourings of a graph G are in the same connected component of $\mathcal{C}_k(G)$. Formally, we have the following decision problem:

k-Colour Path
Instance: Graph G, two k-colourings of G, α and β.
Question: Is there a path between α and β in $\mathcal{C}_k(G)$?

The problem 2-Colour Path is trivial: the 2-colour graph of a connected bipartite graph always consists of two isolated nodes. For 3-colourings, we have:

Theorem 1 ([6]). *The decision problem* 3-Colour Path *is in* P.

The proof of correctness of the polynomial-time algorithm for 3-Colour Path given in [6] can be employed to exhibit a path between the 3-colourings, if such a path exists. Moreover, such a path has length $O(|V(G)|^2)$, proving:

Theorem 2 ([6]). *Let G be a 3-colourable graph with n vertices. Then the diameter of any component of $\mathcal{C}_3(G)$ is $O(n^2)$.*

Our first main result settles the complexity of k-Colour Path:

Theorem 3. *For every $k \geq 4$, the decision problem k-*Colour Path *is* PSPACE*-complete. Moreover, it remains* PSPACE*-complete for the following restricted instances:*

(i) bipartite graphs and any fixed $k \geq 4$;
(ii) planar graphs and any fixed $4 \leq k \leq 6$; *and*
(iii) bipartite planar graphs and $k = 4$.

In our second main result, we show that instances exist where a superpolynomial number of recolourings is needed to go from the first colouring to the second.

Theorem 4. *For every* $k \geq 4$, *there exists a class of graphs* $\{G_{N,k} : N \in \mathbb{N}^*\}$ *with the following properties. The graphs* $G_{N,k}$ *have size* $O(N^2)$, *and for each of them there exist two k-colourings* α *and* β *in the same component of* $\mathcal{C}_k(G_{N,k})$ *which are at distance* $\Omega(2^N)$. *Moreover,*

(i) the graphs $G_{N,k}$ *may be taken to be bipartite;*
(ii) for every $4 \leq k \leq 6$, *the graphs* $G_{N,k}$ *may be taken to be planar (in such a case the graphs have size* $O(N^4)$*); and*
(iii) for $k = 4$, *the graphs* $G_{N,k}$ *may be taken to be planar and bipartite (in such a case the graphs have size* $O(N^4)$*).*

We consider the construction of these instances to be of independent interest, illustrating the complexity of the problem in a different way, particularly with regard to sampling colourings via Glauber dynamics. But also because of its relationship with the well-known NP $\neq$ PSPACE conjecture. A path between α and β in $\mathcal{C}_k(G)$ constitutes a YES-certificate for the k-COLOUR PATH instance G, α, β. So if for any instance the distance between α and β were always polynomial, k-COLOUR PATH would lie in NP. Now, having established that k-COLOUR PATH is PSPACE-complete, asserting that NP $\neq$ PSPACE is *equivalent* to saying that for every possible YES-certificate for k-COLOUR PATH, there exist instances for which the certificate cannot be verified in polynomial time. Hence proving this statement for every possible certificate is a daunting task, but the above theorem proves it for the most natural certificate, in some sense supporting the conjecture.

The rest of the paper is organised as follows. In Section 2 we introduce the notions that will be used in the proofs. In Section 3 we prove Theorem 3 and also show that the values of k in parts (ii) and (iii) of the theorem are tight: for larger values of k, the instance is always a YES instance. Section 4 is devoted to the proof of Theorem 4.

Theorems 1 to 4 together suggest that the computational complexity of k-COLOUR PATH and the possible distance between k-colourings are intimately linked. How strong is this connection between PSPACE-completeness and superpolynomial distances in the colour graph? In particular, bearing in mind the tightness of k in (ii) and (iii) of Theorem 3: is it true that for a planar graph G and $k \geq 7$, or G a bipartite planar graph and $k \geq 5$, the components of $\mathcal{C}_k(G)$ always have polynomial diameter? We formulate this question more generally as a conjecture in Section 3.3, and give a partial answer. For completeness, we remark that graph classes can be constructed for which k-COLOUR PATH is easy, but which still contain instances with colourings at superpolynomial distance; using for example the graphs from Section 4.

Another situation where the results presented here and in [3], [4] find application is that of radio-frequency reassignment. Given that the frequency assignment problem is often modelled as a graph-colouring problem, the task of reassigning frequencies in a network, while avoiding interference and ensuring no connections are lost, can initially be thought of as a graph recolouring problem. See [7] for a discussion of these ideas in the context of cellular phone networks.

It is very interesting to compare the work presented in this paper and [3,4] with [8], which contains remarkably similar results. For a given instance φ of the Boolean satisfiability problem, the authors of [8] define the graph $G(\varphi)$ as the graph with vertex set the satisfying assignments of φ, and assignments adjacent whenever they differ in exactly one bit. They consider the analogous question to the one we address here, finding the same correspondence between PSPACE-complete instance classes of the decision problem and possible superpolynomial distances in the graph of satisfying assignments.

2 Preliminaries: List Colourings and Forbidding Paths

In Sections 3 and 4 we will construct particular k-Colour Path instances G, α, β: first for the PSPACE-hardness proof, and then for the superpolynomial distance proof. In both cases, it is easier to first define *list-colouring instances*: for such instances we give every vertex v a *colour list* $L(v) \subseteq \{1, 2, 3, 4\}$. A proper list-colouring is a proper vertex colouring with the additional constraint that every vertex colour needs to be chosen from the colour list of the vertex. In the same way as that in which we define the colour graph $\mathcal{C}_k(G)$ of G with nodes corresponding to proper k-colourings, we define the *list-colour graph* $\mathcal{C}(G, L)$ of G with nodes corresponding to proper list-colourings, where L represents the colour lists. The problem List-Colour Path is now defined as follows.

List-Colour Path
Instance: Graph G, colour lists $L(v) \subseteq \{1, 2, 3, 4\}$ for all $v \in V(G)$, two proper list-colourings of G, α and β.
Question: Is there a path between α and β in $\mathcal{C}(G, L)$?

Whenever colour lists are given for the vertices of the graph, 'proper list-colouring' should be read when we say 'colouring'.

A List-Colour Path instance can be turned into an equivalent k-Colour Path instance by adding a K_k on vertex set $\{u_1, \ldots, u_k\}$, and extending the colourings with $\alpha(u_i) = \beta(u_i) = i$. Now $\kappa(u_i) = i$ holds for all colourings κ obtainable from α and β. Adding edges vu_i if and only if $i \notin L(v)$ turns the graph into a k-Colour Path instance, where in all k-colourings κ we consider, $\kappa(v) \in L(v)$. By using other frozen graphs that are planar and/or bipartite, and introducing one or multiple copies of such graphs for every vertex, a similar transformation that preserves planarity and/or bipartiteness can be obtained. Figure 1 shows planar graphs with frozen 5- and 6-colourings. By removing a perfect matching from $K_{k,k}$, a bipartite graph that has a frozen k-colouring is obtained. The resulting graph is planar for $k = 4$. Hence we have:

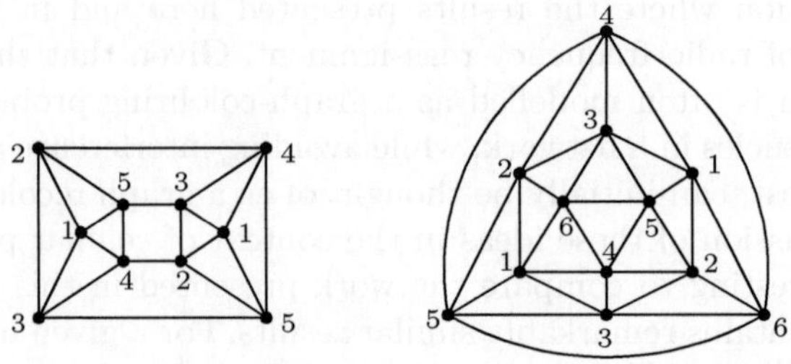

Fig. 1. Planar graphs with respective frozen 5- and 6-colourings

Lemma 5. *For any $k \geq 4$, a* List-Colour Path *instance G, L, α, β with lists $L(v) \subseteq \{1, 2, 3, 4\}$ can be transformed into a k-*Colour Path *instance G', α', β' such that the distance between α and β in $\mathcal{C}(G, L)$ (possibly infinite) is the same as the distance between α' and β' in $\mathcal{C}_k(G')$. Moreover,*

(i) if G is bipartite, this can be done so that G' is also bipartite, for all $k \geq 4$;
(ii) if G is planar, this can be done so that G' is also planar, when $4 \leq k \leq 6$; and
(iii) if G is planar and bipartite, this can be done so that G' is also planar and bipartite, when $k = 4$.

In all cases, this can be done so that $|V(G')| \leq |V(G)|f(k)$ and $|E(G')| \leq |E(G)| + |V(G)|g(k)$, for some functions $f(k)$ and $g(k)$.

Proof. The transformation from a List-Colour Path instance G, L, α, β to a k-Colour Path instance G', α', β' is as follows. Let F be a graph with a frozen k-colouring κ; F may be chosen to be bipartite and/or planar depending on k, as shown above. For every vertex $v \in V(G)$ and colour $c \in \{1, \ldots, k\} \backslash L(v)$, we add a copy of F to G, labelled $F_{v,c}$. We also add an edge between v and a vertex u of $F_{v,c}$ with $\kappa(u) = c$. This yields G'. The colourings α' and β' are obtained by extending α and β using the colouring κ for every $F_{v,c}$.

It is easy to see that every k-colouring obtainable from α' and β' induces the same frozen colouring on every copy of F. Also, because of the way the edges between v and vertices of $F_{v,c}$ are added, all these k-colourings of G' correspond to list colourings of G, and vice versa. This proves that the distance between α and β in $\mathcal{C}(G, L)$ is exactly the same as the distance between α' and β' in $\mathcal{C}_k(G')$.

When G and F are bipartite, the construction of G' starts with a number of bipartite components, and edges are added only between different components. So in this case G' is also bipartite. It can also be seen that G' is planar when G and F are planar: start with a planar embedding of G, and for each copy $F_{v,c}$ of F, consider a planar embedding that has a vertex with colour c on its outer face. These embeddings of $F_{v,c}$ can be inserted into a face of G that is incident with v. Now adding an edge between v and a vertex of $F_{v,c}$ with colour c can be done without destroying planarity.

Since for all $k \geq 4$ we can choose F to be bipartite, for $4 \leq k \leq 6$ we can choose F to be planar, and for $k = 4$ we can choose F to be both planar and bipartite, we are done. □

The next important notion for our proofs is the following. For $a, b \in \{1, \ldots, 4\}$, an (a, b)-*forbidding path* from u to v is a (u, v)-path with colour lists L, such that for $x \in L(u)$ and $y \in L(v)$, a colouring κ with $\kappa(u) = x$ and $\kappa(v) = y$ exists if and only if not both $x = a$ and $y = b$. In addition, for every colouring κ of the path with $\kappa(v) = y$, and every $x \in L(u) \backslash \{\kappa(u)\}$ with $(x, y) \neq (a, b)$, there exists a sequence of recolourings of the vertices of the path that ends by changing the colour of u to x, without earlier changing the colour of u or v. So when we add an (a, b)-forbidding path between vertices u and v of an existing graph, this has no effect on the possible colourings and recolourings of u and v other than restricting them from simultaneously having colours a and b respectively. We may therefore ignore internal vertices of the added paths when proving lower bounds for the distance between colourings, or when asking whether two colourings lie in the same component. The next lemma shows that we do not have to describe (a, b)-forbidding paths in detail every time; as long as $L(u), L(v) \neq \{1, 2, 3, 4\}$, such a path always exists.

Lemma 6. *For any $L_u \subset \{1, 2, 3, 4\}$, $L_v \subset \{1, 2, 3, 4\}$, $a \in L_u$ and $b \in L_v$, there exists an (a, b)-forbidding (u, v)-path P with $L(u) = L_u$, $L(v) = L_v$ and all other colour lists $L(w) \subseteq \{1, 2, 3, 4\}$. Moreover, we can insist P has even length at most six.*

Proof. Let $c \in \{1, 2, 3, 4\} \backslash L(u)$ and $d \in \{1, 2, 3, 4\} \backslash L(v)$. If $c \neq d$ then it can be checked that a path of length four with the following colour lists along the path is an (a, b)-forbidding path: L_u, $\{a, c\}$, $\{c, d\}$, $\{d, b\}$, L_v. If $c = d$, then we choose a path of length six with the following colour lists along the path: L_u, $\{a, c\}$, $\{c, e\}$, $\{e, f\}$, $\{f, c\}$, $\{c, b\}$, L_v, for some $e \in \{1, 2, 3, 4\} \backslash \{a, c\}$ and $f \in \{1, 2, 3, 4\} \backslash \{b, c\}$ with $e \neq f$. □

3 PSPACE-Completeness of k-Colour Path for $k \geq 4$

3.1 A PSPACE-Complete Problem: Sliding Tokens

We first prove that List-Colour Path is PSPACE-complete by giving a reduction from Sliding Tokens, a PSPACE-complete problem described in [9]. We need some definitions. A *token configuration* of a graph G is a set of vertices on which tokens are placed, in such a way that no two tokens are adjacent. (Thus a token configuration can be thought of as an independent set of vertices of G.) A *move* between two token configurations is the displacement of a token from one vertex to an adjacent vertex. (Note that a move must result in a valid token configuration).

Sliding Tokens
Instance: Graph G, two token configurations of G, T_A and T_B.
Question: Is there a sequence of moves transforming T_A into T_B?

The reduction used to prove PSPACE-completeness of SLIDING TOKENS in [9] actually shows that the problem remains PSPACE-complete for very restricted graphs and token configurations. Our reduction to LIST-COLOUR PATH is actually from a slightly wider class of restricted instances for which SLIDING TOKENS remains PSPACE-complete, but we do not give a reduction from the general problem. We proceed to describe the instances G, T_A, T_B of SLIDING TOKENS that we will use for our reduction. For details, see [9].

The graphs G are made up of *token triangles* (copies of K_3) and *token edges* (this involves a slight abuse of terminology: when we say token edge, we actually mean a copy of K_2). Every vertex of G is part of exactly one token triangle or token edge, and token triangles and token edges (disjoint in G) are joined together by *link edges.* In addition, G has a planar embedding where every token triangle bounds a face.

The token configurations T_A and T_B are such that every token triangle and every token edge contains exactly one token on one of its vertices. Such token configurations are called *standard token configurations.* In any sequence of moves from T_A or T_B, a token may never leave its triangle or its edge: the first time a token were to do so, it would become adjacent to another token. Therefore all moves yield standard token configurations.

3.2 The Construction of Equivalent LIST-COLOUR PATH Instances

Given a restricted instance G, T_A, T_B of SLIDING TOKENS as described in Section 3.1, we construct an instance G', L, α, β of LIST-COLOUR PATH such that standard token configurations of G correspond to list-colourings of G', and sliding a token in G corresponds to a sequence of vertex recolourings in G'.

We first label the vertices of G: the token triangles are labelled $1, \ldots, n_t$, and the vertices of triangle i are labelled t_{i1}, t_{i2} and t_{i3}. The token edges are labelled $1, \ldots, n_e$, and the vertices of token edge i are labelled e_{i1} and e_{i2}.

The construction of G' is as follows: for every token triangle i we introduce a vertex t_i, with colour list $L(t_i) = \{1, 2, 3\}$. For every token edge i we introduce a vertex e_i in G', with colour list $L(e_i) = \{1, 2\}$. Whenever a link edge of G joins a vertex t_{ia} with a vertex e_{jb}, we add an (a, b)-forbidding path of even length between t_i and e_j in G'. We do the same for pairs t_{ia} and t_{jb}, and pairs e_{ia} and e_{jb}. Note that this is a polynomial-time transformation that preserves planarity, and that the resulting graph is bipartite.

The two colourings α and β that correspond to T_A and T_B are constructed as follows: if the token of token edge i is on e_{ij} ($j = 1, 2$), we colour e_i with colour j, and analogously for t_{ij}. Because tokens are never adjacent, this does not violate the constraints imposed by the (a, b)-forbidding paths, so we can extend these colour choices to a full colouring.

Lemma 7. *Let G, T_A, T_B be a resticted instance of* SLIDING TOKENS *as described in Section 3.1, and let G', L, α, β be the corresponding instance of* LIST-COLOUR PATH *as constructed above. Then G, T_A, T_B is a YES-instance if and only if G', L, α, β is a YES-instance.*

Proof. Consider a sequence of token moves from T_A to T_B, if it exists. All intermediate token configurations are standard token configurations, and therefore correspond to some choice of colours for the vertices t_i and e_i. Every time a token is moved from e_{ij} to e_{ik}, we recolour e_i from j to k (analogously for token moves from t_{ij} to t_{ik}). Before this can be done, it may be necessary to first recolour internal vertices of incident (a, b)-forbidding paths, but this is always possible.

Similarly, for every sequence of recolourings from α to β we can describe a sequence of token moves from T_A to T_B: whenever a vertex t_i (e_i) is recoloured from colour a to colour b, we move the corresponding token from t_{ia} to t_{ib} (from e_{ia} to e_{ib}). □

Lemma 7 shows that the instance G', L, α, β of LIST-COLOUR PATH we constructed above is equivalent to the given instance of SLIDING TOKENS. In addition, we saw that G' is planar and bipartite. Now by Lemma 5 we can construct equivalent k-COLOUR PATH instances from G', L, α, β. All of these transformations are polynomial-time. The fact that k-COLOUR PATH is in PSPACE follows from the fact that a non-deterministic algorithm for k-COLOUR PATH will use a polynomial amount of space, and Savitch's Theorem, which states that PSPACE = NPSPACE (see [2], p.150 or [10] for details). This proves Theorem 3.

3.3 Tightness of the Hardness Results

Recall that the *colouring number* col(G) of a graph G (also known as the *degeneracy* or the *maximin degree*) is defined as the largest minimum degree of any subgraph of G. That is, $\text{col}(G) = \max_{H \subseteq G} \delta(H)$. The following result was proved in [11]. (The proof is similar to that of Claim 11 below.)

Theorem 8. *For any graph G and integer $k \geq \text{col}(G) + 2$, G is k-mixing.*

Recalling that the colouring number of a planar graph is at most 5, and that the colouring number of a bipartite planar graph is at most 3, Theorems 1, 3 and 8 together yield:

Theorem 9. *Restricted to planar graphs, the decision problem k-*COLOUR PATH *is* PSPACE*-complete for $4 \leq k \leq 6$, and polynomial-time solvable for all other values of k.*
*Restricted to bipartite planar graphs, the decision problem k-*COLOUR PATH *is* PSPACE*-complete for $k = 4$, and polynomial-time solvable for all other values of k.*

We observed in the introduction that the instance classes for which we can prove k-COLOUR PATH to be PSPACE-complete are exactly those for which we can construct graphs with superpolynomial colour graph diameter, and that for one of the classes for which the problem is known to be polynomial-time solvable ($k = 3$), the diameter is always polynomial. Is the diameter also polynomial for the other polynomial-time solvable instance classes mentioned in this section?

We speculate that this is indeed the case, and formulate a stronger conjecture in terms of the colouring number:

Conjecture 10. *For a graph G with n vertices and $k \geq \text{col}(G)+2$, the diameter of $\mathcal{C}_k(G)$ is $O(n^3)$.*

For values of $k \geq 2\,\text{col}(G)+1$, we are able to prove this statement, and even a stronger bound:

Claim 11. *For a graph G with n vertices and $k \geq 2\,\text{col}(G)+1$, the diameter of $\mathcal{C}_k(G)$ is $O(n^2)$.*

Proof. We can label the vertices $v_1, \ldots, v_n$ so that every vertex has at most $\text{col}(G)$ neighbors with a lower index. Using this vertex ordering, we first prove the following statement by induction over n.

Induction hypothesis
Let α and β be distinct k-colourings of G, and let i be the lowest index such that $\alpha(v_i) \neq \beta(v_i)$. There exists a recolouring sequence that starts with α and ends with recolouring v_i to $\beta(v_i)$, where every v_j with $j < i$ is never recoloured, and every v_j with $j \geq i$ is recoloured at most once.

The statement is trivial for $n = 1$. If $i = n$, then v_n can be recoloured to $\beta(v_n)$ because β is a proper colouring that coincides with α on all other vertices. Now suppose $i < n$, and consider $G' = G - v_n$. Let α' be the k-colouring of G' induced by α. By induction we can assume there exists a recolouring sequence starting with α' that ends with recolouring v_i to $\beta(v_i)$, in which vertices v_j with $j < i$ are not recoloured, and vertices v_j with $j \geq i$ are recoloured at most once. So for every vertex we can identify an *old* colour and a *new* colour in this recolouring sequence (they may be the same). Because there are at least $2\,\text{col}(G)+1$ available colours, and v_n has at most $\text{col}(G)$ neighbors, a colour c can be chosen for v_n that is not equal to the old colour or new colour of any of its neighbors. First recolour v_n to c if necessary, and then recolour the rest of the graph according to the recolouring sequence for G'. By the choice of colour c, all intermediate colourings are proper, so this is the desired recolouring sequence for G.

Now, we can keep repeating the above procedure, every time for a new vertex v_i which will have a higher index. This will yield β after at most $O(n^2)$ recolouring steps. □

4 Graphs with Colourings at Superpolynomial Distance

4.1 The Construction of the Graphs

For every integer $N \geq 1$, we construct a graph G_N with colour lists L. (To avoid cluttering the notation, we will denote the colour lists of each G_N by L; which graph these lists belong to will be clear from the context.) The graphs G_N will have size $O(N^2)$ and the $\mathcal{C}(G_N, L)$ will have diameter $\Omega(2^N)$.

For a given N, the graph G_N is constructed as follows. Start with N triangles, each consisting of vertices v_i, v'_i and v^*_i with $L(v_i) = \{1,2\}$, $L(v'_i) = \{1,2,3\}$ and $L(v^*_i) = \{3,4\}$, for $i = 1,\ldots,N$. In a colouring κ where $\kappa(v^*_i) = 3$, triangle i is said to be *locked*, otherwise it is *unlocked*. Now between every pair v^*_i and v^*_j with $i \neq j$ we add a $(4,4)$-forbidding path. So:

Observation 12. *At most one triangle can be unlocked in any colouring.*

We complete the construction of G_N by, for every i, adding a $(4,1)$-forbidding path from v^*_i to v_{i-1}, and adding $(4,2)$-forbidding paths from v^*_i to v_j for all $j \leq i-2$.

Observation 13. *Triangle i can only be unlocked in a colouring κ when $\kappa(v_{i-1}) = 2$ and $\kappa(v_j) = 1$ for all $j \leq i-2$.*

4.2 Bounds on Size and Distance

Observation 14. *The sizes of $V(G_N)$ and $E(G_N)$ are both bounded by $O(N^2)$.*

To show that there exists a pair of colourings of G_N such that exponentially many steps (exponential in N) are needed to go from one to the other, we need only consider the colours of the vertices v_i. These can be seen as N bits with value 1 or 2. We call a colouring κ of G_N a $(c_1, c_2, \ldots, c_N)$*-colouring* if $\kappa(v_i) = c_i$ for all i. All $(c_1, c_2, \ldots, c_N)$-colourings together form the *colour class* $(c_1, c_2, \ldots, c_N)$. Observe that no colour class is empty.

Lemma 15. *Let $(x_1, \ldots, x_N)$ and $(y_1, \ldots, y_N)$ be distinct tuples with all $x_i, y_i \in \{1,2\}$.*

(a) If the tuples differ only on position i, and $x_{i-1} = 2$, and $x_j = 1$ for all $j < i-1$, then from any colouring in class $(x_1, \ldots, x_N)$ we can reach some colouring in class $(y_1, \ldots, y_N)$ via a sequence of recolourings, without ever leaving colour class $(x_1, \ldots, x_N)$ in the intermediate colourings.

(b) Otherwise, there is no colouring in class $(x_1, \ldots, x_N)$ that is adjacent to a colouring in class $(y_1, \ldots, y_N)$.

Proof. Suppose the stated conditions on the tuples hold. We show that any colouring κ in the class $(x_1, \ldots, x_N)$ can be recoloured to a colouring in class $(y_1, \ldots, y_N)$. If in κ, a triangle $j \neq i$ is unlocked, it can be verified that (if necessary) v'_j and internal vertices of incident (a,b)-forbidding paths can be recoloured so that triangle j can be locked, without ever leaving colour class $(x_1, \ldots, x_N)$. At this point, triangle i can be unlocked; none of the (a,b)-forbidding paths incident with v^*_i pose any restriction by our assumption on the tuple, and since all other triangles are locked. Then we can set $\kappa(v'_i) = 3$ and $\kappa(v_i) = y_i$ to obtain a colouring in class $(y_1, \ldots, y_N)$. This proves the first statement.

Now let α be a $(x_1, \ldots, x_N)$-colouring, let β be a $(y_1, \ldots, y_N)$-colouring, and suppose that α and β are adjacent. This means they differ only on one vertex, and because the tuples are distinct, α and β must therefore differ precisely on a vertex v_i, for some i. This means triangle i is unlocked in both colourings. Because of the $(4,1)$- and $(4,2)$-forbidding paths starting at v_i^*, $\alpha(v_{i-1}) = 2$ and $\alpha(v_j) = 1$ for all $j < i-1$. This proves the second statement. □

Theorem 16. *Every graph G_N has two colourings α and β in the same component of $\mathcal{C}(G_N, L)$ which are at distance at least $2^N - 1$.*

Proof. For the colouring α we choose a colouring in class $(1,\ldots,1)$; such a colouring exists since no colour class is empty. Colouring β will be a colouring in class $(1,\ldots,1,2)$. We first prove by induction that such colourings can be obtained from each other by recolourings, using the following induction hypothesis.

Induction hypothesis

There is a path in $\mathcal{C}(G_N, L)$ from any colouring α' in class $(1,\ldots,1,x_0,x_1,\ldots,x_{N-n})$ to some colouring β' in class $(1,\ldots,1,3-x_0,x_1,\ldots,x_{N-n})$.

The colourings differ on vertex v_n: we have $\alpha'(v_n) = x_0$ and $\beta'(v_n) = 3-x_0$, while for all $i \neq n$, we have $\alpha'(v_i) = \beta'(v_i)$. If $n = 1$, the statement follows directly from Lemma 15. If $n > 1$, then from α' we recolour to a $(1,\ldots,1,2,x_0,x_1,\ldots,x_{N-n})$-colouring (which differs from the initial class only in the $(n-1)$-th position), using the induction hypothesis. Then we recolour to a $(1,\ldots,1,2,3-x_0,x_1,\ldots,x_{N-n})$-colouring, using Lemma 15. Finally, using the induction hypothesis again, we can recolour to a $(1,\ldots,1,1,3-x_0,x_1,\ldots,x_{N-n})$-colouring, which proves the statement.

Now we show that to go from a $(1,\ldots,1)$-colouring to a $(1,\ldots,1,2)$-colouring, at least $2^N - 2$ other colour classes need to be visited, using the following induction hypothesis.

Induction hypothesis

To go from a $(1,\ldots,1,1,x_1,\ldots,x_{N-n})$-colouring to a $(1,\ldots,1,2,y_1,\ldots,y_{N-n})$-colouring, at least $2^n - 2$ other colour classes need to be visited.

If $n = 1$, the statement is obvious. If $n > 1$, then consider a shortest path between two colourings in these classes, if it exists. At some point in the sequence of recolourings, the colour of v_n is changed for the first time; before this we must have a $(1,\ldots,1,2,1,z_1,\ldots,z_{N-n})$-colouring, by Lemma 15 (in this colouring, v_{n-1} has colour 2). By the induction hypothesis, at least $2^{n-1}-2$ colour classes have been visited before this colour class was reached. Now changing the colour of v_n to 2 yields a $(1,\ldots,1,2,2,z_1,\ldots,z_{N-n})$-colouring. Using the induction hypothesis again, at least $2^{n-1} - 2$ colour classes need to be visited before class $(1,\ldots,1,2,y_1,\ldots,y_{N-n})$ is reached. This means that in total, at least $2^n - 4 + 2$ intermediate colour classes have been visited in the recolouring procedure. This completes the proof. □

We conclude by remarking that the graphs constructed here can be made bipartite without too much effort, and that they can be made planar using a suitable crossing component. We omit the details because of space constraints. The bipartite graphs thus obtained have size $O(N^2)$ while the planar ones have size $O(N^4)$; for both, the distance between α and β is again at least 2^{N-1}. Lemma 5 now shows that they can be transformed into k-COLOUR PATH instances, completing the proof of Theorem 4.

Acknowledgements. We are indebted to Moshe Vardi for initially suggesting that the decision problem k-COLOUR PATH might be PSPACE-complete for $k \geq 4$.

References

1. Diestel, R.: Graph Theory, 3rd edn. Springer, Heidelberg (2005)
2. Papadimitriou, C.H.: Computational Complexity. Addison-Wesley, Reading (1994)
3. Cereceda, L., van den Heuvel, J., Johnson, M.: Connectedness of the graph of vertex-colourings. CDAM Research Report LSE-CDAM-2005-11(accepted for publication in Discrete Math.) (2005), available from `http://www.cdam.lse.ac.uk/Reports/reports2005.html`
4. Cereceda, L., van den Heuvel, J., Johnson, M.: Mixing 3-colourings in bipartite graphs. CDAM Research Report LSE-CDAM-2007-06 (submitted, 2007), available from `http://www.cdam.lse.ac.uk/Reports/reports2007.html`
5. Jerrum, M.: Counting, Sampling and Integrating: Algorithms and Complexity. Birkhäuser Verlag, Basel (2003)
6. Cereceda, L., van den Heuvel, J., Johnson, M.: Finding paths between 3-colourings (in preparation)
7. Billingham, J., Leese, R., Rajaniemi, H., et al.: Frequency reassignment in cellular phone networks, Smith Institute Study Group Report (2005), available from `http://www.smithinst.ac.uk/Projects/ESGI53/ESGI53-Motorola/index_html`
8. Gopalan, P., Kolaitis, P.G., Maneva, E.N., Papadimitriou, C.H.: The connectivity of Boolean satisfiability: computational and structural dichotomies. In: Bugliesi, M., Preneel, B., Sassone, V., Wegener, I. (eds.) ICALP 2006. LNCS, vol. 4051, pp. 346–357. Springer, Heidelberg (2006), available from `http://arxiv.org/abs/cs.CC/0609072`
9. Hearn, R.A., Demaine, E.D.: PSPACE-completeness of sliding-block puzzles and other problems through the nondeterministic constraint logic model of computation. Theoret. Comput. Sci. 343, 72–96 (2005)
10. Savitch, W.J.: Relationships between nondeterministic and deterministic tape complexities. J. Comput. Syst. Sci. 4(2), 177–192 (1970)
11. Dyer, M., Flaxman, A., Frieze, A., Vigoda, E.: Randomly colouring sparse random graphs with fewer colours than the maximum degree. Random Struct. Algor. 29(4), 450–465 (2006)

Shuffle Expressions and Words with Nested Data

Henrik Björklund[1,*] and Mikołaj Bojańczyk[2,**]

[1] University of Dortmund
[2] Warsaw University

Abstract. In this paper, we develop a theory that studies words with nested data values with the help of shuffle expressions. We study two cases, which we call "ordered" and "unordered". In the unordered case, we show that emptiness (of the two related problems) is decidable. In the ordered case, we prove undecidability. As a proof vehicle for the latter, we introduce the notion of higher-order multicounter automata.

1 Introduction

A data word is a word where each position, in addition to its finite alphabet label, carries a *data value* from an infinite domain. Recent times have seen a flurry of research on data languages, motivated chiefly by applications in XML databases and parameterized verification; see, e.g., [7,9,2,4,11,3,1]. One of the main results is that satisfiability for first-order logic with two variables is decidable over data words, as long as the only operation allowed on the data values is equality testing [2]. The same paper also demonstrates a close connection between data languages, *shuffle expressions*, and *multicounter automata*:

1. Multicounter automata. These are nondeterministic automata with many counters, which can be incremented and decremented, but zero tests are only allowed at the end of the word.
2. Shuffle expressions. These are regular expressions extended with intersection and the shuffle operation.
3. Two-variable data languages. These are properties of data words that can be expressed in certain two-variable logics.

The connection between multicounter automata (or Petri nets) and shuffle expressions was discovered in [5], while the connection between the first two and data languages was discovered in [2].

In this paper, we develop and investigate extensions of the above. We focus on *nested* data values and shuffle expressions. There are two principal motivations.

– When data values are only used to induce an equivalence relation on the positions, such as in the logics mentioned above, one of the chief applications is parameterized verification. A number of processes run in parallel, and in

* Supported by the Deutsche Forschungsgemeinschaft Grant SCHW678/3-1.
** Supported by Polish government grant no. N206 008 32/0810.

L. Kučera and A. Kučera (Eds.): MFCS 2007, LNCS 4708, pp. 750–761, 2007.

the resulting sequence of actions, the individual actions are annotated by the unique id of the process that performed them. This data word can be used to check global properties (by looking at the whole string) and local properties (by considering the sequence of actions of a single process).

To model a system where processes can have subprocesses, and so on, we would want a data word with *nested* data values: each action carries the id of the subprocess which performed it as well as the id of the process that spawned the subprocess, and so on.

- In [5], nesting of the shuffle operation was not considered. This runs contrary to the unrestricted nesting of other operations in regular expressions, and begs the question what happens when unrestricted nesting of shuffles is allowed. We discover that the resulting languages are actually intimately related to languages with nested data. We also discover that this leads to undecidability; and decidability can be recovered only after adding a form of commutativity to the shuffle operation.

We note that our notion of shuffle expressions is different than the one used in some of the literature–see, e.g.,[6]–since we consider four different shuffle operators and allow intersection with regular languages.

Our main topic is logics over words with nested data. We study two two-variable fragments of first order, one with order and one without. The first one is shown to be undecidable, while the second is decidable. Nested shuffle expressions are the main proof vehicle, and an object of independent study. For instance, the expressions we consider capture the language:

$$a^{n_1}\#\cdots\#a^{n_k}\#b^{m_1}\#\cdots\#b^{m_k}\# \quad : \quad n_1,\ldots,n_k \text{ is a permutation of } m_1,\ldots,m_k$$

It is our opinion that the two concepts—logic with nested data and shuffle expressions—shed light on each other in a useful cross-fertilization.

Due to space restrictions, most proofs have been omitted, and will appear in the full version of the paper.

2 A Logic for Words with Nested Data

Let A be a finite alphabet and Δ an infinite set, whose elements will be called *data values*. For $k \in \mathbb{N}$, a *word with k layers of data* is a word where every position, apart from a *label* in A, has k labels $d_1, \ldots, d_k \in \Delta$. The label d_i is called the *i-th data value* of the position. Therefore, such a word is an element $w \in (A \times \Delta^k)^*$. In w, the data values can be seen as inducing k equivalence relations $\sim_1, \ldots, \sim_k$ on the positions of w. That is, two positions are related by $\sim_i$ if they agree on the i-th data value.

We are interested in the data values only insofar as equality is concerned. Therefore, the relations $\sim_1, \ldots, \sim_k$ supply all the information required about a word; and could be used as an alternative definition. Adding more structure—such as a linear order—to the data values leads very quickly to undecidable decision problems, already with one layer, and even for very weak formalisms.

A word with k layers of data is said to have *nested data* if for each $i = 2, \ldots, k$ the relation $\sim_i$ is a *refinement* of $\sim_{i-1}$. In other words, if two positions agree on value i, then they also agree on value $i-1$. For the rest of this section, we fix k and only talk about words and languages with k layers of nested data. Instead of "word with k nested layers of data", we will just write "word with nested data".

In the spirit of Büchi's sequential calculus, we will use logic to express properties of words with nested data. In this setting, a word with nested data is treated as a model for logic, with the logical quantifiers ranging over word positions. The data values are accessed via the relations $\sim_i$. For instance, the formula

$$\forall x \forall y \quad (x \sim_2 y \Rightarrow x \sim_1 y) \wedge \cdots \wedge (x \sim_k y \Rightarrow x \sim_{k-1} y)$$

states that the data values are nested. (This formula is a tautology in our setting, where only models representing words with nested data values are considered.) Each label a from the finite label set A can be accessed by a unary relation; for instance $\forall x \; a(x)$ is true in words where all positions have label a. The linear order on word positions can be accessed via the relation $<$. For instance,

$$\forall x \quad (\forall y \; y \leq x) \Rightarrow (\forall y < x \; y \not\sim_1 x)$$

says that the last position has a different first (and consequently, all others as well) data value than than the other positions. We also use the successor relation $x = y + 1$. Although it can be defined in terms of the order, a third variable z is required, which is too much for the two-variable fragments we considered. We mainly consider satisfiability: given a formula, determine if there is a nested data word that satisfies it. This problem is undecidable in general, but by restricting the formulas, we can obtain decidable fragments.

Satisfiability is undecidable already for the following fragments, see [2]:

- There is only one layer of data. The formulas can use three variables; but the order on word positions can be accessed only via $x = y + 1$ and not $<$.
- There are two layers of data, but these are not necessarily nested. The formulas can use only two variables, $x = y + 1$, $x = y + 2$ and not $<$.

The largest known decidable fragment was presented in [2]:

- There is only one layer of data. The formulas can use only two variables, and the positions can be accessed by both $x = y + 1$ and $<$.

We want to generalize the above result to words with multiple layers of nested data. The result from [2] fails already for two layers:

Theorem 2.1. *With two layers of nested data, satisfiability is undecidable for two-variable first-order logic with the relations $x = y + 1$ and $x < y$.*

Proof (Sketch). We encode computations of a two-counter machine with zero tests. Consider words with two layers of nested data, where the labels are $A = \{start, end, inc, dec\}$ and the positions are labeled according to the regular language $(start(inc{+}dec)^*end)^*$. This is easy to express in our two-variable logic.

Next, we express that each block of the form $start(inc + dec)^*end$ corresponds to a single data value on layer 1:

$$\begin{array}{cc} \forall x\ start(x) \Rightarrow \forall y < x\ x \not\sim_1 y & \forall x\ end(x) \Rightarrow \forall y > x\ x \not\sim_1 y \\ \forall x \exists y \geq x\ x \sim_1 y \wedge end(y) & \forall x \exists y \leq x\ x \sim_1 y \wedge start(y) \end{array} \quad (1)$$

Finally, the we can use equivalence classes (sets of positions with the same data value) on layer two to ensure that the operations *inc* and *dec* are balanced.

Consider a word with nested data that satisfies the above properties. This word can be seen as a computation of a one counter machine without states, where *inc* corresponds to a counter increment, *dec* corresponds to a counter decrement, while *start*, *end* correspond to zero tests.

To get two counters instead of one, we expand the label alphabet from A to $\{1, 2\} \times A$. The regular language we use is now

$$start_1 start_2 (inc_1 + inc_2 + dec_1 + dec_2 + end_1 start_1 + end_2 start_2)^* end_2 end_1.$$

The rest of the construction also generalizes, by duplicating each formula, once for each counters, and adding consistency constraints.

Even with the above result, however, not all hope is lost. Satisfiability is decidable if we lose the order $<$, even with arbitrarily many layers of data.

Definition 1. *$FO^2(+1, \sim_1, \ldots, \sim_k)$ is the fragment of first-order logic that uses only the two variables x and y and the following predicates.*

$a(x)$	*x has label a, where a is a label from A*
$y = x + 1$	*y is the position directly to the right of x*
$x \sim_i y$	*x and y have the same layer i data value, with $i \in \{1, \ldots, k\}$*

Theorem 2.2. *Over words with nested data, satisfiability is decidable for $FO^2(+1, \sim_1, \ldots, \sim_k)$.*

The above result is a consequence of Theorem 3.4, which says that formulas of the logic can be compiled into a type of shuffle expression, and Theorem 5.2, which says that emptiness is decidable for these shuffle expressions.

3 Shuffle Expressions

Recall that a word $w \in A^*$ is called a *shuffle* of words $w_1, \cdots, w_m \in A^*$ if the positions of w can be colored using m colors so that the positions with color $i \in \{1, \ldots, m\}$, when read from left to right, form the word w_i. If $K \subseteq A^*$ is a (possibly infinite) set of words, then $shuffle(K) \subseteq A^*$ is defined as

$$\{w : w \text{ is a shuffle of some } w_1, \ldots, w_m \in K, \text{ for some } m \in \mathbb{N}\}.$$

Note that the words $w_1, \ldots, w_m$ above may include repetitions. For instance, $shuffle(\{a\})$ contains all words a^*. Just as finite automata are connected to regular expressions, multicounter automata are connected to shuffle expressions. This is witnessed by the following result, essentially due to [5]:

Theorem 3.1. *The following language classes are equal, modulo morphisms:*

1. *Languages recognized by multicounter automata;*
2. *Languages of the form $L \cap \mathit{shuffle}(K)$, where L, K are regular.*

The qualification "modulo morphisms" means that any language from class 1 is a morphic image of a language in class 2. (Class 2 is simply contained in class 1, without need for morphisms.) The point of the morphism is to erase bookkeeping, such as annotations with accepting runs.

We can now ask what we get if we add intersection and *shuffle* to regular expressions. The answer is that we get more than we want:

Theorem 3.2. *If regular expressions are extended with shuffle and intersection, all recursively enumerable languages can be defined modulo morphisms.*

In particular, emptiness is undecidable for such extended regular expressions. Disallowing intersection trivializes the emptiness problem — which is the problem we are most interested in here — since $\mathit{shuffle}(K)$ is nonempty if and only if K is. Theorem 3.2 follows directly from Theorems 4.1 and 4.2 below.

In this paper, however, we are most interested in decidability results, especially for logics with data values. It turns out that decidability can be recovered, if we consider a weaker form of shuffling, which is partly commutative.

3.1 Cutting and Combining

To express our modification of the shuffle operation, it is most convenient to decompose $\mathit{shuffle}(L)$ into two operations:

$$\mathit{shuffle}(L) = \mathit{combine}(\mathit{cut}(L)) \ . \tag{2}$$

The first *cut* operation sends a set of words $L \subseteq A^*$ to the set of traces obtained by cutting a word from L into pieces:

$$\mathit{cut}(L) = \{w_1|\cdots|w_k : w_1 \cdots w_k \in L\} \subseteq (A^*)^*.$$

In the above a *trace* is a sequence of finite words, which is written as $w_1|\cdots|w_k$, with $|$ separating consecutive words, called *segments*. Traces are denoted by θ or σ, and will be heavily used later on.

The second operation is called *combine*, and it sends a set of traces $L \subseteq (A^*)^*$ to the set of words that can be combined from these traces:

$$\mathit{combine}(L) = \{w : w \text{ is a combination of } \theta_1, \ldots, \theta_m \in L\ \} \subseteq A^*.$$

By saying that w is a *combination* of traces $\theta_1, \ldots, \theta_m$ we mean that positions of w can be colored with m colors, so that for each color $i = 1, \ldots, m$, the positions with color i give the trace θ_i. In the trace $\theta_i = w_1|\cdots|w_n$, the segments $w_1, \ldots, w_n$ correspond to *maximal* subwords of w that are assigned color i. Consider the following example.

$$\begin{array}{c} a\ b\ c\ c\ c\ b\ a\ c \\ 1\ 1\ 2\ 1\ 3\ 1\ 2\ 2 \end{array} \qquad \begin{array}{l} \theta_1 = ab|c|b \\ \theta_2 = c|ac \\ \theta_3 = c \end{array}$$

By the maximality requirement, the trace θ_2 cannot be replaced by $c|a|c$ Given the above definitions of *cut* and *combine*, it should be fairly clear that equation (2) holds. However, by tweaking the definitions of *cut* and *combine*, we will arrive at variants of the shuffle operation that are decidable and, more importantly, relevant to our investigation of nested data values.

The first modification gives us more control on the way words from $L \subseteq A^*$ are cut into traces. Let $K \subseteq A^*$ be a language. We define

$$cut_K(L) = \{w_1|\cdots|w_k : w_1 \cdots w_k \in L, w_1, \dots, w_k \in K\} \subseteq (A^*)^*.$$

In other words, words from L are cut into traces where each segment belongs to K. Setting $K = A^*$ allows us to recover the standard cut operation, so cut_K is a generalization of *cut*. We only consider the case when K is regular.

The second modification concerns the operation *combine*. An *unordered trace* is a multiset of words, i.e. a trace where the order of segments is not important (however, the ordering of letters inside the segments is). The operation *ucombine*(L) treats the set of traces L as unordered traces:

$$ucombine(L) = \{w : w \text{ is an unordered combination of } \theta_1, \dots, \theta_m \in L\ \} \subseteq A^*.$$

In the above, w is an *unordered combination* of $\theta_1, \dots, \theta_m$ if w is a combination of traces $\sigma_1, \dots, \sigma_m$, such that σ_i is obtained from θ_i by rearranging the order of segments. Thus abc is an unordered combination of traces $\theta_1 = c|a$ and $\theta_2 = b$.

3.2 Four Kinds of Shuffle Expressions

From the above, we obtain four variants of the shuffle operation:

- Shuffle: $shuffle(L) = combine(cut(L))$.
- Controlled shuffle: $cshuffle_K(L) = combine(cut_K(L))$.
- Unordered shuffle: $ushuffle(L) = ucombine(cut(L))$.
- Unordered controlled shuffle: $ucshuffle_K(L) = ucombine(cut_K(L))$.

Each such operation gives rise to its own flavor of extended regular expressions. We will investigate and compare these flavors with respect to decidability and expressive power.

Definition 3.3. Controlled shuffle expressions (CSE) *denote languages obtained by nesting the following operations:*

- *Standard regular expression operations: single letters $a \in A$, the empty word ϵ, concatenation, union and Kleene star.*
- *Intersection with regular languages.*
- *Controlled shuffle $cshuffle_K(L)$, where L is defined by a CSE, but K is a regular word language.*
- *Images under morphisms $f : A^* \to B^*$.*

Shuffle expressions (SE), unordered shuffle expressions (USE) *and* unordered controlled shuffle expressions (UCSE) *are defined analogously, by replacing the type of shuffle operation allowed. All types of operations can be freely nested.*

The point of adding morphic images is to have a form of nondeterministic guessing in the expressions (and therefore more power). This will be illustrated in the following example. Note also that morphic images are necessary due to adding the intersection and shuffle operations; in standard regular expressions the projection operation does not add any power and can be eliminated.

Example 1. In the shuffle operation *shuffle*(L), we have no control over the number of times the language L is used. In this example we show that we can enforce that it is used an even number of times. Let then $L \subseteq A^*$ be defined by an SE. The idea is to expand the alphabet A with a new symbol *start*; each word from L will be prefixed by this symbol. Consider then the expression:

$$K = start \cdot L\,.$$

If we now take the expression *shuffle*(K), we can use the marker *start* to see how many times K was used. By intersecting with the regular language "even number of occurrences of *start*", we can make sure that it was used an even number of times. Finally, the markers can be removed using the erasing morphism $f : (A \cup \{start\})^* \to A^*$ defined by $f(start) = \epsilon$ and $f(a) = a$ for $a \in A$.

Example 2. Unordered shuffling is enough to express some counting properties: *ushuffle*(ab) describes words in $(a+b)^*$ with the same number of a's and b's. Using intersection with the regular language a^*b^*, we get the language $\{a^n b^n\}$.

Example 3. Using the same idea as in the previous example, we can also get the language $L = \{a^n\#b^n\#\}$. Consider now the following expression:

$$(a^*\#)^*(b^*\#)^* \cap ucshuffle_K(L) \qquad \text{where } K = a^*\# + b^*\#$$

This expression defines the set of words

$$a^{n_1}\#\cdots\#a^{n_k}\#b^{m_1}\#\cdots\#b^{m_k}\#,$$

such that $n_1,\ldots,n_k$ is a permutation of $m_1,\ldots,m_k$.

3.3 From Logic to Shuffle Expressions

In this section we state the reduction of satisfiability for $FO^2(+1,\sim_1,\ldots,\sim_k)$ to the emptiness problem for unordered controlled shuffle expressions.

Theorem 3.4. *For every $FO^2(+1,\sim_1,\ldots,\sim_k)$-formula ϕ, a UCSE r can be effectively computed such that the language of r is non-empty if and only if ϕ is satisfiable.*

Proof (Sketch). We can assume that ϕ is in *data normal form*; a normal form very similar to the one used in [2]. This means that ϕ is a set of conjuncts, where each conjunct is fairly simple. The conjuncts express properties of classes by referring to types. When talking of types, these conjuncts only use the expressions "at least one node", "has a node", and "at most one node". Thus, the only relevant information about a class string (where the class can be w.r.t. any of the k equivalence relations), is the number of times each type appears. Furthermore, if a type appears at least twice, the exact number of times is irrelevant.

If we have a footprint mapping $f : T \to \{0, 1, 2\}$, where T is the set of types appearing in ϕ, that tells us if a type appears 0, 1, or 2 or more times, we know everything we need about the class string. We can easily compute the set F of those footprints that are allowed by the conjuncts from ϕ. If we construct, for each $f \in F$, a shuffle expression that accepts exactly those strings that have footprint f, we can combine these expressions (using the $+$ operator) into one that accepts all correct class strings. Using the shuffle operator on this expression gives an expression whose language is such that every word can be extended with (level k) data values in such a fashion that the conjuncts of ϕ that use $\sim_k$ are satisfied.

The idea is to use this construction inductively, starting with level k, until, after using k shuffle operations, all conjuncts of ϕ are taken care of.

The corresondence between r and ϕ is actually stronger: r contains words obtained from models of ϕ by erasing data values. We do not know if the converse translation—from expressions to logic—can be done; possibly the expressions are strictly stronger than the logic. A similar reduction, from logic with order to CSE is possible, but the proof is omitted.

4 Ordered Shuffle Expressions

The following theorem relates CSE, SE and higher-order multicounter automata. The latter, to our best knowledge, are a new model.

Theorem 4.1. *The following language classes are equal:*

1. *Languages defined by controlled shuffle expressions (CSE);*
2. *Languages defined by shuffle expressions (SE);*
3. *Languages defined by higher-order multicounter automata;*
4. *Recursively enumerable languages.*

We define higher-order multicounter automata in Section 4.1, and prove their Turing completeness. The rest of the proof of Theorem 4.1 is omitted.

4.1 Higher-Order Multicounter Automata

A multiset over A is a function $m : A \to \mathbb{N}$. We only consider finite multisets here, where all but a finite number of elements in A are assigned 0. We also consider higher-level multisets (which are also multisets). A level 1 multiset over A is a finite multiset over A. A level $k+1$ multiset over A is a finite multiset of level k multisets over A.

A level k multicounter automaton is defined as follows. It has a state space Q, an input alphabet Σ, and a multiset alphabet A. All of these are finite. The automaton reads an input word $w \in \Sigma^*$ from left to right. At each moment, its memory is a tuple $(q, m_1, m_2, \ldots, m_k)$, where q is one of the states in Q, and each m_i is a level i multiset over A, possibly undefined $\perp$. (We distinguish an empty multiset $\emptyset$ from an undefined one $\perp$.) The initial configuration is $(q_I, \perp, \perp, \ldots, \perp)$, where q_I is some designated initial state.

There is a finite set of transition rules, which say how the machine can modify its memory upon reading an input symbol (or doing an ϵ-transition). Each such transition rule is of the form: when in state q and upon reading the label $a \in \Sigma \cup \{\epsilon\}$, assume state p and do counter operation x. The counter operations are:

- new_i: Change m_i from $\perp$ to $\emptyset$.
- inc_a: Add $a \in A$ to the level 1 multiset m_1.
- dec_a: Remove $a \in A$ from the level 1 multiset m_1.
- $store_i$: Add m_i to the level $i+1$ multiset m_{i+1}; then set m_i to $\perp$.
- $load_i$: Remove nondeterministically some element m from m_{i+1} and store it in m_i. This transition is enabled only when $m_1, \ldots, m_i$ are all $\perp$.

We use *Counterops*$_k$ to denote the possible counter operations in a level k automaton. Note here that the automaton knows which m_i are undefined, since this is controlled by transitions new_i and $store_i$. On the other hand, the automaton does not know if a defined multiset m_i is empty, or not.

What is the accepting condition? We say a level k multiset is *hereditarily empty* if it is empty, or it consists only of hereditarily empty level $k-1$ multisets. The automaton accepts if $m_1, \ldots, m_k$ are all hereditarily empty multisets in all memory cells; and the control state belongs to a designated accepting set.

The above definition is similar, but not identical, to the notion of *nested Petri nets* from, e.g., [8].

Here we show that the machines are Turing complete, already on level 2.

Theorem 4.2. *Level* 2 *multicounter automata recognize all recursively enumerable languages.*

Proof
We show that a level 2 multicounter automaton can simulate a two-counter machine with zero tests. Since the latter type of machine is capable of recognizing all recursively enumerable languages, the statement follows.

A configuration of the two-counter machine, where counter 1 has value i and counter 2 has value j, will be represented by the following level 2 multiset:

$$\{\{\underbrace{x, \ldots, x}_{i \text{ times}}, a\}, \{\underbrace{x, \ldots, x}_{j \text{ times}}, b\}, \underbrace{\emptyset, \ldots, \emptyset}_{k \text{ times}}\} . \quad (3)$$

The occurrences $\emptyset$ are used for bookkeeping; the number k will correspond to the number of zero-tests that have been carried out in the run leading to this configuration. A configuration as above is called *proper.* Our automaton will have the property that improper configurations always lead to improper configurations; furthermore, a failed zero-test will lead to an improper configuration.

We now show how to represent the operations of the simulated machine:

- Zero test on counter 1. We do the following sequence of operations:

$$load_1 \quad dec_a \quad store_1 \quad new_1 \quad inc_a \quad store_1 \ .$$

  If the configuration was improper, it will remain so. If it was proper, the level 2 from (3) multiset will become:

$$\{\{a\}, \{\underbrace{x, \ldots, x}_{j \text{ times}}, b\}, \underbrace{\emptyset, \ldots, \emptyset}_{k \text{ times}}, \{\underbrace{x, \ldots, x}_{i \text{ times}}\}\} \ .$$

  If i was not 0, the above configuration will be improper.
- Increment on counter 1. We do the following sequence of operations:

$$load_1 \quad dec_a \quad inc_x \quad inc_a \quad store_1 \ .$$

  A decrement is done the same way.
- The operations on counter 2 are as above, except b is used instead of a.

One can easily see that the automaton can reach a proper configuration as in (3) if and only if the simulated two-counter machine could have counter values (i, j). Furthermore, the simulating machine can test (once, at the end of its run), if it has reached a proper configuration of the form:

$$\{\{a\}, \{b\}, \underbrace{\emptyset, \ldots, \emptyset}_{k \text{ times}}\} \ .$$

This is done by $load_1 \quad dec_a \quad store_1 \quad load_1 \quad dec_b \quad store_1$ and testing if all memory cells are hereditarily empty. □

5 Unordered Shuffle Expressions

In this section, we state the decidability of the emptiness problems for unordered shuffle expressions, controlled (UCSE) or not (USE). Since $\mathit{ushuffle}(L) = \mathit{ucshuffle}_{A^*}(L)$ if A is the alphabet of L, it is clear that USE is a special case of UCSE. Nevertheless, we chose to state the following independently:

Theorem 5.1. *Emptiness for unordered shuffle expressions is decidable.*

The reason is that the proof is considerably less involved than for the controlled case. It uses a reduction to finite word automata equipped with a Presburger counting condition.

As stated in the introduction, the main goal of this paper is to show decidability of satisfiability for the 2-variable logic from Definition 1 over words with nested data. Theorem 3.4 shows that this problem reduces to emptiness for UCSE. We are now ready to complete the proof of Theorem 2.2, by stating the main combinatorial result of the paper:

Theorem 5.2. *Emptiness for unordered controlled shuffle expressions is decidable.*

The proof is rather involved, and is based on a study of the Parikh images [10] of languages defined by UCSE. Due to space limitations, we can can only present here a very brief outline of the key ideas.

The main technical result is the following Parikh-type theorem:

Theorem 5.3. *The Parikh image of a language defined by a UCSE is semilinear.*

Furthermore, since the semilinear set is effectively obtained, it can be tested for emptiness in a decision procedure for emptiness of UCSE, hence Theorem 5.2 follows. The proof of Theorem 5.3 is by induction, and only one step is nontrivial: when the expression is of the form

$$ucshuffle_M(L) \cap K, \tag{4}$$

where L is defined by a UCSE and K, M are regular languages.

What follows is a very informal description of some of the ideas used in showing that the Parikh image of the language above is semilinear. We first remind the reader how the language (4) is defined. We begin with traces $\theta_1, \dots, \theta_n$ that are obtained from cutting words from L into segments from M. In other words, each trace $\theta_i \in (A^*)^*$ must belong to $cut_M(L)$. Then, the segments of these traces are rearranged and combined to get a word in the regular language K.

The basic idea for computing the Parikh image of (4) is as follows. To θ we assign two vectors: its Parikh image $\pi(\theta) \in \mathbb{N}^A$; and another vector $\rho(\theta) \in \mathbb{N}^B$, called the footprint of θ. The idea behind the footprint is that it contains information on the way θ can be combined with other traces to get a word in K. By using the induction assumption of Theorem 5.3, we can show that these two vectors are related in a semilinear way, i.e. the following vector set is semilinear:

$$Y = \{(\pi(\theta), \rho(\theta)) : \theta \in cut_M(L)\} \subseteq \mathbb{N}^{A \cup B}.$$

Using the mappings π and ρ, the job of calculating the Parikh image of (4) can be split into two phases. In the first phase, the question whether or not traces $\theta_1, \dots, \theta_n$ can be combined into a word from K is rephrased as a condition (*) on the footprints $\rho(\theta_1), \dots, \rho(\theta_n)$. In the second phase, the semilinear set Y is used to go from the from the footprints to the Parikh images. More precisely, we show that the following set is semilinear:

$$\{\pi(\theta_1) + \cdots + \pi(\theta_n) : \theta_1, \dots, \theta_n \text{ are traces such that } \rho(\theta_1), \dots, \rho(\theta_n) \text{ satisfy condition } (*)\}$$

This concludes, since the above set is the Parikh image of (4). Note that in the above, we do not need to quantify over the traces θ_i, since it is enough to verify that two vectors $\pi(\theta_i)$ and $\rho(\theta_i)$ satisfy the semilinear property Y.

We conclude by summarizing the expressive power of the expressions:

$$\text{USE} \subsetneq \text{UCSE} \subsetneq \text{SE} = \text{CSE}.$$

The strictness of the first inequality is not shown due to lack of space. The second inequality follows by undecidability of SE, while the equality was mentioned in Theorem 4.1.

References

1. Björklund, H., Schwentick, T.: On notions of regularity for data languages. In: FCT'07 (to appear, 2007)
2. Bojańczyk, M., Muscholl, A., Schwentick, T., Segoufin, L., David, C.: Two-variable logic on words with data. In: LICS'06, pp. 7–16 (2006)
3. Bouyer, P., Petit, A., Thérien, D.: An algebraic approach to data languages and timed languages. Information and Computation 182(2), 137–162 (2003)
4. Demri, S., Lazić, R.: LTL with the freeze quantifier and register automata. In: LICS'06, pp. 17–26 (2006)
5. Gischer, J.: Shuffle languages, petri nets, and context-sensitive grammars. Communications of the ACM 24(9), 597–605 (1981)
6. Jedrzejowicz, J., Szepietowski, A.: Shuffle languages are in P. TCS 250, 31–53 (2001)
7. Kaminski, M., Francez, N.: Finite-memory automata. TCS 132(2), 329–363 (1994)
8. Lomazova, I.A., Schnoebelen, P.: Some decidability results for nested petri nets. In: PSI'99, pp. 208–220 (2000)
9. Neven, F., Schwentick, T., Vianu, V.: Finite state machines for strings over infinite alphabets. ACM transactions on computational logic 15(3), 403–435 (2004)
10. Parikh, R.: On context-free languages. Journal of the ACM , 570–581 (1966)
11. Segoufin, L.: Automata and logics for words and trees over an infinite alphabet. In: Ésik, Z. (ed.) CSL 2006. LNCS, vol. 4207, pp. 41–57. Springer, Heidelberg (2006)

Author Index

Lecture Notes in Computer Science

For information about Vols. 1–4558

please contact your bookseller or Springer

Vol. 4708: L. Kučera, A. Kučera (Eds.), Mathematical Foundations of Computer Science 2007. XVIII, 764 pages. 2007.

Vol. 4697: L. Choi, Y. Paek, S. Cho (Eds.), Advances in Computer Systems Architecture. XIII, 400 pages. 2007.

Vol. 4682: D.-S. Huang, L. Heutte, M. Loog (Eds.), Advanced Intelligent Computing Theories and Applications. XXVII, 1373 pages. 2007. (Sublibrary LNAI).

Vol. 4681: D.-S. Huang, L. Heutte, M. Loog (Eds.), Advanced Intelligent Computing Theories and Applications. XXVI, 1379 pages. 2007.

Vol. 4679: A.L. Yuille, S.-C. Zhu, D. Cremers, Y. Wang (Eds.), Energy Minimization Methods in Computer Vision and Pattern Recognition. XII, 494 pages. 2007.

Vol. 4673: W.G. Kropatsch, M. Kampel, A. Hanbury (Eds.), Computer Analysis of Images and Patterns. XX, 1006 pages. 2007.

Vol. 4671: V. Malyshkin (Ed.), Parallel Computing Technologies. XIV, 635 pages. 2007.

Vol. 4660: S. Džeroski, J. Todorovski (Eds.), Computational Discovery of Scientific Knowledge. X, 327 pages. 2007. (Sublibrary LNAI).

Vol. 4651: F. Azevedo, P. Barahona, F. Fages, F. Rossi (Eds.), Recent Advances in Constraints. VIII, 185 pages. 2007. (Sublibrary LNAI).

Vol. 4647: R. Martin, M. Sabin, J. Winkler (Eds.), Mathematics of Surfaces XII. IX, 509 pages. 2007.

Vol. 4645: R. Giancarlo, S. Hannenhalli (Eds.), Algorithms in Bioinformatics. XIII, 432 pages. 2007. (Sublibrary LNBI).

Vol. 4643: M.-F. Sagot, M.E.M.T. Walter (Eds.), Advances in Bioinformatics and Computational Biology. XII, 177 pages. 2007. (Sublibrary LNBI).

Vol. 4639: E. Csuhaj-Varj'u, Z. Ésik (Eds.), Fundamentals of Computation Theory. XIV, 508 pages. 2007.

Vol. 4637: C. Kruegel, R. Lippmann, A. Clark (Eds.), Recent Advances in Intrusion Detection. XII, 338 pages. 2007.

Vol. 4635: B. Kokinov, D.C. Richardson, T.R. Roth-Berghofer, L. Vieu (Eds.), Modeling and Using Context. XIV, 574 pages. 2007. (Sublibrary LNAI).

Vol. 4634: H.R. Nielson, G. Filé (Eds.), Static Analysis. XI, 469 pages. 2007.

Vol. 4633: M. Kamel, A. Campilho (Eds.), Image Analysis and Recognition. XII, 1312 pages. 2007.

Vol. 4632: R. Alhajj, H. Gao, X. Li, J. Li, O.R. Zaïane (Eds.), Advanced Data Mining and Applications. XV, 634 pages. 2007. (Sublibrary LNAI).

Vol. 4628: L.N. de Castro, F.J. Von Zuben, H. Knidel (Eds.), Artificial Immune Systems. XII, 438 pages. 2007.

Vol. 4627: M. Charikar, K. Jansen, O. Reingold, J.D.P. Rolim (Eds.), Approximation, Randomization, and Combinatorial Optimization. XII, 626 pages. 2007.

Vol. 4626: R.O. Weber, M.M. Richter (Eds.), Case-Based Reasoning Research and Development. XIII, 534 pages. 2007. (Sublibrary LNAI).

Vol. 4624: T. Mossakowski, U. Montanari, M. Haveraaen (Eds.), Algebra and Coalgebra in Computer Science. XI, 463 pages. 2007.

Vol. 4622: A. Menezes (Ed.), Advances in Cryptology - CRYPTO 2007. XIV, 631 pages. 2007.

Vol. 4619: F. Dehne, J.-R. Sack, N. Zeh (Eds.), Algorithms and Data Structures. XVI, 662 pages. 2007.

Vol. 4618: S.G. Akl, C.S. Calude, M.J. Dinneen, G. Rozenberg, H.T. Wareham (Eds.), Unconventional Computation. X, 243 pages. 2007.

Vol. 4617: V. Torra, Y. Narukawa, Y. Yoshida (Eds.), Modeling Decisions for Artificial Intelligence. XII, 502 pages. 2007. (Sublibrary LNAI).

Vol. 4616: A. Dress, Y. Xu, B. Zhu (Eds.), Combinatorial Optimization and Applications. XI, 390 pages. 2007.

Vol. 4615: R. de Lemos, C. Gacek, A. Romanovsky (Eds.), Architecting Dependable Systems IV. XIV, 435 pages. 2007.

Vol. 4613: F.P. Preparata, Q. Fang (Eds.), Frontiers in Algorithmics. XI, 348 pages. 2007.

Vol. 4612: I. Miguel, W. Ruml (Eds.), Abstraction, Reformulation, and Approximation. XI, 418 pages. 2007. (Sublibrary LNAI).

Vol. 4611: J. Indulska, J. Ma, L.T. Yang, T. Ungerer, J. Cao (Eds.), Ubiquitous Intelligence and Computing. XXIII, 1257 pages. 2007.

Vol. 4610: B. Xiao, L.T. Yang, J. Ma, C. Muller-Schloer, Y. Hua (Eds.), Autonomic and Trusted Computing. XVIII, 571 pages. 2007.

Vol. 4609: E. Ernst (Ed.), ECOOP 2007 – Object-Oriented Programming. XIII, 625 pages. 2007.

Vol. 4608: H.W. Schmidt, I. Crnkovic, G.T. Heineman, J.A. Stafford (Eds.), Component-Based Software Engineering. XII, 283 pages. 2007.

Vol. 4607: L. Baresi, P. Fraternali, G.-J. Houben (Eds.), Web Engineering. XVI, 576 pages. 2007.

Vol. 4606: A. Pras, M. van Sinderen (Eds.), Dependable and Adaptable Networks and Services. XIV, 149 pages. 2007.

Vol. 4605: D. Papadias, D. Zhang, G. Kollios (Eds.), Advances in Spatial and Temporal Databases. X, 479 pages. 2007.

Vol. 4604: U. Priss, S. Polovina, R. Hill (Eds.), Conceptual Structures: Knowledge Architectures for Smart Applications. XII, 514 pages. 2007. (Sublibrary LNAI).

Vol. 4603: F. Pfenning (Ed.), Automated Deduction – CADE-21. XII, 522 pages. 2007. (Sublibrary LNAI).

Vol. 4602: S. Barker, G.-J. Ahn (Eds.), Data and Applications Security XXI. X, 291 pages. 2007.

Vol. 4600: H. Comon-Lundh, C. Kirchner, H. Kirchner (Eds.), Rewriting, Computation and Proof. XVI, 273 pages. 2007.

Vol. 4599: S. Vassiliadis, M. Berekovic, T.D. Hämäläinen (Eds.), Embedded Computer Systems: Architectures, Modeling, and Simulation. XVIII, 466 pages. 2007.

Vol. 4598: G. Lin (Ed.), Computing and Combinatorics. XII, 570 pages. 2007.

Vol. 4597: P. Perner (Ed.), Advances in Data Mining. XI, 353 pages. 2007. (Sublibrary LNAI).

Vol. 4596: L. Arge, C. Cachin, T. Jurdziński, A. Tarlecki (Eds.), Automata, Languages and Programming. XVII, 953 pages. 2007.

Vol. 4595: D. Bošnački, S. Edelkamp (Eds.), Model Checking Software. X, 285 pages. 2007.

Vol. 4594: R. Bellazzi, A. Abu-Hanna, J. Hunter (Eds.), Artificial Intelligence in Medicine. XVI, 509 pages. 2007. (Sublibrary LNAI).

Vol. 4592: Z. Kedad, N. Lammari, E. Métais, F. Meziane, Y. Rezgui (Eds.), Natural Language Processing and Information Systems. XIV, 442 pages. 2007.

Vol. 4591: J. Davies, J. Gibbons (Eds.), Integrated Formal Methods. IX, 660 pages. 2007.

Vol. 4590: W. Damm, H. Hermanns (Eds.), Computer Aided Verification. XV, 562 pages. 2007.

Vol. 4589: J. Münch, P. Abrahamsson (Eds.), Product-Focused Software Process Improvement. XII, 414 pages. 2007.

Vol. 4588: T. Harju, J. Karhumäki, A. Lepistö (Eds.), Developments in Language Theory. XI, 423 pages. 2007.

Vol. 4587: R. Cooper, J. Kennedy (Eds.), Data Management. XIII, 259 pages. 2007.

Vol. 4586: J. Pieprzyk, H. Ghodosi, E. Dawson (Eds.), Information Security and Privacy. XIV, 476 pages. 2007.

Vol. 4585: M. Kryszkiewicz, J.F. Peters, H. Rybinski, A. Skowron (Eds.), Rough Sets and Intelligent Systems Paradigms. XIX, 836 pages. 2007. (Sublibrary LNAI).

Vol. 4584: N. Karssemeijer, B. Lelieveldt (Eds.), Information Processing in Medical Imaging. XX, 777 pages. 2007.

Vol. 4583: S.R. Della Rocca (Ed.), Typed Lambda Calculi and Applications. X, 397 pages. 2007.

Vol. 4582: J. Lopez, P. Samarati, J.L. Ferrer (Eds.), Public Key Infrastructure. XI, 375 pages. 2007.

Vol. 4581: A. Petrenko, M. Veanes, J. Tretmans, W. Grieskamp (Eds.), Testing of Software and Communicating Systems. XII, 379 pages. 2007.

Vol. 4580: B. Ma, K. Zhang (Eds.), Combinatorial Pattern Matching. XII, 366 pages. 2007.

Vol. 4579: B. M. Hämmerli, R. Sommer (Eds.), Detection of Intrusions and Malware, and Vulnerability Assessment. X, 251 pages. 2007.

Vol. 4578: F. Masulli, S. Mitra, G. Pasi (Eds.), Applications of Fuzzy Sets Theory. XVIII, 693 pages. 2007. (Sublibrary LNAI).

Vol. 4577: N. Sebe, Y. Liu, Y.-t. Zhuang, T.S. Huang (Eds.), Multimedia Content Analysis and Mining. XIII, 513 pages. 2007.

Vol. 4576: D. Leivant, R. de Queiroz (Eds.), Logic, Language, Information and Computation. X, 363 pages. 2007.

Vol. 4575: T. Takagi, T. Okamoto, E. Okamoto, T. Okamoto (Eds.), Pairing-Based Cryptography – Pairing 2007. XI, 408 pages. 2007.

Vol. 4574: J. Derrick, J. Vain (Eds.), Formal Techniques for Networked and Distributed Systems – FORTE 2007. XI, 375 pages. 2007.

Vol. 4573: M. Kauers, M. Kerber, R. Miner, W. Windsteiger (Eds.), Towards Mechanized Mathematical Assistants. XIII, 407 pages. 2007. (Sublibrary LNAI).

Vol. 4572: F. Stajano, C. Meadows, S. Capkun, T. Moore (Eds.), Security and Privacy in Ad-hoc and Sensor Networks. X, 247 pages. 2007.

Vol. 4571: P. Perner (Ed.), Machine Learning and Data Mining in Pattern Recognition. XIV, 913 pages. 2007. (Sublibrary LNAI).

Vol. 4570: H.G. Okuno, M. Ali (Eds.), New Trends in Applied Artificial Intelligence. XXI, 1194 pages. 2007. (Sublibrary LNAI).

Vol. 4569: A. Butz, B. Fisher, A. Krüger, P. Olivier, S. Owada (Eds.), Smart Graphics. IX, 237 pages. 2007.

Vol. 4568: T. Ishida, S. R. Fussell, P. T. J. M. Vossen (Eds.), Intercultural Collaboration. XIII, 395 pages. 2007.

Vol. 4566: M.J. Dainoff (Ed.), Ergonomics and Health Aspects of Work with Computers. XVIII, 390 pages. 2007.

Vol. 4565: D.D. Schmorrow, L.M. Reeves (Eds.), Foundations of Augmented Cognition. XIX, 450 pages. 2007. (Sublibrary LNAI).

Vol. 4564: D. Schuler (Ed.), Online Communities and Social Computing. XVII, 520 pages. 2007.

Vol. 4563: R. Shumaker (Ed.), Virtual Reality. XXII, 762 pages. 2007.

Vol. 4562: D. Harris (Ed.), Engineering Psychology and Cognitive Ergonomics. XXIII, 879 pages. 2007. (Sublibrary LNAI).

Vol. 4561: V.G. Duffy (Ed.), Digital Human Modeling. XXIII, 1068 pages. 2007.

Vol. 4560: N. Aykin (Ed.), Usability and Internationalization, Part II. XVIII, 576 pages. 2007.

Vol. 4559: N. Aykin (Ed.), Usability and Internationalization, Part I. XVIII, 661 pages. 2007.